Extended Technology Applications

Extended Technology Applications at the end of each chapter use real applications and real data. They require a step-by-step analysis that encourages group work.

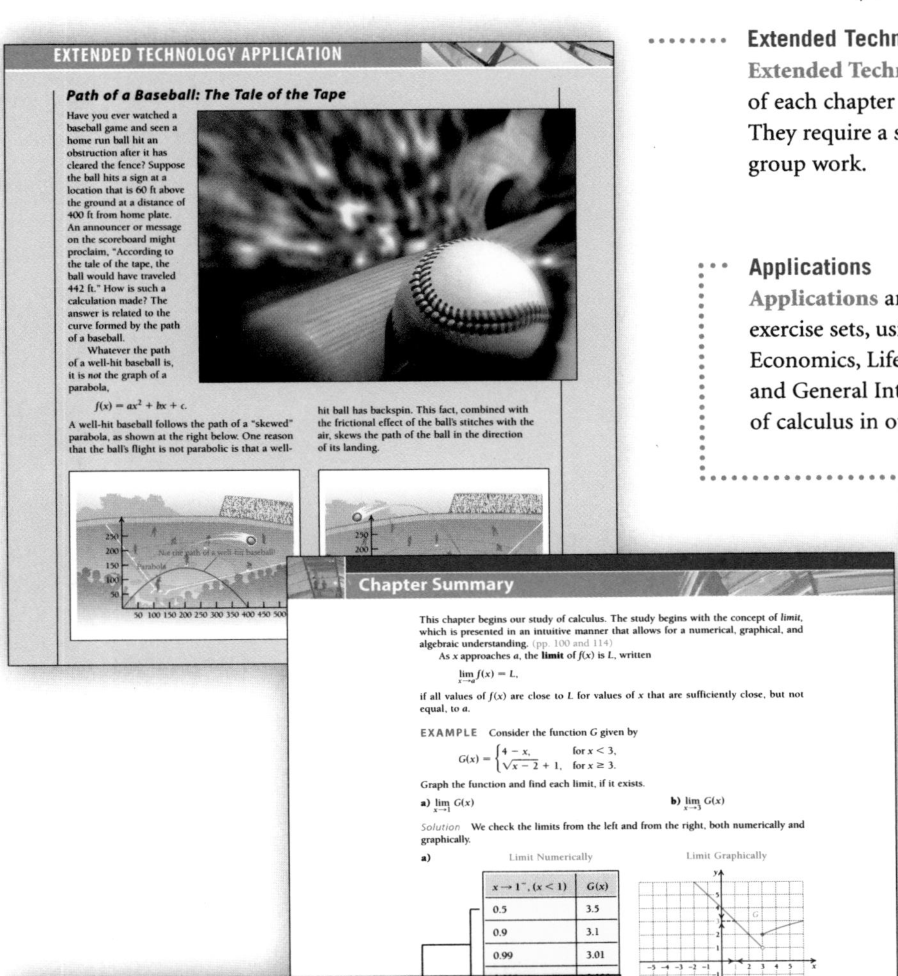

EXTENDED TECHNOLOGY APPLICATION

Path of a Baseball: The Tale of the Tape

Have you ever watched a baseball game and seen a home run ball hit an obstruction after it has cleared the fence? Suppose the ball hits a sign at a location that is 60 ft above the ground at a distance of 400 ft from home plate. An announcer or message on the scoreboard might proclaim, "According to the tale of the tape, the ball would have traveled 442 ft." How is such a calculation made? The answer is related to the curve formed by the path of a baseball.

Whatever the path of a well-hit baseball is, it is *not* the graph of a parabola,

$$f(x) = ax^2 + bx + c.$$

A well-hit baseball follows the path of a "skewed" parabola, as shown at the right below. One reason that the ball's flight is not parabolic is that a well-hit ball has backspin. This fact, combined with the frictional effect of the ball's stitches with the air, skews the path of the ball in the direction of its landing.

Chapter Summary

This chapter begins our study of calculus. The study begins with the concept of *limit*, which is presented in an intuitive manner that allows for a numerical, graphical, and algebraic understanding. (pp. 100 and 114)

As x approaches a, the **limit** of $f(x)$ is L, written

$$\lim_{x \to a} f(x) = L,$$

if all values of $f(x)$ are close to L for values of x that are sufficiently close, but not equal, to a.

EXAMPLE Consider the function G given by

$$G(x) = \begin{cases} 4 - x, & \text{for } x < 3, \\ \sqrt{x-2} + 1, & \text{for } x \geq 3. \end{cases}$$

Graph the function and find each limit, if it exists.

a) $\lim_{x \to 1} G(x)$ **b)** $\lim_{x \to 3} G(x)$

Solution We check the limits from the left and from the right, both numerically and graphically.

a) Limit Numerically Limit Graphically

$x \to 1^-, (x < 1)$	$G(x)$
0.5	3.5
0.9	3.1
0.99	3.01

...ets closer to 1, the outputs $G(x)$ get closer

Applications

Applications are grouped by discipline within the exercise sets, using such headings as Business and Economics, Life/Physical Science, Social Science, and General Interest, to show students the relevance of calculus in other disciplines.

APPLICATIONS

Business and Economics

Taxicab fares. *In New York City, taxicabs charge passengers \$2.50 for entering a cab and then \$0.40 for each one-fifth of a mile (or fraction thereof) traveled. (There are additional charges for slow traffic and idle times, but these are not considered in this problem.) If x represents the distance traveled in miles, then $C(x)$ is the cost of the taxi fare, where*

$C(x) = \$2.50$, if $x = 0$,
$C(x) = \$2.90$, if $0 < x \leq 0.2$,
$C(x) = \$3.30$, if $0.2 < x \leq 0.4$,
$C(x) = \$3.70$, if $0.4 < x \leq 0.6$,

and so on. The graph of C is shown below. (Source: New York City Taxi and Limousine Commission.)

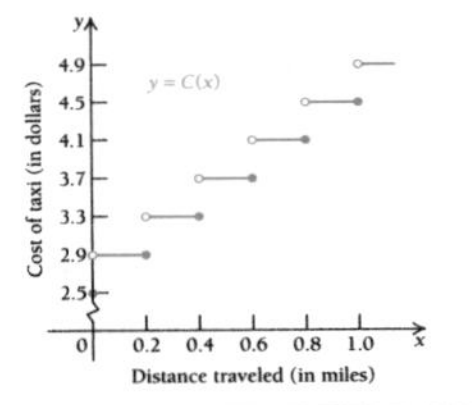

Chapter Summaries

Each chapter ends with a concise **Chapter Summary** that lists important page references and highlights essential examples for students to review.

Chapter Review Exercises

The **Chapter Review Exercises** include a variety of **Concept Reinforcement Exercises** to help build student confidence. Each exercise is accompanied by a bracketed section reference to help students locate the problem type within the chapter.

Chapter Review Exercises

These review exercises are for test preparation. They can also be used as a practice test. Answers are at the back of the book. The bracketed section references tell you what part(s) of the chapter to restudy if your answer is incorrect.

CONCEPT REINFORCEMENT

Classify each statement as either true or false.

1. If $\lim_{x \to 5} f(x)$ exists, then $f(5)$ must exist. [1.1]
2. If $\lim_{x \to 2} f(x) = L$, then $L = f(2)$. [1.1]
3. If f is continuous at $x = 3$, then $\lim_{x \to 3} f(x) = f(3)$. [1.2]
4. A function's average rate of change over the interval [2, 8] is the same as its instantaneous rate of change at $x = 5$. [1.3, 1.4]
5. A function's derivative at a point, if it exists, can be found as the limit of a difference quotient. [1.4]
6. For $f'(5)$ to exist, f must be continuous at 5. [1.4]
7. If f is continuous at 5, then $f'(5)$ must exist. [1.4]
8. The acceleration function is the derivative of the velocity function. [1.8]

Match each function in column A with the rule in column B that would be the most appropriate to use for differentiating the function. [1.5, 1.6]

Column A	Column B
9. $f(x) = x^7$	a) Extended Power Rule
10. $g(x) = x + 9$	b) Product Rule
11. $F(x) = (5x - 3)^4$	c) Sum Rule
12. $G(x) = \frac{2x+1}{3x-4}$	d) Difference Rule
13. $H(x) = f(x) \cdot g(x)$	e) Power Rule
14. $f(x) = 2x - 7$	f) Quotient Rule

For Exercises 15–17, consider

$$\lim_{x \to -7} f(x), \quad \text{where } f(x) = \frac{x^2 + 4x - 21}{x + 7}.$$

15. Limit numerically. [1.1]

a) Complete the following input–output tables.

$x \to -7^-$	$f(x)$
−8	
−7.5	
−7.1	
−7.01	
−7.001	
−7.0001	

$x \to -7^+$	$f(x)$
−6	
−6.5	
−6.9	
−6.99	
−6.999	
−6.9999	

b) Find $\lim_{x \to -7^-} f(x)$, $\lim_{x \to -7^+} f(x)$, and $\lim_{x \to -7} f(x)$, if each exists.

16. Limit graphically. Graph the function, and use the graph to find the limit. [1.1]

17. Limit algebraically. Find the limit algebraically. Show your work. [1.2]

Find each limit, if it exists. If a limit does not exist, state that fact. [1.1, 1.2]

18. $\lim_{x \to -2} \frac{8}{x}$

19. $\lim_{x \to 1} (4x^3 - x^2 + 7x)$

20. $\lim_{x \to -7} \frac{x^2 + 2x - 35}{x + 7}$ 21. $\lim_{x \to \infty} \frac{1}{x} + 3$ [1.1]

From the graphs in Exercises 22 and 23, determine whether each function is continuous and explain why or why not. [1.2]

22. 23.

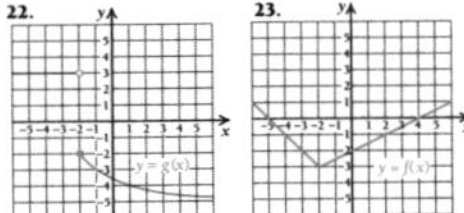

Calculus
AND ITS APPLICATIONS

9 EDITION

MARVIN L. BITTINGER
Indiana University Purdue University Indianapolis

DAVID J. ELLENBOGEN
Community College of Vermont

Boston San Francisco New York
London Toronto Sydney Tokyo Singapore Madrid
Mexico City Munich Paris Cape Town Hong Kong Montreal

Publisher	Greg Tobin
Editor in Chief	Deirdre Lynch
Senior Acquisitions Editor	William Hoffman
Executive Project Manager	Christine O'Brien
Editorial Assistant	Susan Whalen
Senior Managing Editor	Karen Wernholm
Senior Designer/Cover Design	Joyce Cosentino Wells
Digital Assets Manager	Marianne Groth
Media Producer	Michelle Small
Software Development	Mary Durnwald, TestGen and Bob Carroll, MathXL
Marketing Manager	Phyllis Hubbard
Marketing Assistant	Caroline Celano
Senior Technology Specialist	Joe Vetere
Senior Prepress Supervisor	Caroline Fell
Rights and Permissions Advisor	Shannon Barbe
Manufacturing Manager	Evelyn Beaton
Text Design, Art Editing, Photo Research	Geri Davis/The Davis Group, Inc.
Production Coordination	Jane Hoover/Lifland et al., Bookmakers
Composition	Pre-Press PMG
Illustrations	Network Graphics, Inc., and William Melvin

Cover photo: ©2007 Walter Bibikow/Index Stock Imagery. Interior dome view of The Reichstag, Berlin.

Photo Credits

For permission to use copyrighted material, grateful acknowledgment is made to the copyright holders on page 617, which is hereby made part of this copyright page.

Library of Congress Cataloging in Publication Data

Bittinger, Marvin L.

Calculus and its applications / Marvin L. Bittinger, David J. Ellenbogen — 9th ed.

p. cm.

ISBN 0-321-39534-4

1. Calculus. I. Title.

QA303.2.B466 2008

515—dc22 2006046005

6 7 8 9 10—DOW—09

For our children:

Lowell and Chris
Monroe and Zack

Contents

6

Preface

Calculus and Its Applications, Ninth Edition, is the most student-oriented applied calculus text on the market. The authors believe that appealing to students' intuition and speaking in a direct, down-to-earth manner make this text accessible to any student possessing the prerequisite math skills. However, it is not enough for a text to be accessible—it must also provide students with motivation to learn. Tapping into areas of student interest, the authors provide an abundant supply of examples and exercises rich in real-world data from business, economics, environmental studies, health care, and the life sciences. New examples cover applications ranging from charting CEO pay to the melting polar ice cap. Found in every chapter, realistic applications draw students into the discipline and help them to generalize the material and apply it to new and novel situations. To further spark student interest, hundreds of meticulously drawn graphs, illustrations, and tables appear throughout the text, making it a favorite among students who are visual learners.

Appropriate for a one-term course, this text is an introduction to applied calculus. A course in intermediate algebra is a prerequisite, although the Appendix, "Review of Basic Algebra," together with Chapter R, provides a sufficient foundation to unify the diverse backgrounds of most students.

What's New in the Ninth Edition?

New and Revised Content

In response to faculty and student feedback, we have made the following changes to the text's content.

- Because the first chapter in previous editions contained topics typically covered in a preceding algebra course, we have relabeled that chapter Chapter R. Subsequent chapters, corresponding to Chapters 2 through 7 of the 8th edition, are now numbered 1 through 6.
- Within Chapter R, Section R.2 now focuses extra attention on graphs that are piece-wise defined, a topic that many students find challenging.

- Chapter 1 (formerly Chapter 2) has been reorganized—without sacrificing any topics, the chapter now has one fewer section. The coverage of limits and continuity (formerly in Section 2.1) has been divided between two sections: Section 1.1, Limits: A Numerical and Graphical Approach, and Section 1.2, Algebraic Limits and Continuity. This organization allows us to examine limits from the more intuitive numerical and graphical perspectives in Section 1.1, reserving the algebraic perspective for Section 1.2, where it is used for the study of continuity.

  We also now include Instantaneous Rates of Change (formerly Section 2.6) as part of Section 1.4, Differentiation Using Limits of Difference Quotients. This locates the topic of *instantaneous* rates of change immediately after the section in which *average* rates of change are discussed, creating a more natural flow of topics. Instantaneous rates of change are studied again when velocity and acceleration are covered in Section 1.8, Higher-Order Derivatives.
- Chapter 2 has been reconfigured so that marginal cost, revenue, and profit (material formerly found in Section 3.6) is now the first topic covered in Section 2.6, Marginals and Differentials. The concept of using a derivative to approximate nearby function values is thereby made both more intuitive and more application-oriented. Material on differentials closes the section and can now be regarded as a purely mathematical basis for the business example that precedes it.
- The first three sections of Chapter 4 (formerly Chapter 5) have been extensively reorganized. To better emphasize the link between area under a curve, Reimann sums, and antiderivatives, this chapter opens with Section 4.1, The Area under a Graph (in which Reimann sums are introduced), followed by Section 4.2, Area, Antiderivatives, and Integrals (in which the link between antiderivatives and the area under a curve is first developed), and Section 4.3, Area and Definite Integrals (in which the Fundamental Theorem of Calculus is explained and applications in business and science are developed). Coverage of average value (formerly in Section 5.3) now appears with other properties of the integral, in Section 4.4.
- In Chapter 6 (formerly Chapter 7), all coverage of partial derivatives is now in one section (Section 6.2). This better enables instructors to cover (or skip) a topic that previously appeared as two separate and smaller sections.

More Applications and Exercises

In addition to improving the flow of topics within each chapter, we have improved the exercise sets that appear in each section:

- More exercises that make use of real-world data have been included, in applications that students can relate to (health care, finance, environmental studies, etc.).
- At least 20% of the exercises in the 8th edition have been replaced with fresh problems. All applications in need of updating have been made current.
- More exercises have been added to those sections that the authors, with the help of faculty and students, have identified as ones in which students with newly emerging skills can benefit from extra practice problems.

Chapter Summaries

Every chapter now ends with a Chapter Summary in which the authors have compiled notes designed to help students organize their studying. Here, important concepts and definitions and further examples are presented. Page references are included so that students can consult the appropriate section of the chapter for more details.

Concept Reinforcement Exercises

As always, each chapter closes with a set of Chapter Review Exercises, followed by a Chapter Test. New to this edition is the inclusion of 8 to 14 Concept Reinforcement exercises at the beginning of each set of Review Exercises. These are confidence builders for students who have completed their study of the chapter. Presented in matching, true/false, or fill-in-the-blank format, these exercises can also be used in class as oral exercises. Like all review exercises, each concept reinforcement exercise is accompanied by a bracketed section reference to indicate where discussion of the concept appears.

Hallmark Features

Intuitive Approach

Although the word "intuitive" has many meanings and interpretations, its use here means "experience based." Throughout the text, when a particular concept is discussed, its presentation is designed so that the students' learning process is based on their earlier mathematical experience or a new experience presented by the author before the concept is formalized. This is illustrated by the following situations.

- Before the formal definition of continuity is presented, an informal explanation is given, complete with graphs that make use of student intuition into ways in which a function could be discontinuous (see pp. 117–118).
- The definition of the derivative, in Chapter 1, is presented in the context of a discussion of average rates of change (see p. 138). This presentation is more accessible and realistic than the strictly geometric idea of slope.
- When maximization problems involving volume are introduced (see p. 261), a function is derived that is to be maximized. Instead of forging ahead with the standard calculus solution, the student is first asked to stop, make a table of function values, graph the function, and then estimate the maximum value. This experience provides students with more insight into the problem. They recognize that not only do different dimensions yield different volumes, but also that the dimensions yielding the maximum volume may be conjectured or estimated as a result of the calculations.
- Relative maxima and minima (Sections 2.1 and 2.2) and absolute maxima and minima (Section 2.4) are covered in separate sections in Chapter 2, so that students obtain a gradual buildup of these topics as they consider graphing using calculus concepts (see pp. 198–259).
- The explanation underlying the definition of the number e is presented in Chapter 3 both graphically and through a discussion of continuously compounded interest (see pp. 345–346).

Early Coverage of Functions

To minimize the amount of time needed for review, we have an algebra review in an appendix. This allows us to begin Chapter R with coverage of functions, graphs, and models and to incorporate an earlier use of technology. Section R.3 covers finding domain and range. This early and expanded function coverage allows for many uses of functions such as the expansion of graphing applications.

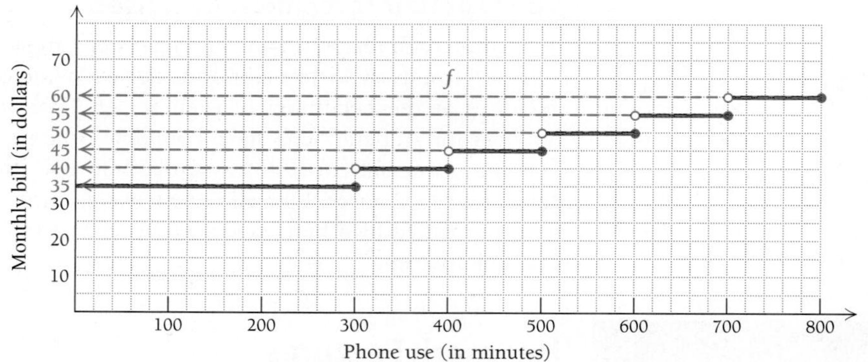

Applications

Relevant and factual applications drawn from a broad spectrum of fields are integrated throughout the text as applied examples and exercises and are also featured in separate application sections of the text. These have been updated and expanded in this edition, to include even more applications using real data. In addition, each chapter opener includes an application that serves as a preview of what students will learn in the chapter.

The applications in the exercise sets are grouped under headings that identify them as reflecting real-life situations: Business and Economics, Life and Physical Sciences, Social Sciences, and General Interest. This organization allows the instructor to gear the assigned exercises to a particular student and also allows the student to know whether a particular exercise applies to his or her major.

The following illustrate some examples of the applications.

- An early discussion in Chapter R of compound interest, total revenue, cost and profit, as well as supply and demand functions sets the stage for subsequent applications throughout the book (see pp. 48–49, 275, and 419).
- The concepts of average rate of change and instantaneous rate of change are introduced with new business applications in Chapter 1. This is prior to the introduction of applications of differentiation in Chapter 2, which include maximum–minimum applications (see pp. 259–274) and implicit differentiation and related rates (see pp. 284–291).
- When the exponential model is studied in Chapter 3, other applications, such as continuously compounded interest and the demand for natural resources, are also considered (see pp. 336–365). Growth and decay are covered in separate sections in Chapter 3 to allow room for the many worthwhile applications that relate to these concepts.
- Applications of integration are covered in a separate chapter (Chapter 5) and include thorough coverage of continuous probability (see pp. 491–514).

- The Extended Technology Application features include interesting and topical situations such as predicting the distance a home run will travel (see pp. 194–196), determining the maximum sustainable harvest for a population (see pp. 301–302) and using total box-office revenue data to determine when video sales should begin (see pp. 387–389).

Mathematical Modeling, Curve Fitting, and Regression

This edition continues to emphasize mathematical modeling, utilizing the advantages of technology as appropriate. For example, the use of the graphing calculator in modeling, as an optional topic, is introduced in Section R.6 and then reinforced many times throughout the text.

$y_1 = -0.49x^2 + 25.95x - 238.49$

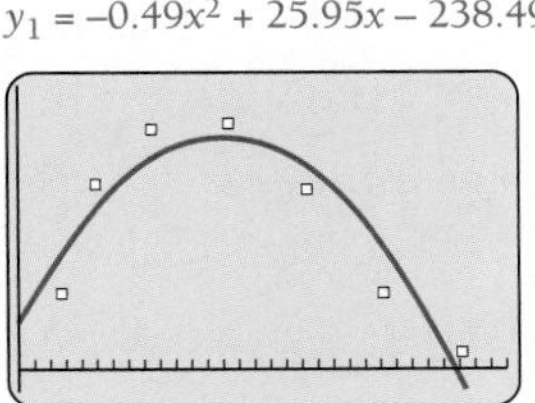

Technology Connection Features. Technology Connection features are included throughout the text to illustrate the use of technology. Whenever appropriate, art that simulates graphs or tables generated by a graphing calculator is included as well.

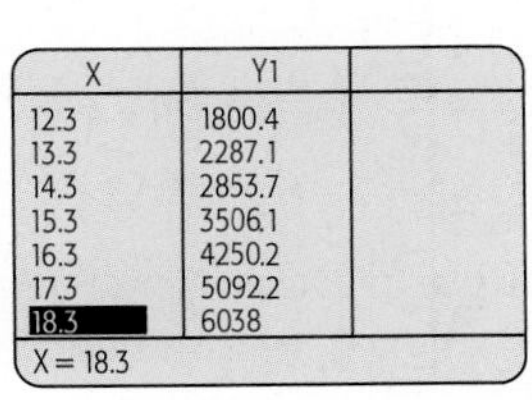

X	Y1
12.3	1800.4
13.3	2287.1
14.3	2853.7
15.3	3506.1
16.3	4250.2
17.3	5092.2
18.3	6038

X = 18.3

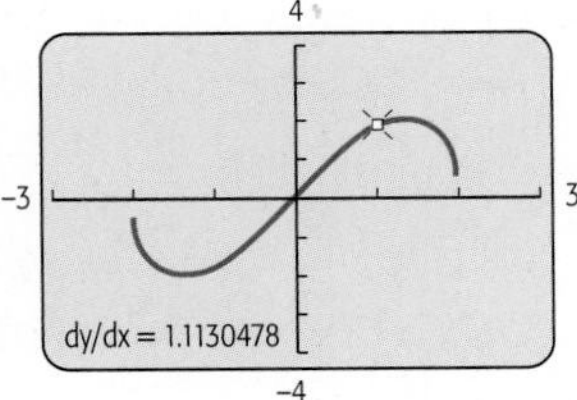

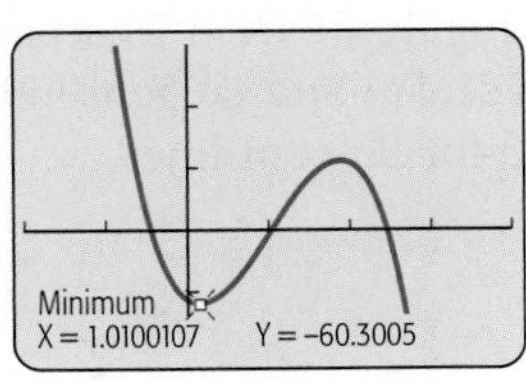

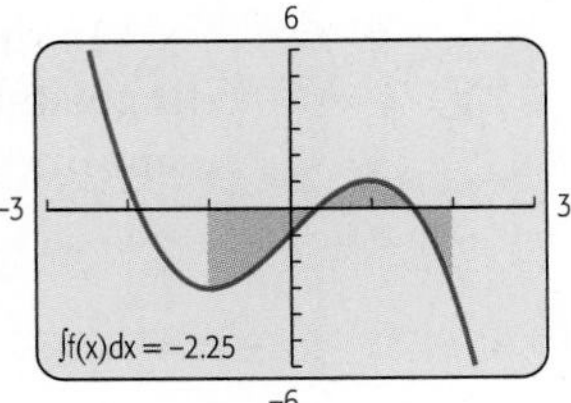

There are four types of Technology Connections for students and instructors to use for exploring key ideas.

- *Lesson/Teaching.* These provide students with an example, followed by exercises to work.
- *Checking.* These tell the students how to verify a solution by using a graphing calculator.
- *Exercises.* These are simply exercises that assume use of a graphing calculator.
- *Exploration/Investigation.* These provide questions to guide students through an investigation.

Technology Connection Exercises. Most exercise sets contain technology-based exercises identified with either a icon or the heading "Technology Connection." These exercises also appear in the Chapter Review Exercises and the Chapter Tests. The Printed Test Bank supplement includes technology-based exercises as well.

Use of Color

The text uses full color as a pedagogical tool. Color is used in a methodical and precise manner so that it enhances the readability of the text for the student and the instructor. When two curves are graphed using the same set of axes, one is usually red and the other blue with the red graph the curve of major importance. This is exemplified in the following graphs from Chapter R (pp. 60 and 61). Note that the equation labels are the same color as the curve. When the instructions say "Graph," the dots match the color of the curve.

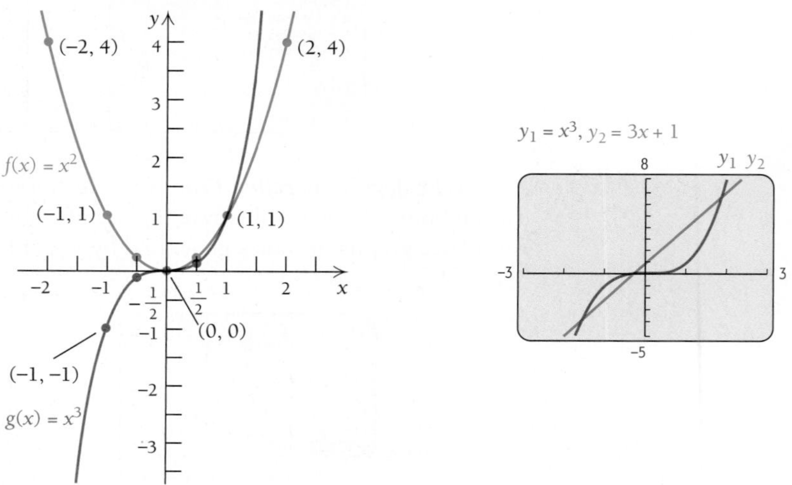

In the following figure from Chapter 1 (p. 137), we see how color is used to distinguish between secant and tangent lines. Throughout the text, blue is used for secant lines and red for tangent lines.

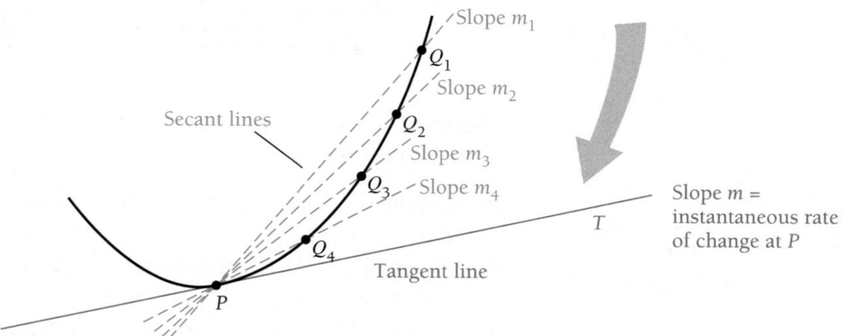

In the following graph and text development from Chapter 2 (pp. 217–218), the color red denotes substitution in equations and blue highlights the corresponding outputs, including maximum and minimum values. This specific use of color is carried out further, as shown in the figure that follows. Note that when dots are used for emphasis other than just merely plotting, they are black.

We next find second coordinates by substituting in the original function:

$$f(-3) = (-3)^3 + 3(-3)^2 - 9(-3) - 13 = 14;$$
$$f(1) = (1)^3 + 3(1)^2 - 9(1) - 13 = -18.$$

Are the points $(-3, 14)$ and $(1, -18)$ relative extrema? Let's look at the second derivative. We use the Second-Derivative Test with the numbers -3 and 1:

$$f''(-3) = 6(-3) + 6 = -12 < 0; \longrightarrow \text{Relative maximum}$$
$$f''(1) = 6(1) + 6 = 12 > 0. \longrightarrow \text{Relative minimum}$$

Thus, $f(-3) = 14$ is a relative maximum and $f(1) = -18$ is a relative minimum. We plot both $(-3, 14)$ and $(1, -18)$, and determine the graph's concavity at each point. Then, by calculating and plotting a few more points, we can make a sketch, as shown below.

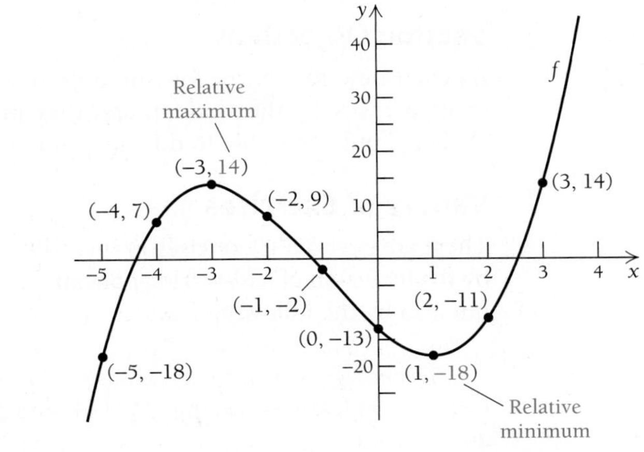

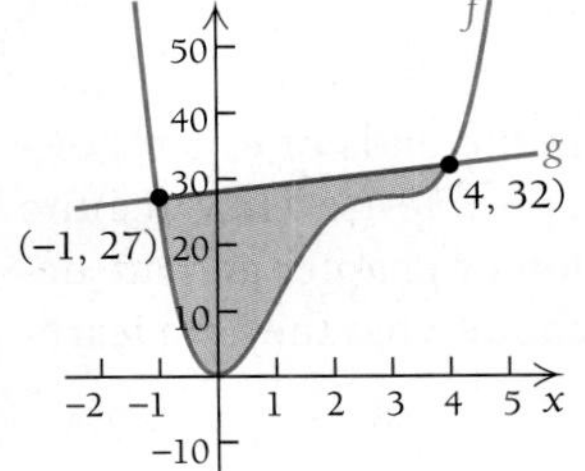

Beginning with integration in Chapter 4, the color amber is used to identify the area, unless certain special concepts are considered in the initial learning process. The figure to the left, from Chapter 4 (p. 434), illustrates the vivid use of blue and red for the curves and labels, and amber for the area. In Chapter 6 (pp. 557–563), we use extra colors for clarity. The top of a three-dimensional surface, as shown in the figure below, is blue, and the bottom of the surface is rose. This color scheme is carried out in all examples in Section 6.3.

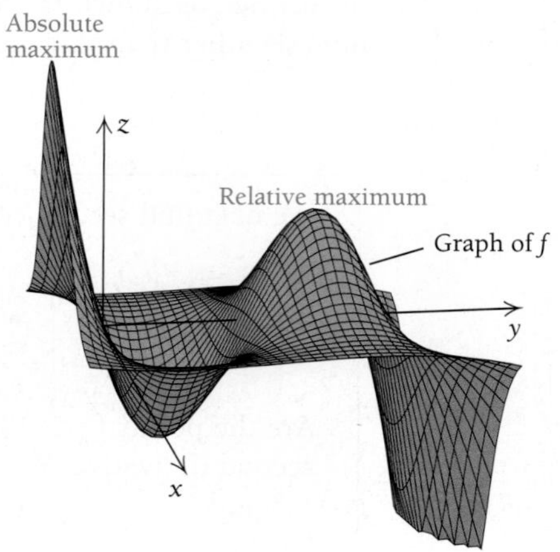

Pedagogical Features

Chapter Openers

Each chapter opener includes an application to whet the student's appetite for the chapter material. This application also provides an intuitive introduction to a key calculus topic.

Section Objectives

As each new section begins, its objectives are stated in the margin. These can be spotted easily by the student, and they provide the answer to the typical question "What should I be able to do after completing this section?"

Variety of Exercises

There are over 3500 exercises in this edition. The exercise sets are enhanced not only by the inclusion of real-world applications, detailed art pieces, and illustrative graphs, but also by the following features.

Technology Connection exercises. These exercises appear in the Technology Connection features (see pp. 34, 144, and 325) and in the exercise sets (see pp. 123, 245, and 437).

Synthesis exercises. Synthesis exercises are included at the ends of every exercise set, including the Chapter Review Exercises and Chapter Tests. They require students to go beyond the immediate objectives of the section or chapter and are designed both to challenge students and to make them think about what they are learning (see pp. 177, 273, 364, and 425).

Thinking and Writing exercises. These exercises appear both within the exercise sets and in the synthesis section at the end of the exercise sets (see pp. 134, 244, and 422). They are denoted by the symbol TW. These ask students to explain mathematical concepts in their own words, thereby strengthening their understanding. The answers to these exercises are given in the *Instructor's Solutions Manual.*

Applications. A section of applied problems is included in most exercise sets. The problems are grouped under headings that identify them as business and economics, life and physical sciences, social sciences, or general interest. Each problem is accompanied by a brief description of its subject matter (see pp. 165–167, 360–364, and 410–411).

Tests and Reviews

Summary and Review. At the end of each chapter is a summary, followed by review exercises and a test. The Chapter Summary (see above, under What's New in the Ninth Edition) provides a set of study notes. Students can use the summary as a basis for solidifying their understanding of the chapter. The Chapter Review Exercises, which include bracketed references to the sections in which the related course content first appears, provide comprehensive coverage of each chapter's material (see pp. 297–299).

Tests. Each chapter ends with a Chapter Test that includes synthesis and technology questions. There is also a Cumulative Review at the end of the text that can serve as a practice final examination. The answers, with section references to the chapter tests and the Cumulative Review, are at the back of the book. Six additional forms of each of the chapter tests and the final examination, with answer keys and ready for classroom use, appear in the *Printed Test Bank.*

Student Supplements

Student's Solutions Manual

- Provides detailed solutions to all odd-numbered exercises, with the exception of the Thinking and Writing exercises
- ISBN: 0-321-45056-6

Graphing Calculator Manual

- Provides instructions and keystroke operations for the TI-83, TI-83/84 Plus, and TI-86
- Includes worked-out examples taken directly from the text
- Topic order corresponds with that of the text
- ISBN: 0-321-45582-7

Video Lectures on CD with optional captioning

- Complete set of digitized videos for student use at home or on campus
- Ideal for distance learning or supplemental instruction
- ISBN: 0-321-45054-X

Addison-Wesley Math Tutor Center

- Provides free tutoring through a registration number that can be packaged with a new textbook or purchased separately

- Staffed by qualified college mathematics instructors
- Accessible via toll-free telephone, toll-free fax, e-mail, and the Internet
- www.aw-bc.com/tutorcenter

Instructor Supplements

Instructor's Edition

- Special edition of the text
- Provides answers to both even and odd-numbered exercises at the back of the text
- ISBN: 0-321-44894-4

Instructor's Solutions Manual

- Provides complete solutions to all text exercises
- ISBN: 0-321-44728-X

Printed Test Bank

- Contains six alternative tests per chapter
- Contains six comprehensive final exams
- Includes answers keys
- ISBN: 0-321-44895-2

TestGen

- Enables instructors to build, edit, print, and administer tests
- Features a computerized bank of questions developed to cover all text objectives
- Available on a dual-platform Windows/Macintosh CD-ROM
- ISBN: 0-321-45584-3

PowerPoint Lecture Presentation

- Classroom presentation software oriented specifically to the text's topic sequence
- Available within MyMathLab or from the Instructor Resource Center

Supplementary Chapters

- Three chapters: Sequences and Series, Differential Equations, and Trigonometric Functions
- Available online at the Instructor Resource Center
- Include many applications and optional technology material

Math XL® MathXL® is a powerful online homework, tutorial, and assessment system that accompanies Addison-Wesley textbooks in mathematics and statistics. With MathXL, instructors can create, edit, and assign online homework and tests using algorithmically generated exercices correlated at the objective level to the textbook. They can also create and assign their own online exercises and import TestGen tests for added flexibility. All student work is tracked in MathXL's online gradebook.

Students can take chapter tests in MathXL and receive personalized study plans based on their test results. The study plan diagnoses weaknesses and links students directly to tutorial exercises for the objectives they need to study and retest. Students can also access supplemental animations and video clips directly from selected exercises. MathXL is available to qualified adopters. For more information, visit the Web site at www.mathxl.com, or contact your Addison-Wesley sales representative.

MyMathLab® MyMathLab is a series of text-specific, easily customizable online courses for Addison-Wesley textbooks in mathematics and statistics. Powered by CourseCompass™ (Pearson Education's online teaching and learning environment) and MathXL® (see above), MyMathLab gives instructors the tools they need to deliver all or a portion of a course online, whether the students are in a lab setting or working from home. MyMathLab provides a rich and flexible set of course materials, featuring free-response exercises that are algorithmically generated for unlimited practice and mastery. Students can also use online tools, such as video lectures, animations, and a multimedia textbook, to independently improve their understanding and performance. Instructors can use MyMathLab's homework and test managers to select and assign online exercises correlated directly to the textbook. They can also create and assign their own online exercises and import TestGen tests for added flexibility. MyMathLab's online gradebook—designed specifically for mathematics and statistics—automatically tracks students' homework and test results and gives the instructor control over how to calculate final grades. Instructors can also add offline (paper-and-pencil) grades to the gradebook. MyMathLab is available to qualified adopters. For more information, visit www.mymathlab.com, or contact your Addison-Wesley sales representative.

InterAct Math Tutorial Web site: www.interactmath.com Get practice and tutorial help online! This interactive tutorial Web site provides algorithmically generated practice exercises that correlate directly to the exercises in the textbook. Students can retry an exercise as many times as they like with new values each time for unlimited practice and mastery. Every exercise is accompanied by an interactive guided solution that provides helpful feedback for incorrect answers, and students can also view a worked-out sample problem that takes them step by step through an exercise similar to the one they're working on.

Acknowledgments

As authors, we have taken many steps to ensure the accuracy of this text. Many devoted individuals comprised the team that was responsible for monitoring the production process in a manner that makes this a work of which we can all be proud. In particular, we would like to thank Jane Hoover for her many helpful suggestions, proofreading, and checking of art. Jane's attention to detail and pleasant demeanor made our work as low in stress as humanly possible, given the demands of the production process. We would also like to thank Christine O'Brien, Executive Project Manager, for her diligence in tending to the countless details and helping with so many of the decisions that a revision of this magnitude requires. Many thanks also to Bill Hoffman, Senior Acquisitions Editor, for his steady hand at the helm in making executive decisions with a perspective that can only be gained from years of experience with this (and other) titles. Carter Fenton merits thanks for working as

Acquisitions Editor and sharing many useful insights when launching this project, and Deirdre Lynch kept the project flowing smoothly during periods of transition. Senior Managing Editor Karen Wernholm also merits thanks for her help in making decisions that helped assure the right person was doing the right job. Susan Whalen provided editorial assistance that made communications run like clockwork. We also wish to thank Michelle Beecher Lanosga for her incredibly helpful data research. Her efforts make the real-world problems in this text as up-to-date as possible, given the production deadlines we faced. Geri Davis deserves credit for both the attractive design of the text and the coordination of the many illustrations, photos, and graphs. She is always a distinct pleasure to work with and sets the standard by which all other art editors are measured. We will forever be grateful to Sybil MacBeth for her tireless proofreading of the answers and pages and her thoughtful insights, suggestions, and corrections. Becky Anderson merits thanks for her help in confirming that our content changes will be well received by users. Many thanks to Chris Burditt, Shannon Travers, and Steve Hansen for compiling the answers. Joyce Wells deserves thanks for designing such a beautiful cover, and Becky Malone gets our thanks for creating the index of applications. Many thanks also to Laurie Hurley, Paul Lorczak, John Morin, and Patricia Nelson, for their careful checking of the typeset pages. Laurie Hurley merits additional thanks for her fine work on the Printed Test Bank.

Most of this edition was class-tested and effectively proofread by the Fall 2006 Calculus I class at Community College of Vermont, in Burlington. We authors are deeply grateful for the useful feedback those dedicated students gave directly to us.

Finally, the following reviewers provided thoughtful and insightful comments that helped immeasurably in the revision of this text.

Kenneth J. Bernard, *Virginia State University*
Melvin D. Lax, *California State University, Long Beach*
John C. Nardo, *Oglethorpe University*
Scott Perkins, *Lake-Sumter Community College*
Abby Roscum, *Marshalltown Community College*
Sounny Slitine, *Palo Alto College*
Jennifer Walsh, *Daytona Beach Community College*

Index of Applications

Business and Economics

Environmental Science

Health Care

Life and Physical Sciences

Social Sciences

Sports

General Interest

Extended Technology Applications

Index of Technology Connections

The following index lists, by topic, 62 optional Technology Connections. These features incorporate in-text comments, including guided explorations and exercises, that explain and encourage the use of technology to explore problem situations. The Technology Connection Features, along with the Technology Connection Exercises in the exercise sets and the Extended Technology Applications, described in the Preface, bolster the optional integration of technology into the course. A *Graphing Calculator Manual*, available for students, features keystroke instructions for several calculator models.

Calculus
AND ITS APPLICATIONS

9TH EDITION

Functions, Graphs, and Models

R

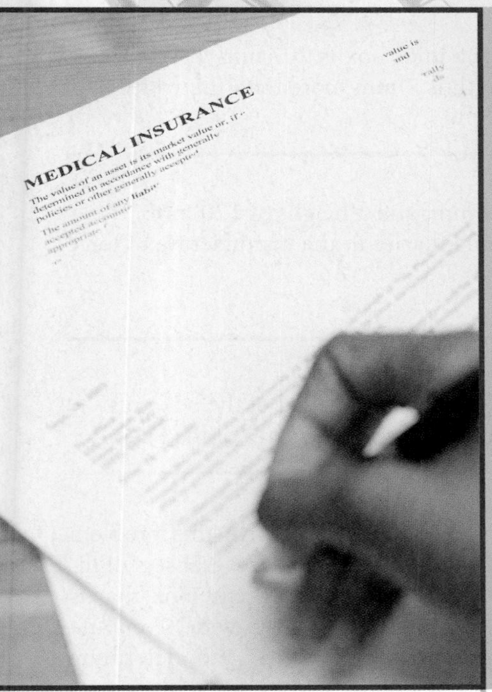

APPLICATION The cost of health insurance continues to rise faster than the rate of inflation. Use the graph below to find the rate at which the annual health insurance premium for a single person was rising between the years 2000 and 2005.

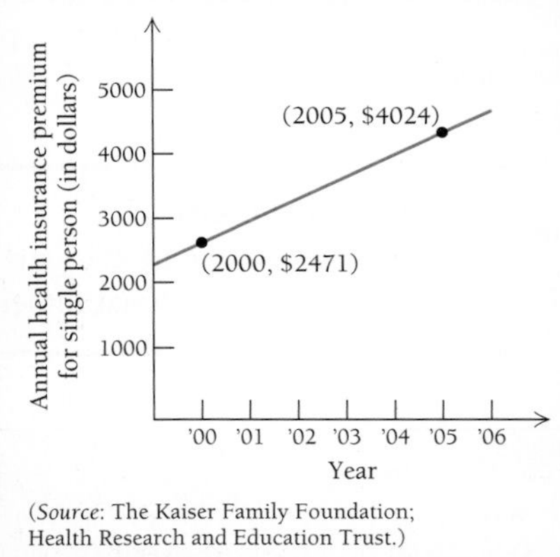

(*Source*: The Kaiser Family Foundation; Health Research and Education Trust.)

This problem appears as Exercise 66 in Exercise Set R.4.

INTRODUCTION

This chapter introduces functions, along with their graphs, notation, and applications. Also presented are many topics that we will consider often throughout the text: supply and demand, total cost, total revenue, total profit, the concept of a mathematical model, and curve fitting.

Skills in using a graphing calculator are also introduced in optional Technology Connections. Details on keystrokes are given in the Graphing Calculator Manual (GCM).

Students needing some algebra review might wish to study the Appendix at the end of the book (on p. 599) along with this chapter.

R.1 Graphs and Equations

OBJECTIVES

- Graph equations.
- Use graphs as mathematical models to make predictions.
- Carry out calculations involving compound interest.

What Is Calculus?

What is calculus? This is a common question at the start of a course like this. Let's consider a simplified answer for now. A more complete answer will evolve throughout the course.

Consider a juice box, as shown at left. The following is a problem that might appear in an algebra course. Try to solve it. (If you need some algebra review, consult the Appendix at the end of the book.)

Algebra Problem

The sum of the height, width, and length of a juice box is 207 mm. If the height is three times the width and the length is 7 mm more than the width, find the dimensions of the box.

The box has a width of 40 mm, a length of 47 mm, and a height of 120 mm.*

The following is a calculus problem that might arise in the manufacture of juice boxes.

Calculus Problem

A juice box is to hold 200 cm^3 (6.75 fl oz) of juice. If the height of the box must be twice the width, what dimensions will minimize the surface area of the box?

Dimensions that assume that the height is twice the width			Total Surface Area,
Width, w	Height, h	Length, l	$2wh + 2lh + 2wl$
5 cm	10 cm	4 cm	220 cm^2
4 cm	8 cm	6.25 cm	214 cm^2 ← Smallest
3 cm	6 cm	$11.\overline{1}$ cm	236 cm^2
2 cm	4 cm	25 cm	316 cm^2

We selected combinations for which $h = 2w$ and the product $w \cdot h \cdot l$ is 200.

One way to solve this problem might be to choose several sets of dimensions for a 200-cm^3 box that is twice as tall as it is wide, compute the resulting areas, and determine which is the least. If you have access to a computer spreadsheet, you might create one and expand the table at left. We let w = width, h = height, l = length, and A = surface area. Then the surface area A is given by

$$A = 2wh + 2lh + 2wl, \quad \text{with } h = 2w.$$

*To find this, let w = width, h = height, and l = length. Then $h = 3w$ and $l = w + 7$, so $w + 3w + w + 7 = 207$. This yields $w = 40$, and thus $h = 120$ and $l = 47$.

From the data in the table, we might conclude that the smallest surface area is 214 cm^2. But how can we be certain that there are no other dimensions that yield a smaller area? We need the tools of calculus to answer this. We will study such maximum–minimum problems in more detail in Chapter 2.

Other topics we will consider in calculus are the slope of a curve at a point, rates of change, area under a curve, accumulations of quantities, and many statistical applications.

Graphs

The study of graphs is an essential aspect of calculus. A graph offers the opportunity to visualize relationships. For instance, the graph below could be used to show how life expectancy has changed over time in the United States. One topic that we consider in calculus is how a change on one axis affects the change on another.

ESTIMATED LIFE EXPECTANCY OF U.S. NEWBORNS BY YEAR OF BIRTH, 1929–2010*

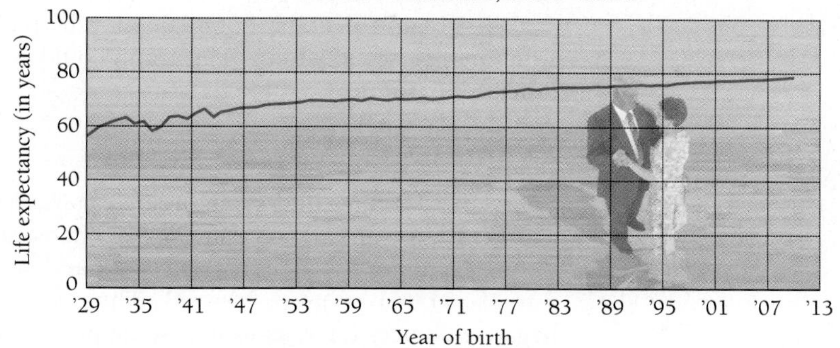

*Figure for 2003 is preliminary; those for 2005 and 2010 are projected.
(*Source*: U.S. National Center for Health Statistics.)

Ordered Pairs and Graphs

Each point in a plane corresponds to an ordered pair of numbers. Note in the figure at the right that the point corresponding to the pair (2, 5) is different from the point corresponding to the pair (5, 2). This is why we call a pair like (2, 5) an **ordered pair.** The first number is called the **first coordinate** of the point, and the second number is called the **second coordinate.** Together these are the *coordinates of the point*. The vertical line is often called the *y-axis*, and the horizontal line is often called the *x-axis*.

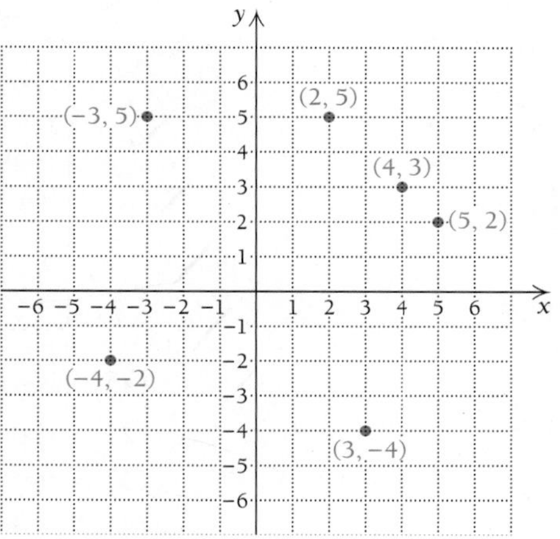

Graphs of Equations

A **solution** of an equation in two variables is an ordered pair of numbers that, when substituted for the variables, forms a true sentence. If not directed otherwise, we usually take the variables in *alphabetical* order. For example, $(-1, 2)$ is a solution of the equation $3x^2 + y = 5$, because when we substitute -1 for x and 2 for y, we get a true sentence:

$$\begin{aligned} 3x^2 + y &= 5 \\ 3(-1)^2 + 2 &\stackrel{?}{=} 5 \\ 3 + 2 &\stackrel{?}{=} 5 \\ 5 &= 5. \quad \text{TRUE} \end{aligned}$$

DEFINITION

The **graph** of an equation is a drawing that represents all ordered pairs that are solutions of the equation.

We obtain the graph of an equation by plotting enough ordered pairs (that are solutions) to see a pattern. The graph could be a line, a curve (or curves), or some other configuration.

EXAMPLE 1 Graph: $y = 2x + 1$.

Solution We first find some ordered pairs that are solutions and arrange them in a table. To find an ordered pair, we can choose any number for x and then determine y. For example, if we choose -2 for x and substitute in $y = 2x + 1$, we find that $y = 2(-2) + 1 = -4 + 1 = -3$. Thus, $(-2, -3)$ is a solution. We make some negative choices for x, as well as some positive choices. If a number takes us off the graph paper, we usually omit the pair from the graph.

x	y	(x, y)
-2	-3	$(-2, -3)$
-1	-1	$(-1, -1)$
0	1	$(0, 1)$
1	3	$(1, 3)$
2	5	$(2, 5)$

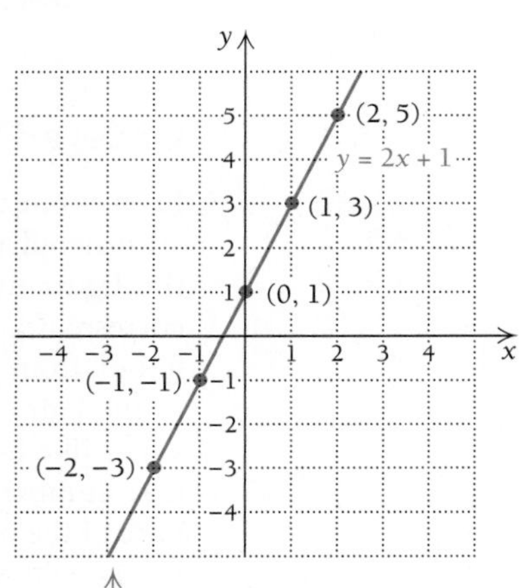

(1) Choose any x.
(2) Compute y.
(3) Form the pair (x, y).
(4) Plot the points.

After we plot the points, we look for a pattern in the graph. If we had enough points, they would make a solid line. We draw the line with a straightedge and label it $y = 2x + 1$. ✦

EXAMPLE 2 Graph: $3x + 5y = 10$.

Solution We could choose x-values, substitute, and solve for y-values, but we first solve for y to ease the calculations.*

$$3x + 5y = 10$$

$$3x + 5y - 3x = 10 - 3x \qquad \text{Subtracting } 3x \text{ from both sides}$$

$$5y = 10 - 3x \qquad \text{Simplifying}$$

$$\tfrac{1}{5} \cdot 5y = \tfrac{1}{5} \cdot (10 - 3x) \qquad \text{Multiplying both sides by } \tfrac{1}{5}\text{, or dividing both sides by 5}$$

$$y = \tfrac{1}{5} \cdot (10) - \tfrac{1}{5} \cdot (3x) \qquad \text{Using the distributive law}$$

$$= 2 - \tfrac{3}{5}x \qquad \text{Simplifying}$$

$$= -\tfrac{3}{5}x + 2$$

Next we use $y = -\frac{3}{5}x + 2$ to find three ordered pairs, choosing multiples of 5 for x to avoid fractions.

x	y	(x, y)
0	2	(0, 2)
5	−1	(5, −1)
−5	5	(−5, 5)

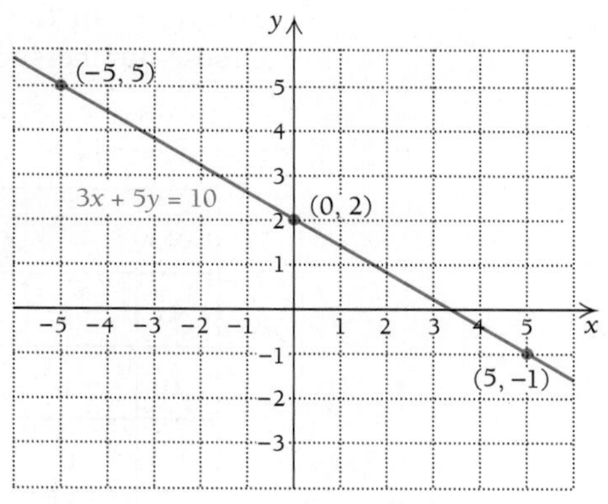

We plot the points, draw the line, and label the graph as shown. ✦

Graphs and properties of linear equations (those whose graphs are straight lines) will be considered in greater detail in Section R.4.

---

*Be sure to consult the Appendix, as needed, for a review of algebra.

EXAMPLE 3 Graph: $y = x^2 - 1$.

Solution

x	y	(x, y)
-2	3	$(-2, 3)$
-1	0	$(-1, 0)$
0	-1	$(0, -1)$
1	0	$(1, 0)$
2	3	$(2, 3)$

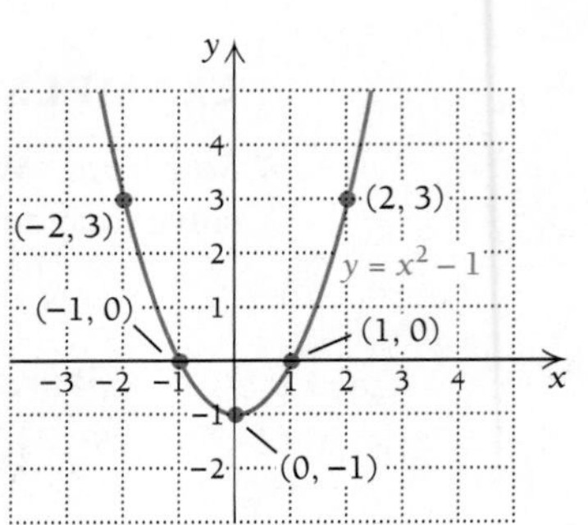

This time the pattern of the points is a curve called a *parabola*. We plot enough points to see a pattern and draw the graph. ✦

EXAMPLE 4 Graph: $x = y^2$.

Solution In this case, x is expressed in terms of the variable y. Thus, we first choose numbers for y and then compute x.

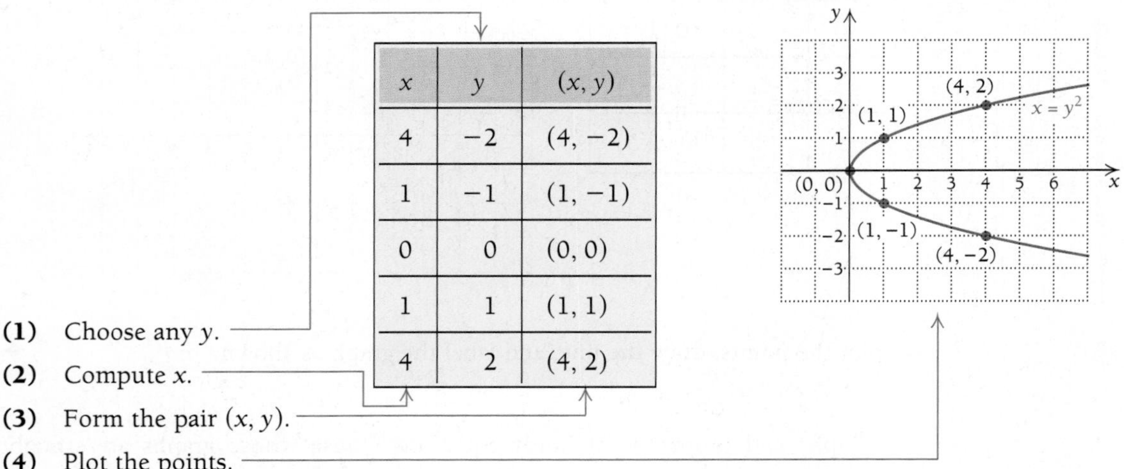

x	y	(x, y)
4	-2	$(4, -2)$
1	-1	$(1, -1)$
0	0	$(0, 0)$
1	1	$(1, 1)$
4	2	$(4, 2)$

We plot these points, keeping in mind that x is still the first coordinate and y the second. We look for a pattern and complete the graph. ✦

TECHNOLOGY CONNECTION

Introduction to the Use of a Graphing Calculator: Windows and Graphs

Viewing Windows

With this first of the optional Technology Connections, we begin to create graphs using a graphing calculator. Most of the coverage will refer to a TI-84 or TI-83 graphing calculator but in a somewhat generic manner, discussing features common to most graphing calculators. Although some keystrokes will be listed, exact keystrokes can be found in the owner's manual for your calculator or in the Graphing Calculator Manual (henceforth referred to as the GCM) that accompanies this text.

One feature common to all graphing calculators is the **viewing window.** This is the rectangular screen in which a graph appears. Windows are described by four numbers, **[L, R, B, T],** which represent the **L**eft and **R**ight endpoints of the x-axis and the **B**ottom and **T**op endpoints of the y-axis. A WINDOW feature can be used to set these dimensions. Below is a window setting of $[-20, 20, -5, 5]$ with axis scaling denoted as Xscl $= 5$ and Yscl $= 1$, which means that there are 5 units between tick marks extending from -20 to 20 on the x-axis and 1 unit between tick marks extending from -5 to 5 on the y-axis.

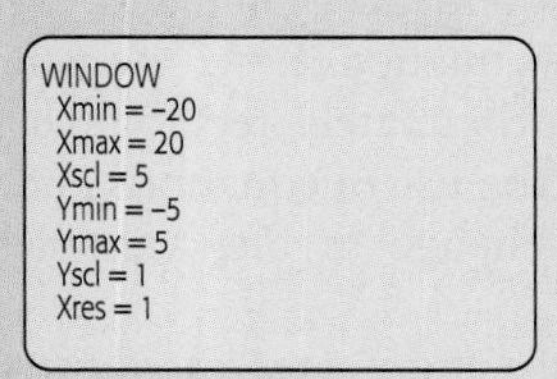
WINDOW
Xmin = –20
Xmax = 20
Xscl = 5
Ymin = –5
Ymax = 5
Yscl = 1
Xres = 1

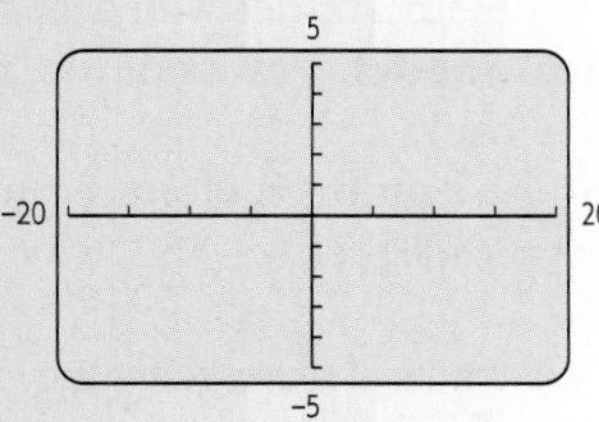

Scales should be chosen with care, since tick marks become blurred and indistinguishable when too many appear. On some graphing calculators, a setting of $[-10, 10, -10, 10]$, Xscl $= 1$, Yscl $= 1$, Xres $= 1$ is considered **standard.**

Graphs are made up of black rectangular dots called **pixels.** The setting Xres allows users to set pixel resolution at 1 through 8 for graphs of equations. At Xres $= 1$, equations are evaluated and graphed at each pixel on the x-axis. At Xres $= 8$, equations are evaluated and graphed at every eighth pixel on the x-axis. The resolution is better for smaller Xres values than for larger values.

Graphs

Let's use a graphing calculator to graph the equation $y = x^3 - 5x + 1$. The equation can be entered using the notation $y = x \wedge 3 - 5x + 1$. We obtain the following graph in the standard viewing window.

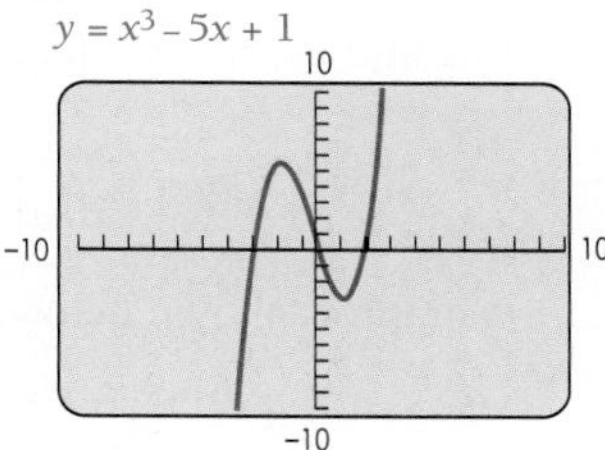

It is often necessary to change viewing windows in order to best reveal the curvature of a graph. For example, each of the following is a graph of $y = 3x^5 - 20x^3$, but with a different viewing window. Which do you think best displays the curvature of the graph?

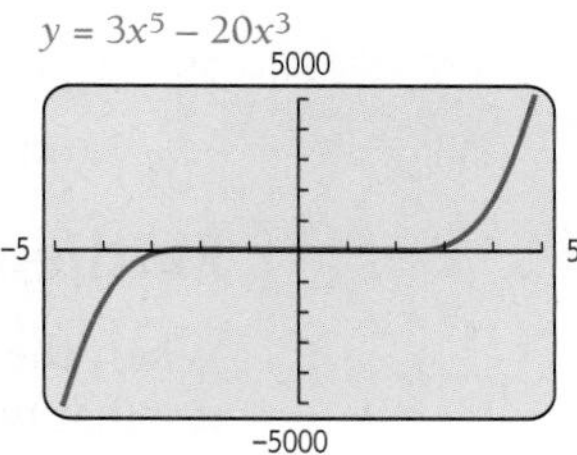

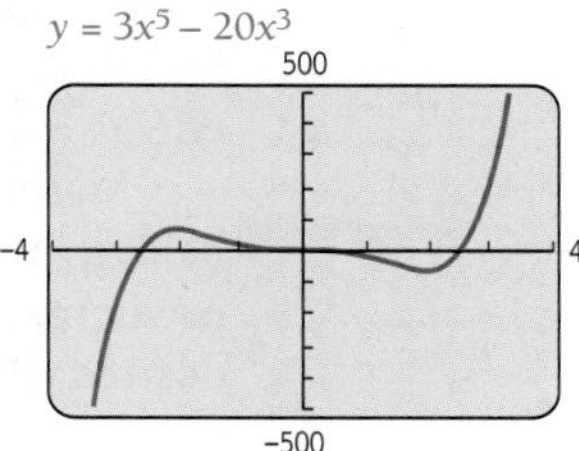

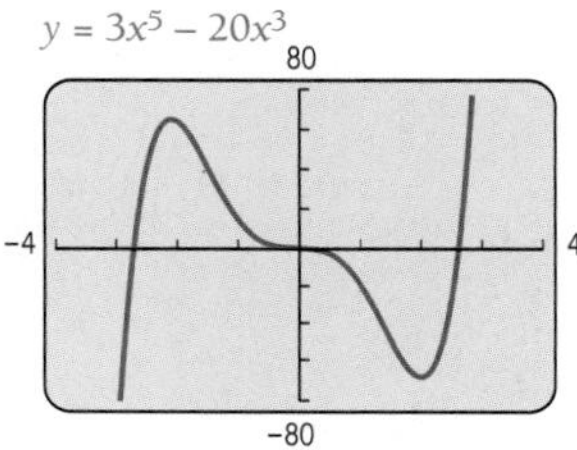

(continued)

Introduction to the Use of a Graphing Calculator: Windows and Graphs (continued)

In general, choosing a window that best reveals a graph's characteristics involves some trial and error and, in some cases, some knowledge about the shape of that graph. We will learn more about the shape of graphs as we continue through the text.

To graph an equation like $3x + 5y = 10$, most calculators require that the equation be solved for y. Thus, we must rewrite and enter the equation as

$$y = \frac{-3x + 10}{5}, \quad \text{or} \quad y = -\frac{3}{5}x + 2.$$

(See Example 2.) Its graph is shown below in the standard window.

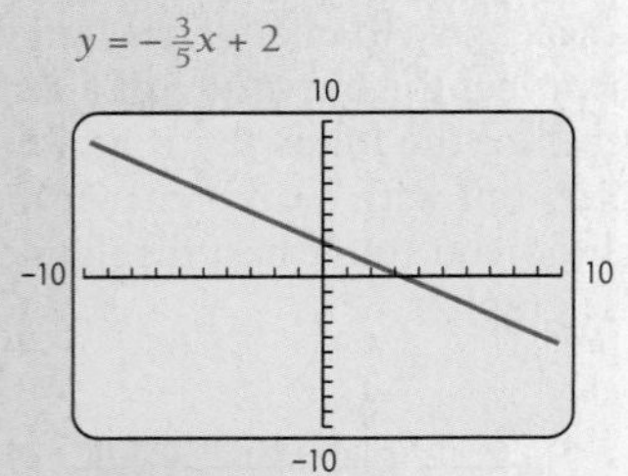

To graph an equation like $x = y^2$, we solve for y and get $y = \sqrt{x}$ or $y = -\sqrt{x}$, which can be written as $y = \pm\sqrt{x}$. We then graph the individual equations $y_1 = \sqrt{x}$ and $y_2 = -\sqrt{x}$.

EXERCISES

Graph each of the following equations. Select the standard window, $[-10, 10, -10, 10]$, with axis scaling Xscl $= 1$ and Yscl $= 1$.

1. $y = x + 3$

2. $y = x - 5$

3. $y = 2x - 1$

4. $y = 3x + 1$

5. $y = -\frac{2}{3}x + 4$

6. $y = -\frac{4}{5}x + 3$

7. $2x - 3y = 18$

8. $5y + 3x = 4$

9. $y = x^2$

10. $y = (x + 4)^2$

11. $y = 8 - x^2$

12. $y = 4 - 3x - x^2$

13. $y + 10 = 5x^2 - 3x$

14. $y - 2 = x^3$

15. $y = x^3 - 7x - 2$

16. $y = x^4 - 3x^2 + x$

17. $y = |x|$ (On most calculators, this is entered as $y = \text{abs}(x)$.)

18. $y = |x - 5|$

19. $y = |x| - 5$

20. $y = 9 - |x|$

Mathematical Models

When the essential parts of a problem are described in mathematical language, we say that we have a **mathematical model.** For example, the equation $A = l \cdot w$ can be used to model the area of a backyard that is only approximately rectangular. Situations in which calculus can be employed often require the use of equations and functions (Section R.2) that typically reflect the way a change in one variable effects a change in another.

Mathematical models are abstracted from real-world situations (see the figure below). Procedures within the mathematical model then give results that allow for predictions in real-world situations. When these predictions are less accurate than desired, the model can be modified or replaced.

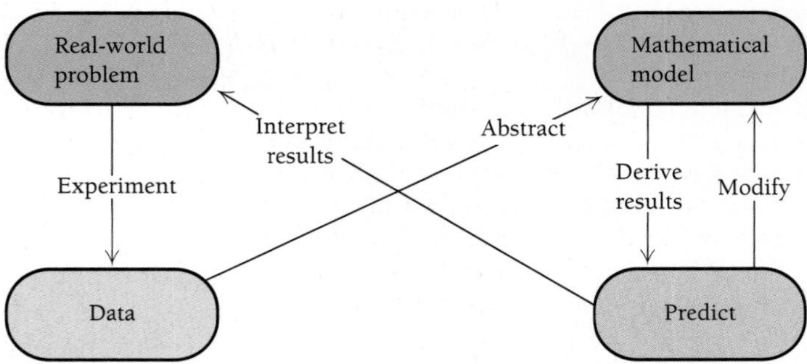

EXAMPLE 5 The graph below shows the numbers of digital photos printed at home from 2000 to 2006.

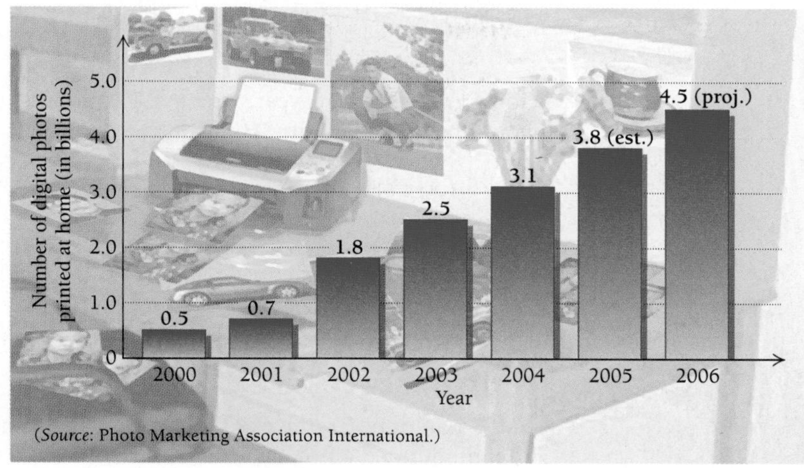

(*Source*: Photo Marketing Association International.)

Use the model $h = 0.7t + 0.3$, where t is the number of years after 2000 and h is the number of digital photos printed at home, in billions, to predict the number of digital photos printed at home in 2008.

Solution Since 2008 is 8 yr after 2000, we substitute, using $t = 8$:

$$h = 0.7t + 0.3 = 0.7(8) + 0.3 = 5.9.$$

According to this model, in 2008, approximately 5.9 billion digital photos will be printed at home. ✦

As is usually the case with models, the model in Example 5 is not perfect. For example, for $t = 1$, we have $h = 1.0$, a value substantially different from the 0.7 given in the original data. To better match the given data, we could use another model or provide an argument in support of using the given model for predictions. For instance, the model $h = 0.013t^4 - 0.166t^3 + 0.676t^2 - 0.199t + 0.476$ can be used to obtain values that nearly match the given data. However, this model (obtained from a graphing calculator in a manner described later in this text) is quite cumbersome and is probably inappropriate, since it predicts the value of h will double between 2006 and 2008.

One important model that is extremely precise involves **compound interest.** Suppose that we invest P dollars at interest rate i, expressed as a decimal and compounded annually. The amount A_1 in the account at the end of the first year is given by

$$A_1 = P + Pi$$ The original amount invested, P, is called the **principal**.

$$= P(1 + i) = Pr,$$

where, for convenience, we let

$$r = 1 + i.$$

Going into the second year, we have Pr dollars, so by the end of the second year, we would have the amount A_2 given by

$$\begin{aligned} A_2 &= A_1 \cdot r = (Pr)r \\ &= Pr^2 = P(1 + i)^2. \end{aligned}$$

Going into the third year, we have Pr^2 dollars, so by the end of the third year, we would have the amount A_3 given by

$$\begin{aligned} A_3 &= A_2 \cdot r = (Pr^2)r \\ &= Pr^3 = P(1 + i)^3. \end{aligned}$$

In general, we have the following theorem.

THEOREM 1

If an amount P is invested at interest rate i, expressed as a decimal and compounded annually, in t years it will grow to the amount A given by

$$A = P(1 + i)^t.$$

EXAMPLE 6 Business: Compound Interest. Suppose that \$1000 is invested at 8%, compounded annually. How much is in the account at the end of 2 yr?

Solution We substitute 1000 for P, 0.08 for i, and 2 for t into the equation $A = P(1 + i)^t$ and get

$$\begin{aligned} A &= 1000(1 + 0.08)^2 \\ &= 1000(1.08)^2 \\ &= 1000(1.1664) \\ &= \$1166.40. \end{aligned}$$

There is \$1166.40 in the account after 2 years. ✦

For interest that is compounded quarterly (four times per year), we can find a formula like the one above, as illustrated in the following diagram.

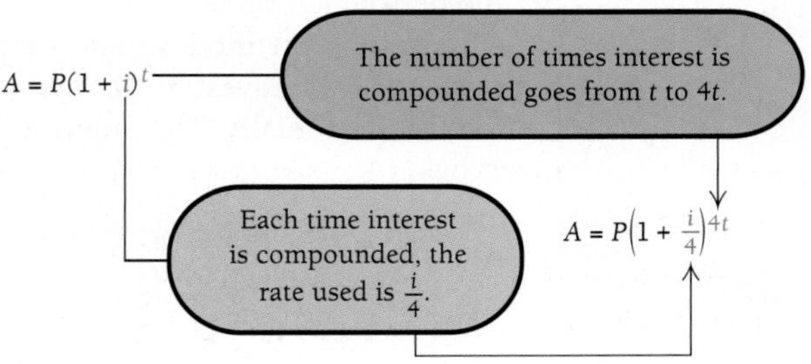

In general, the following theorem applies.

THEOREM 2

If a principal P is invested at interest rate i, expressed as a decimal and compounded n times a year, in t years it will grow to an amount A given by

$$A = P\left(1 + \frac{i}{n}\right)^{nt}.$$

EXAMPLE 7 Business: Compound Interest. Suppose that \$1000 is invested at 8%, compounded quarterly. How much is in the account at the end of 3 yr?

Solution We use the equation $A = P(1 + i/n)^{nt}$, substituting 1000 for P, 0.08 for i, 4 for n (compounding quarterly), and 3 for t. Then we get

$$\begin{aligned} A &= 1000\left(1 + \frac{0.08}{4}\right)^{4\cdot 3} \\ &= 1000(1 + 0.02)^{12} \\ &= 1000(1.02)^{12} \\ &= 1000(1.268241795) && \text{Using a calculator to approximate } (1.02)^{12} \\ &= 1268.241795 \\ &\approx \$1268.24. && \text{The symbol} \approx \text{means "approximately equal to."} \end{aligned}$$

There is \$1268.24 in the account after 3 yr. ◆

A calculator with a y^x or a ^ key and a ten-digit readout was used to find $(1.02)^{12}$ in Example 7. The number of places on a calculator may affect the accuracy of the answer. Thus, you may occasionally find that your answers do not agree with those at the back of the book, which were found on a calculator with a ten-digit readout. In general, when using a calculator, do all computations and round only at the end, as in Example 7. Usually, your answer will agree to at least four digits. It is usually wise to consult with your instructor on the accuracy required.

Exercise Set R.1

Exercises designated by the symbol TW *are Thinking and Writing Exercises. They should be answered using one or two English sentences. Because answers to many such exercises will vary, solutions are not given at the back of the book.*

Graph. (Unless directed otherwise, assume that "Graph" means "Graph by hand.")

1. $y = x + 4$

2. $y = x - 1$

3. $y = -3x$

4. $y = -\frac{1}{4}x$

5. $y = \frac{2}{3}x - 4$

6. $y = -\frac{5}{3}x + 3$

7. $x + y = 5$

8. $x - y = 4$

9. $8y - 2x = 4$

10. $6x + 3y = -9$

11. $5x - 6y = 12$

12. $2x + 5y = 10$

13. $y = x^2 - 5$

14. $y = x^2 - 3$

15. $x = 2 - y^2$

16. $x = y^2 + 2$

17. $y = |x|$

18. $y = |4 - x|$

19. $y = 7 - x^2$

20. $y = 5 - x^2$

21. $y + 1 = x^3$

22. $y - 7 = x^3$

APPLICATIONS

23. Running records. According to at least one study, the world record in any running race can be modeled by a linear equation. In particular, the world record R, in minutes, for the mile run in year x can be modeled by

$$R = -0.00582x + 15.3476.$$

Use this model to estimate the world records for the mile run in 1954 and 2008.

24. Medicine. Ibuprofen is a medication used to relieve pain. The function

$$A = 0.5t^4 + 3.45t^3 - 96.65t^2 + 347.7t, \quad 0 \leq t \leq 6,$$

can be used to estimate the number of milligrams, A, of ibuprofen in the bloodstream t hours after 400 mg of the medication has been swallowed. (*Source:* Based on data from Dr. P. Carey, Burlington, VT.) How many milligrams of ibuprofen are in the bloodstream 2 hr after 400 mg has been swallowed?

25. Freestyle skiing. During the 2004 Winter X-Games, Simon Dumont established a world record by launching himself 22 ft above the edge of a halfpipe. His speed $v(t)$, in miles per hour, upon reentering the pipe can be approximated by $v(t) = 10.9t$, where t is the number of seconds for which he was airborne. Dumont was airborne for approximately 2.34 sec. (*Source:* Based on data from EXPN.com.) How fast was he going when he reentered the halfpipe?

26. Skateboarding. The distance $s(t)$, in feet, traveled by a body falling freely from rest in t seconds is approximated by

$$s(t) = 16t^2.$$

In 2003, Adil Dyani set a record for the highest skateboard drop into a quarterpipe. Dyani was airborne for about 1.36 sec. (*Source:* Based on data from *Guinness Book of World Records 2005.*) How far did he drop?

27. Hearing-impaired Americans. The number N, in millions, of hearing-impaired Americans of age x can be approximated by the graph below. (*Source:* American Speech-Language Hearing Association.)

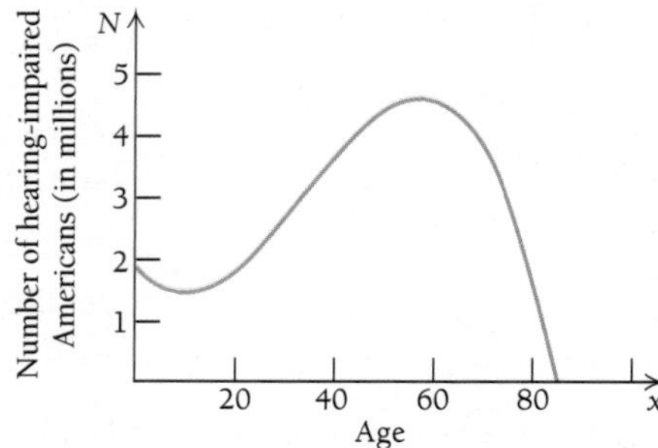

Use the graph to answer the following.

a) Approximate the number of hearing-impaired Americans of ages 20, 40, 50, and 60.

b) For what ages is the number of hearing-impaired Americans approximately 4 million?

c) Examine the graph and try to determine the age at which the greatest number of Americans is hearing-impaired.

TW **d)** What difficulty do you have in making this determination?

28. Life science: incidence of breast cancer. The following graph approximates the incidence of breast cancer y, per 100,000 women, as a function of age x, where x represents ages 25 to 102.

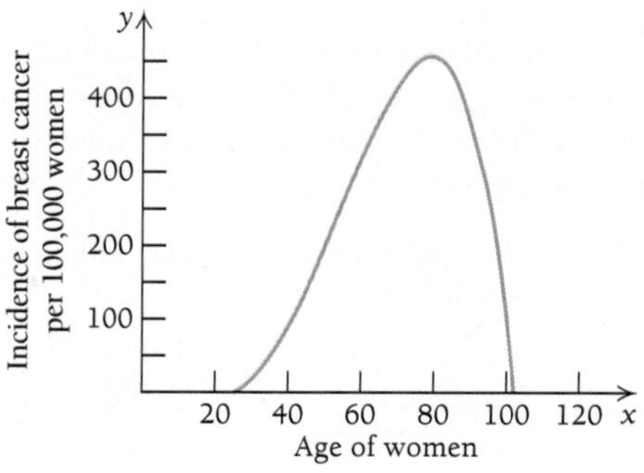

(*Source:* National Cancer Institute.)

a) What is the incidence of breast cancer in 40-yr-old women?

b) For what ages is the incidence of breast cancer about 400 per 100,000 women?

c) Examine the graph and try to determine the age at which the largest incidence of breast cancer occurs.

TW d) What difficulty do you have making this determination?

29. Compound interest. Suppose that \$1000 is invested at 6%. How much is in the account at the end of 1 yr, if interest is compounded:

a) annually? b) semiannually?
c) quarterly? d) daily?
e) hourly?

30. Compound interest. Suppose that \$1000 is invested at $8\frac{1}{2}\%$. How much is in the account at the end of 1 yr, if interest is compounded:

a) annually? b) semiannually?
c) quarterly? d) daily?
e) hourly?

31. Compound interest. Suppose that \$2000 is invested at 4%. How much is in the account at the end of 3 yr, if interest is compounded:

a) annually? b) semiannually?
c) quarterly? d) daily?
e) hourly?

32. Compound interest. Suppose that \$3000 is invested at 5%. How much is in the account at the end of 4 yr, if interest is compounded:

a) annually? b) semiannually?
c) quarterly? d) daily?
e) hourly?

Determining monthly payments on a loan. *If P dollars are borrowed, the monthly payment M, made at the end of each month for n months, is given by*

$$M = P\frac{\frac{i}{12}\left(1 + \frac{i}{12}\right)^n}{\left(1 + \frac{i}{12}\right)^n - 1},$$

where i is the annual interest rate and n is the total number of monthly payments.

33. A car loan is \$18,000, the interest rate is $9\frac{3}{4}\%$, and the loan period is 3 yr. What is the monthly payment?

34. The mortgage on a house is \$100,000, the interest rate is $6\frac{3}{4}\%$, and the loan period is 30 yr. What is the monthly payment?

Annuities. *If P dollars are invested annually into an annuity, after n years, the annuity will be worth*

$$W = P\left[\frac{(1 + i)^n - 1}{i}\right],$$

where i is the annual interest rate.

35. Suppose that \$3000 is invested annually into an annuity that earns 8% interest. How much is the annuity worth after 18 yr? Round to the nearest dollar.

36. Suppose that you establish an annuity that earns $7\frac{1}{4}\%$ interest, and you want it to be worth \$50,000 in 20 yr. How much will you need to invest annually to achieve this goal?

37. Deer population in California. The deer population in California from 1800 to 2000 is approximated in the graph below. The term "abundant" refers to populations above 700,000, "common" refers to populations between 400,000 and 700,000, and "scarce" refers to populations below 400,000. (*Source:* California Department of Fish and Game.)

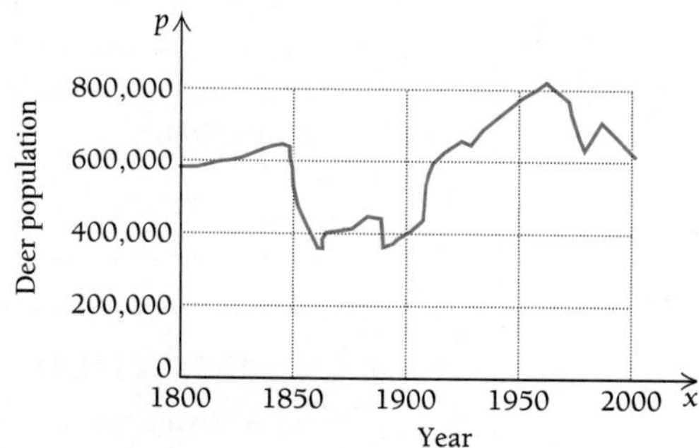

a) For which years were deer common?

b) For which years were deer abundant?

c) For which years were deer scarce?

38. Use the graph in Exercise 37 to answer the following.

a) In approximately what year was the California deer population at its highest?

b) In approximately what year was the California deer population at its lowest?

TW c) In 1848, gold was discovered in California. What does the graph show about the deer population in California in the years just after gold was discovered?

TECHNOLOGY CONNECTION

The Technology Connection heading indicates exercises designed to provide practice using a graphing calculator.

Graph.

39. $y = x - 150$

40. $y = 25 - |x|$

41. $y = x^3 + 2x^2 - 4x - 13$

42. $y = \sqrt{23 - 7x}$

43. $9.6x + 4.2y = -100$

44. $y = -2.3x^2 + 4.8x - 9$

45. $x = 4 + y^2$

46. $x = 8 - y^2$

R.2 Functions and Models

OBJECTIVES

- Determine whether or not a correspondence is a function.
- Find function values.
- Graph functions and determine whether or not a graph is that of a function.
- Graph functions that are piecewise-defined.

Identifying Functions

The idea of a *function* is one of the most important concepts in mathematics. Put simply, a function is a special kind of correspondence between two sets. Let's look at the following.

To each letter on a telephone keypad there corresponds a number.

To each CD in a music store there corresponds its price.

To each real number there corresponds the cube of that number.

In each of these examples, the first set is called the *domain* and the second set is called the *range*. Given a member of the domain, there is *exactly one* member of the range to which it corresponds. This type of correspondence is called a *function*.

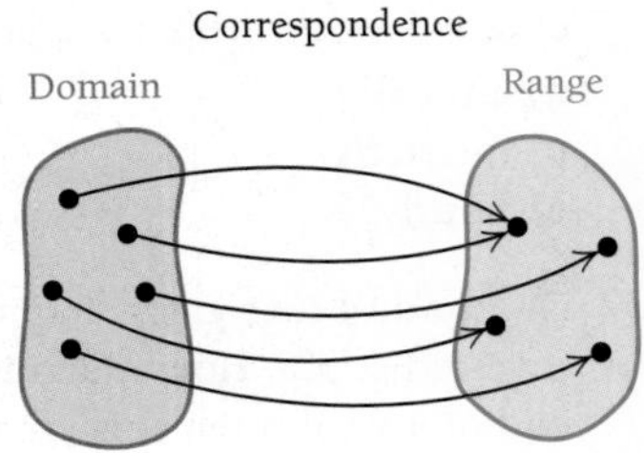

DEFINITION

A **function** is a correspondence between a first set, called the **domain**, and a second set, called the **range**, such that each member of the domain corresponds to *exactly one* member of the range.

EXAMPLE 1 Determine whether or not each correspondence is a function.

a) *Microsoft Stock* (*Source:* finance.yahoo.com.)

Domain		Range
March 14, 2006	⟶	\$27.23
March 15, 2006	⟶	\$27.36
March 16, 2006	⟶	\$27.27
March 17, 2006	⟶	\$27.50

b) *Squaring*

Domain		Range
3	⟶	9
4	⟶	16
5	↘	25
−5	⟶	25

c) *Baseball Teams*

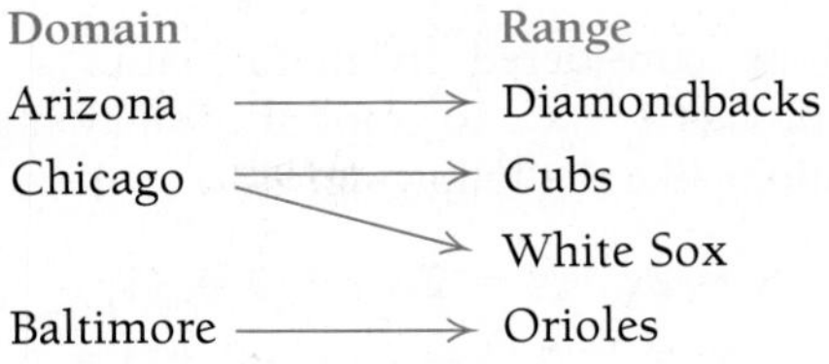

Domain		Range
Arizona	→	Diamondbacks
Chicago	→	Cubs
	↘	White Sox
Baltimore	→	Orioles

d) *Baseball Teams*

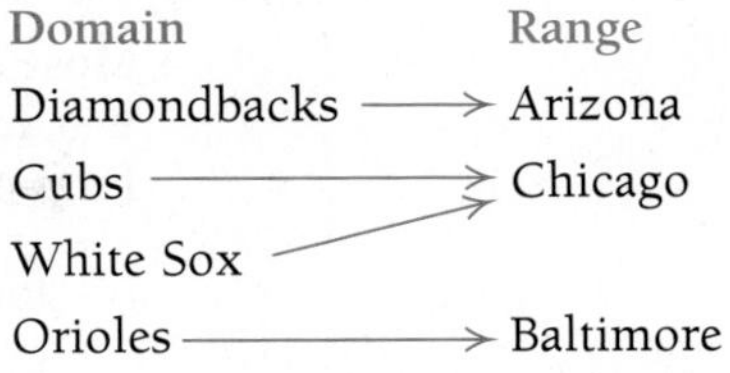

Domain		Range
Diamondbacks	→	Arizona
Cubs	→	Chicago
White Sox	↗	
Orioles	→	Baltimore

Solution

a) The correspondence *is* a function because each member of the domain corresponds (is matched) to only one member of the range.

b) The correspondence *is* a function because each member of the domain corresponds to only one member of the range, even though two members of the domain correspond to 25.

c) The correspondence *is not* a function because one member of the domain, Chicago, corresponds to two members of the range, the Cubs and the White Sox.

d) The correspondence *is* a function because each member of the domain corresponds to only one member of the range, even though two members of the domain correspond to Chicago. ✦

EXAMPLE 2 Determine whether or not each correspondence is a function.

	Domain	Correspondence	Range
a)	A family	Each person's weight	A set of positive numbers
b)	The integers $\{\ldots, -3, -2, -1, 0, 1, 2, 3, \ldots\}$	Each number's square	A set of nonnegative integers: $\{0, 1, 4, 9, 16, 25, \ldots\}$
c)	The set of all states	Each state's members of the U.S. Senate	The set of all U.S. senators

Solution

a) The correspondence is a function because each person has *only one* weight.

b) The correspondence is a function because each integer has *only one* square.

c) The correspondence is *not* a function because each state has *two* U.S. Senators. ✦

Consistent with the definition above, we will regard a function as a set of ordered pairs, such that no two pairs have the same first coordinate paired with different second coordinates. The domain is the set of all first coordinates, and the range is the set of all second coordinates. Function names are usually represented by lowercase letters. Thus, if f represents the function in Example 1(b), we have

$$f = \{(3, 9), (4, 16), (5, 25), (-5, 25)\}$$

and

$$\text{Domain of } f = \{3, 4, 5, -5\}; \quad \text{Range of } f = \{9, 16, 25\}.$$

Finding Function Values

Most functions considered in mathematics are described by equations like $y = 2x + 3$ or $y = 4 - x^2$. To graph the function given by $y = 2x + 3$, we can perform calculations like the following:

for $x = 4$, $y = 2x + 3 = 2 \cdot 4 + 3 = 11$; The graph includes (4, 11).
for $x = -5$, $y = 2x + 3 = 2 \cdot (-5) + 3 = -7$; The graph includes (−5, −7).
for $x = 0$, $y = 2x + 3 = 2 \cdot 0 + 3 = 3$; and so on. The graph includes (0, 3).

For $y = 2x + 3$, the **inputs** (members of the domain) are the values of x substituted into the equation. The **outputs** (members of the range) are the resulting values of y. If we call the function f, we can use x to represent an arbitrary input and $f(x)$, read "f of x" or "f at x" or "the value of f at x," to represent the corresponding output. In this notation, the function given by $y = 2x + 3$ is written as $f(x) = 2x + 3$, and the calculations above can be written more concisely as

$$f(4) = 2 \cdot 4 + 3 = 11;$$
$$f(-5) = 2 \cdot (-5) + 3 = -7;$$
$$f(0) = 2 \cdot 0 + 3 = 3; \text{ and so on.}$$

Thus, instead of writing "when $x = 4$, the value of y is 11," we can simply write "$f(4) = 11$," which is most commonly read as "f of 4 is 11."

It helps to think of a function as a machine. Think of $f(4) = 11$ as the result of putting a member of the domain (an input), 4, into the machine. The machine knows the correspondence $f(x) = 2x + 3$, computes $2 \cdot 4 + 3$, and produces a member of the range (the output), 11.

Function: $f(x) = 2x + 3$	
Input	Output
4	11
−5	−7
0	3
t	$2t + 3$
$a + h$	$2(a + h) + 3$

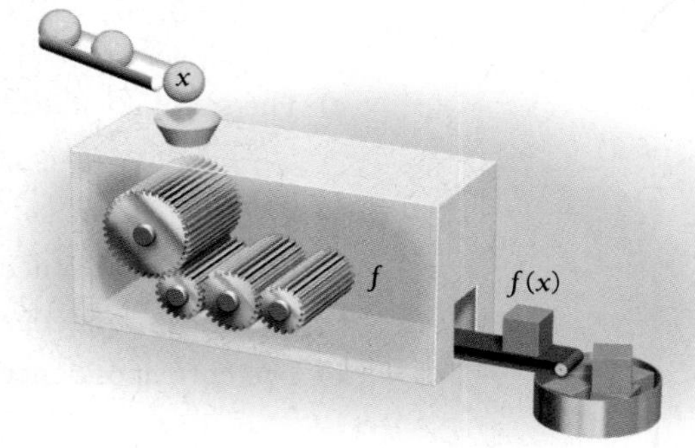

Please remember that $f(x)$ *does not mean* "f times x" and should never be read that way.

EXAMPLE 3 The squaring function f is given by

$$f(x) = x^2.$$

Find $f(-3)$, $f(1)$, $f(k)$, $f(\sqrt{k})$, $f(1 + t)$, and $f(x + h)$.

Solution We have

$$f(-3) = (-3)^2 = 9;$$
$$f(1) = 1^2 = 1;$$
$$f(k) = k^2;$$
$$f(\sqrt{k}) = (\sqrt{k})^2 = k;$$
$$f(1 + t) = (1 + t)^2 = 1 + 2t + t^2;$$

For a review of algebra, see the Appendix on p. 599.

$$f(x + h) = (x + h)^2 = x^2 + 2xh + h^2.$$

To find $f(x + h)$, remember what the function does: It squares the input. Thus, $f(x + h) = (x + h)^2 = x^2 + 2xh + h^2$. This amounts to replacing x on both sides of $f(x) = x^2$ with $x + h$. ✦

EXAMPLE 4 A function f is given by $f(x) = 3x^2 - 2x + 8$. Find $f(0)$, $f(-5)$, and $f(7a)$.

Solution One way to find function values when a formula is given is to think of the formula with blanks, or placeholders, as follows:

$$f(\quad) = 3\quad^2 - 2\quad + 8.$$

To find an output for a given input, we think: "Whatever goes in the blank on the left goes in the blank(s) on the right."

$$f(0) = 3 \cdot 0^2 - 2 \cdot 0 + 8 = 8$$
$$f(-5) = 3(-5)^2 - 2 \cdot (-5) + 8 = 3 \cdot 25 + 10 + 8 = 75 + 10 + 8 = 93$$
$$f(7a) = 3(7a)^2 - 2(7a) + 8 = 3 \cdot 49a^2 - 14a + 8 = 147a^2 - 14a + 8$$ ✦

TECHNOLOGY CONNECTION

The TABLE Feature

One way to find ordered pairs of inputs and outputs of functions is to use the TABLE feature. To see how, consider the function given by $f(x) = x^3 - 5x + 1$. We enter it as $y_1 = x^3 - 5x + 1$. To use the TABLE feature, we access the TABLE SETUP screen and enter the x-value at which the table will start and an increment for the x-value. For this equation, let's set TblStart = 0.3 and ΔTbl = 1. (Other values can be chosen.) This means that the table's x-values will start at 0.3 and increase by 1.

```
TABLE SETUP
  TblStart = .3
  ΔTbl = 1
Indpnt:  Auto  Ask
Depend:  Auto  Ask
```

We next set Indpnt and Depend to Auto and then press TABLE. The result is shown below.

X	Y1
.3	−.473
1.3	−3.303
2.3	1.667
3.31	20.437
4.3	59.007
5.3	123.38
6.3	219.55

X = .3

The arrow keys, ⌃ and ⌄, allow us to scroll up and down the table and extend it to other values not initially shown.

X	Y1
12.3	1800.4
13.3	2287.1
14.3	2853.7
15.3	3506.1
16.3	4250.2
17.3	5092.2
18.3	6038

X = 18.3

(*continued*)

The TABLE Feature (continued)

EXERCISES

Use the function given by $f(x) = x^3 - 5x + 1$ for Exercises 1 and 2.

1. Use the TABLE feature to construct a table starting with $x = 10$ and $\Delta\text{Tbl} = 5$. Find the value of y when x is 10. Then find the value of y when x is 35.
2. Adjust the table settings to Indpnt: Ask. How does the table change? Enter a number of your choice and see what happens. Use this setting to find the value of y when x is 28.

EXAMPLE 5 A function f subtracts the square of an input from the input. A description of f is given by

$$f(x) = x - x^2.$$

Find $f(4)$, $f(x + h)$, and $\dfrac{f(x + h) - f(x)}{h}$.

Solution We have

$$\begin{aligned}
f(4) &= 4 - 4^2 = 4 - 16 = -12; \\
f(x + h) &= (x + h) - (x + h)^2 \\
&= x + h - (x^2 + 2xh + h^2) \\
&= x + h - x^2 - 2xh - h^2; \\
\frac{f(x + h) - f(x)}{h} &= \frac{x + h - x^2 - 2xh - h^2 - (x - x^2)}{h} = \frac{h - 2xh - h^2}{h} \\
&= \frac{h(1 - 2x - h)}{h} \\
&= 1 - 2x - h, \quad \text{for } h \neq 0.
\end{aligned}$$

✦

Graphs of Functions

Consider again the squaring function. The input 3 is associated with the output 9. The input–output pair (3, 9) is one point on the *graph* of this function.

DEFINITION

The **graph** of a function f is a drawing that represents all the input–output pairs $(x, f(x))$. In cases where the function is given by an equation, the graph of a function is the graph of the equation $y = f(x)$.

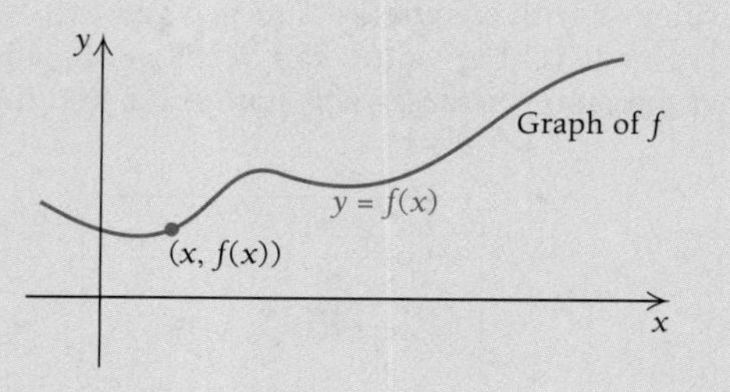

It is customary to locate input values (the domain) on the horizontal axis and output values (the range) on the vertical axis.

EXAMPLE 6 Graph: $f(x) = x^2 - 1$.

Solution

x	$f(x)$	$(x, f(x))$
-2	3	$(-2, 3)$
-1	0	$(-1, 0)$
0	-1	$(0, -1)$
1	0	$(1, 0)$
2	3	$(2, 3)$

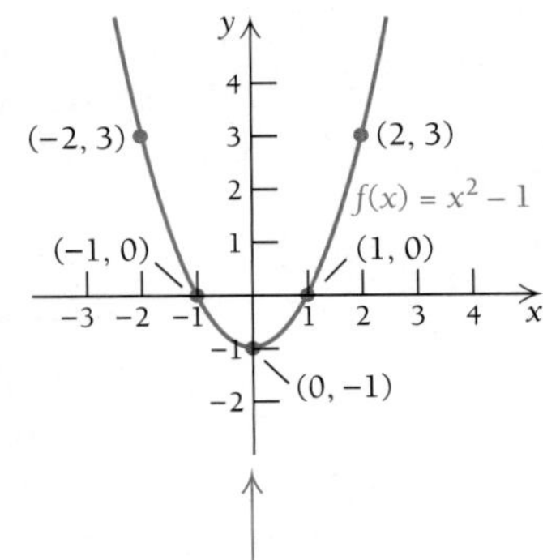

(1) Choose any x.
(2) Compute y.
(3) Form the pair (x, y).
(4) Plot the points.

We plot the input–output pairs from the table and, in this case, draw a curve to complete the graph. ✦

TECHNOLOGY CONNECTION

Graphs and Function Values

We discussed graphing equations in the Technology Connection of Section R.1. Graphing a function makes use of the same procedure. We just change the "$f(x) =$ " notation to "$y =$." Thus, to graph $f(x) = 2x^2 + x$, we key in $y_1 = 2x^2 + x$.

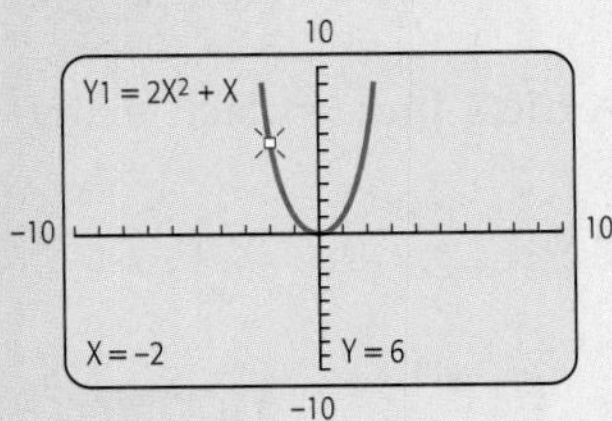

There are several ways in which to find function values. One is to use the TABLE feature, as previously discussed. Another way is to use the TRACE feature. To do so, simply graph the function, press TRACE, and either move the cursor or enter any x-value that is in the window. The corresponding y-value appears automatically. Function values can also be found using the VALUE or Y-VARS feature. Consult an owner's manual or the GCM for details.

EXERCISES

1. Graph $f(x) = x^2 + 3x - 4$. Then find $f(-5)$, $f(-4.7)$, $f(11)$, and $f(2/3)$. (*Hint*: To find $f(11)$, be sure that the window dimensions for the x-values include $x = 11$.)
2. Graph $f(x) = 3.7 - x^2$. Then find $f(-5)$, $f(-4.7)$, $f(11)$, and $f(2/3)$.
3. Graph $f(x) = 4 - 1.2x - 3.4x^2$. Then find $f(-5)$, $f(-4.7)$, $f(11)$, and $f(2/3)$.

The Vertical-Line Test

Let's now determine how we can look at a graph and decide whether it is a graph of a function. In the graph at the right, note that the input x_1 has *two* outputs. Since a function has exactly *one* output for every input, this fact means that the graph does not represent a function. It also means that a vertical line could intersect the graph in more than one place.

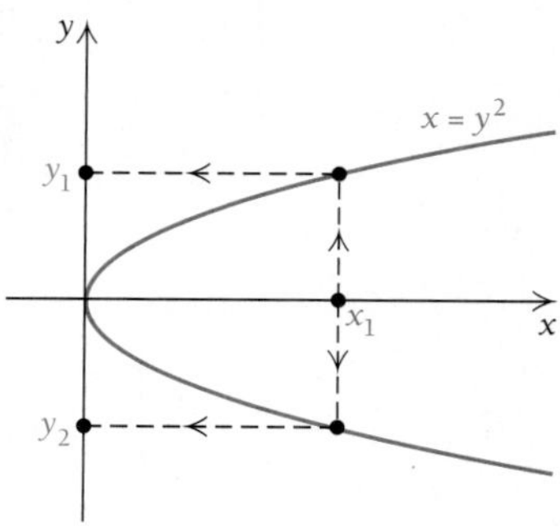

The Vertical-Line Test
A graph represents a function if it is impossible to draw a vertical line that intersects the graph more than once.

EXAMPLE 7 Determine whether each of the following is the graph of a function.

a)

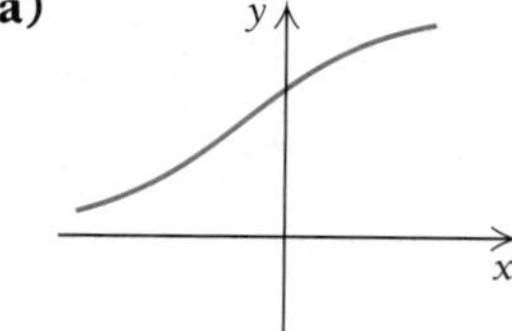

b)

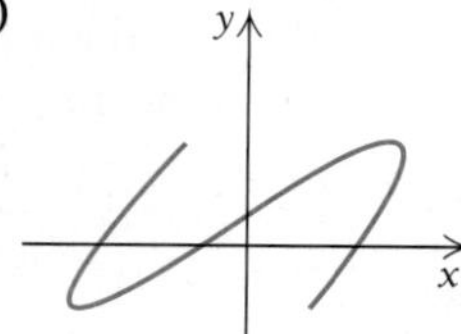

c)

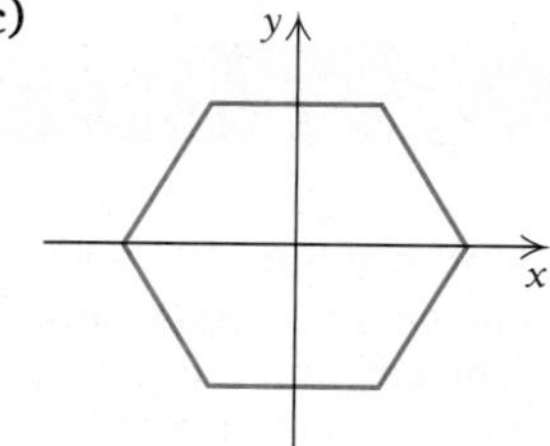

d)

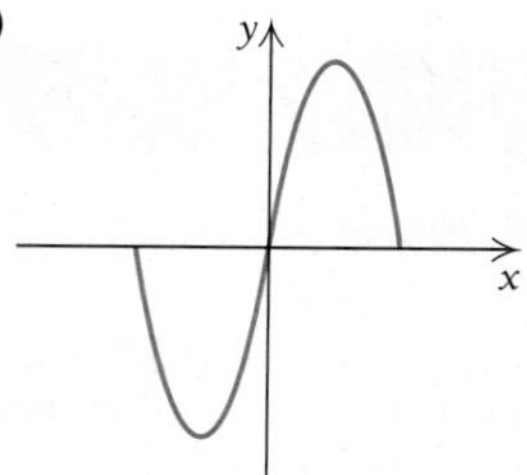

Solution

a) The graph is that of a function. It is impossible to draw a vertical line that intersects the graph more than once.

b) The graph is not that of a function. A vertical line (in fact, many) can intersect the graph more than once.

c) The graph is not that of a function.

d) The graph is that of a function. ✦

Functions Defined Piecewise

Sometimes functions are defined *piecewise.* That is, there are different output formulas for different parts of the domain. You probably have some experience with piecewise-defined functions, although you may not be aware of it. Consider the following real-world application.

EXAMPLE 8 Cellphone Calling Plans. In 2005, Sprint® offered a cellphone calling plan in which a customer's monthly bill can be modeled by the graph below. The amount of the bill is a function f of the number of minutes of phone use.

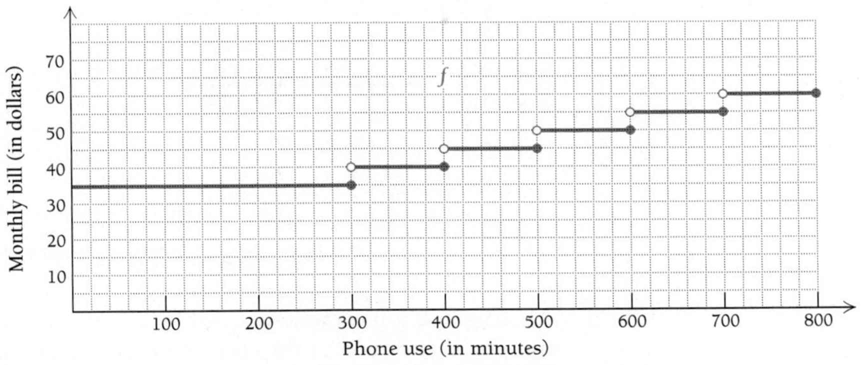

(*Source*: Based on data from *The New York Times*, 9/3/05, p. D8. Bill does not include taxes and fees.)

Each open dot on the graph indicates that the point at that location is not included in the graph.

a) Under this plan, if a customer uses the phone for 360 min, what is his or her monthly bill?

b) If a monthly bill is \$55, for how many minutes did the customer use the phone?

Solution

a) To find the bill for 360 min of use, we locate 360 on the horizontal axis and move directly up to the graph. We then move across to the vertical axis. The bill is

$$f(360) = 40 \quad \text{or} \quad \$40.$$

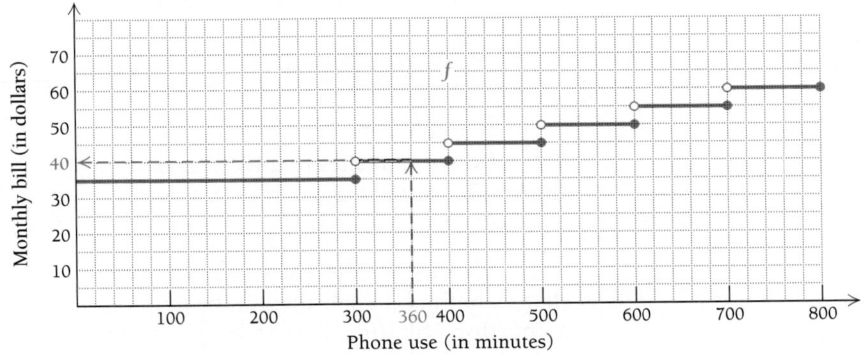

b) To find the number of minutes of use when a monthly bill is \$55, we locate 55 on the vertical axis, move horizontally to the graph, and note that many inputs correspond to 55. For example,

$$f(620) = f(660) = f(700) = 55.$$

A bill of \$55 indicates that the customer used the phone more than 600 min, but not more than 700 min. If t represents the number of minutes of use, we must have $600 < t \leq 700$.

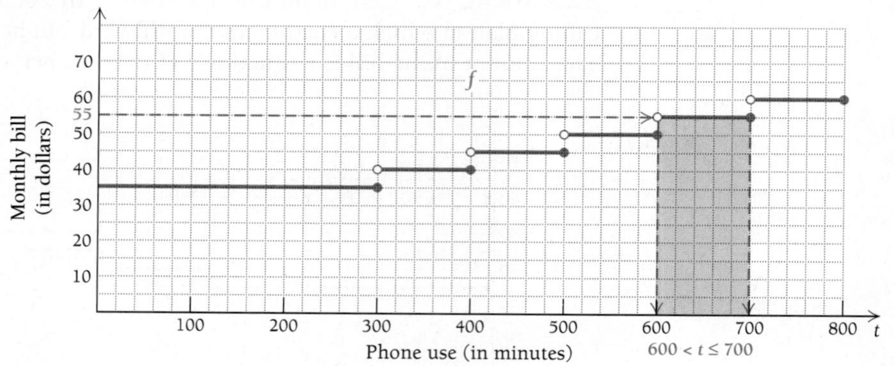

To graph a piecewise-defined function, we usually work from left to right, paying special attention to the correspondence specified for the x-values on each part of the horizontal axis.

EXAMPLE 9 Graph the function defined as follows:

$$f(x) = \begin{cases} 4, & \text{for } x \leq 0, \\ 3 - x^2, & \text{for } 0 < x \leq 2, \\ 2x - 6, & \text{for } x > 2. \end{cases}$$

This means that for any input x less than or equal to 0, the output is 4.

This means that for any input x greater than 0 and less than or equal to 2, the output is $3 - x^2$.

This means that for any input x greater than 2, the output is $2x - 6$.

Solution Working from left to right along the x-axis, we note that for any x-values less than or equal to 0, the graph is the horizontal line $y = 4$. Note that for $f(x) = 4$,

$$f(-2) = 4;$$
$$f(-1) = 4;$$
$$f(0) = 4.$$

The solid dot indicates that $(0, 4)$ is part of the graph.

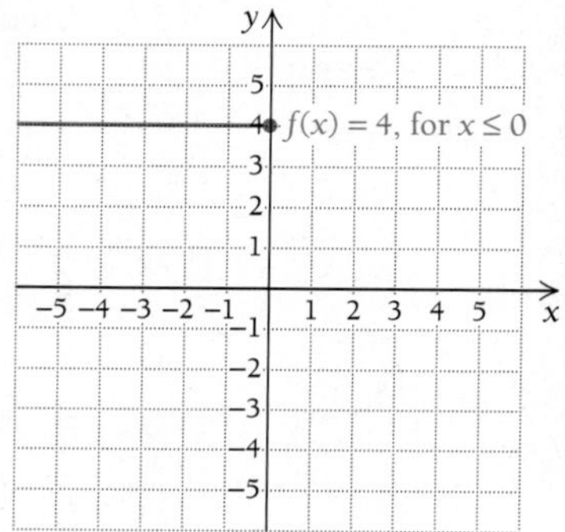

Next, observe that for x-values greater than 0, but not greater than 2, the graph is a portion of the parabola given by $y = 3 - x^2$. Note that for $f(x) = 3 - x^2$,

$$f(0.5) = 3 - 0.5^2 = 2.75;$$
$$f(1) = 2;$$
and $$f(2) = -1.$$

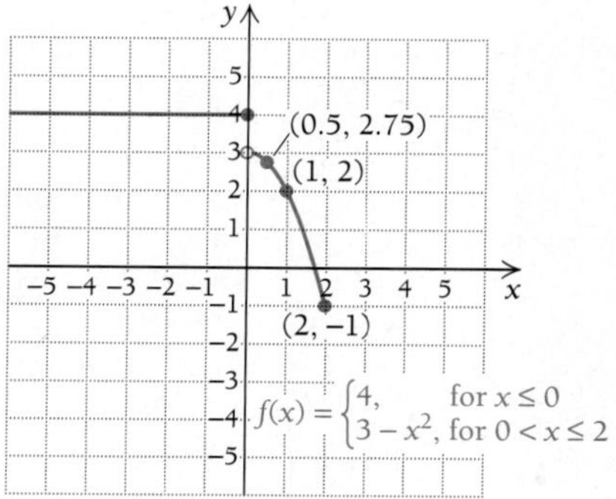

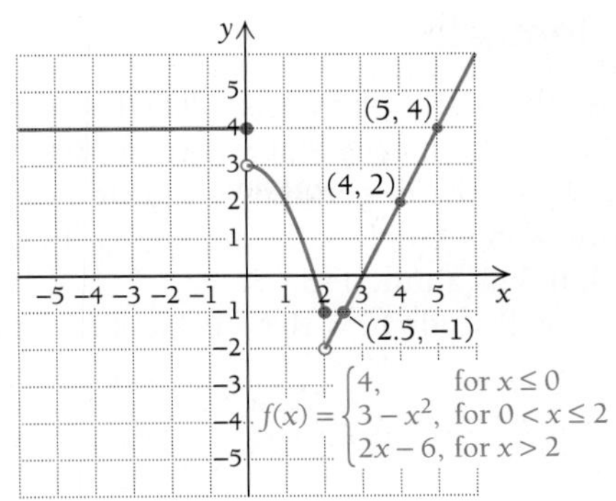

Finally, note that for x-values greater than 2, the graph is the line $y = 2x - 6$. Note that for $f(x) = 2x - 6$,

$$f(2.5) = 2 \cdot 2.5 - 6 = -1;$$
$$f(4) = 2;$$
$$\text{and} \quad f(5) = 4.$$

✦

TECHNOLOGY CONNECTION

Graphing Functions Defined Piecewise

Graphing functions defined piecewise generally involves the use of inequality symbols, which are often accessed using the TEST menu. The function in Example 9 is entered as follows:

```
Plot 1  Plot 2  Plot 3
\Y1 ■ (4)(X ≤ 0) + (3 − X²)(0 <
X)(X ≤ 2) + (2X − 6)(X > 2)■
\Y2 =
\Y3 =
\Y4 =
\Y5 =
```

The graph is shown below.

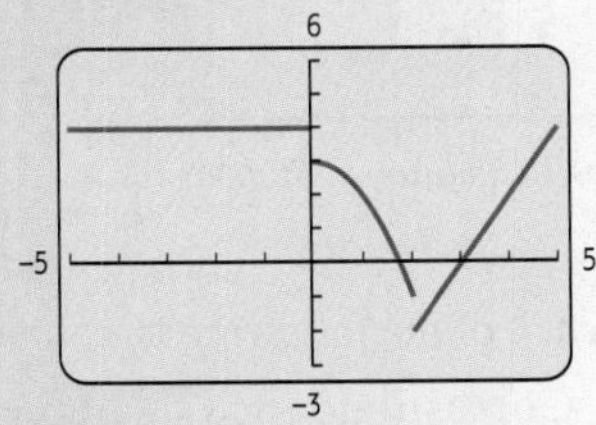

EXERCISES

Graph.

1. $f(x) = \begin{cases} -x - 2, & \text{for } x < -2, \\ 4 - x^2, & \text{for } -2 \le x < 2, \\ x + 3, & \text{for } x \ge 2 \end{cases}$

2. $f(x) = \begin{cases} x^2 - 2, & \text{for } x \le 3, \\ 1, & \text{for } x > 3 \end{cases}$

3. $f(x) = \begin{cases} x + 3, & \text{for } x \le -2, \\ 1, & \text{for } -2 < x \le 3, \\ x^2 - 10, & \text{for } x > 3 \end{cases}$

EXAMPLE 10 Graph the function defined as follows:

$$g(x) = \begin{cases} 3, & \text{for } x = 1, \\ -x + 2, & \text{for } x \neq 1. \end{cases}$$

Solution The function is defined such that $g(1) = 3$ and for all other x-values (that is, for $x \neq 1$), we have $g(x) = -x + 2$. Thus, to graph this function, we graph the line given by $g(x) = -x + 2$, but with an open dot at the point above $x = 1$. To complete the graph, we plot the point $(1, 3)$ since $g(1) = 3$.

x	$g(x)$	$(x, g(x))$
−3	$-(-3) + 2$	$(-3, 5)$
0	$-0 + 2$	$(0, 2)$
1	3	$(1, 3)$
2	$-2 + 2$	$(2, 0)$
3	$-3 + 2$	$(3, -1)$

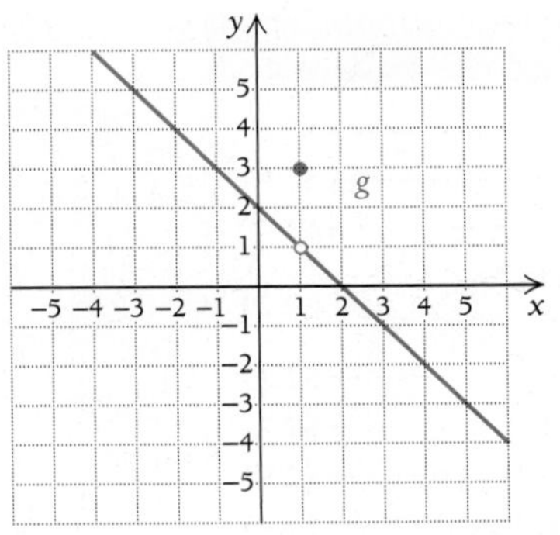

✦

Some Final Remarks

We sometimes use the terminology *y is a function of x*. This means that x is an input and y is an output. It also means that x is the **independent variable** because it represents inputs and y is the **dependent variable** because it represents outputs. We may refer to "a function $y = x^2$" without naming it using a letter f. We may also simply refer to x^2 (alone) as a function.

In calculus we will study how the outputs of a function change when the inputs change.

Note: A review of algebra can be found in the Appendix on p. 599.

Determine whether each correspondence is a function.

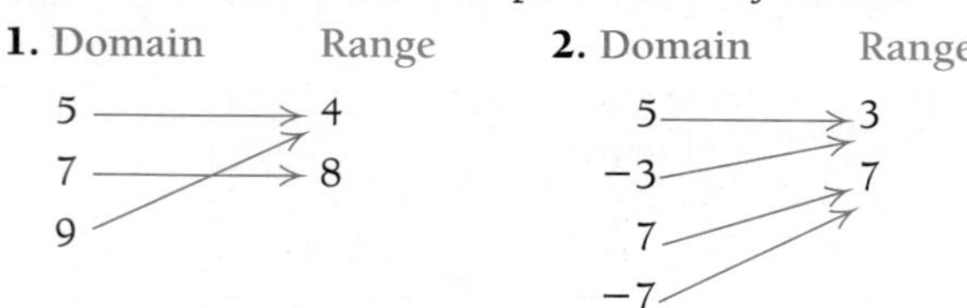

3. Domain → Range

6 → −6
6 → −7
7 → −7
3 → −3

4. Domain → Range

−5 → 1
5 → 1
8 → 1

5. Sandwich prices.

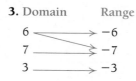

DOMAIN	RANGE
Hamburger	$0.89
Cheeseburger	$0.99
Filet-O-Fish®	$1.99
Quarter Pounder®	$2.29
Big N' Tasty®	$2.29
Big Mac®	$2.39
Crispy Chicken	$2.79
Chicken McGrill®	$2.89
Double Quarter Pounder®	$2.99

(*Source: Nutrition Action Healthletter*, April 2004, p 2.)

6. Sandwich calorie content.

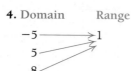

DOMAIN	RANGE
Hamburger	280
Cheeseburger	330
Quarter Pounder®	430
Chicken McGrill®	450
Filet-O-Fish®	470
Big N' Tasty®	540
Crispy Chicken	550
Big Mac®	590
Double Quarter Pounder®	760

(*Source: Nutrition Action Healthletter*, April 2004, p. 2.)

Determine whether each of the following is a function.

	Domain	Correspondence	Range
7.	A class of students	Each student's ID number	A set of numbers
8.	A class of students	Each student's registration form	A set of papers
9.	A class of students	Each student's shoe size	A set of numbers
10.	A class of students	Each student's height	A set of measurements
11.	The set of all real numbers	Square each number and then add 8.	The set of all positive numbers greater than or equal to 8

12. The set of all real numbers	Raise each number to the fourth power.	The set of all nonnegative numbers
13. A set of males	Each person's biological father	A set of males
14. A set of females	Each person's biological mother	A set of females
15. A set of avenues	An intersecting road	A set of cross streets
16. A set of textbooks	An even-numbered page in each book	A set of pages
17. A set of shapes	The area of each shape	A set of area measurements
18. A set of shapes	The perimeter of each shape	A set of length measurements

19. A function f is given by

$f(x) = 3x + 2.$

This function takes a number x, multiplies it by 3, and adds 2.

a) Complete this table.

x	4.1	4.01	4.001	4
$f(x)$				

b) Find $f(5), f(-1), f(k), f(1 + t)$, and $f(x + h)$.

20. A function f is given by

$f(x) = 4x - 3.$

This function takes a number x, multiplies it by 4, and subtracts 3.

a) Complete this table.

x	5.1	5.01	5.001	5
$f(x)$				

b) Find $f(4), f(3), f(-2), f(k), f(1 + t)$, and $f(x + h)$.

21. A function g is given by

$g(x) = x^2 - 3.$

This function takes a number x, squares it, and subtracts 3. Find $g(-1), g(0), g(1), g(5), g(u), g(a + h)$, and $\dfrac{g(a + h) - g(a)}{h}$.

22. A function g is given by

$g(x) = x^2 + 4.$

This function takes a number x, squares it, and adds 4. Find $g(-3), g(0), g(-1), g(7), g(v), g(a + h)$, and $\dfrac{g(a + h) - g(a)}{h}$.

23. A function f is given by

$$f(x) = \frac{1}{(x + 3)^2}.$$

This function takes a number x, adds 3, squares the result, and takes the reciprocal of that result.

a) Find $f(4), f(-3), f(0), f(a), f(t + 4), f(x + h)$, and $\dfrac{f(x + h) - f(x)}{h}$. If an output is undefined, state that fact.

b) Note that f could also be given by

$$f(x) = \frac{1}{x^2 + 6x + 9}.$$

Explain what this does to an input number x.

24. A function f is given by

$$f(x) = \frac{1}{(x - 5)^2}.$$

This function takes a number x, subtracts 5 from it, squares the result, and takes the reciprocal of the square.

a) Find $f(3), f(-1), f(5), f(k), f(t - 1), f(t - 4)$, and $f(x + h)$. If an output is undefined, state that fact.

b) Note that f could also be given by

$$f(x) = \frac{1}{x^2 - 10x + 25}.$$

Explain what this does to an input number x.

Graph each function.

25. $f(x) = 2x - 5$

26. $f(x) = 3x - 1$

27. $g(x) = -4x$

28. $g(x) = -2x$

29. $f(x) = x^2 - 2$

30. $f(x) = x^2 + 4$

31. $f(x) = 6 - x^2$

32. $g(x) = -x^2 + 1$

33. $g(x) = x^3$

34. $g(x) = \frac{1}{2}x^3$

Use the vertical-line test to determine whether each graph is that of a function. (In Exercises 43–46, the vertical dashed lines are not part of the graph.)

35.

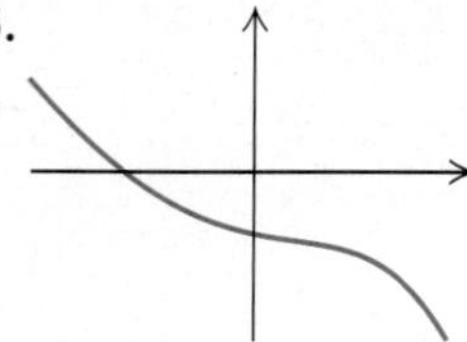

36.

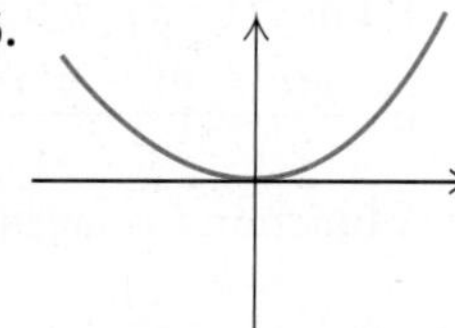

37.

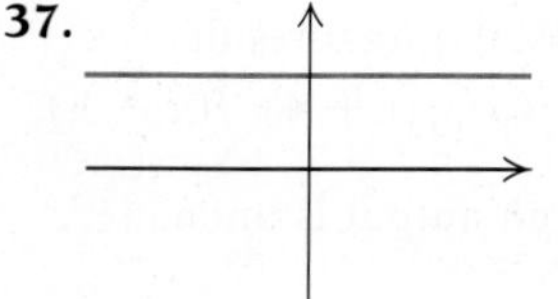

38.

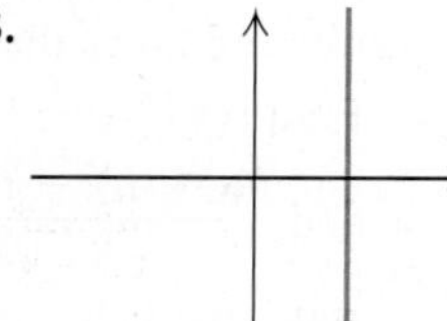

39.

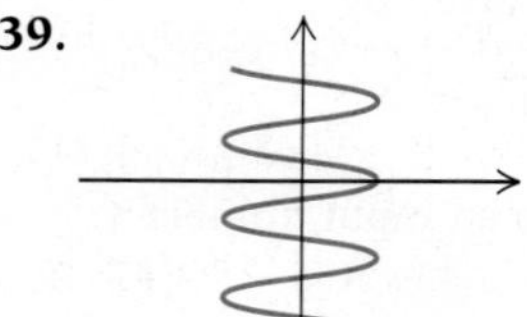

40.

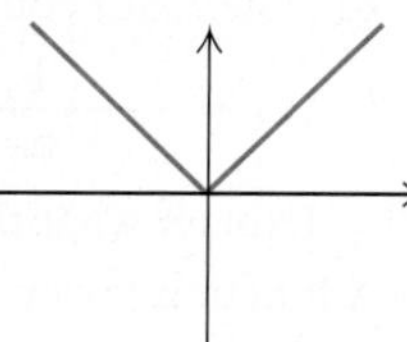

41.

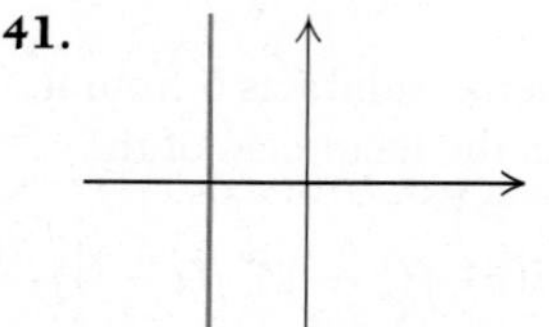

42.

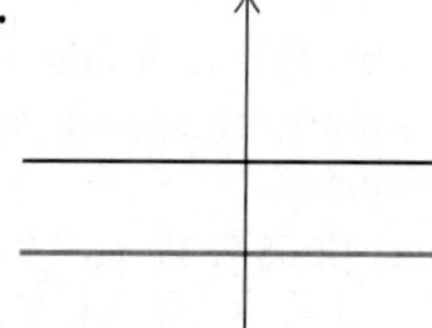

43.

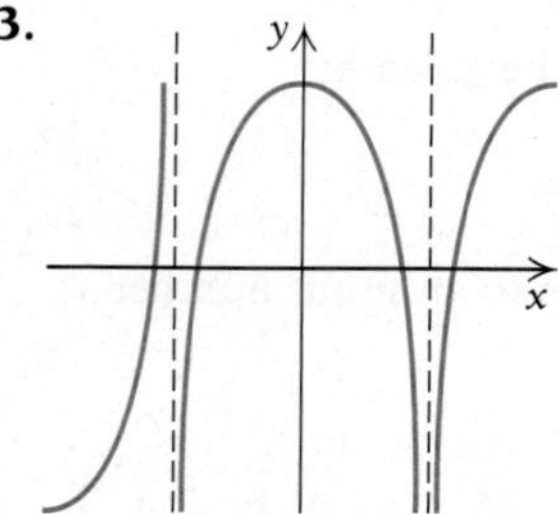

44.

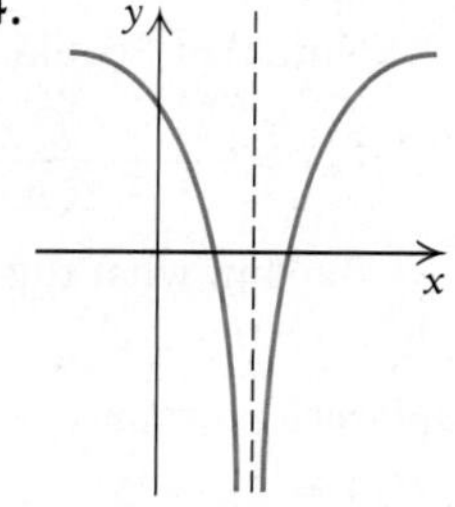

45.

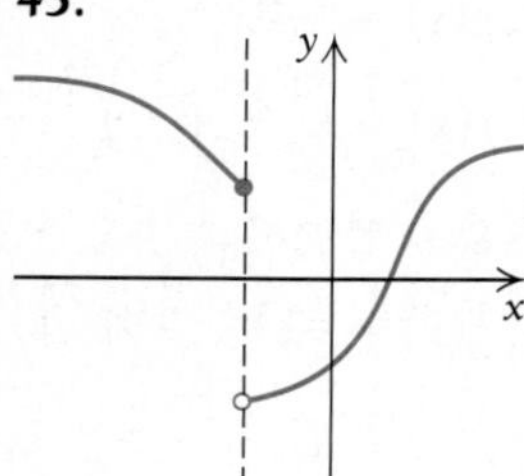

46.

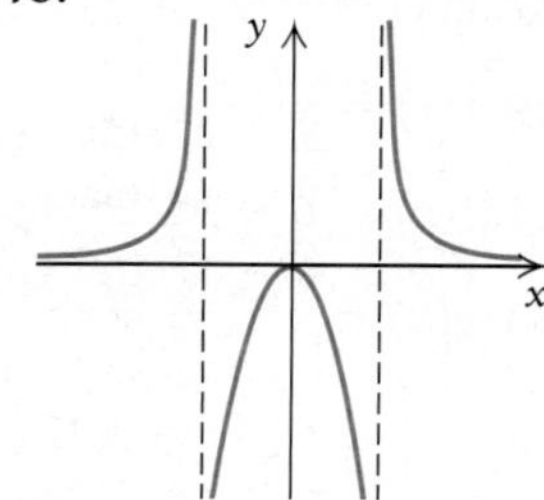

47. a) Graph $x = y^2 - 2$.
b) Is this a function?

48. a) Graph $x = y^2 - 3$.
b) Is this a function?

49. For $f(x) = x^2 - 3x$, find $\dfrac{f(x + h) - f(x)}{h}$.

50. For $f(x) = x^2 + 4x$, find $\dfrac{f(x + h) - f(x)}{h}$.

For Exercises 51–54, consider the function f given by

$$f(x) = \begin{cases} -2x + 1, & \text{for } x < 0, \\ 17, & \text{for } x = 0, \\ x^2 - 3, & \text{for } 0 < x < 4, \\ \frac{1}{2}x + 1, & \text{for } x \geq 4. \end{cases}$$

51. Find $f(-1)$ and $f(1)$.

52. Find $f(-3)$ and $f(3)$.

53. Find $f(0)$ and $f(10)$.

54. Find $f(-5)$ and $f(5)$.

Graph.

55. $f(x) = \begin{cases} 1, & \text{for } x < 0, \\ -1, & \text{for } x \geq 0 \end{cases}$

56. $f(x) = \begin{cases} 2, & \text{for } x \leq 3, \\ -2, & \text{for } x > 3 \end{cases}$

57. $f(x) = \begin{cases} 6, & \text{for } x = -2, \\ x^2, & \text{for } x \neq -2 \end{cases}$

58. $f(x) = \begin{cases} 5, & \text{for } x = 1, \\ x^3, & \text{for } x \neq 1 \end{cases}$

59. $g(x) = \begin{cases} -x, & \text{for } x < 0, \\ 4, & \text{for } x = 0, \\ x + 2, & \text{for } x > 0 \end{cases}$

60. $g(x) = \begin{cases} 2x - 3, & \text{for } x < 1, \\ 5, & \text{for } x = 1, \\ x - 2, & \text{for } x > 1 \end{cases}$

61. $g(x) = \begin{cases} \frac{1}{2}x - 1, & \text{for } x < 2, \\ -4, & \text{for } x = 2, \\ x - 3, & \text{for } x > 2 \end{cases}$

62. $g(x) = \begin{cases} x^2, & \text{for } x < 0, \\ -3, & \text{for } x = 0, \\ -2x + 3, & \text{for } x > 0 \end{cases}$

63. $f(x) = \begin{cases} -7, & \text{for } x = 2, \\ x^2 - 3, & \text{for } x \neq 2 \end{cases}$

64. $f(x) = \begin{cases} -6, & \text{for } x = -3, \\ -x^2 + 5, & \text{for } x \neq -3 \end{cases}$

65. Music revenue. The yearly sales of CDs can be approximated using the graph below.

a) Estimate the number of CDs sold in 2005.

b) In what year were approximately 710 million CDs sold?

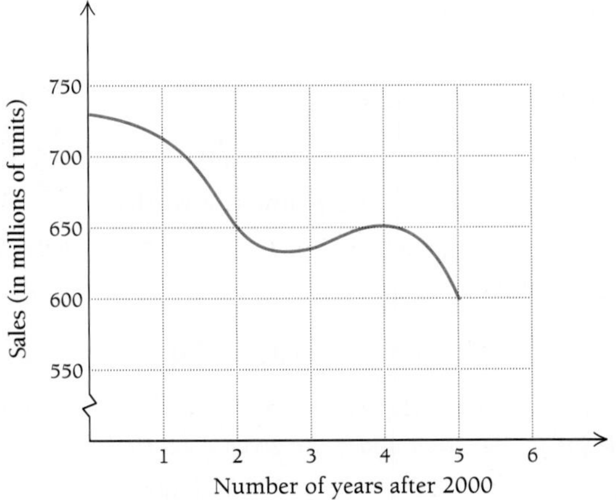

(*Source*: Nielsen.)

66. Major league baseball ticket prices. The average ticket price for a major league baseball game can be approximated using the graph below.

a) Estimate the average ticket price in 2003.

b) In what year did the average price first reach $14?

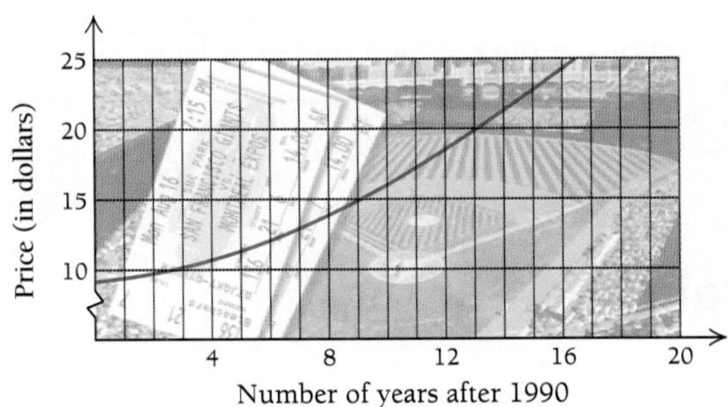

(*Source*: Based on data from www.teammarketing.com.)

67. Federal budget. The following graph shows the gross receipts (blue) and outlays (red), in billions of dollars, of the federal government since 1980.

a) Is each graph that of a function?

b) Estimate the gross receipts of the federal government in 1990.

c) Estimate the year in which gross outlays were $800 billion.

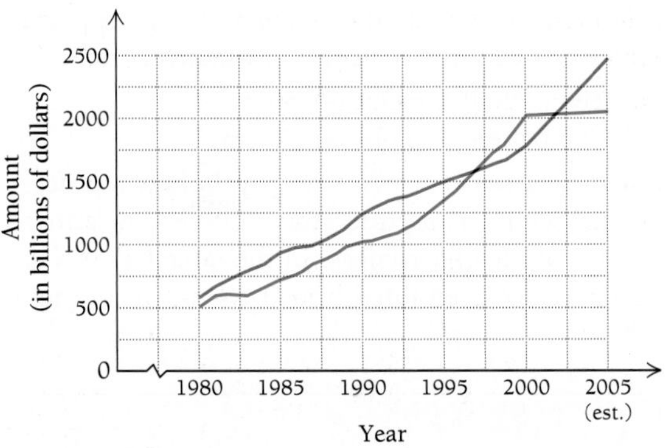

(*Source*: U.S. Census Bureau.)

68. Federal budget. Refer to the graph in Exercise 67.

a) In what two years were the gross receipts and outlays approximately the same?

b) In what year were the gross receipts the least?

c) In what year were the gross outlays the greatest?

Compound interest. *The amount of money, $A(t)$, in a savings account that pays 6% interest, compounded quarterly for t years, with an initial investment of P dollars, is given by*

$$A(t) = P\left(1 + \frac{0.06}{4}\right)^{4t}.$$

69. If $500 is invested at 6%, compounded quarterly, how much will the investment be worth after 2 yr?

70. If $800 is invested at 6%, compounded quarterly, how much will the investment be worth after 3 yr?

Chemotherapy. *In computing the dosage for chemotherapy, a patient's body surface area is needed. A good approximation of a person's surface area s, in square meters (m^2), is given by the formula*

$$s = \sqrt{\frac{hw}{3600}},$$

*where w is the patient's weight in kilograms (kg) and h is the patient's height in centimeters (cm). (**Source:** U.S. Oncology.) Use the preceding information for Exercises 71 and 72.*

71. Assume that a patient's height is 170 cm. Find the patient's approximate surface area assuming that:

a) The patient's weight is 70 kg.

b) The patient's weight is 100 kg.

c) The patient's weight is 50 kg.

72. Assume that a patient's weight is 70 kg. Approximate the patient's surface area assuming that:

a) The patient's height is 150 cm.
b) The patient's height is 180 cm.

73. Scaling stress factors. In psychology a process called *scaling* is used to attach numerical ratings to a group of life experiences. In the table below, various events have been rated from 1 to 100 according to their stress levels.

Event	Scale of Impact
Death of spouse	100
Divorce	73
Jail term	63
Marriage	50
Lost job	47
Pregnancy	40
Death of close friend	37
Loan over $10,000	31
Child leaving home	29
Change in schools	20
Loan less than $10,000	17
Christmas	12

(*Source*: Thomas H. Holmes, University of Washington School of Medicine.)

a) Does the table represent a function? Why or why not?
b) What are the inputs? What are the outputs?

SYNTHESIS

Solve for y in terms of x. Decide whether the resulting equation represents a function.

74. $2x + y - 16 = 4 - 3y + 2x$

75. $2y^2 + 3x = 4x + 5$

76. $(4y^{2/3})^3 = 64x$

77. $(3y^{3/2})^2 = 72x$

TW **78.** Explain why the vertical-line test works.

TW **79.** Is 4 in the domain of f in Exercises 51–54? Explain why or why not.

TECHNOLOGY CONNECTION

In Exercises 80 and 81, use the TABLE *feature to construct a table for the function under the given conditions.*

80. $f(x) = x^3 + 2x^2 - 4x - 13$; TblStart $= -3$; ΔTbl $= 2$

81. $f(x) = \dfrac{3}{x^2 - 4}$; TblStart $= -3$; ΔTbl $= 1$

82. A function f is given by

$$f(x) = |x - 2| + |x + 1| - 5.$$

Find $f(-3)$, $f(-2)$, $f(0)$, and $f(4)$.

83. Graph the function in each of Exercises 80–82.

84. Use the TRACE feature to find several ordered-pair solutions of the function $f(x) = \sqrt{10 - x^2}$.

R.3 Finding Domain and Range

OBJECTIVES

- Write interval notation for a set of points.
- Find the domain and the range of a function.

Set Notation

A **set** is a collection of objects. The set we consider most in calculus is the set of **real numbers,** $\mathbb{R}$. There is a real number for every point on the number line.

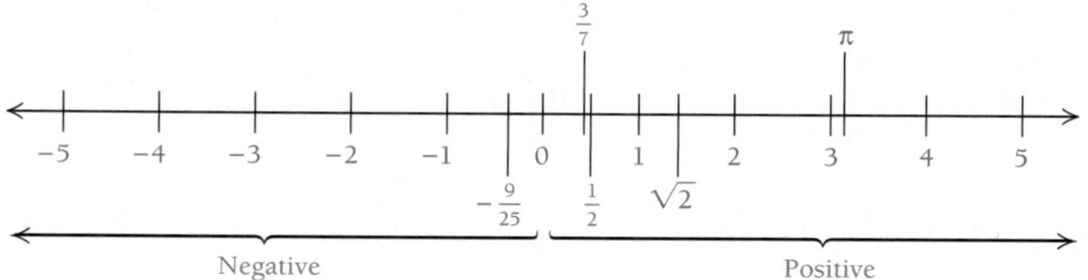

The set consisting of $-\frac{9}{25}$, 0, and $\sqrt{2}$ can be written $\{-\frac{9}{25}, 0, \sqrt{2}\}$. This method of describing sets is known as the **roster method.** It lists every member of the set. To write larger sets often requires **set-builder notation,** which specifies conditions under which an object is in the set. For example, the set of all real numbers less than 4 can be described as follows in set-builder notation:

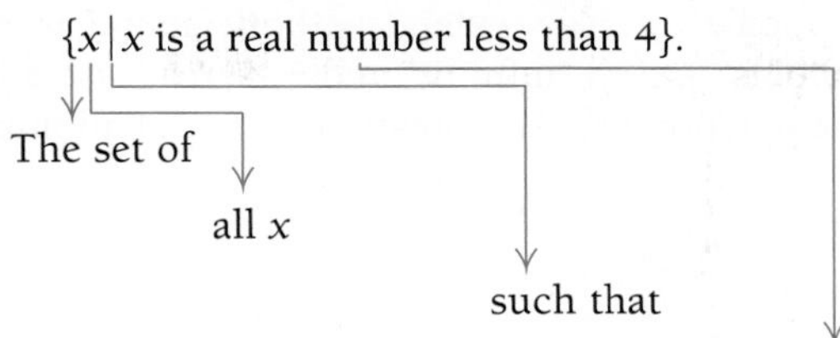

Interval Notation

We can also describe sets using **interval notation.** If a and b are real numbers, with $a < b$, we define the interval (a, b) as the set of all numbers between but not including a and b, that is, the set of all x for which $a < x < b$. Thus,

$$(a, b) = \{x \mid a < x < b\}.$$

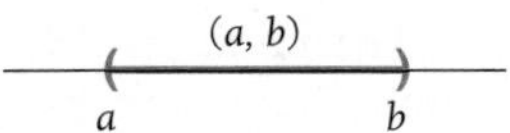

The points a and b are the **endpoints** of the interval. The parentheses indicate that the endpoints are *not* included in the interval.

The interval $[a, b]$ is defined as the set of all x for which $a \leq x \leq b$. Thus,

$$[a, b] = \{x \mid a \leq x \leq b\}.$$

[a, b]

a b

The brackets indicate that the endpoints *are* included in the interval.*

Be careful not to confuse the *interval* (a, b) with the *ordered pair* (a, b) used to represent a point in the plane, as in Section R.1. The context in which the notation appears usually makes the meaning clear.

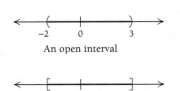

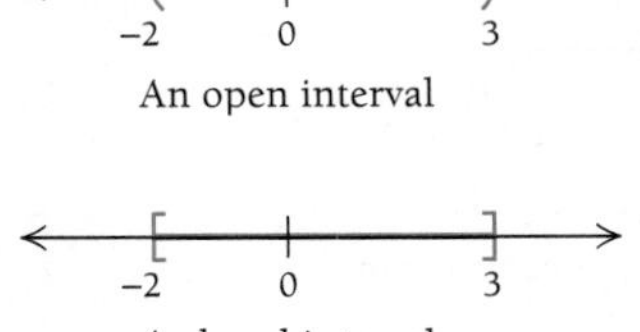

An open interval

-2 0 3

A closed interval

Intervals like $(-2, 3)$, in which neither endpoint is included, are called **open intervals;** intervals like $[-2, 3]$, which include both endpoints, are said to be **closed intervals.** Thus, $[a, b]$ is read "the closed interval a, b" and (a, b) is read "the open interval a, b."

Some intervals are **half-open** and include one endpoint but not the other:

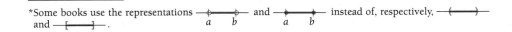

$(a, b] = \{x \mid a < x \leq b\}$. The graph excludes a and includes b.

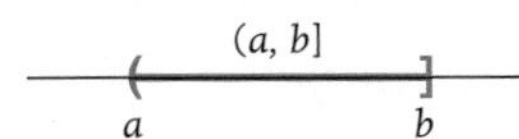

*Some books use the representations and instead of, respectively, and .

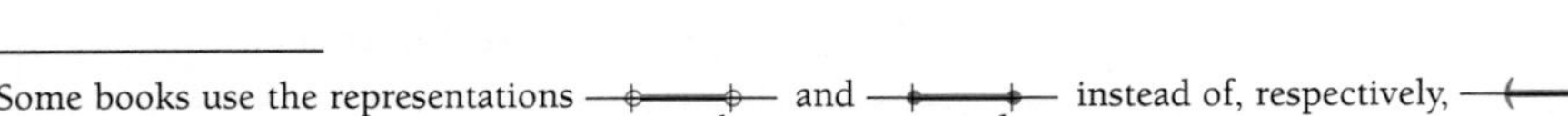

$[a, b) = \{x \mid a \leq x < b\}$. The graph includes a and excludes b.

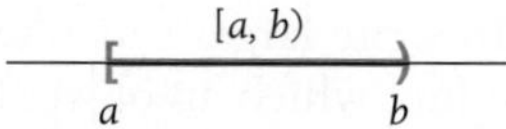

Some intervals extend without bound in one or both directions. We use the symbols ∞, read "infinity," and $-\infty$, read "negative infinity," to name these intervals. The notation $(5, \infty)$ represents the set of all numbers greater than 5. That is,

$$(5, \infty) = \{x \mid x > 5\}.$$

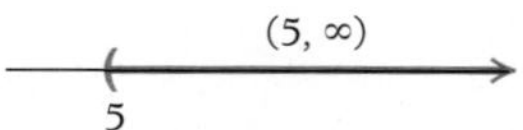

Similarly, the notation $(-\infty, 5)$ represents the set of all numbers less than 5. That is,

$$(-\infty, 5) = \{x \mid x < 5\}.$$

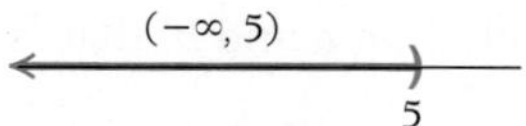

The notations $[5, \infty)$ and $(-\infty, 5]$ are used when we want to include the endpoints. The interval $(-\infty, \infty)$ names the set of all real numbers.

$(-\infty, \infty) = \{x \mid x \text{ is a real number}\}$

Interval notation is summarized in the following table.

Intervals: Notation and Graphs

Interval Notation	Set Notation	Graph
(a, b)	$\{x \mid a < x < b\}$	
$[a, b]$	$\{x \mid a \leq x \leq b\}$	
$[a, b)$	$\{x \mid a \leq x < b\}$	
$(a, b]$	$\{x \mid a < x \leq b\}$	
(a, ∞)	$\{x \mid x > a\}$	
$[a, \infty)$	$\{x \mid x \geq a\}$	
$(-\infty, b)$	$\{x \mid x < b\}$	
$(-\infty, b]$	$\{x \mid x \leq b\}$	
$(-\infty, \infty)$	$\{x \mid x \text{ is a real number}\}$	

EXAMPLE 1 Write interval notation for each set or graph:

a) $\{x \mid -4 < x < 5\}$

b) $\{x \mid x \geq -2\}$

c) 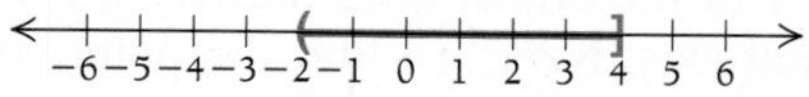

d) 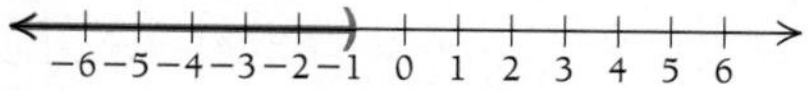

Solution

a) $\{x \mid -4 < x < 5\} = (-4, 5)$

b) $\{x \mid x \geq -2\} = [-2, \infty)$

c) 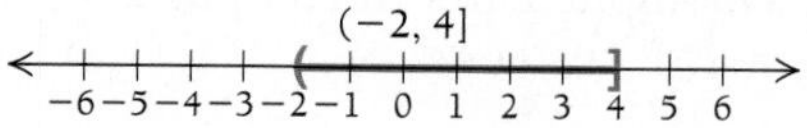

d) 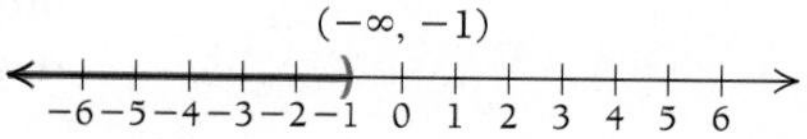

✦

Finding Domain and Range

Recall that when a set of ordered pairs is such that no two different pairs share a common first coordinate, we have a function. The **domain** is the set of all first coordinates, and the **range** is the set of all second coordinates.

EXAMPLE 2 Find the domain and the range of the function f whose graph is shown to the right.

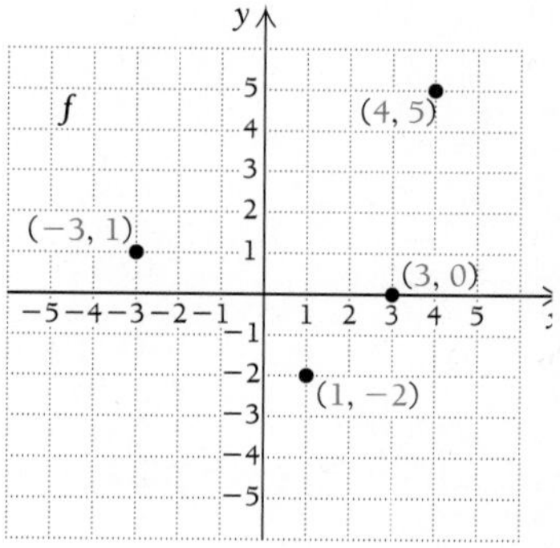

Solution This function consists of just four ordered pairs and can be written as

$$\{(-3, 1), (1, -2), (3, 0), (4, 5)\}.$$

We can determine the domain and the range by reading the x- and the y-values directly from the graph.

The domain is the set of all first coordinates, $\{-3, 1, 3, 4\}$. The range is the set of all second coordinates, $\{1, -2, 0, 5\}$.

✦

EXAMPLE 3 For the function f whose graph is shown to the right, determine each of the following.

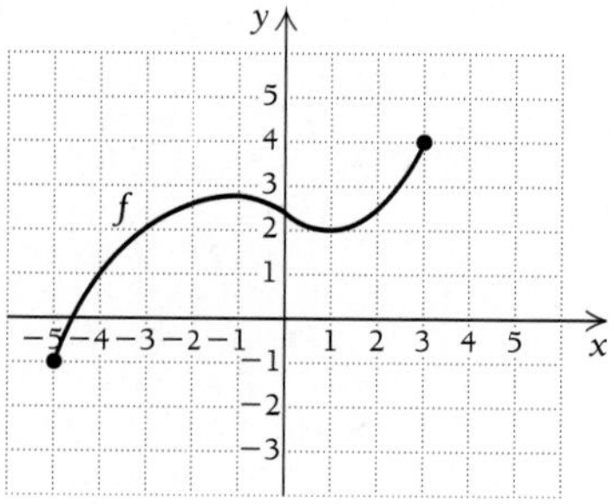

a) The number in the range that is paired with 1 (from the domain). That is, find $f(1)$.

b) The domain of f

c) The number(s) in the domain that is (are) paired with 1 (from the range). That is, find all x-values for which $f(x) = 1$.

d) The range of f

Solution

a) To determine which number in the range is paired with 1 in the domain, we locate 1 on the horizontal axis. Next, we find the point on the graph of f for which 1 is the first coordinate. From that point, we can look to the vertical axis to find the corresponding y-coordinate, 2. The input 1 has the output 2—that is, $f(1) = 2$.

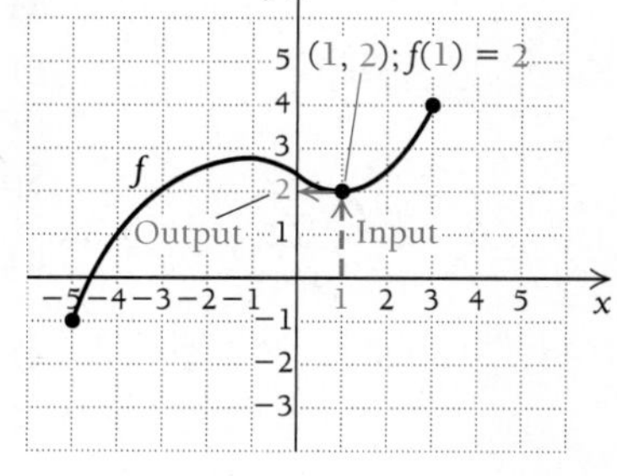

b) The domain of the function is the set of all x-values, or inputs, of the points on the graph. These extend from -5 to 3 and can be viewed as the curve's shadow, or *projection*, onto the x-axis. Thus, the domain is the set $\{x \mid -5 \leq x \leq 3\}$, or, in interval notation, $[-5, 3]$.

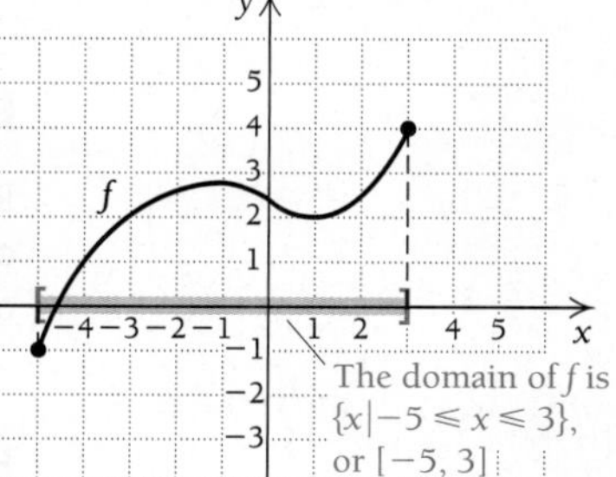

c) To determine which number(s) in the domain is (are) paired with 1 in the range, we locate 1 on the vertical axis (see the graph to the right). From there, we look left and right to the graph of f to find any points for which 1 is the second coordinate. One such point exists: $(-4, 1)$. For this function, we note that $x = -4$ is the only member of the domain paired with 1. For other functions, there might be more than one member of the domain paired with a member of the range.

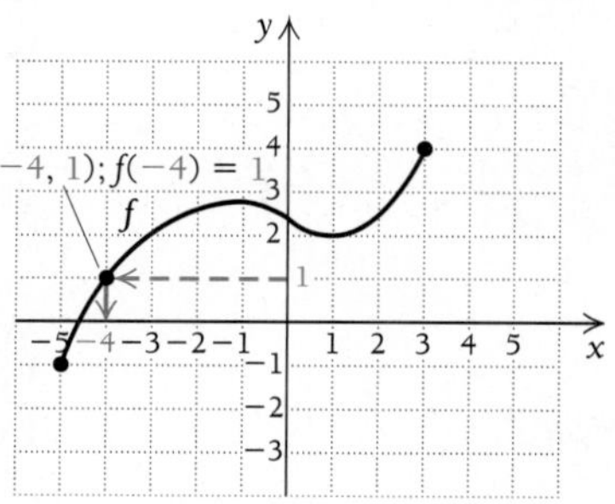

d) The range of the function is the set of all y-values, or outputs, of the points on the graph. These extend from -1 to 4 and can be viewed as the curve's shadow, or projection, onto the y-axis. Thus, the range is the set $\{y \mid -1 \leq y \leq 4\}$, or, in interval notation, $[-1, 4]$.

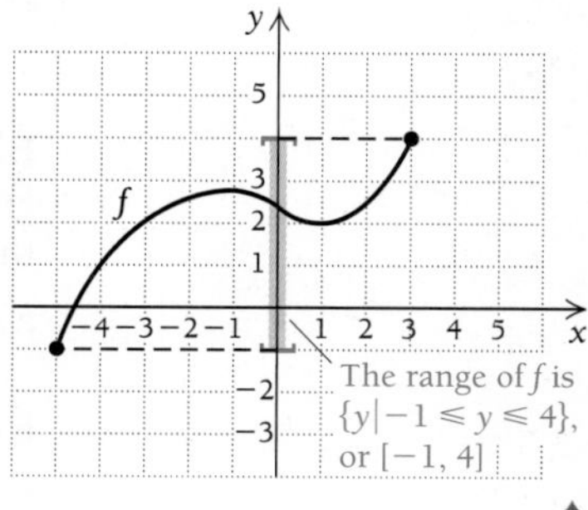

◆

When a function is given by an equation or formula, the domain is understood to be the largest set of real numbers (inputs) for which function values (outputs) can be calculated. That is, the domain is the set of all allowable inputs into the formula. To find the domain, think, "What can we substitute?"

EXAMPLE 4 Find the domain: $f(x) = |x|$.

Solution We ask, "What can we substitute?" Is there any number x for which we cannot calculate $|x|$? The answer is no. Thus, the domain of f is the set of all real numbers. ✦

EXAMPLE 5 Find the domain: $f(x) = \dfrac{3}{2x - 5}$.

Solution We ask, "What can we substitute?" Is there any number x for which we cannot calculate $3/(2x - 5)$? Since $3/(2x - 5)$ cannot be calculated when the denominator $2x - 5$ is 0, we solve the following equation to find those real numbers that must be excluded from the domain of f:

$$2x - 5 = 0 \quad \text{Setting the denominator equal to 0}$$
$$2x = 5 \quad \text{Adding 5 to both sides}$$
$$x = \tfrac{5}{2}. \quad \text{Dividing both sides by 2}$$

Thus, $\frac{5}{2}$ is not in the domain, whereas all other real numbers are. We say that *f is not defined at $\frac{5}{2}$*, or *$f\left(\frac{5}{2}\right)$ does not exist*.

The domain of f is $\left\{x \mid x \text{ is a real number } and\ x \neq \frac{5}{2}\right\}$, or, in interval notation, $\left(-\infty, \frac{5}{2}\right) \cup \left(\frac{5}{2}, \infty\right)$. The symbol $\cup$ indicates the *union* of two sets and means that all elements in both sets are included in the domain. ✦

TECHNOLOGY CONNECTION

Consider the function given by $f(x) = 1/(x - 3)$. The table below was obtained with ΔTbl set at 0.25. Note that the calculator cannot calculate $f(3)$.

ΔTbl = 0.25

X	Y1
2	−1
2.25	−1.333
2.5	−2
2.75	−4
3	ERR:
3.25	4
3.5	2
3.75	1.3333
X = 2	

EXERCISES

1. Make a table for $f(x) = 1/(x^2 - 4)$ from $x = -3$ to $x = 3$ and with ΔTbl set at 0.5.
2. Create a table for the function given in Example 5.

EXAMPLE 6 Find the domain: $f(x) = \sqrt{4 + 3x}$.

Solution We ask, "What can we substitute?" Is there any number x for which we cannot calculate $\sqrt{4 + 3x}$? Since $\sqrt{4 + 3x}$ is not a real number when the radicand $4 + 3x$ is negative, the domain is all real numbers for which $4 + 3x \geq 0$. We find them by solving the inequality. (See the Appendix for a review of inequality solving.)

$$4 + 3x \geq 0$$
$$3x \geq -4 \quad \text{Simplifying}$$
$$x \geq -\tfrac{4}{3} \quad \text{Dividing both sides by 3}$$

The domain is $\left[-\frac{4}{3}, \infty\right)$. ✦

TECHNOLOGY CONNECTION

Determining Domain and Range

Graph each function in the given viewing window. Then determine the domain and the range.

a) $f(x) = 3 - |x|$, $[-10, 10, -10, 10]$
b) $f(x) = x^3 - x$, $[-3, 3, -4, 4]$
c) $f(x) = \frac{12}{x}$, or $12x^{-1}$, $[-14, 14, -14, 14]$
d) $f(x) = x^4 - 2x^2 - 3$, $[-4, 4, -6, 6]$

We have the following.

a)

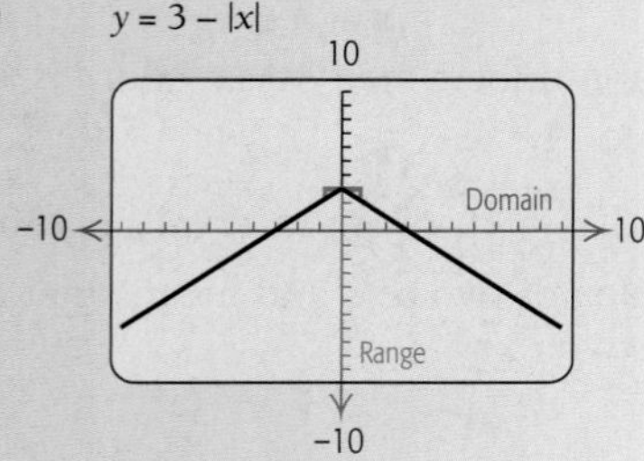

Domain $= \mathbb{R}$
Range $= (-\infty, 3]$

b)

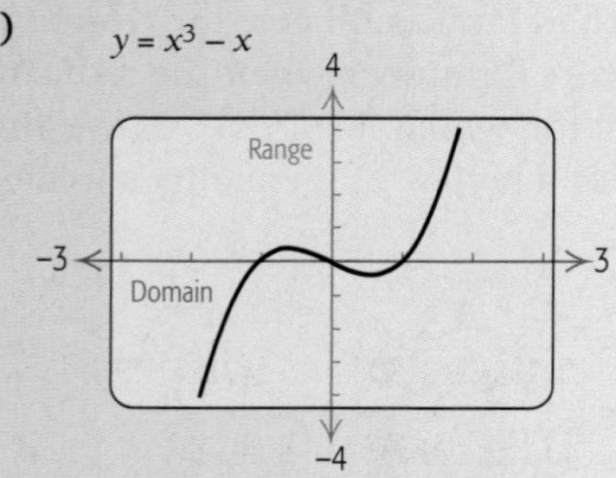

Domain $= \mathbb{R}$
Range $= \mathbb{R}$

c)

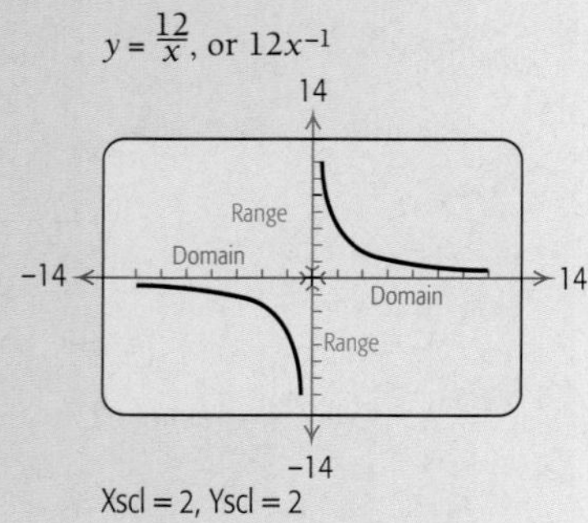

The number 0 is excluded as an input.
Domain $= \{x \mid x$ is a real number *and* $x \neq 0\}$, or $(-\infty, 0) \cup (0, \infty)$;
Range $= \{y \mid y$ is a real number *and* $y \neq 0\}$, or $(-\infty, 0) \cup (0, \infty)$

d)

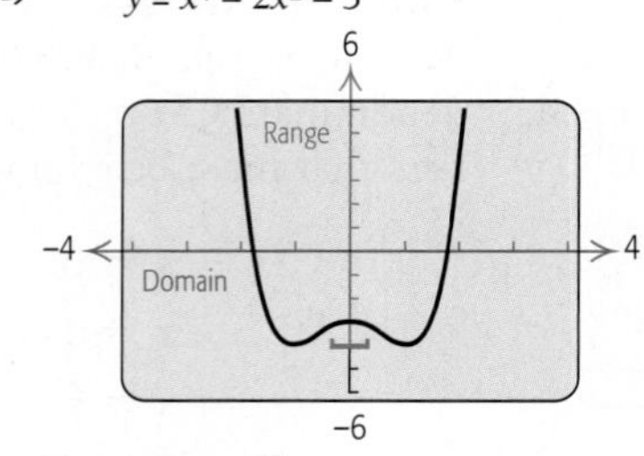

Domain $= \mathbb{R}$
Range $= [-4, \infty)$

We can confirm our results using the TRACE feature, moving the cursor along the curve or entering any x-value in which we have interest. We can also use the TABLE feature. In Example (d), it might not appear as though the domain is all real numbers because the graph seems "thin," but re-examining the formula shows that we can indeed substitute any real number.

EXERCISES

Graph each function in the given viewing window. Then determine the domain and the range.

1. $f(x) = |x| - 4$, $[-10, 10, -10, 10]$
2. $f(x) = 2 + 3x - x^3$, $[-5, 5, -5, 5]$
3. $f(x) = \frac{-3}{x}$, or $-3x^{-1}$, $[-20, 20, -20, 20]$
4. $f(x) = x^4 - 2x^2 - 7$, $[-4, 4, -9, 9]$
5. $f(x) = \sqrt{x + 4}$, $[-8, 8, -8, 8]$
6. $f(x) = \sqrt{9 - x^2}$, $[-5, 5, -5, 5]$
7. $f(x) = -\sqrt{9 - x^2}$, $[-5, 5, -5, 5]$
8. $f(x) = x^3 - 5x^2 + x - 4$, $[-10, 10, -20, 10]$

Domains and Ranges in Applications

Sometimes the domain and the range of a function given by a formula are affected by the context of an application. Let's look again at Exercise 69 in Section R.2.

EXAMPLE 7 Business: Compound Interest. Suppose that \$500 is invested at 6%, compounded quarterly for t years. From Theorem 2 in Section R.1, we know that the amount in the account is given by

$$A(t) = 500\left(1 + \frac{0.06}{4}\right)^{4t}$$
$$= 500(1.015)^{4t}.$$

The amount A is a function of the number of years for which the money is invested. Determine the domain.

Solution We can substitute any real number for t into the formula, but a negative number of years is not meaningful. The context of the application excludes negative numbers. Thus, the domain is the set of all nonnegative numbers, $[0, \infty)$. ✦

EXAMPLE 8 Cellphone Calling Plans. In 2005, Sprint® offered a cellphone calling plan in which a customer's monthly bill could be modeled by the graph below. Find the range of the function shown.

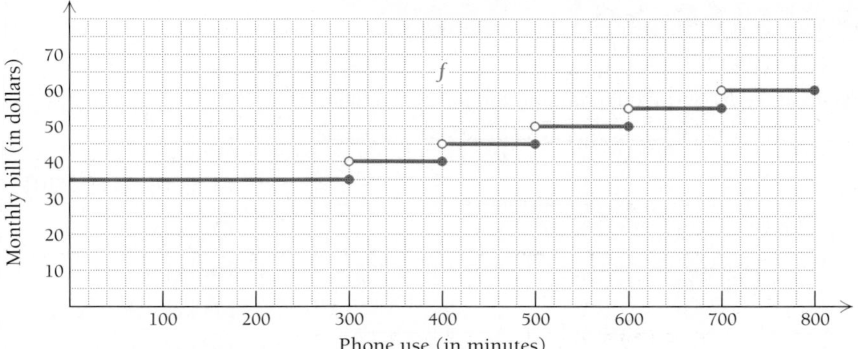

(*Source*: Based on data from *The New York Times*, 9/3/05, p. D8. Bill does not include taxes and fees.)

Solution The range is the set of all outputs—in this case, the different monthly bill amounts—shown in the graph. We see that only six different outputs are used, highlighted in blue. Thus, the range of the function shown is $\{35, 40, 45, 50, 55, 60\}$.

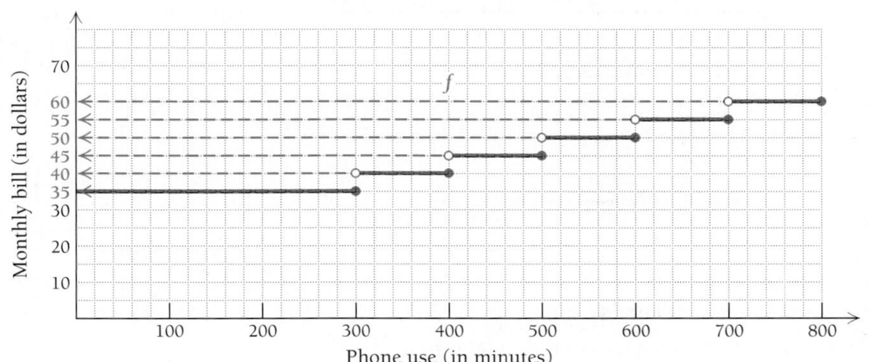

✦

The task of determining the domain and the range of a function is one that we will return to several times as we consider other types of functions in this book.

The following is a review of the function concepts considered in Sections R.1–R.3.

Function Concepts

- Formula for f: $f(x) = x^2 - 7$
- For every input of f, there is exactly one output.
- For the input 1, -6 is the output.
- $f(1) = -6$
- $(1, -6)$ is on the graph.
- Domain = The set of all inputs = The set of all real numbers, $\mathbb{R}$
- Range = The set of all outputs = $[-7, \infty)$

Graph

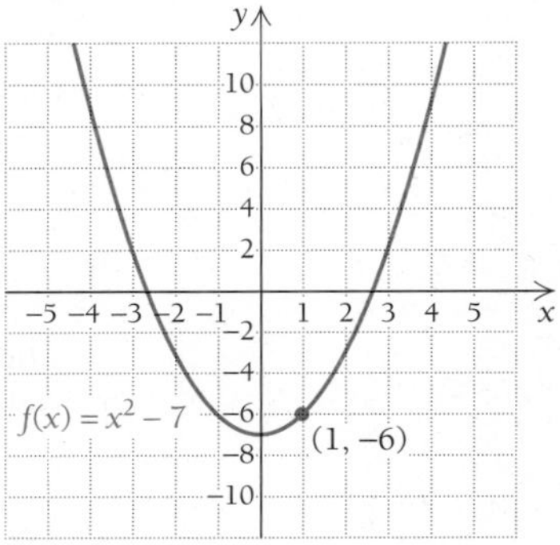

Exercise Set R.3

In Exercises 1–10, write interval notation for each graph.

1. 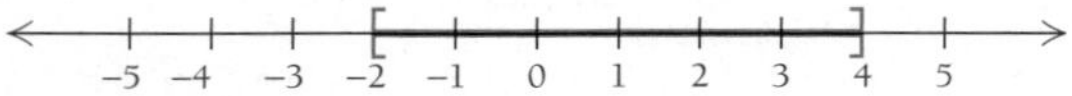

2. 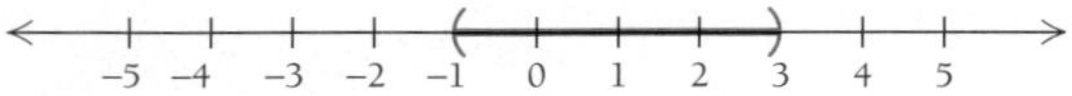

3. 

4. 

5. 

6. 

7. 

8. 

9. 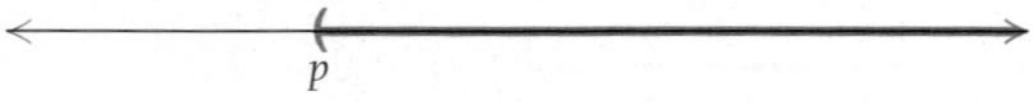

10. 

Write interval notation for each of the following. Then graph the interval on a number line.

11. The set of all numbers x such that $-2 \le x \le 2$

12. The set of all numbers x such that $-5 < x < 5$

13. $\{x \mid -4 \le x < -1\}$

14. $\{x \mid 6 < x \le 20\}$

15. $\{x \mid x \le -2\}$

16. $\{x \mid x > -3\}$

17. $\{x \mid -2 < x \le 3\}$

18. $\{x \mid -10 \le x < 4\}$

19. $\{x \mid x < 12.5\}$

20. $\{x \mid x \ge 12.5\}$

In Exercises 21–32, the graph is that of a function. Determine for each one **(a)** $f(1)$; **(b)** *the domain;* **(c)** *all x-values such that $f(x) = 2$; and* **(d)** *the range.*

21.

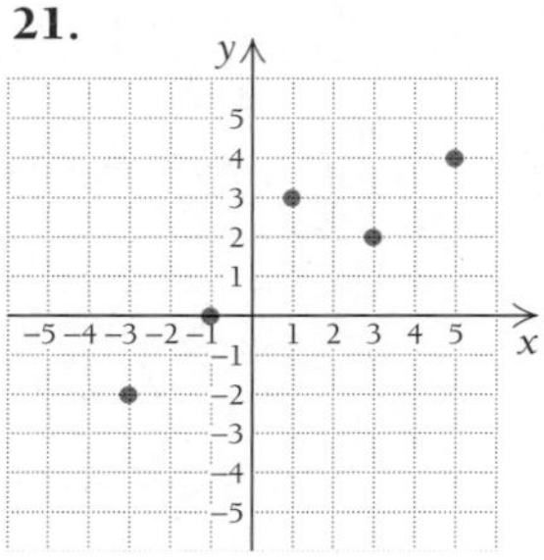

22.

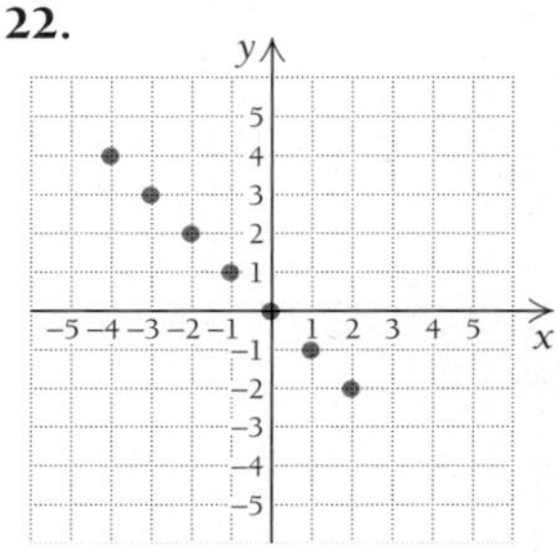

23.

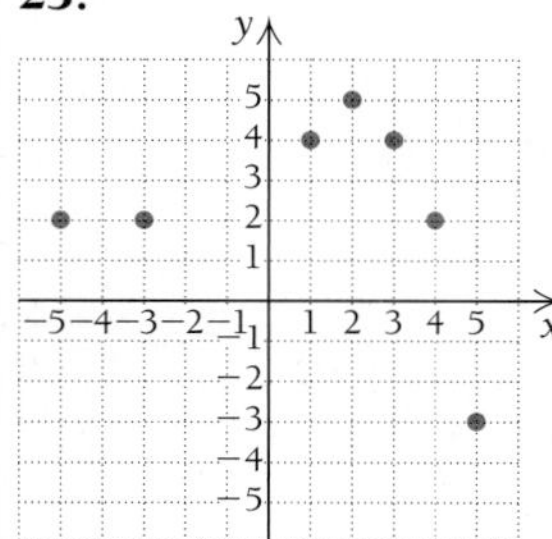

24.

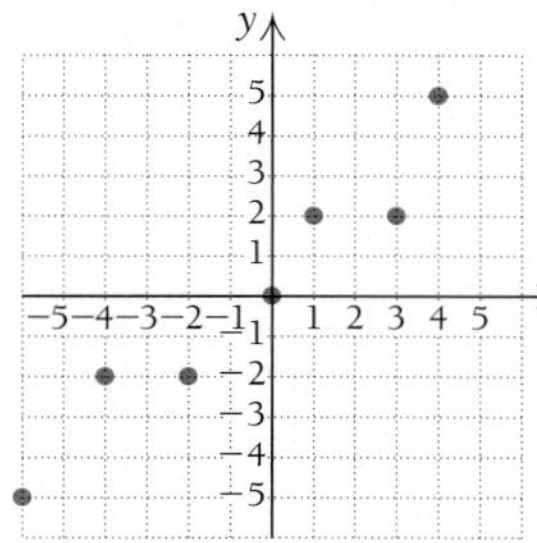

25.

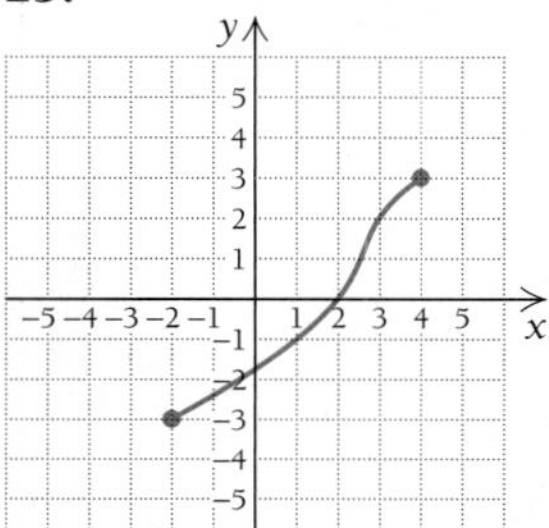

26.

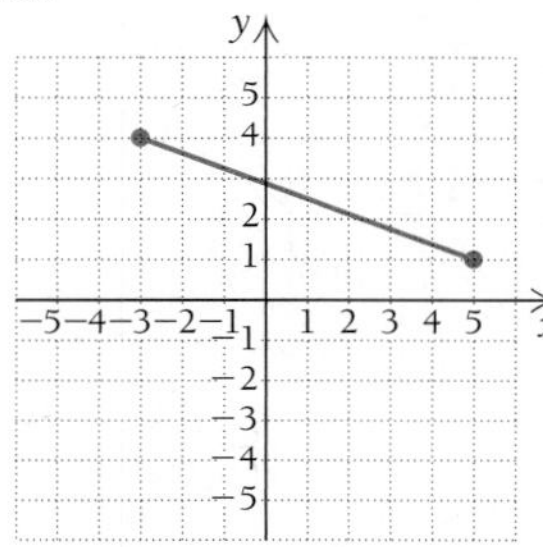

27.

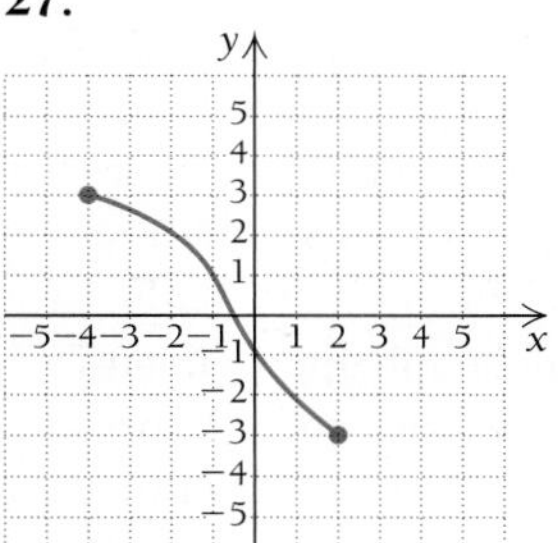

28.

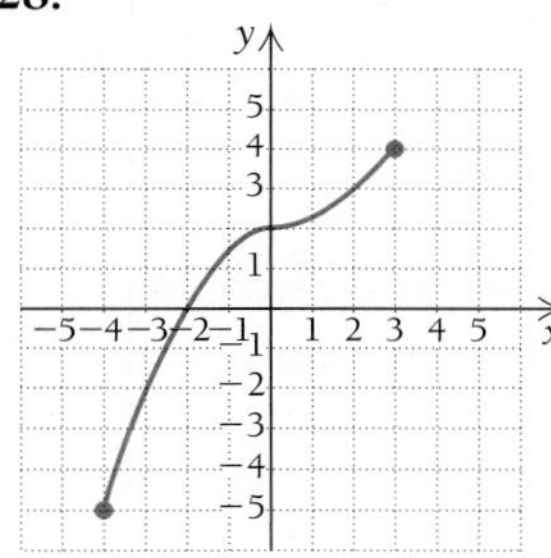

29.

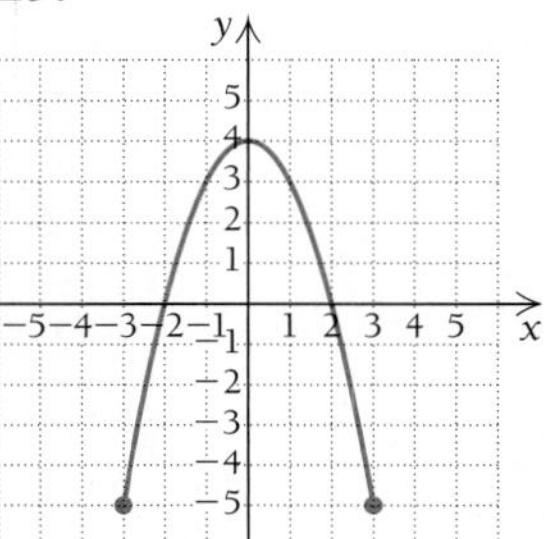

30.

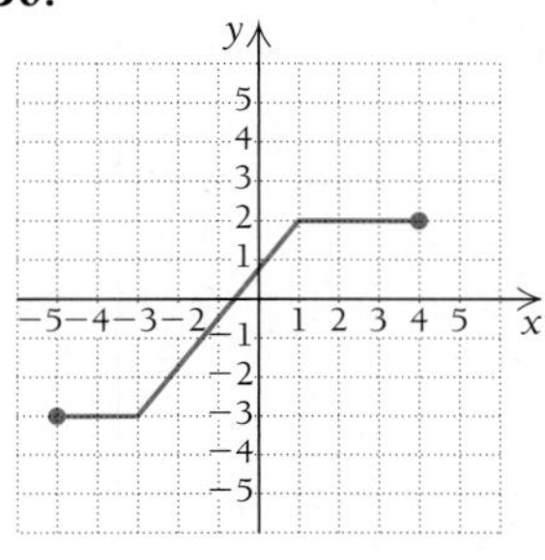

31.

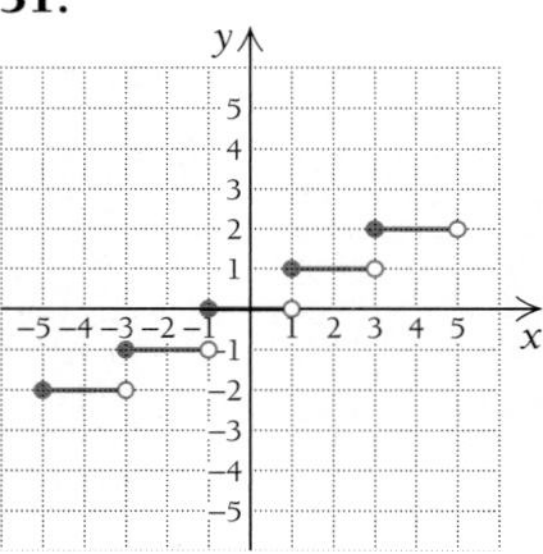

32.

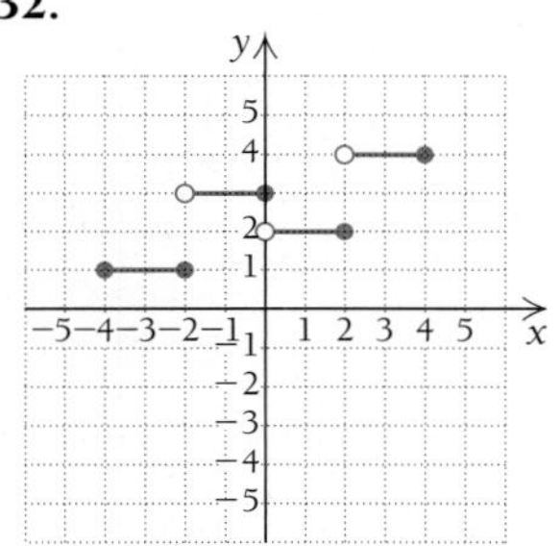

Find the domain of each function given below.

33. $f(x) = \dfrac{6}{2 - x}$

34. $f(x) = \dfrac{2}{x + 3}$

35. $f(x) = \sqrt{2x}$

36. $f(x) = \sqrt{x - 2}$

37. $f(x) = x^2 - 2x + 3$

38. $f(x) = x^2 + 3$

39. $f(x) = \dfrac{x - 2}{6x - 12}$

40. $f(x) = \dfrac{8}{3x - 6}$

41. $f(x) = |x - 4|$

42. $f(x) = |x| - 4$

43. $f(x) = \dfrac{3x - 1}{7 - 2x}$

44. $f(x) = \dfrac{2x - 1}{9 - 2x}$

45. $g(x) = \sqrt{4 + 5x}$

46. $g(x) = \sqrt{2 - 3x}$

47. $g(x) = x^2 - 2x + 1$

48. $g(x) = 4x^3 + 5x^2 - 2x$

49. $g(x) = \dfrac{2x}{x^2 - 25}$

50. $g(x) = \dfrac{x - 1}{x^2 - 36}$

51. $g(x) = |x| + 1$

52. $g(x) = |x + 7|$

53. $g(x) = \dfrac{2x - 6}{x^2 - 6x + 5}$

54. $g(x) = \dfrac{3x - 10}{x^2 - 4x - 5}$

55. For the function f whose graph is shown below, find all x-values for which $f(x) \le 0$.

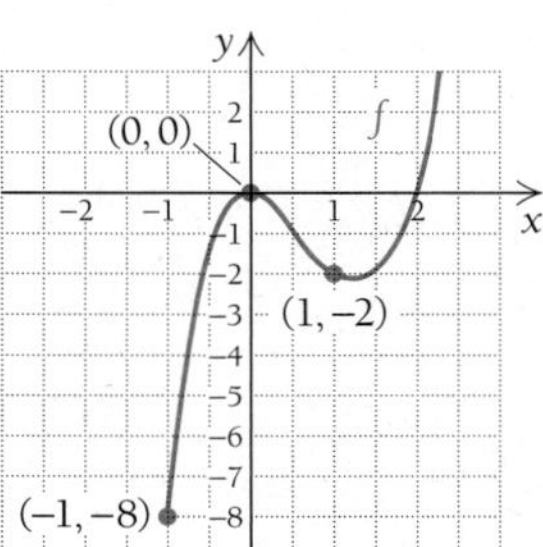

56. For the function g whose graph is shown below, find all x-values for which $g(x) = 1$.

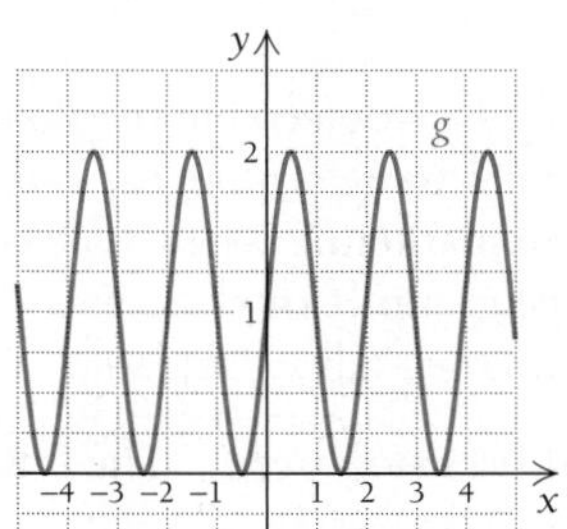

APPLICATIONS

Business and Economics

57. Compound interest. Suppose that \$5000 is invested at 8% interest, compounded semiannually, for t years.

a) The amount A in the account is a function of time. Find an equation for this function.

b) Determine the domain of the function in part (a).

58. Compound interest. Suppose that \$3000 is borrowed as a college loan, at 5% interest, compounded daily, for t years.

a) The amount A that is owed is a function of time. Find an equation for this function.

b) Determine the domain of the function in part (a).

Life and Physical Sciences

59. Hearing-impaired Americans. The following graph (considered in Exercise Set R.1) approximates the number N, in millions, of hearing-impaired Americans as a function of age x. (*Source:* American Speech-Language Hearing Association.) The equation for this graph is the function given by

$$N(x) = -0.00006x^3 + 0.006x^2 - 0.1x + 1.9.$$

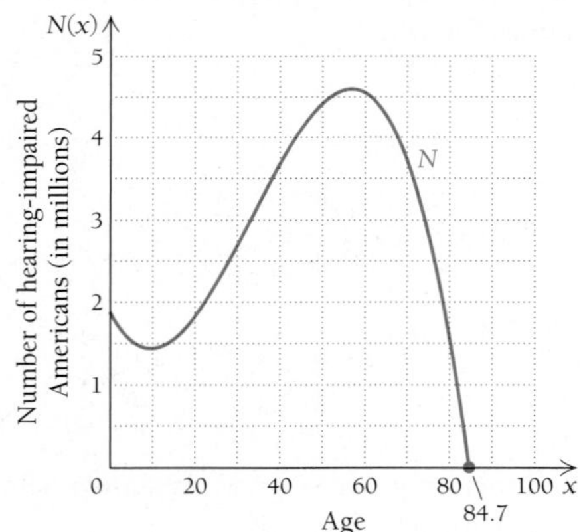

a) Use the graph to determine the domain of N.

b) Use the graph to determine the range of N.

TW **c)** If you were marketing a new type of hearing aid, at what age group (expressed as a 10-yr interval) would you target advertisements? Why?

60. Incidence of breast cancer. The following graph (considered in Exercise Set R.1 without an equation) approximates the incidence of breast cancer I, per 100,000 women, as a function of age x. The equation for this graph is the function given by

$$I(x) = -0.0000554x^4 + 0.0067x^3 - 0.0997x^2 - 0.84x - 0.25.$$

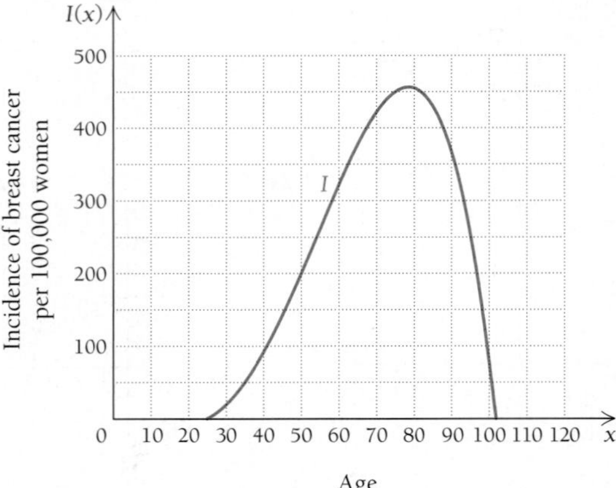

(*Source:* Based on data from the National Cancer Institute.)

a) Use the graph to determine the domain of I.

b) Use the graph to determine the range of I.

TW **c)** What 10-yr age interval sees the greatest increase in the incidence of breast cancer? Explain how you determined this.

61. Lung cancer. The following graph approximates the incidence of lung and bronchus cancer L, per 100,000 males, as a function of t, the number of years since 1940. The equation for this graph is the function given by

$$L(t) = -0.00054t^3 + 0.02917t^2 + 1.2329t + 8.$$

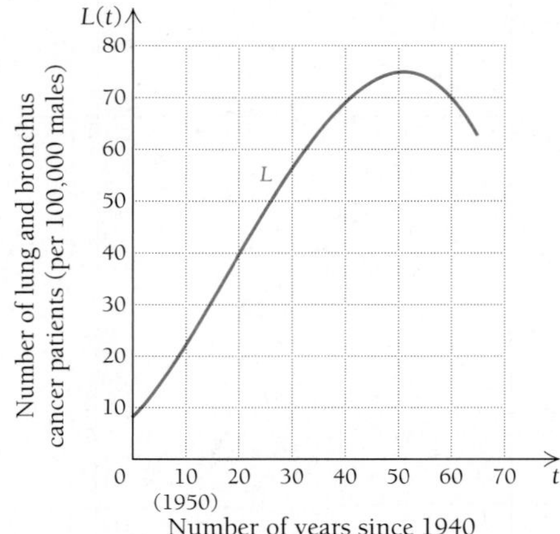

(*Source:* Based on data from the American Cancer Society Surveillance Research, 2005.)

a) Use the graph to estimate the domain of L.

b) Use the graph to estimate the range of L.

62. See Exercise 61.

a) Use the graph to approximate all the x-values where the cancer rate is 50 per 100,000.

b) Use the graph to approximate all the x-values where the cancer rate is 70 per 100,000.

c) Use the formula to approximate the lung and bronchus cancer rate in 2005.

SYNTHESIS

TW **63.** For a given function, $f(2) = -5$. Give as many interpretations of this fact as you can.

TW **64.** Explain how it is possible for the domain and the range of a function to be the same set.

TW **65.** Give an example of a function for which the number 3 is not in the domain, and explain why it is not.

TECHNOLOGY CONNECTION

66. Determine the range of each of the functions in Exercises 33, 35, 39, 40, and 47.

67. Determine the range of each of the functions in Exercises 34, 36, 48, 51, and 54.

R.4 Slope and Linear Functions

OBJECTIVES

- Graph equations of the types $y = f(x) = c$ and $x = a$.
- Graph linear functions.
- Find an equation of a line when given its slope and one point on the line and when given two points on the line.
- Solve applied problems involving slope and linear functions.

Horizontal and Vertical Lines

Let's consider graphs of equations $y = c$ and $x = a$.

EXAMPLE 1

a) Graph $y = 4$.

b) Decide whether the graph represents a function.

Solution

a) The graph consists of all ordered pairs whose second coordinate is 4. To see how a pair such as $(-2, 4)$ could be a solution of $y = 4$, we can consider the equation above in the form

$$y = 0x + 4.$$

Then $(-2, 4)$ is a solution because

$$0(-2) + 4 = 4$$

is true.

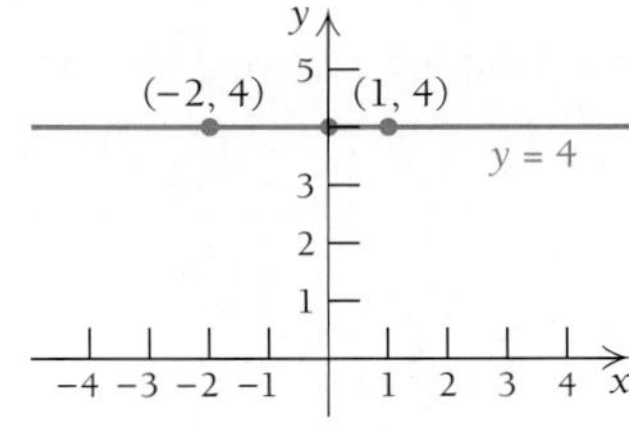

b) The vertical-line test holds. Thus, the graph represents a function. ✦

EXAMPLE 2

a) Graph $x = -3$.

b) Decide whether the graph represents a function.

Solution

a) The graph consists of all ordered pairs whose first coordinate is -3. To see how a pair such as $(-3, 4)$ could be a solution of $x = -3$, we can consider the equation in the form

$$x + 0y = -3.$$

Then $(-3, 4)$ is a solution because

$$(-3) + 0(4) = -3$$

is true.

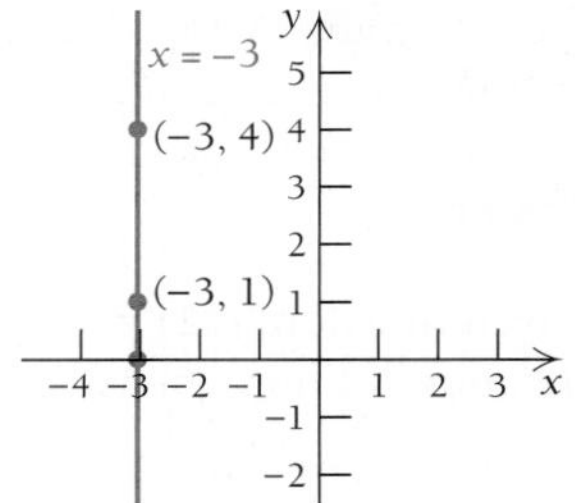

b) This graph does not represent a function because it fails the vertical-line test. The line itself meets the graph more than once—in fact, infinitely many times. ◆

In general, we have the following.

THEOREM 3

The graph of $y = c$, or $f(x) = c$, a horizontal line, is the graph of a function. Such a function is referred to as a **constant function.** The graph of $x = a$, a vertical line, is not the graph of a function.

TECHNOLOGY CONNECTION

Visualizing Slope

Exploratory: Squaring a Viewing Window

The standard $[-10, 10, -10, 10]$ viewing window shown below is not scaled identically on both axes. Note that the intervals on the y-axis are about two-thirds the length of those on the x-axis.

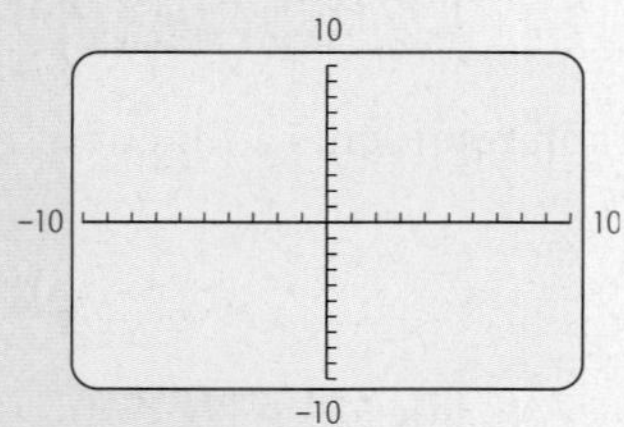

If we change the dimensions of the window to $[-6, 6, -4, 4]$, we get a graph for which the units are visually about the same on both axes.

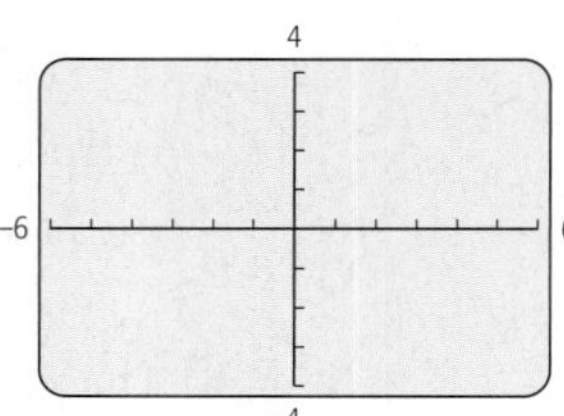

Creating such a window is called **squaring the window.** On many calculators, this is accomplished automatically by selecting the ZSQUARE option of the ZOOM menu.

Visualizing Slope (continued)

Each of the following is a graph of $y = 2x - 3$, but with different viewing windows. When the window is square, as shown in the last graph, we get the most accurate representation of the *slope* of the line.

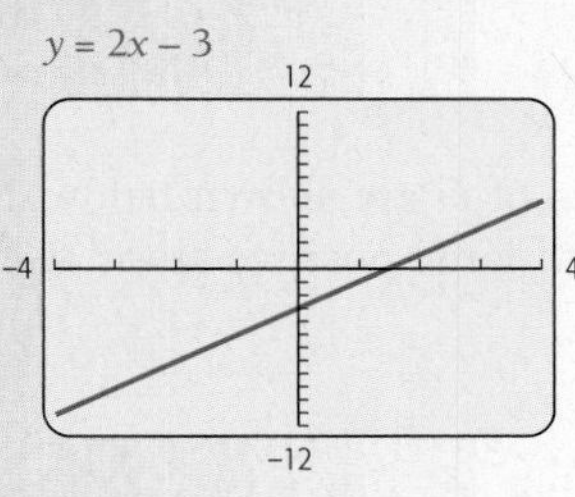

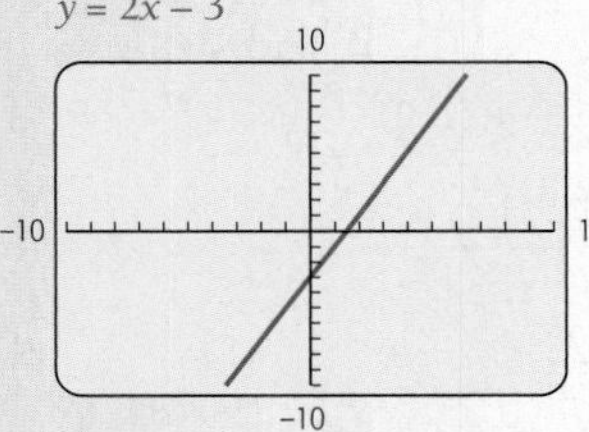

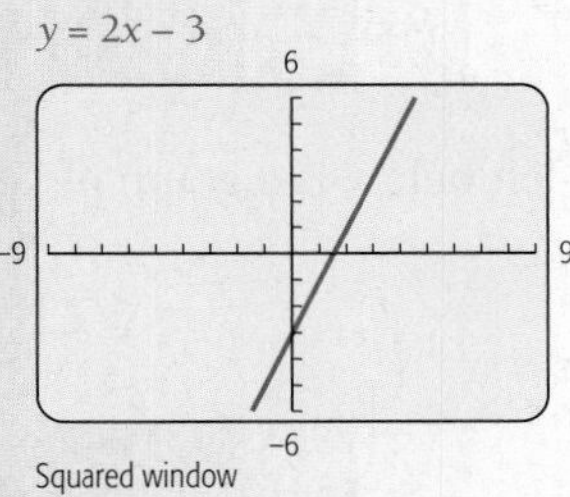

Squared window

EXERCISES

Use a squared viewing window for each of these exercises.

1. Graph $y = x + 1$, $y = 2x + 1$, $y = 3x + 1$, and $y = 10x + 1$. What do you think the graph of $y = 247x + 1$ will look like?
2. Graph $y = x$, $y = \frac{7}{8}x$, $y = 0.47x$, and $y = \frac{2}{31}x$. What do you think the graph of $y = 0.000018x$ will look like?
3. Graph $y = -x$, $y = -2x$, $y = -5x$, and $y = -10x$. What do you think the graph of $y = -247x$ will look like?
4. Graph $y = -x - 1$, $y = -\frac{3}{4}x - 1$, $y = -0.38x - 1$, and $y = -\frac{5}{32}x - 1$. What do you think the graph of $y = -0.000043x - 1$ will look like?

The Equation $y = mx$

Consider the following table of numbers and look for a pattern.

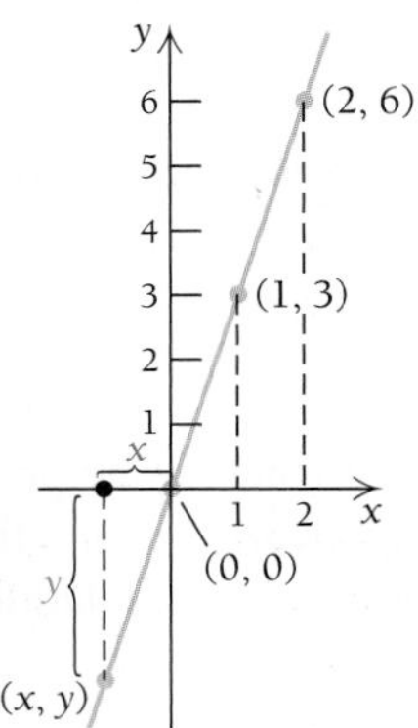

x	1	-1	$-\frac{1}{2}$	2	-2	3	-7	5
y	3	-3	$-\frac{3}{2}$	6	-6	9	-21	15

Note that the ratio of the bottom number to the top one is 3. That is,

$$\frac{y}{x} = 3, \quad \text{or} \quad y = 3x.$$

Ordered pairs from the table can be used to graph the equation $y = 3x$ (see the figure at the left). Note that this is a function.

THEOREM 4

The graph of the function given by

$$y = mx \qquad \text{or} \qquad f(x) = mx$$

is the straight line through the origin $(0, 0)$ and the point $(1, m)$. The constant m is called the **slope** of the line.

Various graphs of $y = mx$ for positive values of m are shown below. Note that such graphs slant up from left to right. A line with large positive slope rises faster than a line with smaller positive slope.

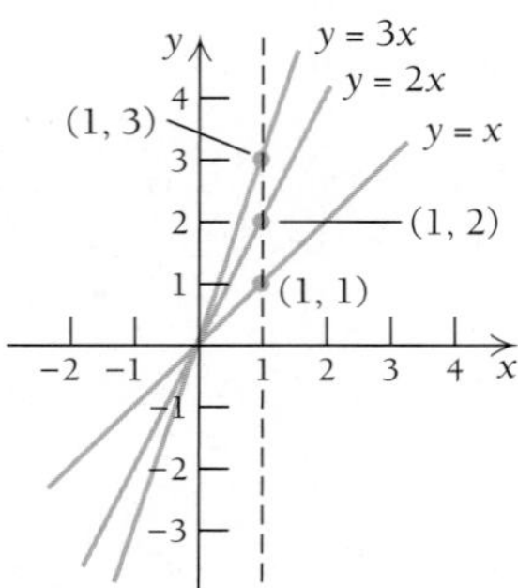

When $m = 0$, $y = 0x$, or $y = 0$. On the left below is a graph of $y = 0$. Note that this is both the x-axis and a horizontal line.

Lines of various slopes.

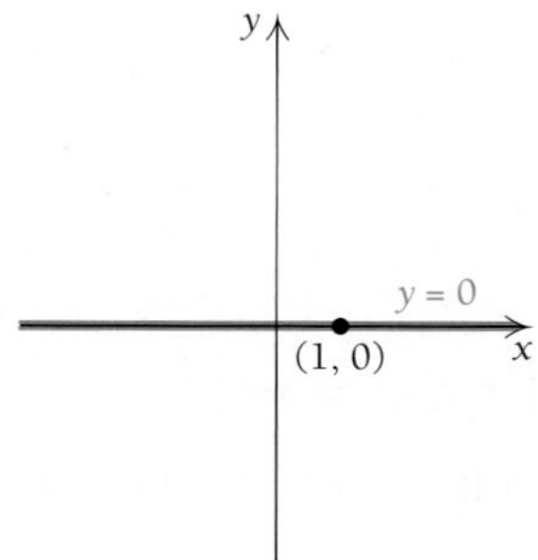

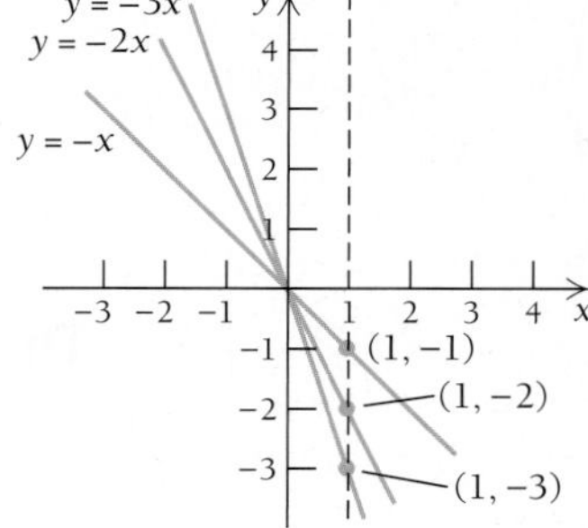

Graphs of $y = mx$ for negative values of m are shown on the right above. Note that such graphs slant down from left to right.

Direct Variation

There are many applications involving equations like $y = mx$, where m is some positive number. In such situations, we say that we have **direct variation,** and m (the slope) is called the **variation constant,** or **constant of proportionality.** Generally, only positive values of x and y are considered.

DEFINITION

The variable y **varies directly** as x if there is some positive constant m such that $y = mx$. We also say that y is **directly proportional** to x.

EXAMPLE 3 Life Science: Weight on Earth and the Moon. The weight M, in pounds, of an object on the moon is directly proportional to the weight E of that object on Earth. An astronaut who weighs 180 lb on Earth will weigh 28.8 lb on the moon.

a) Find an equation of variation.

b) An astronaut weighs 120 lb on Earth. How much will the astronaut weigh on the moon?

Solution

a) The equation has the form $M = mE$. To find m, we substitute:

$$M = mE$$
$$28.8 = m \cdot 180$$
$$\frac{28.8}{180} = m$$
$$0.16 = m.$$

Thus, $M = 0.16E$ is the equation of variation.

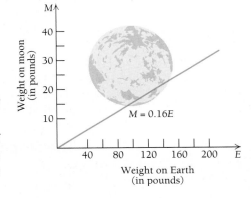

b) To find the weight on the moon of an astronaut who weighs 120 lb on Earth, we substitute 120 for E in the equation of variation,

$$M = 0.16 \cdot 120, \quad \text{Substituting 120 for } E$$

and get

$$M = 19.2.$$

Thus, an astronaut who weighs 120 lb on Earth weighs 19.2 lb on the moon. ✦

TECHNOLOGY CONNECTION

Exploring b

Let's explore the effect of b when graphing $y = mx + b$.

EXERCISES

1. Graph $y_1 = x$. Then, using the same viewing window, compare the graphs of $y_2 = x + 3$ and $y_3 = x - 4$. Next, without drawing them, describe how the graphs of $y = x$ and $y = x - 5$ compare.
2. Use the TABLE feature to compare the values of y_1, y_2, and y_3 when $x = 0$. Then scroll through other values and describe a pattern.

The Equation $y = mx + b$

Compare the graphs of the equations

$$y = 3x \quad \text{and} \quad y = 3x - 2$$

(see the following figure). Note that the graph of $y = 3x - 2$ is a shift 2 units down of the graph of $y = 3x$, and that $y = 3x - 2$ has y-intercept $(0, -2)$. Both graphs represent functions.

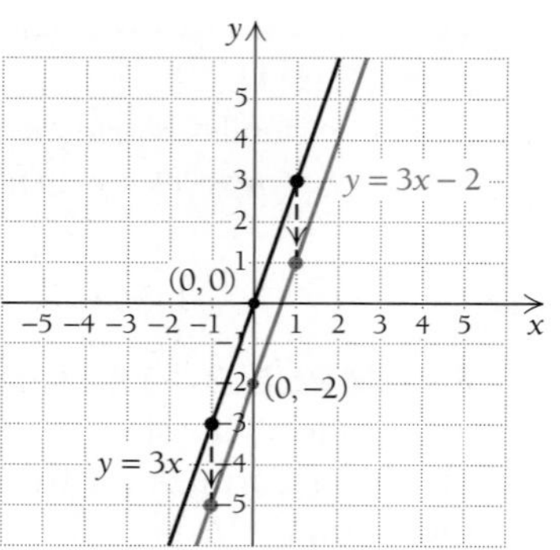

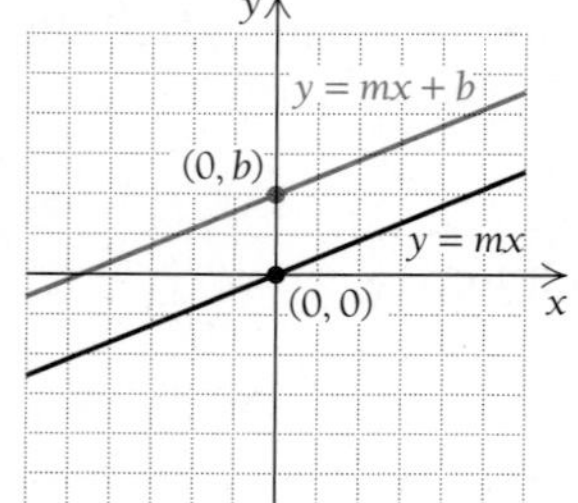

DEFINITION

A **linear function** is given by

$$y = mx + b \quad \text{or} \quad f(x) = mx + b$$

and has a graph that is the straight line parallel to the graph of $y = mx$ and crossing the y-axis at $(0, b)$. The point $(0, b)$ is called the **y-intercept.** (See the figure at the left.)

As before, the constant m is the slope of the line. When $m = 0$, $y = 0x + b = b$, and we have a constant function (see Theorem 3 at the beginning of the section). The graph of such a function is a horizontal line.

The Slope–Intercept Equation

Every nonvertical line l is uniquely determined by its slope m and its y-intercept $(0, b)$. In other words, the slope describes the "slant" of the line, and the y-intercept locates the point at which the line crosses the y-axis. Thus, we have the following definition.

DEFINITION

$y = mx + b$ is called the **slope–intercept equation** of a line.

EXAMPLE 4 Find the slope and the y-intercept of the graph of $2x - 4y - 7 = 0$.

Solution We solve for y:

$$2x - 4y - 7 = 0$$

$$2x - 7 = 4y \qquad \text{Adding } 4y \text{ to both sides}$$

$$\frac{2}{4}x - \frac{7}{4} = y \qquad \text{Dividing both sides by 4}$$

Slope: $\frac{1}{2}$ $\qquad$ y-intercept: $\left(0, -\frac{7}{4}\right)$ ◆

The Point–Slope Equation

Suppose that we know the slope of a line and some point on the line other than the y-intercept. We can still find an equation of the line.

EXAMPLE 5 Find an equation of the line with slope 3 containing the point $(-1, -5)$.

Solution The slope is given as $m = 3$. From the slope–intercept equation, we have

$$y = 3x + b,$$

so we must determine b. Since $(-1, -5)$ is on the line, it follows that

$$-5 = 3(-1) + b$$

$$-5 = -3 + b,$$

so $\quad -2 = b \quad$ and $\quad y = 3x - 2.$ ✦

More generally, if a point (x_1, y_1) is on the line given by

$$y = mx + b, \tag{1}$$

it must follow that

$$y_1 = mx_1 + b. \tag{2}$$

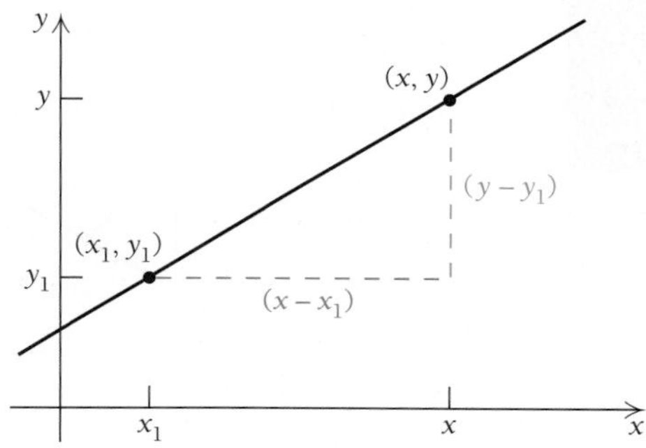

Subtracting the left and right sides of equation (2) from the left and right sides, respectively, of equation (1), we have

$$\begin{aligned} y - y_1 &= (mx + b) - (mx_1 + b) \\ &= mx + b - mx_1 - b \\ &= mx - mx_1 \\ &= m(x - x_1). \end{aligned}$$

DEFINITION

$y - y_1 = m(x - x_1)$ is called the **point–slope equation** of a line.

This definition allows us to write an equation of a line given its slope and the coordinates of *any* point on it.

EXAMPLE 6 Find an equation of the line with slope $\frac{2}{3}$ containing the point $(-1, -5)$.

Solution Substituting in

$$y - y_1 = m(x - x_1),$$

we get

$$\begin{aligned} y - (-5) &= \tfrac{2}{3}[x - (-1)] \\ y + 5 &= \tfrac{2}{3}(x + 1) \\ y + 5 &= \tfrac{2}{3}x + \tfrac{2}{3} \\ y &= \tfrac{2}{3}x + \tfrac{2}{3} - 5 \\ y &= \tfrac{2}{3}x + \tfrac{2}{3} - \tfrac{15}{3} \\ y &= \tfrac{2}{3}x - \tfrac{13}{3}. \end{aligned}$$

✦

Which lines have the same slope?

Computing Slope

We now determine a method of computing the slope of a line when we know the coordinates of two of its points. Suppose that (x_1, y_1) and (x_2, y_2) are the coordinates of two different points, P_1 and P_2, respectively, on a line that is not vertical. Consider a right triangle with legs parallel to the axes, as shown in the following figure.

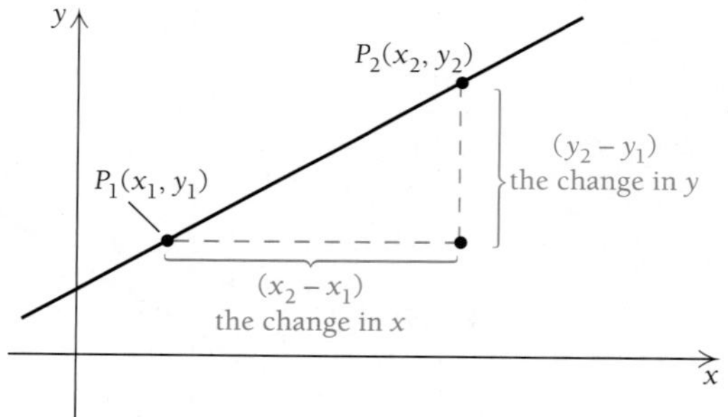

Note that the change in y is $y_2 - y_1$ and the change in x is $x_2 - x_1$. The ratio of these changes is the slope. To see this, consider the point–slope equation,

$$y - y_1 = m(x - x_1).$$

Since (x_2, y_2) is on the line, it must follow that

$$y_2 - y_1 = m(x_2 - x_1). \quad \text{Substituting}$$

Since the line is not vertical, the two x-coordinates must be different; thus, $x_2 - x_1$ is nonzero, and we can divide by it to get the following theorem.

THEOREM 5

$$m = \frac{y_2 - y_1}{x_2 - x_1} = \frac{\text{change in } y}{\text{change in } x} = \text{slope of line containing points } (x_1, y_1) \text{ and } (x_2, y_2)$$

EXAMPLE 7 Find the slope of the line containing the points $(-2, 6)$ and $(-4, 9)$.

Solution We have

$$m = \frac{y_2 - y_1}{x_2 - x_1} = \frac{6 - 9}{-2 - (-4)}$$ We treated $(-2, 6)$ as P_2 and $(-4, 9)$ as P_1.

$$= \frac{-3}{2} = -\frac{3}{2}.$$

Note that it does not matter which point is taken first, so long as we subtract the coordinates in the same order. In this example, we can also find m as follows:

$$m = \frac{9 - 6}{-4 - (-2)} = \frac{3}{-2} = -\frac{3}{2}.$$ Here, $(-4, 9)$ serves as P_2, and $(-2, 6)$ serves as P_1. ✦

If a line is horizontal, the change in y for any two points is 0. Thus, a horizontal line has slope 0. If a line is vertical, the change in x for any two points is 0. Thus, the slope is *not defined* because we cannot divide by 0. A vertical line has undefined slope. Thus, "0 slope" and "undefined slope" are two very different concepts.

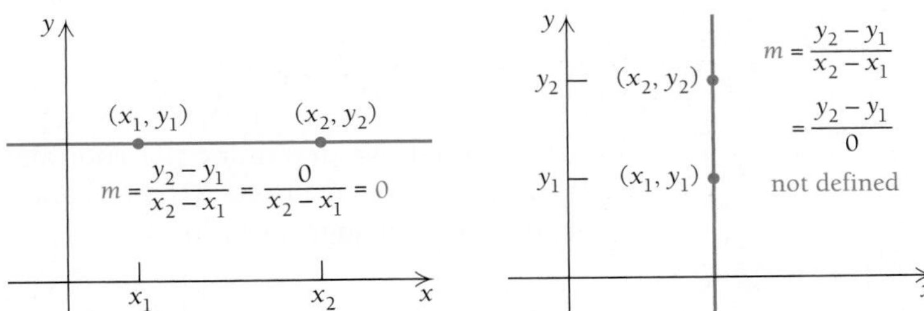

Applications of Slope

Slope has many real-world applications. For example, numbers like 2%, 3%, and 6% are often used to represent the *grade* of a road, a measure of how steep a road on a hill is. A 3% grade $\left(3\% = \frac{3}{100}\right)$ means that for every horizontal distance of 100 ft, the road rises 3 ft. In architecture, the *pitch* of a roof is a measure of how steeply it is angled—a steep pitch sheds more snow than a shallow pitch. Wheelchair-ramp design also involves slope: Building codes rarely allow the steepness of a wheelchair ramp to exceed $\frac{1}{12}$.

Slope can also be considered as an **average rate of change.**

EXAMPLE 8 Life Science: Amount Spent on Cancer Research. The amount appropriated for cancer research has increased steadily over the years and is approximated in the following graph. Find the average rate of change of that amount.

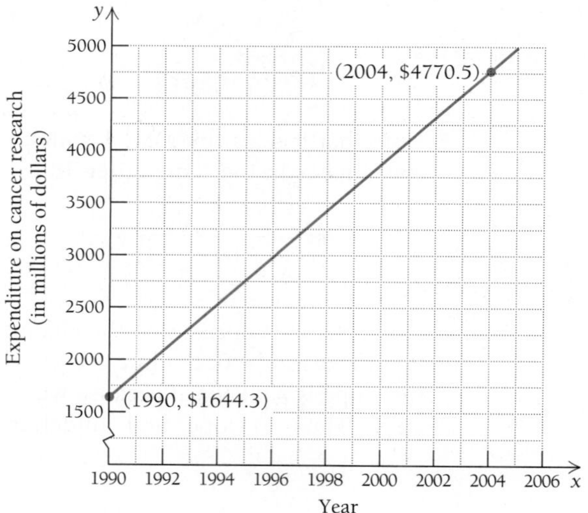

(*Source*: Based on data from the National Cancer Institute.)

Solution First, we determine the coordinates of two points on the graph. In this case, they are given as (1990, \$1644.3) and (2004, \$4770.5). Then we compute the slope, or rate of change, as follows:

$$\text{Slope} = \text{Average rate of change} = \frac{\text{change in } y}{\text{change in } x}$$

$$= \frac{\$4770.5 - \$1644.3}{2004 - 1990} = \frac{\$3126.2}{14} \approx \$223.3 \text{ million/yr.}$$

This result tells us that between 1990 and 2004, the amount appropriated for cancer research increased by about \$223.3 million per year. ✦

Applications of Linear Functions

Many applications are modeled by linear functions.

EXAMPLE 9 Business: Total Cost. Raggs, Ltd., a clothing firm, has **fixed costs** of \$10,000 per year. These costs, such as rent, maintenance, and so on, must be paid no matter how much the company produces. To produce x units of a certain kind of suit, it costs \$20 per suit (unit) in addition to the fixed costs. That is, the **variable costs** for producing x of these suits are $20x$ dollars. These costs are due to the amount produced and stem from items such as material, wages, fuel, and so on. The **total cost** $C(x)$ of producing x suits in a year is given by a function C:

$$C(x) = (\text{Variable costs}) + (\text{Fixed costs}) = 20x + 10{,}000.$$

a) Graph the variable-cost, the fixed-cost, and the total-cost functions.

b) What is the total cost of producing 100 suits? 400 suits?

Solution

a) The variable-cost and fixed-cost functions appear in the graph on the left below. The total-cost function is shown in the graph on the right. From a practical standpoint, the domains of these functions are nonnegative integers 0, 1, 2, 3, and so on, since it does not make sense to make either a negative number or a fractional number of suits. It is common practice to draw the graphs as though the domains were the entire set of nonnegative real numbers.

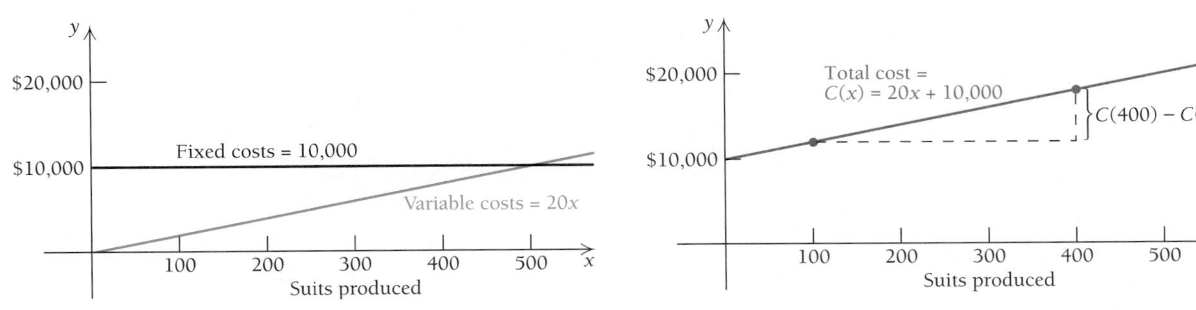

b) The total cost of producing 100 suits is

$$C(100) = 20 \cdot 100 + 10{,}000 = \$12{,}000.$$

The total cost of producing 400 suits is

$$C(400) = 20 \cdot 400 + 10{,}000$$
$$= \$18{,}000.$$

✦

EXAMPLE 10 Business: Profit-and-Loss Analysis. When a business sells an item, it receives the *price* paid by the consumer (this is normally greater than the *cost* to the business of producing the item).

a) The **total revenue** that a business receives is the product of the number of items sold and the price paid per item. Thus, if Raggs, Ltd., sells x suits at \$80 per suit, the total revenue $R(x)$, in dollars, is given by

$$R(x) = \text{Unit price} \cdot \text{Quantity sold} = 80x.$$

If $C(x) = 20x + 10{,}000$ (see Example 9), graph R and C using the same set of axes.

b) The **total profit** that a business receives is the amount left after all costs have been subtracted from the total revenue. Thus, if $P(x)$ represents the total profit when x items are produced and sold, we have

$$P(x) = (\text{Total revenue}) - (\text{Total costs}) = R(x) - C(x).$$

Determine $P(x)$ and draw its graph using the same set of axes as was used for the graph in part (a).

c) The company will *break even* at that value of x for which $P(x) = 0$ (that is, no profit and no loss). This is the point at which $R(x) = C(x)$. Find the **break-even value** of x.

Solution

a) The graphs of $R(x) = 80x$ and $C(x) = 20x + 10{,}000$ are shown below. When $C(x)$ is above $R(x)$, a loss will occur. This is shown by the region shaded red. When $R(x)$ is above $C(x)$, a gain will occur. This is shown by the region shaded gray.

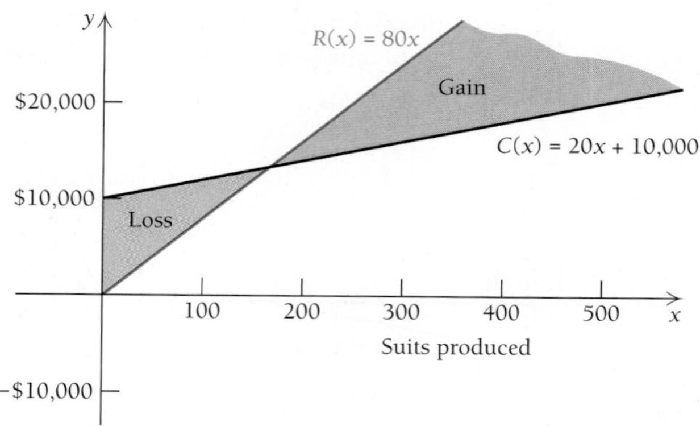

b) To find P, the profit function, we have

$$P(x) = R(x) - C(x) = 80x - (20x + 10{,}000)$$
$$= 60x - 10{,}000.$$

The graph of $P(x)$ is shown by the heavy line. The red portion of the line shows a "negative" profit, or loss. The black portion of the heavy line shows a "positive" profit, or gain.

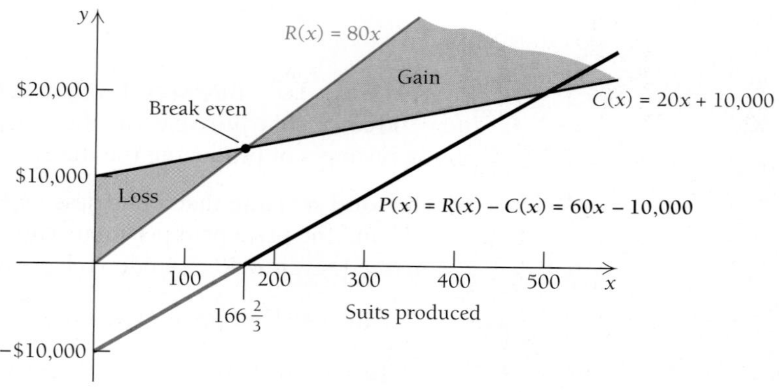

c) To find the break-even value, we solve $R(x) = C(x)$:

$$R(x) = C(x)$$
$$80x = 20x + 10{,}000$$
$$60x = 10{,}000$$
$$x = 166\tfrac{2}{3}.$$

How do we interpret the fractional answer, since it is not possible to produce $\frac{2}{3}$ of a suit? We simply round to 167. Estimates of break-even values are usually sufficient since companies want to operate well away from break-even values in order to maximize profit. ✦

R.4 Exercise Set

Graph.

1. $x = 3$
2. $x = 5$
3. $y = -2$
4. $y = -4$
5. $x = -4.5$
6. $x = -1.5$
7. $y = 3.75$
8. $y = 2.25$

Graph. List the slope and y-intercept.

9. $y = -2x$
10. $y = -3x$
11. $f(x) = 0.5x$
12. $f(x) = -0.5x$
13. $y = 3x - 4$
14. $y = 2x - 5$
15. $g(x) = -x + 3$
16. $g(x) = x - 2.5$
17. $y = 7$
18. $y = -5$

Find the slope and y-intercept.

19. $y - 3x = 6$
20. $y - 4x = 1$
21. $2x + y - 3 = 0$
22. $2x - y + 3 = 0$
23. $2x + 2y + 8 = 0$
24. $3x - 3y + 6 = 0$
25. $x = 3y + 7$
26. $x = -4y + 3$

Find an equation of the line:

27. with $m = -5$, containing $(-2, -3)$.
28. with $m = 7$, containing $(1, 7)$.
29. with $m = -2$, containing $(2, 3)$.
30. with $m = -3$, containing $(5, -2)$.
31. with slope 2, containing $(3, 0)$.
32. with slope -5, containing $(5, 0)$.
33. with y-intercept $(0, -6)$ and slope $\frac{1}{2}$.
34. with y-intercept $(0, 7)$ and slope $\frac{4}{3}$.
35. with slope 0, containing $(2, 3)$.
36. with slope 0, containing $(4, 8)$.

Find the slope of the line containing the given pair of points. If a slope is undefined, state that fact.

37. $(5, -3)$ and $(-2, 1)$
38. $(-2, 1)$ and $(6, 3)$
39. $(2, -3)$ and $(-1, -4)$
40. $(-3, -5)$ and $(1, -6)$
41. $(3, -7)$ and $(3, -9)$
42. $(-4, 2)$ and $(-4, 10)$
43. $\left(\frac{4}{5}, -3\right)$ and $\left(\frac{1}{2}, \frac{2}{5}\right)$
44. $\left(-\frac{3}{16}, -\frac{1}{2}\right)$ and $\left(\frac{5}{8}, -\frac{3}{4}\right)$
45. $(2, 3)$ and $(-1, 3)$
46. $\left(-6, \frac{1}{2}\right)$ and $\left(-7, \frac{1}{2}\right)$
47. $(x, 3x)$ and $(x + h, 3(x + h))$
48. $(x, 4x)$ and $(x + h, 4(x + h))$
49. $(x, 2x + 3)$ and $(x + h, 2(x + h) + 3)$
50. $(x, 3x - 1)$ and $(x + h, 3(x + h) - 1)$

51–60. Find an equation of the line containing the pair of points in each of Exercises 37–46.

61. Find the slope of the skateboard ramp.

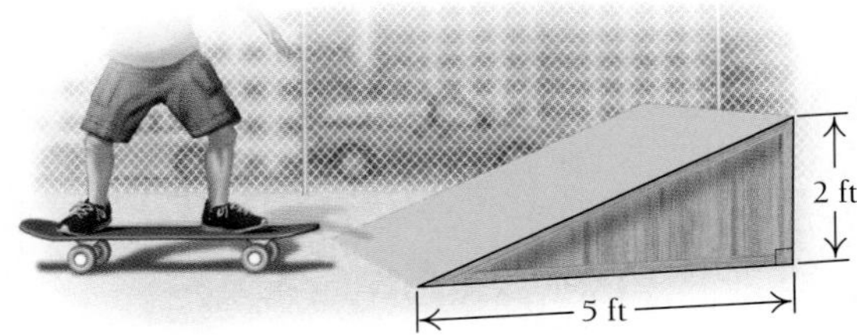

62. Find the slope (or grade) of the treadmill.

63. Find the slope (or head) of the river. Express the answer as a percentage.

64. Stair requirements. A North Carolina state law requires that stairs have minimum treads of 9 in. and maximum risers of 8.25 in. (*Source:* North Carolina Office of the State Fire Marshal.) See the illustration below. According to this law, what is the maximum grade of stairs in North Carolina?

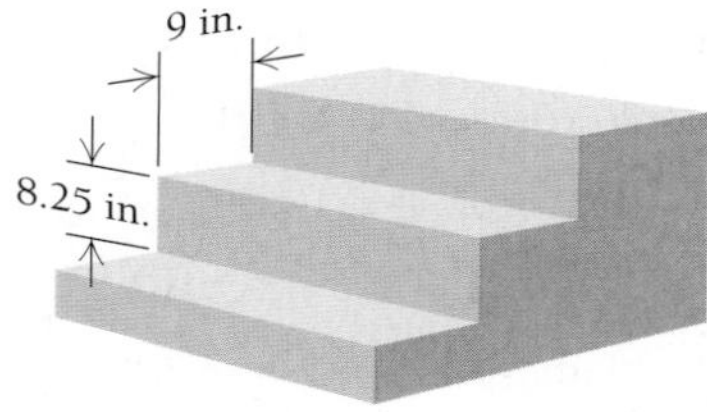

65. Health insurance premiums. Find the average rate of change in the annual premium for a family's health insurance.

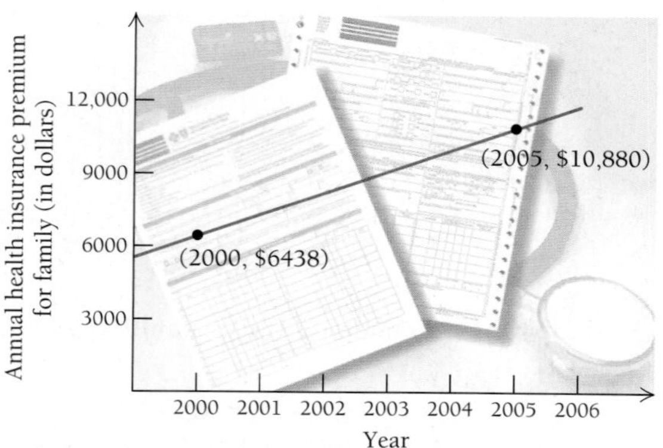

(*Source*: The Kaiser Family Foundation; Health Research and Education Trust.)

66. Health insurance premiums. Find the average rate of change in the annual premium for a single person.

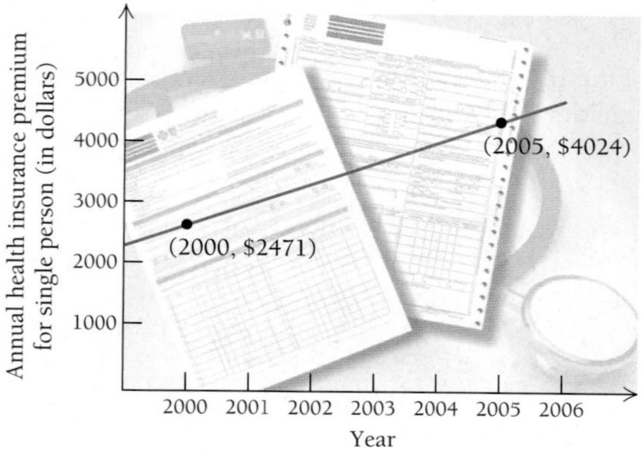

(*Source*: The Kaiser Family Foundation; Health Research and Education Trust.)

67. Find the average rate of change of the tuition and fees at public two-year colleges.

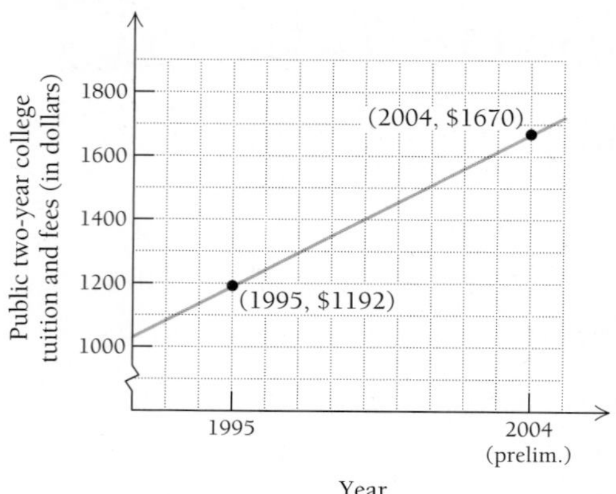

(*Source*: U.S. National Center for Education Statistics, *Digest of Education Statistics*, annual.)

68. Find the average rate of change of the cost of a formal wedding.

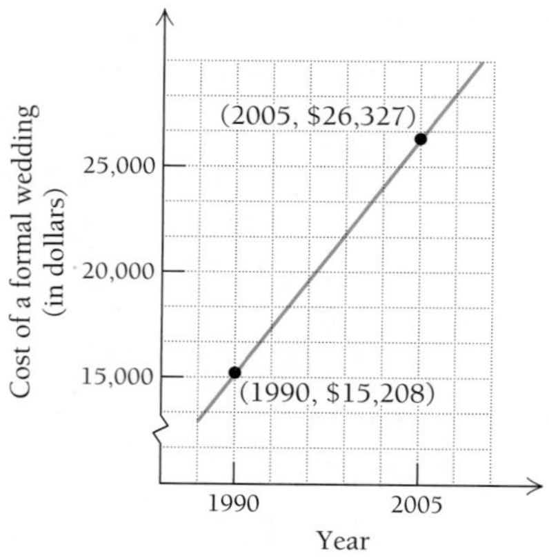

(*Source*: The Fairchild Bridal Group.)

69. Energy conservation. The R-factor of home insulation is directly proportional to its thickness T.

a) Find an equation of variation if $R = 12.51$ when $T = 3$ in.

b) What is the R-factor for insulation that is 6 in. thick?

70. Nerve impulse speed. Impulses in nerve fibers travel at a speed of 293 ft/sec. The distance D, in feet, traveled in t sec is given by $D = 293t$. How long would it take an impulse to travel from the brain to the toes of a person who is 6 ft tall?

71. Muscle weight. The weight M of the muscles in a human is directly proportional to the person's body weight W.

a) It is known that a person who weighs 200 lb has 80 lb of muscles. Find an equation of variation expressing M as a function of W.

b) Express the variation constant as a percent, and interpret the resulting equation.

c) What is the muscle weight of a person who weighs 120 lb?

Muscle weight is directly proportional to body weight.

72. **Brain weight.** The weight B of a human's brain is directly proportional to a person's body weight W.

a) It is known that a person who weighs 120 lb has a brain that weighs 3 lb. Find an equation of variation expressing B as a function of W.

b) Express the variation constant as a percent and interpret the resulting equation.

c) What is the weight of the brain of a person who weighs 160 lb?

73. **Highway tolls.** It has been suggested that since heavier vehicles are responsible for more of the wear and tear on highways, drivers should pay tolls in direct proportion to the weight of their vehicles. Suppose that a Toyota Camry weighing 3350 lb was charged $2.70 for traveling an 80-mile stretch of highway.

a) Find an equation of variation that expresses the amount of the toll T as a function of the vehicle's weight w.

b) What would the toll be if a 3700-lb Jeep Cherokee drove the same stretch of highway?

74. **Inkjet cartridges.** A registrar's office finds that the number of inkjet cartridges, I, required each year for its copiers and printers varies directly with the number of students enrolled, s.

a) Find an equation of variation that expresses I as a function of s, if the office requires 16 cartridges when 2800 students enroll.

b) How many cartridges would be required if 3100 students enrolled?

75. **Profit-and-loss analysis.** Boxowitz, Inc., a computer firm, is planning to sell a new graphing calculator. For the first year, the fixed costs for setting up the new production line are $100,000. The variable costs for producing each calculator are estimated at $20. The sales department projects that 150,000 calculators can be sold during the first year at a price of $45 each.

a) Find and graph $C(x)$, the total cost of producing x calculators.

b) Using the same axes as in part (a), find and graph $R(x)$, the total revenue from the sale of x calculators.

c) Using the same axes as in part (a), find and graph $P(x)$, the total profit from the production and sale of x calculators.

d) What profit or loss will the firm realize if the expected sale of 150,000 calculators occurs?

e) How many calculators must the firm sell in order to break even?

76. **Profit-and-loss analysis.** Red Tide is planning a new line of skis. For the first year, the fixed costs for setting up production are $45,000. The variable costs for producing each pair of skis are estimated at $80, and the selling price will be $255 per pair. It is projected that 3000 pairs will sell the first year.

a) Find and graph $C(x)$, the total cost of producing x pairs of skis.

b) Find and graph $R(x)$, the total revenue from the sale of x pairs of skis. Use the same axes as in part (a).

c) Using the same axes as in part (a), find and graph $P(x)$, the total profit from the production and sale of x pairs of skis.

d) What profit or loss will the company realize if the expected sale of 3000 pairs occurs?

e) How many pairs must the company sell in order to break even?

77. **Straight-line depreciation.** Quick Copy buys an office machine for $5200 on January 1 of a given year. The machine is expected to last for 8 yr, at the end of which time its *salvage value* will be $1100. If the company figures the decline in value to be the same each year, then the *book value*, $V(t)$, after t years, $0 \leq t \leq 8$, is given by

$$V(t) = C - t\left(\frac{C - S}{N}\right),$$

where C is the original cost of the item, N is the number of years of expected life, and S is the salvage value.

a) Find the linear function for the straight-line depreciation of the office machine.
b) Find the book value after 0 yr, 1 yr, 2 yr, 3 yr, 4 yr, 7 yr, and 8 yr.

78. **Profit-and-loss analysis.** Jimmy decides to mow lawns to earn money. The initial cost of his lawnmower is $250. Gasoline and maintenance costs are $4 per lawn.
a) Formulate a function $C(x)$ for the total cost of mowing x lawns.
b) Jimmy determines that the total-profit function for the lawnmowing business is given by $P(x) = 9x - 250$. Find a function for the total revenue from mowing x lawns. How much does Jimmy charge per lawn?
c) How many lawns must Jimmy mow before he begins making a profit?

79. **Straight-line depreciation.** (See Exercise 77.) A business tenant spends $40 per square foot on improvements to a 25,000-ft^2 office space. Under IRS guidelines for straight-line depreciation, these improvements will depreciate completely—that is, have zero salvage value—after 39 yr. Find the depreciated value of the improvements after 10 yr.

80. **Book value.** (See Exercise 77.) The Video Wizard buys a new computer system for $60,000 and projects that its book value will be $2000 after 5 yr. Using straight-line depreciation, find the book value after 3 yr.

81. **Book value.** Tyline Electric uses the function $B(t) = -700t + 3500$ to find the book value, $B(t)$, in dollars, of a photocopier t years after its purchase.

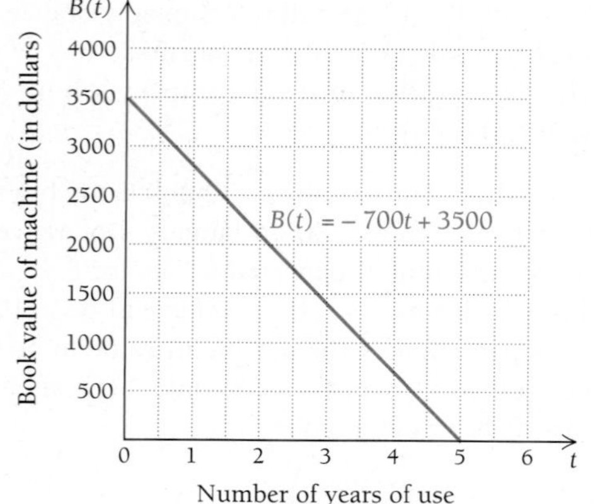

a) What do the numbers -700 and 3500 signify?
b) How long will it take the copier to depreciate completely?
TW c) What is the domain of B? Explain.

82. **Stopping distance on glare ice.** The stopping distance (at some fixed speed) of regular tires on glare ice is given by a linear function of the air temperature F,

$$D(F) = 2F + 115,$$

where $D(F)$ is the stopping distance, in feet, when the air temperature is F, in degrees Fahrenheit.

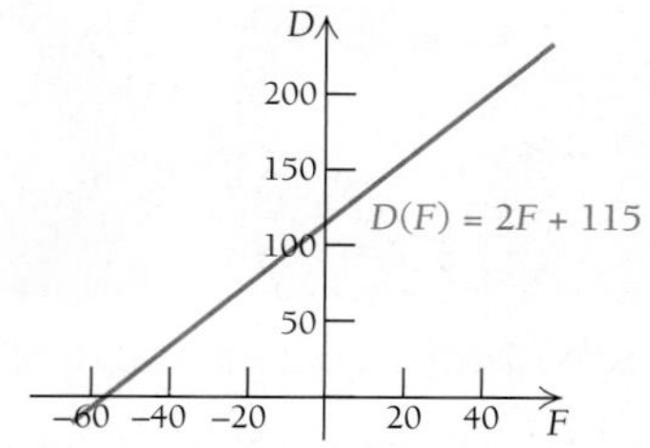

a) Find $D(0°)$, $D(-20°)$, $D(10°)$, and $D(32°)$.
TW b) Explain why the domain should be restricted to the interval $[-57.5°, 32°]$.

83. **Reaction time.** While driving a car, you see a child suddenly crossing the street. Your brain registers the emergency and sends a signal to your foot to hit the brake. The car travels a distance D, in feet, during this time, where D is a function of the speed r, in miles per hour, that the car is traveling when you see the child. That reaction distance is a linear function given by

$$D(r) = \frac{11r + 5}{10}.$$

a) Find $D(5)$, $D(10)$, $D(20)$, $D(50)$, and $D(65)$.
b) Graph $D(r)$.
TW c) What is the domain of the function? Explain.

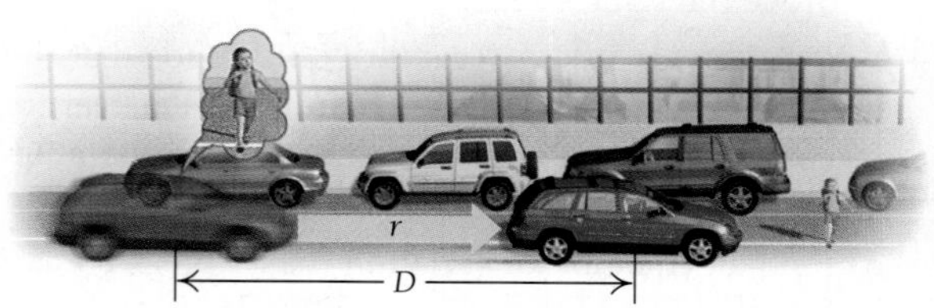

84. Estimating heights. An anthropologist can use certain linear functions to estimate the height of a male or female, given the length of certain bones. The *humerus* is the bone from the elbow to the shoulder. Let $x =$ the length of the humerus, in centimeters. Then the height, in centimeters, of a male with a humerus of length x is given by

$$M(x) = 2.89x + 70.64.$$

The height, in centimeters, of a female with a humerus of length x is given by

$$F(x) = 2.75x + 71.48.$$

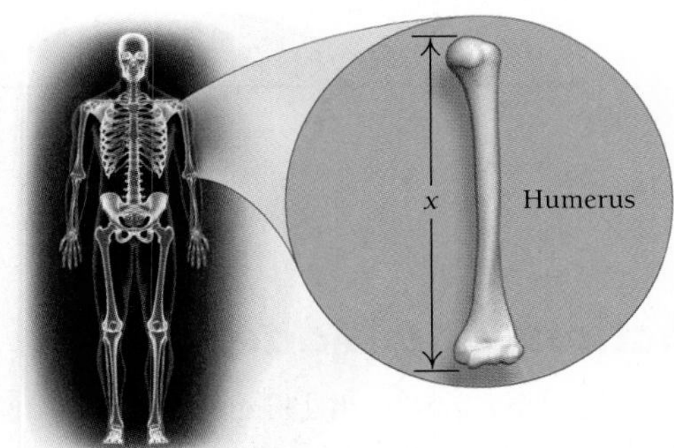

A 26-cm humerus was uncovered in some ruins.

a) If we assume it was from a male, how tall was he?

b) If we assume it was from a female, how tall was she?

85. Manatee population. In January 1999, 1873 manatees were counted in an aerial survey of Florida. In February 2006, 3116 manatees were counted. (*Source:* Florida Fish and Wildlife Conservation Commission.)

a) Using the year as the x-coordinate and the number of manatees as the y-coordinate, find an equation of the line that contains the two data points.

b) Use the equation in part (a) to estimate the number of manatees counted in January 2000.

TW **c)** The actual number counted in January 2000 was 1629. Does the equation found in part (a) give an accurate representation of the number of manatees counted each year?

86. Mallard population. In August 1999, 215 mallards were counted in the annual Mono Lake Fall Shorebird/Waterfowl Count. In August 2003, 543 mallards were counted.

a) Using the year as the x-coordinate and the number of mallards as the y-coordinate, find an equation of the line that contains the two data points.

b) Use the equation in part (a) to estimate the number of mallards counted in August 2002.

TW **c)** The actual number counted in August 2002 was 228. Does the equation found in part (a) give an accurate representation of the number of mallards counted each year?

87. Urban population. The population of Woodland is P. After a growth of 2%, its new population is N.

a) Assuming that N is directly proportional to P, find an equation of variation.

b) Find N when $P = 200{,}000$.

c) Find P when $N = 367{,}200$.

88. Median age of women at first marriage. In general, people in our society are marrying at a later age. The median age, $A(t)$, of women at first marriage can be approximated by the linear function

$$A(t) = 0.08t + 19.7,$$

where t is the number of years after 1950. Thus, $A(0)$ is the median age of women at first marriage in 1950, $A(50)$ is the median age in 2000, and so on.

a) Find $A(0)$, $A(1)$, $A(10)$, $A(30)$, and $A(50)$.

b) What will be the median age of women at first marriage in 2008?

c) Graph $A(t)$.

SYNTHESIS

TW **89.** Explain and compare the situations in which you would use the slope–intercept equation rather than the point–slope equation.

TW **90.** Discuss and relate the concepts of fixed cost, total cost, total revenue, and total profit.

91. Business: daily sales. Match each sentence below with the most appropriate graph (I, II, III, or IV) on the next page.

a) After January 1, daily sales continued to rise, but at a slower rate.

b) After January 1, sales decreased faster than they ever grew.

c) The rate of growth in daily sales doubled after January 1.

d) After January 1, daily sales decreased at half the rate that they grew in December.

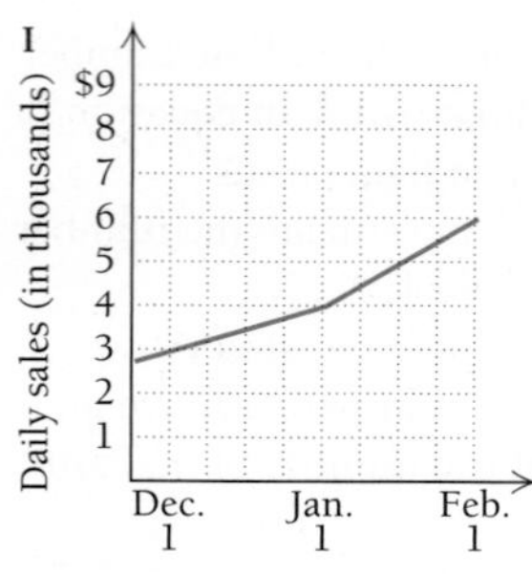

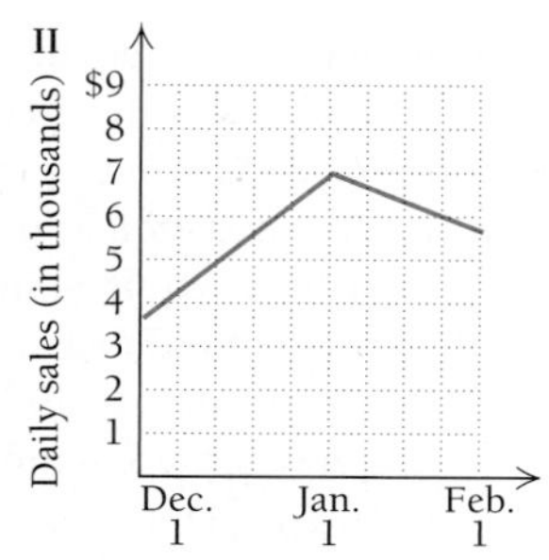

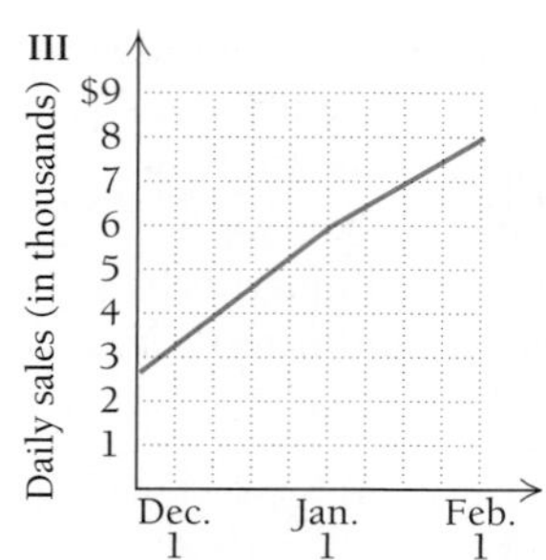

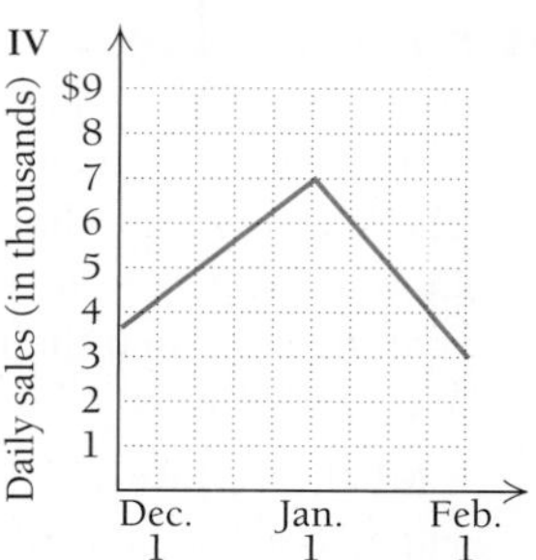

92. Graph some of the total-revenue, total-cost, and total-profit functions in this exercise set using the same set of axes. Identify regions of profit and loss.

R.5 Nonlinear Functions and Models

OBJECTIVES

- Graph functions and solve applied problems.
- Manipulate radical expressions and rational exponents.
- Determine the domain of a rational function and graph certain rational functions.
- Find the equilibrium point given a supply function and a demand function.

There are many functions that have graphs that are not lines. In this section, we study some of these **nonlinear functions** that will frequently appear throughout this course.

Quadratic Functions

DEFINITION

A **quadratic function** f is given by

$$f(x) = ax^2 + bx + c, \quad \text{where } a \neq 0.$$

We have already used quadratic functions—for example, $f(x) = x^2$ and $g(x) = x^2 - 1$. We can create hand-drawn graphs of quadratic functions using the following information.

The graph of a quadratic function $f(x) = ax^2 + bx + c$ is called a **parabola.**

a) It is always a cup-shaped curve, like those in Examples 1 and 2 that follow.
b) It opens upward if $a > 0$ or opens downward if $a < 0$.
c) It has a turning point, or **vertex,** at a point with first coordinate

$$x = -\frac{b}{2a}.$$

d) The vertical line $x = -b/(2a)$ (not part of the graph) serves as the line of symmetry.

EXAMPLE 1 Graph: $f(x) = x^2 - 2x - 3$.

Solution Note that for $f(x) = 1x^2 - 2x - 3$, we have $a = 1$, $b = -2$, and $c = -3$. Since $a > 0$, the graph opens upward. Let's next find the vertex, or turning point. The x-coordinate of the vertex is

$$\begin{aligned} x &= -\frac{b}{2a} \\ &= -\frac{-2}{2(1)} = 1. \end{aligned}$$

Substituting 1 for x, we find the second coordinate of the vertex, $f(1)$:

$$\begin{aligned} f(1) &= 1^2 - 2(1) - 3 \\ &= 1 - 2 - 3 \\ &= -4. \end{aligned}$$

The vertex is $(1, -4)$. The vertical line $x = 1$ is the line of symmetry of the graph. We choose some x-values on each side of the vertex, compute y-values, plot the points, and graph the parabola.

x	$f(x)$
1	−4 ← Vertex
0	−3
2	−3
3	0
4	5
−1	0
−2	5

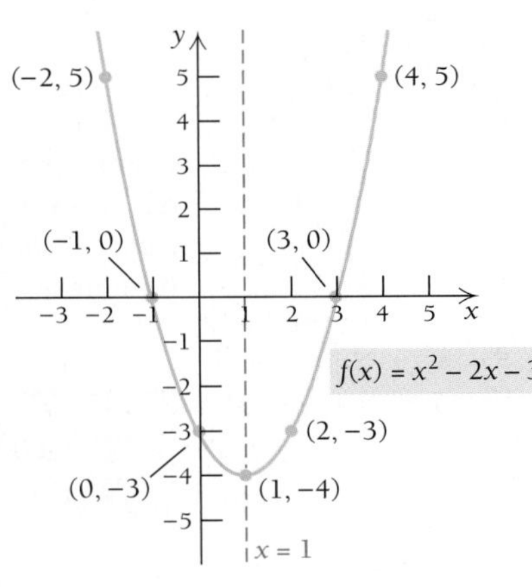

✦

TECHNOLOGY CONNECTION

EXERCISES

Using the procedure of Examples 1 and 2, graph each of the following by hand, using the TABLE feature to create an input–output table for each function. Then press GRAPH to check your sketch.

1. $f(x) = x^2 - 6x + 4$
2. $f(x) = -2x^2 + 4x + 1$

EXAMPLE 2 Graph: $f(x) = -2x^2 + 10x - 7$.

Solution We first note that $a = -2$, and since $a < 0$, the graph will open downward. Let's next find the vertex, or turning point. The x-coordinate of the vertex is

$$\begin{aligned} x &= -\frac{b}{2a} \\ &= -\frac{10}{2(-2)} = \frac{5}{2}. \end{aligned}$$

Substituting $\frac{5}{2}$ for x in the equation, we find the second coordinate of the vertex:

$$\begin{aligned} y = f\left(\tfrac{5}{2}\right) &= -2\left(\tfrac{5}{2}\right)^2 + 10\left(\tfrac{5}{2}\right) - 7 \\ &= -2\left(\tfrac{25}{4}\right) + 25 - 7 \\ &= \tfrac{11}{2}. \end{aligned}$$

The vertex is $\left(\frac{5}{2}, \frac{11}{2}\right)$, and the line of symmetry is $x = \frac{5}{2}$. We choose some x-values on each side of the vertex, compute y-values, plot the points, and graph the parabola:

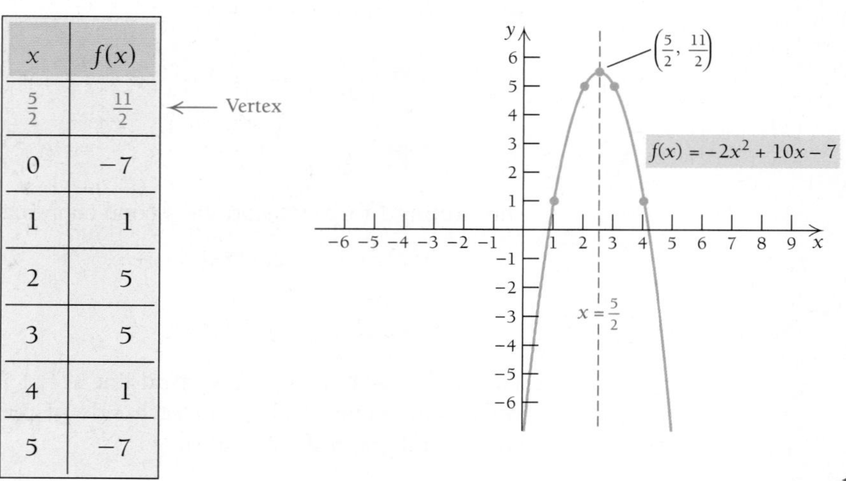

x	$f(x)$
$\frac{5}{2}$	$\frac{11}{2}$
0	−7
1	1
2	5
3	5
4	1
5	−7

✦

First coordinates of points at which a quadratic function intersects the x-axis (x-intercepts), if they exist, can be found by solving the quadratic equation $ax^2 + bx + c = 0$. If real-number solutions exist, they can be found using the *quadratic formula*. See the Appendix at the end of the book for additional review of this important result.

THEOREM 6 The Quadratic Formula

The solutions of any quadratic equation $ax^2 + bx + c = 0$, $a \neq 0$, are given by

$$x = \frac{-b \pm \sqrt{b^2 - 4ac}}{2a}.$$

When solving a quadratic equation, $ax^2 + bx + c = 0$, $a \neq 0$, first try to factor and use the Principle of Zero Products (see the Appendix). When factoring is not possible or seems difficult, use the quadratic formula. It will always give the solutions. When $b^2 - 4ac < 0$, there are no real-number solutions and thus no x-intercepts. There are solutions in an expanded number system called the *complex numbers*. In this text, we will work only with real numbers.

EXAMPLE 3 Solve: $3x^2 - 4x = 2$.

Solution We first find the standard form $ax^2 + bx + c = 0$, and then determine a, b, and c:

$$3x^2 - 4x - 2 = 0,$$
$$a = 3, \quad b = -4, \quad c = -2.$$

We then use the quadratic formula:

$$
\begin{aligned}
x &= \frac{-b \pm \sqrt{b^2 - 4ac}}{2a} \\
&= \frac{-(-4) \pm \sqrt{(-4)^2 - 4(3)(-2)}}{2 \cdot 3} \\
&= \frac{4 \pm \sqrt{16 + 24}}{6} = \frac{4 \pm \sqrt{40}}{6} \\
&= \frac{4 \pm \sqrt{4 \cdot 10}}{6} = \frac{4 \pm 2\sqrt{10}}{6} \\
&= \frac{2(2 \pm \sqrt{10})}{2 \cdot 3} \\
&= \frac{2 \pm \sqrt{10}}{3}.
\end{aligned}
$$

The solutions are $(2 + \sqrt{10})/3$ and $(2 - \sqrt{10})/3$, or approximately 1.721 and -0.387. ✦

TECHNOLOGY CONNECTION

EXERCISE

1. **a)** Below is the graph of

$$f(x) = x^2 - 6x + 8.$$

Using *only* the graph, find the solutions of $x^2 - 6x + 8 = 0$.

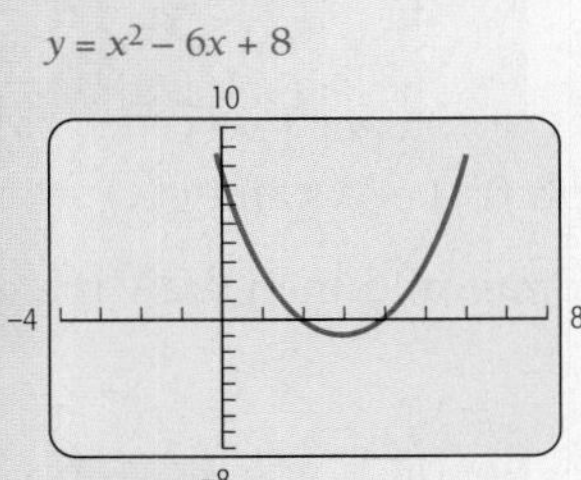

b) Using only a graph, find the solutions of $x^2 + 3x - 10 = 0$.

c) Use the TABLE feature to check your answers to parts (a) and (b).

d) Use a graph and a table to check the solutions of $x^2 + 6x + 8 = 0$, solved algebraically on the right.

Algebraic–Graphical Connection

Let's make an algebraic–graphical connection between the solutions of a quadratic equation and the x-intercepts of a quadratic function.

We just graphed equations of the form $f(x) = ax^2 + bx + c$, $a \neq 0$. Let's look at the graph of $f(x) = x^2 + 6x + 8$ and its x-intercepts, shown below.

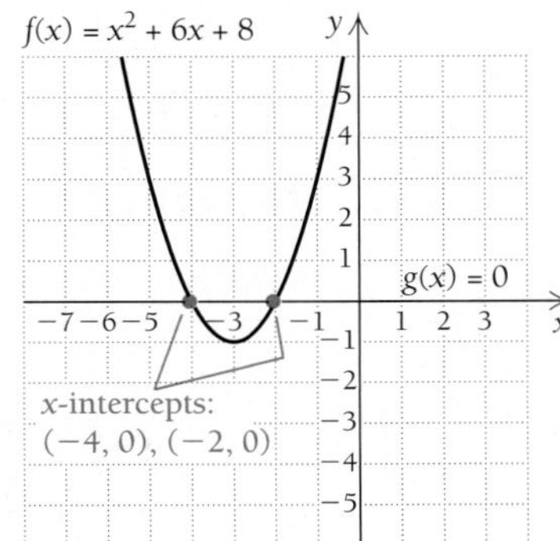

The **x-intercepts,** $(-4, 0)$ and $(-2, 0)$, are the points at which the graph crosses the x-axis. These pairs are also the points of intersection of the graphs of $f(x) = x^2 + 6x + 8$ and $g(x) = 0$ (the x-axis). The x-values, -4 and -2, can be found by solving $f(x) = g(x)$:

$$x^2 + 6x + 8 = 0$$

$$(x + 4)(x + 2) = 0 \qquad \text{Factoring; there is no need for the quadratic formula here.}$$

$$x + 4 = 0 \quad or \quad x + 2 = 0 \qquad \text{Principle of Zero Products}$$

$$x = -4 \quad or \quad x = -2.$$

The solutions of $x^2 + 6x + 8 = 0$ are -4 and -2, which are the first coordinates of the x-intercepts, $(-4, 0)$ and $(-2, 0)$, of the graph of $f(x) = x^2 + 6x + 8$. A brief review of factoring can be found in the Appendix at the end of the book.

Polynomial Functions

Linear and quadratic functions are part of a general class of *polynomial functions.*

DEFINITION

A **polynomial function** f is given by

$$f(x) = a_n x^n + a_{n-1}x^{n-1} + \cdots + a_2x^2 + a_1x^1 + a_0,$$

where n is a nonnegative integer and $a_n, a_{n-1}, \ldots, a_1, a_0$ are real numbers, called the **coefficients.**

The following are examples of polynomial functions:

$f(x) = -5,$ (A constant function)

$f(x) = 4x + 3,$ (A linear function)

$f(x) = -x^2 + 2x + 3,$ (A quadratic function)

$f(x) = 2x^3 - 4x^2 + x + 1.$ (A cubic, or third-degree, function)

In general, creating graphs of polynomial functions other than linear and quadratic functions is difficult without a calculator. We use calculus to sketch such graphs in Chapter 2. Some **power functions,** of the form

$$f(x) = ax^n,$$

are relatively easy to graph.

EXAMPLE 4 Using the same set of axes, graph $f(x) = x^2$ and $g(x) = x^3$.

Solution We set up a table of values, plot the points, and then draw the graphs.

x	x^2	x^3
-2	4	-8
-1	1	-1
$-\frac{1}{2}$	$\frac{1}{4}$	$-\frac{1}{8}$
0	0	0
$\frac{1}{2}$	$\frac{1}{4}$	$\frac{1}{8}$
1	1	1
2	4	8

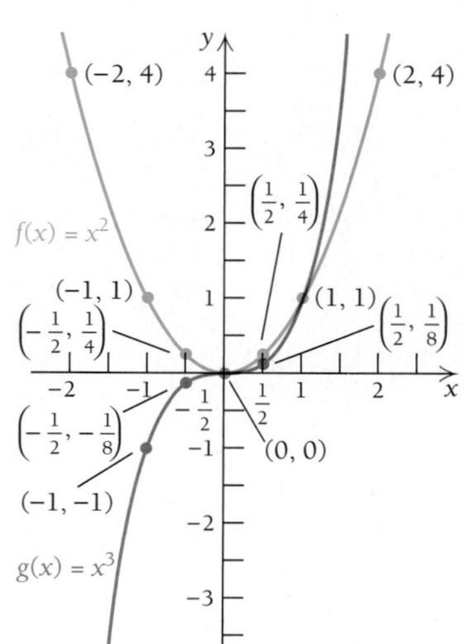

✦

TECHNOLOGY CONNECTION

Solving Polynomial Equations

The INTERSECT Feature

Consider solving the equation

$$x^3 = 3x + 1.$$

Doing so amounts to finding the x-coordinates of the point(s) of intersection of the graphs of

$$f(x) = x^3 \quad \text{and} \quad g(x) = 3x + 1.$$

We enter the functions as

$$y_1 = x^3 \quad \text{and} \quad y_2 = 3x + 1$$

and then graph. We use a $[-3, 3, -5, 8]$ window to see the curvature and possible points of intersection.

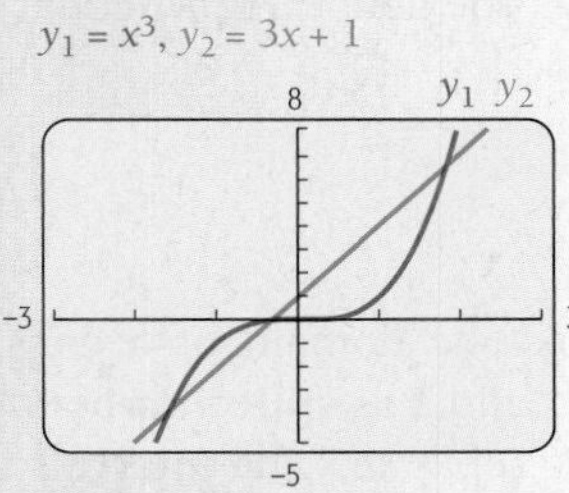

There appear to be at least three points of intersection. Using the INTERSECT feature in the CALC menu, we see that the point of intersection on the left is about $(-1.53, -3.60)$.

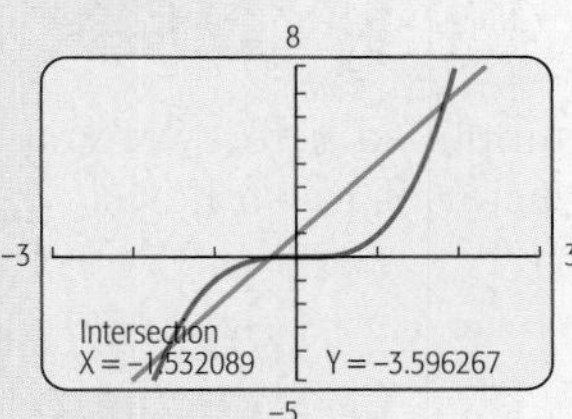

In a similar manner, we find the other points of intersection to be about $(-0.35, -0.04)$ and $(1.88, 6.64)$. The solutions of $x^3 = 3x + 1$ are the x-coordinates of these points, approximately

$$-1.53, -0.35, \text{ and } 1.88.$$

The ZERO Feature

A ZERO, or ROOT, feature can be used to solve an equation. The word "zero" in this context refers to an input, or x-value, for which the output of a function is 0. That is, c is a **zero** of the function f if $f(c) = 0$.

To use such a feature requires a 0 on one side of the equation. Thus, to solve $x^3 = 3x + 1$, we obtain $x^3 - 3x - 1 = 0$ by subtracting $3x + 1$ from both sides. Graphing $y = x^3 - 3x - 1$ and using the ZERO feature to find the zero on the left, we view a screen like the following.

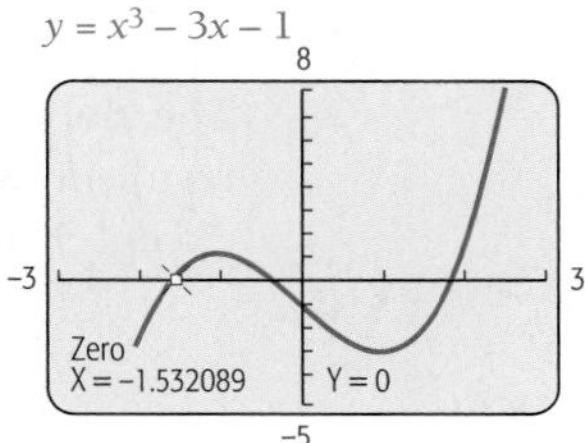

We see that $x^3 - 3x - 1 = 0$ when $x \approx -1.53$, so -1.53 is an approximate solution of the equation $x^3 = 3x + 1$. Proceeding in a similar manner, we can approximate the other solutions as -0.35 and 1.88.

EXERCISES

Using the INTERSECT feature, solve each equation.

1. $x^2 = 10 - 3x$

2. $2x + 24 = x^2$

3. $x^3 = 3x - 2$

4. $x^4 - 2x^2 = 0$

Using the ZERO feature, solve each equation.

5. $0.4x^2 = 280x$
(*Hint:* Use $[-200, 800, -100{,}000, 200{,}000]$.)

6. $\frac{1}{3}x^3 - \frac{1}{2}x^2 = 2x - 1$

7. $x^2 = 0.1x^4 + 0.4$

8. $0 = 2x^4 - 4x^2 + 2$

Find the zeros of each function.

9. $f(x) = 3x^2 - 4x - 2$

10. $f(x) = -x^3 + 6x^2 + 5$

11. $g(x) = x^4 + x^3 - 4x^2 - 2x + 4$

12. $g(x) = -x^4 + x^3 + 11x^2 - 9x - 18$

Rational Functions

DEFINITION

Functions given by the quotient, or ratio, of two polynomials are called **rational functions.**

The following are examples of rational functions:

$$f(x) = \frac{x^2 - 9}{x - 3},$$

$$g(x) = \frac{3x^2 - 4x}{2x + 10},$$

$$h(x) = \frac{x - 3}{x^2 - x - 2}.$$

The domain of a rational function is restricted to those input values that do not result in division by zero. Thus, for f above, the domain consists of all real numbers except 3. To determine the domain of h, we set the denominator equal to 0 and solve:

$$x^2 - x - 2 = 0$$
$$(x + 1)(x - 2) = 0$$
$$x = -1 \quad or \quad x = 2.$$

Therefore, -1 and 2 are not in the domain. The domain of h consists of all real numbers except -1 and 2. We will refer to -1 and 2 as **split numbers** because, in the context of rational functions, they "split," or "separate," the intervals in the domain.

The graphing of most rational functions is rather complicated and is best dealt with using the tools of calculus that we will develop in Chapters 1 and 2. For now we will focus on graphs that are fairly basic and leave the more complicated graphs for Chapter 2.

EXAMPLE 5 Graph: $f(x) = \dfrac{x^2 - 9}{x - 3}$.

Solution This particular function can be simplified before we graph it. We do so by factoring the numerator and removing a factor of 1 as follows:

$$f(x) = \frac{x^2 - 9}{x - 3} \qquad \text{Note that 3 is not in the domain of } f.$$
$$= \frac{(x - 3)(x + 3)}{x - 3}$$
$$= \frac{x - 3}{x - 3} \cdot \frac{x + 3}{1}$$
$$= x + 3, \quad x \neq 3. \qquad \text{We } \textit{must} \text{ specify } x \neq 3.$$

This simplification assumes that x is not 3. By writing $x \neq 3$, we indicate that for any x-value other than 3, the equation $f(x) = x + 3$ is used:

$$f(x) = x + 3, \quad x \neq 3.$$

To find function values, we substitute any value for x other than 3. We make calculations as in the following table and draw the graph. The open circle at (3, 6) indicates that this point is not part of the graph.

x	$f(x)$
-3	0
-2	1
-1	2
0	3
1	4
2	5
4	7

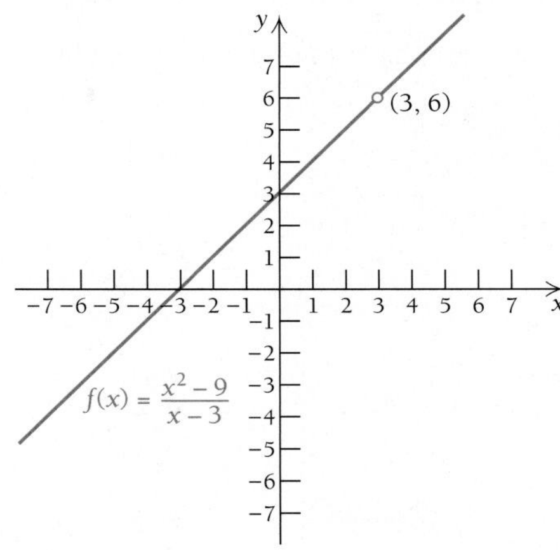

✦

TECHNOLOGY CONNECTION

If $y_1 = (x^2 - 9)/(x - 3)$ and $y_2 = x + 3$ are both graphed, the two graphs appear indistinguishable. To see this, graph both lines and use the arrow keys, ⌃ and ⌄, and the TRACE feature to move the cursor from line to line.

EXERCISES

1. Compare the results of using TRACE and entering the value 3.
2. Use the TABLE feature with TblStart set at -1 and ΔTbl set at 1. How do y_1 and y_2 differ in the resulting table of values?

One important class of rational functions is given by $f(x) = k/x$, where k is a constant.

EXAMPLE 6 Graph: $f(x) = 1/x$.

Solution We make a table of values, plot the points, and then draw the graph.

x	$f(x)$
-3	$-\frac{1}{3}$
-2	$-\frac{1}{2}$
-1	-1
$-\frac{1}{2}$	-2
$-\frac{1}{4}$	-4
$\frac{1}{4}$	4
$\frac{1}{2}$	2
1	1
2	$\frac{1}{2}$
3	$\frac{1}{3}$

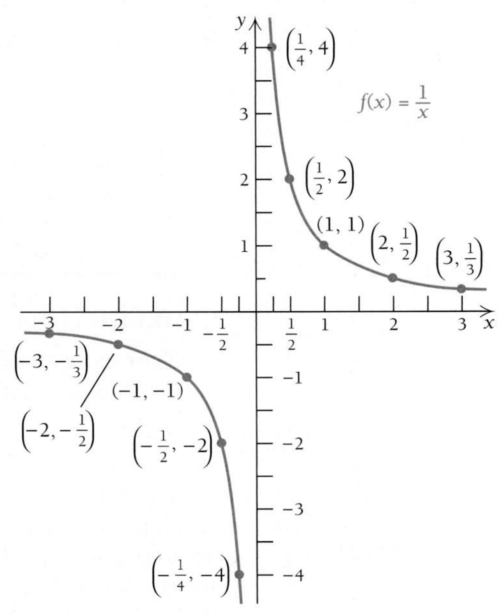

✦

TECHNOLOGY CONNECTION

Graphs of Rational Functions

Consider two graphs of the function given by

$$f(x) = \frac{2x + 1}{x - 3}.$$

CONNECTED mode:

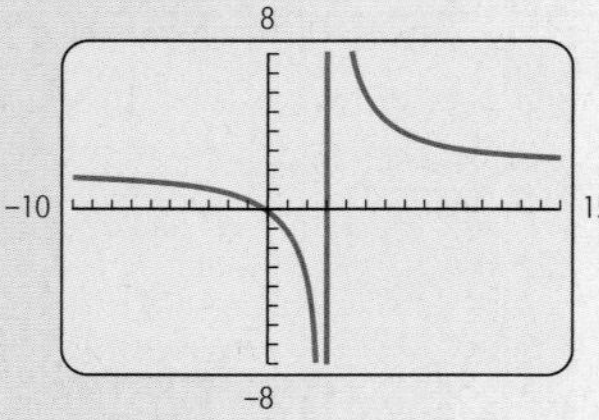

DOT mode:

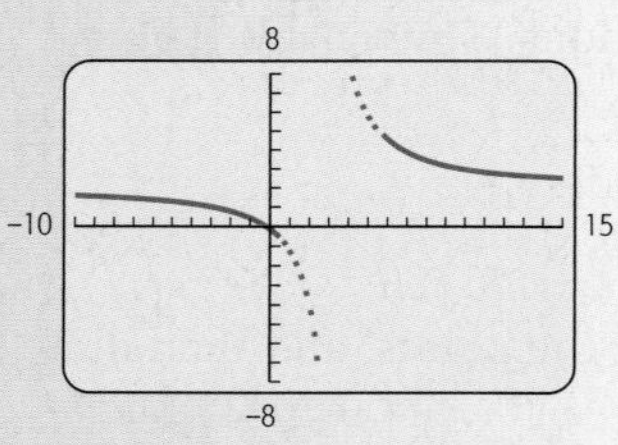

Here use of CONNECTED mode can lead to an *incorrect* graph. Because, in CONNECTED mode, points are joined with line segments, both branches of the graph are connected, making it "appear" as though the vertical line $x = 3$ is part of the graph.

On the other hand, in DOT mode, the calculator simply plots dots representing coordinates of points. When graphing rational functions, it is usually best to use DOT mode.

EXERCISES

Graph each of the following using DOT mode.

1. $f(x) = \dfrac{4}{x - 2}$

2. $f(x) = \dfrac{x}{x + 2}$

3. $f(x) = \dfrac{x^2 - 1}{x^2 + x - 6}$

4. $f(x) = \dfrac{x^2 - 4}{x - 1}$

5. $f(x) = \dfrac{10}{x^2 + 4}$

6. $f(x) = \dfrac{8}{x^2 - 4}$

7. $f(x) = \dfrac{2x + 3}{3x^2 + 7x - 6}$

8. $f(x) = \dfrac{2x^3}{x^2 + 1}$

In Example 6, note that 0 is not in the domain of f because it would yield a denominator of zero. The function is decreasing over the intervals $(-\infty, 0)$ and $(0, \infty)$. The function $f(x) = 1/x$ is an example of **inverse variation.**

DEFINITION

y **varies inversely** as x if there is some positive number k such that $y = k/x$. We also say that y is **inversely proportional** to x.

EXAMPLE 7 Business: Stocks and Gold. Certain economists theorize that stock prices are inversely proportional to the price of gold. That is, when the price of gold goes up, the prices of stocks go down; and when the price of gold goes down, the prices of stocks go up. Let's assume that the Dow Jones Industrial Average D, an index of the overall price of stock, is inversely proportional to the price of gold G, in dollars per ounce. One day the Dow Jones was 9177 and the price of gold was \$364 per ounce. What will the Dow Jones Industrial Average be if the price of gold drops to \$300?

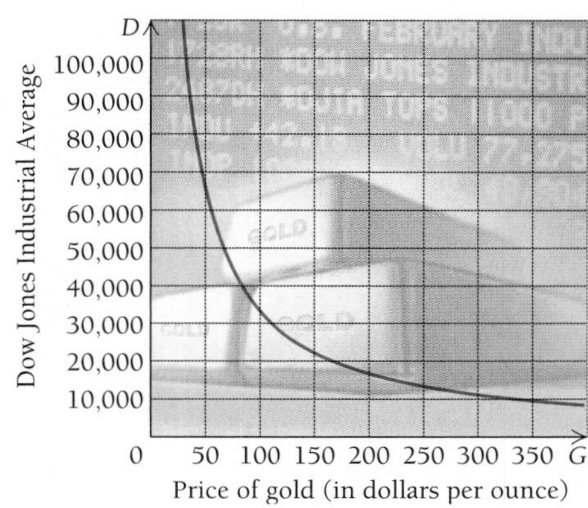

Solution We know that $D = k/G$, so $9177 = k/364$ and $k = 3{,}340{,}428$. Thus,

$$D = \frac{3{,}340{,}428}{G}.$$

We substitute 300 for G and compute D:

$$D = \frac{3{,}340{,}428}{300} \approx 11{,}134.76.$$

Warning! Do not put too much "stock" in the equation of this example. It is meant only to give an idea of economic relationships. An equation to predict the stock market accurately has not been found! ✦

Absolute-Value Functions

The absolute value of a number is its distance from 0 on the number line. We denote the absolute value of a number x as $|x|$. The absolute-value function, given by $f(x) = |x|$, is very important in calculus, and its graph has a distinctive shape.

EXAMPLE 8 Graph: $f(x) = |x|$.

Solution We make a table of values, plot the points, and then draw the graph.

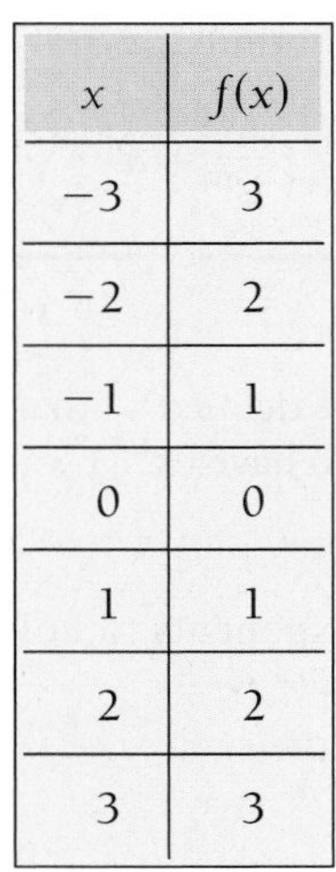

x	$f(x)$
−3	3
−2	2
−1	1
0	0
1	1
2	2
3	3

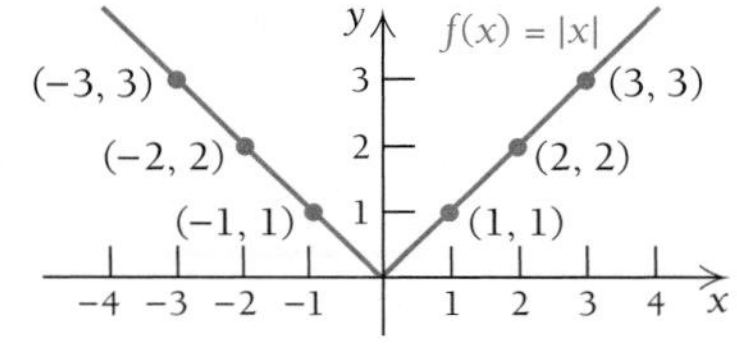

TECHNOLOGY CONNECTION

Absolute value is entered as abs and is often accessed from the NUM option of the MATH menu.

EXERCISES

Graph.

1. $f(x) = \left|\frac{1}{2}x\right|$
2. $g(x) = |x^2 - 4|$

We can think of this function as being defined piecewise by considering the definition of absolute value:

$$f(x) = |x| = \begin{cases} x, & \text{if } x \geq 0, \\ -x, & \text{if } x < 0. \end{cases}$$

✦

Square-Root Functions

The following is an example of a square-root function and its graph.

EXAMPLE 9 Graph: $f(x) = -\sqrt{x}$.

Solution The domain of this function is the set of all nonnegative numbers—the interval $[0, \infty)$. You can find approximate values of square roots on your calculator.

We set up a table of values, plot the points, and then draw the graph.

x	0	1	2	3	4	5
$f(x)$, or $-\sqrt{x}$	0	-1	-1.4	-1.7	-2	-2.2

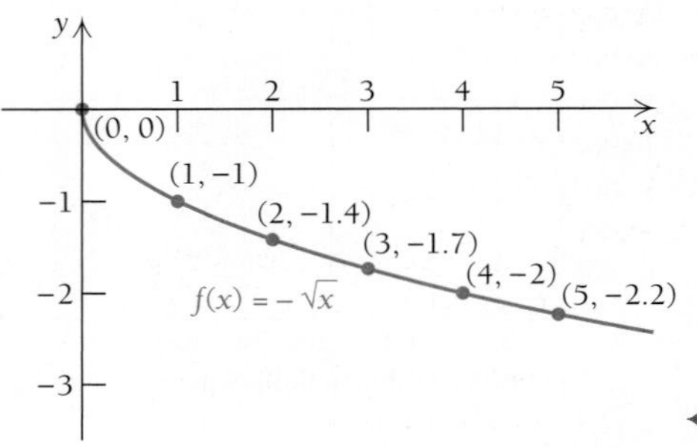

Power Functions with Rational Exponents

We are motivated to define rational exponents so that the following laws of exponents still hold (also see the Appendix):

For any nonzero real number a and any integers n and m,

$$a^n \cdot a^m = a^{n+m}; \quad \frac{a^n}{a^m} = a^{n-m}; \quad (a^n)^m = a^{n \cdot m}; \quad a^{-m} = \frac{1}{a^m}.$$

This suggests that $a^{1/2}$ be defined so that $(a^{1/2})^2 = a^{1/2 \cdot 2} = a^1$. Thus, we define $a^{1/2}$ as $\sqrt{a}$. Similarly, in order to have $(a^{1/3})^3 = a^{1/3 \cdot 3} = a^1$, we define $a^{1/3}$ as $\sqrt[3]{a}$. In general,

$$a^{1/n} = \sqrt[n]{a}, \quad \text{provided } \sqrt[n]{a} \text{ is defined.}$$

Again, for the laws of exponents to hold, we have, assuming that $\sqrt[n]{a}$ exists,

$$a^{m/n} = (a^m)^{1/n} = \sqrt[n]{a^m} = (\sqrt[n]{a})^m,$$

and $a^{-m/n}$ is defined by

$$a^{-m/n} = \frac{1}{a^{m/n}} = \frac{1}{\sqrt[n]{a^m}}.$$

EXAMPLE 10 Rewrite each of the following as an equivalent expression with rational exponents:

a) $\sqrt[4]{x}$

b) $\sqrt[3]{r^2}$

c) $\sqrt{x^{10}}$, for $x \geq 0$

d) $\dfrac{1}{\sqrt[3]{b^5}}$

Solution

a) $\sqrt[4]{x} = x^{1/4}$ **b)** $\sqrt[3]{r^2} = r^{2/3}$

c) $\sqrt{x^{10}} = x^{10/2} = x^5, \quad x \geq 0$ **d)** $\dfrac{1}{\sqrt[3]{b^5}} = \dfrac{1}{b^{5/3}} = b^{-5/3}$ ✦

EXAMPLE 11 Rewrite each of the following as an equivalent expression using radical notation: **a)** $x^{1/3}$; **b)** $t^{6/7}$; **c)** $x^{-2/3}$; **d)** $r^{-1/4}$.

Solution

a) $x^{1/3} = \sqrt[3]{x}$ **b)** $t^{6/7} = \sqrt[7]{t^6}$

c) $x^{-2/3} = \dfrac{1}{x^{2/3}} = \dfrac{1}{\sqrt[3]{x^2}}$ **d)** $r^{-1/4} = \dfrac{1}{r^{1/4}} = \dfrac{1}{\sqrt[4]{r}}$ ✦

EXAMPLE 12 Simplify: **a)** $8^{5/3}$; **b)** $81^{3/4}$.

Solution

a) $8^{5/3} = (8^{1/3})^5 = \left(\sqrt[3]{8}\right)^5 = 2^5 = 32$

b) $81^{3/4} = (81^{1/4})^3 = \left(\sqrt[4]{81}\right)^3 = 3^3 = 27$ ✦

Because even roots (square roots, fourth roots, sixth roots, and so on) of negative numbers are not real numbers, the domain of a radical function may have restrictions.

EXAMPLE 13 Find the domain of the function given by

$$f(x) = \sqrt[4]{2x - 10}.$$

Solution For $f(x)$ to be a real number, $2x - 10$ cannot be negative. Thus, to find the domain of f, we solve the inequality $2x - 10 \geq 0$:

$$2x - 10 \geq 0$$

$$2x \geq 10 \qquad \text{Adding 10 to both sides}$$

$$x \geq 5. \qquad \text{Dividing both sides by 2}$$

The domain of f is $\{x | x \geq 5\}$, or, in interval notation, $[5, \infty)$. ✦

Power functions of the form $f(x) = ax^k$, with k a fraction, do occur in applications.

A fox in its territorial area.

EXAMPLE 14 Life Science: Home Range. The *home range* of an animal is defined as the region to which the animal confines its movements. It has been shown that for carnivorous (meat-eating) mammals the area of that region can be approximated by the function

$$H(w) = 0.11w^{1.36},$$

where w is the mass of the animal, in grams, and $H(w)$ is the area of the home range, in hectares. Graph the function. (*Source:* Based on information in Emlen, J. M., *Ecology: An Evolutionary Approach,* p. 200 (Reading, MA: Addison-Wesley, 1973), and Harestad, A. S., and Bunnel, F. L., "Home Range and Body Weight–A Reevaluation," *Ecology,* Vol. 60, No. 2, pp. 389–402.)

Solution We can approximate function values using a power key, usually labeled (^) or y^x. Note that $w^{1.36} = w^{136/100} = \sqrt[100]{w^{136}}$.

w	0	700	1400	2100	2800	3500
$H(w)$	0	814.2	2089.9	3627.5	5364.5	7266.5

The graph is shown below. Note that the function values increase from left to right. As body weight increases, the area over which the animal moves increases.

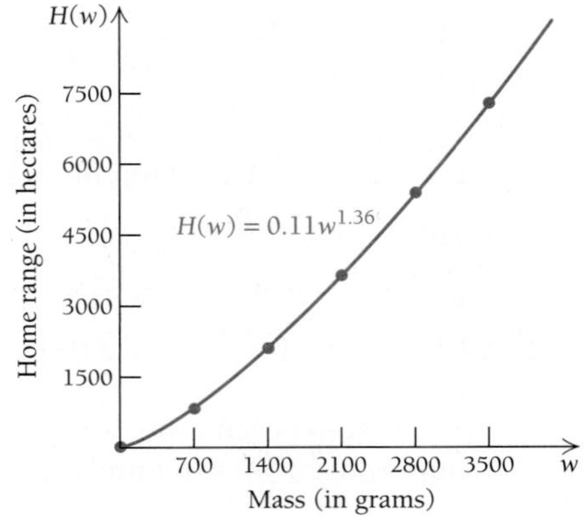

✦

Supply and Demand Functions

Supply and demand in economics are modeled by increasing and decreasing functions.

Demand Functions

The table and graph below show the relationship between the price x per bag of sugar and the quantity q of 5-lb bags that consumers will demand at that price.

DEMAND SCHEDULE

Price, x, per 5-lb Bag	Quantity, q, of 5-lb Bags (in millions)
$5	4
4	5
3	7
2	10
1	15

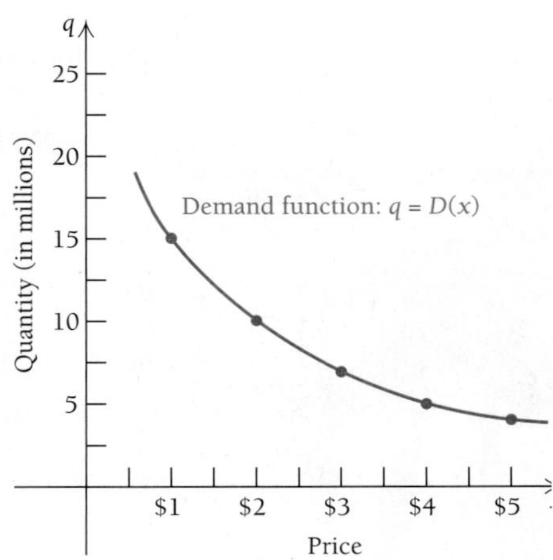

Note that the quantity consumers demand is inversely proportional to the price. As the price goes up, the quantity demanded goes down.

Supply Functions

The next table and graph show the relationship between the price x per bag of sugar and the quantity q of 5-lb bags that sellers are willing to supply, or sell, at that price.

SUPPLY SCHEDULE

Price, x, per 5-lb Bag	Quantity, q, of 5-lb Bags (in millions)
$1	0
2	10
3	16
4	20
5	22

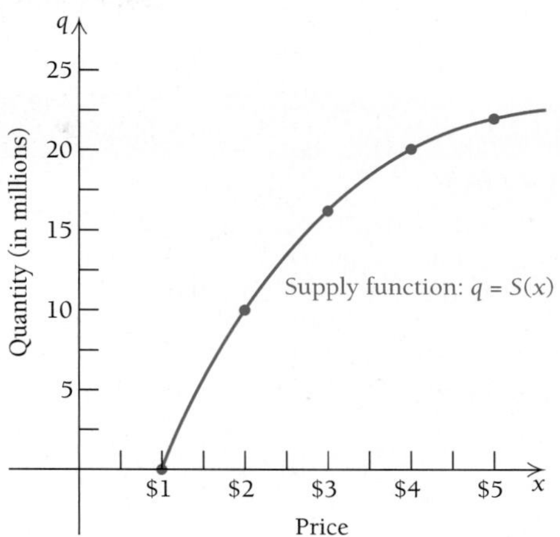

Note that at higher prices, suppliers are willing to supply greater quantities than they are at lower prices.

Let's now look at these curves together. As price increases, supply increases and demand decreases; and as price decreases, demand increases but supply decreases. The point of intersection (x_E, q_E) is called the **equilibrium point.** The equilibrium price x_E (in this case, $2 per bag) corresponds to an equilibrium quantity q_E (in this case, 10 million bags). Sellers are willing to sell 10 million bags at $2/bag, and consumers are willing to buy 10 million bags at that price. The situation is analogous to a buyer and seller haggling over the sale of an item. The equilibrium point, or selling price, is what they finally agree on.

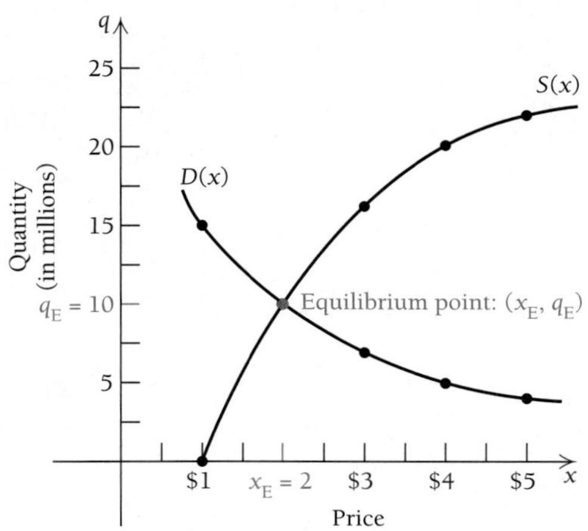

EXAMPLE 15 Economics: Equilibrium Point. Find the equilibrium point for the demand and supply functions for the Ultra-Fine coffee maker. Here q represents the number of coffee makers produced, in hundreds, and x is the price, in dollars.

$$\text{Demand:} \quad q = 50 - \frac{1}{4}x$$

$$\text{Supply:} \quad q = x - 25$$

TECHNOLOGY CONNECTION

EXERCISE

1. Use the INTERSECT feature to find the equilibrium point for the following demand and supply functions.

Demand: $q = 1123.6 - 61.4x$

Supply: $q = 201.8 + 4.6x$

Solution To find the equilibrium point, the quantity demanded must match the quantity produced:

$$50 - \frac{1}{4}x = x - 25$$

$$50 + 25 = x + \frac{1}{4}x \qquad \text{Adding } 25 + \frac{1}{4}x \text{ to each side}$$

$$75 = \frac{5}{4}x$$

$$75 \cdot \frac{4}{5} = x \qquad \text{Multiplying both sides by } \frac{4}{5}$$

$$60 = x.$$

Thus, $x_E = 60$. To find q_E, we substitute x_E into either function. We select the supply function:

$$q_E = x_E - 25 = 60 - 25 = 35.$$

Thus, the equilibrium quantity is 3500 units, and the equilibrium point is (\$60, 3500). ✦

Exercise Set R.5

Graph each pair of equations on one set of axes.

1. $y = \frac{1}{2}x^2$ and $y = -\frac{1}{2}x^2$
2. $y = \frac{1}{4}x^2$ and $y = -\frac{1}{4}x^2$
3. $y = x^2$ and $y = x^2 - 1$
4. $y = x^2$ and $y = x^2 - 3$
5. $y = -2x^2$ and $y = -2x^2 + 1$
6. $y = -3x^2$ and $y = -3x^2 + 2$
7. $y = |x|$ and $y = |x - 3|$
8. $y = |x|$ and $y = |x - 1|$
9. $y = x^3$ and $y = x^3 + 2$
10. $y = x^3$ and $y = x^3 + 1$
11. $y = \sqrt{x}$ and $y = \sqrt{x - 1}$
12. $y = \sqrt{x}$ and $y = \sqrt{x - 2}$

For each of the following, state whether the graph of the function is a parabola. If the graph is a parabola, find the parabola's vertex.

13. $f(x) = x^2 + 4x - 7$
14. $f(x) = x^3 - 2x + 3$
15. $g(x) = 2x^4 - 4x^2 - 3$
16. $g(x) = 3x^2 - 6x$

Graph.

17. $y = x^2 - 4x + 3$
18. $y = x^2 - 6x + 5$
19. $y = -x^2 + 2x - 1$
20. $y = -x^2 - x + 6$
21. $f(x) = 2x^2 - 6x + 1$
22. $f(x) = 3x^2 - 6x + 4$
23. $g(x) = -3x^2 - 4x + 5$
24. $g(x) = -2x^2 - 3x + 7$
25. $y = \frac{2}{x}$
26. $y = \frac{3}{x}$

27. $y = -\dfrac{2}{x}$

28. $y = \dfrac{-3}{x}$

29. $y = \dfrac{1}{x^2}$

30. $y = \dfrac{1}{x - 1}$

31. $y = \sqrt[3]{x}$

32. $y = \dfrac{1}{|x|}$

33. $f(x) = \dfrac{x^2 + 5x + 6}{x + 3}$

34. $g(x) = \dfrac{x^2 + 7x + 10}{x + 2}$

35. $f(x) = \dfrac{x^2 - 1}{x - 1}$

36. $g(x) = \dfrac{x^2 - 25}{x - 5}$

Solve.

37. $x^2 - 2x = 2$

38. $x^2 - 2x + 1 = 5$

39. $x^2 + 6x = 1$

40. $x^2 + 4x = 3$

41. $4x^2 = 4x + 1$

42. $-4x^2 = 4x - 1$

43. $3y^2 + 8y + 2 = 0$

44. $2p^2 - 5p = 1$

45. $x + 7 + \dfrac{9}{x} = 0$ (*Hint:* Multiply both sides by x.)

46. $1 - \dfrac{1}{w} = \dfrac{1}{w^2}$

Rewrite each of the following as an equivalent expression with rational exponents.

47. $\sqrt{x^3}$

48. $\sqrt{x^5}$

49. $\sqrt[5]{a^3}$

50. $\sqrt[4]{b^2}, \quad b \geq 0$

51. $\sqrt[7]{t}$

52. $\sqrt[8]{c}$

53. $\sqrt[4]{x^{12}}, \quad x \geq 0$

54. $\sqrt[3]{t^6}$

55. $\dfrac{1}{\sqrt{t^5}}$

56. $\dfrac{1}{\sqrt{m^4}}$

57. $\dfrac{1}{\sqrt{x^2 + 7}}$

58. $\sqrt{x^3 + 4}$

Rewrite each of the following as an equivalent expression using radical notation.

59. $x^{1/5}$

60. $t^{1/7}$

61. $y^{2/3}$

62. $t^{2/5}$

63. $t^{-2/5}$

64. $y^{-2/3}$

65. $b^{-1/3}$

66. $b^{-1/5}$

67. $e^{-17/6}$

68. $m^{-19/6}$

69. $(x^2 - 3)^{-1/2}$

70. $(y^2 + 7)^{-1/4}$

71. $\dfrac{1}{t^{2/3}}$

72. $\dfrac{1}{w^{-4/5}}$

Simplify.

73. $9^{3/2}$

74. $16^{5/2}$

75. $64^{2/3}$

76. $8^{2/3}$

77. $16^{3/4}$

78. $25^{5/2}$

Determine the domain of each function.

79. $f(x) = \dfrac{x^2 - 25}{x - 5}$

80. $f(x) = \dfrac{x^2 - 4}{x + 2}$

81. $f(x) = \dfrac{x^3}{x^2 - 5x + 6}$

82. $f(x) = \dfrac{x^4 + 7}{x^2 + 6x + 5}$

83. $f(x) = \sqrt{5x + 4}$

84. $f(x) = \sqrt{2x - 6}$

85. $f(x) = \sqrt[4]{7 - x}$

86. $f(x) = \sqrt[6]{5 - x}$

APPLICATIONS

Business and Economics

Find the equilibrium point for each pair of demand and supply functions.

87. Demand: $q = 1000 - 10x$; Supply: $q = 250 + 5x$

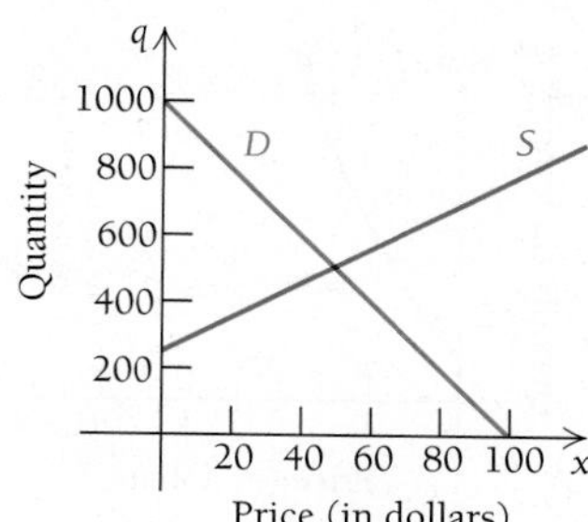

88. Demand: $q = 8800 - 30x$;
Supply: $q = 7000 + 15x$

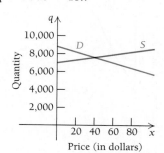

q
10,000
8,000
6,000
4,000
2,000
D
S
Quantity
20 40 60 80
x
Price (in dollars)

89. Demand: $q = \dfrac{5}{x}$; Supply: $q = \dfrac{x}{5}$

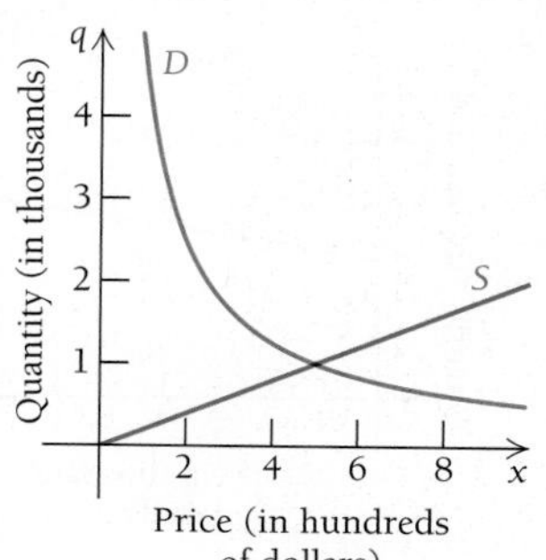

90. Demand: $q = \frac{4}{x}$; Supply: $q = \frac{x}{4}$

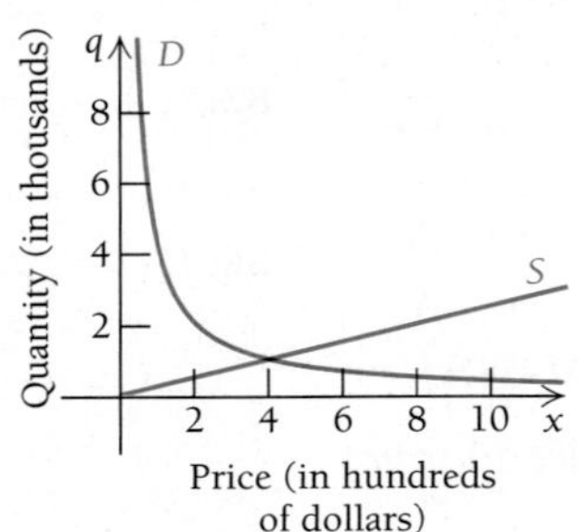

91. Demand: $q = (x - 3)^2$; Supply: $q = x^2 + 2x + 1$ (assume $x \leq 3$)

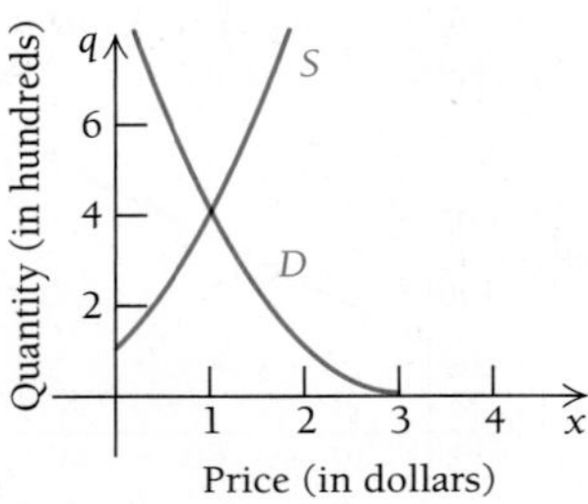

92. Demand: $q = (x - 4)^2$; Supply: $q = x^2 + 2x + 6$ (assume $x \leq 4$)

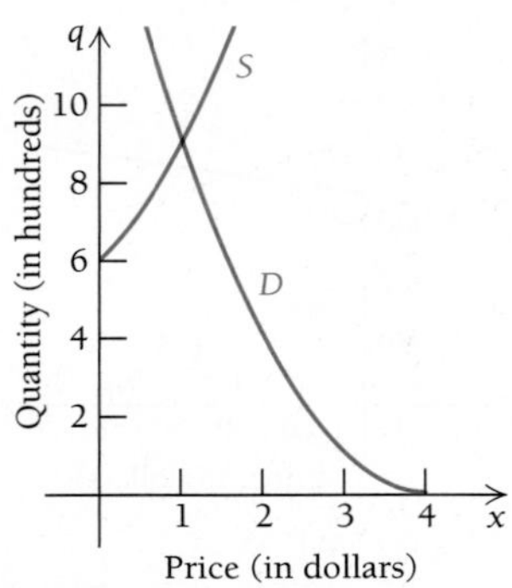

93. Demand: $q = 5 - x$; Supply: $q = \sqrt{x + 7}$

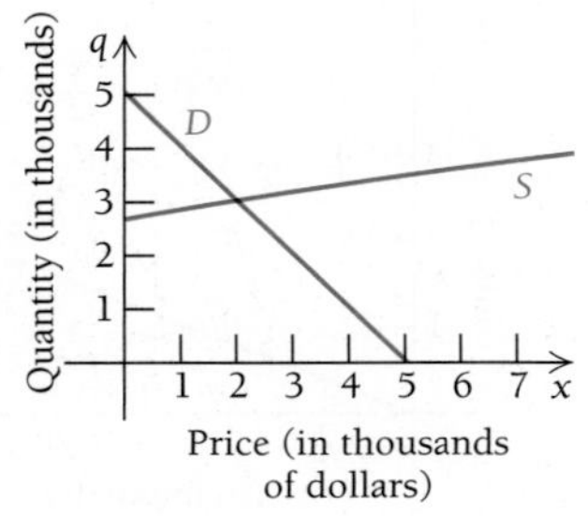

94. Demand: $q = 7 - x$; Supply: $q = 2\sqrt{x + 1}$

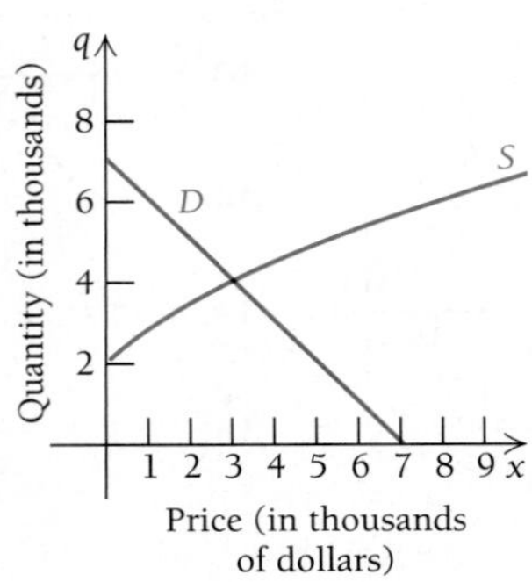

95. Stock prices and prime rate. It is theorized that the price per share of a stock is inversely proportional to the prime (interest) rate. In September 2005, the price per share S of Toyota stock was \$86.89 and the prime rate R was 6.75%. The prime rate rose to 7.50% in March 2006. (*Source:* finance.yahoo.com and Federal Reserve Board.) What would the price per share be if the assumption of inverse proportionality is correct?

96. Demand. The quantity sold x of a plasma television is inversely proportional to the price p. If 85,000 plasma TVs sold for \$2900 each, how many will be sold if the price is \$850 each?

97. Radar range. The function given by

$$R(x) = 11.74x^{0.25}$$

can be used to approximate the maximum range, $R(x)$, in miles, of an ARSR-3 surveillance radar with a peak power of x watts.

a) Determine the maximum radar range when the peak power is 40,000 watts, 50,000 watts, and 60,000 watts.

b) Graph the function.

98. Home range. Refer to Example 14. The home range, in hectares, of an omnivorous mammal (one that eats both plant and meat) of mass w grams is given by

$$H(w) = 0.059w^{0.92}.$$

(*Source:* Harestad, A. S., and Bunnel, F. L., "Home Range and Body Weight—A Reevaluation," *Ecology*, Vol. 10, No. 2 (April, 1979), pp. 389–402.) Complete the table of approximate function values and graph the function.

w	0	1000	2000	3000	4000	5000	6000	7000
$H(w)$	0	34.0						

99. Life science: pollution control. Pollution control has become a very important concern in all countries. If controls are not put in place, it has been predicted that the function

$$P = 1000t^{5/4} + 14{,}000$$

will describe the average pollution, in particles of pollution per cubic centimeter, in most cities at time t, in years, where $t = 0$ corresponds to 1970 and $t = 35$ corresponds to 2005.

a) Predict the pollution in 2005, 2008, and 2014.

b) Graph the function over the interval [0, 50].

100. Surface area and mass. The surface area of a person whose mass is 75 kg can be approximated by the function

$$f(h) = 0.144h^{1/2},$$

where $f(h)$ is measured in square meters and h is the person's height in centimeters. (*Source:* U.S. Oncology.)

a) Find the approximate surface area of a person whose mass is 75 kg and whose height is 180 cm.

b) Find the approximate surface area of a person whose mass is 75 kg and whose height is 170 cm.

c) Graph the function f for $0 \leq h \leq 200$.

SYNTHESIS

101. Zipf's Law. According to Zipf's Law, the number of cities N with a population greater than S is inversely proportional to S. In 2004, there were 51 U.S. cities with a population greater than 350,000. Estimate the number of U.S. cities with a population between 350,000 and 500,000; between 300,000 and 600,000.

TW **102.** At most, how many y-intercepts can a function have? Explain.

TW **103.** Explain the difference between a rational function and a polynomial function. Is every polynomial function a rational function? Why or why not?

TECHNOLOGY CONNECTION

Use the ZERO *feature or the* INTERSECT *feature to approximate the zeros of each function to three decimal places.*

104. $f(x) = x^3 - x$
(Also, use algebra to find the zeros of this function.)

105. $f(x) = 2x^3 - x^2 - 14x - 10$

106. $f(x) = \frac{1}{2}(|x - 4| + |x - 7|) - 4$

107. $f(x) = x^4 + 4x^3 - 36x^2 - 160x + 300$

108. $f(x) = \sqrt{7 - x^2} - 1$

109. $f(x) = |x + 1| + |x - 2| - 5$

110. $f(x) = |x + 1| + |x - 2|$

111. $f(x) = |x + 1| + |x - 2| - 3$

112. $f(x) = x^8 + 8x^7 - 28x^6 - 56x^5 + 70x^4 + 56x^3 - 28x^2 - 8x + 1$

113. Find the equilibrium point for the following demand and supply functions.

Demand: $q = 83 - x$,

Supply: $q = \dfrac{x^2}{576} - 1.9$,

where q is the quantity, in thousands of units, and x is the price per unit, in dollars.

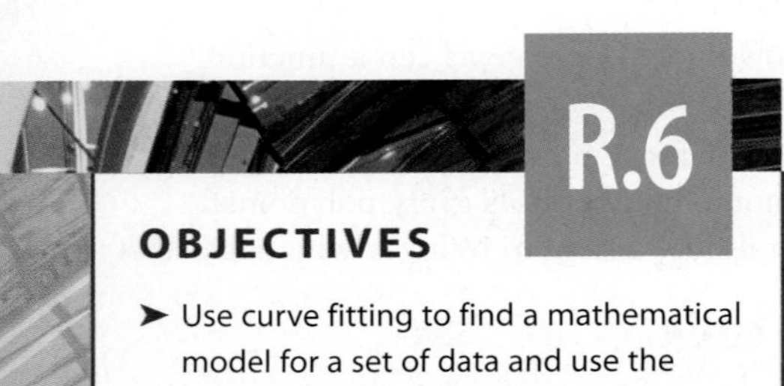

Mathematical Modeling and Curve Fitting

OBJECTIVES

- Use curve fitting to find a mathematical model for a set of data and use the model to make predictions.

Fitting Functions to Data

We have developed a library of functions that can serve as models for many applications. Although others will be introduced later, let's look at those that we have considered. (Cubic and quartic functions are covered in detail in Chapter 2, but we show them for reference.) We will not consider rational functions in this section.

Linear function:
$f(x) = mx + b$

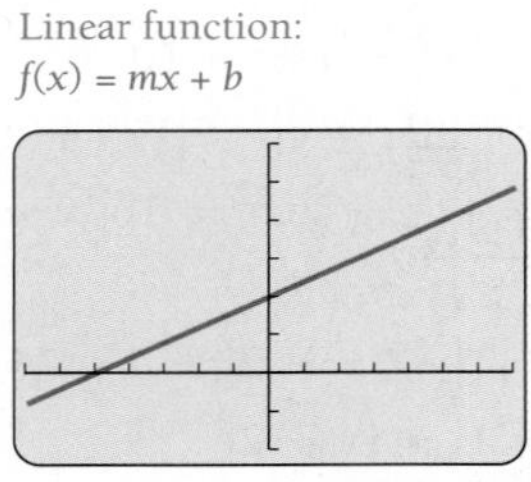

Quadratic function:
$f(x) = ax^2 + bx + c,\ a > 0$

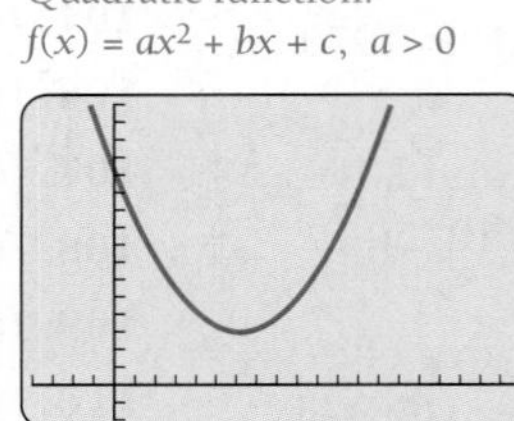

Quadratic function:
$f(x) = ax^2 + bx + c,\ a < 0$

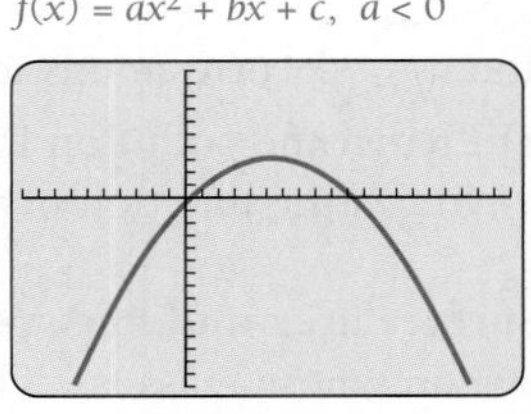

Absolute-value function:
$f(x) = |x|$

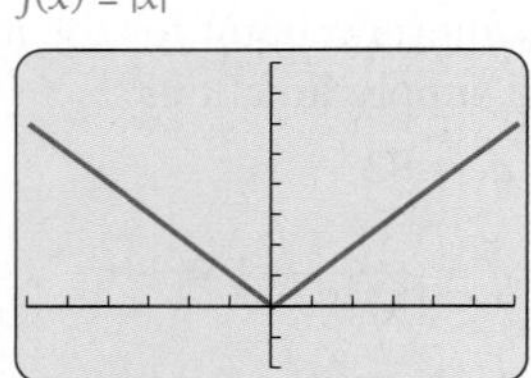

Cubic function:
$f(x) = ax^3 + bx^2 + cx + d,\ a > 0$

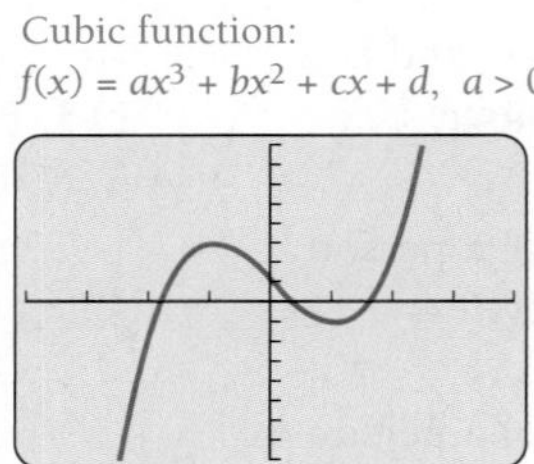

Quartic function:
$f(x) = ax^4 + bx^3 + cx^2 + dx + e,\ a > 0$

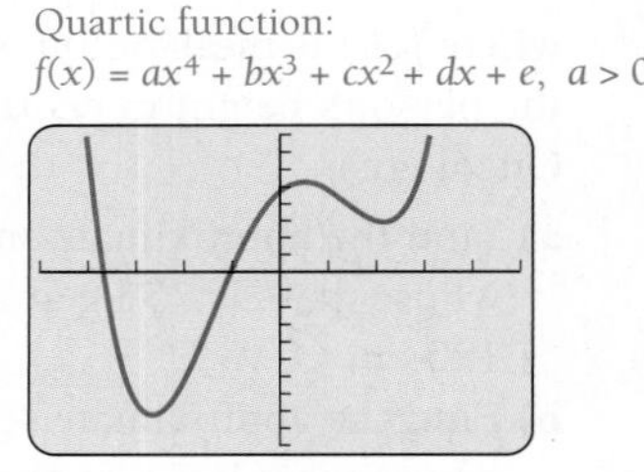

Now let's consider some real-world data. How can we decide which, if any, type of function might fit the data? One simple way is to examine a graph of the data called a **scatterplot.** Then we look for a pattern resembling one of the graphs above. For example, data might be modeled by a linear function if the graph resembles a straight line. The data might be modeled by a quadratic function if the graph rises and then falls, or falls and then rises, in a curved manner resembling a parabola.

Let's now use our library of functions to see which, if any, might fit certain data sets.

EXAMPLE 1 Choosing Models. For the scatterplots and graphs below, determine which, if any, of the following functions might be used as a model for the data.

Linear, $f(x) = mx + b$

Quadratic, $f(x) = ax^2 + bx + c, a > 0$

Quadratic, $f(x) = ax^2 + bx + c, a < 0$

Polynomial, neither quadratic nor linear

a)

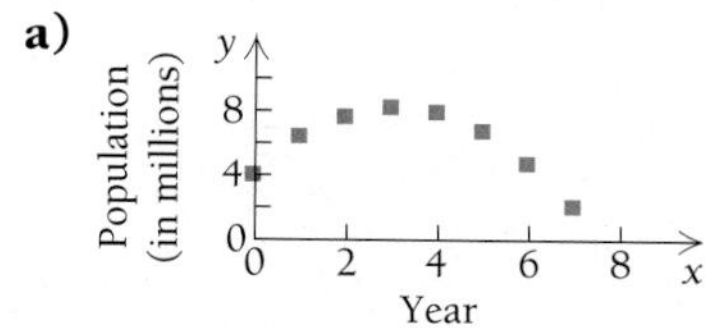

b)

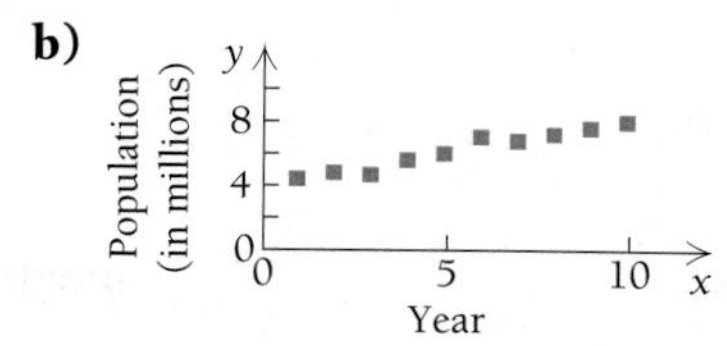

c)

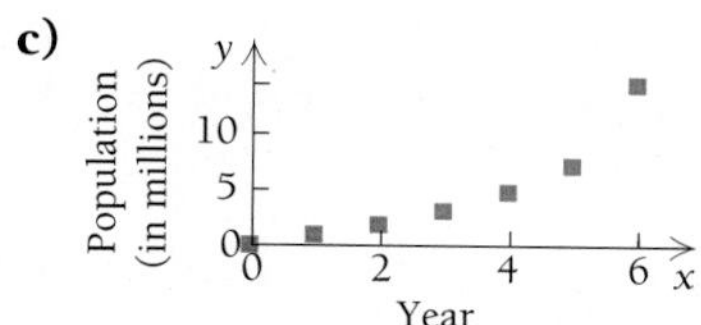

d)

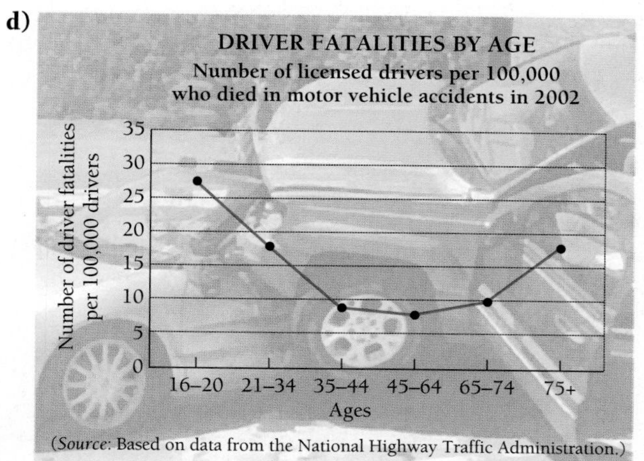

e)

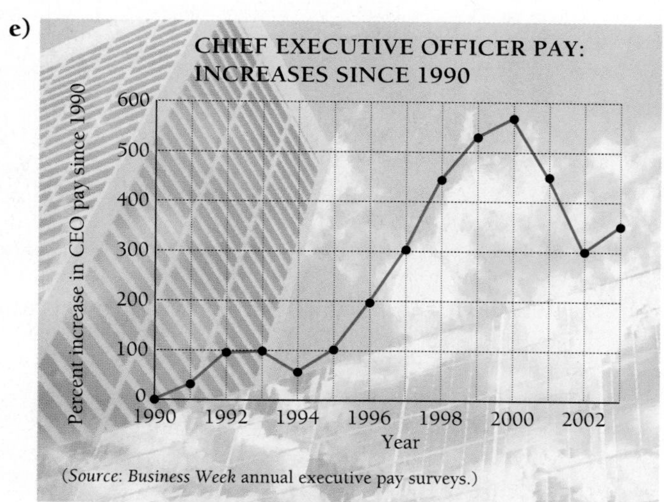

Solution

a)

The data rise and then fall in a curved manner fitting a quadratic function,

$$f(x) = ax^2 + bx + c,\ a < 0.$$

b)

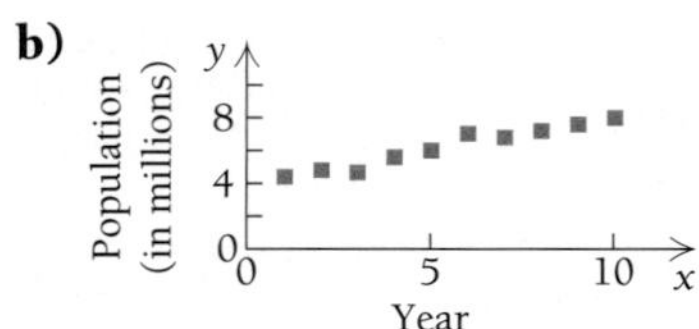

The data seem to fit a linear function,

$$f(x) = mx + b.$$

c)

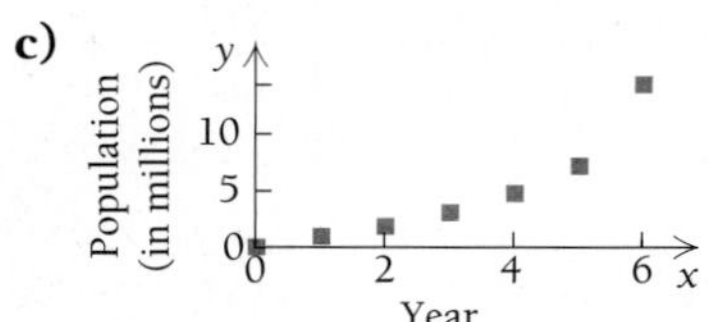

The data rise in a manner fitting the right-hand side of a quadratic function,

$$f(x) = ax^2 + bx + c, a > 0.$$

d)

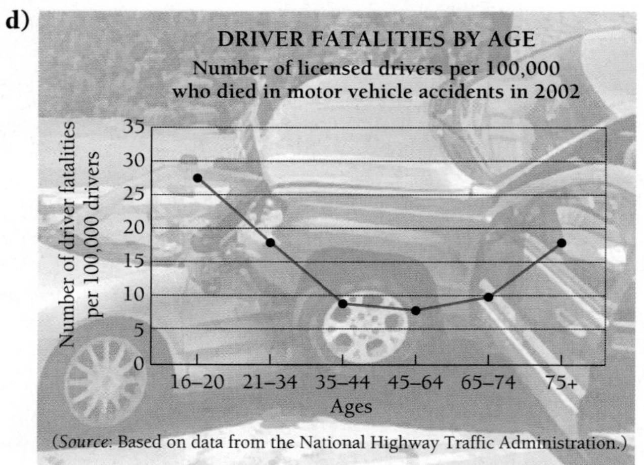

The data fall and then rise in a curved manner fitting a quadratic function,

$$f(x) = ax^2 + bx + c, a > 0.$$

e)

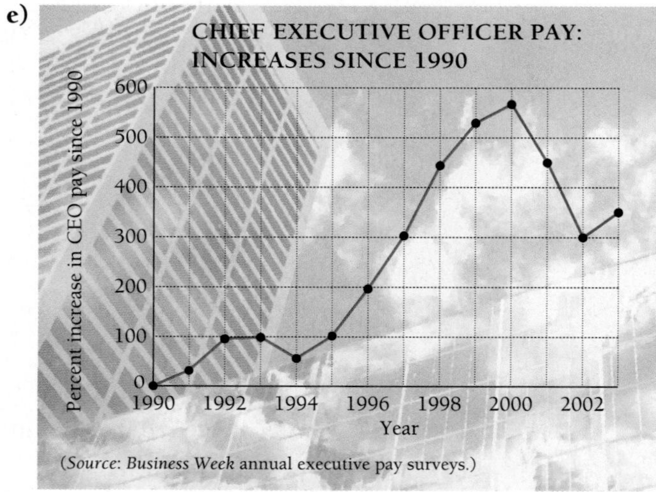

The data rise and fall more than once, so they do not fit a linear or quadratic function but might fit a polynomial function that is neither quadratic nor linear. ✦

It is sometimes possible to find a mathematical model by graphing a set of data as a scatterplot, inspecting the graph to see if a known type of function seems to fit, and then using the data points to derive the equation of a specific function.

EXAMPLE 2 Business: Workers' Wages. The following table shows the annual percent increases in pay since 1996 for a U.S. production worker. (*Source:* Based on data from the *Statistical Abstract of the U.S. 2002* and the Bureau of Labor Statistics.)

Number of years, x, since 1996	1	2	3	4	5	6	7
Percent increase since 1996, P	1.9	7.4	11.7	19.5	28.2	29.7	31.3

a) Make a scatterplot of the data and determine whether the data seem to fit a linear function.

b) Find a linear function that (approximately) fits the data.

c) Use the model to predict the percentage by which 2010 wages will exceed 1996 wages.

Solution

a) The scatterplot is shown at right. The data tend to follow a straight line, although a "perfect" straight line cannot be drawn through the data points.

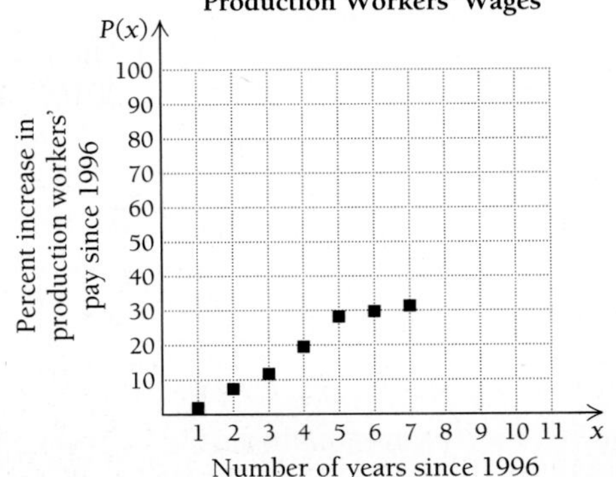

b) We consider the linear function

$$P(x) = mx + b, \qquad (1)$$

where $P(x)$ is the percentage by which the wages x years after 1996 exceed the wages in 1996.

To derive the constants (or parameters) m and b, we choose two data points. Although this procedure is somewhat arbitrary, we try to choose two points that follow the general linear pattern. In this case, we pick $(1, 1.9)$ and $(4, 19.5)$. Since the points are to be solutions of equation (1), it follows that

$$1.9 = m \cdot 1 + b, \quad \text{or} \quad 1.9 = m + b, \qquad (2)$$

$$19.5 = m \cdot 4 + b, \quad \text{or} \quad 19.5 = 4m + b. \qquad (3)$$

We now have a system of equations. We solve by subtracting each side of equation (2) from each side of equation (3) to eliminate b:

$$17.6 = 3m.$$

Then we have

$$m = \frac{17.6}{3}$$
$$\approx 5.87.$$

Substituting 5.87 for m in equation (2), we can solve for b:

$$1.9 = 5.87 + b$$
$$-3.97 = b.$$

Substituting these values of m and b into equation (1), we get the function (model) given by

$$P(x) = 5.87x - 3.97. \qquad (4)$$

We can see the representative data points and the linear function in the graph below.

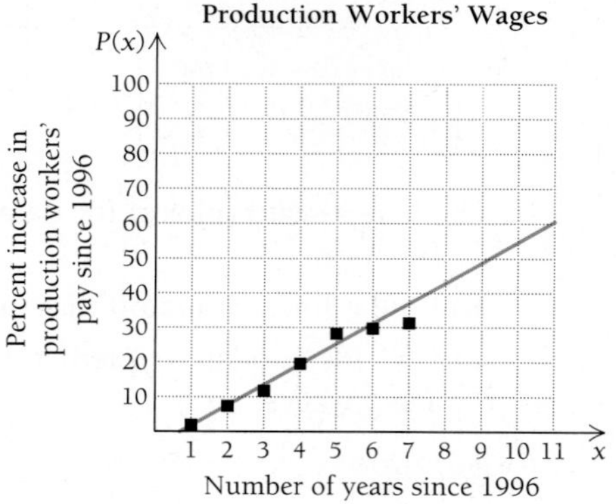

c) The percentage by which U.S. production workers' wages will have increased in 2010, a total of 14 years after 1996, is

$$P(14) = 5.87 \cdot 14 - 3.97 = 78.21\%.$$

Thus, we predict that between 1996 and 2010, the wages paid to U.S. production workers will increase by 78.21%. ◆

TECHNOLOGY CONNECTION

Linear Regression: Fitting a Linear Function to Data

We now consider **linear regression**, the preferred method for fitting a linear function to a set of data. Although the complete basis for this method belongs in Section 6.4, we consider it here because we can carry out the procedure easily using technology. One advantage of linear regression is that it uses *all* data points rather than just two.

EXAMPLE Business: Production Workers' Wages. Consider the data in Example 2.

a) Find the equation of the regression line for the given data. Then graph the regression line with the scatterplot.

b) Use the model to predict the percentage by which wages will increase from 1996 to 2010.

Linear Regression: Fitting a Linear Function to Data (continued)

Solution

a) To fit a linear function to the data using regression, we select the EDIT option of the STAT menu. We then enter the data: (1, 1.9), (2, 7.4), (3, 11.7), (4, 19.5), (5, 28.2), (6, 29.7), and (7, 31.3).

L1	L2	L3 2
1	1.9	
2	7.4	
3	11.7	
4	19.5	
5	28.2	
6	29.7	
7	31.3	

L2(7) = 31.3

To view the data points, we turn on PLOT1 by pressing (▲) ENTER from y_1 at the (Y=) screen and then pressing (GRAPH). To set the window, we could select the ZoomStat option of ZOOM. Instead, we select a $[-1, 15, -2, 100]$ window.

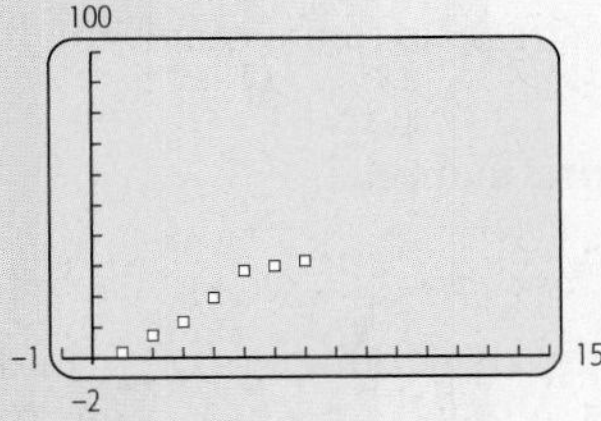

To find the line that best fits the data, we press STAT, select CALC, then LinReg (ax + b), and then press ENTER.

```
LinReg
 y = ax + b
 a = 5.332142857
 b = −2.8
```

The equation of the regression line is approximately $y = 5.33x - 2.8$, or $P(x) = 5.33x - 2.8$. We enter this as y_1 and press (GRAPH).

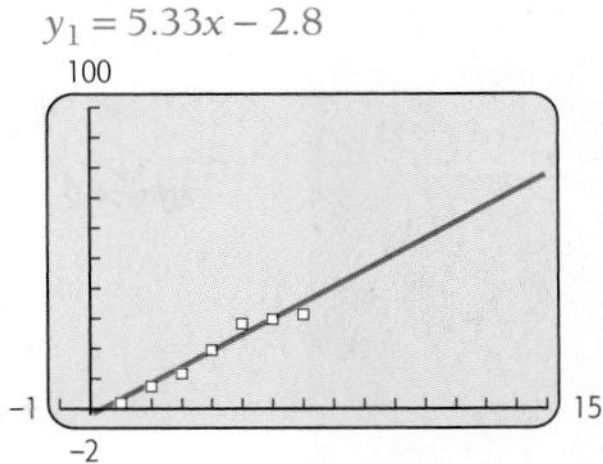

b) To find the percentage increase in production workers' wages from 1996 to 2010, we find $P(14)$; we press (TRACE), move the cursor to the line, and enter 14:

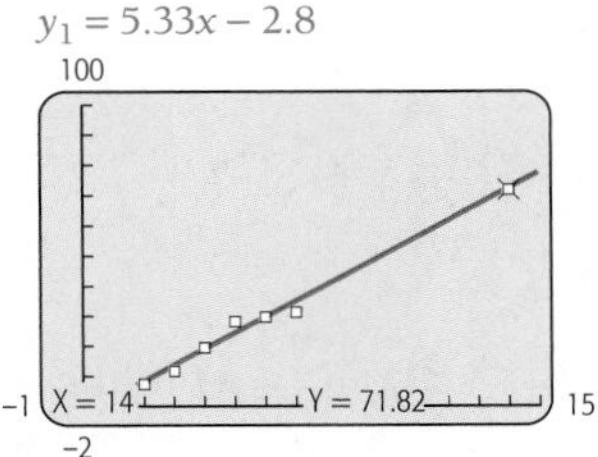

According to the linear regression model, between 1996 and 2010, production workers' wages will rise by 71.82%. Note the discrepancy between this result and the result of 78.21% in Example 2. A different choice of points in part (b) of Example 2 would have yielded a different line and prediction.

EXERCISE

1. *Study time and test scores.* The data in the following table relate study time and test scores.

Study Time (in hours)	Test Grade (in percent)
7	83
8	85
9	88
10	91
11	?

a) Fit a regression line to the data. Then make a scatterplot of the data and graph the regression line with the scatterplot.

b) Use the linear model to predict the test score received when one has studied for 11 hr.

TW **c)** Discuss the appropriateness of a linear model of these data.

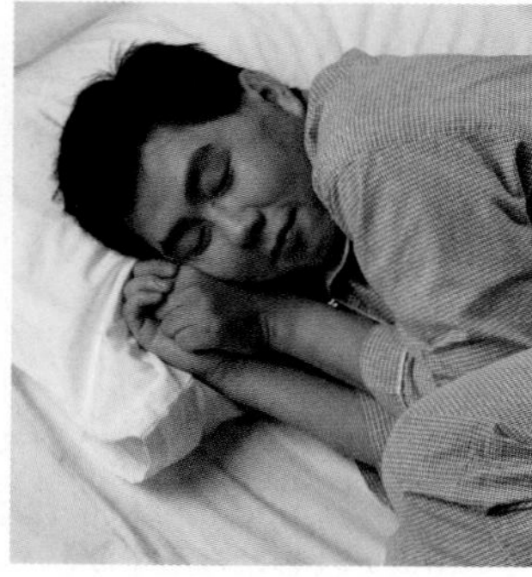

EXAMPLE 3 Life Science: Hours of Sleep and Death Rate. In a study by Dr. Harold J. Morowitz of Yale University, data were gathered that showed the relationship between the death rate of men and the average number of hours per day that the men slept. These data are listed in the following table.

Average Number of Hours of Sleep, x	Death Rate per 100,000 Males, y
5	1121
6	805
7	626
8	813
9	967

(*Source*: Morowitz, Harold J., "Hiding in the Hammond Report," *Hospital Practice*.)

a) Make a scatterplot of the data, and determine whether the data seem to fit a quadratic function.

b) Find a quadratic function that fits the data.

c) Use the model to find the death rate for males who sleep 2 hr, 8 hr, and 10 hr.

Solution

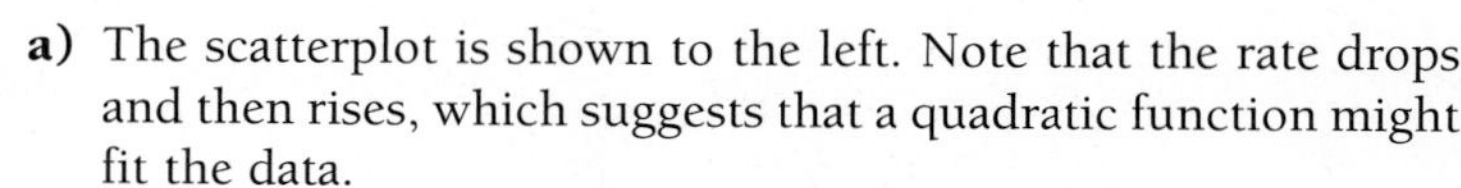

a) The scatterplot is shown to the left. Note that the rate drops and then rises, which suggests that a quadratic function might fit the data.

b) We consider the quadratic model,

$$y = ax^2 + bx + c. \tag{1}$$

To derive the constants (or parameters) a, b, and c, we use the three data points (5, 1121), (7, 626), and (9, 967). Since these points are to be solutions of equation (1), it follows that

$$1121 = a \cdot 5^2 + b \cdot 5 + c, \quad \text{or} \quad 1121 = 25a + 5b + c,$$
$$626 = a \cdot 7^2 + b \cdot 7 + c, \quad \text{or} \quad 626 = 49a + 7b + c,$$
$$967 = a \cdot 9^2 + b \cdot 9 + c, \quad \text{or} \quad 967 = 81a + 9b + c.$$

We solve this system of three equations in three variables using procedures of algebra and get

$$a = 104.5, \quad b = -1501.5, \quad \text{and} \quad c = 6016.$$

Substituting these values into equation (1), we get the function given by

$$y = 104.5x^2 - 1501.5x + 6016.$$

c) The death rate for males who sleep 2 hr is given by

$$y = 104.5(2)^2 - 1501.5(2) + 6016 = 3431.$$

The death rate for males who sleep 8 hr is given by

$$y = 104.5(8)^2 - 1501.5(8) + 6016 = 692.$$

The death rate for males who sleep 10 hr is given by

$$y = 104.5(10)^2 - 1501.5(10) + 6016 = 1451.$$

✦

TECHNOLOGY CONNECTION

Mathematical Modeling Using Regression: Fitting Quadratic and Other Polynomial Functions to Data

Regression can be extended to quadratic, cubic, and quartic polynomial functions.

EXAMPLE Life Science: Live Births to Women of Age x. The following chart relates the average number of live births to women of a particular age.

a) Fit a quadratic function to the data using REGRESSION. Then make a scatterplot of the data and graph the quadratic function with the scatterplot.

b) Fit a cubic function to the data using REGRESSION. Then make a scatterplot of the data and graph the cubic function with the scatterplot.

c) Which function seems to fit the data better?

Age, x	Average Number of Live Births per 1000 Women
16	34
18.5	86.5
22	111.1
27	113.9
32	84.5
37	35.4
42	6.8

(*Source*: Centers for Disease Control and Prevention.)

d) Use the function from part (c) to estimate the average number of live births to women of ages 20 and 30.

Solution We proceed as follows.

a) To fit a quadratic function using REGRESSION, the procedure is similar to what is outlined in the preceding Technology Connection on linear regression. We enter the data but select QuadReg instead of LinReg (ax + b). For the graph below, we used ZoomStat to set the window. We wrote y_1 using approximations of a, b, and c from the QuadReg screen.

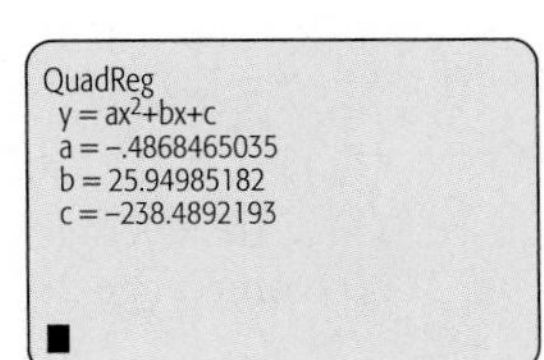

$y_1 = -0.49x^2 + 25.95x - 238.49$

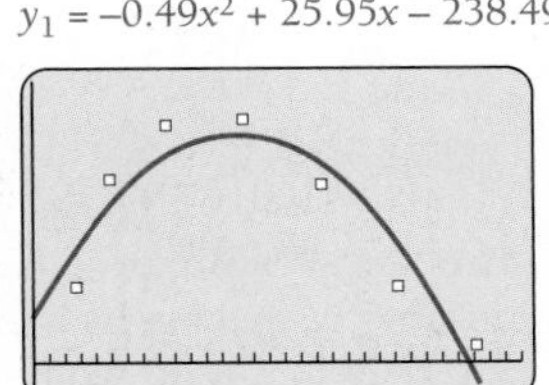

b) To fit a cubic function, we select CubicReg and obtain the following:

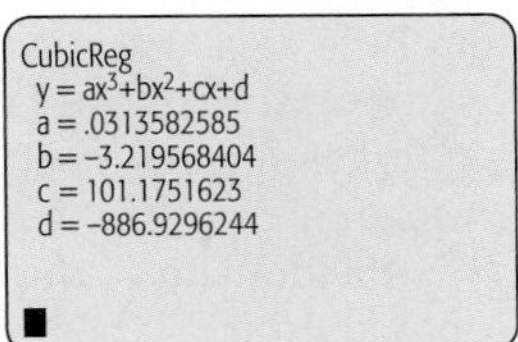

A convenient way to copy a regression function, in full detail, onto the Y= screen is to press (Y=), move the cursor to Y2 (or wherever the function is to appear), press (VARS), and select STATISTICS and then EQ and RegEQ.

$y_2 = 0.031358258459520x^3 - 3.2195684044498x^2 + 101.17516232219x - 886.92962438781$

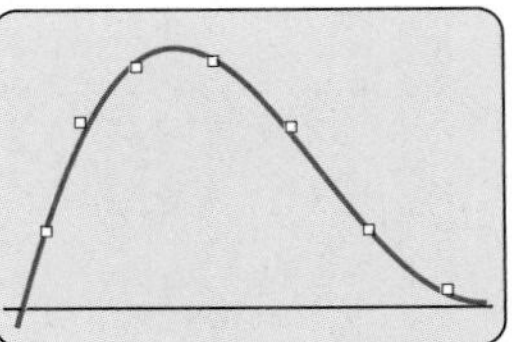

(continued)

Mathematical Modeling Using Regression: Fitting Quadratic and Other Polynomial Functions to Data (continued)

c) The graph of the cubic function seems to fit closer to the data points. Thus we choose it as a model.

d) We press 2ND QUIT to leave the graph screen. Pressing VARS and selecting Y-VARS and then FUNCTION and Y2, we have $Y2(20) \approx 99.6$ and $Y2(30) \approx 97.4$ as shown.

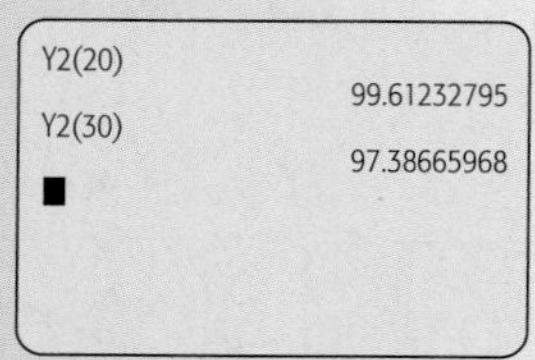

The TRACE feature can also be used. Thus, the average number of live births is 99.6 per 1000 women age 20 and 97.4 per 1000 women age 30.

EXERCISES

1. *Life science: live births.*

a) Use the REGRESSION feature to fit a quartic equation to the live-birth data. Make a scatterplot of the data. Then graph the quartic function with the scatterplot. Decide whether the quartic function gives a better fit than either the quadratic or the cubic function.

TW **b)** Explain why the domain of the cubic live-birth function should probably be restricted to the interval [15, 45].

2. *Business: median household income by age.*

Age, x	Median Income in 2003
19.5	$27,053
29.5	44,779
39.5	55,044
49.5	60,242
59.5	49,215
65	23,787

(*Source*: Based on data in the *Statistical Abstract of the United States, 2005*.)

a) Make a scatterplot of the data and fit a quadratic function to the data using QuadReg. Then graph the quadratic function with the scatterplot.

b) Fit a cubic function to the data using CubicReg. Then graph the cubic function with the scatterplot.

c) Fit a quartic function to the data using QuartReg. Then graph the quartic function with the scatterplot.

d) Which of the quadratic, cubic, or quartic functions seems to best fit the data?

e) Use the function from part (d) to estimate the median household income of people age 25; of people age 45.

3. *Life science: hours of sleep and death rate.* Repeat Example 3 using quadratic regression to fit a function to the data.

Exercise Set R.6

Choosing models. *For the scatterplots and graphs in Exercises 1–9, determine which, if any, of the following functions might be used as a model for the data:*

Linear, $f(x) = mx + b$

Quadratic, $f(x) = ax^2 + bx + c,\ a > 0$

Quadratic, $f(x) = ax^2 + bx + c,\ a < 0$

Polynomial, neither quadratic nor linear

1.

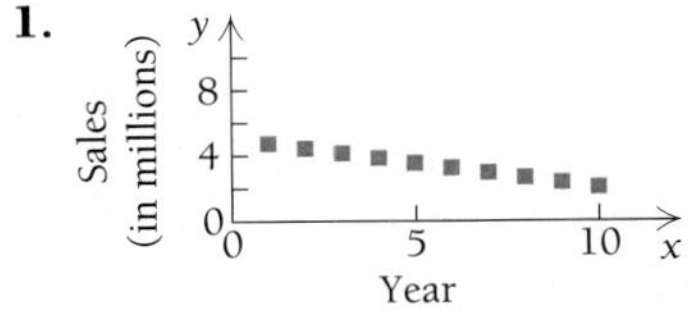

2.

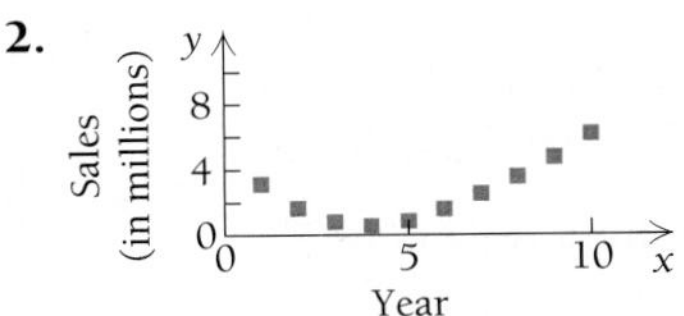

3.

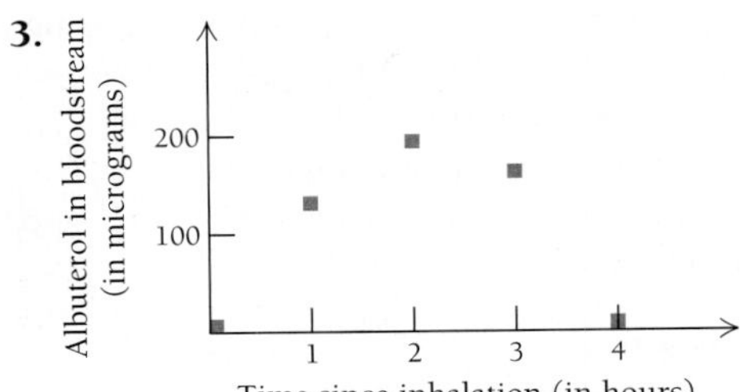

4.

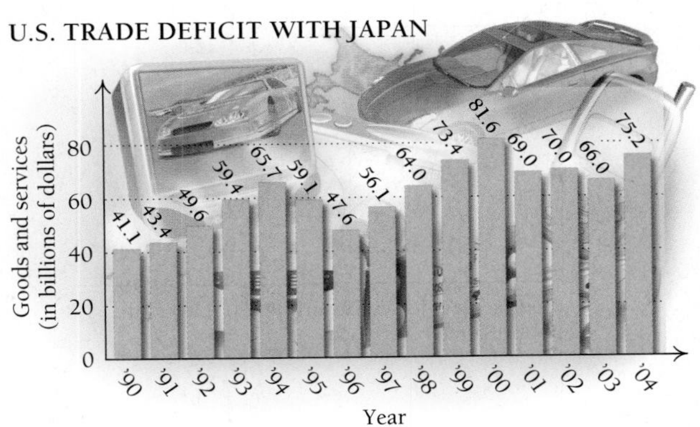

(*Source*: Based on information from U.S. Census Bureau, *Statistical Abstract of the United States, 2006.*)

5.

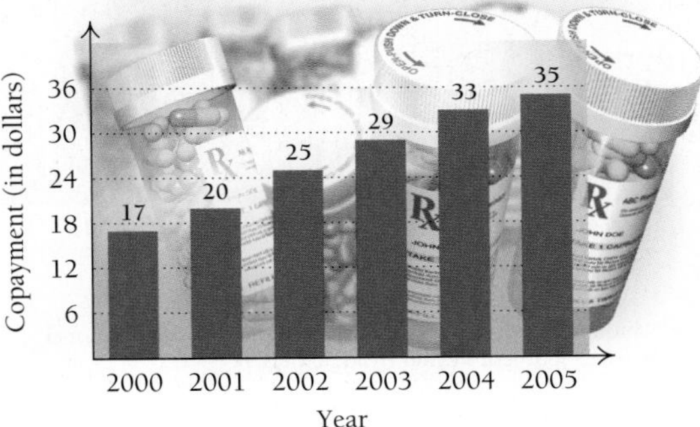

(*Source*: The Kaiser Family Foundation, Health Research and Education Trust.)

6.

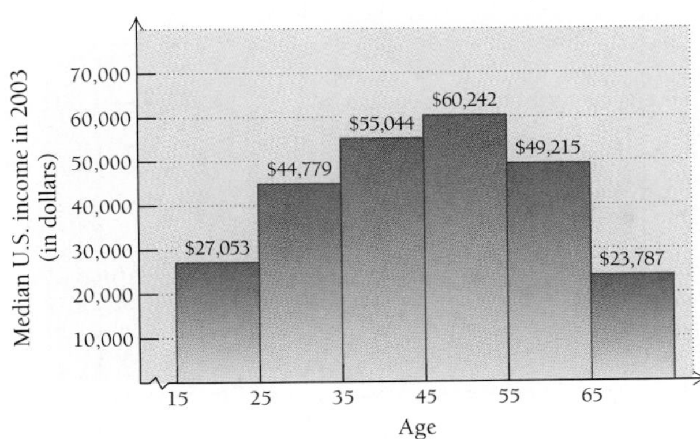

(*Source*: *Statistical Abstract of the United States, 2005.*)

7.

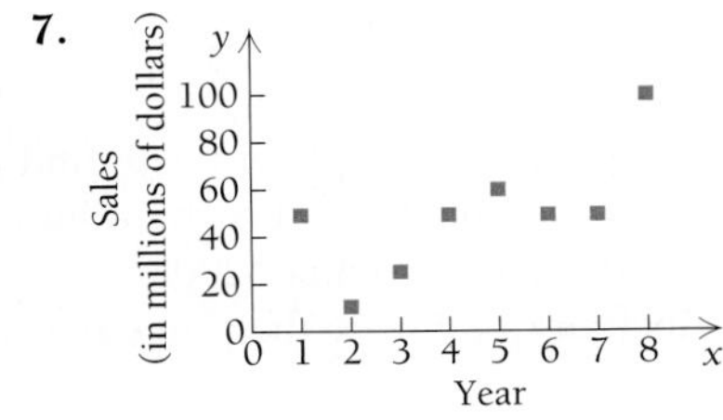

8.

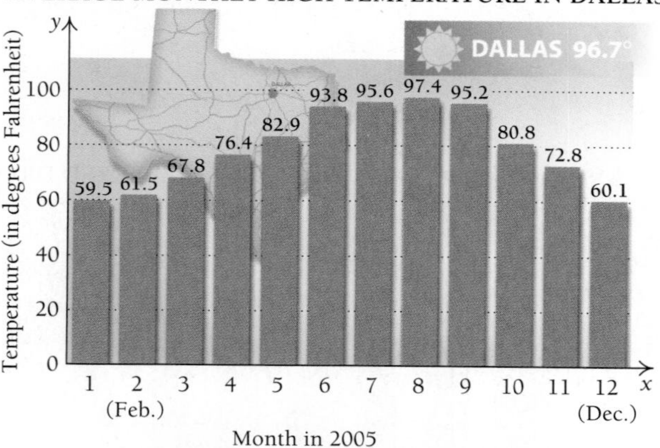

(*Source*: National Oceanic and Atmospheric Administration, National Weather Service, Southern Region Headquarters.)

9.

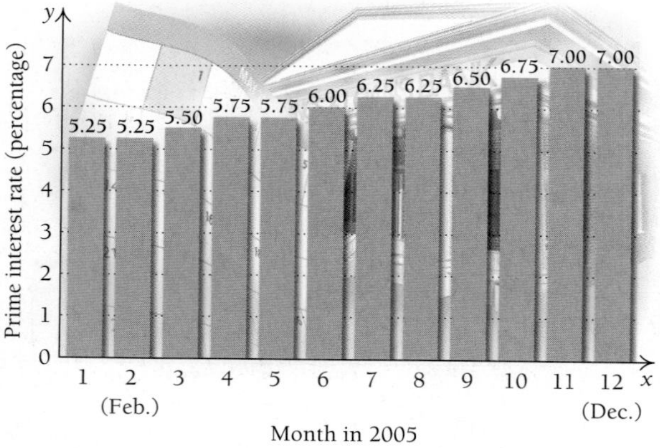

(*Source*: Federal Reserve Board.)

10. Prime interest rate.

a) For the prime interest rate data in Exercise 9, find a linear function that fits the data using the values given for January 2005 and December 2005.

b) Use the linear function to estimate the prime rate in June 2005.

11. Average co-payment for preferred drugs.

Years (since 2000)	Average co-payment
0 (2000)	$13
1 (2001)	15
2 (2002)	17
3 (2003)	19
4 (2004)	21

(*Source*: The Kaiser Family Foundation, Health Research and Education Trust.)

a) Choose two points from the data and find a linear function that fits the data.

b) Graph the scatterplot and the function on the same set of axes.

c) Use the function to predict the average co-payment for preferred drugs in 2010.

12. Absorption of an asthma medication. Use the data from Exercise 3.

a) Find a quadratic function that fits the data using the data points $(0, 0)$, $(2, 200)$, and $(3, 167)$.

b) Use the function to estimate the amount of albuterol in the bloodstream after 4 hr.

TW **c)** Does it make sense to use this function for $t = 6$? Why or why not?

13. Braking distance.

a) Find a quadratic function that fits the following data.

Travel Speed (mph)	Braking Distance (ft)
20	25
40	105
60	300

(*Source*: New Jersey Department of Law and Public Safety.)

b) Use the function to estimate the braking distance of a car that travels at 50 mph.

TW **c)** Does it make sense to use this function when speeds are less than 15 mph? Why or why not?

14. Daytime accidents.

a) Find a quadratic function that fits the following data.

Travel Speed (in km/h)	Number of Daytime Accidents (for every 200 million km driven)
60	100
80	130
100	200

b) Use the function to estimate the number of daytime accidents that occur at 50 km/h for every 200 million km driven.

15. High blood pressure in women.

a) Choose two points from the following data and find a linear function that fits the data.

Age of Female	Percentage of Females with High Blood Pressure
30	1.4
40	8.5
50	19.1
60	31.9
70	53.0

(*Source*: Based on data from Health United States 2005, CDC/NCHS.)

b) Graph the scatterplot and the function on the same set of axes.

c) Use the function to estimate the percent of 55-yr-old women with high blood pressure.

16. High blood pressure in men.

a) Choose two points from the following data and find a linear function that fits the data.

Age of Male	Percentage of Males with High Blood Pressure
30	7.3
40	12.1
50	20.4
60	24.8
70	34.9

(*Source*: Based on data from Health United States 2005, CDC/NCHS.)

b) Graph the scatterplot and the function on the same set of axes.

c) Use the function to estimate the percent of 55-yr-old men with high blood pressure.

SYNTHESIS

TW **17.** Suppose that you have just 3 or 4 data points. Why might it make better sense to use a linear function rather than a quadratic or cubic function that fits these data more closely?

TW **18.** When modeling the number of hours of daylight for the dates April 22 to August 22, which would be a better choice: a linear function or a quadratic function? Explain.

TW **19.** Explain the restrictions that should be placed on the domain of the quadratic function found in Exercise 12 and why such restrictions are needed.

TW **20.** Explain the restrictions that should be placed on the domain of the quadratic function found in Exercise 13 and why such restrictions are needed.

TECHNOLOGY CONNECTION

21. Prime interest rate.

a) Use regression to fit a linear function to the data in Exercise 9.

b) Use the function to estimate the prime rate in June 2005.

c) Compare your answers to those found in Exercise 10. Which is more accurate?

d) Fit a cubic function to the data and use it to estimate the prime rate in June 2005.

TW **e)** Is a linear or cubic model more appropriate for this set of data? Explain.

22. Co-payments for drugs.

a) Use regression to fit a linear function to the data in Exercise 5. Let x be the number of years after 2000.

b) Use the function to estimate the average co-payment in 2005 and 2010.

c) Fit a cubic function to the data and use it to predict the average co-payment in 2005 and 2010.

TW **d)** Is a linear or cubic model more appropriate for this set of data? Explain.

23. Trade deficit with Japan.

a) Use regression to fit a cubic function to the data in Exercise 4. Let x be the number of years after 1990.

b) Use the function to estimate the trade deficit in 2006 and in 2010.

TW **c)** Why might a linear function be a more logical choice than a cubic function for modeling this set of data?

Chapter Summary

This review chapter readies us for the study of calculus with a review of graphs, functions, models, and algebra. For a more detailed review of algebra, see the Appendix on p. 599.

To graph an **ordered pair,** like (4, 3), we locate a point so that the **first coordinate** is the number on the horizontal axis directly above or below the point and the **second coordinate** is the number on the vertical axis directly left or right of the point. (p. 3)

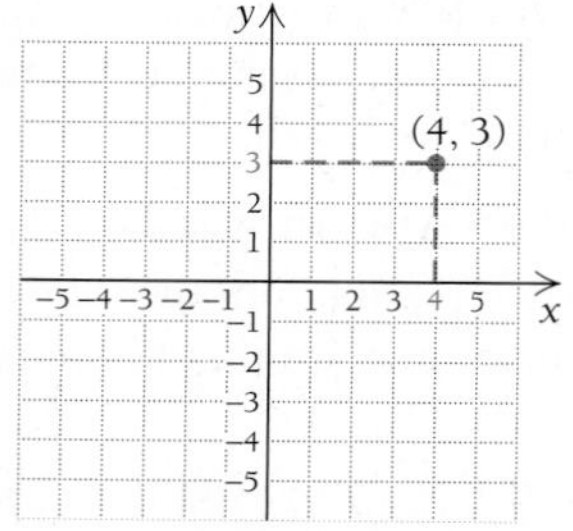

A **solution** of an equation in two variables is an ordered pair of numbers that, when substituted for the variables, forms a true sentence. The **graph** of an equation is a drawing that represents all ordered pairs that are solutions of the equation. (p. 4)

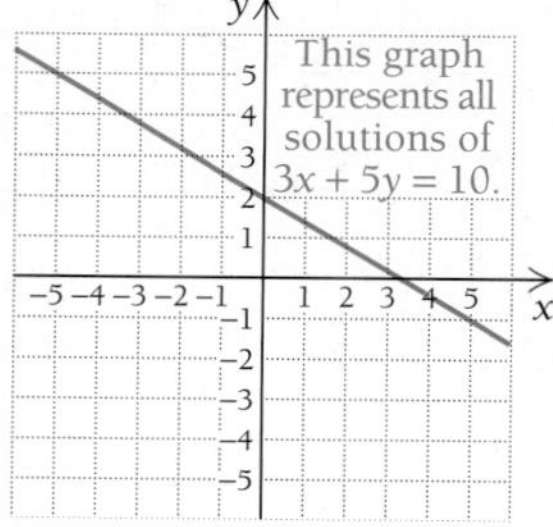

The mathematics used to represent the essential features of a problem comprises a **mathematical model.** For example, the equation $d = 4.9t^2$ can be used to estimate the distance d, in meters, that an object falls in t seconds after it is dropped. (pp. 8, 74)

One of the most important concepts in mathematics, and in calculus in particular, is that of a function:

A **function** is a correspondence between a first set, called the **domain,** and a second set, called the **range,** such that each member of the domain corresponds to *exactly one* member of the range. Thus,

$\{(4, 2), (9, 3), (16, 4), (9, -3)\}$ is *not* a function, but

$\{(4, 2), (9, 3), (16, 4), (25, 5)\}$ is a function. (p. 14)

When graphed, a function will always pass the **vertical-line test:**

A graph represents a function if it is impossible to draw a vertical line that intersects the graph more than once. (p. 20)

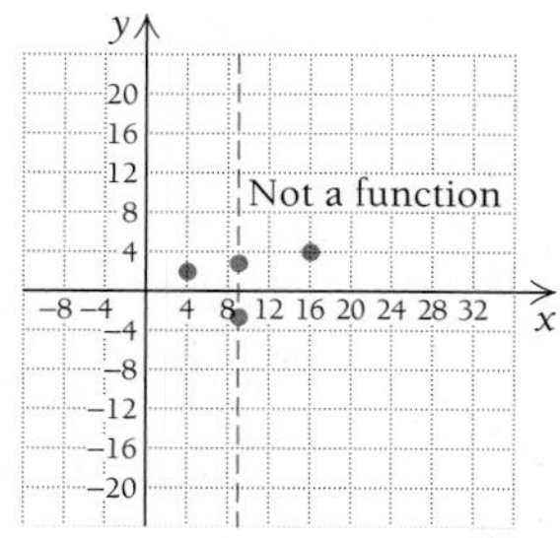

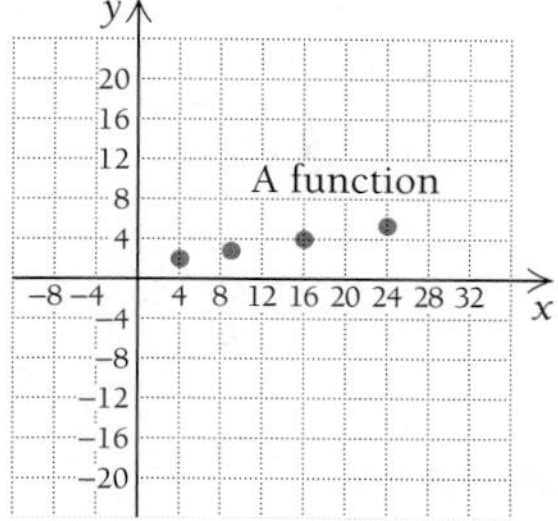

Function notation permits us to easily identify what member of the domain, or **input,** is paired with what member of the range, or **output.** For example, the notation $f(x) = x^2 - 1$ indicates that the input x is paired with the output $x^2 - 1$, and thus $f(4) = 4^2 - 1 = 16 - 1 = 15$. (pp. 16–19)

Like the function f above, most functions in this text are written in the form of an algebraic expression that is evaluated. Unless noted otherwise, the domain of such a function is the set of all real numbers for which the expression can be evaluated. Thus, if

$$f(x) = \frac{x}{x - 7},$$

then the domain of $f = \{x \mid x \text{ is a real number and } x \neq 7\}$. (p. 33)

A function that is **piecewise-defined** specifies different rules for certain members of the domain. For example, the function given by

$$f(x) = \begin{cases} x^2, & \text{for } x < 3, \\ \frac{1}{2}x, & \text{for } x \geq 3, \end{cases}$$

specifies that $f(2)$ and $f(6)$ are found using different rules:

$$f(2) = 2^2 = 4, \quad \text{whereas} \quad f(6) = \tfrac{1}{2} \cdot 6 = 3.$$ (pp. 20–21)

The variable that is used to represent a function's inputs is often called the **independent variable,** and the variable used to represent a function's outputs is the **dependent variable.** Thus, if a function is given by

$$y = 3x^4 - 2x,$$

the independent variable is x and the dependent variable is y. (p. 24)

The domain of any function is a **set,** or collection, of objects. Because there is a **real number** for every point on the number line, usually the domain of a function is a set of real numbers. There are three ways of writing sets. (pp. 29–31)

Roster notation: $\{2, 5, 9, 15\}$

Set-builder notation: $\{x \mid x \text{ is a real number and } x > 3\}$

Interval notation:

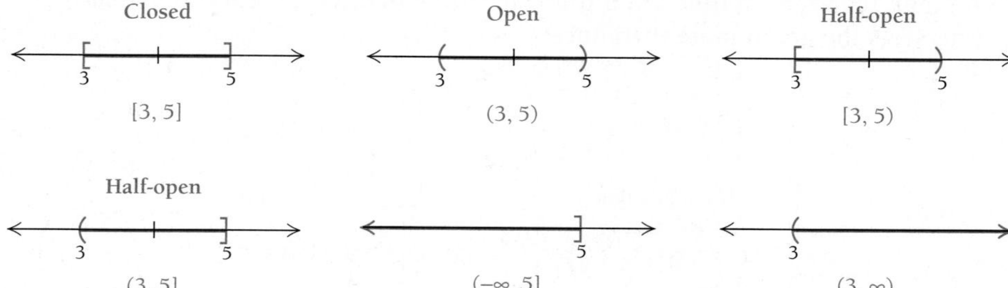

Important Graphs

The graph of an equation written in **slope–intercept form,** $f(x) = mx + b$, is a line with **slope** m and **y-intercept** $(0, b)$. Such a function is said to be **linear.** (pp. 43–45)

The graph of an equation written in **point–slope form,** $y - y_1 = m(x - x_1)$, is a line with slope m passing through the point (x_1, y_1). (p. 45)

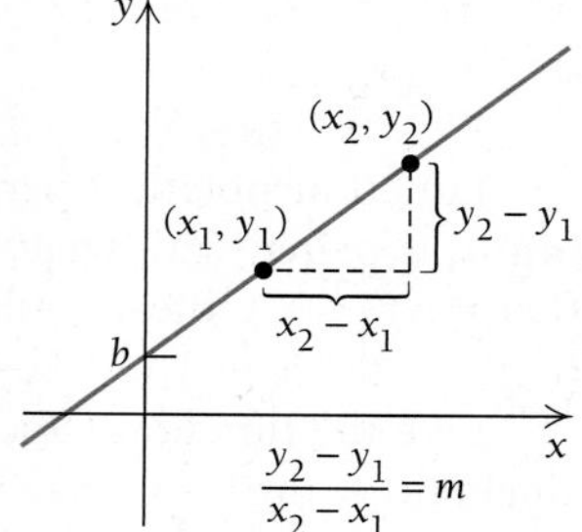

The graph of a **constant function,** given by $f(x) = c$, is a horizontal line. The slope is 0. (pp. 40, 47)

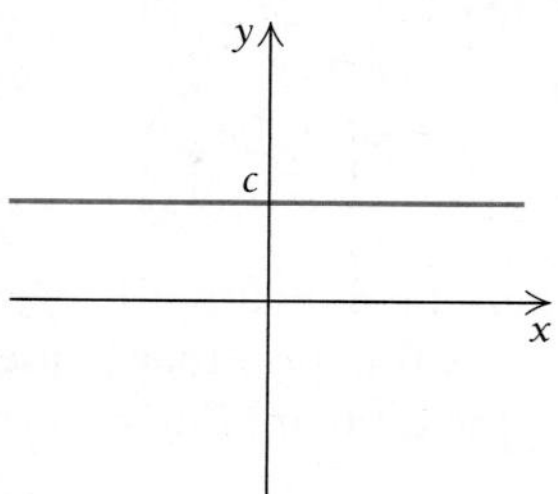

The graph of an equation of the form $x = a$ is a vertical line and is not a function. The slope is undefined. (p. 47)

Slope can also be regarded as **average rate of change.** (p. 47)

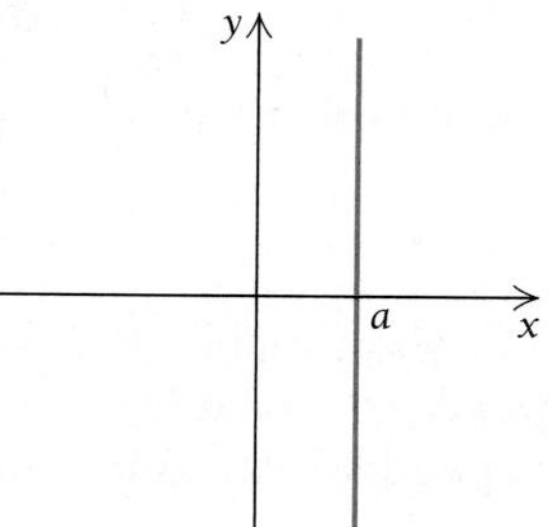

The graph of a **quadratic function,** given by $f(x) = ax^2 + bx + c$, is a **parabola** with a **vertex** at $\left(-\frac{b}{2a}, f\left(-\frac{b}{2a}\right)\right)$. (p. 56)

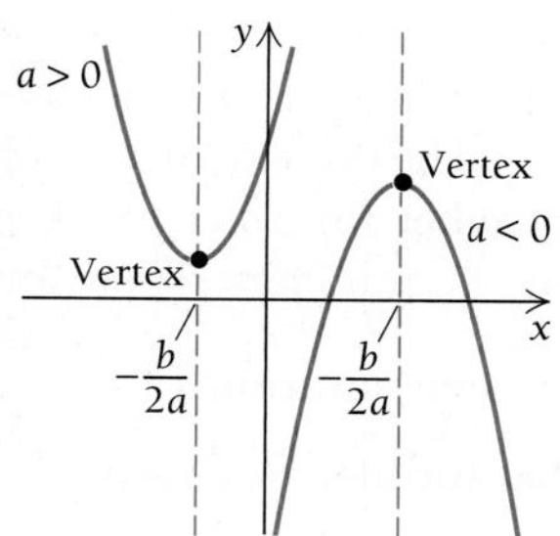

The **x-intercepts** of the graph of a quadratic function are the solutions of $ax^2 + bx + c = 0$ and are given by the **quadratic formula:**

$$x = \frac{-b \pm \sqrt{b^2 - 4ac}}{2a}. \qquad \text{(p. 58)}$$

A **polynomial function** is given by an equation of the form

$$f(x) = a_n x^n + a_{n-1}x^{n-1} + a_{n-2}x^{n-2} + \cdots + a_2x^2 + a_1x^1 + a_0,$$

where n is a nonnegative integer and the a's are real-number **coefficients. Power functions** given by equations of the form $f(x) = ax^n$ are one type of polynomial function that is relatively easy to graph. (p. 60)

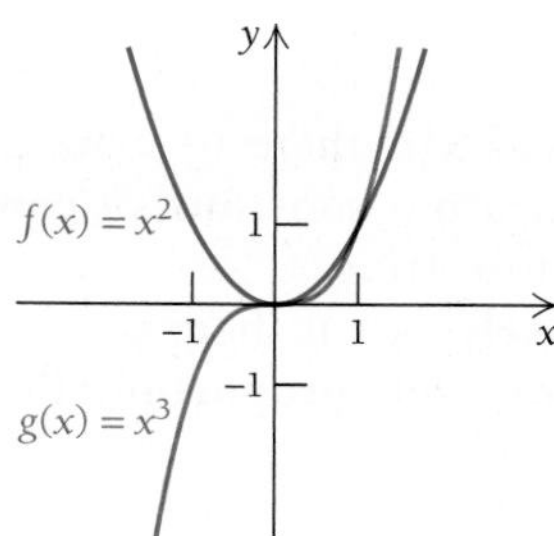

A **rational function** is a function given by the quotient of two polynomials,

$$f(x) = \frac{p(x)}{q(x)}, \quad \text{with } q(x) \neq 0.$$

Numbers for which $q(x)$ is 0 are referred to as **split numbers**. (p. 62)

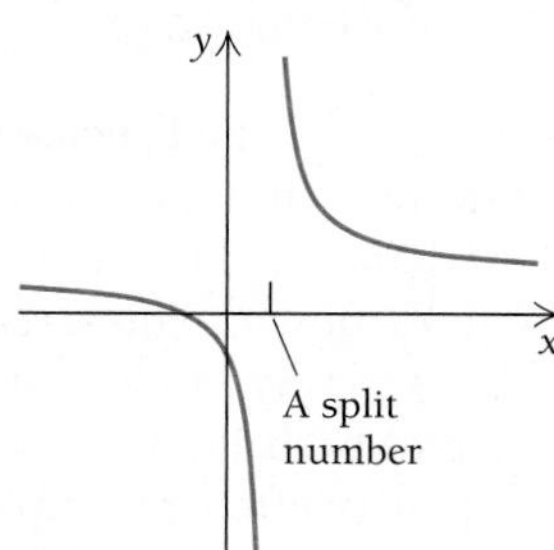

The absolute-value function (p. 65)

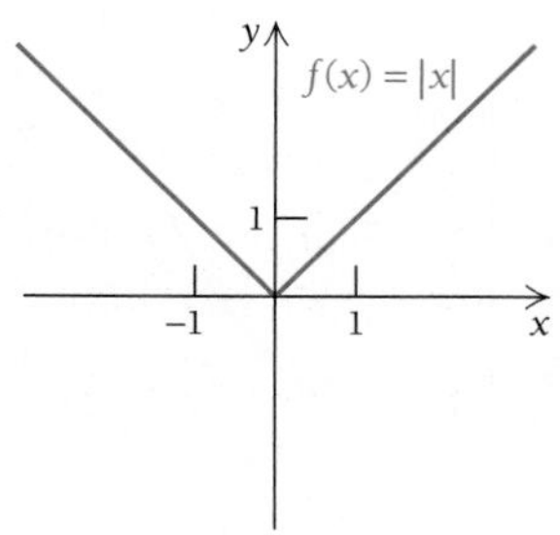

The square-root function (p. 65)

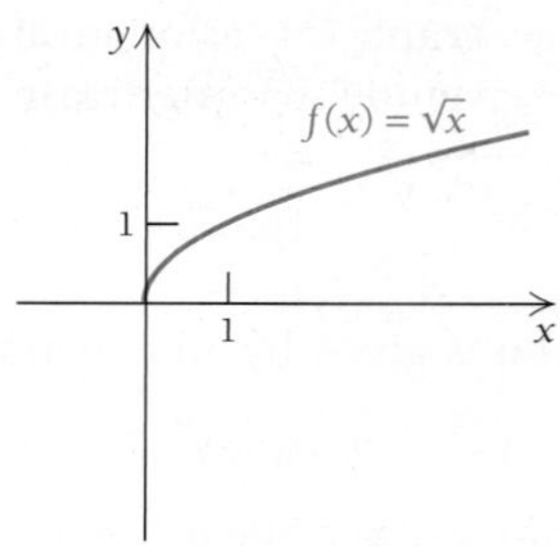

When real-world data are available to graph, we can form a **scatterplot** and then decide which, if any, of the above graphs best models the situation. (p. 74)

Variation

The variable y **varies directly** as x if there is some positive constant m for which $y = mx$. We say that m is the **variation constant,** or **constant of proportionality,** and that y is **directly proportional** to x. (p. 42)

The variable y **varies inversely** as x if there is some positive number k such that $y = k/x$. We also say that y is **inversely proportional** to x. (p. 64)

Business Applications

Total costs are the **fixed costs,** such as rent and maintenance, plus the **variable costs,** such as material and labor, required for production. (p. 48)

Total revenue is the money a business receives and is generally the product of the number of items sold and the price per item.

Total profit is the difference between total revenue and total cost:

$$\begin{matrix}\text{Profit from the sale} \\ \text{of } x \text{ items, } P(x)\end{matrix} = \begin{matrix}\text{Revenue from the sale} \\ \text{of } x \text{ items, } R(x)\end{matrix} - \begin{matrix}\text{Cost of producing} \\ x \text{ items, } C(x).\end{matrix} \qquad \text{(p. 49)}$$

The value of x at which $P(x) = 0$ is called the **break-even value.** (p. 49)

For consumers to purchase more of an item, the price usually needs to drop. If x is the price of each item purchased and $q(x)$ is the quantity of items purchased, then, as x goes up, $q(x)$ generally comes down. This relationship is represented by the **demand function.** For producers to supply more of an item, they often want the price per item to go up. If x is the price of each item purchased and $q(x)$ is the quantity of items produced, then, as x goes up, $q(x)$ often goes up as well. This relationship is represented by the **supply function.** The point at which consumers and producers agree on the price and quantity of items sold is called the **equilibrium point.** (p. 69)

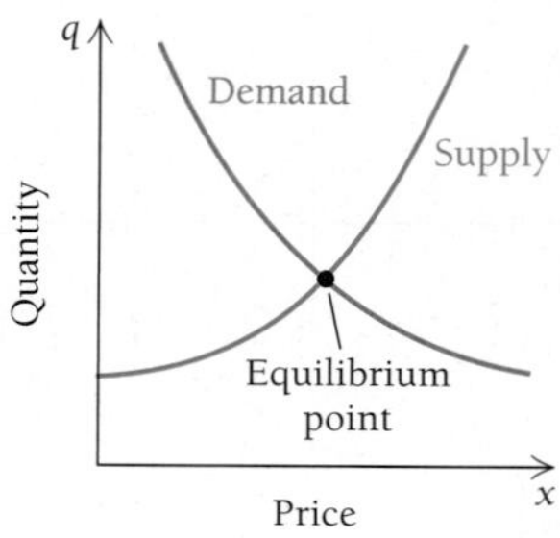

Chapter Review Exercises

These review exercises are for test preparation. They can also be used as a practice test. Answers are at the back of the book. The bracketed section references tell you what part(s) of the chapter to restudy if your answer is incorrect.

CONCEPT REINFORCEMENT

For each equation in column A, select the most appropriate graph in column B. [R.1, R.4, R.5]

Column A | **Column B**

1. $y = |x|$

a)

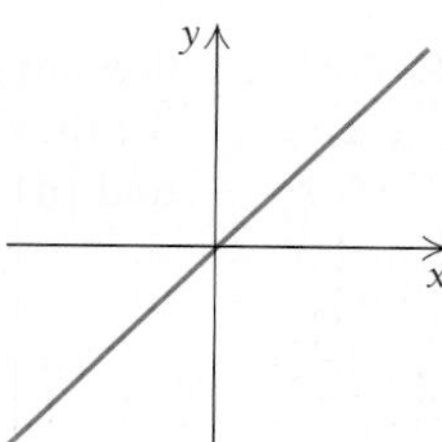

2. $f(x) = x^2 - 1$

b)

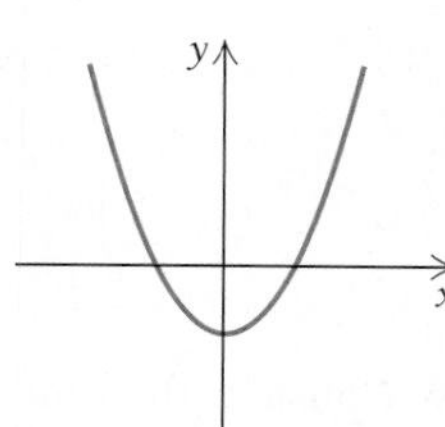

3. $y = -2x - 1$

c)

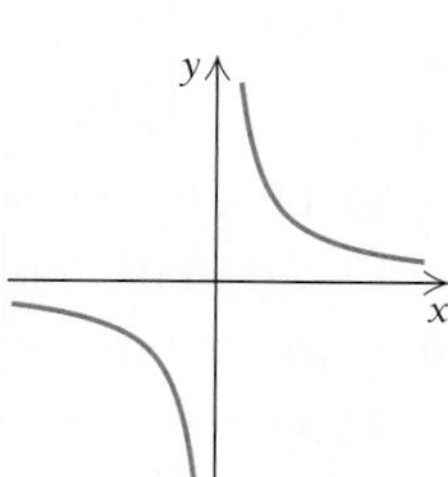

4. $y = x$

d)

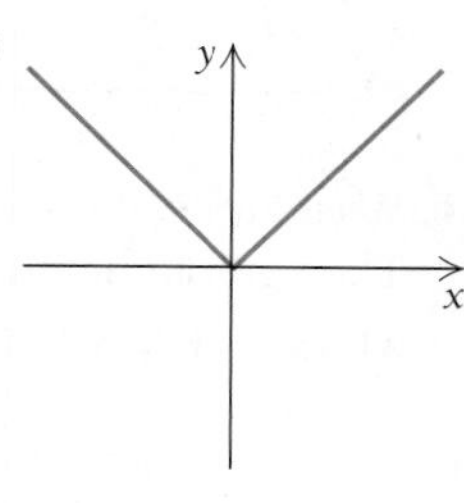

5. $g(x) = \sqrt{x}$

e)

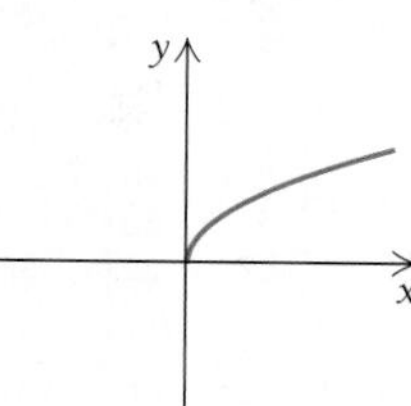

6. $f(x) = \frac{1}{x}$

f)

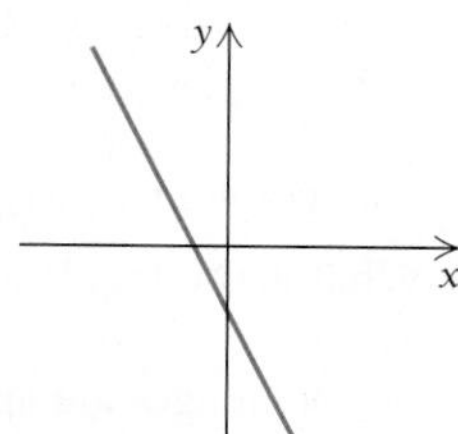

In Exercises 7–14, classify each statement as either true or false.

7. The graph of an equation is a drawing that represents all ordered pairs that are solutions of the equation. [R.1]

8. If $f(-3) = 5$ and $f(3) = 5$, then f cannot be a function. [R.2]

9. The notation $(3, 7)$ can represent a point or an interval. [R.3]

10. An equation of the form $y - y_1 = m(x - x_1)$ has a graph that is a line of slope m passing through (x_1, y_1). [R.4]

11. The graph of an equation of the form $f(x) = ax^2 + bx + c$ has its vertex at $x = b/(2a)$. [R.5]

12. A scatterplot is a random collection of points near a line. [R.6]

13. Unless stated otherwise, the domain of a polynomial function is the set of all real numbers. [R.5]

14. The graph of a constant function has a slope of 0. [R.4]

REVIEW EXERCISES

15. Life science: babies born to women of age x. The following graph relates the number of babies born per 1000 women to the women's age. [R.1, R.3]

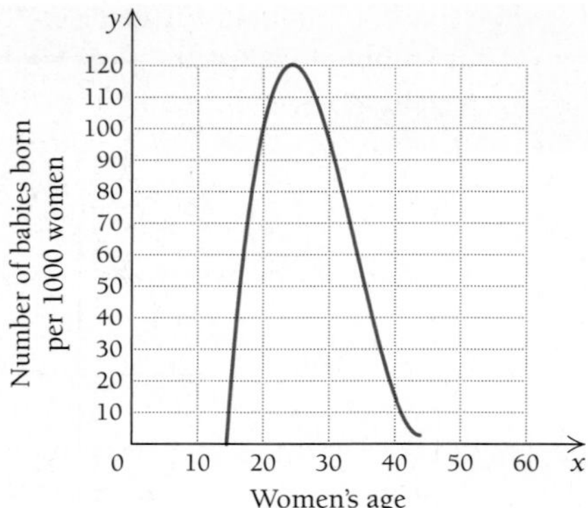

Use the graph to answer the following.

a) What is the incidence of babies born to women of age 35?

b) For what ages are approximately 100 babies born per every 1000 women?

TW **c)** Make an estimate of the domain of the function, and explain why it should be so.

16. Business: compound interest. Suppose that \$1100 is invested at 5%, compounded semiannually. How much is in the account at the end of 4 yr? [R.1]

17. Finance: compound interest. Suppose that \$4000 is borrowed at 12%, compounded annually. How much is owed at the end of 2 yr? [R.1]

18. Is the following correspondence a function? Why or why not? [R.2]

Patricia → Pat
Patrick → Pat
Rebecca → Becky
Richard → Rick
Richard → Rich
Richard → Dick

19. A function is given by $f(x) = -x^2 + x$. Find each of the following. [R.2]

a) $f(3)$ **b)** $f(-5)$
c) $f(a)$ **d)** $f(x + h)$

Graph. [R.5]

20. $y = |x + 1|$

21. $f(x) = (x - 2)^2$

22. $f(x) = \dfrac{x^2 - 16}{x + 4}$

23. $g(x) = \sqrt{x} + 1$

Use the vertical-line test to determine whether each of the following is the graph of a function. [R.2]

24.

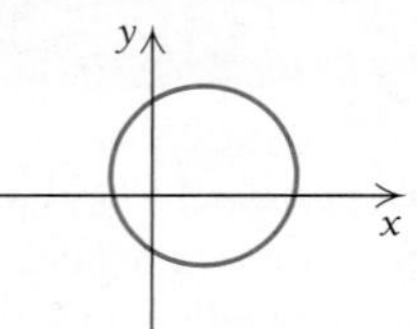

25.

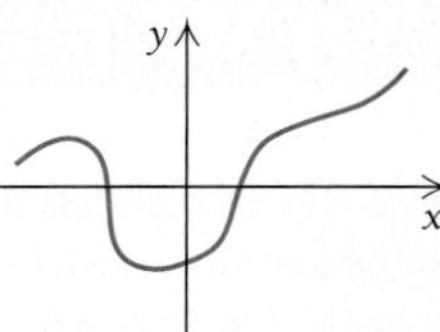

26.

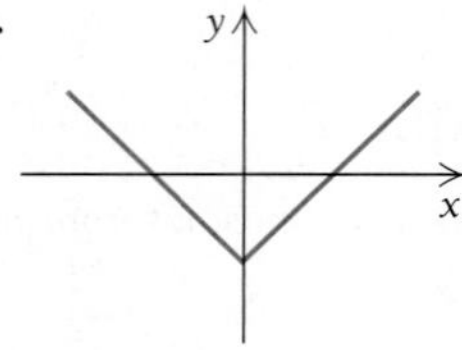

27.

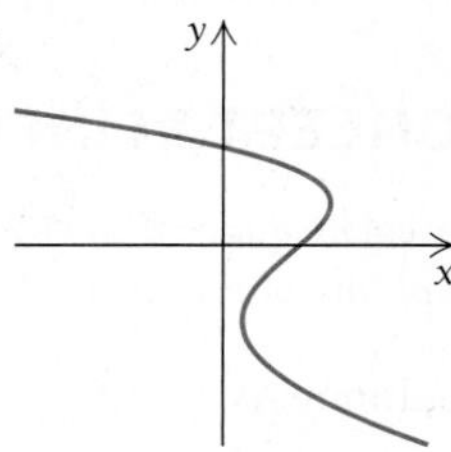

28. For the following graph of function f, determine **(a)** $f(2)$; **(b)** the domain; **(c)** all x-values such that $f(x) = 2$; and **(d)** the range. [R.3]

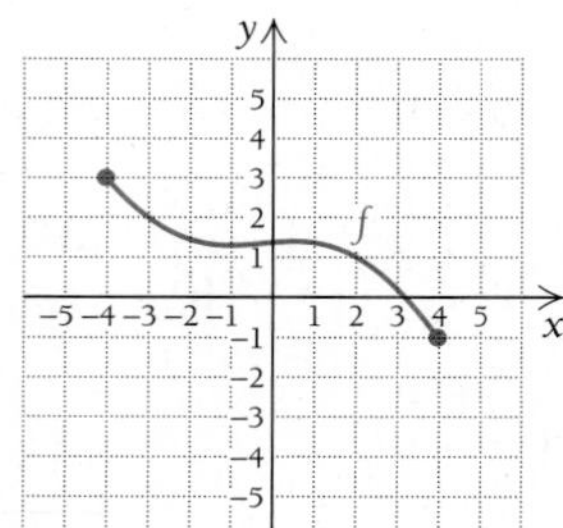

29. Consider the function given by

$$f(x) = \begin{cases} -x^2 + 2, & \text{for } x < 1, \\ 4, & \text{for } 1 \leq x < 2, \\ \frac{1}{2}x, & \text{for } x \geq 2. \end{cases}$$

a) Find $f(-1)$, $f(1.5)$, and $f(6)$. [R.2]

b) Graph the function. [R.2]

30. Write interval notation for each graph. [R.3]

a)

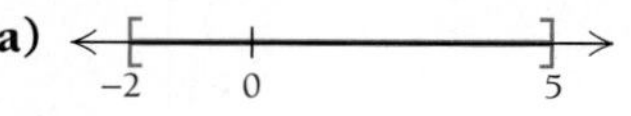

b)

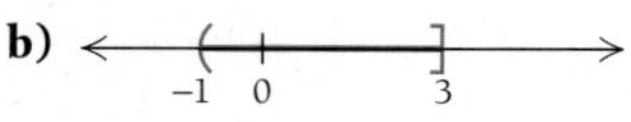

c)

a

31. Write interval notation for each of the following. Then graph the interval on a number line. [R.3]

a) $\{x \mid -4 \leq x < 5\}$ **b)** $\{x \mid x > 2\}$

32. For the function graphed below, determine **(a)** $f(-3)$; **(b)** the domain; **(c)** all x-values for which $f(x) = 4$; **(d)** the range. [R.3]

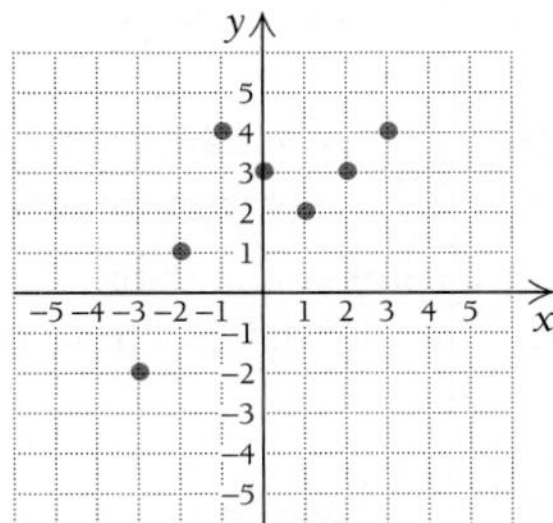

33. Find the domain of f. [R.3, R.5]

a) $f(x) = \dfrac{7}{2x - 10}$

b) $f(x) = \sqrt{x + 6}$

34. What are the slope and the y-intercept of

$y = -3x + 2$? [R.4]

35. Find an equation of the line with slope $\frac{1}{4}$, containing the point $(8, -5)$. [R.4]

36. Find the slope of the line containing the points $(2, -5)$ and $(-3, 10)$. [R.4]

Find the average rate of change. [R.4]

37.

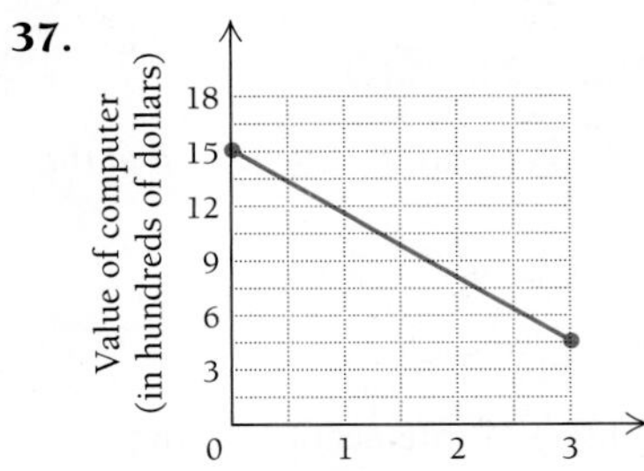

38.

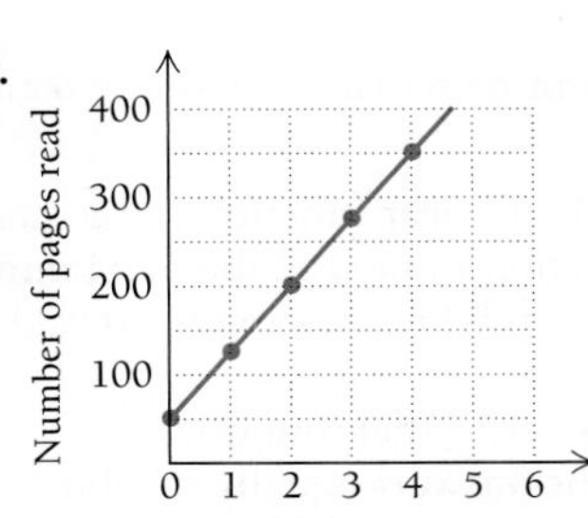

39. Shipping charges. The amount A that Pet-Treats-to-U charges for shipping is directly proportional to the value V of the item(s) being shipped. If the business charges \$2.10 to ship a \$60 gift basket, find an equation of variation expressing A as a function of V. [R.4]

40. Business: profit-and-loss analysis. The band Soul Purpose has fixed costs of \$4000 for producing a new CD. Thereafter, the variable costs are \$0.50 per CD, and the CD will sell for \$10. [R.4]

a) Find and graph $C(x)$, the total cost of producing x CDs.

b) Find and graph $R(x)$, the total revenue from the sale of x CDs. Use the same axes as in part (a).

c) Find and graph $P(x)$, the total profit from the production and sale of x CDs. Use the same axes as in part (b).

d) How many CDs must the band sell in order to break even?

41. Graph each pair of equations on one set of axes. [R.5]

a) $y = \sqrt{x}$ and $y = \sqrt{x} - 3$

b) $y = x^3$ and $y = (x - 1)^3$

42. Graph each of the following. If the graph is a parabola, identify the vertex. [R.5]

a) $f(x) = x^2 - 6x + 8$

b) $g(x) = \sqrt[3]{x} + 2$

c) $y = -\dfrac{1}{x}$

d) $y = \dfrac{x^2 + x - 6}{x - 2}$

43. Solve each of the following. [R.5]

a) $5 + x^2 = 4x + 2$

b) $2x^2 = 4x + 3$

44. Rewrite each of the following as an equivalent expression with rational exponents. [R.5]

a) $\sqrt[5]{x^4}$ **b)** $\sqrt{t^8}$

c) $\dfrac{1}{\sqrt[3]{m^2}}$ **d)** $\dfrac{1}{\sqrt{x^2 - 9}}$

45. Rewrite each of the following as an equivalent expression using radical notation. [R.5]

a) $x^{2/5}$ **b)** $m^{-3/5}$

c) $(x^2 - 5)^{1/2}$ **d)** $\dfrac{1}{t^{-1/3}}$

46. Determine the domain of the function given by

$f(x) = \sqrt[8]{2x - 9}$. [R.5]

47. Economics: equilibrium point. Find the equilibrium point for the given demand and supply functions. [R.5]

Demand: $q = (x - 7)^2$

Supply: $q = x^2 + x + 4$ (assume $x \leq 7$)

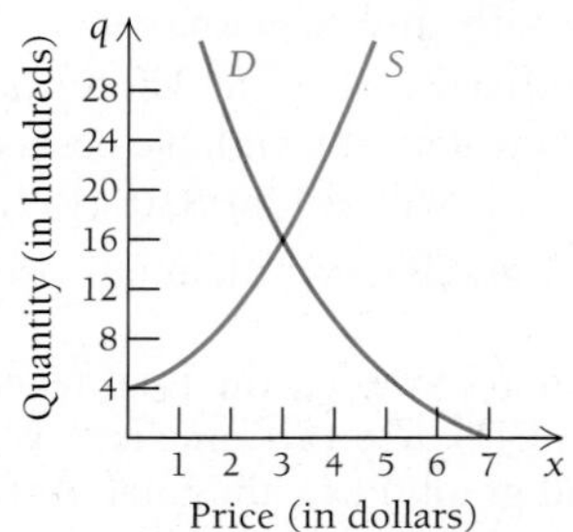

48. Trail maintenance. The amount of time required to maintain a section of the Appalachian Trail varies inversely as the number of volunteers working. If a particular section of trail can be cleared in 4 hr by 9 volunteers, how long would it take 11 volunteers to clear the same section? [R.5]

49. Life science: maximum heart rate. A person exercising should not exceed a maximum heart rate, which depends on his or her gender, age, and resting heart rate. The following table shows data relating resting heart rate and maximum heart rate for a 20-yr-old woman. [R.6]

Resting Heart Rate, r (in beats per minute)	Maximum Heart Rate, M (in beats per minute)
50	170
60	172
70	174
80	176

(*Source*: American Heart Association.)

a) Using the data points (50, 170) and (80, 176), find a linear function that fits the data.
b) Graph the scatterplot and the function on the same set of axes.
c) Use the function to predict the maximum heart rate of a woman whose resting heart rate is 67.

50. Business: ticket profits. The Spring Valley Drama Troupe is performing a new play. Data relating the daily profit P to the number of days after opening night are given below. [R.6]

Days, x	0	9	18	27	36	45
Profit, P (in dollars)	870	548	−100	−100	510	872

a) Make a scatterplot of the data.
b) Decide whether the data seem to fit a quadratic function.
c) Using the data points (0, 870), (18, −100), and (45, 872), find a quadratic function that fits the data.
d) Use the function to estimate the profit made on the 30th day.
TW e) Make an estimate of the domain of this function. Explain its restrictions.

SYNTHESIS

51. Economics: demand. The demand function for Clifton Cheddar Cheese is given by

Demand: $q = 800 - x^3, \quad 0 \leq x \leq 9.28,$

where x is the price per pound and q is in thousands of pounds.

a) Find the number of pounds sold when the price per pound is \$6.50.
b) Find the price per pound when 720,000 lb are sold.

TECHNOLOGY CONNECTION

Graph the function and find the zeros, the domain, and the range. [R.5]

52. $f(x) = x^3 - 9x^2 + 27x + 50$

53. $f(x) = \sqrt[3]{|4 - x^2|} + 1$

54. Approximate the point(s) of intersection of the graphs of the two functions in Exercises 52 and 53. [R.5]

55. Life science: maximum heart rate. Use the data in Exercise 49. [R.6]

a) Use regression to fit a linear function to the data.
b) Use the linear function to predict the maximum heart rate of a woman whose resting heart rate is 67.
TW c) Compare your answer to that found in Exercise 49. Are the answers equally reliable? Why or why not?

56. **Business: ticket profits.** Use the data in Exercise 50. [R.6]

a) Use regression to fit a quadratic function to the data.
b) Use the function to estimate the profit made on the 30th day.
TW c) What factors might cause the Spring Valley Drama Troupe's profit to drop and then rise?

57. **Social sciences: time spent on home computer.** The data in the table to the right relate the average number of minutes spent per month on a home computer, A, to a person's age, x. [R.6]

a) Use regression to fit linear, quadratic, cubic, and quartic functions to the data.
b) Make a scatterplot of the data and graph each function on the scatterplot.
TW c) Which function fits the data best? Why?

Age (in years)	Average Use (in minutes per month)
6.5	363
14.5	645
21	1377
29.5	1727
39.5	1696
49.5	2052
55	2299

(*Source*: Media Matrix; The PC Meter Company.)

Chapter R Test

1. **Social sciences: value of inheritances.** The median value of an inheritance in the United States is shown below.

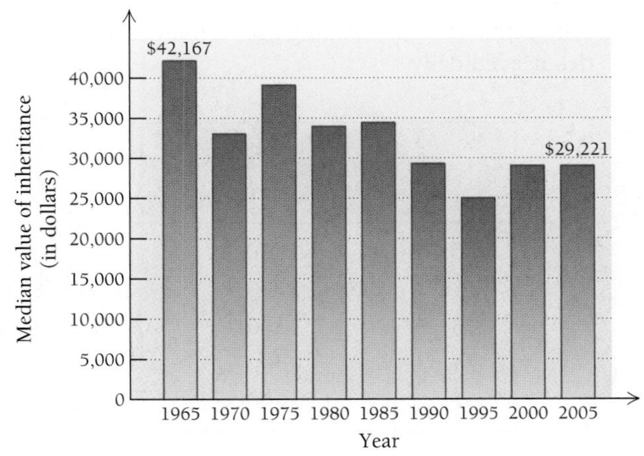

(*Source*: *The New York Times*, 3/26/06, p. 4.)

a) What was the approximate median value of an inheritance in 1990?
b) In what year was the median value of an inheritance approximately $25,000?
c) Estimate the range of the function that expresses median inheritance value as a function of the year.

2. **Business: compound interest.** A person made an investment at 6.5% compounded annually. It has grown to $798.75 in 1 yr. How much was originally invested?

3. A function is given by $f(x) = -x^2 + 5$. Find:

a) $f(-3)$
b) $f(a + h)$.

4. What are the slope and the y-intercept of $y = \frac{4}{5}x - \frac{2}{3}$?

5. Find an equation of the line with slope $\frac{1}{4}$, containing the point $(-3, 7)$.

6. Find the slope of the line containing the points $(-9, 2)$ and $(3, -4)$.

Find the average rate of change.

7.

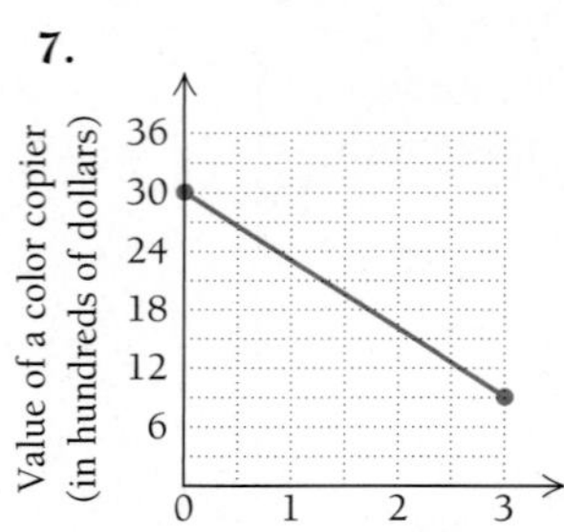

8.

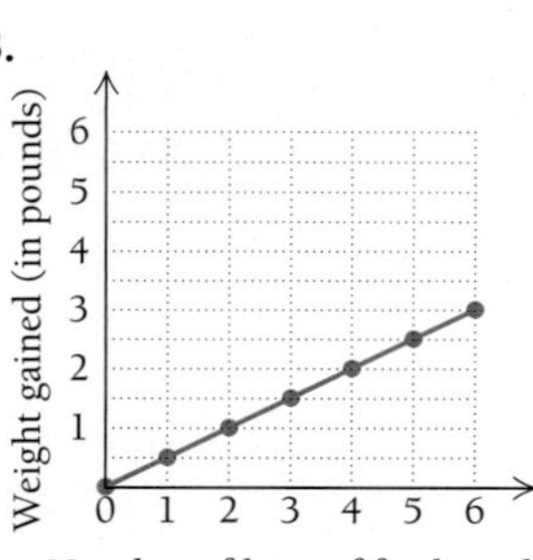

9. Life science: body fluids. The weight F of fluids in a human is directly proportional to body weight W. It is known that a person who weighs 180 lb has 120 lb of fluids. Find an equation of variation expressing F as a function of W.

10. Business: profit-and-loss analysis. A printing shop has fixed costs of \$8000 for producing a newly designed note card. Thereafter, the variable costs are \$0.08 per card. The revenue from each card is expected to be \$0.50.

a) Formulate a function $C(x)$ for the total cost of producing x cards.

b) Formulate a function $R(x)$ for the total revenue from the sale of x cards.

c) Formulate a function $P(x)$ for the total profit from the production and sale of x cards.

d) How many cards must the company sell in order to break even?

11. Economics: equilibrium point. Find the equilibrium point for these demand and supply functions:

Demand: $q = (x - 7)^2, \quad 0 \leq x \leq 7,$

Supply: $q = x^2 + x + 4,$

given that x is the unit price, in dollars, and q is the quantity demanded or supplied, in thousands.

Use the vertical-line test to determine whether each of the following is the graph of a function.

12.

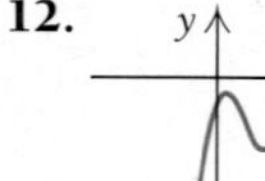

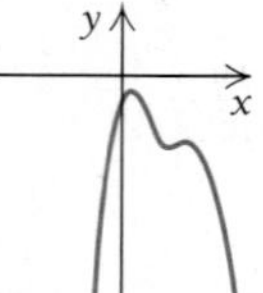

13.

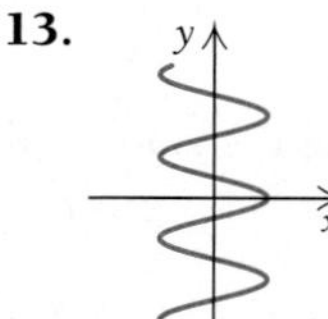

14. For the following graph of a quadratic function f, determine (**a**) $f(1)$; (**b**) the domain; (**c**) all x-values such that $f(x) = 4$; and (**d**) the range.

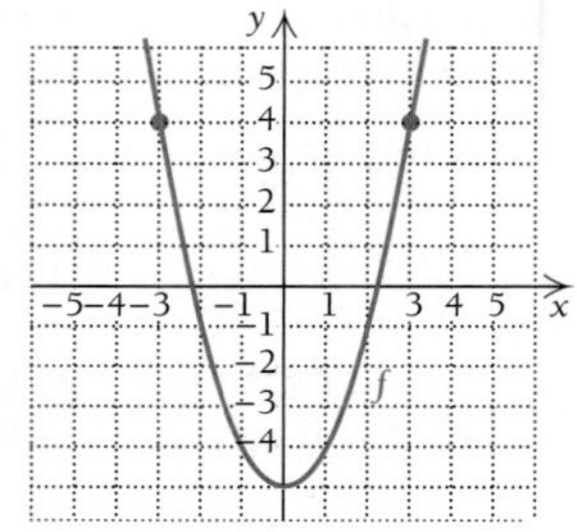

15. Graph: $f(x) = 8/x$.

16. Convert to rational exponents: $1/\sqrt{t}$.

17. Convert to radical notation: $t^{-3/5}$.

18. Graph: $f(x) = \dfrac{x^2 - 1}{x + 1}$.

Determine the domain of each function.

19. $f(x) = \dfrac{x^2 + 20}{(x - 2)(x + 7)}$

20. $f(x) = \dfrac{x}{\sqrt{3x + 6}}$

21. Write interval notation for the following graph.

22. Graph:

$$f(x) = \begin{cases} x^2 + 2, & \text{for } x \geq 0, \\ x^2 - 2, & \text{for } x < 0. \end{cases}$$

23. Nutrition. As people age, their daily caloric needs change. The following table shows data for physically active females, relating age to number of calories needed daily.

Age	Number of Calories Needed Daily
6	1800
11	2200
16	2400
24	2400
41	2200

(*Source*: Based on data from U.S. Department of Agriculture.)

a) Make a scatterplot of the data.

b) Do the data appear to fit a quadratic function?

c) Using the data points (6, 1800), (16, 2400), and (41, 2200), find a quadratic function that fits the data.

d) Use the function from part (c) to predict the number of calories needed daily by a physically active 30-yr-old woman.

TW e) Estimate the domain of the function from part (a). Explain its restrictions.

SYNTHESIS

24. Simplify: $(64^{4/3})^{-1/2}$.

25. Find the domain and the zeros of the function given by

$$f(x) = (5 - 3x)^{1/4} - 7.$$

26. Write an equation that has exactly three solutions: -3, 1, and 4. Answers will vary.

27. A function's average rate of change over the interval $[1, 5]$ is $-\frac{3}{7}$. If $f(1) = 9$, find $f(5)$.

TECHNOLOGY CONNECTION

28. Graph the function and find the zeros and the domain and the range:

$$f(x) = \sqrt[3]{|9 - x^2|} - 1.$$

29. Nutrition. Use the data in Exercise 23.

a) Use REGRESSION to fit a quadratic function to the data.

b) Use the function from part (a) to predict the number of calories needed daily by a physically active 30-yr-old woman.

TW c) Compare your answer from part (b) with that from part (d) of Exercise 23. Which answer do you feel is more accurate? Why?

EXTENDED TECHNOLOGY APPLICATION

Payrolls of Professional Basketball Teams

Extended Technology Applications occur at the end of each chapter. They are designed to consider certain applications in greater depth, make use of calculator skills, and allow for possible group or collaborative learning.

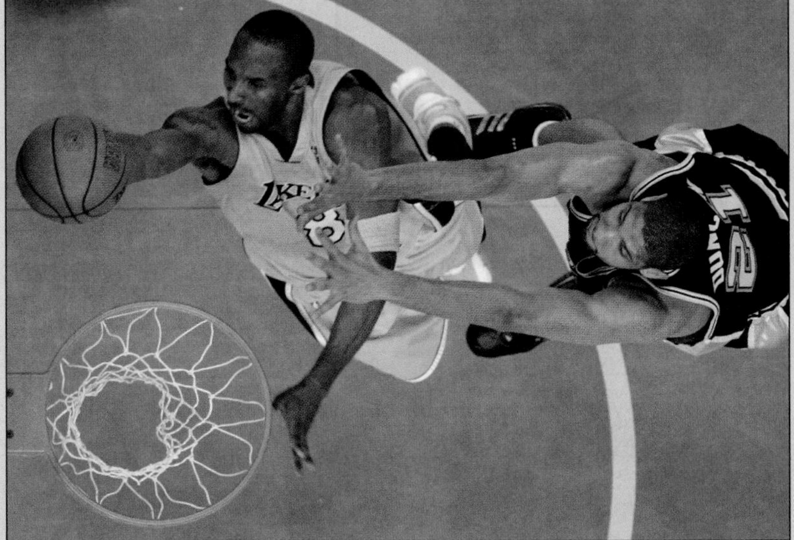

Payrolls in all businesses tend to rise over time, and professional basketball is no exception. The table and the graph on the following page show the average players' payroll for the teams of the National Basketball Association from the 1985–1986 season through the 2004–2005 season. (The data for the 1986–1987, 1989–1990, 2002–2003, and 2003–2004 seasons are omitted.)

EXERCISES

1. a) Using REGRESSION, find a linear function that fits the data.

b) Graph the linear function.

c) Use the function to estimate the average NBA payroll for the 1986–1987 and 2003–2004 seasons. Do these estimates appear reasonable?

(continued)

YEARS AFTER THE 1985–1986 SEASON, t	AVERAGE NBA PAYROLL, P (IN MILLIONS)
0	4.763
2	6.050
3	6.780
5	11.307
6	13.688
7	15.909
8	18.793
9	22.179
10	24.975
11	28.519
12	32.708
13	38.050
14	45.105
15	51.563
16	53.549
19	58.020

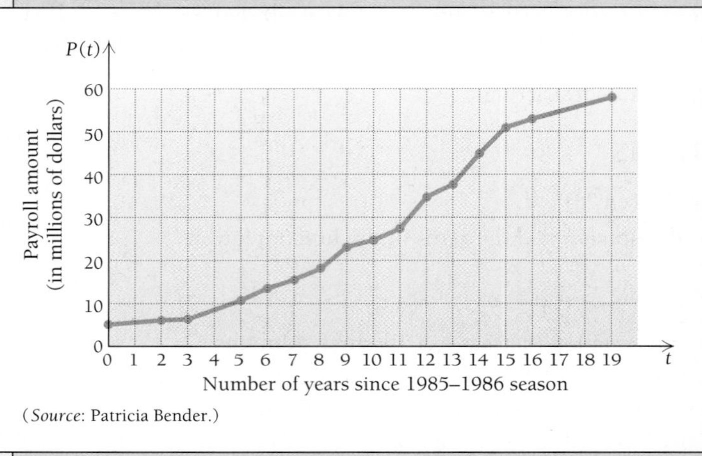

(*Source*: Patricia Bender.)

d) Use the function to predict the average NBA payrolls for the 2009–2010 and 2049–2050 seasons.

e) Use the function to predict when the average NBA payroll will reach $100 million.

2. a) Using REGRESSION, find a quadratic polynomial function,

$$y = ax^2 + bx + c,$$

that fits the data.

b) Graph the quadratic polynomial function.

c) Use the function to estimate the average NBA payroll for the 1986–1987 and 2003–2004 seasons. Do these estimates appear reasonable?

d) Use the function to predict the average NBA payrolls for the 2009–2010 and 2049–2050 seasons.

e) Use the function to predict when the average NBA payroll will reach $100 million.

3. a) Using REGRESSION, find a cubic polynomial function,

$$y = ax^3 + bx^2 + cx + d,$$

that fits the data.

b) Graph the cubic polynomial function.

c) Use the function to estimate the average NBA payroll for the 1986–1987 and 2003–2004 seasons. Do these estimates appear reasonable?

d) Use the function to predict the average NBA payrolls for the 2009–2010 and 2049–2050 seasons.

e) Use the function to predict when the average NBA payroll will reach $100 million.

4. a) Using REGRESSION, find a quartic polynomial function,

$$y = ax^4 + bx^3 + cx^2 + dx + e,$$

that fits the data.

b) Graph the quartic polynomial function.

c) Use the function to estimate the average NBA payroll for the 1986–1987 and 2003–2004 seasons. Do these estimates appear reasonable?

d) Use the function to predict the average NBA payrolls for the 2009–2010 and the 2049–2050 seasons.

e) Use the function to predict when the average NBA payroll will reach $100 million.

5. Discuss the merits and dangers of using each type of function to model the past and future average NBA payroll.

Differentiation

1

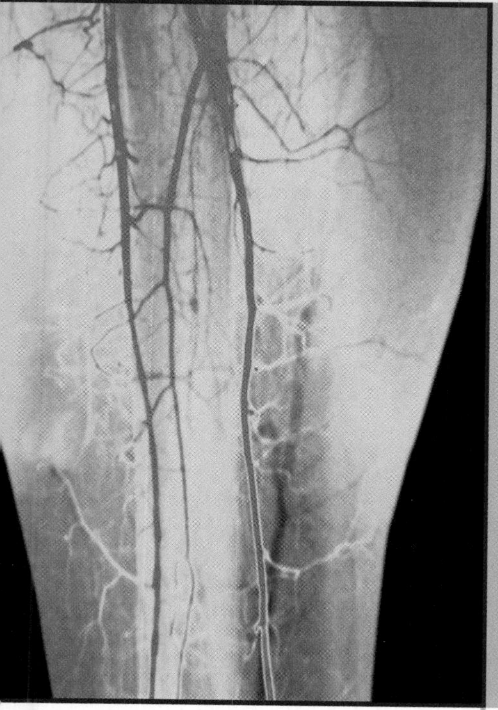

APPLICATION The equation

$$S(r) = \frac{1}{r^4}$$

can be used to determine the resistance to blood flow, S, of a blood vessel that has radius r in millimeters. (*Source*: *Mathematics Teacher*, Vol. 99, No. 4, November 2005.)

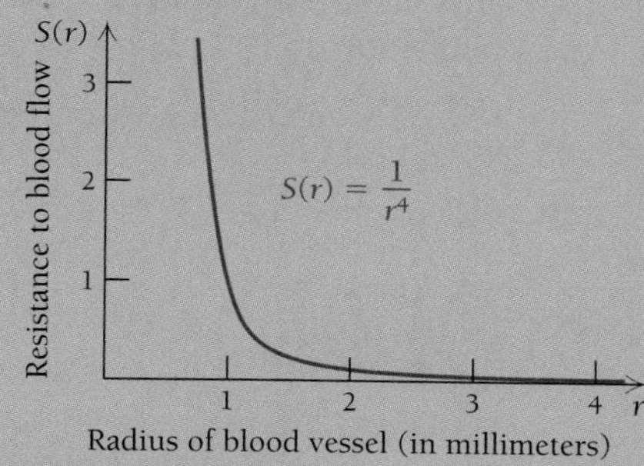

Find the rate of change of resistance with respect to r, the radius of the blood vessel.

This problem appears as Exercise 86 in Section 1.5.

INTRODUCTION

With this chapter, we begin our study of calculus. The first concepts we consider are those of limits and continuity. Then we apply those concepts to establishing the first of the two main building blocks of calculus: differentiation.

Differentiation is a process that takes a formula for a function and derives a formula for another function, called a *derivative,* that allows us to find the slope of the tangent line to a curve at a point. A derivative also represents an instantaneous rate of change. Throughout the chapter, we will learn various techniques for finding derivatives.

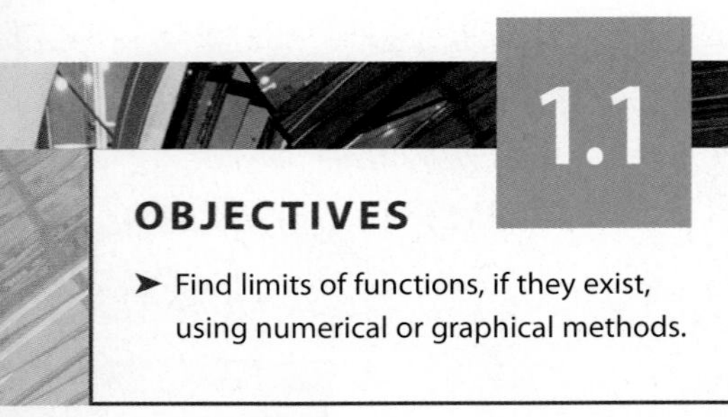

1.1

OBJECTIVES

➤ Find limits of functions, if they exist, using numerical or graphical methods.

Limits: A Numerical and Graphical Approach

In this section, we give an intuitive (meaning "based on prior and present experience") treatment of the concept of a *limit*.

Limits

Suppose that a football is located 5 yd from a goal line, and then because of a penalty, the referee moves the ball half the distance toward the goal. If this happens repeatedly, the ball moves from the 5-yd line to the $2\frac{1}{2}$-yd line, then the $1\frac{1}{4}$-yd line, and so on. The ball never reaches the goal, but its distance to the goal continues to get closer and closer to 0. We say that the *limit* is 0.

One important aspect of the study of calculus is the analysis of how function values, or outputs, change when inputs change. Basic to this study is the notion of limit. Suppose that the inputs get closer and closer to some number. If the corresponding outputs get closer and closer to a number, then that number is called a *limit*. For example, suppose that we select x-values closer and closer to 4 and look at the value of $f(x) = 2x + 3$. In the table and graph below, we see that as input numbers approach 4, but do not equal 4, from the left, output numbers approach 11. Similarly, as input numbers approach 4, but do not equal 4, from the right, output numbers approach 11. Thus, we say:

As x *approaches* 4 from either side, $2x + 3$ *approaches* 11.

Limit Numerically

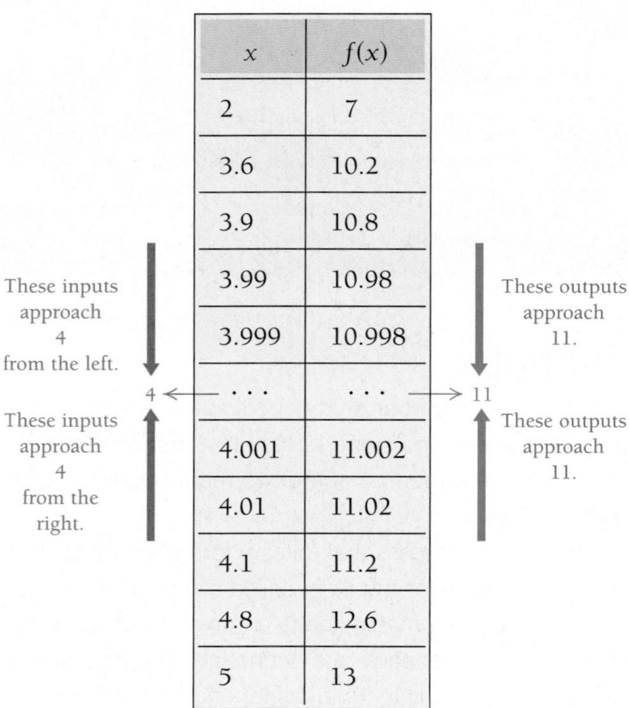

x	$f(x)$
2	7
3.6	10.2
3.9	10.8
3.99	10.98
3.999	10.998
...	...
4.001	11.002
4.01	11.02
4.1	11.2
4.8	12.6
5	13

Limit Graphically

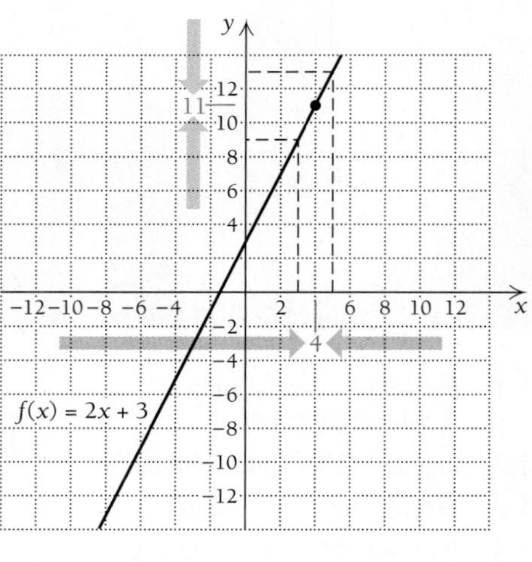

An arrow, $\rightarrow$, is often used to stand for the words "approaches from either side." Thus, the statement above can be written:

$$\text{As } x \rightarrow 4, \quad 2x + 3 \rightarrow 11.$$

The number 11 is said to be the *limit* of $2x + 3$ as x approaches 4 from either side. We can abbreviate this statement as follows:

$$\lim_{x \to 4} (2x + 3) = 11.$$

This is read: "The limit, as x approaches 4, of $2x + 3$ is 11."

DEFINITION

As x approaches a, the **limit** of $f(x)$ is L, written

$$\lim_{x \to a} f(x) = L,$$

if all values of $f(x)$ are close to L for values of x that are sufficiently close, but not equal, to a.

The notation $\lim_{x \to a} f(x)$ indicates that a is being approached from either side. We use the notation

$\lim_{x \to a^+} f(x)$ to indicate the limit from the right

and

$\lim_{x \to a^-} f(x)$ to indicate the limit from the left.

Then, in order for a limit to exist, both of the limits above must exist and be the same. We can rephrase the definition as follows.

THEOREM

As x approaches a, the limit of $f(x)$ is L if the limit from the left exists and the limit from the right exists and both limits are L. That is,

$$\text{if} \quad \lim_{x \to a^+} f(x) = \lim_{x \to a^-} f(x) = L, \quad \text{then} \quad \lim_{x \to a} f(x) = L.$$

TECHNOLOGY CONNECTION

Finding Limits Using the TABLE *and* TRACE *Features*

Exploratory

Consider the function given by $f(x) = 3x - 1$. Let's use the TABLE feature to complete the following table. Note that the inputs do not have the same increment from one to the next, but do approach 6 from either the left or the right. We use TblSet and select Indpnt and ASK mode. Then we enter the inputs shown and use the corresponding outputs to complete the table.

(continued)

Finding Limits Using the TABLE and TRACE Features (continued)

$f(x) = 3x - 1$

x	$f(x)$
5	14
5.8	16.4
5.9	16.7
5.99	16.97
5.999	16.997
6 ←	→ ?
6.001	17.003
6.01	17.03
6.1	17.3
6.4	18.2
7	20

Now we set the table in AUTO mode and starting (TblStart) with a number near 6, we make tables for some increments (ΔTbl) like 0.1, 0.01, −0.1, −0.01, and so on, to determine $\lim_{x \to 6} f(x)$.

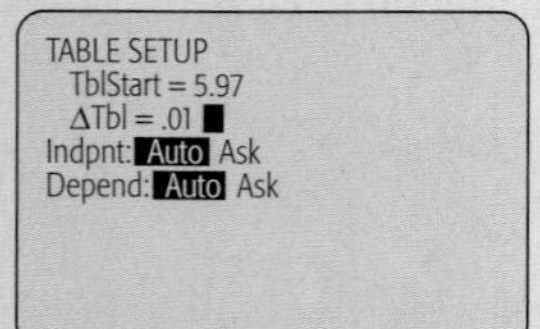

X	Y1
5.97	16.91
5.98	16.94
5.99	16.97
6	17
6.01	17.03
6.02	17.06
6.03	17.09

As an alternative, graphical approach, let's use the TRACE feature with the graph of f. We move the cursor from left to right so that the x-coordinate approaches 6 from the left, changing the window as needed, to see what happens. For example, let's use [5.3, 6.4, 14, 18] and move the cursor from right to left so that the x-coordinate approaches 6 from the right. In general, the TRACE feature is not an efficient way to find limits, but it will help you to visualize the limit process in this early stage of your learning.

Using the TABLE and TRACE features, let's complete the following:

$$\lim_{x \to 6^+} f(x) = 17 \quad \text{and} \quad \lim_{x \to 6^-} f(x) = 17.$$

Thus,

$$\lim_{x \to 6} f(x) = 17.$$

EXERCISES

Consider $f(x) = 3x - 1$. Use the TABLE and TRACE features, making up your own tables, to find each of the following.

1. $\lim_{x \to 2} f(x)$
2. $\lim_{x \to -1} f(x)$

Consider $g(x) = x^3 - 2x - 2$ for Exercises 3–5.

3. Complete the following table.

x	$g(x)$
7	
7.8	
7.9	
7.99	
7.999	
8 ←	→ ?
8.001	
8.01	
8.1	
8.4	
8.9	

Use the TABLE and TRACE features to find each of the following.

4. $\lim_{x \to 8} g(x)$
5. $\lim_{x \to -1} g(x)$

EXAMPLE 1 Consider the function H given by

$$H(x) = \begin{cases} 2x + 2, & \text{for } x < 1, \\ 2x - 4, & \text{for } x \geq 1. \end{cases}$$

Graph the function and find each of the following limits, if they exist. When necessary, state that the limit does not exist.

a) $\lim_{x \to 1} H(x)$ **b)** $\lim_{x \to -3} H(x)$

Solution We check the limits from the left and from the right both numerically, with an input–output table, and graphically.

a) Limit Numerically

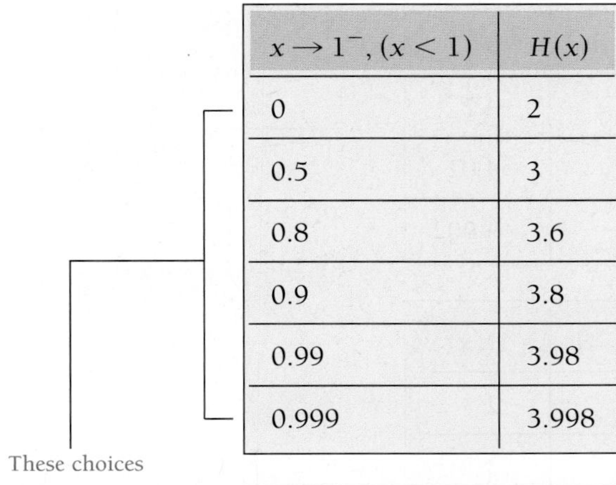

$x \to 1^-$, $(x < 1)$	$H(x)$
0	2
0.5	3
0.8	3.6
0.9	3.8
0.99	3.98
0.999	3.998

These choices can vary.

$x \to 1^+$, $(x > 1)$	$H(x)$
2	0
1.8	−0.4
1.1	−1.8
1.01	−1.98
1.001	−1.998
1.0001	−1.9998

Limit Graphically

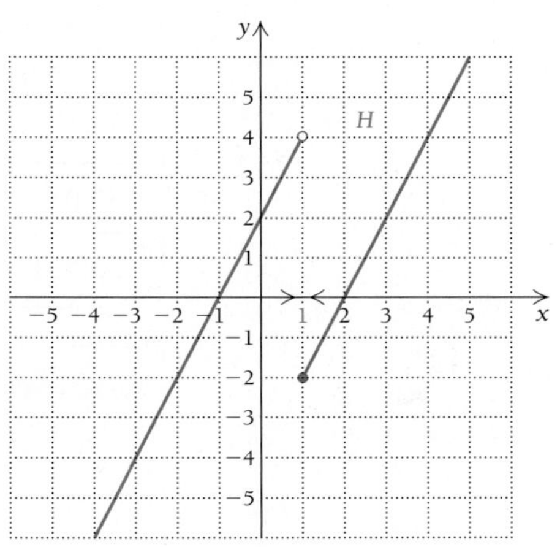

As inputs x approach 1 from the left, outputs $H(x)$ approach 4. Thus, the limit from the left is 4. That is,

$$\lim_{x \to 1^-} H(x) = 4.$$

But as inputs x approach 1 from the right, outputs $H(x)$ approach -2. Thus, the limit from the right is -2. That is,

$$\lim_{x \to 1^+} H(x) = -2.$$

Since the limit from the left, 4, is not the same as the limit from the right, 2, we say that

$$\lim_{x \to 1} H(x) \textit{ does not exist.}$$

We mention in passing that $H(1) = -2$. In this example, the function value exists for $x = 1$, but the limit as x approaches 1 does not exist.

b) Limit Numerically

$x \to -3^-, (x < -3)$	$H(x)$
-4	-6
-3.5	-5
-3.1	-4.2
-3.01	-4.02
-3.001	-4.002

$x \to -3^+, (x > -3)$	$H(x)$
-2	-2
-2.5	-3
-2.9	-3.8
-2.99	-3.98
-2.999	-3.998

Limit Graphically

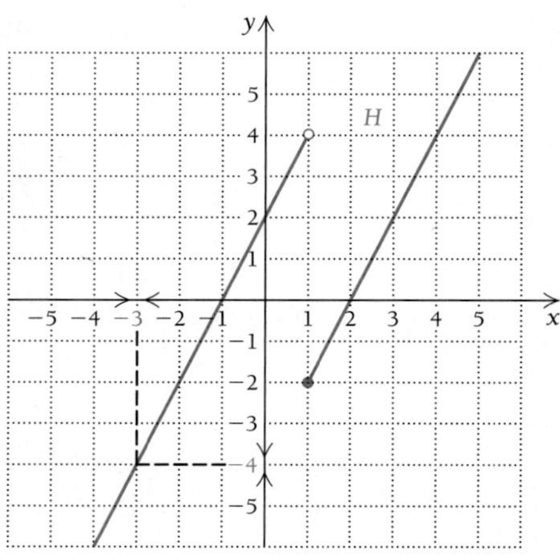

TECHNOLOGY CONNECTION

Check the results of Example 1 using the TABLE feature. See Section R.2 to recall how to graph functions defined piecewise.

As inputs x approach -3 from the left, outputs $H(x)$ approach -4, so the limit from the left is -4. That is,

$$\lim_{x \to -3^-} H(x) = -4.$$

As inputs x approach -3 from the right, outputs $H(x)$ approach -4, so the limit from the right is -4. That is,

$$\lim_{x \to -3^+} H(x) = -4.$$

Since the limits from the left and from the right exist and are the same, we have

$$\lim_{x \to -3} H(x) = -4.$$ ✦

In Example 1, note that $\lim_{x \to 1^-} H(x) = 4$ even though $H(1) \neq 4$.

The limit at a number *a does not depend* on the function value at *a* or even on whether that function value, $f(a)$, exists. That is, whether or not a limit exists at *a* has *nothing* to do with the function value $f(a)$.

The "Wall" Method

As an alternative approach to Example 1, we can draw a "wall" at $x = 1$, as shown in blue on the graph to the left below. We then follow the curve from left to right with a pencil until we hit the wall and mark the location with an ×, assuming it can be determined. Then we follow the curve from right to left until we hit the wall and mark that location with an ×. If the locations are the same, as in the graph to the right below, a limit exists. Thus, for Example 1,

$$\lim_{x \to 1} H(x) \textit{ does not exist}, \quad \text{and} \quad \lim_{x \to -3} H(x) = -4.$$

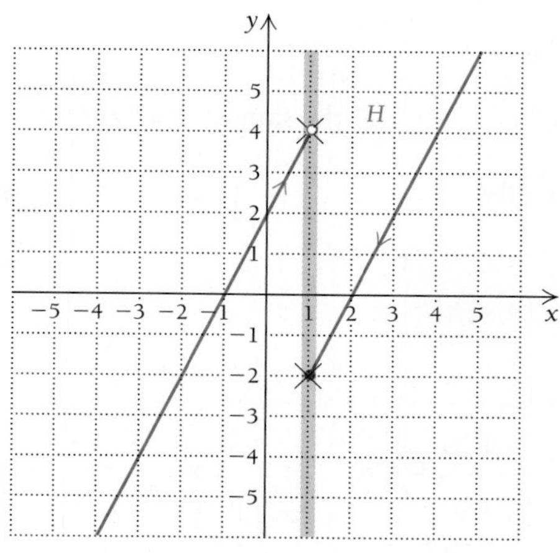

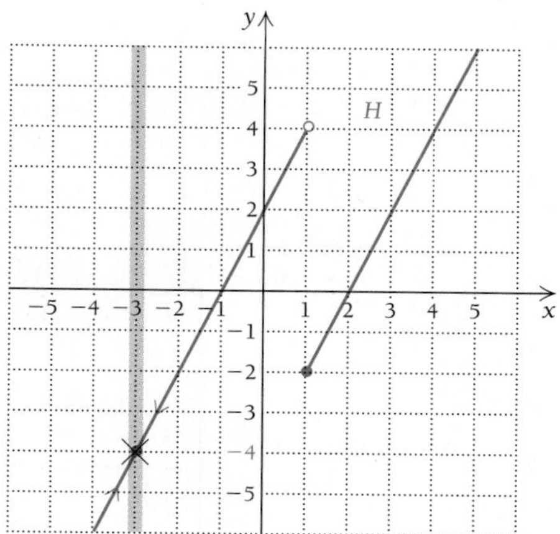

EXAMPLE 2 Consider the function defined as follows:

$$G(x) = \begin{cases} 5, & \text{for } x = 1, \\ x + 1, & \text{for } x \neq 1. \end{cases}$$

Graph the function, and find each of the following limits, if they exist. If necessary, state that the limit does not exist.

a) $\lim_{x \to 1} G(x)$

b) $\lim_{x \to -2} G(x)$

Solution The graph of G follows.

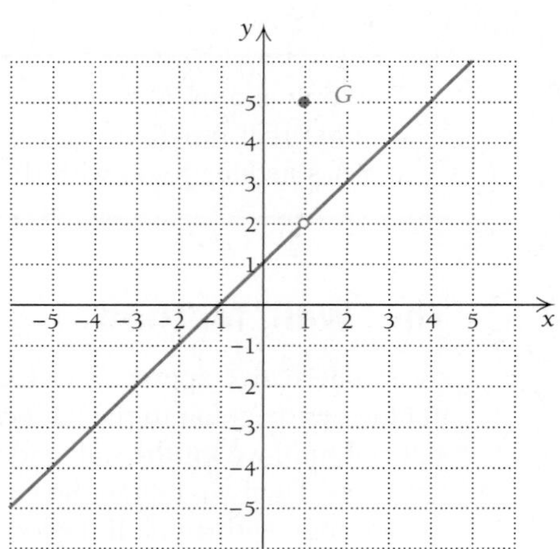

a) As inputs x approach 1 from the left, outputs $G(x)$ approach 2, so the limit from the left is 2. As inputs x approach 1 from the right, outputs $G(x)$ also approach 2, so the limit from the right is 2. Since the limit from the left, 2, is the same as the limit from the right, 2, we have

$$\lim_{x \to 1} G(x) = 2.$$

Note that the limit, 2, is not the same as the function value at 1, which is $G(1) = 5$.

Limit Numerically

$x \to 1^-, (x < 1)$	$G(x)$
0	1
0.5	1.5
0.9	1.9
0.99	1.99

$\lim_{x \to 1} G(x) = 2$

$x \to 1^+, (x > 1)$	$G(x)$
1.5	2.5
1.1	2.1
1.01	2.01
1.001	2.001

Limit Graphically

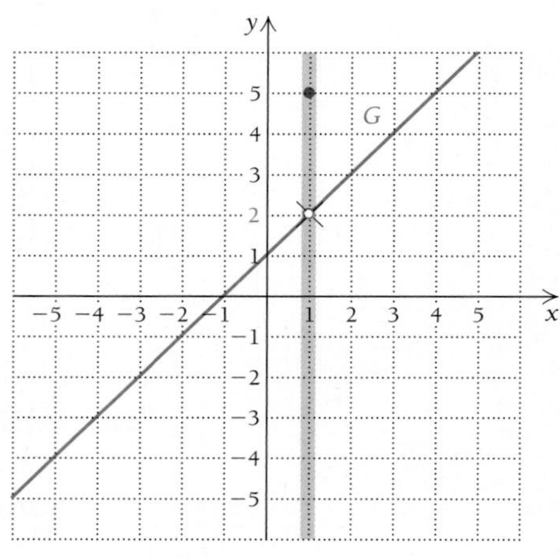

b) Using the same approach as in part (a), we have

$$\lim_{x \to -2} G(x) = -1.$$

Note that in this case, the limit, -1, is the same as the function value at -2, which is $G(-2) = -1$.

Limit Numerically

$x \to -2^-$, $(x < -2)$	$G(x)$
-3	-2
-2.5	-1.5
-2.1	-1.1
-2.01	-1.01

$\lim_{x \to -2} G(x) = -1$

$x \to -2^+$, $(x > -2)$	$G(x)$
-1.5	-0.5
-1.9	-0.9
-1.99	-0.99
-1.999	-0.999

Limit Graphically

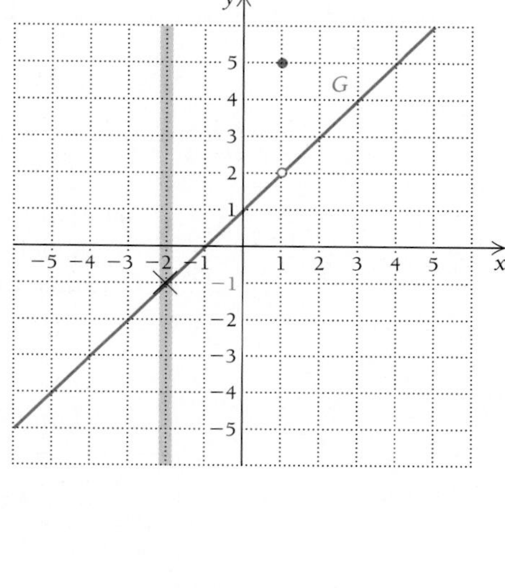

✦

EXAMPLE 3 Consider the function f given by

$$f(x) = \frac{1}{x - 2} + 3.$$

Graph the function, and find each of the following limits, if they exist. If necessary, state that the limit does not exist.

a) $\lim_{x \to 3} f(x)$

b) $\lim_{x \to 2} f(x)$

Solution The graph of f is shown on the next page. Note that it is the same as the graph of $y = \dfrac{1}{x}$ but shifted 2 units to the right and 3 units up.

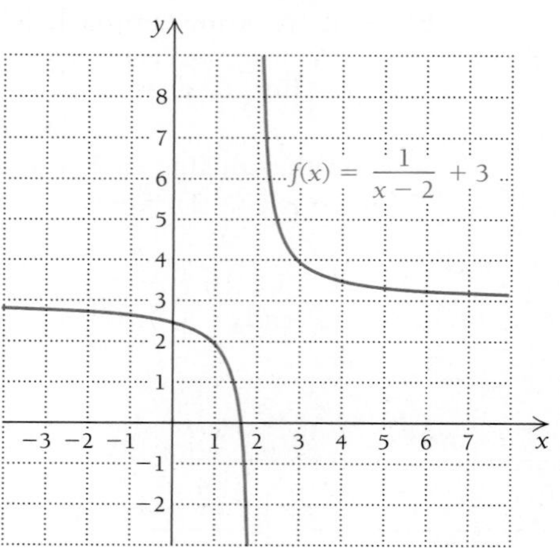

a) As inputs x approach 3 from the left, outputs $f(x)$ approach 4, so the limit from the left is 4. As inputs x approach 3 from the right, outputs $f(x)$ also approach 4. Since the limit from the left, 4, is the same as the limit from the right, we have

$$\lim_{x \to 3} f(x) = 4.$$

Limit Numerically

$x \to 3^-, (x < 3)$	$f(x)$
2.1	13
2.5	5
2.9	$4.\overline{1}$
2.99	$4.\overline{01}$

$\lim_{x \to 3} f(x) = 4$

$x \to 3^+, (x > 3)$	$f(x)$
3.5	$3.\overline{6}$
3.2	$3.8\overline{3}$
3.1	$3.\overline{90}$
3.01	$3.\overline{9900}$

Limit Graphically

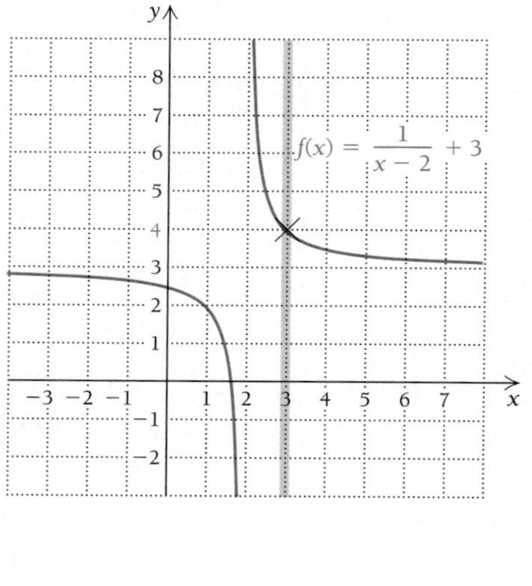

b) As inputs x approach 2 from the left, outputs $f(x)$ become more and more negative, without bound. These numbers do not approach any real number, although it might be said that the limit from the left is negative infinity, $-\infty$. That is,

$$\lim_{x \to 2^-} f(x) = -\infty.$$

As inputs x approach 2 from the right, outputs $f(x)$ become larger and larger, without bound. These numbers do not approach any real number, although it might be said that the limit from the right is infinity, ∞. That is,

$$\lim_{x \to 2^+} f(x) = \infty.$$

Because the left-sided limit differs from the right-sided limit,

$$\lim_{x \to 2} f(x) \text{ does not exist.}$$

Limit Numerically

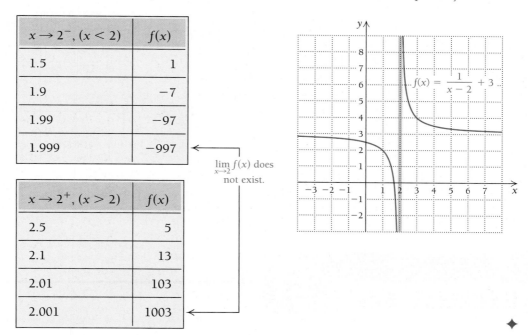

$x \to 2^-, (x < 2)$	$f(x)$
1.5	1
1.9	−7
1.99	−97
1.999	−997

$\lim_{x \to 2} f(x)$ does not exist.

$x \to 2^+, (x > 2)$	$f(x)$
2.5	5
2.1	13
2.01	103
2.001	1003

Limit Graphically

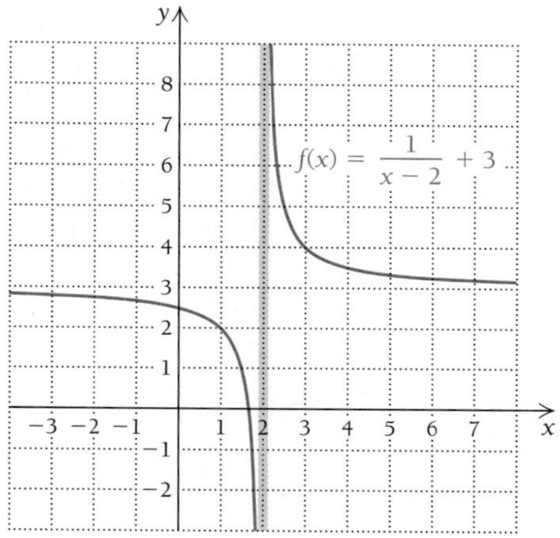

◆

Keep in mind that ∞ and $-\infty$ are not real numbers. We associate ∞ with numbers increasing without bound in a positive direction, as in the interval notation $[5, \infty)$. Similarly, we associate $-\infty$ with numbers decreasing without bound in a negative direction, as in the interval notation $(-\infty, 7]$.

We sometimes need to determine limits when the inputs get larger and larger without bound, that is, when they approach infinity. In such cases, we are finding *limits at infinity*. Such a limit is expressed as

$$\lim_{x \to \infty} f(x).$$

EXAMPLE 4 Consider again the function in Example 3, given by

$$f(x) = \frac{1}{x - 2} + 3.$$

Find $\lim_{x \to \infty} f(x)$.

Solution As inputs x get larger and larger, outputs $f(x)$ get closer and closer to 3. We have

$$\lim_{x \to \infty} f(x) = 3.$$

Limit Numerically

$x \to \infty$	$f(x)$
5	$3.\overline{3}$
10	3.125
100	3.0102
1000	3.0010

← $\lim_{x \to \infty} f(x) = 3$

Limit Graphically

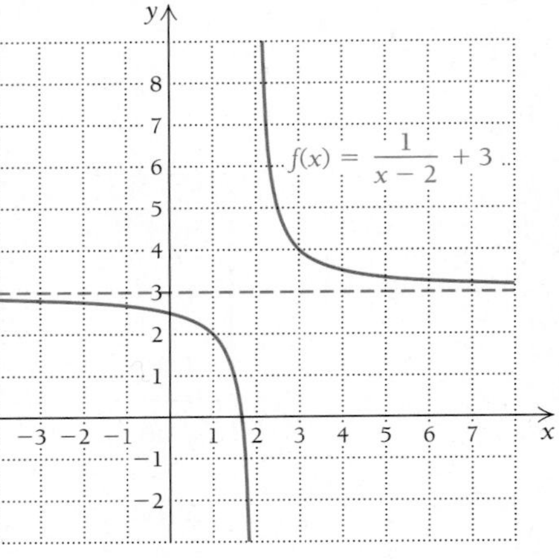

✦

Exercise Set 1.1

Complete each of the following statements.

1. As x approaches 3, the value of $2x + 5$ approaches ________.
2. As x approaches -4, the value of $3x + 7$ approaches ________.
3. As x approaches ________, the value of $-3x$ approaches 6.
4. As x approaches ________, the value of $x - 2$ approaches 5.
5. The notation $\lim_{x \to 4} f(x)$ is read ________.
6. The notation $\lim_{x \to 1} g(x)$ is read ________.
7. The notation $\lim_{x \to 5^-} F(x)$ is read ________.
8. The notation $\lim_{x \to 4^+} G(x)$ is read ________.
9. The notation ________ is read "the limit, as x approaches 2 from the right."
10. The notation ________ is read "the limit, as x approaches 3 from the left."

For Exercises 11–18, consider the function f given by

$$f(x) = \begin{cases} x - 2, & \text{for } x \le 3, \\ x - 1, & \text{for } x > 3. \end{cases}$$

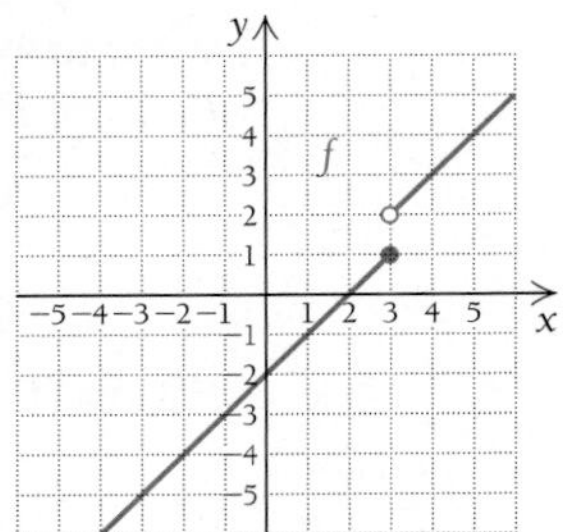

When necessary, state that the limit does not exist.

11. Find $\lim_{x \to 3^+} f(x)$.

12. Find $\lim_{x \to 3^-} f(x)$.

13. Find $\lim_{x \to -1^-} f(x)$.

14. Find $\lim_{x \to -1^+} f(x)$.

15. Find $\lim_{x \to 3} f(x)$.

16. Find $\lim_{x \to -1} f(x)$.

17. Find $\lim_{x \to 4} f(x)$.

18. Find $\lim_{x \to 2} f(x)$.

For Exercises 19–26, consider the function g given by

$$g(x) = \begin{cases} x + 6, & \text{for } x < -2, \\ -\frac{1}{2}x + 1, & \text{for } x \geq -2. \end{cases}$$

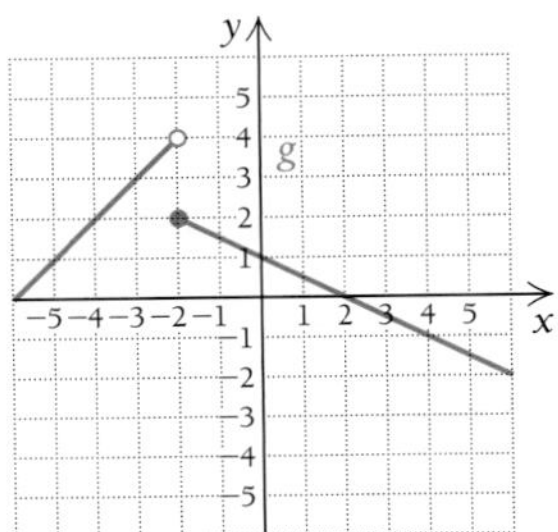

If a limit does not exist, state that fact.

19. $\lim_{x \to -2^-} g(x)$

20. $\lim_{x \to -2^+} g(x)$

21. $\lim_{x \to 4^+} g(x)$

22. $\lim_{x \to 4^-} g(x)$

23. $\lim_{x \to 4} g(x)$

24. $\lim_{x \to -2} g(x)$

25. $\lim_{x \to 2} g(x)$

26. $\lim_{x \to -4} g(x)$

For Exercises 27–34, use the following graph of F to find each limit. When necessary, state that the limit does not exist.

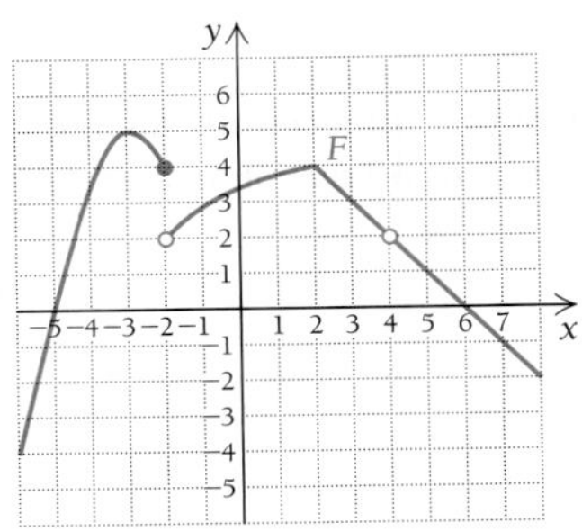

27. $\lim_{x \to -3} F(x)$

28. $\lim_{x \to 2} F(x)$

29. $\lim_{x \to -2} F(x)$

30. $\lim_{x \to -5} F(x)$

31. $\lim_{x \to 4} F(x)$

32. $\lim_{x \to 6} F(x)$

33. $\lim_{x \to -2^+} F(x)$

34. $\lim_{x \to -2^-} F(x)$

For Exercises 35–44, use the following graph of f to find each limit. When necessary, state that the limit does not exist.

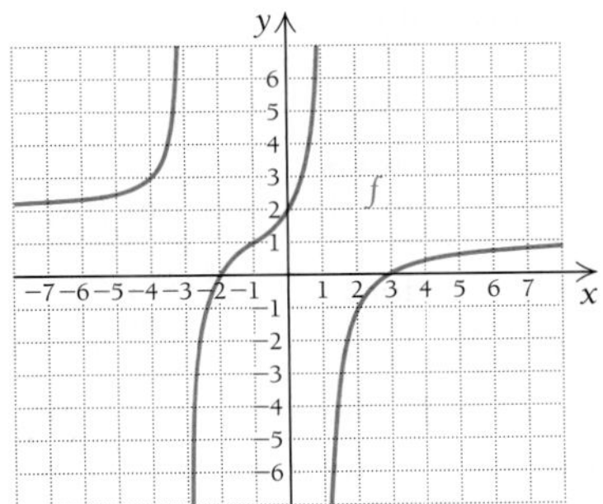

35. $\lim_{x \to -1} f(x)$

36. $\lim_{x \to 2} f(x)$

37. $\lim_{x \to -3} f(x)$

38. $\lim_{x \to 0} f(x)$

39. $\lim_{x \to 3} f(x)$

40. $\lim_{x \to 1} f(x)$

41. $\lim_{x \to -4} f(x)$

42. $\lim_{x \to -2} f(x)$

43. $\lim_{x \to \infty} f(x)$

44. $\lim_{x \to -\infty} f(x)$

For Exercises 45–60, graph each function and then find the specified limits. When necessary, state that the limit does not exist.

45. $f(x) = |x|$; find $\lim_{x \to 0} f(x)$ and $\lim_{x \to -2} f(x)$.

46. $f(x) = x^2$; find $\lim_{x \to -1} f(x)$ and $\lim_{x \to 0} f(x)$.

47. $g(x) = x^2 - 5$; find $\lim_{x \to 0} g(x)$ and $\lim_{x \to -1} g(x)$.

48. $g(x) = |x| + 1$; find $\lim_{x \to -3} g(x)$ and $\lim_{x \to 0} g(x)$.

49. $F(x) = \dfrac{1}{x - 3}$; find $\lim_{x \to 3} F(x)$ and $\lim_{x \to 4} F(x)$.

50. $G(x) = \dfrac{1}{x + 2}$; find $\lim_{x \to -1} G(x)$ and $\lim_{x \to -2} G(x)$.

51. $f(x) = \dfrac{1}{x} - 2$; find $\lim_{x \to \infty} f(x)$ and $\lim_{x \to 0} f(x)$.

52. $f(x) = \dfrac{1}{x} + 3$; find $\lim_{x \to \infty} f(x)$ and $\lim_{x \to 0} f(x)$.

53. $g(x) = \dfrac{1}{x + 2} + 4$; find $\lim_{x \to \infty} g(x)$ and $\lim_{x \to -2} g(x)$.

54. $g(x) = \dfrac{1}{x - 3} + 2$; find $\lim_{x \to \infty} g(x)$ and $\lim_{x \to 3} g(x)$.

55. $F(x) = \begin{cases} 2x + 1, & \text{for } x < 1, \\ x, & \text{for } x \geq 1. \end{cases}$

Find $\lim_{x \to 1^-} F(x)$, $\lim_{x \to 1^+} F(x)$, and $\lim_{x \to 1} F(x)$.

56. $G(x) = \begin{cases} -x + 3, & \text{for } x < 2, \\ x + 1, & \text{for } x \geq 2. \end{cases}$

Find $\lim_{x \to 2^-} G(x)$, $\lim_{x \to 2^+} G(x)$, and $\lim_{x \to 2} G(x)$.

continuous over, or **on, some interval** of the real line if its graph can be traced without lifting the pencil from the paper. If there is any point in an interval where a "jump" or a "hole" occurs, then we say that the function is *not continuous* over that interval. The graphs of functions F, G, and H, which follow, show that these functions are *not* continuous over the whole real line.

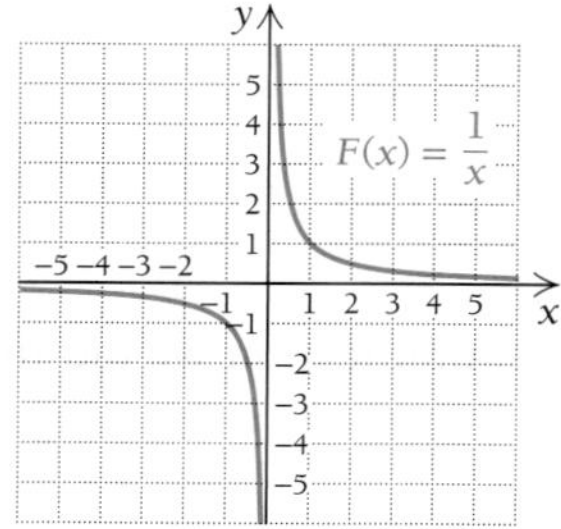

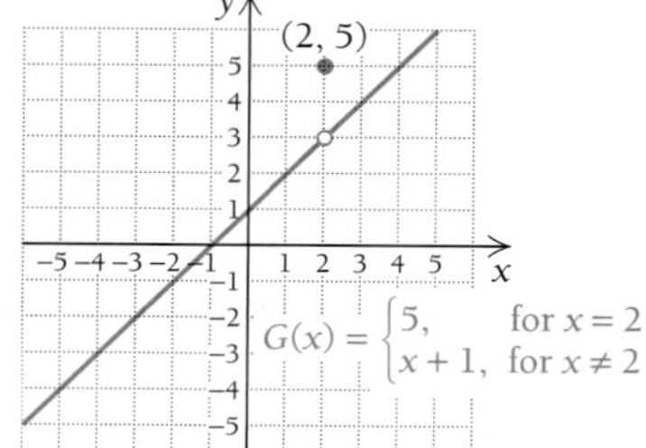

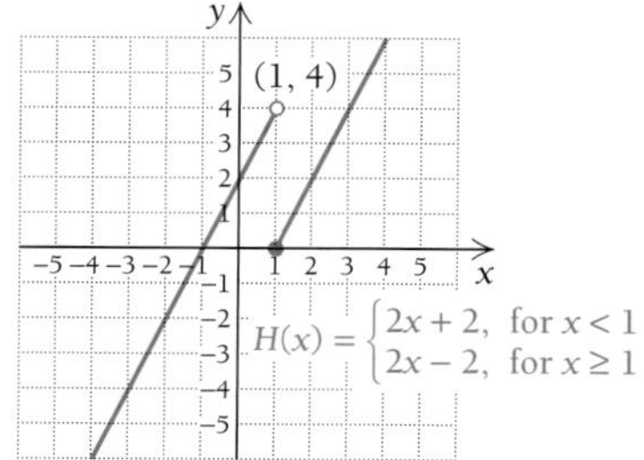

A continuous curve.

In each case, the graph *cannot* be traced without lifting the pencil from the paper. However, each case represents a different situation:

- F is not continuous over $(-\infty, \infty)$ because the point $x = 0$ is not part of the domain. Thus, there is no point to trace at $x = 0$. Note that F *is* continuous over the intervals $(-\infty, 0)$ and $(0, \infty)$.
- G is not continuous over $(-\infty, \infty)$ because it is not continuous at $x = 2$. To see this, trace the graph of G starting to the left of $x = 2$. As x approaches 2 from either side, $G(x)$ approaches 3. However, *at* $x = 2$, $G(x)$ *jumps* up to 5. Note that G is continuous over $(-\infty, 2)$ and $(2, \infty)$.
- H is not continuous over $(-\infty, \infty)$ because it is not continuous at $x = 1$. To see this, trace the graph of H starting to the left of $x = 1$. As x approaches 1 from the left, $H(x)$ approaches 4. However, as x approaches 1 from the right, $H(x)$ is close to 0. Note that H *is* continuous over $(-\infty, 1)$ and $(1, \infty)$.

Each of the above graphs has a *point of discontinuity*. The graph of F is discontinuous at 0, because $F(0)$ does not exist; the graph of G is discontinuous at 2, because $\lim_{x \to 2} G(x) \neq G(2)$; and the graph of H is discontinuous at 1, because $\lim_{x \to 1} H(x)$ does not exist.

DEFINITION

A function f is **continuous** at $x = a$ if:

a) $f(a)$ exists, (The output at a exists.)

b) $\lim_{x \to a} f(x)$ exists, (The limit as $x \to a$ exists.)

and

c) $\lim_{x \to a} f(x) = f(a)$. (The limit is the same as the output.)

A function is **continuous over an interval *I*** if it is continuous at each point in I.

EXAMPLE 6 Determine whether the function given by

$$f(x) = 2x + 3$$

is continuous at $x = 4$.

Solution This function is continuous at $x = 4$ because:

a) $f(4)$ exists, $(f(4) = 11)$

b) $\lim_{x \to 4} f(x)$ exists, $\left(\lim_{x \to 4} f(x) = 11\right.$ was found on pp. 100–101.$\left.\right)$

and

c) $\lim_{x \to 4} f(x) = 11 = f(4)$.

In fact, $f(x) = 2x + 3$ is continuous at any point on the real number line. ✦

EXAMPLE 7 Is the function f given by

$$f(x) = x^2 - 5$$

continuous at $x = 3$? Why or why not?

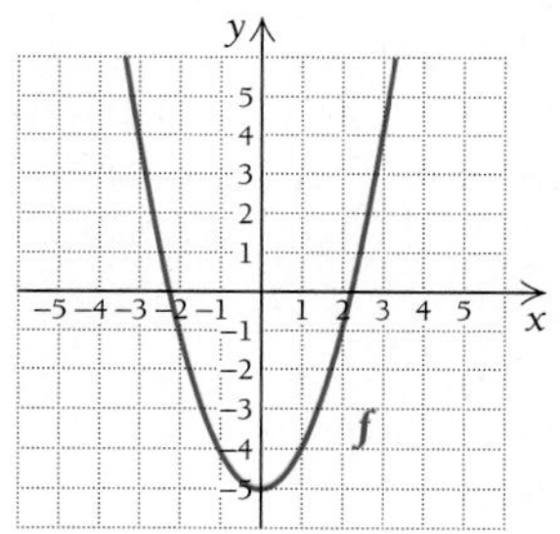

A visualization of the function in Example 7

Solution By the Theorem on Limits of Rational Functions, we have

$$\lim_{x \to 3} f(x) = 3^2 - 5 = 9 - 5 = 4.$$

Since

$$f(3) = 3^2 - 5 = 4,$$

we have

$$\lim_{x \to 3} f(x) = f(3).$$

Thus, f is continuous at $x = 3$. ✦

EXAMPLE 8 Is the function g, given by

$$g(x) = \begin{cases} \frac{1}{2}x + 3, & \text{for } x < -2, \\ x - 1, & \text{for } x \geq -2, \end{cases}$$

continuous at $x = -2$? Why or why not?

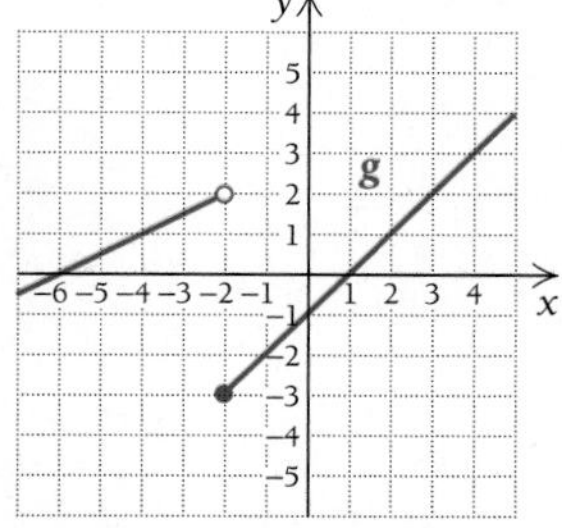

A visualization of the function in Example 8

Solution To find out if g is continuous at −2, we must determine whether $\lim_{x \to -2} g(x) = g(-2)$. Thus, we first note that $g(-2) = -2 - 1 = -3$. To find $\lim_{x \to -2} g(x)$, we look at left- and right-sided limits:

$$\lim_{x \to -2^-} g(x) = \frac{1}{2}(-2) + 3 = -1 + 3 = 2; \lim_{x \to -2^+} g(x) = -2 - 1 = -3.$$

Since $\lim_{x \to -2^-} g(x) \neq \lim_{x \to -2^+} g(x)$, we see that $\lim_{x \to -2} g(x)$ does not exist. Thus, g is not continuous at −2. ✦

EXAMPLE 9 Is the function F, given by

$$F(x) = \begin{cases} \dfrac{x^2 - 16}{x - 4}, & \text{for } x \neq 4, \\ 7, & \text{for } x = 4, \end{cases}$$

continuous at $x = 4$? Why or why not?

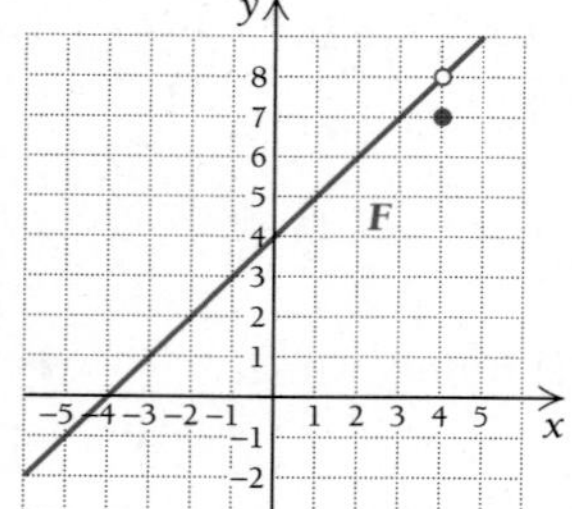

A visualization of the function in Example 9

Solution For F to be continuous at 4, we must have $\lim_{x\to 4} F(x) = F(4)$. Note that $F(4) = 7$. To find $\lim_{x\to 4} F(x)$, we note that, for $x \neq 4$,

$$\frac{x^2 - 16}{x - 4} = \frac{(x - 4)(x + 4)}{x - 4} = x + 4.$$

Thus

$$\lim_{x\to 4} F(x) = 4 + 4 = 8.$$

We see that F is *not* continuous at $x = 4$ since

$$\lim_{x\to 4} F(x) \neq F(4).$$ ✦

The Limit Principles and the Theorem on Limits of Rational Functions can be used to show that if $f(x)$ and $g(x)$ are two arbitrary polynomial functions, then f and g are both continuous. Furthermore, $f + g$, $f - g$, $f \cdot g$, and, assuming $g(x) \neq 0$, f/g are also continuous. We can also use the Limit Principles to show that for n, an integer greater than 1, $\sqrt[n]{f(x)}$ is continuous, provided that $f(x) \geq 0$ when n is even.

EXAMPLE 10 Is the function G, given by

$$G(x) = \begin{cases} -x + 3, & \text{for } x \leq 2, \\ x^2 - 3, & \text{for } x > 2, \end{cases}$$

continuous? Why or why not?

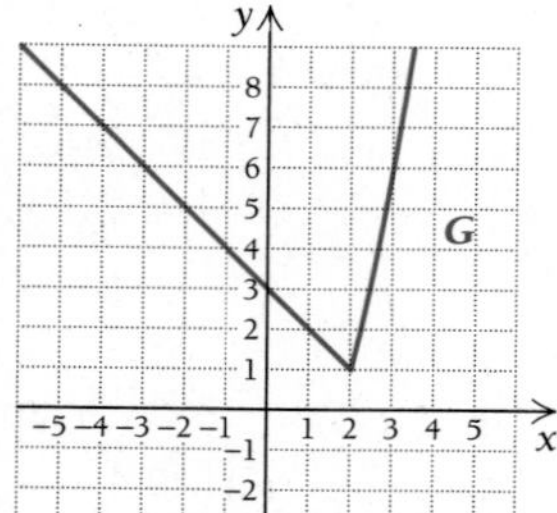

A visualization of the function in Example 10

Solution For G to be continuous, it must be continuous for all real numbers. Since $y = -x + 3$ is continuous on $(-\infty, 2]$ and $y = x^2 - 3$ is continuous on $(2, \infty)$, we need only to determine whether $\lim_{x\to 2} G(x) = G(2)$:

$$G(2) = -2 + 3 = 1;$$

$$\lim_{x\to 2^-} G(x) = -2 + 3 = 1 \quad \text{and} \quad \lim_{x\to 2^+} G(x) = (2)^2 - 3 = 4 - 3 = 1,$$

so $$\lim_{x\to 2} G(x) = 1.$$

Since $\lim_{x\to 2} G(x) = G(2)$, we know that G is continuous. ✦

Exercise Set 1.2

Classify each statement as either true or false.

1. $\lim_{x \to 3} 7 = 7$

2. If $\lim_{x \to 2} f(x) = 9$, then $\lim_{x \to 2} \sqrt{f(x)} = 3$.

3. If $\lim_{x \to 1} g(x) = 5$, then $\lim_{x \to 1} [g(x)]^2 = 10$.

4. If $\lim_{x \to 4} F(x) = 7$, then $\lim_{x \to 4} [c \cdot F(x)] = 7c$.

5. If f is continuous at $x = 2$, then $f(2)$ must exist.

6. If g is discontinuous at $x = 3$, then $g(3)$ must not exist.

7. If $\lim_{x \to 4} F(x)$ exists, then F must be continuous at $x = 4$.

8. If $\lim_{x \to 7} G(x)$ equals $G(7)$, then G must be continuous at $x = 7$.

Use the Theorem on Limits of Rational Functions to find the following limits. When necessary, state that the limit does not exist.

9. $\lim_{x \to 1} (3x + 2)$

10. $\lim_{x \to 2} (4x - 5)$

11. $\lim_{x \to -1} (x^2 - 4)$

12. $\lim_{x \to -2} (x^2 + 3)$

13. $\lim_{x \to 3} (x^2 - 4x + 7)$

14. $\lim_{x \to 5} (x^2 - 6x + 9)$

15. $\lim_{x \to 2} (2x^4 - 3x^3 + 4x - 1)$

16. $\lim_{x \to -1} (3x^5 + 4x^4 - 3x + 6)$

17. $\lim_{x \to 3} \dfrac{x^2 - 8}{x - 2}$

18. $\lim_{x \to 3} \dfrac{x^2 - 25}{x^2 - 5}$

19. $\lim_{x \to 3} \dfrac{x^2 - 9}{x - 3}$

20. $\lim_{x \to 5} \dfrac{x^2 - 25}{x - 5}$

Use the Limit Principles to find the following limits. If a limit does not exist, state that fact.

21. $\lim_{x \to 4} \sqrt{x^2 - 9}$

22. $\lim_{x \to 5} \sqrt{x^2 - 16}$

23. $\lim_{x \to 2} \sqrt{x^2 - 9}$

24. $\lim_{x \to 3} \sqrt{x^2 - 16}$

25. $\lim_{x \to 3^+} \sqrt{x^2 - 9}$

26. $\lim_{x \to -4^-} \sqrt{x^2 - 16}$

Determine whether each of the functions shown in Exercises 27–31 is continuous over the interval $(-6, 6)$.

27.

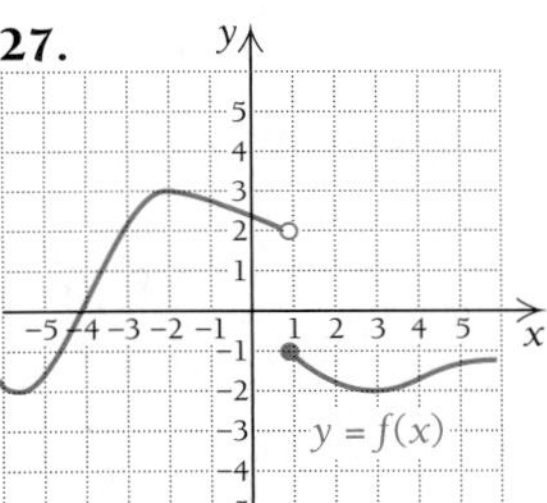

28.

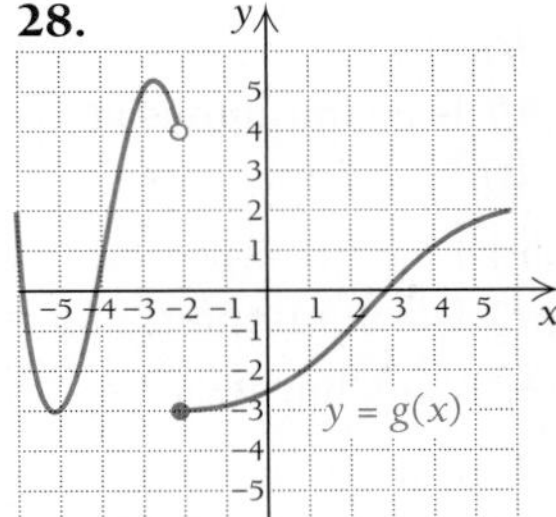

29.

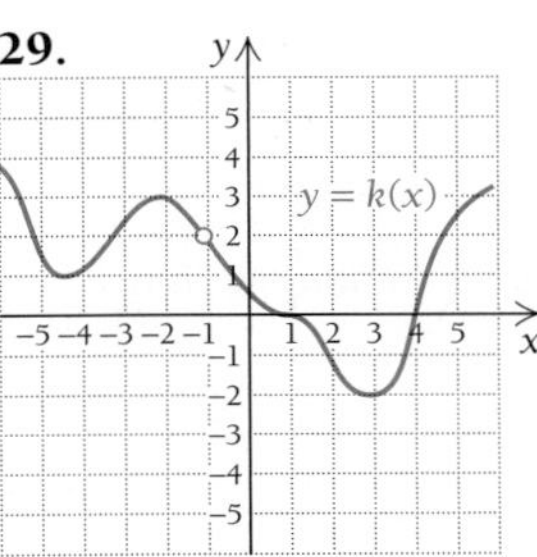

30.

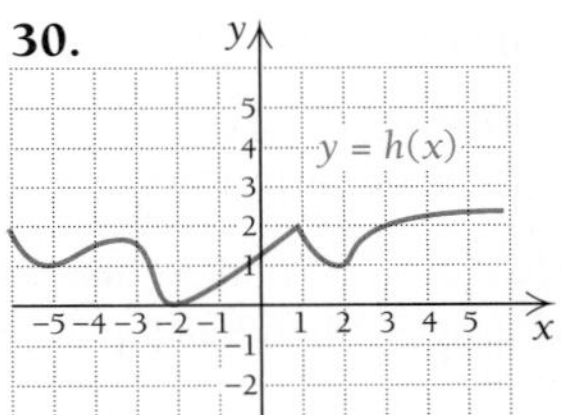

31.

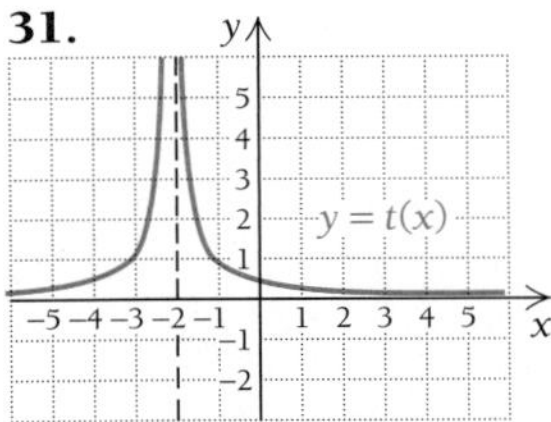

Use the graphs and functions in Exercises 27–31 to answer each of the following. If an expression does not exist, state that fact.

32. a) Find $\lim_{x \to 1^+} f(x)$, $\lim_{x \to 1^-} f(x)$, and $\lim_{x \to 1} f(x)$.

b) Find $f(1)$.

c) Is f continuous at $x = 1$? Why or why not?

d) Find $\lim_{x \to -2} f(x)$.

e) Find $f(-2)$.

f) Is f continuous at $x = -2$? Why or why not?

33. a) Find $\lim_{x \to 1^+} g(x)$, $\lim_{x \to 1^-} g(x)$, and $\lim_{x \to 1} g(x)$.

b) Find $g(1)$.

c) Is g continuous at $x = 1$? Why or why not?

d) Find $\lim_{x \to -2} g(x)$.

e) Find $g(-2)$.

f) Is g continuous at $x = -2$? Why or why not?

34. **a)** Find $\lim_{x\to -1} k(x)$.
b) Find $k(-1)$.
c) Is k continuous at $x = -1$? Why or why not?
d) Find $\lim_{x\to 3} k(x)$.
e) Find $k(3)$.
f) Is k continuous at $x = 3$? Why or why not?

35. **a)** Find $\lim_{x\to 1} h(x)$.
b) Find $h(1)$.
c) Is h continuous at $x = 1$? Why or why not?
d) Find $\lim_{x\to -2} h(x)$.
e) Find $h(-2)$.
f) Is h continuous at $x = -2$? Why or why not?

36. **a)** Find $\lim_{x\to 1} t(x)$.
b) Find $t(1)$.
c) Is t continuous at $x = 1$? Why or why not?
d) Find $\lim_{x\to -2} t(x)$.
e) Find $t(-2)$.
f) Is t continuous at $x = -2$? Why or why not?

In Exercises 37 and 38, use the graphs to find the limits and answer the related questions.

37.

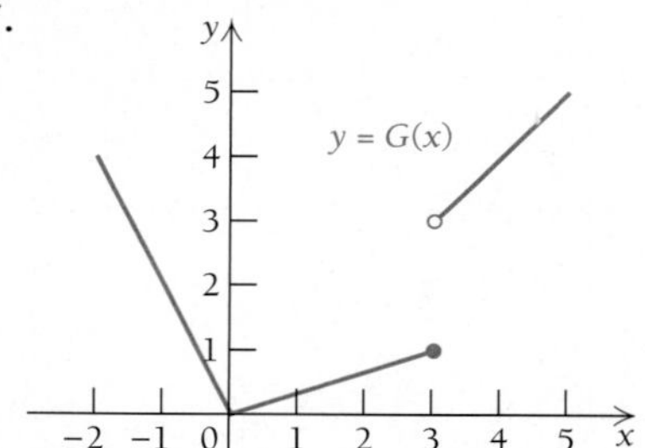

a) Find $\lim_{x\to 3^+} G(x)$. **b)** Find $\lim_{x\to 3^-} G(x)$.
c) Find $\lim_{x\to 3} G(x)$. **d)** Find $G(3)$.
e) Is G continuous at $x = 3$? Why or why not?
f) Is G continuous at $x = 0$? Why or why not?
g) Is G continuous at $x = 2.9$? Why or why not?

38. Consider the function

$$C(x) = \begin{cases} -1, & \text{for } x < 2, \\ 1, & \text{for } x \geq 2. \end{cases}$$

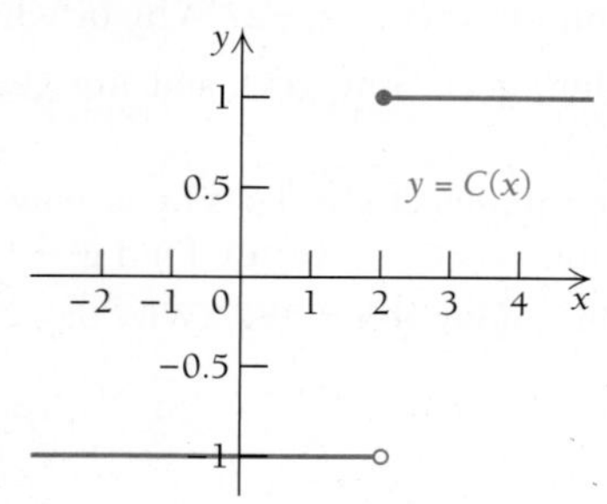

a) Find $\lim_{x\to 2^+} C(x)$.
b) Find $\lim_{x\to 2^-} C(x)$.
c) Find $\lim_{x\to 2} C(x)$.
d) Find $C(2)$.
e) Is C continuous at $x = 2$? Why or why not?
f) Is C continuous at $x = 1.95$? Why or why not?

39. Is the function given by $f(x) = 3x - 2$ continuous at $x = 5$? Why or why not?

40. Is the function given by $g(x) = x^2 - 3x$ continuous at $x = 4$? Why or why not?

41. Is the function given by $G(x) = \frac{1}{x}$ continuous at $x = 0$? Why or why not?

42. Is the function given by $F(x) = \sqrt{x}$ continuous at $x = -1$? Why or why not?

43. Is the function given by

$$g(x) = \begin{cases} \frac{1}{3}x + 4, & \text{for } x \leq 3, \\ 2x - 1, & \text{for } x > 3, \end{cases}$$

continuous at $x = 3$? Why or why not?

44. Is the function given by

$$f(x) = \begin{cases} \frac{1}{2}x + 1, & \text{for } x < 4, \\ -x + 7, & \text{for } x \geq 4, \end{cases}$$

continuous at $x = 4$? Why or why not?

45. Is the function given by

$$F(x) = \begin{cases} \frac{1}{3}x + 4, & \text{for } x \leq 3, \\ 2x - 5, & \text{for } x > 3, \end{cases}$$

continuous at $x = 3$? Why or why not?

46. Is the function given by

$$G(x) = \begin{cases} \frac{1}{2}x + 1, & \text{for } x < 4, \\ -x + 5, & \text{for } x > 4, \end{cases}$$

continuous at $x = 4$? Why or why not?

47. Is the function given by

$$f(x) = \begin{cases} \frac{1}{3}x + 4, & \text{for } x < 3, \\ 2x - 1, & \text{for } x \geq 3, \end{cases}$$

continuous at $x = 3$? Why or why not?

48. Is the function given by

$$g(x) = \begin{cases} \frac{1}{2}x + 1, & \text{for } x < 4, \\ -x + 7, & \text{for } x > 4, \end{cases}$$

continuous at $x = 4$? Why or why not?

49. Is the function given by

$$G(x) = \begin{cases} \dfrac{x^2 - 4}{x - 2}, & \text{for } x \neq 2, \\ 5, & \text{for } x = 2, \end{cases}$$

continuous at $x = 2$? Why or why not?

50. Is the function given by

$$F(x) = \begin{cases} \dfrac{x^2 - 1}{x - 1}, & \text{for } x \neq 1, \\ 4, & \text{for } x = 1, \end{cases}$$

continuous at $x = 1$? Why or why not?

51. Is the function given by

$$f(x) = \begin{cases} \dfrac{x^2 - 4x - 5}{x - 5}, & \text{for } x < 5, \\ x + 1, & \text{for } x \geq 5, \end{cases}$$

continuous at $x = 5$? Why or why not?

52. Is the function given by

$$G(x) = \begin{cases} \dfrac{x^2 - 3x - 4}{x - 4}, & \text{for } x < 4, \\ 2x - 3, & \text{for } x \geq 4, \end{cases}$$

continuous at $x = 4$? Why or why not?

53. Is the function given by $g(x) = \dfrac{1}{x^2 - 7x + 10}$ continuous at $x = 5$? Why or why not?

54. Is the function given by $f(x) = \dfrac{1}{x^2 - 6x + 8}$ continuous at $x = 3$? Why or why not?

55. Is the function given by $F(x) = \dfrac{1}{x^2 - 7x + 10}$ continuous at $x = 4$? Why or why not?

56. Is the function given by $G(x) = \dfrac{1}{x^2 - 6x + 8}$ continuous at $x = 2$? Why or why not?

57. Is the function given by $g(x) = x^2 - 3x + 2$ continuous over the interval $(-4, 4)$? Why or why not?

58. Is the function given by $F(x) = x^2 - 5x + 6$ continuous over the interval $(-5, 5)$? Why or why not?

59. Is the function given by $f(x) = \dfrac{1}{x} + 3$ continuous over the interval $(-7, 7)$? Why or why not?

60. Is the function given by $G(x) = \dfrac{1}{x - 1}$ continuous over the interval $(0, \infty)$? Why or why not?

61. Is the function given by $g(x) = 4x^3 - 6x$ continuous on $\mathbb{R}$?

62. Is the function given by $F(x) = \dfrac{3}{x - 5}$ continuous on $\mathbb{R}$?

SYNTHESIS

Find each limit, if it exists. If a limit does not exist, state that fact.

63. $\lim\limits_{x \to 0} \dfrac{|x|}{x}$

64. $\lim\limits_{x \to -2} \dfrac{x^3 + 8}{x^2 - 4}$

TECHNOLOGY CONNECTION

In Section 1.1, we discussed how to use the TABLE feature to find limits. Consider

$$\lim_{x \to 0} \frac{\sqrt{1 + x} - 1}{x}.$$

Input–output tables for this function are shown below. The table on the left uses TblStart $= -1$ and ΔTbl $= 0.5$. By using smaller and smaller step values and beginning closer to 0, we can refine the table and obtain a better estimate of the limit. On the right is an input–output table with TblStart $= -0.03$ and ΔTbl $= 0.01$.

X	Y1
-1	1
-0.5	0.58579
0	ERR
0.5	0.44949
1	0.41421
1.5	0.38743
2	0.36603

X=-1

X	Y1
-0.03	0.50381
-0.02	0.50253
-0.01	0.50126
0	ERR
0.01	0.49876
0.02	0.49752
0.03	0.49631

X=-0.03

It appears that the limit is 0.5. We can verify this by graphing

$$y = \frac{\sqrt{1 + x} - 1}{x}$$

and tracing the curve near $x = 0$, zooming in on that portion of the curve.

We see that

$$\lim_{x \to 0} \frac{\sqrt{1 + x} - 1}{x} = 0.5.$$

To verify this algebraically, multiply $\dfrac{\sqrt{1 + x} - 1}{x}$ by 1, using $\dfrac{\sqrt{1 + x} + 1}{\sqrt{1 + x} + 1}$. Then simplify the result and find the limit.

In Exercises 65–72, find each limit. Use the TABLE *feature and start with* $\Delta\text{Tbl} = 0.1$. *Then use* 0.01, 0.001, *and* 0.0001. *When you think you know the limit, graph and use the* TRACE *feature to verify your assertion. Then try to verify it algebraically.*

65. $\lim_{a \to -2} \dfrac{a^2 - 4}{\sqrt{a^2 + 5} - 3}$

66. $\lim_{x \to 1} \dfrac{\sqrt{x} - 1}{x - 1}$

67. $\lim_{x \to 0} \dfrac{\sqrt{3 - x} - \sqrt{3}}{x}$

68. $\lim_{x \to 0} \dfrac{\sqrt{4 + x} - \sqrt{4 - x}}{x}$

69. $\lim_{x \to 1} \dfrac{x - \sqrt[4]{x}}{x - 1}$

70. $\lim_{x \to 0} \dfrac{\sqrt{7 + 2x} - \sqrt{7}}{x}$

71. $\lim_{x \to 4} \dfrac{2 - \sqrt{x}}{4 - x}$

72. $\lim_{x \to 0} \dfrac{7 - \sqrt{49 - x^2}}{x}$

1.3 Average Rates of Change

OBJECTIVES

- Compute an average rate of change.
- Find a simplified difference quotient.

Let's say that a car travels 110 mi in 2 hr. Its *average rate of change (speed)* is 110 mi/2 hr, or 55 mi/hr (55 mph). Suppose that you are on a freeway and you begin accelerating. Glancing at the speedometer, you see that at that instant your *instantaneous rate of change* is 55 mph. These are two quite different concepts. The first you are probably familiar with. The second involves ideas of limits and calculus. To understand instantaneous rate of change, we first need to develop a solid understanding of average rate of change.

The following graph shows the total production of suits by Raggs, Ltd., during one morning of work. Industrial psychologists have found curves like this typical of the production of factory workers.

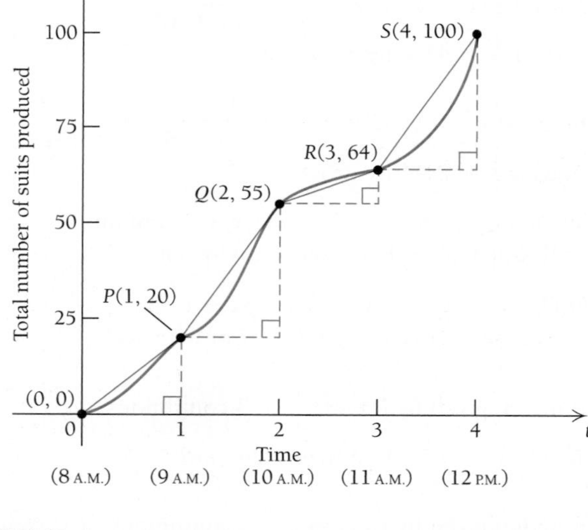

EXAMPLE 1 Business: Production. What was the number of suits produced at Raggs, Ltd., from 9 A.M. to 10 A.M.?

Solution At 9 A.M., 20 suits had been produced. At 10 A.M., 55 suits had been produced. In the hour from 9 A.M. to 10 A.M., the number of suits produced was

55 suits − 20 suits, or 35 suits.

Note that 35 is the slope of the line from P to Q. ✦

EXAMPLE 2 Business: Average Rate of Change. What was the average number of suits produced per hour from 9 A.M. to 11 A.M.?

Solution We have

$$\frac{64 \text{ suits} - 20 \text{ suits}}{11 \text{ A.M.} - 9 \text{ A.M.}} = \frac{44 \text{ suits}}{2 \text{ hr}}$$

$$= 22 \frac{\text{suits}}{\text{hr}}.$$

Note that 22 is the slope of the line from P to R.

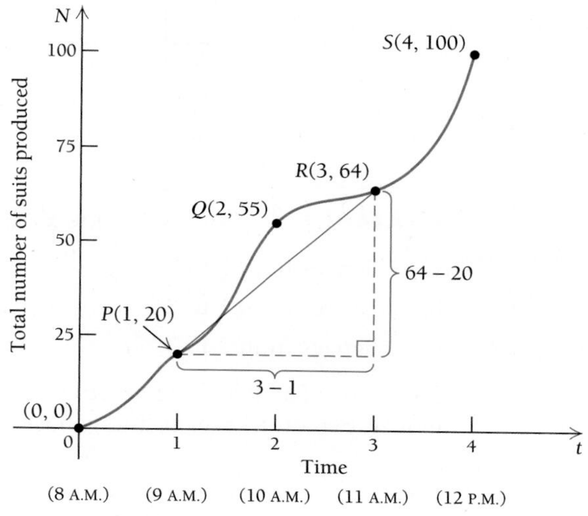

✦

Let's consider a function $y = f(x)$ and two inputs x_1 and x_2. The *change in input*, or the *change in x*, is

$$x_2 - x_1.$$

The *change in output*, or the *change in y*, is

$$y_2 - y_1,$$

where $y_1 = f(x_1)$ and $y_2 = f(x_2)$.

DEFINITION

The **average rate of change of y with respect to x,** as x changes from x_1 to x_2, is the ratio of the change in output to the change in input:

$$\frac{y_2 - y_1}{x_2 - x_1}, \quad \text{where } x_2 \neq x_1.$$

If we look at a graph of the function, we see that

$$\frac{y_2 - y_1}{x_2 - x_1} = \frac{f(x_2) - f(x_1)}{x_2 - x_1},$$

which is both the average rate of change and the slope of the line from $P(x_1, y_1)$ to $Q(x_2, y_2)$.* The line passing through P and Q, denoted $\overleftrightarrow{PQ}$, is called a **secant line.**

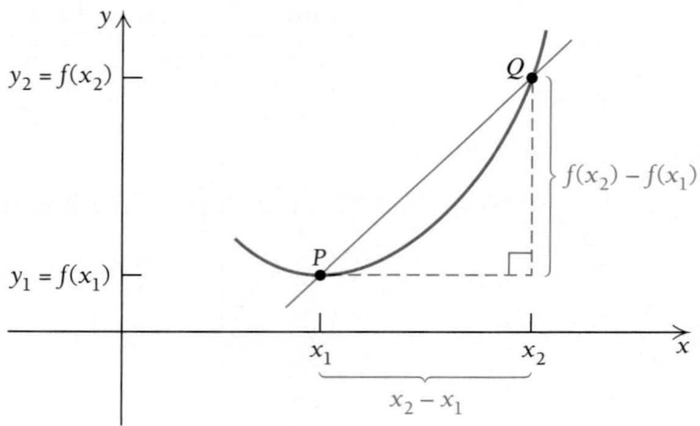

EXAMPLE 3 For $y = f(x) = x^2$, find the average rate of change as:

a) x changes from 1 to 3.

b) x changes from 1 to 2.

c) x changes from 2 to 3.

Solution The following graph is not necessary to the computations but gives us a look at two of the secant lines whose slopes are being computed.

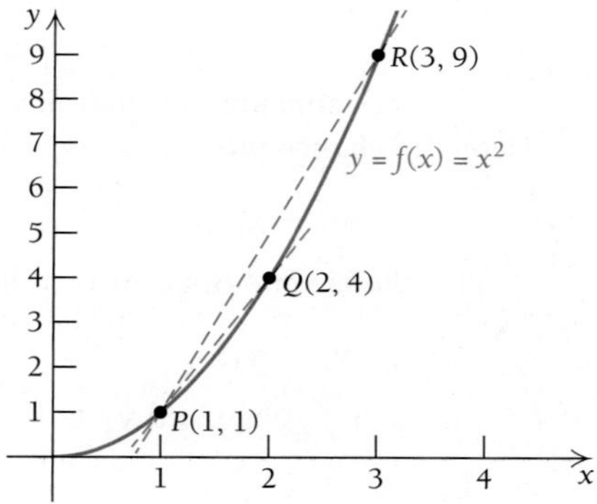

a) When $x_1 = 1$,

$$y_1 = f(x_1) = f(1) = 1^2 = 1;$$

and when $x_2 = 3$,

$$y_2 = f(x_2) = f(3) = 3^2 = 9.$$

*The notation $P(x_1, y_1)$ simply means that point P has coordinates (x_1, y_1).

The average rate of change is

$$\frac{y_2 - y_1}{x_2 - x_1} = \frac{f(x_2) - f(x_1)}{x_2 - x_1}$$
$$= \frac{9 - 1}{3 - 1}$$
$$= \frac{8}{2} = 4.$$

b) When $x_1 = 1$,

$$y_1 = f(x_1) = f(1) = 1^2 = 1;$$

and when $x_2 = 2$,

$$y_2 = f(x_2) = f(2) = 2^2 = 4.$$

The average rate of change is

$$\frac{4 - 1}{2 - 1} = \frac{3}{1} = 3.$$

c) When $x_1 = 2$,

$$y_1 = f(x_1) = f(2) = 2^2 = 4;$$

and when $x_2 = 3$,

$$y_2 = f(x_2) = f(3) = 3^2 = 9.$$

The average rate of change is

$$\frac{9 - 4}{3 - 2} = \frac{5}{1} = 5.$$ ✦

For a linear function, the average rate of change is the same for any choice of x_1 and x_2. As we saw in Example 3, a function that is not linear has average rates of change that vary with the choice of x_1 and x_2.

Difference Quotients as Average Rates of Change

We now develop a notation for average rates of change that does not require subscripts. Instead of x_1, we will write simply x; in place of x_2, we will write $x + h$.

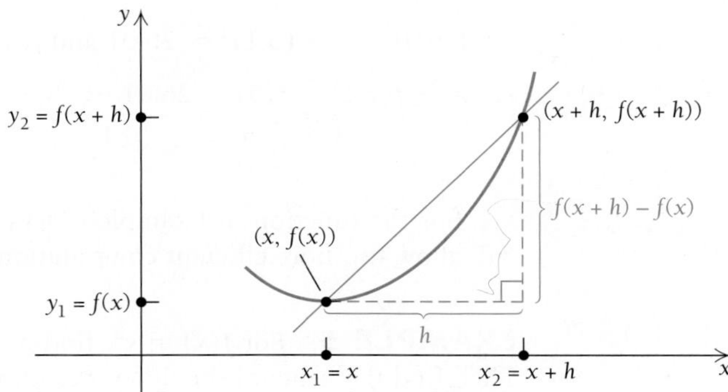

To get from x_1, or x, to x_2, we move a distance h. Thus, $x_2 = x + h$. Then the average rate of change, also called a *difference quotient,* is given by

$$\frac{y_2 - y_1}{x_2 - x_1} = \frac{f(x_2) - f(x_1)}{x_2 - x_1} = \frac{f(x + h) - f(x)}{(x + h) - x} = \frac{f(x + h) - f(x)}{h}.$$

DEFINITION

The average rate of change of f with respect to x is also called the **difference quotient.** It is given by

$$\frac{f(x + h) - f(x)}{h}, \quad \text{where } h \neq 0.$$

The difference quotient is equal to the slope of the line from $(x, f(x))$ to $(x + h, f(x + h))$.

EXERCISES

Use a calculator to show that $f(x + h) \neq f(x) + f(h)$ for each of the following functions.

1. $f(x) = x^4 + x^2$; let $x = 6$ and $h = 0.02$.
2. $f(x) = x^3 - 2x^2 + 4$; let $x = 6$ and $h = 0.1$.

Keep in mind that, in general, $f(x + h) \neq f(x) + f(h)$. (You can check this using $f(x) = x^2$, as in the Technology Connection at the left.)

EXAMPLE 4 For $f(x) = x^2$, find the difference quotient when:

a) $x = 5$ and $h = 3$.

b) $x = 5$ and $h = 0.1$.

Solution

a) We substitute $x = 5$ and $h = 3$ into the formula:

$$\frac{f(x + h) - f(x)}{h} = \frac{f(5 + 3) - f(5)}{3} = \frac{f(8) - f(5)}{3}.$$

Now $f(8) = 8^2 = 64$ and $f(5) = 5^2 = 25$, and we have

$$\frac{f(8) - f(5)}{3} = \frac{64 - 25}{3} = \frac{39}{3} = 13.$$

The difference quotient is 13. It is also the slope of the line from (5, 25) to (8, 64).

b) We substitute $x = 5$ and $h = 0.1$ into the formula:

$$\frac{f(x + h) - f(x)}{h} = \frac{f(5 + 0.1) - f(5)}{0.1} = \frac{f(5.1) - f(5)}{0.1}.$$

Now $f(5.1) = (5.1)^2 = 26.01$ and $f(5) = 25$, and we have

$$\frac{f(5.1) - f(5)}{0.1} = \frac{26.01 - 25}{0.1} = \frac{1.01}{0.1} = 10.1.$$

✦

For the function in Example 4, let's find a form of the difference quotient that will allow for more efficient computations.

EXAMPLE 5 For $f(x) = x^2$, find a simplified form of the difference quotient. Then find the value of the difference quotient when $x = 5$ and $h = 0.1$.

Solution We have

$$f(x) = x^2,$$

so

$$f(x + h) = (x + h)^2 = x^2 + 2xh + h^2.$$

Then

$$f(x + h) - f(x) = (x^2 + 2xh + h^2) - x^2 = 2xh + h^2.$$

Thus,

$$\frac{f(x + h) - f(x)}{h} = \frac{2xh + h^2}{h} = \frac{h(2x + h)}{h} = 2x + h, \quad h \neq 0.$$

This is a simplified form of this difference quotient. It is important to note that any difference quotient is defined only when $h \neq 0$. The simplification above is valid only for nonzero values of h.

When $x = 5$ and $h = 0.1$,

$$\frac{f(x + h) - f(x)}{h} = 2x + h = 2 \cdot 5 + 0.1 = 10 + 0.1 = 10.1.$$

✦

Compare the results of Example 4(b) and Example 5. In general, computations are easier when a simplified form of a difference quotient is found before any specific calculations are performed.

EXAMPLE 6 For $f(x) = x^3$, find a simplified form of the difference quotient.

Solution For $f(x) = x^3$,

$$f(x + h) = (x + h)^3 = x^3 + 3x^2h + 3xh^2 + h^3.$$

(This is shown in the Appendix at the end of this book.) Then

$$f(x + h) - f(x) = (x^3 + 3x^2h + 3xh^2 + h^3) - x^3 = 3x^2h + 3xh^2 + h^3.$$

Thus,

$$\frac{f(x + h) - f(x)}{h} = \frac{3x^2h + 3xh^2 + h^3}{h}$$

It is understood that $h \neq 0$.

$$= \frac{h(3x^2 + 3xh + h^2)}{h}$$

$$= 3x^2 + 3xh + h^2, \quad h \neq 0.$$

Again, this is true *only* for $h \neq 0$.

✦

EXAMPLE 7 For $f(x) = 3/x$, find a simplified form of the difference quotient.

Solution For $f(x) = 3/x$,

$$f(x + h) = \frac{3}{x + h}.$$

Then

$$f(x+h) - f(x) = \frac{3}{x+h} - \frac{3}{x}$$

$$= \frac{3}{x+h} \cdot \frac{x}{x} - \frac{3}{x} \cdot \frac{x+h}{x+h}$$

Here we are multiplying by 1 to get a common denominator.

$$= \frac{3x - 3(x+h)}{x(x+h)}$$

$$= \frac{3x - 3x - 3h}{x(x+h)}$$

$$= \frac{-3h}{x(x+h)}.$$

Thus,

$$\frac{f(x+h) - f(x)}{h} = \frac{\frac{-3h}{x(x+h)}}{h}$$

$$= \frac{-3h}{x(x+h)} \cdot \frac{1}{h} = \frac{-3}{x(x+h)}, \quad h \neq 0.$$

This is true *only* for $h \neq 0$. ✦

Exercise Set 1.3

For the function in each of Exercises 1–16, **(a)** *find a simplified form of the difference quotient and then* **(b)** *complete the following table.*

x	h	$\frac{f(x+h) - f(x)}{h}$
5	2	
5	1	
5	0.1	
5	0.01	

1. $f(x) = 4x^2$

2. $f(x) = 5x^2$

3. $f(x) = -4x^2$

4. $f(x) = -5x^2$

5. $f(x) = x^2 + x$

6. $f(x) = x^2 - x$

7. $f(x) = \frac{2}{x}$

8. $f(x) = \frac{9}{x}$

9. $f(x) = -2x + 5$

10. $f(x) = 2x + 3$

11. $f(x) = 1 - x^3$

12. $f(x) = 12x^3$

13. $f(x) = x^2 - 3x$

14. $f(x) = x^2 - 4x$

15. $f(x) = x^2 + 4x - 3$

16. $f(x) = x^2 - 3x + 5$

APPLICATIONS

Business and Economics

*For Exercises 17–24, use each graph to estimate the average rate of change of the percentage of new employees in that type of employment from 1994 to 2000, from 2000 to 2005, and from 1994 to 2005. (**Source:** The New York Times, 10/15/05, p. C3.)*

17. Total employment.

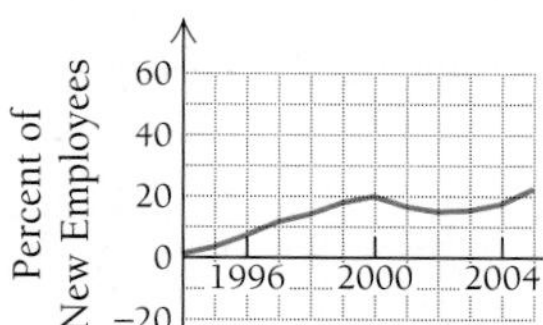

18. Construction.

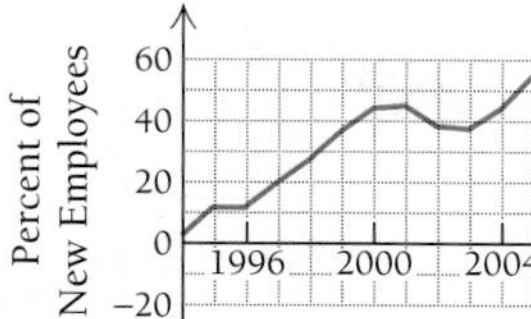

19. Professional services.

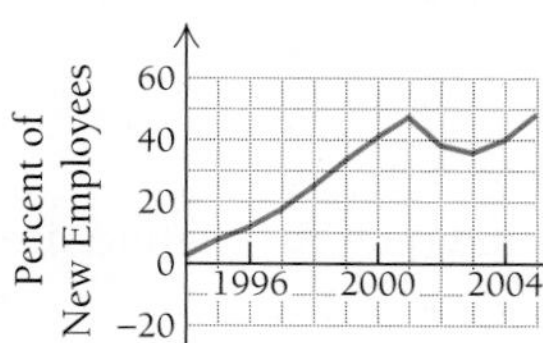

20. Health care.

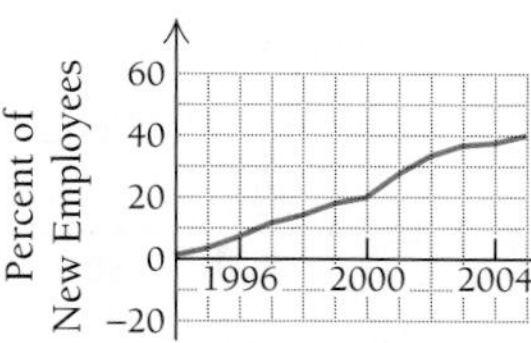

21. Education.

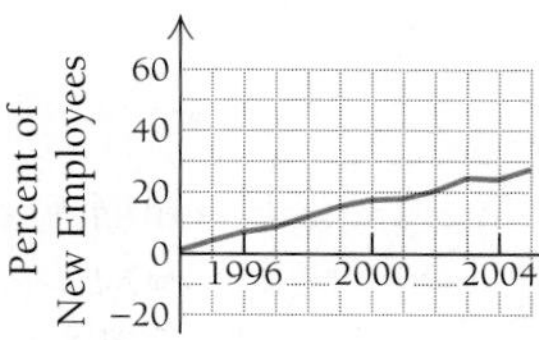

22. Government.

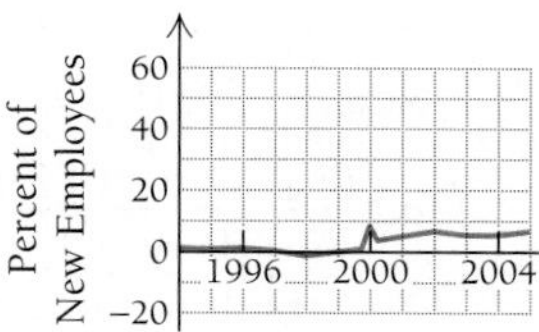

23. Natural resources.

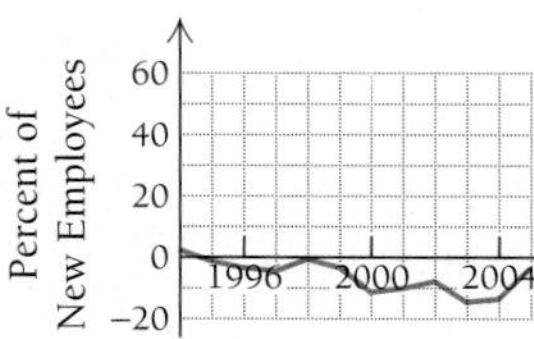

24. Manufacturing.

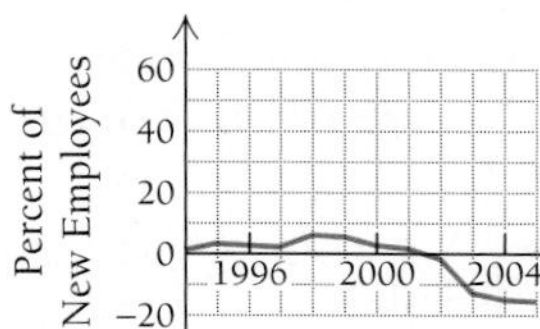

25. Use the following graph to find the average rate of change in U.S. energy consumption from 1960 to 1970, from 1980 to 1990, and from 1990 to 2005.

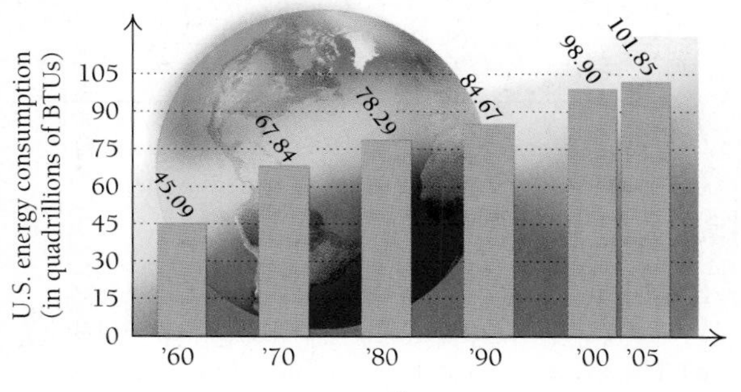

Source: U.S. Energy Information Administration.

26. Use the following graph to find the average rate of change of the U.S. trade deficit with Japan from 1990 to 1995, from 1995 to 2000, and from 2000 to 2004.

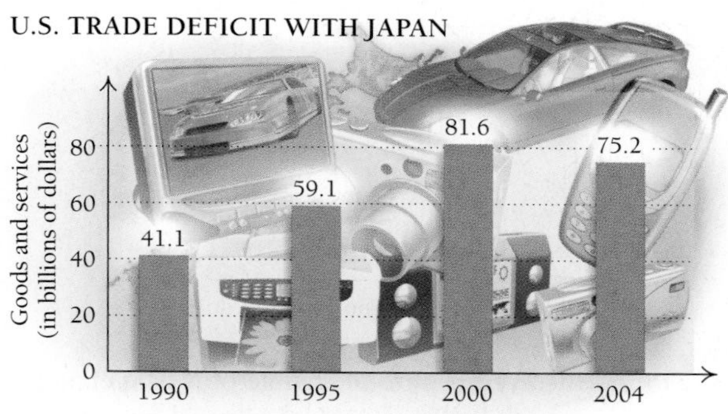

(*Source*: Based on information from U.S. Census Bureau, *Statistical Abstract of the United States, 2006*.)

27. Utility. Utility is a type of function that occurs in economics. When a consumer receives x units of a certain product, a certain amount of pleasure, or utility U, is derived. The following is a graph of a typical utility function.

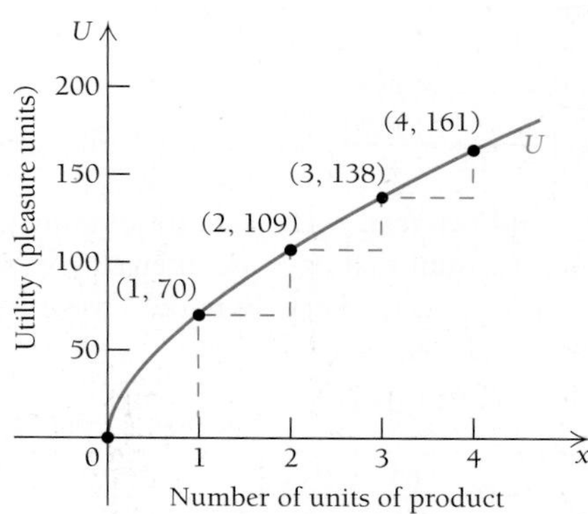

a) Find the average rate of change of U as x changes from 0 to 1; from 1 to 2; from 2 to 3; from 3 to 4.

TW b) Why do you think the average rates of change are decreasing as x increases?

28. Advertising results. The graph on the next page shows a typical response to advertising. After an amount a is spent on advertising, the company sells $N(a)$ units of a product.

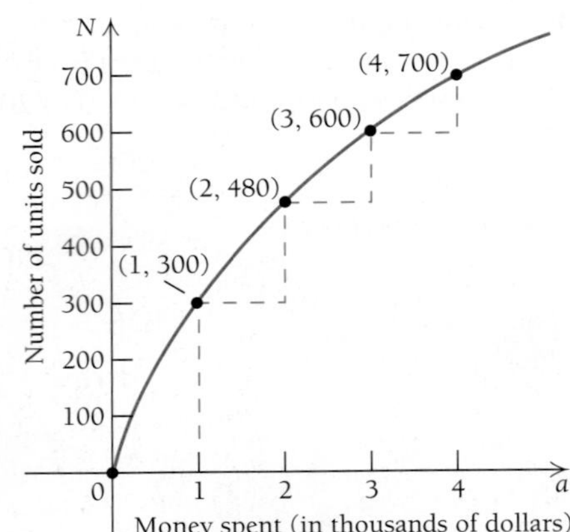

a) Find the average rate of change of N as a changes from 0 to 1; from 1 to 2; from 2 to 3; from 3 to 4.

TW b) Why do you think the average rates of change are decreasing as x increases?

29. **Baseball ticket prices.** Based on data from Major League Baseball, the average price of a ticket to a major league game can be approximated by

$$p(x) = 0.03x^2 + 0.56x + 8.63,$$

where x is the number of years after 1991 and $p(x)$ is in dollars. (*Source*: Based on data from www.teammarketing.com.)

a) Find $p(4)$.
b) Find $p(14)$.
c) Find $p(14) - p(4)$.
d) Find $\dfrac{p(14) - p(4)}{14 - 4}$, and interpret this result.

30. **Compound interest.** The amount of money, $A(t)$, in a savings account that pays 6% interest, compounded quarterly for t years, when an initial investment of \$2000 is made, is given by

$$A(t) = 2000(1.015)^{4t}.$$

a) Find $A(3)$.
b) Find $A(5)$.
c) Find $A(5) - A(3)$.
d) Find $\dfrac{A(5) - A(3)}{5 - 3}$, and interpret this result.

31. **Credit card debt.** When a balance of \$5000 is owed on a credit card and interest is being charged at a rate of 14% per year, the total amount owed after t years, $A(t)$, is given by

$$A(t) = 5000(1.14)^t.$$

Find $\dfrac{A(3) - A(2)}{3 - 2}$, and interpret this result.

32. **Credit card debt.** When a balance of \$3000 is owed on a credit card and interest is charged at a rate of 17% per year, the total amount owed after t years, $A(t)$, is given by

$$A(t) = 3000(1.17)^t.$$

Find $\dfrac{A(4) - A(3)}{4 - 3}$, and interpret this result.

33. **Total cost.** Suppose that Sport Stylz Inc. determines that the cost, in dollars, of producing x cellphone-sunglasses is given by

$$C(x) = -0.05x^2 + 50x.$$

Find $\dfrac{C(301) - C(300)}{301 - 300}$, and interpret the significance of this result to the company.

34. **Total revenue.** Suppose that Sports Stylz Inc. determines that the revenue, in dollars, from the sale of x cellphone-sunglasses is given by

$$R(x) = -0.01x^2 + 1000x.$$

Find $\dfrac{R(301) - R(300)}{301 - 300}$, and interpret the significance of this result to the company.

Life and Physical Sciences

35. **Growth of a baby.** The median weights of babies at age t months are graphed below.

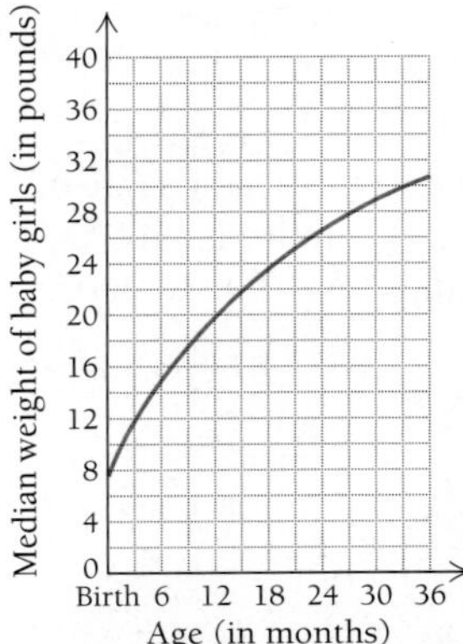

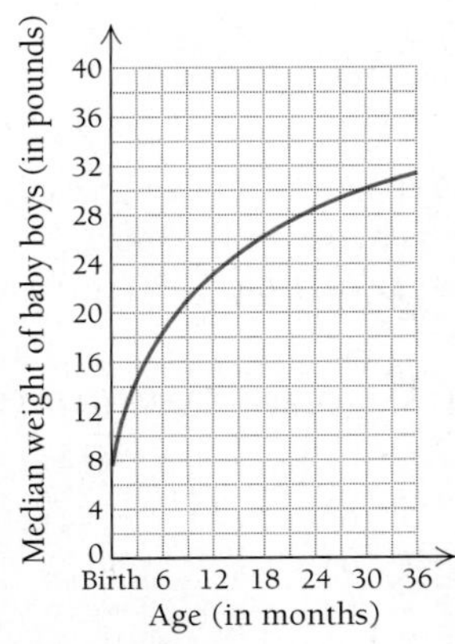

(*Source*: Developed by the National Center for Health Statistics in collaboration with the National Center for Chronic Disease Prevention and Health Promotion, 2000.)

Use the graph of girls' median weight to estimate:

a) The average growth rate of a girl during her first 12 months. (Your answer should be in pounds per month.)

b) The average growth rate of a girl during her second 12 months.

c) The average growth rate of a girl during her first 24 months.

d) Based on your answers in parts (a)–(c) and the graph, estimate the growth rate of a typical 12-month-old girl. Use a straightedge.

e) When does the graph indicate that a baby girl's growth rate is greatest?

36. Growth of a baby. Use the graph of boys' median weight in Exercise 35 to estimate:

a) The average growth rate of a boy during his first 15 months. (Your answer should be in pounds per month.)

b) The average growth rate of a boy during his second 15 months. (Your answer should be in pounds per month.)

c) The average growth rate of a boy during his first 30 months. (Your answer should be in pounds per month.)

TW **d)** Based on your answers in parts (a)–(c) and the graph, estimate the growth rate of a typical boy at exactly 15 months, and explain how you arrived at this figure.

37. Home range. It has been shown that the home range, in hectares, of a carnivorous mammal weighing w grams can be approximated by

$$H(w) = 0.11w^{1.36}.$$

(*Source*: Based on information in Emlen, J. M., *Ecology: An Evolutionary Approach*, p. 200, Reading, MA: Addison-Wesley, 1973; and Harestad, A. S., and Bunnel, F. L., "Home Range and Body Weight—A Reevaluation," *Ecology*, Vol. 60, No. 2, pp. 389–402.)

a) Find the average rate at which a carnivorous mammal's home range increases as the animal's weight grows from 500 g to 700 g.

b) Find $\dfrac{H(300) - H(200)}{300 - 200}$, and interpret this result.

38. Radar range. The function given by $R(x) = 11.74x^{1/4}$ can be used to approximate the maximum range $R(x)$, in miles, of an ARSR-3 surveillance radar with a peak power of x watts (W). (*Source*: *Introduction to RADAR Techniques*, Federal Aviation Administration, 1988.)

a) Find the rate at which the maximum radar range changes as peak power increases from 40,000 W to 60,000 W.

b) Find $\dfrac{R(60{,}000) - R(50{,}000)}{60{,}000 - 50{,}000}$, and interpret this result.

39. Memory. The total number of words, $M(t)$, that a person can memorize in t minutes is shown in the following graph.

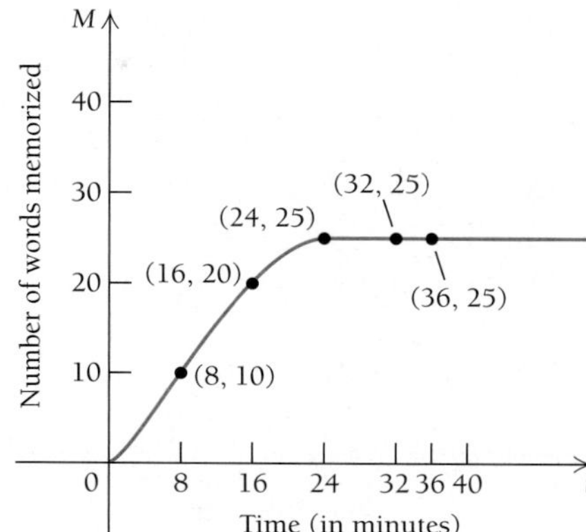

a) Find the average rate of change of M as t changes from 0 to 8; from 8 to 16; from 16 to 24; from 24 to 32; from 32 to 36.

TW **b)** Why do the average rates of change become 0 after 24 min?

40. Average velocity. Suppose that in t hours, a truck travels $s(t)$ miles, where

$$s(t) = 10t^2.$$

a) Find $s(5) - s(2)$. What does this represent?

b) Find the average rate of change of distance with respect to time as t changes from $t_1 = 2$ to $t_2 = 5$. This is known as **average velocity,** or **speed.**

41. Average velocity. In t seconds, an object dropped from a certain height will fall $s(t)$ feet, where

$$s(t) = 16t^2.$$

a) Find $s(5) - s(3)$.

b) What is the average rate of change of distance with respect to time during the period from 3 to 5 sec? This is also *average velocity*.

42. Gas mileage. At the beginning of a trip, the odometer on a car reads 30,680, and the car has a full tank of gas. At the end of the trip, the odometer reads 31,077. It takes 13.5 gal of gas to refill the tank.

a) What is the average rate at which the car was traveling, in miles per gallon?

b) What is the average rate of gas consumption in gallons per mile?

Social Sciences

43. Population growth. The two curves below describe the numbers of people in two countries at time t, in years.

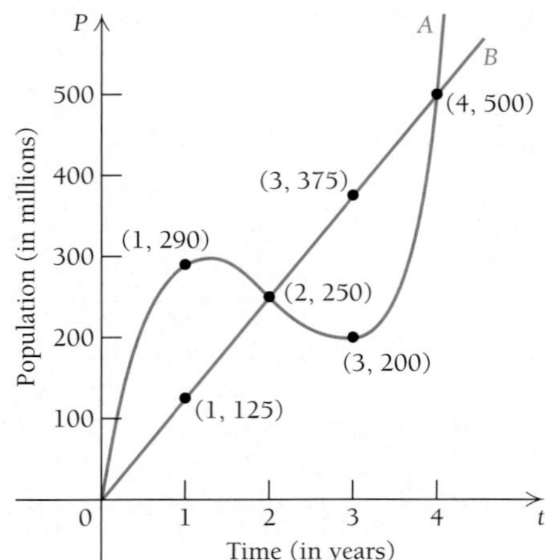

a) Find the average rate of change of each population with respect to time t as t changes from 0 to 4. This is often called the **average growth rate.**

TW b) If the calculation in part (a) were the only one made, would we detect the fact that the populations were growing differently? Explain.

c) Find the average rates of change of each population as t changes from 0 to 1; from 1 to 2; from 2 to 3; from 3 to 4.

TW d) For which population does the statement "the population grew consistently at a rate of 125 million per year" convey accurate information? Why?

SYNTHESIS

TW **44. Business: comparing rates of change.** The following two graphs show the number of federally insured banks and the Nasdaq Composite Stock Index over a 6-month period.

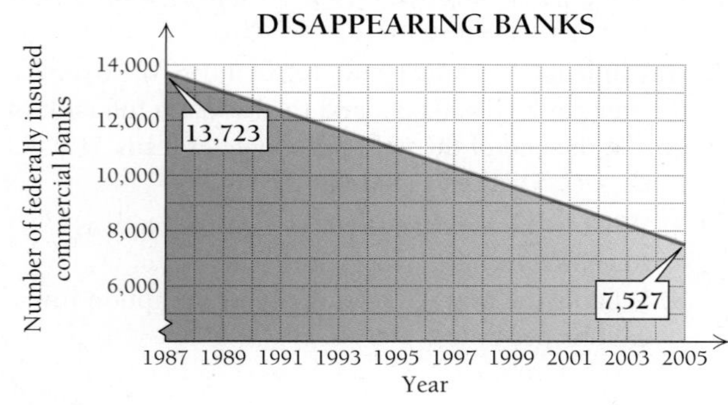

(*Source*: Based on data from the Federal Deposit Insurance Corp.)

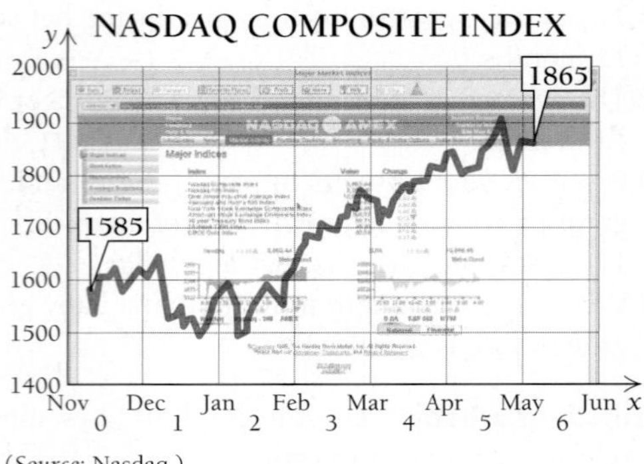

(*Source*: Nasdaq.)

Explain the difference between these graphs in as many ways as you can. Be sure to mention average rates of change.

45. Rising cost of college. Like the cost of most things, the cost of a college education has gone up over the past 30 years. The graphs below display the yearly costs of 4-year colleges in 2005 dollars—indicating that the costs prior to 2005 have been adjusted for inflation.

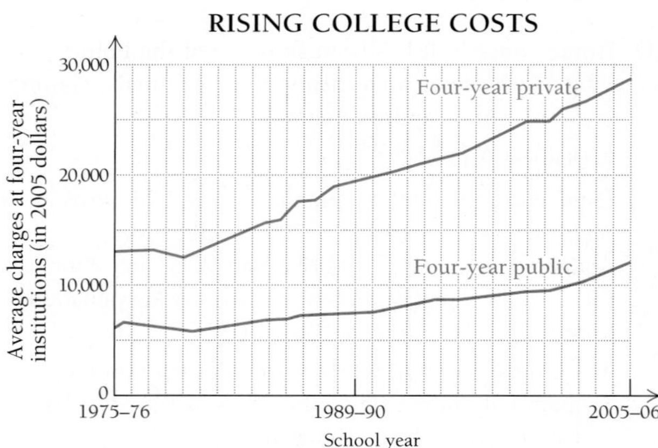

(*Source*: *The New York Times*, Oct. 19, 2005, p. A12.)

a) In what school year did the cost of a private 4-year college increase the most?

b) In what school year(s) did the cost of a public 4-year college increase the most?

c) Assuming an annual inflation rate of 3%, calculate the cost of a year at a public and at a private 4-year college in 1975. Express the costs in 1975 dollars.

Find the simplified difference quotient for each function listed.

46. $f(x) = mx + b$

47. $f(x) = ax^2 + bx + c$

48. $f(x) = ax^3 + bx^2$

49. $f(x) = \sqrt{x}$

$\left(\textit{Hint}: \text{Multiply by 1 using } \dfrac{\sqrt{x+h}+\sqrt{x}}{\sqrt{x+h}+\sqrt{x}}.\right)$

50. $f(x) = x^3 + x$

51. $f(x) = \dfrac{1}{x^2}$

52. $f(x) = \dfrac{1}{1-x}$

53. $f(x) = \dfrac{x}{1+x}$

54. $f(x) = \sqrt{3-2x}$

55. $f(x) = \dfrac{1}{\sqrt{x}}$

1.4 Differentiation Using Limits of Difference Quotients

OBJECTIVES

- Find derivatives and values of derivatives.
- Find equations of tangent lines.

We have seen that the *slope of a secant line* connecting the points $(x, f(x))$ and $(x + h, f(x + h))$ represents the *average rate of change* of $f(x)$ over the interval $[x, x + h]$. This rate is represented by the difference quotient

$$\frac{f(x+h) - f(x)}{h}.$$

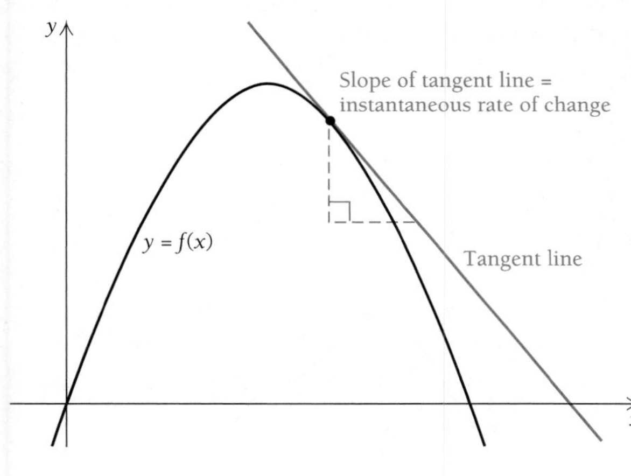

As h gets smaller, the slope of this secant line more closely approximates the slope of the tangent line at $(x, f(x))$. By finding the *slope of the tangent line*, we can find the exact rate of change at this point, a value called the *instantaneous rate of change*.

Tangent Lines

A line *tangent* to a circle is a line that touches the circle exactly once.

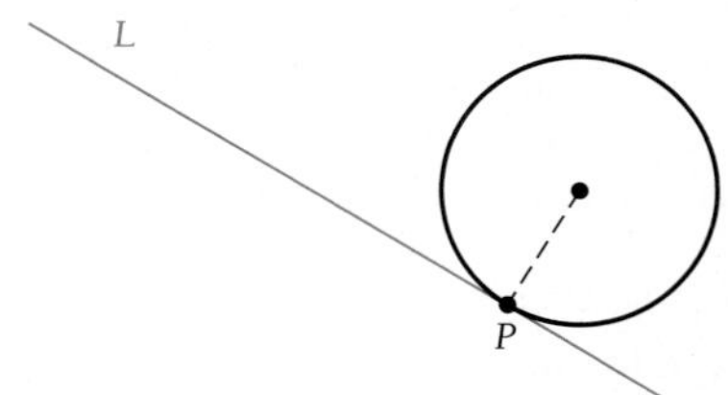

This definition becomes unworkable with other curves. For example, consider the curve shown in Fig. 1. Line L touches the curve at point P but meets the curve at other places as well. It is considered a tangent line, but "touching at one point" cannot be its definition.

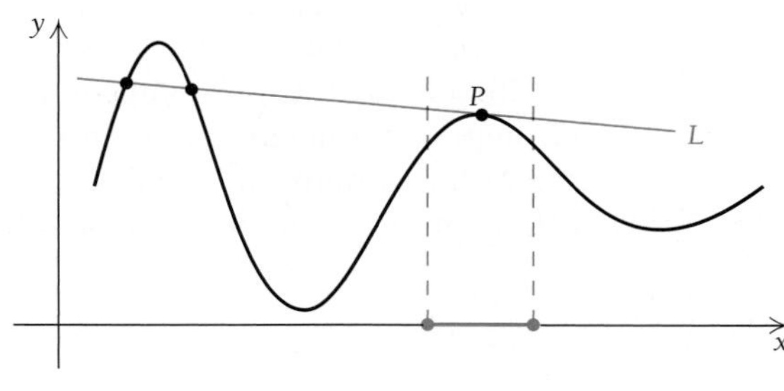

FIGURE 1

The cable of a chairlift runs tangent to the rollers on each tower.

Note in Fig. 1 that over a small interval containing P, line L does touch the curve exactly once. This is still not a suitable definition of a tangent line because it would allow line M in Fig. 2 to be a tangent line, which is not the case.

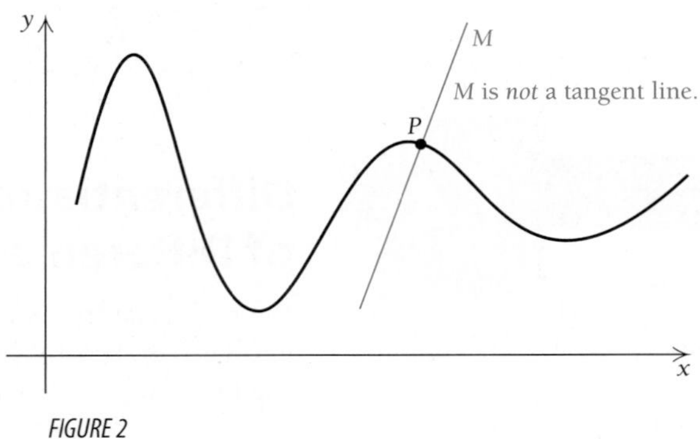

FIGURE 2

Soon we will give a definition of a tangent line, but for now we will rely on intuition. In Fig. 3, all of the lines except for L_1 and L_2 are tangent lines.

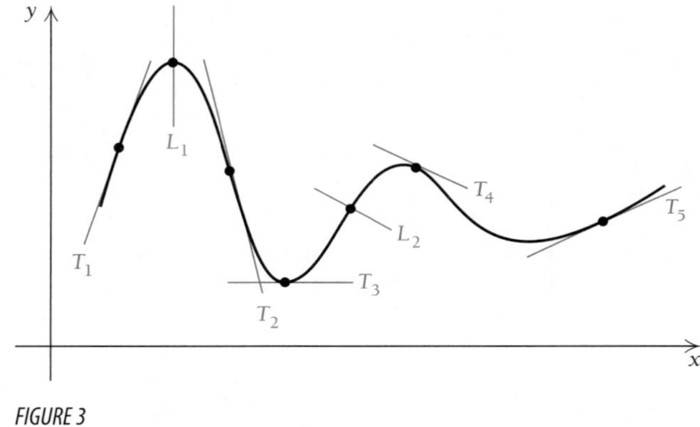

FIGURE 3

TECHNOLOGY CONNECTION

Exploratory

Note: Each Technology Connection labeled "Exploratory" is designed to lead students through a discovery process. Because of this, answers are not provided.

Graph $y_1 = 3x^5 - 20x^3$ with the viewing window $[-3, 3, -80, 80]$, with Xscl = 1 and Yscl = 10. Then also graph the lines $y_2 = -7x - 10$, $y_3 = -30x + 13$, and $y_4 = -45x + 28$. Which line appears to be tangent to the graph of y_1 at $(1, -17)$? If necessary, zoom in near $(1, -17)$ to refine your guess.

Differentiation Using Limits

We now define *tangent line* so that it makes sense for *any* curve. To do this, we use the notion of limit.

To obtain the line tangent to the curve at point P, consider secant lines through P and neighboring points Q_1, Q_2, and so on. As the Q's approach P, the secant lines approach line T. Each secant line has a slope. The slopes m_1, m_2, m_3, and so on, of the secant lines approach the slope m of line T. We *define* line T as the **tangent line,** the line that contains point P and has slope m, where m is the limit of the slopes of the secant lines as the points Q approach P.

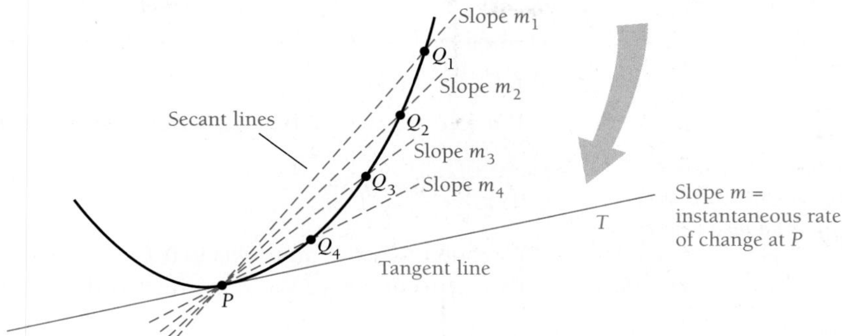

How might we calculate the limit m? Suppose that P has coordinates $(x, f(x))$. Then the first coordinate of Q is x plus some number h, or $x + h$. The coordinates of Q are $(x + h, f(x + h))$, as shown in Fig. 4.

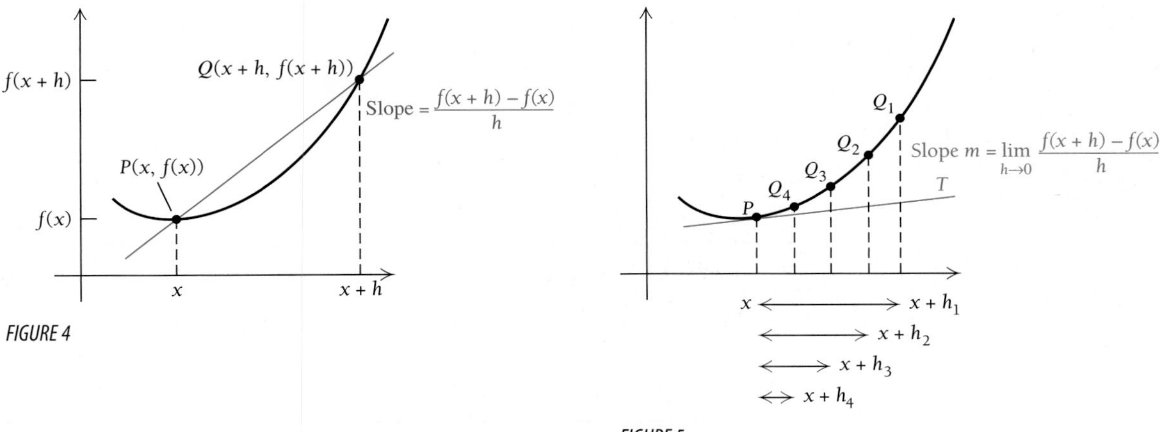

FIGURE 4

FIGURE 5

From Section 1.3, we know that the slope of the secant $\overleftrightarrow{PQ}$ is given by

$$\frac{f(x + h) - f(x)}{h}.$$

Now, as we see in Fig. 5, as the Q's approach P, the x-values $x + h$ approach x. That is, h approaches 0. Thus, we have the following.

The slope of the tangent line at $(x, f(x)) = m = \lim_{h \to 0} \frac{f(x + h) - f(x)}{h}$.

This limit is also the **instantaneous rate of change** of $f(x)$ at x.

The formal definition of the *derivative of a function* f can now be given. We will designate the derivative at x as $f'(x)$, rather than m. The notation $f'(x)$ is read "the derivative of f at x," "f prime at x," or "f prime of x."

> *"Nothing in this world is so powerful as an idea whose time has come."*
> *Victor Hugo*

DEFINITION

For a function $y = f(x)$, its **derivative** at x is the function f' defined by

$$f'(x) = \lim_{h \to 0} \frac{f(x+h) - f(x)}{h},$$

provided that the limit exists. If $f'(x)$ exists, then we say that f is **differentiable** at x. We sometimes call f' the **derived function.**

Let's now calculate some formulas for derivatives. That is, given a formula for a function f, we will attempt to find a formula for f'.

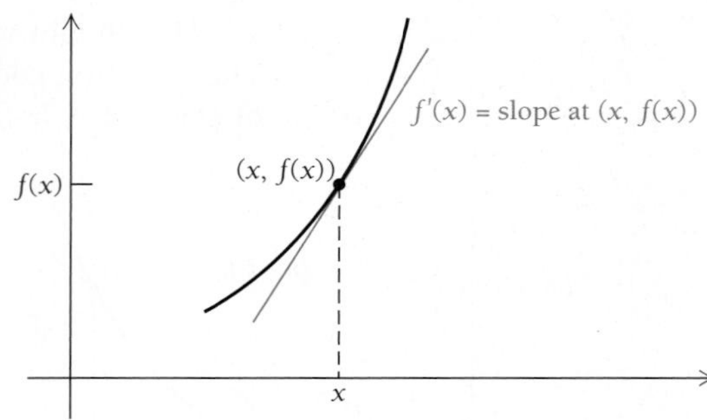

There are three steps in calculating a derivative.

1. Write down the difference quotient, $(f(x+h) - f(x))/h$.
2. Simplify the difference quotient.
3. Find the limit as h approaches 0.

EXAMPLE 1 For $f(x) = x^2$, find $f'(x)$. Then find $f'(-3)$ and $f'(4)$.

Solution We have

1. $$\frac{f(x+h) - f(x)}{h} = \frac{(x+h)^2 - x^2}{h}$$

2. $$\frac{f(x+h) - f(x)}{h} = \frac{x^2 + 2xh + h^2 - x^2}{h}$$ Evaluating $f(x+h)$ and $f(x)$

$$= \frac{2xh + h^2}{h} = \frac{h(2x+h)}{h}$$
$$= 2x + h, \quad h \neq 0$$ Simplifying

3. We want to find

$$\lim_{h \to 0} \frac{f(x+h) - f(x)}{h} = \lim_{h \to 0} (2x + h).$$

As $h \to 0$, we see that $(2x + h) \to 2x$. Thus,

$$\lim_{h \to 0} (2x + h) = 2x,$$

and we have

$$f'(x) = 2x.$$

Using the fact that $f'(x) = 2x$, it follows that

$$f'(-3) = 2 \cdot (-3) = -6, \quad \text{and} \quad f'(4) = 2 \cdot 4 = 8.$$

This tells us that at $x = -3$, the curve has a tangent line whose slope is

$$f'(-3) = -6,$$

and at $x = 4$, the tangent line has slope

$$f'(4) = 8.$$

We can also say:

- The tangent line to the curve at the point $(-3, 9)$ has slope -6.
- The tangent line to the curve at the point $(4, 16)$ has slope 8.
- The instantaneous rate of change at $x = -3$ is -6.
- The instantaneous rate of change at $x = 4$ is 8.

TECHNOLOGY CONNECTION

Exploratory

To see exactly how the definition of derivative works and to check the results of Example 1, let $y_1 = ((-3 + x)^2 - (-3)^2)/x$ and note that we are using x in place of h. Use the TABLE feature and enter smaller and smaller values for x. Then repeat the procedure for $y_2 = ((4 + x)^2 - 4^2)/x$. How do these results confirm those in Example 1?

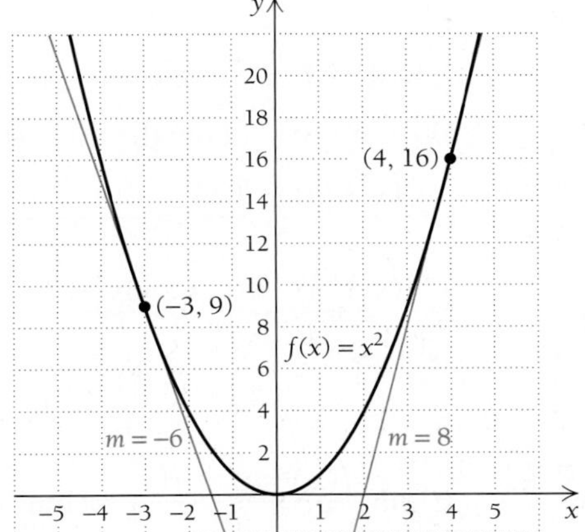

✦

EXAMPLE 2 For $f(x) = x^3$, find $f'(x)$. Then find $f'(-1)$ and $f'(1.5)$.

Solution

1. We have

$$\frac{f(x + h) - f(x)}{h} = \frac{(x + h)^3 - x^3}{h}.$$

2. In Example 6 of Section 1.3 (on p. 129), we showed how this difference quotient can be simplified to

$$\frac{f(x + h) - f(x)}{h} = 3x^2 + 3xh + h^2.$$

3. We then have

$$\lim_{h \to 0} \frac{f(x+h) - f(x)}{h} = \lim_{h \to 0} (3x^2 + 3xh + h^2) = 3x^2.$$

Thus,

$$f'(-1) = 3(-1)^2 = 3 \quad \text{and} \quad f'(1.5) = 3(1.5)^2 = 6.75.$$

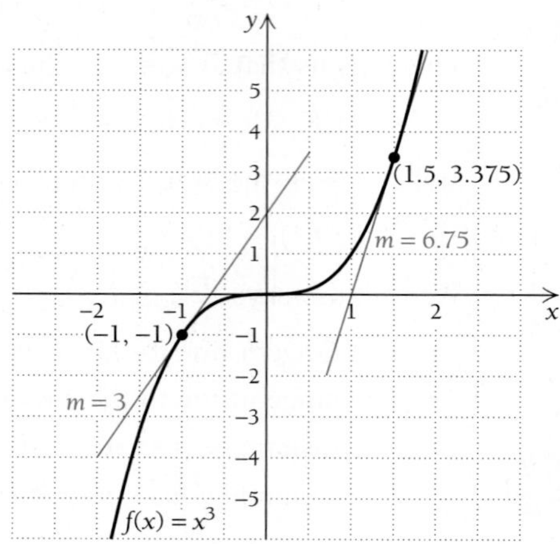

✦

EXAMPLE 3 For $f(x) = 3x - 4$, find $f'(x)$ and $f'(2)$.

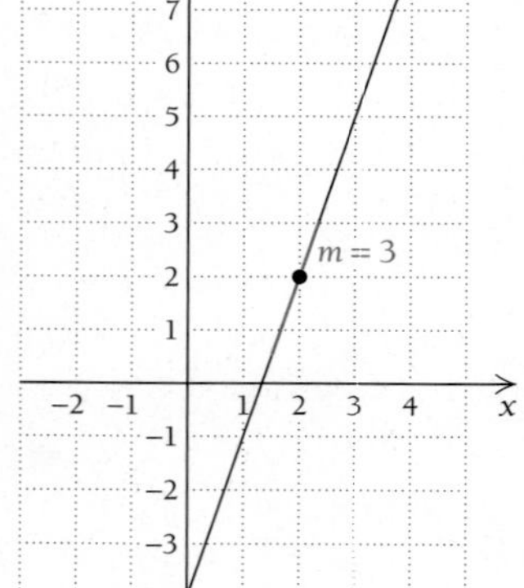

Solution We follow the three steps given above.

1. $\dfrac{f(x+h) - f(x)}{h} = \dfrac{3(x+h) - 4 - (3x - 4)}{h}$ The parentheses are important.

2. $\dfrac{f(x+h) - f(x)}{h} = \dfrac{3x + 3h - 4 - 3x + 4}{h}$ Using the distributive law

$= \dfrac{3h}{h} = 3, \quad h \neq 0;$ Simplifying

3. $\displaystyle\lim_{h \to 0} \frac{f(x+h) - f(x)}{h} = \lim_{h \to 0} 3 = 3$, since 3 is a constant.

Thus, if $f(x) = 3x - 4$, then $f'(x) = 3$ and $f'(2) = 3$. ✦

The result of Example 3 suggests that, for a straight line, the slope of a tangent line is the slope of the straight line itself. That is, a general formula for the derivative of a linear function

$$f(x) = mx + b$$

is

$$f'(x) = m.$$

The formula can be verified in a manner similar to that used in Example 3.

Examples 1–3 and Example 4, which follows, involve a somewhat lengthy process, but in Section 1.5 we will develop some faster techniques. It is very important in this section, however, to fully understand the concept of derivative.

EXAMPLE 4 For $f(x) = 3/x$:

a) Find $f'(x)$.

b) Find $f'(2)$.

c) Find an equation of the tangent line to the curve at $x = 2$.

Solution

a) 1. We have

$$\frac{f(x+h) - f(x)}{h} = \frac{[3/(x+h)] - (3/x)}{h}.$$

2. In Example 7 of Section 1.3 (on pp. 129–130), we showed that this difference quotient simplifies to

$$\frac{f(x+h) - f(x)}{h} = \frac{-3}{x(x+h)}, \quad h \neq 0.$$

3. We want to find

$$\lim_{h \to 0} \frac{f(x+h) - f(x)}{h} = \lim_{h \to 0} \frac{-3}{x(x+h)}.$$

As $h \to 0$, we have $x + h \to x$. Thus,

$$f'(x) = \lim_{h \to 0} \frac{-3}{x(x+h)} = \frac{-3}{x \cdot x} = \frac{-3}{x^2}.$$

b) Since $f'(x) = -3/x^2$, we have

$$f'(2) = \frac{-3}{2^2} = -\frac{3}{4}.$$ This is the slope of the tangent line at $x = 2$.

c) We can find an equation of the tangent line at $x = 2$ if we know the line's slope and a point that is on the line. In part (b), we found that the slope at $x = 2$ is $-\frac{3}{4}$. To find a point on the line, we compute $f(2)$:

$$f(2) = \frac{3}{2}.$$ CAUTION! Be careful to use f when computing y-values and f' when computing slope.

We have

Point: $\left(2, \frac{3}{2}\right)$, This is $(x_1, f(x_1))$.

Slope: $-\frac{3}{4}$. This is $f'(x_1)$.

We substitute into the point–slope equation (see Section R.4):

$$\begin{aligned} y - y_1 &= m(x - x_1) \\ y - \tfrac{3}{2} &= -\tfrac{3}{4}(x - 2) \\ y &= -\tfrac{3}{4}x + \tfrac{3}{2} + \tfrac{3}{2} \\ &= -\tfrac{3}{4}x + 3. \end{aligned}$$ Rewriting in slope–intercept form

The equation of the tangent line to the curve at $x = 2$ is

$$y = -\tfrac{3}{4}x + 3.$$

TECHNOLOGY CONNECTION

Curves and Tangent Lines

EXERCISES

1. For $f(x) = 3/x$, find $f'(x)$, $f'(-2)$, and $f'\left(-\frac{1}{2}\right)$.
2. Find an equation of the tangent line to the graph of $f(x) = 3/x$ at $\left(-2, -\frac{3}{2}\right)$ and an equation of the tangent line to the curve at $\left(-\frac{1}{2}, -6\right)$. Then graph the curve $f(x) = 3/x$ and both tangent lines. Use ZOOM to view the graphs near the points of tangency.
3. To check your equations of tangent lines, first graph $f(x) = 3/x$. Then use the TRACE feature to move the cursor to the point of tangency. Finally, select Tangent from the DRAW menu.

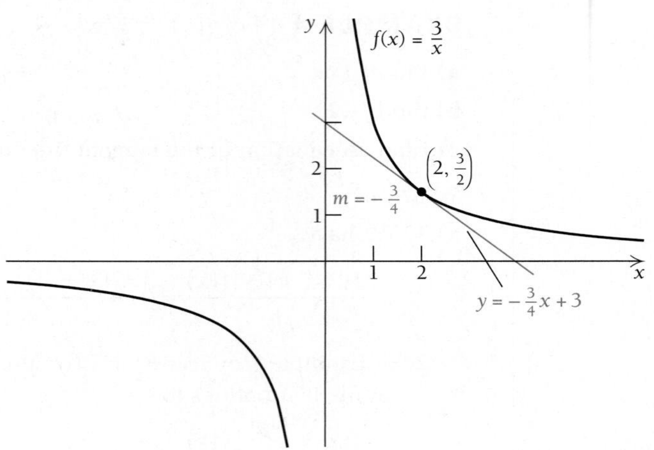

✦

In Example 4, note that since $f(0)$ does not exist for $f(x) = 3/x$, we cannot evaluate the difference quotient

$$\frac{f(0 + h) - f(0)}{h}.$$

Thus, $f'(0)$ does not exist. We say that "f is not differentiable at 0."

When a function is not defined at a point, it is not differentiable at that point. In general, if a function is discontinuous at a point, it is not differentiable at that point.

It can happen that a function f is defined and continuous at a point but its derivative f' is not defined. The function f given by

$$f(x) = |x|$$

is an example. Note that

$$\lim_{x \to 0} f(x) = 0 = f(0),$$

so the function is both defined and continuous at $x = 0$.

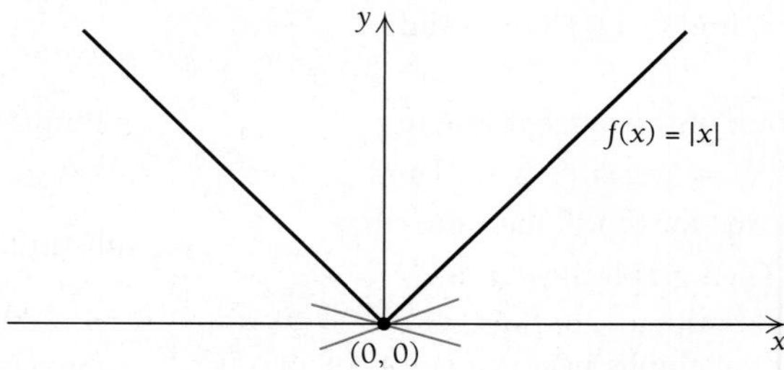

Suppose that we try to draw a tangent line at $(0, 0)$. A function like this with a corner (not smooth) would seem to have many tangent lines at $(0, 0)$, and thus many slopes. The derivative at such a point would not be unique. Let's try to calculate the derivative at 0.

Since

$$f(x) = |x| = \begin{cases} x, & \text{for } x \geq 0, \\ -x, & \text{for } x < 0, \end{cases}$$

it follows from our earlier work with lines that

$$f'(x) = \begin{cases} 1, & \text{for } x > 0, \\ -1, & \text{for } x < 0. \end{cases}$$

Then, since

$$\lim_{x \to 0^+} f'(x) \neq \lim_{x \to 0^-} f'(x),$$

it follows that $f'(0)$ does not exist.

In general, if a function has a "sharp point" or "corner," it will not have a derivative at that point.

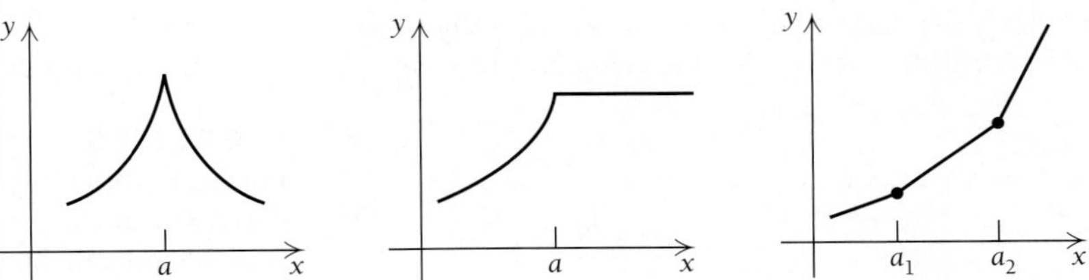

A function will also fail to be differentiable at a point if it has a vertical tangent at that point. For example, the function shown below has a vertical tangent at point a. Recall that since the slope of a vertical line is undefined, there is no derivative at such a point.

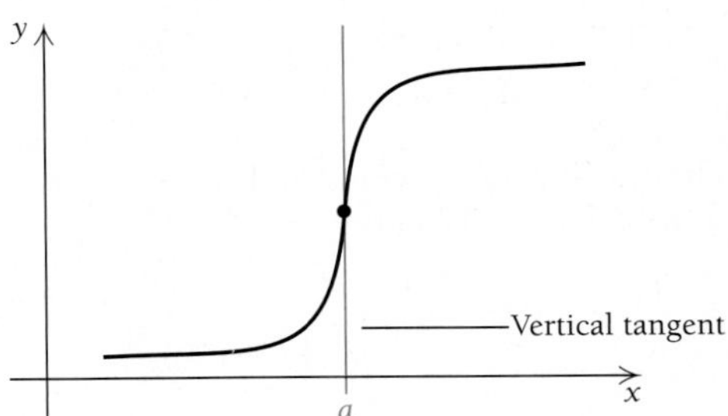

The function given by $f(x) = |x|$ illustrates the fact that although a function may be continuous at each point in an interval I, it may not be differentiable at each point in I. That is, continuity does not imply differentiability. However, differentiability *does* guarantee continuity. That is, if $f'(a)$ exists, then f is continuous at a. The function $f(x) = x^2$ is an example of a function that is differentiable over the interval $(-\infty, \infty)$ and is thus continuous everywhere. Thus, when a function is differentiable over an interval, it is not just continuous, but is also *smooth* in the sense that there are no "sharp points," or "corners," in the graph.

TECHNOLOGY CONNECTION

Numerical Differentiation and Drawing Tangent Lines

Calculators have the capability of finding slopes of tangent lines, that is, of taking the derivative at a specific x-value. Let's consider

$$f(x) = x(100 - x),$$

and find the value of the derivative at $x = 70$.

Calculating the Derivative

Select nDeriv (numerical derivative) from the MATH menu (option 8) and enter the function, the variable, and the value at which the derivative is to be evaluated. When we enter nDeriv($x(100 - x)$, x, 70), the calculator returns -40.

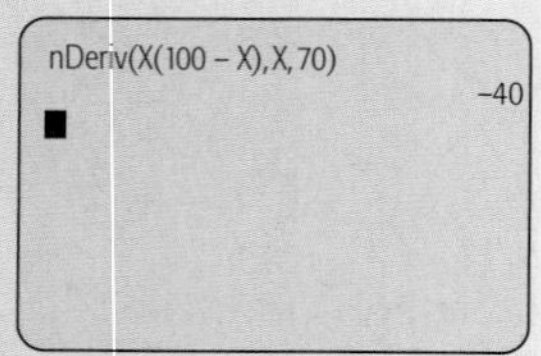

We see that the derivative of the function at $x = 70$ is -40. That is, the slope of the tangent line to the curve at $x = 70$ is -40. Note that if we wish to graph the derivative of y_1, we can simply graph $y_2 =$ nDeriv(y_1, x, x).

Drawing the Tangent Line

We first graph the function using the viewing window $[-10, 100, -10, 3000]$, with Xscl $= 10$ and Yscl $= 1000$.

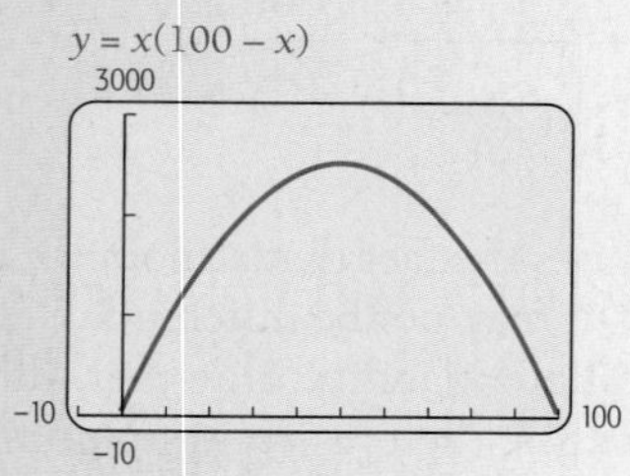

To draw a tangent line at $x = 70$, we graph the function and select Tangent from the DRAW menu. Then we enter 70 and view both the graph of $f(x)$ and the tangent line at $x = 70$.

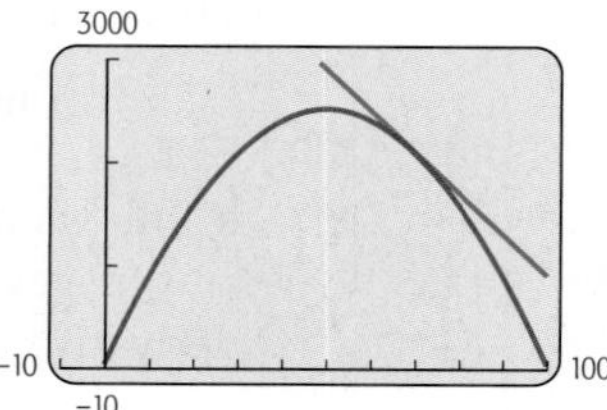

For more on derivatives and tangent lines, see p. 149.

EXERCISES

For each of the following functions, evaluate the derivative at the given points. Then draw the graph and the tangent line. If a derivative does not exist, state that fact. When appropriate, round to two decimal places.

1. $f(x) = x(100 - x)$;
   $x = 20, x = 37, x = 50, x = 90$
2. $f(x) = -\frac{1}{3}x^3 + 6x^2 - 11x - 50$;
   $x = -5, x = 0, x = 7, x = 12, x = 15$
3. $f(x) = 6x^2 - x^3$;
   $x = -2, x = 0, x = 2, x = 4, x = 6.3$
4. $f(x) = x\sqrt{4 - x^2}$;
   $x = -2, x = -1.3, x = -0.5, x = 0, x = 1, x = 2$
5. TW For the function in Exercise 4, try to draw a tangent line at $x = 3$ and estimate the derivative. What goes wrong? Explain.

Exercise Set 1.4

In Exercises 1–16:

a) *Graph the function.*

b) *Draw tangent lines to the graph at points whose x-coordinates are* -2, 0, *and* 1.

c) *Find* $f'(x)$ *by determining* $\lim_{h\to 0} \frac{f(x+h)-f(x)}{h}$.

d) *Find* $f'(-2)$, $f'(0)$, *and* $f'(1)$. *These slopes should match those of the lines you drew in part (b).*

1. $f(x) = \frac{3}{2}x^2$

2. $f(x) = \frac{1}{2}x^2$

3. $f(x) = -2x^2$

4. $f(x) = -3x^2$

5. $f(x) = x^3$

6. $f(x) = -x^3$

7. $f(x) = 2x + 3$

8. $f(x) = -2x + 5$

9. $f(x) = \frac{1}{2}x - 3$

10. $f(x) = \frac{3}{4}x - 2$

11. $f(x) = x^2 + x$

12. $f(x) = x^2 - x$

13. $f(x) = 2x^2 + 3x - 2$

14. $f(x) = 5x^2 - 2x + 7$

15. $f(x) = \frac{1}{x}$

16. $f(x) = \frac{2}{x}$

17. Find an equation of the tangent line to the graph of $f(x) = x^2$ at **(a)** $(3, 9)$; **(b)** $(-1, 1)$; **(c)** $(10, 100)$. See Example 1.

18. Find an equation of the tangent line to the graph of $f(x) = x^3$ at **(a)** $(-2, -8)$; **(b)** $(0, 0)$; **(c)** $(4, 64)$. See Example 2.

19. Find an equation of the tangent line to the graph of $f(x) = 2/x$ at **(a)** $(1, 2)$; **(b)** $(-1, -2)$; **(c)** $(100, 0.02)$. See Exercise 16.

20. Find an equation of the tangent line to the graph of $f(x) = -1/x$ at **(a)** $(-1, 1)$; **(b)** $\left(2, -\frac{1}{2}\right)$; **(c)** $\left(-5, \frac{1}{5}\right)$.

21. Find an equation of the tangent line to the graph of $f(x) = 4 - x^2$ at **(a)** $(-1, 3)$; **(b)** $(0, 4)$; **(c)** $(5, -21)$.

22. Find an equation of the tangent line to the graph of $f(x) = x^2 - 2x$ at **(a)** $(-2, 8)$; **(b)** $(1, -1)$; **(c)** $(4, 8)$.

23. Find $f'(x)$ for $f(x) = mx + b$.

24. Find $f'(x)$ for $f(x) = ax^2 + bx$.

List the points in the graph at which each function is not differentiable.

25.

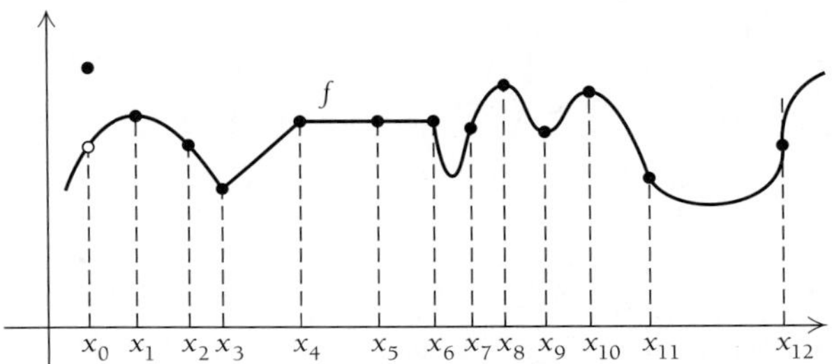

26.

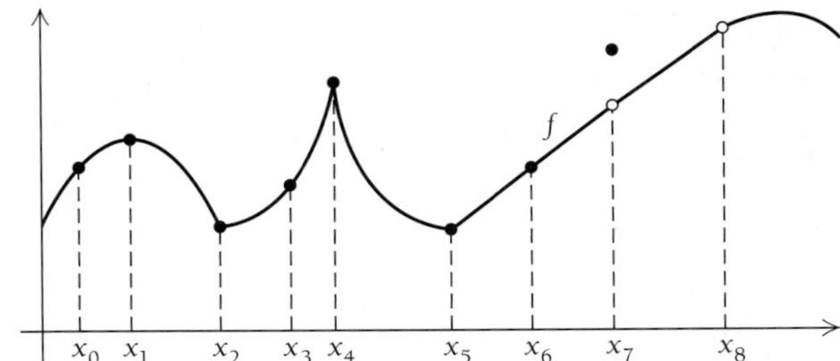

27. Draw a graph that is continuous, but not differentiable, at $x = 3$.

28. Draw a graph that has a horizontal tangent line at $x = 5$.

APPLICATIONS

Business and Economics

29. The postage function. Consider the postage function defined in Exercises 64–68 of Exercise Set 1.1, on p. 112. At what values in the domain is the function not differentiable?

30. **The taxicab fare function.** Consider the taxicab fare function defined in Exercises 61–63 of Exercise Set 1.1, on p. 112. At what values is the function not differentiable?

31. **Baseball ticket prices.** Consider the model for average Major League Baseball ticket prices in Exercise 29 of Exercise Set 1.3, on p. 132. At what value is the function not differentiable?

SYNTHESIS

TW 32. Which of the lines in the following graph appear to be tangent lines? Try to explain why or why not.

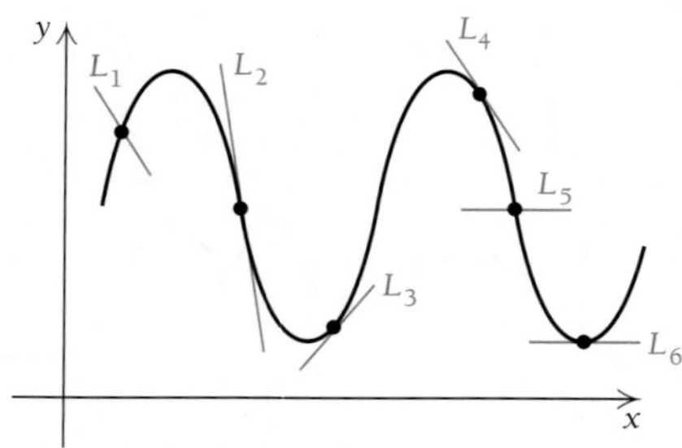

TW 33. On the following graph, use a blue colored pencil to draw each secant line from point P to the points Q. Then use a red colored pencil to draw a tangent line to the curve at P. Describe what happens.

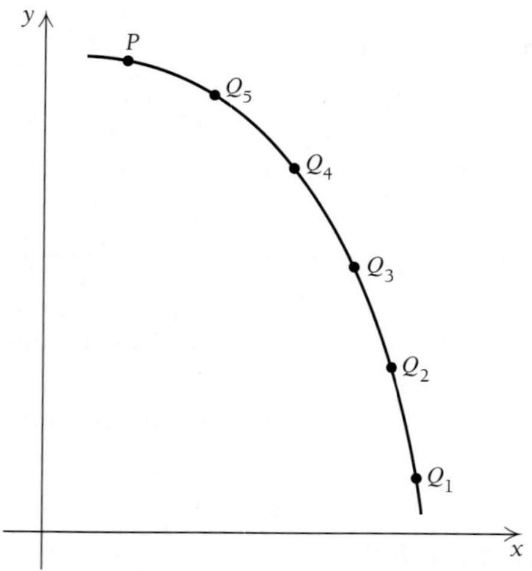

Find $f'(x)$ for each function.

34. $f(x) = x^4$

35. $f(x) = \dfrac{1}{x^2}$

36. $f(x) = \dfrac{1}{1 - x}$

37. $f(x) = \dfrac{x}{1 + x}$

38. $f(x) = \dfrac{2x}{x + 1}$

39. $f(x) = \sqrt{x}$

$\left(\textit{Hint}: \text{Multiply the difference quotient by } \dfrac{\sqrt{x+h} + \sqrt{x}}{\sqrt{x+h} + \sqrt{x}}.\right)$

40. $f(x) = \dfrac{1}{\sqrt{x}}$

41. Consider the function f given by

$$f(x) = \frac{x^2 - 9}{x + 3}.$$

a) For what x-value(s) is this function not differentiable?

TW b) Describe the simplest way to find $f'(4)$.

TECHNOLOGY CONNECTION

42–47. Use a calculator to check your answers to Exercises 17–22.

48. **Business: growth of an investment.** A company determines that the value of an investment is V, in millions of dollars, after time t, in years, where V is given by

$$V(t) = 5t^3 - 30t^2 + 45t + 5\sqrt{t}.$$

Note: Calculators often use only the variables y and x, so you may need to change the variables when entering this function.

a) Graph V over the interval $[0, 5]$.

b) Find the equation of the secant line passing through the points $(1, V(1))$ and $(5, V(5))$. Then graph this secant line using the same axes as in part (a).

c) Find the average rate of change of the investment between year 1 and year 5.

d) Repeat parts (b) and (c) for the following pairs of points: $(1, V(1))$ and $(4, V(4))$; $(1, V(1))$ and $(3, V(3))$; $(1, V(1))$ and $(1.5, V(1.5))$.

e) What *appears* to be the slope of the tangent line to the graph at the point $(1, V(1))$?

f) Approximate the rate at which the value of the investment is changing after 1 yr.

49. Use a calculator to determine where $f'(x)$ does not exist, if $f(x) = \sqrt[3]{x - 5}$.

1.5 Differentiation Techniques: The Power and Sum–Difference Rules

OBJECTIVES

- Differentiate using the Power Rule or the Sum–Difference Rule.
- Differentiate a constant or a constant times a function.
- Determine points at which a tangent line has a specified slope.

Leibniz Notation

When y is a function of x, we will also designate the derivative, $f'(x)$, as

$$\frac{dy}{dx},$$

which is read "the derivative of y with respect to x." This notation was invented by the German mathematician Leibniz. It does *not* mean dy divided by dx, nor does it mean y/x, with the d's having been removed. Think of dy/dx as a single entity made up of many parts. That is, we cannot interpret dy/dx as a quotient until meanings are given to dy and dx, which we will not do here. For example, if $y = x^2$, then

$$f'(x) = \frac{dy}{dx} = 2x.$$

Historical Note: The German mathematician and philosopher Gottfried Wilhelm von Leibniz (1646–1716) and the English mathematician, philosopher, and physicist Sir Isaac Newton (1642–1727) are both credited with the invention of calculus, though each performed his work independently. Newton used the dot notation $\dot{y}$ for dy/dt, where y is a function of time; this notation is still used, though it is not as common as Leibniz notation.

We can also write

$$\frac{d}{dx} f(x)$$

to denote the derivative of f with respect to x. For example,

$$\frac{d}{dx}x^2 = 2x.$$ This is read "the derivative, with respect to x, of x^2 is $2x$."

The value of dy/dx when $x = 5$ can be denoted by

$$\left.\frac{dy}{dx}\right|_{x=5}.$$ This is read "the derivative of y with respect to x, evaluated at $x = 5$."

Thus, for $dy/dx = 2x$,

$$\left.\frac{dy}{dx}\right|_{x=5} = 2 \cdot 5, \quad \text{or} \quad 10.$$

In general, for $y = f(x)$,

$$\left.\frac{dy}{dx}\right|_{x=a} = f'(a).$$

Sometimes the notation y' (read "y prime") or $D_x y$ is used in place of dy/dx or $f'(x)$.

The Power Rule

In the remainder of this section, we will develop rules and techniques to streamline differentiation.

Look for a pattern in the following table, which contains functions and derivatives that we have found in previous work.

Function	Derivative
x^2	$2x^1$
x^3	$3x^2$
x^4	$4x^3$
$\frac{1}{x} = x^{-1}$	$-1 \cdot x^{-2} = \frac{-1}{x^2}$
$\frac{1}{x^2} = x^{-2}$	$-2 \cdot x^{-3} = \frac{-2}{x^3}$

Perhaps you have discovered the following theorem.

THEOREM 1

The Power Rule

For any real number k,

$$\frac{d}{dx}x^k = k \cdot x^{k-1}.$$

We proved this theorem for the cases $k = 2, 3$, and -1 in Examples 1 and 2 and Exercise 15, respectively, of Section 1.4. The proof of the general case, for k any real number, is beyond the scope of this text. Note that the Power Rule holds no matter what the exponent. That is, to differentiate x^k, we write the exponent k as the coefficient, followed by x with an exponent 1 less than k.

$$\frac{d}{dx}x^k = k \cdot x^{k-1}$$

1. Write the exponent as the coefficient.
2. Subtract 1 from the exponent.

EXAMPLE 1 Differentiate each of the following:

a) $y = x^5$; **b)** $y = x$; **c)** $y = x^{-4}$.

Solution

a) $\frac{d}{dx}x^5 = 5 \cdot x^{5-1} = 5x^4$ Using the Power Rule

b) $\frac{d}{dx}x = 1 \cdot x^{1-1} = 1 \cdot x^0 = 1$

c) $\frac{d}{dx}x^{-4} = -4 \cdot x^{-4-1} = -4x^{-5}$, or $-4 \cdot \frac{1}{x^5}$, or $-\frac{4}{x^5}$ ✦

The Power Rule also allows us to differentiate expressions with rational exponents.

EXAMPLE 2 Differentiate:

a) $y = \sqrt{x}$; **b)** $y = x^{0.7}$.

Solution

a) $\frac{d}{dx}\sqrt{x} = \frac{d}{dx}x^{1/2} = \frac{1}{2} \cdot x^{(1/2)-1}$

$= \frac{1}{2}x^{-1/2}$, or $\frac{1}{2} \cdot \frac{1}{x^{1/2}}$, or $\frac{1}{2} \cdot \frac{1}{\sqrt{x}}$, or $\frac{1}{2\sqrt{x}}$

b) $\frac{d}{dx}x^{0.7} = 0.7x^{(0.7)-1} = 0.7x^{-0.3}$ ✦

TECHNOLOGY CONNECTION

More on Numerical Differentiation and Tangent Lines

Consider $f(x) = x\sqrt{4 - x^2}$, graphed below.

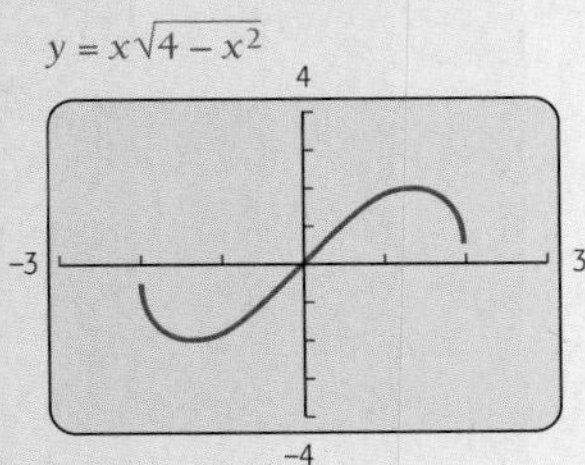

To find the value of dy/dx at a point, we select dy/dx from the CALC menu.

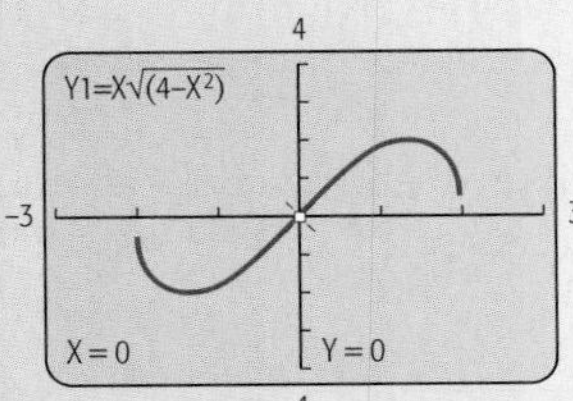

Next we key in the desired x-value or use the arrow keys to move the cursor to the desired point. We then press ENTER to obtain the value of the derivative at the given x-value.

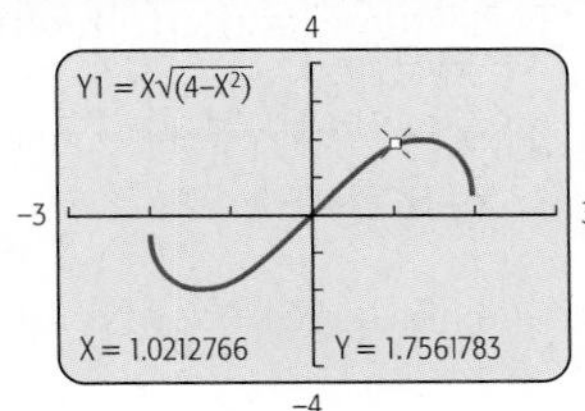

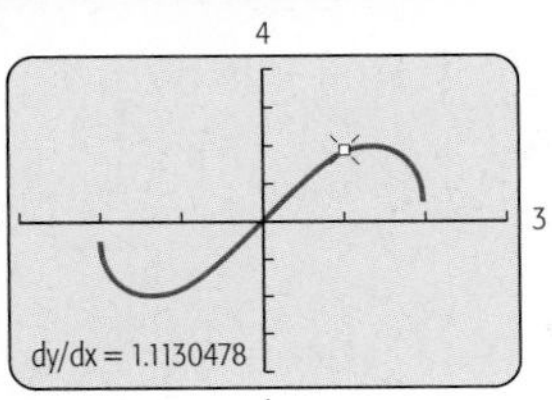

We can also use the Tangent feature from the DRAW menu to draw the tangent line at the point where the derivative was found. Both the line and its equation will appear on the calculator screen.

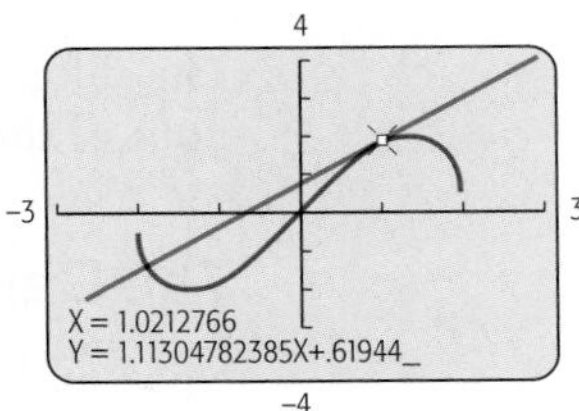

EXERCISES

For each of the following functions, use dy/dx to find the derivative, and then draw the tangent line at the given point. When selecting the viewing window, be sure to include the specified x-values.

1. $f(x) = x(200 - x)$;
$x = 24, x = 138, x = 150, x = 190$

2. $f(x) = x^3 - 6x^2$;
$x = -2, x = 0, x = 2, x = 4, x = 6.3$

3. $f(x) = -4.32 + 1.44x + 3x^2 - x^3$;
$x = -0.5, x = 0.5, x = 2.1$

TECHNOLOGY CONNECTION

Exploratory

Graph the constant function $y = -3$. Then find the derivative of this function at $x = -6$, $x = 0$, and $x = 8$. What do you conclude about the derivative of a constant function?

The Derivative of a Constant Function

Consider the constant function given by $F(x) = c$. Note that the slope at each point on its graph is 0.

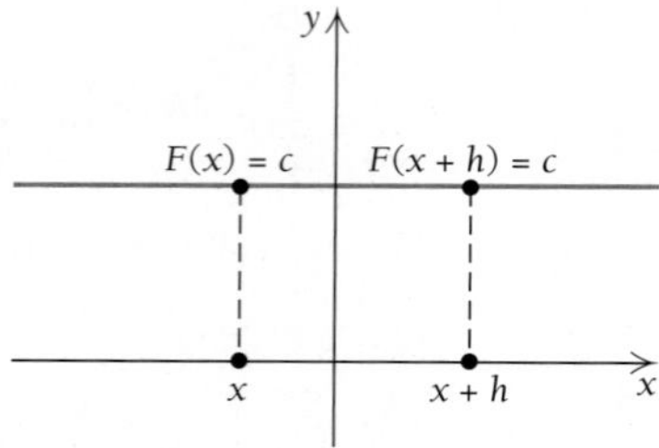

This suggests the following theorem.

THEOREM 2

The derivative of a constant function is 0. That is, $\frac{d}{dx}c = 0$.

Proof. Let F be the function given by $F(x) = c$. Then

$$\frac{F(x+h) - F(x)}{h} = \frac{c - c}{h}$$
$$= \frac{0}{h} = 0.$$

The difference quotient for this function is always 0. Thus, as h approaches 0, the limit of the difference quotient is 0, so $F'(x) = 0$. ✦

The Derivative of a Constant Times a Function

Now let's consider differentiating functions such as

$$f(x) = 5x^2 \quad \text{and} \quad g(x) = -7x^4.$$

Note that we already know how to differentiate x^2 and x^4. Let's again look for a pattern in the results of Section 1.4 and its exercise set.

Function	Derivative
$5x^2$	$10x$
$3x^{-1}$	$-3x^{-2}$
$-7x^2$	$-14x$
$1 \cdot x^3$	$3x^2$

Perhaps you have discovered the following theorem.

THEOREM 3

The derivative of a constant times a function is the constant times the derivative of the function. Using derivative notation, we can write this as

$$\frac{d}{dx}[c \cdot f(x)] = c \cdot \frac{d}{dx} f(x).$$

Proof. Let $F(x) = cf(x)$. Then

$$\begin{aligned} F'(x) &= \lim_{h \to 0} \frac{F(x+h) - F(x)}{h} && \text{Using the definition of derivative} \\ &= \lim_{h \to 0} \frac{cf(x+h) - cf(x)}{h} && \text{Substituting} \\ &= c \cdot \lim_{h \to 0} \left[\frac{f(x+h) - f(x)}{h} \right] && \text{Factoring and using the Limit Principles} \\ &= c \cdot f'(x). && \text{Using the definition of derivative} \end{aligned}$$

✦

Combining this rule with the Power Rule allows us to find many derivatives.

EXAMPLE 3 Find each of the following derivatives:

a) $\frac{d}{dx} 7x^4$; **b)** $\frac{d}{dx}(-9x)$; **c)** $\frac{d}{dx}\left(\frac{1}{5x^2}\right)$.

Solution

a) $\frac{d}{dx} 7x^4 = 7\frac{d}{dx}x^4 = 7 \cdot 4 \cdot x^{4-1} = 28x^3$ With practice, this may be done in one step.

b) $\frac{d}{dx}(-9x) = -9\frac{d}{dx}x = -9 \cdot 1 = -9$

c)
$$\begin{aligned} \frac{d}{dx}\left(\frac{1}{5x^2}\right) = \frac{d}{dx}\left(\frac{1}{5}x^{-2}\right) &= \frac{1}{5} \cdot \frac{d}{dx}x^{-2} \\ &= \frac{1}{5}(-2)x^{-2-1} \\ &= -\frac{2}{5}x^{-3}, \quad \text{or} \quad -\frac{2}{5x^3} \end{aligned}$$

✦

Recall from Section 1.4 that a function's derivative at x is also its instantaneous rate of change at x.

EXAMPLE 4 Life Science: Volume of a Tumor. The spherical volume V of a tumor can be approximated by

$$V(r) = \tfrac{4}{3}\pi r^3,$$

where r is the radius of the tumor, in centimeters.

a) Find the rate of change of the volume with respect to the radius.

b) Find the rate of change of the volume at $r = 1.2$ cm.

Solution

a) $\frac{dV}{dr} = V'(r) = 3 \cdot \frac{4}{3} \cdot \pi r^2 = 4\pi r^2$

(This expression turns out to be equal to the tumor's surface area.)

b) $V'(1.2) = 4\pi(1.2)^2 = 5.76\pi \approx 18 \frac{\text{cm}^3}{\text{cm}} = 18 \text{ cm}^2$

When the radius is 1.2 cm, the volume is changing at the rate of 18 cm^3 for every change of 1 cm in the radius. ✦

The Derivative of a Sum or a Difference

In Exercise 11 of Exercise Set 1.4, you found that the derivative of

$$f(x) = x^2 + x$$

is

$$f'(x) = 2x + 1.$$

Note that the derivative of x^2 is $2x$, the derivative of x is 1, and the sum of these derivatives is $f'(x)$. This illustrates the following.

TECHNOLOGY CONNECTION

Exploratory

Let $y_1 = x(100 - x)$, and $y_2 = x\sqrt{100 - x^2}$. Using the Y-VARS option from the VARS menu, enter Y3 as Y1 + Y2. Find the derivative of each of the three functions at $x = 8$ using the numerical differentiation feature.

Compare your answers. How do you think you can find the derivative of a sum?

THEOREM 4

The Sum–Difference Rule

Sum. The derivative of a sum is the sum of the derivatives:

$$\frac{d}{dx}[f(x) + g(x)] = \frac{d}{dx} f(x) + \frac{d}{dx} g(x).$$

Difference. The derivative of a difference is the difference of the derivatives:

$$\frac{d}{dx}[f(x) - g(x)] = \frac{d}{dx} f(x) - \frac{d}{dx} g(x).$$

Proof. The proof of the Sum Rule relies on the fact that the limit of a sum is the sum of the limits. Let $F(x) = f(x) + g(x)$. Then

$$\lim_{h \to 0} \frac{F(x+h) - F(x)}{h} = \lim_{h \to 0} \frac{[f(x+h) + g(x+h)] - [f(x) + g(x)]}{h}$$

$$= \lim_{h \to 0} \left[\frac{f(x+h) - f(x)}{h} + \frac{g(x+h) - g(x)}{h} \right] = f'(x) + g'(x).$$

To prove the Difference Rule, we note that

$$\frac{d}{dx}(f(x) - g(x)) = \frac{d}{dx}(f(x) + (-1)g(x)) = \frac{d}{dx} f(x) + \frac{d}{dx}(-1)g(x)$$

$$= f'(x) + (-1)\frac{d}{dx} g(x) = f'(x) - g'(x).$$ ✦

Any function that is a sum or difference of several terms can be differentiated term by term.

EXAMPLE 5 Find each of the following derivatives:

a) $\frac{d}{dx}(5x^3 - 7)$; **b)** $\frac{d}{dx}\left(24x - \sqrt{x} + \frac{5}{x}\right)$.

Solution

a)
$$\begin{aligned}\frac{d}{dx}(5x^3 - 7) &= \frac{d}{dx}(5x^3) - \frac{d}{dx}(7)\\ &= 5\frac{d}{dx}x^3 - 0\\ &= 5 \cdot 3x^2\\ &= 15x^2\end{aligned}$$

b)
$$\begin{aligned}\frac{d}{dx}\left(24x - \sqrt{x} + \frac{5}{x}\right) &= \frac{d}{dx}(24x) - \frac{d}{dx}(\sqrt{x}) + \frac{d}{dx}\left(\frac{5}{x}\right)\\ &= 24 \cdot \frac{d}{dx}x - \frac{d}{dx}x^{1/2} + 5 \cdot \frac{d}{dx}x^{-1}\\ &= 24 \cdot 1 - \frac{1}{2}x^{(1/2)-1} + 5(-1)x^{-1-1}\\ &= 24 - \frac{1}{2}x^{-1/2} - 5x^{-2}\\ &= 24 - \frac{1}{2\sqrt{x}} - \frac{5}{x^2}\end{aligned}$$ ✦

A word of caution! The derivative of

$$f(x) + c,$$

a function plus a constant, is just the derivative of the function,

$$f'(x).$$

The derivative of

$$c \cdot f(x),$$

a function times a constant, is the constant times the derivative

$$c \cdot f'(x).$$

That is, the constant is retained for a product, but not for a sum.

Slopes of Tangent Lines

It is important to be able to determine points at which the tangent line to a curve has a certain slope, that is, points at which the derivative attains a certain value.

EXAMPLE 6 Find the points on the graph of $f(x) = -x^3 + 6x^2$ at which the tangent line is horizontal.

Solution The derivative is used to find the slope of a tangent line, and a horizontal tangent line has slope 0. Therefore, we are seeking all x for which $f'(x) = 0$:

$$f'(x) = 0 \qquad \text{Setting the derivative equal to 0}$$

$$\frac{d}{dx}(-x^3 + 6x^2) = 0$$

$$-3x^2 + 12x = 0. \qquad \text{Differentiating}$$

We factor and solve:

$$-3x(x - 4) = 0$$

$$-3x = 0 \quad \text{or} \quad x - 4 = 0$$

$$x = 0 \quad \text{or} \quad x = 4.$$

We are to find the points *on the graph*, so we must determine the second coordinates from the original equation, $f(x) = -x^3 + 6x^2$.

$$f(0) = -0^3 + 6 \cdot 0^2 = 0.$$

$$f(4) = -4^3 + 6 \cdot 4^2 = -64 + 96 = 32.$$

Thus, the points we are seeking are (0, 0) and (4, 32), as shown on the graph.

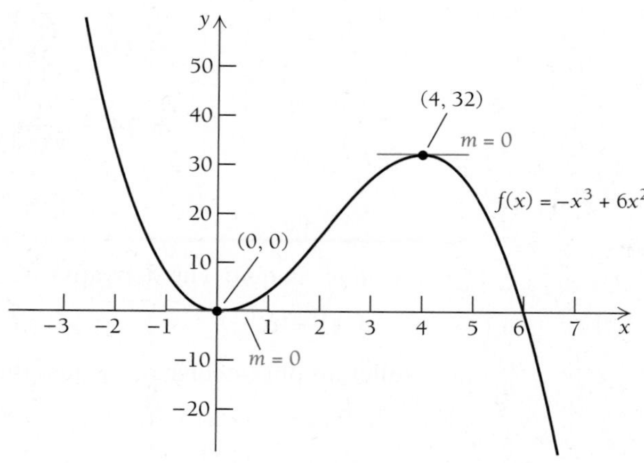

✦

EXAMPLE 7 Find the points on the graph of $f(x) = -x^3 + 6x^2$ at which the tangent line has slope 6.

Solution We want to find values of x for which $f'(x) = 6$. That is, we want to find x such that

$$-3x^2 + 12x = 6. \qquad \text{As in Example 6, note that } \frac{d}{dx}(-x^3 + 6x^2) = -3x^2 + 12x.$$

To solve, we add -6 on both sides and get

$$-3x^2 + 12x - 6 = 0.$$

We can simplify this equation by multiplying by $-\frac{1}{3}$, since each term has a common factor of -3. This gives us

$$x^2 - 4x + 2 = 0.$$

This is a quadratic equation, not readily factorable, so we use the quadratic formula, where $a = 1$, $b = -4$, and $c = 2$:

$$\begin{aligned} x &= \frac{-b \pm \sqrt{b^2 - 4ac}}{2a} \\ &= \frac{-(-4) \pm \sqrt{(-4)^2 - 4 \cdot 1 \cdot 2}}{2 \cdot 1} \\ &= \frac{4 \pm \sqrt{8}}{2} \\ &= \frac{2 \cdot 2 \pm 2\sqrt{2}}{2 \cdot 1} \\ &= \frac{2}{2} \cdot \frac{2 \pm \sqrt{2}}{1} \\ &= 2 \pm \sqrt{2}. \end{aligned}$$

The solutions are $2 + \sqrt{2}$ and $2 - \sqrt{2}$.

We determine the second coordinates from the original equation:

$$\begin{aligned} f(2 + \sqrt{2}) &= -(2 + \sqrt{2})^3 + 6(2 + \sqrt{2})^2 \\ &= -[(2 + \sqrt{2})^2(2 + \sqrt{2})] + 6(4 + 4\sqrt{2} + 2) \\ &= -[(6 + 4\sqrt{2})(2 + \sqrt{2})] + 6(6 + 4\sqrt{2}) \\ &= -[12 + 6\sqrt{2} + 8\sqrt{2} + 8] + 36 + 24\sqrt{2} \\ &= -[20 + 14\sqrt{2}] + 36 + 24\sqrt{2} \\ &= -20 - 14\sqrt{2} + 36 + 24\sqrt{2} \\ &= 16 + 10\sqrt{2}. \end{aligned}$$

It can be similarly shown that

$$f(2 - \sqrt{2}) = 16 - 10\sqrt{2}.$$

Thus, the points we are seeking are $(2 + \sqrt{2}, 16 + 10\sqrt{2})$ and $(2 - \sqrt{2}, 16 - 10\sqrt{2})$, shown on the following graph.

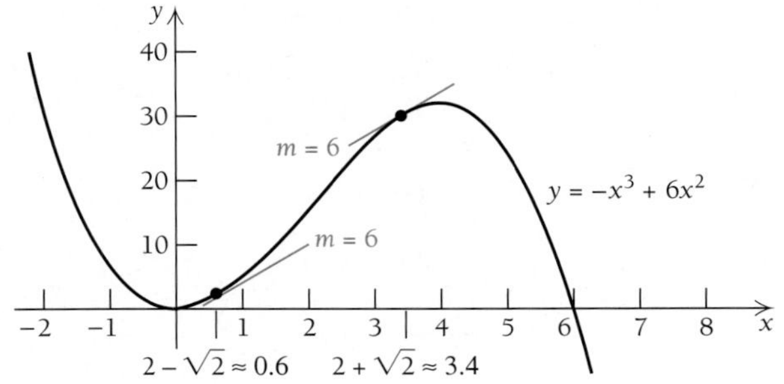

✦

TECHNOLOGY CONNECTION

EXERCISE

1. Graph $y = \frac{1}{3}x^3 - 2x^2 + 4x$, and draw tangent lines at various points. Estimate points at which the tangent line is horizontal. Then use calculus, as in Examples 6 and 7, to find the exact results.

Exercise Set 1.5

Find $\frac{dy}{dx}$.

1. $y = x^7$
2. $y = x^8$
3. $y = -3x$
4. $y = -0.5x$
5. $y = 12$
6. $y = 7$
7. $y = 2x^{15}$
8. $y = 3x^{10}$
9. $y = x^{-6}$
10. $y = x^{-8}$
11. $y = 4x^{-2}$
12. $y = 3x^{-5}$
13. $y = x^3 + 3x^2$
14. $y = x^4 - 7x$
15. $y = 8\sqrt{x}$
16. $y = 4\sqrt{x}$
17. $y = x^{0.9}$
18. $y = x^{0.7}$
19. $y = \frac{1}{2}x^{4/5}$
20. $y = -4.8x^{1/3}$
21. $y = \frac{7}{x^3}$
22. $y = \frac{6}{x^4}$
23. $y = \frac{4x}{5}$
24. $y = \frac{3x}{4}$

Find each derivative.

25. $\frac{d}{dx}\left(\sqrt[4]{x} - \frac{3}{x}\right)$
26. $\frac{d}{dx}\left(\sqrt[5]{x} - \frac{2}{x}\right)$
27. $\frac{d}{dx}\left(\sqrt{x} - \frac{2}{\sqrt{x}}\right)$
28. $\frac{d}{dx}\left(\sqrt[3]{x} + \frac{4}{\sqrt{x}}\right)$
29. $\frac{d}{dx}\left(-2\sqrt[3]{x^5}\right)$
30. $\frac{d}{dx}\left(-\sqrt[4]{x^3}\right)$
31. $\frac{d}{dx}(5x^2 - 7x + 3)$
32. $\frac{d}{dx}(6x^2 - 5x + 9)$

Find $f'(x)$.

33. $f(x) = 0.6x^{1.5}$
34. $f(x) = 0.3x^{1.2}$
35. $f(x) = \frac{2x}{3}$
36. $f(x) = \frac{3x}{4}$
37. $f(x) = \frac{4}{7x^3}$
38. $f(x) = \frac{2}{5x^6}$
39. $f(x) = \frac{5}{x} - x^{2/3}$
40. $f(x) = \frac{4}{x} - x^{3/5}$
41. $f(x) = 4x - 7$
42. $f(x) = 7x - 14$
43. $f(x) = \frac{x^{4/3}}{4}$
44. $f(x) = \frac{x^{3/2}}{3}$
45. $f(x) = -0.01x^2 - 0.5x + 70$
46. $f(x) = -0.01x^2 + 0.4x + 50$

Find y'.

47. $y = 3x^{-2/3} + x^{3/4} + x^{6/5} + \frac{8}{x^3}$
48. $y = x^{-3/4} - 3x^{2/3} + x^{5/4} + \frac{2}{x^4}$
49. $y = \frac{2}{x} - \frac{x}{2}$
50. $y = \frac{x}{7} + \frac{7}{x}$
51. If $f(x) = x^2 + 4x - 5$, find $f'(10)$.
52. If $f(x) = \sqrt{x}$, find $f'(4)$.
53. If $y = \frac{4}{x^2}$, find $\left.\frac{dy}{dx}\right|_{x=-2}$
54. If $y = x + \frac{2}{x^3}$, find $\left.\frac{dy}{dx}\right|_{x=1}$
55. Find an equation of the tangent line to the graph of $f(x) = x^3 - 2x + 1$
    a) at $(2, 5)$; **b)** at $(-1, 2)$; **c)** at $(0, 1)$.
56. Find an equation of the tangent line to the graph of $f(x) = x^2 - \sqrt{x}$
    a) at $(1, 0)$; **b)** at $(4, 14)$; **c)** at $(9, 78)$.

For each function, find the points on the graph at which the tangent line is horizontal. If none exist, state that fact.

57. $y = x^2 - 3$
58. $y = -x^2 + 4$
59. $y = -x^3 + 1$
60. $y = x^3 - 2$
61. $y = 3x^2 - 5x + 4$
62. $y = 5x^2 - 3x + 8$
63. $y = -0.01x^2 - 0.5x + 70$
64. $y = -0.01x^2 + 0.4x + 50$
65. $y = 2x + 4$
66. $y = -2x + 5$
67. $y = 4$
68. $y = -3$
69. $y = -x^3 + x^2 + 5x - 1$
70. $y = -\frac{1}{3}x^3 + 6x^2 - 11x - 50$
71. $y = \frac{1}{3}x^3 - 3x + 2$
72. $y = x^3 - 6x + 1$
73. $f(x) = \frac{1}{3}x^3 + \frac{1}{2}x^2 - 2$
74. $f(x) = \frac{1}{3}x^3 - 3x^2 + 9x - 9$

For each function, find the points on the graph at which the tangent line has slope 1.

75. $y = 20x - x^2$

76. $y = 6x - x^2$

77. $y = -0.025x^2 + 4x$

78. $y = -0.01x^2 + 2x$

79. $y = \frac{1}{3}x^3 + 2x^2 + 2x$

80. $y = \frac{1}{3}x^3 - x^2 - 4x + 1$

APPLICATIONS

Life and Natural Sciences

81. Healing wound. The circular area A, in square centimeters, of a healing wound is approximated by

$$A(r) = 3.14r^2,$$

where r is the wound's radius, in centimeters.

a) Find the rate of change of the area with respect to the radius.

TW **b)** Explain the meaning of your answer to part (a).

82. Healing wound. The circumference C, in centimeters, of a healing wound is approximated by

$$C(r) = 6.28r,$$

where r is the wound's radius, in centimeters.

a) Find the rate of change of the circumference with respect to the radius.

TW **b)** Explain the meaning of your answer to part (a).

83. Growth of a baby. The median weight of a boy whose age is between 0 and 36 months can be approximated by the function

$$w(t) = 8.15 + 1.82t - 0.0596t^2 + 0.000758t^3,$$

where t is measured in months and w is measured in pounds.

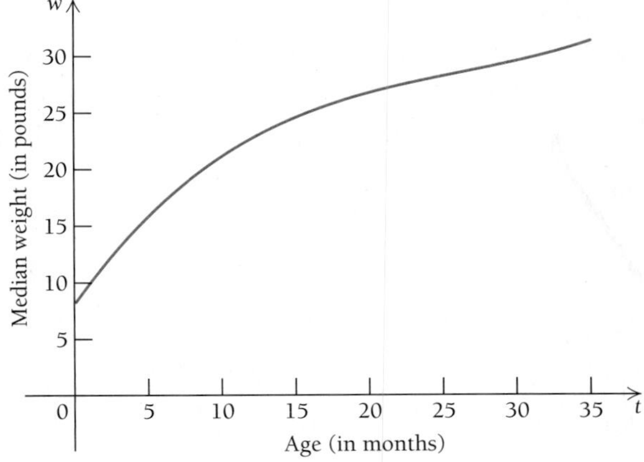

(*Source*: Centers for Disease Control. Developed by the National Center for Health Statistics in collaboration with the National Center for Chronic Disease Prevention and Health Promotion, 2000.)

Use this approximation to find the following for a boy with median weight:

a) The rate of change of weight with respect to time.

b) The weight of the baby at age 10 months.

c) The rate of change of the baby's weight with respect to time at age 10 months.

84. Temperature during an illness. The temperature T of a person during an illness is given by

$$T(t) = -0.1t^2 + 1.2t + 98.6,$$

where T is the temperature, in degrees Fahrenheit, at time t, in days.

a) Find the rate of change of the temperature with respect to time.

b) Find the temperature at $t = 1.5$ days.

c) Find the rate of change at $t = 1.5$ days.

85. Heart rate. The equation

$$R(v) = \frac{6000}{v}$$

can be used to determine the heart rate, R, of a person whose heart pumps 6000 milliliters (mL) of blood per minute and v milliliters of blood per beat. (*Source*: *Mathematics Teacher*, Vol. 99, No. 4, November 2005.)

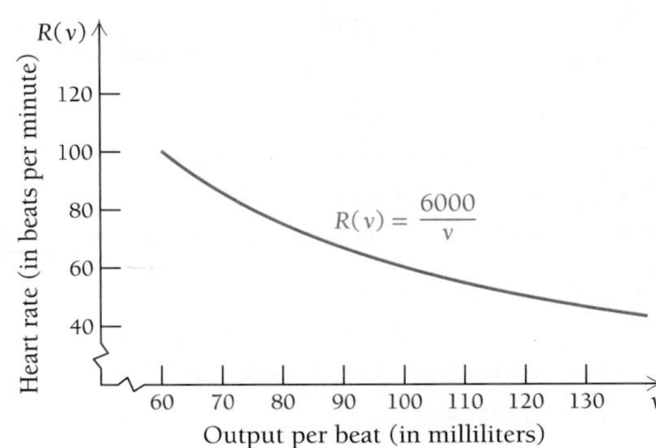

a) Find the rate of change of heart rate with respect to v, the output per beat.

b) Find the heart rate at $v = 80$ mL per beat.

c) Find the rate of change at $v = 80$ mL per beat.

86. Blood flow resistance. The equation

$$S(r) = \frac{1}{r^4}$$

can be used to determine the resistance to blood flow, S, of a blood vessel that has radius r, in millimeters (mm). (*Source*: *Mathematics Teacher*, Vol. 99, No. 4, November 2005.)

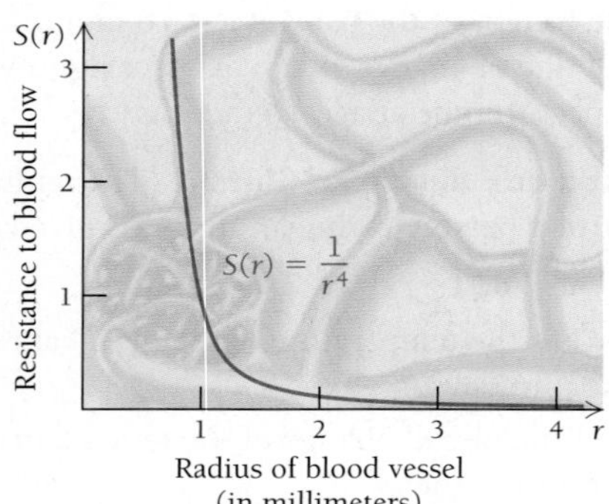

a) Find the rate of change of resistance with respect to r, the radius of the blood vessel.
b) Find the resistance at $r = 1.2$ mm.
c) Find the rate of change of S with respect to r when $r = 0.8$ mm.

Social Sciences

87. Population growth rate. The population of a city grows from an initial size of 100,000 to a size P given by

$$P(t) = 100{,}000 + 2000t^2,$$

where t is in years.

a) Find the **growth rate,** dP/dt.
b) Find the population after 10 yr.
c) Find the growth rate at $t = 10$.
TW **d)** Explain the meaning of your answer to part (c).

88. Median age of women at first marriage. The median age of women at first marriage can be approximated by the linear function

$$A(t) = 0.08t + 19.7,$$

where $A(t)$ is the median age of women marrying for the first time at t years after 1950.

a) Find the rate of change of the median age A with respect to time t.
TW **b)** Explain the meaning of your answer to part (a).

General Interest

89. View to the horizon. The view V, or distance in miles, that one can see to the horizon from a height h, in feet, is given by

$$V = 1.22\sqrt{h}.$$

a) Find the rate of change of V with respect to h.
b) How far can one see to the horizon from an airplane window at a height of 40,000 ft?
c) Find the rate of change at $h = 40{,}000$.
TW **d)** Explain the meaning of your answers to parts (a) and (c).

90. Baseball ticket prices. The average price, in dollars, of a ticket for a Major League baseball game x years after 1990 can be estimated by

$$p(x) = 9.41 - 0.19x + 0.09x^2.$$

a) Find the rate of change of the average ticket price with respect to the year, dp/dx.
b) What is the average ticket price in 2007?
c) What is the rate of change of the average ticket price in 2007?

SYNTHESIS

For each function, find the interval(s) for which $f'(x)$ is positive.

91. $f(x) = x^2 - 4x + 1$

92. $f(x) = x^2 + 7x + 2$

93. $f(x) = \frac{1}{3}x^3 - x^2 - 3x + 5$

94. Find the points on the graph of

$$y = x^4 - \tfrac{4}{3}x^2 - 4$$

at which the tangent line is horizontal.

95. Find the points on the graph of

$$y = 2x^6 - x^4 - 2$$

at which the tangent line is horizontal.

Find dy/dx. Each function can be differentiated using the rules developed in this section, but some algebra may be required beforehand.

96. $y = (x + 3)(x - 2)$

97. $y = (x - 1)(x + 1)$

98. $y = \dfrac{x^5 - x^3}{x^2}$

99. $y = \dfrac{5x^2 - 8x + 3}{8}$

100. $y = \dfrac{x^5 + x}{x^2}$

101. $y = \dfrac{x^5 - 3x^4 + 2x + 4}{x^2}$

102. $y = (-4x)^3$

103. $y = \sqrt{7x}$

104. $y = \sqrt[3]{8x}$

105. $y = (x - 3)^2$

106. $y = \left(\sqrt{x} - \dfrac{1}{\sqrt{x}}\right)^2$

107. $y = (\sqrt{x} + \sqrt[3]{x})^2$

108. $y = (x + 1)^3$

109. Use Theorem 1 to prove that the derivative of 1 is 0.

TW **110.** When might Leibniz notation be more convenient than function notation?

TW **111.** Write a short biographical paper on Leibniz and/or Newton. Emphasize the contributions each man made to many areas of science and society.

TECHNOLOGY CONNECTION

Graph each of the following. Then estimate the x-values at which tangent lines are horizontal.

112. $f(x) = x^4 - 3x^2 + 1$

113. $f(x) = 1.6x^3 - 2.3x - 3.7$

114. $f(x) = 10.2x^4 - 6.9x^3$

115. $f(x) = \dfrac{5x^2 + 8x - 3}{3x^2 + 2}$

For each of the following, graph f and f′ and then determine f′(1). Use the nDeriv *option from the* MATH *menu, if available* (*see p. 144*)*. For Exercises 120 and 121,* nDeriv *must be used.*

116. $f(x) = 20x^3 - 3x^5$

117. $f(x) = x^4 - 3x^2 + 1$

118. $f(x) = x^3 - 2x - 2$

119. $f(x) = x^4 - x^3$

120. $f(x) = \dfrac{4x}{x^2 + 1}$

121. $f(x) = \dfrac{5x^2 + 8x - 3}{3x^2 + 2}$

1.6 Differentiation Techniques: The Product and Quotient Rules

OBJECTIVES

- Differentiate using the Product and Quotient Rules.
- Use the Quotient Rule to differentiate the average cost, revenue, and profit functions.

The Product Rule

The derivative of a sum is the sum of the derivatives, but the derivative of a product is *not* the product of the derivatives. To see this, consider x^2 and x^5. The product is x^7, and the derivative of this product is $7x^6$. The individual derivatives are $2x$ and $5x^4$, and the product of these derivatives is $10x^5$, which is not $7x^6$.

The following is the rule for finding the derivative of a product.

THEOREM 5

The Product Rule

Let $F(x) = f(x) \cdot g(x)$. Then

$$F'(x) = \frac{d}{dx}[f(x) \cdot g(x)] = f(x) \cdot \left[\frac{d}{dx} g(x)\right] + g(x) \cdot \left[\frac{d}{dx} f(x)\right].$$

The derivative of a product is the first factor times the derivative of the second factor, plus the second factor times the derivative of the first factor.

The proof of the Product Rule is outlined in Exercise 123 at the end of this section.

Let's check the Product Rule for $x^2 \cdot x^5$. There are five steps.

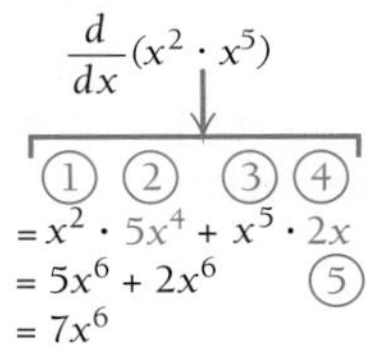

1. Write down the first factor.
2. Multiply it by the derivative of the second factor.
3. Write down the second factor.
4. Multiply it by the derivative of the first factor.
5. Add the result of steps (1) and (2) to the result of steps (3) and (4).

Usually we try to write the results in simplified form. In Examples 1 and 2, we do not simplify in order to better emphasize the steps being performed.

EXAMPLE 1 Find $\frac{d}{dx}[(x^4 - 2x^3 - 7)(3x^2 - 5x)]$. Do not simplify.

Solution We have

$$\frac{d}{dx}(x^4 - 2x^3 - 7)(3x^2 - 5x) = (x^4 - 2x^3 - 7)(6x - 5) + (3x^2 - 5x)(4x^3 - 6x^2).$$

Note that we could have multiplied the polynomials and then differentiated, thereby avoiding the use of the Product Rule, but this would have been more work. ✦

EXAMPLE 2 For $F(x) = (x^2 + 4x - 11)(7x^3 - \sqrt{x})$, find $F'(x)$. Do not simplify.

Solution We rewrite this as

$$F(x) = (x^2 + 4x - 11)(7x^3 - x^{1/2}).$$

Then, using the Product Rule, we have

$$F'(x) = (x^2 + 4x - 11)\left(21x^2 - \tfrac{1}{2}x^{-1/2}\right) + (7x^3 - x^{1/2})(2x + 4).$$ ✦

The Quotient Rule

The derivative of a quotient is *not* the quotient of the derivatives. To see why, consider x^5 and x^2. The quotient x^5/x^2 is x^3, and the derivative of this quotient is $3x^2$. The individual derivatives are $5x^4$ and $2x$, and the quotient of these derivatives, $5x^4/(2x)$, is $(5/2)x^3$, which is not $3x^2$.

The rule for differentiating quotients is as follows.

THEOREM 6

The Quotient Rule

If $Q(x) = \dfrac{N(x)}{D(x)}$, then $Q'(x) = \dfrac{D(x) \cdot N'(x) - N(x) \cdot D'(x)}{[D(x)]^2}$.

The derivative of a quotient is the denominator times the derivative of the numerator, minus the numerator times the derivative of the denominator, all divided by the square of the denominator.

(If we think of the function in the numerator as the first function and the function in the denominator as the second function, then we can reword the Quotient Rule as "the derivative of a quotient is the second function times the derivative of the first function minus the first function times the derivative of the second function, all divided by the square of the second function.")

A proof of this result is outlined in Exercise 93 of Section 1.7 (on p. 177).

The Quotient Rule is illustrated below.

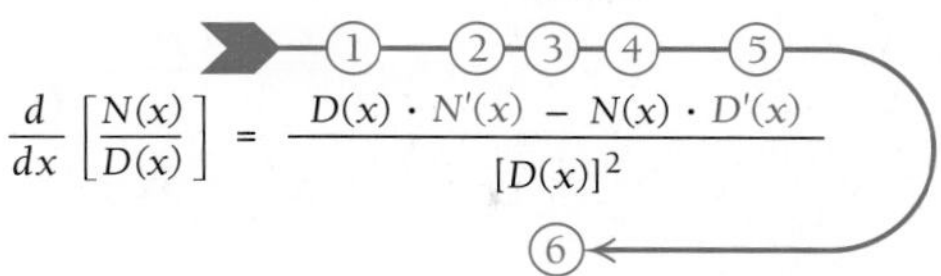

1. Write down the denominator.
2. Multiply the denominator by the derivative of the numerator.
3. Write a minus sign.
4. Write down the numerator.
5. Multiply it by the derivative of the denominator.
6. Divide by the square of the denominator.

EXAMPLE 3 For $Q(x) = x^5/x^2$, find $Q'(x)$.

Solution We have already seen that $x^5/x^2 = x^3$ and $\frac{d}{dx}x^3 = 3x^2$, but we wish to practice using the Quotient Rule:

$$Q'(x) = \frac{x^2 \cdot 5x^4 - x^5 \cdot 2x}{(x^2)^2}$$

$$= \frac{5x^6 - 2x^6}{x^4} = \frac{3x^6}{x^4} = 3x^2. \qquad \text{This checks with the result above.}$$

✦

EXAMPLE 4 Differentiate: $f(x) = \dfrac{1 + x^2}{x^3}$.

Solution

$$f'(x) = \frac{x^3 \cdot 2x - (1 + x^2) \cdot 3x^2}{(x^3)^2} \qquad \text{Using the Quotient Rule}$$

$$= \frac{2x^4 - 3x^2 - 3x^4}{x^6} = \frac{-x^4 - 3x^2}{x^6}$$

$$= \frac{x^2(-x^2 - 3)}{x^2 \cdot x^4} \qquad \text{Factoring}$$

$$= \frac{-x^2 - 3}{x^4} \qquad \text{Removing a factor equal to 1: } \frac{x^2}{x^2} = 1$$

✦

EXAMPLE 5 Differentiate: $f(x) = \dfrac{x^2 - 3x}{x - 1}$.

Solution We have

$$f'(x) = \frac{(x-1)(2x-3) - (x^2-3x)\cdot 1}{(x-1)^2} \quad \text{Using the Quotient Rule}$$

$$= \frac{2x^2 - 5x + 3 - x^2 + 3x}{(x-1)^2} \quad \text{Using the distributive law}$$

$$= \frac{x^2 - 2x + 3}{(x-1)^2}. \quad \text{Simplifying}$$

It is not necessary to multiply out $(x - 1)^2$. ✦

TECHNOLOGY CONNECTION

Checking Derivatives Graphically

To check Example 5, we first enter the function:

$$y_1 = \frac{x^2 - 3x}{x - 1}.$$

Then we enter the possible derivative:

$$y_2 = \frac{x^2 - 2x + 3}{(x-1)^2}.$$

For the third function, we enter

$$y_3 = \text{nDeriv}(y_1, x, x).$$

Next, we deselect y_1 and graph y_2 and y_3. We use different graph styles and the Sequential mode to see each graph as it appears on the screen.

$y_2 = \dfrac{x^2 - 2x + 3}{(x-1)^2}$, $y_3 = \text{nDeriv}(y_1, x, x)$

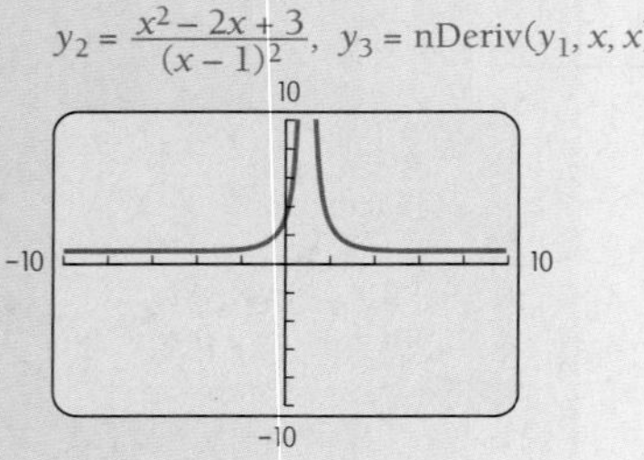

Since the graphs appear to coincide, it appears that $y_2 = y_3$ and we have a check. This is considered a partial check, however, because the graphs might not coincide at a point not in the viewing window.

We can also use a table to check that $y_2 = y_3$.

X	Y2	Y3
5.97	1.081	1.081
5.98	1.0806	1.0806
5.99	1.0803	1.0803
6	1.08	1.08
6.01	1.0797	1.0797
6.02	1.0794	1.0794
6.03	1.079	1.079

X = 5.97

You should verify that had we miscalculated the derivative as, say, $y_2 = (x^2 - 2x - 8)/(x - 1)^2$, neither the tables nor the graphs of y_2 and y_3 would agree.

$y_2 = \dfrac{x^2 - 2x - 8}{(x-1)^2}$

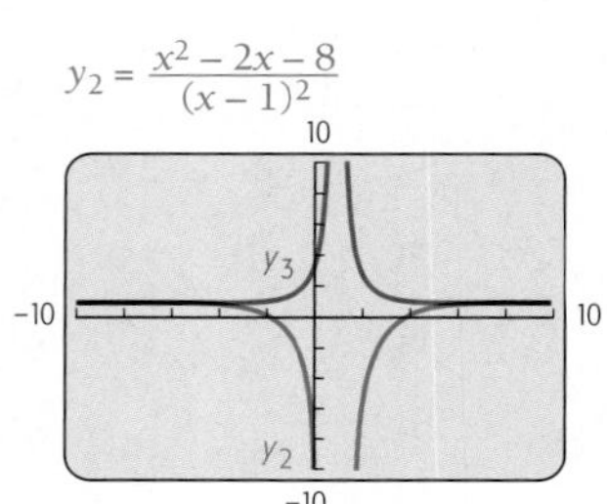

Checking Derivatives Graphically (continued)

EXERCISES

1. For the function

$$f(x) = \frac{x^2 - 4x}{x + 2},$$

use graphs and tables to determine which of the following seems to be the correct derivative.

a) $f'(x) = \dfrac{-x^2 - 4x - 8}{(x + 2)^2}$

b) $f'(x) = \dfrac{x^2 - 4x + 8}{(x + 2)^2}$

c) $f'(x) = \dfrac{x^2 + 4x - 8}{(x + 2)^2}$

2–5. Check the results of Examples 1–4 in this section.

Application of the Quotient Rule

The total cost, total revenue, and total profit functions, discussed in Section R.4, pertain to the accumulated cost, revenue, and profit when x items are produced. Because of economy of scale and other factors, it is common for the cost, revenue (price), and profit for, say, the 10th item to differ from those for the 1000th item. For this reason, a business is often interested in the *average* cost, revenue, and profit associated with the production and sale of x items.

DEFINITION

If $C(x)$ is the cost of producing x items, then the **average cost** of producing x items is $\dfrac{C(x)}{x}$;

if $R(x)$ is the revenue from the sale of x items, then the **average revenue** from selling x items is $\dfrac{R(x)}{x}$;

if $P(x)$ is the profit from the sale of x items, then the **average profit** from selling x items is $\dfrac{P(x)}{x}$.

EXAMPLE 6 Business. Paulsen's Greenhouse finds that the cost, in dollars, of growing x hundred geraniums is given by

$$C(x) = 200 + 100\sqrt[4]{x}.$$

If the revenue from the sale of x hundred geraniums is given by

$$R(x) = 120 + 90\sqrt{x},$$

find each of the following.

a) The average cost, the average revenue, and the average profit when x hundred geraniums are grown and sold.

b) The rate at which average profit is changing when 300 geraniums are being grown and sold.

Solution

a) We let A_C, A_R, and A_P represent average cost, average revenue, and average profit, respectively. Then

$$A_C(x) = \frac{C(x)}{x} = \frac{200 + 100\sqrt[4]{x}}{x};$$

$$A_R(x) = \frac{R(x)}{x} = \frac{120 + 90\sqrt{x}}{x};$$

$$A_P(x) = \frac{P(x)}{x} = \frac{R(x) - C(x)}{x} = \frac{-80 + 90\sqrt{x} - 100\sqrt[4]{x}}{x}.$$

b) To find the rate at which average profit is changing when 300 geraniums are being grown, we calculate $A_P'(300)$:

$$A_P'(x) = \frac{d}{dx}\left[\frac{-80 + 90x^{1/2} - 100x^{1/4}}{x}\right]$$

$$= \frac{x(\frac{1}{2} \cdot 90x^{1/2-1} - \frac{1}{4} \cdot 100x^{1/4-1}) - (-80 + 90x^{1/2} - 100x^{1/4}) \cdot 1}{x^2}$$

$$= \frac{45x^{1/2} - 25x^{1/4} + 80 - 90x^{1/2} + 100x^{1/4}}{x^2} = \frac{75x^{1/4} - 45x^{1/2} + 80}{x^2}$$

$$A_P'(300) = \frac{75\sqrt[4]{300} - 45\sqrt{300} + 80}{300^2} \approx -0.0043.$$

When 300 geraniums are being grown, the average profit is decreasing at a rate of about 0.4¢ per plant. ✦

TECHNOLOGY CONNECTION

Using Y-VARS

One way to save keystrokes on most calculators is to use the Y-VARS option on the VARS menu.

To check Example 6, we let $y_1 = 200 + 100x^{0.25}$ and $y_2 = 120 + 90x^{0.5}$. To express the profit function as y_3, we press Y= and move the cursor to enter y_3. Next we press VARS and select Y-VARS and then FUNCTION. From the FUNCTION menu we select Y2, which then appears on the Y= screen. After pressing −, we repeat the procedure to get Y1 on the Y= screen.

```
Plot 1   Plot 2   Plot 3
\Y1=200+100X^0.25
\Y2=120+90X^0.5
\Y3=Y2-Y1
\Y4=
\Y5=
\Y6=
\Y7=
```

EXERCISES

1. Use the Y-VARS option to enter $y_4 = y_1/x$, $y_5 = y_2/x$, and $y_6 = y_3/x$, and explain what each of the functions represents.
2. Use nDeriv from the MATH menu or dy/dx from the CALC menu to check part (b) of Example 6.

Exercise Set 1.6

Differentiate two ways: first, by using the Product Rule; then, by multiplying the expressions before differentiating. Compare your results as a check.

1. $y = x^5 \cdot x^6$

2. $y = x^9 \cdot x^4$

3. $f(x) = (2x + 5)(3x - 4)$

4. $g(x) = (3x - 2)(4x + 1)$

5. $G(x) = 4x^2(x^3 + 5x)$

6. $F(x) = 3x^4(x^2 - 4x)$

7. $y = \left(3\sqrt{x} + 2\right)x^2$

8. $y = \left(4\sqrt{x} + 3\right)x^3$

9. $g(x) = (4x - 3)(2x^2 + 3x + 5)$

10. $f(x) = (2x + 5)(3x^2 - 4x + 1)$

11. $F(t) = \left(\sqrt{t} + 2\right)\left(3t - 4\sqrt{t} + 7\right)$

12. $G(t) = \left(2t + 3\sqrt{t} + 5\right)\left(\sqrt{t} + 4\right)$

Differentiate two ways: first, by using the Quotient Rule; then, by dividing the expressions before differentiating. Compare your results as a check.

13. $y = \dfrac{x^7}{x^3}$

14. $y = \dfrac{x^6}{x^4}$

15. $f(x) = \dfrac{2x^5 + x^2}{x}$

16. $g(x) = \dfrac{3x^7 - x^3}{x}$

17. $G(x) = \dfrac{8x^3 - 1}{2x - 1}$

18. $F(x) = \dfrac{x^3 + 27}{x + 3}$

19. $y = \dfrac{t^2 - 16}{t + 4}$

20. $y = \dfrac{t^2 - 25}{t - 5}$

Differentiate each function.

21. $f(x) = (3x^2 - 2x + 5)(4x^2 + 3x - 1)$

22. $g(x) = (5x^2 + 4x - 3)(2x^2 - 3x + 1)$

23. $y = \dfrac{5x^2 - 1}{2x^3 + 3}$

24. $y = \dfrac{3x^4 + 2x}{x^3 - 1}$

25. $G(x) = \left(8x + \sqrt{x}\right)\left(5x^2 + 3\right)$

26. $F(x) = (-3x^2 + 4x)\left(7\sqrt{x} + 1\right)$

27. $g(t) = \dfrac{t}{3 - t} + 5t^3$

28. $f(t) = \dfrac{t}{5 + 2t} - 2t^4$

29. $F(x) = (x + 3)^2$
[*Hint:* $(x + 3)^2 = (x + 3)(x + 3)$.]

30. $G(x) = (5x - 4)^2$

31. $y = (x^3 - 4x)^2$

32. $y = (3x^2 - 4x + 5)^2$

33. $g(x) = 5x^{-3}(x^4 - 5x^3 + 10x - 2)$

34. $f(x) = 6x^{-4}(6x^3 + 10x^2 - 8x + 3)$

35. $F(t) = \left(t + \dfrac{2}{t}\right)(t^2 - 3)$

36. $G(t) = (3t^5 - t^2)\left(t - \dfrac{5}{t}\right)$

37. $y = \dfrac{x^2 + 1}{x^3 - 1} - 5x^2$

38. $y = \dfrac{x^3 - 1}{x^2 + 1} + 4x^3$

39. $y = \dfrac{\sqrt[3]{x} - 7}{\sqrt{x} + 3}$

40. $y = \dfrac{\sqrt{x} + 4}{\sqrt[3]{x} - 5}$

41. $f(x) = \dfrac{x}{x^{-1} + 1}$

42. $f(x) = \dfrac{x^{-1}}{x + x^{-1}}$

43. $F(t) = \dfrac{1}{t - 4}$

44. $G(t) = \dfrac{1}{t + 2}$

45. $f(x) = \dfrac{3x^2 + 2x}{x^2 + 1}$

46. $f(x) = \dfrac{3x^2 - 5x}{x^2 - 1}$

47. $g(t) = \dfrac{-t^2 + 3t + 5}{t^2 - 2t + 4}$

48. $f(t) = \dfrac{3t^2 + 2t - 1}{-t^2 + 4t + 1}$

49–96. Use a graphing calculator to check the results of Exercises 1–48.

97. Find an equation of the tangent line to the graph of $y = 8/(x^2 + 4)$ at **(a)** $(0, 2)$; **(b)** $(-2, 1)$.

98. Find an equation of the tangent line to the graph of $y = \sqrt{x}/(x + 1)$ at **(a)** $x = 1$; **(b)** $x = \frac{1}{4}$.

99. Find an equation of the tangent line to the graph of $y = x^2 + 3/(x - 1)$ at **(a)** $x = 2$; **(b)** $x = 3$.

100. Find an equation of the tangent line to the graph of $y = 4x/(1 + x^2)$ at **(a)** $(0, 0)$; **(b)** $(-1, -2)$.

APPLICATIONS

Business and Economics

101. Average cost. Summertime Fabrics finds that the cost, in dollars, of producing x jackets is given by $C(x) = 950 + 15\sqrt{x}$. Find the rate at which the average cost is changing when 400 jackets have been produced.

102. Average cost. Tongue-Tied Sauces, Inc., finds that the cost, in dollars, of producing x bottles of barbecue sauce is given by $C(x) = 375 + 0.75x^{3/4}$. Find the rate at which the average cost is changing when 81 bottles of barbecue sauce have been produced.

103. Average revenue. Summertime Fabrics finds that the revenue, in dollars, from the sale of x jackets is given by $R(x) = 85\sqrt{x}$. Find the rate at which average revenue is changing when 400 jackets have been produced.

104. Average revenue. Tongue-Tied Sauces, Inc., finds that the revenue, in dollars, from the sale of x bottles of barbecue sauce is given by $R(x) = 7.5x^{0.7}$. Find the rate at which average revenue is changing when 81 bottles of barbecue sauce have been produced.

105. Average profit. Use the information in Exercises 101 and 103 to determine the rate at which Summertime Fabrics' average profit per jacket is changing when 400 jackets have been produced and sold.

106. Average profit. Use the information in Exercises 102 and 104 to determine the rate at which Tongue-Tied Sauces' average profit per bottle of barbecue sauce is changing when 81 bottles have been produced and sold.

107. Average profit. Sparkle Pottery has determined that the cost, in dollars, of producing x vases is given by

$$C(x) = 4300 + 2.1x^{0.6}.$$

If the revenue from the sale of x vases is given by $R(x) = 65x^{0.9}$, find the rate at which the average profit per vase is changing when 50 vases have been made and sold.

108. Average profit. Cruzin' Boards has found that the cost, in dollars, of producing x skateboards is given by

$$C(x) = 900 + 18x^{0.7}.$$

If the revenue from the sale of x skateboards is given by $R(x) = 75x^{0.8}$, find the rate at which the average profit per skateboard is changing when 20 skateboards have been built and sold.

109. Gross domestic product. The U.S. gross domestic product (in billions of dollars) can be approximated using the function

$$P(t) = 567 + t(36t^{0.6} - 104),$$

where t is the number of the years since 1960.

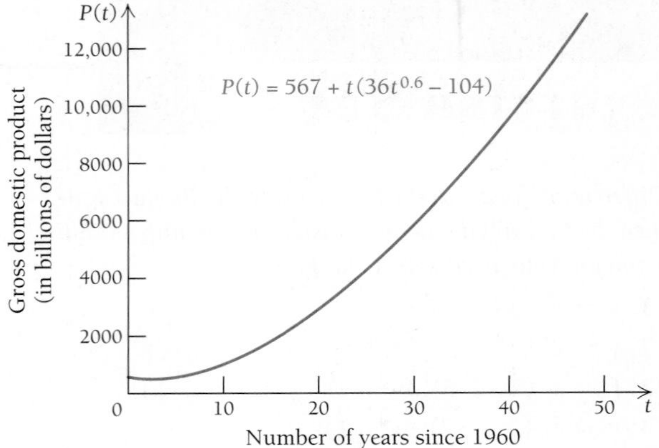

(*Source*: U.S. Bureau of Economic Analysis.)

a) Find $P'(t)$.
b) Find $P'(45)$.
TW c) In words, explain what $P'(45)$ means.

Social Sciences

110. Population growth. The population P, in thousands, of a small city is given by

$$P(t) = \frac{500t}{2t^2 + 9},$$

where t is the time, in months.

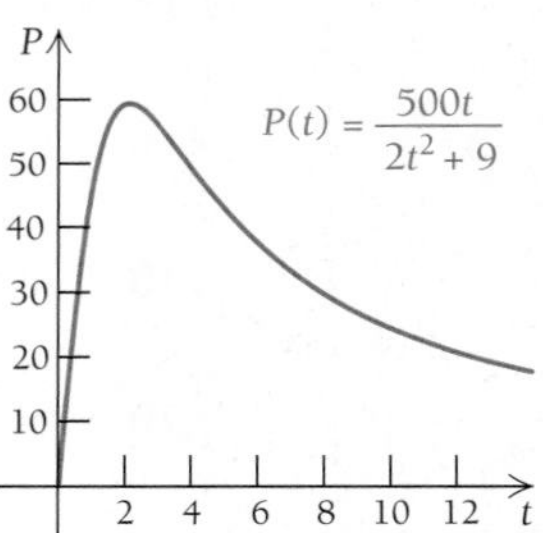

a) Find the growth rate.
b) Find the population after 12 months.
c) Find the growth rate at $t = 12$ months.

Life and Physical Sciences

111. Temperature during an illness. The temperature T of a person during an illness is given by

$$T(t) = \frac{4t}{t^2 + 1} + 98.6,$$

where T is the temperature, in degrees Fahrenheit, at time t, in hours.

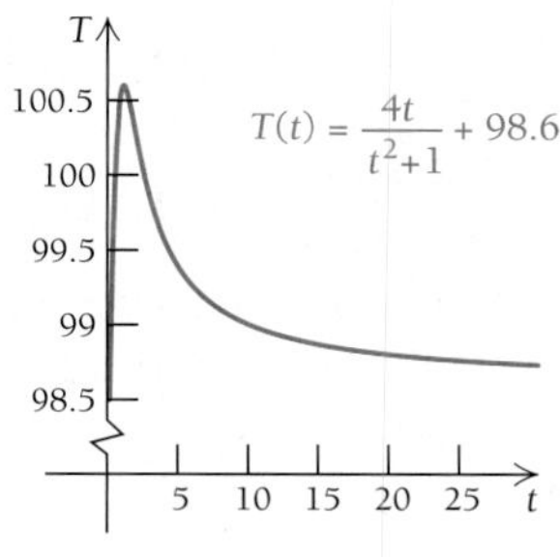

a) Find the rate of change of the temperature with respect to time.

b) Find the temperature at $t = 2$ hr.

c) Find the rate of change of the temperature at $t = 2$ hr.

SYNTHESIS

Differentiate each function.

112. $f(x) = \dfrac{7 - \dfrac{3}{2x}}{\dfrac{4}{x^2} + 5}$ (*Hint:* Simplify before differentiating.)

113. $y(t) = 5t(t - 1)(2t + 3)$

114. $f(x) = x(3x^3 + 6x - 2)(3x^4 + 7)$

115. $g(x) = (x^3 - 8) \cdot \dfrac{x^2 + 1}{x^2 - 1}$

116. $f(t) = (t^5 + 3) \cdot \dfrac{t^3 - 1}{t^3 + 1}$

117. $f(x) = \dfrac{(x - 1)(x^2 + x + 1)}{x^4 - 3x^3 - 5}$

118. Let $f(x) = \dfrac{x}{x + 1}$ and $g(x) = \dfrac{-1}{x + 1}$.

a) Compute $f'(x)$.

b) Compute $g'(x)$.

TW c) What can you conclude about f and g on the basis of your results from parts (a) and (b)?

119. Let $f(x) = \dfrac{x^2}{x^2 - 1}$ and $g(x) = \dfrac{1}{x^2 - 1}$.

a) Compute $f'(x)$.

b) Compute $g'(x)$.

TW c) What can you conclude about the graphs of f and g on the basis of your results from parts (a) and (b)?

TW **120.** Write a rule for finding the derivative of $f(x) \cdot g(x) \cdot h(x)$. Describe the rule in words.

TW **121.** Is the derivative of the reciprocal of $f(x)$ the reciprocal of the derivative of $f'(x)$? Why or why not?

122. Sensitivity. The reaction R of the body to a dose Q of medication is often represented by the general function

$$R(Q) = Q^2\left(\frac{k}{2} - \frac{Q}{3}\right),$$

where k is a constant and R is in millimeters of mercury (mmHg) if the reaction is a change in blood pressure or in degrees Fahrenheit (°F) if the reaction is a change in temperature. The rate of change dR/dQ is defined to be the body's *sensitivity* to the medication.

a) Find a formula for the sensitivity.

TW b) Explain the meaning of your answer to part (a).

123. A proof of the Product Rule appears below. Provide a justification for each step.

a) $\dfrac{d}{dx}[f(x) \cdot g(x)] = \lim\limits_{h \to 0} \dfrac{f(x + h)g(x + h) - f(x)g(x)}{h}$

b) $= \lim\limits_{h \to 0} \dfrac{f(x + h)g(x + h) - f(x + h)g(x) + f(x + h)g(x) - f(x)g(x)}{h}$

c) $= \lim\limits_{h \to 0} \dfrac{f(x + h)g(x + h) - f(x + h)g(x)}{h} + \lim\limits_{h \to 0} \dfrac{f(x + h)g(x) - f(x)g(x)}{h}$

d) $= \lim\limits_{h \to 0} \left[f(x + h) \cdot \dfrac{g(x + h) - g(x)}{h}\right] + \lim\limits_{h \to 0} \left[g(x) \cdot \dfrac{f(x + h) - f(x)}{h}\right]$

e) $= f(x) \cdot \lim\limits_{h \to 0} \dfrac{g(x + h) - g(x)}{h} + g(x) \cdot \lim\limits_{h \to 0} \dfrac{f(x + h) - f(x)}{h}$

f) $= f(x) \cdot g'(x) + g(x) \cdot f'(x)$

g) $= f(x) \cdot \left[\dfrac{d}{dx} g(x)\right] + g(x) \cdot \left[\dfrac{d}{dx} f(x)\right]$

TECHNOLOGY CONNECTION

124. Business. Refer to Exercises 102, 104, and 106. At what rate is Tongue-Tied Sauces' profit changing at the break-even point? At what rate is the average profit per bottle of barbecue sauce changing at that point?

125. Business. Refer to Exercises 101, 103, and 105. At what rate is Summertime Fabrics' profit changing at the break-even point? At what rate is the average profit per jacket changing at that point?

For the function in each of Exercises 126–131, graph f and f′. Then estimate points at which the tangent line to f is horizontal. If no such point exists, state that fact.

126. $f(x) = x^2(x - 2)(x + 2)$

127. $f(x) = \left(x + \dfrac{2}{x}\right)(x^2 - 3)$

128. $f(x) = \dfrac{x^3 - 1}{x^2 + 1}$

129. $f(x) = \dfrac{0.3x}{0.04 + x^2}$

130. $f(x) = \dfrac{0.01x^2}{x^4 + 0.0256}$

131. $f(x) = \dfrac{4x}{x^2 + 1}$

132. Use a graph to decide which of the following seems to be the correct derivative of the function in Exercise 131.

$$y_1 = \frac{2}{x}$$

$$y_2 = \frac{4 - 4x}{x^2 + 1}$$

$$y_3 = \frac{4 - 4x^2}{(x^2 + 1)^2}$$

$$y_4 = \frac{4x^2 - 4}{(x^2 + 1)^2}$$

1.7 The Chain Rule

OBJECTIVES

- Find the composition of two functions.
- Differentiate using the Extended Power Rule or the Chain Rule.

The Extended Power Rule

How can we differentiate more complicated functions such as

$$y = (1 + x^2)^3,$$
$$y = (1 + x^2)^{89},$$

or

$$y = (1 + x^2)^{1/3}?$$

For $(1 + x^2)^3$, we can expand and then differentiate. Although this could be done for $(1 + x^2)^{89}$, it would certainly be time-consuming, and such an expansion would not work for $(1 + x^2)^{1/3}$. Not knowing a rule, we might suspect that the derivative of the function $y = (1 + x^2)^3$ is

$$\frac{dy}{dx} = 3(1 + x^2)^2. \tag{1}$$

To check this, we first expand $(1 + x^2)^3$:

$$\begin{aligned}(1 + x^2)^3 &= (1 + x^2)(1 + x^2)(1 + x^2)\\ &= (1 + 2x^2 + x^4)(1 + x^2)\\ &= 1 + 2x^2 + x^4 + x^2 + 2x^4 + x^6\\ &= 1 + 3x^2 + 3x^4 + x^6\end{aligned}$$

We then differentiate the result:

$$\frac{dy}{dx} = 6x + 12x^3 + 6x^5 \qquad \text{Differentiating}$$

$$= (1 + 2x^2 + x^4)6x$$
$$= 3(1 + x^2)^2 \cdot 2x. \qquad (2)$$

Attempting to match equation (1)

Comparing this with equation (1), we see that the Power Rule is not sufficient for such a differentiation. Note that the factor $2x$ in the actual derivative, equation (2), is the derivative of the "inside" function, $1 + x^2$. This is consistent with the following new rule (explained in more detail on p. 173).

THEOREM 7

The Extended Power Rule

Suppose that $g(x)$ is a differentiable function of x. Then, for any real number k,

$$\frac{d}{dx}[g(x)]^k = k[g(x)]^{k-1} \cdot \frac{d}{dx}g(x).$$

Let's differentiate $(1 + x^3)^5$. There are three steps to carry out.

$(1 + x^3)^5$	**1.** Mentally block out the "inside" function, $1 + x^3$.
$5(1 + x^3)^4$	**2.** Differentiate the "outside" function, $(1 + x^3)^5$.
$5(1 + x^3)^4 \cdot 3x^2$	**3.** Multiply by the derivative of the "inside" function.
$= 15x^2(1 + x^3)^4$	

Step (3) is quite commonly overlooked. Try not to forget it!

TECHNOLOGY CONNECTION

One way to check your differentiation of y_1 is to enter your derivative as y_2 and see if the graph of y_2 coincides with the graph of $y_2 = \text{nDeriv}(y_1, x, x)$. Use this approach to check Example 1. Be sure to use the Sequential mode and different graph styles for the two curves.

EXAMPLE 1 Differentiate: $f(x) = (1 + x^3)^{1/2}$.

Solution

$$\frac{d}{dx}(1 + x^3)^{1/2} = \frac{1}{2}(1 + x^3)^{1/2-1} \cdot 3x^2$$
$$= \frac{3x^2}{2}(1 + x^3)^{-1/2}$$
$$= \frac{3x^2}{2\sqrt{1 + x^3}}$$

✦

EXAMPLE 2 Differentiate: $y = (1 - x^2)^3 + (5 + 4x)^2$.

Solution Here we combine the Sum–Difference Rule and the Extended Power Rule:

$$\frac{dy}{dx} = 3(1 - x^2)^2(-2x) + 2(5 + 4x)^1 \cdot 4.$$

We differentiate each term using the Extended Power Rule.

Since $dy/dx = 3(1 - x^2)^2(-2x) + 2(5 + 4x) \cdot 4$, it follows that

$$\begin{aligned}\frac{dy}{dx} &= -6x(1 - x^2)^2 + 8(5 + 4x) \\ &= -6x(1 - 2x^2 + x^4) + 40 + 32x \\ &= -6x + 12x^3 - 6x^5 + 40 + 32x \\ &= 40 + 26x + 12x^3 - 6x^5.\end{aligned}$$

✦

EXAMPLE 3 Differentiate: $f(x) = (3x - 5)^4(7 - x)^{10}$.

Solution Here we combine the Product Rule and the Extended Power Rule:

$$\begin{aligned}f'(x) &= (3x - 5)^4 \cdot 10(7 - x)^9(-1) + (7 - x)^{10}\, 4(3x - 5)^3(3) \\ &= -10(3x - 5)^4(7 - x)^9 + (7 - x)^{10}\, 12(3x - 5)^3 \\ &= 2(3x - 5)^3(7 - x)^9[-5(3x - 5) + 6(7 - x)] \\ &= 2(3x - 5)^3(7 - x)^9(-15x + 25 + 42 - 6x) \\ &= 2(3x - 5)^3(7 - x)^9(67 - 21x).\end{aligned}$$

We factor out $2(3x - 5)^3(7 - x)^9$.

✦

EXAMPLE 4 Differentiate: $f(x) = \sqrt[4]{\dfrac{x + 3}{x - 2}}$.

Solution Here we use the Quotient Rule to differentiate the inside function:

$$\begin{aligned}\frac{d}{dx}\sqrt[4]{\frac{x + 3}{x - 2}} = \frac{d}{dx}\left(\frac{x + 3}{x - 2}\right)^{1/4} &= \frac{1}{4}\left(\frac{x + 3}{x - 2}\right)^{1/4-1}\left[\frac{(x - 2)1 - 1(x + 3)}{(x - 2)^2}\right] \\ &= \frac{1}{4}\left(\frac{x + 3}{x - 2}\right)^{-3/4}\left[\frac{x - 2 - x - 3}{(x - 2)^2}\right] \\ &= \frac{1}{4}\left(\frac{x + 3}{x - 2}\right)^{-3/4}\left[\frac{-5}{(x - 2)^2}\right], \quad \text{or} \quad \frac{-5}{4(x + 3)^{3/4}(x - 2)^{5/4}}.\end{aligned}$$

✦

Composition of Functions and the Chain Rule

The Extended Power Rule is a special case of a more general rule called the *Chain Rule*. Before discussing the Chain Rule, let's consider *composition of functions*.

One author of this text exercises three times a week at a local YMCA. When he recently bought a pair of running shoes, he found a label on which the numbers at the bottom indicate equivalent shoe sizes in five countries.

Author Marv Bittinger and his size-$11\frac{1}{2}$ running shoes

This label suggests that there are functions that convert one country's shoe sizes to those used in another country. There is, indeed, a function g that gives a correspondence between shoe sizes in the United States and those in France:

$$g(x) = \frac{4x + 92}{3},$$

where x is the U.S. size and $g(x)$ is the French size. Thus, a U.S. size $11\frac{1}{2}$ corresponds to a French size

$$g\left(11\tfrac{1}{2}\right) = \frac{4 \cdot 11\frac{1}{2} + 92}{3}, \quad \text{or } 46.$$

There is also a function f that gives a correspondence between shoe sizes in France and those in Japan. The function is given by

$$f(x) = \frac{15x - 100}{2},$$

where x is the French size and $f(x)$ is the corresponding Japanese size. Thus, a French size 46 corresponds to a Japanese size

$$f(46) = \frac{15 \cdot 46 - 100}{2}, \quad \text{or } 295.$$

It seems reasonable to conclude that a shoe size of $11\frac{1}{2}$ in the United States corresponds to a size of 295 in Japan and that some function h describes this correspondence. Can we find a formula for h?

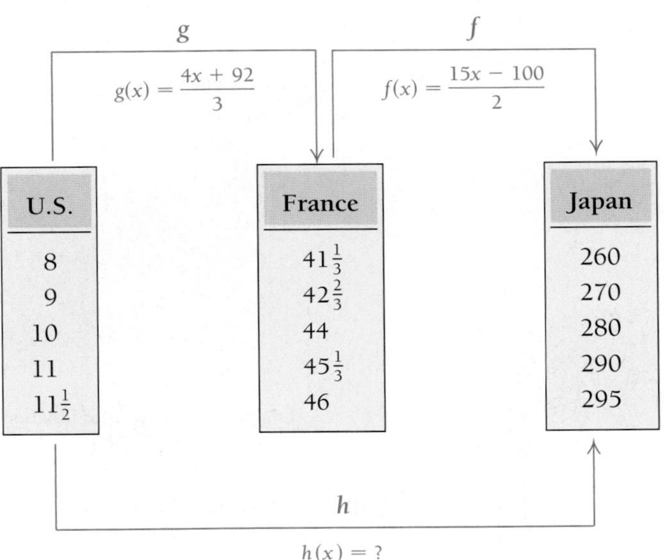

A shoe size x in the United States corresponds to a shoe size $g(x)$ in France, where

$$g(x) = \frac{4x + 92}{3}.$$

Thus, $(4x + 92)/3$ represents a shoe size in France. If we replace x in $f(x)$ with $(4x + 92)/3$, we can find the corresponding shoe size in Japan:

$$f(g(x)) = \frac{15\left(\dfrac{4x + 92}{3}\right) - 100}{2}$$

$$= \frac{5(4x + 92) - 100}{2} = \frac{20x + 460 - 100}{2}$$

$$= \frac{20x + 360}{2} = 10x + 180.$$

This gives a formula for h: $h(x) = 10x + 180$. As a check, a shoe size of $11\frac{1}{2}$ in the United States corresponds to a shoe size of $h\left(11\frac{1}{2}\right) = 10\left(11\frac{1}{2}\right) + 180 = 295$ in Japan. The function h is the *composition* of f and g, symbolized by $f \circ g$ and read as "f composed with g," or simply "f circle g."

DEFINITION

The **composed** function $f \circ g$, the **composition** of f and g, is defined as

$$(f \circ g)(x) = f(g(x)).$$

We can visualize the composition of functions as shown below.

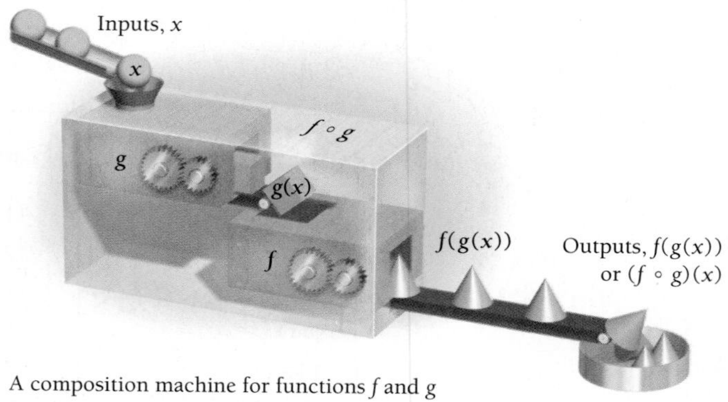

A composition machine for functions f and g

To find $(f \circ g)(x)$, we substitute $g(x)$ for x in $f(x)$.

EXAMPLE 5 For $f(x) = x^3$ and $g(x) = 1 + x^2$, find $(f \circ g)(x)$ and $(g \circ f)(x)$.

Solution Consider each function separately:

$f(x) = x^3$ — This function cubes each input.

and

$g(x) = 1 + x^2.$ — This function adds 1 to the square of each input.

a) The function $f \circ g$ first does what g does (adds 1 to the square) and then does what f does (cubes). We find $f(g(x))$ by substituting $g(x)$ for x:

$$\begin{aligned}(f \circ g)(x) = f(g(x)) &= f(1 + x^2) && \text{Using } g(x) \text{ as an input}\\ &= (1 + x^2)^3\\ &= 1 + 3x^2 + 3x^4 + x^6.\end{aligned}$$

b) The function $g \circ f$ first does what f does (cubes) and then does what g does (adds 1 to the square). We find $g(f(x))$ by substituting $f(x)$ for x:

$$\begin{aligned}(g \circ f)(x) = g(f(x)) &= g(x^3) && \text{Using } f(x) \text{ as an input}\\ &= 1 + (x^3)^2 = 1 + x^6.\end{aligned}$$

✦

EXAMPLE 6 For $f(x) = \sqrt{x}$ and $g(x) = x - 1$, find $(f \circ g)(x)$ and $(g \circ f)(x)$.

Solution

$$(f \circ g)(x) = f(g(x)) = f(x - 1) = \sqrt{x - 1}$$
$$(g \circ f)(x) = g(f(x)) = g(\sqrt{x}) = \sqrt{x} - 1$$

✦

Keep in mind that, in general, $(f \circ g)(x) \neq (g \circ f)(x)$. We see this fact demonstrated in Examples 5 and 6.

How do we differentiate a composition of functions? The following theorem tells us.

THEOREM 8

The Chain Rule

The derivative of the composition $f \circ g$ is given by

$$\frac{d}{dx}[(f \circ g)(x)] = \frac{d}{dx}[f(g(x))] = f'(g(x)) \cdot g'(x).$$

As we noted earlier, the Extended Power Rule is a special case of the Chain Rule. Consider $f(x) = x^k$. For any other function $g(x)$, we have $(f \circ g)(x) = [g(x)]^k$, and the derivative of the composition is

$$\frac{d}{dx}[g(x)]^k = k[g(x)]^{k-1} \cdot g'(x).$$

The Chain Rule often appears in another form. Suppose that $y = f(u)$ and $u = g(x)$. Then

$$\frac{dy}{dx} = \frac{dy}{du} \cdot \frac{du}{dx}.$$

To better understand the Chain Rule, suppose that a video game manufacturer wished to determine its rate of profit, in *dollars* per *minute*. One way to find this rate would be to multiply the rate of profit, in *dollars* per *item*, by the production rate, in *items* per *minute*. That is,

$$\begin{bmatrix}\text{Change in profits}\\ \text{with respect to time}\end{bmatrix} = \begin{bmatrix}\text{Change in profits with respect}\\ \text{to number of games produced}\end{bmatrix} \cdot \begin{bmatrix}\text{Change in number of games}\\ \text{produced with respect to time.}\end{bmatrix}$$

EXAMPLE 7 For $y = 2 + \sqrt{u}$ and $u = x^3 + 1$, find dy/du, du/dx, and dy/dx.

Solution First we find dy/du and du/dx:

$$\frac{dy}{du} = \frac{1}{2}u^{-1/2} \quad \text{and} \quad \frac{du}{dx} = 3x^2.$$

Then

$$\begin{aligned}\frac{dy}{dx} &= \frac{dy}{du} \cdot \frac{du}{dx}\\ &= \frac{1}{2\sqrt{u}} \cdot 3x^2\\ &= \frac{3x^2}{2\sqrt{x^3 + 1}}. \quad \text{Substituting } x^3 + 1 \text{ for } u\end{aligned}$$

EXAMPLE 8 For $y = u^2 - 3u$ and $u = 5t - 1$, find dy/dt.

Solution We have $\frac{dy}{dt} = \frac{dy}{du} \cdot \frac{du}{dt}$, with $\frac{dy}{du} = 2u - 3$ and $\frac{du}{dt} = 5$.

Thus,

$$\begin{aligned}\frac{dy}{dt} &= (2u - 3) \cdot 5 = 10u - 15\\ &= 10(5t - 1) - 15 \quad \text{Substituting } 5t - 1 \text{ for } u\\ &= 50t - 10 - 15\\ &= 50t - 25.\end{aligned}$$

Exercise Set 1.7

Differentiate each function.

1. $y = (2x + 1)^2$
2. $y = (3 - 2x)^2$

(Exercises 1–2: Check by expanding and then differentiating.)

3. $y = (7 - x)^{55}$
4. $y = (8 - x)^{100}$
5. $y = \sqrt{1 + 8x}$
6. $y = \sqrt{1 - x}$
7. $y = \sqrt{3x^2 - 4}$
8. $y = \sqrt{4x^2 + 1}$
9. $y = (8x^2 - 6)^{-40}$
10. $y = (4x^2 + 1)^{-50}$
11. $y = (x - 4)^8(2x + 3)^6$
12. $y = (x + 5)^7(4x - 1)^{10}$
13. $y = \frac{1}{(3x + 8)^2}$
14. $y = \frac{1}{(4x + 5)^2}$
15. $y = \frac{4x^2}{(7 - 5x)^3}$
16. $y = \frac{7x^3}{(4 - 9x)^5}$
17. $f(x) = (1 + x^3)^3 - (2 + x^8)^4$
18. $f(x) = (3 + x^3)^5 - (1 + x^7)^4$
19. $f(x) = x^2 + (200 - x)^2$
20. $f(x) = x^2 + (100 - x)^2$
21. $g(x) = \sqrt{x} + (x - 3)^3$
22. $G(x) = \sqrt[3]{2x - 1} + (4 - x)^2$
23. $f(x) = -5x(2x - 3)^4$
24. $f(x) = -3x(5x + 4)^6$
25. $g(x) = (3x - 1)^7(2x + 1)^5$
26. $F(x) = (5x + 2)^4(2x - 3)^8$
27. $f(x) = x^2\sqrt{4x - 1}$
28. $f(x) = x^3\sqrt{5x + 2}$
29. $G(x) = \sqrt[3]{x^5 + 6x}$

30. $F(x) = \sqrt[4]{x^2 - 5x + 2}$

31. $f(x) = \left(\frac{3x - 1}{5x + 2}\right)^4$ **32.** $f(x) = \left(\frac{2x}{x^2 + 1}\right)^3$

33. $g(x) = \sqrt{\frac{4 - x}{3 + x}}$ **34.** $g(x) = \sqrt{\frac{3 + 2x}{5 - x}}$

35. $f(x) = (2x^3 - 3x^2 + 4x + 1)^{100}$

36. $f(x) = (7x^4 + 6x^3 - x)^{204}$

37. $g(x) = \left(\frac{2x + 3}{5x - 1}\right)^{-4}$ **38.** $h(x) = \left(\frac{1 - 3x}{2 - 7x}\right)^{-5}$

39. $f(x) = \sqrt{\frac{x^2 + x}{x^2 - x}}$ **40.** $f(x) = \sqrt[3]{\frac{4 - x^3}{x - x^2}}$

41. $f(x) = \frac{(2x + 3)^4}{(3x - 2)^5}$ **42.** $f(x) = \frac{(5x - 4)^7}{(6x + 1)^3}$

43. $f(x) = 12(2x + 1)^{2/3}(3x - 4)^{5/4}$

44. $y = 6\sqrt[3]{x^2 + x}(x^4 - 6x)^3$

Find $\frac{dy}{du}$, $\frac{du}{dx}$, *and* $\frac{dy}{dx}$.

45. $y = \sqrt{u}$ and $u = x^2 - 1$

46. $y = \frac{15}{u^3}$ and $u = 2x + 1$

47. $y = u^{50}$ and $u = 4x^3 - 2x^2$

48. $y = \frac{u + 1}{u - 1}$ and $u = 1 + \sqrt{x}$

49. $y = u(u + 1)$ and $u = x^3 - 2x$

50. $y = (u + 1)(u - 1)$ and $u = x^3 + 1$

Find $\frac{dy}{dx}$ *for each pair of functions.*

51. $y = 5u^2 + 3u$ and $u = x^3 + 1$

52. $y = u^3 - 7u^2$ and $u = x^2 + 3$

53. $y = \sqrt[3]{2u + 5}$ and $u = x^2 - x$

54. $y = \sqrt{7 - 3u}$ and $u = x^2 - 9$

55. Find $\frac{dy}{dt}$ if $y = \frac{1}{u^2 + u}$ and $u = 5 + 3t$.

56. Find $\frac{dy}{dt}$ if $y = \frac{1}{3u^5 - 7}$ and $u = 7t^2 + 1$.

57. Find an equation for the tangent line to the graph of $y = \sqrt{x^2 + 3x}$ at the point $(1, 2)$.

58. Find an equation for the tangent line to the graph of $y = (x^3 - 4x)^{10}$ at the point $(2, 0)$.

59. Find an equation for the tangent line to the graph of $y = x\sqrt{2x + 3}$ at the point $(3, 9)$.

60. Find an equation for the tangent line to the graph of $y = \left(\frac{2x + 3}{x - 1}\right)^3$ at the point $(2, 343)$.

61. Consider

$$f(x) = \frac{x^2}{(1 + x)^5}.$$

a) Find $f'(x)$ using the Quotient Rule and the Extended Power Rule.

b) Note that $f(x) = x^2(1 + x)^{-5}$. Find $f'(x)$ using the Product Rule and the Extended Power Rule.

c) Compare your answers to parts (a) and (b).

62. Consider

$$g(x) = \left(\frac{6x + 1}{2x - 5}\right)^2.$$

a) Find $g'(x)$ using the Extended Power Rule.

b) Note that

$$g(x) = \frac{36x^2 + 12x + 1}{4x^2 - 20x + 25}.$$

Find $g'(x)$ using the Quotient Rule.

TW **c)** Compare your answers to parts (a) and (b). Which approach was easier, and why?

Find $f(x)$ *and* $g(x)$ *such that* $h(x) = (f \circ g)(x)$. *Answers may vary.*

63. $h(x) = (3x^2 - 7)^5$

64. $h(x) = \frac{1}{\sqrt{7x + 2}}$

65. $h(x) = \frac{x^3 + 1}{x^3 - 1}$

66. $h(x) = (\sqrt{x} + 5)^4$

Do Exercises 67–70 in two ways. First, use the Chain Rule to find the answer. Next, check your answer by finding $f(g(x))$*, taking the derivative, and substituting.*

67. $f(u) = u^3$, $g(x) = u = 2x^4 + 1$
Find $(f \circ g)'(-1)$.

68. $f(u) = \frac{u + 1}{u - 1}$, $g(x) = u = \sqrt{x}$
Find $(f \circ g)'(4)$.

69. $f(u) = \sqrt[3]{u}$, $g(x) = u = 1 + 3x^2$
Find $(f \circ g)'(2)$.

70. $f(u) = 2u^5$, $g(x) = u = \frac{3 - x}{4 + x}$
Find $(f \circ g)'(-10)$.

APPLICATIONS

Business and Economics

71. Total revenue. A total-revenue function is given by

$$R(x) = 1000\sqrt{x^2 - 0.1x},$$

where $R(x)$ is the total revenue, in thousands of dollars, from the sale of x items. Find the rate at which total revenue is changing when 20 items have been sold.

72. Total cost. A total-cost function is given by

$$C(x) = 2000(x^2 + 2)^{1/3} + 700,$$

where $C(x)$ is the total cost, in thousands of dollars, of producing x items. Find the rate at which total cost is changing when 20 items have been produced.

73. Total profit. Use the total-cost and total-revenue functions in Exercises 71 and 72 to find the rate at which total profit is changing when x items have been produced and sold.

74. Total cost. A company determines that its total cost, in thousands of dollars, for producing x items is

$$C(x) = \sqrt{5x^2 + 60},$$

and it plans to boost production t months from now according to the function

$$x(t) = 20t + 40.$$

How fast will costs be rising 4 months from now?

75. Consumer credit. The total outstanding consumer credit of the United States (in billions of dollars) can be modeled by the function

$$C(x) = 0.21x^4 - 5.92x^3 + 50.53x^2 - 18.92x + 1114.93$$

where x is the number of years since 1995.

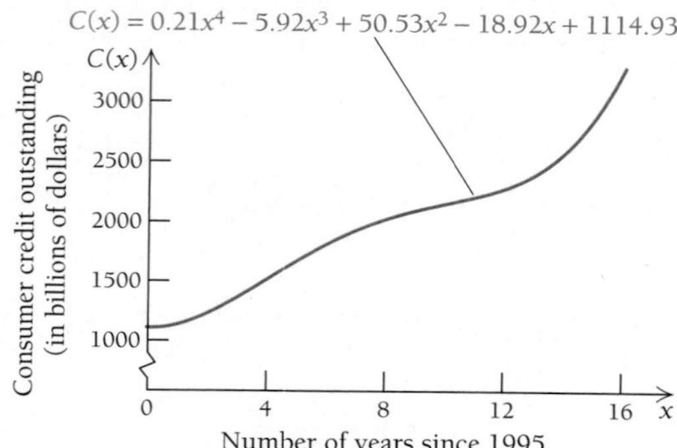

(*Source*: Federal Reserve Board.)

a) Find dC/dx.

TW b) Interpret the meaning of dC/dx.

c) Using this model, predict how quickly outstanding consumer credit will be rising in 2010.

76. Utility. Utility is a type of function that occurs in economics. When a consumer receives x units of a product, a certain amount of pleasure, or utility, U, is derived. Suppose that the utility related to the number of tickets x for a ride at a county fair is

$$U(x) = 80\sqrt{\frac{2x + 1}{3x + 4}}.$$

Find the rate at which the utility changes with respect to the number of tickets bought.

77. Compound interest. If \$1000 is invested at interest rate i, compounded annually, in 3 yr it will grow to an amount A given by (see Section R.1)

$$A = \$1000(1 + i)^3.$$

a) Find the rate of change, dA/di.

TW b) Interpret the meaning of dA/di.

78. Compound interest. If \$1000 is invested at interest rate i, compounded quarterly, in 5 yr it will grow to an amount, A, given by

$$A = \$1000\left(1 + \frac{i}{4}\right)^{20}.$$

a) Find the rate of change, dA/di.

TW b) Interpret the meaning of dA/di.

79. Consumer demand. Suppose that the demand function for a product is given by

$$D(p) = \frac{80{,}000}{p}$$

and that price p is a function of time given by $p = 1.6t + 9$, where t is in days.

a) Find the demand as a function of time t.

b) Find the rate of change of the quantity demanded when $t = 100$ days.

80. Business profit. A company is selling microcomputers. It determines that its total profit, in dollars, is given by

$$P(x) = 0.08x^2 + 80x,$$

where x is the number of units produced and sold. Suppose that x is a function of time, in months, where $x = 5t + 1$.

a) Find the total profit as a function of time t.

b) Find the rate of change of total profit when $t = 48$ months.

Life and Physical Sciences

81. Chemotherapy. The dosage for Carboplatin Chemotherapy drugs depends on several parameters of the particular drug as well as the age, weight, and sex of the patient. For female patients, the formulas giving the dosage for such drugs are

$$D = 0.85A(c + 25)$$

and

$$c = (140 - y)\frac{w}{72x},$$

where A and x depend on which drug is used, D is the dosage in milligrams (mg), c is called the creatine clearance, y is the patient's age in years, and w is the patient's weight in kg. (*Source:* US Oncology.)

a) Suppose that a patient is a 45-year-old woman and the drug has parameters $A = 5$ and $x = 0.6$. Use this information to find formulas for D and c that give D as a function of c and c as a function of w.

b) Use your formulas from part (a) to compute dD/dc.

c) Use your formulas from part (a) to compute dc/dw.

d) Compute dD/dw.

TW **e)** Interpret the meaning of the derivative dD/dw.

SYNTHESIS

Differentiate.

82. $y = \sqrt{(2x - 3)^2 + 1}$

83. $y = \sqrt[3]{x^3 + 6x + 1} \cdot x^5$

84. $s = \sqrt[4]{t^4 + 3t^2 + 8} \cdot 3t$

85. $y = \left(\dfrac{x}{\sqrt{x - 1}}\right)^3$

86. $y = \left(x\sqrt{1 + x^2}\right)^3$

87. $y = \dfrac{\sqrt{1 - x^2}}{1 - x}$

88. $w = \dfrac{u}{\sqrt{1 + u^2}}$

89. $y = \left(\dfrac{x^2 - x - 1}{x^2 + 1}\right)^3$

90. $g(x) = \sqrt{\dfrac{x^2 - 4x}{2x + 1}}$

91. $f(t) = \sqrt{3t + \sqrt{t}}$

92. $F(x) = [6x(3 - x)^5 + 2]^4$

TW **93.** The following is the beginning of an alternative proof of the Quotient Rule that uses the Product Rule and the Power Rule. Complete the proof, giving reasons for each step.

Proof. Let

$$Q(x) = \frac{N(x)}{D(x)}.$$

Then

$$Q(x) = N(x) \cdot [D(x)]^{-1}.$$

Therefore, ...

TECHNOLOGY CONNECTION

For the function in each of Exercises 94 and 95, graph f and f' over the given interval. Then estimate points at which the tangent line is horizontal.

94. $f(x) = 1.68x\sqrt{9.2 - x^2}$; $[-3, 3]$

95. $f(x) = \sqrt{6x^3 - 3x^2 - 48x + 45}$; $[-5, 5]$

Find the derivative of each of the following functions analytically. Then use a calculator to check the results.

96. $f(x) = x\sqrt{4 - x^2}$

97. $g(x) = \dfrac{4x}{\sqrt{x - 10}}$

98. $f(x) = \left(\sqrt{2x - 1} + x^3\right)^5$

1.8 Higher-Order Derivatives

OBJECTIVES

- Find derivatives of higher order.
- Given a formula for distance, find velocity and acceleration.

Consider the function given by

$$y = f(x) = x^5 - 3x^4 + x.$$

Its derivative f' is given by

$$y' = f'(x) = 5x^4 - 12x^3 + 1.$$

The derivative function f' can also be differentiated. We can think of the derivative of f' as the rate of change of the slope of the tangent lines of f. It can also be regarded as the rate at which $f'(x)$ is changing. We use the notation f'' for the derivative $(f')'$. That is,

$$f''(x) = \frac{d}{dx} f'(x).$$

We call f'' the *second derivative* of f. For $f(x) = x^5 - 3x^4 + x$, the second derivative is given by

$$y'' = f''(x) = 20x^3 - 36x^2.$$

Continuing in this manner, we have

$$f'''(x) = 60x^2 - 72x, \quad \text{The third derivative of } f$$
$$f''''(x) = 120x - 72, \quad \text{The fourth derivative of } f$$
$$f'''''(x) = 120. \quad \text{The fifth derivative of } f$$

When notation like $f'''(x)$ gets lengthy, we abbreviate it using a symbol in parentheses. Thus, $f^{(n)}(x)$ is the nth derivative. For the function above,

$$f^{(4)}(x) = 120x - 72,$$
$$f^{(5)}(x) = 120,$$
$$f^{(6)}(x) = 0, \quad \text{and}$$
$$f^{(n)}(x) = 0, \quad \text{for any integer } n \geq 6.$$

Leibniz notation for the second derivative of a function given by $y = f(x)$ is

$$\frac{d^2y}{dx^2}, \quad \text{or} \quad \frac{d}{dx}\left(\frac{dy}{dx}\right),$$

read "the second derivative of y with respect to x." The 2's in this notation are *not* exponents. If $y = x^5 - 3x^4 + x$, then

$$\frac{d^2y}{dx^2} = 20x^3 - 36x^2.$$

Leibniz notation for the third derivative is d^3y/dx^3; for the fourth derivative, d^4y/dx^4; and so on:

$$\frac{d^3y}{dx^3} = 60x^2 - 72x,$$
$$\frac{d^4y}{dx^4} = 120x - 72,$$
$$\frac{d^5y}{dx^5} = 120.$$

EXAMPLE 1 For $y = 1/x$, find d^2y/dx^2.

Solution We have $y = x^{-1}$, so

$$\frac{dy}{dx} = -1 \cdot x^{-1-1} = -x^{-2}, \quad \text{or} \quad -\frac{1}{x^2}.$$

Then

$$\frac{d^2y}{dx^2} = (-2)(-1)x^{-2-1} = 2x^{-3}, \quad \text{or} \quad \frac{2}{x^3}.$$ ✦

EXAMPLE 2 For $y = (x^2 + 10x)^{20}$, find y' and y''.

Solution To find y', we use the Extended Power Rule:

$$\begin{aligned} y' &= 20(x^2 + 10x)^{19}(2x + 10) \\ &= 20(x^2 + 10x)^{19} \cdot 2(x + 5) \\ &= 40(x^2 + 10x)^{19}(x + 5). \end{aligned}$$

To find y'', we use the Product Rule and the Extended Power Rule:

$$\begin{aligned} y'' &= 40(x^2 + 10x)^{19}(1) + (x + 5)19 \cdot 40(x^2 + 10x)^{18}(2x + 10) \\ &= 40(x^2 + 10x)^{19} + 760(x + 5)(x^2 + 10x)^{18}\,2(x + 5) \\ &= 40(x^2 + 10x)^{19} + 1520(x + 5)^2(x^2 + 10x)^{18} \\ &= 40(x^2 + 10x)^{18}[(x^2 + 10x) + 38(x + 5)^2] \qquad \text{Factoring} \\ &= 40(x^2 + 10x)^{18}[x^2 + 10x + 38(x^2 + 10x + 25)] \\ &= 40(x^2 + 10x)^{18}(39x^2 + 390x + 950). \end{aligned}$$ ✦

Velocity and Acceleration

We have already seen that a function's derivative represents its instantaneous rate of change. When the function relates a change in distance to a change in time, the instantaneous rate of change is called *speed*, or *velocity*.* The letter v is generally used to stand for velocity.

DEFINITION

The **velocity** of an object that is $s(t)$ units from a starting point at time t is given by

$$\text{Velocity} = v(t) = s'(t) = \lim_{h \to 0} \frac{s(t + h) - s(t)}{h}$$

*In this text, the words "speed" and "velocity" are used interchangeably. In physics and engineering, this is not done, since velocity requires direction and speed does not.

EXAMPLE 3 Physical Science: Velocity. Suppose that an object travels so that its distance s, in miles, from its starting point is a function of time t, in hours, as follows:

$$s(t) = 10t^2.$$

a) Find the average velocity between the times $t = 2$ hr and $t = 5$ hr.

b) Find the (instantaneous) velocity when $t = 4$ hr.

Solution

a) From $t = 2$ hr to $t = 5$ hr, we have

$$\frac{\text{Difference in miles}}{\text{Difference in hours}} = \frac{s(5) - s(2)}{3}$$
$$= \frac{10 \cdot 5^2 \text{ mi} - 10 \cdot 2^2 \text{ mi}}{3 \text{ hr}} = 70\,\frac{\text{mi}}{\text{hr}}.$$

b) The instantaneous velocity is given by

$$\lim_{h \to 0} \frac{s(t + h) - s(t)}{h} = s'(t).$$

We know how to find this limit quickly from the special techniques learned in Section 1.5. Thus, $s'(t) = 20t$, and

$$s'(4) = 20 \cdot 4 = 80\,\frac{\text{mi}}{\text{hr}}.$$

✦

Often velocity itself is a function of time. When a jet takes off or a vehicle comes to a sudden stop, the change in velocity is easily felt by passengers. The rate at which velocity changes is called *acceleration*. Suppose that Car A reaches a speed of 65 mi/hr in 8.4 sec and Car B reaches a speed of 65 mi/hr in 8 sec; then B has a *faster acceleration* than A. We generally use the letter a for acceleration. It is useful to think of acceleration as the rate at which velocity is changing.

DEFINITION

Acceleration $= a(t) = v'(t) = s''(t)$.

EXAMPLE 4 Physical Science: Distance, Velocity, and Acceleration. For $s(t) = 10t^2$, find $v(t)$ and $a(t)$, where s is the distance from the starting point, in miles, and t is in hours. Then find the distance, velocity, and acceleration when $t = 4$ hr.

Solution We have $s(t) = 10t^2$. Thus,

$$v(t) = s'(t) = 20t \quad \text{and} \quad a(t) = v'(t) = s''(t) = 20.$$

It follows that

$$s(4) = 10(4)^2 = 160 \text{ mi},$$
$$v(4) = 20(4) = 80 \text{ mi/hr},$$
and
$$a(4) = 20 \text{ mi/hr}^2.$$

If this distance function applies to motion of a vehicle, then at time $t = 4$ hr, the vehicle has traveled 160 mi, the velocity is 80 mi/hr, and the acceleration is 20 miles per hour per hour, which we abbreviate as 20 mi/hr^2. ✦

Note from Example 4 that since acceleration represents the *rate* at which velocity is changing, the units in which it is measured involve a unit of time squared:

$$\frac{\text{Change in velocity}}{\text{Change in time}} = \frac{\text{mi/hr}}{\text{hr}} = \frac{\text{mi}}{\text{hr}} \cdot \frac{1}{\text{hr}} = \text{mi/hr}^2.$$

EXAMPLE 5 Free Fall. When an object is dropped, the distance it falls in t seconds, assuming that air resistance is negligible, is given by

$$s(t) = 4.905t^2,$$

where $s(t)$ is in meters (m). If a stone is dropped from a cliff, find each of the following, assuming that air resistance is negligible: **(a)** how far it has traveled 5 sec after being dropped, **(b)** how fast it is traveling 5 sec after being dropped, and **(c)** the stone's acceleration after it has been falling for 5 sec.

Solution

a) After 5 sec, the stone has traveled

$$s(5) = 4.905(5)^2 = 4.905(25) = 122.625 \text{ m}.$$

b) The speed at which the stone falls is given by

$$v(t) = s'(t) = 9.81t.$$

Thus,

$$v(5) = 9.81 \cdot 5 = 49.05 \text{ m/sec}.$$

c) The stone's acceleration after t sec is constant:

$$a(t) = v'(t) = s''(t) = 9.81 \text{ m/sec}^2.$$

Thus, $s''(5) = 9.81 \text{ m/sec}^2$. ✦

There are many important real-world applications that make use of the second derivative. In Chapter 2, we will examine many of these, including applications in the fields of economics, health care, and the natural and physical sciences.

Exercise Set 1.8

Find d^2y/dx^2.

1. $y = x^5 + 9$

2. $y = x^4 - 7$

3. $y = 2x^4 - 5x$

4. $y = 5x^3 + 4x$

5. $y = 4x^2 + 3x - 1$

6. $y = 4x^2 - 5x + 7$

7. $y = 7x + 2$

8. $y = 6x - 3$

9. $y = \dfrac{1}{x^2}$

10. $y = \dfrac{1}{x^3}$

11. $y = \sqrt{x}$

12. $y = \sqrt[4]{x}$

Find $f''(x)$.

13. $f(x) = x^4 + \frac{3}{x}$

14. $f(x) = x^3 - \frac{5}{x}$

15. $f(x) = x^{1/5}$

16. $f(x) = x^{1/3}$

17. $f(x) = 4x^{-3}$

18. $f(x) = 2x^{-2}$

19. $f(x) = (x^2 + 3x)^7$

20. $f(x) = (x^3 + 2x)^6$

21. $f(x) = (2x^2 - 3x + 1)^{10}$

22. $f(x) = (3x^2 + 2x + 1)^5$

23. $f(x) = \sqrt[4]{(x^2 + 1)^3}$

24. $f(x) = \sqrt[3]{(x^2 - 1)^2}$

Find y''.

25. $y = x^{2/3} + 4x$

26. $y = x^{3/2} - 5x$

27. $y = (x^3 - x)^{3/4}$

28. $y = (x^4 + x)^{2/3}$

29. $y = 2x^{5/4} + x^{1/2}$

30. $y = 3x^{4/3} - x^{1/2}$

31. $y = \frac{2}{x^3} + \frac{1}{x^2}$

32. $y = \frac{3}{x^4} - \frac{1}{x}$

33. $y = (x^2 + 3)(4x - 1)$

34. $y = (x^3 - 2)(5x + 1)$

35. $y = \frac{3x + 1}{2x - 3}$

36. $y = \frac{2x + 3}{5x - 1}$

37. For $y = x^4$, find d^4y/dx^4.

38. For $y = x^5$, find d^4y/dx^4.

39. For $y = x^6 - x^3 + 2x$, find d^5y/dx^5.

40. For $y = x^7 - 8x^2 + 2$, find d^6y/dx^6.

41. For $f(x) = x^{-2} - x^{1/2}$, find $f^{(4)}(x)$.

42. For $f(x) = x^{-3} + 2x^{1/3}$, find $f^{(5)}(x)$.

43. For $g(x) = x^4 - 3x^3 - 7x^2 - 6x + 9$, find $g^{(6)}(x)$.

44. For $g(x) = 6x^5 + 2x^4 - 4x^3 + 7x^2 - 8x + 3$, find $g^{(7)}(x)$.

APPLICATIONS

Life and Physical Sciences

45. Given

$$s(t) = t^3 + t,$$

where s is in feet and t is in seconds, find each of the following.

a) $v(t)$ **b)** $a(t)$

c) The velocity and acceleration when $t = 4$ sec

46. Given

$$s(t) = -10t^2 + 2t + 5,$$

where s is in meters and t is in seconds, find each of the following.

a) $v(t)$ **b)** $a(t)$

c) The velocity and acceleration when $t = 1$ sec

47. Given

$$s(t) = 3t + 10,$$

where s is in miles and t is in hours, find each of the following.

a) $v(t)$ **b)** $a(t)$

c) The velocity and acceleration when $t = 2$ hr

TW **d)** When the distance function is given by a linear function, we have *uniform motion.* What does uniform motion mean in terms of velocity and acceleration?

48. Given

$$s(t) = t^2 - \frac{1}{2}t + 3,$$

where s is in meters and t is in seconds, find each of the following.

a) $v(t)$ **b)** $a(t)$

c) The velocity and acceleration when $t = 1$ sec

49. Free fall. When an object is dropped, the distance it falls in t seconds, assuming that air resistance is negligible, is given by

$$s(t) = 16t^2,$$

where $s(t)$ is in feet. Suppose that a medic's reflex hammer is dropped from a hovering helicopter. Find **(a)** how far the hammer falls in 3 sec, **(b)** how fast the hammer is traveling 3 sec after being dropped, and **(c)** the hammer's acceleration after it has been falling for 3 sec.

50. Free fall. (See Exercise 49.) Suppose a worker drops a bolt from a scaffold high above a work site. Assuming that air resistance is negligible, find **(a)** how far the bolt falls in 2 sec, **(b)** how fast the bolt is traveling 2 sec after being dropped, and **(c)** the bolt's acceleration after it has been falling for 2 sec.

51. Free fall. Find the velocity and acceleration of the stone in Example 5 after it has been falling for 2 sec.

52. Free fall. Find the velocity and acceleration of the stone in Example 5 after it has been falling for 3 sec.

53. The following graph describes an airplane's distance from its last point of rest.

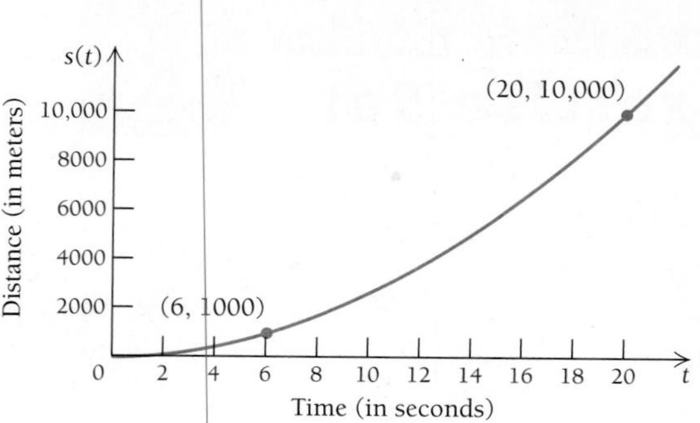

a) Is the plane's velocity greater at $t = 6$ sec or $t = 20$ sec? How can you tell?

b) Is the plane's acceleration positive or negative? How can you tell?

54. The following graph describes a bicycle racer's distance from a roadside television camera.

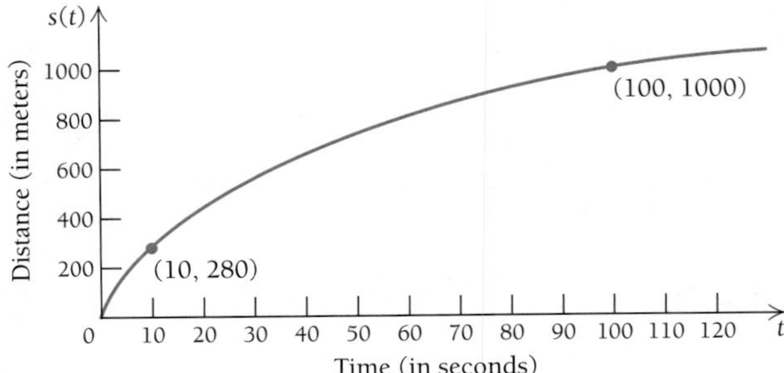

a) When is the bicyclist's velocity the greatest? How can you tell?

b) Is the bicyclist's acceleration positive or negative? How can you tell?

55. Sales. A company determines that monthly sales S, in thousands of dollars, after t months of marketing a product is given by

$$S(t) = 2t^3 - 40t^2 + 220t + 160.$$

a) Find $S'(1)$, $S'(2)$, and $S'(4)$.

b) Find $S''(1)$, $S''(2)$, and $S''(4)$.

TW **c)** Interpret the meaning of your answers to parts (a) and (b).

56. Sales. A business discovers that the number of items sold t days after launching a new sales promotion is given by

$$N(t) = 2t^3 - 3t^2 + 2t.$$

a) Find $N'(1)$, $N'(2)$, and $N'(4)$.

b) Find $N''(1)$, $N''(2)$, and $N''(4)$.

TW **c)** Interpret the meaning of your answers to parts (a) and (b).

SYNTHESIS

Find y''' for each function.

57. $y = \dfrac{1}{1 - x}$

58. $y = x\sqrt{1 + x^2}$

59. $y = \dfrac{1}{\sqrt{2x + 1}}$

60. $y = \dfrac{3x - 1}{2x + 3}$

Find y'' for each function.

61. $y = \dfrac{\sqrt{x} + 1}{\sqrt{x} - 1}$

62. $y = \dfrac{x}{\sqrt{x - 1}}$

63. For $y = x^k$, find d^5y/dx^5.

64. For $y = ax^3 + bx^2 + cx + d$, find d^3y/dx^3.

Find the first through the fourth derivatives. Be sure to simplify at each stage before continuing.

65. $f(x) = \dfrac{x - 1}{x + 2}$

66. $f(x) = \dfrac{x + 3}{x - 2}$

67. Baseball. A batted baseball is dropping from a height of 180 ft. For how many seconds must it fall to reach a speed of 50 mi/hr? (*Hint*: See Exercise 49.)

TECHNOLOGY CONNECTION

For the distance function in each of Exercises 68–71, graph s, v, and a over the given interval. Then use the graphs to determine the point(s) at which the velocity will switch from increasing to decreasing or from decreasing to increasing.

68. $s(t) = 0.1t^4 - t^2 + 0.4;\quad [-5, 5]$

69. $s(t) = -t^3 + 3t;\quad [-3, 3]$

70. $s(t) = t^4 + t^3 - 4t^2 - 2t + 4;\quad [-3, 3]$

71. $s(t) = t^3 - 3t^2 + 2;\quad [-2, 4]$

Chapter Summary

This chapter begins our study of calculus. The study begins with the concept of *limit*, which is presented in an intuitive manner that allows for a numerical, graphical, and algebraic understanding. (pp. 100 and 114)

As x approaches a, the **limit** of $f(x)$ is L, written

$$\lim_{x \to a} f(x) = L,$$

if all values of $f(x)$ are close to L for values of x that are sufficiently close, but not equal, to a.

EXAMPLE Consider the function G given by

$$G(x) = \begin{cases} 4 - x, & \text{for } x < 3, \\ \sqrt{x - 2} + 1, & \text{for } x \geq 3. \end{cases}$$

Graph the function and find each limit, if it exists.

a) $\lim_{x \to 1} G(x)$

b) $\lim_{x \to 3} G(x)$

Solution We check the limits from the left and from the right, both numerically and graphically.

a)

Limit Numerically

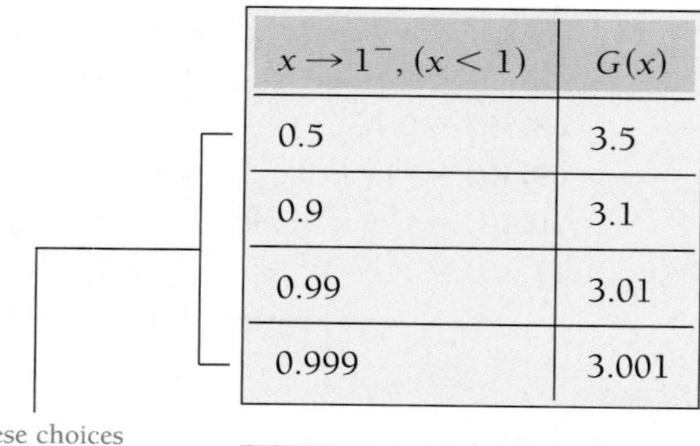

$x \to 1^-, (x < 1)$	$G(x)$
0.5	3.5
0.9	3.1
0.99	3.01
0.999	3.001

These choices can vary.

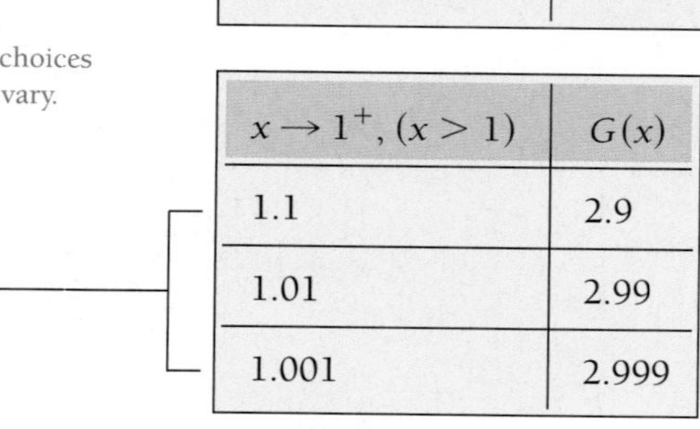

$x \to 1^+, (x > 1)$	$G(x)$
1.1	2.9
1.01	2.99
1.001	2.999

Limit Graphically

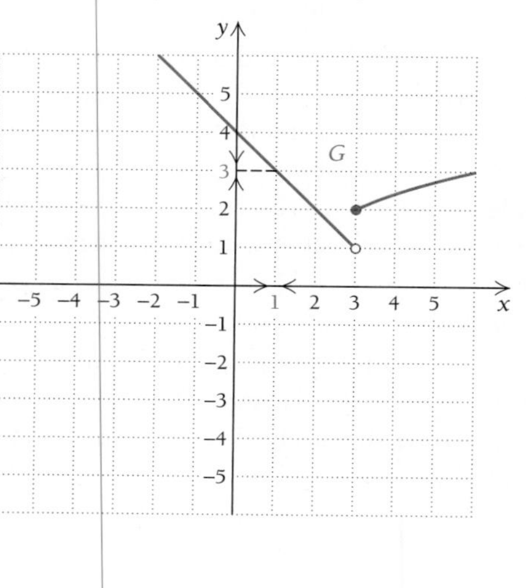

Both the tables and the graph show that as x gets closer to 1, the outputs $G(x)$ get closer to 3. Thus, $\lim_{x \to 1} G(x) = 3$.

b) Limit Numerically

$x \to 3^-$, $(x < 3)$	$G(x)$
2.5	1.5
2.9	1.1
2.99	1.01
2.999	1.001

$x \to 3^+$, $(x > 3)$	$G(x)$
3.5	2.2247
3.1	2.0488
3.01	2.0050
3.001	2.0005

Limit Graphically

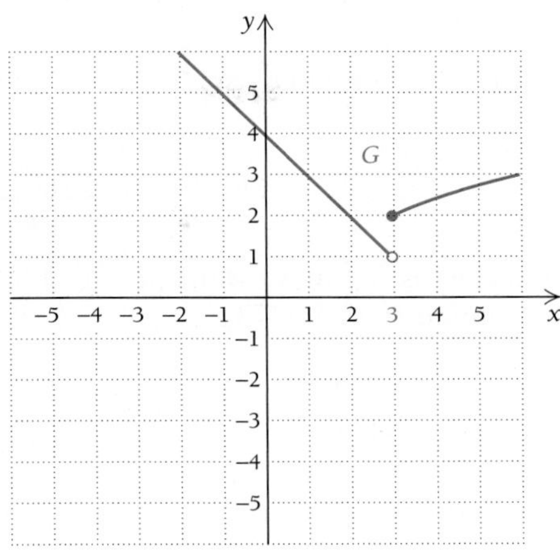

$\lim_{x\to 3} G(x)$ does not exist.

Both the tables and the graph indicate that $\lim_{x\to 3^-} G(x) \neq \lim_{x\to 3^+} G(x)$. Since the left-sided and right-sided limits differ, $\lim_{x\to 3} G(x)$ does not exist. ✦

Algebraic methods allow us to streamline our work. In particular, for any rational function F, with a in the domain of F,

$$\lim_{x\to a} F(x) = F(a).$$ (p. 116)

EXAMPLE Find $\lim_{x\to 3} (x^3 - 2x^2 + 7)$.

Solution Every polynomial function is a rational function; thus,

$$\begin{aligned}\lim_{x\to 3} (x^3 - 2x^2 + 7) &= 3^3 - 2 \cdot 3^2 + 7 \\ &= 27 - 18 + 7 \\ &= 16.\end{aligned}$$ ✦

The concept of limit is central to the definition of continuity. A function f is continuous at $x = a$ if:

a) $f(a)$ exists. (The output at a exists.)

b) $\lim_{x\to a} f(x)$ exists. (The limit as $x \to a$ exists.)

and

c) $\lim_{x\to a} f(x) = f(a)$. (The limit is the same as the output.) (p. 118)

EXAMPLE Is the function g given by $g(x) = \dfrac{x^2 - 3x - 4}{x + 1}$ continuous over $[-3, 3]$?

Solution For g to be continuous over $[-3, 3]$, it must be continuous at each point in $[-3, 3]$. Note that

$$g(x) = \frac{x^2 - 3x - 4}{x + 1} = \frac{(x + 1)(x - 4)}{x + 1} = x - 4, \quad \text{provided } x \neq -1.$$

Since -1 is not in the domain of g, it follows that $g(-1)$ does not exist. Thus, g is not continuous over $[-3, 3]$.

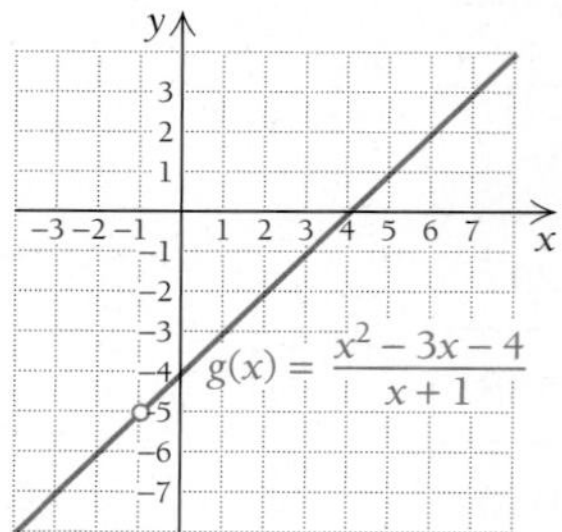

✦

When a function is continuous, it is possible to discuss the rate at which its second coordinates are changing with respect to its first coordinates.

The **average rate of change** as x changes from x_1 to x_2 is the slope of the line passing through (x_1, y_1) and (x_2, y_2). (p. 125)

EXAMPLE Find the average rate of change of $y = f(x) = \sqrt{x - 1} + 2$ as x changes from 1 to 5.

Solution We have $f(1) = 2$ and $f(5) = 4$, so the average rate of change is

$$\frac{y_2 - y_1}{x_2 - x_1} = \frac{4 - 2}{5 - 1} = \frac{2}{4} = \frac{1}{2}.$$

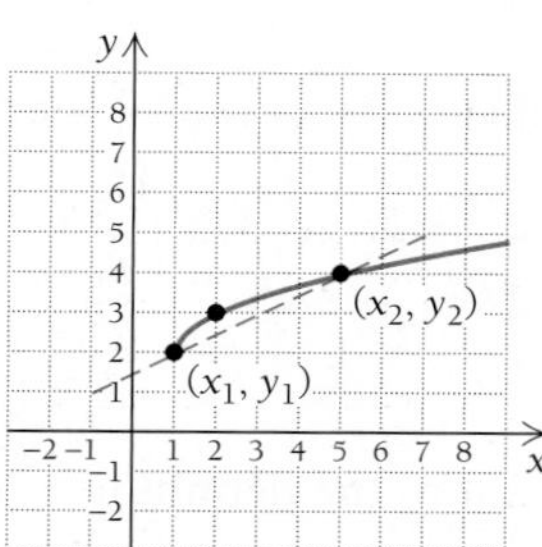

✦

The rate of change in the preceding example can be expressed using the **difference quotient:**

$$\frac{f(x_1 + h) - f(x_1)}{h},$$

where $x_1 = 1$ and $h = 4$. (p. 128)

EXAMPLE For $f(x) = x^2 + 3$, find a simplified form of the difference quotient.

Solution We have $f(x + h) = x^2 + 2xh + h^2 + 3$, so

$$\begin{aligned}\frac{f(x+h) - f(x)}{h} &= \frac{x^2 + 2xh + h^2 + 3 - (x^2 + 3)}{h} \\ &= \frac{2xh + h^2}{h} = \frac{h(2x + h)}{h} \\ &= 2x + h, \quad h \neq 0.\end{aligned}$$

✦

By taking the limit as $h \to 0$, we can use a difference quotient to find a function's **instantaneous rate of change** at a point $(x, f(x))$. This number is the slope of the tangent line at that point and is most commonly referred to as the **derivative** of f at x. (pp. 137–138)

EXAMPLE For $f(x) = -x^2 + 5$, find $f'(x)$ and $f'(2)$.

Solution We have

$$\begin{aligned}\frac{f(x+h) - f(x)}{h} &= \frac{-(x+h)^2 + 5 - (-x^2 + 5)}{h} \\ &= \frac{-(x^2 + 2xh + h^2) + 5 + x^2 - 5}{h} \\ &= \frac{-2xh - h^2}{h} \\ &= -2x - h, \quad h \neq 0.\end{aligned}$$

Since

$$f'(x) = \lim_{h \to 0} \frac{f(x+h) - f(x)}{h} = \lim_{h \to 0}(-2x - h),$$

we have

$$f'(x) = -2x.$$

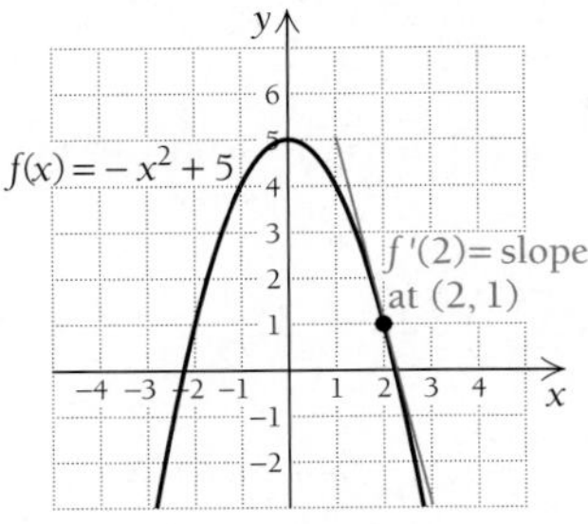

It follows that $f'(2) = -2 \cdot 2 = -4$. ✦

If $f(x) = y$, we often replace the notation $f'(x)$ with **Leibniz notation,** $\frac{dy}{dx}$. We can also write $\frac{d}{dx}f(x)$ to indicate the derivative, with respect to x, of $f(x)$. (p. 147)

Several rules permit us to avoid using difference quotients when calculating certain derivatives.

The Power Rule: $\frac{d}{dx}x^k = k \cdot x^{k-1}$ (p. 148)

EXAMPLE $\frac{d}{dx}x^7 = 7x^6$ ✦

The derivative of a constant: $\frac{d}{dx}c = 0$ (p. 150)

EXAMPLE $\frac{d}{dx}34 = 0$ ✦

The derivative of a constant times a function: $\frac{d}{dx}[c \cdot f(x)] = c \cdot f'(x)$ (p. 151)

EXAMPLE $\frac{d}{dx}3x^8 = 3 \cdot 8x^7 = 24x^7$ ✦

The Sum Rule: $\frac{d}{dx}[f(x) + g(x)] = f'(x) + g'(x)$ (p. 152)

EXAMPLE $\frac{d}{dx}[x^7 + 3x] = 7x^6 + 3$ ✦

The Difference Rule: $\frac{d}{dx}[f(x) - g(x)] = f'(x) - g'(x)$ (p. 152)

EXAMPLE $\frac{d}{dx}\left[5x - x^4\right] = 5 - 4x^3$ ✦

The Product Rule: $\frac{d}{dx}[f(x) \cdot g(x)] = f(x) \cdot g'(x) + g(x) \cdot f'(x)$ (p. 159)

EXAMPLE

$$\frac{d}{dx}\left[(2x + 3)\sqrt{x}\right] = (2x + 3)\frac{1}{2}x^{-1/2} + \sqrt{x} \cdot 2$$

$$= \frac{2x + 3}{2\sqrt{x}} + 2\sqrt{x}$$ ✦

The Quotient Rule: $\frac{d}{dx}\left[\frac{f(x)}{g(x)}\right] = \frac{g(x) \cdot f'(x) - f(x) \cdot g'(x)}{[g(x)]^2}$ (p. 160)

EXAMPLE $\frac{d}{dx}\left(\frac{3x - 1}{2x + 5}\right) = \frac{(2x + 5)3 - (3x - 1)2}{(2x + 5)^2} = \frac{17}{(2x + 5)^2}$ ✦

The Extended Power Rule: $\frac{d}{dx}[g(x)]^k = k[g(x)]^{k-1} \cdot \frac{d}{dx}g(x)$ (p. 169)

EXAMPLE $\frac{d}{dx}(3x + 5)^7 = 7(3x + 5)^6 \cdot 3 = 21(3x + 5)^6$ ✦

The Chain Rule: $\frac{d}{dx}[(f \circ g)(x)] = \frac{d}{dx}[f(g(x))] = f'(g(x)) \cdot g'(x)$ (p. 173)

EXAMPLE For $y = u^3 + u^2$ and $u = 4t + 5$, find $\frac{dy}{dt}$.

Solution

$$\begin{aligned}
\frac{dy}{dt} = \frac{dy}{du} \cdot \frac{du}{dt} &= (3u^2 + 2u)4 \\
&= (3(4t + 5)^2 + 2(4t + 5))4 \\
&= 12(4t + 5)^2 + 8(4t + 5) \\
&= 12(16t^2 + 40t + 25) + 32t + 40 \\
&= 192t^2 + 512t + 340
\end{aligned}$$ ✦

In many cases, it is useful to find the derivative of a derivative. This is called a **second derivative** and notated as f'' or y'' or $\frac{d^2y}{dx^2}$. (p. 178)

EXAMPLE If $f(x) = x^5$, find $f''(x)$.

Solution We have $f'(x) = 5x^4$, so $f''(x) = 20x^3$. ✦

One of the most common uses of the second derivative is in physics, where acceleration is the derivative of velocity (with respect to time) and velocity is the derivative of distance, or position (again with respect to time). Thus, acceleration is the second derivative of the distance function with respect to time. (pp. 179–180)

EXAMPLE An object is $s(t) = 3t^2 + 1$ miles from a starting point t hours after a clock was started. Find the object's velocity and acceleration after 2 hr.

Solution Velocity $= v(t) = s'(t) = 6t$, so $v(2) = 12$ mph.
Acceleration $= a(t) = s''(t) = 6$, so $a(2) = 6\,\text{mi/hr}^2$. ✦

Chapter Review Exercises

These review exercises are for test preparation. They can also be used as a practice test. Answers are at the back of the book. The bracketed section references tell you what part(s) of the chapter to restudy if your answer is incorrect.

CONCEPT REINFORCEMENT

Classify each statement as either true or false.

1. If $\lim_{x\to 5} f(x)$ exists, then $f(5)$ must exist. [1.1]

2. If $\lim_{x\to 2} f(x) = L$, then $L = f(2)$. [1.1]

3. If f is continuous at $x = 3$, then $\lim_{x\to 3} f(x) = f(3)$. [1.2]

4. A function's average rate of change over the interval [2, 8] is the same as its instantaneous rate of change at $x = 5$. [1.3, 1.4]

5. A function's derivative at a point, if it exists, can be found as the limit of a difference quotient. [1.4]

6. For $f'(5)$ to exist, f must be continuous at 5. [1.4]

7. If f is continuous at 5, then $f'(5)$ must exist. [1.4]

8. The acceleration function is the derivative of the velocity function. [1.8]

Match each function in column A with the rule in column B that would be the most appropriate to use for differentiating the function. [1.5, 1.6]

Column A	Column B
9. $f(x) = x^7$	**a)** Extended Power Rule
10. $g(x) = x + 9$	**b)** Product Rule
11. $F(x) = (5x - 3)^4$	**c)** Sum Rule
12. $G(x) = \dfrac{2x + 1}{3x - 4}$	**d)** Difference Rule
13. $H(x) = f(x) \cdot g(x)$	**e)** Power Rule
14. $f(x) = 2x - 7$	**f)** Quotient Rule

For Exercises 15–17, consider

$$\lim_{x\to -7} f(x), \quad \text{where } f(x) = \frac{x^2 + 4x - 21}{x + 7}.$$

15. Limit numerically. [1.1]

a) Complete the following input–output tables.

$x \to -7^-$	$f(x)$
-8	
-7.5	
-7.1	
-7.01	
-7.001	
-7.0001	

$x \to -7^+$	$f(x)$
-6	
-6.5	
-6.9	
-6.99	
-6.999	
-6.9999	

b) Find $\lim_{x\to -7^-} f(x)$, $\lim_{x\to -7^+} f(x)$, and $\lim_{x\to -7} f(x)$, if each exists.

16. Limit graphically. Graph the function, and use the graph to find the limit. [1.1]

17. Limit algebraically. Find the limit algebraically. Show your work. [1.2]

Find each limit, if it exists. If a limit does not exist, state that fact. [1.1, 1.2]

18. $\lim_{x\to -2} \dfrac{8}{x}$

19. $\lim_{x\to 1} (4x^3 - x^2 + 7x)$

20. $\lim_{x\to -7} \dfrac{x^2 + 2x - 35}{x + 7}$

21. $\lim_{x\to \infty} \dfrac{1}{x} + 3$ [1.1]

From the graphs in Exercises 22 and 23, determine whether each function is continuous and explain why or why not. [1.2]

22.

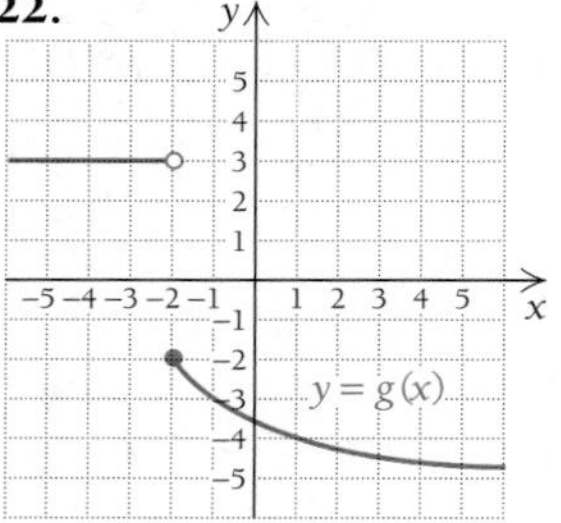

23.

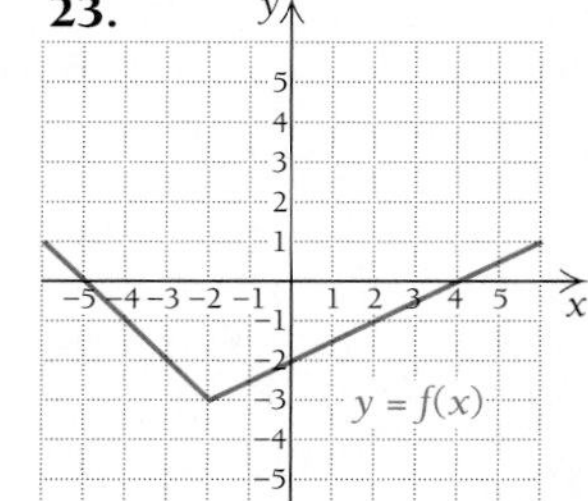

For the function graphed in Exercise 22, answer the following.

24. Find $\lim_{x \to 1} g(x)$. [1.2]

25. Find $g(1)$. [1.2]

26. Is g continuous at 1? Why or why not? [1.2]

27. Find $\lim_{x \to -2} g(x)$. [1.2]

28. Find $g(-2)$. [1.2]

29. Is g continuous at -2? Why or why not? [1.2]

30. For $f(x) = x^3 + x^2 - 2x$, find the average rate of change as x changes from -1 to 2. [1.3]

31. Find a simplified difference quotient for $g(x) = -3x + 2$. [1.3]

32. Find a simplified difference quotient for $f(x) = 2x^2 - 3$. [1.3]

33. Find an equation of the tangent line to the graph of $y = x^2 + 3x$ at the point $(-1, -2)$. [1.4]

34. Find the points on the graph of $y = -x^2 + 8x - 11$ at which the tangent line is horizontal. [1.5]

35. Find the points on the graph of $y = 5x^2 - 49x + 12$ at which the tangent line has slope 1. [1.5]

Find dy/dx.

36. $y = 9x^5$ [1.5]

37. $y = 8\sqrt[3]{x}$ [1.5]

38. $y = \dfrac{-3}{x^8}$ [1.5]

39. $y = 15x^{2/5}$ [1.5]

40. $y = 0.1x^7 - 3x^4 - x^3 + 6$ [1.5]

Differentiate.

41. $f(x) = \dfrac{5}{12}x^6 + 8x^4 - 2x$ [1.5]

42. $y = \dfrac{x^3 + x}{x}$ [1.5, 1.6]

43. $y = \dfrac{x^2 + 8}{8 - x}$ [1.6]

44. $g(x) = (5 - x)^2(2x - 1)^5$ [1.6]

45. $f(x) = (x^5 - 3)^7$ [1.7]

46. $f(x) = x^2(4x + 2)^{3/4}$ [1.7]

47. For $y = x^3 - \dfrac{2}{x}$, find $\dfrac{d^4y}{dx^4}$. [1.8]

48. For $y = \dfrac{3}{42}x^7 - 10x^3 + 13x^2 + 28x - 2$, find y''. [1.8]

49. For $s(t) = t + t^4$, with t in seconds and $s(t)$ in feet, find each of the following. [1.8]

a) $v(t)$
b) $a(t)$
c) The velocity and the acceleration when $t = 2$ sec

50. Business: average revenue, cost, and profit. Given revenue and cost functions $R(x) = 40x$ and $C(x) = 5\sqrt{x} + 100$, find each of the following. Assume $R(x)$ and $C(x)$ are in dollars and x is the number of items produced. [1.6]

a) The average cost, the average revenue, and the average profit when x items are produced and sold
b) The rate at which average cost is changing when 9 items are produced

51. Life science: growth rate. The population of a city grows from an initial size of 10,000 to a size P, given by $P = 10{,}000 + 50t^2$, where t is in years. [1.5]

a) Find the growth rate.
b) Find the number of people in the city after 20 yr (at $t = 20$).
c) Find the growth rate at $t = 20$.

52. Find $(f \circ g)(x)$ and $(g \circ f)(x)$, given that $f(x) = x^2 + 5$ and $g(x) = 1 - 2x$. [1.7]

SYNTHESIS

53. Differentiate $y = \dfrac{x\sqrt{1 + 3x}}{1 + x^3}$. [1.7]

TECHNOLOGY CONNECTION

Create an input–output table that includes each of the following limits. Start with ΔTbl = 0.1 and then go to 0.01, 0.001, and 0.0001. When you think you know the limit, graph the functions, and use the TRACE *feature to verify your assertion.*

54. $\lim_{x \to 1} \dfrac{2 - \sqrt{x + 3}}{x - 1}$ [1.1, 1.5]

55. $\lim_{x \to 11} \dfrac{\sqrt{x - 2} - 3}{x - 11}$ [1.1, 1.5]

56. Graph f and f' over the given interval. Then estimate points at which the tangent line to f is horizontal.[1.5]

$f(x) = 3.8x^5 - 18.6x^3$; $[-3, 3]$

Chapter 1 Test

For Exercises 1–3, consider

$$\lim_{x\to 6} f(x), \quad \text{where } f(x) = \frac{x^2 - 36}{x - 6}.$$

1. Numerical limits.

a) Complete the following input–output tables.

$x \to 6^-$	$f(x)$
5	
5.7	
5.9	
5.99	
5.999	
5.9999	

$x \to 6^+$	$f(x)$
7	
6.5	
6.1	
6.01	
6.001	
6.0001	

b) Find $\lim_{x\to 6^-} f(x)$, $\lim_{x\to 6^+} f(x)$, and $\lim_{x\to 6} f(x)$, if each exists.

2. Graphical limits. Graph the function, and use the graph to find the limit.

3. Algebraic limits. Find the limit algebraically. Show your work.

Graphical limits. *Consider the following graph of function f for Exercises 4–11.*

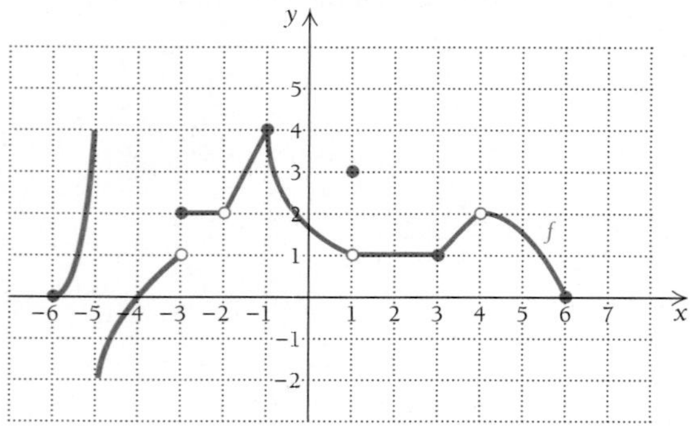

Find each limit, if it exists. If a limit does not exist, state that fact.

4. $\lim_{x\to -5} f(x)$

5. $\lim_{x\to -4} f(x)$

6. $\lim_{x\to -3} f(x)$

7. $\lim_{x\to -2} f(x)$

8. $\lim_{x\to -1} f(x)$

9. $\lim_{x\to 1} f(x)$

10. $\lim_{x\to 2} f(x)$

11. $\lim_{x\to 3} f(x)$

Determine whether each function is continuous. If a function is not continuous, state why.

12.

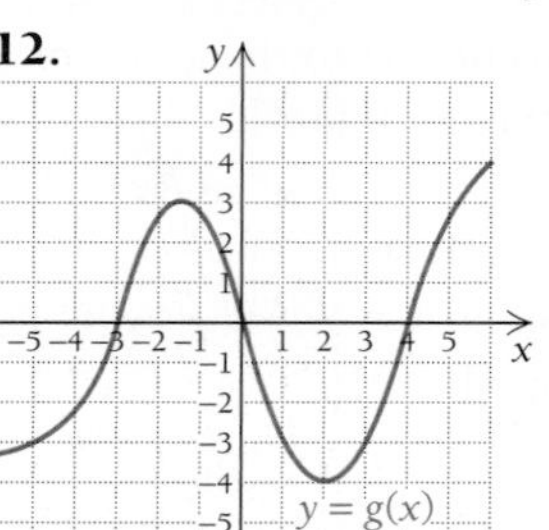

13.

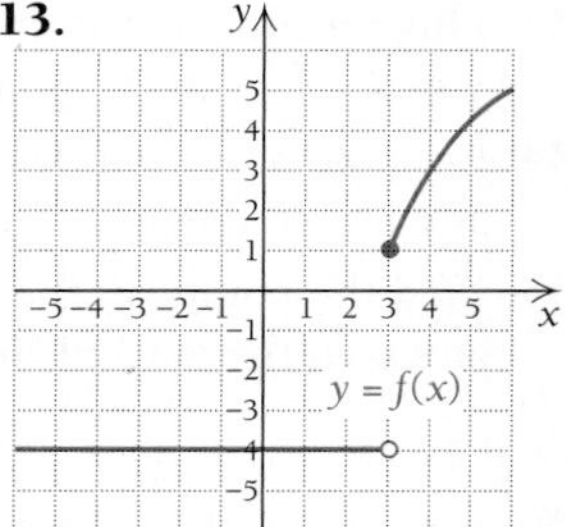

Consider the function shown in Exercise 13.

14. Find $\lim_{x\to 3} f(x)$.

15. Find $f(3)$.

16. Is f continuous at 3?

17. Find $\lim_{x\to 4} f(x)$.

18. Find $f(4)$.

19. Is f continuous at 4?

Find each limit, if it exists. If a limit does not exist, state why.

20. $\lim_{x\to 1} (3x^4 - 2x^2 + 5)$

21. $\lim_{x\to 2^+} \frac{x - 2}{x(x^2 - 4)}$

22. $\lim_{x\to 0} \frac{7}{x}$

23. Find a simplified difference quotient for

$f(x) = 2x^2 + 3x - 9$.

24. Find an equation of the tangent line to the graph of $y = x + (4/x)$ at the point $(4, 5)$.

25. Find the point(s) on the graph of $y = x^3 - 3x^2$ at which the tangent line is horizontal.

Find dy/dx.

26. $y = x^{23}$

27. $y = 4\sqrt[3]{x} + 5\sqrt{x}$

28. $y = \frac{-10}{x}$

29. $y = x^{5/4}$

30. $y = -0.5x^2 + 0.61x + 90$

Differentiate.

31. $y = \frac{1}{3}x^3 - x^2 + 2x + 4$

32. $y = \frac{3x - 4}{x^3}$

33. $f(x) = \frac{x}{5 - x}$

34. $f(x) = (x + 3)^4(7 - x)^5$

35. $y = (x^5 - 4x^3 + x)^{-5}$

36. $f(x) = x\sqrt{x^2 + 5}$

37. For $y = x^4 - 3x^2$, find $\frac{d^3y}{dx^3}$.

38. Business: average revenue, cost, and profit. Given revenue and cost functions $R(x) = 50x$ and $C(x) = x^{2/3} + 750$, where x is the number of items produced and $R(x)$ and $C(x)$ are in dollars, find the following:

a) The average revenue, the average cost, and the average profit when x items are produced

b) The rate at which average cost is changing when 8 items are produced

39. Social sciences: memory. In a certain memory experiment, a person is able to memorize M words after t minutes, where $M = -0.001t^3 + 0.1t^2$.

a) Find the rate of change of the number of words memorized with respect to time.

b) How many words are memorized during the first 10 min (at $t = 10$)?

c) At what rate are words being memorized after 10 min (at $t = 10$)?

40. Find $(f \circ g)(x)$ and $(g \circ f)(x)$ for $f(x) = x^2 - x$ and $g(x) = 2x^3$.

SYNTHESIS

41. Differentiate $y = \sqrt{(1 - 3x)^{2/3}(1 + 3x)^{1/3}}$.

42. Find $\lim_{x \to 3} \frac{x^3 - 27}{x - 3}$.

TECHNOLOGY CONNECTION

43. Graph f and f' over the interval $[0, 5]$. Then estimate points at which the tangent line to f is horizontal.

$$f(x) = 5x^3 - 30x^2 + 45x + 5\sqrt{x}; \quad [0, 5]$$

44. Find the following limit by creating a table of values:

$$\lim_{x \to 0} \frac{\sqrt{5x + 25} - 5}{x}.$$

Start with ΔTbl $= 0.1$ and then go to 0.01 and 0.001. When you think you know the limit, graph

$$y = \frac{\sqrt{5x + 25} - 5}{x},$$

and use the TRACE feature to verify your assertion.

EXTENDED TECHNOLOGY APPLICATION

Path of a Baseball: The Tale of the Tape

Have you ever watched a baseball game and seen a home run ball hit an obstruction after it has cleared the fence? Suppose the ball hits a sign at a location that is 60 ft above the ground at a distance of 400 ft from home plate. An announcer or message on the scoreboard might proclaim, "According to the tale of the tape, the ball would have traveled 442 ft." How is such a calculation made? The answer is related to the curve formed by the path of a baseball.

Whatever the path of a well-hit baseball is, it is *not* the graph of a parabola,

$$f(x) = ax^2 + bx + c.$$

A well-hit baseball follows the path of a "skewed" parabola, as shown at the right below. One reason that the ball's flight is not parabolic is that a well-hit ball has backspin. This fact, combined with the frictional effect of the ball's stitches with the air, skews the path of the ball in the direction of its landing.

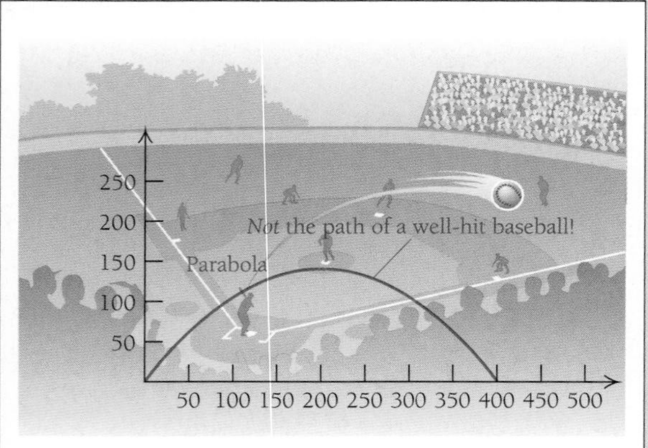

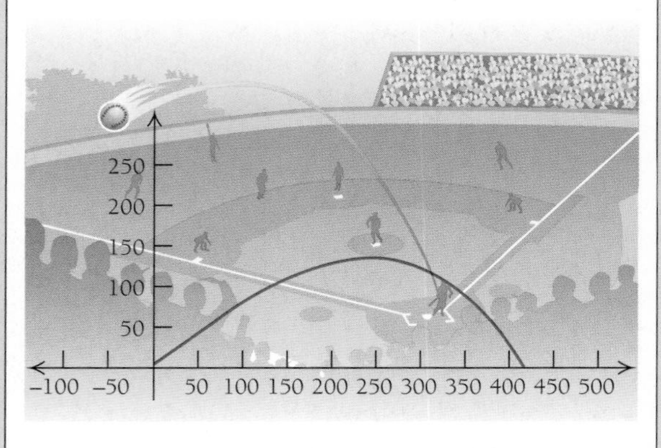

Let's see if we can model the path of a baseball. Consider the following data.

HORIZONTAL DISTANCE, x (IN FEET)	0	50	100	200	285	300	360	400
VERTICAL DISTANCE, y (IN FEET)	4.5	43	82	130	142	134	100	60

Assume for the given data that (0, 4.5) is the point at home plate at which the ball is hit, roughly 4.5 ft off the ground. Also, assume that the ball has hit a billboard 60 ft above the ground and 400 ft from home plate.

EXERCISES

1. Plot the points and connect them with line segments. This can be done on many calculators by pressing STAT PLOT, turning on PLOT, and selecting the appropriate type.

2. a) Use REGRESSION to find a cubic function

$$y = ax^3 + bx^2 + cx + d$$

that fits the data.

b) Graph the function over the interval [0, 500].

c) Does the function closely model the given data?

d) Predict the horizontal distance from home plate at which the ball would have hit the ground had it not hit the billboard.

e) Find the rate of change of the ball's height with respect to its horizontal distance from home plate.

f) Find the point(s) at which the graph has a horizontal tangent line. Explain the significance of the point(s).

3. a) Use REGRESSION to find a quartic function

$$y = ax^4 + bx^3 + cx^2 + dx + e$$

that fits the data.

b) Graph the function over the interval [0, 500].

c) Does the function closely model the given data?

d) Predict the horizontal distance from home plate at which the ball would have hit the ground had it not hit the billboard.

e) Find the rate of change of the ball's height with respect to its horizontal distance from home plate.

f) Find the point(s) at which the graph has a horizontal tangent line. Explain the significance of the point(s).

4. a) Although most calculators cannot fit such a function to the data, assume that the equation

$$y = 0.0015x\sqrt{202{,}500 - x^2}$$

has been found using some type of curve-fitting technique. Graph the function over the interval [0, 500].

b) Predict the horizontal distance from home plate at which the ball would have hit the ground had it not hit the billboard.

c) Find the rate of change of the ball's height with respect to its horizontal distance from home plate.

d) Find the point(s) at which the graph has a horizontal tangent line. Explain the significance of the point(s).

5. Compare the answers in Exercises 2(d), 3(d), and 4(b). Discuss the relative merits of using the quartic model in Exercise 3 with the model in Exercise 4 to make the prediction.

(continued)

Tale of the tape. Actually, scoreboard operators in the major leagues use different models to predict the distance that a home run ball would have traveled. The models are linear and are related to the trajectory of the ball, that is, how high the ball is hit. See the following graph.

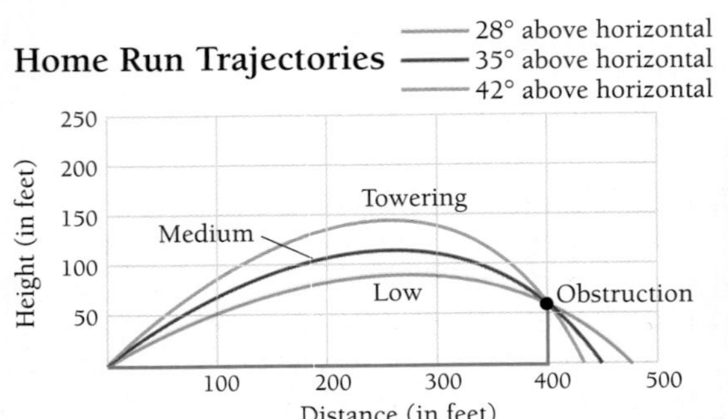

Suppose that a ball hits an obstruction d feet horizontally from home plate at a height of H feet. Then the estimated horizontal distance D that the ball would have traveled, depending on its trajectory type, is

Low trajectory:	$D = 1.1H + d$,
Medium trajectory:	$D = 0.7H + d$,
Towering trajectory:	$D = 0.5H + d$.

6. For a ball striking an obstacle at $d = 400$ ft and $H = 60$ ft, estimate how far the ball would have traveled if it were following a low trajectory, or a medium trajectory, or a towering trajectory.
7. In 1953, Hall-of-Famer Mickey Mantle hit a towering home run in old Griffith Stadium in Washington, D.C., that hit an obstruction 60 ft high and 460 ft from home plate. Reporters asserted at the time that the ball would have traveled 565 ft. Is this estimate valid?

Many thanks to Robert K. Adair, professor of physics at Yale University, for many of the ideas presented in this section.

Applications of Differentiation

2

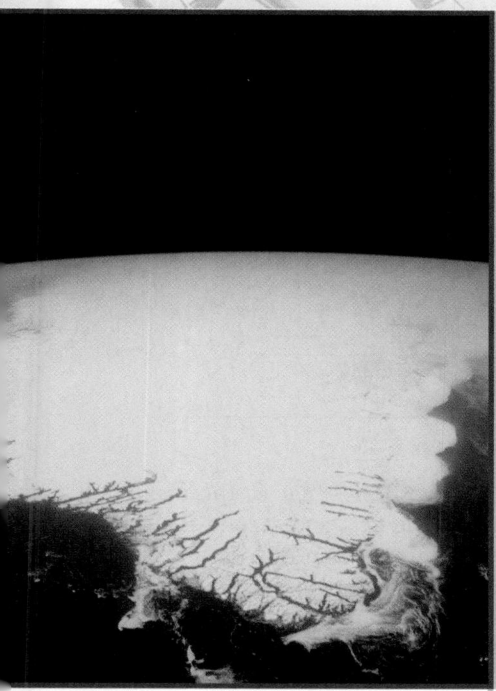

APPLICATION In a trend that scientists attribute, at least in part, to global warming, the floating cap of sea ice on the Arctic Ocean has been shrinking since 1950. The ice cap always shrinks in summer and grows in winter. Average minimum size of the ice cap, in square miles, can be approximated by

$$A = \pi r^2.$$

In 2005, the radius of the ice cap was approximately 808 mi and was shrinking at the rate of approximately 4.3 mi/yr. (*Source*: Based on data for 2005 and information from gsfc.nasa.gov and the *New York Times*, 9/29/05.) How fast was the area changing at that time?

This problem appears as Exercise 39 in Exercise Set 2.7.

INTRODUCTION

In this chapter, we explore many applications of differentiation. We learn to find maximum and minimum values of functions, and that skill allows us to solve many kinds of problems in which we need to find the largest and/or smallest value in a real-world situation. We also apply our differentiation skills to graphing and to approximating function values.

OBJECTIVES

- Find relative extrema of a continuous function using the First-Derivative Test.
- Sketch graphs of continuous functions.

2.1 Using First Derivatives to Find Maximum and Minimum Values and Sketch Graphs

The graph below illustrates a typical life cycle of a retail product. Note that the number of items sold varies with respect to time. Sales begin at a small level and increase to a point of maximum sales, after which they taper off to a low level, where the decline is probably due to the effect of new competitive products. The company then rejuvenates the product by making improvements. Think about products such as TVs, records, compact discs, and CD-ROMs. Where might each be in a typical product life cycle, and is that curve appropriate?

The graph of a typical product life cycle can be used to illustrate many of the ideas that we will consider in this chapter.

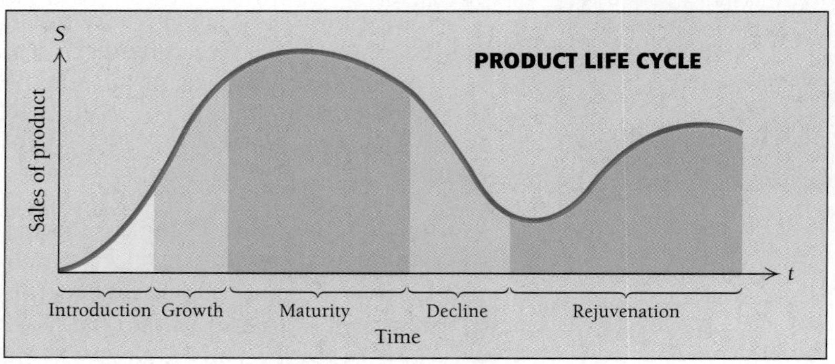

Finding the largest and smallest values of a function—that is, the maximum and minimum values—has extensive application. The first and second derivatives of a function are tools of calculus that give us information about the shape of a graph that is often helpful in finding maximum and minimum values and in graphing functions. Throughout this section, we will assume that the functions f are continuous, but this does not necessarily imply that f' and f'' are continuous.

Increasing and Decreasing Functions

If the graph of a function rises from left to right over an interval I, the function is said to be **increasing** on, or over, I.

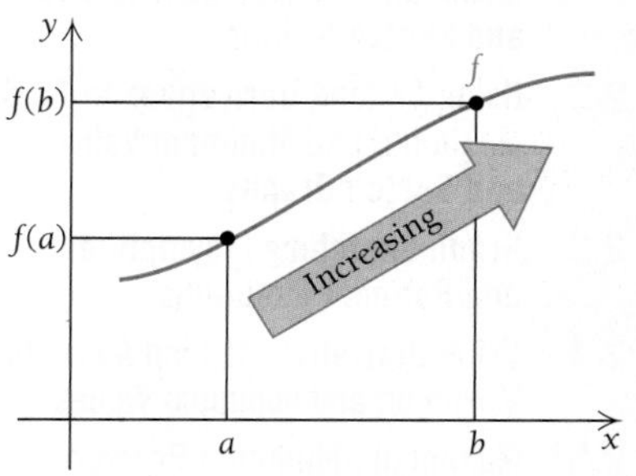

f is an increasing function over I: for all a, b in I, if $a < b$, then $f(a) < f(b)$.

TECHNOLOGY CONNECTION

Exploratory

Graph the function

$$y = -\tfrac{1}{3}x^3 + 6x^2 - 11x - 50$$

and its derivative

$$y' = -x^2 + 12x - 11$$

using the window $[-10, 25, -100, 150]$, with Xscl = 5 and Yscl = 25. Then TRACE from left to right along each graph. As you move the cursor from left to right, note that the x-coordinate always increases. If a function is increasing over an interval, the y-coordinate will increase as well. If a function is decreasing over an interval, the y-coordinate will decrease.

Over what intervals is the function increasing?
Over what intervals is the function decreasing?
Over what intervals is the derivative positive?
Over what intervals is the derivative negative?

What rules can you propose relating the sign of y' to the behavior of y?

If the graph drops from left to right, the function is said to be **decreasing** on, or over, I.

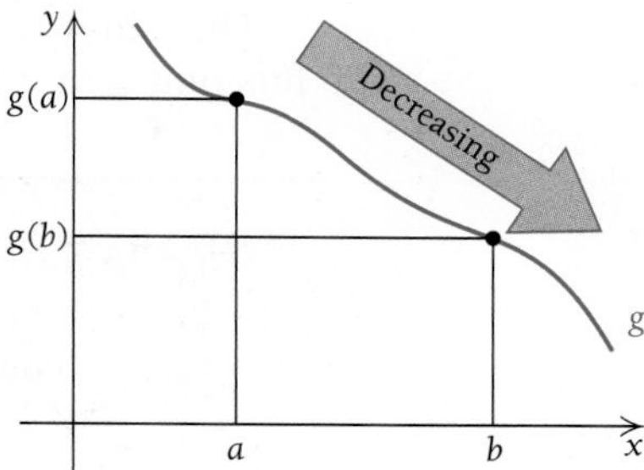

g is a decreasing function over I:
for all a, b in I, if $a < b$, then $g(a) > g(b)$.

We can describe these phenomena mathematically as follows.

DEFINITIONS

A function f is **increasing** over I if, for every a and b in I,

$$\text{if } a < b, \quad \text{then } f(a) < f(b).$$

(If the input a is less than the input b, then the output for a is less than the output for b.)

A function f is **decreasing** over I if, for every a and b in I,

$$\text{if } a < b, \quad \text{then } f(a) > f(b).$$

(If the input a is less than the input b, then the output for a is greater than the output for b.)

Note that the directions of the inequalities stay the same for an increasing function, but they differ for a decreasing function.

In Chapter R, we saw how the slope of a linear function determines whether that function is increasing or decreasing (or neither). For a general function, the derivative yields similar information.

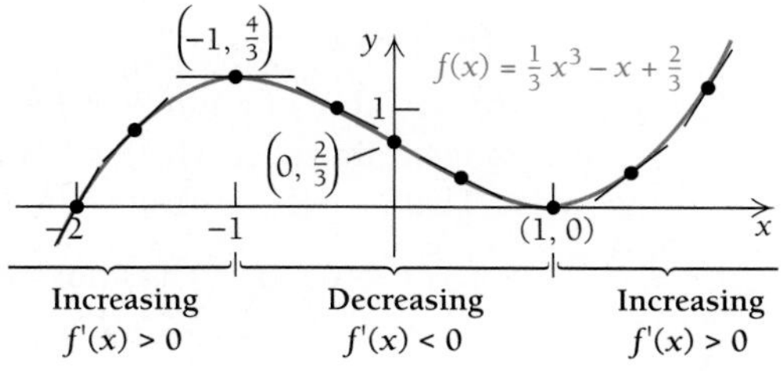

f is increasing over the intervals $(-\infty, -1]$ and $[1, \infty)$; slopes of tangent lines are positive.

f is decreasing over the interval $[-1, 1]$; slopes of tangent lines are negative.

Also note that the definitions of increasing and decreasing allow endpoints to be included in the intervals under consideration. Thus, for example, the function shown at the bottom of page 199 is decreasing not just on $(-1, 1)$, but on $[-1, 1]$.

The following theorem shows how we can use derivatives to determine whether a function is increasing or decreasing.

THEOREM 1

If $f'(x) > 0$ for all x in an interval I, then f is increasing over I.
If $f'(x) < 0$ for all x in an interval I, then f is decreasing over I.

Critical Values

Consider the following graph of a continuous function.

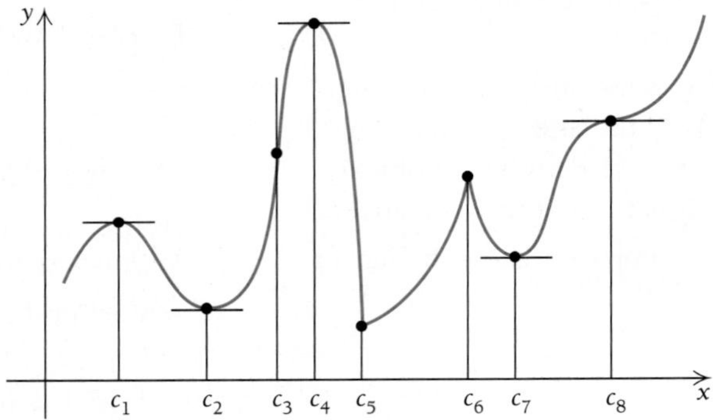

Note the following:

1. $f'(c) = 0$ at c_1, c_2, c_4, c_7, and c_8. That is, the tangent line to the graph is horizontal for these values.
2. $f'(c)$ does not exist at c_3, c_5, and c_6. The tangent line is vertical at c_3 and there is a corner point, or sharp point, at both c_5 and c_6. (See also the discussion at the end of Section 1.4.)

DEFINITION

A **critical value** of a function f is any number c, in the domain of f, for which the tangent line at $(c, f(c))$ is horizontal or for which the derivative does not exist. That is, c is a critical value if $f(c)$ exists and

$$f'(c) = 0 \quad \text{or} \quad f'(c) \text{ does not exist.}$$

Thus, in the preceding graph:

1. c_1, c_2, c_4, c_7, and c_8 are critical values because $f'(c) = 0$ for each value.
2. c_3, c_5, and c_6 are critical values because $f'(c)$ does not exist for each value.

Also note that a continuous function can change from increasing to decreasing or from decreasing to increasing *only* at a critical value. In the graph above, c_1, c_2, c_4, c_5, c_6, and c_7 separate the intervals over which the function changes from increasing to decreasing or from decreasing to increasing. Although c_3 and c_8 are critical values, they do not separate intervals over which the function changes from increasing to decreasing or from decreasing to increasing.

Finding Relative Maximum and Minimum Values

Consider the graph below. Note the "peaks" and "valleys" at the interior points c_1, c_2, and c_3.

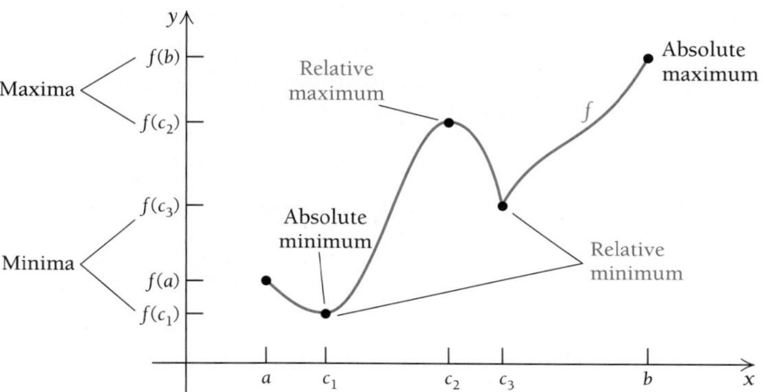

Here $f(c_2)$ is an example of a **relative maximum** (plural: **maxima**). Each of $f(c_1)$ and $f(c_3)$ is called a **relative minimum** (plural: **minima**). The terms **local maximum** and **local minimum** are also used.

DEFINITIONS

Let I be the domain of f.

$f(c)$ is a **relative minimum** if there exists within I an open interval I_1 containing c such that $f(c) \leq f(x)$, for all x in I_1;

and

$f(c)$ is a **relative maximum** if there exists within I an open interval I_2 containing c such that $f(c) \geq f(x)$, for all x in I_2.

A relative maximum can be thought of loosely as the second coordinate of a "peak" that may or may not be the highest point over all of I. Similarly, a relative minimum can be thought of as the second coordinate of a "valley" that may or may not be the lowest point on I. The second coordinates of the points that are the highest and the lowest on the interval are, respectively, the **absolute maximum** and the **absolute minimum**. For now, we focus on finding relative maximum or minimum values, collectively referred to as **relative extrema** (singular: **extremum**).

Look again at the preceding graph. The x-values at which a continuous function has relative extrema are those values for which the derivative is 0 or for which the derivative does not exist—the critical values.

THEOREM 2

If a function f has a relative extreme value $f(c)$ on an open interval, then c is a critical value, so

$$f'(c) = 0 \quad \text{or} \quad f'(c) \text{ does not exist.}$$

Theorem 2 is very useful, but it is important to understand it precisely. What it says is that to find relative extrema, we need only consider those inputs for which the derivative is 0 or for which it does not exist. We can think of a critical value as a *candidate* for a value where a relative extremum *might* occur. That is, Theorem 2 does not say that every critical value will yield a relative maximum or minimum. Consider, for example, the graph of

$$f(x) = (x - 1)^3 + 2,$$

shown at the right. Note that

$$f'(x) = 3(x - 1)^2,$$

and

$$f'(1) = 3(1 - 1)^2 = 0.$$

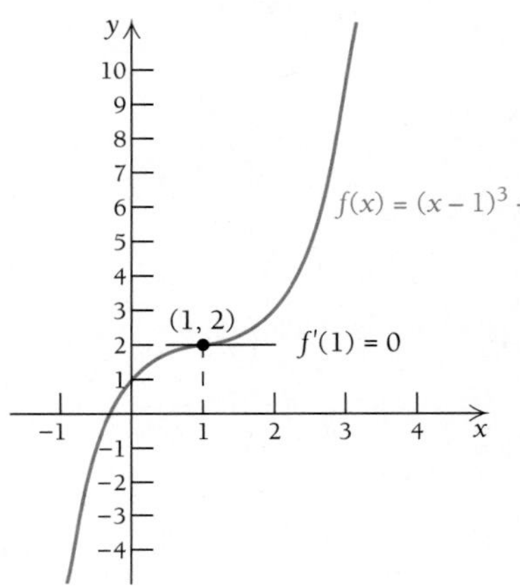

The function has $c = 1$ as a critical value, but has no relative maximum or minimum at that value.

Theorem 2 does say that if a relative maximum or minimum occurs, then the first coordinate of that extremum will be a critical value. How can we tell when the existence of a critical value leads us to a relative extremum? The following graph leads us to a test.

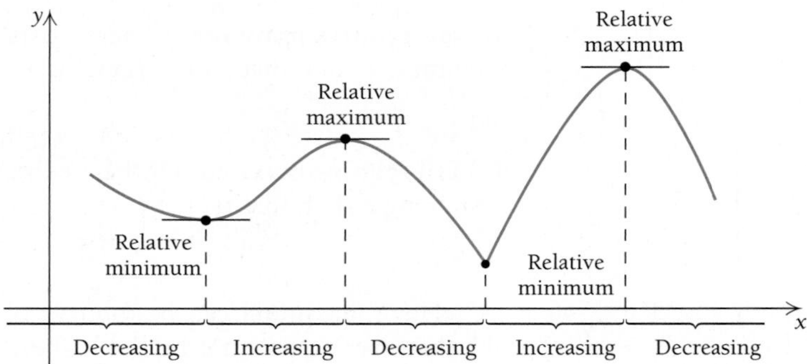

Note that at a critical value for which there is a relative minimum, the function is decreasing to the left of the critical value and increasing to the right. At a critical value for which there is a relative maximum, the function is increasing to the left of the critical value and decreasing to the right. In both cases, the derivative changes signs on either side of the critical value.

Graph over the interval (a, b)	$f(c)$	Sign of $f'(x)$ for x in (a, c)	Sign of $f'(x)$ for x in (c, b)	Increasing or decreasing
	Relative minimum	$-$	$+$	Decreasing on $(a, c]$; increasing on $[c, b)$
	Relative maximum	$+$	$-$	Increasing on $(a, c]$; decreasing on $[c, b)$
	No relative maxima or minima	$-$	$-$	Decreasing on (a, b)
	No relative maxima or minima	$+$	$+$	Increasing on (a, b)

Derivatives tell us when a function is increasing or decreasing. This leads us to the First-Derivative Test.

THEOREM 3 The First-Derivative Test for Relative Extrema

For any continuous function f that has exactly one critical value c in an open interval (a, b):

F1. f has a relative minimum at c if $f'(x) < 0$ on (a, c) and $f'(x) > 0$ on (c, b). That is, f is decreasing to the left of c and increasing to the right of c.

F2. f has a relative maximum at c if $f'(x) > 0$ on (a, c) and $f'(x) < 0$ on (c, b). That is, f is increasing to the left of c and decreasing to the right of c.

F3. f has neither a relative maximum nor a relative minimum at c if $f'(x)$ has the same sign on (a, c) as on (c, b).

Now let's see how we can use the First-Derivative Test to create and understand graphs and to find relative extrema.

EXAMPLE 1 Graph the function f given by

$$f(x) = 2x^3 - 3x^2 - 12x + 12,$$

and find the relative extrema.

Solution Suppose that we are trying to graph this function but don't know any calculus. What can we do? We could plot several points to determine in which direction the graph seems to be turning. Let's pick some x-values and see what happens.

x	$f(x)$
−3	−33
−2	8
−1	19
0	12
1	−1
2	−8
3	3
4	44

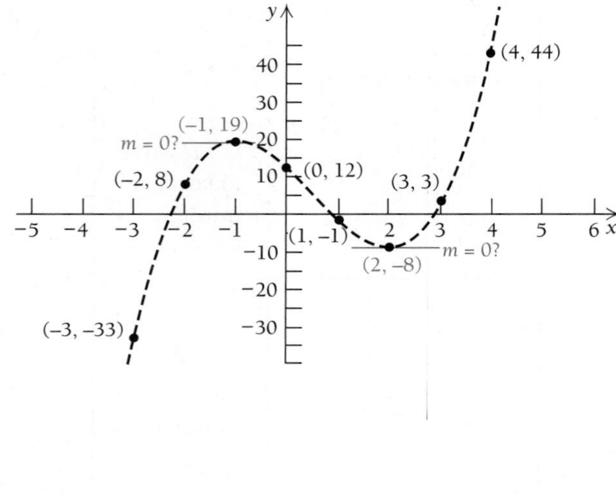

We plot the points and use them to sketch a "best guess" of the graph, shown as the dashed line in the figure above. According to this rough sketch, it would seem that the graph has a tangent line with slope 0 somewhere around $x = -1$ and $x = 2$. But how do we know for sure? We can begin by finding a general expression for the derivative:

$$f'(x) = 6x^2 - 6x - 12.$$

We next determine where $f'(x)$ does not exist or where $f'(x) = 0$. Since we can evaluate $f'(x) = 6x^2 - 6x - 12$ for any real number, there is no value for which $f'(x)$ does not exist. So the only possibilities for critical values are those where $f'(x) = 0$, locations at which there are horizontal tangents. To find such values, we solve $f'(x) = 0$:

$$6x^2 - 6x - 12 = 0$$

$$x^2 - x - 2 = 0 \qquad \text{Dividing both sides by 6}$$

$$(x + 1)(x - 2) = 0 \qquad \text{Factoring}$$

$$x + 1 = 0 \quad \text{or} \quad x - 2 = 0 \qquad \text{Using the Principle of Zero Products}$$

$$x = -1 \quad \text{or} \quad x = 2.$$

The critical values are -1 and 2. Since it is at these values that a relative maximum or minimum might exist, we examine the intervals on each side of the critical values: $A(-\infty, -1)$, $B(-1, 2)$, and $C(2, \infty)$, as shown below.

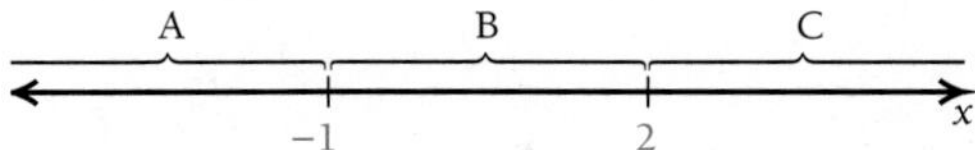

Next, we analyze the sign of the derivative on each interval. If $f'(x)$ is positive for one value in the interval, then it will be positive for all numbers in the interval. Similarly, if it is negative for one value, it will be negative for all values in the interval. This is because in order for the derivative to change signs, it would have to be 0 or undefined at some value. Such a value would be a critical value. Thus, we merely choose a test value in each interval and make a substitution. The test values we choose are -2, 0, and 4.

A: Test -2, $f'(-2) = 6(-2)^2 - 6(-2) - 12$
$= 24 + 12 - 12 = 24 > 0;$

B: Test 0, $f'(0) = 6(0)^2 - 6(0) - 12 = -12 < 0;$

C: Test 4, $f'(4) = 6(4)^2 - 6(4) - 12$
$= 96 - 24 - 12 = 60 > 0.$

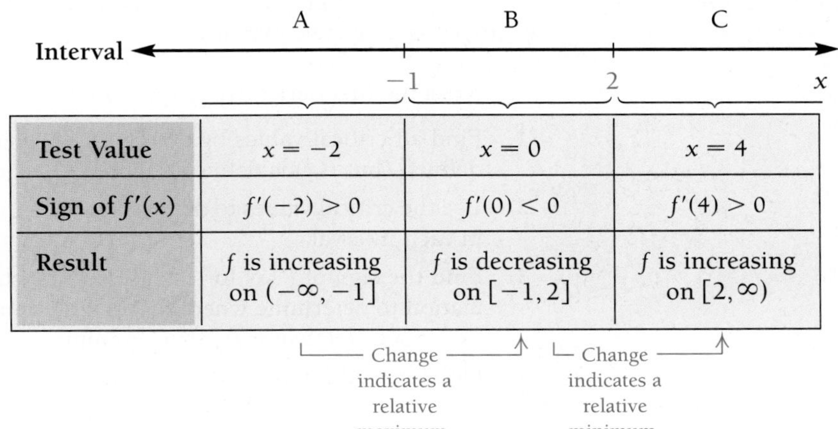

Interval	A	B	C
Test Value	$x = -2$	$x = 0$	$x = 4$
Sign of $f'(x)$	$f'(-2) > 0$	$f'(0) < 0$	$f'(4) > 0$
Result	f is increasing on $(-\infty, -1]$	f is decreasing on $[-1, 2]$	f is increasing on $[2, \infty)$

Therefore, by the First-Derivative Test,

f has a relative maximum at $x = -1$ given by

$$f(-1) = 2(-1)^3 - 3(-1)^2 - 12(-1) + 12$$ Substituting into the *original* function

$$= 19$$ This is a relative maximum.

and f has a relative minimum at $x = 2$ given by

$$f(2) = 2(2)^3 - 3(2)^2 - 12(2) + 12 = -8.$$ This is a relative minimum.

Thus, there is a relative maximum at $(-1, 19)$ and a relative minimum at $(2, -8)$, as we suspected from the sketch of the graph.

The information we have obtained can be very useful in sketching a graph of the function. We know that this polynomial is continuous, and we know where the function is increasing, where it is decreasing, and where it has relative extrema. We complete the graph by using a calculator to generate some additional function values. The graph of the function, shown below in red, has been scaled to clearly show its curving nature.

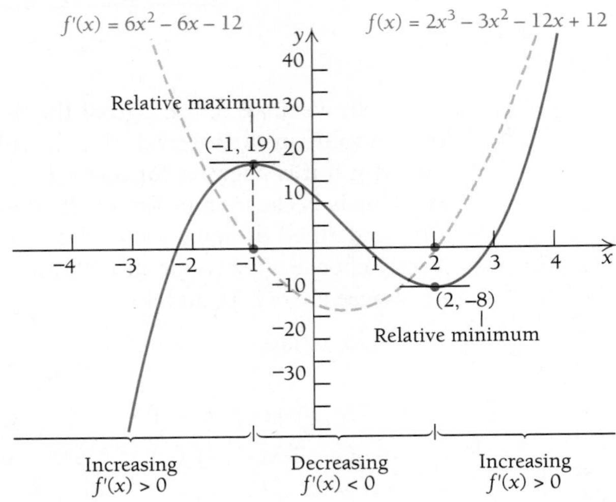

TECHNOLOGY CONNECTION

Exploratory

Consider the function f given by

$$f(x) = x^3 - 3x + 2.$$

Graph both f and f' using the same set of axes. Examine the graphs using the TABLE and TRACE features. Where do you think the relative extrema of $f(x)$ occur? Where is the derivative equal to 0? Where does $f(x)$ have critical values?

For reference, the graph of the derivative is shown in blue. Note that $f'(x) = 0$ where $f(x)$ has relative extrema. ✦

To use the first derivative for graphing a function f:

1. Find all critical values by determining where $f'(x)$ is 0 and where $f'(x)$ is undefined (but $f(x)$ is defined). Find $f(x)$ for each critical value.
2. Use the critical values to divide the x-axis into intervals and choose a test value in each interval.
3. Find the sign of $f'(x)$ for each test value chosen in step 2, and use this information to determine where $f(x)$ is increasing or decreasing and to classify any extrema as relative maxima or minima.
4. Plot some additional points and sketch the graph.

The *derivative* f' is used to find the critical values of f. The test values are substituted into the *derivative* f', and the function values are found using the *original* function f.

EXAMPLE 2 Find the relative extrema of the function f given by

$$f(x) = 2x^3 - x^4.$$

Then sketch the graph.

Solution First, we must determine the critical values. To do so, we find $f'(x)$:

$$f'(x) = 6x^2 - 4x^3.$$

Next, we find where $f'(x)$ does not exist or where $f'(x) = 0$. We can replace x in $f'(x) = 6x^2 - 4x^3$ with any real number. Thus, $f'(x)$ exists for all real numbers. So the only possibilities for critical values are where $f'(x) = 0$, that is, where the tangent line is horizontal:

$$\begin{aligned} 6x^2 - 4x^3 &= 0 && \text{Setting } f'(x) \text{ equal to } 0 \\ 2x^2(3 - 2x) &= 0 && \text{Factoring} \\ 2x^2 = 0 \quad &\text{or} \quad 3 - 2x = 0 \\ x^2 = 0 \quad &\text{or} \quad 3 = 2x \\ x = 0 \quad &\text{or} \quad x = \tfrac{3}{2}. \end{aligned}$$

The critical values are 0 and $\frac{3}{2}$. We use these values to divide the x-axis into three intervals: $A(-\infty, 0)$, $B\left(0, \frac{3}{2}\right)$, and $C\left(\frac{3}{2}, \infty\right)$, as shown below.

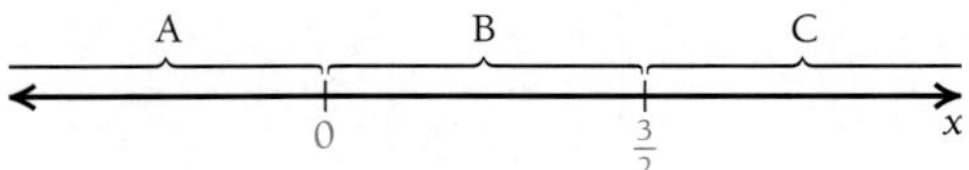

Note that $f\left(\frac{3}{2}\right) = 2\left(\frac{3}{2}\right)^3 - \left(\frac{3}{2}\right)^4 = \frac{27}{16}$ and $f(0) = 2 \cdot 0^3 - 0^4 = 0$ are possible extrema.

We now determine the sign of the derivative on each interval by choosing a test value in each interval and substituting. We generally choose test values for which it is easy to compute $f'(x)$.

$$\begin{aligned} &\text{A:} \quad \text{Test } -1, \quad f'(-1) = 6(-1)^2 - 4(-1)^3 \\ &\qquad\qquad\qquad\qquad\quad = 6 + 4 = 10 > 0; \\ &\text{B:} \quad \text{Test } 1, \quad f'(1) = 6(1)^2 - 4(1)^3 \\ &\qquad\qquad\qquad\qquad\quad = 6 - 4 = 2 > 0; \\ &\text{C:} \quad \text{Test } 2, \quad f'(2) = 6(2)^2 - 4(2)^3 \\ &\qquad\qquad\qquad\qquad\quad = 24 - 32 = -8 < 0. \end{aligned}$$

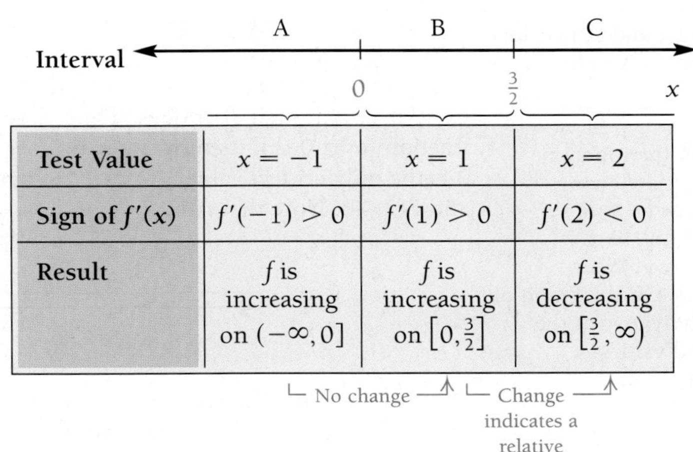

Interval	A	B	C
Test Value	$x = -1$	$x = 1$	$x = 2$
Sign of $f'(x)$	$f'(-1) > 0$	$f'(1) > 0$	$f'(2) < 0$
Result	f is increasing on $(-\infty, 0]$	f is increasing on $\left[0, \frac{3}{2}\right]$	f is decreasing on $\left[\frac{3}{2}, \infty\right)$

Therefore, by the First-Derivative Test, f has no extremum at $x = 0$, since $f(x)$ is increasing on both sides of 0, and f has a relative maximum at $x = \frac{3}{2}$. Thus, $f\left(\frac{3}{2}\right)$, or $\frac{27}{16}$, is a relative maximum.

We use the information obtained to sketch the graph below. Other function values are listed in the table. (More can be generated by the student.)

x	$f(x)$, approximately
-1	-3
-0.5	-0.31
0	0
0.5	0.19
1	1
1.25	1.46
2	0

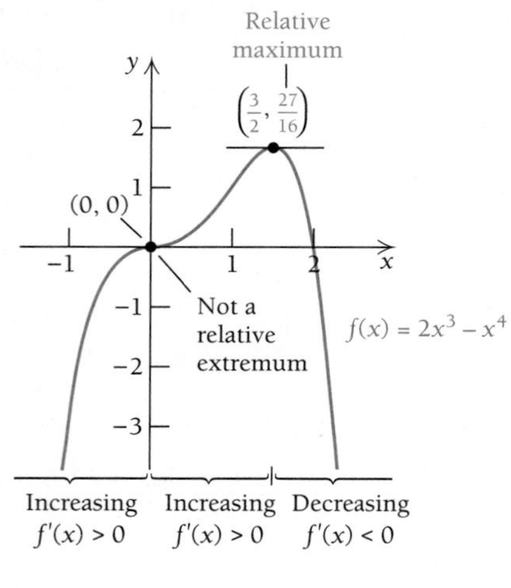

✦

EXAMPLE 3 Find the relative extrema of the function f given by

$$f(x) = (x - 2)^{2/3} + 1.$$

Then sketch the graph.

Solution First, we determine the critical values. To do so, we find $f'(x)$:

$$f'(x) = \frac{2}{3}(x - 2)^{-1/3}$$

$$= \frac{2}{3\sqrt[3]{x - 2}}.$$

Next, we find where $f'(x)$ does not exist or where $f'(x) = 0$. Note that $f'(x)$ does not exist at 2, although $f(x)$ does. Thus, 2 is a critical value. Since the only way for a fraction to be 0 is if its numerator is 0, we see that $f'(x) = 0$ has no solution. Thus, 2 is the only critical value. We use 2 to divide the x-axis into the intervals $A(-\infty, 2)$ and $B(2, \infty)$. Note that $f(2) = (2 - 2)^{2/3} + 1 = 1$.

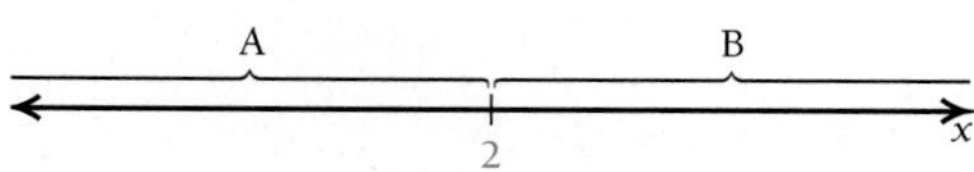

TECHNOLOGY CONNECTION

EXERCISES

In Exercises 1 and 2, consider the function f given by

$$f(x) = 2 - (x - 1)^{2/3}.$$

1. Graph the function using the viewing window $[-4, 6, -2, 4]$.
2. Graph the first derivative. What happens to the graph of the derivative at the critical values?

To determine the sign of the derivative, we choose a test value in each interval and substitute. We choose test values 0 and 3. It is not necessary to find an exact value of the derivative; we need only determine the sign. Sometimes we can do this by just examining the formula for the derivative:

$$\text{A: Test } 0, \quad f'(0) = \frac{2}{3\sqrt[3]{0 - 2}} < 0;$$

B: Test 3, $f'(3) = \dfrac{2}{3\sqrt[3]{3-2}} > 0.$

Interval ← 2 → x

Test Value	$x = 0$	$x = 3$
Sign of $f'(x)$	$f'(0) < 0$	$f'(3) > 0$
Result	f is decreasing on $(-\infty, 2]$	f is increasing on $[2, \infty)$

Change indicates a relative minimum.

Since we have a change from decreasing to increasing, we conclude from the First-Derivative Test that a relative minimum occurs at $(2, f(2))$, or $(2, 1)$. The graph has *no* tangent line at $(2, 1)$, since $f'(2)$ does not exist.

We use the information obtained to sketch the graph below. Other function values are listed in the table.

x	$f(x)$, approximately
-1	3.08
-0.5	2.84
0	2.59
0.5	2.31
1	2
1.5	1.63
2	1
2.5	1.63
3	2
3.5	2.31
4	2.59

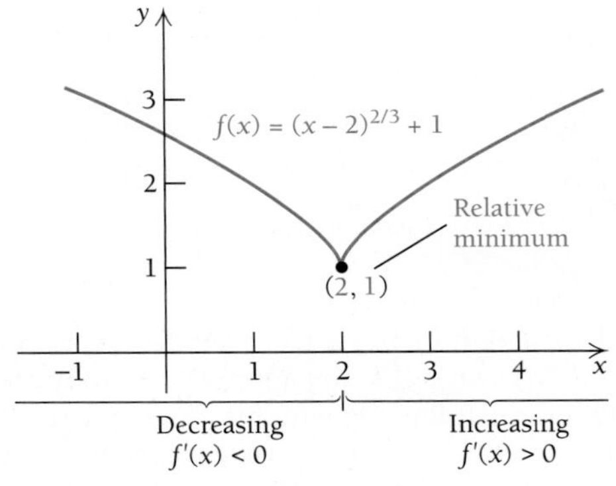

✦

TECHNOLOGY CONNECTION

Finding Relative Extrema

To explore some methods for approximating relative extrema, let's find the relative extrema of

$$f(x) = -0.4x^3 + 6.2x^2 - 11.3x - 54.8.$$

We first graph the function, using a window that reveals the curvature.

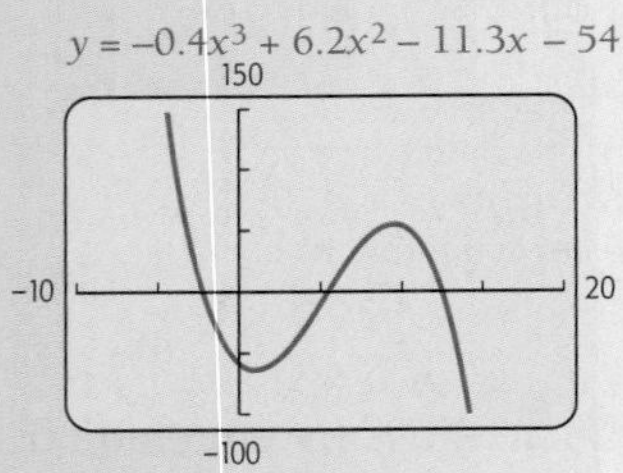

Method 1: TRACE

Beginning with the window shown above, we press TRACE and move the cursor along the curve, noting where relative extrema might occur.

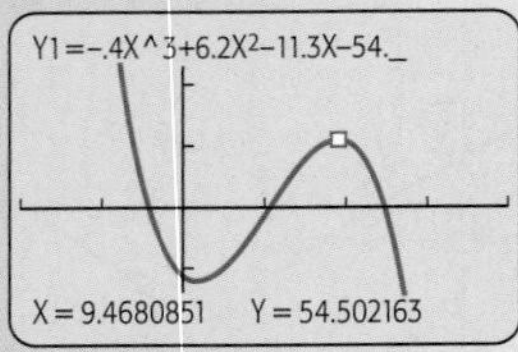

A relative maximum seems to be about $y = 54.5$ at $x = 9.47$. We can refine the approximation by zooming in to obtain the following window. We press TRACE and move the cursor along the curve, again noting where the y-value is largest. The approximation is about $y = 54.61$ at $x = 9.31$.

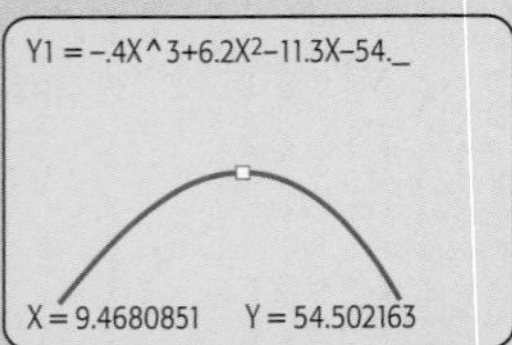

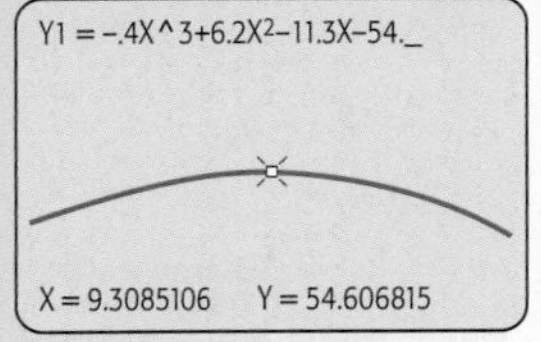

We can continue in this manner until the desired accuracy is achieved.

Method 2: TABLE

We can also use the TABLE feature, adjusting starting points and step values to improve accuracy:

TblStart = 9.3 ΔTbl = .01

X	Y1
9.3	54.605
9.31	54.607
9.32	54.608
9.33	54.608
9.34	54.607
9.35	54.604
9.36	54.601

X=9.32

The approximation seems to be nearly $y = 54.61$ at an x-value between 9.32 and 9.33. We could next set up a new table showing function values between $f(9.32)$ and $f(9.33)$ to refine the approximation.

Method 3: MAXIMUM, MINIMUM

Using the MAXIMUM option from the CALC menu, we find that a relative maximum of about 54.61 occurs at $x \approx 9.32$.

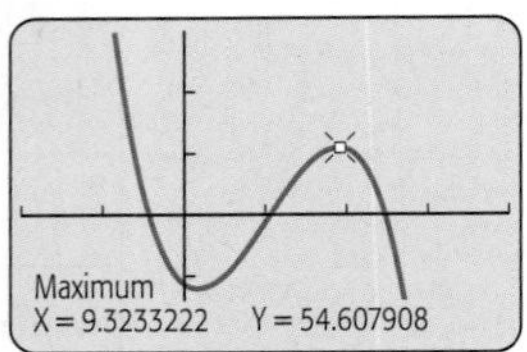

Method 4: fMax or fMin

This feature calculates a relative maximum or minimum value over any specified closed interval. We see from the initial graph that a relative maximum occurs in the interval $[-10, 20]$. Using the fMax option from the MATH menu, we see that a relative maximum occurs on $[-10, 20]$ when $x \approx 9.32$.

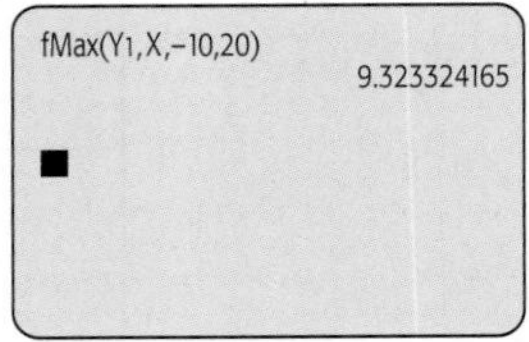

Finding Relative Extrema (continued)

To obtain the maximum value, we evaluate the function at the given x-value, obtaining the following.

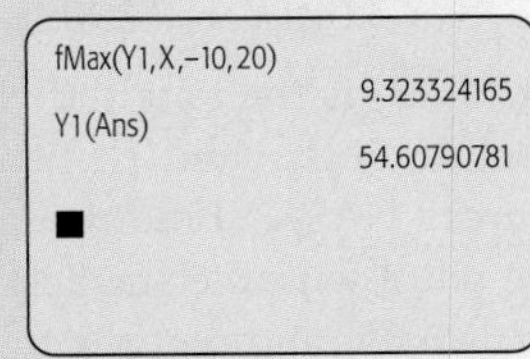

The approximation is about $y = 54.61$ at $x = 9.32$.

Using any of these methods, we find the relative minimum to be about $y = -60.30$ at $x = 1.01$.

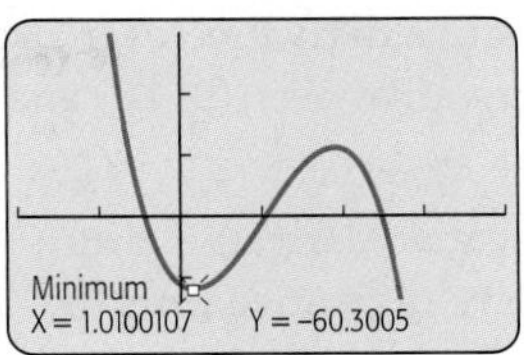

EXERCISE

1. Using one of the methods just described, approximate the relative extrema of the function in Example 1.

Exercise Set 2.1

Find the relative extrema of each function, if they exist. List each extremum along with the x-value at which it occurs. Then sketch a graph of the function.

1. $f(x) = x^2 + 4x + 5$

2. $f(x) = x^2 + 6x - 3$

3. $f(x) = 5 - x - x^2$

4. $f(x) = 2 - 3x - 2x^2$

5. $g(x) = 1 + 6x + 3x^2$

6. $F(x) = 0.5x^2 + 2x - 11$

7. $G(x) = x^3 - x^2 - x + 2$

8. $g(x) = x^3 + \frac{1}{2}x^2 - 2x + 5$

9. $f(x) = x^3 - 3x + 6$

10. $f(x) = x^3 - 3x^2$

11. $f(x) = 3x^2 + 2x^3$

12. $f(x) = x^3 + 3x$

13. $g(x) = 2x^3 - 16$

14. $F(x) = 1 - x^3$

15. $G(x) = x^3 - 6x^2 + 10$

16. $f(x) = 12 + 9x - 3x^2 - x^3$

17. $g(x) = x^3 - x^4$

18. $f(x) = x^4 - 2x^3$

19. $f(x) = \frac{1}{3}x^3 - 2x^2 + 4x - 1$

20. $F(x) = -\frac{1}{3}x^3 + 3x^2 - 9x + 2$

21. $g(x) = 2x^4 - 20x^2 + 18$

22. $f(x) = 3x^4 - 15x^2 + 12$

23. $F(x) = \sqrt[3]{x - 1}$

24. $G(x) = \sqrt[3]{x + 2}$

25. $f(x) = 1 - x^{2/3}$

26. $f(x) = (x + 3)^{2/3} - 5$

27. $G(x) = \dfrac{-8}{x^2 + 1}$

28. $F(x) = \dfrac{5}{x^2 + 1}$

29. $g(x) = \dfrac{4x}{x^2 + 1}$

30. $g(x) = \dfrac{x^2}{x^2 + 1}$

31. $f(x) = \sqrt[3]{x}$

32. $f(x) = (x + 1)^{1/3}$

33. $g(x) = \sqrt{x^2 + 2x + 5}$

34. $F(x) = \dfrac{1}{\sqrt{x^2 + 1}}$

35–68. Check the results of Exercises 1–34 using a calculator.

For Exercises 69–78, draw a graph to match the description given. Answers will vary.

69. $f(x)$ is increasing over $(-\infty, 2]$ and decreasing over $[2, \infty)$.

70. $g(x)$ is decreasing over $(-\infty, -3]$ and increasing over $[-3, \infty)$.

71. $G(x)$ is decreasing over $(-\infty, 4]$ and $[9, \infty)$ and increasing over $[4, 9]$.

72. $F(x)$ is increasing over $(-\infty, 5]$ and $[12, \infty)$ and decreasing over $[5, 12]$.

73. $g(x)$ has a positive derivative over $(-\infty, -3)$ and a negative derivative over $(-3, \infty)$.

74. $f(x)$ has a negative derivative over $(-\infty, 1)$ and a positive derivative over $(1, \infty)$.

75. $F(x)$ has a negative derivative over $(-\infty, 2)$ and $(5, 9)$ and a positive derivative over $(2, 5)$ and $(9, \infty)$.

76. $G(x)$ has a positive derivative over $(-\infty, -2)$ and $(4, 7)$ and a negative derivative over $(-2, 4)$ and $(7, \infty)$.

77. $f(x)$ has a positive derivative over $(-\infty, 3)$ and $(3, 9)$, a negative derivative over $(9, \infty)$, and a derivative equal to 0 at $x = 3$.

78. $g(x)$ has a negative derivative over $(-\infty, 5)$ and $(5, 8)$, a positive derivative over $(8, \infty)$, and a derivative equal to 0 at $x = 5$.

APPLICATIONS

79. Temperature during an illness. The temperature of a person during an intestinal illness is given by

$$T(t) = -0.1t^2 + 1.2t + 98.6, \quad 0 \le t \le 12,$$

where T is the temperature (°F) at time t, in days. Find the relative extrema and sketch a graph of the function.

80. Advertising. Brody Electronics estimates that it will sell N units of a new toy after spending a thousands of dollars on advertising, where

$$N(a) = -a^2 + 300a + 6, \quad 0 \le a \le 300.$$

Find the relative extrema and sketch a graph of the function.

81. Investment in oil exploration. "Oil companies have been finding less oil than they are pumping out, in part because they have been investing less in exploration" (*New York Times*, 10/28/05, p. C2).

The function given by

$$A(t) = 0.0265t^3 - 0.453t^2 + 1.796t + 7.47, \quad 0 \le t \le 10,$$

can be used to estimate the amount, in billions of dollars, invested by large oil companies in exploration for new reserves t years after 1995. (***Source:*** Based on data from Wood MacKenzie, John S. Herold Inc., and Prof. Lloyd Heinze, Texas Tech, as printed in *New York Times*, 10/28/05.) Find the relative extrema, and sketch a graph of the function.

82. Solar eclipse. On June 21, 2001, there was a total eclipse of the sun in Africa. Between the times 12:00 noon and 1:00 P.M. Universal Time, the center of the eclipse followed a path along the earth described approximately by

$$f(x) = \sqrt{138.1 - 5.025x + 0.2902x^2},$$

where x $(0 \le x \le 22)$ is the longitude east and $f(x)$ is the latitude south. Both x and $f(x)$ are measured in degrees. (***Source:*** NASA.) Find the latitude and longitude of the point with the smallest southern latitude along this path, and plot the function f.

TW **83.** Consider this graph.

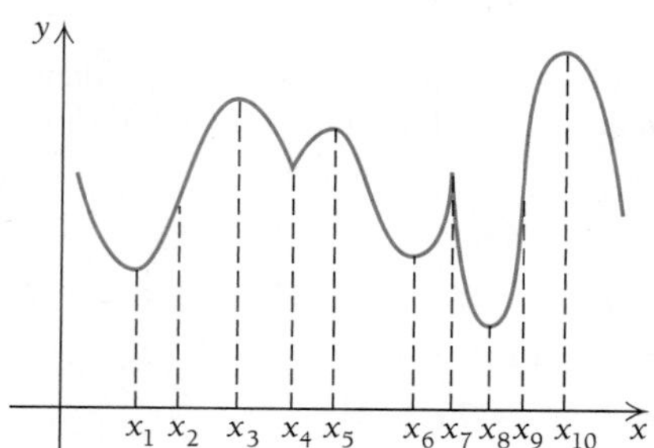

Explain the idea of a critical value. Then determine which x-values are critical values, and state why.

TW **84.** Consider this graph.

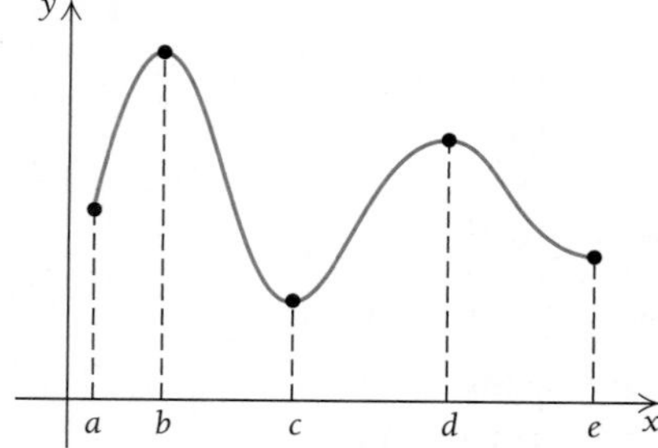

Using the graph and the intervals noted, explain how to relate the concept of the function being increasing or decreasing to the first derivative.

SYNTHESIS

In Exercises 85–90, the graph of a derivative f' is shown. Use the information in each graph to determine where f is increasing or decreasing and the x-values of any extrema. Then sketch a possible graph of f.

85.

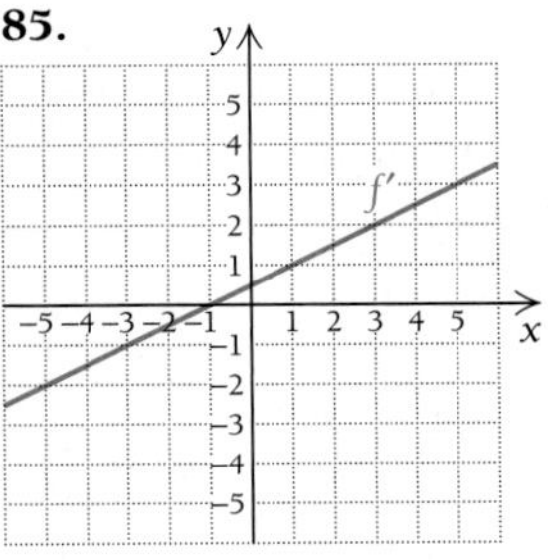

86.

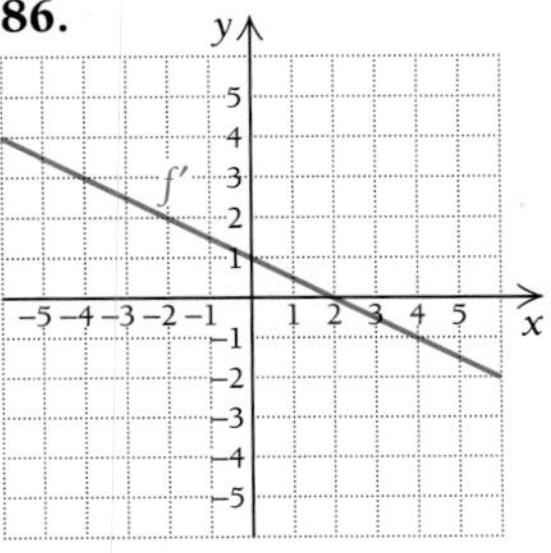

87.

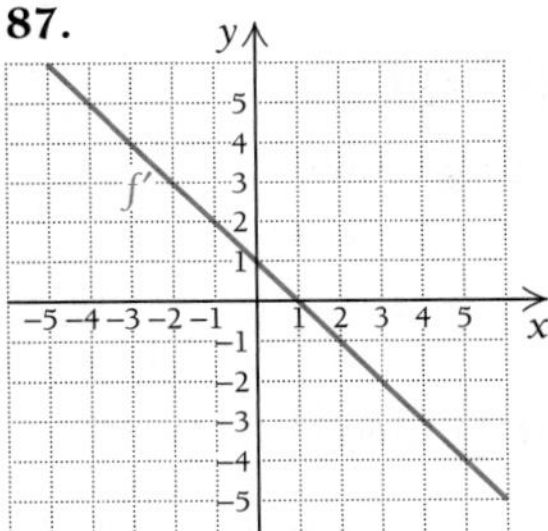

88.

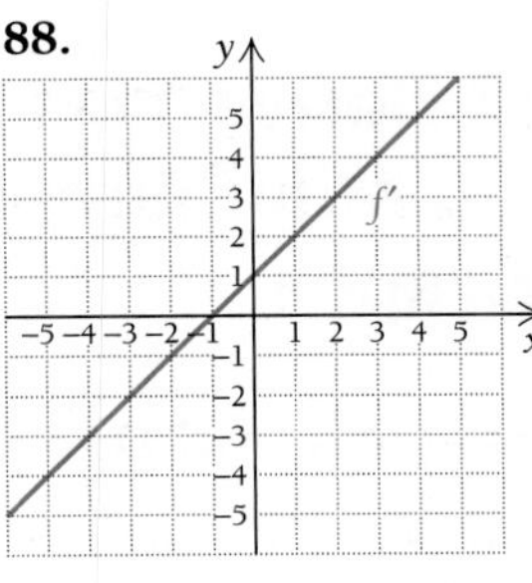

89.

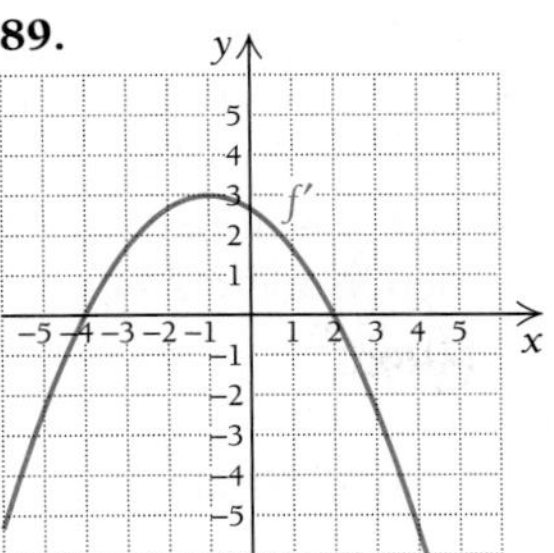

90.

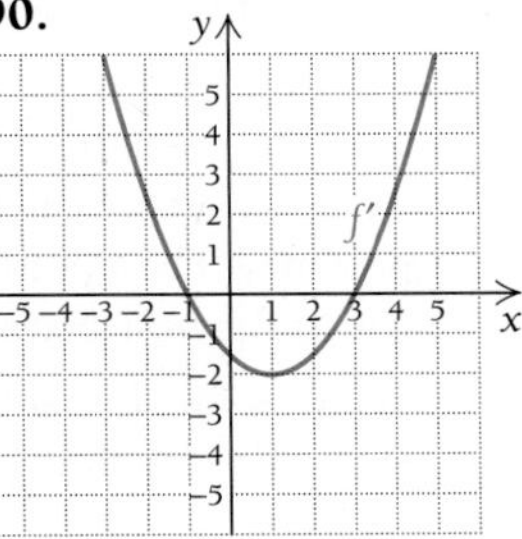

TECHNOLOGY CONNECTION

Graph each function. Then estimate any relative extrema.

91. $f(x) = -x^6 - 4x^5 + 54x^4 + 160x^3 - 641x^2 - 828x + 1200$

92. $f(x) = x^4 + 4x^3 - 36x^2 - 160x + 400$

93. $f(x) = \sqrt[3]{|4 - x^2|} + 1$

94. $f(x) = x\sqrt{9 - x^2}$

Life science: caloric intake and life expectancy. *The data in the following table give, for various countries, daily caloric intake, projected life expectancy, and infant mortality. Use the data for Exercises 95 and 96 on p. 214.*

Country	Daily Caloric Intake (2003)	Life Expectancy (2004)	Infant Mortality (in deaths per 1000 births, in 2004)
Argentina	2959	75	16
Australia	3135	81	5
Bolivia	2219	64	54
Canada*	3605	79	5
Dominican Republic	2281	68	27
Germany*	3484	78	4
Haiti	2109	52	74
Mexico	3171	75	23
United States	3754	78	7
Venezuela	2272	73	16

(*Source*: Based on data from UNICEF and the United Nations Food and Agricultural Organization.)
*All data for Canada and Germany are from 2003.

TW **95. Life expectancy and daily caloric intake.**

a) Use the regression procedures of Section R.6 to fit a cubic function $y = f(x)$ to the data on p. 213, where x is daily caloric intake and y is life expectancy. Then fit a quartic function and decide which fits best. Explain.

b) What is the domain of the function?

c) Does the function have any relative extrema? Explain.

TW **96. Infant mortality and daily caloric intake.**

a) Use the regression procedures of Section R.6 to fit a cubic function $y = f(x)$ to the data on p. 213, where x is daily caloric intake and y is infant mortality. Then fit a quartic function and decide which fits best. Explain.

b) What is the domain of the function?

c) Does the function have any relative extrema? Explain.

TW **97.** Describe a procedure that can be used to select an appropriate viewing window for the functions given in **(a)** Exercises 1–16 and **(b)** Exercises 91–94.

2.2 Using Second Derivatives to Find Maximum and Minimum Values and Sketch Graphs

OBJECTIVES

- Classify the relative extrema of a function using the Second-Derivative Test.
- Sketch the graph of a continuous function.

Concavity: Increasing and Decreasing Derivatives

The graphs of two functions are shown below. The graph of f is turning up and the graph of g is turning down. Let's see if we can relate these observations to the functions' derivatives.

Consider first the graph of f. Take a ruler, or straightedge, and draw tangent lines as you move along the curve from left to right. What happens to the slopes of the tangent lines? Do the same for the graph of g. Look at the curvature and decide whether you see a pattern.

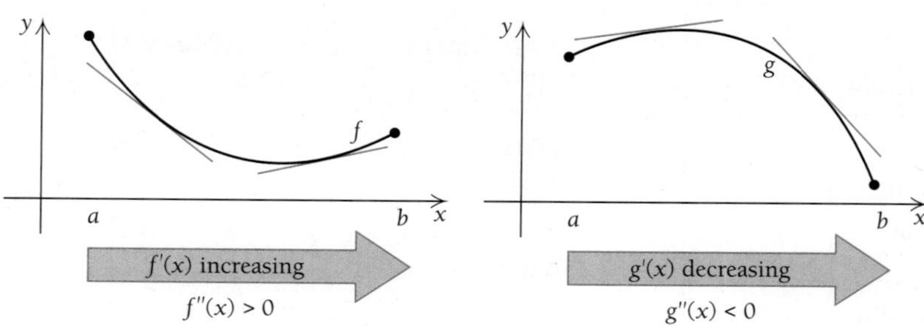

For the graph of f, the slopes of the tangent lines are increasing. That is, f' is increasing over the interval. This can be determined by noting that $f''(x)$ is positive, since the relationship between f' and f'' is like the relationship between f and f'. Note also that all the tangent lines for f are below the graph. For the graph of g, the slopes are decreasing. This can be determined by noting that g' is decreasing at x any time that $g''(x)$ is negative. For g, all tangent lines are above the graph.

DEFINITION

Suppose that f is a function whose derivative f' exists at every point in an open interval I. Then

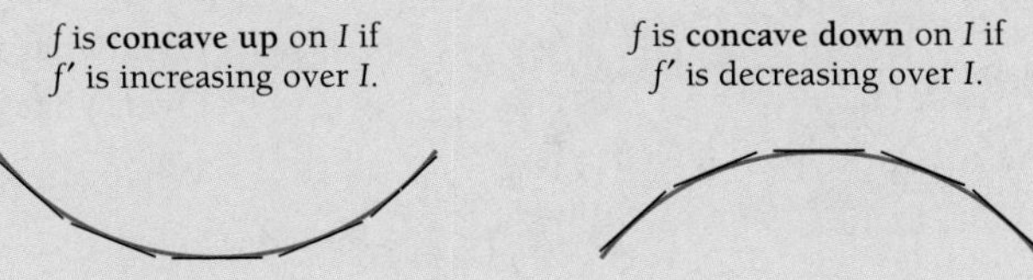
f is **concave up** on I if
f' is increasing over I.
f is **concave down** on I if
f' is decreasing over I.

The following theorem states how the concavity of a function's graph and the second derivative of the function are related.

THEOREM 4 A Test for Concavity

1. If $f''(x) > 0$ on an interval I, then the graph of f is concave up. (f' is increasing, so f is turning up on I.)
2. If $f''(x) < 0$ on an interval I, then the graph of f is concave down. (f' is decreasing, so f is turning down on I.)

Keep in mind that a function can be decreasing and concave up, decreasing and concave down, increasing and concave up, or increasing and concave down. That is, *concavity* and *increasing/decreasing* are independent concepts. It is the increasing or decreasing aspect of the *derivative* that tells us about the function's concavity.

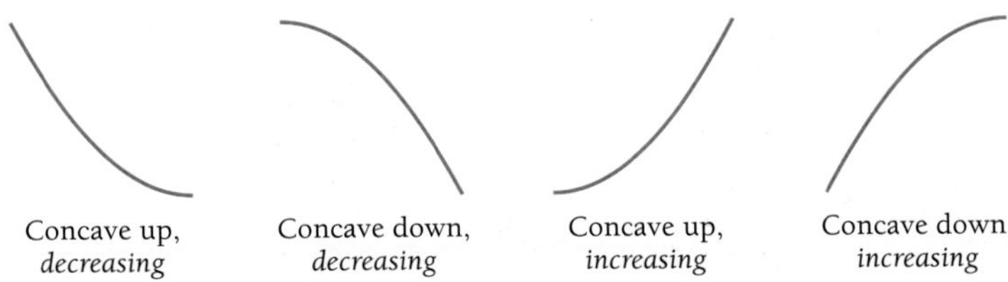
Concave up, *decreasing*
Concave down, *decreasing*
Concave up, *increasing*
Concave down, *increasing*

TECHNOLOGY CONNECTION

Exploratory

Graph the function

$$f(x) = -\tfrac{1}{3}x^3 + 6x^2 - 11x - 50$$

and its second derivative,

$$f''(x) = -2x + 12,$$

using the viewing window $[-10, 25, -100, 150]$, with Xscl = 5 and Yscl = 25.

Over what intervals is the graph of f concave up?
Over what intervals is the graph of f concave down?
Over what intervals is the graph of f'' positive?
Over what intervals is the graph of f'' negative?

What can you conjecture?

(continued)

Technology Connection, Exploratory (continued)

Now graph the first derivative

$$f'(x) = -x^2 + 12x - 11$$

and the second derivative

$$f''(x) = -2x + 12$$

using the viewing window $[-10, 25, -200, 50]$, with Xscl = 5 and Yscl = 25.

Over what intervals is the first derivative f' increasing?
Over what intervals is the first derivative f' decreasing?
Over what intervals is the graph of f'' positive?
Over what intervals is the graph of f'' negative?

What can you conjecture?

Classifying Relative Extrema Using Second Derivatives

In the following discussion, we see how we can use second derivatives to determine whether a function has a relative extremum on an open interval.

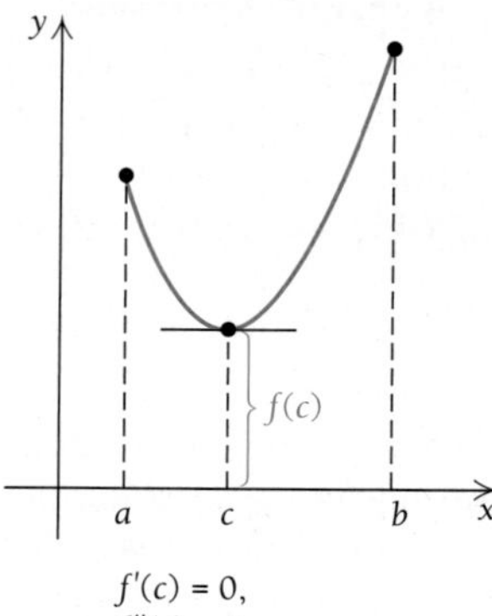

$f'(c) = 0$,
$f''(c) > 0$,
The graph is concave up around c.
Therefore, $f(c)$ is a relative minimum.

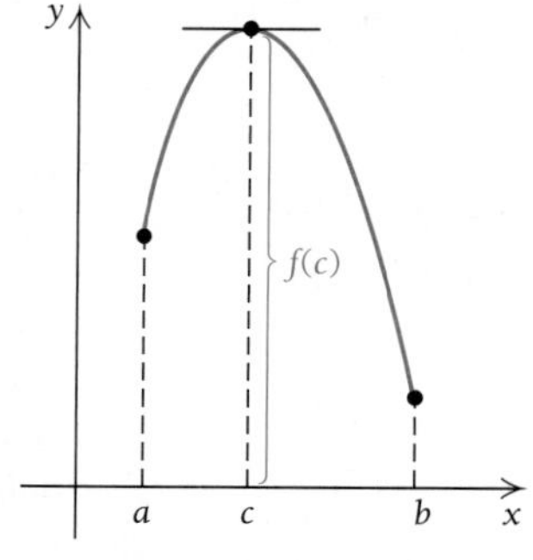

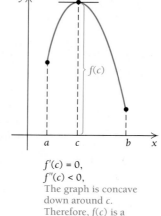

$f'(c) = 0$,
$f''(c) < 0$,
The graph is concave down around c.
Therefore, $f(c)$ is a relative maximum.

A helpful memory aid consists of happy and sad faces.

THEOREM 5 The Second-Derivative Test for Relative Extrema

Suppose that f is differentiable for every x in an open interval (a, b) and that there is a critical value c in (a, b) for which $f'(c) = 0$. Then:

1. $f(c)$ is a relative minimum if $f''(c) > 0$.
2. $f(c)$ is a relative maximum if $f''(c) < 0$.

For $f''(c) = 0$, the First-Derivative Test can be used to determine whether $f(x)$ is a relative extremum.

Consider the following graphs. In each one, f' and f'' are both 0 at $c = 2$, but the first function has an extremum and the second function does not. When c is a critical value and $f''(c) = 0$, an extremum may or may not exist at c. Note too that if $f'(c)$ does not exist and c is a critical value, then $f''(c)$ also does not exist. Again, an approach other than the Second-Derivative Test must be used to determine whether $f(c)$ is an extremum.

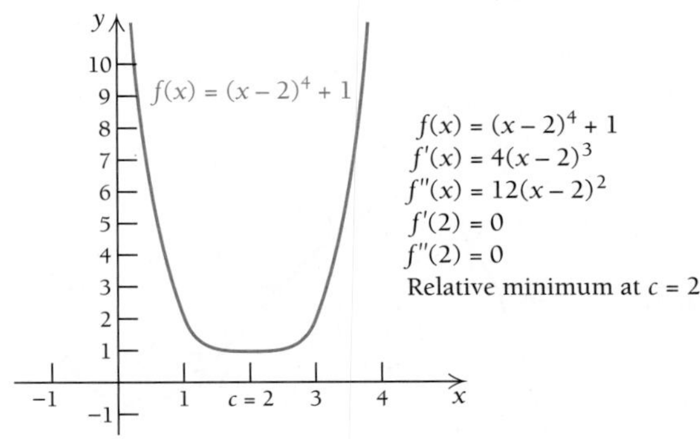

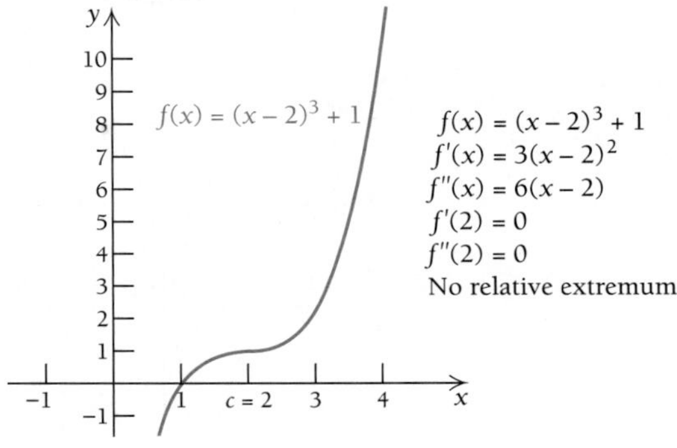

EXAMPLE 1 Graph the function f given by

$$f(x) = x^3 + 3x^2 - 9x - 13,$$

and find the relative extrema.

TECHNOLOGY CONNECTION

Exploratory

Consider the function given by

$$y = x^3 - 3x^2 - 9x - 1.$$

Use a graph to estimate the relative extrema. Then find the first and second derivatives. Graph both in the same window. Use the ZERO feature to determine where the first derivative is zero. Verify that relative extrema occur at those x-values by checking the sign of the second derivative. Then check your work using the approach of Example 1.

Solution To find any critical values, we determine $f'(x)$. To determine whether any critical values lead to extrema, we also find $f''(x)$:

$$f'(x) = 3x^2 + 6x - 9,$$
$$f''(x) = 6x + 6.$$

Then we solve $f'(x) = 0$:

$$3x^2 + 6x - 9 = 0$$
$$x^2 + 2x - 3 = 0 \quad \text{Dividing both sides by 3}$$
$$(x+3)(x-1) = 0 \quad \text{Factoring}$$
$$x + 3 = 0 \quad \text{or} \quad x - 1 = 0 \quad \text{Using the Principle of Zero Products}$$
$$x = -3 \quad \text{or} \quad x = 1.$$

We next find second coordinates by substituting in the original function:

$$f(-3) = (-3)^3 + 3(-3)^2 - 9(-3) - 13 = 14;$$
$$f(1) = (1)^3 + 3(1)^2 - 9(1) - 13 = -18.$$

Are the points $(-3, 14)$ and $(1, -18)$ relative extrema? Let's look at the second derivative. We use the Second-Derivative Test with the numbers -3 and 1:

$$f''(-3) = 6(-3) + 6 = -12 < 0; \longrightarrow \text{Relative maximum}$$
$$f''(1) = 6(1) + 6 = 12 > 0. \longrightarrow \text{Relative minimum}$$

Thus, $f(-3) = 14$ is a relative maximum and $f(1) = -18$ is a relative minimum. We plot both $(-3, 14)$ and $(1, -18)$, including short arcs at each point to indicate the graph's concavity. By calculating and plotting a few more points, we can make a sketch, as shown below.

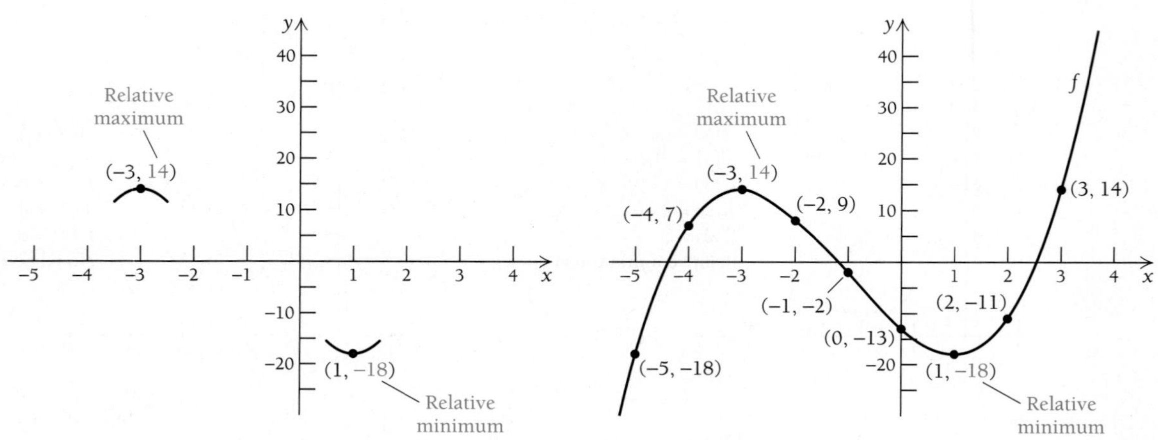

✦

EXAMPLE 2 Find the relative extrema of the function f given by

$$f(x) = 3x^5 - 20x^3,$$

and sketch the graph.

Solution We find both the first and second derivatives:

$$f'(x) = 15x^4 - 60x^2,$$
$$f''(x) = 60x^3 - 120x.$$

Then we solve $f'(x) = 0$ to find any critical values:

$$15x^4 - 60x^2 = 0$$
$$15x^2(x^2 - 4) = 0$$
$$15x^2(x + 2)(x - 2) = 0 \quad \text{Factoring}$$
$$15x^2 = 0 \quad \text{or} \quad x + 2 = 0 \quad \text{or} \quad x - 2 = 0 \quad \text{Using the Principle of Zero Products}$$
$$x = 0 \quad \text{or} \quad x = -2 \quad \text{or} \quad x = 2.$$

We next find second coordinates by substituting in the original function:

$$f(-2) = 3(-2)^5 - 20(-2)^3 = 64;$$
$$f(2) = 3(2)^5 - 20(2)^3 = -64;$$
$$f(0) = 3(0)^5 - 20(0)^3 = 0.$$

All three of these y-values are candidates for relative extrema.

We now use the Second-Derivative Test with the numbers -2, 2, and 0:

$$f''(-2) = 60(-2)^3 - 120(-2) = -240 < 0; \rightarrow \text{Relative maximum}$$
$$f''(2) = 60(2)^3 - 120(2) = 240 > 0; \rightarrow \text{Relative minimum}$$
$$f''(0) = 60(0)^3 - 120(0) = 0. \rightarrow \text{The Second-Derivative Test fails. Use the First-Derivative Test.}$$

Thus, $f(-2) = 64$ is a relative maximum and $f(2) = -64$ is a relative minimum. Since $f'(-1) < 0$ and $f'(1) < 0$, we know that f is decreasing on both $[-2, 0]$ and $[0, 2]$. Thus, we know by the First-Derivative Test that f has no relative extremum at $(0, 0)$. We complete the graph, plotting other points as needed. The extrema are shown in the graph below.

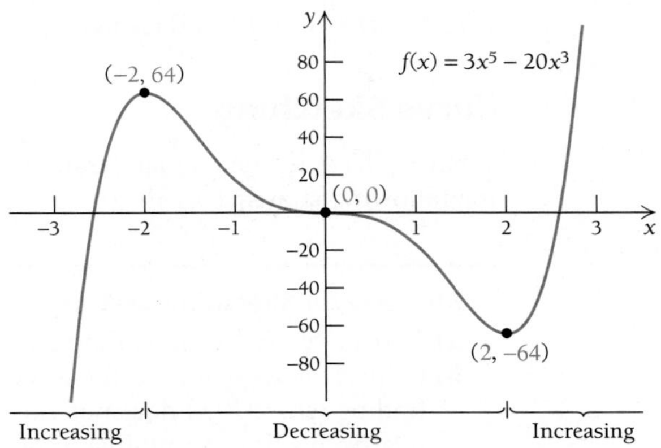

✦

Points of Inflection

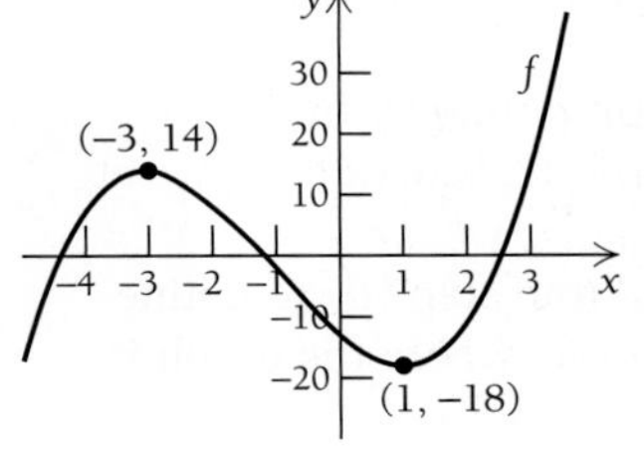

Let's reexamine the graph of f in Example 1. Because the graph is concave down at $(-3, 14)$ and concave up at $(1, -18)$, there exists some *point of inflection* at which the concavity changes.

A **point of inflection,** or an **inflection point,** is a point across which the direction of concavity changes. For example, in Figs. 1–3, point P is an inflection point. The figures display the sign of $f''(x)$ to indicate the concavity on either side of P.

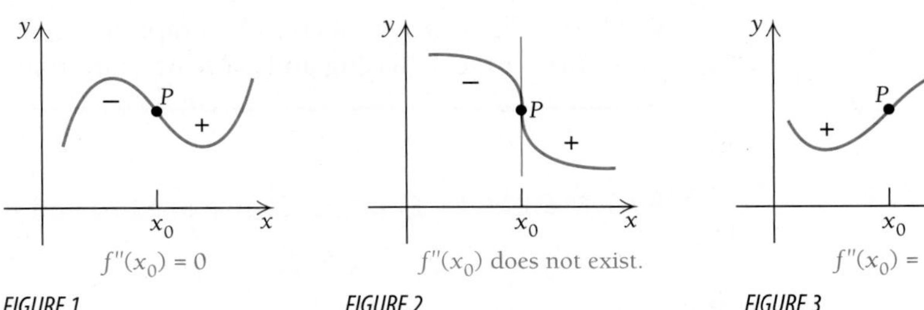

FIGURE 1 FIGURE 2 FIGURE 3

As we move to the right along each curve, the concavity changes at P. Since, as we move through P, the sign of $f''(x)$ changes, either the value of $f''(x_0)$ at P must be 0, as in Figs. 1 and 3, or $f''(x_0)$ must not exist, as in Fig. 2.

THEOREM 6 Finding Points of Inflection

If a function f has a point of inflection, it must occur at a point x_0, where

$$f''(x_0) = 0 \quad \text{or} \quad f''(x_0) \text{ does not exist.}$$

Thus, to find candidates for points of inflection, we look for numbers x_0 for which $f''(x_0) = 0$ or for which $f''(x_0)$ does not exist. Then, if $f''(x)$ changes sign as x moves through x_0 (see Figs. 1–3), we have a point of inflection at x_0.

Theorem 6, about points of inflection, is completely analogous to Theorem 2 about relative extrema. Theorem 2 tells us that relative extrema occur when $f'(x) = 0$ or $f'(x)$ does not exist. Theorem 6 tells us that points of inflection occur when $f''(x) = 0$ or $f''(x)$ does not exist.

Curve Sketching

What we have learned so far greatly enhances our ability to sketch curves. We use the following strategy:

Strategy for Sketching Graphs*

a) *Derivatives and domain.* Find $f'(x)$ and $f''(x)$. Note the domain of f.

b) *Critical values of f.* Find the critical values by solving $f'(x) = 0$ and finding where $f'(x)$ does not exist. These numbers yield candidates for relative maxima or minima. Find the function values at these points.

c) *Increasing and/or decreasing; relative extrema.* Substitute each critical value, x_0, from step (b) into $f''(x)$. If $f''(x_0) < 0$, then $f(x_0)$ is a relative maximum and f is increasing to the left of x_0 and decreasing to the right. If $f''(x_0) > 0$, then $f(x_0)$ is a relative minimum and f is decreasing to the left of x_0 and increasing to the right.

d) *Inflection points.* Determine candidates for inflection points by finding where $f''(x) = 0$ or where $f''(x)$ does not exist. Find the function values at these points.

e) *Concavity.* Use the candidates for inflection points from step (d) to define intervals. Substitute test values into $f''(x)$ to determine where the graph is concave up ($f''(x) > 0$) and where it is concave down ($f''(x) < 0$).

f) *Sketch the graph.* Sketch the graph using the information from steps (a)–(e), calculating and plotting extra points as needed.

EXAMPLE 3 Find the relative maxima and minima of the function f given by

$$f(x) = x^3 - 3x + 2,$$

and sketch the graph.

Solution

a) *Derivatives and domain.* Find $f'(x)$ and $f''(x)$:

$$f'(x) = 3x^2 - 3,$$
$$f''(x) = 6x.$$

The domain of f (and of any polynomial function) is $\mathbb{R}$, the set of all real numbers.

---

*This strategy is refined further, for rational functions, in Section 2.3.

b) *Critical values of f.* Find the critical values by finding where $f'(x)$ does not exist and by solving $f'(x) = 0$. We know that $f'(x) = 3x^2 - 3$ exists for all values of x, so the only critical values are where $f'(x)$ is 0:

$$3x^2 - 3 = 0 \qquad \text{Setting } f'(x) \text{ equal to 0}$$
$$3x^2 = 3$$
$$x^2 = 1$$
$$x = \pm 1.$$

We have $f(-1) = 4$ and $f(1) = 0$, so $(-1, 4)$ and $(1, 0)$ are on the graph.

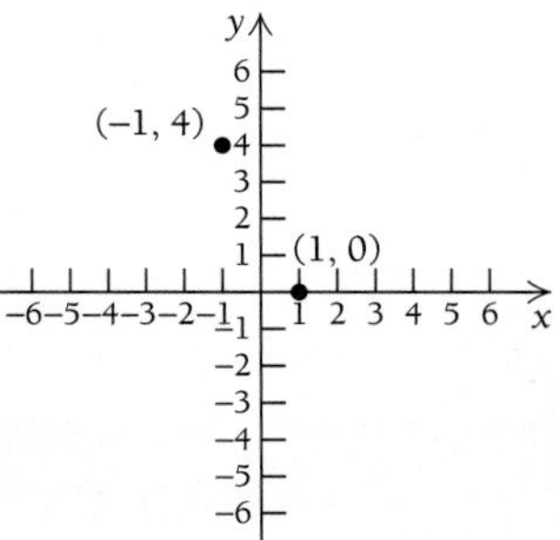

c) *Increasing and/or decreasing; relative extrema.* Substitute the critical values into $f''(x)$:

$$f''(-1) = 6(-1) = -6 < 0,$$

so $f(-1) = 4$ is a relative maximum, with f increasing on $(-\infty, -1]$ and decreasing on $[-1, 1]$. The graph is concave down at the point $(-1, 4)$.

$$f''(1) = 6 \cdot 1 = 6 > 0,$$

so $f(1) = 0$ is a relative minimum, with f decreasing on $[-1, 1]$ and increasing on $[1, \infty)$. The graph is concave up at the point $(1, 0)$.

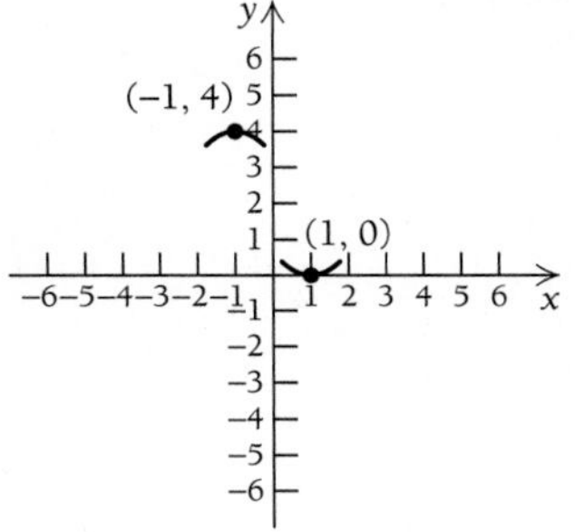

d) *Inflection points.* Find possible inflection points by finding where $f''(x)$ does not exist and by solving $f''(x) = 0$. We know that $f''(x) = 6x$ exists for all values of x, so we try to solve $f''(x) = 0$:

$$6x = 0 \qquad \text{Setting } f''(x) \text{ equal to 0}$$
$$x = 0. \qquad \text{Dividing both sides by 6}$$

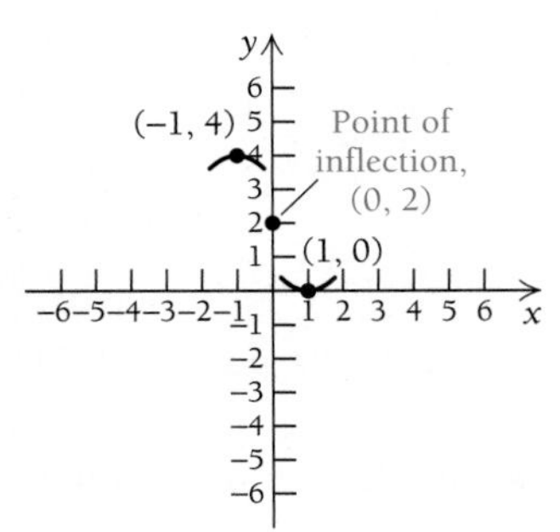

We have $f(0) = 2$, which gives us another point, $(0, 2)$, that lies on the graph.

e) *Concavity.* Find the intervals on which f is concave up or concave down, using the point $(0, 2)$ from step (d). From step (c), we can conclude that f is concave down over the interval $(-\infty, 0)$ and concave up over $(0, \infty)$.

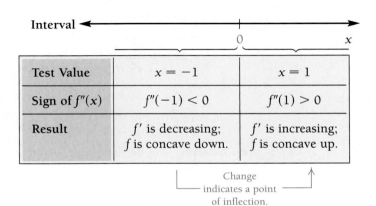

Point of inflection, (0, 2)

Interval — 0 — x

Test Value	$x = -1$	$x = 1$
Sign of $f''(x)$	$f''(-1) < 0$	$f''(1) > 0$
Result	f' is decreasing; f is concave down.	f' is increasing; f is concave up.

Change indicates a point of inflection.

f) *Sketch the graph.* Sketch the graph using the information in steps (a)–(e). Calculate some extra function values if desired. The graph follows.

x	$f(x)$
-3	-16
-2	0
-1	4
0	2
1	0
2	4
3	20

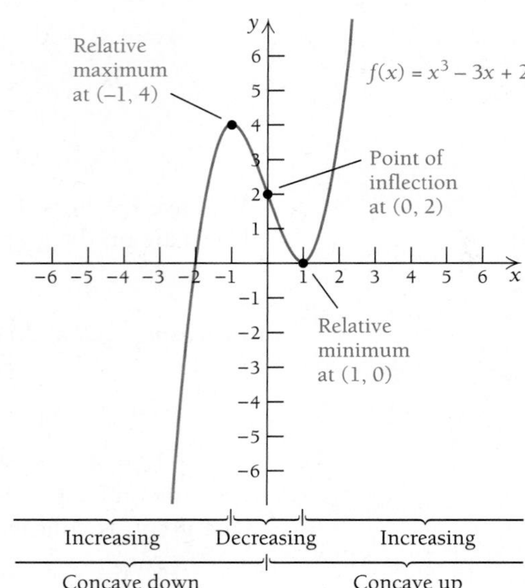

TECHNOLOGY CONNECTION

Check the results of Example 3 using a calculator.

EXAMPLE 4 Find the relative maxima and minima of the function f given by

$$f(x) = x^4 - 2x^2,$$

and sketch the graph.

Solution

a) *Derivatives and domain.* Find $f'(x)$ and $f''(x)$:

$$f'(x) = 4x^3 - 4x,$$
$$f''(x) = 12x^2 - 4.$$

The domain of f is $\mathbb{R}$.

b) *Critical values.* Since $f'(x) = 4x^3 - 4x$ exists for all values of x, the only critical values are where $f'(x) = 0$:

$$4x^3 - 4x = 0 \quad \text{Setting } f'(x) \text{ equal to 0}$$
$$4x(x^2 - 1) = 0$$
$$4x = 0 \quad \text{or} \quad x^2 - 1 = 0$$
$$x = 0 \quad \text{or} \quad x^2 = 1$$
$$x = \pm 1.$$

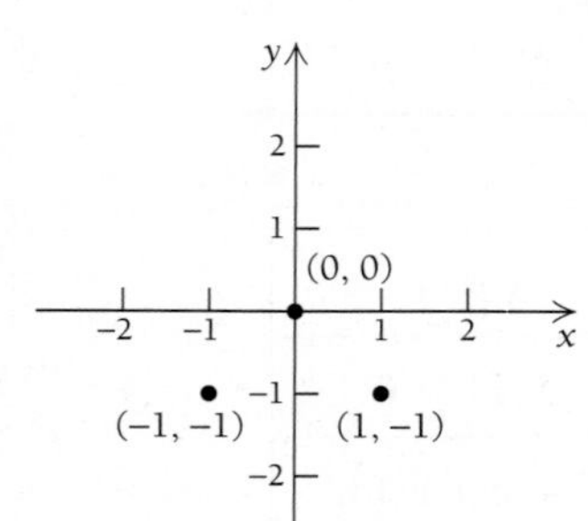

We have $f(0) = 0$, $f(-1) = -1$, and $f(1) = -1$, which gives the points $(0, 0)$, $(-1, -1)$, and $(1, -1)$ on the graph.

c) *Increasing and/or decreasing; relative extrema.* Substitute the critical values into $f''(x)$:

$$f''(0) = 12 \cdot 0^2 - 4 = -4 < 0,$$

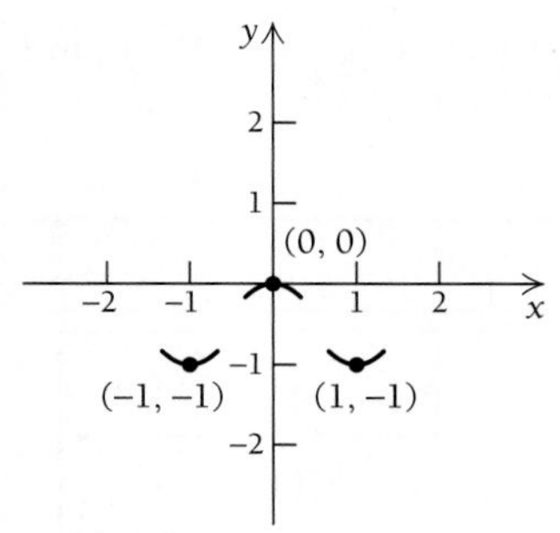

so $f(0) = 0$ is a relative maximum, with f increasing on $[-1, 0]$ and decreasing on $[0, 1]$. The graph is concave down at the point $(0, 0)$.

$$f''(-1) = 12(-1)^2 - 4 = 8 > 0,$$

so $f(-1) = -1$ is a relative minimum, with f decreasing on $(-\infty, -1]$ and increasing on $[-1, 0]$. The graph is concave up at the point $(-1, -1)$.

$$f''(1) = 12 \cdot 1^2 - 4 = 8 > 0,$$

so $f(1) = -1$ is also a relative minimum, with f decreasing on $[0, 1]$ and increasing on $[1, \infty)$. The graph is concave up at the point $(1, -1)$.

d) *Inflection points.* Find where $f''(x)$ does not exist and where $f''(x) = 0$. Since $f''(x)$ exists for all real numbers, we just solve $f''(x) = 0$:

$$\begin{aligned} 12x^2 - 4 &= 0 \qquad \text{Setting } f''(x) \text{ equal to } 0 \\ 4(3x^2 - 1) &= 0 \\ 3x^2 - 1 &= 0 \\ 3x^2 &= 1 \\ x^2 &= \frac{1}{3} \\ x &= \pm\sqrt{\frac{1}{3}} \\ &= \pm\frac{1}{\sqrt{3}}. \end{aligned}$$

We have

$$\begin{aligned} f\left(\frac{1}{\sqrt{3}}\right) &= \left(\frac{1}{\sqrt{3}}\right)^4 - 2\left(\frac{1}{\sqrt{3}}\right)^2 \\ &= \frac{1}{9} - \frac{2}{3} = -\frac{5}{9} \end{aligned}$$

and

$$f\left(-\frac{1}{\sqrt{3}}\right) = -\frac{5}{9}.$$

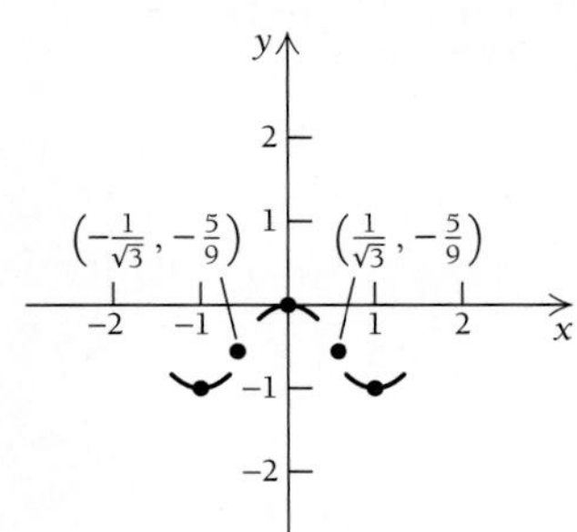

These values give

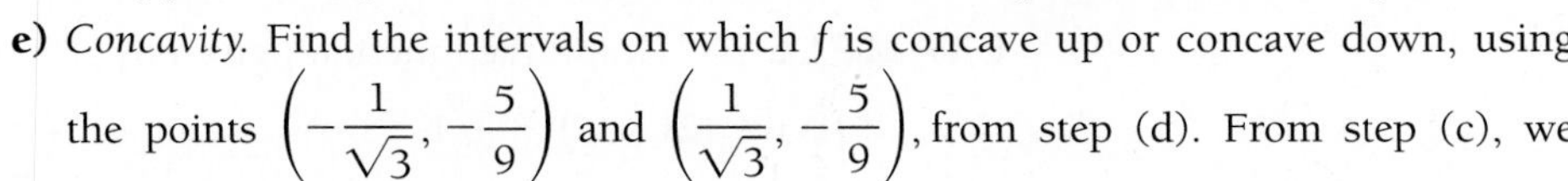

$$\left(-\frac{1}{\sqrt{3}}, -\frac{5}{9}\right) \quad \text{and} \quad \left(\frac{1}{\sqrt{3}}, -\frac{5}{9}\right),$$

or approximately $(-0.6, -0.6)$ and $(0.6, -0.6)$ as possible inflection points.

e) *Concavity.* Find the intervals on which f is concave up or concave down, using the points $\left(-\frac{1}{\sqrt{3}}, -\frac{5}{9}\right)$ and $\left(\frac{1}{\sqrt{3}}, -\frac{5}{9}\right)$, from step (d). From step (c), we can conclude that f is concave up over the intervals $(-\infty, -1/\sqrt{3})$ and $(1/\sqrt{3}, \infty)$ and concave down over the interval $(-1/\sqrt{3}, 1/\sqrt{3})$.

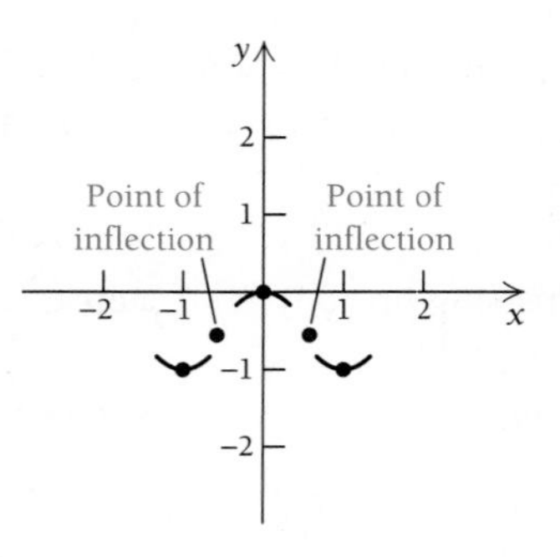

Interval: $-1/\sqrt{3}$, $1/\sqrt{3}$, x

Test Value	$x = -1$	$x = 0$	$x = 1$
Sign of $f''(x)$	$f''(-1) > 0$	$f''(0) < 0$	$f'(1) > 0$
Result	f' is increasing; f is concave up.	f' is decreasing; f is concave down.	f' is increasing; f is concave up.

Change indicates a point of inflection. Change indicates a point of inflection.

f) *Sketch the graph.* Sketch the graph using the information in steps (a)–(e). By solving $x^4 - 2x^2 = 0$, we can find the x-intercepts easily. They are $(-\sqrt{2}, 0)$, $(0, 0)$, and $(\sqrt{2}, 0)$. This also aids with graphing. Extra function values can be calculated if desired. The graph is shown below.

x	$f(x)$, approximately
-2	8
-1.5	0.56
-1	-1
-0.5	-0.44
0	0
0.5	-0.44
1	-1
1.5	0.56
2	8

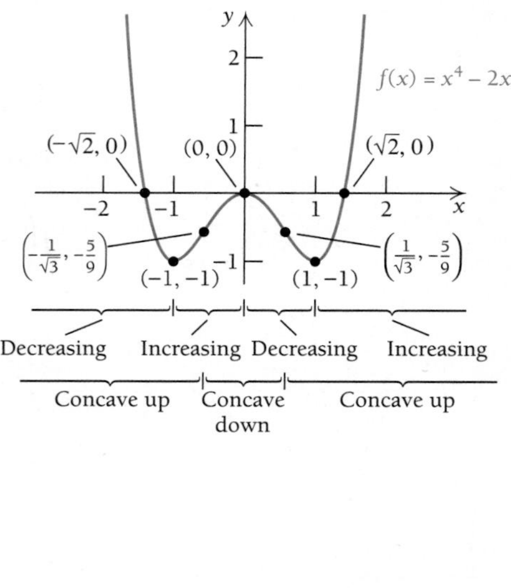

✦

TECHNOLOGY CONNECTION

EXERCISE

1. Consider $f(x) = x^3(x - 2)^3$. How many relative extrema do you anticipate finding? Where do you think they will be?

Graph f, f', and f'' using $[-1, 3, -2, 6]$ as a viewing window. Estimate the relative extrema and the inflection points of f. Then check your work using the analytic methods of Examples 3 and 4.

EXAMPLE 5 Graph the function f given by

$$f(x) = (2x - 5)^{1/3} + 1.$$

List the coordinates of any extrema and points of inflection. State where the function is increasing or decreasing, as well as where it is concave up or concave down.

Solution

a) *Derivatives and domain.* Find $f'(x)$ and $f''(x)$:

$$f'(x) = \tfrac{1}{3}(2x - 5)^{-2/3} \cdot 2 = \tfrac{2}{3}(2x - 5)^{-2/3}, \text{ or } \frac{2}{3(2x - 5)^{2/3}};$$

$$f''(x) = -\tfrac{4}{9}(2x - 5)^{-5/3} \cdot 2 = -\tfrac{8}{9}(2x - 5)^{-5/3}, \text{ or } \frac{-8}{9(2x - 5)^{5/3}}.$$

The domain of f is $\mathbb{R}$.

b) *Critical values.* Since

$$f'(x) = \frac{2}{3(2x - 5)^{2/3}}$$

is never 0 (a fraction equals 0 only when its numerator is 0), the only critical value is when $f'(x)$ does not exist. The only time $f'(x)$ does not exist is when its denominator is 0:

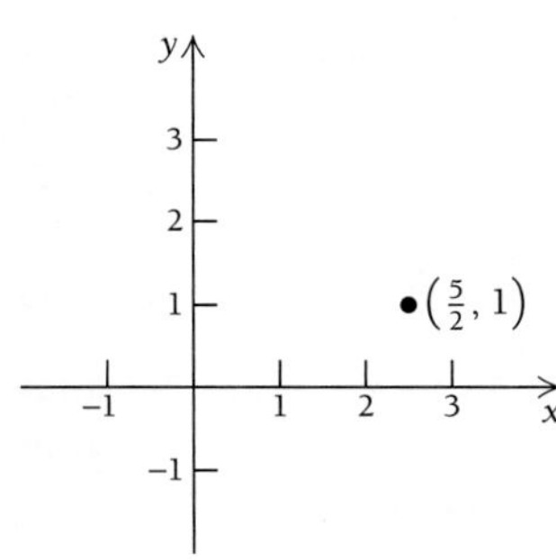

$$\begin{aligned} 3(2x - 5)^{2/3} &= 0 \\ (2x - 5)^{2/3} &= 0 && \text{Dividing both sides by 3} \\ (2x - 5)^2 &= 0 && \text{Cubing both sides} \\ 2x - 5 &= 0 \\ 2x &= 5 \\ x &= \tfrac{5}{2} \end{aligned}$$

We now have $f(\frac{5}{2}) = (2 \cdot \frac{5}{2} - 5)^{1/3} + 1 = 0 + 1 = 1$, so the point $(\frac{5}{2}, 1)$ is on the graph.

c) *Increasing and/or decreasing; relative extrema.* Substitute the critical value into $f''(x)$:

$$f''\left(\tfrac{5}{2}\right) = \frac{-8}{9\left(2 \cdot \frac{5}{2} - 5\right)^{5/3}} = \frac{-8}{9 \cdot 0} = \frac{-8}{0}.$$

Since $f''(\frac{5}{2})$ does not exist, the Second-Derivative Test cannot be used at $x = \frac{5}{2}$. Instead, we use the First-Derivative Test, selecting 2 and 3 as test values on either side of $\frac{5}{2}$:

$$f'(2) = \frac{2}{3(2 \cdot 2 - 5)^{2/3}} = \frac{2}{3(-1)^{2/3}} = \frac{2}{3 \cdot 1} = \frac{2}{3},$$

$$\text{and } f'(3) = \frac{2}{3(2 \cdot 3 - 5)^{2/3}} = \frac{2}{3 \cdot 1^{2/3}} = \frac{2}{3 \cdot 1} = \frac{2}{3}.$$

Since $f'(x) > 0$ on either side of $x = \frac{5}{2}$, we know that f is increasing on both $(-\infty, \frac{5}{2}]$ and $[\frac{5}{2}, \infty)$; thus, $f(\frac{5}{2}) = 1$ is not an extremum.

d) *Inflection points.* Find where $f''(x)$ does not exist and where $f''(x) = 0$. Since $f''(x)$ is never 0 (why?), we only need to find where $f''(x)$ does not exist. Since $f''(x)$ cannot exist where $f'(x)$ does not exist, we know from step (b) that a possible inflection point is $(\frac{5}{2}, 1)$.

e) *Concavity.* We check the concavity on either side of $x = \frac{5}{2}$:

$$f''(2) = \frac{-8}{9(2 \cdot 2 - 5)^{5/3}} = \frac{-8}{9(-1)} > 0,$$

so f is concave up at $x = 2$;

$$f''(3) = \frac{-8}{9(2 \cdot 3 - 5)^{5/3}} = \frac{-8}{9 \cdot 1} < 0,$$

so f is concave down at $x = 3$.

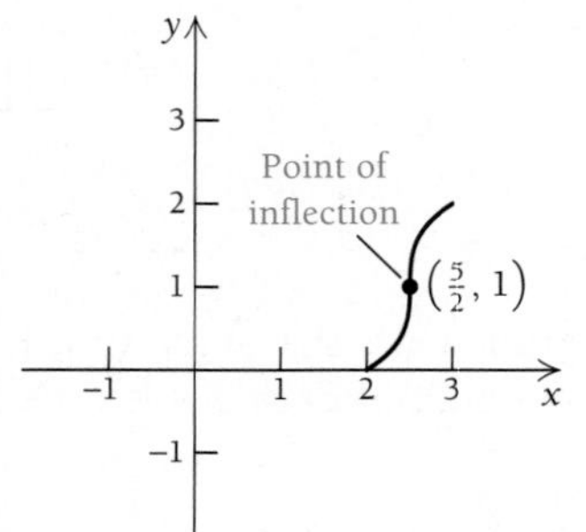

Interval — x, divided at $\frac{5}{2}$

Test Value	$x = 2$	$x = 3$
Sign of $f''(x)$	$f''(2) > 0$	$f''(3) < 0$
Result	f' is increasing; f is concave up.	f' is decreasing; f is concave down.

Change indicates a point of inflection.

f) *Sketch the graph.* Sketch the graph using the information in steps (a)–(e). By solving $(2x - 5)^{1/3} + 1 = 0$, we can find the x-intercept—it is $(2, 0)$. Extra function values can be calculated, if desired. The graph is shown below.

x	$f(x)$, approximately
0	−0.71
1	−0.44
2	0
$\frac{5}{2}$	1
3	2
4	2.44
5	2.71

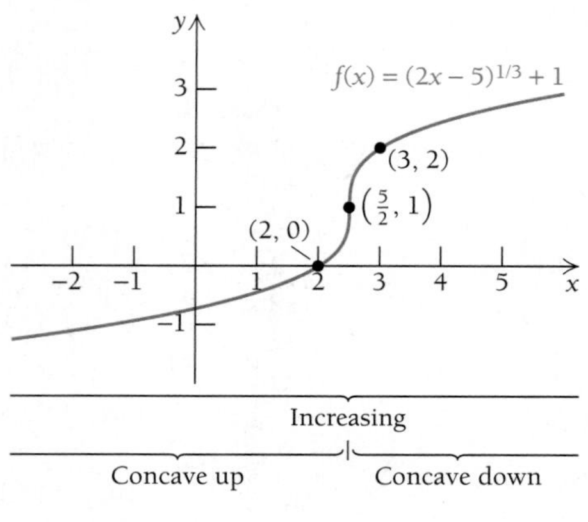

✦

The following figures illustrate some information concerning Example 1 that can be found from the first and second derivatives of f. The relative extrema are shown in Figs. 4 and 5. In Fig. 5, we see that the x-coordinates of the x-intercepts of f' are the critical values of f. Note that the intervals over which f

is increasing or decreasing are those intervals for which f' is positive or negative, respectively.

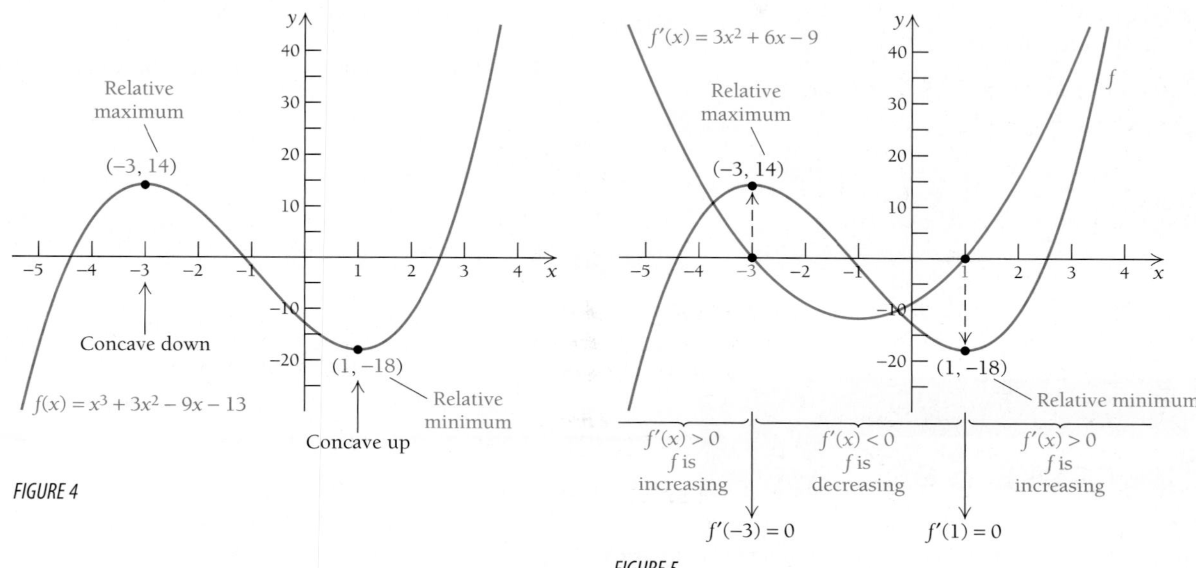

FIGURE 4

FIGURE 5

In Fig. 6, the intervals over which f' is increasing or decreasing are, respectively, those intervals over which f'' is positive or negative. And finally in Fig. 7, we note that when $f''(x) < 0$, the graph of f is concave down, and when $f''(x) > 0$, the graph of f is concave up.

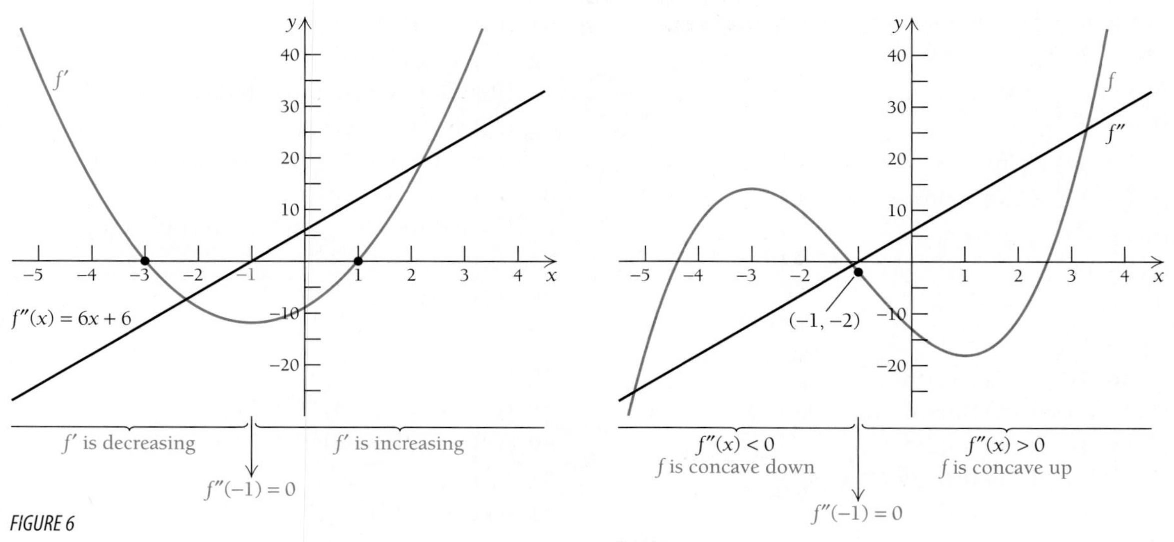

FIGURE 6

FIGURE 7

TECHNOLOGY CONNECTION

EXERCISES

Graph the following:

$$f(x) = 3x^5 - 5x^3, \quad f'(x) = 15x^4 - 15x^2,$$

and

$$f''(x) = 60x^3 - 30x,$$

using the window $[-3, 3, -10, 10]$.

1. From the graph of f', estimate the critical values of f.
2. From the graph of f'', estimate the x-values of any inflection points of f.

Exercise Set 2.2

For each function, find all relative extrema and classify each as a maximum or minimum. Use the Second-Derivative Test where possible.

1. $f(x) = 5 - x^2$

2. $f(x) = 4 - x^2$

3. $f(x) = x^2 - x$

4. $f(x) = x^2 + x - 1$

5. $f(x) = -5x^2 + 8x - 7$

6. $f(x) = -4x^2 + 3x - 1$

7. $f(x) = 8x^3 - 6x + 1$

8. $f(x) = x^3 - 12x - 1$

Sketch the graph of each function. List the coordinates of where extrema or points of inflection occur. State where the function is increasing or decreasing, as well as where it is concave up or concave down.

9. $f(x) = x^3 - 12x$

10. $f(x) = x^3 - 27x$

11. $f(x) = 3x^3 - 36x - 3$

12. $f(x) = 2x^3 - 3x^2 - 36x + 28$

13. $f(x) = \frac{8}{3}x^3 - 2x + \frac{1}{3}$

14. $f(x) = 80 - 9x^2 - x^3$

15. $f(x) = -x^3 + 3x^2 - 4$

16. $f(x) = -x^3 + 3x - 2$

17. $f(x) = 3x^4 - 16x^3 + 18x^2$
(Round results to three decimal places.)

18. $f(x) = 3x^4 + 4x^3 - 12x^2 + 5$
(Round results to three decimal places.)

19. $f(x) = x^4 - 6x^2$

20. $f(x) = 2x^2 - x^4$

21. $f(x) = x^3 - 2x^2 - 4x + 3$

22. $f(x) = x^3 - 6x^2 + 9x + 1$

23. $f(x) = 3x^4 + 4x^3$

24. $f(x) = x^4 - 2x^3$

25. $f(x) = x^3 - 6x^2 - 135x$

26. $f(x) = x^3 - 3x^2 - 144x - 140$

27. $f(x) = x^4 - 4x^3 + 10$

28. $f(x) = \frac{4}{3}x^3 - 2x^2 + x$

29. $f(x) = x^3 - 6x^2 + 12x - 6$

30. $f(x) = x^3 + 3x + 1$

31. $f(x) = 5x^3 - 3x^5$

32. $f(x) = 20x^3 - 3x^5$
(Round results to three decimal places.)

33. $f(x) = x^2(3 - x)^2$
(Round results to three decimal places.)

34. $f(x) = x^2(1 - x)^2$
(Round results to three decimal places.)

35. $f(x) = (x + 1)^{2/3}$

36. $f(x) = (x - 1)^{2/3}$

37. $f(x) = (x - 3)^{1/3} - 1$

38. $f(x) = (x - 2)^{1/3} + 3$

39. $f(x) = -2(x - 4)^{2/3} + 5$

40. $f(x) = -3(x - 2)^{2/3} + 3$

41. $f(x) = x\sqrt{4 - x^2}$

42. $f(x) = -x\sqrt{1 - x^2}$

43. $f(x) = \dfrac{x}{x^2 + 1}$

44. $f(x) = \dfrac{8x}{x^2 + 1}$

45. $f(x) = \dfrac{3}{x^2 + 1}$

46. $f(x) = \dfrac{-4}{x^2 + 1}$

For Exercises 47–54, sketch a graph that possesses the characteristics listed. Answers may vary.

47. f is increasing and concave up on $(-\infty, 4)$, f is increasing and concave down on $(4, \infty)$.

48. f is decreasing and concave up on $(-\infty, 2)$, f is decreasing and concave down on $(2, \infty)$.

49. f is increasing and concave down on $(-\infty, 1)$, f is increasing and concave up on $(1, \infty)$.

50. f is decreasing and concave down on $(-\infty, 3)$, f is decreasing and concave up on $(3, \infty)$.

51. f is concave down at $(1, 5)$, concave up at $(7, -2)$, and has an inflection point at $(4, 1)$.

52. f is concave up at $(1, -3)$, concave down at $(8, 7)$, and has an inflection point at $(5, 4)$.

53. $f'(-1) = 0$, $f''(-1) > 0$, $f(-1) = -5$; $f'(7) = 0$, $f''(7) < 0$, $f(7) = 10$; $f''(3) = 0$, and $f(3) = 2$

54. $f'(-3) = 0$, $f''(-3) < 0$, $f(-3) = 8$; $f'(9) = 0$, $f''(9) > 0$, $f(9) = -6$; $f''(2) = 0$, and $f(2) = 1$

55–100. Check the results of Exercises 1–46 with a graphing claculator.

APPLICATIONS

Business and Economics

Total revenue, cost, and profit. *Using the same set of axes, sketch the graphs of the total-revenue, total-cost, and total-profit functions.*

101. $R(x) = 50x - 0.5x^2$, $C(x) = 4x + 10$

102. $R(x) = 50x - 0.5x^2$, $C(x) = 10x + 3$

103. Small business. The percentage of the U.S. national income generated by nonfarm proprietors may be modeled by the function

$$p(x) = \frac{13x^3 - 240x^2 - 2460x + 585{,}000}{75{,}000},$$

where x is the number of years since 1970. Sketch the graph of this function for $0 \le x \le 40$.

104. Labor force. The percentage of the U.S. civilian labor force aged 45–54 may be modeled by the function

$$f(x) = 0.025x^2 - 0.71x + 20.44,$$

where x is the number of years after 1970. Sketch the graph of this function for $0 \le x \le 30$.

Life and Physical Sciences

105. Coughing velocity. A person coughs when a foreign object is in the windpipe. The velocity of the cough depends on the size of the object. Suppose a person has a windpipe with a 20-mm radius. If a foreign object has a radius r, in millimeters, then the velocity V, in millimeters/second, needed to remove the object by a cough is given by

$$V(r) = k(20r^2 - r^3), \quad 0 \le r \le 20,$$

where k is some positive constant. For what size object is the maximum velocity needed to remove the object?

106. New York temperatures. The average temperature in New York can be approximated by the function

$$T(x) = 43.5 - 18.4x + 8.57x^2 - 0.996x^3 + 0.0338x^4,$$

where T represents the temperature, in degrees Fahrenheit, $x = 1$ represents the middle of January, $x = 2$ represents the middle of February, and so on. (*Source*: www.WorldClimate.com.)

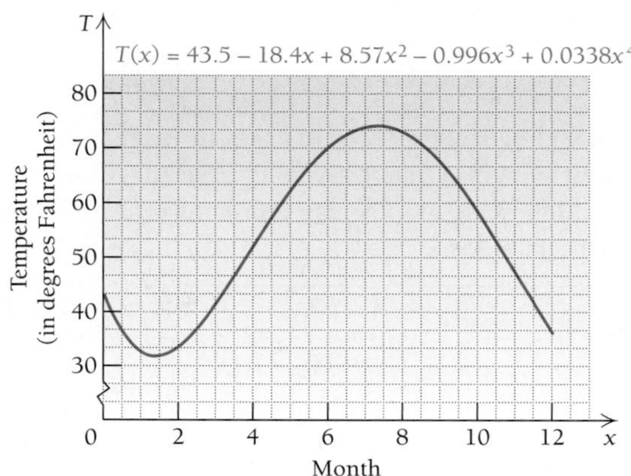

a) Based on the graph, when would you expect the highest temperature to occur in New York?

b) Based on the graph, when would you expect the lowest temperature to occur?

TW **c)** Use the Second-Derivative Test to estimate the points of inflection for the function $T(x)$. What is the significance of these points?

TW **107. Hours of daylight.** The number of hours of daylight in Chicago is represented in the graph below. On what dates is the number of hours of daylight changing most rapidly? How can you tell?

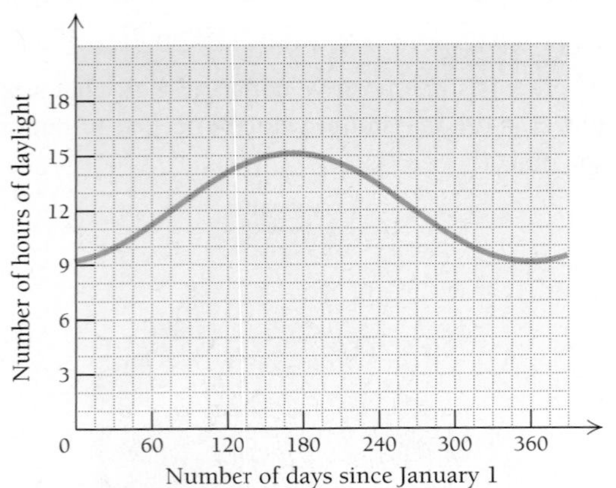

(*Source*: Based on data from Astronomical Applications Dept., U.S. Naval Observatory.)

SYNTHESIS

TW *In each of Exercises 108 and 109, determine which graph is the derivative of the other and explain why.*

108.

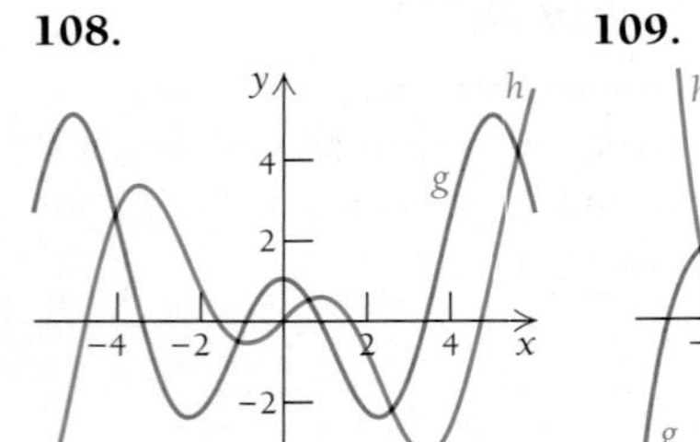

109.

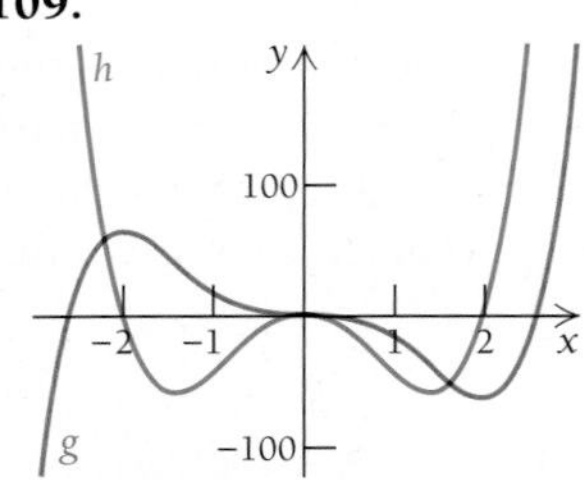

TW **110. Social sciences: three aspects of love.** Researchers at Yale University have suggested that the following graphs may represent three different aspects of love.

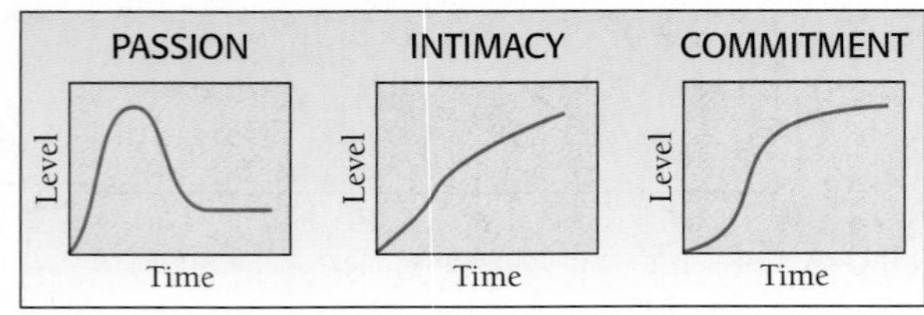

(*Source*: From "A Triangular Theory of Love," by R. J. Sternberg, 1986, *Psychological Review*, 93(2), 119–135. Copyright 1986 by the American Psychological Association, Inc. Reprinted by permission.)

Analyze each of these graphs in terms of the concepts you have learned: relative extrema, concavity, increasing, decreasing, and so on. Do you agree with the researchers regarding the shapes of these graphs? Why or why not?

111. Use calculus to prove that the relative minimum or maximum for any function f for which

$f(x) = ax^2 + bx + c, \quad a \neq 0,$

occurs at $x = -b/(2a)$.

TECHNOLOGY CONNECTION

Graph each function. Then estimate any relative extrema. Where appropriate, round to three decimal places.

112. $f(x) = 3x^{2/3} - 2x$

113. $f(x) = 4x - 6x^{2/3}$

114. $f(x) = x^2(x - 2)^3$

115. $f(x) = x^2(1 - x)^3$

116. $f(x) = x - \sqrt{x}$

117. $f(x) = (x - 1)^{2/3} - (x + 1)^{2/3}$

TW **118. Social sciences: time spent on home computer.** The following data relate the average number of minutes spent per month on a home computer to a person's age.

Age (in years)	Average Use (in minutes per month)
6.5	363
14.5	645
21	1377
29.5	1727
39.5	1696
49.5	2052
55 and up	2299

(*Source*: Media Matrix; The PC Meter Company.)

a) Use the regression procedures of Section R.6 to fit linear, cubic, and quartic functions $y = f(x)$ to the data, where x is age and y is average use per month. Decide which function best fits the data. Explain.

b) What is the domain of the function?

c) Does the function have any relative extrema? Explain.

2.3

OBJECTIVES

- Find limits involving infinity.
- Determine the asymptotes of a function's graph.
- Graph rational functions.

Graph Sketching: Asymptotes and Rational Functions

Rational Functions

Thus far we have considered a strategy for graphing a continuous function using the tools of calculus. We now want to consider some discontinuous functions, most of which are rational functions. Our graphing skills must now allow for discontinuities as well as certain lines called *asymptotes.*

Let's review the definition of a rational function.

DEFINITION

A **rational function** is a function f that can be described by

$$f(x) = \frac{P(x)}{Q(x)},$$

where $P(x)$ and $Q(x)$ are polynomials, with $Q(x)$ not the zero polynomial. The domain of f consists of all inputs x for which $Q(x) \neq 0$.

Polynomials are themselves a special kind of rational function, since $Q(x)$ can be the polynomial 1. Here we are considering graphs of rational functions in which the denominator is not a constant. Before we do so, however, we need to reconsider limits.

Vertical and Horizontal Asymptotes

Figure 1 shows the graph of the rational function

$$f(x) = \frac{x^2 - 1}{x^2 + x - 6} = \frac{(x-1)(x+1)}{(x-2)(x+3)}.$$

FIGURE 1

Note that as x gets closer to 2 from the left, the function values get smaller and smaller negatively, approaching $-\infty$. As x gets closer to 2 from the right, the function values get larger and larger positively. Thus,

$$\lim_{x\to 2^-} f(x) = -\infty \quad \text{and} \quad \lim_{x\to 2^+} f(x) = \infty.$$

For this graph, we can think of the line $x = 2$ as a "limiting line" called a *vertical asymptote*. Similarly, the line $x = -3$ is another vertical asymptote.

DEFINITION

The line $x = a$ is a **vertical asymptote** if any of the following limit statements is true:

$$\lim_{x\to a^-} f(x) = \infty, \quad \lim_{x\to a^-} f(x) = -\infty, \quad \lim_{x\to a^+} f(x) = \infty, \quad \text{or} \quad \lim_{x\to a^+} f(x) = -\infty.$$

The graph of a rational function *never* crosses a vertical asymptote. If the expression that defines the rational function f is simplified, meaning that it has no common factor other than -1 or 1, then if a is an input that makes the denominator 0, the line $x = a$ is a vertical asymptote.

For example,

$$f(x) = \frac{x^2 - 9}{x - 3} = \frac{(x-3)(x+3)}{x-3}$$

does not have a vertical asymptote at $x = 3$, even though 3 is an input that makes the denominator 0. This is because when $(x^2 - 9)/(x - 3)$ is simplified, it has $x - 3$ as a common factor of the numerator and the denominator. In contrast,

$$g(x) = \frac{x^2 - 4}{x^2 + x - 12} = \frac{(x+2)(x-2)}{(x-3)(x+4)}$$

is simplified and has $x = 3$ and $x = -4$ as vertical asymptotes.

Figure 2 shows the four ways in which a vertical asymptote can occur.

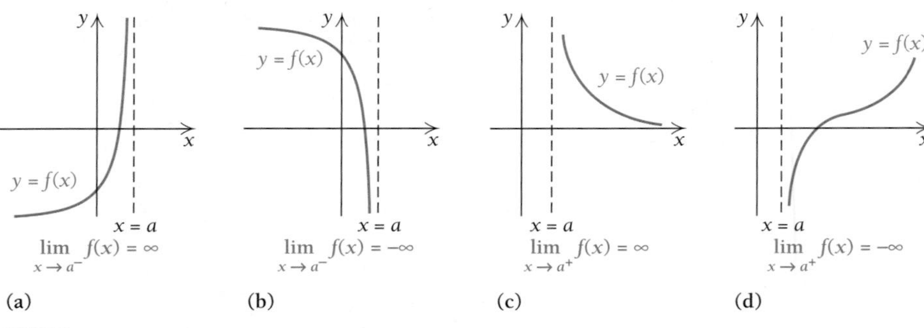

FIGURE 2

EXAMPLE 1 Determine the vertical asymptotes: $f(x) = \dfrac{3x - 2}{x(x-5)(x+3)}$.

Solution The expression is in simplified form. The vertical asymptotes are the lines $x = 0, x = 5,$ and $x = -3$. ◆

TECHNOLOGY CONNECTION

Asymptotes

Our discussion here allows us to attach the term "vertical asymptote" to those mysterious vertical lines that appear when graphing rational functions in the CONNECTED mode. For example, consider the graph of $f(x) = 8/(x^2 - 4)$, using the window $[-6, 6, -8, 8]$. Vertical asymptotes occur at $x = -2$ and $x = 2$. These lines are not part of the graph.

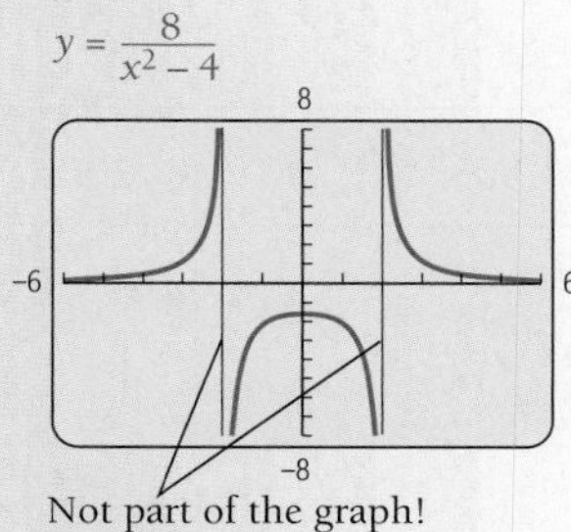

EXERCISES

Graph each of the following in both DOT and CONNECTED modes. Try to locate the vertical asymptotes visually. Then verify your results using the method of Examples 1 and 2. You may need to try different viewing windows.

1. $f(x) = \dfrac{x^2 + 7x + 10}{x^2 + 3x - 28}$

2. $f(x) = \dfrac{x^2 + 5}{x^3 - x^2 - 6x}$

EXAMPLE 2 Determine the vertical asymptotes of the function given by

$$f(x) = \frac{x^2 - 2x}{x^3 - x}.$$

Solution We write the expression in simplified form:

$$f(x) = \frac{x^2 - 2x}{x^3 - x} = \frac{x(x - 2)}{x(x - 1)(x + 1)}$$
$$= \frac{x - 2}{(x - 1)(x + 1)}, \quad x \neq 0.$$

The expression is now in simplified form. The vertical asymptotes are the lines $x = -1$ and $x = 1$. ✦

Look again at the graph in Fig. 1. Note that function values get closer and closer to 1 as x approaches $-\infty$, meaning that $f(x) \to 1$ as $x \to -\infty$. Also, function values get closer and closer to 1 as x approaches ∞, meaning that $f(x) \to 1$ as $x \to \infty$. Thus,

$$\lim_{x \to -\infty} f(x) = 1 \quad \text{and} \quad \lim_{x \to \infty} f(x) = 1.$$

The line $y = 1$ is called a *horizontal asymptote.*

DEFINITION

The line $y = b$ is a **horizontal asymptote** if either or both of the following limit statements is true:

$$\lim_{x \to -\infty} f(x) = b \quad \text{or} \quad \lim_{x \to \infty} f(x) = b.$$

The graph of a rational function may or may not cross a horizontal asymptote. Horizontal asymptotes occur when the degree of the numerator is less than or equal to the degree of the denominator. (The degree of a polynomial in one variable is the highest power of that variable.)

In Figs. 3–5, we see three ways in which horizontal asymptotes can occur.

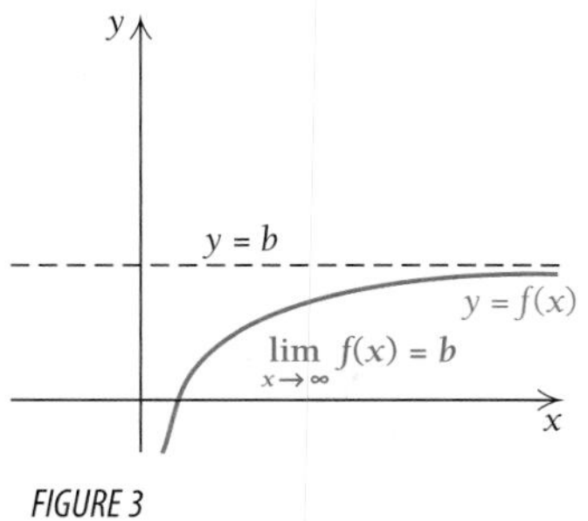

FIGURE 3

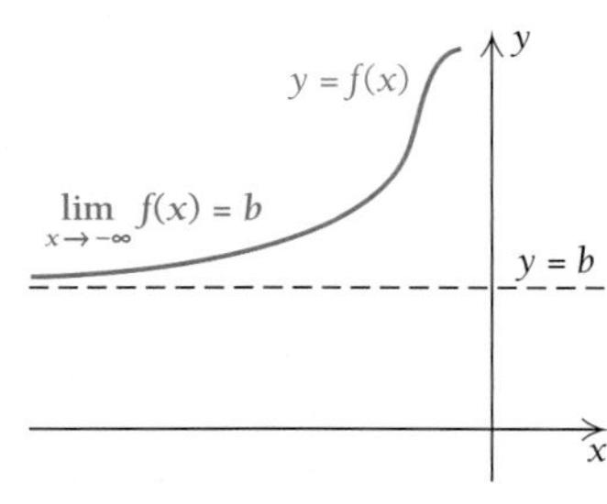

FIGURE 4

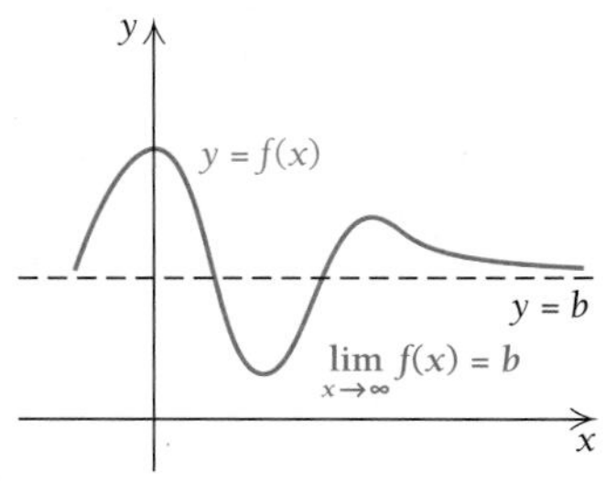

FIGURE 5

Horizontal asymptotes are found by determining the limit of a rational function as inputs approach $-\infty$ or ∞.

EXAMPLE 3 Determine the horizontal asymptote of the function given by

$$f(x) = \frac{3x - 4}{x}.$$

Solution To find the horizontal asymptote, we consider

$$\lim_{x\to\infty} f(x) = \lim_{x\to\infty} \frac{3x - 4}{x}.$$

One way to find such a limit is to use an input–output table, as follows, using progressively larger x-values.

Inputs, x	1	10	50	100	2000
Outputs, $\frac{3x-4}{x}$	-1	2.6	2.92	2.96	2.998

As the inputs get larger and larger without bound, the outputs get closer and closer to 3. Thus,

$$\lim_{x\to\infty} \frac{3x - 4}{x} = 3.$$

Another way to find this limit is to use some algebra and the fact that

$$\text{as } x \to \infty, \text{ we have } \frac{1}{x} \to 0, \text{ and more generally, } \frac{b}{ax^n} \to 0,$$

for any positive integer n and any constants a and b, $a \neq 0$. We multiply by 1, using $(1/x) \div (1/x)$. This amounts to dividing both the numerator and the denominator by x:

$$\begin{aligned}
\lim_{x\to\infty} \frac{3x - 4}{x} &= \lim_{x\to\infty} \frac{3x - 4}{x} \cdot \frac{(1/x)}{(1/x)} \\
&= \lim_{x\to\infty} \frac{\frac{3x}{x} - \frac{4}{x}}{\frac{x}{x}} \\
&= \lim_{x\to\infty} \frac{3 - \frac{4}{x}}{1} \\
&= \lim_{x\to\infty} \left(3 - \frac{4}{x}\right) \\
&= 3 - 0 = 3.
\end{aligned}$$

TECHNOLOGY CONNECTION

EXERCISES

1. Verify the limit
$$\lim_{x\to\infty} \frac{3x - 4}{x} = 3$$
by using the TABLE feature with larger and larger x-values.

X	Y1
50	2.92
150	2.9733
250	2.984
350	2.9886
450	2.9911
550	2.9927
650	2.9938

X = 50

X	Y1
500	2.992
1500	2.9973
2500	2.9984
3500	2.9989
4500	2.9991
5500	2.9993
6500	2.9994

X = 500

2. Graph the function
$$f(x) = \frac{3x - 4}{x}$$
in DOT mode. Then use TRACE, moving the cursor along the graph from left to right, and observe the behavior of the y-coordinates.

For Exercises 3 and 4, consider $\lim_{x\to\infty} \frac{2x + 5}{x}$.

3. Use the TABLE feature to find the limit.
4. Graph the function in DOT mode, and use TRACE to find the limit.

In a similar manner, it can be shown that

$$\lim_{x \to -\infty} f(x) = 3.$$

The horizontal asymptote is the line $y = 3$. ✦

EXAMPLE 4 Determine the horizontal asymptote of the function given by

$$f(x) = \frac{3x^2 + 2x - 4}{2x^2 - x + 1}.$$

Solution As in Example 3, the degree of the numerator is the same as the degree of the denominator. Let's adapt the algebraic approach used in that example.

To do so, we divide the numerator and the denominator by x^2 and find the limit as $|x|$ gets larger and larger:

$$f(x) = \frac{3x^2 + 2x - 4}{2x^2 - x + 1} = \frac{3 + \frac{2}{x} - \frac{4}{x^2}}{2 - \frac{1}{x} + \frac{1}{x^2}}.$$

As $|x|$ gets very large, the numerator approaches 3 and the denominator approaches 2. Therefore, the function gets very close to $\frac{3}{2}$. Thus,

$$\lim_{x \to -\infty} f(x) = \frac{3}{2} \quad \text{and} \quad \lim_{x \to \infty} f(x) = \frac{3}{2}.$$

The line $y = \frac{3}{2}$ is a horizontal asymptote. ✦

Examples 3 and 4 lead to the following result.

When the degree of the numerator is the same as the degree of the denominator, the line $y = a/b$ is a horizontal asymptote, where a is the leading coefficient of the numerator and b is the leading coefficient of the denominator.

TECHNOLOGY CONNECTION

To see why a horizontal asymptote is determined by the leading terms in the expression's numerator and denominator, consider the terms $3x^2$, $2x$, and -4 in Example 4. If $y_1 = 3x^2$ and $y_2 = 3x^2 + 2x - 4$, a table reveals that as x gets larger, the difference between y_2 and y_1 becomes less significant. This is because $3x^2$ grows far more rapidly than does $2x - 4$.

X	Y1	Y2
20	1200	1236
120	43200	43436
220	145200	145636
320	307200	307836
420	529200	530036
520	811200	812236
620	1.15 E6	1.15 E6

X = 20

EXERCISES

1. Let $y_1 = 2x^2$ and $y_2 = 2x^2 - x + 1$. Use a table to show that as x gets large, $y_1 \approx y_2$.
2. Let $y_1 = (3x^2)/(2x^2)$ and $y_2 = (3x^2 + 2x - 1)/(2x^2 - x + 1)$. Show that for large x, we have $y_1 \approx y_2$.

EXAMPLE 5 Determine the horizontal asymptote:

$$f(x) = \frac{2x + 3}{x^3 - 2x^2 + 4}.$$

Solution Since the degree of the numerator is less than the degree of the denominator, there is a horizontal asymptote. To identify that asymptote, we divide both the numerator and denominator by the highest power of x in the denominator, just as in Examples 3 and 4, and find the limits as $|x| \to \infty$:

$$f(x) = \frac{2x + 3}{x^3 - 2x^2 + 4} = \frac{\frac{2}{x^2} + \frac{3}{x^3}}{1 - \frac{2}{x} + \frac{4}{x^3}}.$$

TECHNOLOGY CONNECTION

EXERCISES

Graph each of the following. Try to locate the horizontal asymptotes using the TABLE and TRACE features. Verify your results using the methods of Examples 3–5.

1. $f(x) = \dfrac{x^2 + 5}{x^3 - x^2 - 6x}$

2. $f(x) = \dfrac{9x^4 - 7x^2 - 9}{3x^4 + 7x^2 + 9}$

3. $f(x) = \dfrac{135x^5 - x^2}{x^7}$

4. $f(x) = \dfrac{3x^2 - 4x + 3}{6x^2 + 2x - 5}$

As x gets smaller and smaller negatively, $|x|$ gets larger and larger. Similarly, as x gets larger and larger positively, $|x|$ gets larger and larger. Thus, as $|x|$ becomes very large, every expression with a denominator that is a power of x comes ever closer to 0. Thus, the numerator of $f(x)$ approaches 0 as its denominator approaches 1; hence the entire expression takes on values ever closer to 0. That is, for $x \to -\infty$ or $x \to \infty$, we have

$$f(x) \approx \frac{0 + 0}{1 - 0 + 0},$$

so $$\lim_{x \to -\infty} f(x) = 0 \quad \text{and} \quad \lim_{x \to \infty} f(x) = 0,$$

and the x-axis, the line $y = 0$, is a horizontal asymptote. ✦

When the degree of the numerator is less than the degree of the denominator, the x-axis, or the line $y = 0$, is a horizontal asymptote.

Slant Asymptotes

Some asymptotes are neither vertical nor horizontal. For example, in the graph of

$$f(x) = \frac{x^2 - 4}{x - 1},$$

shown at right, as $|x|$ gets larger and larger, the curve gets closer and closer to $y = x + 1$. The line $y = x + 1$ is called a *slant asymptote*, or *oblique asymptote*. In Example 6, we will see how the line $y = x + 1$ was determined.

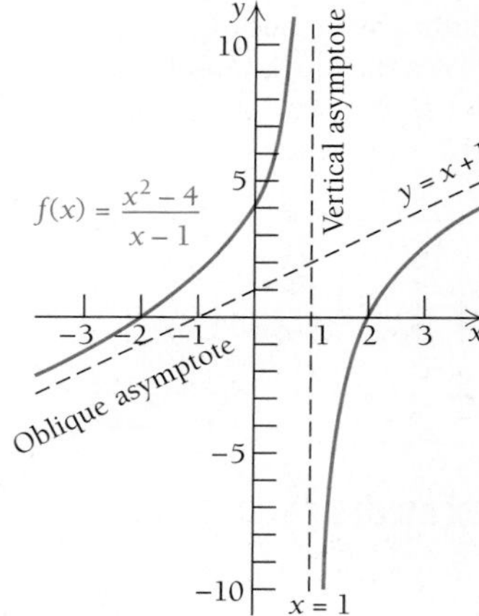

DEFINITION

A linear asymptote that is neither vertical nor horizontal is called a **slant**, or **oblique, asymptote**. For any rational function of the form $f(x) = p(x)/q(x)$, a slant asymptote occurs when the degree of $p(x)$ is exactly 1 more than the degree of $q(x)$. A graph can cross a slant asymptote.

How can we find a slant asymptote? One way is by division.

EXAMPLE 6 Find the slant asymptote:

$$f(x) = \frac{x^2 - 4}{x - 1}.$$

Solution When we divide the numerator by the denominator, we obtain a quotient of $x + 1$ and a remainder of -3:

$$\begin{array}{r} x+1 \\ x-1\overline{)x^2 \qquad -4} \\ \underline{x^2-x} \\ x-4 \\ \underline{x-1} \\ -3 \end{array} \qquad f(x) = \frac{x^2-4}{x-1} = (x+1) + \frac{-3}{x-1}.$$

Now we can see that when $|x|$ gets very large, $-3/(x-1)$ approaches 0. Thus, for very large $|x|$, the expression $x + 1$ is the dominant part of

$$(x+1) + \frac{-3}{x-1}.$$

Thus, $y = x + 1$ is a slant asymptote. ✦

TECHNOLOGY CONNECTION

EXERCISES

Graph each of the following. Try to visually locate the slant asymptotes. Then use the method in Example 6 to find each slant asymptote and graph it along with the original function.

1. $f(x) = \dfrac{3x^2 - 7x + 8}{x - 2}$

2. $f(x) = \dfrac{5x^3 + 2x + 1}{x^2 - 4}$

Intercepts

If they exist, the **x-intercepts** of a function occur at those values of x for which $y = f(x) = 0$, and they give us points at which the graph crosses the x-axis. If it exists, the **y-intercept** of a function occurs at the value of y for which $x = 0$, and it gives us the point at which the graph crosses the y-axis.

EXAMPLE 7 Find the intercepts of

$$f(x) = \frac{x^3 - x^2 - 6x}{x^2 - 3x + 2}.$$

Solution We factor the numerator and the denominator:

$$f(x) = \frac{x(x+2)(x-3)}{(x-1)(x-2)}.$$

To find the x-intercepts, we solve the equation $f(x) = 0$. Such values occur when the numerator is 0 and the denominator is not. Thus, we solve the equation

$$x(x+2)(x-3) = 0.$$

The x-values that make the numerator 0 are 0, -2, and 3. Since none of these make the denominator 0, they yield the x-intercepts $(0, 0)$, $(-2, 0)$, and $(3, 0)$.

To find the y-intercept, we let $x = 0$:

$$f(0) = \frac{0^3 - 0^2 - 6(0)}{0^2 - 3(0) + 2} = 0.$$

In this case, the y-intercept is also an x-intercept, $(0, 0)$. ✦

TECHNOLOGY CONNECTION

EXERCISES

Graph each of the following. Use the ZERO feature and a table in ASK mode to find the x- and y-intercepts.

1. $f(x) = \dfrac{x(x-3)(x+5)}{(x+2)(x-4)}$

2. $f(x) = \dfrac{x^3 + 2x^2 - 3x}{x^2 + 5}$

Sketching Graphs

We can now refine our analytic strategy for graphing.

Strategy for Sketching Graphs

a) *Intercepts.* Find the x-intercept(s) and the y-intercept of the graph.
b) *Asymptotes.* Find any vertical, horizontal, or slant asymptotes.
c) *Derivatives and domain.* Find $f'(x)$ and $f''(x)$. Find the domain of f.
d) *Critical values of f.* Find any inputs for which $f'(x)$ is not defined or for which $f'(x) = 0$.
e) *Increasing and/or decreasing; relative extrema.* Substitute each critical value, x_0, from step (d) into $f''(x)$. If $f''(x_0) < 0$, then x_0 yields a relative maximum and f is increasing to the left of x_0 and decreasing to the right. If $f''(x_0) > 0$, then x_0 yields a relative minimum and f is decreasing to the left of x_0 and increasing to the right. On intervals where no critical value exists, use f' and test values to find where f is increasing or decreasing.
f) *Inflection points.* Determine candidates for inflection points by finding x-values for which $f''(x)$ does not exist or for which $f''(x) = 0$. Find the function values at these points. If a function value $f(x)$ does not exist, then the function does not have an inflection point at x.
g) *Concavity.* Use the values from step (f) as endpoints of intervals. Determine the concavity over each interval by checking to see where f' is increasing—that is, $f''(x) > 0$—and where f' is decreasing—that is, $f''(x) < 0$. Do this by substituting a test value from each interval into $f''(x)$. Use the results of step (d).
h) *Sketch the graph.* Use the information from steps (a)–(g) to sketch the graph, plotting extra points as needed.

EXAMPLE 8 Sketch the graph of $f(x) = \dfrac{8}{x^2 - 4}$.

Solution

a) *Intercepts.* The x-intercepts occur at values for which the numerator is 0 but the denominator is not. Since in this case the numerator is the constant 8, there are no x-intercepts. To find the y-intercept, we compute $f(0)$:

$$f(0) = \frac{8}{0^2 - 4} = \frac{8}{-4} = -2.$$

This gives us one point on the graph, $(0, -2)$.

b) *Asymptotes.*

Vertical: The denominator, $x^2 - 4 = (x + 2)(x - 2)$, is 0 for x-values of -2 and 2. Thus, the graph has the lines $x = -2$ and $x = 2$ as vertical asymptotes. We draw them using dashed lines (they are *not* part of the actual graph, just guidelines).

Horizontal: The degree of the numerator is less than the degree of the denominator, so the x-axis, $y = 0$, is the horizontal asymptote.

Slant: There is no slant asymptote since the degree of the numerator is not 1 more than the degree of the denominator.

c) *Derivatives and domain.* Find $f'(x)$ and $f''(x)$. Using the Quotient Rule, we get

$$f'(x) = \frac{-16x}{(x^2 - 4)^2} \quad \text{and} \quad f''(x) = \frac{16(3x^2 + 4)}{(x^2 - 4)^3}.$$

The domain of f is $(-\infty, -2) \cup (-2, 2) \cup (2, \infty)$ as determined in step (b).

d) *Critical values of f.* We look for values of x for which $f'(x) = 0$ or for which $f'(x)$ does not exist. From step (c), we see that $f'(x) = 0$ for values of x for which $-16x = 0$, but the denominator is not 0. The only such number is 0 itself. The derivative $f'(x)$ does not exist at -2 and 2, but neither value is in the domain of f. Thus, the only critical value is 0.

e) *Increasing and/or decreasing; relative extrema.* Use the undefined values and the critical values to determine the intervals over which f is increasing and the intervals over which f is decreasing. The values to consider are $-2, 0$, and 2.

Since

$$f''(0) = \frac{16(3 \cdot 0^2 + 4)}{(0^2 - 4)^3} = \frac{64}{-64} < 0,$$

we know that a relative maximum exists at $(0, f(0))$, or $(0, -2)$. Thus, f is increasing on the interval $(-2, 0]$ and decreasing on $[0, 2)$.

Since $f''(x)$ does not exist for the x-values -2 and 2, we use $f'(x)$ and test values to see if f is increasing or decreasing on $(-\infty, -2)$ and $(2, \infty)$:

$$\text{Test } -3, \quad f'(-3) = \frac{-16(-3)}{[(-3)^2 - 4]^2} = \frac{48}{25} > 0, \text{ so } f \text{ is increasing on } (-\infty, -2);$$

$$\text{Test } 3, \quad f'(3) = \frac{-16(3)}{[(3)^2 - 4]^2} = \frac{-48}{25} < 0, \text{ so } f \text{ is decreasing on } (2, \infty).$$

f) *Inflection points.* Determine candidates for inflection points by finding where $f''(x)$ does not exist and where $f''(x) = 0$. The only values for which $f''(x)$ does not exist are where $x^2 - 4 = 0$, or -2 and 2. Neither value is in the domain of f, so we focus solely on where $f''(x) = 0$, or

$$16(3x^2 + 4) = 0.$$

Since $16(3x^2 + 4) > 0$ for all real numbers x, there are no points of inflection.

g) *Concavity.* Since no values were found in step (f), the only place where concavity could change is on either side of the vertical asymptotes, $x = -2$ and $x = 2$. To determine the concavity, we check to see where $f''(x)$ is positive or negative. The numbers -2 and 2 divide the x-axis into three intervals. We choose test values in each interval and make a substitution into f'':

$$\text{Test } -3, \quad f''(-3) = \frac{16[3(-3)^2 + 4]}{[(-3)^2 - 4]^3} > 0;$$

$$\text{Test } 0, \quad f''(0) = \frac{16[3(0)^2 + 4]}{[(0)^2 - 4]^3} < 0;$$ We already knew this from step (e).

$$\text{Test } 3, \quad f''(3) = \frac{16[3(3)^2 + 4]}{[(3)^2 - 4]^3} > 0.$$

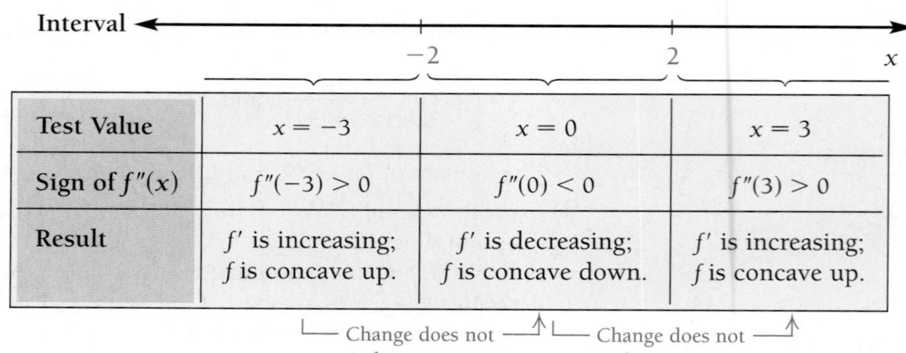

Test Value	$x = -3$	$x = 0$	$x = 3$
Sign of $f''(x)$	$f''(-3) > 0$	$f''(0) < 0$	$f''(3) > 0$
Result	f' is increasing; f is concave up.	f' is decreasing; f is concave down.	f' is increasing; f is concave up.

The function is concave up over the intervals $(-\infty, -2)$ and $(2, \infty)$. The function is concave down over the interval $(-2, 2)$.

h) *Sketch the graph.* Sketch the graph using the information in the following table, plotting extra points as needed. The graph is shown below.

x	$f(x)$ approximately
−5	0.38
−4	0.67
−3	1.6
−1	−2.67
0	−2
1	−2.67
3	1.6
4	0.67
5	0.38

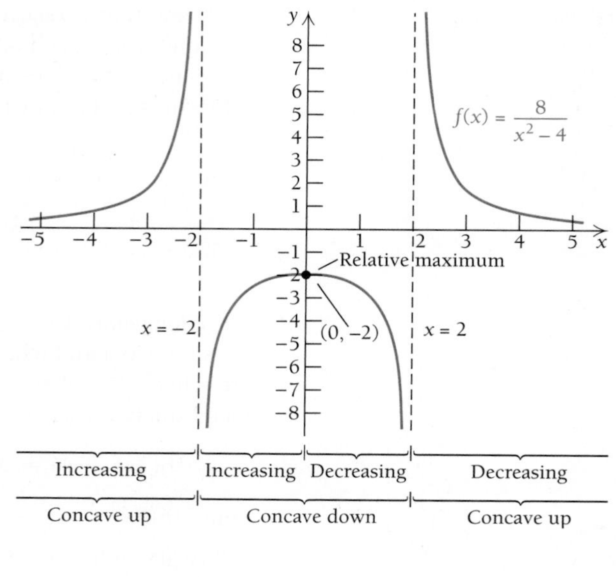

EXAMPLE 9 Sketch the graph of $f(x) = \dfrac{x^2 + 4}{x}$.

Solution

a) *Intercepts.* The equation $f(x) = 0$ has no real-number solution. Thus, there are no x-intercepts. The number 0 is not in the domain of the function. Thus, there is no y-intercept.

b) *Asymptotes.*

Vertical: Since replacing x with 0 makes the denominator 0, the line $x = 0$ is a vertical asymptote.

Horizontal: The degree of the numerator is greater than the degree of the denominator, so there is no horizontal asymptote.

Slant: The degree of the numerator is 1 greater than the degree of the denominator, so there is a slant asymptote. We do the division

$$\begin{array}{r} x\phantom{{}+4} \\ x\overline{)x^2 + 4} \\ \underline{x^2\phantom{{}+4}} \\ 4 \end{array}$$

and express the function in the form

$$f(x) = x + \frac{4}{x}.$$

As $|x|$ gets larger, $4/x$ approaches 0, so the line $y = x$ is a slant asymptote.

c) *Derivatives and domain.* Find $f'(x)$ and $f''(x)$:

$$f'(x) = 1 - 4x^{-2} = 1 - \frac{4}{x^2};$$

$$f''(x) = 8x^{-3} = \frac{8}{x^3}.$$

The domain of f is $(-\infty, 0) \cup (0, \infty)$ or all real numbers except 0.

d) *Critical values of f.* We see from step (c) that $f'(x)$ is undefined at $x = 0$, but 0 is not in the domain of f. Thus, to find critical values, we solve $f'(x) = 0$, looking for solutions other than 0:

$$1 - \frac{4}{x^2} = 0 \qquad \text{Setting } f'(x) \text{ equal to } 0$$

$$1 = \frac{4}{x^2}$$

$$x^2 = 4 \qquad \text{Multiplying both sides by } x^2$$

$$x = \pm 2.$$

Thus, -2 and 2 are critical values.

e) *Increasing and/or decreasing; relative extrema.* Use the points found in step (d) to find intervals over which f is increasing and intervals over which f is decreasing. The points to consider are -2, 0, and 2.

Since

$$f''(-2) = \frac{8}{(-2)^3} = -1 < 0,$$

we know that a relative maximum exists at $(-2, f(-2))$, or $(-2, -4)$. Thus, f is increasing on $(-\infty, -2]$ and decreasing on $[-2, 0)$.

Since

$$f''(2) = \frac{8}{(2)^3} = 1 > 0,$$

we know that a relative minimum exists at $(2, f(2))$, or $(2, 4)$. Thus, f is decreasing on $(0, 2]$ and increasing on $[2, \infty)$.

f) *Inflection points.* Determine candidates for inflection points by finding where $f''(x)$ does not exist or where $f''(x) = 0$. The only value for which $f''(x)$ does not exist is 0, but 0 is not in the domain of f. Thus, the only place an inflection point could occur is where $f''(x) = 0$:

$$\frac{8}{x^3} = 0.$$

But this equation has no solution. Thus, there are no points of inflection.

g) *Concavity.* Since no values were found in step (f), the only place where concavity could change would be on either side of the vertical asymptote $x = 0$. In step (e), we used the Second-Derivative Test to determine relative extrema. From that work, we know that f is concave down over the interval $(-\infty, 0)$ and concave up over $(0, \infty)$.

Interval ← 0 → x

Test Value	$x = -2$	$x = 2$
Sign of $f''(x)$	$f''(-2) < 0$	$f''(2) > 0$
Result	f' is decreasing; f is concave down	f' is increasing; f is concave up

Change does not indicate a point of inflection since $f(0)$ does not exist.

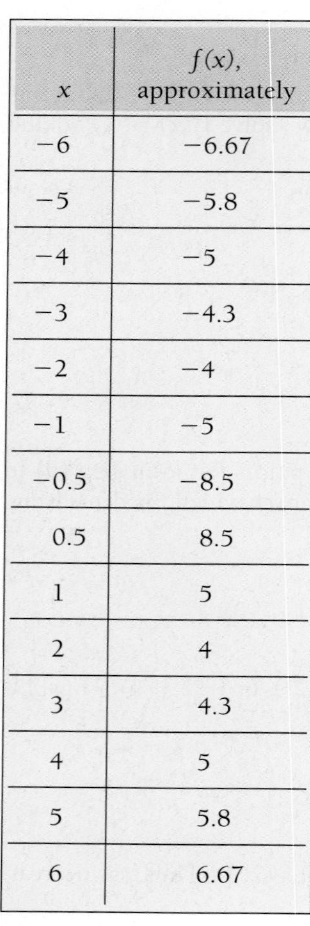

x	$f(x)$, approximately
−6	−6.67
−5	−5.8
−4	−5
−3	−4.3
−2	−4
−1	−5
−0.5	−8.5
0.5	8.5
1	5
2	4
3	4.3
4	5
5	5.8
6	6.67

h) *Sketch the graph.* Sketch the graph using the preceding information and additional computed values of f, as needed. The graph follows.

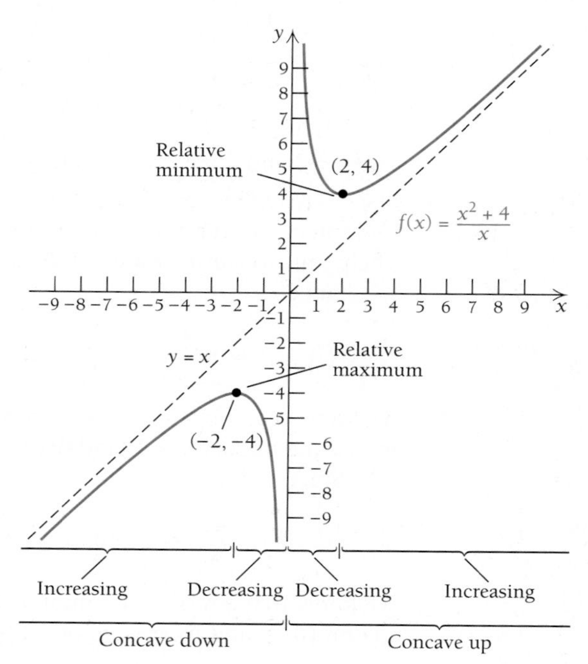

Exercise Set 2.3

Determine the vertical asymptote(s) of each function. If none exists, state that fact.

1. $f(x) = \dfrac{2x - 3}{x - 5}$
2. $f(x) = \dfrac{x + 4}{x - 2}$
3. $f(x) = \dfrac{3x}{x^2 - 9}$
4. $f(x) = \dfrac{5x}{x^2 - 25}$
5. $f(x) = \dfrac{x + 2}{x^3 - 6x^2 + 8x}$
6. $f(x) = \dfrac{x + 3}{x^3 - x}$
7. $f(x) = \dfrac{x + 6}{x^2 + 7x + 6}$
8. $f(x) = \dfrac{x + 2}{x^2 + 6x + 8}$
9. $f(x) = \dfrac{6}{x^2 + 36}$
10. $f(x) = \dfrac{7}{x^2 + 49}$

Determine the horizontal asymptote of each function. If none exists, state that fact.

11. $f(x) = \dfrac{6x}{8x + 3}$
12. $f(x) = \dfrac{3x^2}{6x^2 + x}$
13. $f(x) = \dfrac{4x}{x^2 - 3x}$
14. $f(x) = \dfrac{2x}{3x^3 - x^2}$
15. $f(x) = 5 - \dfrac{3}{x}$
16. $f(x) = 4 + \dfrac{2}{x}$
17. $f(x) = \dfrac{8x^4 - 5x^2}{2x^3 + x^2}$
18. $f(x) = \dfrac{6x^3 + 4x}{3x^2 - x}$
19. $f(x) = \dfrac{6x^4 + 4x^2 - 7}{2x^5 - x + 3}$
20. $f(x) = \dfrac{4x^3 - 3x + 2}{x^3 + 2x - 4}$
21. $f(x) = \dfrac{2x^3 - 4x + 1}{4x^3 + 2x - 3}$
22. $f(x) = \dfrac{5x^4 - 2x^3 + x}{x^5 - x^3 + 8}$

Sketch the graph of each function. Indicate where each function is increasing or decreasing, where any relative extrema occur, where asymptotes occur, where the graph is concave up or concave down, where any points of inflection occur, and where any intercepts occur.

23. $f(x) = -\dfrac{5}{x}$
24. $f(x) = \dfrac{4}{x}$
25. $f(x) = \dfrac{1}{x - 5}$
26. $f(x) = \dfrac{-2}{x - 5}$
27. $f(x) = \dfrac{1}{x + 2}$
28. $f(x) = \dfrac{1}{x - 3}$
29. $f(x) = \dfrac{-3}{x - 3}$
30. $f(x) = \dfrac{-2}{x + 5}$
31. $f(x) = \dfrac{3x - 1}{x}$
32. $f(x) = \dfrac{2x + 1}{x}$
33. $f(x) = x + \dfrac{2}{x}$
34. $f(x) = x + \dfrac{9}{x}$
35. $f(x) = \dfrac{-1}{x^2}$
36. $f(x) = \dfrac{2}{x^2}$
37. $f(x) = \dfrac{x}{x + 2}$
38. $f(x) = \dfrac{x}{x - 3}$
39. $f(x) = \dfrac{-1}{x^2 + 2}$
40. $f(x) = \dfrac{1}{x^2 + 3}$
41. $f(x) = \dfrac{x + 3}{x^2 - 9}$ (*Hint:* Simplify.)
42. $f(x) = \dfrac{x - 1}{x^2 - 1}$
43. $f(x) = \dfrac{x - 1}{x + 2}$
44. $f(x) = \dfrac{x - 2}{x + 1}$
45. $f(x) = \dfrac{x^2 - 4}{x + 3}$
46. $f(x) = \dfrac{x^2 - 9}{x + 1}$
47. $f(x) = \dfrac{x + 1}{x^2 - 2x - 3}$
48. $f(x) = \dfrac{x - 3}{x^2 + 2x - 15}$
49. $f(x) = \dfrac{2x^2}{x^2 - 16}$
50. $f(x) = \dfrac{x^2 + x - 2}{2x^2 - 2}$
51. $f(x) = \dfrac{1}{x^2 - 1}$
52. $f(x) = \dfrac{10}{x^2 + 4}$
53. $f(x) = \dfrac{x^2 + 1}{x}$
54. $f(x) = \dfrac{x^3}{x^2 - 1}$
55. $f(x) = \dfrac{x^2 - 9}{x - 3}$
56. $f(x) = \dfrac{x^2 - 16}{x + 4}$

APPLICATIONS

Business and Economics

57. **Depreciation.** Suppose that the value V of the inventory at Fido's Pet Supply decreases, or depreciates, with time t, in months, where

$$V(t) = 50 - \frac{25t^2}{(t + 2)^2}.$$

a) Find $V(0)$, $V(5)$, $V(10)$, and $V(70)$.

b) Find the maximum value of the inventory over the interval $[0, \infty)$.

c) Sketch a graph of V.

TW d) Does there seem to be a value below which $V(t)$ will never fall? Explain.

58. **Average cost.** The total-cost function for Acme, Inc., to produce x units of a product is given by

$$C(x) = 3x^2 + 80.$$

a) The *average cost* is given by $A(x) = C(x)/x$. Find $A(x)$.

b) Graph the average cost.

c) Find the slant asymptote for the graph of $y = A(x)$, and interpret its significance.

59. **Cost of pollution control.** Cities and companies find that the cost of pollution control increases with respect to the percentage of pollutants to be removed from a situation. Suppose that the cost C, in dollars, of removing $p\%$ of the pollutants from a chemical spill is given by

$$C(p) = \frac{48{,}000}{100 - p}.$$

a) Find $C(0)$, $C(20)$, $C(80)$, and $C(90)$.

b) Find the domain of C.

c) Sketch a graph of C.

TW d) Can the company or city afford to remove 100% of the pollutants from this spill? Explain.

60. **Total cost and revenue.** The total cost and total revenue, in dollars, from producing x couches are given by

$$C(x) = 5000 + 600x \quad \text{and} \quad R(x) = -\tfrac{1}{2}x^2 + 1000x.$$

a) Find the total-profit function, $P(x)$.

b) The *average profit* is given by $A(x) = P(x)/x$. Find $A(x)$.

c) Graph the average profit.

d) Find the slant asymptote for the graph of $y = A(x)$.

61. **Purchasing price.** Since 1970, the purchasing power of the dollar, as measured by consumer prices, can be modeled by the function

$$P(x) = \frac{2.632}{1 + 0.116x},$$

where x is the number of years since 1970. (*Source*: U.S. Bureau of Economic Analysis.)

a) Find $P(10)$, $P(20)$, and $P(40)$.

b) When will the purchasing power be \$0.50?

c) Find $\lim_{x\to\infty} P(x)$.

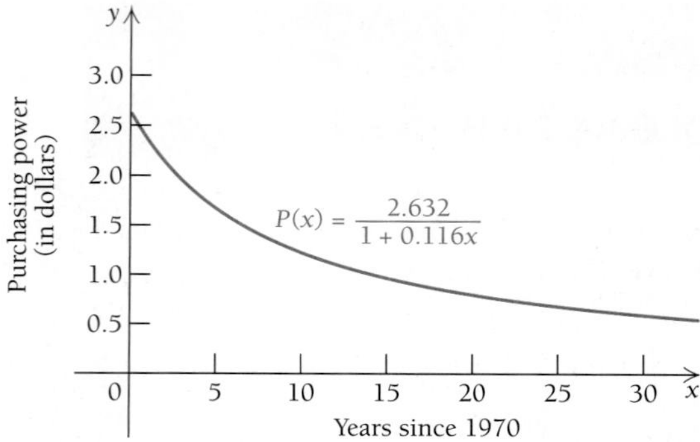

Life and Physical Sciences

62. **Medication in the bloodstream.** After an injection, the amount of a medication A, in cubic centimeters (cc), in the bloodstream decreases with time t, in hours. Suppose that under certain conditions A is given by

$$A(t) = \frac{A_0}{t^2 + 1},$$

where A_0 is the initial amount of the medication. Assume that an initial amount of 100 cc is injected.

a) Find $A(0)$, $A(1)$, $A(2)$, $A(7)$, and $A(10)$.

b) Find the maximum amount of medication in the bloodstream over the interval $[0, \infty)$.

c) Sketch a graph of the function.

TW d) According to this function, does the medication ever completely leave the bloodstream? Explain your answer.

General Interest

63. **Baseball: earned-run average.** A pitcher's *earned-run average* (the average number of runs given up every 9 innings, or 1 game) is given by

$$E = 9 \cdot \frac{r}{n},$$

where r is the number of earned runs allowed in n innings. Suppose that we fix the number of earned runs allowed at 4 and let n vary. We get a function given by

$$E(n) = 9 \cdot \frac{4}{n}.$$

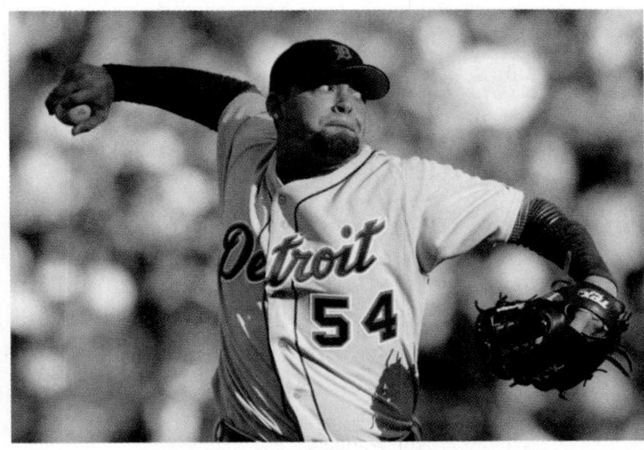

a) Complete the following table, rounding to two decimal places.

Innings Pitched (n)	Earned-Run Average (E)
9	
8	
7	
6	
5	
4	
3	
2	
1	
$\frac{2}{3}$	
$\frac{1}{3}$	

b) Find $\lim_{n \to 0} E(n)$.

SYNTHESIS

TW **64.** Explain why a vertical asymptote is only a guide and is not part of the graph of a function.

TW **65.** Using graphs and limits, explain the idea of an asymptote to the graph of a function. Describe three types of asymptotes.

Find each limit, if it exists.

66. $\lim_{x \to -\infty} \frac{-3x^2 + 5}{2 - x}$

67. $\lim_{x \to 0} \frac{|x|}{x}$

68. $\lim_{x \to -2} \frac{x^3 + 8}{x^2 - 4}$

69. $\lim_{x \to \infty} \frac{-6x^3 + 7x}{2x^2 - 3x - 10}$

70. $\lim_{x \to -\infty} \frac{-6x^3 + 7x}{2x^2 - 3x - 10}$

71. $\lim_{x \to 1} \frac{x^3 - 1}{x^2 - 1}$

72. $\lim_{x \to -\infty} \frac{7x^5 + x - 9}{6x + x^3}$

73. $\lim_{x \to -\infty} \frac{2x^4 + x}{x + 1}$

TECHNOLOGY CONNECTION

Graph each function.

74. $f(x) = x^2 + \frac{1}{x^2}$

75. $f(x) = \frac{x}{\sqrt{x^2 + 1}}$

76. $f(x) = \frac{x^3 + 4x^2 + x - 6}{x^2 - x - 2}$

77. $f(x) = \frac{x^3 + 2x^2 - 15x}{x^2 - 5x - 14}$

78. $f(x) = \frac{x^3 + 2x^2 - 3x}{x^2 - 25}$

79. $f(x) = \left|\frac{1}{x} - 2\right|$

80. Graph the function

$$f(x) = \frac{x^2 - 3}{2x - 4}.$$

Using only the TRACE and ZOOM features:

a) Find all the x-intercepts.

b) Find the y-intercept.

c) Find all the asymptotes.

TW **81.** Graph the function given by

$$f(x) = \frac{\sqrt{x^2 + 3x + 2}}{x - 3}.$$

a) Estimate $\lim_{x \to \infty} f(x)$ and $\lim_{x \to -\infty} f(x)$ using the graph and input–output tables as needed to refine your estimates.

b) Describe the outputs of the function over the interval $(-2, -1)$.

c) What appears to be the domain of the function? Explain.

d) Find $\lim_{x \to -2^-} f(x)$ and $\lim_{x \to -1^+} f(x)$.

82. Not all asymptotes are linear. Use long division to find an equation for the nonlinear asymptote that is approached by the graph of

$$f(x) = \frac{x^5 + x - 9}{x^3 + 6x}.$$

Then graph the function and its asymptote.

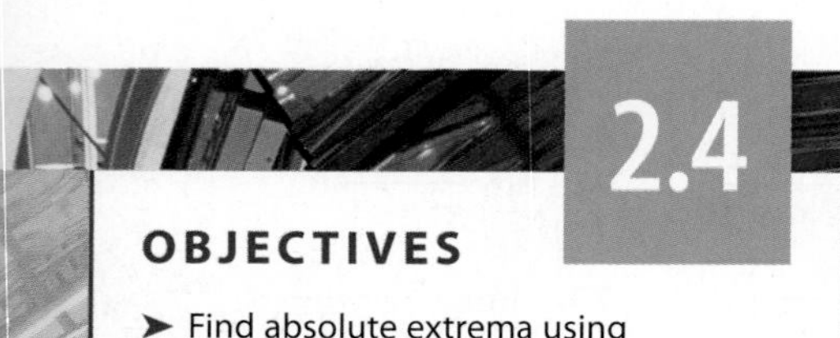

2.4 Using Derivatives to Find Absolute Maximum and Minimum Values

OBJECTIVES

- Find absolute extrema using Maximum–Minimum Principle 1.
- Find absolute extrema using Maximum–Minimum Principle 2.

Absolute Maximum and Minimum Values

A relative minimum may or may not be an absolute minimum, meaning the smallest value of the function over its entire domain. Similarly, a relative maximum may or may not be an absolute maximum, meaning the greatest value of a function over its entire domain.

The function in the following graph has relative minima at interior points c_1 and c_3 of the closed interval $[a, b]$.

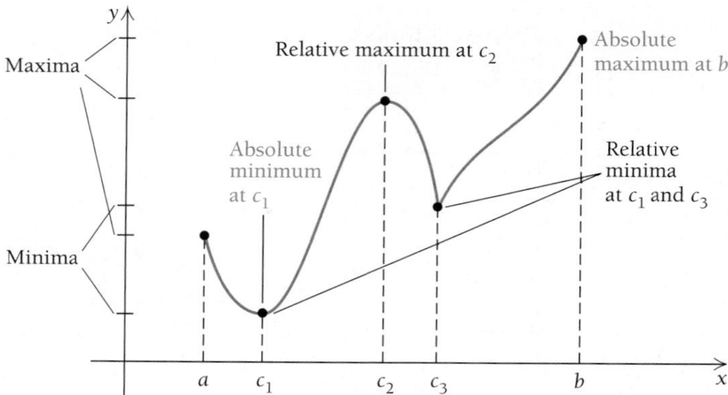

The relative minimum at c_1 is also the absolute minimum. On the other hand, the relative maximum at c_2 is *not* the absolute maximum. The absolute maximum occurs at the endpoint b.

DEFINITION

Suppose that f is a function with domain I.

$f(c)$ is an **absolute minimum** if $f(c) \leq f(x)$ for all x in I.

$f(c)$ is an **absolute maximum** if $f(c) \geq f(x)$ for all x in I.

Finding Absolute Maximum and Minimum Values Over Closed Intervals

We first consider a continuous function for which the domain is a closed interval. Look at the graphs in Figs. 1 and 2 and try to determine where the absolute maxima and minima (extrema) occur for each interval.

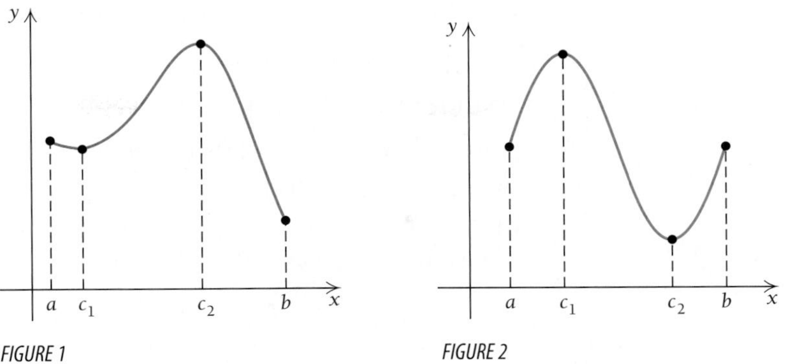

FIGURE 1

FIGURE 2

Note that each of the functions does indeed have an absolute maximum value and an absolute minimum value. This leads us to the following theorem.

THEOREM 7 The Extreme-Value Theorem

A continuous function f defined over a closed interval $[a, b]$ must have an absolute maximum value and an absolute minimum value over $[a, b]$.

Look again at the preceding graphs and consider the critical values and the endpoints. In Fig. 1, the graph starts at $f(a)$ and falls to $f(c_1)$. Then it rises from $f(c_1)$ to $f(c_2)$. From there it falls to $f(b)$. In Fig. 2, the graph starts at $f(a)$ and rises to $f(c_1)$. Then it falls from $f(c_1)$ to $f(c_2)$. From there it rises to $f(b)$. It seems reasonable that whatever the maximum and minimum values are, they occur among the function values $f(a)$, $f(c_1)$, $f(c_2)$, and $f(b)$. This leads us to a procedure for determining *absolute extrema.*

THEOREM 8 Maximum–Minimum Principle 1

Suppose that f is a continuous function defined over a closed interval $[a, b]$. To find the absolute maximum and minimum values over $[a, b]$:

a) First find $f'(x)$.

b) Then determine all critical values in $[a, b]$. That is, find all c in $[a, b]$ for which

$$f'(c) = 0 \quad \text{or} \quad f'(c) \text{ does not exist.}$$

c) List the values from step (b) and the endpoints of the interval:

$$a, c_1, c_2, \ldots, c_n, b.$$

d) Evaluate $f(x)$ for each value in step (c):

$$f(a), f(c_1), f(c_2), \ldots, f(c_n), f(b).$$

The largest of these is the **absolute maximum of f over $[a, b]$**. The smallest of these is the **absolute minimum of f over $[a, b]$**.

EXAMPLE 1 Find the absolute maximum and minimum values of

$$f(x) = x^3 - 3x + 2$$

over the interval $\left[-2, \frac{3}{2}\right]$.

Solution Keep in mind that we are considering only the interval $\left[-2, \frac{3}{2}\right]$.

a) Find $f'(x)$: $\quad f'(x) = 3x^2 - 3.$

b) Find the critical values. The derivative exists for all real numbers. Thus, we merely solve $f'(x) = 0$:

$$\begin{aligned} 3x^2 - 3 &= 0 \\ 3x^2 &= 3 \\ x^2 &= 1 \\ x &= \pm 1. \end{aligned}$$

c) List the critical values and the endpoints: $-2, -1, 1,$ and $\frac{3}{2}$.

d) Evaluate f for each value in step (c):

$$f(-2) = (-2)^3 - 3(-2) + 2 = -8 + 6 + 2 = 0; \longrightarrow \text{Minimum}$$
$$f(-1) = (-1)^3 - 3(-1) + 2 = -1 + 3 + 2 = 4; \longrightarrow \text{Maximum}$$
$$f(1) = (1)^3 - 3(1) + 2 = 1 - 3 + 2 = 0; \longrightarrow \text{Minimum}$$
$$f\left(\tfrac{3}{2}\right) = \left(\tfrac{3}{2}\right)^3 - 3\left(\tfrac{3}{2}\right) + 2 = \tfrac{27}{8} - \tfrac{9}{2} + 2 = \tfrac{7}{8}.$$

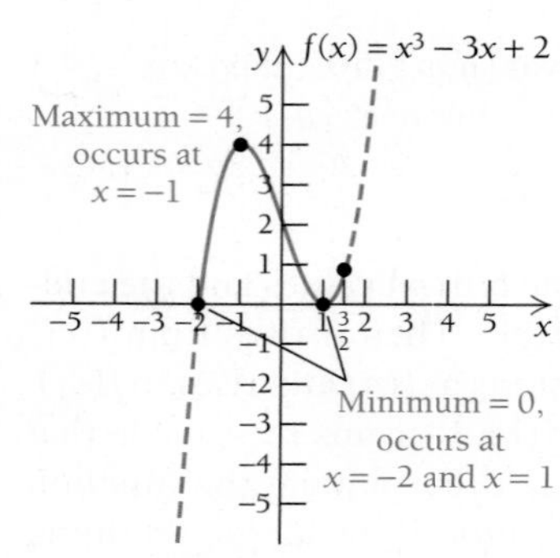

A visualization of Example 1

The largest of these values, 4, is the maximum. It occurs at $x = -1$. The smallest of these values is 0. It occurs twice: at $x = -2$ and $x = 1$. Thus, over the interval $\left[-2, \frac{3}{2}\right]$, the

$$\text{absolute maximum} = 4 \text{ at } x = -1$$

and the

$$\text{absolute minimum} = 0 \text{ at } x = -2 \text{ and } x = 1.$$

✦

Note that an absolute maximum or minimum value can occur at more than one point.

TECHNOLOGY CONNECTION

Finding Absolute Extrema

To find the absolute extrema of Example 1, we can use any of the methods described in the Technology Connection on p. 210. In this case, we adapt Methods 3 and 4.

Method 3

Method 3 is selected because there are relative extrema in the interval $\left[-2, \frac{3}{2}\right]$. This method gives us approximations for the relative extrema.

$y = x^3 - 3x + 2$

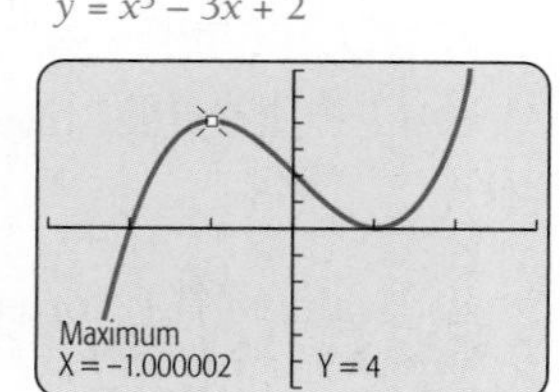

$y = x^3 - 3x + 2$

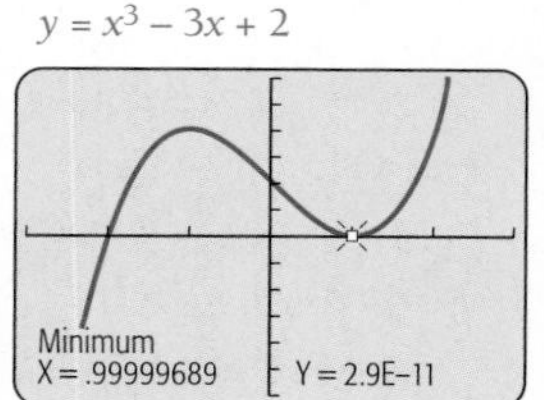

Finding Absolute Extrema (continued)

Next, we check function values at these x-values and at the endpoints, using Maximum–Minimum Principle 1 to determine the absolute maximum and minimum values over $\left[-2, \frac{3}{2}\right]$.

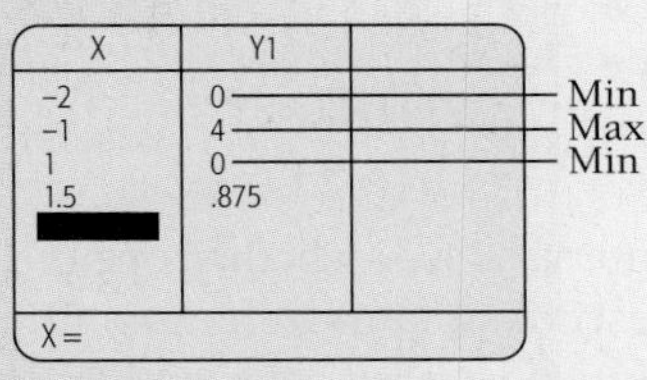

Method 4

Example 2 (to follow) considers the same function as in Example 1, but over a different interval. This time we use fMax and fMin features from the MATH menu.

Because there are no relative extrema, we can use fMax and fMin. The minimum and maximum values occur at the endpoints.

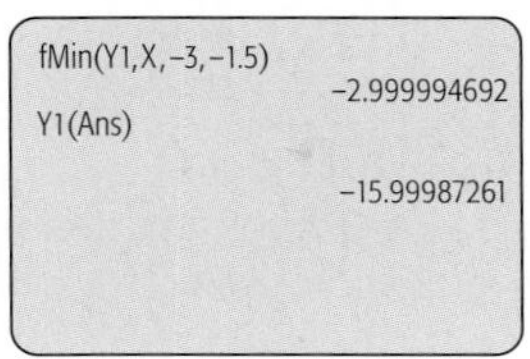

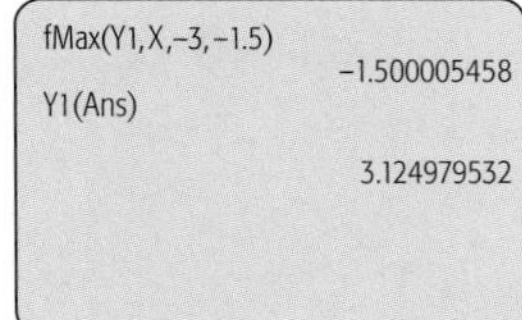

EXERCISE

1. Use a graph to estimate the absolute maximum and minimum values of $f(x) = x^3 - x^2 - x + 2$, first over the interval $[-2, 1]$ and then over the interval $[-1, 2]$. Then check your work using the methods of Examples 1 and 2.

EXAMPLE 2 Find the absolute maximum and minimum values of

$$f(x) = x^3 - 3x + 2$$

over the interval $\left[-3, -\frac{3}{2}\right]$.

Solution As in Example 1, the derivative is 0 at -1 and 1. But neither -1 nor 1 is in the interval $\left[-3, -\frac{3}{2}\right]$, so there are no critical values in this interval. Thus, the maximum and minimum values occur at the endpoints:

$$\begin{aligned} f(-3) &= (-3)^3 - 3(-3) + 2 \\ &= -27 + 9 + 2 = -16; \longrightarrow \text{Minimum} \\ f\left(-\frac{3}{2}\right) &= \left(-\frac{3}{2}\right)^3 - 3\left(-\frac{3}{2}\right) + 2 \\ &= -\frac{27}{8} + \frac{9}{2} + 2 = \frac{25}{8} = 3\frac{1}{8}. \longrightarrow \text{Maximum} \end{aligned}$$

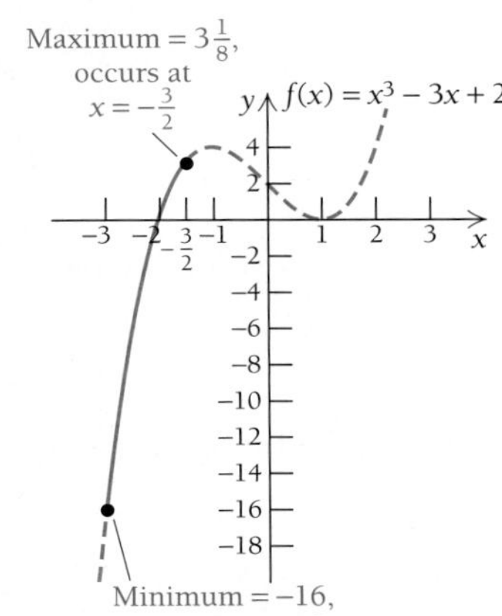

A visualization of Example 2

Thus, the absolute maximum over the interval $\left[-3, -\frac{3}{2}\right]$, is $3\frac{1}{8}$, which occurs at $x = -\frac{3}{2}$, and the absolute minimum over $\left[-3, -\frac{3}{2}\right]$ is -16, which occurs at $x = -3$. ✦

Finding Absolute Maximum and Minimum Values Over Other Intervals

When there is only one critical value c in I, we may not need to check endpoint values to determine whether the function has an absolute maximum or minimum value at that point.

THEOREM 9 Maximum–Minimum Principle 2

Suppose that f is a function such that $f'(x)$ exists for every x in an interval I, and that there is *exactly one* (critical) value c in I, for which $f'(c) = 0$. Then

$f(c)$ is the absolute maximum value over I if $f''(c) < 0$

or

$f(c)$ is the absolute minimum value over I if $f''(c) > 0$.

Theorem 9 holds no matter what the interval I is—whether open, closed, or infinite in length. If $f''(c) = 0$, either we must use Maximum–Minimum Principle 1 or we must know more about the behavior of the function over the given interval.

TECHNOLOGY CONNECTION

Finding Absolute Extrema

Let's do Example 3 graphically, by adapting Methods 1 and 2 of the Technology Connection on p. 210. Strictly speaking, we cannot use the fMin or fMax options of the MATH menu or the MAXIMUM or MINIMUM options from the CALC menu since we do not have a closed interval.

Methods 1 and 2

We create a graph, examine its shape, and use TRACE and/or TABLE. This procedure leads us to see that there is indeed no absolute minimum. We do get an absolute maximum $f(x) = 4$ at $x = 2$.

EXERCISE

1. Use a graph to estimate the absolute maximum and minimum values of $f(x) = x^2 - 4x$. Then check your work using the method of Example 3.

EXAMPLE 3 Find the absolute maximum and minimum values of

$$f(x) = 4x - x^2.$$

Solution When no interval is specified, we consider the entire domain of the function. In this case, the domain is the set of all real numbers.

a) Find $f'(x)$:

$$f'(x) = 4 - 2x.$$

b) Find the critical values. The derivative exists for all real numbers. Thus, we merely solve $f'(x) = 0$:

$$4 - 2x = 0$$
$$-2x = -4$$
$$x = 2.$$

c) Since there is only one critical value, we can apply Maximum–Minimum Principle 2 using the second derivative:

$$f''(x) = -2.$$

The second derivative is constant. Thus, $f''(2) = -2$, and since this is negative, we have the

$$\text{absolute maximum} = f(2) = 4 \cdot 2 - 2^2 = 8 - 4 = 4 \text{ at } x = 2.$$

The function has no minimum, as the graph, shown below, indicates.

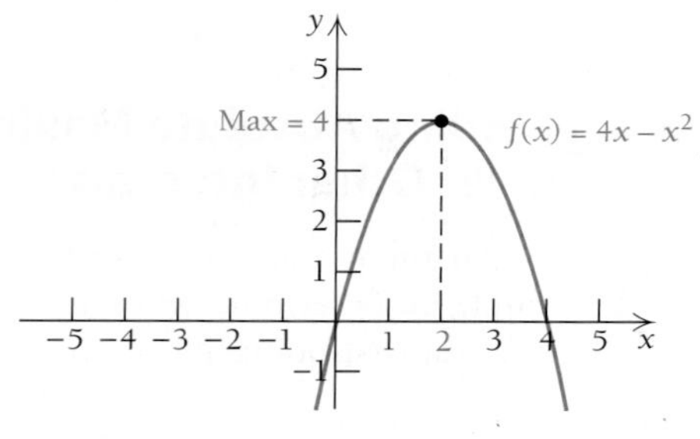

✦

EXAMPLE 4 Find the absolute maximum and minimum values of $f(x) = 4x - x^2$ over the interval $[1, 4]$.

Solution By the reasoning in Example 3, we know that the absolute maximum of f on $(-\infty, \infty)$ is $f(2)$, or 4. Since 2 is in the interval $[1, 4]$, we know that the absolute maximum of f over $[1, 4]$ will occur at 2. To find the absolute minimum, we need to check the endpoints:

$$f(1) = 4 \cdot 1 - 1^2 = 3$$

and

$$f(4) = 4 \cdot 4 - 4^2 = 0.$$

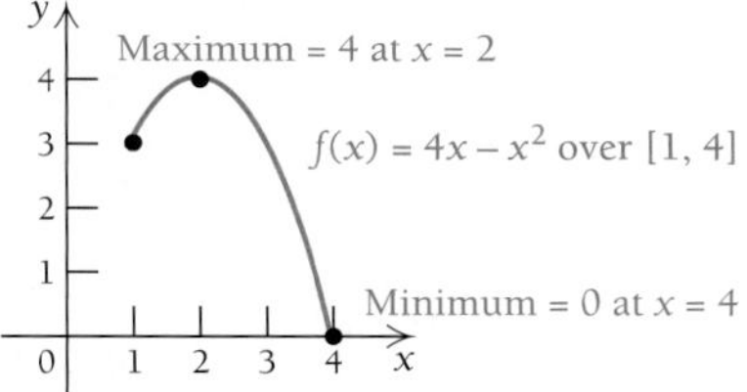

We see from the graph above that the minimum is 0. It occurs at $x = 4$. Thus, the

$$\text{absolute maximum} = 4 \text{ at } x = 2$$

and

$$\text{absolute minimum} = 0 \text{ at } x = 4.$$

✦

We have thus far restricted the use of Maximum–Minimum Principle 2 to intervals with one critical value. Suppose that a closed interval contains two critical values. Then we could break the interval up into two subintervals, consider maximum and minimum values over those subintervals, and compare. But we would need to consider values at the endpoints, and since we would, in effect, be using Maximum–Minimum Principle 1, we may as well use it at the outset.

A Strategy for Finding Maximum and Minimum Values

The following general strategy can be used when finding maximum and minimum values of continuous functions.

A Strategy for Finding Absolute Maximum and Minimum Values

To find absolute maximum and minimum values of a continuous function over an interval:

a) Find $f'(x)$.
b) Find the critical values.
c) If the interval is closed and there is more than one critical value, use Maximum–Minimum Principle 1.
d) If the interval is closed and there is exactly one critical value, use either Maximum–Minimum Principle 1 or Maximum–Minimum Principle 2. If it is easy to find $f''(x)$, use Maximum–Minimum Principle 2.
e) If the interval is not closed, such as $(-\infty, \infty)$, $(0, \infty)$, or (a, b), and the function has only one critical value, use Maximum–Minimum Principle 2. In such a case, if the function has a maximum, it will have no minimum; and if it has a minimum, it will have no maximum.

The case of finding absolute maximum and minimum values when more than one critical value occurs in an interval that is not closed, such as those listed in step (e) above, must be dealt with by a detailed graph or by techniques beyond the scope of this book.

EXAMPLE 5 Find the absolute maximum and minimum values of

$$f(x) = (x - 2)^3 + 1.$$

Solution

a) Find $f'(x)$.

$$f'(x) = 3(x - 2)^2.$$

b) Find the critical values. The derivative exists for all real numbers. Thus, we solve $f'(x) = 0$:

$$\begin{aligned} 3(x - 2)^2 &= 0 \\ (x - 2)^2 &= 0 \\ x - 2 &= 0 \\ x &= 2. \end{aligned}$$

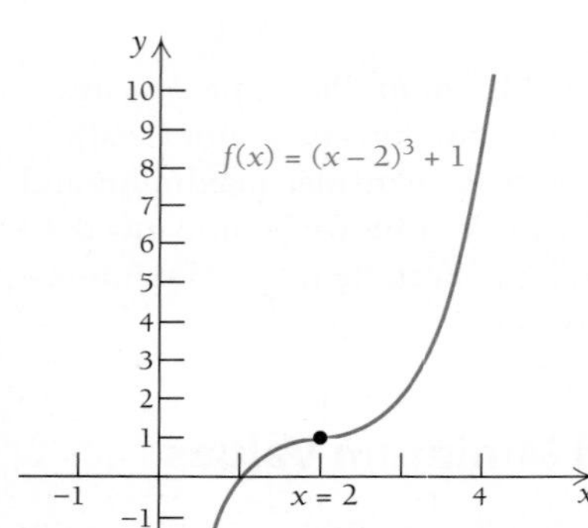

A visualization of Example 5

c) Since there is only one critical value and there are no endpoints, we can try to apply Maximum–Minimum Principle 2 using the second derivative:

$$f''(x) = 6(x - 2).$$

We have

$$f''(2) = 6(2 - 2) = 0,$$

so Maximum–Minimum Principle 2 does not apply. We cannot use Maximum–Minimum Principle 1 because there are no endpoints. But note that $f'(x) = 3(x - 2)^2$ is never negative. Thus, $f(x)$ is increasing everywhere except at $x = 2$, so there is no maximum and no minimum. For $x < 2$, say $x = 1$, we have $f''(1) = -6 < 0$. Similarly, for $x > 2$, say $x = 3$, we have $f''(3) = 6 > 0$. Thus, at $x = 2$, the function has a *point of inflection.* ✦

EXAMPLE 6 Find the absolute maximum and minimum values of

$$f(x) = 5x + \frac{35}{x}$$

over the interval $(0, \infty)$.

Solution

a) Find $f'(x)$. We first express $f(x)$ as

$$f(x) = 5x + 35x^{-1}.$$

Then

$$\begin{aligned} f'(x) &= 5 - 35x^{-2} \\ &= 5 - \frac{35}{x^2}. \end{aligned}$$

b) Find the critical values. Since $f'(x)$ exists for all values of x in $(0, \infty)$, the only critical values are those for which $f'(x) = 0$:

$$5 - \frac{35}{x^2} = 0$$

$$5 = \frac{35}{x^2}$$

$$5x^2 = 35 \qquad \text{Multiplying both sides by } x^2, \text{ since } x \neq 0$$

$$x^2 = 7$$

$$x = \pm\sqrt{7} \approx \pm 2.646.$$

c) The interval is not closed and is $(0, \infty)$. The only critical value is $\sqrt{7}$. Therefore, we can apply Maximum–Minimum Principle 2 using the second derivative,

$$f''(x) = 70x^{-3} = \frac{70}{x^3},$$

to determine whether we have a maximum or a minimum. Since

$$f''(\sqrt{7}) = \frac{70}{(\sqrt{7})^3} > 0,$$

an absolute minimum occurs at $x = \sqrt{7}$:

$$\begin{aligned}
\text{Absolute minimum} &= f(\sqrt{7}) \\
&= 5 \cdot \sqrt{7} + \frac{35}{\sqrt{7}} \\
&= 5\sqrt{7} + \frac{35}{\sqrt{7}} \cdot \frac{\sqrt{7}}{\sqrt{7}} \\
&= 5\sqrt{7} + \frac{35\sqrt{7}}{7} \\
&= 5\sqrt{7} + 5\sqrt{7} \\
&= 10\sqrt{7} \approx 26.458 \quad \text{at } x = \sqrt{7}.
\end{aligned}$$

The function has no maximum value, which can happen since the interval $(0, \infty)$ is *not* closed.

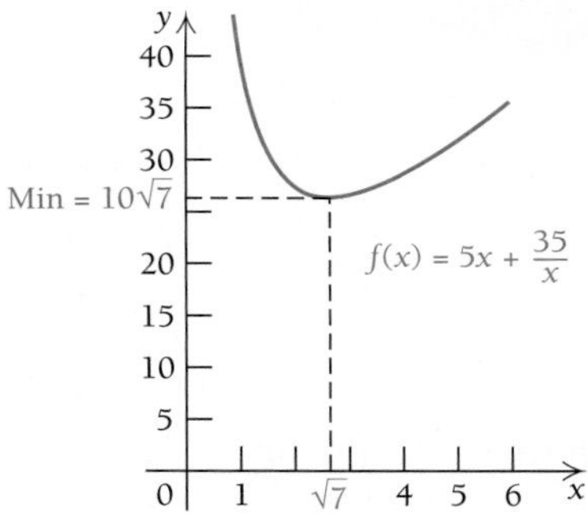

TECHNOLOGY CONNECTION

Finding Absolute Extrema

Let's do Example 6 using MAXIMUM and MINIMUM from the CALC menu. The shape of the graph leads us to see that there is no absolute maximum, but there is an absolute minimum.

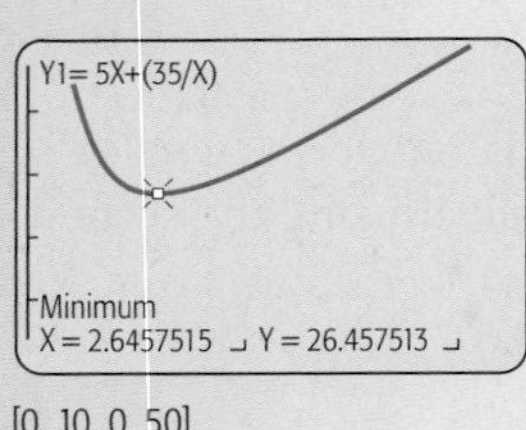

[0, 10, 0, 50]

Note that

$$\sqrt{7} \approx 2.6458 \quad \text{and} \quad 10\sqrt{7} \approx 26.458,$$

which confirms the analytic solution.

EXERCISE

1. Use a graph to estimate the absolute maximum and minimum values of $f(x) = 10x + 1/x$ over the interval $(0, \infty)$. Then check your work using the analytic method of Example 6.

Exercise Set 2.4

1. Fuel economy. According to the U.S. Department of Energy, a vehicle's fuel economy, in miles per gallon (mpg), decreases rapidly for speeds over 60 mph.

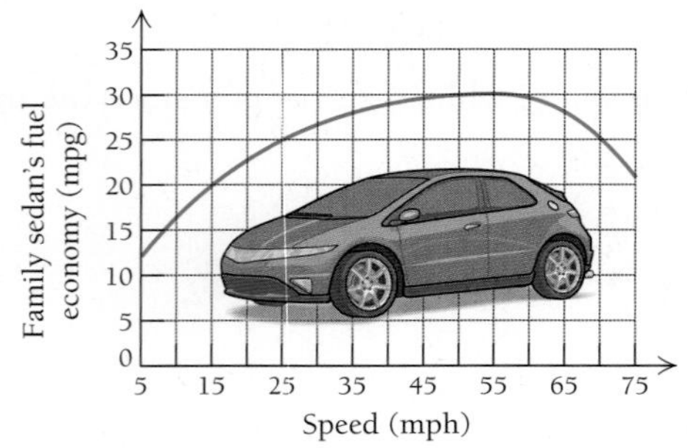

(*Sources*: Based on information from U.S. Dept. of Energy and a study by West, B.H., McGill, R.N., Hodgson, J.W., Sluder, S.S., and Smith, D.E., Oak Ridge National Laboratory, 1999.)

a) Estimate the speed at which the absolute maximum gasoline mileage is obtained.

b) Estimate the speed at which the absolute minimum gasoline mileage is obtained.

c) What is the mileage obtained at 70 mph?

2. Fuel economy. Using the graph in Exercise 1, estimate the absolute maximum and the absolute minimum fuel economy over the interval [30, 70].

Find the absolute maximum and minimum values of each function over the indicated interval, and indicate the x-values at which they occur.

3. $f(x) = 5 + x - x^2$; $[0, 2]$

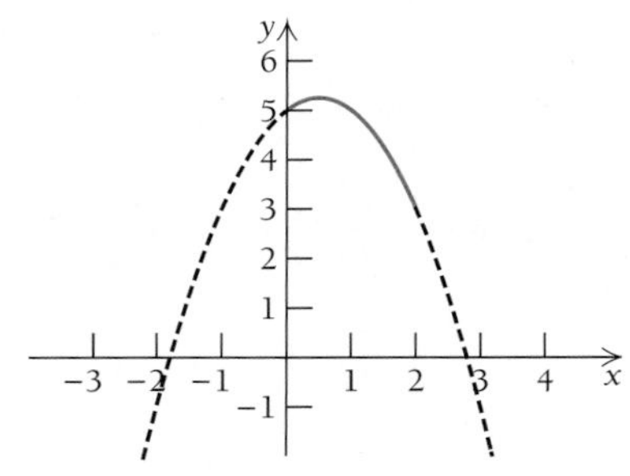

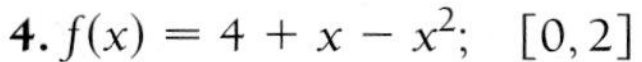

4. $f(x) = 4 + x - x^2$; $[0, 2]$

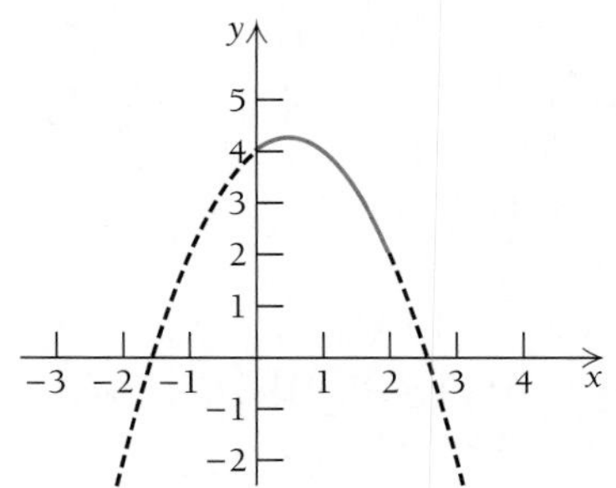

5. $f(x) = x^3 - x^2 - x + 2$; $[-1, 2]$

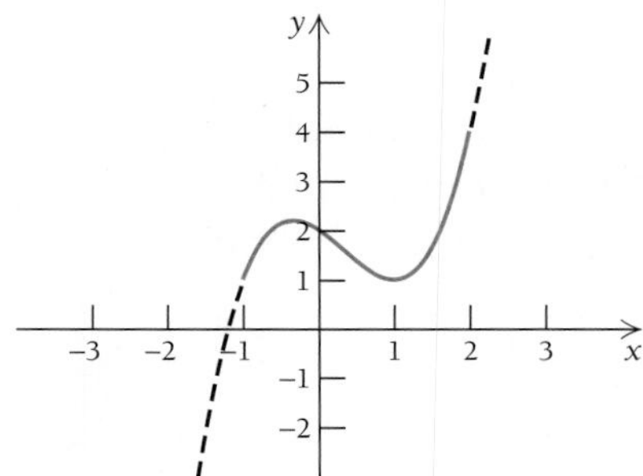

6. $f(x) = x^3 + \frac{1}{2}x^2 - 2x + 5$; $[-2, 1]$

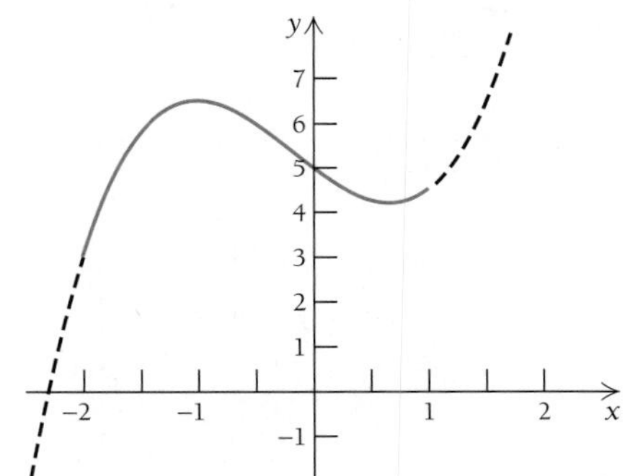

7. $f(x) = x^3 - x^2 - x + 3$; $[-1, 0]$

8. $f(x) = x^3 + \frac{1}{2}x^2 - 2x + 4$; $[-2, 0]$

9. $f(x) = 5x - 7$; $[-2, 3]$

10. $f(x) = 2x + 4$; $[-1, 1]$

11. $f(x) = 7 - 4x$; $[-2, 5]$

12. $f(x) = -2 - 3x$; $[-10, 10]$

13. $f(x) = -5$; $[-1, 1]$

14. $g(x) = 24$; $[4, 13]$

15. $f(x) = x^2 - 6x - 3$; $[-1, 5]$

16. $f(x) = x^2 - 4x + 5$; $[-1, 3]$

17. $f(x) = 3 - 2x - 5x^2$; $[-3, 3]$

18. $f(x) = 1 + 6x - 3x^2$; $[0, 4]$

19. $f(x) = x^3 - 3x^2$; $[0, 5]$

20. $f(x) = x^3 - 3x + 6$; $[-1, 3]$

21. $f(x) = x^3 - 3x$; $[-5, 1]$

22. $f(x) = 3x^2 - 2x^3$; $[-5, 1]$

23. $f(x) = 1 - x^3$; $[-8, 8]$

24. $f(x) = 2x^3$; $[-10, 10]$

25. $f(x) = 12 + 9x - 3x^2 - x^3$; $[-3, 1]$

26. $f(x) = x^3 - 6x^2 + 10$; $[0, 4]$

27. $f(x) = x^4 - 2x^3$; $[-2, 2]$

28. $f(x) = x^3 - x^4$; $[-1, 1]$

29. $f(x) = x^4 - 2x^2 + 5$; $[-2, 2]$

30. $f(x) = x^4 - 8x^2 + 3$; $[-3, 3]$

31. $f(x) = (x + 3)^{2/3} - 5$; $[-4, 5]$

32. $f(x) = 1 - x^{2/3}$; $[-8, 8]$

33. $f(x) = x + \dfrac{1}{x}$; $[1, 20]$

34. $f(x) = x + \dfrac{4}{x}$; $[-8, -1]$

35. $f(x) = \dfrac{x^2}{x^2 + 1}$; $[-2, 2]$

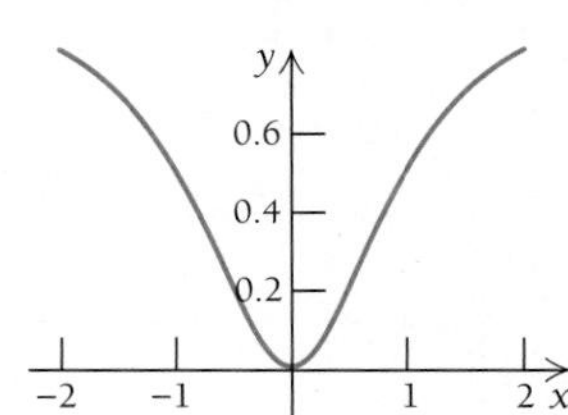

36. $f(x) = \dfrac{4x}{x^2 + 1}$, $[-3, 3]$

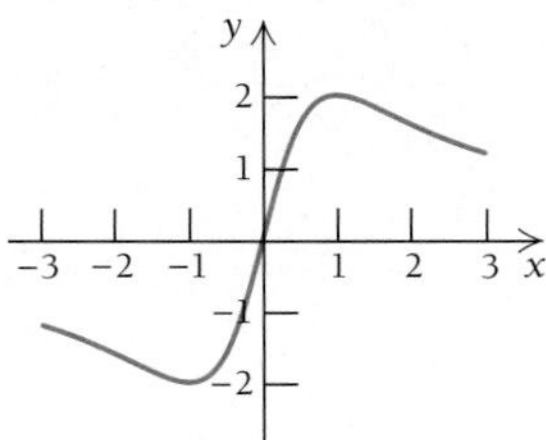

37. $f(x) = (x + 1)^{1/3}$; $[-2, 26]$

38. $f(x) = \sqrt[3]{x}$; $[8, 64]$

39–48. Check Exercises 3, 5, 9, 13, 19, 23, 33, 35, 37, and 38 with a graphing calculator.

Find the absolute maximum and minimum values of each function, if they exist, over the indicated interval. Also indicate the x-value at which each extremum occurs. When no interval is specified, use the real line, $(-\infty, \infty)$.

49. $f(x) = 12x - x^2$

50. $f(x) = 30x - x^2$

51. $f(x) = 2x^2 - 40x + 270$

52. $f(x) = 2x^2 - 20x + 340$

53. $f(x) = x - \frac{4}{3}x^3;\quad (0, \infty)$

54. $f(x) = 16x - \frac{4}{3}x^3;\quad (0, \infty)$

55. $f(x) = x(60 - x)$

56. $f(x) = x(25 - x)$

57. $f(x) = \frac{1}{3}x^3 - 3x;\quad [-2, 2]$

58. $f(x) = \frac{1}{3}x^3 - 5x;\quad [-3, 3]$

59. $f(x) = -0.001x^2 + 4.8x - 60$

60. $f(x) = -0.01x^2 + 1.4x - 30$

61. $f(x) = -\frac{1}{3}x^3 + 6x^2 - 11x - 50;\quad (0, 3)$

62. $f(x) = -x^3 + x^2 + 5x - 1;\quad (0, \infty)$

63. $f(x) = 15x^2 - \frac{1}{2}x^3;\quad [0, 30]$

64. $f(x) = 4x^2 - \frac{1}{2}x^3;\quad [0, 8]$

65. $f(x) = 2x + \dfrac{72}{x};\quad (0, \infty)$

66. $f(x) = x + \dfrac{3600}{x};\quad (0, \infty)$

67. $f(x) = x^2 + \dfrac{432}{x};\quad (0, \infty)$

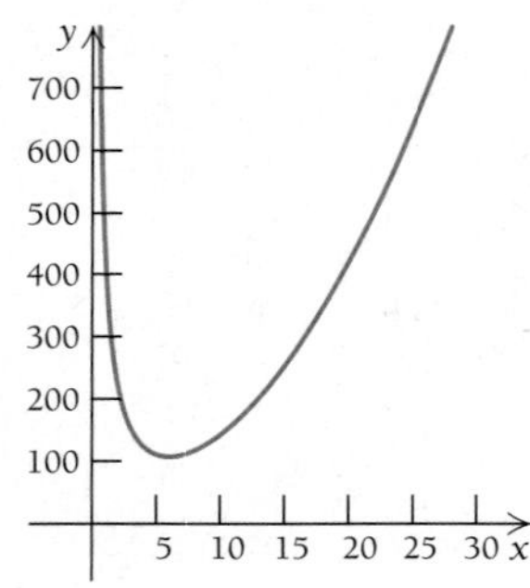

68. $f(x) = x^2 + \dfrac{250}{x};\quad (0, \infty)$

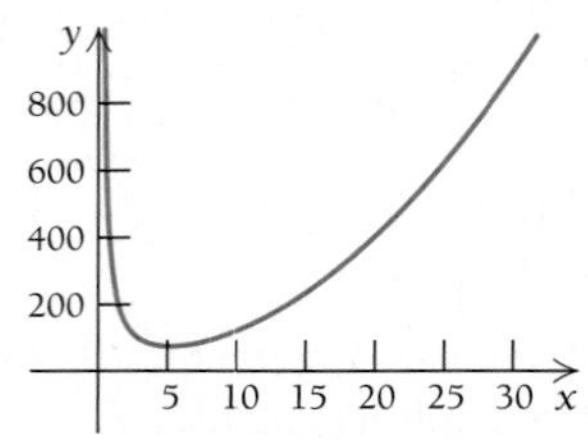

69. $f(x) = 2x^4 - x;\quad [-1, 1]$

70. $f(x) = 2x^4 + x;\quad [-1, 1]$

71. $f(x) = \sqrt[3]{x};\quad [0, 8]$

72. $f(x) = \sqrt{x};\quad [0, 4]$

73. $f(x) = (x + 1)^3$

74. $f(x) = (x - 1)^3$

75. $f(x) = 2x - 3;\quad [-1, 1]$

76. $f(x) = 9 - 5x;\quad [-10, 10]$

77. $f(x) = 2x - 3;\quad [-1, 5)$

78. $f(x) = 9 - 5x;\quad [-2, 3)$

79. $f(x) = x^{2/3};\quad [-1, 1]$

80. $g(x) = x^{2/3}$

81. $f(x) = \frac{1}{3}x^3 - x + \frac{2}{3}$

82. $f(x) = \frac{1}{3}x^3 - \frac{1}{2}x^2 - 2x + 1$

83. $f(x) = \frac{1}{3}x^3 - 2x^2 + x;\quad [0, 4]$

84. $g(x) = \frac{1}{3}x^3 + 2x^2 + x;\quad [-4, 0]$

85. $t(x) = x^4 - 2x^2$

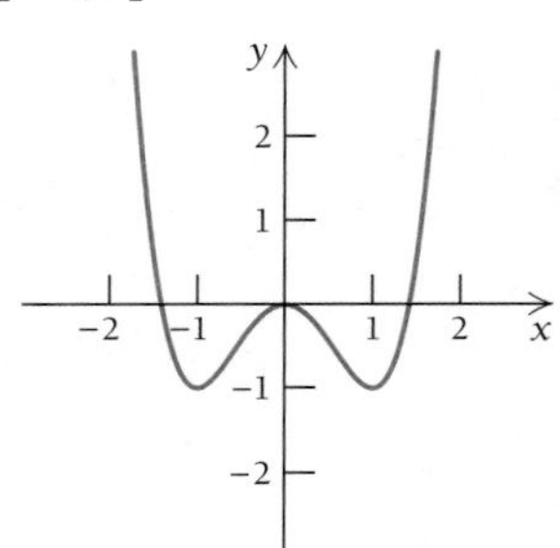

86. $f(x) = 2x^4 - 4x^2 + 2$

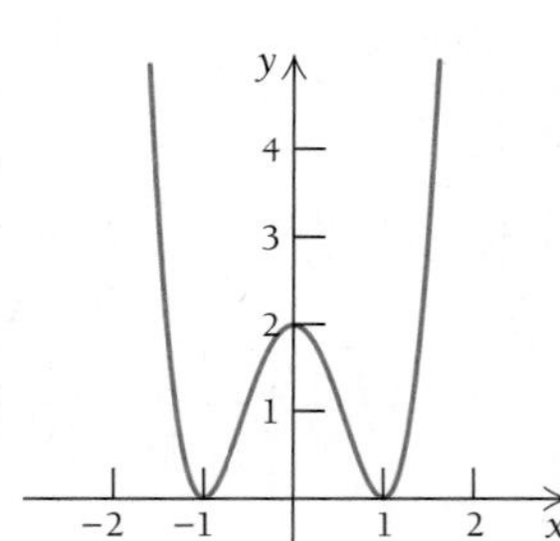

87–96. Check Exercises 49, 51, 53, 57, 61, 65, 67, 69, 73, and 85 with a graphing calculator.

APPLICATIONS

Business and Economics

97. Monthly productivity. An employee's monthly productivity M, in number of units produced, is found to be a function of the number t of years of service. For a certain product, a productivity function is given by

$$M(t) = -2t^2 + 100t + 180, \quad 0 \leq t \leq 40.$$

Find the maximum productivity and the year in which it is achieved.

98. Advertising. Sound Software estimates that it will sell N units of a program after spending a dollars on advertising, where

$$N(a) = -a^2 + 300a + 6, \quad 0 \leq a \leq 300,$$

and a is in thousands of dollars. Find the maximum number of units that can be sold and the amount that must be spent on advertising in order to achieve that maximum.

99. Small business. The percentage of the U.S. national income generated by nonfarm proprietors may be modeled by the function

$$p(x) = \frac{13x^3 - 240x^2 - 2460x + 585{,}000}{75{,}000},$$

where x is the number of years since 1970. (***Source*:** U.S. Census Bureau.) According to this model, in what year from 1970 through 2000 was this percentage a minimum? Calculate the answer, and then check it on the graph.

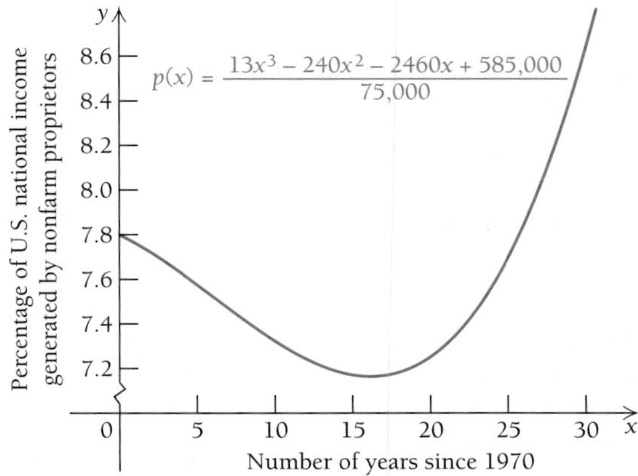

100. Labor force. The percentage of the U.S. civilian labor force aged 45–54 may be modeled by the function

$$f(x) = 0.025x^2 - 0.71x + 20.44,$$

where x is the number of years since 1970. (***Source*:** U.S. Census Bureau.) According to this model, in what year from 1970 through 2000 was this percentage a minimum?

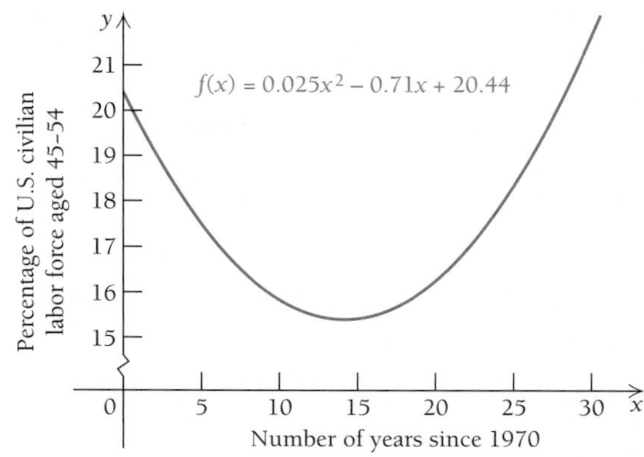

101. Worldwide oil production. One model of worldwide oil production is the function

$$P(t) = 0.000008533t^4 - 0.001685t^3 + 0.090t^2 - 0.687t + 4.00, \quad 0 \leq t \leq 90,$$

where $P(t)$ is the number of barrels, in billions, produced in a year, t years after 1950. (***Source*:** Based on data from *Beyond Oil*, by Kenneth S. Deffeyes, p. xii, Hill and Wang, New York, 2005.) According to this model, in what year did worldwide oil production achieve an absolute maximum? What was that maximum? (*Hint:* Do not solve $P'(t) = 0$ algebraically.)

102. Maximizing profit. Corner Stone Electronics determines that its total weekly profit, in dollars, from the production and sale of x amplifiers is given by

$$P(x) = \frac{1500}{x^2 - 6x + 10}.$$

Find the number of amplifiers, x, for which the total weekly profit is a maximum.

Maximizing profit. *The total-cost and total-revenue functions for producing x items are*

$$C(x) = 5000 + 600x \quad \text{and} \quad R(x) = -\frac{1}{2}x^2 + 1000x,$$

where $0 \le x \le 600$. Use this function for Exercises 103 and 104.

103. a) Find the total-profit function $P(x)$.

b) Find the number of items, x, for which the total profit is a maximum.

104. a) The *average profit* is given by $A(x) = P(x)/x$. Find $A(x)$.

b) Find the number of items, x, for which the average profit is a maximum.

Life and Physical Sciences

105. Blood pressure. For a dosage of x cubic centimeters (cc) of a certain drug, the resulting blood pressure B is approximated by

$$B(x) = 305x^2 - 1830x^3, \quad 0 \le x \le 0.16.$$

Find the maximum blood pressure and the dosage at which it occurs.

SYNTHESIS

TW **106.** Explain the usefulness of the second derivative in finding the absolute extrema of a function.

Find the absolute maximum and minimum values of the function, if they exist, over the indicated interval.

107. $g(x) = x\sqrt{x+3}$; $[-3, 3]$

108. $h(x) = x\sqrt{1-x}$; $[0, 1]$

109. Business: total cost. Certain costs in a business environment can be separated into two components: those that increase with volume and those that decrease with volume. For example, customer service becomes more expensive as its quality increases, but part of the increased cost is offset by fewer customer complaints. A firm has determined that its cost of service, $C(x)$, in thousands of dollars, is given by the following function, where x represents the number of "quality units":

$$C(x) = (2x + 4) + \left(\frac{2}{x-6}\right), \quad x > 6.$$

Find the number of "quality units" that the firm should use in order to minimize its total cost of service.

110. Let

$$y = (x-a)^2 + (x-b)^2.$$

For what value of x is y a minimum?

TW **111.** Explain the usefulness of the first derivative in finding the absolute extrema of a function.

TECHNOLOGY CONNECTION

112. Worldwide oil production. Refer to Exercise 101. In what year was worldwide oil production increasing most rapidly and at what rate was it increasing?

113. U.S. oil production. One model of oil production in the United States is given by

$$P(t) = 0.0000000219t^4 - 0.0000167t^3 + 0.00155t^2 + 0.002t + 0.22, \quad 0 \le t \le 110,$$

where $P(t)$ is the number of barrels of oil, in billions, produced in a year, t years after 1910. (***Source***: Based on data from *Beyond Oil*, by Kenneth S. Deffeyes, p. 41, Hill and Wang, New York, 2005.)

a) According to this model, what is the absolute maximum amount of oil produced in the United States and in what year did that production occur?

b) According to this model, at what rate was United States oil production declining in 2004?

Graph each function over the given interval. Visually estimate where absolute maximum and minimum values occur. Then use the TABLE *feature to refine your estimate.*

114. $f(x) = x^{2/3}(x-5)$; $[1, 4]$

115. $f(x) = \frac{3}{4}(x^2 - 1)^{2/3}$; $\left[\frac{1}{2}, \infty\right)$

116. $f(x) = x\left(\frac{x}{2} - 5\right)^4$; $\mathbb{R}$

117. Life and physical sciences: contractions during pregnancy. The following table and graph give the pressure of a pregnant woman's contractions as a function of time.

Time, t (in minutes)	Pressure (in millimeters of mercury)
0	10
1	8
2	9.5
3	15
4	12
5	14
6	14.5

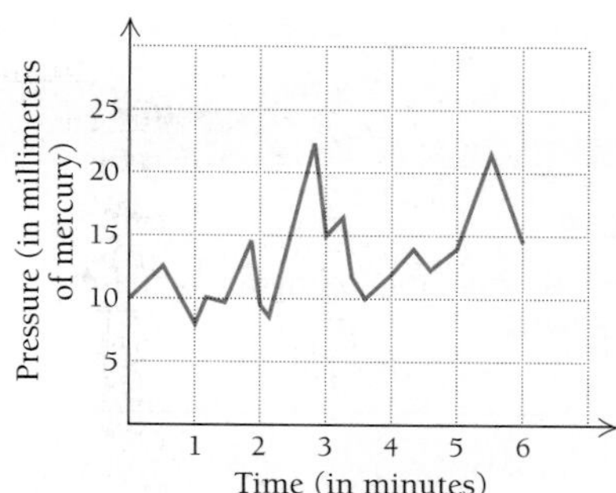

Use a calculator that has the REGRESSION option.

a) Fit a linear equation to the data. Predict the pressure of the contractions after 7 min.

b) Fit a quartic polynomial to the data. Predict the pressure of the contractions after 7 min. Find the smallest contraction over the interval $[0, 10]$.

OBJECTIVE

- Solve maximum – minimum problems using calculus.

Maximum – Minimum Problems; Business and Economics Applications

An important use of calculus is the solving of maximum–minimum problems, that is, finding the absolute maximum or minimum value of some varying quantity Q and the point at which that maximum or minimum occurs.

EXAMPLE 1 Maximizing Area. A hobby store has 20 ft of fencing to fence off a rectangular area for an electric train in one corner of its display room. The two sides up against the wall require no fence. What dimensions of the rectangle will maximize the area? What is the maximum area?

Solution At first glance, we might think that it does not matter what dimensions we use: They will all yield the same area. This is not the case. Let's first make a drawing and express the area in terms of one variable. If we let $x =$ the length, in feet, of one side and $y =$ the length, in feet, of the other, then, since the sum of the lengths must be 20 ft, we have

$$x + y = 20 \quad \text{and} \quad y = 20 - x.$$

Thus, the area is given by

$$\begin{aligned} A &= xy \\ &= x(20 - x) \\ &= 20x - x^2. \end{aligned}$$

TECHNOLOGY CONNECTION

EXERCISES

1. Complete this table using a calculator as needed.

x	$y = 20 - x$	$A = x(20 - x)$
0		
4		
6.5		
8		
10		
12		
13.2		
20		

2. Graph $A(x) = x(20 - x)$ over the interval $[0, 20]$.
3. Estimate the maximum value, and state where it occurs.

We are trying to find the maximum value of

$$A(x) = 20x - x^2$$

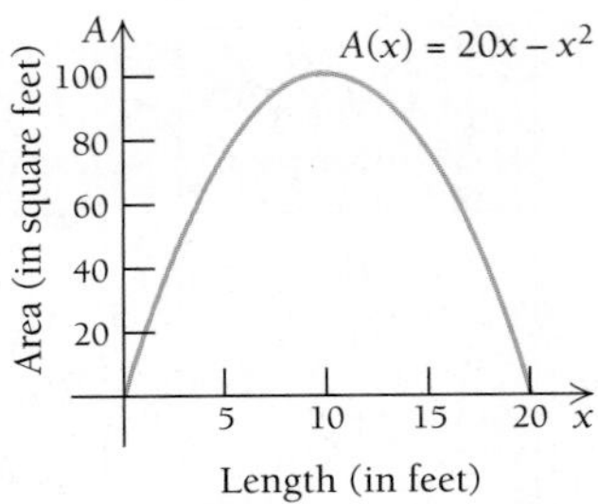

over the interval $(0, 20)$. We consider the interval $(0, 20)$ because x is a length and cannot be negative or 0. Since there is only 20 ft of fencing, x cannot be greater than 20. Also, x cannot be 20 because then the length of y would be 0.

a) We first find $A'(x)$: $\quad A'(x) = 20 - 2x.$

b) This derivative exists for all values of x in $(0, 20)$. Thus, the only critical values are where

$$A'(x) = 20 - 2x = 0$$
$$-2x = -20$$
$$x = 10.$$

Since there is only one critical value, we can use the second derivative to determine whether we have a maximum. Note that

$$A''(x) = -2,$$

which is a constant. Thus, $A''(10)$ is negative, so $A(10)$ is a maximum. Now

$$A(10) = 10(20 - 10)$$
$$= 10 \cdot 10$$
$$= 100.$$

Thus, the maximum area of 100 ft^2 is obtained using 10 ft for the length of one side and $20 - 10$, or 10 ft for the other. Note that $A(5) = 75$, $A(16) = 64$, and $A(12) = 96$; so length does affect area. ✦

Here is a general strategy for solving maximum–minimum problems. Although it may not guarantee success, it should certainly improve your chances.

A Strategy for Solving Maximum–Minimum Problems

1. Read the problem carefully. If relevant, make a drawing.
2. Make a list of appropriate variables and constants, noting what varies, what stays fixed, and what units are used. Label the measurements on your drawing, if one exists.
3. Translate the problem to an equation involving a quantity Q to be maximized or minimized. Try to represent Q in terms of the variables of step (2).
4. Try to express Q as a function of one variable. Use the procedures developed in Sections 2.1–2.4 to determine the maximum or minimum values and the points at which they occur.

EXAMPLE 2 Maximizing Volume. From a thin piece of cardboard 8 in. by 8 in., square corners are cut out so that the sides can be folded up to make a box. What dimensions will yield a box of maximum volume? What is the maximum volume?

Solution We might again think at first that it does not matter what the dimensions are, but our experience with Example 1 suggests otherwise. We make a drawing in which x is the length, in inches, of each square to be cut. It is important to note that since the original square is 8 in. by 8 in., after the smaller squares are removed, the lengths of the sides of the box will be $(8 - 2x)$ in. by $(8 - 2x)$ in.

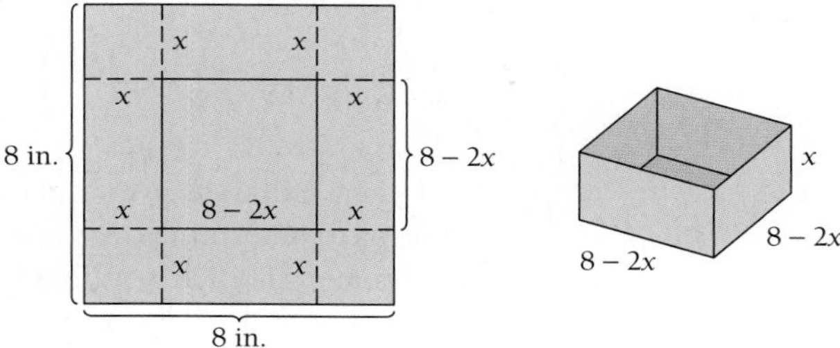

After the four small squares are removed and the sides are folded up, the volume V of the resulting box is

$$V = l \cdot w \cdot h = (8 - 2x) \cdot (8 - 2x) \cdot x,$$

or $$V(x) = (64 - 32x + 4x^2)x = 4x^3 - 32x^2 + 64x.$$

From the drawing, we see that x must be less than 4 in. Thus, we need to maximize

$$V(x) = 4x^3 - 32x^2 + 64x \quad \text{over the interval } (0, 4).$$

To do so, we first find $V'(x)$:

$$V'(x) = 12x^2 - 64x + 64.$$

Since $V'(x)$ exists for all x in the interval $(0, 4)$, we can set it equal to 0 to find the critical values:

$$\begin{aligned} V'(x) = 12x^2 - 64x + 64 &= 0 \\ 4(3x^2 - 16x + 16) &= 0 \\ 4(3x - 4)(x - 4) &= 0 \\ 3x - 4 = 0 \quad &\text{or} \quad x - 4 = 0 \\ 3x = 4 \quad &\text{or} \quad x = 4 \\ x = \tfrac{4}{3} \quad &\text{or} \quad x = 4. \end{aligned}$$

The only critical value in $(0, 4)$ is $\frac{4}{3}$. Thus, we can use the second derivative,

$$V''(x) = 24x - 64,$$

TECHNOLOGY CONNECTION

EXERCISES

1. Complete this table to help visualize Example 2.

x	$8 - 2x$	$4x^3 - 32x^2 + 64x$
0		
0.5		
1.0		
1.5		
2.0		
2.5		
3.0		
3.5		
4.0		

2. Graph $V(x) = 4x^3 - 32x^2 + 64x$ over the interval $(0, 4)$.
3. Estimate a maximum value, and state where it occurs.

to determine whether we have a maximum. Since

$$V''\left(\tfrac{4}{3}\right) = 24 \cdot \tfrac{4}{3} - 64 = 32 - 64 < 0,$$

we know that $V\left(\tfrac{4}{3}\right)$ is a maximum.

Thus, to maximize the box's volume, small squares with edges measuring $\tfrac{4}{3}$ in., or $1\tfrac{1}{3}$ in., should be cut from each corner of the original 8 in. by 8 in. piece of cardboard. When the sides are folded up, the resulting box will have sides of length

$$8 - 2x = 8 - 2 \cdot \frac{4}{3} = \frac{24}{3} - \frac{8}{3} = \frac{16}{3} = 5\frac{1}{3}\text{ in.}$$

and a height of $1\tfrac{1}{3}$ in. The maximum volume is

$$V\left(\frac{4}{3}\right) = 4\left(\frac{4}{3}\right)^3 - 32\left(\frac{4}{3}\right)^2 + 64\left(\frac{4}{3}\right) = \frac{1024}{27} = 37\frac{25}{27}\text{ in}^3.$$ ✦

In Example 3, an open-top container of fixed volume is to be constructed. We want to determine the dimensions that will allow it to be built with the least amount of material. Such a problem could be important from an ecological standpoint.

EXAMPLE 3 Minimizing Surface Area. A container firm is designing an open-top rectangular box, with a square base, that will hold 108 cubic centimeters (cm^3). What dimensions yield the minimum surface area? What is the minimum surface area?

Solution We first make a drawing. The surface area, S, of the box is given by the area of the base plus the area of the four sides. Thus,

$$S = x^2 + 4xy,$$

where x and y are, respectively, the length and height, in centimeters (cm), of the completed box. The volume must be 108 cm^3, and is given by

$$V = x^2y = 108.$$

To express S in terms of one variable, we solve $x^2y = 108$ for y:

$$y = \frac{108}{x^2}.$$

Then

$$S(x) = x^2 + 4x\left(\frac{108}{x^2}\right)$$
$$= x^2 + \frac{432}{x}.$$

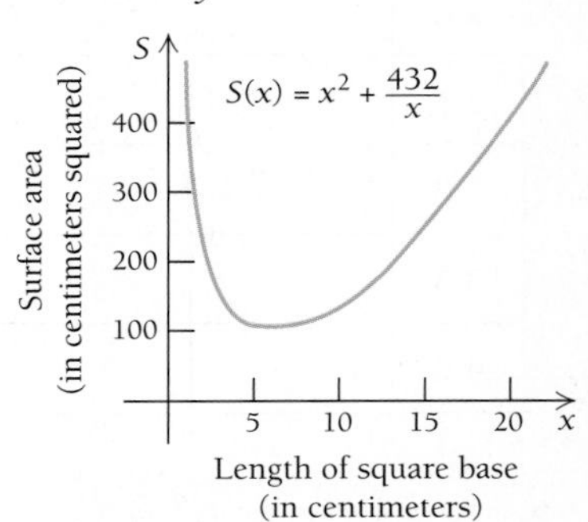

The nature of the application requires $x > 0$. Thus, we are minimizing S over the interval $(0, \infty)$. We first find dS/dx:

$$\frac{dS}{dx} = 2x - \frac{432}{x^2}.$$ The notation $S'(x)$ can also be used.

Since dS/dx exists for all x in $(0, \infty)$, the only critical values occur where $dS/dx = 0$:

$$2x - \frac{432}{x^2} = 0 \qquad \text{Setting the first derivative equal to 0}$$

$$x^2\left(2x - \frac{432}{x^2}\right) = x^2 \cdot 0 \qquad \text{We multiply by } x^2 \text{ to clear the fractions.}$$

$$2x^3 - 432 = 0$$

$$2x^3 = 432$$

$$x^3 = 216$$

$$x = 6. \qquad \text{Finding the cube root of each side}$$

This is the only critical value, so we can use the second derivative to determine whether we have a minimum:

$$\frac{d^2S}{dx^2} = 2 + \frac{864}{x^3}.$$

Note that this is positive for all $x > 0$. Thus, we have a minimum at $x = 6$. When $x = 6$, it follows that $y = 3$:

$$y = \frac{108}{6^2} = \frac{108}{36} = 3.$$

Thus, the surface area is minimized when $x = 6$ cm and $y = 3$ cm. The minimum surface area is

$$S = 6^2 + 4 \cdot 6 \cdot 3$$
$$= 108 \text{ cm}^2.$$

By coincidence, this is the same number as the fixed volume. ✦

Business and Economics Applications

EXAMPLE 4 Business: Maximizing Revenue. A stereo manufacturer determines that in order to sell x units of a new stereo, the price per unit, in dollars, must be

$$p(x) = 1000 - x.$$

The manufacturer also determines that the total cost of producing x units is given by

$$C(x) = 3000 + 20x.$$

a) Find the total revenue $R(x)$.

b) Find the total profit $P(x)$.

c) How many units must the company produce and sell in order to maximize profit?

d) What is the maximum profit?

e) What price per unit must be charged in order to make this maximum profit?

Solution

a) $R(x) =$ Total revenue

$= (\text{Number of units}) \cdot (\text{Price per unit})$

$= \quad x \quad \cdot \quad p$

$= x(1000 - x) = 1000x - x^2$

b) $P(x) = \text{Total revenue} - \text{Total cost}$

$$= R(x) - C(x)$$
$$= (1000x - x^2) - (3000 + 20x)$$
$$= -x^2 + 980x - 3000$$

c) To find the maximum value of $P(x)$, we first find $P'(x)$:

$$P'(x) = -2x + 980.$$

This is defined for all real numbers, so the only critical values will come from solving $P'(x) = 0$:

$$P'(x) = -2x + 980 = 0$$
$$-2x = -980$$
$$x = 490.$$

There is only one critical value. We can therefore try to use the second derivative to determine whether we have an absolute maximum. Note that

$$P''(x) = -2, \quad \text{a constant.}$$

Thus, $P''(490)$ is negative, and so profit is maximized when 490 units are produced and sold.

d) The maximum profit is given by

$$P(490) = -(490)^2 + 980 \cdot 490 - 3000$$
$$= \$237{,}100.$$

Thus, the stereo manufacturer makes a maximum profit of \$237,100 by producing and selling 490 stereos.

e) The price per unit needed to make the maximum profit is

$$p = 1000 - 490 = \$510.$$ ✦

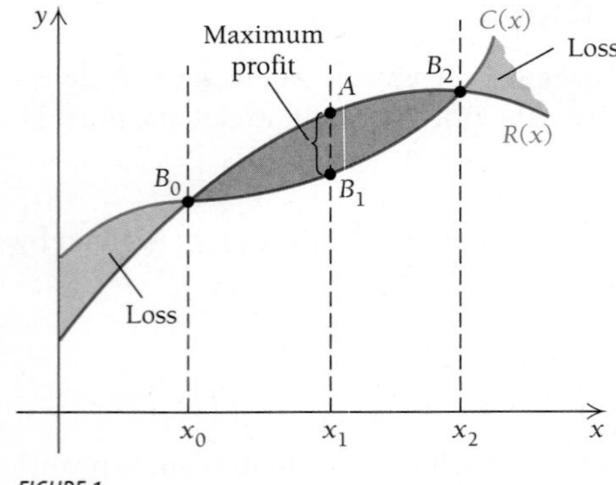

FIGURE 1

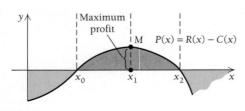

FIGURE 2

Let's take a general look at the total-profit function and its related functions.

Figure 1 shows an example of total-cost and total-revenue functions. We can estimate what the maximum profit might be by looking for the widest gap between $R(x)$ and $C(x)$, when $R(x) > C(x)$. Points B_0 and B_2 are break-even points.

Figure 2 shows the related total-profit function. Note that when production is too low ($< x_0$), there is a loss, perhaps due to high fixed or initial costs and low revenue. When production is too high ($> x_2$), there is also a loss, perhaps due to the increased cost of overtime pay or expansion.

The business operates at a profit everywhere between x_0 and x_2. Note that maximum profit occurs at a critical value x_1 of $P(x)$. If we assume that $P'(x)$ exists for all x in some interval, usually $[0, \infty)$, this critical value occurs at some number x such that

$$P'(x) = 0 \quad \text{and} \quad P''(x) < 0.$$

Since $P(x) = R(x) - C(x)$, it follows that

$$P'(x) = R'(x) - C'(x) \quad \text{and} \quad P''(x) = R''(x) - C''(x).$$

Thus, the maximum profit occurs at some number x such that

$$P'(x) = R'(x) - C'(x) = 0 \quad \text{and} \quad P''(x) = R''(x) - C''(x) < 0,$$

or

$$R'(x) = C'(x) \quad \text{and} \quad R''(x) < C''(x).$$

In summary, we have the following theorem.

THEOREM 10

Maximum profit occurs at those x-values for which

$$R'(x) = C'(x) \quad \text{and} \quad R''(x) < C''(x).^*$$

You can check that the results in parts (c) and (d) of Example 4 can be easily found using Theorem 10.

EXAMPLE 5 Business: Determining a Ticket Price. Promoters of international fund-raising concerts must walk a fine line between profit and loss, especially when determining the price to charge for admission to closed-circuit TV showings in local theaters. By keeping records, a theater determines that at an admission price of \$26, it averages 1000 people in attendance. For every drop in price of \$1, it gains 50 customers. Each customer spends an average of \$4 on concessions. What admission price should the theater charge in order to maximize total revenue?

Solution Let x be the number of dollars by which the price of \$26 should be decreased. (If x is negative, the price is increased.) We first express the total revenue R as a function of x. Note that the increase in ticket sales is $50x$ when the price drops x dollars:

$$\begin{aligned} R(x) &= (\text{Revenue from tickets}) + (\text{Revenue from concessions}) \\ &= (\text{Number of people}) \cdot (\text{Ticket price}) + (\text{Number of people}) \cdot 4 \\ &= (1000 + 50x)(26 - x) + (1000 + 50x) \cdot 4 \\ &= 26{,}000 - 1000x + 1300x - 50x^2 + 4000 + 200x, \end{aligned}$$

or $\quad R(x) = -50x^2 + 500x + 30{,}000.$

To find x such that $R(x)$ is a maximum, we first find $R'(x)$:

$$R'(x) = -100x + 500.$$

This derivative exists for all real numbers x. Thus, the only critical values are where $R'(x) = 0$; so we solve that equation:

$$\begin{aligned} -100x + 500 &= 0 \\ -100x &= -500 \\ x &= 5 \end{aligned}$$ This corresponds to lowering the price by \$5.

*In Section 2.6, the concepts of *marginal revenue* and *marginal cost* are introduced, allowing $R'(x) = C'(x)$ to be regarded as Marginal revenue = Marginal cost.

Since this is the only critical value, we can use the second derivative,

$$R''(x) = -100,$$

to determine whether we have a maximum. Since $R''(5)$ is negative, $R(5)$ is a maximum. Therefore, in order to maximize revenue, the theater should charge

$$\$26 - \$5, \quad \text{or} \quad \$21 \text{ per ticket.}$$

✦

Minimizing Inventory Costs

A retail business outlet needs to be concerned about inventory costs. Suppose, for example, that an appliance store sells 2500 television sets per year. It *could* operate by ordering all the sets at once. But then the owners would face the carrying costs (insurance, building space, and so on) of storing them all. Thus, they might make several, say 5, smaller orders, so that the largest number they would ever have to store is 500. However, each time they reorder, there are costs for paperwork, delivery charges, labor, and so on. It seems, therefore, that there must be some balance between carrying costs and reorder costs. Let's see how calculus can help determine what that balance might be. We are trying to minimize the following function:

$$\text{Total inventory costs} = \begin{pmatrix}\text{Yearly carrying}\\ \text{costs}\end{pmatrix} + \begin{pmatrix}\text{Yearly reorder}\\ \text{costs}\end{pmatrix}.$$

The *lot size* x is the largest number ordered each reordering period. If x units are ordered each period, then during that time somewhere between 0 and x units are in stock. To have a representative expression for the amount in stock at any one time in the period, we can use the average, $x/2$. This represents the average amount held in stock over the course of each time period.

Refer to the graphs shown below. If the lot size is 2500, then during the period between orders, there are somewhere between 0 and 2500 units in stock. On average, there are 2500/2, or 1250 units in stock. If the lot size is 1250, then during the period between orders, there are somewhere between 0 and 1250 units in stock. On average, there are 1250/2, or 625 units in stock. In general, if the lot size is x, the average inventory is $x/2$.

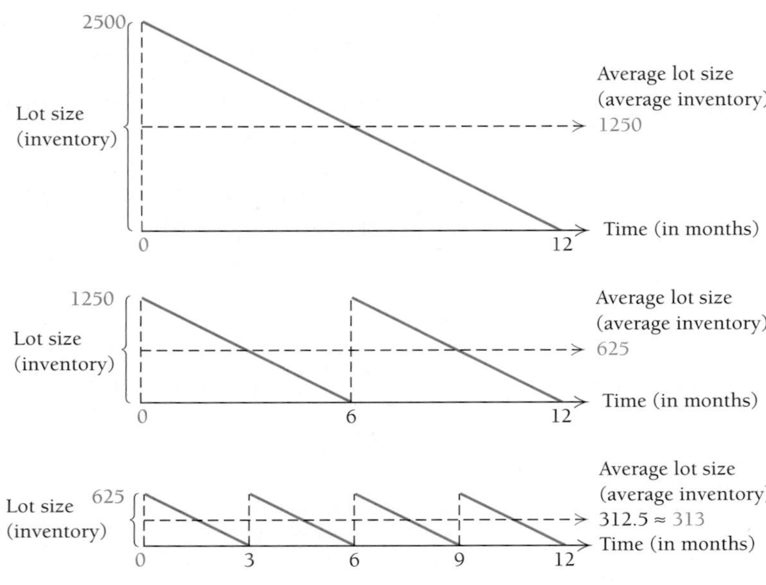

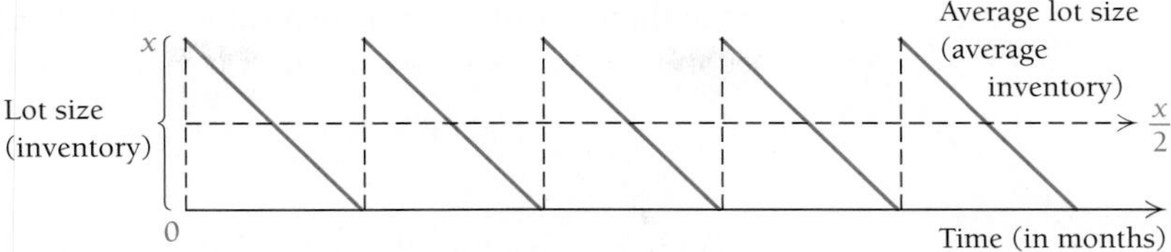

EXAMPLE 6 Business: Minimizing Inventory Costs. A retail appliance store sells 2500 television sets per year. It costs \$10 to store one set for a year. To reorder, there is a fixed cost of \$20, plus a fee of \$9 per set. How many times per year should the store reorder, and in what lot size, to minimize inventory costs?

Solution Let x = the lot size. Inventory costs are given by

$$C(x) = (\text{Yearly carrying costs}) + (\text{Yearly reorder costs}).$$

We consider each component of inventory costs separately.

a) *Yearly carrying costs.* The average amount held in stock is $x/2$, and it costs \$10 per set for storage. Thus,

$$\begin{aligned}\text{Yearly carrying costs} &= \begin{pmatrix}\text{Yearly cost}\\ \text{per item}\end{pmatrix} \cdot \begin{pmatrix}\text{Average number}\\ \text{of items}\end{pmatrix}\\ &= 10 \cdot \frac{x}{2}.\end{aligned}$$

b) *Yearly reorder costs.* We know that x is the lot size, and we let N be the number of reorders each year. Then $Nx = 2500$, and $N = 2500/x$. Thus,

$$\begin{aligned}\text{Yearly reorder costs} &= \begin{pmatrix}\text{Cost of each}\\ \text{order}\end{pmatrix} \cdot \begin{pmatrix}\text{Number of}\\ \text{reorders}\end{pmatrix}\\ &= (20 + 9x)\frac{2500}{x}.\end{aligned}$$

c) Thus, we have

$$\begin{aligned}C(x) &= 10 \cdot \frac{x}{2} + (20 + 9x)\frac{2500}{x}\\ &= 5x + \frac{50{,}000}{x} + 22{,}500 = 5x + 50{,}000x^{-1} + 22{,}500.\end{aligned}$$

d) To find a minimum value of C over $[1, 2500]$, we first find $C'(x)$:

$$C'(x) = 5 - \frac{50{,}000}{x^2}.$$

e) $C'(x)$ exists for all x in $[1, 2500]$, so the only critical values are those x-values such that $C'(x) = 0$. We solve $C'(x) = 0$:

$$\begin{aligned}5 - \frac{50{,}000}{x^2} &= 0\\ 5 &= \frac{50{,}000}{x^2}\\ 5x^2 &= 50{,}000\\ x^2 &= 10{,}000\\ x &= \pm 100.\end{aligned}$$

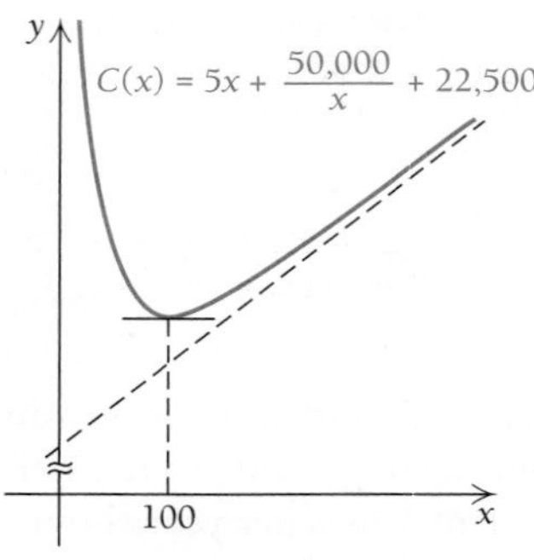

Since there is only one critical value in $[1, 2500]$, that is, $x = 100$, we can use the second derivative to see whether it yields a maximum or a minimum:

$$C''(x) = \frac{100{,}000}{x^3}.$$

$C''(x)$ is positive for all x in $[1, 2500]$, so we have a minimum at $x = 100$. Thus, to minimize inventory costs, the store should order sets $(2500/100)$, or 25 times per year. The lot size is 100. ✦

TECHNOLOGY CONNECTION

Exploratory

Many calculators can make tables and/or spreadsheets of function values. In reference to Example 6, without using calculus, one might make an estimate of the lot size that will minimize total inventory costs by using a table like the one below. Complete the table, and estimate the solution of Example 6.

EXERCISES

1. Graph $C(x)$ over the interval $[1, 2500]$.
2. Graphically estimate the minimum value, and note where it occurs. Does the table confirm the graph?

Lot Size, x	Number of Reorders, $\frac{2500}{x}$	Average Inventory, $\frac{x}{2}$	Carrying Costs, $10 \cdot \frac{x}{2}$	Cost of Each Order, $20 + 9x$	Reorder Costs, $(20 + 9x)\frac{2500}{x}$	Total Inventory Costs, $C(x) = 10 \cdot \frac{x}{2} + (20 + 9x)\frac{2500}{x}$
2500	1	1250	\$12,500	\$22,520	\$22,520	\$35,020
1250	2	625	6,250	11,270	22,540	
500	5	250	2,500	4,520		
250	10	125				
167	15	84				
125	20					
100	25					
90	28					
50	50					

What happens in problems like Example 6 if the answer is not a whole number? For those cases, we consider the two whole numbers closest to the answer and substitute them into $C(x)$. The value that yields the smaller $C(x)$ is the lot size.

EXAMPLE 7 **Business: Minimizing Inventory Costs.** Reconsider Example 6, but change the \$10 storage cost to \$20. How many times per year should the store reorder television sets, and in what lot size, in order to minimize inventory costs?

Solution Comparing this situation with that in Example 6, we find that the inventory cost function becomes

$$C(x) = 20 \cdot \frac{x}{2} + (20 + 9x)\frac{2500}{x}$$

$$= 10x + \frac{50{,}000}{x} + 22{,}500 = 10x + 50{,}000x^{-1} + 22{,}500.$$

Then we find $C'(x)$, set it equal to 0, and solve for x:

$$\begin{aligned} C'(x) = 10 - \frac{50{,}000}{x^2} &= 0 \\ 10 &= \frac{50{,}000}{x^2} \\ 10x^2 &= 50{,}000 \\ x^2 &= 5000 \\ x &= \sqrt{5000} \\ &\approx 70.7. \end{aligned}$$

It is impossible to reorder 70.7 sets each time, so we consider the two numbers closest to 70.7, which are 70 and 71. Since

$$C(70) \approx \$23{,}914.29 \quad \text{and} \quad C(71) \approx \$23{,}914.23,$$

it follows that the lot size that will minimize cost is 71, although the difference, \$0.06, is not much. (*Note*: Such a procedure will not work for all functions but will work for the type we are considering here.) The number of times an order should be placed is 2500/71 with a remainder of 15, indicating that 35 orders should be placed. Of those, $35 - 15 = 20$ will be for 71 items and 15 will be for 72 items. ✦

The lot size that minimizes total inventory costs is often referred to as the *economic ordering quantity*. Three assumptions are made in using the preceding method to determine the economic ordering quantity. First, the demand for the product is the same year round. For television sets this may be reasonable, but for seasonal items such as clothing or skis, this assumption is unrealistic. Second, the time between the placing of an order and its receipt remains consistent throughout the year. Finally, the various costs involved, such as storage, shipping charges, and so on, do not vary. This may not be reasonable in a time of inflation, although one may account for variation by anticipating what they might be and using average costs. Regardless, the model described above is useful, allowing us to analyze a seemingly difficult problem using calculus.

Exercise Set 2.5

1. Of all numbers whose sum is 50, find the two that have the maximum product. That is, maximize $Q = xy$, where $x + y = 50$.

2. Of all numbers whose sum is 70, find the two that have the maximum product. That is, maximize $Q = xy$, where $x + y = 70$.

TW 3. In Exercise 1, can there be a minimum product? Why or why not?

TW 4. In Exercise 2, can there be a minimum product? Why or why not?

5. Of all numbers whose difference is 4, find the two that have the minimum product.

6. Of all numbers whose difference is 6, find the two that have the minimum product.

7. Maximize $Q = xy^2$, where x and y are positive numbers such that $x + y^2 = 1$.

8. Maximize $Q = xy^2$, where x and y are positive numbers such that $x + y^2 = 4$.

9. Minimize $Q = 2x^2 + 3y^2$, where $x + y = 5$.

10. Minimize $Q = x^2 + 2y^2$, where $x + y = 3$.

11. Maximize $Q = xy$, where x and y are positive numbers such that $\frac{4}{3}x^2 + y = 16$.

12. Maximize $Q = xy$, where x and y are positive numbers such that $x + \frac{4}{3}y^2 = 1$.

13. **Maximizing area.** A lifeguard needs to rope off a rectangular swimming area in front of Long Lake Beach, using 180 yd of rope and floats. What dimensions of the rectangle will maximize the area? What is the maximum area? (Note that the beach is one side of the rectangle.)

14. **Maximizing area.** A rancher wants to enclose two rectangular areas near a river, one for sheep and one for cattle. There are 240 yd of fencing available. What is the largest total area that can be enclosed?

15. **Maximizing area.** A carpenter is building a rectangular shed with a fixed perimeter of 54 ft. What are the dimensions of the largest shed that can be built? What is its area?

16. **Maximizing area.** Of all rectangles that have a perimeter of 42 ft, find the dimensions of the one with the largest area. What is its area?

17. **Maximizing volume.** From a 50-cm-by-50-cm sheet of aluminum, square corners are cut out so that the sides can be folded up to make a box. What dimensions will yield a box of maximum volume? What is the maximum volume?

18. **Maximizing volume.** From a thin piece of cardboard 20 in. by 20 in., square corners are cut out so that the sides can be folded up to make a box. What dimensions will yield a box of maximum volume? What is the maximum volume?

19. **Minimizing surface area.** Drum Tight Containers is designing an open-top, square-based, rectangular box that will have a volume of 62.5 in^3. What dimensions will minimize surface area? What is the minimum surface area?

20. **Minimizing surface area.** A soup company is constructing an open-top, square-based, rectangular metal tank that will have a volume of 32 ft^3. What dimensions will minimize surface area? What is the minimum surface area?

21. Minimizing surface area. Open Air Waste Management is designing a rectangular construction dumpster that will be twice as long as it is wide and must hold 12 yd^3 of debris. Find the dimensions of the dumpster that will minimize its surface area.

22. Minimizing surface area. Ever Green Gardening is designing a rectangular compost container that will be twice as tall as it is wide and must hold 18 ft^3 of composted food scraps. Find the dimensions of the compost container with minimal surface area (include the bottom and top).

APPLICATIONS

Business and Economics

Maximizing profit. *Find the maximum profit and the number of units that must be produced and sold in order to yield the maximum profit. Assume that revenue, $R(x)$, and cost, $C(x)$, are in dollars for Exercises 23–26.*

23. $R(x) = 50x - 0.5x^2$, $C(x) = 4x + 10$

24. $R(x) = 50x - 0.5x^2$, $C(x) = 10x + 3$

25. $R(x) = 2x$, $C(x) = 0.01x^2 + 0.6x + 30$

26. $R(x) = 5x$, $C(x) = 0.001x^2 + 1.2x + 60$

27. $R(x) = 9x - 2x^2$, $C(x) = x^3 - 3x^2 + 4x + 1$; assume that $R(x)$ and $C(x)$ are in thousands of dollars, and x is in thousands of units.

28. $R(x) = 100x - x^2$,
$C(x) = \frac{1}{3}x^3 - 6x^2 + 89x + 100$;
assume that $R(x)$ and $C(x)$ are in thousands of dollars, and x is in thousands of units.

29. Maximizing profit. Raggs, Ltd., a clothing firm, determines that in order to sell x suits, the price per suit must be

$$p = 150 - 0.5x.$$

It also determines that the total cost of producing x suits is given by

$$C(x) = 4000 + 0.25x^2.$$

a) Find the total revenue, $R(x)$.
b) Find the total profit, $P(x)$.
c) How many suits must the company produce and sell in order to maximize profit?
d) What is the maximum profit?
e) What price per suit must be charged in order to maximize profit?

30. Maximizing profit. Riverside Appliances is marketing a new refrigerator. It determines that in order to sell x refrigerators, the price per refrigerator must be

$$p = 280 - 0.4x.$$

It also determines that the total cost of producing x refrigerators is given by

$$C(x) = 5000 + 0.6x^2.$$

a) Find the total revenue, $R(x)$.
b) Find the total profit, $P(x)$.
c) How many refrigerators must the company produce and sell in order to maximize profit?
d) What is the maximum profit?
e) What price per refrigerator must be charged in order to maximize profit?

31. Maximizing revenue. A university is trying to determine what price to charge for tickets to football games. At a price of \$18 per ticket, attendance averages 40,000 people per game. Every decrease of \$3 adds 10,000 people to the average number. Every person at the game spends an average of \$4.50 on concessions. What price per ticket should be charged in order to maximize revenue? How many people will attend at that price?

32. Maximizing profit. Gritz-Charlston is a 300-unit luxury hotel. All rooms are occupied when the hotel charges \$80 per day for a room. For every increase of x dollars in the daily room rate, there are x rooms vacant. Each occupied room costs \$22 per day to service and maintain. What should the hotel charge per day in order to maximize profit?

33. Maximizing yield. An apple farm yields an average of 30 bushels of apples per tree when 20 trees are planted on an acre of ground. Each time 1 more tree is planted per acre, the yield decreases by 1 bushel (bu) per tree as a result of crowding. How many trees should be planted on an acre in order to get the highest yield?

34. Nitrogen prices. During 2001, nitrogen prices fell by 41%. Over the same period of time, nitrogen demand went up by 12%. (**Source**: *Chemical Week*.)

a) Assuming a linear change in demand, find the demand function, $q(x)$, by finding the equation of the line that passes through the points $(1, 1)$ and $(0.59, 1.12)$. Here x is the price as a fraction of the January 2001 price, and $q(x)$ is the demand as a fraction of the demand in January.

b) As a percentage of the January 2001 price, what should the price of nitrogen be to maximize revenue?

35. Vanity license plates. According to a pricing model, increasing the fee for vanity license plates by \$1 decreases the percentage of a state's population that will request them by 0.04%. (*Source*: E. D. Craft, "The demand for vanity (plates): Elasticities, net revenue maximization, and deadweight loss," *Contemporary Economic Policy*, Vol. 20, 133–144 (2002).)

a) Recently, the fee for vanity license plates in Maryland was \$25, and the percentage of the state's population that had vanity plates was 2.13%. Use this information to construct the demand function, $q(x)$, for the percentage of Maryland's population that will request vanity license plates for a fee of x dollars.

b) Find the fee, x, that will maximize revenue from vanity plates.

36. Maximizing revenue. When a theater owner charges \$5 for admission, there is an average attendance of 180 people. For every \$0.10 increase in admission, there is a loss of 1 customer from the average number. What admission should be charged in order to maximize revenue?

37. Minimizing costs. A rectangular box with a volume of 320 ft^3 is to be constructed with a square base and top. The cost per square foot for the bottom is 15¢, for the top is 10¢, and for the sides is 2.5¢. What dimensions will minimize the cost?

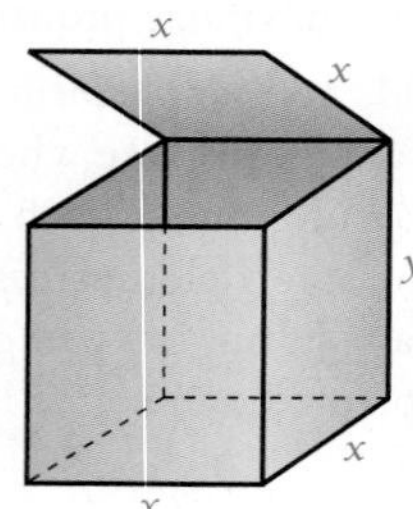

38. Maximizing area. A publisher decides that each page in a new book must have an area of 73.125 in^2, a 0.75-in. margin at the top and at the bottom of each page, and a 0.5-in. margin on each of the sides. What should the outside dimensions of each page be so that the printed area is a maximum?

39. Minimizing inventory costs. A sporting goods store sells 100 pool tables per year. It costs \$20 to store one pool table for a year. To reorder, there is a fixed cost of \$40 per shipment plus \$16 for each pool table. How many times per year should the store order pool tables, and in what lot size, in order to minimize inventory costs?

40. Minimizing inventory costs. A pro shop in a bowling center sells 200 bowling balls per year. It costs \$4 to store one bowling ball for a year. To reorder, there is a fixed cost of \$1, plus \$0.50 for each bowling ball. How many times per year should the shop order bowling balls, and in what lot size, in order to minimize inventory costs?

41. Minimizing inventory costs. A retail outlet for Boxowitz Calculators sells 720 calculators per year. It costs \$2 to store one calculator for a year. To reorder, there is a fixed cost of \$5, plus \$2.50 for each calculator. How many times per year should the store order calculators, and in what lot size, in order to minimize inventory costs?

42. Minimizing inventory costs. Bon Temps Surf and SCUBA Shop sells 360 surfboards per year. It costs \$8 to store one surfboard for a year. Each reorder costs \$10, plus an additional \$5 for each surfboard ordered. How many times per year should the store order surfboards, and in what lot size, in order to minimize inventory costs?

43. Minimizing inventory costs. Repeat Exercise 41 using the same data, but assume yearly sales of 256 calculators with the fixed cost of each reorder set at \$4.

44. Minimizing inventory costs. Repeat Exercise 42 using the same data, but change the reorder costs from an additional \$5 per surfboard to \$6 per surfboard.

General Interest

45. Maximizing volume. The postal service places a limit of 84 in. on the combined length and girth of (distance around) a package to be sent parcel post. What dimensions of a rectangular box with square cross-section will contain the largest volume that can be mailed? (*Hint*: There are two different girths.)

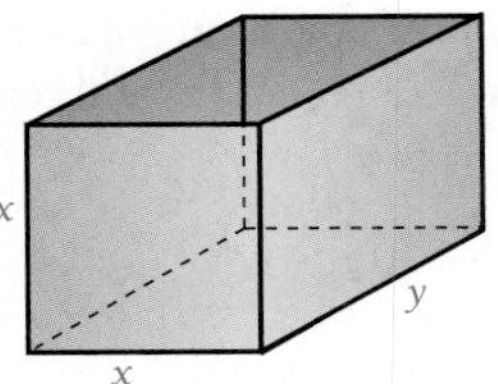

46. Minimizing cost. A rectangular play area is to be fenced off in a person's yard and is to contain 48 yd^2. The next-door neighbor agrees to pay half the cost of the fence on the side of the play area that lies along the property line. What dimensions will minimize the cost of the fence?

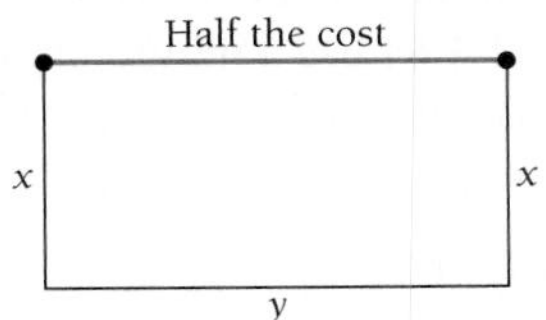

47. Maximizing light. A Norman window is a rectangle with a semicircle on top. Suppose that the perimeter of a particular Norman window is to be 24 ft. What should its dimensions be in order to allow the maximum amount of light to enter through the window?

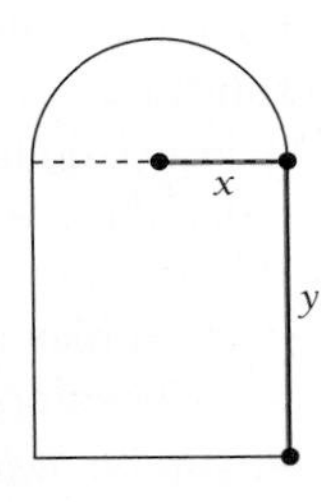

48. Maximizing light. Repeat Exercise 47, but assume that the semicircle is to be stained glass, which transmits only half as much light as clear glass does.

SYNTHESIS

49. For what positive number is the sum of its reciprocal and five times its square a minimum?

50. For what positive number is the sum of its reciprocal and four times its square a minimum?

51. Maximizing profit. The amount of money that customers deposit in a bank in savings accounts is directly proportional to the interest rate that the bank pays on that money. Suppose that a bank was able to turn around and loan out all the money deposited in its savings accounts at an interest rate of 18%. What interest rate should it pay on its savings accounts in order to maximize profit?

52. A 24-in. piece of wire is cut in two pieces. One piece is used to form a circle and the other to form a square. How should the wire be cut so that the sum of the areas is a minimum? A maximum?

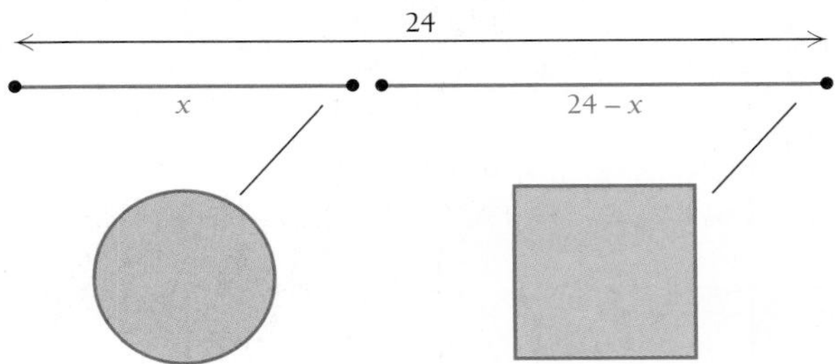

53. Business: minimizing costs. A power line is to be constructed from a power station at point A to an island at point C, which is 1 mi directly out in the water from a point B on the shore. Point B is 4 mi downshore from the power station at A. It costs \$5000 per mile to lay the power line under water and \$3000 per mile to lay the line under ground. At what point S downshore from A should the line come to the shore in order to minimize cost? Note that S could very well be B or A. (*Hint*: The length of CS is $\sqrt{1 + x^2}$.)

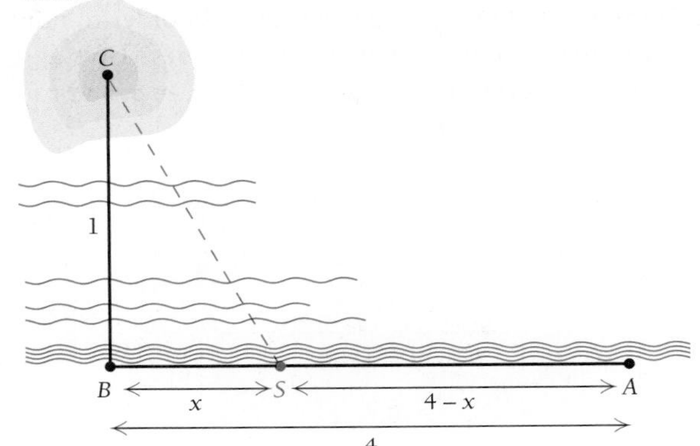

54. Life science: flights of homing pigeons. It is known that homing pigeons tend to avoid flying over water in the daytime, perhaps because the downdrafts of air over water make flying difficult. Suppose that a homing pigeon is released on an island at point C, which is 3 mi directly out in the water from a point B on shore. Point B is 8 mi downshore from the pigeon's home loft at point A. Assume that a pigeon flying over water uses energy at a rate 1.28 times the rate over land. Toward what point S downshore from A should the pigeon fly in order to minimize the total energy required to get to the home loft at A? Assume that

Total energy =
(Energy rate over water) · (Distance over water)
+ (Energy rate over land) · (Distance over land).

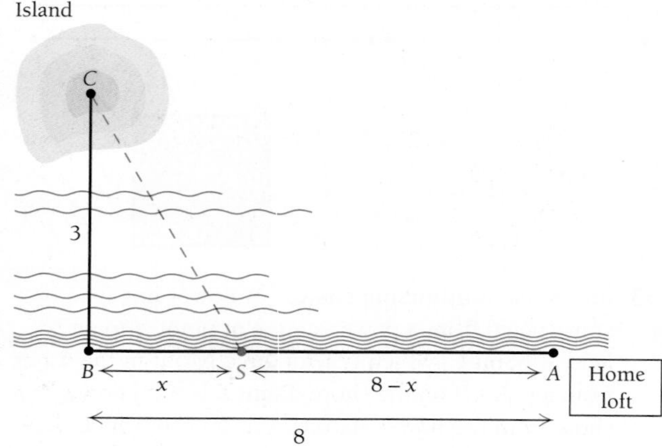

55. Business: minimizing distance. A road is to be built between two cities C_1 and C_2, which are on opposite sides of a river of uniform width r. Because of the river, a bridge must be built. C_1 is a units from the river, and C_2 is b units from the river, with $a \leq b$. Where should the bridge be located in order to minimize the total distance between the cities? Give a general solution using the constants a, b, p, and r as shown in the figure.

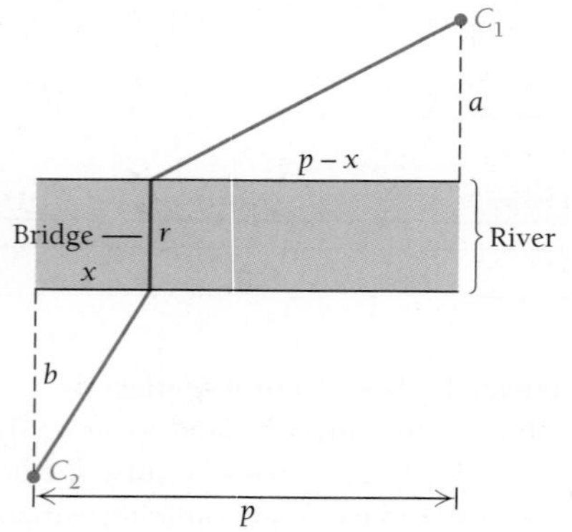

56. Business: minimizing cost. The total cost, in dollars, of producing x units of a certain product is given by

$$C(x) = 8x + 20 + \frac{x^3}{100}.$$

a) Find the average cost, $A(x) = C(x)/x$.
b) Find $C'(x)$ and $A'(x)$.
c) Find the minimum of $A(x)$ and the value x_0 at which it occurs. Find $C'(x_0)$.
d) Compare $A(x_0)$ and $C'(x_0)$.

57. Business: minimizing cost. Consider

$$A(x) = C(x)/x.$$

a) Find $A'(x)$ in terms of $C'(x)$ and $C(x)$.
b) Show that if $A(x)$ has a minimum, then it will occur at that value of x_0 for which

$$C'(x_0) = A(x_0) = \frac{C(x_0)}{x_0}.$$

This result shows that if average cost can be minimized, such a minimum will occur when marginal cost equals average cost.

58. Minimize $Q = x^3 + 2y^3$, where x and y are positive numbers, such that $x + y = 1$.

59. Minimize $Q = 3x + y^3$, where $x^2 + y^2 = 2$.

60. Business: minimizing inventory costs—a general solution. A store sells Q units of a product per year. It costs a dollars to store one unit for a year. To reorder, there is a fixed cost of b dollars, plus c dollars for each unit. How many times per year should the store reorder, and in what lot size, in order to minimize inventory costs?

61. Business: minimizing inventory costs. Use the general solution found in Exercise 60 to find how many times per year a store should reorder, and in what lot size, when $Q = 2500$, $a = \$10$, $b = \$20$, and $c = \$9$.

OBJECTIVES

- Find marginal cost, revenue, and profit.
- Find Δy and dy.
- Use differentials for approximations.

Marginals and Differentials

In this section, we consider ways of using calculus to make linear approximations. Suppose, for example, that a company is considering an increase in production. Usually the company wants at least an approximation of what the resulting changes in *cost*, *revenue*, and *profit* will be.

Marginal Cost, Revenue, and Profit

Suppose that a band is self-producing its own CD and considering an increase in monthly production from 12 cartons to 13. To estimate the resulting increase in cost, it would be reasonable to find the rate at which cost is increasing when 12 cartons are produced and add that to the cost of producing 12 cartons. That is,

$$C(13) \approx C(12) + C'(12).$$

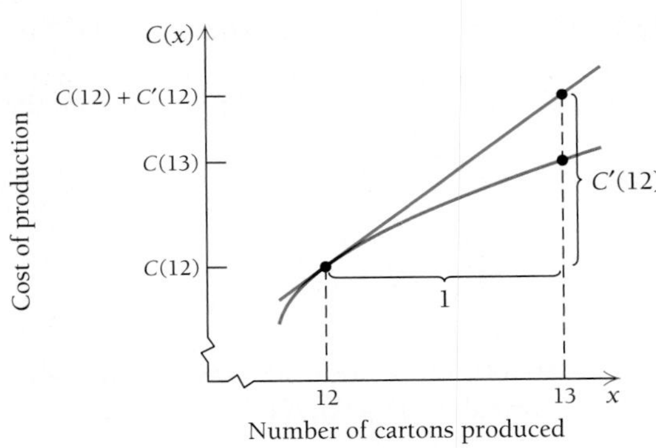

The number $C'(12)$ is called the *marginal cost at 12*. Remember that $C'(12)$ is the slope of the tangent line at the point $(12, C(12))$. If, for example, this slope is $\frac{3}{4}$, we can regard it as a vertical change of 3 with a horizontal change of 4, or a vertical change of $\frac{3}{4}$ with a horizontal change of 1. It is this latter interpretation that we use for estimating. Graphically, this interpretation can be viewed as shown at the left. Note in the figure that $C'(12)$ is slightly more than the difference between $C(13)$ and $C(12)$, or $C(13) - C(12)$. For other curves, $C'(12)$ may be slightly less than $C(13) - C(12)$. Almost always, however, it is simpler to compute $C'(12)$ than it is to compute $C(13) - C(12)$.

Generalizing, we have the following.

DEFINITIONS

Let $C(x)$, $R(x)$, and $P(x)$ represent, respectively, the total cost, revenue, and profit from the production and sale of x items.

The **marginal cost*** at x, given by $C'(x)$, is the approximate cost of the $(x + 1)$st item:

$$C'(x) \approx C(x + 1) - C(x), \text{ or } C(x + 1) \approx C(x) + C'(x).$$

The **marginal revenue** at x, given by $R'(x)$, is the approximate revenue from the $(x + 1)$st item:

$$R'(x) \approx R(x + 1) - R(x), \text{ or } R(x + 1) \approx R(x) + R'(x).$$

The **marginal profit** at x, given by $P'(x)$, is the approximate profit from the $(x + 1)$st item:

$$P'(x) \approx P(x + 1) - P(x), \text{ or } P(x + 1) \approx P(x) + P'(x).$$

---

*The term "marginal" comes from the Marginalist School of Economic Thought, which originated in Austria for the purpose of applying mathematics and statistics to the study of economics.

You can confirm that $P'(x) = R'(x) - C'(x)$.

EXAMPLE 1 Business: Marginal Cost, Revenue, and Profit. Given

$$C(x) = 62x^2 + 27{,}500 \quad \text{and}$$
$$R(x) = x^3 - 12x^2 + 40x + 10,$$

find each of the following.

a) Total profit, $P(x)$

b) Total cost, revenue, and profit from the production and sale of 50 units of the product

c) The marginal cost, revenue, and profit when 50 units are produced and sold

Solution

a) Total profit $= P(x) = R(x) - C(x)$
$$= x^3 - 12x^2 + 40x + 10 - (62x^2 + 27{,}500)$$
$$= x^3 - 74x^2 + 40x - 27{,}490$$

b) $C(50) = 62 \cdot 50^2 + 27{,}500 = \$182{,}500$ (the total cost of producing the first 50 units);
$R(50) = 50^3 - 12 \cdot 50^2 + 40 \cdot 50 + 10 = \$97{,}010$ (the total revenue from the sale of the first 50 units);

$$P(50) = R(50) - C(50)$$
$$= \$97{,}010 - \$182{,}500 \qquad \text{We could also use } P(x) \text{ from part (a).}$$
$$= -\$85{,}490 \qquad \text{There is a } \textit{loss} \text{ of \$85,490 when 50 units are produced and sold.}$$

c) $C'(x) = 124x$, so $C'(50) = 124 \cdot 50 = \$6200$. Once 50 units have been made, the approximate cost of the 51st unit (marginal cost) is \$6200.

$$R'(x) = 3x^2 - 24x + 40, \text{ so } R'(50) = 3 \cdot 50^2 - 24 \cdot 50 + 40 = \$6340.$$

Once 50 units have been sold, the approximate revenue from the 51st unit (marginal revenue) is \$6340.

$$P'(x) = 3x^2 - 148x + 40, \text{ so } P'(50) = 3 \cdot 50^2 - 148 \cdot 50 + 40 = \$140.$$

Once 50 units have been produced and sold, the approximate profit from the sale of the 51st item (marginal profit) is \$140. ✦

TECHNOLOGY CONNECTION

Business: Marginal Revenue, Cost, and Profit

EXERCISE

1. Using the viewing window [0, 100, 0, 2000], graph these total-revenue and total-cost functions:

$$R(x) = 50x - 0.5x^2$$

and

$$C(x) = 10x + 3.$$

Then find $P(x)$ and graph it using the same viewing window. Find $R'(x)$, $C'(x)$, and $P'(x)$, and graph them using [0, 60, 0, 60]. Then find $R(40)$, $C(40)$, $P(40)$, $R'(40)$, $C'(40)$, and $P'(40)$. Which marginal function is constant?

Often, in business, a formula for $C(x)$, $R(x)$, or $P(x)$ does not exist. In such cases, marginals are especially useful—by knowing $C(a)$ and $C'(a)$, we can make a reasonable prediction of $C(a + 1)$. In Example 1, formulas *do* exist, so it is possible to see how accurate our predictions were. We check $C(51) - C(50)$ and leave the checks of $R(51) - R(50)$ and $P(51) - P(50)$ for you (see the Technology Connection on p. 277):

$$C(51) - C(50) = 62 \cdot 51^2 + 27{,}500 - (62 \cdot 50^2 + 27{,}500)$$
$$= 6262,$$

whereas $\qquad C'(50) = 6200.$

In this case, $C'(50)$ provides an approximation of $C(51)$ that is within 1% of the actual value.

TECHNOLOGY CONNECTION

To check the accuracy of $R'(50)$ as an estimate of $R(51) - R(50)$, let $y_1 = x^3 - 12x^2 + 40x + 10$, $y_2 = y_1(x + 1) - y_1(x)$, and $y_3 = \text{nDeriv}(y_1, x, x)$. By using TABLE with Indpnt: Ask, we can display a table in which y_2 (the difference between $y_1(x + 1)$ and $y_1(x)$) can be compared with $y_1'(x)$.

X	Y2	Y3
40	3989	3880
48	5933	5800
50	6479	6340

X=

EXERCISE

1. Create a table to check the effectiveness of using $P'(50)$ to approximate $P(51) - P(50)$.

Note that marginal cost is different from *average* cost:

$$\text{Average cost per unit for 50 units} = \frac{C(50)}{50} \begin{array}{l}\longleftarrow \text{Total cost of 50 units} \\ \longleftarrow \text{The number of units, 50}\end{array}$$

$$= \frac{182{,}500}{50} = \$3650,$$

whereas

$$\text{Marginal cost when 50 units are produced} = \$6200$$
$$\approx \text{cost of the 51st unit.}$$

Differentials and Delta Notation

Just as the marginal cost $C'(x_0)$ can be used to estimate $C(x_0 + 1)$, the value of the derivative of any continuous function, $f'(x_0)$, can be used to estimate values of $f(x)$ for x-values near x_0. Before we do so, however, we need to develop some notation.

Recall the difference quotient

$$\frac{f(x + h) - f(x)}{h},$$

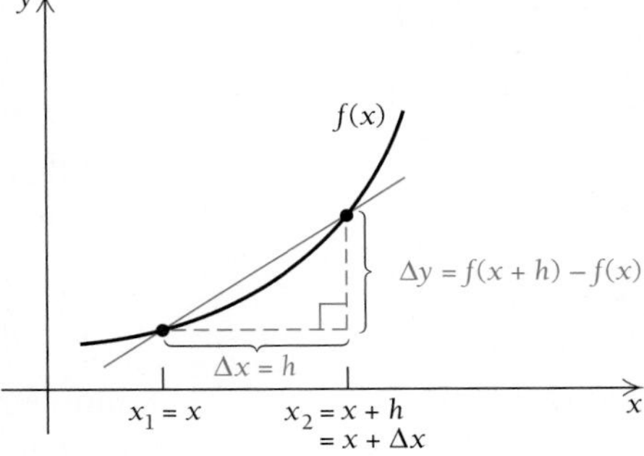

illustrated in this graph. The difference quotient is used to define the derivative of a function at x. The number h is considered to be a *change* in x. Another notation for such a change is Δx, read "delta x" and called **delta notation.** The expression Δx is *not* the product of Δ and x, but is its own entity; that is, it is a new type of variable that represents the *change* in the value of x from a *first* value to a *second*. Thus,

$$\Delta x = (x + h) - x = h.$$

If subscripts are used for the first and second values of x, we have

$$\Delta x = x_2 - x_1, \quad \text{or} \quad x_2 = x_1 + \Delta x.$$

Note that the value of Δx can be positive or negative. For example,

if $x_1 = 4$ and $\Delta x = 0.7$, then $x_2 = 4.7$,

and if $x_1 = 4$ and $\Delta x = -0.7$, then $x_2 = 3.3$.

We generally omit the subscripts and use x and $x + \Delta x$. Now suppose that we have a function given by $y = f(x)$. A change in x from x to $x + \Delta x$ yields a change in y from $f(x)$ to $f(x + \Delta x)$. The change in y is given by

$$\Delta y = f(x + \Delta x) - f(x).$$

EXAMPLE 2 For $y = x^2$, $x = 4$, and $\Delta x = 0.1$, find Δy.

Solution We have

$$\begin{aligned}\Delta y &= (4 + 0.1)^2 - 4^2 \\ &= (4.1)^2 - 4^2 = 16.81 - 16 = 0.81.\end{aligned}$$

✦

EXAMPLE 3 For $y = x^3$, $x = 2$, and $\Delta x = -0.1$, find Δy.

Solution We have

$$\begin{aligned}\Delta y &= [2 + (-0.1)]^3 - 2^3 \\ &= (1.9)^3 - 2^3 = 6.859 - 8 = -1.141.\end{aligned}$$

✦

Let's now use calculus to predict function values. If delta notation is used, the difference quotient

$$\frac{f(x + h) - f(x)}{h}$$

becomes

$$\frac{f(x + \Delta x) - f(x)}{\Delta x} = \frac{\Delta y}{\Delta x}.$$

We can then express the derivative as

$$\frac{dy}{dx} = \lim_{\Delta x \to 0} \frac{\Delta y}{\Delta x}.$$

Note that the delta notation resembles Leibniz notation (see Section 1.5).

For values of Δx close to 0, we have the approximation

$$\frac{dy}{dx} \approx \frac{\Delta y}{\Delta x}, \quad \text{or} \quad f'(x) \approx \frac{\Delta y}{\Delta x}.$$

Multiplying both sides of the second expression by Δx gives us

$$\Delta y \approx f'(x)\,\Delta x.$$

We can see this in the graph at right.

From this graph, it seems reasonable to assume that, for small values of Δx, the y-values on the tangent line can be used to estimate function values on the curve.

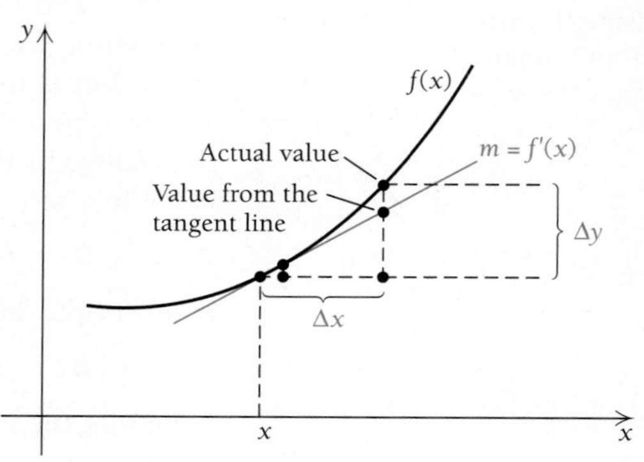

For f, a continuous, differentiable function, and small Δx,

$$f'(x) \approx \frac{\Delta y}{\Delta x} \quad \text{and} \quad \Delta y \approx f'(x) \cdot \Delta x.$$

Let's illustrate this idea by considering the square-root function, $f(x) = \sqrt{x}$. We know how to approximate $\sqrt{27}$ using a calculator. But suppose we didn't. We could begin with $\sqrt{25}$ and use as a change in input $\Delta x = 2$. We would use the corresponding change in y, that is, $\Delta y \approx f'(x)\,\Delta x$, to estimate $\sqrt{27}$.

EXAMPLE 4 Approximate $\sqrt{27}$ using $\Delta y \approx f'(x)\,\Delta x$.

Solution We first think of the number closest to 27 that is a perfect square. This is 25. What we will do is approximate how $y = \sqrt{x}$ changes when x changes from 25 to 27.

From the box above, we have

$$\left.\begin{aligned}\Delta y &\approx f'(x)\cdot\Delta x\\ &\approx \tfrac{1}{2}x^{-1/2}\cdot\Delta x\end{aligned}\right\}\quad \text{Using } y = \sqrt{x} = x^{1/2} \text{ as } f(x)$$

We are interested in Δy as x changes from 25 to 27, so

$$\begin{aligned}\Delta y &\approx \frac{1}{2\sqrt{25}}\cdot 2 \qquad \text{Replacing } x \text{ with 25 and } \Delta x \text{ with 2}\\ &\approx \frac{1}{5} = 0.2.\end{aligned}$$

We can now approximate $\sqrt{27}$:

$$\begin{aligned}\sqrt{27} &= \sqrt{25} + \Delta y\\ &= 5 + \Delta y\\ &\approx 5 + 0.2\\ &\approx 5.2.\end{aligned}$$

To five decimal places, $\sqrt{27} = 5.19615$. Thus, our approximation is fairly accurate. ✦

Up to now we have not defined the symbols dy and dx as separate entities, and we have treated dy/dx as one symbol. We now define dy and dx. These symbols are called **differentials.**

DEFINITION

For $y = f(x)$, we define

dx, called the **differential of x,** by $dx = \Delta x$

and

dy, called the **differential of y,** by $dy = f'(x)\,dx$.

We can illustrate dx and dy as shown at the right. Note that $dx = \Delta x$, but $dy \neq \Delta y$, though $dy \approx \Delta y$, for small values of dx.

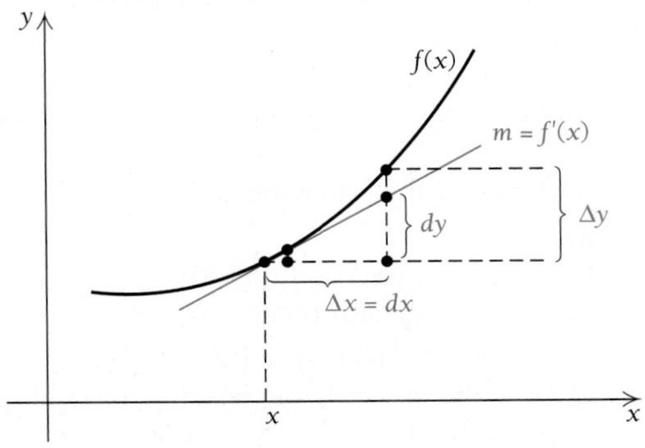

EXAMPLE 5 For $y = x(4 - x)^3$:

a) Find dy.

b) Find dy when $x = 5$ and $dx = 0.2$.

Solution

a) First, we find dy/dx:

$$\frac{dy}{dx} = -x[3(4 - x)^2] + (4 - x)^3$$ Using the Product Rule and the Extended Power Rule

$$= (4 - x)^2[-3x + (4 - x)]$$

$$= (4 - x)^2[-4x + 4]$$ Simplifying

$$= -4(4 - x)^2(x - 1).$$

Then

$$dy = -4(4 - x)^2(x - 1)\, dx.$$

Note that the expression for dy contains two variables, x and dx.

b) When $x = 5$ and $dx = 0.2$,

$$dy = -4(4 - 5)^2(5 - 1)0.2 = -4(-1)^2(4)(0.2) = -3.2.$$ ✦

Historically, differentials were quite valuable when used to make approximations. However, with the advent of computers and graphing calculators, such use has diminished considerably. The use of marginals remains important in the study of business and economics.

Exercise Set 2.6

APPLICATIONS

Business and Economics

1. Marginal revenue, cost, and profit. Let $R(x)$, $C(x)$, and $P(x)$ be, respectively, the revenue, cost, and profit, in dollars, from the production and sale of x items. If

$$R(x) = 5x \quad \text{and} \quad C(x) = 0.001x^2 + 1.2x + 60,$$

find each of the following.

a) $P(x)$

b) $R(100)$, $C(100)$, and $P(100)$

c) $R'(x)$, $C'(x)$, and $P'(x)$

d) $R'(100)$, $C'(100)$, and $P'(100)$

TW **e)** Describe in words the meaning of each quantity in parts (b) and (d).

2. Marginal revenue, cost, and profit. Let $R(x)$, $C(x)$, and $P(x)$ be, respectively, the revenue, cost, and profit, in dollars, from the production and sale of x items. If

$$R(x) = 50x - 0.5x^2 \quad \text{and} \quad C(x) = 4x + 10,$$

find each of the following.

a) $P(x)$

b) $R(20)$, $C(20)$, and $P(20)$

c) $R'(x)$, $C'(x)$, and $P'(x)$

d) $R'(20)$, $C'(20)$, and $P'(20)$

3. Marginal cost. Suppose that the monthly cost, in dollars, of producing x chairs is

$$C(x) = 0.001x^3 + 0.07x^2 + 19x + 700,$$

and currently 25 chairs are produced monthly.

a) What is the current monthly cost?

b) What would be the additional cost of increasing production to 26 chairs monthly?

c) What is the marginal cost when $x = 25$?
d) Use marginal cost to estimate the difference in cost between producing 25 and 27 chairs per month.
e) Use the answer from part (d) to predict $C(27)$.

4. Marginal cost. Suppose that the daily cost, in dollars, of producing x radios is

$$C(x) = 0.002x^3 + 0.1x^2 + 42x + 300,$$

and currently 40 radios are produced daily.

a) What is the current daily cost?
b) What would be the additional daily cost of increasing production to 41 radios daily?
c) What is the marginal cost when $x = 40$?
d) Use marginal cost to estimate the daily cost of increasing production to 42 radios daily.

5. Marginal revenue. Pierce Manufacturing determines that the daily revenue, in dollars, from the sale of x lawn chairs is

$$R(x) = 0.005x^3 + 0.01x^2 + 0.5x.$$

Currently, Pierce sells 70 lawn chairs daily.

a) What is the current daily revenue?
b) How much would revenue increase if 73 lawn chairs were sold each day?
c) What is the marginal revenue when 70 lawn chairs are sold daily?
d) Use the answer from part (c) to estimate $R(71)$, $R(72)$, and $R(73)$.

6. Marginal profit. For Sunshine Motors, the weekly profit, in dollars, of selling x cars is

$$P(x) = -0.006x^3 - 0.2x^2 + 900x - 1200,$$

and currently 60 cars are sold weekly.

a) What is the current weekly profit?
b) How much profit would be lost if the dealership were able to sell only 59 cars weekly?
c) What is the marginal profit when $x = 60$?
d) Use marginal profit to estimate the weekly profit if sales increase to 61 cars weekly.

7. Marginal profit. Crawford Computing finds that its weekly profit, in dollars, from the production and sale of x laptop computers is

$$P(x) = -0.004x^3 - 0.3x^2 + 600x - 800.$$

Currently Crawford builds and sells 9 laptops weekly.

a) What is the current weekly profit?
b) How much profit would be lost if production and sales dropped to 8 laptops weekly?
c) What is the marginal profit when $x = 9$?
d) Use the answers from parts (a)–(c) to estimate the profit resulting from the production and sale of 10 laptops weekly.

8. Marginal revenue. Solano Carriers finds that its monthly revenue, in dollars, from the sale of x carry-on suitcases is

$$R(x) = 0.007x^3 - 0.5x^2 + 150x.$$

Currently Solano is selling 26 carry-on suitcases monthly.

a) What is the current monthly revenue?
b) How much would revenue increase if sales increased from 26 to 28 suitcases?
c) What is the marginal revenue when 26 suitcases are sold?
d) Use the answers from parts (a)–(c) to estimate the revenue resulting from selling 27 suitcases per month.

9. Sales. Let $N(x)$ be the number of computers sold annually when the price is x dollars per computer. Explain in words what occurs if $N(1000) = 500{,}000$ and $N'(1000) = -100$.

10. Sales. Estimate the number of computers sold in Exercise 9 if the price is raised to \$1025.

For Exercises 11–16, assume that $C(x)$ and $R(x)$ are in dollars and x is the number of units produced and sold.

11. For the total-cost function

$$C(x) = 0.01x^2 + 0.6x + 30,$$

find ΔC and $C'(x)$ when $x = 70$ and $\Delta x = 1$.

12. For the total-cost function

$$C(x) = 0.01x^2 + 1.6x + 100,$$

find ΔC and $C'(x)$ when $x = 80$ and $\Delta x = 1$.

13. For the total-revenue function

$$R(x) = 2x,$$

find ΔR and $R'(x)$ when $x = 70$ and $\Delta x = 1$.

14. For the total-revenue function

$$R(x) = 3x,$$

find ΔR and $R'(x)$ when $x = 80$ and $\Delta x = 1$.

15. a) Using $C(x)$ from Exercise 11 and $R(x)$ from Exercise 13, find the total profit, $P(x)$.
b) Find ΔP and $P'(x)$ when $x = 70$ and $\Delta x = 1$.

16. a) Using $C(x)$ from Exercise 12 and $R(x)$ from Exercise 14, find the total profit, $P(x)$.
b) Find ΔP and $P'(x)$ when $x = 80$ and $\Delta x = 1$.

17. Marginal demand. The demand, D, for a new rollerball pen is given by

$$D = 0.007p^3 - 0.5p^2 + 150p,$$

where p is the price in dollars.

a) Find the rate of change of quantity with respect to price, dD/dp.

b) How many units will consumers want to buy when the price is \$25 per unit?

TW **c)** Find the rate of change at $p = 25$, and interpret this result.

TW **d)** Would you expect dD/dp to be positive or negative? Why?

18. Marginal productivity. An employee's monthly productivity, M, in number of units produced, is found to be a function of the number of years of service, t. For a certain product, the productivity function is given by

$$M(t) = -2t^2 + 100t + 180.$$

a) Find the productivity of an employee after 5 yr, 10 yr, 25 yr, and 45 yr of service.

b) Find the marginal productivity.

TW **c)** Find the marginal productivity at $t = 5$; $t = 10$; $t = 25$; $t = 45$; and interpret the results.

TW **d)** Explain how the employee's marginal productivity might be related to experience and to age.

19. Average cost. The average cost for a company to produce x units of a product is given by the function

$$A(x) = \frac{13x + 100}{x}.$$

Use $A'(x)$ to estimate the change in average cost as production goes from 100 units to 101 units.

20. Supply. A supply function for a certain product is given by

$$S(p) = 0.08p^3 + 2p^2 + 10p + 11,$$

where $S(p)$ is the number of items produced when the price is p dollars. Use $S'(p)$ to estimate how many more units a producer will supply when the price changes from \$18.00 per unit to \$18.20 per unit.

21. Gross domestic product. The U.S. gross domestic product, in billions of current dollars, may be modeled by the function

$$P(x) = 567 + x(36x^{0.6} - 104),$$

where x is the number of years since 1960. (*Source*: U.S. Bureau for Economic Analysis.) Use $P'(x)$ to estimate how much the gross domestic product will increase from 2009 to 2010.

22. Advertising. Norris Inc. finds that it sells N units of a product after spending x thousands of dollars on advertising, where

$$N(x) = -x^2 + 300x + 6.$$

Use $N'(x)$ to estimate how many more units Norris will sell by increasing its advertising expenditure from \$100,000 to \$101,000.

Marginal tax rate. *Businesses and individuals are frequently concerned about their marginal tax rate, or the rate at which the next dollar earned is taxed. In progressive taxation, the 80,001st dollar earned is taxed at a higher rate than the 25,001st dollar earned and at a lower rate than the 140,001st dollar earned. Use the graph below, showing the marginal tax rate for 2005, to answer Exercises 23–26.*

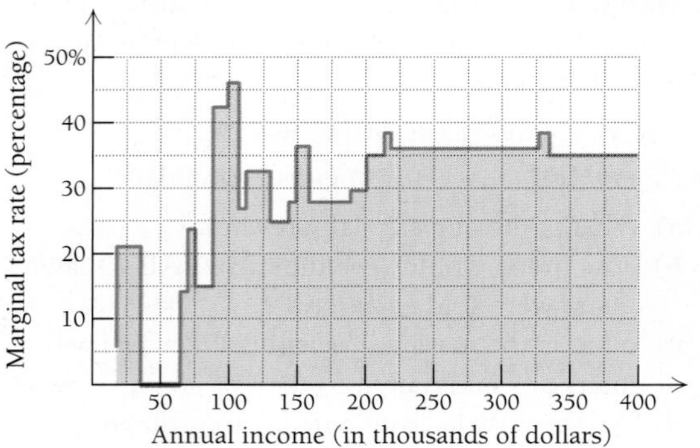

(*Source*: "Towards Fundamental Tax Reform" by Alan Auerbach and Kevin Hassett, *New York Times*, 5/5/05, p. C2.)

TW **23.** Was the taxation in 2005 progressive? Why or why not?

24. Marcy and Tyrone work for the same marketing agency. Because she is not yet a partner, Marcy's year-end income is approximately \$95,000; Tyrone's year-end income is approximately \$150,000. Suppose that one of them is to receive another \$5000 in income for the year. Which one would keep more of that \$5000 after taxes? Why?

25. Alan earns \$25,000 per year and is considering a second job that would earn him another \$2000 annually. At what rate will his tax liability (the amount he must pay in taxes) change if he takes the extra job? Express your answer in tax dollars paid per dollar earned.

26. Iris earns \$50,000 per year and is considering extra work that would earn her an extra \$3000 annually. At what rate will her tax liability grow if she takes the extra work (see Exercise 25)?

Find Δy and $f'(x)\,\Delta x$. Round to four and two decimal places, respectively.

27. For $y = f(x) = x^2$, $x = 2$, and $\Delta x = 0.01$

28. For $y = f(x) = x^3$, $x = 2$, and $\Delta x = 0.01$

29. For $y = f(x) = x + x^2$, $x = 3$, and $\Delta x = 0.04$

30. For $y = f(x) = x - x^2$, $x = 3$, and $\Delta x = 0.02$

31. For $y = f(x) = 1/x^2$, $x = 1$, and $\Delta x = 0.5$

32. For $y = f(x) = 1/x$, $x = 1$, and $\Delta x = 0.2$

33. For $y = f(x) = 3x - 1$, $x = 4$, and $\Delta x = 2$

34. For $y = f(x) = 2x - 3$, $x = 8$, and $\Delta x = 0.5$

Use $\Delta y \approx f'(x)\,\Delta x$ to find a decimal approximation of each radical expression. Round to three decimal places.

35. $\sqrt{26}$

36. $\sqrt{10}$

37. $\sqrt{102}$

38. $\sqrt{103}$

39. $\sqrt[3]{1005}$

40. $\sqrt[3]{28}$

Find dy.

41. $y = \sqrt{x + 1}$

42. $y = \sqrt{3x - 2}$

43. $y = (2x^3 + 1)^{3/2}$

44. $y = x^3(2x + 5)^2$

45. $y = \sqrt[5]{x + 27}$

46. $y = \dfrac{x^3 + x + 2}{x^2 + 3}$

47. $y = x^4 - 2x^3 + 5x^2 + 3x - 4$

48. $y = (7 - x)^8$

49. In Exercise 47, find dy when $x = 2$ and $dx = 0.1$.

50. In Exercise 48, find dy when $x = 1$ and $dx = 0.01$.

51. For $y = (3x - 10)^5$, find dy when $x = 4$ and $dx = 0.03$.

52. For $y = x^5 - 2x^3 - 7x$, find dy when $x = 3$ and $dx = 0.02$.

53. For $f(x) = x^4 - x^2 + 8$, use a differential to approximate $f(5.1)$.

54. For $f(x) = x^3 - 5x + 9$, use a differential to approximate $f(3.2)$.

SYNTHESIS

Life and Physical Sciences

55. Body surface area. Certain chemotherapy dosages depend on a patient's surface area. According to the Gehan and George model,

$$S = 0.02235h^{0.42246}w^{0.51456},$$

where h is the patient's height in centimeters, w is his or her weight in kilograms, and S is the approximation to his or her surface area in square meters. (*Source*: www.halls.md.) Joanne is 160 cm tall and weighs 60 kg. Use a differential to estimate how much her surface area changes after her weight decreases by 1 kg.

56. Healing wound. The circular area of a healing wound is given by $A = \pi r^2$, where r is the radius, in centimeters. By approximately how much does the area decrease when the radius is decreased from 2 cm to 1.9 cm? Use 3.14 for π.

57. Medical dosage. The function

$$N(t) = \frac{0.8t + 1000}{5t + 4}$$

gives the bodily concentration $N(t)$, in parts per million, of a dosage of medication after time t, in hours. Use differentials to determine whether the concentration changes more from 1.0 hr to 1.1 hr or from 2.8 hr to 2.9 hr.

General Interest

58. Major League ticket prices. The average ticket price of a major league baseball game can be modeled by the function

$$p(x) = 0.09x^2 - 0.19x + 9.41,$$

where x is the number of years after 1990. (*Source*: Major League Baseball.) Use differentials to predict whether ticket price will increase more between 2010 and 2012 or between 2030 and 2031.

59. Suppose that a rope surrounds the earth at the equator. The rope is lengthened by 10 ft. By about how much is the rope raised above the earth?

Business and Economics

60. Marginal average cost. In Section R.6, we defined the average cost of producing x units of a product in terms of the total cost $C(x)$ by $A(x) = C(x)/x$. Find a general expression for *marginal average cost*, $A'(x)$.

Marginal revenue. *In each of Exercises 61–65, a demand function, $p = D(x)$, expresses price, in dollars, as a function of the number of items produced and sold. Find the marginal revenue.*

61. $p = 100 - \sqrt{x}$

62. $p = 400 - x$

63. $p = 500 - x$

64. $p = \dfrac{4000}{x} + 3$

65. $p = \dfrac{3000}{x} + 5$

TW **66.** Look up "differential" in a book or Web site devoted to math history. In a short paragraph, describe your findings.

TW **67.** Explain the uses of the differential.

2.7 Implicit Differentiation and Related Rates*

OBJECTIVES

- Differentiate implicitly.
- Solve related-rate problems.

Implicit Differentiation

Consider the equation

$$y^3 = x.$$

This equation *implies* that y is a function of x, for if we solve for y, we get

$$y = \sqrt[3]{x} = x^{1/3}.$$

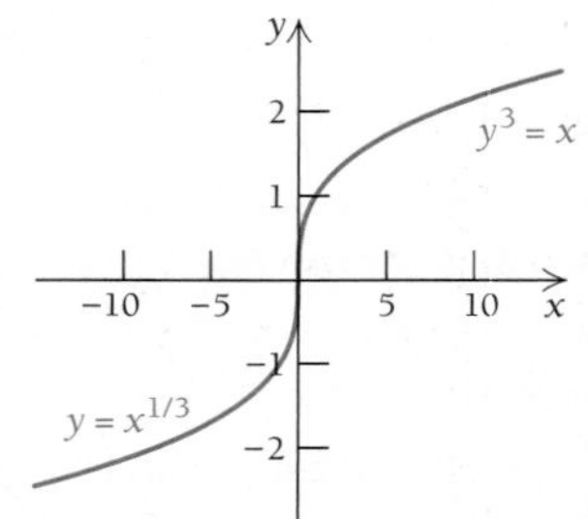

We know from our earlier work that

$$\frac{dy}{dx} = \frac{1}{3}x^{-2/3}. \tag{1}$$

A method known as **implicit differentiation** allows us to find dy/dx *without* solving for y. To do so, we use the Chain Rule, treating y as a function of x, and differentiate both sides of

$$y^3 = x$$

with respect to x:

$$\frac{d}{dx}y^3 = \frac{d}{dx}x.$$

The derivative on the left side is found using the Extended Power Rule:

$$3y^2\frac{dy}{dx} = 1.$$

Finally, we solve for dy/dx by dividing both sides by $3y^2$:

$$\frac{dy}{dx} = \frac{1}{3y^2}, \quad \text{or} \quad \frac{1}{3}y^{-2}.$$

We can show that this indeed gives us the same answer as equation (1) by replacing y with $x^{1/3}$:

$$\frac{dy}{dx} = \frac{1}{3}y^{-2} = \frac{1}{3}(x^{1/3})^{-2} = \frac{1}{3}x^{-2/3}.$$

Often, it is difficult or impossible to solve for y and to express dy/dx solely in terms of x. For example, the equation

$$y^3 + x^2y^5 - x^4 = 27$$

determines y as a function of x, but it would be difficult to solve for y. We can nevertheless find a formula for the derivative of y *without* solving for y. To do so usually

*This section can be omitted without loss of continuity.

involves computing $\frac{d}{dx}y^n$ for various integers n, and hence involves the Extended Power Rule in the form

$$\frac{d}{dx}y^n = ny^{n-1} \cdot \frac{dy}{dx}.$$

EXAMPLE 1 For $y^3 + x^2y^5 - x^4 = 27$:

a) Find dy/dx using implicit differentiation.

b) Find the slope of the tangent line to the curve at the point $(0, 3)$.

Solution

a) We differentiate the term x^2y^5 using the Product Rule. Because y is a function of x, it is critical that dy/dx is included as a factor in the result any time a term involving y is differentiated. When an expression involving just x is differentiated, there is no factor dy/dx.

$$\frac{d}{dx}(y^3 + x^2y^5 - x^4) = \frac{d}{dx}(27) \qquad \text{Differentiating both sides with respect to } x$$

$$\frac{d}{dx}y^3 + \frac{d}{dx}x^2y^5 - \frac{d}{dx}x^4 = 0$$

$$3y^2 \cdot \frac{dy}{dx} + x^2 \cdot 5y^4 \cdot \frac{dy}{dx} + y^5 \cdot 2x - 4x^3 = 0. \qquad \text{Using the Extended Power Rule and the Product Rule}$$

We next isolate those terms with dy/dx as a factor on one side:

$$3y^2 \cdot \frac{dy}{dx} + 5x^2y^4 \cdot \frac{dy}{dx} = 4x^3 - 2xy^5 \qquad \text{Adding } 4x^3 - 2xy^5 \text{ to both sides}$$

$$(3y^2 + 5x^2y^4)\frac{dy}{dx} = 4x^3 - 2xy^5 \qquad \text{Factoring out } dy/dx$$

$$\frac{dy}{dx} = \frac{4x^3 - 2xy^5}{3y^2 + 5x^2y^4}. \qquad \text{Solving for } dy/dx\text{. Leave the answer in terms of } x \text{ and } y.$$

b) To find the slope of the tangent line to the curve at $(0, 3)$, we replace x with 0 and y with 3:

$$\frac{dy}{dx} = \frac{4 \cdot 0^3 - 2 \cdot 0 \cdot 3^5}{3 \cdot 3^2 + 5 \cdot 0^2 \cdot 3^4} = 0.$$

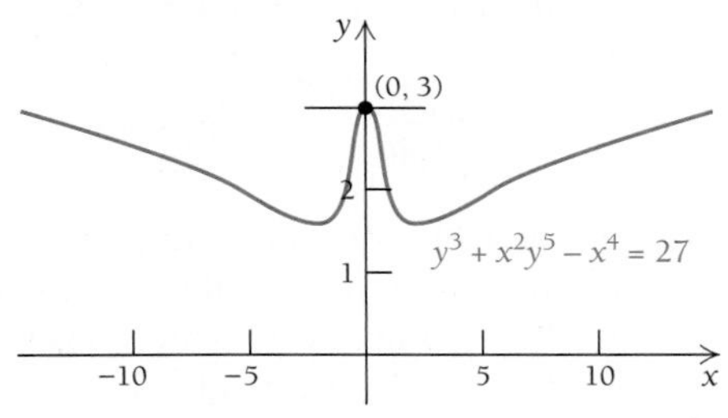

✦

The steps in Example 1 are typical of those used when differentiating implicitly.

To differentiate implicitly:

a) Differentiate both sides of the equation with respect to x (or whatever variable you are differentiating with respect to).
b) Apply the rules for differentiation (the Power, Product, Quotient, and Chain Rules) as necessary. Any time an expression involving y is differentiated, dy/dx will be a factor in the result.
c) Isolate all terms with dy/dx as a factor on one side of the equation.
d) If necessary, factor out dy/dx.
e) If necessary, divide both sides of the equation to isolate dy/dx.

The demand function for a product (see Sections R.5 and 1.7) is often given implicitly.

EXAMPLE 2 For the demand equation $x = \sqrt{200 - p^3}$, differentiate implicitly to find dp/dx.

Solution

$$\frac{d}{dx}x = \frac{d}{dx}\sqrt{200 - p^3}$$

$$1 = \frac{1}{2}(200 - p^3)^{-1/2} \cdot (-3p^2) \cdot \frac{dp}{dx}$$ Using the Extended Power Rule twice

$$1 = \frac{-3p^2}{2\sqrt{200 - p^3}} \cdot \frac{dp}{dx}$$

$$\frac{2\sqrt{200 - p^3}}{-3p^2} = \frac{dp}{dx}$$ ✦

Related Rates

Suppose that y is a function of x, say

$$y = f(x),$$

and x is a function of time, t. Since y depends on x and x depends on t, it follows that y depends on t. The Chain Rule gives the following:

$$\frac{dy}{dt} = \frac{dy}{dx} \cdot \frac{dx}{dt}.$$

Thus, the rate of change of y is *related* to the rate of change of x. Let's see how this comes up in problems. It helps to keep in mind that any variable can be thought of as a function of time t, even though a specific expression in terms of t may not be given or its rate of change with respect to t may be 0.

EXAMPLE 3 Business: Service Area. A restaurant supplier services the restaurants in a circular area in such a way that the radius r is increasing at the rate of 2 mi per year at the moment when $r = 5$ mi. At that moment, how fast is the area increasing?

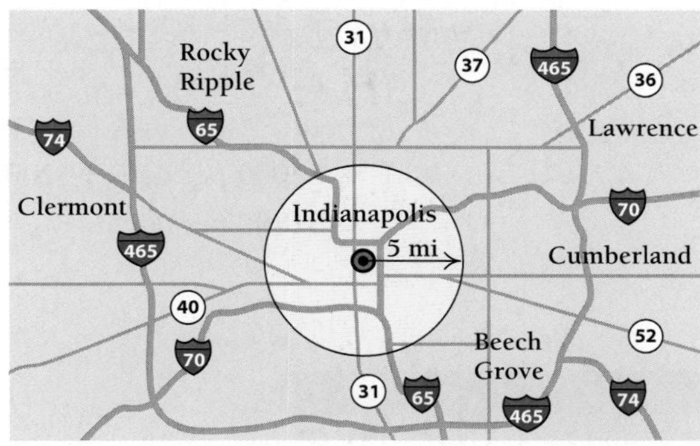

Solution The area A and the radius r are always related by the equation for the area of a circle:

$$A = \pi r^2.$$

We take the derivative of both sides with respect to t:

$$\frac{dA}{dt} = 2\pi r \cdot \frac{dr}{dt}.$$ The factor dr/dt results from the Chain Rule and the fact that r is assumed to be a function of t.

At the moment in question, $dr/dt = 2$ mi/yr and $r = 5$ mi, so

$$\frac{dA}{dt} = 2\pi(5 \text{ mi})\left(2\frac{\text{mi}}{\text{yr}}\right)$$
$$= 20\pi \frac{\text{mi}^2}{\text{yr}} \approx 63 \text{ square miles per year.}$$ ✦

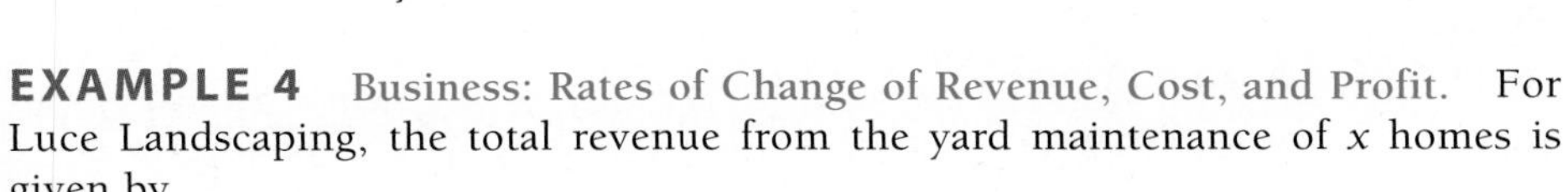

EXAMPLE 4 Business: Rates of Change of Revenue, Cost, and Profit. For Luce Landscaping, the total revenue from the yard maintenance of x homes is given by

$$R(x) = 1000x - x^2,$$

and the total cost is given by

$$C(x) = 3000 + 20x.$$

Suppose that Luce is adding 10 homes per day at the moment when the 400th customer is signed. At that moment, what is the rate of change of (a) total revenue, (b) total cost, and (c) total profit?

Solution

a) $\frac{dR}{dt} = 1000 \cdot \frac{dx}{dt} - 2x \cdot \frac{dx}{dt}$ Differentiating both sides with respect to time

$= 1000 \cdot 10 - 2(400)10$ Substituting 10 for dx/dt and 400 for x

$= \$2000$ per day

b) $\frac{dC}{dt} = 20 \cdot \frac{dx}{dt}$ Differentiating both sides with respect to time

$= 20(10)$

$= \$200$ per day

c) Since $P = R - C$,

$$\frac{dP}{dt} = \frac{dR}{dt} - \frac{dC}{dt}$$

$= \$2000$ per day $- \$200$ per day

$= \$1800$ per day.

Exercise Set 2.7

Differentiate implicitly to find dy/dx. Then find the slope of the curve at the given point.

1. $x^3 + 2y^3 = 6;\quad (2, -1)$

2. $3x^3 - y^2 = 8;\quad (2, 4)$

3. $2x^2 - 3y^3 = 5;\quad (-2, 1)$

4. $2x^3 + 4y^2 = -12;\quad (-2, -1)$

5. $x^2 - y^2 = 1;\quad (\sqrt{3}, \sqrt{2})$

6. $x^2 + y^2 = 1;\quad \left(\frac{1}{2}, \frac{\sqrt{3}}{2}\right)$

7. $3x^2y^4 = 12;\quad (2, -1)$

8. $2x^3y^2 = -18;\quad (-1, 3)$

9. $x^3 - x^2y^2 = -9;\quad (3, -2)$

10. $x^4 - x^2y^3 = 12;\quad (-2, 1)$

11. $xy - x + 2y = 3;\quad \left(-5, \frac{2}{3}\right)$

12. $xy + y^2 - 2x = 0;\quad (1, -2)$

13. $x^2y - 2x^3 - y^3 + 1 = 0;\quad (2, -3)$

14. $4x^3 - y^4 - 3y + 5x + 1 = 0;\quad (1, -2)$

Differentiate implicitly to find dy/dx.

15. $2xy + 3 = 0$

16. $x^2 + 2xy = 3y^2$

17. $x^2 - y^2 = 16$

18. $x^2 + y^2 = 25$

19. $y^5 = x^3$

20. $y^3 = x^5$

21. $x^2y^3 + x^3y^4 = 11$

22. $x^3y^2 - x^5y^3 = -19$

For each demand equation in Exercises 23–30, differentiate implicitly to find dp/dx.

23. $p^3 + p - 3x = 50$

24. $p^2 + p + 2x = 40$

25. $xp^3 = 24$

26. $x^3p^2 = 108$

27. $\frac{xp}{x + p} = 2$ (*Hint*: Clear the fraction first.)

28. $\frac{x^2p + xp + 1}{2x + p} = 1$ (*Hint*: Clear the fraction first.)

29. $(p + 4)(x + 3) = 48$

30. $1000 - 300p + 25p^2 = x$

31. Two variable quantities A and B are found to be related by the equation

$$A^3 + B^3 = 9.$$

What is the rate of change dA/dt at the moment when $A = 2$ and $dB/dt = 3$?

32. Two variable quantities G and H, nonnegative, are found to be related by the equation

$$G^2 + H^2 = 25.$$

What is the rate of change dH/dt when $dG/dt = 3$ and $G = 0$? $G = 1$? $G = 3$?

APPLICATIONS

Business and Economics

Rates of change of total revenue, cost, and profit. *In Exercises 33–36, find the rates of change of total revenue, cost, and profit with respect to time. Assume that $R(x)$ and $C(x)$ are in dollars.*

33. $R(x) = 50x - 0.5x^2$,
$C(x) = 4x + 10$,
when $x = 30$ and $dx/dt = 20$ units per day

34. $R(x) = 50x - 0.5x^2$,
$C(x) = 10x + 3$,
when $x = 10$ and $dx/dt = 5$ units per day

35. $R(x) = 2x$,
$C(x) = 0.01x^2 + 0.6x + 30$,
when $x = 20$ and $dx/dt = 8$ units per day

36. $R(x) = 280x - 0.4x^2$,
$C(x) = 5000 + 0.6x^2$,
when $x = 200$ and $dx/dt = 300$ units per day

37. Change of sales. Suppose that the price p, in dollars, and number of sales, x, of a certain item follow the equation

$$5p + 4x + 2px = 60.$$

Suppose also that p and x are both functions of time, measured in days. Find the rate at which x is changing when

$$x = 3, p = 5, \text{ and } \frac{dp}{dt} = 1.5.$$

38. Change of revenue. For x and p as described in Exercise 37, find the rate at which the total revenue $R = xp$ is changing when

$$x = 3, p = 5, \text{ and } \frac{dp}{dt} = 1.5.$$

Life and Natural Sciences

39. Rate of change of the Arctic ice cap. In a trend that scientists attribute, at least in part, to global warming, the floating cap of sea ice on the Arctic Ocean has been shrinking since 1950. The ice cap always shrinks in summer and grows in winter. Average minimum size of the ice cap, in square miles, can be approximated by

$$A = \pi r^2.$$

In 2005, the radius of the ice cap was approximately 808 mi and was shrinking at a rate of approximately 4.3 mi/yr. (*Source*: Based on data for 2005 and information from gsfc.nasa.gov and the *New York Times*, 9/29/05.) How fast was the area changing at that time?

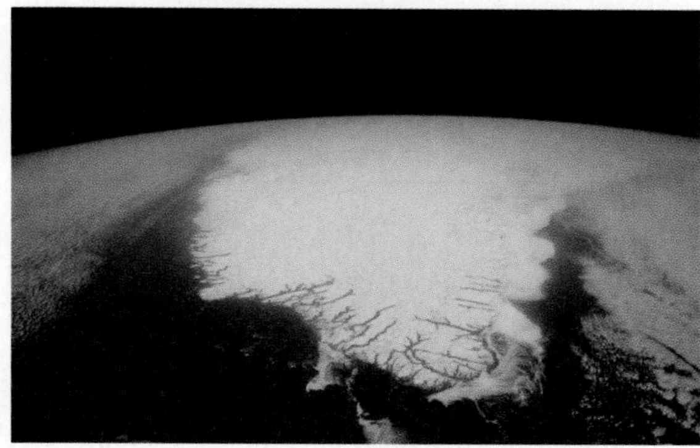

40. Rate of change of a healing wound. The area of a healing wound is given by

$$A = \pi r^2.$$

The radius is decreasing at the rate of 1 millimeter per day (-1 mm/day) at the moment when $r = 25$ mm. How fast is the area decreasing at that moment?

41. Body surface area. Certain chemotherapy dosages depend on a patient's surface area. According to the Mosteller model,

$$S = \frac{\sqrt{hw}}{60},$$

where h is the patient's height in centimeters, w is the patient's weight in kilograms, and S is the approximation to the patient's surface area in square meters. (*Source*: www.halls.md.) Assume that Tom's height is a constant 165 cm, but he is on a diet. If he loses 2 kg per month, how fast is his surface area decreasing at the instant he weighs 70 kg?

Poiseuille's Law. *The flow of blood in a blood vessel is faster toward the center of the vessel and slower toward the outside. The speed of the blood V, in millimeters per second (mm/sec), is given by*

$$V = \frac{p}{4Lv}(R^2 - r^2),$$

where R is the radius of the blood vessel, r is the distance of the blood from the center of the vessel, and p, L, and v are physical constants related to pressure, length, and viscosity of the blood vessels, respectively. Assume that dV/dt is measured in millimeters per second squared (mm / sec²). Use this formula for Exercises 42 and 43.

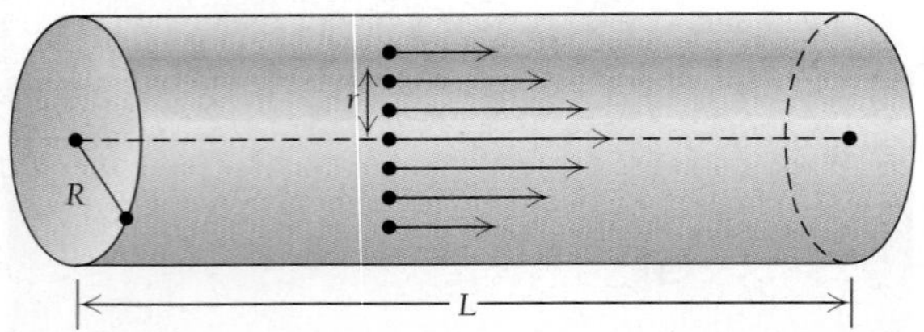

42. Assume that r is a constant as well as p, L, and v.

a) Find the rate of change dV/dt in terms of R and dR/dt when $L = 80$ mm, $p = 500$, and $v = 0.003$.

b) A person goes out into the cold to shovel snow. Cold air has the effect of contracting blood vessels far from the heart. Suppose that a blood vessel contracts at a rate of

$$\frac{dR}{dt} = -0.0002 \text{ mm/sec}$$

at a place in the blood vessel where the radius $R = 0.075$ mm. Find the rate of change, dV/dt, at that location.

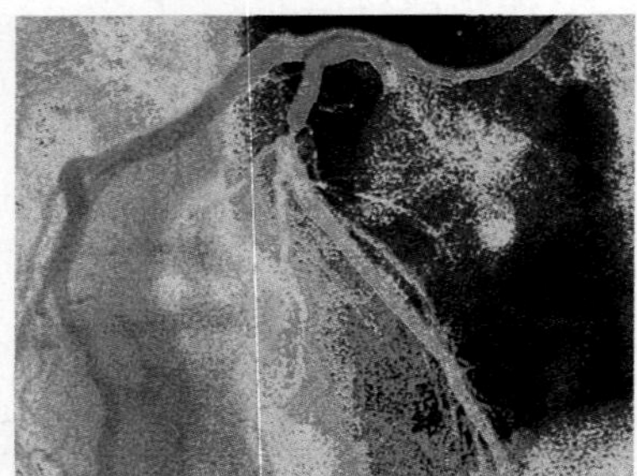

The flow of blood in a blood vessel can be modeled by Poiseuille's Law.

43. Assume that r is a constant as well as p, L, and v.

a) Find the rate of change dV/dt in terms of R and dR/dt when $L = 70$ mm, $p = 400$, and $v = 0.003$.

b) When shoveling snow in cold air, a person with a history of heart trouble can develop angina (chest pains) due to contracting blood vessels. To counteract this, he or she may take a nitroglycerin tablet, which dilates the blood vessels. Suppose that after a nitroglycerin tablet is taken, a blood vessel dilates at a rate of

$$\frac{dR}{dt} = 0.00015 \text{ mm/sec}$$

at a place in the blood vessel where the radius $R = 0.1$ mm. Find the rate of change, dV/dt.

General Interest

44. Two cars start from the same point at the same time. One travels north at 25 mph, and the other travels east at 60 mph. How fast is the distance between them increasing at the end of 1 hr? (*Hint*: $D^2 = x^2 + y^2$. To find D after 1 hr, solve $D^2 = 25^2 + 60^2$.)

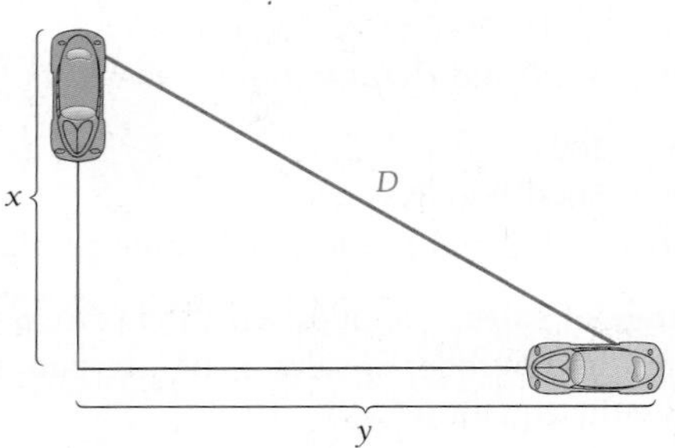

45. A ladder 26 ft long leans against a vertical wall. If the lower end is being moved away from the wall at the rate of 5 ft/s, how fast is the height of the top changing (this will be a negative rate) when the lower end is 10 ft from the wall?

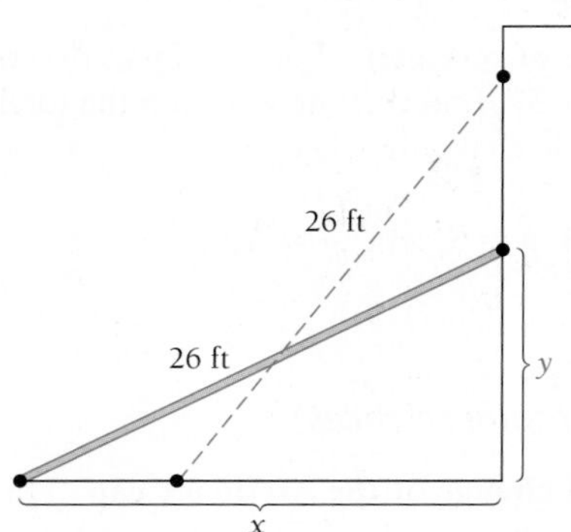

46. An inner city revitalization zone is a rectangle that is twice as long as it is wide. A diagonal through the region is growing at a rate of 90 m per year at a time when the region is 440 m wide. How fast is the area changing at that point in time?

47. The volume of a cantaloupe is given by
$V = \frac{4}{3}\pi r^3$.
The radius is growing at the rate of 0.7 cm/week, at a time when the radius is 7.5 cm. How fast is the volume changing at that moment?

SYNTHESIS

Differentiate implicitly to find dy/dx.

48. $\sqrt{x} + \sqrt{y} = 1$

49. $\frac{1}{x^2} + \frac{1}{y^2} = 5$

50. $y^3 = \frac{x-1}{x+1}$

51. $y^2 = \frac{x^2-1}{x^2+1}$

52. $x^{3/2} + y^{2/3} = 1$

53. $(x - y)^3 + (x + y)^3 = x^5 + y^5$

Differentiate implicitly to find d^2y/dx^2.

54. $xy + x - 2y = 4$

55. $y^2 - xy + x^2 = 5$

56. $x^2 - y^2 = 5$

57. $x^3 - y^3 = 8$

TW 58. Explain the usefulness of implicit differentiation.

TW 59. Look up the word "implicit" in a dictionary. Explain how that definition can be related to the concept of a function that is defined "implicitly."

TECHNOLOGY CONNECTION

Graph each of the following equations. On most calculators, equations must be solved for y before they can be entered.

60. $x^2 + y^2 = 4$
Note: You will probably need to sketch the graph in two parts: $y = \sqrt{4 - x^2}$ and $y = -\sqrt{4 - x^2}$. Then graph the tangent line to the graph at the point $(-1, \sqrt{3})$.

61. $x^4 = y^2 + x^6$
Then graph the tangent line to the graph at the point $(-0.8, 0.384)$.

62. $y^4 = y^2 - x^2$

63. $x^3 = y^2(2 - x)$

64. $y^2 = x^3$

Chapter Summary

This chapter explored the many ways in which derivatives can be used in both curve sketching and real-world situations. Special attention was given to using the derivative as a way of identifying when and where a maximum or minimum value occurs.

One important use of the derivative is to determine where a function is increasing or decreasing:

If $f'(x) > 0$ for all x in an interval I, then f is **increasing** over I.
If $f'(x) < 0$ for all x in an interval I, then f is **decreasing** over I. (p. 200)

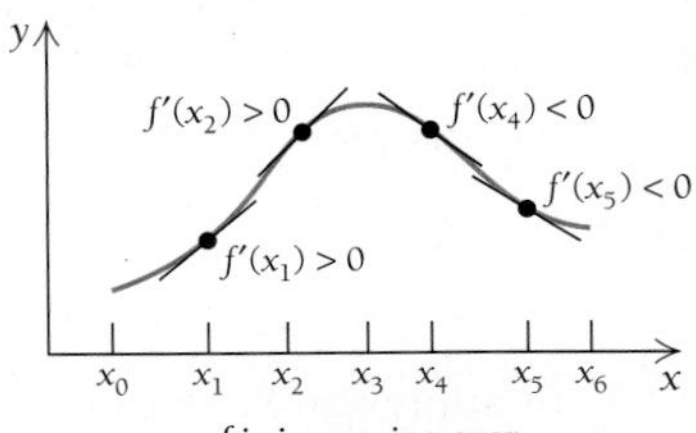

f is increasing over the interval $[x_0, x_3]$ and decreasing over $[x_3, x_6]$.

To identify where maximum or minimum values occur, we look for points where the sign of $f'(x)$ changes. On a continuous function, this occurs at **critical values.** A critical value of f is any number c for which $f(c)$ exists and

$$f'(c) = 0 \quad \text{or} \quad f'(c) \text{ does not exist.}$$ (p. 200)

For a continuous function f, any **relative** (or **local**) **minimum** or **maximum** or any **absolute extremum** will occur at a critical value. (p. 201)

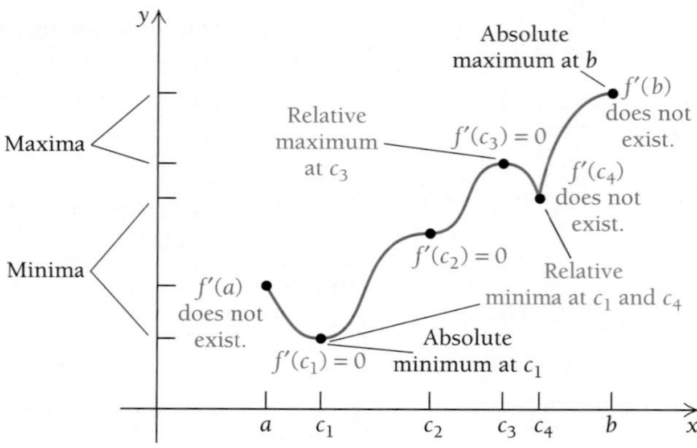

For the above figure, over the interval $[a, b]$, the relative minima are $f(c_1)$ and $f(c_4)$. The relative maximum is $f(c_3)$. The absolute minimum is $f(c_1)$, and the absolute maximum is $f(b)$. Note that $f'(c_2) = 0$, yet $f(c_2)$ is neither a relative maximum nor a relative minimum.

To determine where extrema occur, we look for those x-values for which $f'(x) = 0$ or $f'(x)$ does not exist. To classify the associated y-values as maxima or minima, we can use the **First-Derivative Test.** (p. 203)

EXAMPLE Find the relative extrema of the function given by

$$f(x) = \tfrac{1}{3}x^3 - \tfrac{1}{2}x^2 - 20x + 7.$$

Solution Critical values occur where $f'(x) = 0$ or where $f'(x)$ does not exist:

$$f'(x) = x^2 - x - 20.$$

Since $f'(x)$ exists for all real numbers, the only critical values occur where $f'(x) = 0$:

$$x^2 - x - 20 = 0$$
$$(x + 4)(x - 5) = 0$$
$$x + 4 = 0 \quad \text{or} \quad x - 5 = 0$$
$$x = -4 \quad \text{or} \quad x = 5$$

The critical values are -4 and 5. To apply the First-Derivative Test, we check the sign of $f'(x)$ to the left and the right of each critical value, using test values.

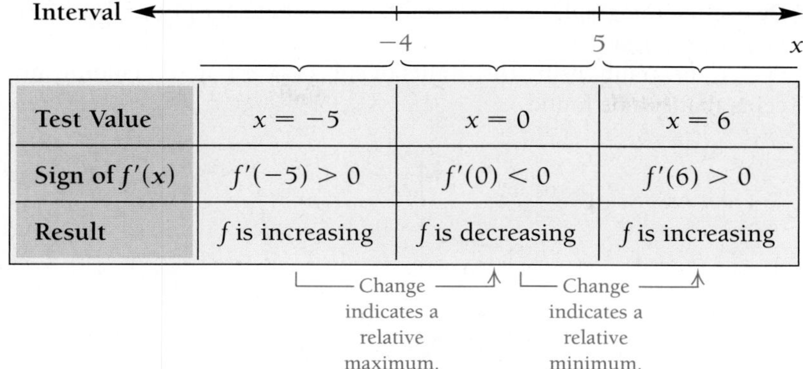

Test Value	$x = -5$	$x = 0$	$x = 6$
Sign of $f'(x)$	$f'(-5) > 0$	$f'(0) < 0$	$f'(6) > 0$
Result	f is increasing	f is decreasing	f is increasing

We see that

$$\begin{aligned} f(-4) &= \tfrac{1}{3}(-4)^3 - \tfrac{1}{2}(-4)^2 - 20(-4) + 7 \\ &= -\tfrac{64}{3} - \tfrac{16}{2} + 80 + 7 = -21\tfrac{1}{3} + 79 \\ &= 57\tfrac{2}{3} \text{ is a relative maximum,} \end{aligned}$$

and

$$\begin{aligned} f(5) &= \tfrac{1}{3} \cdot 5^3 - \tfrac{1}{2} \cdot 5^2 - 20 \cdot 5 + 7 \\ &= \tfrac{125}{3} - \tfrac{25}{2} - 100 + 7 = 41\tfrac{2}{3} - 12\tfrac{1}{2} - 93 \\ &= -63\tfrac{5}{6} \text{ is a relative minimum.} \end{aligned}$$

✦

Another way to classify extrema as maxima or minima is to use the **Second-Derivative Test.** To use this test on the function in the preceding example, we first find

$$f''(x) = 2x - 1.$$

Next, rather than use test values, we simply note that

$f''(-4) = 2(-4) - 1 < 0$, indicating that f is *concave down* at $x = -4$,

and $f''(5) = 2 \cdot 5 - 1 > 0$, indicating that f is *concave up* at $x = 5$. (p. 216)

These results indicate that a relative maximum occurs at $x = -4$ and a relative minimum occurs at $x = 5$.

Points at which a function's concavity changes from concave up to concave down or from concave down to concave up are known as **points of inflection.** If such a point exists, it will occur at one of the points where $f''(x) = 0$ or where $f''(x)$ does not exist. (p. 219)

The preceding information allows us to develop a procedure for graphing continuous functions. To sketch the graph of a continuous function f:

a) Find $f'(x)$ and $f''(x)$, and note the domain of f.
b) Find all critical values and their corresponding y-values.
c) Apply the Second-Derivative Test to determine any relative extrema, using the results of step (b). This will also identify where the function is increasing or decreasing.
d) Locate any points of inflection by checking the concavity on either side of those points for which $f''(x) = 0$ or for which $f''(x)$ does not exist.

e) Determine the graph's concavity and where it changes from the information in steps (c) and (d).

f) Sketch the graph, using the results of steps (a)–(e) and including any intercepts that can be found.

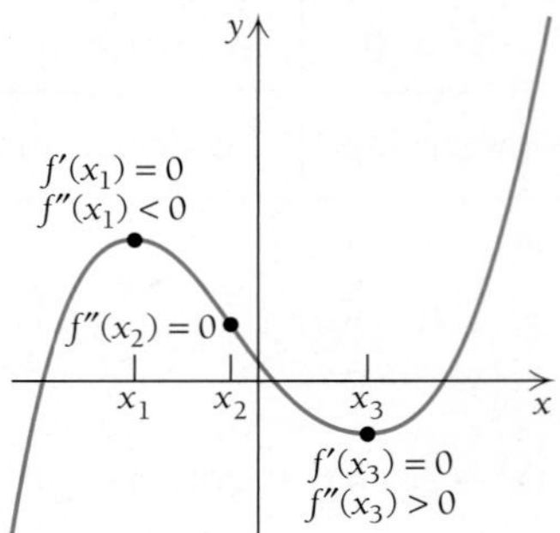

Rational functions of the form $f(x) = P(x)/Q(x)$, where $P(x)$ and $Q(x)$ are polynomials, with $Q(x)$ not the zero polynomial, often have graphs with **vertical, horizontal,** or **slant (oblique) asymptotes.** (pp. 232–236)

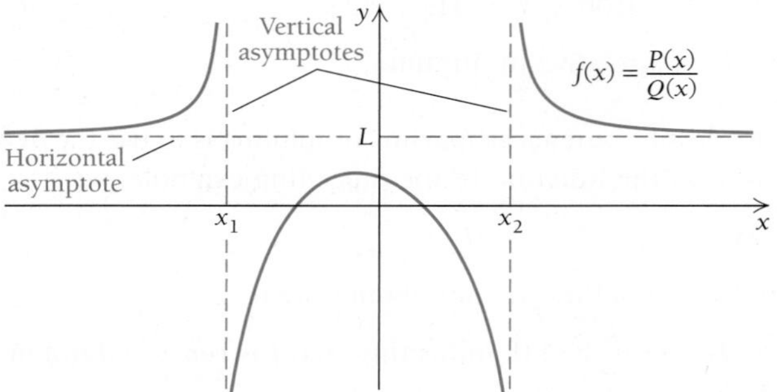

Rational functions are graphed in much the same way as continuous functions are (see above). The primary difference is the existence of asymptotes:

To locate any vertical asymptote(s), assuming that $f(x)$ is simplified, we find all x-values for which $Q(x) = 0$. In the figure above, we have $Q(x_1) = 0$ and $Q(x_2) = 0$.

To locate any horizontal or slant asymptote, we find $\lim_{x\to\infty} f(x)$ and $\lim_{x\to-\infty} f(x)$.

- If the degree of $P(x)$ is less than that of $Q(x)$, then the horizontal asymptote is $y = 0$, or the x-axis.
- If the degree of $P(x)$ is the same as the degree of $Q(x)$, then the horizontal asymptote is $y = L$, where L is the ratio of the leading coefficient of $P(x)$ to the leading coefficient of $Q(x)$.
- If the degree of $P(x)$ is one greater than that of $Q(x)$, then the asymptote is a slant asymptote and can be found by dividing $P(x)$ by $Q(x)$ and noting that, as $|x| \to \infty$, the remainder, divided by $Q(x)$, approaches 0.

To find the absolute extrema of a continuous function f over a closed interval $[a, b]$, it suffices to find all critical values, $c_1, c_2, \ldots, c_n$, and then evaluate

$$f(a), f(c_1), f(c_2), f(c_3), \ldots, f(c_n), f(b).$$

The largest of these values is the **absolute maximum** and the smallest is the **absolute minimum** over $[a, b]$. (p. 247)

To find the absolute extrema of a continuous function f with only one critical value over an open interval, such as $(-\infty, \infty)$ or $(0, \infty)$, the First- or Second-Derivative Test can be used to determine whether an absolute maximum or minimum exists at that x-value. (p. 250)

Applications in which it is necessary to identify when extrema occur are numerous and include many situations in which cost is to be minimized or profit is to be maximized. In general, this type of problem is solved by first translating the problem into an equation in which some quantity Q, that is to be maximized or minimized, is a function of a predetermined variable. We then find critical values and extrema as outlined above. (pp. 260–261)

The final use of derivatives covered in this chapter was in the calculation of marginals and differentials. Simply put, **marginal cost,** when x items are currently being produced, is the approximate cost of the $(x + 1)$st item. This is the same as the rate at which cost is changing at the point $(x, C(x))$, or $C'(x)$. Similar descriptions apply to **marginal revenue** and **marginal profit.** (p. 275)

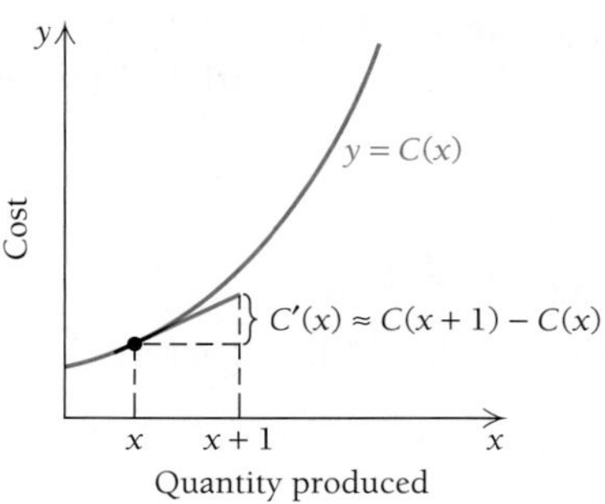

The discussion of differentials used the same concepts as used for marginal cost, profit, and revenue. The chief difference is the use of **delta notation** to indicate change. Specifically, we have

$$\Delta x = x_2 - x_1 \quad \text{and} \quad \Delta y = y_2 - y_1.$$

Thus, for small Δx, we have $\Delta y \approx f'(x)\,\Delta x$. (p. 277)

Differentials are defined as follows:

For $y = f(x)$, dx, the *differential of* x, is Δx,

and dy, the *differential of* y, is $f'(x)\,dx$.

Thus, $dx = \Delta x$, but $dy \neq \Delta y$. Instead, $dy \approx \Delta y$ for small values of Δx. (p. 279)

EXAMPLE For $y = \sqrt[3]{x}$, find dy when $x = 27$ and $dx = 2$.

Solution Note that

$$\frac{dy}{dx} = \frac{d}{dx}x^{1/3} = \frac{1}{3}x^{-2/3} = \frac{1}{3x^{2/3}}.$$

Thus,

$$dy = \frac{1}{3x^{2/3}} \cdot dx,$$

and for $x = 27$ and $dx = 2$, we have

$$dy = \frac{1}{3 \cdot 27^{2/3}} \cdot 2 = \frac{2}{27}.$$

✦

The result of the preceding example can be used to approximate $\sqrt[3]{29}$, using the fact that $\sqrt[3]{29} = \sqrt[3]{27} + \Delta y$, with $\Delta x = 2$:

$$\begin{aligned}\sqrt[3]{29} &\approx \sqrt[3]{27} + dy \\ &\approx 3 + \tfrac{2}{27} \\ &\approx 3.074.\end{aligned}$$

The actual value of $\sqrt[3]{29}$ is closer to 3.072.

Sometimes it is difficult or impossible to solve an equation for y, despite the fact that y is a function of x. When this occurs, **implicit differentiation** affords us another way to find dy/dx. (p. 284)

EXAMPLE Find dy/dx if $y^5 = x^3 + 7$.

Solution We differentiate both sides with respect to x:

$$\begin{aligned}\frac{d}{dx}y^5 &= \frac{d}{dx}(x^3 + 7) \\ 5y^4 \cdot \frac{dy}{dx} &= 3x^2 \\ \frac{dy}{dx} &= \frac{3x^2}{5y^4}.\end{aligned}$$

✦

In some applied problems, one variable is a function of another variable, and both change with respect to time. These problems of *related rates* are solved by differentiating both sides with respect to t, in much the same way that $\frac{d}{dx}(y^5)$ was found in the preceding example. (p. 286)

Chapter Review Exercises

These review exercises are for test preparation. They can also be used as a practice test. Answers are at the back of the book. The bracketed section references tell you what part(s) of the chapter to restudy if your answer is incorrect.

CONCEPT REINFORCEMENT

Match each description in column A with the most appropriate graph in column B. [2.1–2.4]

Column A

Column B

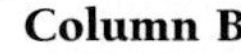

1. A function with a relative maximum but no absolute extrema

a)

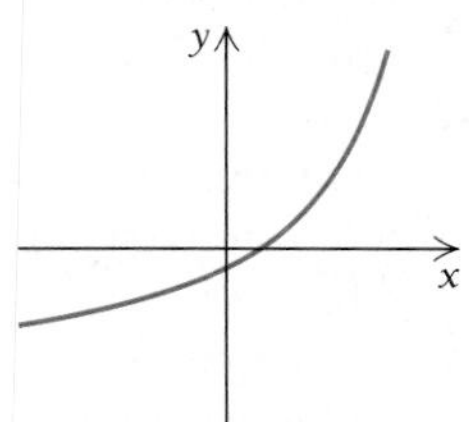

2. A function with both a vertical asymptote and a horizontal asymptote

b)

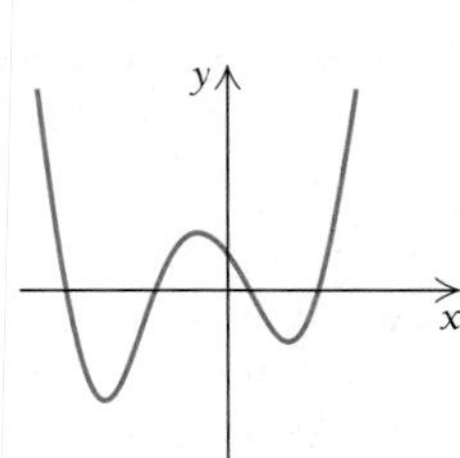

3. A function that is concave up and decreasing

c)

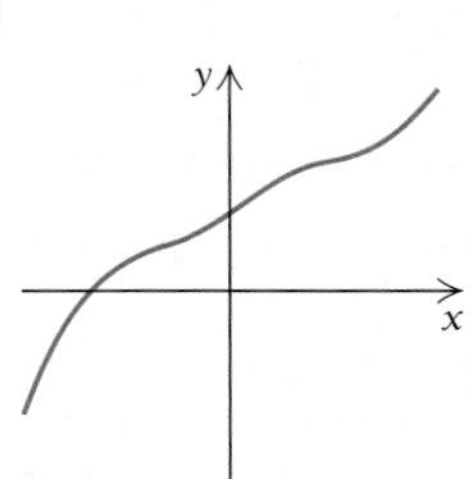

4. A function that is concave up and increasing

d)

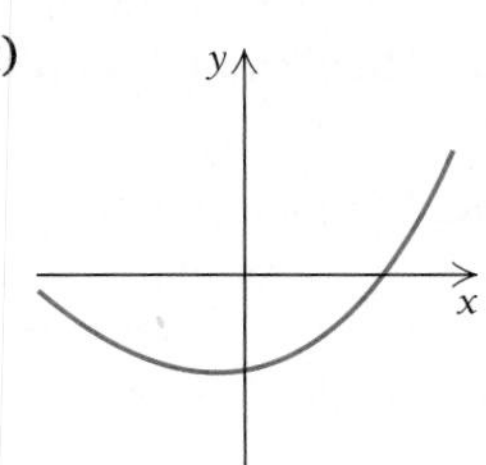

5. A function with three critical values

e)

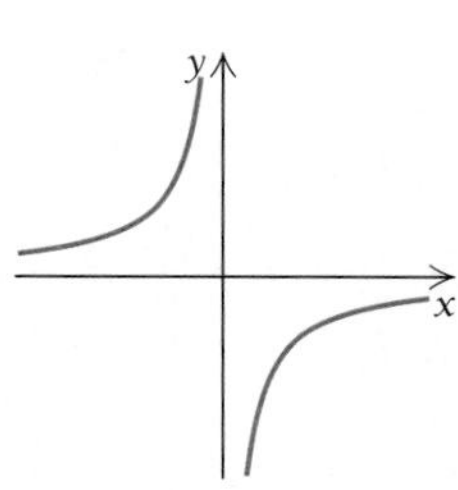

6. A function with one critical value and a second derivative that is always positive

f)

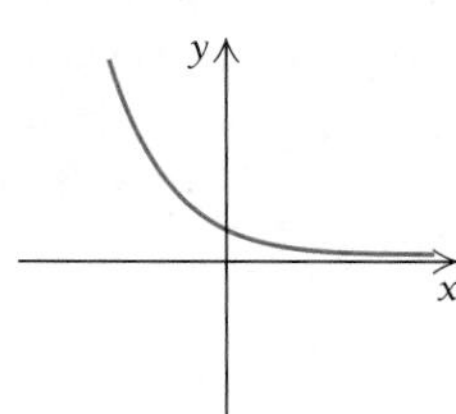

7. A function with a first derivative that is always positive

g)

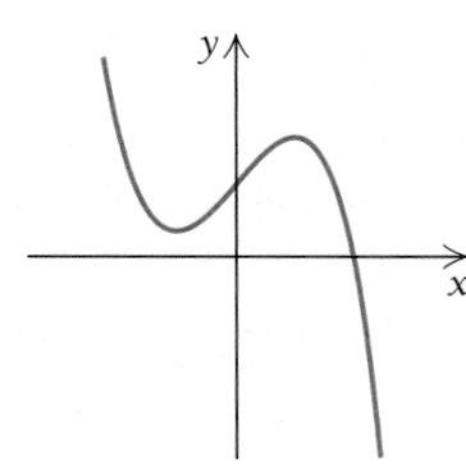

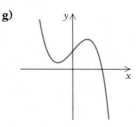

In Exercises 8–13, classify each statement as either true or false.

8. Every continuous function has at least one critical value. [2.1]
9. If a continuous function $y = f(x)$ has extrema, they will occur where $f'(x) = 0$. [2.1]
10. If $f'(c) = 0$ and $f''(c) > 0$, then $f(c)$ is a relative minimum. [2.2]
11. If $f'(c) = 0$ and $f''(c) = 0$, then $f(c)$ cannot be a relative minimum. [2.2]
12. If the graph of $f(x) = P(x)/Q(x)$ has a horizontal asymptote, then the degree of the polynomial $P(x)$ must be the same as that of the polynomial $Q(x)$. [2.3]
13. Unless a check such as the First- or Second-Derivative Test is performed, the steps used to maximize a function could actually minimize that same function. [2.4]

REVIEW EXERCISES

For each function given, find any extrema, along with the x-value at which they occur. Then sketch a graph of the function. [2.1]

14. $f(x) = 4 - 3x - x^2$

15. $f(x) = x^4 - 2x^2 + 3$

16. $f(x) = \dfrac{-8x}{x^2 + 1}$

17. $f(x) = 4 + (x - 1)^3$

18. $f(x) = x^3 + x^2 - x + 3$

19. $f(x) = 3x^{2/3}$

20. $f(x) = 2x^3 - 3x^2 - 12x + 10$

21. $f(x) = x^3 - 3x + 2$

Sketch the graph of each function. List any minimum or maximum values and where they occur, as well as any points of inflection. State where the function is increasing or decreasing, as well as where it is concave up or concave down. [2.2]

22. $f(x) = \dfrac{1}{3}x^3 + 3x^2 + 9x + 2$

23. $f(x) = x^2 - 10x + 8$

24. $f(x) = 4x^3 - 6x^2 - 24x + 5$

25. $f(x) = x^4 - 2x^2$

26. $f(x) = 3x^4 + 2x^3 - 3x^2 + 1$ (Round to three decimal places where appropriate.)

27. $f(x) = \frac{1}{5}x^5 + \frac{3}{4}x^4 - \frac{4}{3}x^3 + 8$ (Round to three decimal places where appropriate.)

Sketch the graph of each function. Indicate where each function is increasing or decreasing, the coordinates at which relative extrema occur, where any asymptotes occur, where the graph is concave up or concave down, and where any intercepts occur. [2.3]

28. $f(x) = \dfrac{2x + 5}{x + 1}$

29. $f(x) = \dfrac{x}{x - 2}$

30. $f(x) = \dfrac{5}{x^2 - 16}$

31. $f(x) = -\dfrac{x + 1}{x^2 - x - 2}$

32. $f(x) = \dfrac{x^2 - 2x + 2}{x - 1}$

33. $f(x) = \dfrac{x^2 + 3}{x}$

Find the absolute maximum and minimum values of each function, if they exist, over the indicated interval. Indicate the x-value at which each extremum occurs. Where no interval is specified, use the real line. [2.4]

34. $f(x) = x^4 - 2x^2 + 3;\ [0, 3]$

35. $f(x) = 8x^2 - x^3;\ [-1, 8]$

36. $f(x) = x + \dfrac{50}{x};\ (0, \infty)$

37. $f(x) = x^4 - 2x^2 + 1$

38. Of all numbers whose sum is 60, find the two that have the maximum product. [2.5]

39. Find the minimum value of $Q = x^2 - 2y^2$, where $x - 2y = 1$. [2.5]

40. Business: maximizing profit. If

$$R(x) = 52x - 0.5x^2 \quad \text{and} \quad C(x) = 22x - 1,$$

find the maximum profit and the number of units that must be produced and sold in order to yield this maximum profit. Assume that $R(x)$ and $C(x)$ are in dollars. [2.5]

41. A rectangular box with a square base and a cover is to have a volume of 2500 ft^3. If the cost per square foot for the bottom is \$2, for the top is \$3, and for the sides is \$1, what should the dimensions be in order to minimize the cost? [2.5]

42. Business: minimizing inventory cost. A store in California sells 360 hybrid bicycles per year. It costs \$8 to store one bicycle for a year. To reorder, there is a fixed cost of \$10, plus \$2 for each bicycle. How many times per year should the store order bicycles, and in what lot size, in order to minimize inventory costs? [2.5]

43. Business: marginal revenue. Crane Foods determines that its daily revenue, $R(x)$, in dollars, from the sale of x frozen dinners is

$$R(x) = 4x^{3/4}.$$

a) What is Crane's daily revenue when 81 frozen dinners are sold?

b) What is Crane's marginal revenue when 81 frozen dinners are sold?

c) Use the answers from parts (a) and (b) to estimate $R(82)$. [2.6]

For Exercises 44 and 45, $y = f(x) = 2x^3 + x$. [2.6]

44. Find Δy and $f'(x)\,\Delta x$, given that $x = 1$ and $\Delta x = -0.05$.

45. a) Find dy.
b) Find dy when $x = -2$ and $dx = 0.01$.

46. Approximate $\sqrt{83}$ using $\Delta y \approx f'(x)\,\Delta x$. [2.6]

47. Differentiate the following implicitly to find dy/dx. Then find the slope of the curve at the given point.

$$2x^3 + 2y^3 = -9xy; \quad (-1, -2) \qquad [2.7]$$

48. A ladder 25 ft long leans against a vertical wall. If the lower end is being moved away from the wall at the rate of 6 ft/sec, how fast is the height of the top decreasing when the lower end is 7 ft from the wall? [2.7]

49. Business: total revenue, cost, and profit. Find the rates of change, with respect to time, of total revenue, cost, and profit for

$$R(x) = 120x - 0.5x^2 \quad \text{and} \quad C(x) = 15x + 6,$$

when $x = 100$ and $dx/dt = 30$ units per day. Assume that $R(x)$ and $C(x)$ are in dollars. [2.7]

SYNTHESIS

50. Find the absolute maximum and minimum values, if they exist, over the indicated interval.

$$f(x) = (x - 3)^{2/5}; \quad (-\infty, \infty) \qquad [2.4]$$

51. Differentiate implicitly to find dy/dx:

$$(x - y)^4 + (x + y)^4 = x^6 + y^6. \qquad [2.7]$$

52. Find the relative maxima and minima of

$$y = x^4 - 8x^3 - 270x^2. \qquad [2.1 \text{ and } 2.2]$$

TECHNOLOGY CONNECTION

Use a calculator to estimate the relative extrema of each function. [2.1 and 2.2]

53. $f(x) = 3.8x^5 - 18.6x^3$

54. $f(x) = \sqrt[3]{|9 - x^2|} - 1$

55. Life and physical sciences: incidence of breast cancer. The following table provides data relating the incidence of breast cancer per 100,000 women of various ages.

a) Use REGRESSION to fit linear, quadratic, cubic, and quartic functions to the data.

Age	Incidence per 100,000
0	0
27	10
32	25
37	60
42	125
47	187
52	224
57	270
62	340
67	408
72	437
77	475
82	460
87	420

(*Source*: National Cancer Institute.)

b) Which function best fits the data?

c) Determine the domain of the function on the basis of the function and the problem situation, and explain.

d) Determine the maximum value of the function on the domain. At what age is the incidence of breast cancer the greatest? [2.1 and 2.2]

Note: The function used in Exercise 28 of Section R.1 was found in this manner.

Chapter 2 Test

Find all relative minimum or maximum values as well as the x-values at which they occur. State where each function is increasing or decreasing. Then sketch a graph of the function.

1. $f(x) = x^2 - 4x - 5$

2. $f(x) = 4 + 3x - x^3$

3. $f(x) = (x - 2)^{2/3} - 4$

4. $f(x) = \dfrac{16}{x^2 + 4}$

Sketch a graph of each function. List any extrema, and indicate any asymptotes or points of inflection.

5. $f(x) = x^3 + x^2 - x + 1$

6. $f(x) = 2x^4 - 4x^2 + 1$

7. $f(x) = (x - 2)^3 + 3$

8. $f(x) = x\sqrt{9 - x^2}$

9. $f(x) = \dfrac{2}{x - 1}$

10. $f(x) = \dfrac{-8}{x^2 - 4}$

11. $f(x) = \dfrac{x^2 - 1}{x}$

12. $f(x) = \dfrac{x - 3}{x + 2}$

Find the absolute maximum and minimum values, if they exist, of each function over the indicated interval. Where no interval is specified, use the real line.

13. $f(x) = x(6 - x)$

14. $f(x) = x^3 + x^2 - x + 1;\ \left[-2, \frac{1}{2}\right]$
15. $f(x) = -x^2 + 8.6x + 10$
16. $f(x) = -2x + 5;\ [-1, 1]$
17. $f(x) = -2x + 5$
18. $f(x) = 3x^2 - x - 1$
19. $f(x) = x^2 + \dfrac{128}{x};\ (0, \infty)$
20. Of all numbers whose difference is 8, find the two that have the minimum product.
21. Minimize $Q = x^2 + y^2$, where $x - y = 10$.
22. **Business: maximum profit.** Find the maximum profit and the number of units, x, that must be produced and sold in order to yield the maximum profit. Assume that $R(x)$ and $C(x)$ are the revenue and cost, in dollars, when x units are produced.

$$R(x) = x^2 + 110x + 60,$$
$$C(x) = 1.1x^2 + 10x + 80$$

23. From a thin piece of cardboard 60 in. by 60 in., square corners are cut out so that the sides can be folded up to make an open box. What dimensions will yield a box of maximum volume? What is the maximum volume?
24. **Business: minimizing inventory costs.** Ironside Sports sells 1225 tennis rackets per year. It costs \$2 to store one tennis racket for a year. To reorder, there is a fixed cost of \$1, plus \$0.50 for each tennis racket. How many times per year should Ironside order tennis rackets, and in what lot size, in order to minimize inventory costs?
25. For $y = f(x) = x^2 - 3$, $x = 5$, and $\Delta x = 0.1$, find Δy and $f'(x)\,\Delta x$.
26. Approximate $\sqrt{50}$ using $\Delta y \approx f'(x)\,\Delta x$.
27. For $y = \sqrt{x^2 + 3}$:
    a) Find dy.
    b) Find dy when $x = 2$ and $dx = 0.01$.
28. Differentiate the following implicitly to find dy/dx. Then find the slope of the curve at $(1, 2)$:

$$x^3 + y^3 = 9.$$

29. A pole 13 ft long leans against a vertical wall. If the lower end is moving away from the wall at the rate of 0.4 ft/sec, how fast is the upper end coming down when the lower end is 12 ft from the wall?

SYNTHESIS

30. Find the absolute maximum and minimum values of the following function, if they exist, over $[0, \infty)$:

$$f(x) = \frac{x^2}{1 + x^3}.$$

31. **Business: minimizing average cost.** The total cost in dollars of producing x units of a product is given by

$$C(x) = 100x + 100\sqrt{x} + \frac{\sqrt{x^3}}{100}.$$

How many units should be produced to minimize the average cost?

TECHNOLOGY CONNECTION

32. Use a calculator to estimate any extrema of this function:

$$f(x) = 5x^3 - 30x^2 + 45x + 5\sqrt{x}.$$

33. **Business: advertising.** The business of manufacturing and selling bowling balls is one of frequent changes. Companies introduce new models to the market about every 3 to 4 months. Typically, a new model is created because of advances in technology such as new surface stock or a new way to place weight blocks in a ball. To decide how to best use advertising dollars, companies track the sales in relation to the amount spent on advertising. Suppose that a company has the following data from past sales.

Amount Spent on Advertising (in thousands)	Number of Bowling Balls Sold, N
\$ 0	8
50	13,115
100	19,780
150	22,612
200	20,083
250	12,430
300	4

a) Use REGRESSION to fit linear, quadratic, cubic, and quartic functions to the data.
TW **b)** Determine the domain of the function on the basis of the function in part (a) that fits best and the problem situation. Justify your answer.
c) Determine the maximum value of the function on the domain. How much should the company spend on advertising its next new model in order to maximize the number of bowling balls sold?

EXTENDED TECHNOLOGY APPLICATION

Maximum Sustainable Harvest

In certain situations, biologists are able to determine what is called a **reproduction curve.** This is a function

$$y = f(P)$$

such that if P is the population after P years, then $f(P)$ is the population a year later, at time $t + 1$. Such a curve is shown below.

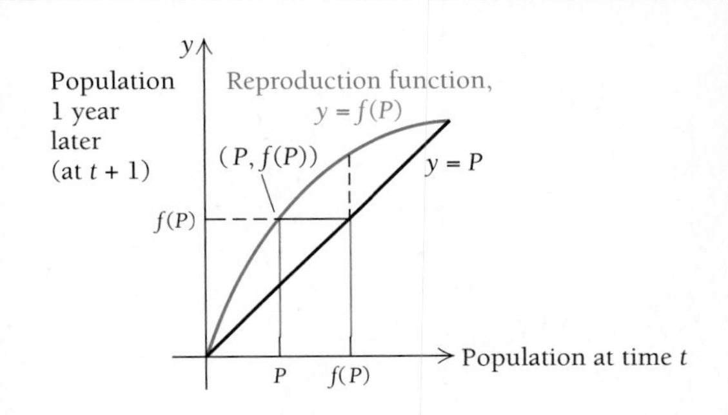

The line $y = P$ is significant because if it ever coincides with the curve $y = f(P)$, then we know that the population stays the same from year to year. Here the graph of f lies mostly above the line, indicating that the population is increasing.

Too many deer in a forest can deplete the food supply and eventually cause the population to decrease for lack of food. Often in such cases and with some controversy, hunters are allowed to "harvest" some of the deer. Then with a greater food supply, the remaining deer population might actually prosper and increase.

We know that a population P will grow to a population $f(P)$ in a year. If this were a population of fur-bearing animals and the population were increasing, then hunters could "harvest" the amount

$$f(P) - P$$

each year without shrinking the initial population P. If the population were remaining the same or decreasing, then such a harvest would deplete the population.

Suppose that we want to know the value of P_0 that would allow the harvest to be the largest. If we could determine that P_0, we could let the population grow until it reached that level and then begin harvesting year after year the amount $f(P_0) - P_0$.

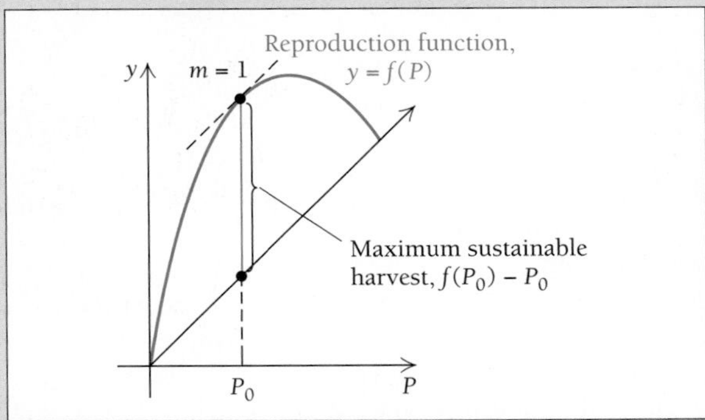

Let the harvest function H be given by

$$H(P) = f(P) - P.$$

Then $H'(P) = f'(P) - 1.$

Now, if we assume that $H'(P)$ exists for all values of P and that there is only one critical value, it follows that the *maximum sustainable harvest* occurs at that value P_0 such that

$$H'(P_0) = f'(P_0) - 1 = 0$$

and $H''(P_0) = f''(P_0) < 0.$

Or, equivalently, we have the following.

THEOREM

The **maximum sustainable harvest** occurs at P_0 such that

$$f'(P_0) = 1 \quad \text{and} \quad f''(P_0) < 0,$$

and is given by

$$H(P_0) = f(P_0) - P_0.$$

EXERCISES

For each reproduction curve in Exercises 1–3, do the following.

a) Graph the reproduction function, the line $y = P$, and the harvest function using the same viewing window.

b) Find the population at which the maximum sustainable harvest occurs. Use both a graphical solution and a calculus solution.

c) Find the maximum sustainable harvest.

1. $f(P) = P(10 - P)$, where P is measured in thousands.

2. $f(P) = -0.025P^2 + 4P$, where P is measured in thousands. This is the reproduction curve in the Hudson Bay area for the snowshoe hare, a fur-bearing animal.

3. $f(P) = -0.01P^2 + 2P$, where P is measured in thousands. This is the reproduction curve in the Hudson Bay area for the lynx, a fur-bearing animal.

For each reproduction curve in Exercises 4 and 5, do the following.

a) Graph the reproduction function, the line $y = P$, and the harvest function using the same viewing window.

b) Graphically determine the population at which the maximum sustainable harvest occurs.

c) Find the maximum sustainable harvest.

4. $f(P) = 40\sqrt{P}$, where P is measured in thousands. Assume that this is the reproduction curve for the brown trout population in a large lake.

5. $f(P) = 0.237P\sqrt{2000 - P^2}$, where P is measured in thousands.

6. The table below lists data regarding the reproduction of a certain animal.

a) Use REGRESSION to fit a cubic polynomial to these data.

b) Graph the reproduction function, the line $y = P$, and the harvest function using the same viewing window.

c) Graphically determine the population at which the maximum sustainable harvest occurs.

POPULATION, P (IN THOUSANDS)	POPULATION $f(P)$, 1 YEAR LATER
10	9.7
20	23.1
30	37.4
40	46.2
50	42.6

Exponential and Logarithmic Functions

3

APPLICATION More Americans are buying organic fruit and vegetables and products made with organic ingredients. The amount $A(t)$, in billions of dollars, spent on organic food and beverages t years after 1995 can be approximated by

$$A(t) = 2.43e^{0.18t}, \quad 0 \leq t \leq 15.$$

(*Source*: Based on data and projections from *Nutrition Business Journal*, 2004.) Estimate the rate at which spending on organic food and beverages was growing in 2006.

This problem appears as Exercise 81 in Section 3.1.

INTRODUCTION

In this chapter, we consider two types of functions that are closely related: *exponential functions* and *logarithmic functions.* After learning to find derivatives of such functions, we will study applications in the areas of population growth, decay, interest compounded continuously, spread of disease, and carbon dating.

3.1

OBJECTIVES

- Graph exponential functions.
- Differentiate exponential functions.

Exponential Functions

Graphs of Exponential Functions

Consider the following graph. The rapid rise of the graph indicates that it approximates an *exponential function*. We now consider such functions and many of their applications.

World Population Growth

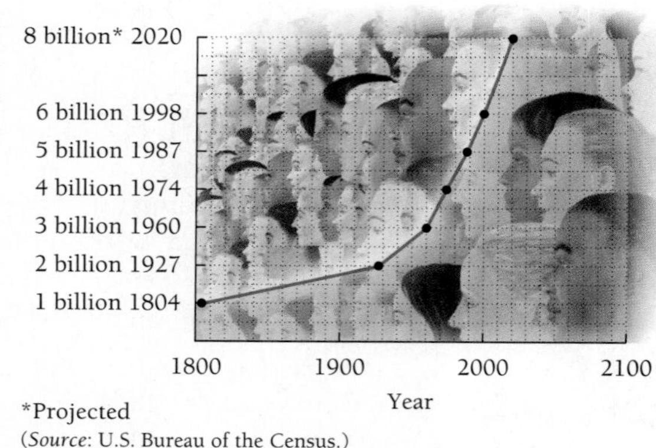

*Projected
(*Source*: U.S. Bureau of the Census.)

Let's review definitions of expressions of the form a^x, where x is a rational number. For example,

$$a^{2.34} \quad \text{or} \quad a^{234/100}$$

means "raise a to the 234th power and then take the 100th root $\left(\sqrt[100]{a^{234}}\right)$."

What about expressions with irrational exponents, such as $2^{\sqrt{2}}$, 2^{π}, or $2^{-\sqrt{3}}$? An *irrational number* is a number named by an infinite, nonrepeating decimal. Let's consider 2^{π}. We know that π is irrational with an infinite, nonrepeating decimal expansion:

$$3.141592653\ldots.$$

This means that π is approached as a limit by the rational numbers

$$3,\ 3.1,\ 3.14,\ 3.141,\ 3.1415, \ldots,$$

so it seems reasonable that 2^{π} should be approached as a limit by the rational powers

$$2^3,\ 2^{3.1},\ 2^{3.14},\ 2^{3.141},\ 2^{3.1415}, \ldots.$$

Estimating each power with a calculator, we get the following:

$$8,\ 8.574188,\ 8.815241,\ 8.821353,\ 8.824411, \ldots.$$

TECHNOLOGY CONNECTION

We can approximate 2^{π} using the exponential key: 2 [^] π [ENTER]. We obtain

$$2^{\pi} \approx 8.824977827.$$

EXERCISES

Approximate.

1. 5^{π} **2.** $5^{\sqrt{3}}$
3. $7^{-\sqrt{2}}$ **4.** $18^{-\pi}$

In general, a^x is approximated by the values of a^r for rational numbers r near x; a^x is the limit of a^r as r approaches x through rational values. Thus, for $a > 0$, the usual laws of exponents, such as

$$a^x \cdot a^y = a^{x+y}, \qquad a^x \div a^y = a^{x-y}, \qquad (a^x)^y = a^{xy}, \quad \text{and} \quad a^{-x} = \frac{1}{a^x},$$

can be applied to real number exponents. Moreover, the function so obtained, $f(x) = a^x$, is continuous.

DEFINITION

An **exponential function** f is given by

$$f(x) = a^x,$$

where x is any real number, $a > 0$, and $a \neq 1$. The number a is called the **base.**

The following are examples of exponential functions:

$$f(x) = 2^x, \qquad f(x) = \left(\tfrac{1}{2}\right)^x, \qquad f(x) = (0.4)^x.$$

Note that in contrast to power functions like $y = x^2$ and $y = x^3$, in an exponential function, the variable is in the exponent, not the base. Exponential functions have countless applications. For now, however, let's consider their graphs.

EXAMPLE 1 Graph: $y = f(x) = 2^x$.

Solution

a) First, we find some function values. Note that 2^x is always positive:

$$x = 0, \quad y = 2^0 = 1;$$
$$x = \frac{1}{2}, \quad y = 2^{1/2} = \sqrt{2} \approx 1.4;$$
$$x = 1, \quad y = 2^1 = 2;$$
$$x = 2, \quad y = 2^2 = 4;$$
$$x = 3, \quad y = 2^3 = 8;$$
$$x = -1, \quad y = 2^{-1} = \frac{1}{2};$$
$$x = -2, \quad y = 2^{-2} = \frac{1}{2^2} = \frac{1}{4}.$$

x	$y = f(x) = 2^x$
0	1
$\frac{1}{2}$	1.4
1	2
2	4
3	8
-1	$\frac{1}{2}$
-2	$\frac{1}{4}$

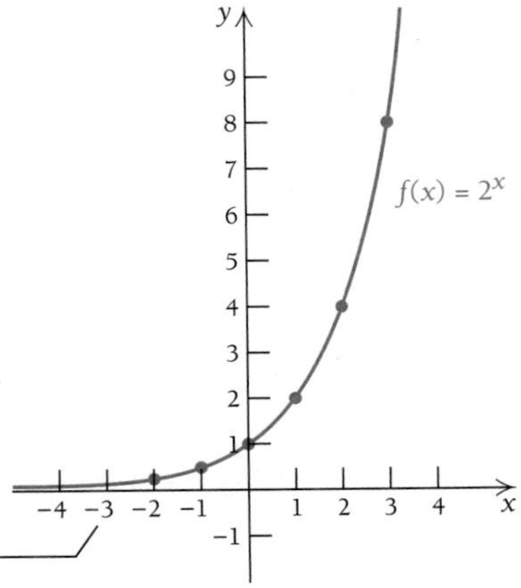

The curve comes very close to the x-axis, but does not touch or cross it. The x-axis is a horizontal asymptote.

b) Next, we plot the points and connect them with a smooth curve, as shown on p. 305. The graph is continuous, increasing without bound, and concave up. We see too that the x-axis is a horizontal asymptote (see Section 2.3), that is,

$$\lim_{x \to -\infty} f(x) = 0 \quad \text{and} \quad \lim_{x \to \infty} f(x) = \infty.$$ ✦

EXAMPLE 2 Graph: $y = f(x) = \left(\frac{1}{2}\right)^x$.

Solution

a) First, we note that

$$y = f(x) = \left(\tfrac{1}{2}\right)^x$$
$$= (2^{-1})^x$$
$$= 2^{-x}.$$

This will ease our work in calculating function values:

x	$y = f(x) = \left(\frac{1}{2}\right)^x$
0	1
$\frac{1}{2}$	0.7
1	$\frac{1}{2}$
2	$\frac{1}{4}$
-1	2
-2	4
-3	8

$x = 0,\quad y = 2^{-0} = 1;$

$x = \frac{1}{2},\quad y = 2^{-1/2} = \frac{1}{2^{1/2}}$

$= \frac{1}{\sqrt{2}} \approx \frac{1}{1.4} \approx 0.7;$

$x = 1,\quad y = 2^{-1} = \frac{1}{2};$

$x = 2,\quad y = 2^{-2} = \frac{1}{4};$

$x = -1,\quad y = 2^{-(-1)} = 2;$

$x = -2,\quad y = 2^{-(-2)} = 4;$

$x = -3,\quad y = 2^{-(-3)} = 8.$

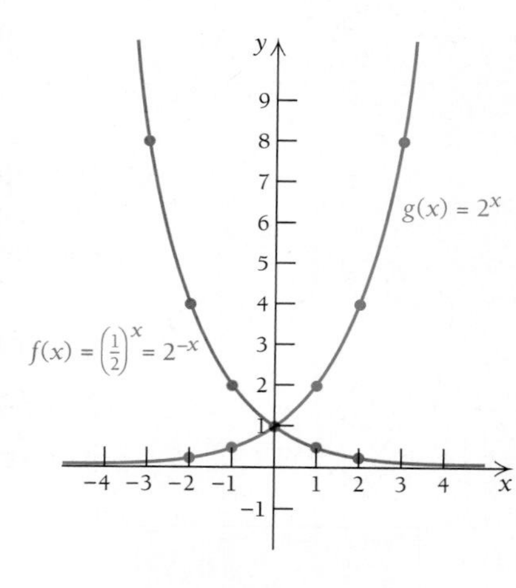

Note that the graph of $y = f(-x)$ is the reflection of the graph of $y = f(x)$ across the y-axis. Thus, we *expect* the graphs of $y = \left(\frac{1}{2}\right)^x = 2^{-x}$ and $y = 2^x$ to be symmetric with respect to the y-axis.

b) Next, we plot these points and connect them with a smooth curve, as shown by the red curve in the figure. The graph is continuous, decreasing, and concave up. We see too that the x-axis is a horizontal asymptote, that is,

$$\lim_{x \to \infty} f(x) = 0 \quad \text{and} \quad \lim_{x \to -\infty} f(x) = \infty.$$

The graph of $g(x) = 2^x$ is shown as a blue curve, for comparison. ✦

TECHNOLOGY CONNECTION

Exploratory Exercise: Growth

Take a sheet of $8\frac{1}{2}$-in.-by-11-in. paper and cut it into two equal pieces. Then cut these again to obtain four equal pieces. Then cut these to get eight equal pieces, and so on, performing five cutting steps.

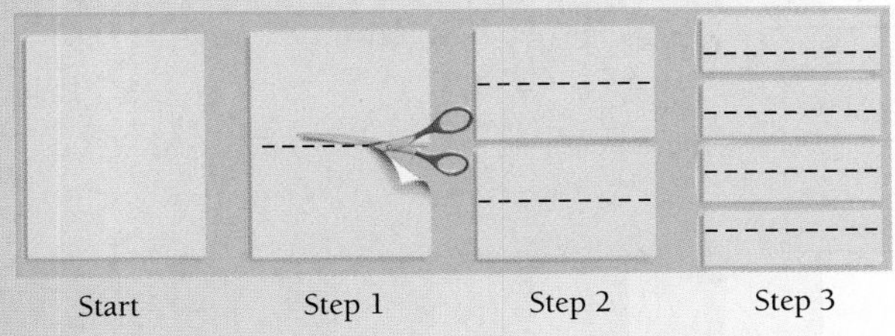

	t	$0.004 \cdot 2^t$
Start	0	$0.004 \cdot 2^0$, or 0.004
Step 1	1	$0.004 \cdot 2^1$, or 0.008
Step 2	2	$0.004 \cdot 2^2$, or 0.016
Step 3	3	
Step 4	4	
Step 5	5	

a) Place all the pieces in a stack and measure the thickness.

b) A piece of paper is typically 0.004 in. thick. Check the measurement in part (a) by completing the table.

c) Graph the function $f(t) = 0.004(2)^t$.

d) Compute the thickness of the stack (in miles) after 25 steps.

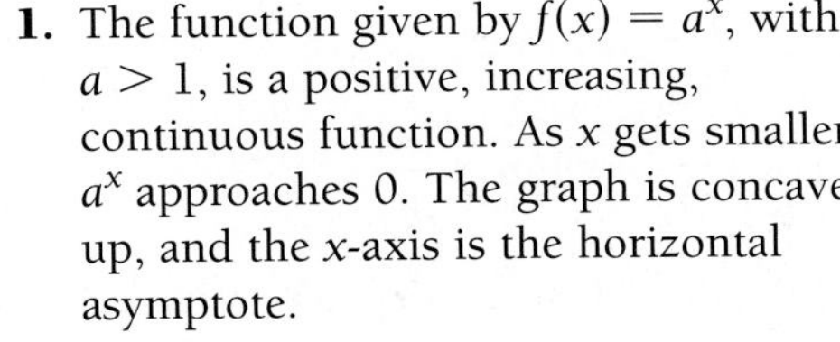

TECHNOLOGY CONNECTION

Exploratory

Check the graphs of the functions in Examples 1 and 2. Then graph

$$f(x) = 3^x \quad \text{and}$$
$$g(x) = \left(\tfrac{1}{3}\right)^x$$

and look for patterns.

1. The function given by $f(x) = a^x$, with $a > 1$, is a positive, increasing, continuous function. As x gets smaller, a^x approaches 0. The graph is concave up, and the x-axis is the horizontal asymptote.

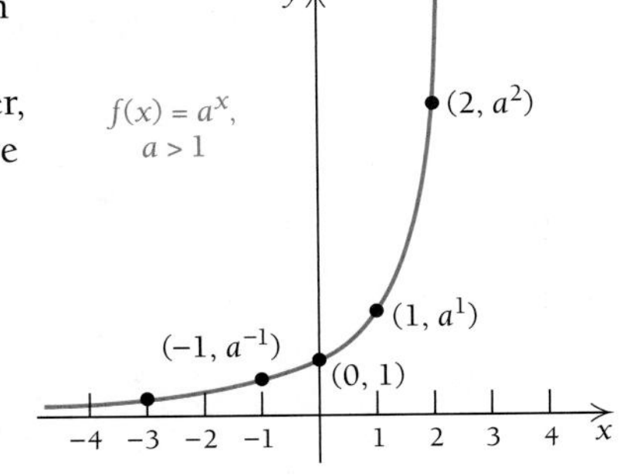

2. The function given by $f(x) = a^x$, with $0 < a < 1$, is a positive, decreasing, continuous function. As x gets larger, a^x approaches 0. The graph is concave up, and the x-axis is the horizontal asymptote.

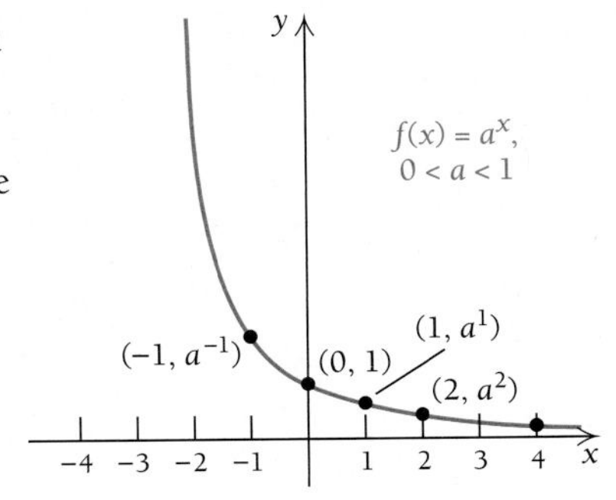

For $a = 1$, we have $f(x) = a^x = 1^x = 1$; so, in this case, f is a constant function. This is why we do not allow 1 to be the base of an exponential function.

The Derivative of a^x; the Number e

Let's consider finding the derivative of the exponential function

$$f(x) = a^x.$$

The derivative is given by

$$\begin{aligned} f'(x) &= \lim_{h \to 0} \frac{f(x+h) - f(x)}{h} && \text{Definition of the derivative} \\ &= \lim_{h \to 0} \frac{a^{x+h} - a^x}{h} && \text{Substituting } a^{x+h} \text{ for } f(x+h) \text{ and } a^x \text{ for } f(x) \\ &= \lim_{h \to 0} \frac{a^x \cdot a^h - a^x \cdot 1}{h} && \text{Using the laws for exponents} \\ &= \lim_{h \to 0} \left(a^x \cdot \frac{a^h - 1}{h} \right) && \text{Factoring} \\ &= a^x \cdot \lim_{h \to 0} \frac{a^h - 1}{h}. && \end{aligned}$$

a^x is constant with respect to h, and the limit of a constant times a function is the constant times the limit of that function.

We get

$$f'(x) = a^x \cdot \lim_{h \to 0} \frac{a^h - 1}{h}.$$

In particular, for $g(x) = 2^x$,

$$g'(x) = 2^x \cdot \lim_{h \to 0} \frac{2^h - 1}{h}. \quad \text{Substituting 2 for } a$$

Note that the limit does not depend on the value of x at which we are evaluating the derivative. In order for $g'(x)$ to exist, we must determine whether

$$\lim_{h \to 0} \frac{2^h - 1}{h} \quad \text{exists.}$$

Let's investigate this question.

We choose a sequence of numbers h approaching 0 and compute $(2^h - 1)/h$, listing the results in a table, as shown at left. It seems reasonable to assume that $(2^h - 1)/h$ has a limit as h approaches 0 and that its approximate value is 0.7; thus,

h	$\frac{2^h - 1}{h}$
0.5	0.8284
0.25	0.7568
0.175	0.7369
0.0625	0.7084
0.03125	0.7007
0.00111	0.6934
0.000001	0.6931

$$g'(x) \approx (0.7)2^x.$$

In other words, the derivative is a constant times 2^x. Similarly, for $t(x) = 3^x$,

$$t'(x) = 3^x \cdot \lim_{h \to 0} \frac{3^h - 1}{h}. \quad \text{Substituting 3 for } a$$

Again, we can find an approximation for the limit that does not depend on the value of x at which we are evaluating the derivative. Consider the table on p. 309. Again,

h	$\frac{3^h - 1}{h}$
0.5	1.4641
0.25	1.2643
0.175	1.2113
0.0625	1.1372
0.03125	1.1177
0.00111	1.0993
0.000001	1.0986

TECHNOLOGY CONNECTION

EXERCISE

Using TBLSET in ASK mode, produce the tables shown on p. 308 and above.

it seems reasonable to conclude that $(3^h - 1)/h$ has a limit as h approaches 0. This time, the approximate value is 1.1; thus,

$$t'(x) \approx (1.1)3^x.$$

In other words, the derivative is a constant times 3^x.

Let's now observe and analyze what we have done. We proved that

$$\text{if } f(x) = a^x, \quad \text{then } f'(x) = a^x \cdot \lim_{h \to 0} \frac{a^h - 1}{h}.$$

Consider $\lim_{h \to 0} \frac{a^h - 1}{h}$.

1. For $a = 2$,

$$\lim_{h \to 0} \frac{a^h - 1}{h} = \lim_{h \to 0} \frac{2^h - 1}{h} \approx 0.7.$$

2. For $a = 3$,

$$\lim_{h \to 0} \frac{a^h - 1}{h} = \lim_{h \to 0} \frac{3^h - 1}{h} \approx 1.1.$$

It seems reasonable to conclude that, for some choice of a between 2 and 3, we have

$$\lim_{h \to 0} \frac{a^h - 1}{h} = 1.$$

To find that a, it suffices to look for a value such that

$$\frac{a^h - 1}{h} = 1. \qquad (1)$$

Multiplying both sides by h and then adding 1 to both sides, we have

$$a^h = 1 + h.$$

Raising both sides to the power $1/h$, we have

$$a = (1 + h)^{1/h}. \qquad (2)$$

Since equations (1) and (2) are equivalent, it follows that

$$\lim_{h \to 0} \frac{a^h - 1}{h} = \lim_{h \to 0} 1 \quad \text{and} \quad \lim_{h \to 0} a = \lim_{h \to 0} (1 + h)^{1/h}$$

are also equivalent. Thus,

$$\lim_{h \to 0} \frac{a^h - 1}{h} = 1 \quad \text{and} \quad a = \lim_{h \to 0} (1 + h)^{1/h}$$

are equivalent. Thus, for there to be a number a for which $\lim_{h \to 0} \frac{a^h - 1}{h} = 1$, we must have $a = \lim_{h \to 0} (1 + h)^{1/h}$. This last equation gives us the special number we are searching for.* The number is named e, in honor of Leonhard Euler (pronounced "Oiler"), the great Swiss mathematician (1707–1783) who did groundbreaking work with it.

h	$(1 + h)^{1/h}$
0.5	2.25
0.1	2.5937
0.01	2.7048
0.001	2.7169
−0.01	2.732
−0.001	2.7196
0.0001	2.7181

DEFINITION

$$e = \lim_{h \to 0} (1 + h)^{1/h} \approx 2.718281828459$$

It follows that, for the exponential function $f(x) = e^x$,

$$f'(x) = e^x \cdot \lim_{h \to 0} \frac{e^h - 1}{h}$$
$$= e^x \cdot 1$$
$$= e^x.$$

That is, the derivative of e^x is e^x.

We have shown that if $f(x) = e^x$, it follows that $f'(x) = e^x$.

THEOREM 1

$$\frac{d}{dx} e^x = e^x$$

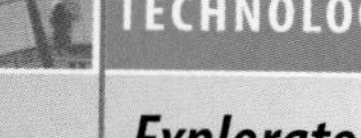

TECHNOLOGY CONNECTION

Exploratory

Graph $f(x) = e^x$ using the viewing window $[-5, 5, -1, 10]$. Trace to a point and note the y-value. Then find dy/dx at that point, using the CALC key, and compare y and dy/dx.

Repeat this process for three other values of x. What do you observe?

Theorem 1 says that for the function $f(x) = e^x$, the derivative at x (the slope of the tangent line) is the same as the function value at x.

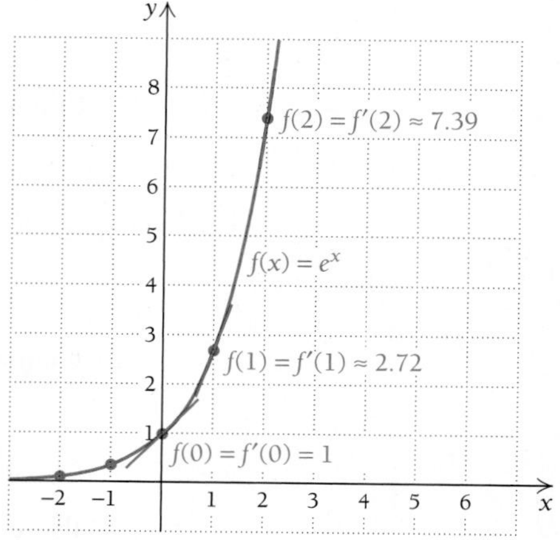

*This derivation is based on one presented in Appendix 4 of *e: The Story of a Number*, by Eli Maor (Princeton University Press, 1998).

Finding Derivatives of Functions Involving *e*

We can use Theorem 1 in combination with other theorems derived earlier to differentiate a variety of functions.

TECHNOLOGY CONNECTION

Exploratory

Check the results of Example 3 by entering each function as y_1 and letting $y_2 = \text{nDeriv}(y_1, x, x)$. Then enter the derivatives from Example 3 as y_3 and use graphs or a table to compare y_2 and y_3.

EXAMPLE 3 Find dy/dx: **a)** $y = 3e^x$; **b)** $y = x^2e^x$; **c)** $y = \dfrac{e^x}{x^3}$.

Solution

a) $\dfrac{d}{dx}(3e^x) = 3\dfrac{d}{dx}e^x$ Recall that $\frac{d}{dx}[c \cdot f(x)] = c \cdot f'(x)$.

$= 3e^x$

b) $\dfrac{d}{dx}(x^2e^x) = x^2 \cdot e^x + e^x \cdot 2x$ Using the Product Rule

$= e^x(x^2 + 2x), \quad \text{or} \quad xe^x(x + 2)$ Factoring

c) $\dfrac{d}{dx}\left(\dfrac{e^x}{x^3}\right) = \dfrac{x^3 \cdot e^x - e^x \cdot 3x^2}{x^6}$ Using the Quotient Rule

$= \dfrac{x^2e^x(x - 3)}{x^6}$ Factoring

$= \dfrac{e^x(x - 3)}{x^4}$ Simplifying ✦

Suppose that we have a more complicated function in the exponent, such as

$$h(x) = e^{x^2-5x}.$$

This is a composition of functions. For such a function, we have

$$h(x) = e^{f(x)} = g(f(x)), \quad \text{where} \quad g(x) = e^x.$$

Now $g'(x) = e^x$. Then by the Chain Rule (Section 1.7), we have

$$\begin{aligned} h'(x) &= g'(f(x)) \cdot f'(x) \\ &= e^{f(x)} \cdot f'(x). \end{aligned}$$

For the case above, $f(x) = x^2 - 5x$, so $f'(x) = 2x - 5$. Then

$$\begin{aligned} h'(x) &= g'(f(x)) \cdot f'(x) \\ &= e^{f(x)} \cdot f'(x) \\ &= e^{x^2-5x}(2x - 5). \end{aligned}$$

The next theorem, which we have proven using the Chain Rule, allows us to find derivatives of functions like the one above.

THEOREM 2

$$\frac{d}{dx}e^{f(x)} = e^{f(x)} \cdot f'(x)$$

or

$$\frac{d}{dx}e^{u} = e^{u} \cdot \frac{du}{dx}$$

The derivative of e to some power is the product of e to that power and the derivative of the power.

The following gives us a way to remember this rule.

$$h(x) = e^{x^2-5x}$$

Rewrite the original function.

$$h'(x) = e^{x^2-5x}(2x - 5)$$

Multiply by the derivative of the exponent.

EXAMPLE 4 Differentiate each of the following with respect to x:

a) $y = e^{8x}$; **b)** $y = e^{-x^2+4x-7}$; **c)** $e^{\sqrt{x^2-3}}$.

Solution

a) $\frac{d}{dx}e^{8x} = e^{8x} \cdot 8$, or $8e^{8x}$

b) $\frac{d}{dx}e^{-x^2+4x-7} = e^{-x^2+4x-7}(-2x + 4)$

c)
$$\begin{aligned}\frac{d}{dx}e^{\sqrt{x^2-3}} &= \frac{d}{dx}e^{(x^2-3)^{1/2}} \\ &= e^{(x^2-3)^{1/2}} \cdot \tfrac{1}{2}(x^2 - 3)^{-1/2} \cdot 2x \qquad \text{Using the Chain Rule twice} \\ &= e^{\sqrt{x^2-3}} \cdot x \cdot (x^2 - 3)^{-1/2} \\ &= \frac{e^{\sqrt{x^2-3}} \cdot x}{\sqrt{x^2 - 3}}\end{aligned}$$

✦

Graphs of e^x, e^{-x}, and $1 - e^{-kx}$

Now that we know how to find the derivative of $f(x) = e^x$, let's look at the graph of $f(x) = e^x$ from the standpoint of calculus concepts and the curve-sketching techniques discussed in Section 2.2.

EXAMPLE 5 Graph: $f(x) = e^x$. Analyze the graph using calculus.

Solution We simply find some function values using a calculator, plot the points, and sketch the graph as shown below.

x	$f(x)$
-2	0.135
-1	0.368
0	1
1	2.718
2	7.389

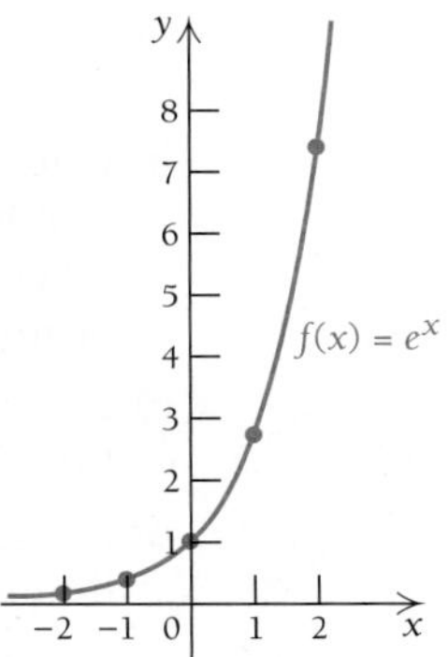

We analyze the graph using calculus as follows.

a) *Derivatives.* Since $f(x) = e^x$, it follows that $f'(x) = e^x$, so $f''(x) = e^x$.

b) *Critical values of f.* Since $f'(x) = e^x > 0$ for all real numbers x, we know that the derivative exists for all real numbers and there is no solution of the equation $f'(x) = 0$. There are no critical values and therefore no maximum or minimum values.

c) *Increasing.* We have $f'(x) = e^x > 0$ for all real numbers x, so the function f is increasing over the entire real line, $(-\infty, \infty)$.

d) *Inflection points.* We have $f''(x) = e^x > 0$ for all real numbers x, so the equation $f''(x) = 0$ has no solution and there are no points of inflection.

e) *Concavity.* Since $f''(x) = e^x > 0$ for all real numbers x, the function f' is increasing and the graph is concave up over the entire real line. ✦

EXAMPLE 6 Graph: $g(x) = e^{-x}$. Analyze the graph using calculus.

Solution First, we find some function values, plot the points, and sketch the graph.

x	$g(x)$
-2	7.389
-1	2.718
0	1
1	0.368
2	0.135

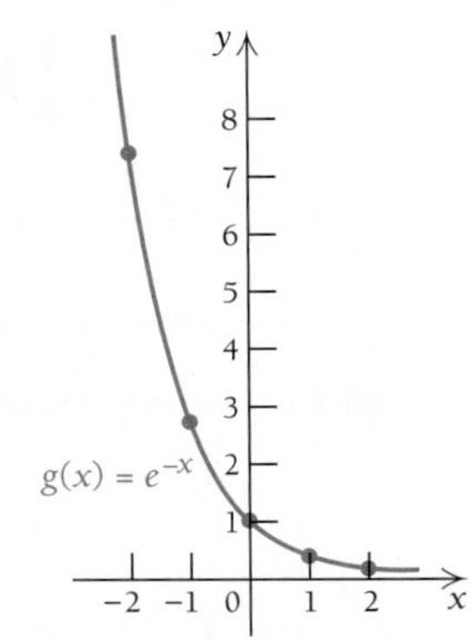

We can then analyze the graph using calculus as follows.

a) *Derivatives.* Since $g(x) = e^{-x}$, we have

$$g'(x) = e^{-x}(-1) = -e^{-x},$$

so

$$g''(x) = -e^{-x}(-1) = e^{-x}.$$

b) *Critical values of g.* Since $e^{-x} = 1/e^x > 0$, we have $g'(x) = -e^{-x} < 0$ for all real numbers x. Thus, the derivative exists for all real numbers, and the equation $g'(x) = 0$ has no solution. There are no critical values and therefore no maximum or minimum values.

c) *Decreasing.* Since the derivative $g'(x) = -e^{-x} < 0$ for all real numbers x, the function g is decreasing over the entire real line.

d) *Inflection points.* We have $g''(x) = e^{-x} > 0$, so the equation $g''(x) = 0$ has no solution and there are no points of inflection.

e) *Concavity.* We also know that since $g''(x) = e^{-x} > 0$ for all real numbers x, the function g' is increasing and the graph is concave up over the entire real line.

✦

Functions of the type $f(x) = 1 - e^{-kx}$, with $x \geq 0$, have important applications.

EXAMPLE 7 Graph: $h(x) = 1 - e^{-2x}$, with $x \geq 0$. Analyze the graph using calculus.

Solution First, we find some function values, plot the points, and sketch the graph.

x	$h(x)$
0	0
0.5	0.63212
1	0.86466
2	0.98168
3	0.99752
4	0.99966
5	0.99995

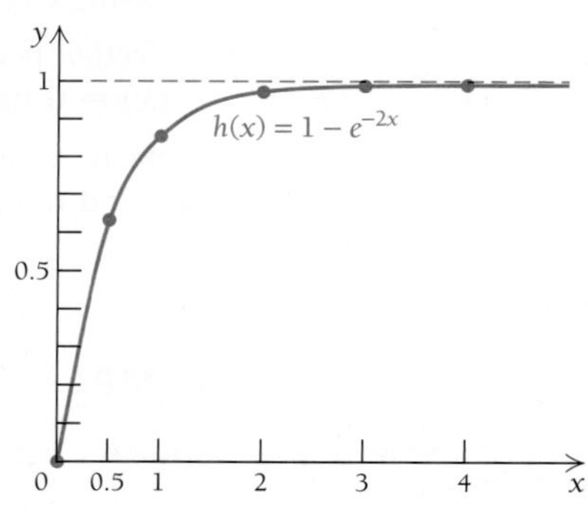

We can analyze the graph using calculus as follows.

a) *Derivatives.* Since $h(x) = 1 - e^{-2x}$,

$$h'(x) = -e^{-2x}(-2) = 2e^{-2x}$$

and

$$h''(x) = 2e^{-2x}(-2) = -4e^{-2x}.$$

b) *Critical values.* Since $e^{-2x} = 1/e^{2x} > 0$, we have $2e^{-2x} > 0$. Thus, $h'(x) = 0$ has no solution, and since $h'(x)$ exists for all $x > 0$, it follows that there are no critical values on the interval $(0, \infty)$

c) *Increasing.* Since $2e^{-2x} > 0$ for all real numbers x, we know that h is increasing over the interval $[0, \infty)$.

d) *Inflection points.* Since $h''(x) = -4e^{-2x} < 0$, we know that the equation $h''(x) = 0$ has no solution; thus, there are no points of inflection.

e) *Concavity.* Since $h''(x) = -4e^{-2x} < 0$, we know that h' is decreasing and the graph is concave down over the interval $(0, \infty)$.

In general, for $k > 0$, the graph of $h(x) = 1 - e^{-kx}$ is increasing, which we expect since $h'(x) = ke^{-kx}$ is always positive. Note that $h(x)$ approaches 1 as x approaches ∞.

A word of caution! Functions of the type a^x (for example, 2^x, 3^x, and e^x) are different from functions of the type x^a (for example, x^2, x^3, $x^{1/2}$). For a^x, the variable is in the exponent. For x^a, the variable is in the base. The derivative of a^x is not xa^{x-1}. In particular, we have the following:

$$\frac{d}{dx}e^x \neq xe^{x-1}, \quad \text{but} \quad \frac{d}{dx}e^x = e^x.$$

Exercise Set 3.1

Graph.

1. $y = 4^x$
2. $y = 5^x$
3. $y = (0.25)^x$
4. $y = (0.2)^x$
5. $f(x) = \left(\frac{3}{2}\right)^x$
6. $f(x) = \left(\frac{4}{3}\right)^x$
7. $g(x) = \left(\frac{2}{3}\right)^x$
8. $g(x) = \left(\frac{3}{4}\right)^x$
9. $f(x) = (2.5)^x$
10. $f(x) = (1.2)^x$

Differentiate.

11. $f(x) = e^x$
12. $f(x) = e^{-x}$
13. $g(x) = e^{2x}$
14. $g(x) = e^{3x}$
15. $f(x) = 6e^x$
16. $f(x) = 4e^x$
17. $F(x) = e^{-7x}$
18. $F(x) = e^{-4x}$
19. $G(x) = 2e^{4x}$
20. $g(x) = 3e^{5x}$
21. $f(x) = -3e^{-x}$
22. $G(x) = -7e^{-x}$
23. $g(x) = \frac{1}{2}e^{-5x}$
24. $f(x) = \frac{1}{3}e^{-4x}$
25. $F(x) = -\frac{2}{3}e^{x^2}$
26. $g(x) = -\frac{4}{5}e^{x^3}$
27. $G(x) = 7 + 3e^{5x}$
28. $F(x) = 4 - e^{2x}$
29. $f(x) = x^5 - 2e^{6x}$
30. $G(x) = x^3 - 5e^{2x}$
31. $g(x) = x^5e^{2x}$
32. $f(x) = x^7e^{4x}$
33. $F(x) = \dfrac{e^{2x}}{x^4}$
34. $g(x) = \dfrac{e^{3x}}{x^6}$
35. $f(x) = (x^2 + 3x - 9)e^x$
36. $f(x) = (x^2 - 2x + 2)e^x$
37. $f(x) = \dfrac{e^x}{x^4}$
38. $f(x) = \dfrac{e^x}{x^5}$
39. $f(x) = e^{-x^2+7x}$
40. $f(x) = e^{-x^2+8x}$
41. $f(x) = e^{-x^2/2}$
42. $f(x) = e^{x^2/2}$
43. $y = e^{\sqrt{x-7}}$
44. $y = e^{\sqrt{x-4}}$
45. $y = \sqrt{e^x - 1}$
46. $y = \sqrt{e^x + 1}$
47. $y = xe^{-2x} + e^{-x} + x^3$
48. $y = e^x + x^3 - xe^x$
49. $y = 1 - e^{-x}$
50. $y = 1 - e^{-3x}$
51. $y = 1 - e^{-kx}$
52. $y = 1 - e^{-mx}$
53. $g(x) = (4x^2 + 3x)e^{x^2-7x}$
54. $g(x) = (5x^2 - 8x)e^{x^2-4x}$

Graph each function. Then determine critical values, inflection points, regions over which the function is increasing or decreasing, and the concavity.

55. $f(x) = e^{2x}$

56. $g(x) = e^{-2x}$

57. $g(x) = e^{(1/2)x}$

58. $f(x) = e^{(1/3)x}$

59. $f(x) = \frac{1}{2}e^{-x}$

60. $g(x) = \frac{1}{3}e^{-x}$

61. $F(x) = -e^{(1/3)x}$

62. $G(x) = -e^{(1/2)x}$

63. $g(x) = 2(1 - e^{-x})$, for $x \geq 0$

64. $f(x) = 3 - e^{-x}$, for $x \geq 0$

65–74. For each function given in Exercises 55–64, graph the function and its first and second derivatives using a calculator.

75. Find the slope of the line tangent to the graph of $f(x) = e^x$ at the point $(0, 1)$.

76. Find the slope of the line tangent to the graph of $f(x) = 2e^{-3x}$ at the point $(0, 2)$.

77. Find an equation of the line tangent to the graph of $g(x) = e^{-x}$ at the point $(0, 1)$.

78. Find an equation of the line tangent to the graph of $f(x) = e^{2x}$ at the point $(0, 1)$.

79. and 80. For each of Exercises 77 and 78, graph the function and the tangent line using a calculator.

APPLICATIONS

Business and Economics

81. Organic food. More Americans are buying organic fruit and vegetables and products made with organic ingredients. The amount $A(t)$, in billions of dollars, spent on organic food and beverages t years after 1995 can be approximated by

$$A(t) = 2.43e^{0.18t}, \quad 0 \leq t \leq 15.$$

(**Source**: Based on data and projections from *Nutrition Business Journal*, 2004.)

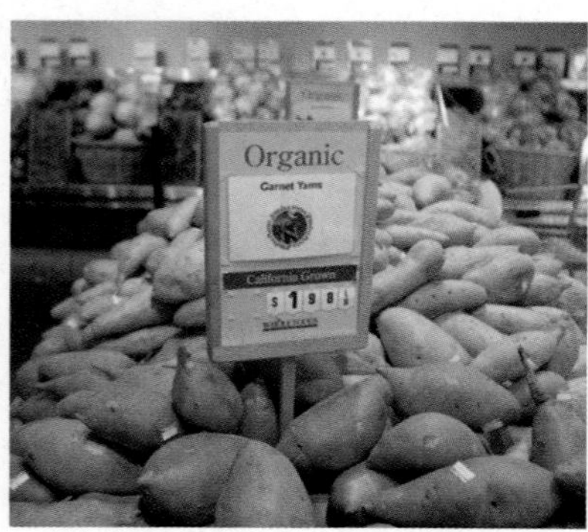

a) Estimate the amount that Americans will spend on organic food and beverages in 2009.

b) Estimate the rate at which spending on organic food and beverages was growing in 2006.

82. Organic food. Refer to Exercise 81.

a) By how much did spending on organic food and beverages increase between 2006 and 2007?

TW **b)** Compare your answer for part (a) with the answer for part (b) of Exercise 81, and explain why the values are similar.

Cellular phones. *The total number of cellular phones, in thousands, in use in Israel approximately follows the function*

$$T(x) = \frac{5200}{1 + 16.3e^{-0.7x}},$$

where x is the number of years since 1994. (**Source**: *A. Tischler, R. Ventura, and J. Watters, "Cellular Telephones in the Israeli Market: The Demand, the Choice of Provider and Potential Revenues," Applied Economics, Vol. 33 (2001), 1479–1492.) Use this function for Exercises 83 and 84.*

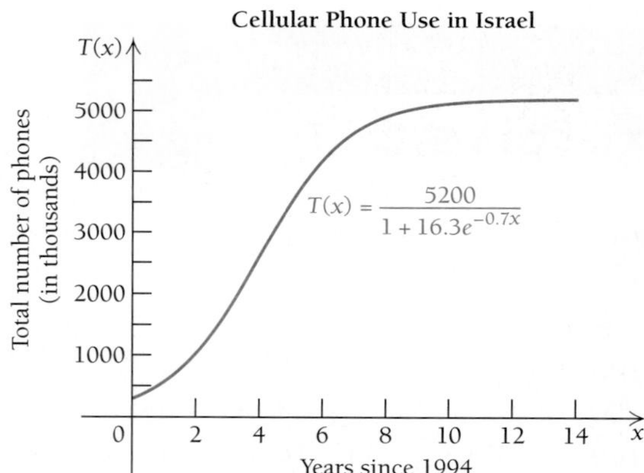

83. a) Estimate the total number of cellular phones in use in Israel in 2005 and 2010.

b) Find $\lim_{x \to \infty} T(x)$. Interpret this limit.

84. a) Find $T'(x)$.

b) Use $T'(x)$ to estimate the number of additional cell phones that will be put into use in Israel between 2009 and 2010.

85. Marginal cost. A company's total cost, in millions of dollars, is given by

$$C(t) = 100 - 50e^{-t},$$

where t is the time in years since the start-up date.

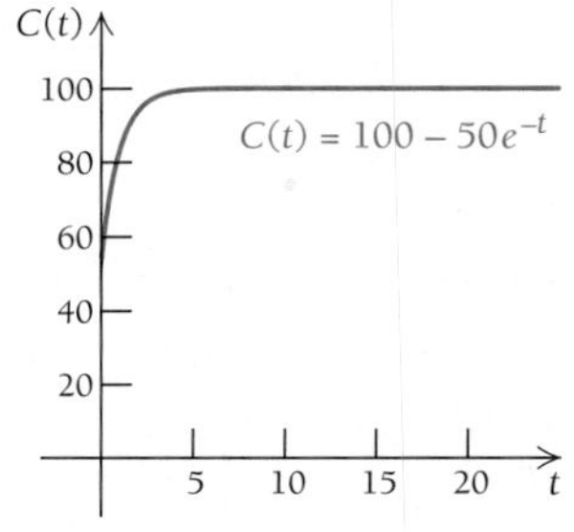

Find each of the following.

a) The marginal cost, $C'(t)$

b) $C'(0)$

c) $C'(4)$ (Round to the nearest thousand.)

TW **d)** Find $\lim_{t\to\infty} C(t)$ and $\lim_{t\to\infty} C'(t)$. Why do you think the company's costs tend to level off as time increases?

86. Marginal cost. A company's total cost, in millions of dollars, is given by

$$C(t) = 200 - 40e^{-t},$$

where t is the time in years since the start-up date.

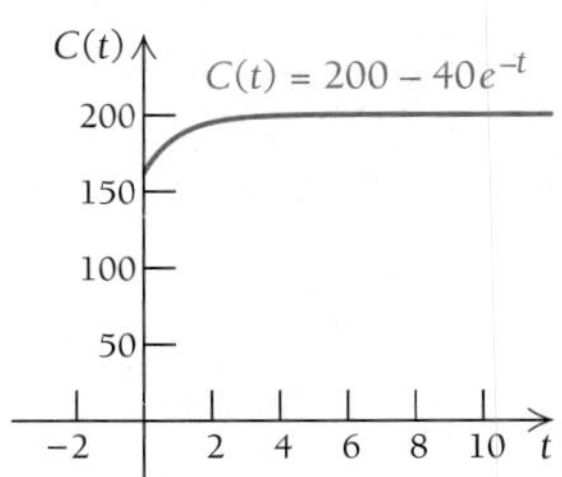

Find each of the following.

a) The marginal cost $C'(t)$

b) $C'(0)$

c) $C'(5)$ (Round to the nearest thousand.)

TW **d)** Find $\lim_{t\to\infty} C(t)$ and $\lim_{t\to\infty} C'(t)$. Why do you think the company's costs tend to level off as time increases?

87. Marginal demand. At a price of x dollars, the demand, in thousands of units, for a certain MP3 player is given by the demand function

$$q = 240e^{-0.003x}.$$

a) How many MP3 players will be bought at a price of \$250? Round to the nearest thousand.

b) Graph the demand function for $0 \le x \le 400$.

c) Find the marginal demand, $q'(x)$.

TW **d)** Interpret the meaning of the derivative.

88. Marginal supply. At a price of x dollars, the supply function for the MP3 player in Exercise 87 is given by

$$q = 75e^{0.004x},$$

where q is in thousands of units.

a) How many MP3 players will be supplied at a price of \$250? Round to the nearest thousand.

b) Graph the supply function for $0 \le x \le 400$.

c) Find the marginal supply, $q'(x)$.

TW **d)** Interpret the meaning of the derivative.

Life and Physical Sciences

89. Medication concentration. The concentration C, in parts per million, of a medication in the body t hours after ingestion is given by the function

$$C(t) = 10t^2e^{-t}.$$

a) Find the concentration after 0 hr; 1 hr; 2 hr; 3 hr; 10 hr.

b) Sketch a graph of the function for $0 \le t \le 10$.

c) Find the rate of change of the concentration, $C'(t)$.

d) Find the maximum value of the concentration and the time at which it occurs.

TW **e)** Interpret the meaning of the derivative.

Death rate. The annual death rate (per thousand people) in Mexico can be modeled by the function

$$D(t) = 34.4 - \frac{30.48}{1 + 29.44e^{-0.072t}},$$

where t is the number of years since 1900. (***Source***: B. J. L. Berry, L. S. Hall, R. Hernandez-Guerrero, and P. H. Martin, "México's Demographic Transition: Public Policy and Spatial Process," *Population and Environment*, Vol. 21, 363–383 (2000).) Use this function for Exercises 90 and 91.

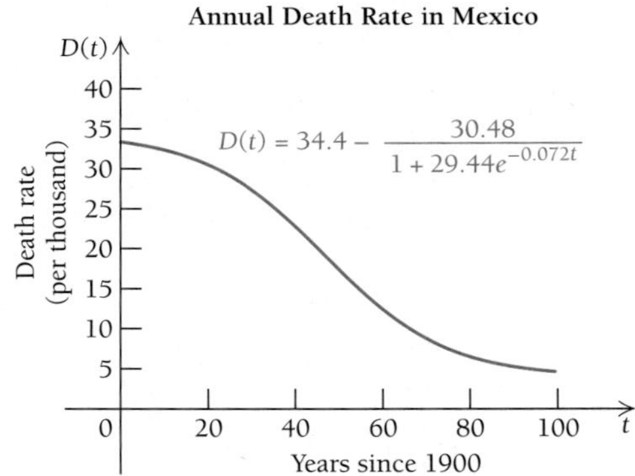

90. **a)** According to this model, what will Mexico's death rate be in 2010? In 2050?

b) Find and interpret $\lim_{t\to\infty} D(t)$.

91. **a)** Find $D'(t)$, and interpret its meaning.

b) Use $D'(t)$ to estimate how much Mexico's death rate will decline between 2009 and 2010.

Social Sciences

92. **Ebbinghaus learning model.** Suppose that you are given the task of learning 100% of a block of knowledge. Human nature tells us that we would retain only a percentage P of the knowledge t weeks after we have learned it. The *Ebbinghaus learning model* asserts that P is given by

$$P(t) = Q + (100 - Q)e^{-kt},$$

where Q is the percentage that we would never forget and k is a constant that depends on the knowledge learned. Suppose that $Q = 40$ and $k = 0.7$.

a) Find the percentage retained after 0 weeks; 1 week; 2 weeks; 6 weeks; 10 weeks.

b) Find $\lim_{t\to\infty} P(t)$.

c) Sketch a graph of P.

d) Find the rate of change of P with respect to time t.

TW **e)** Interpret the meaning of the derivative.

SYNTHESIS

Differentiate.

93. $y = (e^{3x} + 1)^5$

94. $y = (e^{x^2} - 2)^4$

95. $y = \dfrac{e^{3t} - e^{7t}}{e^{4t}}$

96. $y = \sqrt[3]{e^{3t} + t}$

97. $y = \dfrac{e^x}{x^2 + 1}$

98. $y = \dfrac{e^x}{1 - e^x}$

99. $f(x) = e^{\sqrt{x}} + \sqrt{e^x}$

100. $f(x) = \dfrac{1}{e^x} + e^{1/x}$

101. $f(x) = e^{x/2} \cdot \sqrt{x - 1}$

102. $f(x) = \dfrac{xe^{-x}}{1 + x^2}$

103. $f(x) = \dfrac{e^x - e^{-x}}{e^x + e^{-x}}$

104. $f(x) = e^{e^x}$

Exercises 105 and 106 each give an expression for e. Find the function values that are approximations for e. Round to five decimal places.

105. For $f(t) = (1 + t)^{1/t}$ we have $e = \lim_{t\to 0} f(t)$. Find $f(1), f(0.5), f(0.2), f(0.1)$, and $f(0.001)$.

106. For $g(t) = t^{1/(t-1)}$, we have $e = \lim_{t\to 1} g(t)$. Find $g(0.5)$, $g(0.9)$, $g(0.99)$, $g(0.999)$, and $g(0.9998)$.

107. Find the maximum value of $f(x) = x^2e^{-x}$ over $[0, 4]$.

108. Find the minimum value of $f(x) = xe^x$ over $[-2, 0]$.

TW **109.** A student made the following error on a test:

$$\frac{d}{dx}e^x = xe^{x-1}.$$

Explain the error and how to correct it.

TW **110.** Describe the differences in the graphs of $f(x) = 3^x$ and $g(x) = x^3$.

TECHNOLOGY CONNECTION

Graph each function in Exercises 111 and 112, and find all relative extrema.

111. $f(x) = x^2e^{-x}$

112. $f(x) = e^{-x^2}$

For each of the functions in Exercises 113–116, graph f, f′, and f″.

113. $f(x) = e^x$

114. $f(x) = e^{-x}$

115. $f(x) = 2e^{0.3x}$

116. $f(x) = 1000e^{-0.08x}$

117. Graph

$$f(x) = \left(1 + \frac{1}{x}\right)^x.$$

Use the TABLE feature and very large values of x to confirm that e is approached as a limit.

3.2 Logarithmic Functions

OBJECTIVES

- Convert between exponential and logarithmic equations.
- Solve exponential equations.
- Solve problems involving exponential and logarithmic functions.
- Differentiate functions involving natural logarithms.

Graphs of Logarithmic Functions

Suppose that we want to solve the equation

$$10^y = 1000.$$

We are trying to find the power of 10 that will give 1000. Since $10^3 = 1000$, the answer is 3. The number 3 is called the "logarithm, base 10, of 1000."

DEFINITION

A **logarithm** is defined as follows:

$$\log_a x = y \quad \text{means} \quad a^y = x, \quad a > 0, a \neq 1.$$

The number $\log_a x$ is the power y to which we raise a to get x. The number a is called the *logarithmic base.* We read $\log_a x$ as "the logarithm, base a, of x."

For logarithms base 10, $\log_{10} x$ is the power y such that $10^y = x$. Therefore, a logarithm can be thought of as an exponent. We can convert from a logarithmic equation to an exponential equation, and conversely, as follows.

Logarithmic Equation	Exponential Equation
$\log_a M = N$	$a^N = M$
$\log_{10} 100 = 2$ $\log_5 \frac{1}{25} = -2$ $\log_{49} 7 = \frac{1}{2}$	$10^2 = 100$ $5^{-2} = \frac{1}{25}$ $49^{1/2} = 7$

TECHNOLOGY CONNECTION

Exploratory

Graph $f(x) = 2^x$. Then use the TABLE feature to find the coordinates of points on the graph. How can each ordered pair help you make a hand drawing of a graph of $g(x) = \log_2 x$?

In order to graph a logarithmic equation, we can graph its equivalent exponential equation.

EXAMPLE 1 Graph: $y = \log_2 x$.

Solution We first write the equivalent exponential equation:

$$2^y = x.$$

We select values for y and find the corresponding values of 2^y. Then we plot points, remembering that x is still the first coordinate, and connect the points with a smooth curve.

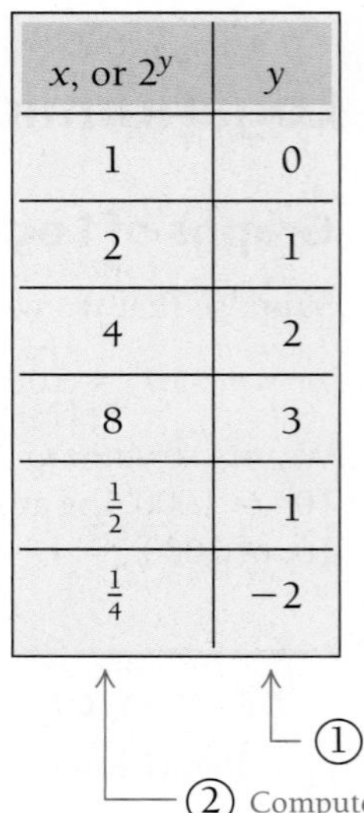

x, or 2^y	y
1	0
2	1
4	2
8	3
$\frac{1}{2}$	-1
$\frac{1}{4}$	-2

① Select y.

② Compute x.

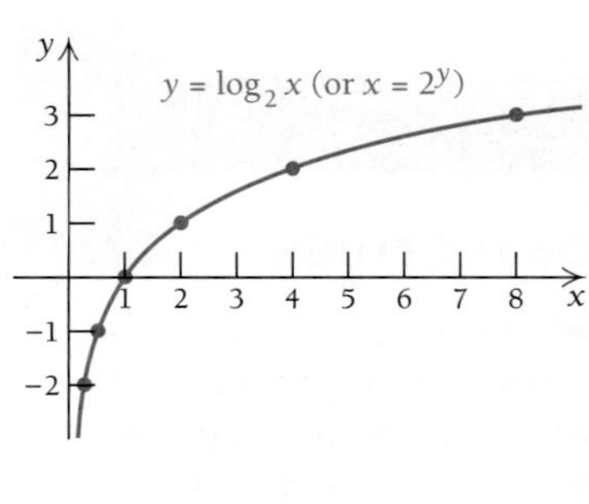

✦

The graphs of $f(x) = 2^x$ and $g(x) = \log_2 x$ are shown below on the same set of axes. Note that we can obtain the graph of g by reflecting the graph of f across the line $y = x$. Graphs that can be obtained in this manner are known as *inverses* of each other.

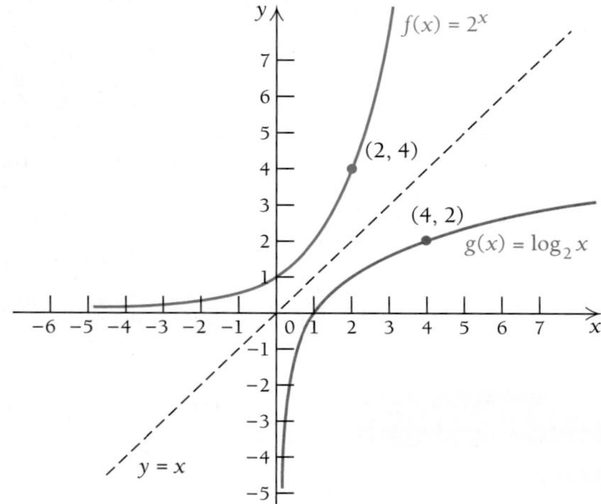

Although we cannot develop inverses in detail here, it is important to note that they "undo" each other. For example,

$$f(3) = 2^3 = 8 \qquad \text{The input 3 gives the output 8.}$$

and $$g(8) = \log_2 8 = 3. \qquad \text{The input 8 gets us back to 3.}$$

TECHNOLOGY CONNECTION

Graphing Logarithmic Functions

To graph $y = \log_2 x$, we first graph $y_1 = 2^x$. We next select the DrawInv option from the DRAW menu and then the Y-VARS option from the VARS menu, followed by 1, 1, and ENTER to draw the inverse of y_1. Both graphs are drawn together.

$y_1 = 2^x, \; y_2 = \log_2 x$

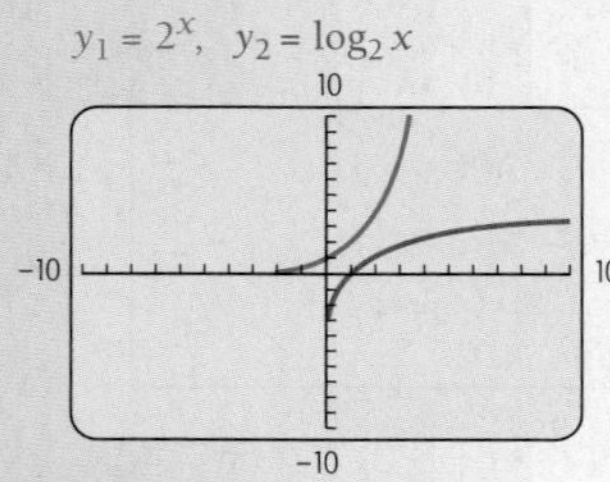

EXERCISES

Graph.

1. $y = \log_3 x$ **2.** $y = \log_5 x$

3. $f(x) = \log_e x$ **4.** $f(x) = \log_{10} x$

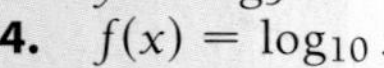

Basic Properties of Logarithms

The following are some basic properties of logarithms. The proofs of P1–P3 follow from properties of exponents and are outlined in Exercises 103–105 at the end of this section. Properties P4–P6 follow directly from the definition of logarithm, and a proof of P7 is outlined in Exercise 106.

THEOREM 3 Properties of Logarithms

For any positive numbers M, N, a, and b, with $a, b \neq 1$, and any real number k:

P1. $\log_a (MN) = \log_a M + \log_a N$

P2. $\log_a \frac{M}{N} = \log_a M - \log_a N$

P3. $\log_a (M^k) = k \cdot \log_a M$

P4. $\log_a a = 1$

P5. $\log_a (a^k) = k$

P6. $\log_a 1 = 0$

P7. $\log_b M = \frac{\log_a M}{\log_a b}$ (The change-of-base formula)

Let's illustrate these properties.

EXAMPLE 2 Given

$$\log_a 2 = 0.301 \quad \text{and} \quad \log_a 3 = 0.477,$$

find each of the following: **a)** $\log_a 6$; **b)** $\log_a \frac{2}{3}$; **c)** $\log_a 81$; **d)** $\log_a \frac{1}{3}$; **e)** $\log_a \sqrt{a}$; **f)** $\log_a (2a)$; **g)** $\log_a 5$; **h)** $\frac{\log_a 3}{\log_a 2}$.

Solution

a) $\log_a 6 = \log_a (2 \cdot 3)$
$= \log_a 2 + \log_a 3$ By P1
$= 0.301 + 0.477$
$= 0.778$

b) $\log_a \frac{2}{3} = \log_a 2 - \log_a 3$ By P2
$= 0.301 - 0.477$
$= -0.176$

c) $\log_a 81 = \log_a 3^4$
$= 4 \log_a 3$ By P3
$= 4(0.477)$
$= 1.908$

d) $\log_a \frac{1}{3} = \log_a 1 - \log_a 3$ By P2

$= 0 - 0.477$ By P6

$= -0.477$

e) $\log_a \sqrt{a} = \log_a (a^{1/2}) = \frac{1}{2}$ By P5

f) $\log_a (2a) = \log_a 2 + \log_a a$ By P1

$= 0.301 + 1$ By P4

$= 1.301$

g) *There is no way to find* $\log_a 5$ *using the properties of logarithms.*
$(\log_a 5 \neq \log_a 2 + \log_a 3)$

h) $\dfrac{\log_a 3}{\log_a 2} = \dfrac{0.477}{0.301} \approx 1.58$

We simply divided and used none of the properties. ✦

Common Logarithms

The number $\log_{10} x$ is the **common logarithm** of x and is abbreviated $\log x$; that is:

DEFINITION

For any positive number x, $\log x = \log_{10} x$.

Thus, when we write "$\log x$" with no base indicated, base 10 is understood. Note the following comparison of common logarithms and powers of 10.

$1000 = 10^3$	The common logarithms at the right follow from the powers at the left.	$\log 1000 = 3$
$100 = 10^2$		$\log 100 = 2$
$10 = 10^1$		$\log 10 = 1$
$1 = 10^0$		$\log 1 = 0$
$0.1 = 10^{-1}$		$\log 0.1 = -1$
$0.01 = 10^{-2}$		$\log 0.01 = -2$
$0.001 = 10^{-3}$		$\log 0.001 = -3$

Since $\log 100 = 2$ and $\log 1000 = 3$, it seems reasonable that $\log 500$ is somewhere between 2 and 3. Tables were originally used for such approximations, but since the advent of the calculator, logarithm tables are rarely used. Using a calculator with a LOG key, we find that $\log 500 \approx 2.6990$.

Before calculators became so readily available, common logarithms were used extensively to do certain computations. In fact, computation is the reason logarithms were developed. Since standard notation for numbers is based on 10, it was logical to use base-10, or common, logarithms for computations. Today, computations with common logarithms are mainly of historical interest; the logarithmic functions, base e, are far more important.

Natural Logarithms

The number e, which is approximately 2.718282, was developed in Section 3.1, and has extensive application in many fields. The number $\log_e x$ is the **natural logarithm** of x and is abbreviated $\ln x$.

DEFINITION

For any positive number x, $\ln x = \log_e x$.

The following basic properties of natural logarithms parallel those given earlier for logarithms in general.

THEOREM 4 Properties of Natural Logarithms

P1. $\ln (MN) = \ln M + \ln N$

P2. $\ln \dfrac{M}{N} = \ln M - \ln N$

P3. $\ln (a^k) = k \cdot \ln a$

P4. $\ln e = 1$

P5. $\ln (e^k) = k$

P6. $\ln 1 = 0$

P7. $\log_b M = \dfrac{\ln M}{\ln b}$ and $\ln M = \dfrac{\log M}{\log e}$

TECHNOLOGY CONNECTION

To enter $y = \log_{10} x$, the key labeled LOG can be used.

EXERCISES

1. Graph $f(x) = 10^x$, $y = x$, and $g(x) = \log_{10} x$ using the same set of axes. Then find $f(3)$, $f(0.699)$, $g(5)$, and $g(1000)$.
2. Use the LOG key and P7 of Theorem 4 to graph $y = \log_2 x$.

Let's illustrate the properties of Theorem 4.

EXAMPLE 3 Given

$$\ln 2 = 0.6931 \quad \text{and} \quad \ln 3 = 1.0986,$$

find each of the following: **a)** $\ln 6$; **b)** $\ln 81$; **c)** $\ln \frac{1}{3}$; **d)** $\ln (2e^5)$; **e)** $\log_2 3$.

Solution

a) $\ln 6 = \ln (2 \cdot 3) = \ln 2 + \ln 3$ By P1
$= 0.6931 + 1.0986$
$= 1.7917$

b) $\ln 81 = \ln (3^4)$
$= 4 \ln 3$ By P3
$= 4(1.0986)$
$= 4.3944$

c) $\ln \frac{1}{3} = \ln 1 - \ln 3$ By P2
$= 0 - 1.0986$ By P6
$= -1.0986$

d) $\ln(2e^5) = \ln 2 + \ln(e^5)$ By P1

$= 0.6931 + 5$ By P5

$= 5.6931$

e) $\log_2 3 = \dfrac{\ln 3}{\ln 2} = \dfrac{1.0986}{0.6931} = 1.5851$ By P7 ✦

Finding Natural Logarithms Using a Calculator

You should have a calculator with an **LN** key. You can find natural logarithms directly using this key.

EXAMPLE 4 Approximate each of the following to six decimal places:

a) $\ln 5.24$;   **b)** $\ln 0.001278$.

Solution We use a calculator with an **LN** key.

a) $\ln 5.24 \approx 1.656321$ **b)** $\ln 0.001278 \approx -6.662459$ ✦

Exponential Equations

If an equation contains a variable in an exponent, the equation is **exponential.** We can use logarithms to manipulate or solve exponential equations.

EXAMPLE 5 Solve $e^t = 40$ for t.

Solution We have

$\ln e^t = \ln 40$ Taking the natural logarithm on both sides

$t = \ln 40$ By P5; remember that $\ln e^t$ means $\log_e e^t$.

$t \approx 3.688879$ Using a calculator

$t \approx 3.7$

Note that this is an approximation for t even though an equals sign is often used. ✦

EXAMPLE 6 Solve $e^{-0.04t} = 0.05$ for t.

Solution We have

$\ln e^{-0.04t} = \ln 0.05$ Taking the natural logarithm on both sides

$-0.04t = \ln 0.05$ By P5

$$t = \frac{\ln 0.05}{-0.04}$$

$$t \approx \frac{-2.995732}{-0.04}$$ Using a calculator

$t \approx 75.$ ✦

In Example 6, we rounded $\ln 0.05$ to -2.995732 in an intermediate step. When using a calculator, you should find

$$\frac{\ln 0.05}{-0.04}$$

by keying in

pressing ENTER, and rounding at the end. Answers at the back of this book have been found in this manner. Remember, the number of places in a table or on a calculator affects the accuracy of the answer. Usually, your answer should agree with that in the Answers section to at least three digits.

TECHNOLOGY CONNECTION

Solving Exponential Equations

Let's solve the equation of Example 5, $e^t = 40$, graphically.

Method 1: The INTERSECT Feature

We change the variable to x and consider the system of equations $y_1 = e^x$ and $y_2 = 40$. We graph the equations in the window $[-1, 8, -10, 70]$ to see the curvature and point of intersection.

$y_1 = e^x, \quad y_2 = 40$

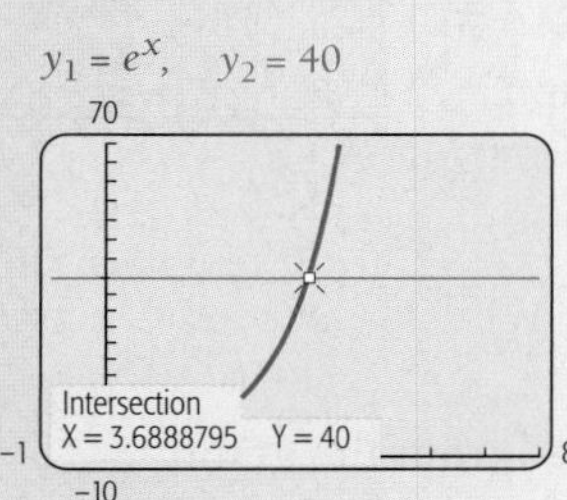

Then we use the INTERSECT option from the CALC menu to find the point of intersection, about $(3.7, 40)$. The x-coordinate, 3.7, is the solution of $e^t = 40$.

Method 2: The ZERO Feature

We change the variable to x and get a 0 on one side of the equation: $e^x - 40 = 0$. Then we graph $y = e^x - 40$ in the window $[-1, 8, -10, 10]$.

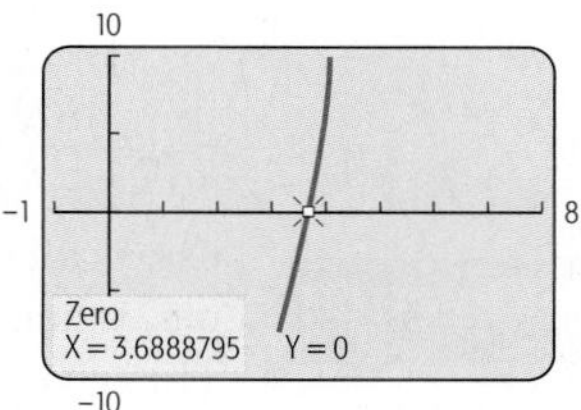

Using the ZERO option from the CALC menu, we see that the x-intercept is about $(3.7, 0)$, so 3.7 is the solution of $e^t = 40$.

EXERCISES

Solve using a calculator.

1. $e^t = 1000$
2. $e^{-x} = 60$
3. $e^{-0.04t} = 0.05$
4. $e^{0.23x} = 41{,}378$
5. $15e^{0.2x} = 34{,}785.13$

Graphs of Natural Logarithmic Functions

There are two ways in which we might graph $y = f(x) = \ln x$. One is to graph the equivalent equation $x = e^y$ by selecting values for y and calculating the corresponding values of e^y. We then plot points, remembering that x is still the first coordinate.

x, or e^y	y
0.1	−2
0.4	−1
1.0	0
2.7	1
7.4	2
20.1	3

① Select y.
② Compute x.

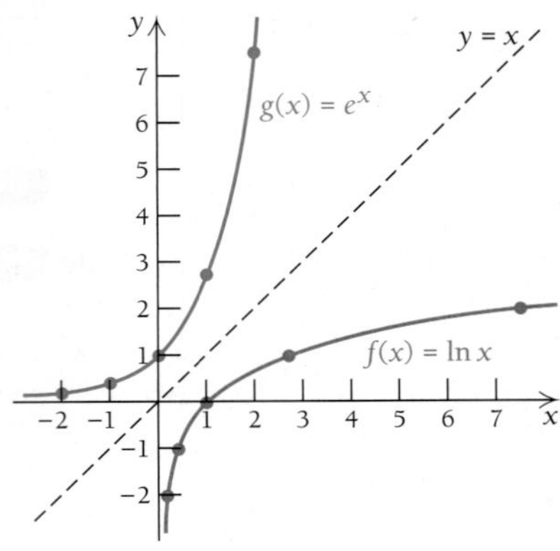

The graph above shows the graph of $g(x) = e^x$ for comparison. Note again that the functions are inverses of each other. That is, the graph of $y = \ln x$, or $x = e^y$, is a reflection, or mirror image, across the line $y = x$ of the graph of $y = e^x$. Any ordered pair (a, b) on the graph of g yields an ordered pair (b, a) on f. Note too that $\lim_{x \to 0^+} \ln x = -\infty$ and the y-axis is a vertical asymptote.

The second method of graphing $y = \ln x$ is to use a calculator to find function values. For example, when $x = 2$, then $y = \ln 2 \approx 0.6931 \approx 0.7$. This gives the pair $(2, 0.7)$ shown on the graph.

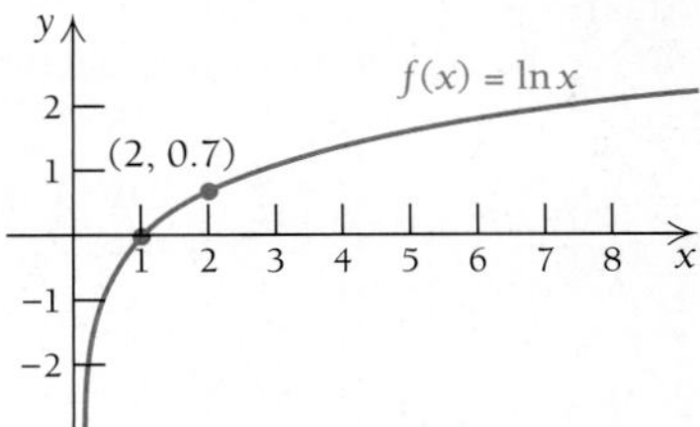

TECHNOLOGY CONNECTION

Graphing Logarithmic Functions

To graph $y = \ln x$, we can use the LN key and enter the function as $y_1 = \ln(x)$.

The following properties can be observed from the graph.

THEOREM 5

$\ln x$ exists only for positive numbers x. The domain is $(0, \infty)$.

$\ln x < 0$ for $0 < x < 1$.

$\ln x = 0$ when $x = 1$.

$\ln x > 0$ for $x > 1$.

The function given by $f(x) = \ln x$ is always increasing. The range is the entire real line, $(-\infty, \infty)$.

Derivatives of Natural Logarithmic Functions

Let's find the derivative of

$$f(x) = \ln x. \tag{1}$$

We first write its equivalent exponential equation:

$$e^{f(x)} = x. \quad \text{$\ln x = \log_e x = f(x)$, so $e^{f(x)} = x$, by the definition of logarithms.} \tag{2}$$

Now we differentiate on both sides of this equation:

$$\frac{d}{dx}e^{f(x)} = \frac{d}{dx}x$$

$$e^{f(x)} \cdot f'(x) = 1 \quad \text{By the Chain Rule}$$

$$x \cdot f'(x) = 1 \quad \text{Substituting x for $e^{f(x)}$ from equation (2)}$$

$$f'(x) = \frac{1}{x}.$$

Thus, we have the following.

THEOREM 6

For any positive number x,

$$\frac{d}{dx}\ln x = \frac{1}{x}.$$

Theorem 6 asserts that for the function $f(x) = \ln x$, to find the slope of the tangent line at x, we need only take the reciprocal of x. This is true only for positive values of x, since $\ln x$ is defined only for positive numbers. (For negative numbers x, this derivative formula becomes

$$\frac{d}{dx}\ln |x| = \frac{1}{x},$$

but we will seldom consider such a case in this text.)

Let's find some derivatives.

EXAMPLE 7 Differentiate: **a)** $y = 3 \ln x$; **b)** $y = x^2 \ln x + 5x$; **c)** $y = \dfrac{\ln x}{x^3}$.

Solution

a) $$\frac{d}{dx}(3 \ln x) = 3\frac{d}{dx}\ln x = \frac{3}{x}$$

b) $\frac{d}{dx}(x^2 \ln x + 5x) = x^2 \cdot \frac{1}{x} + \ln x \cdot 2x + 5$ Using the Product Rule on $x^2 \ln x$

$= x + 2x \cdot \ln x + 5$ Simplifying

c) $\frac{d}{dx}\frac{\ln x}{x^3} = \frac{x^3 \cdot (1/x) - (\ln x)(3x^2)}{x^6}$ By the Quotient Rule

$= \frac{x^2 - 3x^2 \ln x}{x^6}$

$= \frac{x^2(1 - 3 \ln x)}{x^6}$ Factoring

$= \frac{1 - 3 \ln x}{x^4}$ Simplifying ✦

Suppose that we want to differentiate a more complicated function that is of the form $h(x) = \ln f(x)$, such as

$$h(x) = \ln (x^2 - 8x).$$

This can be regarded as

$$h(x) = g(f(x)), \quad \text{where} \quad g(x) = \ln x \quad \text{and} \quad f(x) = x^2 - 8x.$$

Now $g'(x) = 1/x$, so by the Chain Rule (Section 1.7), we have

$$h'(x) = g'(f(x)) \cdot f'(x)$$
$$= \frac{1}{f(x)} \cdot f'(x).$$

For the above case, $f(x) = x^2 - 8x$, so $f'(x) = 2x - 8$. Then

$$h'(x) = \frac{1}{x^2 - 8x} \cdot (2x - 8) = \frac{2x - 8}{x^2 - 8x}.$$

The following rule, which we have proven using the Chain Rule, allows us to find derivatives of functions like the one above.

THEOREM 7

$$\frac{d}{dx} \ln f(x) = \frac{1}{f(x)} \cdot f'(x) = \frac{f'(x)}{f(x)},$$

or

$$\frac{d}{dx} \ln u = \frac{1}{u} \cdot \frac{du}{dx}.$$

The derivative of the natural logarithm of a function is the derivative of the function divided by the function.

The following gives us a way of remembering this rule.

$$h(x) = \ln (x^2 - 8x)$$

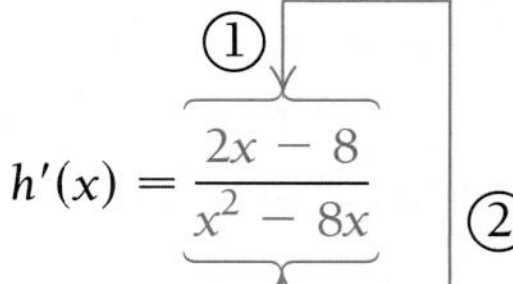

$$h'(x) = \frac{2x - 8}{x^2 - 8x}$$

① Differentiate the "inside" function.

② Divide by the "inside" function.

EXAMPLE 8 Differentiate: **a)** $y = \ln (3x)$; **b)** $y = \ln (x^2 - 5)$; **c)** $f(x) = \ln (\ln x)$; **d)** $f(x) = \ln \left(\dfrac{x^3 + 4}{x}\right)$.

Solution

a) If $y = \ln (3x)$, then

$$\frac{dy}{dx} = \frac{3}{3x} = \frac{1}{x}.$$

Note that we could have done this using the fact that $\ln (MN) = \ln M + \ln N$:

$$\ln (3x) = \ln 3 + \ln x;$$

then, since ln 3 is a constant, we have

$$\frac{d}{dx}\ln (3x) = \frac{d}{dx}\ln 3 + \frac{d}{dx}\ln x = 0 + \frac{1}{x} = \frac{1}{x}.$$

b) If $y = \ln (x^2 - 5)$, then

$$\frac{dy}{dx} = \frac{2x}{x^2 - 5}.$$

c) If $f(x) = \ln (\ln x)$, then

$$f'(x) = \frac{1}{\ln x} \cdot \frac{d}{dx}\ln x = \frac{1}{\ln x} \cdot \frac{1}{x} = \frac{1}{x \ln x}.$$

d) If $f(x) = \ln \left(\dfrac{x^3 + 4}{x}\right)$, then, since $\ln \dfrac{M}{N} = \ln M - \ln N$, we have

$$f'(x) = \frac{d}{dx}[\ln (x^3 + 4) - \ln x]$$ By P2. This avoids use of the Quotient Rule.

$$= \frac{3x^2}{x^3 + 4} - \frac{1}{x}$$

$$= \frac{3x^2}{x^3 + 4} \cdot \frac{x}{x} - \frac{1}{x} \cdot \frac{x^3 + 4}{x^3 + 4}$$ Finding a common denominator

$$= \frac{(3x^2)x - (x^3 + 4)}{x(x^3 + 4)}$$

$$= \frac{3x^3 - x^3 - 4}{x(x^3 + 4)} = \frac{2x^3 - 4}{x(x^3 + 4)}$$ Simplifying ♦

TECHNOLOGY CONNECTION

Exploratory

To check part (a) of Example 8, we let $y_1 = \ln (3x)$, $y_2 = \text{nDeriv}(y_1, x, x)$, and $y_3 = 1/x$. Either GRAPH or TABLE can then be used to show that $y_2 = y_3$. Use this approach to check parts (b), (c), and (d) of Example 8.

Applications

EXAMPLE 9 Social Science: Forgetting. In a psychological experiment, students were shown a set of nonsense syllables, such as POK, RIZ, DEQ, and so on, and asked to recall them every second thereafter. The percentage $R(t)$ who retained the syllables after t seconds was found to be given by the logarithmic learning model

$$R(t) = 80 - 27 \ln t, \quad \text{for } t \geq 1.$$

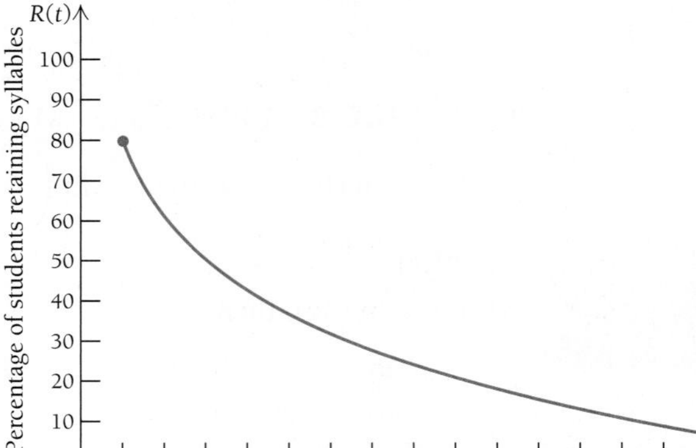

TECHNOLOGY CONNECTION

Exploratory

Graph $y = 80 - 27 \ln x$, from Example 9, using the viewing window $[1, 14, -1, 100]$. Trace along the graph. Describe the meaning of each coordinate in an ordered pair.

a) What percentage of students retained the syllables after 1 sec?

b) Find $R'(2)$, and explain what it represents.

Solution

a) $R(1) = 80 - 27 \cdot \ln 1 = 80 - 27 \cdot 0 = 80\%$

b) $\dfrac{d}{dt}(80 - 27 \ln t) = 0 - 27 \cdot \dfrac{1}{t} = -\dfrac{27}{t}$,

so $\quad R'(2) = -\dfrac{27}{2} = -13.5.$

This result indicates that 2 sec after students have been shown the syllables, the percentage of them who remember the syllables is shrinking at the rate of 13.5% per second. ✦

EXAMPLE 10 Business: An Advertising Model. A company begins a radio advertising campaign in New York City to market a new product. The percentage of the "target market" that buys a product is normally a function of the duration of the advertising campaign. The radio station estimates this percentage, as a decimal, by using $f(t) = 1 - e^{-0.04t}$ for this type of product, where t is the number of days of the campaign. The target market is approximately 1,000,000 people and the price per unit is \$0.50. If the campaign costs \$1000 per day, how long should it last in order to maximize profit?

Solution That the percentage of the target market that buys the product, expressed as a decimal, can be modeled using $f(t) = 1 - e^{-0.04t}$ is justified if we graph f. The function increases from 0 (0%) toward 1 (100%). The longer the advertising campaign, the larger the percentage of the market that has bought the product.

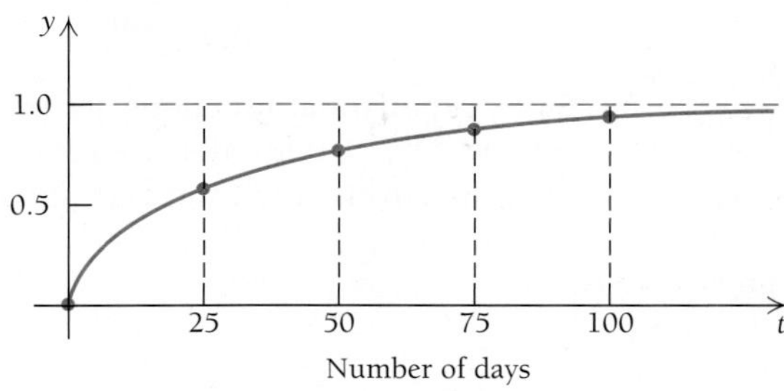

The total-profit function, here expressed in terms of time t, is given by

$$\text{Profit} = \text{Revenue} - \text{Cost}$$
$$P(t) = R(t) - C(t).$$

We find $R(t)$ and $C(t)$:

$$R(t) = (\text{Percentage buying}) \cdot (\text{Target market}) \cdot (\text{Price per unit})$$
$$= (1 - e^{-0.04t})(1{,}000{,}000)(0.5) = 500{,}000 - 500{,}000e^{-0.04t},$$

and

$$C(t) = (\text{Advertising costs per day}) \cdot (\text{Number of days}) = 1000t.$$

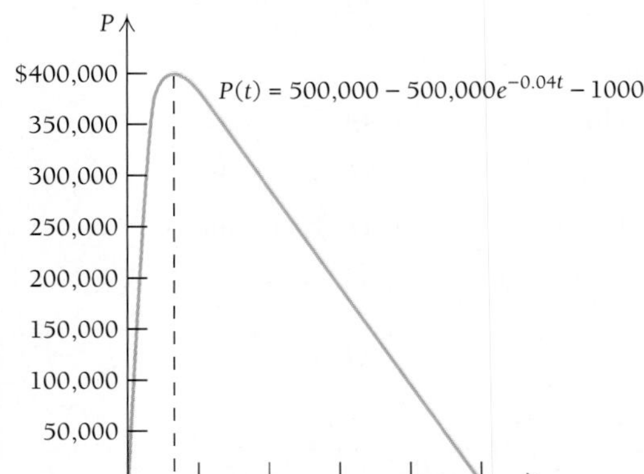

Next, we find $P(t)$ and take its derivative:

$$P(t) = R(t) - C(t)$$
$$= 500{,}000 - 500{,}000e^{-0.04t} - 1000t$$
$$P'(t) = -500{,}000e^{-0.04t}(-0.04) - 1000$$
$$= 20{,}000e^{-0.04t} - 1000.$$

We then set the first derivative equal to 0 and solve:

$$20{,}000e^{-0.04t} - 1000 = 0$$
$$20{,}000e^{-0.04t} = 1000$$
$$e^{-0.04t} = \frac{1000}{20{,}000} = 0.05$$
$$\ln e^{-0.04t} = \ln 0.05$$
$$-0.04t = \ln 0.05$$
$$t = \frac{\ln 0.05}{-0.04}$$
$$t \approx 75.$$

TECHNOLOGY CONNECTION

Exploratory

Graph the function P in Example 10, and verify that there is maximum profit if the length of the advertising campaign is about 75 days.

We have only one critical value, so we can use the second derivative to determine whether we have a maximum:

$$P''(t) = 20{,}000e^{-0.04t}(-0.04)$$
$$= -800e^{-0.04t}.$$

Since exponential functions are positive, $e^{-0.04t} > 0$ for all numbers t. Thus, since $-800e^{-0.04t} < 0$ for all t, we have $P''(75) < 0$, and we have a maximum.

The advertising campaign should run for 75 days in order to maximize profit. ◆

Exercise Set 3.2

Write an equivalent exponential equation.

1. $\log_2 8 = 3$
2. $\log_3 81 = 4$
3. $\log_8 2 = \frac{1}{3}$
4. $\log_{27} 3 = \frac{1}{3}$
5. $\log_a K = J$
6. $\log_a J = K$
7. $-\log_{10} h = p$
8. $-\log_b V = w$

Write an equivalent logarithmic equation.

9. $e^M = b$
10. $e^t = p$
11. $10^2 = 100$
12. $10^3 = 1000$
13. $10^{-1} = 0.1$
14. $10^{-2} = 0.01$
15. $M^p = V$
16. $Q^n = T$

Given $\log_b 3 = 1.099$ and $\log_b 5 = 1.609$, find each value.

17. $\log_b \frac{5}{3}$
18. $\log_b \frac{1}{5}$
19. $\log_b 15$
20. $\log_b \sqrt{b^3}$
21. $\log_b (5b)$
22. $\log_b 75$

Given $\ln 4 = 1.3863$ and $\ln 5 = 1.6094$, find each value. Do not use a calculator.

23. $\ln 20$
24. $\ln \frac{5}{4}$
25. $\ln \frac{1}{5}$
26. $\ln (4e)$
27. $\ln \sqrt{e^8}$
28. $\ln 80$

Find each logarithm. Round to six decimal places.

29. $\ln 5894$
30. $\ln 99{,}999$
31. $\ln 0.0182$
32. $\ln 0.00087$
33. $\ln 8100$
34. $\ln 0.011$

Solve for t.

35. $e^t = 80$
36. $e^t = 10$
37. $e^{2t} = 1000$
38. $e^{3t} = 900$
39. $e^{-t} = 0.1$
40. $e^{-t} = 0.01$
41. $e^{-0.02t} = 0.06$
42. $e^{0.07t} = 2$

Differentiate.

43. $y = -8 \ln x$
44. $y = -9 \ln x$
45. $y = x^4 \ln x - \frac{1}{2}x^2$
46. $y = x^6 \ln x - \frac{1}{4}x^4$
47. $f(x) = \ln (6x)$
48. $f(x) = \ln (9x)$
49. $g(x) = x^2 \ln (7x)$
50. $g(x) = x^5 \ln (3x)$
51. $y = \dfrac{\ln x}{x^4}$
52. $y = \dfrac{\ln x}{x^5}$
53. $y = \ln \dfrac{x^2}{4}$ $\left(\textit{Hint: } \ln \dfrac{A}{B} = \ln A - \ln B\right)$
54. $y = \ln \dfrac{x^4}{2}$
55. $y = \ln (3x^2 + 2x - 1)$
56. $y = \ln (7x^2 + 5x + 2)$
57. $f(x) = \ln \left(\dfrac{x^2 - 7}{x}\right)$
58. $f(x) = \ln \left(\dfrac{x^2 + 5}{x}\right)$
59. $g(x) = e^x \ln x^2$
60. $g(x) = e^{2x} \ln x$
61. $f(x) = \ln (e^x + 1)$
62. $f(x) = \ln (e^x - 2)$

63. $g(x) = (\ln x)^4$ (*Hint*: Use the Extended Power Rule.)

64. $g(x) = (\ln x)^3$

65. $f(x) = \ln (\ln (8x))$

66. $f(x) = \ln (\ln (3x))$

67. $g(x) = \ln (5x) \cdot \ln (3x)$

68. $g(x) = \ln (2x) \cdot \ln (7x)$

69. Find the equation of the line tangent to the graph of $y = (x^2 - x) \ln (6x)$ at $x = 2$.

70. Find the equation of the line tangent to the graph of $y = e^{3x} \cdot \ln (4x)$ at $x = 1$.

71. Find the equation of the line tangent to the graph of $y = (\ln x)^2$ at $x = 3$.

72. Find the equation of the line tangent to the graph of $y = \ln (4x^2 - 7)$ at $x = 2$.

APPLICATIONS

Business and Economics

73. Advertising. A model for consumers' response to advertising is given by

$$N(a) = 2000 + 500 \ln a, \quad a \geq 1,$$

where $N(a)$ is the number of units sold and a is the amount spent on advertising, in thousands of dollars.

a) How many units were sold after spending \$1000 on advertising?

b) Find $N'(a)$ and $N'(10)$.

c) Find the maximum and minimum values, if they exist.

TW **d)** Find $\lim_{a \to \infty} N'(a)$. Discuss whether it makes sense to continue to spend more and more dollars on advertising.

74. Advertising. A model for consumers' response to advertising is given by

$$N(a) = 1000 + 200 \ln a, \quad a \geq 1,$$

where $N(a)$ is the number of units sold and a is the amount spent on advertising, in thousands of dollars.

a) How many units were sold after spending \$1000 on advertising?

b) Find $N'(a)$ and $N'(10)$.

c) Find the maximum and minimum values, if they exist.

TW **d)** Find $N'(a)$. Discuss $\lim_{a \to \infty} N'(a)$. Does it make sense to spend more and more dollars on advertising? Why or why not?

75. An advertising model. Solve Example 10 given that the advertising campaign costs \$2000 per day.

76. An advertising model. Solve Example 10 given that the advertising campaign costs \$4000 per day.

77. Growth of a stock. The value, $V(t)$, in dollars, of a stock t months after it is purchased is modeled by

$$V(t) = 58(1 - e^{-1.1t}) + 20.$$

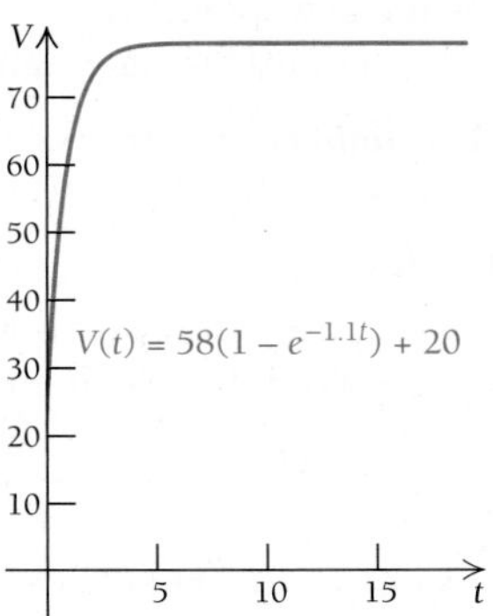

a) Find $V(1)$ and $V(12)$.

b) Find $V'(t)$.

c) After how many months will the value of the stock first reach \$75?

TW **d)** Find $\lim_{t \to \infty} V(t)$. Discuss the value of the stock over a long period of time. Is this trend typical?

78. Marginal revenue. The demand for a new computer game can be modeled by

$$p(x) = 53.5 - 8 \ln x,$$

where $p(x)$ is the price consumers will pay, in dollars, and x is the number of games sold, in thousands. Recall that total revenue is given by $R(x) = x \cdot p(x)$.

a) Find $R(x)$.

b) Find the marginal revenue, $R'(x)$.

TW **c)** Is there any price at which revenue will be maximized? Why or why not?

79. Marginal profit. The profit, in thousands of dollars, from the sale of x thousand mechanical pencils, can be estimated by

$$P(x) = 2x - 0.3x \ln x.$$

a) Find the marginal profit, $P'(x)$.

TW **b)** Find $P'(150)$, and explain what this number represents.

c) How many thousands of mechanical pencils should be sold to maximize profit?

Life and Physical Sciences

80. Acceptance of a new medicine. The percentage P of doctors who prescribe a certain new medicine is

$$P(t) = 100(1 - e^{-0.2t}),$$

where t is the time, in months.

a) Find $P(1)$ and $P(6)$.

b) Find $P'(t)$.

c) How many months will it take for 90% of doctors to prescribe the new medicine?

TW **d)** Find $\lim_{t\to\infty} P(t)$, and discuss its meaning.

81. The Reynolds number. For many kinds of animals, the Reynolds number, R, is given by

$$R = A \ln r - Br,$$

where A and B are positive constants and r is the radius of the aorta. Find the maximum value of R.

Social Sciences

82. Forgetting. Students in a botany class took a final exam. They took equivalent forms of the exam at monthly intervals thereafter. After t months, the average score $S(t)$, as a percentage, was found to be

$$S(t) = 68 - 20 \ln (t + 1), \quad t \geq 0.$$

a) What was the average score when the students initially took the test?

b) What was the average score after 4 months?

c) What was the average score after 24 months?

d) What percentage of their original answers did the students retain after 2 years (24 months)?

e) Find $S'(t)$.

f) Find the maximum value, if one exists.

TW **g)** Find $\lim_{t\to\infty} S(t)$, and discuss its meaning.

83. Forgetting. Students in a zoology class took a final exam. They took equivalent forms of the exam at monthly intervals thereafter. After t months, the average score $S(t)$, as a percentage, was found to be given by

$$S(t) = 78 - 15 \ln (t + 1), \quad t \geq 0.$$

a) What was the average score when they initially took the test, $t = 0$?

b) What was the average score after 4 months?

c) What was the average score after 24 months?

d) What percentage of their original answers did the students retain after 2 years (24 months)?

e) Find $S'(t)$.

f) Find the maximum and minimum values, if they exist.

TW **g)** Find $\lim_{t\to\infty} S(t)$ and discuss its meaning.

84. Walking speed. Bornstein and Bornstein found in a study that the average walking speed v, in feet per second, of a person living in a city of population p, in thousands, is

$$v(p) = 0.37 \ln p + 0.05.$$

(*Source*: *International Journal of Psychology*.)

a) The population of Seattle is 571,000 ($p = 571$). What is the average walking speed of a person living in Seattle?

b) The population of New York is 8,100,000. What is the average walking speed of a person living in New York?

c) Find $v'(p)$.

TW **d)** Interpret $v'(p)$ found in part (c).

85. The Hullian learning model. A keyboarder learns to type W words per minute after t weeks of practice, where W is given by

$$W(t) = 100(1 - e^{-0.3t}).$$

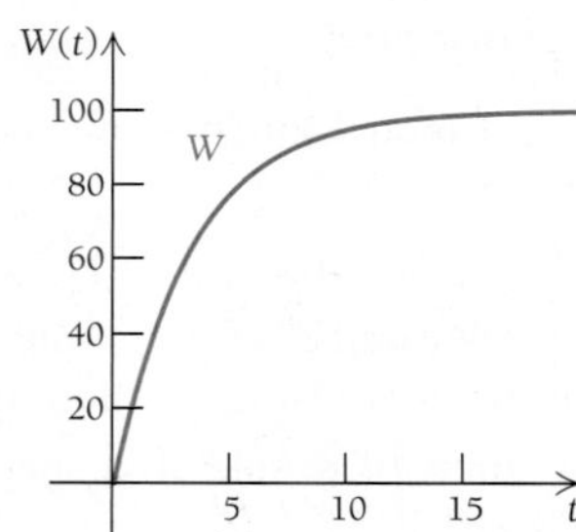

a) Find $W(1)$ and $W(8)$.

b) Find $W'(t)$.

c) After how many weeks will the keyboarder's speed be 95 words per minute?

TW **d)** Find $\lim_{t\to\infty} W(t)$, and discuss its meaning.

SYNTHESIS

Solve for t.

86. $P = P_0 e^{-kt}$

87. $P = P_0 e^{kt}$

Differentiate.

88. $f(x) = \ln (x^3 + 1)^5$

89. $f(t) = \ln (t^2 - t)^7$

90. $g(x) = [\ln (x + 5)]^4$

91. $f(x) = \ln [\ln (\ln (3x))]$

92. $f(t) = \ln [(t^3 + 3)(t^2 - 1)]$

93. $f(t) = \ln \frac{1-t}{1+t}$

94. $y = \ln \frac{x^5}{(8x+5)^2}$

95. $f(x) = \log_5 x$

96. $f(x) = \log_7 x$

97. $y = \ln \sqrt{5 + x^2}$

98. $f(t) = \frac{\ln t^2}{t^2}$

99. $f(x) = \frac{1}{5}x^5\left(\ln x - \frac{1}{5}\right)$

100. $y = \frac{x^{n+1}}{n+1}\left(\ln x - \frac{1}{n+1}\right)$

101. $f(x) = \ln \frac{1+\sqrt{x}}{1-\sqrt{x}}$

102. $f(x) = \ln (\ln x)^3$

To prove Properties P1, P2, P3, and P7 of Theorem 3, let $X = \log_a M$ and $Y = \log_a N$, and give reasons for the steps listed in Exercises 103–106.

103. Proof of P1 of Theorem 3.

$M = a^X$ and $N = a^Y$, ______
so $MN = a^X \cdot a^Y = a^{X+Y}$. ______
Thus, $\log_a (MN) = X + Y$ ______
$= \log_a M + \log_a N$. ______

104. Proof of P2 of Theorem 3.

$M = a^X$ and $N = a^Y$, ______
so $\frac{M}{N} = \frac{a^X}{a^Y} = a^{X-Y}$. ______
Thus, $\log_a \frac{M}{N} = X - Y$ ______
$= \log_a M - \log_a N$. ______

105. Proof of P3 of Theorem 3.

$M = a^X$, ______
so $M^k = (a^X)^k$ ______
$= a^{Xk}$. ______
Thus, $\log_a M^k = Xk$ ______
$= k \cdot \log_a M$. ______

106. Proof of P7 of Theorem 3.

Let $\log_b M = R$.
Then $b^R = M$, ______
and $\log_a (b^R) = \log_a M$. ______
Thus, $R \cdot \log_a b = \log_a M$, ______
and $R = \frac{\log_a M}{\log_a b}$. ______
It follows that
$\log_b M = \frac{\log_a M}{\log_a b}$. ______

107. Find $\lim_{h \to 0} \frac{\ln (1+h)}{h}$.

108. For any $k > 0$, $\ln (kx) = \ln k + \ln x$. Use this fact to show graphically why

$$\frac{d}{dx} \ln (kx) = \frac{d}{dx} \ln x = \frac{1}{x}.$$

TECHNOLOGY CONNECTION

109. Use natural logarithms to determine which is larger, e^π or π^e. (*Hint*: $y = \ln x$ is an increasing function.)

TW **110.** Find $\sqrt[e]{e}$. Compare it to other expressions of the type $\sqrt[x]{x}$, with $x > 0$. What can you conclude?

Use input–output tables to find each limit.

111. $\lim_{x \to 1} \ln x$

112. $\lim_{x \to \infty} \ln x$

Graph each function f and its derivative f'.

113. $f(x) = \ln x$

114. $f(x) = x \ln x$

115. $f(x) = x^2 \ln x$

116. $f(x) = \frac{\ln x}{x^2}$

Find the minimum value of each function.

117. $f(x) = x \ln x$

118. $f(x) = x^2 \ln x$

3.3 Applications: Uninhibited and Limited Growth Models

OBJECTIVES

- Find functions that satisfy $dP/dt = kP$.
- Convert between growth rate and doubling time.
- Solve application problems using exponential growth and limited growth models.

Exponential Growth

Consider the function

$$f(x) = 2e^{3x}.$$

Differentiating, we get

$$\begin{aligned} f'(x) &= 2e^{3x} \cdot 3 \\ &= f(x) \cdot 3. \end{aligned}$$

Graphically, this says that the derivative, or slope of the tangent line, is simply the constant 3 times the function value.

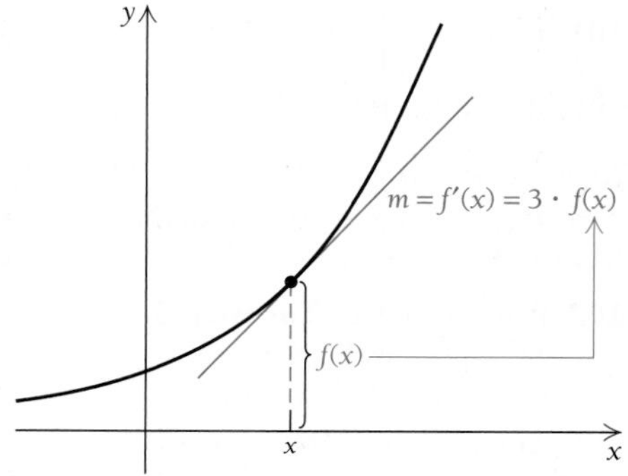

Although we do not prove it here, the exponential function $f(x) = ce^{kx}$ is the only function for which the derivative is a constant times the function itself.

THEOREM 8

A function $y = f(x)$ satisfies the equation

$$\frac{dy}{dx} = ky \quad \text{or} \quad f'(x) = k \cdot f(x)$$

if and only if

$$y = ce^{kx} \quad \text{or} \quad f(x) = ce^{kx}$$

for some constant c.

EXAMPLE 1 Find the general form of the function that satisfies the equation

$$\frac{dA}{dt} = 5A.$$

Solution The function is $A = ce^{5t}$, or $A(t) = ce^{5t}$, where c is an arbitrary constant. As a check, note that

$$A'(t) = ce^{5t} \cdot 5 = 5 \cdot A(t).$$ ✦

EXAMPLE 2 Find the general form of the function that satisfies the equation

$$\frac{dP}{dt} = kP.$$

Solution The function is $P = ce^{kt}$, or $P(t) = ce^{kt}$, where c is an arbitrary constant. *Check*:

$$\frac{dP}{dt} = ce^{kt} \cdot k = kP.$$ ✦

Whereas the solution of an algebraic equation is a number, the solutions of the equations in Examples 1 and 2 are functions. For example, the solution of $2x + 5 = 11$ is the number 3, and the solution of the equation $dP/dt = kP$ is the function $P(t) = ce^{kt}$. An equation like $dP/dt = kP$, which includes a derivative and which has a function as a solution, is called a *differential equation*.

EXAMPLE 3 Solve the differential equation

$$f'(z) = k \cdot f(z).$$

Solution The solution is $f(z) = ce^{kz}$. *Check*: $f'(z) = ce^{kz} \cdot k = f(z) \cdot k$. ✦

Uninhibited Population Growth

The equation

$$\frac{dP}{dt} = kP \quad \text{or} \quad P'(t) = kP(t), \quad \text{with } k > 0,$$

is the basic model of uninhibited (unrestrained) population growth, whether the population is comprised of humans, bacteria in a culture, or dollars invested with interest compounded continuously. Neglecting special inhibiting or stimulating factors, a population normally reproduces at a rate proportional to its size, and this is exactly what $dP/dt = kP$ says. The only function that satisfies this differential equation is given by

$$P(t) = ce^{kt}, \quad \text{where } t \text{ is time.}$$

Note that

$$P(0) = ce^{k \cdot 0} = ce^{0} = c \cdot 1 = c,$$

so c represents the initial population, which we denote P_0:

$$P(t) = P_0e^{kt}.$$

What will the world population be in 2010?

The graph of $P(t) = P_0e^{kt}$, for $k > 0$, shows how uninhibited growth produces a "population explosion."

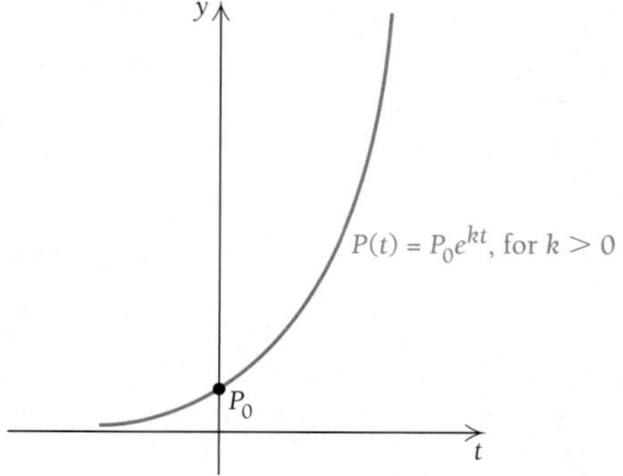

The constant k is called the **rate of exponential growth,** or simply the **growth rate.** This is not the rate of change of the population size, which varies according to

$$\frac{dP}{dt} = kP,$$

but the constant by which P must be multiplied in order to get the instantaneous rate of change at any point in time. It is similar to the daily interest rate paid by a bank. If the daily interest rate is $0.07/365$, then any given balance P is growing at the rate of $0.07/365 \cdot P$ dollars per day. Because of the compounding, after 1 year, the interest earned will exceed 7% of P. When interest is compounded continuously, the interest rate is a true exponential growth rate. A detailed explanation of this is presented at the end of this section.

EXAMPLE 4 Business: Interest Compounded Continuously. Suppose that an amount P_0, in dollars, is invested in a savings account where interest is compounded continuously at 7% per year. That is, the balance P grows at the rate given by

$$\frac{dP}{dt} = 0.07P.$$

a) Find the function that satisfies the equation. Write it in terms of P_0 and 0.07.

b) Suppose that \$100 is invested. What is the balance after 1 yr?

c) In what period of time will an investment of \$100 double itself?

Solution

a) $P(t) = P_0e^{0.07t}$ Note that $P(0) = P_0$.

b) $P(1) = 100e^{0.07(1)} = 100e^{0.07}$

$\approx 100(1.072508)$ It is best to skip this step when using a calculator.

$\approx \$107.25$

c) We are looking for a number T such that $P(T) = \$200$. The number T is called the **doubling time.** To find T, we solve the equation

$$200 = 100e^{0.07 \cdot T}$$

$$2 = e^{0.07T}.$$

We use natural logarithms to solve this equation:

$$\ln 2 = \ln e^{0.07T} \quad \text{Finding the natural logarithm of both sides}$$
$$\ln 2 = 0.07T \quad \text{By P5: } \ln e^k = k$$
$$\frac{\ln 2}{0.07} = T$$
$$9.9 \approx T.$$

Thus, \$100 will double itself in approximately 9.9 yr. ✦

To find a general expression relating growth rate k and doubling time T, we solve the following:

$$2P_0 = P_0 e^{kT}$$
$$2 = e^{kT} \quad \text{Dividing by } P_0$$
$$\ln 2 = \ln e^{kT}$$
$$\ln 2 = kT.$$

Note that this relationship between k and T does not depend on P_0. We now have the following theorem.

THEOREM 9

The *growth rate* k and the *doubling time* T are related by

$$kT = \ln 2 \approx 0.693147,$$

or $$k = \frac{\ln 2}{T} \approx \frac{0.693147}{T},$$

and $$T = \frac{\ln 2}{k} \approx \frac{0.693147}{k}.$$

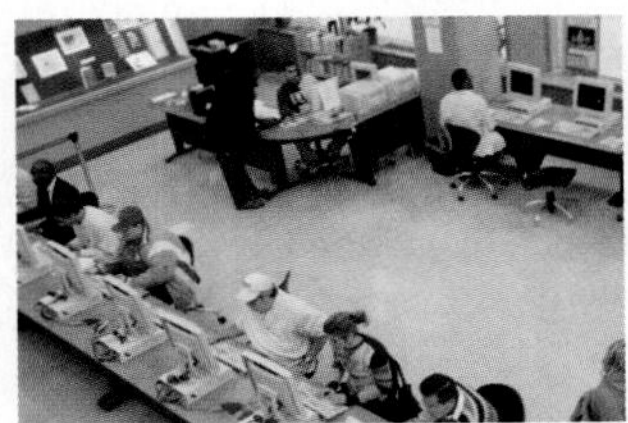

EXAMPLE 5 Business: Internet Use. Worldwide use of the Internet is increasing at an exponential rate, with traffic doubling every 100 days. What is the exponential growth rate?

Solution We have

$$k = \frac{\ln 2}{T} = \frac{0.693147}{100 \text{ days}} \quad \text{If possible, enter the calculation as } (\ln 2)/100 \text{ without approximating the logarithmic value.}$$
$$\approx 0.006931 \cdot \frac{1}{\text{day}}$$

The exponential growth rate is approximately 0.69% per day. ✦

The Rule of 70

The relationship between doubling time T and interest rate k is the basis of a rule often used in business, called the **Rule of 70.** To estimate how long it takes to double your money, divide 70 by the rate of return:

$$T = \frac{\ln 2}{k} \approx \frac{0.693147}{k}$$

$$\approx \frac{69.3147}{100k} \approx \frac{70}{100k}.$$

Remember that k is the interest rate written as a decimal.

EXAMPLE 6 Life Science: World Population Growth. The world population was approximately 6.0400 billion at the beginning of 2000. It has been estimated that the population is growing exponentially at the rate of 0.016, or 1.6% per year. (How was this estimate determined? The answer is in the model we develop in the following Technology Connection.) Thus,

$$\frac{dP}{dt} = 0.016P,$$

where t is the time, in years, after 2000. (*Source*: Based on data from the U.S. Census Bureau.)

a) Find the function that satisfies the equation. Assume that $P_0 = 6.0400$ and $k = 0.016$.

b) Estimate the world population at the beginning of 2020 ($t = 20$).

c) After what period of time will the population be double that in 2000?

Solution

a) $P(t) = 6.0400e^{0.016t}$

b) $P(20) = 6.0400e^{0.016(20)} = 6.0400e^{0.32} \approx 8.3179$ billion

c) $T = \frac{\ln 2}{k} = \frac{\ln 2}{0.016} = 43.3$ yr

Thus, according to this model, the population in 2000 will double itself by 2043. (No wonder ecologists are alarmed!) ✦

Under ideal conditions, the growth rate of this rapidly growing population of rabbits might be 11.7% per day. When will this population of rabbits double?

TECHNOLOGY CONNECTION

Exponential Models Using Regression

Projecting World Population Growth

The table below shows data regarding world population growth. A graph illustrating these data, along with the projected population in 2020, appeared in Section 3.1.

Year	World Population (in billions)
1927	2
1960	3
1974	4
1987	5
1998	6

How was the population projected for 2020? The graph shows a rapidly growing population that can be modeled with an exponential function. We carry out the regression procedure very much as we did in Section R.6, but here we choose ExpReg rather than LinReg.

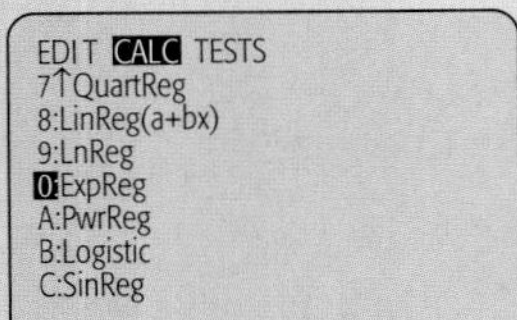

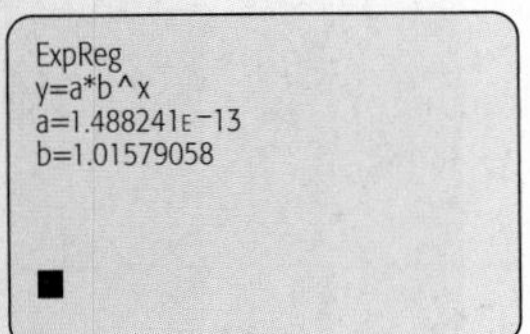

Note that this gives us an exponential model of the type $y = a \cdot b^x$, where y is the population, in billions, in year x.

$$y = (1.488241 \cdot 10^{-13})(1.01579058)^x. \qquad (1)$$

The base here is not e, but we can make a conversion to an exponential function, base e, using the fact that $b = e^{\ln b}$ and then multiplying exponents:

$$\begin{aligned} b^x &= (1.01579058)^x = (e^{\ln 1.01579058})^x \\ &= e^{(\ln 1.01579058)x} \\ &\approx e^{0.0156672059x}. \end{aligned}$$

We can now write equation (1) as

$$y = (1.488241 \cdot 10^{-13})e^{0.0156672059x}. \qquad (2)$$

The advantage of this form is that we see the growth rate. Here the world population growth rate is about 0.016, or 1.6%. To find world population in 2008, we can substitute 2008 for x in either equation (1) or (2). We choose equation (2):

$$\begin{aligned} y &= (1.488241 \cdot 10^{-13})e^{0.0156672059(2008)} \\ &\approx 6.8465 \text{ billion.} \end{aligned}$$

EXERCISES

Use equation (1) or equation (2) to estimate world population in each year.

1. 2007 **2.** 2012 **3.** 2020 **4.** 2050

Projecting College Costs

For Exercises 5 and 6, use the data regarding projected college costs listed in the table below.

School Year	Cost of Attending a Public 4-yr College or University
1987–1988	$ 4199
1989–1990	4715
1991–1992	5452
1993–1994	6212
1995–1996	6743
1997–1998	7469
1999–2000	8080
2001–2002	9032
2003–2004	10,530
2005–2006	12,127

(*Source*: College Board, New York, NY.)

EXERCISES

5. Use REGRESSION to fit an exponential function $y = a \cdot b^x$ to the data. Let 1987–1988 be represented by $x = 0$ and let $y =$ the cost, in dollars. Then convert that formula to an exponential function, base e, and determine the exponential growth rate.

6. Use either of the exponential functions found in Exercise 5 to estimate college costs in 2007, 2018, and 2040.

Some myths about alcohol. *It's a fact—the blood alcohol concentration (BAC) in the human body is measurable. And there's no cure for its effect on the central nervous system except time. It takes time for the body's metabolism to recover. That means a cup of coffee, a cold shower, and fresh air can't erase the effect of several drinks.*

There are variables, of course: a person's body weight, how many drinks have been consumed in a given time, how much has been eaten, and so on. These account for different BAC levels. But the myth that some people can "handle their liquor" better than others is a gross rationalization—especially when it comes to driving. Some people can act more sober than others. But an automobile doesn't act; it reacts.

In the preceding Technology Connection, we used *regression* to create an exponential model. There is another way to create such a model if regression is not an option. As shown in Example 7, two representative data points are sufficient to determine P_0 and k in $P(t) = P_0e^{kt}$.

EXAMPLE 7 Life Science: Alcohol Absorption and the Risk of Having an Accident. Extensive research has provided data linking the risk R (in percent) of having an automobile accident to the blood alcohol level b (in percent). Using two representative points, (0, 1%) and (0.14, 20%), we can approximate the data with an exponential function. The modeling assumption is that the rate of change of the risk R with respect to the blood alcohol level b is given by

$$\frac{dR}{db} = kR.$$

a) Find the function that satisfies this equation. Assume that $R_0 = 1\%$.

b) At what blood alcohol level will the risk of having an accident be 100%? Round to the nearest hundredth.

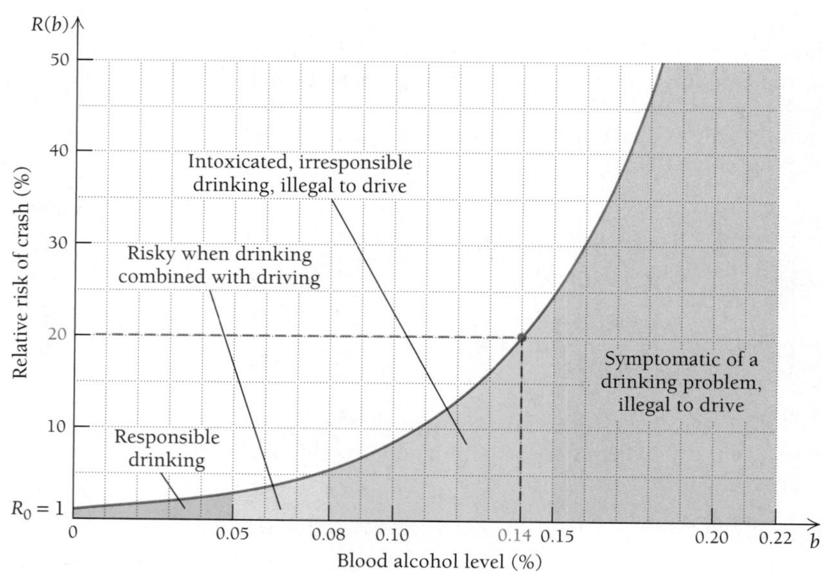

Solution Both R and b are percentages, but we omit the % symbol in the computations.

a) Because of the modeling assumption, we have $R(b) = R_0e^{kb}$. Since $R_0 = 1$, it follows that

$$R(b) = 1e^{kb} = e^{kb}.$$

We have made use of the data point (0, 1).

Next, we use the other data point, (0.14, 20), to help determine k. We solve

$$R(b) = e^{kb}, \quad \text{or} \quad 20 = e^{k(0.14)}$$

for k, using natural logarithms:

$$20 = e^{k(0.14)} = e^{0.14k}$$
$$\ln 20 = \ln e^{0.14k}$$
$$\ln 20 = 0.14k$$
$$\frac{\ln 20}{0.14} = k$$
$$\frac{2.995732}{0.14} \approx k \qquad \text{Skip this step when using a calculator.}$$
$$21.4 \approx k. \qquad \text{Rounded to the nearest tenth}$$

The desired function is $R(b) = e^{21.4b}$.

b) We substitute 100 for $R(b)$ and solve the equation for b:

$$100 = e^{21.4b}$$
$$\ln 100 = \ln e^{21.4b}$$
$$\ln 100 = 21.4b$$
$$\frac{\ln 100}{21.4} = b$$
$$\frac{4.605170}{21.4} \approx b \qquad \text{Skip this step when using a calculator.}$$
$$0.22 \approx b. \qquad \text{Rounded to the nearest hundredth}$$

Note that in part (a), we find ln 20 and divide by 0.14, obtaining approximately 21.4. We then use that value for k in part (b). Answers are found this way in the exercises. You may note some variance in the last one or two decimal places of your answers if you round as you go.

Thus, according to this model, the risk of an accident is 100% when the blood alcohol level is 0.22%. ✦

Models of Limited Growth

The growth model $P(t) = P_0e^{kt}$ has many applications to unlimited population growth, as we have seen in this section. However, there are often factors that prevent a population from exceeding some limiting value L—perhaps a limitation on food, living space, or other natural resources. One model of such growth is

$$P(t) = \frac{L}{1 + be^{-kt}}, \quad \text{for } k > 0,$$

which is called the *logistic equation.*

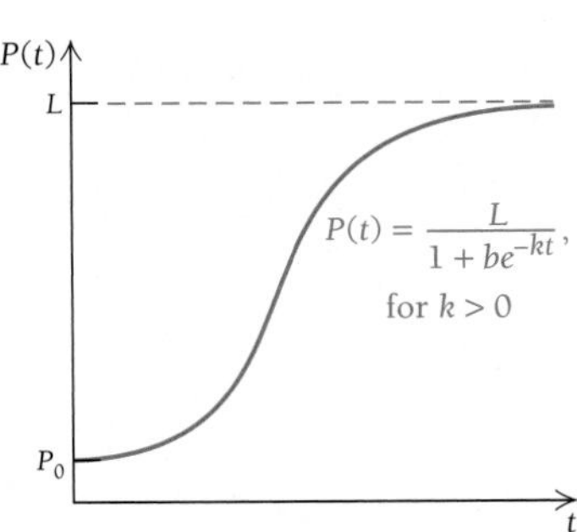

EXAMPLE 8 Spread of Infection. Spread by skin-to-skin contact or via shared towels or clothing, methicillin-resistant *Staphylococcus aureus* (MRSA) can easily spread a staph infection throughout a university. Left unchecked, the number of cases of MRSA on a university campus t weeks after the first 9 cases occur can be modeled by

$$N(t) = \frac{568.803}{1 + 62.200e^{-0.092t}}.$$

(*Source*: Based on information from the Vermont Department of Health, Epidemiology Division.)

a) Find the number of infected students after 3 weeks; 40 weeks; 80 weeks.

b) Find the rate at which the disease is spreading after 20 weeks.

c) Explain why an unrestricted growth model is inappropriate but a logistic equation is appropriate for this situation. Then use a calculator to graph the equation.

Solution

a) We use a calculator to find the function values:

$$N(3) = 11.8; \quad N(40) = 221.4; \quad N(80) = 547.2.$$

After 3 weeks, approximately 12 students are infected.

After 40 weeks, approximately 221 students are infected.

After 80 weeks, approximately 547 students are infected.

b) We find the rate of change, $N'(t)$, using the Quotient Rule:

$$N'(t) = \frac{(1 + 62.200e^{-0.092t}) \cdot 0 - 568.803(62.200e^{-0.092t})(-0.092)}{(1 + 62.200e^{-0.092t})^2}$$

$$= \frac{3254.918e^{-0.092t}}{(1 + 62.200e^{-0.092t})^2}.$$

Next, we use a calculator to evaluate the derivative:

$$N'(20) = 4.368.$$

After 20 weeks, the disease is spreading through the campus at a rate of about 4 new cases per week.

c) Unrestricted growth is inappropriate for modeling this situation because as more students become infected, fewer are left to be newly infected. The logistic equation, when graphed as shown below, displays the rapid spread of the disease initially, as well as the slower growth in later weeks when there are fewer students left to be newly infected.

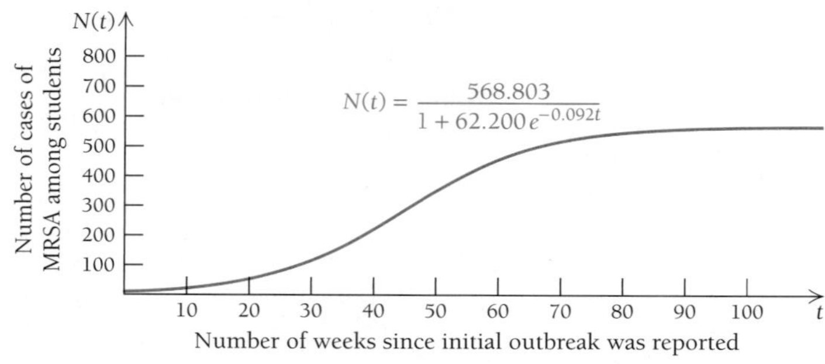

◆

Another model of limited growth is provided by the function

$$P(t) = L(1 - e^{-kt}), \quad \text{for } k > 0,$$

which is shown graphed below. This function also increases over the entire interval $[0, \infty)$, but increases most rapidly at the beginning, unlike the logistic equation.

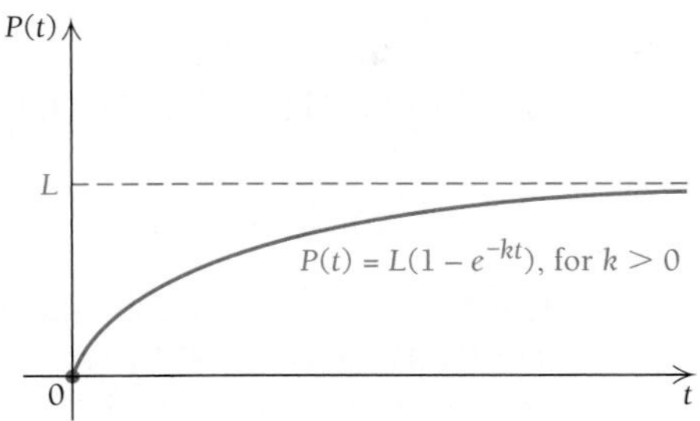

An Alternative Derivation of e and $P(t) = P_0e^{kt}$

The number e can also be found using the *compound-interest formula* (which was developed in Chapter R),

$$A = P\left(1 + \frac{i}{n}\right)^{nt},$$

where A is the amount that an initial investment P will be worth after t years at interest rate i, expressed as a decimal, compounded n times per year.

Suppose that \$1 is invested at 100% interest for 1 yr (though obviously no bank would pay this). The formula becomes

$$A = \left(1 + \frac{1}{n}\right)^{n}.$$

Suppose that the number of compounding periods, n, increases indefinitely. Let's investigate the behavior of the function. We obtain the following table of values and graph.

n	$A = \left(1 + \frac{1}{n}\right)^{n}$
1	\$2.00000
2	\$2.25000
3	\$2.37037
4	\$2.44141
12	\$2.61304
52	\$2.69260
365	\$2.71457
8,760	\$2.71813
525,600	\$2.71828

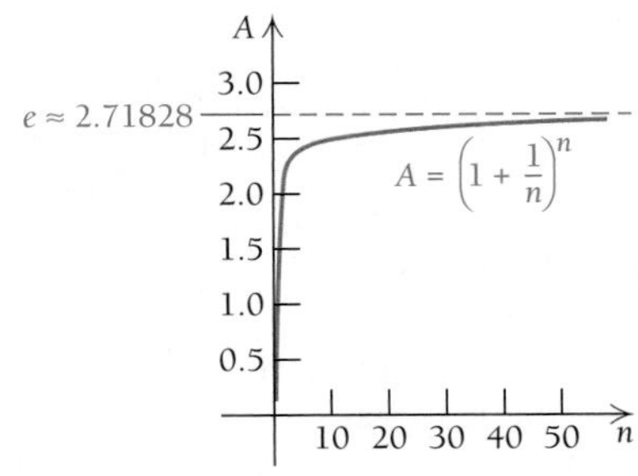

If interest is compounded *continuously*, we have $A = \lim_{n\to\infty}\left(1 + \frac{1}{n}\right)^n$, or equivalently,

$$A = \lim_{h\to 0}\left(1 + \frac{1}{\frac{1}{h}}\right)^{1/h}, \quad \text{or} \quad A = \lim_{h\to 0}(1 + h)^{1/h}.$$

Recall from Section 3.1 that $\lim_{h\to 0}(1 + h)^{1/h} = e$. Thus,

$$\lim_{n\to\infty}\left(1 + \frac{1}{n}\right)^n = e.$$

This result is confirmed by the graph and table on p. 345. For \$1, invested at an interest rate of 100% with increasingly frequent compounding periods, the greatest value it could grow to in 1 yr is about to \$2.7183.

To develop the formula

$$P(t) = P_0e^{kt},$$

we again start with the compound-interest formula,

$$A = P\left(1 + \frac{i}{n}\right)^{nt},$$

and assume that interest will be compounded continuously. Let $P = P_0$ and $i = k$ to obtain

$$P(t) = P_0\left(1 + \frac{k}{n}\right)^{nt}.$$

We are interested in what happens as n approaches ∞. To find this limit, we first let

$$\frac{k}{n} = \frac{1}{q}, \quad \text{so that} \quad qk = n.$$

Note that since k is a positive constant, as n gets large, so must q. Thus,

$$P(t) = \lim_{n\to\infty}\left[P_0\left(1 + \frac{k}{n}\right)^{nt}\right]$$ Letting the number of compounding periods become infinite

$$= P_0 \lim_{q\to\infty}\left[\left(1 + \frac{1}{q}\right)^{qkt}\right]$$ The limit of a constant times a function is the constant times the limit. We also substitute $1/q$ for k/n and qk for n. Also, $q \to \infty$ because $n \to \infty$.

$$= P_0\left[\lim_{q\to\infty}\left(1 + \frac{1}{q}\right)^q\right]^{kt}$$ The limit of a power is the power of the limit: a form of L2 in Section 1.2.

$$= P_0[e]^{kt}.$$

TECHNOLOGY CONNECTION

Exploratory

Graph

$$y = \left(1 + \frac{1}{x}\right)^x$$

using the viewing window [0, 5000, 0, 5], with Xscl = 1000 and Yscl = 1. Trace along the graph. Why does the graph appear to be horizontal? As you trace to the right, note the value of the y-coordinate. Is it approaching a constant? What seems to be its limiting value?

Exercise Set 3.3

1. Find the general form of f if $f'(x) = 4f(x)$.
2. Find the general form of g if $g'(x) = 6g(x)$.
3. Find the general form of the function that satisfies $dA/dt = -9A$.
4. Find the general form of the function that satisfies $dP/dt = -3P(t)$.
5. Find the general form of the function that satisfies $dQ/dt = kQ$.
6. Find the general form of the function that satisfies $dR/dt = kR$.

APPLICATIONS

Business and Economics

7. **U.S. patents.** The number of applications for patents, N, has grown dramatically in recent years, with growth averaging about 4.6% per year. That is,

$$N'(t) = 0.046N(t).$$

(*Source*: Based on data in the *New York Times*, 11/13/05, p. C1.)

   a) Find the function that satisfies this equation. Assume that $t = 0$ corresponds to 1980, when approximately 112,000 patent applications were received.
   b) Estimate the number of patent applications in 2010.
   c) Estimate the doubling time for $N(t)$.

8. **Franchise expansion.** Pete Zah's, Inc., is selling franchises for pizza shops throughout the country. The marketing manager estimates that the number of franchises, N, will increase at the rate of 10% per year, that is,

$$\frac{dN}{dt} = 0.10N.$$

   a) Find the function that satisfies this equation. Assume that the number of franchises at $t = 0$ is 50.
   b) How many franchises will there be in 20 yr?
   c) In what period of time will the initial number of 50 franchises double?

9. **Compound interest.** Suppose that P_0 is invested in a savings account in which interest is compounded continuously at 6.5% per year. That is, the balance P grows at the rate given by

$$\frac{dP}{dt} = 0.065P.$$

   a) Find the function that satisfies the equation. Write it in terms of P_0 and 0.065.
   b) Suppose that \$1000 is invested. What is the balance after 1 yr? After 2 yr?
   c) When will an investment of \$1000 double itself?

10. **Compound interest.** Suppose that P_0 is invested in a savings account in which interest is compounded continuously at 8% per year. That is, the balance P grows at the rate given by

$$\frac{dP}{dt} = 0.08P.$$

   a) Find the function that satisfies the equation. Write it in terms of P_0 and 0.08.
   b) Suppose that \$20,000 is invested. What is the balance after 1 yr? After 2 yr?
   c) When will an investment of \$20,000 double itself?

11. **Bottled water sales.** Since 2000, sales of bottled water have increased at the rate of approximately 9.3% per year. That is, the volume of bottled water sold, G, in billions of gallons, t years after 2000 is growing at the rate given by

$$\frac{dG}{dt} = 0.093G.$$

(*Source*: Based on data from the Beverage Marketing Corporation.)

   a) Find the function that satisfies the equation, given that approximately 4.7 billion gallons of bottled water were sold in 2000.

b) Predict the number of gallons of water that will be sold in 2010.
c) What is the doubling time for $G(t)$?

12. **Immigration prosecutions.** Since 1994, immigration violations in the United States have grown at the rate of approximately 18.2% per year, surpassing drug violations as the crime category with the most prosecutions. That is, the number of immigration prosecutions, P, occurring t years after 1994 can be approximated by

$$\frac{dP}{dt} = 0.182P.$$

(*Source*: Based on data from the Transactional Records Access Clearinghouse, as published in the *New York Times*, 9/29/05, p. A27.)

a) Find the function that satisfies the equation, given that approximately 6100 immigration prosecutions occurred in 1994.
b) Estimate the number of immigration cases prosecuted in 2007.
c) What is the doubling time for $P(t)$?

13. **Annual interest rate.** A bank advertises that it compounds interest continuously and that it will double your money in 15 yr. What is its annual interest rate?

14. **Annual interest rate.** A bank advertises that it compounds interest continuously and that it will double your money in 12 yr. What is its annual interest rate?

15. **Oil demand.** The growth rate of the demand for oil in the United States is 10% per year. When will the demand be double that of 2006?

16. **Coal demand.** The growth rate of the demand for coal in the world is 4% per year. When will the demand be double that of 2006?

Interest compounded continuously. *For Exercises 17–20, complete the following.*

	Initial Investment at $t = 0$, P_0	Interest Rate, k	Doubling Time, T (in years)	Amount After 5 yr
17.	\$75,000	6.2%		
18.	\$5,000			\$7,130.90
19.		8.4%		\$11,414.71
20.			11	\$17,539.32

21. **Art masterpieces.** In 2004, a collector paid \$104,168,000 for Pablo Picasso's "Garçon à la Pipe." The same painting sold for \$30,000 in 1950. (*Source*: BBC News, 5/6/04.)

a) Find the exponential growth rate k, to three decimal places, and determine the exponential growth function V, for which $V(t)$ is the painting's value, in dollars, t years after 1950.
b) Predict the value of the painting in 2010.
c) What is the doubling time for the value of the painting?
d) How long after 1950 will the value of the painting be \$1 billion?

22. **Federal budget.** In 1950, the U.S. federal budget was \$39.4 billion. In 2006, the federal budget was \$2153.9 billion. (*Source*: U.S. Office of Management and Budget.) Assume that the growth of the U.S. federal budget follows an exponential model and use 1950 as the base ($t = 0$).

a) Find the value of k to five decimal places, and write the function, with $B(t)$, the budget, in billions of dollars.
b) Estimate what the U.S. federal budget will be in 2010.
c) When will the U.S. federal budget be \$10 trillion, that is, \$10,000 billion?

23. **Per capita income.** In 1960, the U.S. per capita personal income was \$2277. In 2004, it was \$32,907. (*Source*: U.S. Census Bureau, *Statistical Abstract of the United States*.) Assume that the growth of U.S. per capita personal income follows an exponential model.

a) Using $t = 0$ for 1960, write the function.
b) Predict what the U.S. per capita personal income will be in 2020.
c) When will the U.S. per capita personal income be \$100,000?
d) In what year was the U.S. per capita personal income double that of 1960?

24. Consumer price index. The *consumer price index* compares the costs of goods and services over various years, where 1967 is used as a base ($t = 0$). The same goods and services that cost \$100 in 1967 cost \$184.50 in 1977. Assuming an exponential model:

a) Write the function, rounding k to four decimal places.
b) Estimate what the goods and services costing \$100 in 1967 will cost in 2010.
c) In what year did the same goods and services cost twice the 1967 price?

Sales of paper shredders. *Data in the following bar graph show paper shredder sales in recent years. Use these data for Exercises 25 and 26.*

Shredder Boom: Estimated Paper Shredder Sales

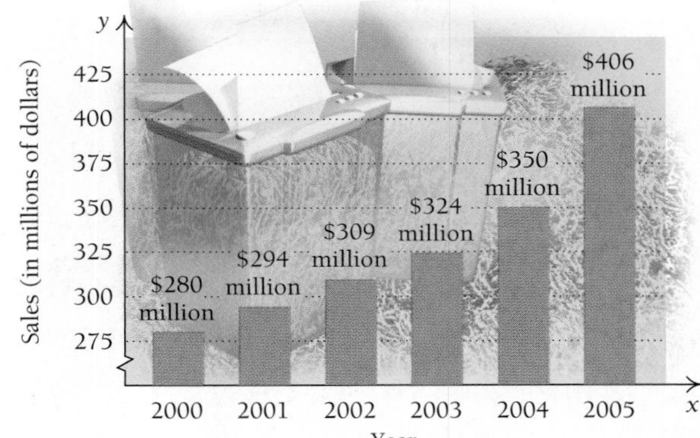

(*Source*: Based on data from www.sfgate.com.)

25. a) Use REGRESSION to fit an exponential function $y = a \cdot b^x$ to the data. Let y be in millions of dollars. Then convert that formula to an exponential function, base e, where x is the number of years after 1990, and determine the exponential growth rate. (See the Technology Connection on p. 341.)
b) Estimate the total sales of paper shredders in 2007 and in 2012.
c) After what amount of time will sales be \$500 million?
d) What is the doubling time for sales of shredders?

26. a) To find an exponential function, base e, that fits the data, find k using the points (10, 280) and (15, 406). Then write the function. (Assume that x is the number of years after 1990.)
b) Estimate the total sales of paper shredders in 2007 and in 2012.
c) After what amount of time will total sales be \$500 million?
d) What is the doubling time for sales of shredders?
TW **e)** Compare your answers to parts (a)–(d) with those from Exercise 25. Decide which exponential function seems better to you, and explain why.

27. Value of Manhattan Island. Peter Minuit of the Dutch West India Company purchased Manhattan Island from the natives living there in 1626 for \$24 worth of merchandise. Assuming an exponential rate of inflation of 5%, how much will Manhattan be worth in 2010?

28. Total revenue. Intel, a computer chip manufacturer, reported \$1265 million in total revenue in 1986. In 2005, the total revenue was \$38.8 billion. (*Source*: U.S. Securities and Exchange Commission.) Assuming an exponential model, find the growth rate k, to four decimal places, and write the revenue function R, with $R(t)$ in billions of dollars. Then predict the company's total revenue for 2012.

29. Average salary of Major League baseball players. In 1970, the average salary of Major League baseball players was \$29,303. In 2005, the average salary was \$2,632,655. (*Source*: *Baseball Almanac*.) Assuming exponential growth occurred, what was the growth rate to the nearest hundredth of a percent? What will the average salary be in 2010? In 2020?

30. Cost of a first-class postage stamp. The cost of a first-class postage stamp in 1962 was 4¢. In 2007, it was 39¢. This was exponential growth. What was the growth rate? Predict the cost of a first-class postage stamp in 2010 and in 2020.

31. Effect of advertising. Suppose that SpryBorg Inc. introduces a new computer game in Houston using television advertisements. Surveys show that $P\%$ of the target audience buy the game after x ads are broadcast, satisfying

$$P(x) = \frac{100}{1 + 49e^{-0.13x}}.$$

a) What percentage buy the game without seeing a TV ad ($x = 0$)?
b) What percentage buy the game after the ad is run 5 times? 10 times? 20 times? 30 times? 50 times? 60 times?
c) Find the rate of change, $P'(x)$.
d) Sketch a graph of the function.

32. Cost of a Hershey bar. The cost of a Hershey bar was \$0.05 in 1962 and was increasing at an exponential growth rate of 9.7%.

a) Predict the cost of a Hershey bar in 2010 and in 2015.
TW **b)** Are the prices realistic? Explain.

Life and Physical Sciences

Population growth. *For Exercises 33–38, complete the following.*

Population	*Exponential Growth Rate, k*	*Doubling Time, T (in years)*
33. Mexico	3.5%/yr	
34. Europe		69.31
35. Oil reserves		6.931
36. Coal reserves		17.3
37. Alaska	2.794%/yr	
38. Central America		19.8

39. Yellowstone grizzly bears. In 1972, the population of grizzly bears in Yellowstone National Park had shrunk to approximately 190. In 2005, the number of Yellowstone grizzlies had grown to about 610. (*Source*: Based on data in the *New York Times*, 9/26/05.) Find an exponential function that fits the data, and then predict Yellowstone's grizzly bear population in 2012. Round k to three decimal places.

40. Bicentennial growth of the United States. The population of the United States in 1776 was about 2,508,000. In the country's bicentennial year, the population was about 216,000,000.

a) Assuming an exponential model, what was the growth rate of the United States through its bicentennial year?

TW **b)** Is exponential growth a reasonable assumption? Explain.

41. Blood alcohol level. Refer to Example 7 (on alcohol absorption). At what blood alcohol level will the risk of an accident be 80%?

42. Blood alcohol level. Refer to Example 7. At what blood alcohol level will the risk of an accident be 90%?

43. Limited population growth. A ship carrying 1000 passengers has the misfortune to be wrecked on a small island from which the passengers are never rescued. The natural resources of the island restrict the growth of the population to a *limiting value* of 5780, to which the population gets closer and closer but which it never reaches. The population of the island after time t, in years, is approximated by the logistic equation

$$P(t) = \frac{5780}{1 + 4.78e^{-0.4t}}.$$

a) Find the population after 0 yr; 1 yr; 2 yr; 5 yr; 10 yr; 20 yr.

b) Find the rate of change, $P'(t)$.

c) Sketch a graph of the function.

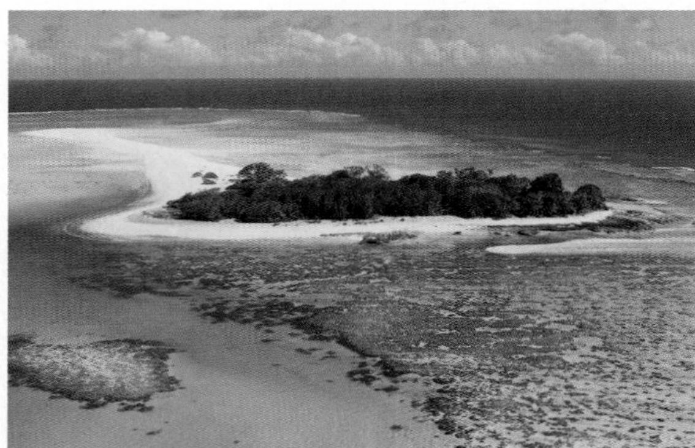

This island may have achieved its limited population growth.

44. Limited population growth. A lake is stocked with 400 rainbow trout. The size of the lake, the availability of food, and the number of other fish restrict growth in the lake to a *limiting value* of 2500. (See Exercise 43.) The population of trout in the lake after time t, in months, is approximated by

$$P(t) = \frac{2500}{1 + 5.25e^{-0.32t}}.$$

a) Find the population after 0 months; 1 month; 5 months; 10 months; 15 months; 20 months.

b) Find the rate of change, $P'(t)$.

c) Sketch a graph of the function.

Social Sciences

45. Women college graduates. The number of women graduating from 4-yr colleges in the United States has grown from 1930, when 48,869 women earned a bachelor's degree, to 2005, when approximately 832,000 women received such a degree. (*Source*: National Center for Education Statistics.) Find an exponential function that fits the data, and the exponential growth rate, rounded to the nearest hundredth of a percent.

46. Hullian learning model. The Hullian learning model asserts that the probability p of mastering a task after t learning trials is approximated by

$$p(t) = 1 - e^{-kt},$$

where k is a constant that depends on the task to be learned. Suppose that a new dance is being taught to an aerobics class. For this particular dance, the constant $k = 0.28$.

a) What is the probability of mastering the dance's steps in 1 trial? 2 trials? 5 trials? 11 trials? 16 trials? 20 trials?

b) Find the rate of change, $p'(t)$.

c) Sketch a graph of the function.

47. Diffusion of information. Pharmaceutical firms invest significant money in testing any new medication. After the drug is approved by the Federal Drug Administration, it still takes time for physicians to fully accept and start prescribing the medication. The acceptance by physicians approaches a *limiting value* of 100%, or 1, after time t, in months. Suppose that the percentage P of physicians prescribing a new cancer medication after t months is approximated by

$$P(t) = 100(1 - e^{-0.4t}).$$

a) What percentage of doctors are prescribing the medication after 0 months? 1 month? 2 months? 3 months? 5 months? 12 months? 16 months?

b) Find $P'(7)$, and interpret its meaning.

c) Sketch a graph of the function.

48. Spread of a rumor. The rumor "People who study math all get scholarships" spreads across a college campus. Data in the following table show the number of students N who have heard the rumor after time t, in days.

a) Use REGRESSION to fit a logistic equation,

$$N(t) = \frac{c}{1 + ae^{-bt}},$$

to the data.

b) Estimate the limiting value of the function. At most, how many students will hear the rumor?

Time, t (in days)	Number, N, Who Have Heard the Rumor
1	1
2	2
3	4
4	7
5	12
6	18
7	24
8	26
9	28
10	28
11	29
12	30

c) Graph the function.

d) Find the rate of change, $N'(t)$.

TW **e)** Find $\lim_{t \to \infty} N'(t)$, and explain its meaning.

SYNTHESIS

TW *We have now studied models for linear, quadratic, exponential, and logistic growth. In the real world, understanding which is the most appropriate type of model for a given situation is an important skill. For each situation in Exercises 49–56, identify the most appropriate type of model and explain why you chose that model. List any restrictions you would place on the domain of the function.*

49. The growth in value of a U.S. savings bond

50. The growth in the length of Zachary's hair following a haircut

51. The increasing number of computers infected with a virus on a college campus

52. The drop and rise of a lake's water level during and after a drought

53. The rapidly growing sales of organic foods

54. The rapidly growing percentage of produce that has been genetically modified

55. The number of manufacturing jobs that have left the United States since 1995

56. The life expectancy of the average American

57. Find an expression relating the exponential growth rate k and the *quadrupling time* T_4.

58. Find an expression relating the exponential growth rate k and the *tripling time* T_3.

59. A quantity Q_1 grows exponentially with a doubling time of 1 yr. A quantity Q_2 grows exponentially with a doubling time of 2 yr. If the initial amounts of Q_1 and Q_2 are the same, how long will it take for Q_1 to be twice the size of Q_2?

60. To what exponential growth rate per hour does a growth rate of 100% per day correspond?

Business: effective annual yield. *Suppose that \$100 is invested at 7%, compounded continuously, for 1 yr. We know from Example 4 that the ending balance will be \$107.25. This would also be the ending balance if \$100 were invested at 7.25%, compounded once a year (simple interest). The rate of 7.25% is called the* effective annual yield. *In general, if P_0 is invested at interest rate k, compounded continuously, then the effective annual yield is that number i satisfying $P_0(1 + i) = P_0e^k$. Then, $1 + i = e^k$, or*

$$\text{Effective annual yield} = i = e^k - 1.$$

61. An amount is invested at 7.3% per year compounded continuously. What is the effective annual yield?

62. An amount is invested at 8% per year compounded continuously. What is the effective annual yield?

63. The effective annual yield on an investment compounded continuously is 9.42%. At what rate was it invested?

64. The effective annual yield on an investment compounded continuously is 6.61%. At what rate was it invested?

65. To show that the exponential growth rate can be determined using any two points, let

$$y_1 = Ce^{kt_1}, \quad \text{and} \quad y_2 = Ce^{kt_2}.$$

Solve this system of equations for k to show that k can be calculated directly, using (t_1, y_1) and (t_2, y_2).

TW **66.** Complete the table below, which relates growth rate k and doubling time T.

Growth Rate, k (per year)	1%	2%			14%
Doubling Time, T (in years)			15	10	

Graph $T = (\ln 2)/k$. Is this a linear relationship? Explain.

TW **67.** Describe the differences in the graphs of an exponential function and a logistic function.

TW **68.** Explain how the Rule of 70 could be useful to someone studying inflation.

3.4 Applications: Decay

OBJECTIVES

- Find a function that satisfies $dP/dt = -kP$.
- Convert between decay rate and half-life.
- Solve applied problems involving exponential decay.

In the equation of population growth, $dP/dt = kP$, the constant k is actually given by

$$k = (\text{Birth rate}) - (\text{Death rate}).$$

Thus, a population "grows" only when the *birth rate* is greater than the *death rate.* When the birth rate is less than the death rate, k will be negative, and the population will be decreasing, or "decaying," at a rate proportional to its size. For convenience in our computations, we will express such a negative value as $-k$, where $k > 0$. The equation

$$\frac{dP}{dt} = -kP, \quad \text{where } k > 0,$$

shows P to be *decreasing* as a function of time, and the solution

$$P(t) = P_0e^{-kt}$$

TECHNOLOGY CONNECTION

Exploratory

Using the same set of axes, graph $y_1 = e^{2x}$ and $y_2 = e^{-2x}$. Compare the graphs.

Then, using the same set of axes, graph $y_1 = 100e^{-0.06x}$ and $y_2 = 100e^{0.06x}$. Compare these graphs.

shows it to be decreasing exponentially. This is called **exponential decay.** The amount present initially at $t = 0$ is again P_0.

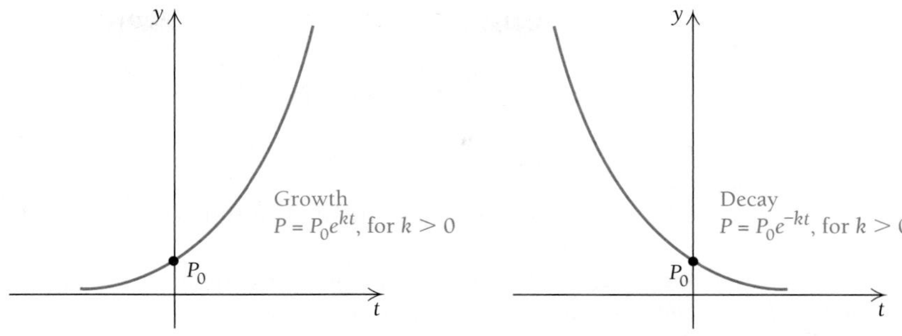

Radioactive Decay

Radioactive elements decay exponentially; that is, they disintegrate at a rate that is proportional to the amount present.

EXAMPLE 1 Life Science: Decay. Strontium-90 has a decay rate of 2.8% per year. The rate of change of an amount N of this radioactive element is given by

$$\frac{dN}{dt} = -0.028N.$$

a) Find the function that satisfies the equation. Let N_0 represent the amount present at $t = 0$.

b) Suppose that 1000 grams (g) of strontium-90 is present at $t = 0$. How much will remain after 70 yr?

c) After how long will half of the 1000 g remain?

Solution

a) $N(t) = N_0e^{-0.028t}$

b)
$$\begin{aligned} N(70) &= 1000e^{-0.028(70)} \\ &= 1000e^{-1.96} \\ &\approx 1000(0.1408584209) \\ &\approx 140.8584209. \end{aligned}$$

After 70 yr, about 140.9 g of the strontium-90 remains.

c) We are asking, "At what time T will $N(T)$ be half of N_0, or $\frac{1}{2} \cdot 1000$?" The number T is called the **half-life.** To find T, we solve the equation

$$500 = 1000e^{-0.028T} \qquad \text{We use 500 because } 500 = \tfrac{1}{2} \cdot 1000.$$

$$\frac{1}{2} = e^{-0.028T} \qquad \text{Dividing both sides by 1000}$$

$$\ln\frac{1}{2} = \ln e^{-0.028T} \qquad \text{Taking the natural logarithm of both sides}$$

$$\ln 1 - \ln 2 = -0.028T \qquad \text{Using the properties of logarithms}$$

$$0 - \ln 2 = -0.028T$$

$$\frac{-\ln 2}{-0.028} = T \qquad \text{Dividing both sides by } -0.028$$

$$\frac{\ln 2}{0.028} = T$$

$$\frac{0.693147}{0.028} \approx T$$

$$25 \approx T.$$

Thus, the half-life of strontium-90 is about 25 yr. ✦

How can scientists determine that the remains of an animal or plant have lost 30% of the carbon-14? The assumption is that the percentage of carbon-14 in the atmosphere and in living plants and animals is the same. When a plant or animal dies, the amount of carbon-14 decays exponentially. The scientist burns the remains and uses a Geiger counter to determine the percentage of carbon-14 in the smoke. The amount by which this varies from the percentage in the atmosphere indicates how much carbon-14 has been lost through decay.

The process of carbon-14 dating was developed by the American chemist Willard F. Libby in 1952. It is known that the radioactivity of a living plant measures 16 disintegrations per gram per minute. Since the half-life of carbon-14 is 5750 years, a dead plant with an activity of 8 disintegrations per gram per minute is 5750 years old, one with an activity of 4 disintegrations per gram per minute is 11,500 years old, and so on. Carbon-14 dating can be used to determine the age of organic objects from 30,000 to 40,000 years old. Beyond such an age, it is too difficult to measure the radioactivity, and other methods are used.

We can find a general expression relating the decay rate k and the half-life T by solving the equation

$$\tfrac{1}{2}P_0 = P_0e^{-kT}$$

$$\tfrac{1}{2} = e^{-kT} \qquad \text{Note that these steps correspond to those above.}$$

$$\ln \tfrac{1}{2} = \ln e^{-kT}$$

$$\ln 1 - \ln 2 = -kT$$

$$0 - \ln 2 = -kT$$

$$-\ln 2 = -kT$$

$$\ln 2 = kT.$$

Again, we have the following.

THEOREM 10

The *decay rate*, k, and the *half-life*, T, are related by

$$kT = \ln 2 = 0.693147,$$

or

$$k = \frac{\ln 2}{T} \quad \text{and} \quad T = \frac{\ln 2}{k}.$$

Thus, the half-life, T, depends only on the decay rate, k. In particular, it is independent of the initial population size.

The effect of half-life is shown in the radioactive decay curve to the right. Note that the exponential function gets close to, but never reaches, 0 as t gets larger. Thus, in theory, a radioactive substance never completely decays.

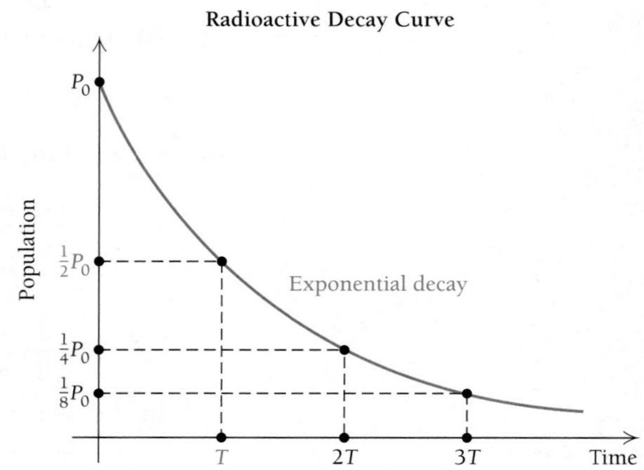

EXAMPLE 2 Life Science: Half-life. Plutonium-239, a common product of a functioning nuclear reactor, can be deadly to people exposed to it. Its decay rate is about 0.0028% per year. What is its half-life?

Solution We have

$$\begin{aligned} T &= \frac{\ln 2}{k} \\ &= \frac{\ln 2}{0.000028} \qquad \text{Converting the percentage to decimal notation} \\ &\approx 24{,}755. \end{aligned}$$

Thus, the half-life of plutonium is about 24,755 yr. ✦

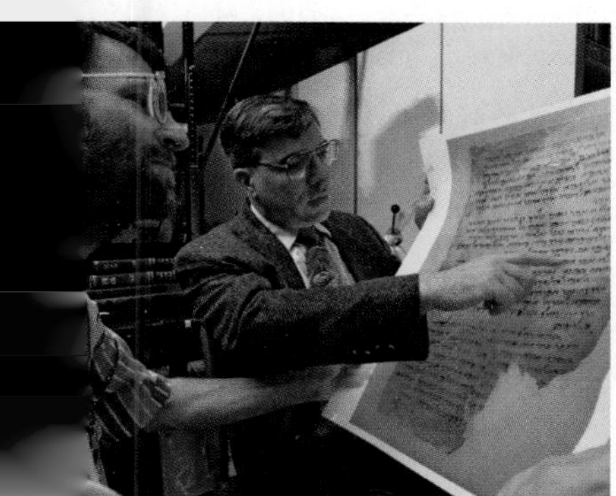

In 1947, a Bedouin youth looking for a stray goat climbed into a cave at Kirbet Qumran on the shores of the Dead Sea near Jericho and came upon earthenware jars containing an incalculable treasure of ancient manuscripts. Shown here are fragments of those so-called Dead Sea Scrolls, a portion of some 600 or so texts found so far and which concern the Jewish books of the Bible. Officials date them before 70 A.D.*, making them the oldest Biblical manuscripts by 1000 years.*

EXAMPLE 3 Life Science: Carbon Dating. The radioactive element carbon-14 has a half-life of 5750 yr. The percentage of carbon-14 present in the remains of plants and animals can be used to determine age. Archaeologists found that the linen wrapping from one of the Dead Sea Scrolls had lost 22.3% of its carbon-14. How old was the linen wrapping?

Solution Our plan is to find the exponential equation of the form $N(t) = N_0e^{-kt}$, replace $N(t)$ with $(1 - 0.223)N_0$, and solve for t. First, however, we must find the decay rate, k:

$$k = \frac{\ln 2}{T} = \frac{0.693147}{5750} \approx 0.0001205, \quad \text{or} \quad 0.01205\% \text{ per year.}$$

Thus, the amount $N(t)$ that remains from an initial amount N_0 after t years is given by:

$$N(t) = N_0e^{-0.0001205t}. \qquad \text{Remember: } k \text{ is positive, so } -k \text{ is negative.}$$

(*Note:* This equation can be used for all subsequent carbon-dating problems.) If the Dead Sea Scrolls lost 22.3% of their carbon-14 from an initial amount P_0, then $77.7\% \cdot P_0$ remains. To find the age t of the scrolls, we solve the following equation for t:

$$\begin{aligned} 77.7\%\, N_0 &= N_0e^{-0.0001205t} \\ 0.777 &= e^{-0.0001205t} \\ \ln 0.777 &= \ln e^{-0.0001205t} \\ \ln 0.777 &= -0.0001205t \\ \frac{\ln 0.777}{-0.0001205} &= t \\ 2094 &\approx t. \end{aligned}$$

Thus, the linen wrapping of the Dead Sea Scrolls is about 2094 yr old. ✦

A Business Application: Present Value

A representative of a financial institution is often asked to solve a problem like the following.

EXAMPLE 4 Business: Present Value. Following the birth of their granddaughter, two grandparents want to make an initial investment of P_0 that will grow to \$10,000 by the child's 20th birthday. Interest is compounded continuously at 6%. What should the initial investment be?

Solution Using the equation $P = P_0e^{kt}$, we find P_0 such that

$$10{,}000 = P_0e^{0.06 \cdot 20},$$

or

$$10{,}000 = P_0e^{1.2}.$$

Now

$$\frac{10{,}000}{e^{1.2}} = P_0,$$

or

$$10{,}000e^{-1.2} = P_0,$$

and, using a calculator, we have

$$P_0 = 10{,}000e^{-1.2}$$
$$\approx \$3011.94$$

Thus, the grandparents must deposit \$3011.94, which will grow to \$10,000 by the child's 20th birthday. ✦

Economists call \$3011.94 the *present value* of \$10,000 due 20 yr from now at 6%, compounded continuously. The process of computing present value is called **discounting.** Another way to pose this problem is to ask "What must I invest now, at 6%, compounded continuously, in order to have \$10,000 in 20 years?" The answer is \$3011.94, and it is the present value of \$10,000.

Computing present value can be interpreted as exponential decay from the future back to the present.

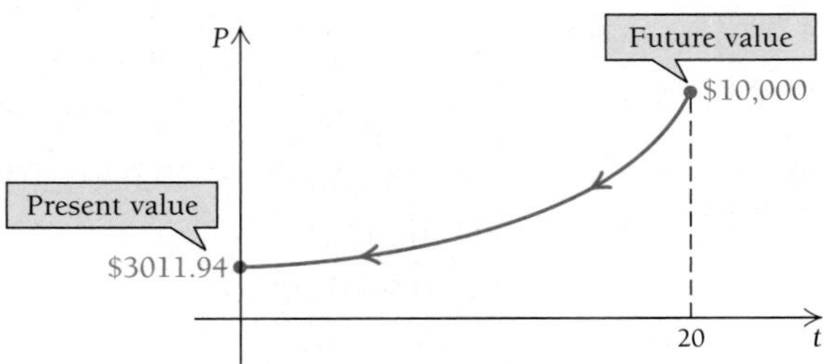

In general, the present value P_0 of an amount P due t years later is found by solving the following equation for P_0:

$$P_0e^{kt} = P$$

$$P_0 = \frac{P}{e^{kt}} = Pe^{-kt}.$$

THEOREM 11

The **present value** P_0 of an amount P due t years later, at interest rate k, compounded continuously, is given by

$$P_0 = Pe^{-kt}.$$

Newton's Law of Cooling

Consider the following situation. A hot cup of soup, at a temperature of 200°, is placed in a 70° room.* The temperature of the soup decreases over time t, in minutes, according to the model known as **Newton's Law of Cooling.**

Newton's Law of Cooling

The temperature T of a cooling object drops at a rate that is proportional to the difference $T - C$, where C is the constant temperature of the surrounding medium. Thus,

$$\frac{dT}{dt} = -k(T - C). \quad (1)$$

The function that satisfies equation (1) is

$$T = T(t) = ae^{-kt} + C. \quad (2)$$

To check that $T(t) = ae^{-kt} + C$ is the solution, simply find dT/dt and substitute dT/dt and $T(t)$ into equation (1). This check is left to the student.

EXAMPLE 5 Life Science: Scalding Coffee. McDivett's Pie Shoppes, a national restaurant firm, finds that the temperature of its freshly brewed coffee is 130°. The company fears that if customers spill hot coffee on themselves, lawsuits might result. Room temperature in the restaurants is generally 72°. The temperature of the coffee cools to 120° after 4.3 min. The company determines that it is safer to serve the coffee at a temperature of 105°. How long does it take a cup of coffee to cool to 105°?

Solution Note that C, the surrounding air temperature, is 72°. To find the value of a in equation (2) for Newton's Law of Cooling, we observe that at $t = 0$, we have $T(0) = 130°$. We solve for a as follows:

$$130 = ae^{-k \cdot 0} + 72$$

$$130 = a + 72$$

$$58 = a. \quad \text{Note that } a = 130 - 72, \text{ the difference between the original temperatures.}$$

*Assume throughout this section that all temperatures are in degrees Fahrenheit unless noted otherwise.

Next, we find k using the fact that $T(4.3) = 120$:

$$\begin{aligned}
120 &= 58e^{-k\cdot(4.3)} + 72 \\
48 &= 58e^{-4.3k} \\
\frac{48}{58} &= e^{-4.3k} \\
\ln\frac{48}{58} &= \ln e^{-4.3k} \\
-0.1892420 &\approx -4.3k \\
k &\approx 0.044.
\end{aligned}$$

We now have $T(t) = 58e^{-0.044t}$. To see how long it will take the coffee to cool to 105°, we set $T(t) = 105$ and solve for t:

$$\begin{aligned}
105 &= 58e^{-0.044t} + 72 \\
33 &= 58e^{-0.044t} \\
\frac{33}{58} &= e^{-0.044t} \\
\ln\frac{33}{58} &= \ln e^{-0.044t} \\
-0.5639354 &\approx -0.044t \\
t &\approx 12.8 \text{ min.}
\end{aligned}$$

Thus, to cool to 105° the coffee should be allowed to cool for about 13 min. ✦

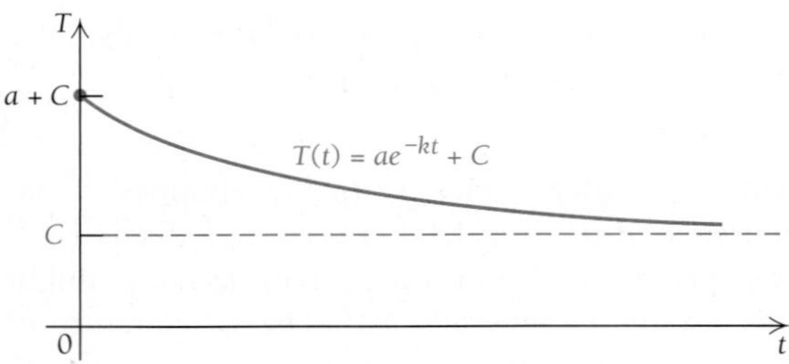

The graph of $T(t) = ae^{-kt} + C$ shows that $\lim_{t\to\infty} T(t) = C$. The temperature of the object decreases toward the temperature of the surrounding medium.

Mathematically, this model tells us that the object's temperature never quite reaches C. In practice, the temperature of the cooling object will get so close to that of the surrounding medium that no device could detect a difference. Let's now see how Newton's Law of Cooling can be used in solving a crime.

EXAMPLE 6 Life Science: When Was the Murder Committed? Found stabbed by a collection of number 2 pencils, Prof. Cal Kulice's body was slumped over a stack of exams with plenty of red marks on them. A coroner arrives at noon, immediately takes the temperature of the body, and finds it to be 94.6°. She waits 1 hr, takes the temperature again, and finds it to be 93.4°. She also notes that the temperature of the room is 70°. When was the murder committed?

Solution Note that C, the surrounding air temperature, is 70°. To find a in $T(t) = ae^{-kt} + C$, we assume that the temperature of the body was normal when the murder occurred. Thus, $T = 98.6°$ at $t = 0$:

$$\begin{aligned}
98.6 &= ae^{-k\cdot 0} + 70, \\
a &= 28.6.
\end{aligned}$$

This gives $T(t) = 28.6e^{-kt} + 70$.

To find the number of hours N since the murder was committed, we must first determine k. From the two temperature readings the coroner made, we have

$$94.6 = 28.6e^{-kN} + 70, \quad \text{or} \quad 24.6 = 28.6e^{-kN}; \qquad (3)$$
$$93.4 = 28.6e^{-k(N+1)} + 70, \quad \text{or} \quad 23.4 = 28.6e^{-k(N+1)}. \qquad (4)$$

Dividing equation (3) by equation (4), we get

$$\begin{aligned} \frac{24.6}{23.4} &= \frac{28.6e^{-kN}}{28.6e^{-k(N+1)}} \\ &= e^{-kN+k(N+1)} \\ &= e^{-kN+kN+k} = e^k. \end{aligned}$$

We solve this equation for k:

$$\ln \frac{24.6}{23.4} = \ln e^k \qquad \text{Taking the natural logarithm on both sides}$$
$$0.05 \approx k.$$

Next, we substitute back into equation (3) and solve for N:

$$\begin{aligned} 24.6 &= 28.6e^{-0.05N} \\ \frac{24.6}{28.6} &= e^{-0.05N} \\ \ln \frac{24.6}{28.6} &= \ln e^{-0.05N} \\ -0.150660 &\approx -0.05N \\ 3 &\approx N. \end{aligned}$$

Since the coroner arrived at noon, or 12 o'clock, the murder occurred at about 9:00 A.M. ✦

In summary, we have added several functions to our candidates for curve fitting. Let's review them.

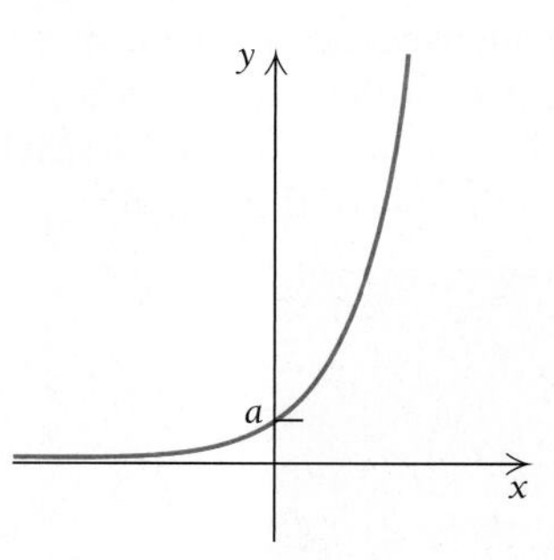

Exponential:
$f(x) = ab^x$, or ae^{kx}
$a, b, k > 0, b \neq 1$

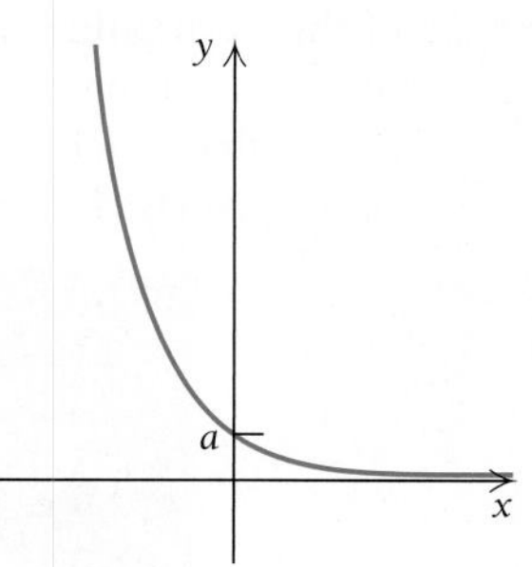

Exponential:
$f(x) = ab^{-x}$, or ae^{-kx}
$a, b, k > 0, b \neq 1$

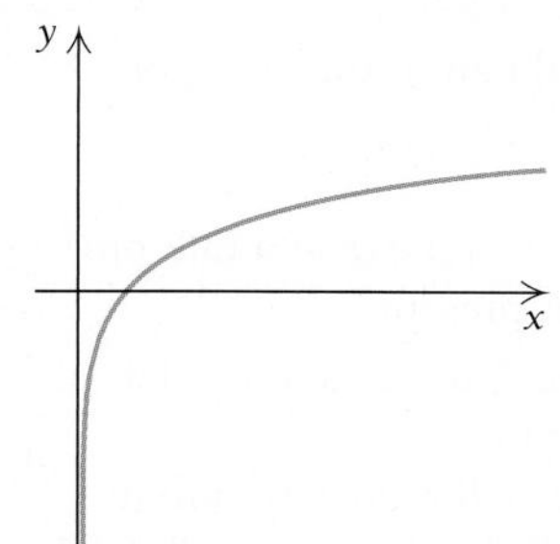

Logarithmic:
$f(x) = a + b \ln x$

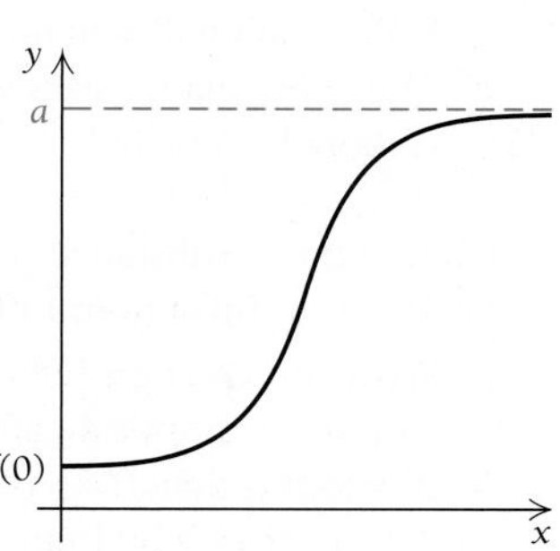

Logistic:
$f(x) = \dfrac{a}{1 + be^{-kx}}$

Now, when we analyze a set of data, we can consider these models, as well as linear, quadratic, polynomial, and rational functions, for curve fitting.

Exercise Set 3.4

APPLICATIONS

Life and Physical Sciences

1. **Radioactive decay.** Iodine-131 has a decay rate of 9.6% per day. The rate of change of an amount N of iodine-131 is given by

$$\frac{dN}{dt} = -0.096\,N,$$

where t is the number of days since the decay began.

a) Let N_0 represent the amount of iodine-131 present at $t = 0$. Find the exponential function that models the situation.

b) Suppose that 500 g of iodine-131 is present at $t = 0$. How much will remain after 4 days?

c) After how many days will half of the 500 g of iodine-131 remain?

2. **Radioactive decay.** Carbon-14 has a decay rate of 0.01205% per year. The rate of change of an amount N of carbon-14 is given by

$$\frac{dN}{dt} = -0.0001205N,$$

where t is the number of years since the decay began.

a) Let N_0 represent the amount of carbon-14 present at $t = 0$. Find the exponential function that models the situation.

b) Suppose 200 g of carbon-14 is present at $t = 0$. How much will remain after 800 yr?

c) After how many years will half of the 200 g of carbon-14 remain?

3. **Chemistry.** Substance A decomposes at a rate proportional to the amount of A present.

a) Write an equation relating A to the amount left of an initial amount A_0 after time t.

b) It is found that 10 lb of A will reduce to 5 lb in 3.3 hr. After how long will there be only 1 lb left?

4. **Chemistry.** Substance A decomposes at a rate proportional to the amount of A present.

a) Write an equation relating A to the amount left of an initial amount A_0 after time t.

b) It is found that 8 g of A will reduce to 4 g in 3 hr. After how long will there be only 1 g left?

Radioactive decay. *For Exercises 5–8, complete the following.*

	Radioactive Substance	*Decay Rate, k*	*Half-life, T*
5.	Polonium-218		3 min
6.	Uranium-238		4560 yr
7.	Lead-210	3.15%/yr	
8.	Strontium-90	2.77%/yr	

9. **Half-life.** Of an initial amount of 1000 g of lead-210, how much will remain after 100 yr? See Exercise 7 for the value of k.

10. **Half-life.** Of an initial amount of 1000 g of polonium-218, how much will remain after 20 min? See Exercise 5 for the value of k.

11. **Carbon dating.** How old is an ivory tusk that has lost 40% of its carbon-14?

12. **Carbon dating.** How old is a piece of wood that has lost 90% of its carbon-14?

13. **Cancer treatment.** Iodine-125 is often used to treat cancer and has a half-life of 60.1 days. In a sample, the amount of iodine-125 decreased by 25% while in storage. How long was the sample sitting on the shelf?

14. **Carbon dating.** How old is a Chinese artifact that has lost 60% of its carbon-14?

15. **Carbon dating.** Recently, while digging in Chaco Canyon, New Mexico, archaeologists found corn pollen that had lost 38.1% of its carbon-14. The age of this corn pollen was evidence that Indians had been cultivating crops in the Southwest centuries earlier than scientists had thought. (*Source*: *American Anthropologist*.) What was the age of the pollen?

Chaco Canyon, New Mexico

Business and Economics

16. Present value. Following the birth of a child, a parent wants to make an initial investment P_0 that will grow to \$30,000 by the child's 20th birthday. Interest is compounded continuously at 6%. What should the initial investment be?

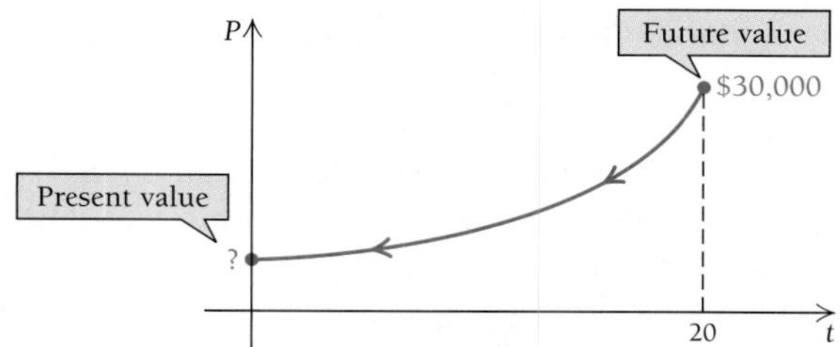

17. Present value. Following the birth of a child, a parent wants to make an initial investment P_0 that will grow to \$40,000 by the child's 20th birthday. Interest is compounded continuously at 5.3%. What should the initial investment be?

18. Present value. A homeowner wants to have \$15,000 available in 5 yr to pay for new siding. Interest is 6.1%, compounded continuously. How much money should be invested?

19. Sports salaries. An athlete signs a contract that guarantees a \$9-million salary 6 yr from now. Assuming that money can be invested at 5.7%, with interest compounded continuously, what is the present value of that year's salary?

20. Actors' salaries. An actor signs a film contract that will pay \$12 million when the film is completed 3 yr from now. Assuming that money can be invested at 6.2%, with interest compounded continuously, what is the present value of that payment?

21. Estate planning. A person has a trust fund that will yield \$80,000 in 13 yr. A CPA is preparing a financial statement for this client and wants to take into account the present value of the trust fund in computing the client's net worth. Interest is compounded continuously at 8.3%. What is the present value of the trust fund?

22. Supply and demand. The supply and demand for stereos produced by a sound company are given by

$$S(x) = \ln x \quad \text{and} \quad D(x) = \ln \frac{163{,}000}{x},$$

where $S(x)$ is the number of stereos that the company is willing to sell at price x and $D(x)$ is the quantity that the public is willing to buy at price x. Find the equilibrium point. (See Section R.5.)

23. Salvage value. A business estimates that the salvage value $V(t)$, in dollars, of a piece of machinery after t years is given by

$$V(t) = 40{,}000e^{-t}.$$

a) What did the machinery cost initially?

b) What is the salvage value after 2 yr?

TW **c)** Find the rate of change of the salvage value, and explain its meaning.

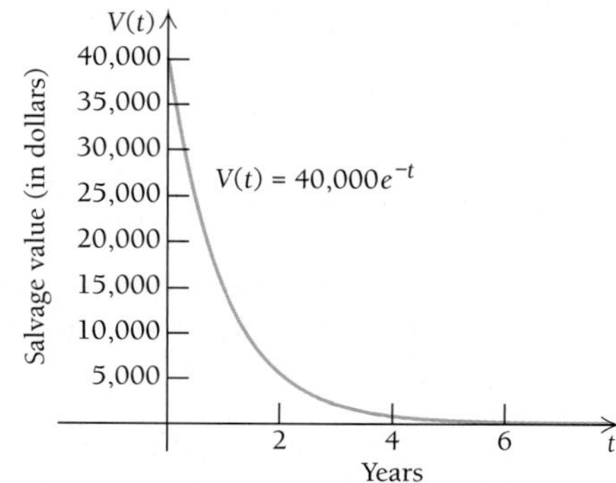

24. Consumer price index. The consumer price index compares the costs of goods and services over various years, where 1967 is used as a base. The same goods and services that cost \$100 in 1967 cost \$42 in 1940. Assuming the exponential-decay model:

a) Find the value k, and write the equation. Round k to the nearest thousandth.

b) Estimate what the goods and services that cost \$100 in 1967 cost in 1900.

25. Salvage value. A company tracks the value of a particular photocopier over a period of years. The data in the table below show the value of the copier at time t, in years, after the date of purchase.

Time, t (in years)	Salvage Value
0	\$34,000
1	22,791
2	15,277
3	10,241
4	6,865
5	4,600
6	3,084

(*Source*: International Data Corporation.)

a) Use REGRESSION to fit an exponential function $y = a \cdot b^x$ to the data. Then convert that formula to $V(t) = V_0e^{-kt}$, where V_0 is the value when the copier is purchased and t is the time, in years, from the date of purchase. (See the Technology Connection on p. 341.)

b) Estimate the salvage value of the copier after 7 yr; 10 yr.

c) After what amount of time will the salvage value be \$1000?

d) After how long will the copier be worth half of its original value?

TW e) Find the rate of change of the salvage value, and interpret its meaning.

Social Sciences

26. **Forgetting.** In an art history class, students took a final exam. They were subsequently retested with an equivalent test at monthly intervals. Their average retest scores t months later are given in the following table.

Time, t (in months)	Score, y
1	84.9%
2	84.6%
3	84.4%
4	84.2%
5	84.1%
6	83.9%

a) Use REGRESSION to fit a logarithmic function $y = a + b \ln x$ to the data.

b) Use the function to predict the average test score after 8 months; 10 months; 24 months; 36 months.

c) After how long will the test scores fall below 82%?

TW d) Find the rate of change of the scores, and interpret its meaning.

27. **Decline in beef consumption.** The annual consumption of beef per person was about 80 lb in 1985 and about 67 lb in 1996. Assuming that $B(t)$, the annual beef consumption t years after 1985, is decreasing according to the exponential decay model:

a) Find the value of k, and write the equation.

b) Estimate the consumption of beef in 2000.

c) In what year (theoretically) will the consumption of beef be 20 lb per person?

28. **Population decrease of Russia.** The population of Russia dropped from 150 million in 1995 to 142.9 million in 2006. (*Source*: CIA–*The World Factbook.*) Assume that $P(t)$, the population, in millions, t years after 1995, is decreasing according to the exponential decay model.

a) Find the value of k, and write the equation.

b) Estimate the population of Russia in 2012.

c) When will the population of Russia be 100 million?

29. **Population decrease of Ukraine.** The population of Ukraine dropped from 51.9 million in 1995 to 46.7 million in 2006. (*Source*: CIA–*The World Factbook.*) Assume that $P(t)$, the population, in millions, t years after 1995, is decreasing according to the exponential decay model.

a) Find the value of k, and write the equation.

b) Estimate the population of Ukraine in 2015.

c) After how many years will the population of Ukraine be 1, according to this model?

Life and Natural Sciences

30. **Cooling.** After warming the water in the Salisbury-Tulis' hot tub to 100°, the heating element fails. The surrounding air temperature is 40°, and in 5 min the water temperature drops to 95°.

a) Find the value of the constant a in Newton's Law of Cooling.

b) Find the value of the constant k. Round to five decimal places.

c) What is the water temperature after 10 min?

d) How long does it take the water to cool to 41°?

TW e) Find the rate of change of the water temperature, and interpret its meaning.

31. **Cooling.** The temperature in a whirlpool bath is 102°, and the room temperature is 75°. The water cools to 90° in 10 min.

a) Find the value of the constant a in Newton's Law of Cooling.

b) Find the value of the constant k. Round to five decimal places.

c) What is the water temperature after 20 min?

d) How long does it take the water to cool to 80°?

TW e) Find the rate of change of the water temperature, and interpret its meaning.

32. **Criminology.** A coroner arrives at a murder scene at 2 A.M. He takes the temperature of the body and finds it to be 61.6°. He waits 1 hr, takes the temperature again, and finds it to be 57.2°. The body is in a meat freezer, where the temperature is 10°. When was the murder committed?

33. Criminology. A coroner arrives at a murder scene at 11 P.M. She finds the temperature of the body to be 85.9°. She waits 1 hr, takes the temperature again, and finds it to be 83.4°. She notes that the room temperature is 60°. When was the murder committed?

34. Prisoner-of-war protest. The initial weight of a prisoner of war is 140 lb. To protest the conditions of her imprisonment, she begins a fast. Her weight t days after her last meal is approximated by

$$W = 140e^{-0.009t}.$$

a) How much does the prisoner weigh after 25 days?
b) At what rate is the prisoner's weight changing after 25 days?

35. Political protest. A monk weighing 170 lb begins a fast to protest a war. His weight after t days is given by

$$W = 170e^{-0.008t}.$$

a) When the war ends 20 days later, how much does the monk weigh?
b) At what rate is the monk losing weight after 20 days (before any food is consumed)?

36. Atmospheric pressure. Atmospheric pressure P at altitude a is given by

$$P = P_0e^{-0.00005a},$$

where P_0 is the pressure at sea level. Assume that $P_0 = 14.7 \text{ lb/in}^2$ (pounds per square inch).

a) Find the pressure at an altitude of 1000 ft.
b) Find the pressure at an altitude of 20,000 ft.
c) At what altitude is the pressure 14.7 lb/in^2?
TW **d)** Find the rate of change of the pressure, and interpret its meaning.

37. Satellite power. The power supply of a satellite is a radioisotope (radioactive substance). The power output P, in watts (W), decreases at a rate proportional to the amount present; P is given by

$$P = 50e^{-0.004t},$$

where t is the time, in days.

a) How much power will be available after 375 days?
b) What is the half-life of the power supply?
c) The satellite's equipment cannot operate on fewer than 10 watts of power. How long can the satellite stay in operation?
d) How much power did the satellite have to begin with?
TW **e)** Find the rate of change of the power output, and interpret its meaning.

Modeling

For each of the following scatterplots, determine which, if any, of these functions might be used as a model for the data:

a) Quadratic: $f(x) = ax^2 + bx + c$
b) Polynomial, not quadratic
c) Exponential: $f(x) = ae^{kx}$, $k > 0$
d) Exponential: $f(x) = ae^{-kx}$, $k > 0$
e) Logarithmic: $f(x) = a + b \ln x$
f) Logistic: $f(x) = \dfrac{a}{1 + be^{-kx}}$

38.

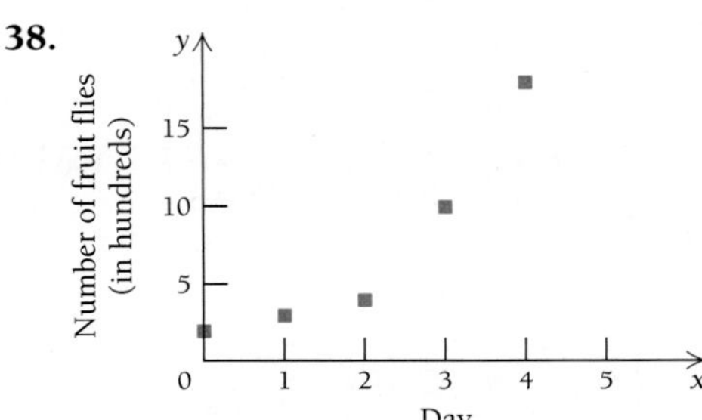

39.

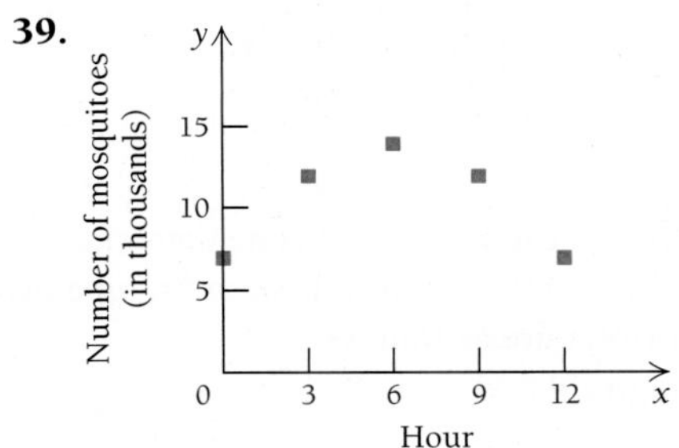

40. **41.**

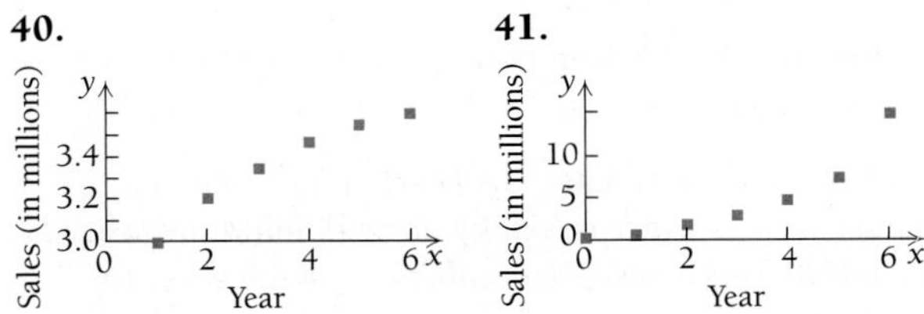

42.

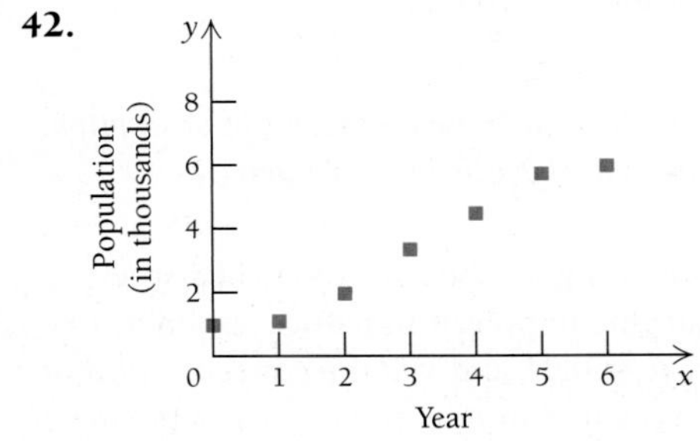

43.

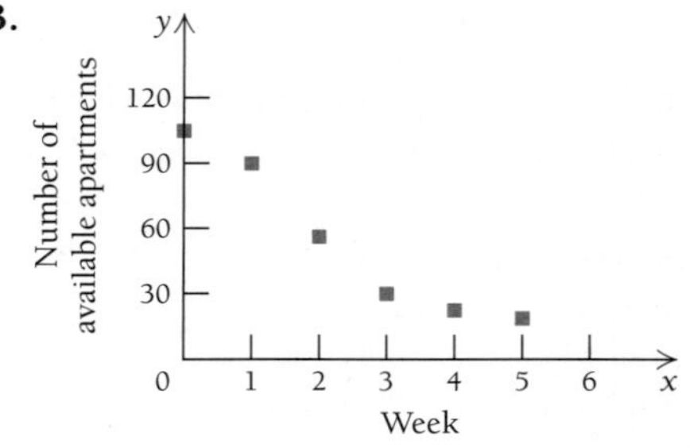

44.

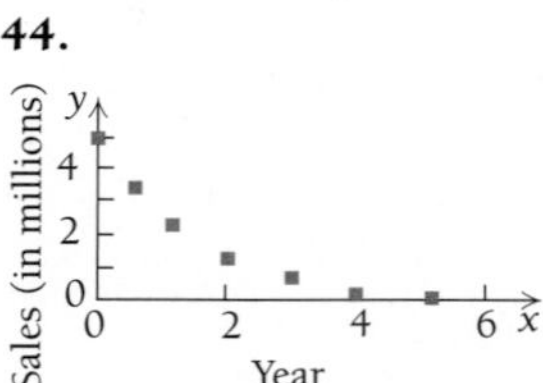

45.

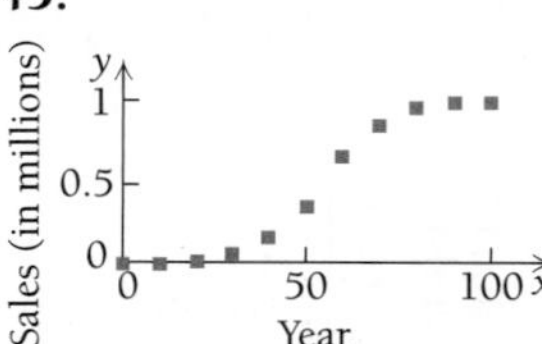

46.

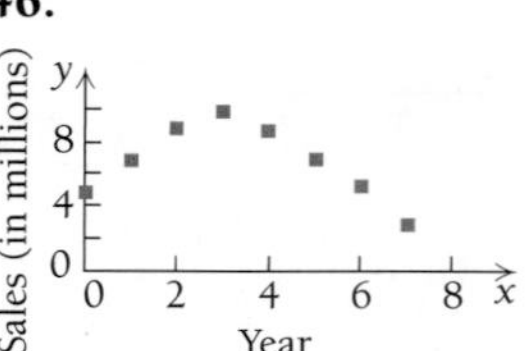

47.

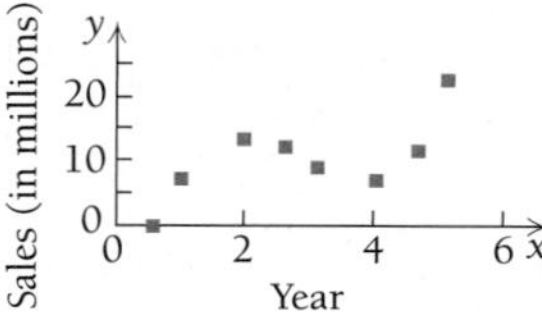

SYNTHESIS

48. **Economics: supply and demand.** The demand, $D(x)$, and supply, $S(x)$, functions for a certain type of multipurpose printer are as follows:

$$D(x) = q = 480e^{-0.003x}$$

and

$$S(x) = q = 150e^{0.004x}.$$

Find the equilibrium point. Assume that x is the price in dollars.

The Beer–Lambert Law. *A beam of light enters a medium such as water or smoky air with initial intensity I_0. Its intensity is decreased depending on the thickness (or concentration) of the medium. The intensity I at a depth (or concentration) of x units is given by*

$$I = I_0e^{-\mu x}.$$

The constant μ ("mu"), called the coefficient of absorption, varies with the medium. Use this law for Exercises 49 and 50.

TW 49. **Light through smog.** Concentrations of particulates in the air due to pollution reduce sunlight. In a smoggy area, $\mu = 0.01$ and x is the concentration of particulates measured in micrograms per cubic meter (mcg/m^3). What change is more significant—dropping pollution levels from 100 mcg/m^3 to 90 mcg/m^3 or dropping them from 60 mcg/m^3 to 50 mcg/m^3? Why?

TW 50. **Light through sea water.** Sea water has $\mu = 1.4$ and x is measured in meters. What would increase cloudiness more—dropping x from 2 m to 5 m or dropping x from 7 m to 10 m? Explain.

51. **Newton's Law of Cooling.** Consider the following exploratory situation. Fill a glass with hot tap water. Place a thermometer in the glass and measure the temperature. Check the temperature every 30 min thereafter. Plot your data on this graph, and connect the points with a smooth curve.

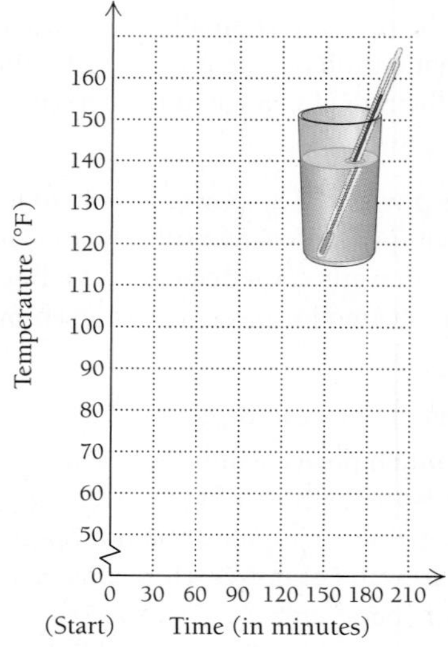

a) What was the temperature of the water when you began?

b) At what temperature does there seem to be a leveling off of the graph?

c) What is the difference between your answers to parts (a) and (b)?

d) How does the water temperature in part (b) compare with the room temperature?

e) Find an equation that fits the data. Use this equation to check values of other data points. How do they compare?

f) Is it ever "theoretically" possible for the temperature of the water to be the same as the room temperature? Explain.

TW **g)** Find the rate of change of the temperature, and interpret its meaning.

TW **52.** An interest rate decreases from 8% to 7.2%. Explain why this increases the present value of an amount due 10 yr later.

53. Population growth. The population of the United States was 281 million in 2000. At that time, it was estimated that the population P was growing exponentially at the rate of 0.9% per year; that is,

$$\frac{dP}{dt} = 0.009P.$$

(*Source*: U.S. Census Bureau.) In what year will the U.S. population of 2007 be doubled?

3.5 The Derivatives of a^x and $\log_a x$

OBJECTIVES

- Differentiate functions involving a^x.
- Differentiate functions involving $\log_a x$.

The Derivative of a^x

To find the derivative of a^x, for any base a, we first express a^x as a power of e. To do this, we recall that $\log_b x$ is the power to which b is raised in order to get x. Thus,

$$b^{\log_b x} = x.$$

In particular, it follows that

$$e^{\log_e A} = A, \quad \text{or } e^{\ln A} = A.$$

If we replace A with a^x, we have

$$e^{\ln a^x} = a^x, \quad \text{or } a^x = e^{\ln a^x}. \tag{1}$$

To find the derivative of a^x, we differentiate both sides:

$$\begin{aligned}
\frac{d}{dx}a^x &= \frac{d}{dx}e^{\ln a^x} \\
&= \frac{d}{dx}e^{x \ln a} && \text{Using a property of logarithms} \\
&= \frac{d}{dx}e^{(\ln a)x} \\
&= e^{(\ln a)x} \cdot \ln a && \text{Differentiating } e^{kx} \text{ with respect to } x \\
&= e^{\ln a^x} \cdot \ln a && \text{Using a property of logarithms} \\
&= a^x \cdot \ln a. && \text{Using equation (1)}
\end{aligned}$$

Thus, we have the following theorem.

THEOREM 12

$$\frac{d}{dx}a^x = (\ln a)\, a^x$$

EXAMPLE 1 Differentiate: **a)** $y = 2^x$; **b)** $y = (1.4)^x$; **c)** $f(x) = 3^{2x}$.

Solution

a) $\frac{d}{dx}2^x = (\ln 2)2^x$ Using Theorem 12

Note that $\ln 2 \approx 0.7$, so this equation verifies our earlier approximation of the derivative of 2^x as $(0.7)2^x$ in Section 3.1.

b) $\frac{d}{dx}(1.4)^x = (\ln 1.4)(1.4)^x$

c) Since $f(x) = 3^{2x}$ is of the form $f(x) = 3^{g(x)}$, the Chain Rule applies:

$$f'(x) = (\ln 3)3^{2x} \cdot \frac{d}{dx}(2x)$$
$$= \ln 3 \cdot 3^{2x} \cdot 2 = 2 \ln 3 \cdot 3^{2x}$$ ✦

Compare these formulas:

$$\frac{d}{dx}a^x = (\ln a)\, a^x \quad \text{and} \quad \frac{d}{dx}e^x = e^x.$$

It is the simplicity of the latter formula that is a reason for the use of base e in calculus. The many applications of e in natural phenomena provide other reasons.

One other result also follows from what we have done. If

$$f(x) = a^x,$$

we now know that

$$f'(x) = a^x (\ln a).$$

Alternatively, in Section 3.1, we showed that if $f(x) = a^x$, then

$$f'(x) = a^x \cdot \lim_{h \to 0} \frac{a^h - 1}{h}.$$

Thus,

$$a^x (\ln a) = a^x \cdot \lim_{h \to 0} \frac{a^h - 1}{h}.$$

Dividing both sides by a^x, we have the following.

THEOREM 13

$$\ln a = \lim_{h \to 0} \frac{a^h - 1}{h}$$

The Derivative of $\log_a x$

Just as the derivative of a^x is expressed in terms of $\ln a$, so too is the derivative of $\log_a x$. To find this derivative, we first express $\log_a x$ in terms of $\ln a$ using the change-of-base formula (P7 of Theorem 4) from Section 3.2:

$$\frac{d}{dx}\log_a x = \frac{d}{dx}\left(\frac{\log_e x}{\log_e a}\right) \qquad \text{Using the change-of-base formula}$$

$$= \frac{d}{dx}\left(\frac{\ln x}{\ln a}\right)$$

$$= \frac{1}{\ln a}\cdot\frac{d}{dx}(\ln x) \qquad \frac{1}{\ln a} \text{ is a constant.}$$

$$= \frac{1}{\ln a}\cdot\frac{1}{x}.$$

THEOREM 14

$$\frac{d}{dx}\log_a x = \frac{1}{\ln a}\cdot\frac{1}{x}$$

Comparing this equation with

$$\frac{d}{dx}\ln x = \frac{1}{x},$$

we see another reason for the use of base e in calculus.

EXAMPLE 2 Differentiate: **a)** $y = \log_8 x$; **b)** $y = \log x$; **c)** $f(x) = \log_3 (x^2 + 1)$; **d)** $f(x) = x^3 \log_5 x$.

Solution

a) $\dfrac{d}{dx}\log_8 x = \dfrac{1}{\ln 8}\cdot\dfrac{1}{x}$ Using Theorem 14

b) $\dfrac{d}{dx}\log x = \log_{10} x$ $\log x$ means $\log_{10} x$.

$$= \frac{1}{\ln 10}\cdot\frac{1}{x}$$

c) Note that $f(x) = \log_3 (x^2 + 1)$ is of the form $f(x) = \log_3 (g(x))$, so the Chain Rule is required:

$$f'(x) = \frac{1}{\ln 3}\cdot\frac{1}{x^2+1}\cdot\frac{d}{dx}(x^2+1) \qquad \text{Using the Chain Rule}$$

$$= \frac{1}{\ln 3}\cdot\frac{1}{x^2+1}\cdot 2x$$

$$= \frac{2x}{(\ln 3)(x^2+1)}.$$

TECHNOLOGY CONNECTION

Exploratory

Using the nDeriv feature, check the results of Examples 1 and 2 graphically. Then differentiate $y = \log_2 x$, and check the result with your calculator.

d) Since $f(x) = x^3 \log_5 x$ is of the form $f(x) = g(x) \cdot h(x)$, the Product Rule is applied:

$$f'(x) = x^3 \cdot \frac{d}{dx} \log_5 x + \log_5 x \cdot 3x^2 \qquad \text{Using the Product Rule}$$

$$= x^3 \cdot \frac{1}{\ln 5} \cdot \frac{1}{x} + \log_5 x \cdot 3x^2$$

$$= \frac{x^2}{\ln 5} + 3x^2 \log_5 x, \quad \text{or} \quad x^2\left(\frac{1}{\ln 5} + 3\log_5 x\right). \quad ✦$$

Exercise Set 3.5

Differentiate.

1. $y = 7^x$
2. $y = 6^x$
3. $f(x) = 8^x$
4. $f(x) = 15^x$
5. $g(x) = x^3(5.4)^x$
6. $g(x) = x^5(3.7)^x$
7. $y = 7^{x^4+2}$
8. $y = 4^{x^2+5}$
9. $y = e^{8x}$
10. $y = e^{x^2}$
11. $f(x) = 3^{x^4+1}$
12. $f(x) = 12^{7x-4}$
13. $y = \log_4 x$
14. $y = \log_8 x$
15. $y = \log_{17} x$
16. $y = \log_{23} x$
17. $g(x) = \log_6 (5x + 1)$
18. $g(x) = \log_{32} (9x - 2)$
19. $F(x) = \log (6x - 7)$
20. $G(x) = \log (5x + 4)$
21. $y = \log_8 (x^3 + x)$
22. $y = \log_9 (x^4 - x)$
23. $f(x) = 4 \log_7 (\sqrt{x} - 2)$
24. $g(x) = -\log_6 (\sqrt[3]{x} + 5)$
25. $y = 6^x \cdot \log_7 x$
26. $y = 5^x \cdot \log_2 x$
27. $G(x) = (\log_{12} x)^5$
28. $F(x) = (\log_9 x)^7$
29. $g(x) = \dfrac{7^x}{4x + 1}$
30. $f(x) = \dfrac{6^x}{5x - 1}$
31. $y = 5^{2x^3-1} \cdot \log (6x + 5)$
32. $y = \log (7x + 3) \cdot 4^{2x^4+8}$
33. $F(x) = 7^x \cdot (\log_4 x)^9$
34. $G(x) = \log_9 x \cdot (4^x)^6$
35. $f(x) = (3x^5 + x)^5 \log_3 x$
36. $g(x) = \sqrt{x^3 - x}\,(\log_5 x)$

APPLICATIONS

Business and Economics

37. **Double declining balance depreciation.** An office machine is purchased for \$5200. Under certain assumptions, its salvage value, V, in dollars, is depreciated according to a method called *double declining balance*, by basically 80% each year, and is given by

$$V(t) = 5200(0.80)^t,$$

where t is the time, in years after purchase.

a) Find $V'(t)$.

TW **b)** Interpret the meaning of $V'(t)$.

38. **Recycling aluminum cans.** It is known that 45% of all aluminum cans distributed will be recycled each year. A beverage company uses 250,000 lb of aluminum cans. After recycling, the amount of aluminum, in pounds, still in use after t years is given by

$$N(t) = 250{,}000(0.45)^t.$$

(*Source*: Based on data from The Container Recycling Institute.)

a) Find $N'(t)$.

TW **b)** Interpret the meaning of $N'(t)$.

39. Household liability. The total financial liability, in billions of dollars, of U.S. households can be modeled by the function

$$L(t) = 1547(1.083)^t,$$

where t is the number of years after 1980. The graph of this function is given below.

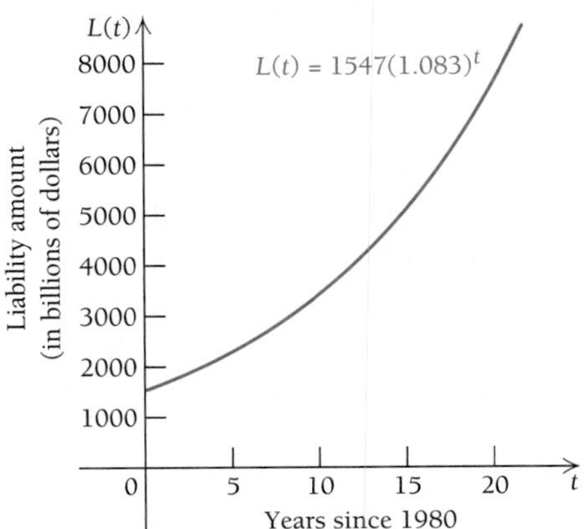

a) Using this model, predict the total financial liability of U.S. households in 2012.

b) Find $L'(25)$.

TW **c)** Interpret the meaning of $L'(25)$.

40. Small business. The number of nonfarm proprietorships, in thousands, in the United States can be modeled by the function

$$N(t) = 8400 \ln t - 10{,}500,$$

where t is the number of years after 1970. The graph of this function is given below.

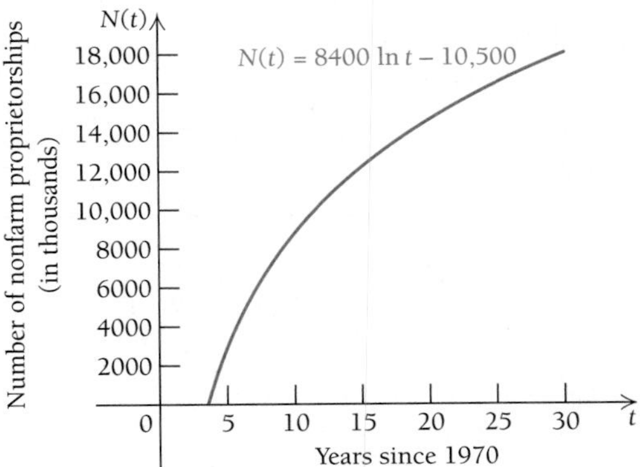

a) Using this model, predict the number of nonfarm proprietorships in the United States in 2014.

b) Find $N'(45)$.

TW **c)** Interpret the meaning of $N'(45)$.

Life and Physical Sciences

41. Agriculture. Farmers wishing to avoid the use of genetically modified (GMO) seeds are increasingly concerned about inadvertently growing GMO plants as a result of pollen drifting from nearby farms. Assuming that these farmers raise their own seeds, the fractional portion of their crop that remains free of GMO plants t years later can be approximated by

$$P(t) = (0.98)^t.$$

a) Using this model, predict the fractional portion of the crop that will be GMO-free 10 yr after a neighboring farm begins to use GMO seeds.

b) Find $P'(15)$.

TW **c)** Interpret the meaning of $P'(15)$.

Earthquake magnitude. *The magnitude R (measured on the Richter scale) of an earthquake of intensity I is defined as*

$$R = \log \frac{I}{I_0},$$

where I_0 is a minimum intensity used for comparison. When one earthquake is 10 times as intense as another, its magnitude on the Richter scale is 1 higher. If one earthquake is 100 times as intense as another, its magnitude on the Richter scale is 2 higher, and so on. Thus, an earthquake whose magnitude is 7 on the Richter scale is 10 times as intense as an earthquake whose magnitude is 6. Earthquakes can be interpreted as multiples of the minimum intensity I_0. Use this information for Exercises 42 and 43.

42. In 1986, there was an earthquake near Cleveland, Ohio. It had an intensity of $10^5 \cdot I_0$. What was its magnitude on the Richter scale?

43. On October 17, 1989, there was an earthquake in San Francisco, California, during the World Series. It had an intensity of $10^{6.9} \cdot I_0$. What was its magnitude on the Richter scale?

This photograph shows part of the damage in the San Francisco, California, area earthquake in 1989.

44. Earthquake intensity. The intensity of an earthquake is given by

$$I = I_0 10^R,$$

where R is the magnitude on the Richter scale and I_0 is the minimum intensity, at which $R = 0$, used for comparison.

a) Find I, in terms of I_0, for an earthquake of magnitude 7 on the Richter scale.

b) Find I, in terms of I_0, for an earthquake of magnitude 8 on the Richter scale.

c) Compare your answers to parts (a) and (b).

d) Find the rate of change dI/dR.

TW e) Interpret the meaning of dI/dR.

45. Intensity of sound. The intensity of a sound is given by

$$I = I_0 10^{0.1L},$$

where L is the loudness of the sound as measured in decibels and I_0 is the minimum intensity detectable by the human ear.

a) Find I, in terms of I_0, for the loudness of a power mower, which is 100 decibels.

b) Find I, in terms of I_0, for the loudness of a just audible sound, which is 10 decibels.

c) Compare your answers to parts (a) and (b).

d) Find the rate of change dI/dL.

TW e) Interpret the meaning of dI/dL.

46. Earthquake magnitude. The magnitude R (measured on the Richter scale) of an earthquake of intensity I is defined as

$$R = \log \frac{I}{I_0},$$

where I_0 is the minimum intensity (used for comparison). (The exponential form of this definition is given in Exercise 44.)

a) Find the rate of change dR/dI.

TW b) Interpret the meaning of dR/dI.

47. Loudness of sound. The loudness L of a sound of intensity I is defined as

$$L = 10 \log \frac{I}{I_0},$$

where I_0 is the minimum intensity detectable by the human ear and L is the loudness measured in decibels. (The exponential form of this definition is given in Exercise 45.)

a) Find the rate of change dL/dI.

TW b) Interpret the meaning of dL/dI.

48. Response to drug dosage. The response y to a dosage x of a drug is given by

$$y = m \log x + b,$$

where m and b are constants. The response may be hard to measure with a number. The patient might perspire more, have an increase in temperature, or faint.

a) Find the rate of change dy/dx.

TW b) Interpret the meaning of dy/dx.

SYNTHESIS

49. Find $\lim\limits_{h \to 0} \dfrac{3^h - 1}{h}$. (*Hint*: See p. 366.)

Use the Chain Rule, implicit differentiation, and other formulas in this section to differentiate each function given in Exercises 50–57.

50. $f(x) = 3^{(2^x)}$

51. $y = 2^{x^4}$

52. $y = x^x$, for $x > 0$

53. $y = \log_3 (\log x)$

54. $f(x) = x^{e^x}$, for $x > 0$

55. $y = a^{f(x)}$

56. $y = \log_a f(x)$, for $f(x)$ positive

57. $y = [f(x)]^{g(x)}$, for $f(x)$ positive

TW **58.** In your own words, prove the formula for finding the derivative of $f(x) = a^x$.

TW **59.** In your own words, prove the formula for finding the derivative of $f(x) = \log_a x$.

3.6 An Economics Application: Elasticity of Demand

OBJECTIVES

- Find the elasticity of a demand function.
- Find the maximum of a total-revenue function.
- Characterize demand in terms of elasticity.

Retailers and manufacturers often need to know how a small change in price will affect the demand for a product. If a small increase in price produces no change in demand, a price increase may make sense; if a small increase in price creates a large drop in demand, the increase is probably ill advised. To measure the sensitivity of demand to a small percent increase in price, economists calculate the *elasticity of demand.*

Suppose that q represents a quantity of goods purchased and x is the price per unit of the goods. Recall that q and x are related by the demand function

$$q = D(x).$$

Suppose that there is a change Δx in the price per unit. The percent change in price is given by

$$\frac{\Delta x}{x} = \frac{\Delta x}{x} \cdot \frac{100}{100} = \frac{\Delta x \cdot 100}{x}\%.$$

A change in the price produces a change Δq in the quantity sold. The percent change in quantity is given by

$$\frac{\Delta q}{q} = \frac{\Delta q \cdot 100}{q}\%.$$

The ratio of the percent change in quantity to the percent change in price is

$$\frac{\Delta q/q}{\Delta x/x},$$

which can be expressed as

$$\frac{x}{q} \cdot \frac{\Delta q}{\Delta x}. \qquad (1)$$

Note that for differentiable functions,

$$\lim_{\Delta x \to 0} \frac{\Delta q}{\Delta x} = \frac{dq}{dx},$$

so the limit as Δx approaches 0 of the expression in equation (1) becomes

$$\lim_{\Delta x \to 0} \frac{x}{q} \cdot \frac{\Delta q}{\Delta x} = \frac{x}{q} \cdot \frac{dq}{dx} = \frac{x}{q} \cdot D'(x) = \frac{x}{D(x)} \cdot D'(x).$$

This result is the basis of the following definition.

DEFINITION

The **elasticity of demand** E is given as a function of price x by

$$E(x) = -\frac{x \cdot D'(x)}{D(x)}.$$

To understand the purpose of the negative sign in the preceding definition, note that the price, x, and the demand, $D(x)$, are both nonnegative. Since $D(x)$ is normally decreasing, $D'(x)$ is usually negative. By inserting a negative sign in the definition, economists make $E(x)$ nonnegative and easier to work with.

EXAMPLE 1 Economics: Demand for DVD Rentals. Lake Shore Video has found that demand for rentals of its DVDs is given by

$$q = D(x) = 120 - 20x,$$

where q is the number of DVDs rented per day at x dollars per rental. Find each of the following.

a) The quantity demanded when the price is \$2 per rental

b) The elasticity as a function of x

c) The elasticity at $x = 2$ and at $x = 4$. Interpret the meaning of these values of the elasticity.

d) The value of x for which $E(x) = 1$. Interpret the meaning of this price.

e) The total-revenue function, $R(x) = x \cdot D(x)$

f) The price x at which total revenue is a maximum

Solution

a) For $x = 2$, we have $D(2) = 120 - 20(2) = 80$. Thus, 80 DVDs per day will be rented at a price of \$2 per rental.

b) To find the elasticity, we first find the derivative $D'(x)$:

$$D'(x) = -20.$$

Then we substitute -20 for $D'(x)$ and $120 - 20x$ for $D(x)$ in the expression for elasticity:

$$E(x) = -\frac{x \cdot D'(x)}{D(x)} = -\frac{x \cdot (-20)}{120 - 20x} = \frac{20x}{120 - 20x} = \frac{x}{6 - x}.$$

c) $E(2) = \dfrac{2}{6 - 2} = \dfrac{1}{2}$

At $x = 2$, the elasticity is $\frac{1}{2}$, which is less than 1. Thus, the ratio of the percent change in quantity to the percent change in price is less than 1. A small percentage increase in price will cause an even smaller percentage decrease in the quantity sold.

$$E(4) = \frac{4}{6 - 4} = 2$$

At $x = 4$, the elasticity is 2, which is greater than 1. Thus, the ratio of the percent change in quantity to the percent change in price is greater than 1. A small percentage increase in price will cause a larger percentage decrease in the quantity sold.

TECHNOLOGY CONNECTION

Economics: Demand for In-Ear Radios

A company determines that the demand function for in-ear radios is

$$q = D(x) = 300 - x,$$

where q is the number sold per day when the price is x dollars per radio.

EXERCISES

1. Find the elasticity E and the total revenue R.
2. Using only the first quadrant, graph the demand, elasticity, and total-revenue functions on the same set of axes.
3. Find the price x for which the total revenue is a maximum. Use the method of Example 1. Check the answer graphically.

d) We set $E(x) = 1$ and solve for p:

$$\frac{x}{6 - x} = 1$$

$$x = 6 - x \qquad \text{We multiply both sides by } 6 - x\text{, assuming that } x \neq 6.$$

$$2x = 6$$

$$x = 3.$$

Thus, when the price is \$3 per rental, the ratio of the percent change in quantity to the percent change in price is 1.

e) Recall that the total revenue $R(x)$ is given by $x \cdot D(x)$. Then

$$R(x) = x \cdot D(x) = x(120 - 20x) = 120x - 20x^2.$$

f) To find the price x that maximizes total revenue, we find $R'(x)$:

$$R'(x) = 120 - 40x.$$

We see that $R'(x)$ exists for all x in the interval $[0, \infty)$. Thus, we solve:

$$R'(x) = 120 - 40x = 0$$

$$-40x = -120$$

$$x = 3.$$

Since there is only one critical value, we can try to use the second derivative to see if we have a maximum:

$$R''(x) = -40 < 0.$$

Thus, $R''(3)$ is negative, so $R(3)$ is a maximum. That is, total revenue is a maximum at \$3 per rental. ✦

Note in parts (d) and (f) of Example 1 that the value of x for which $E(x) = 1$ is the same as the value of x for which total revenue is a maximum. The following theorem states that this is always the case.

THEOREM 15

Total revenue is increasing at those x-values for which $E(x) < 1$.

Total revenue is decreasing at those x-values for which $E(x) > 1$.

Total revenue is maximized at the value(s) of x for which $E(x) = 1$.

Proof. We know that

$$R(x) = x \cdot D(x),$$

so

$$R'(x) = x \cdot D'(x) + D(x) \cdot 1 \qquad \text{Using the product rule}$$

$$= D(x)\left[\frac{x \cdot D'(x)}{D(x)} + 1\right] \qquad \text{Check this by multiplying.}$$

$$= D(x)[-E(x) + 1]$$

$$= D(x)[1 - E(x)]$$

Since we can assume that $D(x) > 0$, it follows that $R'(x)$ is positive for $E(x) < 1$, is negative for $E(x) > 1$, and is 0 when $E(x) = 1$. Thus, total revenue is increasing for $E(x) < 1$, is decreasing for $E(x) > 1$, and is maximized when $E(x) = 1$. ✦

Elasticity and Revenue

For a particular value of the price x:

1. The demand is *inelastic* if $E(x) < 1$. An increase in price will bring an increase in revenue. If demand is inelastic, then revenue is increasing.
2. The demand has *unit elasticity* if $E(x) = 1$. The demand has unit elasticity when revenue is at a maximum.
3. The demand is *elastic* if $E(x) > 1$. An increase in price will bring a decrease in revenue. If demand is elastic, then revenue is decreasing.

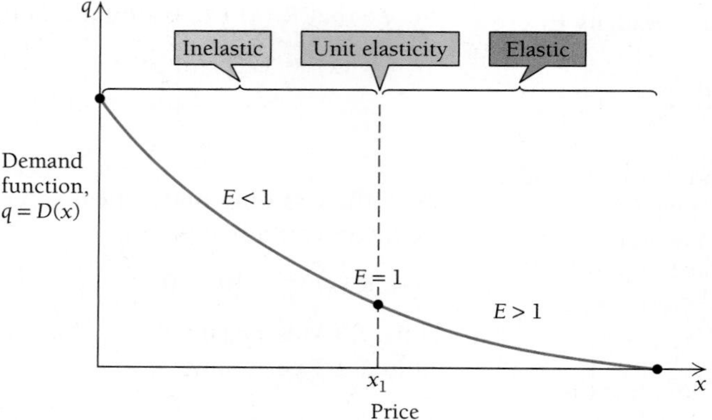

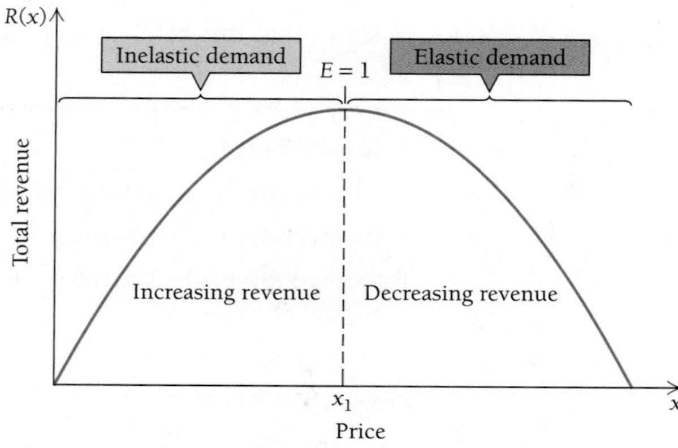

In summary, suppose that Lake Shore Video in Example 1 raises the price per rental and that the total revenue increases. Then we say the demand is *inelastic*. If the total revenue decreases, we say the demand is *elastic*. Some price elasticities in the U.S. economy are listed in the following table.

Price Elasticities in the U.S. Economy

Industry	Elasticity
Elastic Demands	
Metals	1.52
Electrical engineering products	1.39
Mechanical engineering products	1.30
Furniture	1.26
Motor vehicles	1.14
Instrument engineering products	1.10
Professional services	1.09
Transportation services	1.03
Inelastic Demands	
Gas, electricity, and water	0.92
Oil	0.91
Chemicals	0.89
Beverages (all types)	0.78
Tobacco	0.61
Food	0.58
Banking and insurance services	0.56
Housing services	0.55
Clothing	0.49
Agricultural and fish products	0.42
Books, magazines, and newspapers	0.34
Coal	0.32

(*Source*: Ahsan Mansur and John Whalley, "Numerical specification of applied general equilibrium models: Estimation, calibration, and data." In H. E. Scarf and J. B. Shoven (eds.), *Applied General Equilibrium Analysis*. (New York: Cambridge University Press, 1984), p. 109.)

Exercise Set 3.6

For the demand function given in each of Exercises 1–12, find the following.

a) *The elasticity*

b) *The elasticity at the given price, stating whether the demand is elastic or inelastic*

c) *The value(s) of x for which total revenue is a maximum (assume that x is in dollars)*

1. $q = D(x) = 400 - x; \quad x = 125$

2. $q = D(x) = 500 - x; \quad x = 38$

3. $q = D(x) = 200 - 4x; \quad x = 46$

4. $q = D(x) = 500 - 2x; \quad x = 57$

5. $q = D(x) = \dfrac{400}{x}; \quad x = 50$

6. $q = D(x) = \dfrac{3000}{x}; \quad x = 60$

7. $q = D(x) = \sqrt{600 - x}; \quad x = 100$

8. $q = D(x) = \sqrt{300 - x}; \quad x = 250$

9. $q = D(x) = 100e^{-0.25x}$; $x = 10$

10. $q = D(x) = 200e^{-0.05x}$; $x = 80$

11. $q = D(x) = \dfrac{100}{(x + 3)^2}$; $x = 1$

12. $q = D(x) = \dfrac{500}{(2x + 12)^2}$; $x = 8$

APPLICATIONS

Business and Economics

13. Demand for chocolate chip cookies. Good Times Bakers works out a demand function for its chocolate chip cookies and finds it to be

$$q = D(x) = 967 - 25x,$$

where q is the quantity of cookies sold when the price per cookie, in cents, is x.

a) Find the elasticity.
b) At what price is the elasticity of demand equal to 1?
c) At what prices is the elasticity of demand elastic?
d) At what prices is the elasticity of demand inelastic?
e) At what price is the revenue a maximum?
f) At a price of 20¢ per cookie, will a small increase in price cause the total revenue to increase or decrease?

14. Demand for oil. Suppose that you have been hired by OPEC as an economic consultant concerning the world demand for oil. The demand function is

$$q = D(x) = 63{,}000 + 50x - 25x^2, \quad 0 \le x \le 50,$$

where q is measured in millions of barrels of oil per day at a price of x dollars per barrel.

a) Find the elasticity.
b) Find the elasticity at a price of \$10 per barrel, stating whether the demand is elastic or inelastic at that price.
c) Find the elasticity at a price of \$20 per barrel, stating whether the demand is elastic or inelastic at that price.
d) Find the elasticity at a price of \$30 per barrel, stating whether the demand is elastic or inelastic at that price.
e) At what price is the revenue a maximum?
f) What quantity of oil will be sold at the price that maximizes revenue? Check current world prices to see how this compares with your answer.
g) At a price of \$30 per barrel, will a small increase in price cause the total revenue to increase or decrease?

15. Demand for computer games. High Wire Electronics determines the following demand function for a new game:

$$q = D(x) = \sqrt{200 - x^3},$$

where q is the number of games sold per day when the price is x dollars per game.

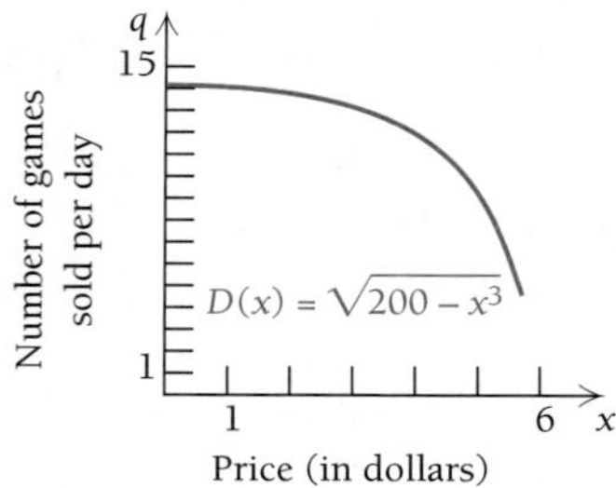

a) Find the elasticity.
b) Find the elasticity when $x = 3$.
c) At $x = 3$, will a small increase in price cause the total revenue to increase or decrease?

16. Demand for tomato plants. Sunshine Gardens determines the following demand function during early summer for tomato plants:

$$q = D(x) = \frac{2x + 300}{10x + 11},$$

where q is the number of plants sold per day when the price is x dollars per plant.

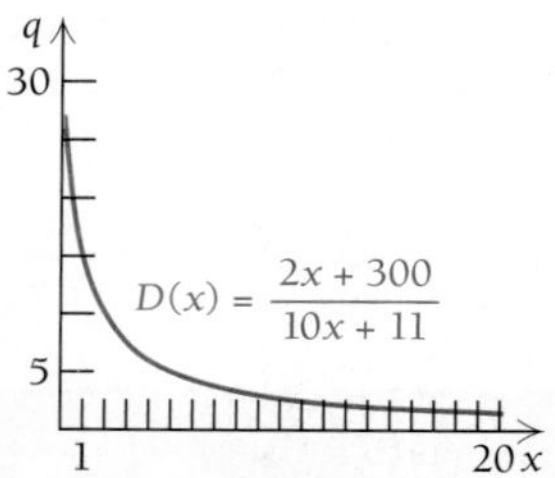

a) Find the elasticity.
b) Find the elasticity when $x = 3$.
c) At \$3 per plant, will a small increase in price cause the total revenue to increase or decrease?

SYNTHESIS

17. Economics: constant elasticity curve.

a) Find the elasticity of the demand function

$$q = D(x) = \frac{k}{x^n},$$

where k is a positive constant and n is an integer greater than 0.

b) Is the value of the elasticity dependent on the price per unit?

c) Does the total revenue have a maximum? When?

18. Economics: exponential demand curve.

a) Find the elasticity of the demand function

$$q = D(x) = Ae^{-kx},$$

where A and k are positive constants.

b) Is the value of the elasticity dependent on the price per unit?

c) Does the total revenue have a maximum? At what value of x?

19. Let

$$L(x) = \ln D(x).$$

Describe the elasticity in terms of $L'(x)$.

TW **20.** Explain in your own words the concept of elasticity and its usefulness to economists. Do some library or online research or consult an economist in order to determine when and how this concept was first developed.

TW **21.** Explain how the elasticity of demand for a product can be affected by the availability of substitutes for the product.

Chapter Summary

This chapter examined exponential and logarithmic functions, showing how such functions are differentiated and how they are used in real-world situations.

An **exponential function**, f, is a function of the form $f(x) = a^x$, where x is any real number and a, the **base,** is any positive number other than 1. For example, the functions given by $f(x) = \left(\frac{1}{2}\right)^x$ and $g(x) = 3^x$ are both exponential. (p. 305)

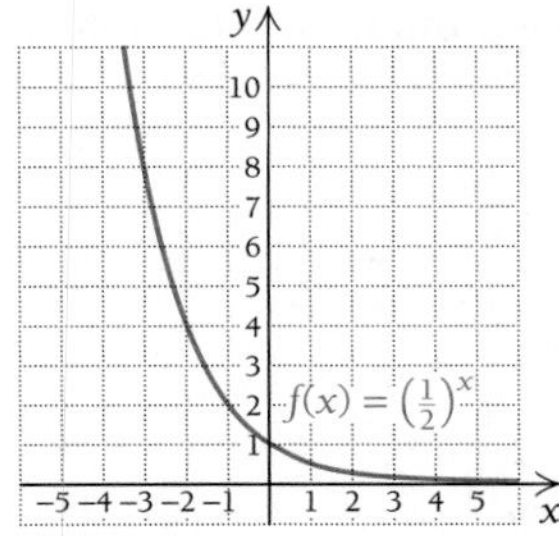

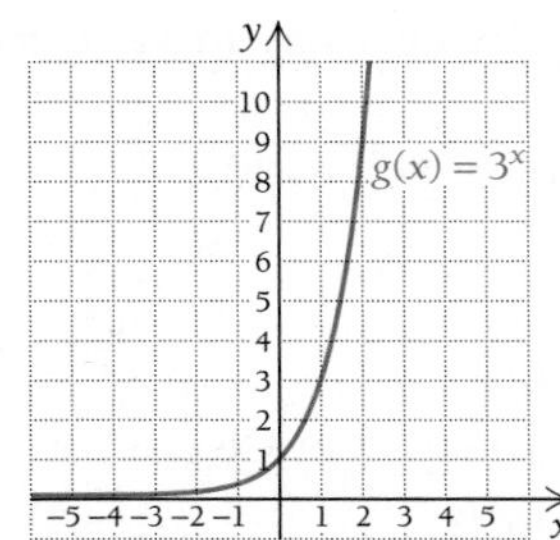

Only one base exists for which $\frac{d}{dx}a^x = a^x$ and that number is e. That is,

$$\frac{d}{dx}e^x = e^x,$$

where $e \approx 2.718$ and $e = \lim_{h \to 0} (1 + h)^{1/h}$. (p. 310)

Functions like $y = e^x$ can appear in a variety of mathematical settings. To differentiate $f(x) = (5x^2 + 3)e^{7x}$, for example, both the Product and the Chain Rules are needed:

$$\frac{d}{dx}[(5x^2 + 3)e^{7x}] = (5x^2 + 3)e^{7x} \cdot 7 + e^{7x} \cdot 10x.$$ (p. 312)

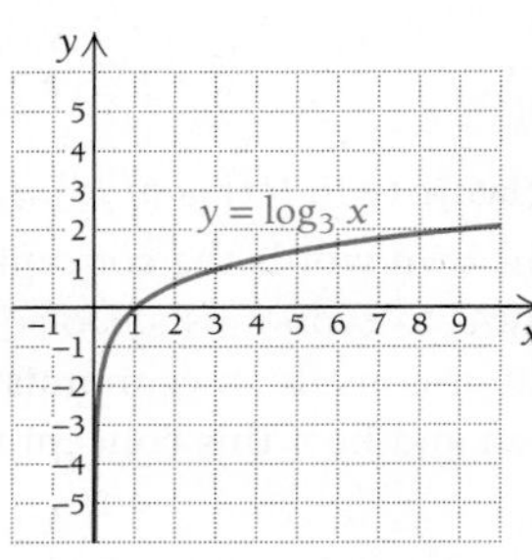

A *logarithmic function*, g, is any function of the form $y = \log_a x$, where a, the *base*, is a positive number other than 1. Logarithmic functions are inverses of exponential functions. Thus, for

$$f(x) = 4.3^x, \quad \text{we have} \quad f(2) = 4.3^2. \qquad \text{(p. 319)}$$

Since

$$g(x) = \log_{4.3} x, \quad \text{we have} \quad g(4.3^2) = \log_{4.3} 4.3^2 = 2.$$

It is important to remember that $\log_b c$ *is the exponent to which* b *is raised to get* c. It is understood that $\log c$ means $\log_{10} c$ and that $\ln c$ means $\log_e c$.

Thus, $\log 1000 = 3$ and $\ln e^{\sqrt{5}} = \sqrt{5}$. (pp. 321–323)

Important properties of logarithms follow. We assume that M, N, and a are positive, with $a \neq 1$, and k is a real number:

$$\log_a (MN) = \log_a M + \log_a N$$
$$\log_a \frac{M}{N} = \log_a M - \log_a N$$
$$\log_a (M^k) = k \log_a M$$
$$\log_a a = 1$$
$$\log_a a^k = k$$
$$\log_a 1 = 0$$
$$\log_b M = \frac{\log_a M}{\log_a b} \qquad \text{(p. 321)}$$

Logarithms provide a way to solve certain exponential equations.

EXAMPLE Solve: $5e^{2t} = 80$.

Solution We have

$$5e^{2t} = 80$$
$$e^{2t} = 16 \qquad \text{Dividing both sides by 5}$$
$$\ln e^{2t} = \ln 16 \qquad \text{Taking the natural log of both sides}$$
$$2t = \ln 16 \qquad \text{Using a property of logarithms}$$
$$t = \frac{\ln 16}{2}$$
$$t \approx 1.386$$

✦

The derivative of the natural logarithm function is easy to calculate:

$$\frac{d}{dx} \ln x = \frac{1}{x}. \qquad \text{(p. 327)}$$

The function $y = \ln x$ appears in many mathematical settings, some of which require the rules for differentiation developed in earlier chapters. In the following, we use both Product and Chain Rules:

$$\frac{d}{dx}[\ln(5x)\cdot(x^3-7x)] = \ln(5x)\cdot(3x^2-7) + (x^3-7x)\cdot\frac{1}{5x}\cdot 5$$
$$= \ln(5x)\cdot(3x^2-7) + x^2 - 7.$$

Because exponential functions are the only functions for which the derivative (rate of change) is directly proportional to the function value at any point in time, they are used to model many real-world situations, as the following examples illustrate.

EXAMPLE The balance P in a savings account grows at a rate given by

$$\frac{dP}{dt} = 0.04P,$$

where t is time, in years.

a) Find the function that satisfies the equation (let $P(0) = P_0$).

b) After what period of time will an initial investment, P_0, double itself?

Solution

a) $P(t) = P_0e^{0.04t}$

Check:

$$\frac{d}{dt}P_0e^{0.04t} = P_0e^{0.04t}\cdot 0.04$$
$$= 0.04P_0e^{0.04t}$$
$$= 0.04P(t)$$

b) We set $P(t) = 2P_0$ and solve for t:

$$2P_0 = P_0e^{0.04t}$$
$$2 = e^{0.04t} \quad \text{Dividing both sides by } P_0$$
$$\ln 2 = \ln(e^{0.04t}) \quad \text{Taking the natural log of both sides}$$
$$\ln 2 = 0.04t$$
$$\frac{\ln 2}{0.04} = t, \quad \text{or} \quad t \approx 17.3 \text{ yr}.$$

✦

In general, the **growth rate,** k, and the **doubling time,** T, are related by

$$kT = \ln 2 \approx 0.693147,$$

or

$$k = \frac{\ln 2}{T} \approx \frac{0.693147}{T} \quad \text{and} \quad T = \frac{\ln 2}{k} \approx \frac{0.693147}{k}.$$ (p. 339)

The **Rule of 70** rewrites the last of these equations as

$$\text{Doubling time} = T \approx \frac{70}{100k},$$

which allows us to simply write k as a percentage when calculating an approximation for the doubling time. (p. 340)

Two models for limited, or restricted, growth were also studied and are illustrated by the following graphs. (pp. 343, 345)

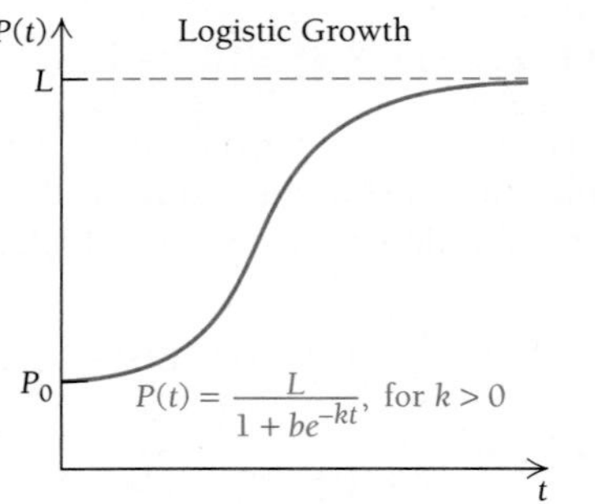

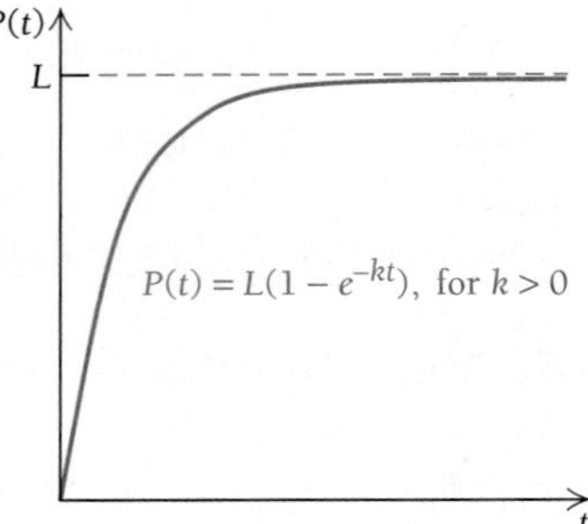

Exponential growth, modeled by $P(t) = P_0e^{kt}$, with $k > 0$, becomes **exponential decay** when the exponent is $-kt$. (pp. 352–353)

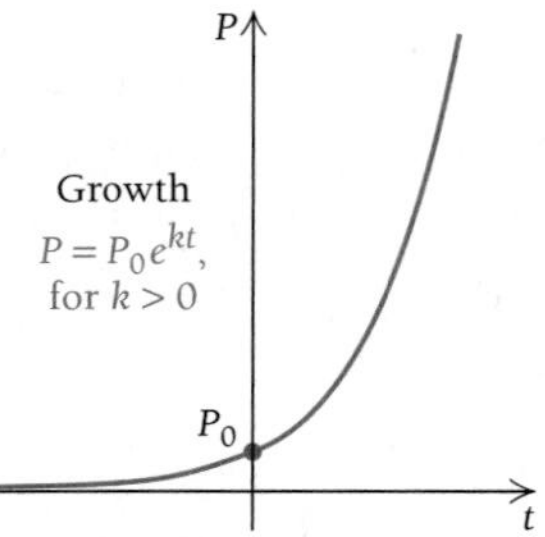

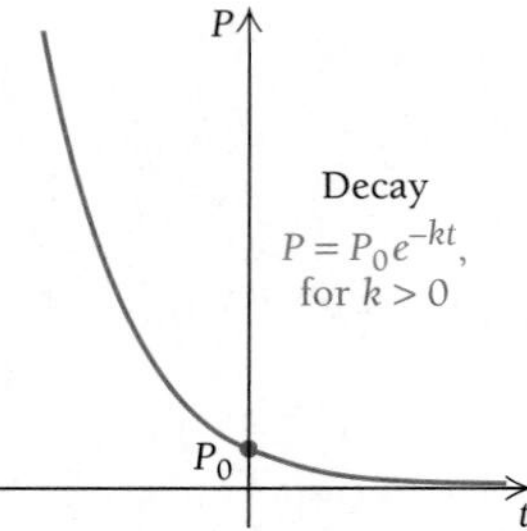

Exponential decay occurs in many real-world situations. One of the most common is in *radioactive decay.* (p. 353)

EXAMPLE Lead-210 has a decay rate of 3.15% per year. The rate of change of an amount N of lead-210 is given by

$$\frac{dN}{dt} = -0.0315N.$$

a) Find the function that satisfies the equation.

b) How much of an 80-g sample of lead-210 will remain after 20 yr?

Solution

a) $N(t) = N_0e^{-0.0315t}$

Check:

$$\frac{d}{dt}(N_0e^{-0.0315t}) = N_0e^{-0.0315t}(-0.0315)$$

$$= -0.0315 \cdot N(t).$$

b) $N(20) = 80e^{-0.0315(20)} \approx 42.6 \text{ g}$ ✦

Half-life, T, and *decay rate*, k, are related according to

$$kT \approx 0.693147. \qquad \text{(p. 354)}$$

The decay equation, $P_0 = Pe^{-kt}$, is also used to calculate **present value.** (p. 356)

It is not difficult to differentiate exponential and logarithmic functions with bases other than e:

$$\frac{d}{dx}a^x = (\ln a)a^x \quad \text{and} \quad \frac{d}{dx}\log_a x = \frac{1}{\ln a}\cdot\frac{1}{x}. \qquad \text{(pp. 365–367)}$$

EXAMPLE Differentiate: **a)** $y = 7^{0.3x}$; **b)** $y = \log_6 x$.

Solution

a) $\frac{d}{dx}7^{0.3x} = \ln 7 \cdot 7^{0.3x} \cdot 0.3$

$\approx 0.58 \cdot 7^{0.3x}$

b) $\frac{d}{dx}\log_6 x = \frac{1}{\ln 6}\cdot\frac{1}{x} \approx \frac{0.558}{x}$ ✦

Exponential functions play a key role in certain calculations for economics and business.

The **elasticity of demand,** E, is a function of price, x:

$$E(x) = -\frac{x \cdot D'(x)}{D(x)}. \qquad \text{(p. 371)}$$

When $E(x)$ exceeds 1, revenue is decreasing;
when $E(x)$ is less than 1, revenue is increasing; and
when $E(x)$ is 1, revenue is maximized. (p. 373)

EXAMPLE The Leslie Davis Band finds that demand for their CD at their performances is given by

$$q = D(x) = 50 - 2x,$$

where x is the price, in dollars, of each CD sold and q is the number of CDs sold at a performance.

a) Find the elasticity when the price is \$10 per CD, and interpret the result.

b) Find the price at which revenue is maximized.

Solution

a) Elasticity at $x = E(x) = -\frac{x \cdot D'(x)}{D(x)}$

$= -\frac{x(-2)}{50 - 2x}$

$= -\frac{-2x}{50 - 2x} = \frac{x}{25 - x}.$

Thus,

$$E(10) = \frac{10}{25 - 10} = \frac{2}{3}.$$

Since $E(10)$ is less than 1, the demand for the CD is *inelastic*, and an increase in price will increase revenue.

b) Revenue is maximized when $E(x) = 1$:

$$-\frac{-2x}{50 - 2x} = 1$$

$$\frac{2x}{50 - 2x} = 1$$

$$\frac{x}{25 - x} = 1$$

$$x = 25 - x$$

$$2x = 25$$

$$x = 12.5.$$

At a price of \$12.50 per CD, revenue will be maximized. ✦

Chapter Review Exercises

These review exercises are for test preparation. They can also be used as a practice test. Answers are at the back of the book. The bracketed section references tell you what part(s) of the chapter to restudy if your answer is incorrect.

CONCEPT REINFORCEMENT

Match each equation in column A with the most appropriate graph in column B. [3.1–3.4]

Column A

1. $P(t) = 50e^{0.03t}$

2. $P(t) = \frac{50}{1 + 2e^{-0.02t}}$

3. $P(t) = 50e^{-0.02t}$

Column B

a)

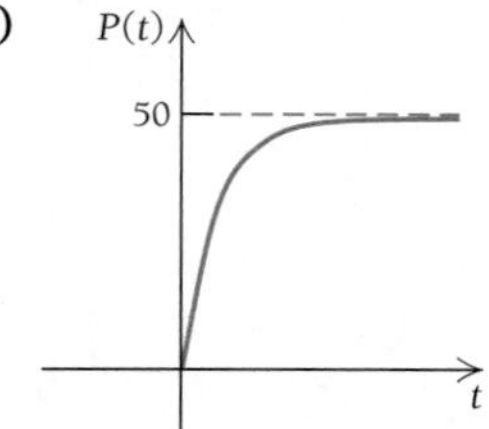

b)

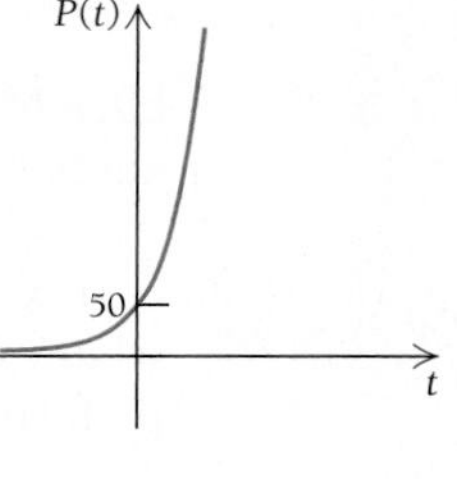

c)

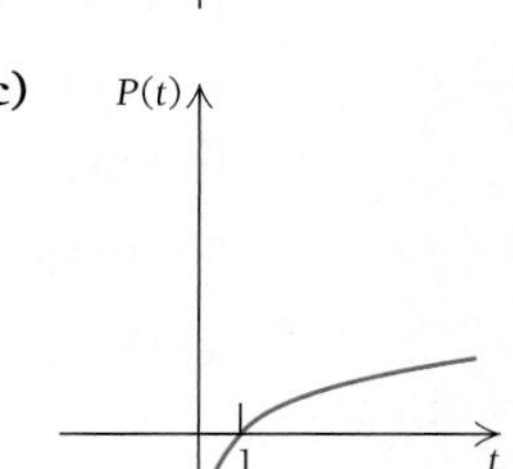

Column A	Column B
4. $P(t) = \ln t$	d) 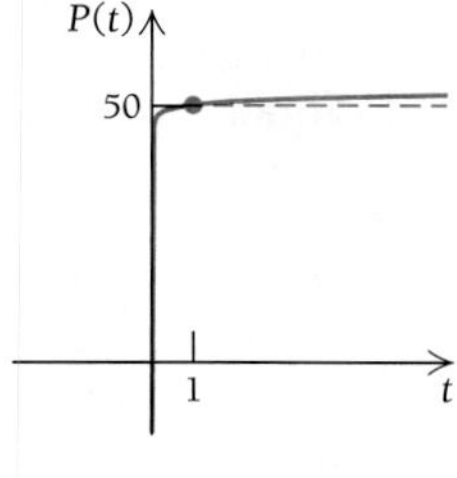
5. $P(t) = 50(1 - e^{-0.04t})$	e) 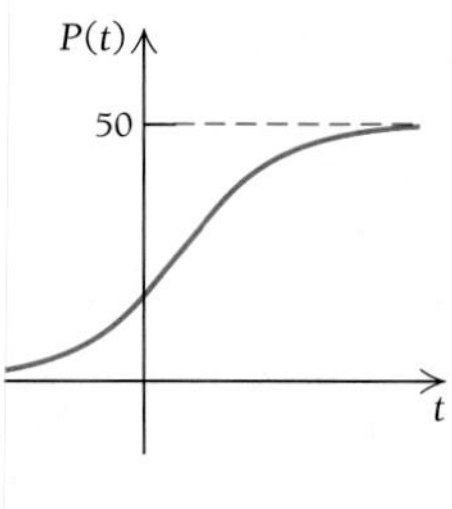
6. $P(t) = 50 + \ln t$	f) 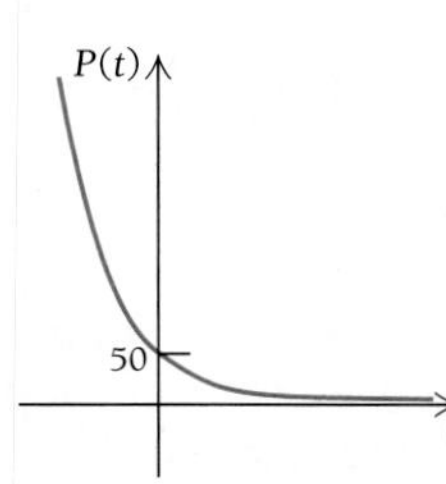

Classify each statement as either true or false.

7. The base a in the exponential function given by $f(x) = a^x$ must be greater than 1. [3.1]

8. The base a in the logarithmic function given by $g(x) = \log_a x$ must be greater than 0. [3.2]

9. If $f'(x) = c \cdot f(x)$ for $c \neq 0$ and $f(x) \neq 0$, then f must be an exponential function. [3.3]

10. With exponential growth, the doubling time depends on the size of the original population. [3.3]

11. A radioactive element's half-life determines what its decay constant is. [3.4]

12. A radioactive element's half-life depends on how much of the element is initially present. [3.4]

13. For any exponential function of the form $f(x) = a^x$, it follows that $f'(x) = \ln a \cdot a^x$. [3.5]

14. For any logarithmic function of the form $g(x) = \log_a x$, it follows that $g'(x) = \frac{1}{a} \cdot \frac{1}{x}$. [3.5]

15. Revenue is maximized when an elasticity of 1 is achieved. [3.6]

REVIEW EXERCISES

Differentiate each function.

16. $y = \ln x$ [3.2]

17. $y = e^x$ [3.1]

18. $y = \ln (x^4 + 5)$ [3.2]

19. $y = e^{2\sqrt{x}}$ [3.1]

20. $f(x) = \ln\sqrt{x}$ [3.2]

21. $f(x) = x^4 e^{3x}$ [3.1]

22. $f(x) = \dfrac{\ln x}{x^3}$ [3.2]

23. $f(x) = e^{x^2} \cdot \ln 4x$ [3.1, 3.2]

24. $f(x) = e^{4x} - \ln \dfrac{x}{4}$ [3.1, 3.2]

25. $g(x) = x^8 - 8 \ln x$ [3.2]

26. $y = \dfrac{\ln e^x}{e^x}$ [3.1, 3.2]

27. $F(x) = 9^x$ [3.5]

28. $g(x) = \log_2 x$ [3.5]

29. $y = 3^x \cdot \log_4 (2x + 1)$ [3.5]

Graph each function. [3.1]

30. $f(x) = 4^x$

31. $g(x) = \left(\frac{1}{3}\right)^x$

Given $\log_a 2 = 1.8301$ *and* $\log_a 7 = 5.0999$, *find each logarithm.* [3.2]

32. $\log_a 14$

33. $\log_a \frac{2}{7}$

34. $\log_a 28$

35. $\log_a 3.5$

36. $\log_a \sqrt{7}$

37. $\log_a \frac{1}{4}$

38. Find the function that satisfies $dQ/dt = 7Q$, given that $Q(0) = 25$. [3.3]

39. **Life science: population growth.** The population of Boomtown doubled in 16 yr. What was the growth rate of the city? Round to the nearest tenth of a percent. [3.3]

40. **Business: interest compounded continuously.** Suppose that \$8300 is deposited in a savings and loan association, where the interest rate is 6.8%, compounded continuously. How long will it take for the \$8300 to double itself? Round to the nearest tenth of a year. [3.3]

41. **Business: cost of a prime-rib dinner.** The average cost C of a prime-rib dinner was \$4.65 in 1962. In 1986, it was \$15.81. Assuming that the exponential growth model applies: [3.3]

a) Find the exponential growth rate to three decimal places, and write the function that models the situation.

b) What will the cost of such a dinner be in 2007? In 2012?

42. Business: franchise growth. A clothing firm is selling franchises throughout the United States and Canada. It is estimated that the number of franchises N will increase at the rate of 12% per year, that is,

$$\frac{dN}{dt} = 0.12N,$$

where t is the time, in years. [3.3]

a) Find the function that satisfies the equation, assuming that the number of franchises in 2007 ($t = 0$) is 60.

b) How many franchises will there be in 2013?

c) After how long will the number of franchises be 120? Round to the nearest tenth of a year.

43. Life science: decay rate. The decay rate of a certain radioactive substance is 13% per year. What is its half-life? Round to the nearest tenth of a year. [3.4]

44. Life science: half-life. The half-life of radon-222 is 3.8 days. What is its decay rate? Round to the nearest tenth of a percent. [3.4]

45. Life science: decay rate. A certain radioactive element has a decay rate of 7% per day, that is,

$$\frac{dA}{dt} = -0.07A,$$

where A is the amount of the element present at time t, in days. [3.4]

a) Find a function that satisfies the equation if the amount of the element present at $t = 0$ is 800 g.

b) After 20 days, how much of the 800 g will remain? Round to the nearest gram.

c) After how long will half of the original amount remain?

46. Social science: The Hullian Learning Model. The probability p of mastering a certain assembly-line task after t learning trials is given by

$$p(t) = 1 - e^{-0.7t}. \quad [3.3]$$

a) What is the probability of learning the task after 1 trial? 2 trials? 5 trials? 10 trials? 14 trials?

b) Find the rate of change, $p'(t)$.

TW **c)** Interpret the meaning of $p'(t)$.

d) Sketch a graph of the function.

47. Business: present value. Find the present value of \$1,000,000 due 40 yr later at 8.4%, compounded continuously. [3.4]

48. Economics: elasticity of demand. Consider the demand function

$$q = D(x) = \frac{600}{(x + 4)^2}. \quad [3.6]$$

a) Find the elasticity.

b) Find the elasticity at $x = \$1$, stating whether the demand is elastic or inelastic.

c) Find the elasticity at $x = \$12$, stating whether the demand is elastic or inelastic.

d) At a price of \$12, will a small increase in price cause the total revenue to increase or decrease?

e) Find the value of x for which the total revenue is a maximum.

SYNTHESIS

49. Differentiate: $y = \dfrac{e^{2x} + e^{-2x}}{e^{2x} - e^{-2x}}$. [3.1]

50. Find the minimum value of $f(x) = x^4 \ln(4x)$. [3.2]

TECHNOLOGY CONNECTION

51. Graph: $f(x) = \dfrac{e^{1/x}}{(1 + e^{1/x})^2}$. [3.1]

52. Find $\lim\limits_{x \to 0} \dfrac{e^{1/x}}{(1 + e^{1/x})^2}$. [3.1]

53. Business: shopping on the Internet. Online sales of all types of consumer products have been increasing at an exponential rate. Data in the following table show online retail sales, in billions of dollars. [3.3]

Years, t, after 1998	U.S. Online Retail Sales (in billions)
0	\$ 4.9
1	14.7
2	28.0
3	34.3
4	44.7
5	55.7
6	69.2
7	86.3

(*Source*: U.S. Census Bureau.)

a) Use REGRESSION to fit an exponential function $y = a \cdot b^x$ to the data. Then convert that formula to an exponential function, base e, where t is the number of years after 1998, and determine the exponential growth rate.

b) Estimate online sales in 2010; in 2020.

c) After what amount of time will online sales be \$400 billion?

d) What is the doubling time of online sales?

Chapter 3 Test

Differentiate.

1. $y = 2e^{3x}$
2. $y = (\ln x)^4$
3. $f(x) = e^{-x^2}$
4. $f(x) = \ln \frac{x}{7}$
5. $f(x) = e^x - 5x^3$
6. $f(x) = 3e^x \ln x$
7. $y = 7^x + 3^x$
8. $y = \log_{14} x$

Given $\log_b 2 = 0.2560$ *and* $\log_b 9 = 0.8114$, *find each of the following.*

9. $\log_b 18$
10. $\log_b 4.5$
11. $\log_b 3$

12. Find the function that satisfies $dM/dt = 6M$, with $M(0) = 2$.

13. The doubling time for a certain bacteria population is 3 hr. What is the growth rate? Round to the nearest tenth of a percent.

APPLICATIONS

14. **Business: interest compounded continuously.** An investment is made at 6.931% per year, compounded continuously. What is the doubling time? Round to the nearest tenth of a year.

15. **Business: cost of milk.** The cost C of a gallon of milk was \$0.54 in 1941. In 2006, it was \$3.22. (*Source*: U.S. Department of Labor, Bureau of Labor Statistics.) Assuming that the exponential growth model applies:

a) Find the exponential growth rate to the nearest tenth of a percent, and write the equation.
b) Find the cost of a gallon of milk in 2008; in 2015.

16. **Life science: drug dosage.** A dose of a drug is injected into the body of a patient. The drug amount in the body decreases at the rate of 10% per hour, that is,

$$\frac{dA}{dt} = -0.1A,$$

where A is the amount in the body and t is the time, in hours.

a) A dose of 3 cubic centimeters (cc) is administered. Assuming $A_0 = 3$, find the function that satisfies the equation.
b) How much of the initial dose of 3 cc will remain after 10 hr?
c) After how long does half of the original dose remain?

17. **Life science: decay rate.** The decay rate of radium-226 is 4.209% per century. What is its half-life?

18. **Life science: half-life.** The half-life of bohrium-267 is 17 sec. What is its decay rate? Express the rate as a percentage rounded to four decimal places.

19. **Business: effect of advertising.** Twin City Roasters introduced a new coffee in a trial run. The firm advertised the coffee on television and found that the percentage P of people who bought the coffee after t ads had been run satisfied the function

$$P(t) = \frac{100}{1 + 24e^{-0.28t}}.$$

a) What percentage of the people bought the coffee before seeing the ad ($t = 0$)?
b) What percentage bought the coffee after the ad had been run 1 time? 5 times? 10 times? 15 times? 20 times? 30 times? 35 times?
c) Find the rate of change, $P'(t)$.
TW d) Interpret the meaning of $P'(t)$.
e) Sketch a graph of the function.

20. In 2004, the New York Mets signed Pedro Martinez to a contract paying him \$13 million in 2008. (*Source*: *Newsday*, Dec 17, 2004.) Find the value of that amount in 2004, assuming 4.3% interest, compounded continuously.

21. **Economics: elasticity of demand.** Consider the demand function

$$q = D(x) = 400e^{-0.2x}.$$

a) Find the elasticity.
b) Find the elasticity at $x = 3$, and state whether the demand is elastic or inelastic.
c) Find the elasticity at $x = 18$, and state whether the demand is elastic or inelastic.
d) At a price of \$3, will a small increase in price cause the total revenue to increase or decrease?
e) Find the price for which the total revenue is a maximum.

SYNTHESIS

22. Differentiate: $y = x(\ln x)^2 - 2x \ln x + 2x$.

23. Find the maximum and minimum values of $f(x) = x^4e^{-x}$ over $[0, 10]$.

TECHNOLOGY CONNECTION

24. Graph: $f(x) = \dfrac{e^x - e^{-x}}{e^x + e^{-x}}$.

25. Find $\lim_{x \to 0} \dfrac{e^x - e^{-x}}{e^x + e^{-x}}$.

26. Business: average price of a television commercial. The cost of a 30-sec television commercial that runs during the Super Bowl has been increasing exponentially over the past several years. Data in the table below show costs for years after 1990.

Years, t, after 1990	Cost of commercial
1	\$ 800,000
3	850,000
5	1,000,000
8	1,300,000
13	2,100,000
16	2,600,000

(*Source*: National Football League.)

a) Use REGRESSION to fit an exponential function $y = a \cdot b^x$ to the data. Then convert that formula to an exponential function, base e, where t is the number of years after 1990.

b) Estimate the cost of a commercial run during the Super Bowl in 2010; in 2020.

c) After what amount of time will the cost be \$1 billion?

d) What is the doubling time of the cost of a commercial run during the Super Bowl?

EXTENDED TECHNOLOGY APPLICATION

The Business of Motion-Picture Box-Office Revenue

It is possible to use a logistic growth equation to predict the total U.S. box-office revenue of a motion picture.

EXERCISES

Consider the following data on a movie for Exercises 1–5.

MOVIE REVENUE: *CRASH* WEEK 1 = MAY 6, 2005		
WEEK IN RELEASE, t	GROSS REVENUE, G (in millions, current week)	TOTAL REVENUE, R (in millions)
1	\$ 9.11	\$ 9.11
2	10.50	19.61
3	8.04	27.65
4	8.55	36.20
5	4.73	40.93
6	3.37	44.30
7	2.12	46.42
8	1.78	48.20
9	1.49	49.69

(*Source*: www.boxofficereport.com.)

1. Make a scatterplot of the data points (t, R). Does it seem to fit a logistic equation? Why or why not?

2. Use REGRESSION to fit a logistic function,

$$R(t) = \frac{c}{1 + ae^{-bt}},$$

to the data.

3. Find the rate of change $R'(t)$, and explain its meaning. Find $\lim_{t \to \infty} R'(t)$, and explain its meaning.

4. Make a scatterplot of the data points (t, G). Select a model for regression that you think best fits the data, and make a case for why you have chosen it. Find the domain of the function.

5. Typically, the distributors of a movie wait until its weekly U.S. box-office revenue is 0 before it begins home video sales. Use the function found in Exercise 4 to predict when the distributor should begin home video sales. Why might you wait longer than the predicted time to begin such sales?

(continued)

Now let's consider data on another movie for Exercises 6–10.

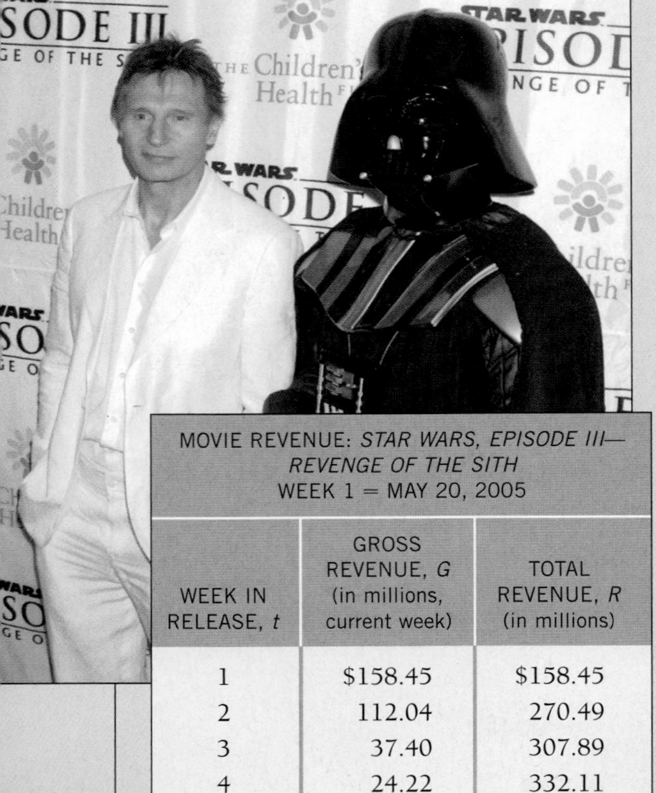

MOVIE REVENUE: *STAR WARS, EPISODE III—REVENGE OF THE SITH* WEEK 1 = MAY 20, 2005		
WEEK IN RELEASE, t	GROSS REVENUE, G (in millions, current week)	TOTAL REVENUE, R (in millions)
1	\$158.45	\$158.45
2	112.04	270.49
3	37.40	307.89
4	24.22	332.11
5	16.03	348.14
6	10.33	358.47
7	8.03	366.50
8	4.32	370.82
9	3.11	373.93

(*Source*: www.boxofficereport.com.)

6. Make a scatterplot of the data points (t, R). Does it seem to fit a logistic equation? Why or why not?
7. Use REGRESSION to fit a logistic function,

$$R(t) = \frac{c}{1 + ae^{-bt}},$$

to the data.

8. Find the rate of change $R'(t)$, and explain its meaning. Find $\lim_{t \to \infty} R'(t)$, and explain its meaning.
9. Next, make a scatterplot of the data points (t, G). Select a model for regression that you think best fits the data, and make a case for why you have chosen it. Find the domain of the function.
10. Use the function found in Exercise 9 to predict when the distributor should begin home video sales. (See Exercise 5.)

Next, consider data on a third movie for Exercises 11–15.

MOVIE REVENUE: *CURIOUS GEORGE* WEEK 1 = FEBRUARY 10, 2006		
WEEK IN RELEASE, t	GROSS REVENUE, G (in millions, current week)	TOTAL REVENUE, R (in millions)
1	\$14.70	\$14.70
2	18.82	33.52
3	9.83	43.35
4	5.95	49.30
5	3.28	52.58
6	2.81	55.39
7	1.50	56.89

(*Source*: www.boxofficereport.com.)

11. Make a scatterplot of the data points (t, R). Does a logistic equation appear to fit the data? Why or why not?

12. Use REGRESSION to fit a logistic function,

$$R(t) = \frac{c}{1 + ae^{-bt}},$$

to the data.

13. Find the rate of change $R'(t)$, and explain its meaning. Find $\lim_{t \to \infty} R'(t)$, and explain its meaning.

14. Make a scatterplot of the data points (t, G). Select a model for regression that you think best fits the data, and make a case for why you have chosen it. Find the domain of the function.

15. Use the function found in Exercise 14 to predict when the distributor should begin home video sales. (See Exercise 5.)

Consider the data on a fourth movie for Exercises 16–20.

MOVIE REVENUE: *REMEMBER THE TITANS* WEEK 1 = SEPTEMBER 29, 2000		
WEEK IN RELEASE, t	GROSS REVENUE, G (in millions, current week)	TOTAL REVENUE, R (in millions)
1	\$26.65	\$ 26.65
2	24.50	51.15
3	16.29	67.44
4	12.30	79.74
5	10.04	89.78
6	8.84	98.62
7	6.13	104.75
8	3.53	108.28
9	2.31	110.59

(*Source*: www.boxofficereport.com.)

16. Make a scatterplot of the data points (t, R). Does it seem to fit a logistic equation? Why or why not?

17. Use REGRESSION to fit a logistic function,

$$R(t) = \frac{c}{1 + ae^{-bt}},$$

to the data.

18. Find the rate of change $R'(t)$, and explain its meaning. Find $\lim_{t \to \infty} R'(t)$, and explain its meaning.

19. Make a scatterplot of the data points (t, G). Select a model for regression that you think best fits the data, and make a case for why you have chosen it. Find the domain of the function.

20. Use the function found in Exercise 19 to predict when the distributor should begin home video sales. (See Exercise 5.)

Now let's consider some questions regarding all the preceding data on the four movies.

21. Does a logistic function seem to fit all graphs of the data points (t, R)? Why do you think this happens?

22. What type of function best fits the data points (t, G)? Explain why.

Integration

4

APPLICATION For the years 1994 through 2000, the annual rate of change in the national credit market debt, in billions of dollars per year, could be modeled by the function

$$D'(t) = 1975 - 1190t + 597t^2 - 71.3t^3,$$

where t is the number of years since 1994. (*Source:* Based on data from the Federal Reserve System.) Find the national credit market debt in 2000, given that $D(0) = 17{,}198$.

This problem appears as Exercise 60 in Exercise Set 4.2.

INTRODUCTION

Suppose that we do the reverse of differentiating—that is, we try to find a function whose derivative is a given function. This process is called *antidifferentiation*, or *integration*. It is the main topic of this chapter and the second main branch of calculus, the first being differentiation. We will see that we can use integration to find the area under a curve over a closed interval, to find the accumulation of a certain quantity over an interval, and to calculate the average value of a function.

We first consider the meaning of integration and then learn several techniques for integrating.

4.1 The Area under a Graph

OBJECTIVES

- Use the area under a graph to find total cost.
- Use rectangles to approximate the area under a graph.

Total Cost and Area

There are many situations in which it is useful to determine the area under a graph. For instance, suppose that Green Leaf Skateboards determines that, for the first 50 skateboards produced, its cost is \$40 per board. The total cost of producing those boards is $\$40 \cdot 50$, or \$2000, represented by the area shaded red in the graph below. For the next 75 skateboards, Green Leaf's cost drops to \$30 per board, so the production cost of those boards is $\$30 \cdot 75$, or \$2250. This amount is represented by the area shaded blue.

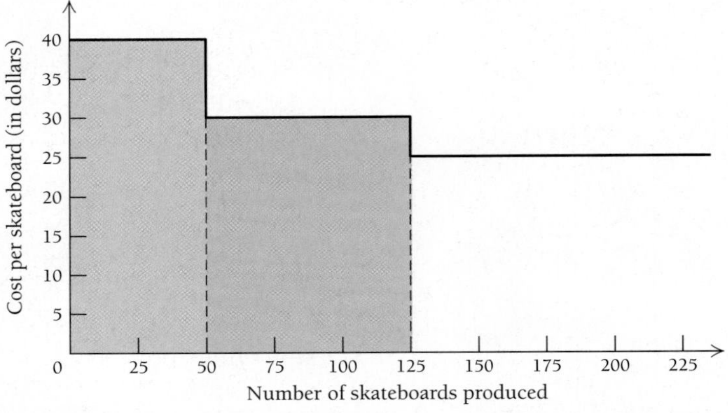

Next, suppose that Green Leaf's production cost drops to \$25 per board after 125 boards are built. If Green Leaf produces 150 skateboards, we can determine the total cost of production by finding the area under the graph, as shown below.

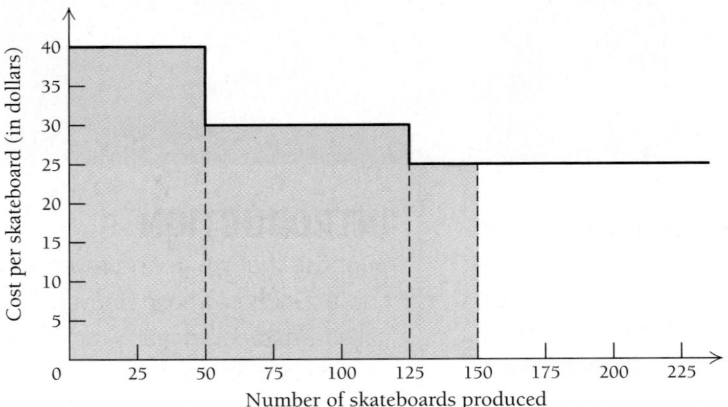

$$
\begin{aligned}
\text{Total cost of 150 skateboards} &= \text{Cost of skateboards 1 through 50} + \text{Cost of skateboards 51 through 125} + \text{Cost of skateboards 126 through 150} \\
&= 50 \cdot \$40 + 75 \cdot \$30 + 25 \cdot \$25 \\
&= \$2000 + \$2250 + \$625 = \$4875
\end{aligned}
$$

This is the sum of the areas.

In many real-world situations, the cost per unit produced changes more gradually than it did in the preceding application.

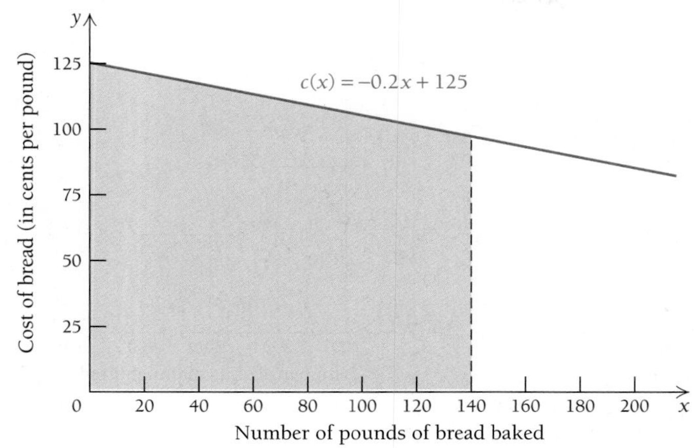

EXAMPLE 1 Business: Total Cost. Covington Bakers determines that the cost, in cents per pound, of its sourdough bread is

$$c(x) = -0.2x + 125, \quad x < 500$$

where x is the number of pounds of bread baked. Find the total cost of baking 140 lb of bread. Disregard fixed costs.

Solution Note that the cost per pound of bread baked goes down as the number of pounds goes up. Just as we did to find the total cost of producing 150 skateboards, we use the area under the graph to find the total cost of baking 140 lb of bread. This time, however, we must find the area of a trapezoid.

$$\begin{aligned} \text{Total cost of baking 140 lb of bread} &= \text{Area of trapezopid} \\ &= \frac{1}{2}h(b_1 + b_2) \\ &= \frac{1}{2} \cdot 140(c(0) + c(140)) \\ &= 70(125 + 97) \\ &= 15{,}540¢, \text{ or } \$155.40 \end{aligned}$$

h is height and b_1 and b_2 are the lengths of the bases.
View the trapezoid sideways.

As a check, we can examine the units involved:

$$\text{Cost per pound} \cdot \text{Number of pounds}$$
$$¢/\text{lb} \cdot \text{lb} = ¢$$

Covington's total cost of baking 140 lb of sourdough bread is \$155.40. ✦

In Example 1, the cost function corresponds to a changing unit cost (the cost per pound). For a continuous function, this is equivalent to the marginal cost. In the next example, we are given a nonlinear equation for marginal cost. To find the area under such a curve, we need to use a different approach.

EXAMPLE 2 Business: Total Cost. Raggs, Ltd., determines that the marginal cost, in dollars, when x suits are manufactured is

$$C'(x) = 0.0008x^2 - 0.45x + 100.$$

Ignoring fixed costs, approximate the total cost of producing 400 suits by using four rectangles.

Solution We divide the interval $[0, 400]$ into four subintervals, each of width $\Delta x = 400/4 = 100$. To determine the height of each rectangle, we use the left endpoint of each subinterval.

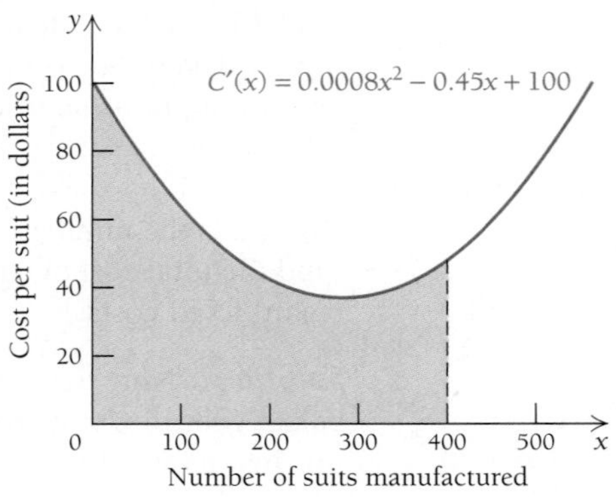

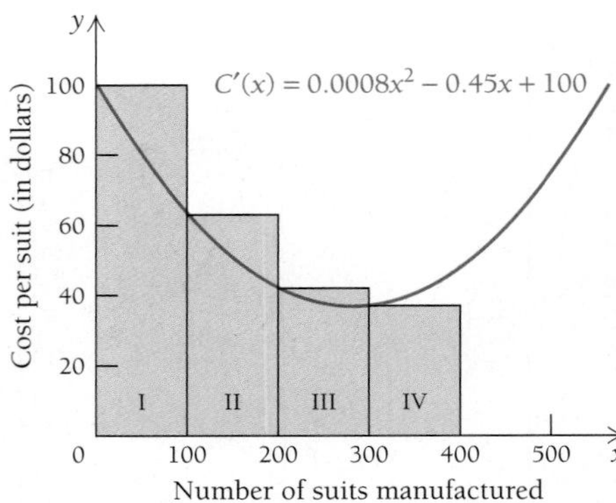

Then, we have

$$\begin{aligned}\text{Total cost of producing 400 suits} &\approx \text{Area I} + \text{Area II} + \text{Area III} + \text{Area IV}\\ &\approx C'(0)\cdot 100 + C'(100)\cdot 100\\ &\quad + C'(200)\cdot 100 + C'(300)\cdot 100\\ &\approx \$100\cdot 100 + \$63\cdot 100 + \$42\cdot 100 + \$37\cdot 100\\ &\approx \$10{,}000 + \$6300 + \$4200 + \$3700\\ &\approx \$24{,}200.\end{aligned}$$

The total cost of producing 400 suits is approximately \$24,200. ✦

Riemann Sums

In Example 2, we approximated the total cost of manufacturing 400 suits. How might we make that approximation more exact? We do so by using more rectangles that have a narrower width. This type of calculation is most easily accomplished with the use of *summation notation*, which is introduced below.

In the following figure, $[a, b]$ is divided into four subintervals, each having width $\Delta x = (b - a)/4$.

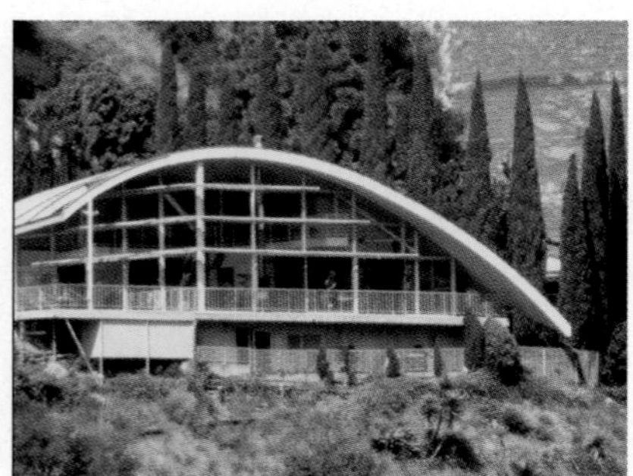

The area under a curve can be approximated by a sum of rectangular areas.

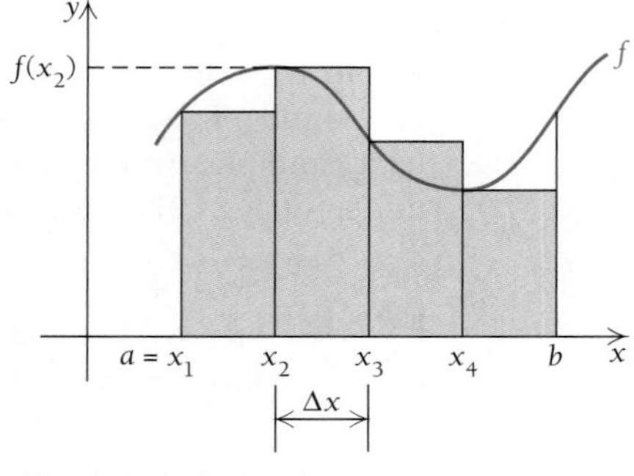

The heights of the rectangles shown are

$f(x_1)$, $f(x_2)$, $f(x_3)$, and $f(x_4)$.

The area of the region under the curve is approximately the sum of the areas of the four rectangles:

$$f(x_1)\,\Delta x + f(x_2)\,\Delta x + f(x_3)\,\Delta x + f(x_4)\,\Delta x.$$

We can denote this sum with **summation**, or **sigma**, **notation**, which uses the Greek capital letter sigma, Σ:

$$\sum_{i=1}^{4} f(x_i)\,\Delta x, \quad \text{or} \quad \textstyle\sum_{i=1}^{4} f(x_i)\,\Delta x.$$

This is read "the sum of the product $f(x_i)\,\Delta x$ from $i = 1$ to $i = 4$." To recover the original expression, we substitute the numbers 1 through 4 successively for i in $f(x_i)\,\Delta x$ and write plus signs between the results.

Before we continue, let's consider some examples involving summation notation.

EXAMPLE 3 Write summation notation for $2 + 4 + 6 + 8 + 10$.

Solution Note that we are adding consecutive multiples of 2:

$$2 + 4 + 6 + 8 + 10 = \sum_{i=1}^{5} 2i.$$ ✦

EXAMPLE 4 Write summation notation for

$$g(x_1)\,\Delta x + g(x_2)\,\Delta x + \cdots + g(x_{19})\,\Delta x.$$

Solution

$$g(x_1)\,\Delta x + g(x_2)\,\Delta x + \cdots + g(x_{19})\,\Delta x = \sum_{i=1}^{19} g(x_i)\,\Delta x$$ ✦

EXAMPLE 5 Express

$$\sum_{i=1}^{4} 3^i$$

without using summation notation.

Solution

$$\sum_{i=1}^{4} 3^i = 3^1 + 3^2 + 3^3 + 3^4, \quad \text{or} \quad 120$$ ✦

EXAMPLE 6 Express

$$\sum_{i=1}^{30} h(x_i)\,\Delta x$$

without using summation notation.

Solution

$$\sum_{i=1}^{30} h(x_i)\,\Delta x = h(x_1)\,\Delta x + h(x_2)\,\Delta x + \cdots + h(x_{30})\,\Delta x$$ ✦

Approximation of area by rectangles becomes more accurate as we use smaller subintervals and hence more rectangles, as shown in the following figures.

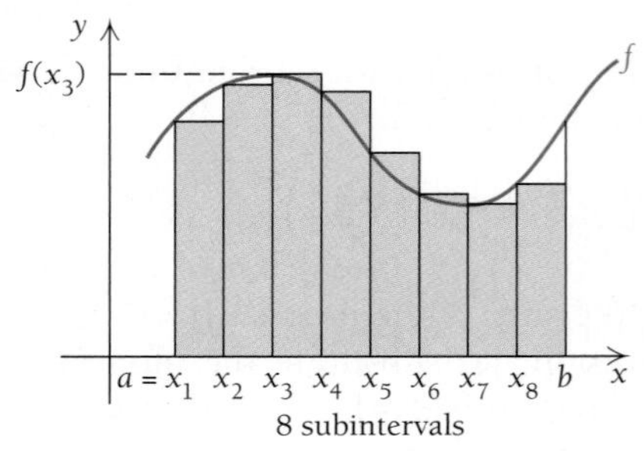

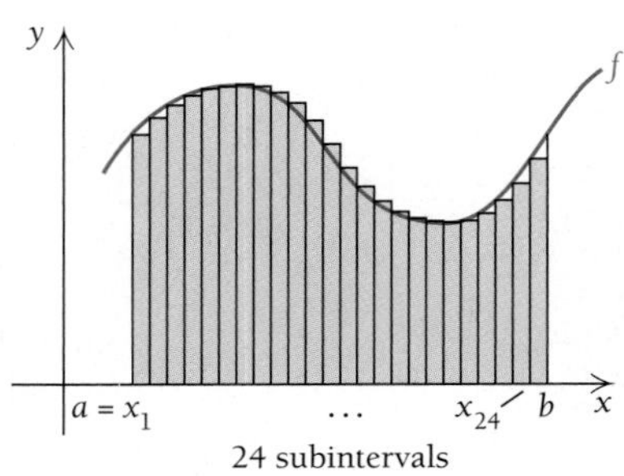

In general, suppose that the interval $[a, b]$ is divided into n equally sized subintervals, each of width $\Delta x = (b - a)/n$. We construct rectangles with heights

$$f(x_1), f(x_2), \ldots, f(x_n).$$

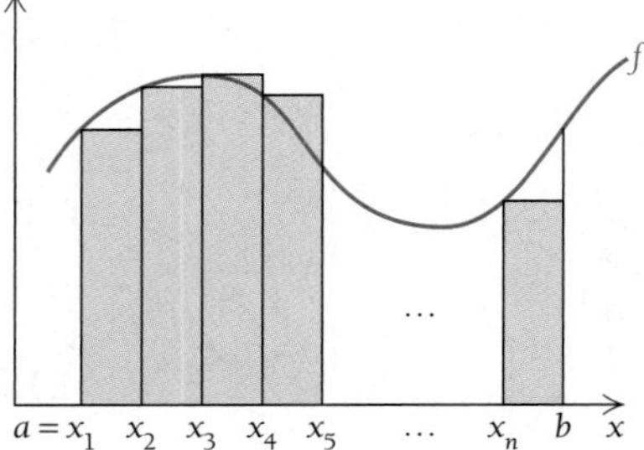

The width of each rectangle is Δx, so the first rectangle has an area of $f(x_1)\,\Delta x$, the second rectangle has an area of $f(x_2)\,\Delta x$, and so on. The area of the region under the curve is approximated by the sum of the areas of the rectangles:

$$\sum_{i=1}^{n} f(x_i)\,\Delta x.$$

EXAMPLE 7 Consider the graph of

$$f(x) = 600x - x^2$$

over the interval $[0, 600]$.

a) Approximate the area by dividing the interval into 6 subintervals.

b) Approximate the area by dividing the interval into 12 subintervals.

Solution

a) We divide $[0, 600]$ into 6 subintervals of size

$$\Delta x = \frac{600 - 0}{6} = 100,$$

with x_i ranging from $x_1 = 0$ to $x_6 = 500$.

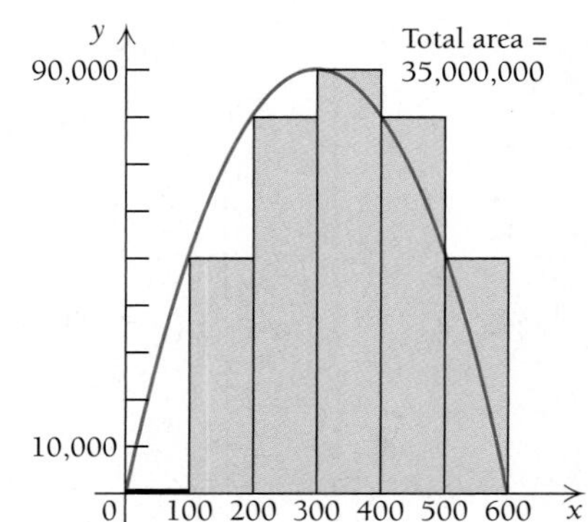

Thus, the area under the curve is approximately

$$\begin{aligned}\sum_{i=1}^{6} f(x_i)\,\Delta x &= f(0)\cdot 100 + f(100)\cdot 100 + f(200)\cdot 100 \\ &\quad + f(300)\cdot 100 + f(400)\cdot 100 + f(500)\cdot 100 \\ &= 0\cdot 100 + 50{,}000\cdot 100 + 80{,}000\cdot 100 \\ &\quad + 90{,}000\cdot 100 + 80{,}000\cdot 100 + 50{,}000\cdot 100 \\ &= 35{,}000{,}000.\end{aligned}$$

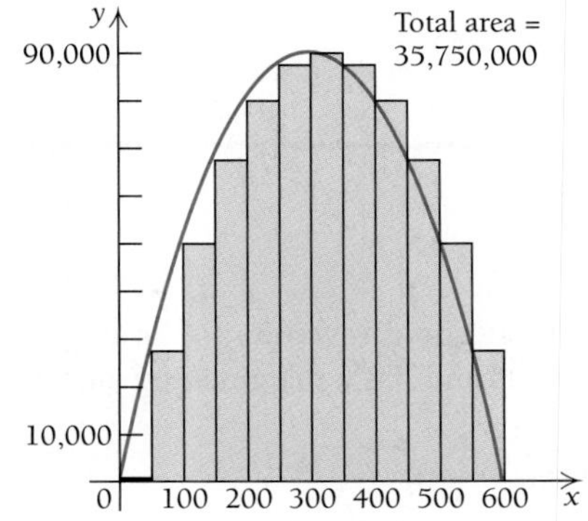

b) We divide the interval $[0, 600]$ into 12 subintervals, all of size $\Delta x = (600 - 0)/12 = 50$, with x_i ranging from $x_1 = 0$ to $x_{12} = 550$. Thus, we have another approximation of the area under the curve:

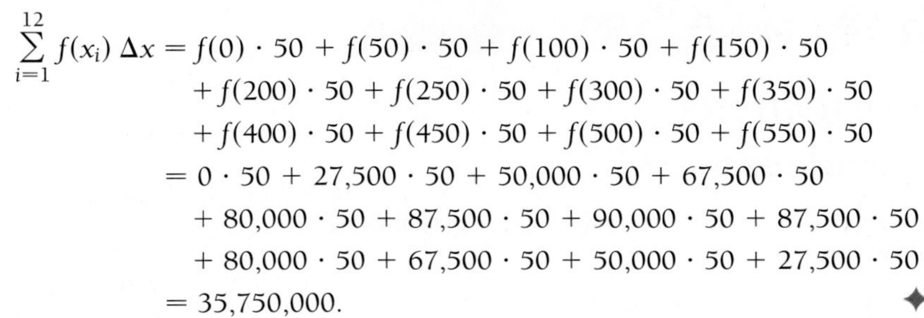

$$\begin{aligned}\sum_{i=1}^{12} f(x_i)\,\Delta x &= f(0)\cdot 50 + f(50)\cdot 50 + f(100)\cdot 50 + f(150)\cdot 50 \\ &\quad + f(200)\cdot 50 + f(250)\cdot 50 + f(300)\cdot 50 + f(350)\cdot 50 \\ &\quad + f(400)\cdot 50 + f(450)\cdot 50 + f(500)\cdot 50 + f(550)\cdot 50 \\ &= 0\cdot 50 + 27{,}500\cdot 50 + 50{,}000\cdot 50 + 67{,}500\cdot 50 \\ &\quad + 80{,}000\cdot 50 + 87{,}500\cdot 50 + 90{,}000\cdot 50 + 87{,}500\cdot 50 \\ &\quad + 80{,}000\cdot 50 + 67{,}500\cdot 50 + 50{,}000\cdot 50 + 27{,}500\cdot 50 \\ &= 35{,}750{,}000.\end{aligned}$$

✦

Note that in Example 7 the approximation using $n = 12$ is closer to the exact value than the one using $n = 6$.

The sums used in Examples 2 and 7 to approximate the area under a curve are called *Riemann sums* (pronounced "Ree-mahn") in honor of the great German mathematician G. F. Bernhard Riemann (1826–1866). Riemann sums can be calculated using any x-value within each subinterval. *For simplicity, in this text we will always use the left endpoint of each subinterval.*

EXAMPLE 8 Use 5 subintervals to approximate the area under the graph of $f(x) = 0.1x^3 - 2.3x^2 + 12x + 25$ over the interval $[1, 16]$.

Solution We divide $[1, 16]$ into 5 subintervals of size $\Delta x = (16 - 1)/5 = 3$, with x_i ranging from $x_1 = 1$ to $x_5 = 13$. Although a drawing is not required, we can make one to help visualize the area.

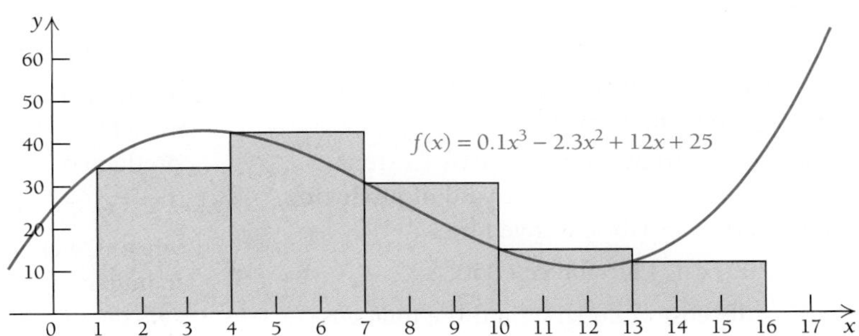

The area under the curve from 1 to 16 is approximately

$$\begin{aligned}\sum_{i=1}^{5} f(x_i)\,\Delta x &= f(1)\cdot 3 + f(4)\cdot 3 + f(7)\cdot 3 + f(10)\cdot 3 + f(13)\cdot 3\\ &= 34.8\cdot 3 + 42.6\cdot 3 + 30.6\cdot 3 + 15\cdot 3 + 12\cdot 3\\ &= 405.\end{aligned}$$

✦

In Section 4.2, we will see how Riemann sums can lead to an *exact* answer for the area under a curve.

Exercise Set 4.1

APPLICATIONS

Business and Economics

In Exercises 1–8, calculate total cost, disregarding any fixed costs.

1. Total cost from marginal cost. Redline Roasting has found that the cost, in dollars per pound, of the coffee it roasts is

$$c(x) = -0.012x + 6.50, \quad \text{for } x \le 300,$$

where x is the number of pounds of coffee roasted. Find the total cost of roasting 200 lb of coffee.

2. Total cost from marginal cost. Sylvie's Old World Cheeses has found that the cost, in dollars per kilogram, of the cheese it produces is

$$c(x) = -0.003x + 4.25, \quad \text{for } x \le 500,$$

where x is the number of kilograms of cheese produced. Find the total cost of producing 400 kg of cheese.

3. Total cost from marginal cost. Photos From Nature has found that the cost per card of producing x note cards is given by

$$c(x) = -0.04x + 85, \quad \text{for } x \le 1000,$$

where $c(x)$ is the cost, in cents, per card. Find the total cost of producing 650 cards.

4. Total cost from marginal cost. Cleo's Custom Fabrics has found that the cost per yard of producing x yards of a particular fabric is given by

$$c(x) = -0.007x + 12, \quad \text{for } x \le 350,$$

where $c(x)$ is the cost in dollars. Find the total cost of producing 200 yd of this material.

5. Total revenue from marginal revenue. A concert promoter sells x tickets and has a marginal-revenue function given by

$$R'(x) = 2x - 1150,$$

where $R'(x)$ is in dollars per ticket. This means that the rate of change of total revenue with respect to the number of tickets sold, x, is $R'(x)$. Find the total revenue from the sale of the first 300 tickets.

6. Total profit from marginal profit. Poyse Inc. has a marginal-profit function given by

$$P'(x) = -2x + 80,$$

where $P'(x)$ is in dollars per unit. This means that the rate of change of total profit with respect to the number of units produced, x, is $P'(x)$. Find the total profit from the production and sale of the first 40 units.

7. Total cost from marginal costs. Raggs, Ltd., determines that its marginal cost, in dollars per dress, is given by

$$C'(x) = -\frac{2}{25}x + 50, \quad \text{for } x \le 450.$$

Find the total cost of producing the first 200 dresses.

8. Total cost from marginal cost. Using the information and answer for Exercise 7, find the cost of producing the 201st dress through the 400th dress.

9. Total cost from marginal cost. Ship Shape Woodworkers has found that the marginal cost of producing x feet of custom molding is given by

$$C'(x) = -0.00002x^2 - 0.04x + 45, \quad \text{for } x \le 800,$$

where $C'(x)$ is in cents. Approximate the total cost of manufacturing 800 ft of molding, using 5 subintervals over $[0, 800]$ and the left endpoint of each subinterval.

10. Total cost from marginal cost. Soulful Scents has found that the marginal cost of producing x ounces of a new fragrance is given by

$$C'(x) = 0.0005x^2 - 0.1x + 30, \quad \text{for } x \leq 125,$$

where $C'(x)$ is in dollars. Use 5 subintervals over $[0, 100]$ and the left endpoint of each subinterval to approximate the total cost of producing 100 oz of the fragrance.

11. Total cost from marginal cost. Shelly's Roadside Fruit has found that the marginal cost of producing x pints of fresh-squeezed orange juice is given by

$$C'(x) = 0.000008x^2 - 0.004x + 2, \quad \text{for } x \leq 350,$$

where $C'(x)$ is in dollars. Approximate the total cost of manufacturing 270 pt of juice, using 3 subintervals over $[0, 270]$ and the left endpoint of each subinterval.

12. Total cost from marginal cost. Mangianello Paving, Inc., has found that the marginal cost, in dollars, of paving a road surface with asphalt is given by

$$C'(x) = \frac{1}{6}x^2 - 20x + 1800, \quad \text{for } x \leq 80,$$

where x is measured in hundreds of feet. Use 4 subintervals over $[0, 40]$ and the left endpoint of each subinterval to approximate the total cost of paving 4000 ft of road surface.

In Exercises 13–18, write summation notation for each expression.

13. $3 + 6 + 9 + 12 + 15 + 18$

14. $5 + 10 + 15 + 20 + 25 + 30 + 35$

15. $f(x_1) + f(x_2) + f(x_3) + f(x_4)$

16. $g(x_1) + g(x_2) + g(x_3) + g(x_4) + g(x_5)$

17. $G(x_1) + G(x_2) + \cdots + G(x_{15})$

18. $F(x_1) + F(x_2) + \cdots + F(x_{17})$

19. Express $\sum_{i=1}^{4} 2^i$ without using summation notation.

20. Express $\sum_{i=0}^{5} (-2)^i$ without using summation notation.

21. Express $\sum_{i=1}^{5} f(x_i)$ without using summation notation.

22. Express $\sum_{i=1}^{4} g(x_i)$ without using summation notation.

23. a) Approximate the area under the following graph of $f(x) = \dfrac{1}{x^2}$ over the interval $[1, 7]$ by computing the area of each rectangle to four decimal places and then adding.

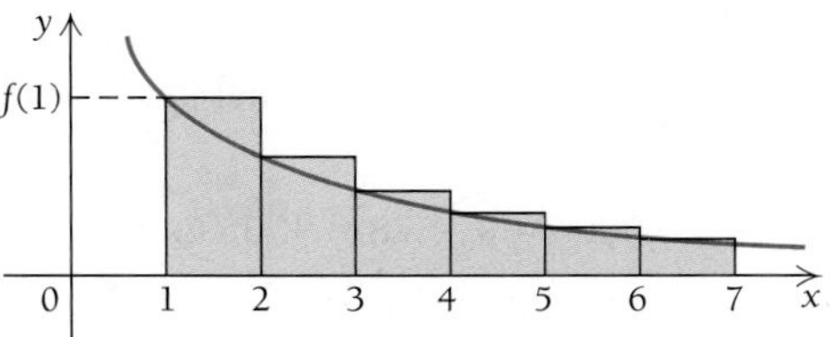

b) Approximate the area under the graph of $f(x) = \dfrac{1}{x^2}$ over the interval $[1, 7]$ by computing the area of each rectangle to four decimal places and then adding. Compare your answer to that for part (a).

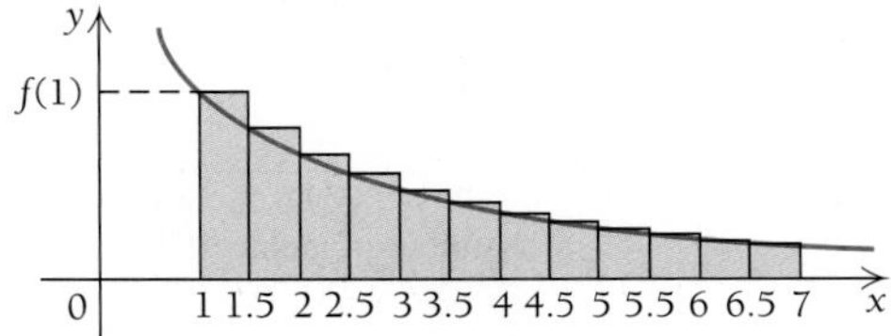

24. a) Approximate the area under the graph of $f(x) = x^2 + 1$ over the interval $[0, 5]$ by computing the area of each rectangle and then adding.

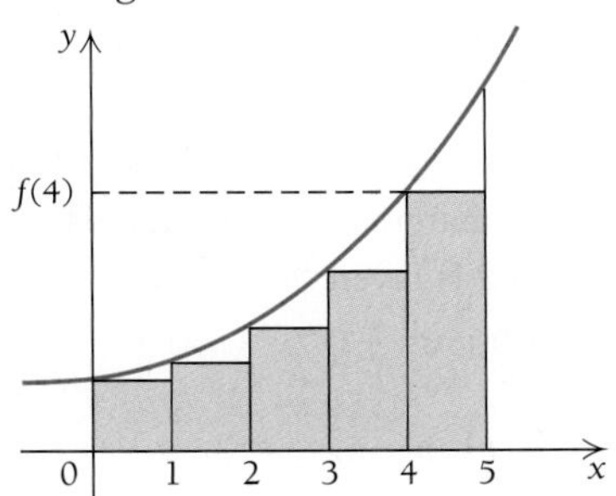

b) Approximate the area under the graph of $f(x) = x^2 + 1$ over the interval $[0, 5]$ by computing the area of each rectangle and then adding. Compare your answer to that for part (a).

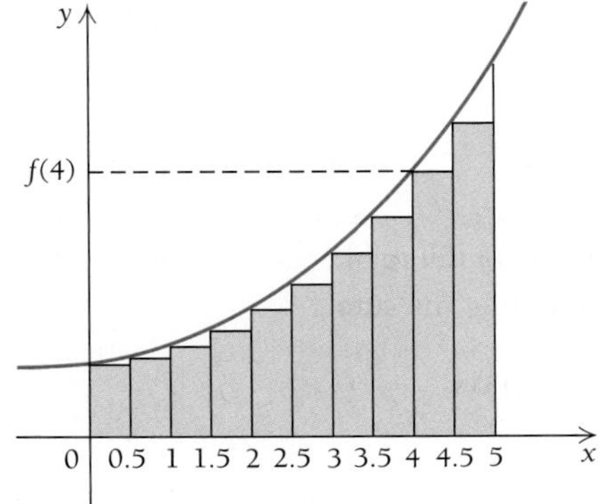

25. Total profit from marginal profit. Holcomb Hill Fitness has found that the marginal profit, $P'(x)$, in cents, is given by

$$P'(x) = -0.0006x^3 + 0.28x^2 + 55.6x, \quad \text{for } x \leq 500,$$

where x is the number of members currently enrolled at the health club.

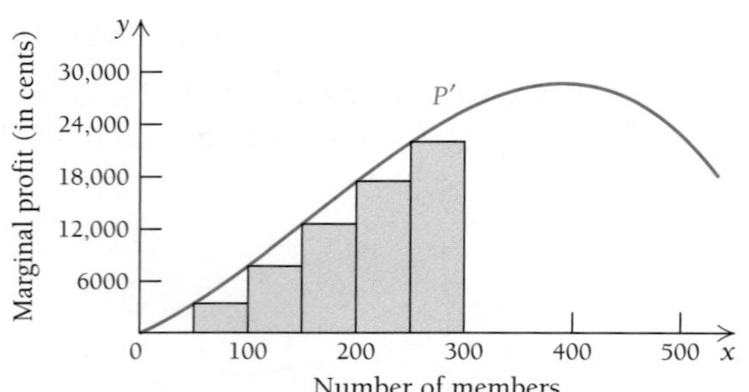

Approximate the total profit when 300 members are enrolled by computing the sum

$$\sum_{i=1}^{6} P'(x_i)\,\Delta x,$$

with $\Delta x = 50$.

26. Total cost from marginal cost. Raggs, Ltd., has found that the marginal cost, in dollars, for the xth jacket produced is given by

$$C'(x) = 0.0003x^2 - 0.2x + 50.$$

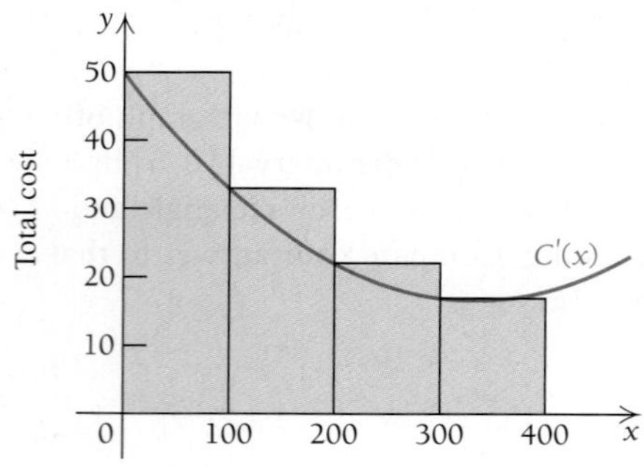

Approximate the total cost of producing 400 jackets by computing the sum

$$\sum_{i=1}^{4} C'(x_i)\,\Delta x,$$

with $\Delta x = 100$.

27. Approximate the area under the graph of

$$f(x) = 0.01x^4 - 1.44x^2 + 60$$

over the interval $[2, 10]$ by dividing the interval into 4 subintervals.

28. Approximate the area under the graph of

$$g(x) = -0.02x^4 + 0.28x^3 - 0.3x^2 + 20$$

over the interval $[3, 12]$ by dividing the interval into 4 subintervals.

29. Approximate the area under the graph of

$$F(x) = 0.2x^3 + 2x^2 - 0.2x - 2$$

over the interval $[-8, -3]$ using 5 subintervals.

30. Approximate the area under the graph of

$$G(x) = 0.1x^3 + 1.2x^2 - 0.4x - 4.8$$

over the interval $[-10, -4]$ using 6 subintervals.

SYNTHESIS

31. Show that, for any function f defined for all x_i's, and any constant k, we have

$$\sum_{i=1}^{4} kf(x_i) = k\sum_{i=1}^{4} f(x_i).$$

Then show that, in general,

$$\sum_{i=1}^{n} kf(x_i) = k\sum_{i=1}^{n} f(x_i),$$

for any constant k and any function f defined for all x_i's.

The Trapezoidal Rule. *Another way to approximate an integral is to replace each rectangle in the sum with a trapezoid (see Figs. 1 and 2). The area of a trapezoid is $h(c_1 + c_2)/2$, where c_1 and c_2 are the lengths of the parallel sides. Thus, in Fig. 2,*

Area under f over $[a, b]$

$$\approx \Delta x\frac{f(a) + f(m)}{2} + \Delta x\frac{f(m) + f(b)}{2}$$

$$\approx \Delta x\left[\frac{f(a)}{2} + f(m) + \frac{f(b)}{2}\right].$$

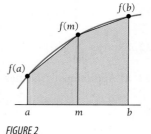

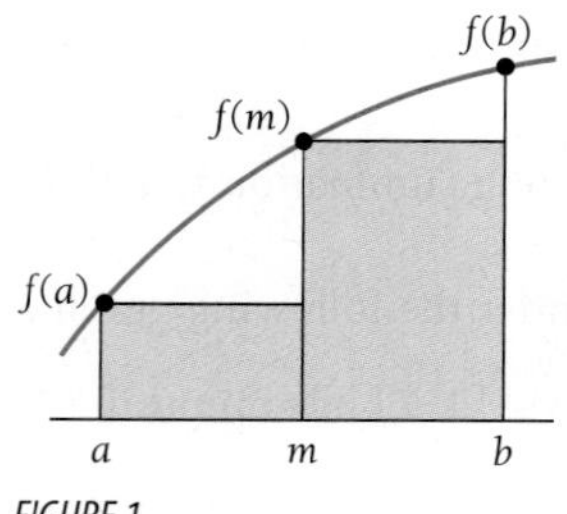

FIGURE 1

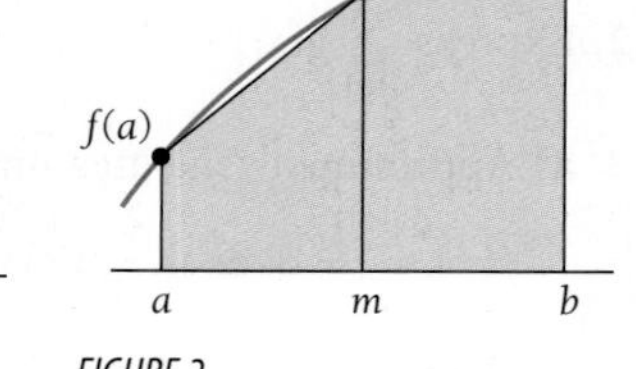

FIGURE 2

For an interval $[a, b]$ subdivided into n equal subintervals of length $\Delta x = (b - a)/n$, we get the approximation

Area under f over $[a, b]$

$$\approx \Delta x\left[\frac{f(a)}{2} + f(x_2) + f(x_3) + \cdots + f(x_n) + \frac{f(b)}{2}\right],$$

where $x_1 = a$ and

$$x_n = x_{n-1} + \Delta x \quad \text{or} \quad x_n = a + (n - 1)\Delta x.$$

This is called the **Trapezoidal Rule**.

32. Use the Trapezoidal Rule and the interval subdivision of Exercise 23(a) to approximate the area under the graph of $f(x) = 1/x^2$ over the interval $[1, 7]$.

33. Use the Trapezoidal Rule and the interval subdivision of Exercise 24(a) to approximate the area under the graph of $f(x) = x^2 + 1$ over the interval $[0, 5]$.

TECHNOLOGY CONNECTION

The exact area of a semicircle of radius r can be found using the formula $A = \frac{1}{2}\pi r^2$. Using this equation, compare the answers to Exercises 34 and 35 with the exact area. Note that most calculators do not show entire semicircles.

34. Approximate the area under the graph of $f(x) = \sqrt{25 - x^2}$ using 10 rectangles.

35. Approximate the area under the graph of $g(x) = \sqrt{49 - x^2}$ using 14 rectangles.

4.2

OBJECTIVES

- Find an antiderivative of a function.
- Evaluate indefinite integrals using the basic integration formulas.
- Use initial conditions, or boundary conditions, to determine an antiderivative.

Area, Antiderivatives, and Integrals

Is it possible to derive a formula for the area under a curve? In Section 4.1, we found that the areas of rectangles can be used to approximate such an area and that the more rectangles we use, the more closely they approximate the area. To determine the actual area, we find the sum of the areas, as the number of rectangles, n, increases without bound. The exact area, A, is thus defined by the Riemann sum

$$A = \lim_{n \to \infty} \sum_{i=1}^{n} f(x_i)\,\Delta x.$$

For certain functions, it is unnecessary to use Riemann sums. Consider, for example, a constant function $f(x) = b$ over the interval from 0 to some arbitrary value, x, that is, $[0, x]$.

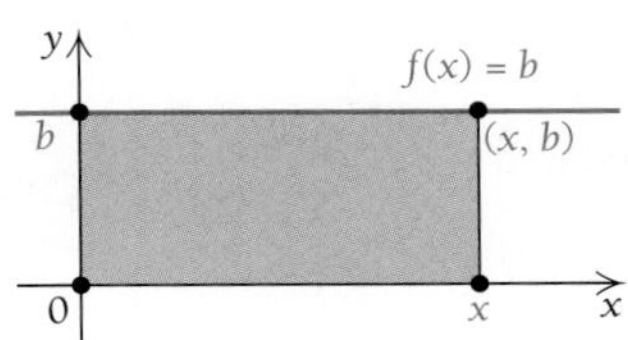

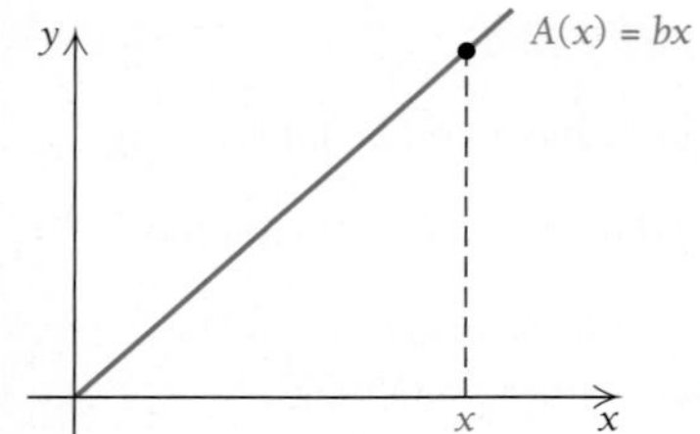

Refer to the figure on the previous page, which shows a rectangle whose area is the base times the height, $x \cdot b$, or bx. Suppose that we allow x to vary, yielding rectangles of differing widths. The area of each rectangle is still bx. We now have an area function:

$$A(x) = bx.$$

Its graph is shown at the left.

Let's next consider the linear function $f(x) = mx$ over the interval from 0 to x, or $[0, x]$, as shown below.

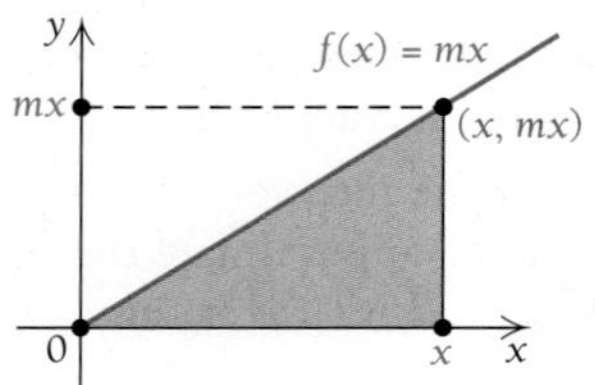

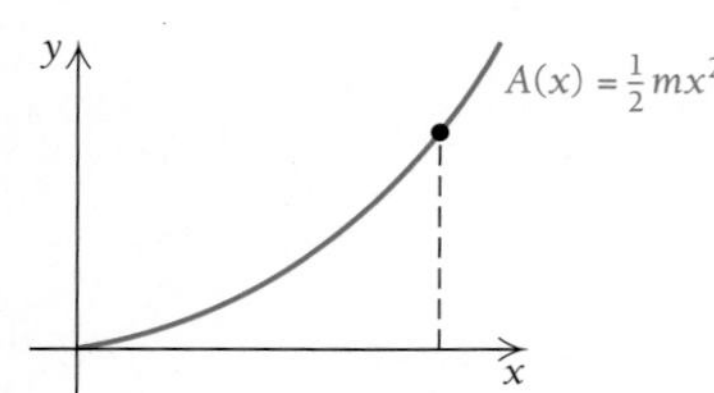

The figure formed this time is a triangle, and its area is one-half the base times the height, $\frac{1}{2} \cdot x \cdot (mx)$, or $\frac{1}{2}mx^2$. If we allow x to vary, we again get an area function:

$$A(x) = \tfrac{1}{2}mx^2.$$

Its graph is shown at the left.

Exploratory Exercises

FINDING AREAS

1. Consider the constant function $f(x) = 5$.

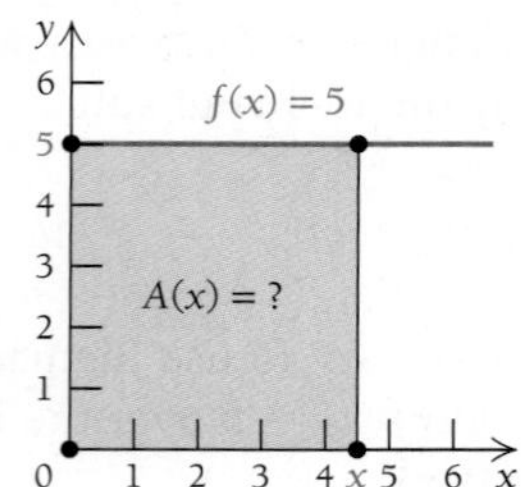

a) Find $A(x)$.
b) Find $A(1)$, $A(2)$, and $A(7)$.
c) How are $f(x)$ and $A(x)$ related?

2. Consider the function $f(x) = 3x$.

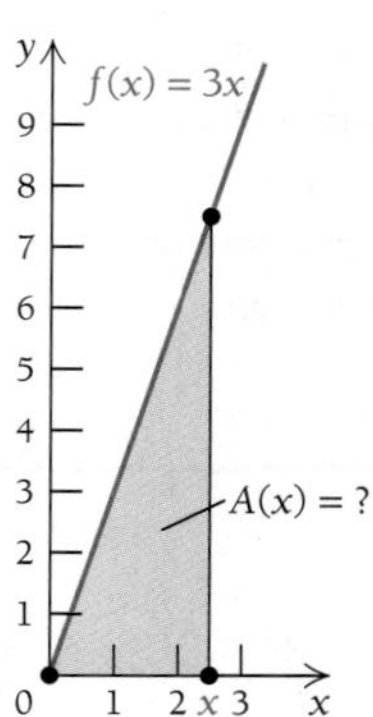

a) Find $A(x)$.
b) Find $A(1)$, $A(2)$, and $A(3.5)$.
c) How are $f(x)$ and $A(x)$ related?

Next, let's consider the linear function $f(x) = mx + b$ over the interval from 0 to x, or $[0, x]$, as shown below.

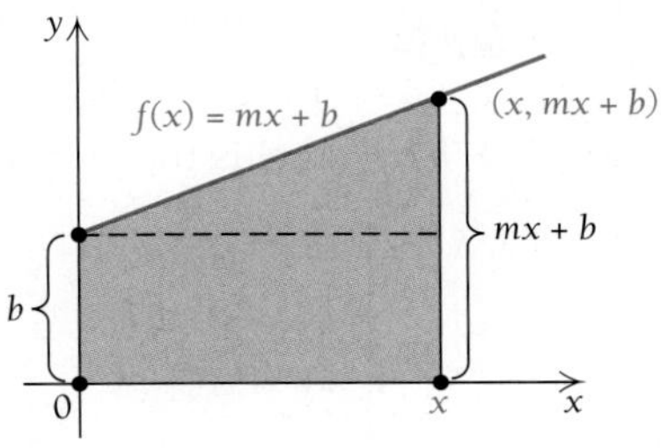

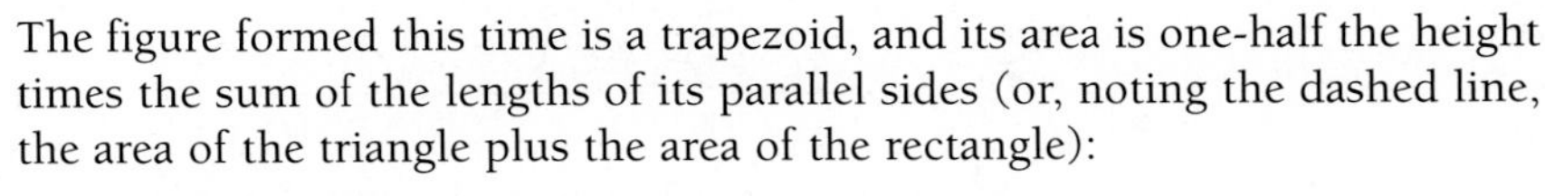

The figure formed this time is a trapezoid, and its area is one-half the height times the sum of the lengths of its parallel sides (or, noting the dashed line, the area of the triangle plus the area of the rectangle):

$$\tfrac{1}{2} \cdot x \cdot [b + (mx + b)],$$

or $$\tfrac{1}{2} \cdot x \cdot (mx + 2b),$$

or $$\tfrac{1}{2}mx^2 + bx.$$

If we allow x to vary, we again get an area function:

$$A(x) = \tfrac{1}{2}mx^2 + bx.$$

Its graph is shown at the left.

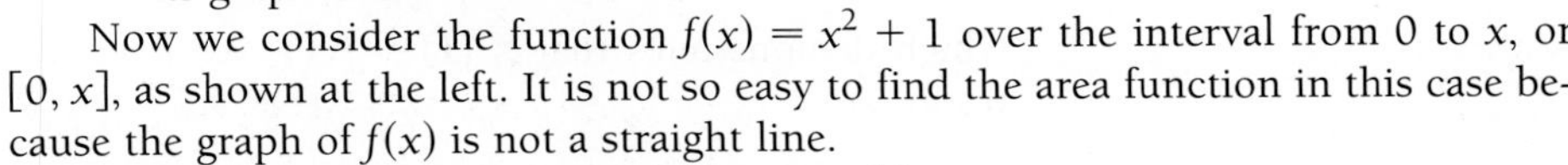

Now we consider the function $f(x) = x^2 + 1$ over the interval from 0 to x, or $[0, x]$, as shown at the left. It is not so easy to find the area function in this case because the graph of $f(x)$ is not a straight line.

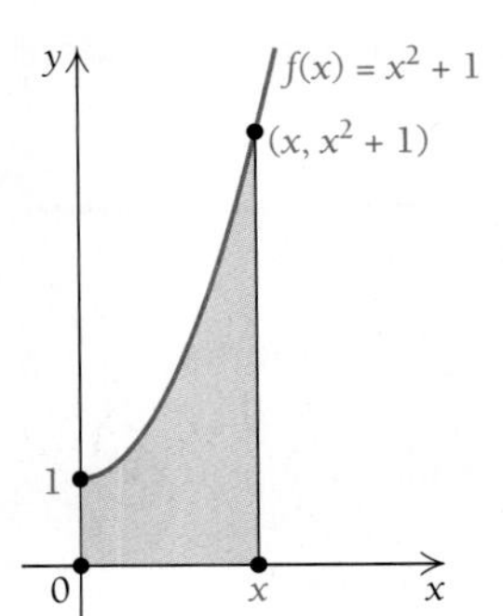

Let's look for a pattern in the following table.

$f(x)$	$A(x)$
$f(x) = 5$	$A(x) = 5x$
$f(x) = b$	$A(x) = bx$
$f(x) = 3x$	$A(x) = \frac{3}{2}x^2$
$f(x) = mx$	$A(x) = \frac{1}{2}mx^2$
$f(x) = mx + b$	$A(x) = \frac{1}{2}mx^2 + bx$

You may have noticed that each time the derivative of the area function, $A(x)$, is $f(x)$. Is this always the case?

We answer this by letting $A(x)$ represent the area under a nonnegative continuous function f over the interval $[0, x]$.

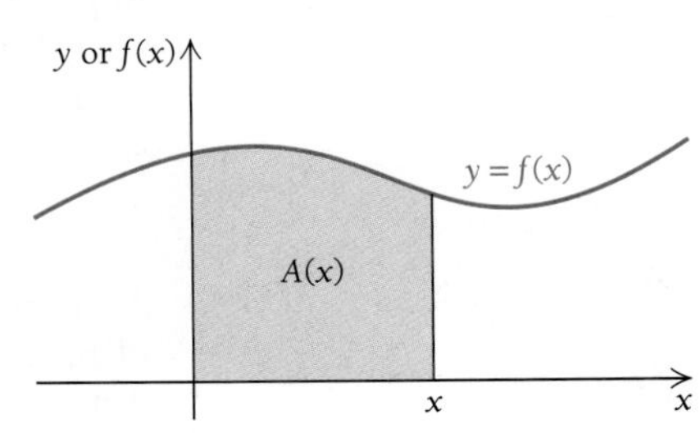

To find $A'(x)$, we use the definition of derivative:

$$A'(x) = \lim_{h \to 0} \frac{A(x + h) - A(x)}{h}.$$

Since $A(x + h)$ is the area under f over the interval $[0, x + h]$, it follows that $A(x + h) - A(x)$ is the area under f between x and $x + h$.

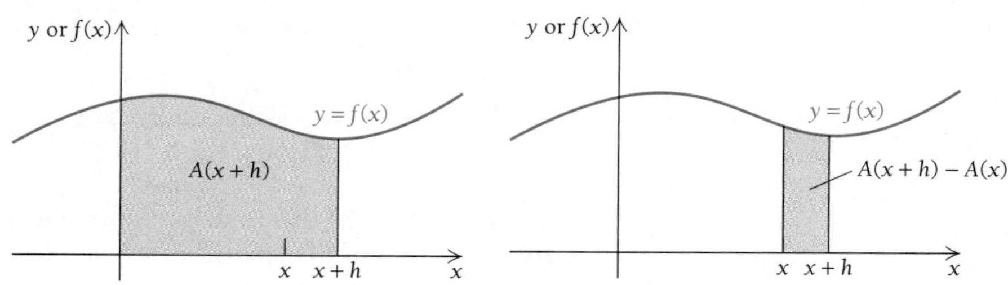

As h approaches zero, the area $A(x + h) - A(x)$ approaches the area of a rectangle with width h and height $f(x)$. That is,

$$A(x + h) - A(x) \approx h \cdot f(x).$$

Thus, $$\frac{A(x + h) - A(x)}{h} \approx f(x), \qquad \text{Dividing both sides by } h$$

and $$\lim_{h \to 0} \frac{A(x + h) - A(x)}{h} = \lim_{h \to 0} f(x) = f(x),$$

which demonstrates that $A'(x) = f(x)$. We have proved the following remarkable result.

THEOREM 1

Let f be a nonnegative continuous function over an interval $[0, b]$, and let $A(x)$ be the area between the graph of f and the x-axis over the interval $[0, x]$, with $0 < x < b$. Then $A(x)$ is a differentiable function of x and $A'(x) = f(x)$.

Theorem 1 answers the question posed earlier: Yes, the derivative of the area function is always the function under which the area is being calculated. Before discussing how to find particular formulas for $A(x)$, we need some new terminology.

The Antiderivative

Suppose that $g(x)$ is such that $g'(x) = 8$. Can we find $g(x)$? It is easy to see that one such function is $g(x) = 8x$. That is, $g(x) = 8x$ is a function whose derivative is 8. Is there another function whose derivative is 8? In fact, there are many. Here are some examples:

$$g(x) = 8x + 3, \quad g(x) = 8x - 10, \quad g(x) = 8x + 7.4, \quad g(x) = 8x + \sqrt{2}.$$

All these functions are $8x$ plus some constant. No other functions have a derivative of 8. Another way of saying this is that any two functions having a derivative of 8 differ by, at most, a constant. This is generalized in the following theorem.

THEOREM 2

If two functions F and G have the same derivative over an interval, then

$$F(x) = G(x) + C, \quad \text{where } C \text{ is a constant.}$$

The reverse of differentiating is *antidifferentiating*, and the result of antidifferentiating is called an **antiderivative.** Above, we found antiderivatives of the function $g'(x) = 8$. There are many, but they are all $8x$ plus some constant.

EXAMPLE 1 Antidifferentiate (find the antiderivatives of) x^2.

Solution One antiderivative is $x^3/3$. All other antiderivatives differ from this by a constant, so we can represent any one of them as follows:

$$\frac{x^3}{3} + C.$$

To check this, we differentiate $x^3/3 + C$:

$$\frac{d}{dx}\left(\frac{x^3}{3} + C\right) = \frac{d}{dx}\left(\frac{x^3}{3}\right) + \frac{d}{dx}(C) = 3 \cdot \frac{x^2}{3} + 0 = x^2.$$

Thus, $x^3/3 + C$ is an antiderivative of x^2. ✦

The solution to Example 1, which includes the constant term C, is the *general form* of the antiderivative of x^2.

Theorems 1 and 2 allow us to transform the problem of finding a formula for the area under a curve into one of finding an antiderivative. We will return to the topic of area in Section 4.3.

Integrals and Integration

The process of antidifferentiating is often called **integration.** To indicate that the antiderivative of x^2 is $x^3/3 + C$, we write

$$\int x^2\, dx = \frac{x^3}{3} + C, \tag{1}$$

where the notation

$$\int f(x)\, dx$$

is used to represent the antiderivative of a function $f(x)$. More generally, we can write

$$\int f(x)\, dx = F(x) + C, \tag{2}$$

where $F(x) + C$ is the general form of the antiderivative of $f(x)$. Equation (2) is read "the *indefinite integral* of $f(x)\, dx$ is $F(x) + C$." The constant C is called the *constant of integration* and can have any fixed value; hence the use of the word "indefinite."

The notation $\int f(x)\, dx$, from Leibniz, is called an *integral*, or more precisely, an **indefinite integral.** The symbol $\int$ is called an *integral sign*. The left side of equation (2) is often read "the integral of $f(x)$, with respect to x" or more briefly, "the integral of $f(x)$." In this context, $f(x)$ is called the *integrand*. For now, think of dx as indicating that the variable involved in the integration is x. More will be said about dx later.

EXAMPLE 2 Evaluate: $\int x^9\,dx$.

Solution We have

$$\int x^9\,dx = \frac{x^{10}}{10} + C. \qquad \text{Check: } \frac{d}{dx}\left(\frac{x^{10}}{10} + C\right) = \frac{10x^9}{10} + 0 = x^9.$$

The left side of this equation is read "the integral of x^9, dx" (though "dx" is often omitted in reading). In this case, the integrand is x^9. Note that the exponent in the antiderivative, $x^{10}/10$, is 1 more than the exponent in the integrand. Dividing by 10 anticipates the multiplication by 10 brought on by differentiation. ✦

Be careful to note that

$$\int e^x\,dx \neq \frac{e^{x+1}}{x+1} + C.$$

EXAMPLE 3 Evaluate: $\int 5e^{4x}\,dx$.

Solution We have

$$\int 5e^{4x}\,dx = \tfrac{5}{4}e^{4x} + C.$$

We can check by differentiating the antiderivative. The derivative of $\frac{5}{4}e^{4x} + C$ is $\frac{5}{4}e^{4x} \cdot 4$, or $5e^{4x}$, which is the integrand. ✦

To integrate (or antidifferentiate), we make use of differentiation formulas, in effect, stating them in reverse. Below are some of these integration formulas. Each can be verified by differentiating the right-hand side and noting that the result is the integrand.

THEOREM 3 Basic Integration Formulas

1. $\int k\,dx = kx + C$ (k is a constant)

2. $\int x^r\,dx = \dfrac{x^{r+1}}{r+1} + C$, provided $r \neq -1$

(To integrate a power of x other than -1, increase the power by 1 and divide by the increased power.)

3. $\int x^{-1}\,dx = \int \dfrac{1}{x}\,dx = \int \dfrac{dx}{x} = \ln x + C, \quad x > 0$

$\int x^{-1}\,dx = \ln |x| + C, \quad x < 0$

(We will generally assume that $x > 0$.)

4. $\int be^{ax}\,dx = \dfrac{b}{a}e^{ax} + C$

The following rules, along with the preceding formulas, allow us to find many integrals. These rules are derived by reversing two familiar differentiation rules.

THEOREM 4

Rule A. $\int kf(x)\,dx = k\int f(x)\,dx$

(The integral of a constant times a function is the constant times the integral of the function.)

Rule B. $\int [f(x) \pm g(x)]\,dx = \int f(x)\,dx \pm \int g(x)\,dx$

(The integral of a sum or a difference is the sum or the difference of the integrals.)

EXAMPLE 4 Evaluate $\int (5x + 4x^3)\,dx$.

Solution

$$\begin{aligned}\int (5x + 4x^3)\,dx &= \int 5x\,dx + \int 4x^3\,dx && \text{Rule B}\\ &= 5\int x\,dx + 4\int x^3\,dx && \text{Rule A}\end{aligned}$$

Then, using integration formula 2 of Theorem 3 for $\int x\,dx$ and $\int x^3\,dx$, we have

$$\int (5x + 4x^3)\,dx = 5 \cdot \frac{x^2}{2} + 4 \cdot \frac{x^4}{4} + C = \frac{5}{2}x^2 + x^4 + C.$$

Don't forget the constant of integration. Note that it is not necessary to write two constants of integration. If we did, we could add them and regard C as the sum.

We can check this result as follows:

$$\frac{d}{dx}\left(\frac{5}{2}x^2 + x^4 + C\right) = 2 \cdot \frac{5}{2} \cdot x + 4x^3 = 5x + 4x^3.$$ ✦

We can always check an integration by differentiating.

Do not forget to include C, the constant of integration.

EXAMPLE 5 Evaluate and then check by differentiation:

a) $\int (7e^{6x} - \sqrt{x})\,dx$; **b)** $\int \left(1 - \frac{3}{x} + \frac{1}{x^4}\right)dx$.

Solution

a)
$$\begin{aligned}\int \left(7e^{6x} - \sqrt{x}\right)dx &= \int 7e^{6x}\,dx - \int \sqrt{x}\,dx\\ &= \int 7e^{6x}\,dx - \int x^{1/2}dx\\ &= \frac{7}{6}e^{6x} - \frac{x^{(1/2)+1}}{\frac{1}{2}+1} + C\\ &= \frac{7}{6}e^{6x} - \frac{x^{3/2}}{\frac{3}{2}} + C\\ &= \frac{7}{6}e^{6x} - \frac{2}{3}x^{3/2} + C\end{aligned}$$

To check this, we differentiate:

$$\frac{d}{dx}\left(\frac{7}{6}e^{6x} - \frac{2}{3}x^{3/2} + C\right) = \frac{7}{6}e^{6x} \cdot 6 - \frac{3}{2} \cdot \frac{2}{3}x^{1/2} + 0 = 7e^{6x} - \sqrt{x}.$$

b) $$\int \left(1 - \frac{3}{x} + \frac{1}{x^4}\right) dx = \int 1\, dx - 3\int \frac{dx}{x} + \int x^{-4}\, dx$$

$$= x - 3\ln x + \frac{x^{-4+1}}{-4+1} + C \qquad \text{We assume } x > 0.$$

$$= x - 3\ln x - \frac{x^{-3}}{3} + C$$

To check this, we differentiate:

$$\frac{d}{dx}\left(x - 3\ln x - \frac{x^{-3}}{3} + C\right) = 1 - 3 \cdot \frac{1}{x} - (-3) \cdot \frac{1}{3}x^{-4} + 0 = 1 - \frac{3}{x} + \frac{1}{x^4}.$$

✦

TECHNOLOGY CONNECTION

Exploratory

Using the same set of axes, graph

$$y = x^4 - 4x^3$$

and

$$y = x^4 - 4x^3 + 7.$$

Compare the graphs. Draw a tangent line to each graph at $x = -1$, $x = 0$, and $x = 1$ and find the slope of each line. Compare the slopes. Then find the derivative of each function. Compare the derivatives. Is either function an antiderivative of $y = 4x^3 - 12x^2$?

Initial Conditions and Antiderivatives

The graphs of the antiderivatives of $y' = x^2$ are the graphs of the functions

$$y = \int x^2\, dx = \frac{x^3}{3} + C$$

for the various values of the constant C.

As we can see in the figures below, x^2 is the derivative of each function; that is, the tangent line at the point

$$\left(a, \frac{a^3}{3} + C\right)$$

has slope a^2. The curves $y = x^3/3 + C$ fill up the plane, with exactly one curve going through any given point (x_0, y_0).

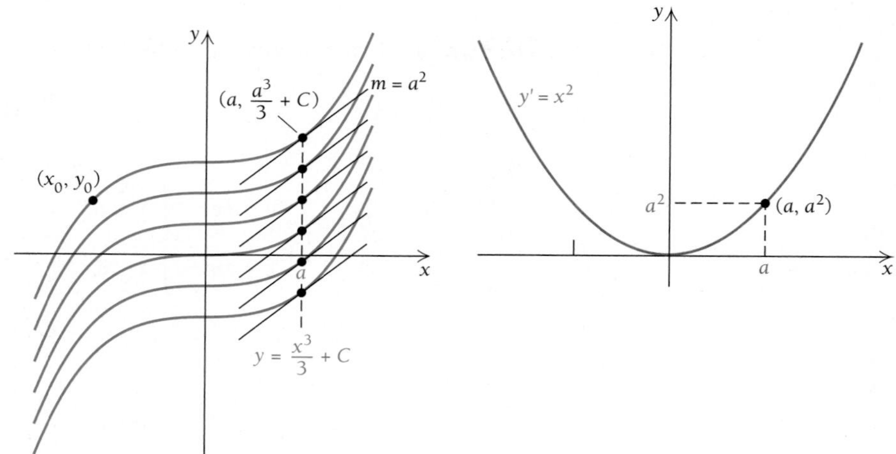

Suppose that we look for an antiderivative of x^2 with a specified value at a certain point—say, $f(-1) = 2$. We will find that there is only one such function.

EXAMPLE 6 Find the function f such that

$$f'(x) = x^2 \quad \text{and} \quad f(-1) = 2.$$

Solution

We first find $f(x)$ by integrating:

$$f(x) = \int x^2\,dx = \frac{x^3}{3} + C.$$

Then the condition $f(-1) = 2$ allows us to find C:

$$f(-1) = \frac{(-1)^3}{3} + C = 2. \qquad \text{Setting } f(-1) \text{ equal to 2}$$

Solving for C, we get

$$-\tfrac{1}{3} + C = 2$$
$$C = 2 + \tfrac{1}{3}, \text{ or } \tfrac{7}{3}.$$

Thus, $f(x) = \dfrac{x^3}{3} + \dfrac{7}{3}$. ✦

We are often given further information about a function, such as, in Example 6, $f(-1) = 2$. The statement that $f(-1) = 2$ is called an *initial condition*, or a *boundary condition*. Sometimes the term *initial value*, or *boundary value*, is used. In Section 4.3, we will see how initial conditions appear in applications and return to the problem of calculating the area under a curve.

Exercise Set 4.2

Evaluate.

1. $\int x^6\,dx$

2. $\int x^7\,dx$

3. $\int 2\,dx$

4. $\int 4\,dx$

5. $\int x^{1/4}\,dx$

6. $\int x^{1/3}\,dx$

7. $\int (x^2 + x - 1)\,dx$

8. $\int (x^2 - x + 2)\,dx$

9. $\int (2t^2 + 5t - 3)\,dt$

10. $\int (3t^2 - 4t + 7)\,dt$

11. $\int \frac{1}{x^3}\,dx$

12. $\int \frac{1}{x^5}\,dx$

13. $\int \sqrt[3]{x}\,dx$

14. $\int \sqrt{x}\,dx$

15. $\int \sqrt{x^5}\,dx$

16. $\int \sqrt[3]{x^2}\,dx$

17. $\int \frac{dx}{x^4}$

18. $\int \frac{dx}{x^2}$

19. $\int \frac{1}{x}\,dx$

20. $\int \frac{2}{x}\,dx$

21. $\int \left(\frac{3}{x} + \frac{5}{x^2}\right) dx$

22. $\int \left(\frac{4}{x^3} + \frac{7}{x}\right) dx$

23. $\int \frac{-7}{\sqrt[3]{x^2}}\,dx$

24. $\int \frac{5}{\sqrt[4]{x^3}}\,dx$

25. $\int 2e^{2x}\,dx$

26. $\int 4e^{4x}\,dx$

27. $\int e^{3x}\,dx$

28. $\int e^{5x}\,dx$

29. $\int e^{7x}\,dx$

30. $\int e^{6x}\,dx$

31. $\int 5e^{3x}\,dx$

32. $\int 2e^{5x}\,dx$

33. $\int 6e^{8x}\,dx$

34. $\int 12e^{3x}\,dx$

35. $\int \frac{2}{3}e^{-9x}\,dx$

36. $\int \frac{4}{5}e^{-10x}\,dx$

37. $\int (5x^2 - 2e^{7x})\,dx$

38. $\int (2x^5 - 4e^{3x})\,dx$

39. $\int \left(x^2 - \frac{3}{2}\sqrt{x} + x^{-4/3}\right)dx$

40. $\int \left(x^4 + \frac{1}{8\sqrt{x}} - \frac{4}{5}x^{-2/5}\right)dx$

41. $\int (3x + 2)^2\,dx$ (*Hint*: Expand first.)

42. $\int (x + 4)^2\,dx$

43. $\int \left(\frac{3}{x} - 5e^{2x} + \sqrt{x^7}\right)dx$

44. $\int \left(2e^{6x} - \frac{3}{x} + \sqrt[3]{x^4}\right)dx$

45. $\int \left(\frac{7}{\sqrt{x}} - \frac{2}{3}e^{5x} - \frac{8}{x}\right)dx$

46. $\int \left(\frac{4}{\sqrt[5]{x}} + \frac{3}{4}e^{6x} - \frac{7}{x}\right)dx$

Find f such that:

47. $f'(x) = x - 3, \quad f(2) = 9$

48. $f'(x) = x - 5, \quad f(1) = 6$

49. $f'(x) = x^2 - 4, \quad f(0) = 7$

50. $f'(x) = x^2 + 1, \quad f(0) = 8$

51. $f'(x) = 5x^2 + 3x - 7, \quad f(0) = 9$

52. $f'(x) = 8x^2 + 4x - 2, \quad f(0) = 6$

53. $f'(x) = 3x^2 - 5x + 1, \quad f(1) = \frac{7}{2}$

54. $f'(x) = 6x^2 - 4x + 2, \quad f(1) = 9$

55. $f'(x) = 5e^{2x}, \quad f(0) = \frac{1}{2}$

56. $f'(x) = 3e^{4x}, \quad f(0) = \frac{7}{4}$

57. $f'(x) = \frac{4}{\sqrt{x}}, \quad f(1) = -5$

58. $f'(x) = \frac{2}{\sqrt[3]{x}}, \quad f(1) = 1$

APPLICATIONS

Business and Economics

Credit market debt. *From 1994 to 2000, the annual rate of change in the national credit market debt, in billions of dollars per year, could be modeled by the function*

$$D'(t) = 1975 - 1190t + 597t^2 - 71.3t^3,$$

*where t is the number of years since 1994. (**Source:** Federal Reserve System.) Use the preceding information for Exercises 59 and 60.*

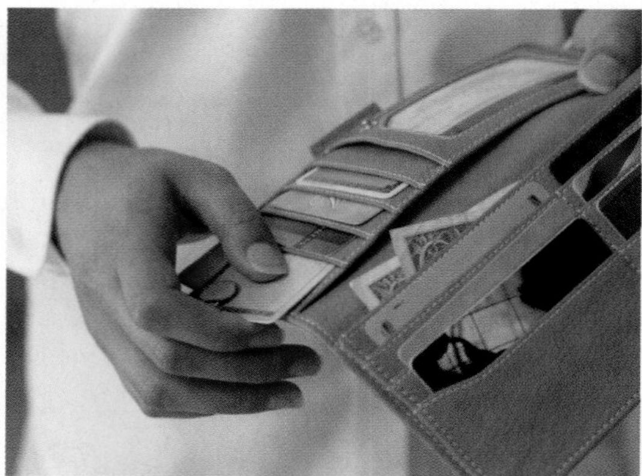

59. Find the national credit market debt, $D(t)$, during the years 1994 through 2000 given that $D(0) = 17{,}198$.

60. What was the national credit market debt in 2000, given that $D(0) = 17{,}198$?

61. **Total cost from marginal cost.** A company determines that the marginal cost, C', of producing the xth unit of a product is given by

$$C'(x) = x^3 - 2x.$$

Find the total-cost function, C, assuming that $C(x)$ is in dollars and that fixed costs are \$7000.

62. **Total cost from marginal cost.** A company determines that the marginal cost, C', of producing the xth unit of a product is given by

$$C'(x) = x^3 - x.$$

Find the total-cost function, C, assuming that $C(x)$ is in dollars and that fixed costs are \$6500.

63. **Total revenue from marginal revenue.** A company determines that the marginal revenue, R', in dollars, from selling the xth unit of a product is given by

$$R'(x) = x^2 - 3.$$

a) Find the total-revenue function, R, assuming that $R(0) = 0$.

TW **b)** Why is $R(0) = 0$ a reasonable assumption?

64. **Total revenue from marginal revenue.** A company determines that the marginal revenue, R', in dollars, from selling the xth unit of a product is given by

$$R'(x) = x^2 - 1.$$

a) Find the total-revenue function, R, assuming that $R(0) = 0$.

TW **b)** Why is $R(0) = 0$ a reasonable assumption?

65. Demand from marginal demand. A company finds that the rate at which the quantity of a product that consumers demand changes with respect to price is given by the marginal-demand function

$$D'(x) = -\frac{4000}{x^2},$$

where x is the price per unit, in dollars. Find the demand function if it is known that 1003 units of the product are demanded by consumers when the price is \$4 per unit.

66. Supply from marginal supply. A company finds that the rate at which a seller's quantity supplied changes with respect to price is given by the marginal-supply function

$$S'(x) = 0.24x^2 + 4x + 10,$$

where x is the price per unit, in dollars. Find the supply function if it is known that the seller will sell 121 units of the product when the price is \$5 per unit.

67. Efficiency of a machine operator. The rate at which a machine operator's efficiency, E (expressed as a percentage), changes with respect to time t is given by

$$\frac{dE}{dt} = 30 - 10t,$$

A machine operator's efficiency changes with respect to time.

where t is the number of hours the operator has been at work.

a) Find $E(t)$, given that the operator's efficiency after working 2 hr is 72%; that is, $E(2) = 72$.

b) Use the answer to part (a) to find the operator's efficiency after 3 hr; after 5 hr.

68. Efficiency of a machine operator. The rate at which a machine operator's efficiency, E (expressed as a percentage), changes with respect to time t is given by

$$\frac{dE}{dt} = 40 - 10t,$$

where t is the number of hours the operator has been at work.

a) Find $E(t)$, given that the operator's efficiency after working 2 hr is 72%; that is, $E(2) = 72$.

b) Use the answer to part (a) to find the operator's efficiency after 4 hr; after 8 hr.

Social and Natural Sciences

69. Spread of influenza. During 34 weeks in the 2000–2001 flu season, the rate at which the number of cases of influenza per 100,000 people in Ireland changed could be approximated by

$$I'(t) = 3.389e^{0.1049t},$$

where I is the total number of people per 100,000 who have contracted influenza and t is time measured in weeks. (*Source*: Departments of Public Health.)

a) Estimate $I(t)$, the total number per 100,000 who have contracted influenza by time t. Assume that $I(0) = 0$.

b) Approximately how many people per 100,000 contracted influenza during the first 27 weeks?

c) Approximately how many people per 100,000 contracted influenza during the whole 34 weeks?

d) Approximately how many people per 100,000 contracted influenza during the last 7 of the 34 weeks?

70. Memory. In a memory experiment, the rate at which students memorize Spanish vocabulary is found to be given by

$$M'(t) = 0.2t - 0.003t^2,$$

where $M(t)$ is the number of words memorized in t minutes.

a) Find $M(t)$ if it is known that $M(0) = 0$.

b) How many words are memorized in 8 min?

SYNTHESIS

Find f.

71. $f'(t) = \sqrt{t} + \frac{1}{\sqrt{t}}, \quad f(4) = 0$

72. $f'(t) = t^{\sqrt{3}}, \quad f(0) = 8$

Evaluate. Each of the following can be integrated using the rules developed in this section, but some algebra may be required beforehand.

73. $\int (5t + 4)^2 t^4 \, dt$

74. $\int (x - 1)^2 x^3 \, dx$

75. $\int (1 - t)\sqrt{t} \, dt$

76. $\int \frac{(t + 3)^2}{\sqrt{t}} \, dt$

77. $\int \frac{x^4 - 6x^2 - 7}{x^3} \, dx$

78. $\int (t + 1)^3 \, dt$

79. $\int \frac{1}{\ln 10} \frac{dx}{x}$

80. $\int be^{ax} \, dx$

81. $\int (3x - 5)(2x + 1)^2 \, dx$

82. $\int \sqrt[3]{64x^4} \, dx$

83. $\int \frac{x^2 - 1}{x + 1} \, dx$

84. $\int \frac{t^3 + 8}{t + 2} \, dt$

TW **85.** On a test, a student makes this statement: "The function $f(x) = x^2$ has a unique integral." Is this a true statement? Why or why not?

TW **86.** Describe the graphical interpretation of an antiderivative.

4.3 Area and Definite Integrals

OBJECTIVES

- Find the area under a curve over a given closed interval.
- Evaluate a definite integral.
- Interpret an area below the horizontal axis.
- Solve applied problems involving definite integrals.

We now use integration to find areas of certain regions. In many cases, such an area represents the solution of an important real-world problem.

Definite Integrals

In Section 4.2, we found that if $A(x)$ represents the area under the graph of a nonnegative continuous function $y = f(x)$ over the interval $[0, x]$, then $A(x)$ is an antiderivative of $f(x)$. Referring to the graph below, we see that the area over $[a, b]$ is the same as the area over $[0, b]$ minus the area over $[0, a]$, or $A(b) - A(a)$.

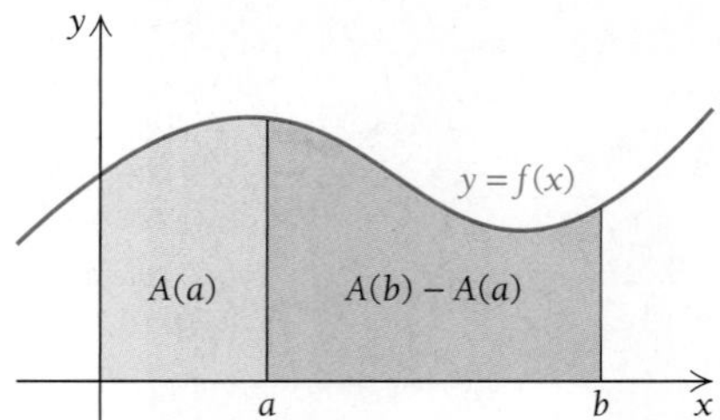

In Section 4.2, we also found that a function's antiderivatives can differ only in their constant terms. Thus, if $F(x)$ is another antiderivative of $f(x)$, then $A(x) = F(x) + C$, for some constant C, and

$$A(b) - A(a) = F(b) + C - (F(a) + C) = F(b) - F(a).$$

This result tells us that as long as an area is computed by substituting an interval's endpoints into an antiderivative and then subtracting, *any* antiderivative—and any choice of C—can be used. It generally simplifies computations to choose 0 as the value of C.

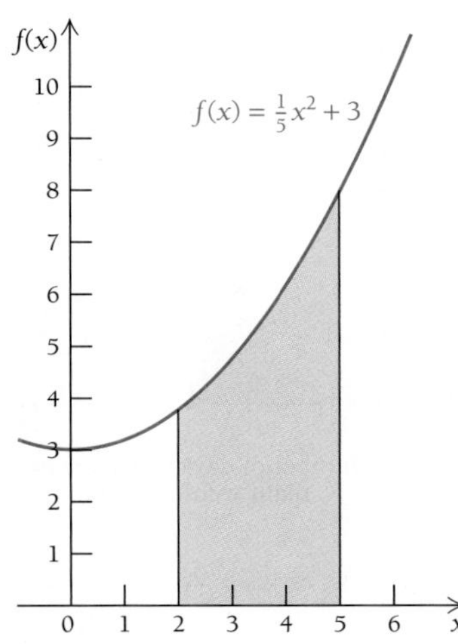

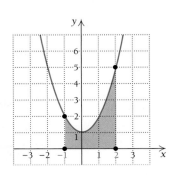

EXAMPLE 1 Find the area under the graph of $f(x) = \frac{1}{5}x^2 + 3$ over the interval $[2, 5]$.

Solution Although making a drawing is not required, doing so helps us visualize the problem.

Note that every antiderivative of $f(x) = \frac{1}{5}x^2 + 3$ is of the form

$$F(x) = \tfrac{1}{15}x^3 + 3x + C.$$

For simplicity, we set $C = 0$, so

$$\begin{aligned}\text{Area over } [2, 5] &= F(5) - F(2)\\ &= \tfrac{125}{15} + 15 - \left(\tfrac{8}{15} + 6\right)\\ &= 16\tfrac{4}{5}.\end{aligned}$$

✦

Although it is possible to express the area under a curve as the limit of a Riemann sum, as we did in Section 4.1, antiderivatives are usually much easier to work with.

To find the area under the graph of a nonnegative, continuous function f over the interval $[a, b]$:

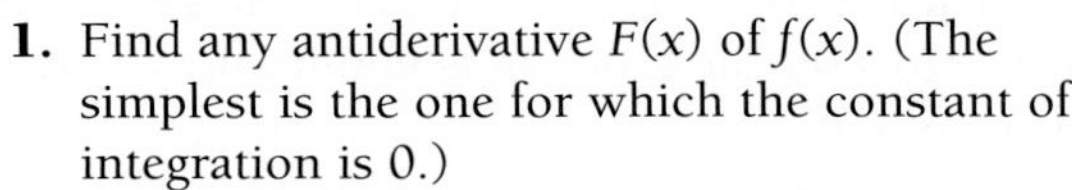

1. Find any antiderivative $F(x)$ of $f(x)$. (The simplest is the one for which the constant of integration is 0.)
2. Evaluate $F(x)$ using b and a, and compute $F(b) - F(a)$. The result is the area under the graph over the interval $[a, b]$.

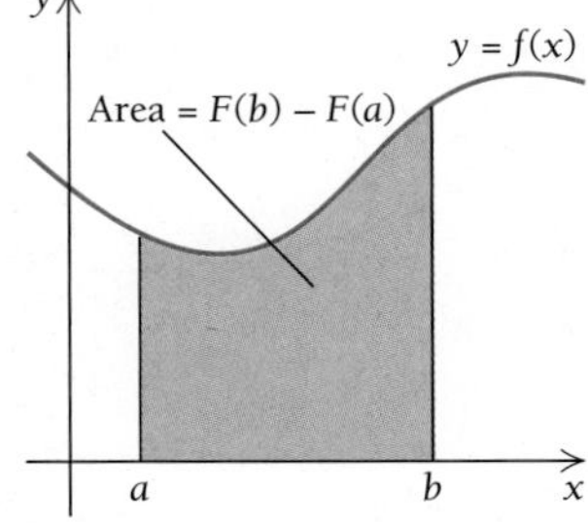

EXAMPLE 2 Find the area under the graph of $y = x^2 + 1$ over the interval $[-1, 2]$.

Solution In this case, $f(x) = x^2 + 1$, with $a = -1$ and $b = 2$.

1. Find any antiderivative $F(x)$ of $f(x)$. We choose the simplest one:

$$F(x) = \frac{x^3}{3} + x.$$

2. Substitute 2 and -1, and find the difference $F(2) - F(-1)$:

$$\begin{aligned}F(2) - F(-1) &= \left[\frac{2^3}{3} + 2\right] - \left[\frac{(-1)^3}{3} + (-1)\right]\\ &= \frac{8}{3} + 2 - \left[\frac{-1}{3} - 1\right]\\ &= \frac{8}{3} + 2 + \frac{1}{3} + 1\\ &= 6.\end{aligned}$$

We can make a partial check by counting the squares and parts of squares shaded on the graph to the left. ✦

EXAMPLE 3 Let $y = x^3$ represent the number of kilowatts (kW) generated by a new power plant each day, x days after going on line. Find the area under the graph of $y = x^3$ over the interval $[0, 5]$ and interpret the significance of the area.

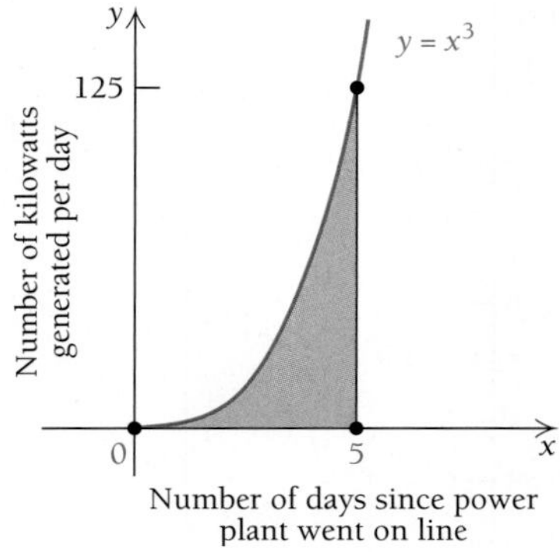

Solution In this case, $f(x) = x^3$, $a = 0$, and $b = 5$.

1. Find any antiderivative $F(x)$ of $f(x)$. We choose the simplest one:

$$F(x) = \frac{x^4}{4}.$$

2. Substitute 5 and 0, and find the difference $F(5) - F(0)$:

$$F(5) - F(0) = \frac{5^4}{4} - \frac{0^4}{4} = \frac{625}{4} = 156\tfrac{1}{4}.$$

The area represents the total number of kilowatts generated during the first 5 days. Note that kW/day · days = kW. ✦

The difference $F(b) - F(a)$ has the same value for all antiderivatives of a function f whether the function is nonnegative or not. It is called the *definite integral* of f from a to b.

DEFINITION

Let f be any continuous function over the interval $[a, b]$ and F be any antiderivative of f. Then the **definite integral** of f from a to b is

$$\int_a^b f(x)\,dx = F(b) - F(a).$$

Evaluating definite integrals is called *integrating*. The numbers a and b are known as the **limits of integration.** Note that this use of the word *limit* indicates an endpoint of an interval, not a value that is being approached, as you learned in Chapter 1.

EXAMPLE 4 Evaluate: $\int_a^b x^2\,dx$.

Solution Using the antiderivative $F(x) = x^3/3$, we have

$$\int_a^b x^2\,dx = \frac{b^3}{3} - \frac{a^3}{3}.$$ ✦

It is convenient to use an intermediate notation:

$$\int_a^b f(x)\,dx = \big[F(x)\big]_a^b = F(b) - F(a),$$

where $F(x)$ is an antiderivative of $f(x)$.

EXAMPLE 5 Evaluate each of the following:

a) $\int_{-1}^{4} (x^2 - x)\, dx$;

b) $\int_{0}^{3} e^x\, dx$;

c) $\int_{1}^{e} \left(1 + 2x - \frac{1}{x}\right) dx$ (assume $x > 0$).

Solution

a)
$$\begin{aligned}\int_{-1}^{4} (x^2 - x)\, dx &= \left[\frac{x^3}{3} - \frac{x^2}{2}\right]_{-1}^{4} = \left(\frac{4^3}{3} - \frac{4^2}{2}\right) - \left(\frac{(-1)^3}{3} - \frac{(-1)^2}{2}\right)\\ &= \left(\frac{64}{3} - \frac{16}{2}\right) - \left(\frac{-1}{3} - \frac{1}{2}\right)\\ &= \frac{64}{3} - 8 + \frac{1}{3} + \frac{1}{2} = 14\tfrac{1}{6}\end{aligned}$$

b)
$$\int_{0}^{3} e^x\, dx = \left[e^x\right]_0^3 = e^3 - e^0 = e^3 - 1$$

c)
$$\begin{aligned}\int_{1}^{e} \left(1 + 2x - \frac{1}{x}\right) dx &= \left[x + x^2 - \ln x\right]_1^e \qquad \text{We assume } x > 0.\\ &= (e + e^2 - \ln e) - (1 + 1^2 - \ln 1)\\ &= (e + e^2 - 1) - (1 + 1 - 0)\\ &= e + e^2 - 1 - 1 - 1\\ &= e + e^2 - 3\end{aligned}$$

✦

The fact that we can express the integral of a function either as a limit of a sum or in terms of an antiderivative is so important that it has a name: the *Fundamental Theorem of Integral Calculus.*

THE FUNDAMENTAL THEOREM OF INTEGRAL CALCULUS

If a continuous function f has an antiderivative F over $[a, b]$, then

$$\lim_{n\to\infty} \sum_{i=1}^{n} f(x_i)\,\Delta x = \int_a^b f(x)\, dx = F(b) - F(a).$$

It is interesting to envision taking the limit as stretching the summation sign, Σ, into something resembling an S (the integral sign) and redefining Δx as dx. Because Δx is used in the limit, dx appears in the integral notation.

More on Area

When we evaluate the definite integral of a nonnegative function, we get the area under the graph over an interval.

EXAMPLE 6 Suppose that y is the profit per mile traveled and x is the number of miles traveled, in thousands. Find the area under $y = 1/x$ over the interval $[1, 4]$ and interpret the significance of this area.

Solution

$$\int_1^4 \frac{dx}{x} = \left[\ln x\right]_1^4 = \ln 4 - \ln 1$$
$$= \ln 4 - 0 \approx 1.3863$$

Considering the units, dollars/mile · miles = dollars, we see that the area represents a total profit of $1386.30 when the miles traveled increase from 1000 to 4000 miles.

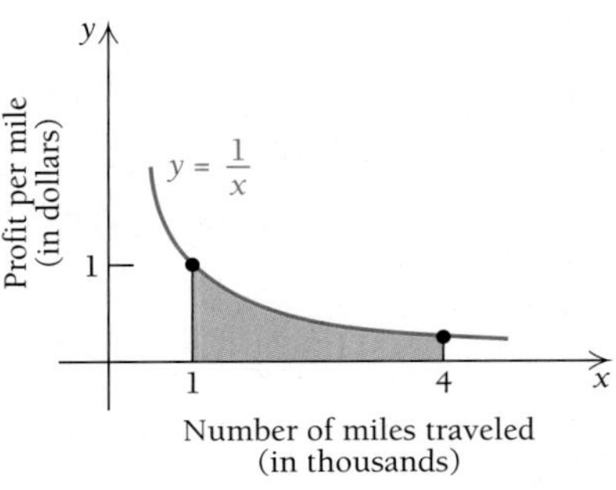

◆

EXAMPLE 7 Find the area under $y = 1/x^2$ over the interval $[1, b]$.

Solution

$$\int_1^b \frac{dx}{x^2} = \int_1^b x^{-2}\,dx$$
$$= \left[\frac{x^{-2+1}}{-2+1}\right]_1^b$$
$$= \left[\frac{x^{-1}}{-1}\right]_1^b = \left[-\frac{1}{x}\right]_1^b$$
$$= \left(-\frac{1}{b}\right) - \left(-\frac{1}{1}\right)$$
$$= 1 - \frac{1}{b}$$

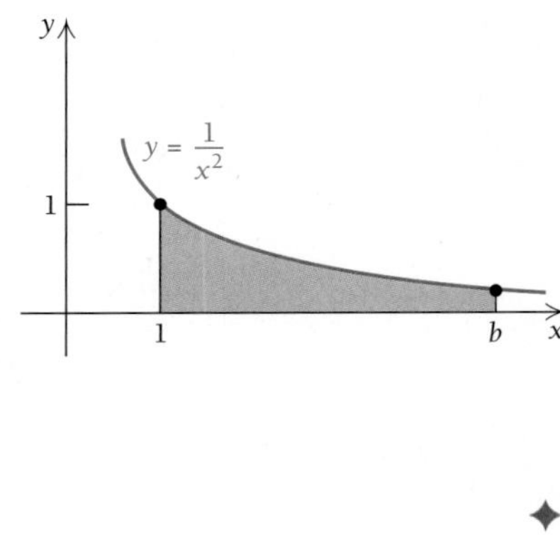

◆

Now let's compare the definite integrals of the functions $y = x^2$ and $y = -x^2$:

$$\int_0^2 x^2\,dx = \left[\frac{x^3}{3}\right]_0^2 = \frac{2^3}{3} - \frac{0^3}{3} = \frac{8}{3}$$

$$\int_0^2 -x^2\,dx = \left[-\frac{x^3}{3}\right]_0^2 = -\frac{2^3}{3} + \frac{0^3}{3} = -\frac{8}{3}$$

The graphs of the functions $y = x^2$ and $y = -x^2$ are reflections of each other across the x-axis. Thus, the shaded areas are the same, $\frac{8}{3}$. The evaluation procedure for $y = -x^2$ gave us $-\frac{8}{3}$. This illustrates that for negative-valued functions, the definite integral gives us the opposite of the area between the curve and the x-axis.

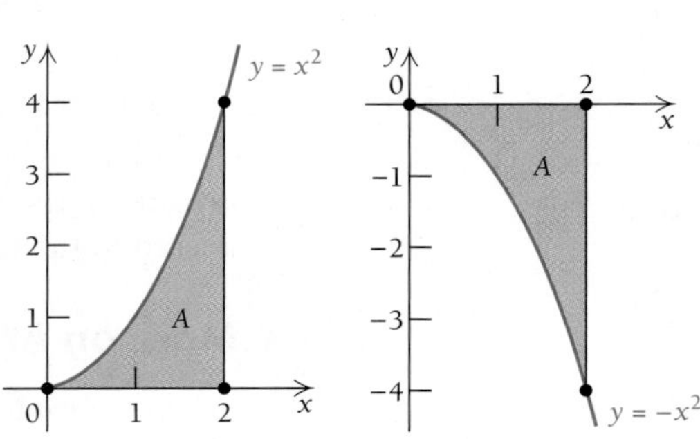

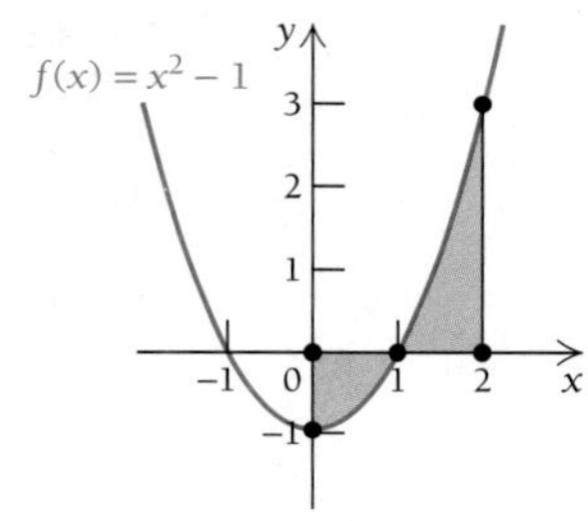

Now let's consider $f(x) = x^2 - 1$ over the interval $[0, 2]$. It has both positive and negative values. We apply the preceding evaluation procedure, even though function values are not all nonnegative. We do so in two ways.

First, let's use the fact that for any a, b, c, if $a < b < c$, then

$$\int_a^c f(x)\,dx = \int_a^b f(x)\,dx + \int_b^c f(x)\,dx.$$

The area from a to b plus the area from b to c is the area from a to c.

(We will consider this property of integrals again in Section 4.4.) Note that 1 is an x-intercept in $[0, 2]$.

$$\begin{aligned}
\int_0^2 (x^2 - 1)\,dx &= \int_0^1 (x^2 - 1)\,dx + \int_1^2 (x^2 - 1)\,dx \\
&= \left[\frac{x^3}{3} - x\right]_0^1 + \left[\frac{x^3}{3} - x\right]_1^2 \\
&= \underbrace{\left[\left(\frac{1^3}{3} - 1\right) - \left(\frac{0^3}{3} - 0\right)\right]} + \underbrace{\left[\left(\frac{2^3}{3} - 2\right) - \left(\frac{1^3}{3} - 1\right)\right]} \\
&= \left[\frac{1}{3} - 1\right] + \left[\frac{8}{3} - 2 - \frac{1}{3} + 1\right] \\
&= -\frac{2}{3} + \frac{4}{3} \\
&= \frac{2}{3}.
\end{aligned}$$

This shows that the area above the x-axis exceeds the area below the x-axis by $\frac{2}{3}$ units.

Now let's evaluate the original integral in another, more direct, way:

$$\begin{aligned}
\int_0^2 (x^2 - 1)\,dx &= \left[\frac{x^3}{3} - x\right]_0^2 \\
&= \left(\frac{2^3}{3} - 2\right) - \left(\frac{0^3}{3} - 0\right) \\
&= \left(\frac{8}{3} - 2\right) - 0 \\
&= \frac{2}{3}.
\end{aligned}$$

The definite integral of a continuous function over an interval is the sum of the areas above the x-axis minus the sum of the areas below the x-axis.

EXAMPLE 8 Consider $\int_{-1}^{2} (-x^3 + 3x - 1)\,dx$. Predict the sign of the integral, by examining the graph, and then evaluate the integral.

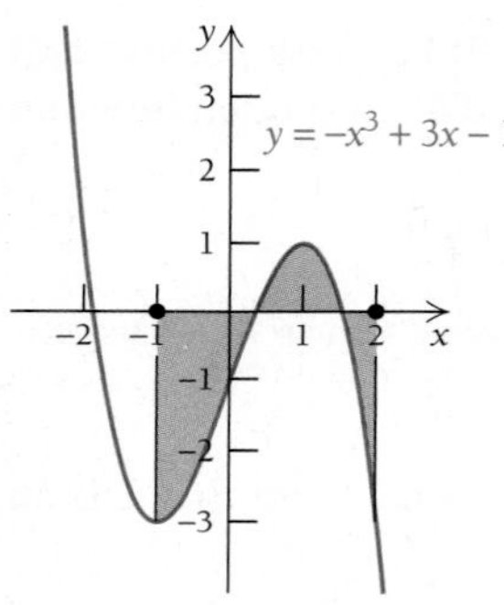

Solution From the graph, it appears that there is considerably more area below the x-axis than above. Thus, we expect that

$$\int_{-1}^{2} (-x^3 + 3x - 1)\, dx < 0.$$

Evaluating the integral, we have

$$\begin{aligned}\int_{-1}^{2} (-x^3 + 3x - 1)\, dx &= \left[-\frac{x^4}{4} + \frac{3}{2}x^2 - x\right]_{-1}^{2} \\ &= \left(-\frac{2^4}{4} + \frac{3}{2}\cdot 2^2 - 2\right) - \left(-\frac{(-1)^4}{4} + \frac{3}{2}(-1)^2 - (-1)\right) \\ &= (-4 + 6 - 2) - \left(-\frac{1}{4} + \frac{3}{2} + 1\right) = 0 - 2\tfrac{1}{4} \\ &= -2\tfrac{1}{4}.\end{aligned}$$

As a partial check, we note that the result is negative, as predicted. ✦

TECHNOLOGY CONNECTION

Approximating Definite Integrals

There are two methods for evaluating definite integrals with a calculator. Let's consider the function from Example 8: $f(x) = -x^3 + 3x - 1$.

Method 1: fnInt

First, we select fnInt from the MATH menu. Next, we enter the function, the variable, and the endpoints of the interval over which we are integrating. The calculator returns the same value for the definite integral as we found in Example 8.

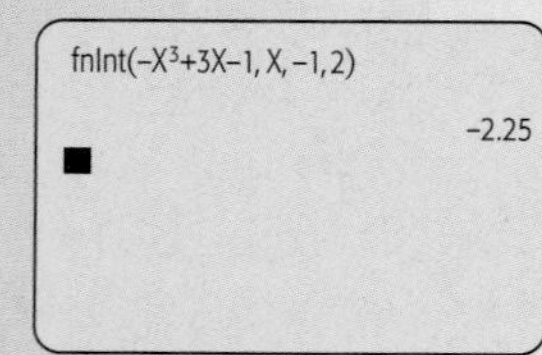

Method 2: ∫ f(x) dx

We first graph $y_1 = -x^3 + 3x - 1$. Then we select ∫f(x) dx from the CALC menu and enter the lower and upper limits of integration. The calculator shades the area and returns the same value for the definite integral as found in Example 8.

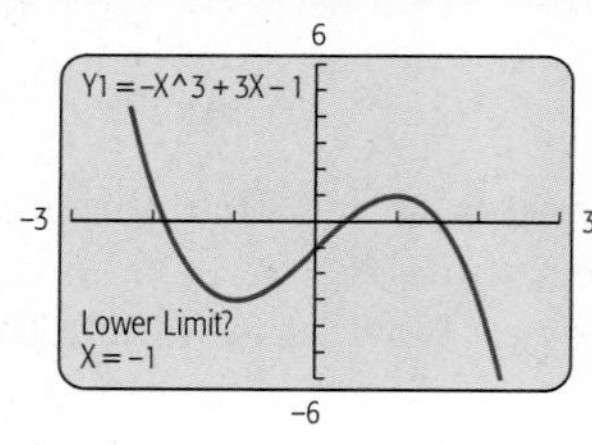

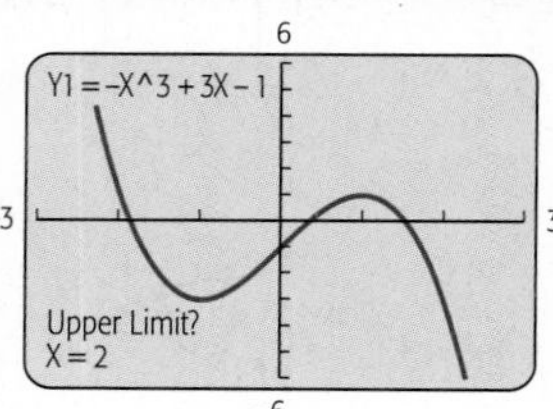

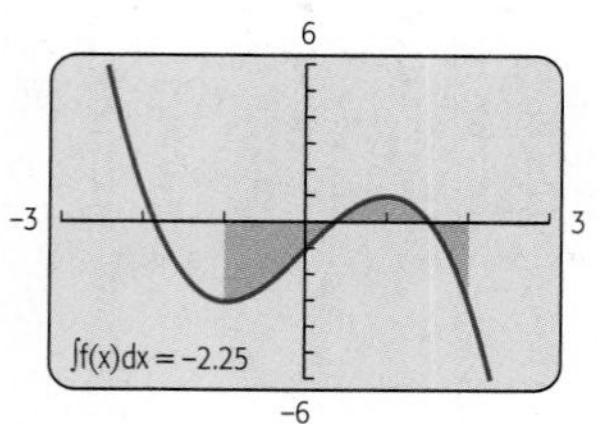

EXERCISES

Evaluate each definite integral.

1. $\int_{-1}^{2} (x^2 - 1)\, dx$

2. $\int_{-2}^{3} (x^3 - 3x + 1)\, dx$

3. $\int_{1}^{6} \frac{\ln x}{x^2}\, dx$

4. $\int_{-8}^{2} \frac{4}{(1 + e^x)^2}\, dx$

5. $\int_{-10}^{10} (0.002x^4 - 0.3x^2 + 4x - 7)\, dx$

Applications Involving Definite Integrals

Determining Total Profit

EXAMPLE 9 Business: Total Profit from Marginal Profit. Northeast Airlines determines that the marginal profit resulting from the sale of x seats on a jet traveling from Atlanta to Kansas City, in hundreds of dollars, is given by

$$P'(x) = \sqrt{x} - 6.$$

Find the total profit when 60 seats are sold.

Solution We integrate to find $P(60)$:

$$\begin{aligned} P(60) &= \int_0^{60} P'(x)\,dx \\ &= \int_0^{60} (\sqrt{x} - 6)\,dx \\ &= \left[\frac{2}{3}x^{3/2} - 6x\right]_0^{60} \\ &\approx -50.1613. \quad \text{Using a calculator} \end{aligned}$$

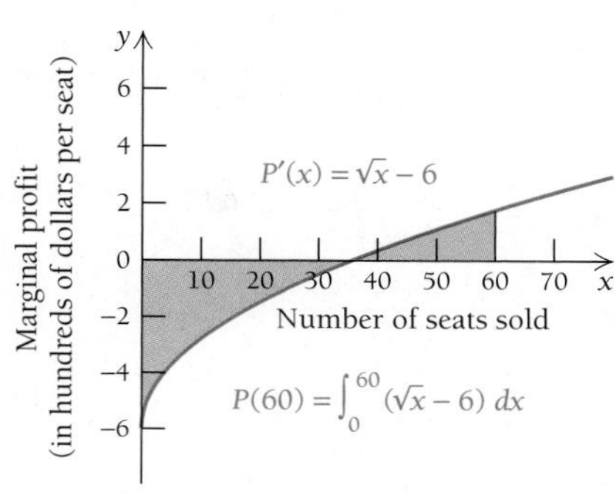

When 60 seats are sold, Northeast's profit is $-\$5016.13$. That is, the airline will lose \$5016.13 on the flight. ✦

Finding Velocity and Distance from Acceleration

Recall that the position coordinate at time t of an object moving along a number line is $s(t)$. Then

$s'(t) = v(t) =$ the **velocity** at time t,

$s''(t) = v'(t) = a(t) =$ the **acceleration** at time t.

EXAMPLE 10 Physical Science: Distance. Suppose that $v(t) = 5t^4$ and $s(0) = 9$. Find $s(t)$. Assume that $s(t)$ is in feet and $v(t)$ is in feet per second (ft/sec).

Solution We first find $s(t)$ by integrating:

$$s(t) = \int v(t)\,dt = \int 5t^4\,dt = t^5 + C.$$

Next we determine C by using the initial condition $s(0) = 9$, which is the starting position for s at time $t = 0$:

$$\begin{aligned} s(0) = 0^5 + C &= 9 \\ C &= 9. \end{aligned}$$

Thus, $s(t) = t^5 + 9$. ✦

EXAMPLE 11 Physical Science: Distance. Suppose that $a(t) = 12t^2 - 6$, with $v(0) =$ the initial velocity $= 5$, and $s(0) =$ the initial position $= 10$. Find $s(t)$.

Solution

1. We first find $v(t)$ by integrating $a(t)$:

$$\begin{aligned} v(t) &= \int a(t)\,dt \\ &= \int (12t^2 - 6)\,dt \\ &= 4t^3 - 6t + C_1. \end{aligned}$$

The condition $v(0) = 5$ allows us to find C_1:

$$\begin{aligned} v(0) = 4 \cdot 0^3 - 6 \cdot 0 + C_1 &= 5 \\ C_1 &= 5. \end{aligned}$$

Thus, $v(t) = 4t^3 - 6t + 5$.

2. Next we find $s(t)$ by integrating $v(t)$:

$$\begin{aligned} s(t) &= \int v(t)\,dt \\ &= \int (4t^3 - 6t + 5)\,dt \\ &= t^4 - 3t^2 + 5t + C_2. \end{aligned}$$

The condition $s(0) = 10$ allows us to find C_2:

$$\begin{aligned} s(0) = 0^4 - 3 \cdot 0^2 + 5 \cdot 0 + C_2 &= 10 \\ C_2 &= 10. \end{aligned}$$

Thus, $s(t) = t^4 - 3t^2 + 5t + 10$.

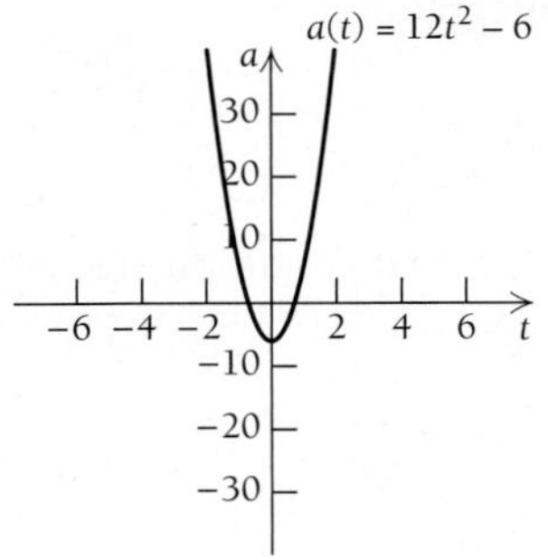

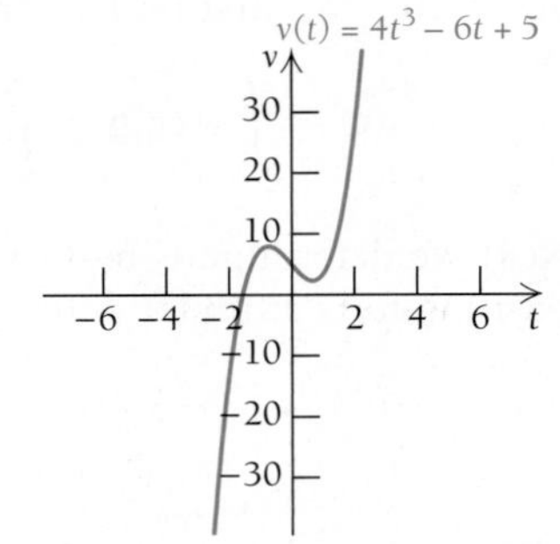

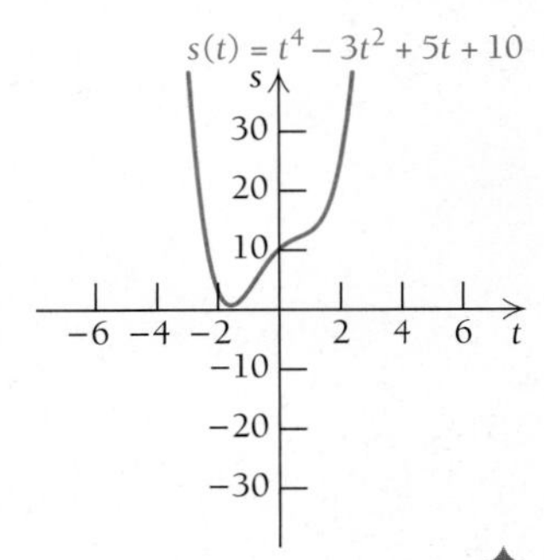

✦

EXAMPLE 12 Physical Science: Total Distance Traveled. A particle starts out from some origin. Its velocity, in miles per hour, is given by

$$v(t) = \sqrt{t} + t,$$

where t is the number of hours since the particle left the origin. How far does the particle travel during the second, third, and fourth hours (from $t = 1$ to $t = 4$)?

Solution Recall that velocity, or speed, is the rate of change of distance with respect to time. In other words, velocity is the derivative of the distance function, and the distance function is an antiderivative of the velocity function. To find the total distance traveled from $t = 1$ to $t = 4$, we evaluate the integral

$$\int_1^4 (\sqrt{t} + t)\, dt.$$

We have

$$\begin{aligned}
\int_1^4 (\sqrt{t} + t)\, dt &= \int_1^4 (t^{1/2} + t)\, dt \\
&= \left[\tfrac{2}{3} t^{3/2} + \tfrac{1}{2} t^2\right]_1^4 \\
&= \tfrac{2}{3} \cdot 4^{3/2} + \tfrac{1}{2} \cdot 4^2 - \left(\tfrac{2}{3} \cdot 1^{3/2} + \tfrac{1}{2} \cdot 1^2\right) \\
&= \tfrac{16}{3} + \tfrac{16}{2} - \tfrac{2}{3} - \tfrac{1}{2} \\
&= \tfrac{14}{3} + \tfrac{15}{2} \\
&= \tfrac{73}{6} \\
&= 12\tfrac{1}{6} \text{ mi.}
\end{aligned}$$

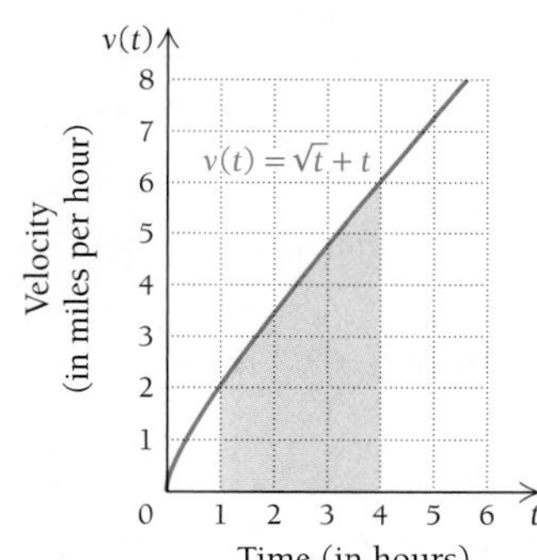

As a check, we can count shaded squares and parts of squares on the graph at the left. As another partial check, we can observe that the units used are mi/hr · hr = mi. ✦

Exercise Set 4.3

Find the area under the given curve over the indicated interval.

1. $y = 4$; $[1, 3]$

2. $y = 5$; $[1, 3]$

3. $y = 2x$; $[1, 3]$

4. $y = x^2$; $[0, 3]$

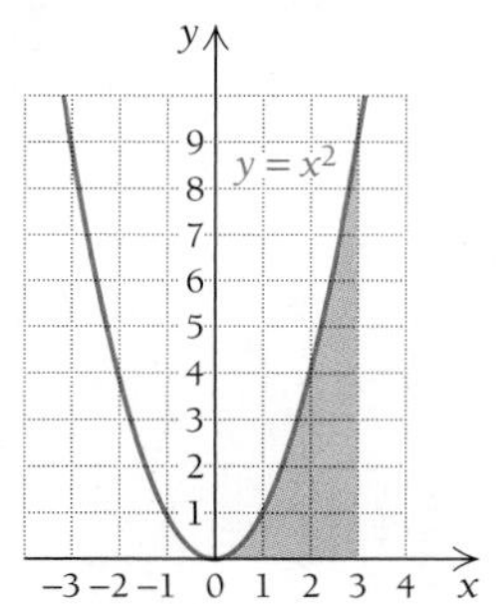

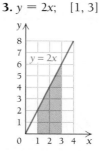

5. $y = x^2$; $[0, 5]$

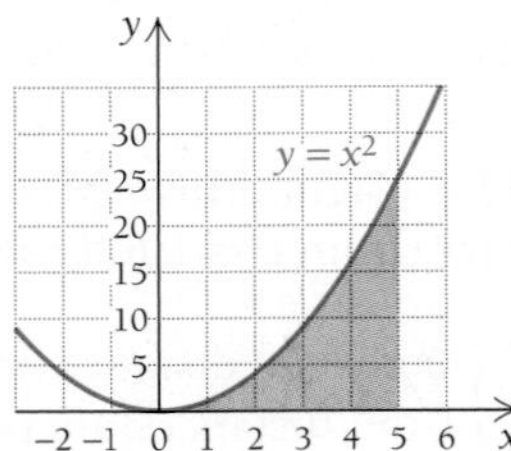

6. $y = x^3$; $[0, 2]$

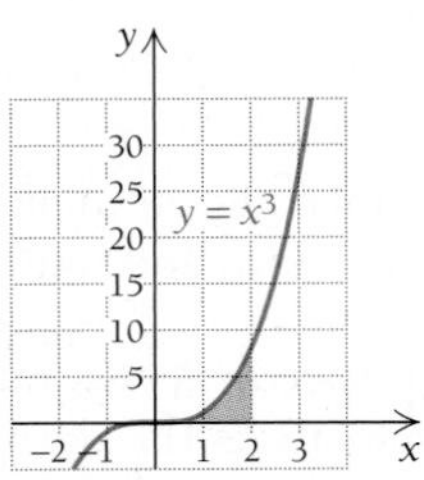

7. $y = x^3$; $[0, 1]$

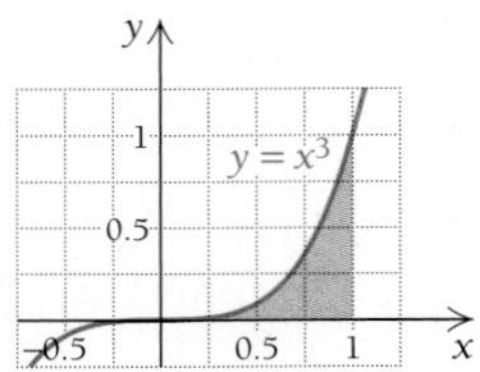

8. $y = 1 - x^2$; $[-1, 1]$

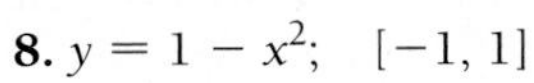

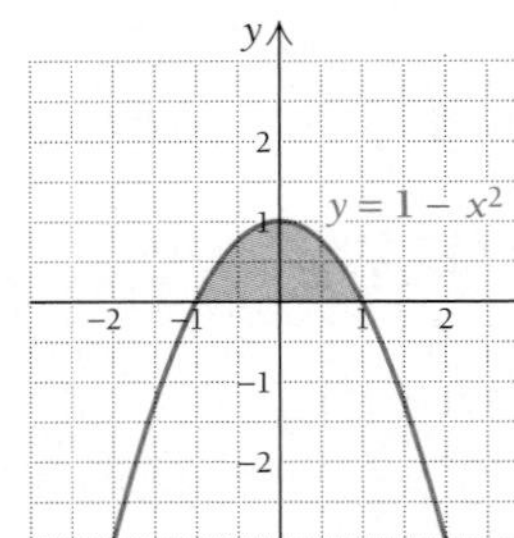

9. $y = 4 - x^2$; $[-2, 2]$

10. $y = e^x$; $[0, 2]$

11. $y = e^x$; $[0, 3]$

12. $y = \frac{2}{x}$; $[1, 4]$

13. $y = \frac{3}{x}$; $[1, 6]$

14. $y = x^2 - 4x$; $[-4, -2]$

In each of Exercises 15–24, explain what the shaded area represents.

15.

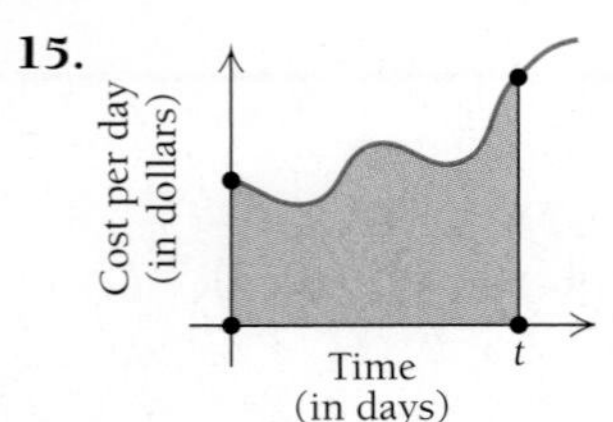

16.

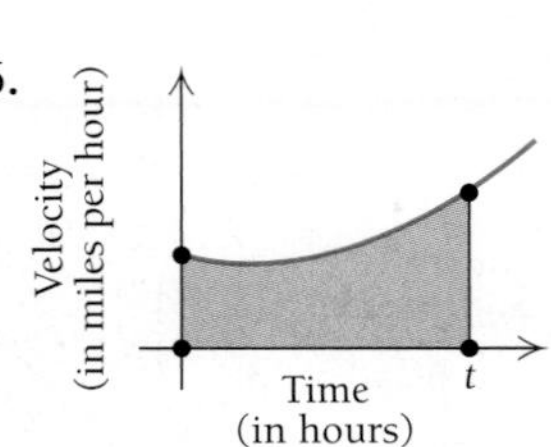

17.

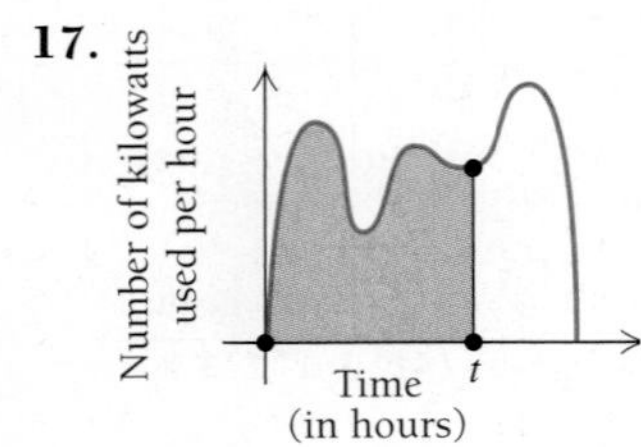

18.

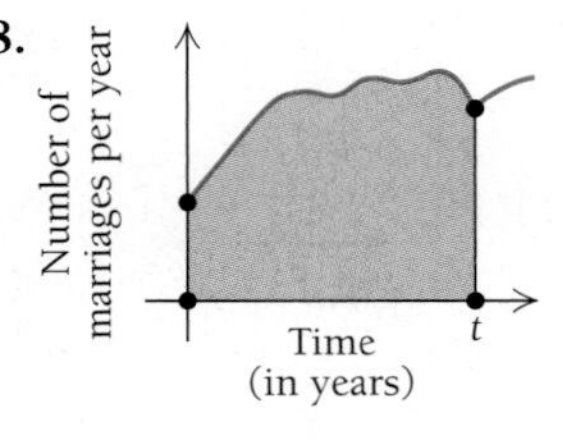

19.

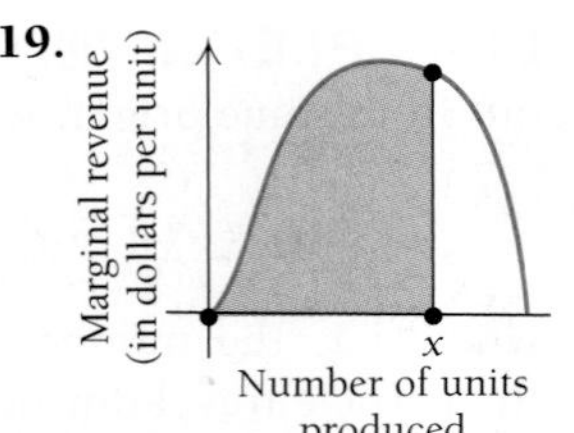

20.

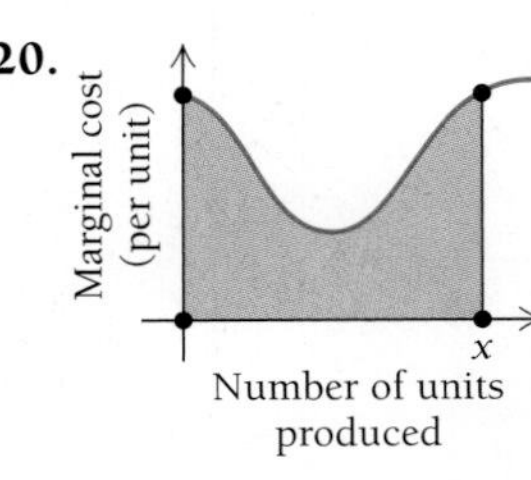

21.

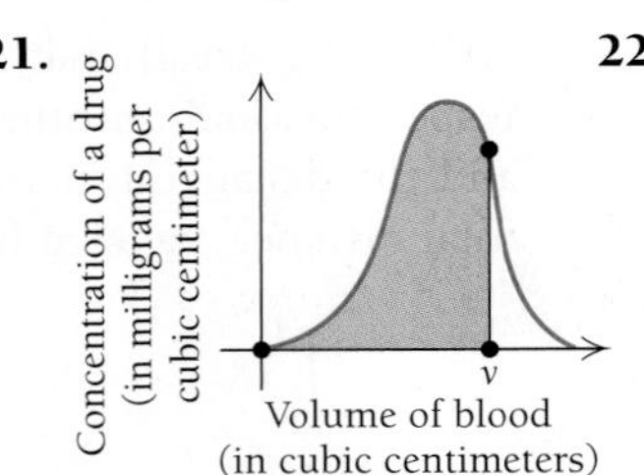

22.

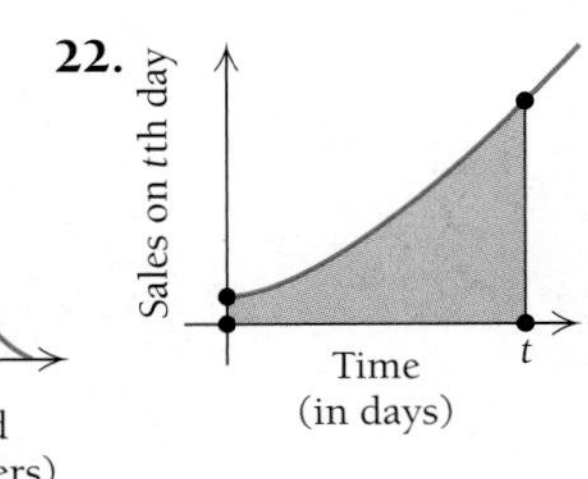

23. **24.**

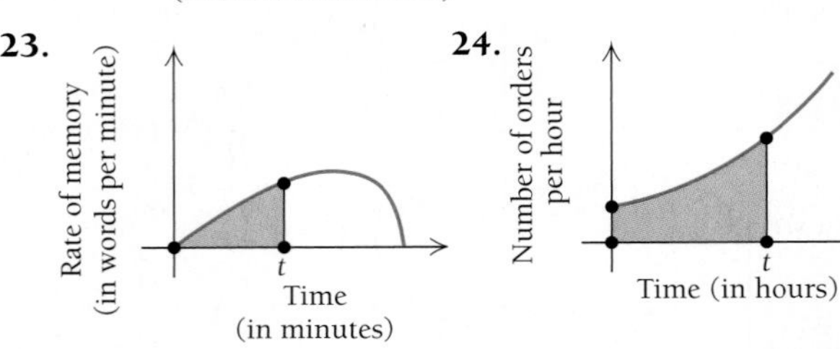

Find the area under the graph of each function over the interval given.

25. $y = x^3$; $[0, 2]$

26. $y = x^4$; $[0, 1]$

27. $y = x^2 + x + 1$; $[2, 3]$

28. $y = 2 - x - x^2$; $[-2, 1]$

29. $y = 5 - x^2$; $[-1, 2]$

30. $y = e^x$; $[-2, 3]$

31. $y = e^x$; $[-1, 5]$

32. $y = 2x + \frac{1}{x^2}$; $[1, 4]$

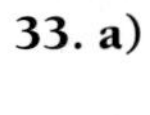

In Exercises 33 and 34, determine visually whether $\int_a^b f(x)\,dx$ is positive, negative, or zero, and express $\int_a^b f(x)\,dx$ in terms of the area A. Explain your result.

33. a)

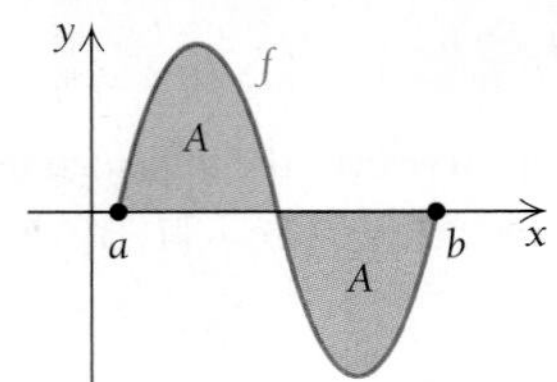

b)

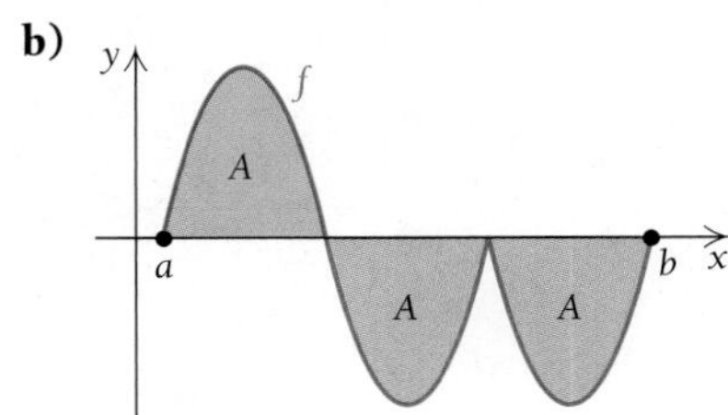

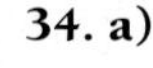

34. a)

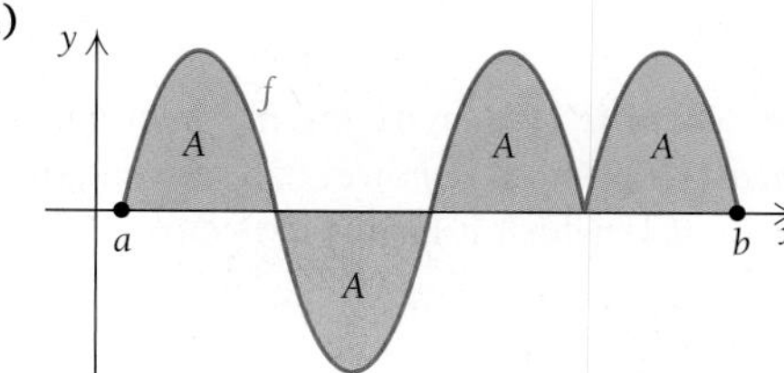

b)

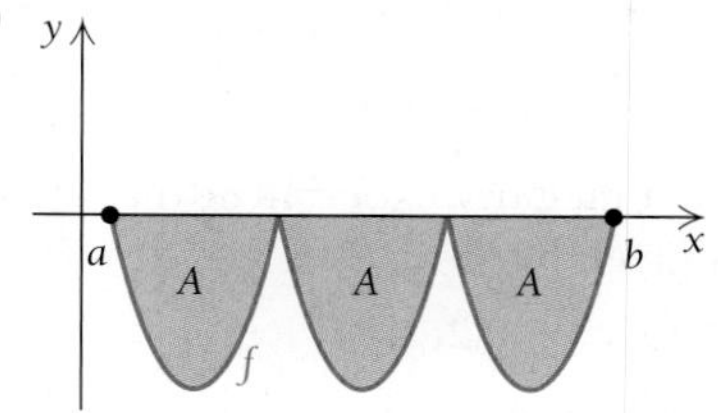

Evaluate. Then interpret the result in terms of the area above or below the x-axis.

35. $\int_0^{1.5} (x - x^2)\, dx$

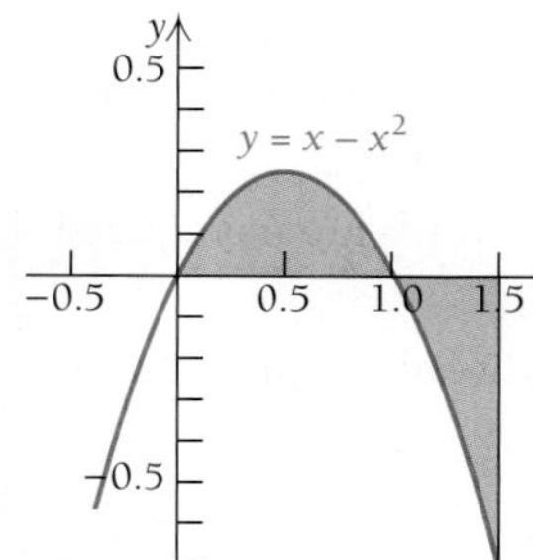

36. $\int_0^{2} (x^2 - x)\, dx$

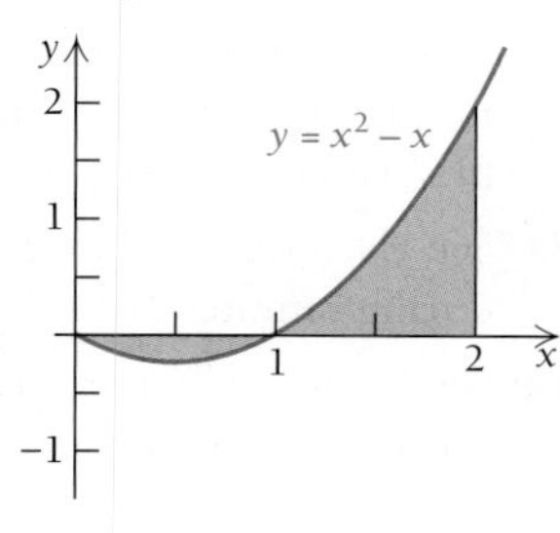

37. $\int_{-1}^{1} (x^4 - x^2)\, dx$

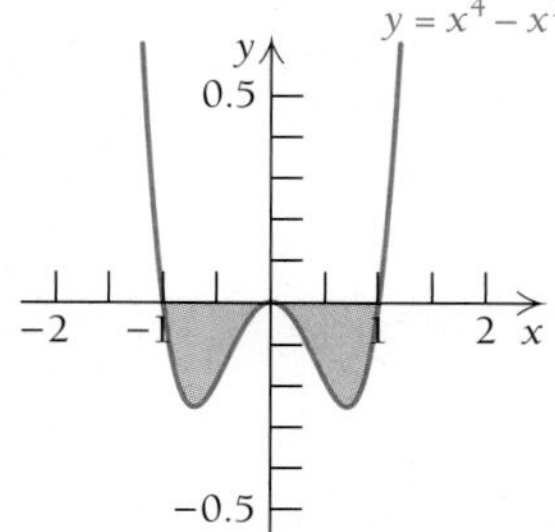

38. $\int_0^{b} -2e^{3x}\, dx$

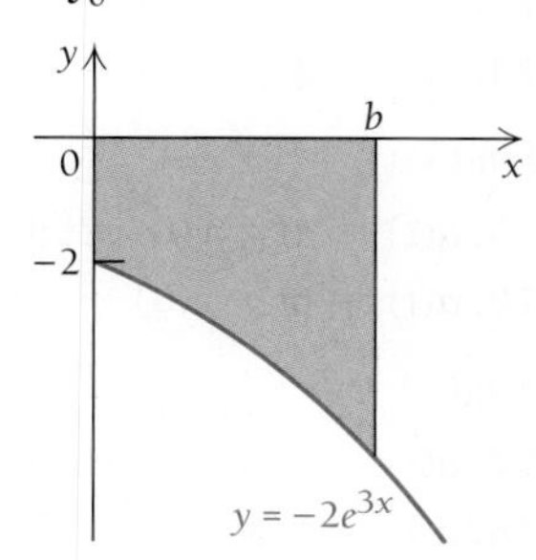

39–42. Check the results of each of Exercises 35–38 using a graphing calculator.

Evaluate.

43. $\int_1^{3} (3t^2 + 7)\, dt$

44. $\int_1^{2} (4t^3 - 1)\, dt$

45. $\int_1^{4} (\sqrt{x} - 1)\, dx$

46. $\int_1^{8} (\sqrt[3]{x} - 2)\, dx$

47. $\int_{-2}^{5} (2x^2 - 3x + 7)\, dx$

48. $\int_{-2}^{3} (-x^2 + 4x - 5)\, dx$

49. $\int_{-5}^{2} e^t\, dt$

50. $\int_{-2}^{3} e^{-t}\, dt$

51. $\int_a^{b} \frac{1}{2}x^2\, dx$

52. $\int_a^{b} \frac{1}{5}x^3\, dx$

53. $\int_a^{b} e^{2t}\, dt$

54. $\int_a^{b} -e^t\, dt$

55. $\int_1^{e} \left(x + \frac{1}{x}\right) dx$

56. $\int_1^{e} \left(x - \frac{1}{x}\right) dx$

57. $\int_0^{2} \sqrt{2x}\, dx$ (*Hint:* Simplify first.)

58. $\int_0^{27} \sqrt{3x}\, dx$

APPLICATIONS

Business and Economics

59. Business: total profit. Pure Water Enterprises finds that the marginal profit, in dollars, from drilling a well that is x feet deep is given by

$$P'(x) = \sqrt[5]{x}.$$

Find the profit when a well 250 ft deep is drilled.

60. Business: total revenue. Sally's Sweets finds that the marginal revenue, in dollars, from the sale of x pounds of maple-coated pecans is given by

$$R'(x) = 6x^{-1/6}.$$

Find the revenue when 300 lb of maple-coated pecans are produced.

61. Business: increasing total cost. Kitchens-to-Please Contracting determines that the marginal cost, in dollars per foot, of installing x feet of kitchen countertop is given by

$$C'(x) = 8x^{-1/3}.$$

Find the cost of installing an extra 14 ft of countertop after 50 ft have already been ordered.

62. Business: increasing total profit. Laso Industries finds that the marginal profit, in dollars, from the sale of x digital control boards is given by

$$P'(x) = 2.6x^{0.1}.$$

A customer orders 1200 digital control boards and later increases the order to 1500. Find the extra profit resulting from the increase in order size.

63. Accumulated sales. A company estimates that its sales will grow continuously at a rate given by the function

$$S'(t) = 20e^t,$$

where $S'(t)$ is the rate at which sales are increasing, in dollars per day, on day t.

a) Find the accumulated sales for the first 5 days.

b) Find the sales from the 2nd day through the 5th day. (This is the integral from 1 to 5.)

64. Accumulated sales. Raggs, Ltd., estimates that its sales will grow continuously at a rate given by the function

$$S'(t) = 10e^t,$$

where $S'(t)$ is the rate at which sales are increasing, in dollars per day, on day t.

a) Find the accumulated sales for the first 5 days.

b) Find the sales from the 2nd day through the 5th day. (This is the integral from 1 to 5.)

Credit market debt. *The annual rate of change in the national credit market debt (in billions of dollars per year) can be modeled by the function*

$$D'(t) = 857.98 + 829.66t - 197.34t^2 + 15.36t^3,$$

where t is the number of years since 1995. (**Source:** *Based on data from the Federal Reserve System.) Use the preceding information for Exercises 65 and 66.*

65. By how much did the credit market debt increase between 1996 and 2000?

66. By how much did the credit market debt increase between 1999 and 2005?

Industrial learning curve. *A company is producing a new product. Due to the nature of the product, the time required to produce each unit decreases as workers become more familiar with the production procedure. It is determined that the function for the learning process is*

$$T(x) = 2 + 0.3\left(\frac{1}{x}\right),$$

where T(x) is the time, in hours, required to produce the xth unit. Use this information for Exercises 67 and 68.

67. Find the total time required for a new worker to produce units 1 through 10; units 20 through 30.

68. Find the total time required for a new worker to produce units 1 through 20; units 20 through 40.

Social Sciences

Memorizing. *The rate of memorizing information initially increases. Eventually, however, a maximum rate is reached, after which it begins to decrease.*

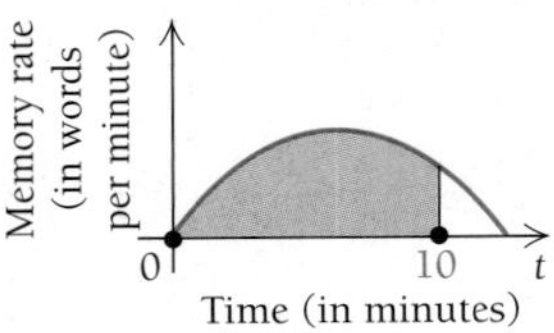

69. Suppose that in a memory experiment the rate of memorizing is given by

$$M'(t) = -0.009t^2 + 0.2t,$$

where $M'(t)$ is the memory rate, in words per minute. How many words are memorized in the first 10 min (from $t = 0$ to $t = 10$)?

70. Suppose that in another memory experiment the rate of memorizing is given by

$$M'(t) = -0.003t^2 + 0.2t,$$

where $M'(t)$ is the memory rate, in words per minute. How many words are memorized in the first 10 min (from $t = 0$ to $t = 10$)?

71. See Exercise 69. How many words are memorized during minutes 10–15?

72. See Exercise 70. How many words are memorized during minutes 10–17?

Life and Physical Sciences

Find s(t).

73. $v(t) = 3t^2, \quad s(0) = 4$

74. $v(t) = 2t, \quad s(0) = 10$

Find v(t).

75. $a(t) = 4t, \quad v(0) = 20$

76. $a(t) = 6t, \quad v(0) = 30$

Find s(t).

77. $a(t) = -2t + 6$, with $v(0) = 6$ and $s(0) = 10$

78. $a(t) = -6t + 7$, with $v(0) = 10$ and $s(0) = 20$

79. Physics. A particle is released as part of an experiment. Its speed t seconds after release is given by $v(t) = -0.5t^2 + 10t$, where $v(t)$ is in meters per second.

a) How far does the particle travel during the first 5 sec?

b) How far does it travel during the second 5 sec?

80. Physics. A particle is released during an experiment. Its speed t minutes after release is given by $v(t) = -0.3t^2 + 9t$, where $v(t)$ is in kilometers per minute.

a) How far does the particle travel during the first 10 min?
b) How far does it travel during the second 10 min?

81. Distance and speed. A motorcycle accelerates at a constant rate from 0 mph ($v(0) = 0$) to 60 mph in 15 sec.

a) How fast is it traveling after 15 sec?
b) How far has it traveled after 15 sec? (*Hint:* Convert seconds to hours.)

82. Distance and speed. A car accelerates at a constant rate from 0 mph to 60 mph in 30 sec.

a) How fast is it traveling after 30 sec?
b) How far has it traveled after 30 sec?

83. Distance and speed. A bicyclist accelerates at a constant rate from a standstill to 30 km/hr in 45 sec.

a) How fast is the bicyclist traveling after 20 sec?
b) How far has the bicyclist traveled after 45 sec?

84. Distance and speed. A cheetah accelerates at a constant rate from 0 to 50 km/hr in 20 sec.

a) How fast is the cheetah moving after 10 sec?
b) How far has the cheetah traveled after 20 sec?

85. Distance. For a freely falling object, $a(t) = -32$ ft/sec^2, $v(0) =$ initial velocity $= v_0$ (in ft/sec), and $s(0) =$ initial height $= s_0$ (in ft). Find a general expression for $s(t)$ in terms of v_0 and s_0.

86. Time. A ball is thrown upward from a height of 10 ft, that is, $s(0) = 10$, at an initial velocity of 80 ft/sec, or $v(0) = 80$. How long will it take before the ball hits the ground? (See Exercise 85.)

87. Distance. A car accelerates at a constant rate from 0 to 60 mph in $\frac{1}{2}$ min. How far does the car travel during that time?

88. Distance. A motorcycle accelerates at a constant rate from 0 to 50 mph in 15 sec. How far does it travel during that time?

89. Physics. A particle starts out from the origin. Its velocity, in miles per hour, after t hours is given by

$$v(t) = 3t^2 + 2t.$$

How far does it travel from the 2nd hour through the 5th hour (from $t = 1$ to $t = 5$)?

90. Physics. A particle starts out from the origin. Its velocity, in miles per hour, after t hours is given by

$$v(t) = 4t^3 + 2t.$$

How far does it travel from the start through the 3rd hour (from $t = 0$ to $t = 3$)?

SYNTHESIS

91. Accumulated sales. Bluetape, Inc. estimates that its sales will grow continuously at a rate given by

$$S'(t) = 0.5e^t,$$

where $S'(t)$ is the rate at which sales are increasing, in dollars per day, on day t. On what day will accumulated sales first exceed $10,000?

92. Total pollution. A factory is polluting a lake in such a way that the rate of pollutants entering the lake at time t, in months, is given by

$$N'(t) = 280t^{3/2},$$

where N is the total number of pounds of pollutants in the lake at time t.

a) How many pounds of pollutants enter the lake in 16 months?
b) An environmental board tells the factory that it must begin cleanup procedures after 50,000 lb of pollutants have entered the lake. After what length of time will this occur?

Evaluate.

93. $\displaystyle\int_2^3 \frac{x^2 - 1}{x - 1}\,dx$

94. $\displaystyle\int_1^5 \frac{x^5 - x^{-1}}{x^2}\,dx$

95. $\displaystyle\int_4^{16} (x - 1)\sqrt{x}\,dx$

96. $\displaystyle\int_0^1 (x + 2)^3\,dx$

97. $\displaystyle\int_1^8 \frac{\sqrt[3]{x^2} - 1}{\sqrt[3]{x}}\,dx$

98. $\displaystyle\int_0^1 \frac{x^3 + 8}{x + 2}\,dx$

99. $\displaystyle\int_2^5 (t + \sqrt{3})(t - \sqrt{3})\,dt$

100. $\int_0^1 (t+1)^3\, dt$

101. $\int_1^3 \left(x - \frac{1}{x}\right)^2 dx$

102. $\int_1^3 \frac{t^5 - t}{t^3}\, dt$

103. $\int_4^9 \frac{t+1}{\sqrt{t}}\, dt$

TW *Find the error in each of the following. Explain.*

104. $$\int_1^2 (x^2 + x + 1)\, dx = \left[\frac{1}{3}x^3 + \frac{1}{2}x^2 + x\right]_1^2$$
$$= \left(\frac{1}{3} \cdot 2^3 + \frac{1}{2} \cdot 2^2 + 2\right)$$
$$= \frac{20}{3}$$

105. $$\int_1^2 (\ln x - e^x)\, dx = \left[\frac{1}{x} - e^x\right]_1^2$$
$$= \left(\frac{1}{2} - e^2\right) - (1 - e^1)$$
$$= e - e^2 - \frac{1}{2}$$

TECHNOLOGY CONNECTION

Evaluate.

106. $\int_{-1.2}^{6.3} (x^3 - 9x^2 + 27x + 50)\, dx$

107. $\int_{-8}^{1.4} (x^4 + 4x^3 - 36x^2 - 160x + 300)\, dx$

108. $\int_{-2}^{2} \sqrt{4 - x^2}\, dx$

109. $\int_{-1}^{1} (3 + \sqrt{1 - x^2})\, dx$

110. $\int_0^8 x(x-5)^4\, dx$

111. $\int_{-2}^{2} x^{2/3}\left(\frac{5}{2} - x\right) dx$

112. $\int_2^4 \frac{x^2 - 4}{x^2 - 3}\, dx$

113. $\int_{-10}^{10} \frac{8}{x^2 + 4}\, dx$

TW 114. Prove that $\int_a^b f(x)\, dx = -\int_b^a f(x)\, dx$.

4.4 Properties of Definite Integrals

OBJECTIVES

- Use properties of definite integrals to find the area between curves.
- Solve applied problems involving definite integrals.
- Determine the average value of a function.

The Area under a Curve

We have seen that the definite integral

$$\int_a^c f(x)\, dx$$

can be regarded as the area under the graph of $y = f(x)$ over the interval $[a, c]$. Thus, if b is such that $a < b < c$, the above integral can be expressed as a sum, as stated in the following theorem.

THEOREM 5

For $a < b < c$,

$$\int_a^c f(x)\, dx = \int_a^b f(x)\, dx + \int_b^c f(x)\, dx.$$

For any number b between a and c, the integral from a to c is the integral from a to b plus the integral from b to c.

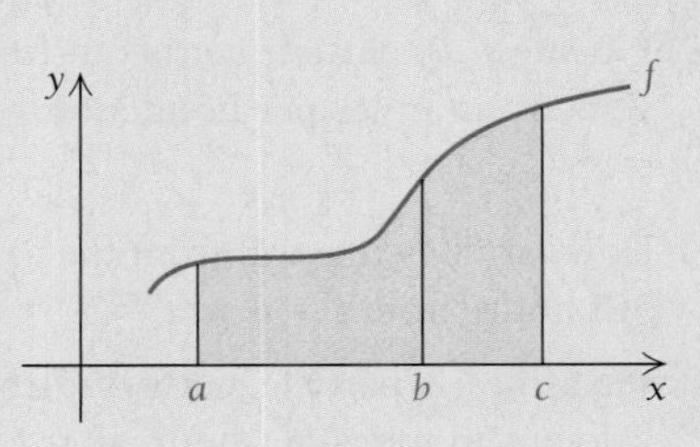

Theorem 5 is especially useful when a function is defined piecewise, in different ways over subintervals.

EXAMPLE 1 Find the area under the graph of $y = f(x)$ from -4 to 5, where

$$f(x) = \begin{cases} 9, & \text{for } x < 3, \\ x^2, & \text{for } x \geq 3. \end{cases}$$

Solution

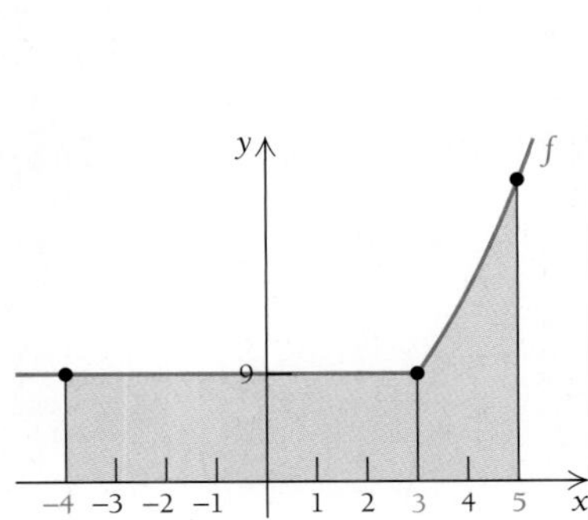

$$\begin{aligned}\int_{-4}^{5} f(x)\,dx &= \int_{-4}^{3} f(x)\,dx + \int_{3}^{5} f(x)\,dx \\ &= \int_{-4}^{3} 9\,dx + \int_{3}^{5} x^2\,dx \\ &= 9[x]_{-4}^{3} + \left[\frac{x^3}{3}\right]_3^5 \\ &= 9(3-(-4)) + \left(\frac{5^3}{3} - \frac{3^3}{3}\right) \\ &= 95\tfrac{2}{3}\end{aligned}$$

✦

The Area of a Region Bounded by Two Graphs

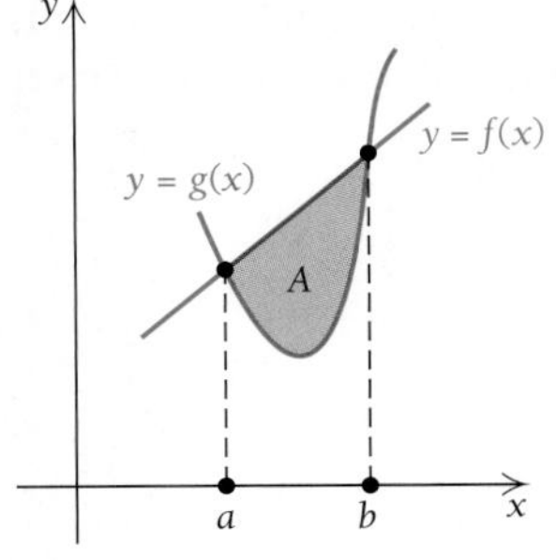

Suppose that we want to find the area of a region bounded by the graphs of two functions, $y = f(x)$ and $y = g(x)$, as shown at the left.

Note that the area of the desired region A is the area of A_2 minus that of A_1.

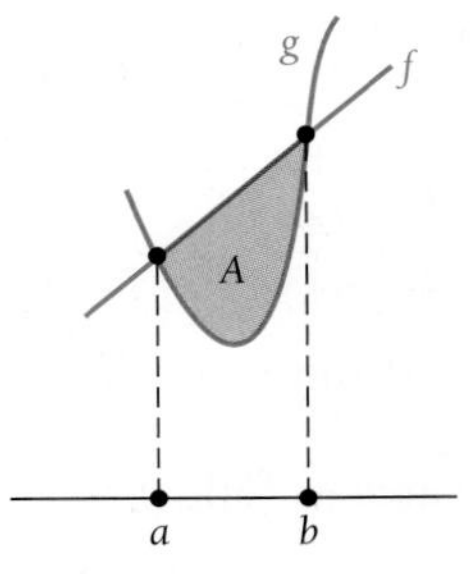

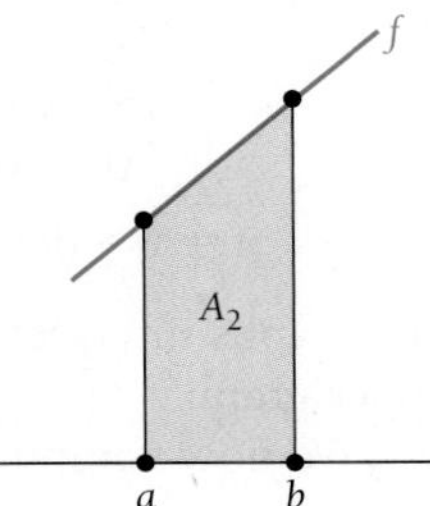

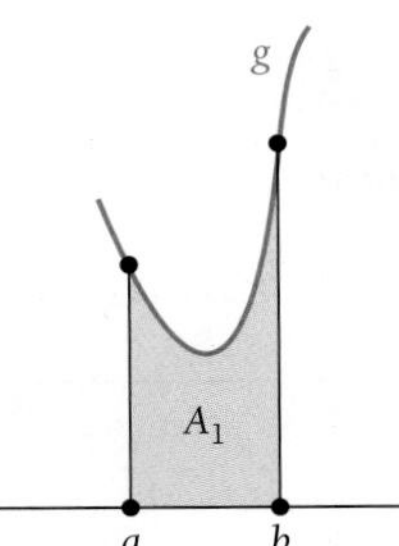

Thus,

$$A = \int_a^b f(x)\,dx - \int_a^b g(x)\,dx,$$

or

$$A = \int_a^b [f(x) - g(x)]\,dx.$$

In general, we have the following.

THEOREM 6

Let f and g be continuous functions and suppose that $f(x) \geq g(x)$ over the interval $[a, b]$. Then the area of the region between the two curves, from $x = a$ to $x = b$, is

$$\int_a^b [f(x) - g(x)]\, dx.$$

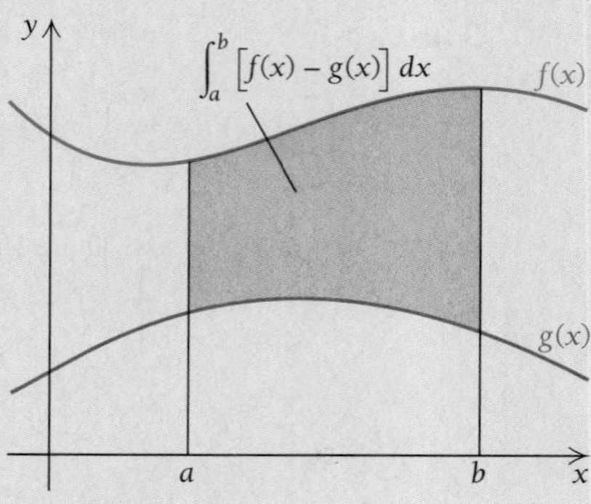

EXAMPLE 2 Find the area of the region that is bounded by the graphs of $f(x) = 2x + 1$ and $g(x) = x^2 + 1$.

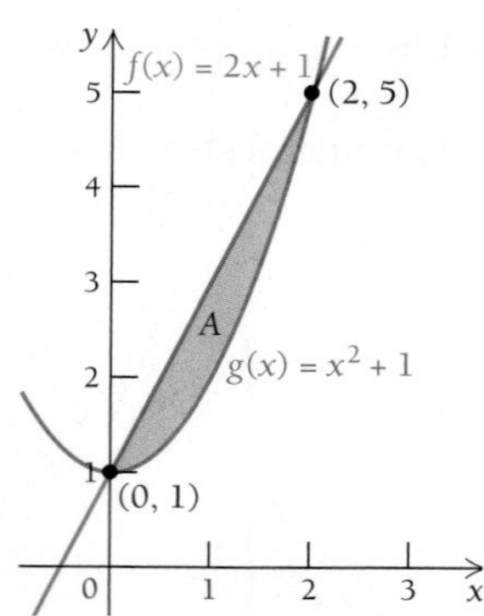

Solution First, we make a reasonably accurate sketch, as in the figure at the left, to determine which is the upper graph. To calculate the points of intersection, we set $f(x)$ equal to $g(x)$ and solve.

$$\begin{aligned} f(x) &= g(x) \\ 2x + 1 &= x^2 + 1 \\ 0 &= x^2 - 2x \\ 0 &= x(x - 2) \\ x = 0 \quad &\text{or} \quad x = 2 \end{aligned}$$

The graphs intersect at $x = 0$ and $x = 2$. We see that, over the interval $[0, 2]$, f is the upper graph.

We now compute the area as follows:

$$\begin{aligned} \int_0^2 [(2x + 1) - (x^2 + 1)]\, dx &= \int_0^2 (2x - x^2)\, dx \\ &= \left[x^2 - \frac{x^3}{3}\right]_0^2 \\ &= \left(2^2 - \frac{2^3}{3}\right) - \left(0^2 - \frac{0^3}{3}\right) \\ &= 4 - \frac{8}{3} \\ &= \frac{4}{3}. \end{aligned}$$

✦

TECHNOLOGY CONNECTION

To find the area bounded by two graphs, such as $y_1 = -2x - 7$ and $y_2 = -x^2 - 4$, we graph each function and use INTERSECT from the CALC menu to determine the points of intersection.

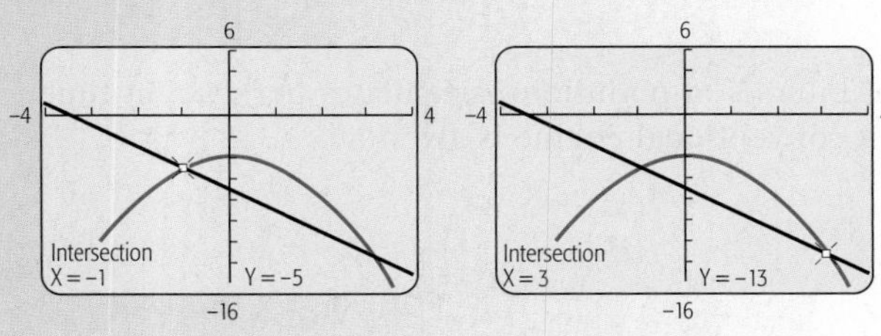

Next, we use fnInt from the MATH menu.

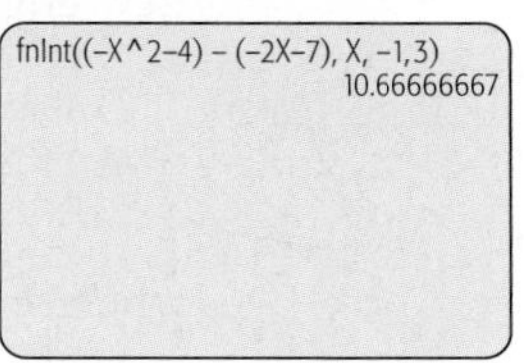

The area bounded by the two curves is about 10.7.

EXERCISE

1. Find the area of the region in Example 2 using this approach.

EXAMPLE 3 Find the area of the region bounded by

$$y = x^4 - 3x^3 - 4x^2 + 10, \qquad y = 40 - x^2, \qquad x = 1, \quad \text{and} \quad x = 3.$$

Solution First, we make a reasonably accurate sketch, as in the figure below, to ensure that we have the correct configuration. Note that over $[1, 3]$, the upper graph is $y = 40 - x^2$. Thus, $40 - x^2 \geq x^4 - 3x^3 - 4x^2 + 10$ over $[1, 3]$.

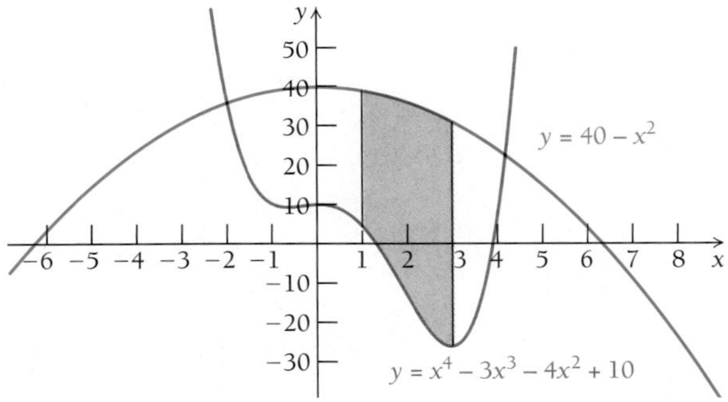

By subtracting a negative number (the area between the red curve and the x-axis), we are adding areas. The limits of integration are stated, so we can compute the area as follows:

$$\int_1^3 [(40 - x^2) - (x^4 - 3x^3 - 4x^2 + 10)]\,dx = \int_1^3 (-x^4 + 3x^3 + 3x^2 + 30)\,dx$$

$$= \left[-\frac{x^5}{5} + \frac{3}{4}x^4 + x^3 + 30x\right]_1^3$$

$$= \left(-\frac{3^5}{5} + \frac{3}{4}\cdot 3^4 + 3^3 + 30\cdot 3\right)$$

$$-\left(-\frac{1^5}{5} + \frac{3}{4}\cdot 1^4 + 1^3 + 30\cdot 1\right)$$

$$= 97.6.$$

✦

An Application

EXAMPLE 4 Life Science: Emission Control. A clever college student develops an engine that is believed to meet all state standards for emission control. The new engine's rate of emission is given by

$$E(t) = 2t^2,$$

where $E(t)$ is the emissions, in billions of pollution particulates per year, at time t, in years. The emission rate of a conventional engine is given by

$$C(t) = 9 + t^2.$$

The graphs of both curves are shown at the right.

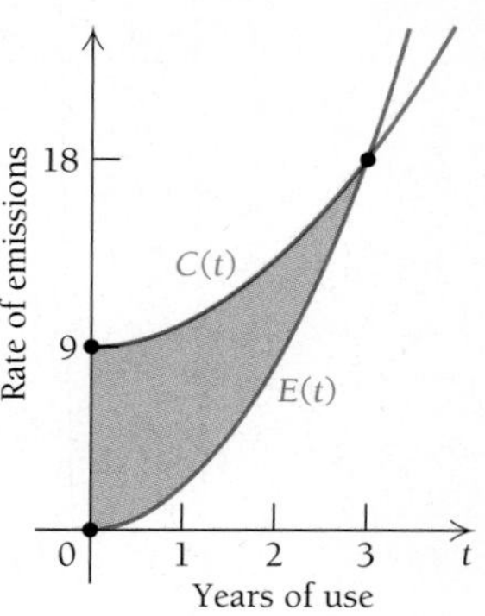

a) At what point in time will the emission rates be the same?

b) What reduction in emissions results from using the student's engine?

Solution

a) The rate of emission will be the same when $E(t) = C(t)$, or

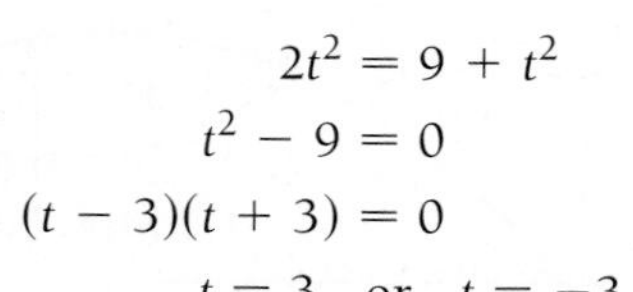

$$\begin{aligned} 2t^2 &= 9 + t^2 \\ t^2 - 9 &= 0 \\ (t - 3)(t + 3) &= 0 \\ t = 3 \quad &\text{or} \quad t = -3. \end{aligned}$$

Since negative time has no meaning in this problem, the emission rates will be the same when $t = 3$ yr.

b) The reduction in emissions is represented by the area of the shaded region in the figure above. It is the area between $C(t) = 9 + t^2$ and $E(t) = 2t^2$, from $t = 0$ to $t = 3$, and is computed as follows:

$$\begin{aligned} \int_0^3 [(9 + t^2) - 2t^2]\,dt &= \int_0^3 (9 - t^2)\,dt \\ &= \left[9t - \frac{t^3}{3}\right]_0^3 \\ &= \left(9 \cdot 3 - \frac{3^3}{3}\right) - \left(9 \cdot 0 - \frac{0^3}{3}\right) \\ &= 27 - 9 \\ &= 18 \text{ billion pollution particulates.} \end{aligned}$$

✦

Average Value of a Continuous Function

Another important use of the area under a curve is in finding the average value of a continuous function over a closed interval.

Suppose that

$$T = f(t)$$

is the temperature at time t recorded at a weather station on a certain day. The station uses a 24-hr clock, so the domain of the temperature function is the interval $[0, 24]$. The function is continuous, as shown below.

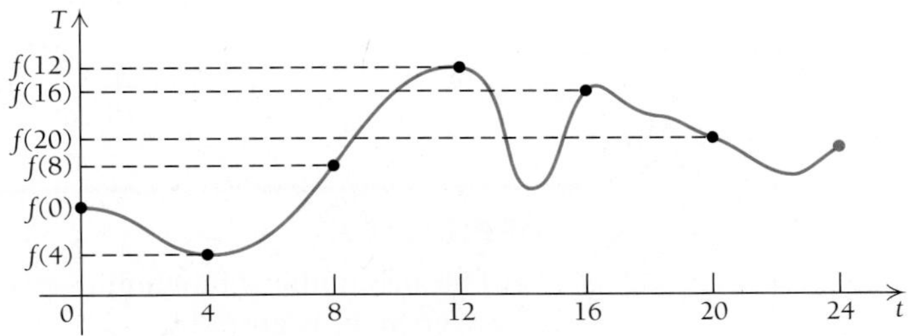

To find the average temperature for the day, we might take six temperature readings at 4-hr intervals, starting at midnight:

$$T_0 = f(0), \qquad T_1 = f(4),$$
$$T_2 = f(8), \qquad T_3 = f(12),$$
$$T_4 = f(16), \qquad T_5 = f(20).$$

The average reading would then be the sum of these six readings divided by 6:

$$T_{av} = \frac{T_0 + T_1 + T_2 + T_3 + T_4 + T_5}{6}.$$

This computation of the average temperature has limitations. For example, suppose that it is a hot summer day, and at 2:00 in the afternoon (hour 14 on the 24-hr clock), there is a short thunderstorm that cools the air for an hour between our readings. This temporary dip would not show up in the average computed above.

What can we do? We could take 48 readings at half-hour intervals. This should give us a better result. In fact, the shorter the time between readings, the better the result should be. It seems reasonable that we might define the **average value** of T over $[0, 24]$ to be the limit, as n approaches ∞, of the average of n values:

$$\text{Average value of } T = \lim_{n\to\infty}\left(\frac{1}{n}\sum_{i=1}^{n} T_i\right)$$
$$= \lim_{n\to\infty}\left(\frac{1}{n}\sum_{i=1}^{n} f(t_i)\right).$$

Note that this is not too far from our definition of an integral. All we need is to get Δt, which is $(24 - 0)/n$, or $24/n$, into the summation. We accomplish this by multiplying by 1, writing 1 as $\frac{1}{\Delta t} \cdot \Delta t$:

$$\begin{aligned}
\text{Average value of } T &= \lim_{n\to\infty}\left(\frac{1}{\Delta t}\cdot\frac{1}{n}\sum_{i=1}^{n} f(t_i)\,\Delta t\right) \\
&= \lim_{n\to\infty}\left(\frac{n}{24}\cdot\frac{1}{n}\sum_{i=1}^{n} f(t_i)\,\Delta t\right) \qquad \Delta t = \frac{24}{n}, \text{ so } \frac{1}{\Delta t} = \frac{n}{24} \\
&= \frac{1}{24}\lim_{n\to\infty}\sum_{i=1}^{n} f(t_i)\,\Delta t \\
&= \frac{1}{24}\int_0^{24} f(t)\,dt.
\end{aligned}$$

DEFINITION

Let f be a continuous function over a closed interval $[a, b]$. Its **average value, y_{av},** over $[a, b]$ is given by

$$y_{av} = \frac{1}{b-a}\int_a^b f(x)\,dx.$$

Let's consider average value in another way. If we multiply both sides of

$$y_{av} = \frac{1}{b-a}\int_a^b f(x)\,dx$$

by $b - a$, we get

$$(b-a)y_{av} = \int_a^b f(x)\,dx.$$

Now the expression on the left side is the area of a rectangle of length $b - a$ and height y_{av}. The area of such a rectangle is the same as the area under the graph of $y = f(x)$ over the interval $[a, b]$, as shown in the figure.

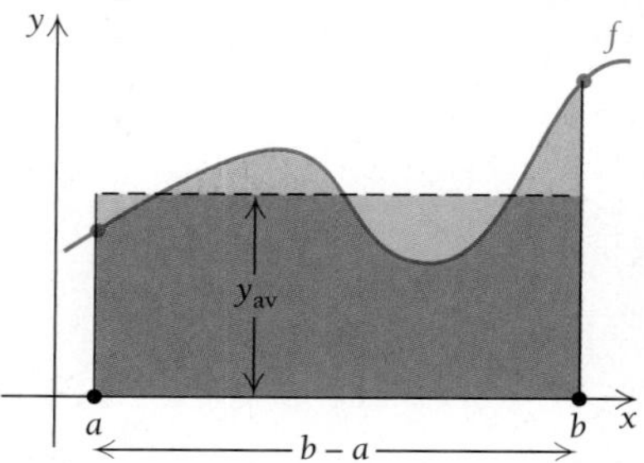

TECHNOLOGY CONNECTION

EXERCISE

1. Graph $f(x) = x^4$. Compute the average value of the function over the interval $[0, 2]$, using the method of Example 5. Then use that value, y_{av}, and draw a graph of it as a horizontal line using the same set of axes. What does this line represent in comparison to the graph of $f(x) = x^4$ and its associated area?

EXAMPLE 5 Find the average value of $f(x) = x^2$ over the interval $[0, 2]$.

Solution The average value is

$$\begin{aligned}
\frac{1}{2-0}\int_0^2 x^2\,dx &= \frac{1}{2}\left[\frac{x^3}{3}\right]_0^2 \\
&= \frac{1}{2}\left(\frac{2^3}{3} - \frac{0^3}{3}\right) \\
&= \frac{1}{2}\cdot\frac{8}{3} = \frac{4}{3}, \quad \text{or} \quad 1\frac{1}{3}.
\end{aligned}$$

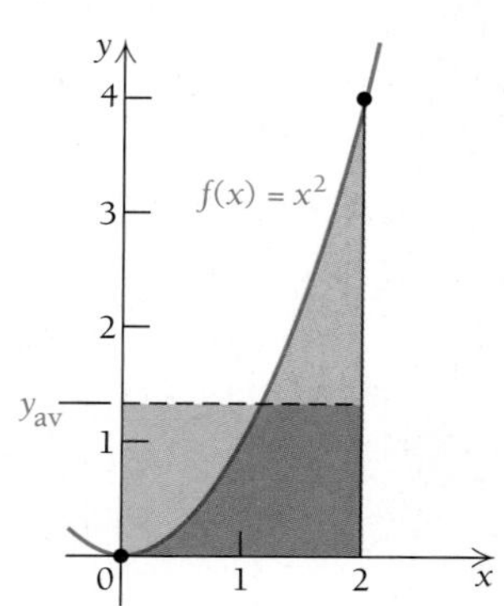

Note that although the values of $f(x)$ increase from 0 to 4 over $[0, 2]$, we do not expect the average value to be 2 (which is half of 4), because we see from the graph that $f(x)$ is less than 2 over more than half the interval. ✦

EXAMPLE 6 Rico's speed, in miles per hour, t minutes after entering the freeway, is given by

$$v(t) = -\frac{1}{200}t^3 + \frac{3}{20}t^2 - \frac{3}{8}t + 60, \quad t \leq 30.$$

From 5 min after entering the freeway to 25 min, what was Rico's average speed? How far did he travel over that time interval?

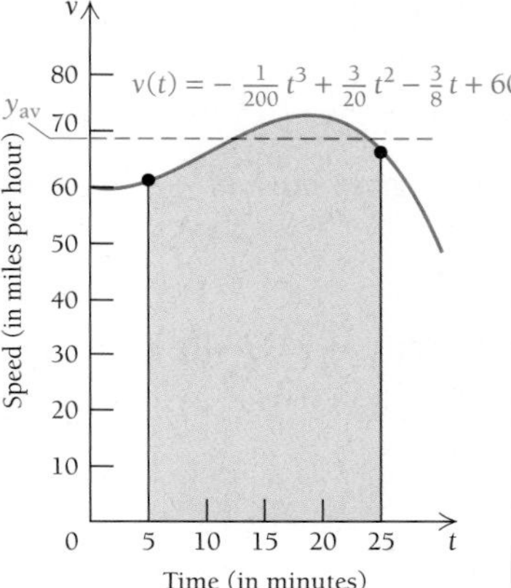

Solution The average speed is

$$\begin{aligned}
&\frac{1}{25-5}\int_5^{25}\left(-\frac{1}{200}t^3 + \frac{3}{20}t^2 - \frac{3}{8}t + 60\right)dt \\
&= \frac{1}{20}\left[-\frac{1}{800}t^4 + \frac{1}{20}t^3 - \frac{3}{16}t^2 + 60t\right]_5^{25} \\
&= \frac{1}{20}\left[\left(-\frac{1}{800}\cdot 25^4 + \frac{1}{20}\cdot 25^3 - \frac{3}{16}\cdot 25^2 + 60\cdot 25\right)\right. \\
&\qquad \left. -\left(-\frac{1}{800}\cdot 5^4 + \frac{1}{20}\cdot 5^3 - \frac{3}{16}\cdot 5^2 + 60\cdot 5\right)\right] \\
&= \frac{1}{20}\left(\frac{53{,}625}{32} - \frac{9625}{32}\right) \\
&= 68\tfrac{3}{4} \text{ mph.}
\end{aligned}$$

To find how far Rico traveled over the time interval $[5, 25]$, we first note that t is given in minutes, not hours. Since 25 min − 5 min = 20 min is $\frac{1}{3}$ hr, the distance traveled over $[5, 25]$ is

$$\tfrac{1}{3}\cdot 68\tfrac{3}{4} = 22\tfrac{11}{12} \text{ mi.}$$ ✦

Exercise Set 4.4

Find the area under the graph of f over the interval $[1, 5]$.

1. $f(x) = \begin{cases} 2x + 1, & \text{for } x \leq 3, \\ 10 - x, & \text{for } x > 3 \end{cases}$

2. $f(x) = \begin{cases} x + 5, & \text{for } x \leq 4, \\ 11 - \frac{1}{2}x, & \text{for } x > 4 \end{cases}$

Find the area under the graph of g over the interval $[-2, 3]$.

3. $g(x) = \begin{cases} x^2 + 4, & \text{for } x \leq 0, \\ 4 - x, & \text{for } x > 0 \end{cases}$

4. $g(x) = \begin{cases} -x^2 + 5, & \text{for } x \leq 0, \\ x + 5, & \text{for } x > 0 \end{cases}$

Find the area under the graph of f over the interval $[-6, 4]$.

5. $f(x) = \begin{cases} -x^2 - 6x + 7, & \text{for } x < 1, \\ \frac{3}{2}x - 1, & \text{for } x \geq 1 \end{cases}$

6. $f(x) = \begin{cases} -x - 1, & \text{for } x < -1, \\ -x^2 + 4x + 5, & \text{for } x \geq -1 \end{cases}$

In Exercises 7–12, determine the x-values at which the graphs of f and g cross. If no such x-values exist, state that fact.

7. $f(x) = 9, \quad g(x) = x^2$

8. $f(x) = 8, \quad g(x) = \frac{1}{2}x^2$

9. $f(x) = 7, \quad g(x) = x^2 - 3x + 2$

10. $f(x) = -6, \quad g(x) = x^2 + 3x + 13$

11. $f(x) = x^2 - x - 5, \quad g(x) = x + 10$

12. $f(x) = x^2 - 7x + 20, \quad g(x) = 2x + 6$

Find the area of the shaded region.

13. $f(x) = 2x + x^2 - x^3, \quad g(x) = 0$

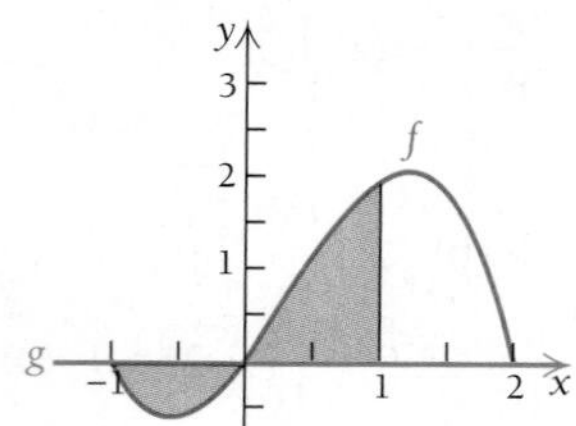

14. $f(x) = x^3 + 3x^2 - 9x - 12, \quad g(x) = 4x + 3$

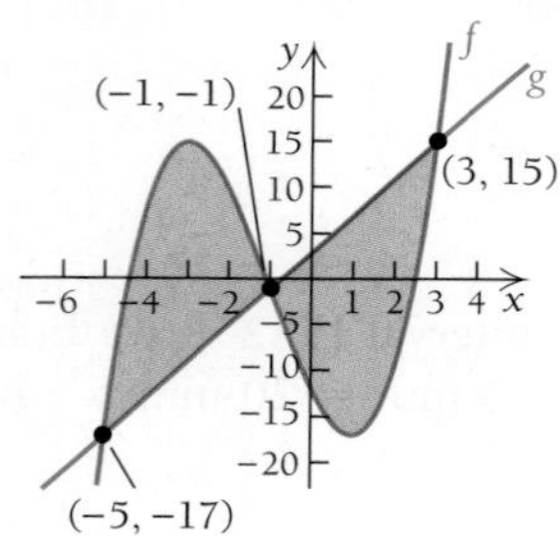

15. $f(x) = x^4 - 8x^3 + 18x^2, \quad g(x) = x + 28$

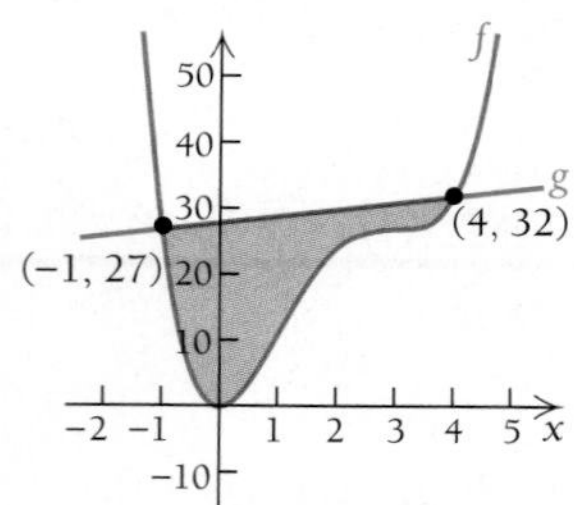

16. $f(x) = 4x - x^2, \quad g(x) = x^2 - 6x + 8$

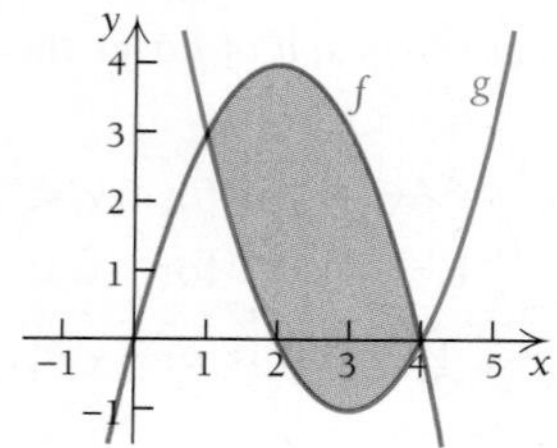

Find the area of the region bounded by the graphs of the given equations.

17. $y = x, y = x^3, x = 0, x = 1$

18. $y = x, y = x^4$

19. $y = x + 2, y = x^2$

20. $y = x^2 - 2x, y = x$

21. $y = 6x - x^2, y = x$

22. $y = x^2 - 6x, y = -x$

23. $y = 2x - x^2, y = -x$

24. $y = x^2, y = \sqrt{x}$

25. $y = x, y = \sqrt[4]{x}$

26. $y = 3, y = x, x = 0$

27. $y = 5, y = \sqrt{x}, x = 0$

28. $y = x^2, y = x^3$

29. $y = 4 - x^2, y = 4 - 4x$

30. $y = x^2 + 1, y = x^2, x = 1, x = 3$

31. $y = x^2 + 3, y = x^2, x = 1, x = 2$

32. $y = 2x^2 - x - 3, y = x^2 + x$

33. $y = 2x^2 - 6x + 5, y = x^2 + 6x - 15$

Find the average value over the given interval.

34. $y = 2x^3; \quad [-1, 1]$

35. $y = 4 - x^2; \quad [-2, 2]$

36. $y = e^x; \quad [0, 1]$

37. $y = e^{-x}; \quad [0, 1]$

38. $y = x^2 - x + 1; \quad [0, 2]$

39. $f(x) = x^2 + x - 2; \quad [0, 4]$

40. $f(x) = mx + 1; \quad [0, 2]$

41. $f(x) = 4x + 5; \quad [0, a]$

42. $f(x) = x^n, n \neq 0; \quad [0, 1]$

43. $f(x) = x^n, n \neq 0; \quad [1, 2]$

44. $f(x) = \dfrac{n}{x}; \quad [1, 5]$

APPLICATIONS

Business and Economics

45. Total and average daily profit. Shylls, Inc., determines that its marginal revenue per day is given by

$$R'(t) = 100e^t, \quad R(0) = 0,$$

where $R(t)$ is the total accumulated revenue, in dollars, on the tth day. The company's marginal cost per day is given by

$$C'(t) = 100 - 0.2t, \quad C(0) = 0,$$

where $C(t)$ is the total accumulated cost, in dollars, on the tth day.

a) Find the total profit from $t = 0$ to $t = 10$ (the first 10 days). *Note:*

$$P(T) = R(T) - C(T) = \int_0^T [R'(t) - C'(t)]\, dt.$$

b) Find the average daily profit for the first 10 days (from $t = 0$ to $t = 10$).

46. Total and average daily profit. Great Green, Inc., determines that its marginal revenue per day is given by

$$R'(t) = 75e^t - 2t, \quad R(0) = 0,$$

where $R(t)$ is the total accumulated revenue, in dollars, on the tth day. The company's marginal cost per day is given by

$$C'(t) = 75 - 3t, \quad C(0) = 0,$$

where $C(t)$ is the total accumulated cost, in dollars, on the tth day.

a) Find the total profit from $t = 0$ to $t = 10$ (see Exercise 45).
b) Find the average daily profit for the first 10 days.

47. Accumulated sales. ProArt, Inc., determines that its weekly online sales, $S(t)$, in hundreds of dollars, t weeks after online sales began, can be estimated by

$$S(t) = 9e^t.$$

Find the average weekly sales for the first 5 weeks after online sales began.

48. Accumulated sales. Music Manager, Ltd., estimates that monthly revenue, $R(t)$, in thousands of dollars, attributable to its Web site t months after the Web site was completed, is given by

$$R(t) = 0.5e^t.$$

Find the average monthly revenue attributable to the Web site for its first 4 months of operation.

49. Refer to Exercise 47. Find ProArt's average weekly online sales for weeks 2 through 5 ($t = 1$ to $t = 5$).

50. Refer to Exercise 48. Find the average monthly revenue from Music Manager's Web site for months 3 through 5 ($t = 2$ to $t = 5$).

Social Sciences

51. Memorizing. In a memory experiment, Alice is able to memorize words at the rate given by

$$m'(t) = -0.009t^2 + 0.2t \quad \text{(words per minute)}.$$

In the same memory experiment, Ben is able to memorize words at the rate given by

$$M'(t) = -0.003t^2 + 0.2t \quad \text{(words per minute)}.$$

a) Who has the higher rate of memorization?
b) How many more words does that person memorize from $t = 0$ to $t = 10$ (during the first 10 min of the experiment)?
c) Over the first 10 min of the experiment, on average, how many words per minute did Alice memorize?
d) Over the first 10 min of the experiment, on average, how many words per minute did Ben memorize?

52. Results of studying. Antonio's score on a test is given by

$$s(t) = t^2, \quad 0 \le t \le 10,$$

where $s(t)$ is his score after t hours of studying. Bonnie's score on the same test is given by

$$S(t) = 10t, \quad 0 \le t \le 10,$$

where $S(t)$ is her score after t hours of studying.

a) For $0 < t < 10$, who will have the higher test score?
b) Find the average value of $s(t)$ over the interval $[7, 10]$, and explain what it represents.
c) Find the average value of $S(t)$ over the interval $[6, 10]$, and explain what it represents.
d) Assuming that both students have the same study habits and are equally likely to study for any number of hours, t, in $[0, 10]$, on average, how far apart will their test scores be?

53. Results of practice. A keyboarder's speed over a 5-min interval is given by

$$W(t) = -6t^2 + 12t + 90, \quad t \text{ in } [0, 5],$$

where $W(t)$ is the speed, in words per minute, at time t.

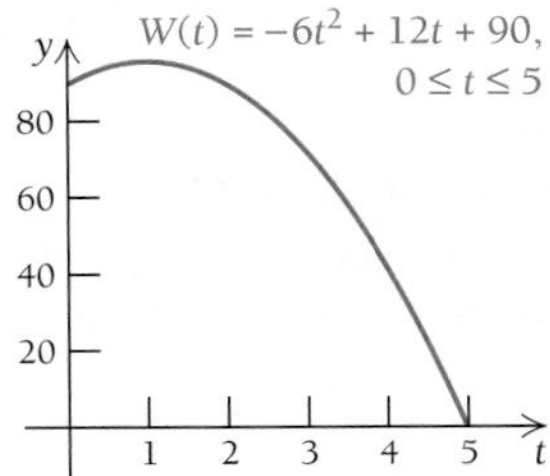

a) Find the speed at the beginning of the interval.
b) Find the maximum speed and when it occurs.
c) Find the average speed over the 5-min interval.

54. Average population. The population of the United States can be approximated by

$$P(t) = 282.3e^{0.01t},$$

where P is in millions and t is the number of years since 2000. (*Source:* Based on data from Population Division, U.S. Census Bureau.) Find the average value of the population from 2001 to 2005.

Natural and Life Sciences

55. Average drug dose. The concentration, C, of phenylbutazone, in micrograms per milliliter (μg/mL), in the plasma of a calf injected with this anti-inflammatory agent is given approximately by

$$C(t) = 42.03e^{-0.01050t},$$

where t is the number of hours after the injection and $0 \le t \le 120$. (*Source:* A. K. Arifah and P. Lees, "Pharmacodynamics and Pharmacokinetics of Phenylbutazone in Calves," *Journal of Veterinary Pharmacology and Therapeutics*, Vol. 25, 299–309 (2002).)

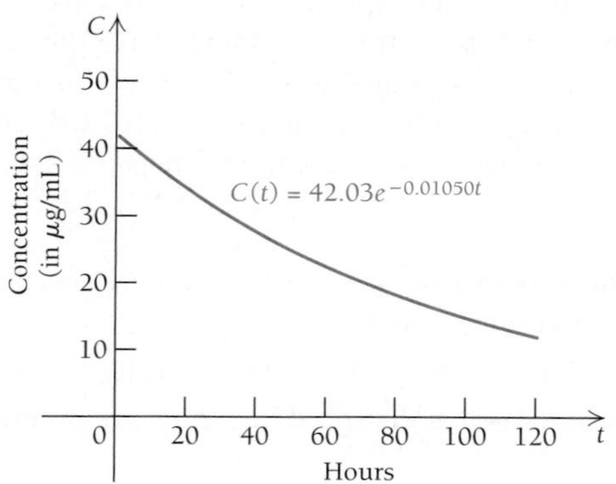

a) Given that this model is accurate for $0 \le t \le 120$, what is the initial dosage?

b) What is the average amount of phenylbutazone in the calf's body for the time between 10 and 120 hours?

56. New York temperature. For any date, the average temperature on that date in New York can be approximated by the function

$$T(x) = 43.5 - 18.4x + 8.57x^2 - 0.996x^3 + 0.0338x^4,$$

where T represents the temperature in degrees Fahrenheit, $x = 1$ represents the middle of January, $x = 2$ represents the middle of February, and so on. (*Source:* www.worldclimate.com.) Compute the average temperature in New York over the whole year to the nearest degree.

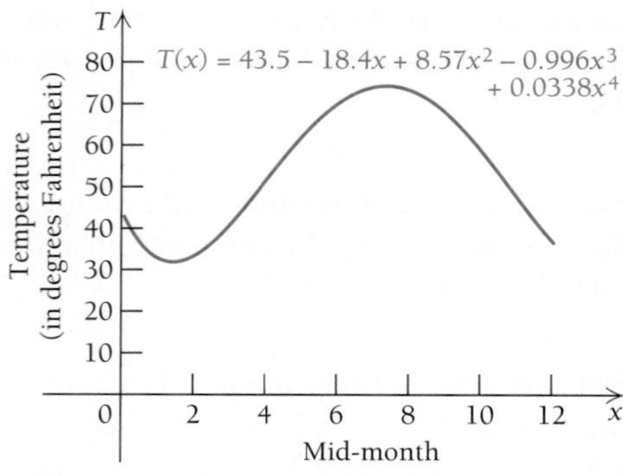

57. Outside temperature. The temperature over a 10-hr period is given by

$$f(t) = -t^2 + 5t + 40, \quad 0 \le t \le 10.$$

a) Find the average temperature.

b) Find the minimum temperature.

c) Find the maximum temperature.

58. Engine emissions. The emissions of an engine are given by

$$E(t) = 2t^2,$$

where $E(t)$ is the engine's rate of emission, in billions of pollution particulates per year, at time t, in years. Find the average emissions from $t = 1$ to $t = 5$.

SYNTHESIS

Find the area of the region bounded by the given graphs.

59. $y = x^2$, $y = x^{-2}$, $x = 5$

60. $y = e^x$, $y = e^{-x}$, $x = -2$

61. $y = x + 6$, $y = -2x$, $y = x^3$

62. $y = x^2$, $y = x^3$, $x = -1$

63. $x + 2y = 2$, $y - x = 1$, $2x + y = 7$

64. Find the area bounded by $y = 3x^5 - 20x^3$, the x-axis, and the first coordinates of the relative maximum and minimum values of the function.

65. Find the area bounded by $y = x^3 - 3x + 2$, the x-axis, and the first coordinates of the relative maximum and minimum values of the function.

66. Life science: Poiseuille's Law. The flow of blood in a blood vessel is faster toward the center of the vessel and slower toward the outside. The speed of the blood is given by

$$V = \frac{p}{4Lv}(R^2 - r^2),$$

where R is the radius of the blood vessel, r is the distance of the blood from the center of the vessel, and p, v, and L are physical constants related to the pressure and viscosity of the blood and the length of the blood vessel. If R is constant, we can think of V as a function of r:

$$V(r) = \frac{p}{4Lv}(R^2 - r^2).$$

The total blood flow, Q, is given by

$$Q = \int_0^R 2\pi \cdot V(r) \cdot r \cdot dr.$$

Find Q.

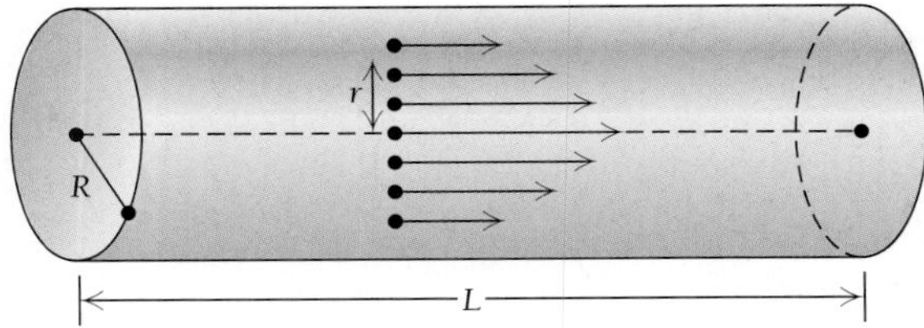

67. Solve for K, given that

$$\int_1^2 [(3x^2 + 5x) - (3x + K)]\, dx = 6.$$

TECHNOLOGY CONNECTION

Find the area of the region enclosed by the given graphs.

68. $y = x^2 + 4x,\ y = \sqrt{16 - x^2}$

69. $y = x\sqrt{4 - x^2},\ y = \dfrac{-4x}{x^2 + 1},\ x = 0,\ x = 2$

70. $y = 2x^2 + x - 4,\ y = 1 - x + 8x^2 - 4x^4$

71. $y = \sqrt{1 - x^2},\ y = 1 - x^2,\ x = -1,\ x = 1$

72. Consider the following functions:

$f(x) = 3.8x^5 - 18.6x^3,$

$g(x) = 19x^4 - 55.8x^2.$

a) Graph these functions in the window $[-3, 3, -80, 80]$, with Yscl $= 10$.

b) Estimate the first coordinates a, b, and c of the three points of intersection of the two graphs.

c) Find the area between the curves on the interval $[a, b]$.

d) Find the area between the curves on the interval $[b, c]$.

4.5 Integration Techniques: Substitution

OBJECTIVES

- Evaluate integrals using substitution.
- Solve applied problems involving integration by substitution.

The following formulas provide a basis for an integration technique called *substitution*.

A. $\displaystyle\int u^r\, du = \frac{u^{r+1}}{r + 1} + C,$ assuming $r \neq -1$

B. $\displaystyle\int e^u\, du = e^u + C$

C. $\displaystyle\int \frac{1}{u}\, du = \ln|u| + C;$ or

$\displaystyle\int \frac{1}{u}\, du = \ln u + C, \quad u > 0$

(Unless noted otherwise, we will assume $u > 0$.)

Recall the Leibniz notation, dy/dx, for a derivative. We gave specific definitions of the differentials dy and dx in Section 2.6. Recall that

$$\frac{dy}{dx} = f'(x)$$

and

$$dy = f'(x)\,dx.$$

We will make extensive use of this notation in this section.

EXAMPLE 1 For $y = f(x) = x^3$, find dy.

Solution We have

$$\frac{dy}{dx} = f'(x) = 3x^2,$$

so

$$dy = f'(x)\,dx = 3x^2\,dx.$$ ✦

EXAMPLE 2 For $u = F(x) = x^{2/3}$, find du.

Solution We have

$$\frac{du}{dx} = F'(x) = \tfrac{2}{3}x^{-1/3},$$

so

$$du = F'(x)\,dx = \tfrac{2}{3}x^{-1/3}\,dx.$$ ✦

EXAMPLE 3 For $u = g(x) = \ln x$, find du.

Solution We have

$$\frac{du}{dx} = g'(x) = \frac{1}{x},$$

so

$$du = g'(x)\,dx = \frac{1}{x}\,dx, \quad \text{or} \quad \frac{dx}{x}.$$ ✦

EXAMPLE 4 For $y = f(x) = e^{x^2}$, find dy.

Solution Using the Chain Rule, we have

$$\frac{dy}{dx} = f'(x) = e^{x^2} \cdot 2x,$$

so

$$dy = f'(x)\,dx = e^{x^2} \cdot 2x\,dx.$$ ✦

So far, the dx in

$$\int f(x)\,dx$$

has played no role in integrating other than to indicate the variable of integration. Now it becomes convenient to make use of dx. Consider the integral

$$\int 2xe^{x^2}\,dx.$$

Finding an antiderivative may seem impossible. Yet, if we note that $2xe^{x^2} = e^{x^2} \cdot 2x$, we see in Example 4 that $f(x) = e^{x^2}$ is an antiderivative of $f'(x) = 2xe^{x^2}$. How might we find such an antiderivative directly? Suppose that we let $u = x^2$. Then

$$\frac{du}{dx} = 2x, \quad \text{and} \quad du = 2x\,dx.$$

If we substitute u for x^2 and du for $2x\,dx$, we have

$$\int 2xe^{x^2}\,dx = \int e^{x^2}\,2x\,dx = \int e^u\,du.$$

Since

$$\int e^u\,du = e^u + C,$$

it follows that

$$\begin{aligned}\int 2xe^{x^2}\,dx &= \int e^u\,du \\ &= e^u + C \\ &= e^{x^2} + C.\end{aligned}$$

In effect, we have used the Chain Rule in reverse. We can check the result by differentiating. The procedure is referred to as *substitution*, or *change of variable*. It can involve *trial-and-error*, but you will become more proficient the more you practice. If you try a substitution that doesn't result in an integrand that can be easily integrated, try again. Many integrations cannot be carried out using substitution, but any integration that fits formula A, B, or C on p. 437 can be done this way.

Let's consider some additional examples.

EXAMPLE 5 Evaluate: $\int 3x^2\,(x^3 + 1)^{10}\,dx.$

Solution Note that $3x^2$ is the derivative of x^3. Thus,

$$\begin{aligned}\int 3x^2\,(x^3 + 1)^{10}\,dx &= \int (x^3 + 1)^{10}\,3x^2\,dx \qquad \text{Substitution} \quad u = x^3 + 1,\ du = 3x^2\,dx \\ &= \int u^{10}\,du \\ &= \frac{u^{11}}{11} + C \\ &= \tfrac{1}{11}(x^3 + 1)^{11} + C. \qquad \text{"Reversing" the substitution}\end{aligned}$$

As a check, we can always differentiate:

$$\frac{d}{dx}\left(\tfrac{1}{11}(x^3 + 1)^{11} + C\right) = \tfrac{11}{11}(x^3 + 1)^{10} \cdot 3x^2 + 0$$
$$= (x^3 + 1)^{10} \cdot 3x^2$$
$$= 3x^2 (x^3 + 1)^{10}.$$

✦

EXAMPLE 6 Evaluate: $\displaystyle\int \frac{2x\,dx}{1 + x^2}$.

Solution

$$\int \frac{2x\,dx}{1 + x^2} = \int \frac{du}{u}$$ Substitution — Let $u = 1 + x^2$; then $du = 2x\,dx$.

$$= \ln u + C$$
$$= \ln (1 + x^2) + C$$

✦

EXAMPLE 7 Evaluate: $\displaystyle\int \frac{2x\,dx}{(1 + x^2)^2}$.

Solution

$$\int \frac{2x\,dx}{(1 + x^2)^2} = \int \frac{du}{u^2}$$ Substitution — $u = 1 + x^2$, $du = 2x\,dx$

$$= \int u^{-2}\,du$$
$$= -u^{-1} + C$$
$$= -\frac{1}{u} + C$$
$$= \frac{-1}{1 + x^2} + C$$

Don't forget to reverse the substitution after integrating.

✦

EXAMPLE 8 Evaluate: $\displaystyle\int \frac{\ln (3x)\,dx}{x}$.

Solution

$$\int \frac{\ln (3x)\,dx}{x} = \int u\,du$$ Substitution — $u = \ln (3x)$, $du = \frac{1}{x}\,dx$

$$= \frac{u^2}{2} + C$$
$$= \frac{(\ln (3x))^2}{2} + C$$

✦

EXAMPLE 9 Evaluate: $\int xe^{x^2}\,dx$.

Solution If we try $u = x^2$, we have $du = 2x\,dx$. We don't have $2x\,dx$ in $\int xe^{x^2}\,dx$. We do have $x\,dx$, so we need a factor of 2. To provide this factor, we multiply by 1, in the form $\frac{1}{2}\cdot 2$:

$$\int xe^{x^2}\,dx = \frac{1}{2}\int 2xe^{x^2}\,dx \qquad a\int f(x)\,dx = \int af(x)\,dx$$
$$= \frac{1}{2}\int e^{x^2}(2x\,dx) = \frac{1}{2}\int e^u\,du$$
$$= \frac{1}{2}e^u + C = \frac{1}{2}e^{x^2} + C.$$

With practice, you will be able to make certain substitutions mentally and just write down the answer. Example 10 illustrates one such case.

EXAMPLE 10 Evaluate: $\int \frac{dx}{x+3}$.

Solution

$$\int \frac{dx}{x+3} = \int \frac{du}{u} \qquad \text{Substitution} \quad u = x + 3,\ du = 1\,dx = dx$$
$$= \ln u + C$$
$$= \ln(x+3) + C \qquad \text{We assume } x + 3 > 0.$$

EXAMPLE 11 Evaluate: $\int_0^1 5x\sqrt{x^2+3}\,dx$. Round to the nearest thousandth.

TECHNOLOGY CONNECTION

EXERCISE

1. Use a calculator to evaluate $\int_0^1 5x\sqrt{x^2+3}\,dx$.

Solution We first find the indefinite integral and then evaluate that integral over [0, 1]:

$$\int 5x\sqrt{x^2+3}\,dx = 5\int x\sqrt{x^2+3}\,dx$$
$$= \frac{5}{2}\int 2x\sqrt{x^2+3}\,dx \qquad \text{Substitution} \quad u = x^2 + 3,\ du = 2x\,dx$$
$$= \frac{5}{2}\int \sqrt{x^2+3}\,2x\,dx$$
$$= \frac{5}{2}\int \sqrt{u}\,du$$
$$= \frac{5}{2}\int u^{1/2}\,du$$
$$= \frac{5}{2}\cdot\frac{2}{3}u^{3/2} + C$$
$$= \frac{5}{3}(x^2+3)^{3/2} + C.$$

Using 0 for the constant C, we have

$$\int_0^1 5x\sqrt{x^2+3}\,dx = \left[\frac{5}{3}(x^2+3)^{3/2}\right]_0^1$$
$$= \frac{5}{3}\left[(x^2+3)^{3/2}\right]_0^1$$
$$= \frac{5}{3}\left[4^{3/2} - 3^{3/2}\right]$$
$$\approx 4.673.$$

Exercise Set 4.5

Evaluate. Assume $u > 0$ when $\ln u$ appears. (Be sure to check by differentiating!)

1. $\int (8 + x^3)^5\, 3x^2\, dx$

2. $\int (x^2 - 7)^6\, 2x\, dx$

3. $\int (x^2 - 6)^7\, x\, dx$

4. $\int (x^3 + 1)^4\, x^2\, dx$

5. $\int (3t^4 + 2)t^3\, dt$

6. $\int (2t^5 - 3)t^4\, dt$

7. $\int \frac{2}{1 + 2x}\, dx$

8. $\int \frac{5}{5x + 7}\, dx$

9. $\int (\ln x)^3 \frac{1}{x}\, dx$

10. $\int (\ln x)^7 \frac{1}{x}\, dx$

11. $\int e^{3x}\, dx$

12. $\int e^{7x}\, dx$

13. $\int e^{x/3}\, dx$

14. $\int e^{x/2}\, dx$

15. $\int x^4 e^{x^5}\, dx$

16. $\int x^3 e^{x^4}\, dx$

17. $\int te^{-t^2}\, dt$

18. $\int t^2 e^{-t^3}\, dt$

19. $\int \frac{1}{5 + 2x}\, dx$

20. $\int \frac{1}{2 + 8x}\, dx$

21. $\int \frac{dx}{12 + 3x}$

22. $\int \frac{dx}{1 + 7x}$

23. $\int \frac{dx}{1 - x}$

24. $\int \frac{dx}{4 - x}$

25. $\int t(t^2 - 1)^5\, dt$

26. $\int t^2(t^3 - 1)^7\, dt$

27. $\int (x^4 + x^3 + x^2)^7(4x^3 + 3x^2 + 2x)\, dx$

28. $\int (x^3 - x^2 - x)^9(3x^2 - 2x - 1)\, dx$

29. $\int \frac{e^x\, dx}{4 + e^x}$

30. $\int \frac{e^t\, dt}{3 + e^t}$

31. $\int \frac{\ln x^2}{x}\, dx$ (*Hint*: Use the properties of logarithms.)

32. $\int \frac{(\ln x)^2}{x}\, dx$

33. $\int \frac{dx}{x \ln x}$

34. $\int \frac{dx}{x \ln x^2}$

35. $\int x\sqrt{ax^2 + b}\, dx$

36. $\int \sqrt{ax + b}\, dx$

37. $\int P_0 e^{kt}\, dt$

38. $\int be^{ax}\, dx$

39. $\int \frac{x^3\, dx}{(2 - x^4)^7}$

40. $\int \frac{3x^2\, dx}{(1 + x^3)^5}$

41. $\int 12x\sqrt[5]{1 + 6x^2}\, dx$

42. $\int 5x\sqrt[4]{1 - x^2}\, dx$

Evaluate.

43. $\int_0^1 2xe^{x^2}\, dx$

44. $\int_0^1 3x^2 e^{x^3}\, dx$

45. $\int_0^1 x(x^2 + 1)^5\, dx$

46. $\int_1^2 x(x^2 - 1)^7\, dx$

47. $\int_0^4 \frac{dt}{1 + t}$

48. $\int_0^2 e^{4x}\, dx$

49. $\int_1^4 \frac{2x + 1}{x^2 + x - 1}\, dx$

50. $\int_1^3 \frac{2x + 3}{x^2 + 3x}\, dx$

51. $\int_0^b e^{-x}\, dx$

52. $\int_0^b 2e^{-2x}\, dx$

53. $\int_0^b me^{-mx}\, dx$

54. $\int_0^b ke^{-kx}\, dx$

55. $\int_0^4 (x-6)^2\,dx$

56. $\int_0^3 (x-5)^2\,dx$

57. $\int_0^2 \frac{3x^2\,dx}{(1+x^3)^5}$

58. $\int_{-1}^0 \frac{x^3\,dx}{(2-x^4)^7}$

59. $\int_0^{\sqrt{7}} 7x\sqrt[3]{1+x^2}\,dx$

60. $\int_0^1 12x\sqrt[5]{1-x^2}\,dx$

61. Use a graphing calculator to check the results of any of Exercises 43–60.

APPLICATIONS

Business and Economics

62. Demand from marginal demand. A firm has the marginal-demand function

$$D'(x) = \frac{-2000x}{\sqrt{25-x^2}}.$$

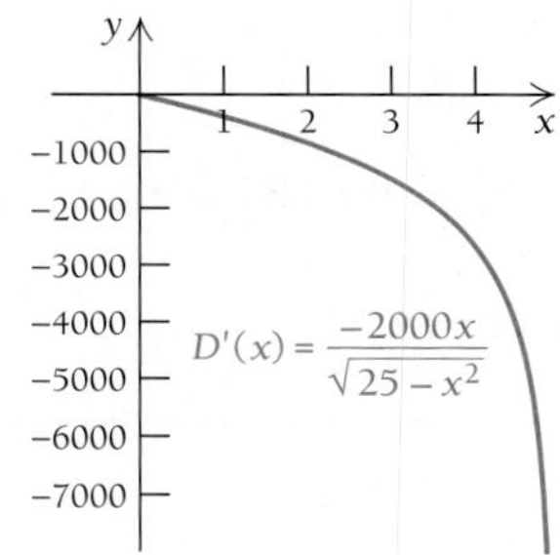

Find the demand function given that $D = 13{,}000$ when $x = \$3$ per unit.

63. Value of an investment. V. King Manufacturing buys a new machine for \$250,000. The marginal revenue from the sale of products produced by the machine after t years is given by

$$R'(t) = 4000t.$$

The salvage value of the machine, in dollars, after t years is given by

$$V(t) = 200{,}000 - 25{,}000e^{0.1t}.$$

The total profit from the machine, in dollars, after t years is given by

$$P(t) = \begin{pmatrix}\text{Revenue}\\ \text{from}\\ \text{sale of}\\ \text{product}\end{pmatrix} + \begin{pmatrix}\text{Revenue}\\ \text{from}\\ \text{sale of}\\ \text{machine}\end{pmatrix} - \begin{pmatrix}\text{Cost}\\ \text{of}\\ \text{machine}\end{pmatrix}.$$

The company knows that $R(0) = 0$.

a) Find $P(t)$.

b) Find $P(10)$.

64. Profit from marginal profit. A firm has the marginal-profit function

$$\frac{dP}{dx} = \frac{9000-3000x}{(x^2-6x+10)^2}.$$

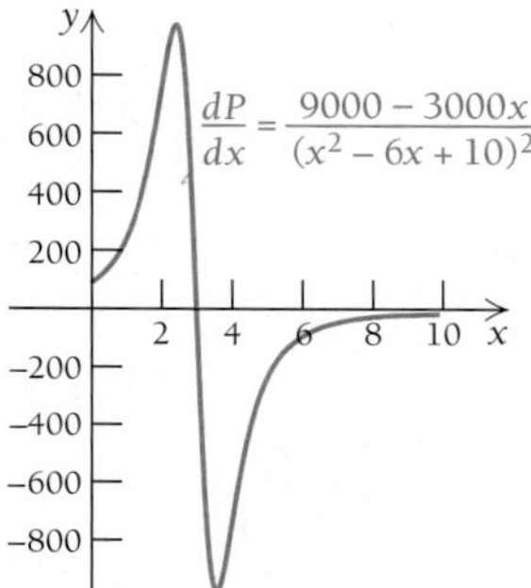

Find the total-profit function given that $P = \$1500$ at $x = 3$.

Social Sciences

65. Divorce rate. The divorce rate in the United States is approximated by

$$D(t) = 100{,}000e^{0.025t},$$

where $D(t)$ is the number of divorces occurring at time t and t is the number of years measured from 1900. That is, $t = 0$ corresponds to 1900, $t = 98\frac{9}{365}$ corresponds to January 9, 1998, and so on.

a) Find the total number of divorces from 1900 to 2005. Note that this is

$$\int_0^{105} D(t)\,dt.$$

b) Find the total number of divorces from 1980 to 2006. Note that this is

$$\int_{80}^{106} D(t)\,dt.$$

SYNTHESIS

Find the area of the shaded region.

66.

67.

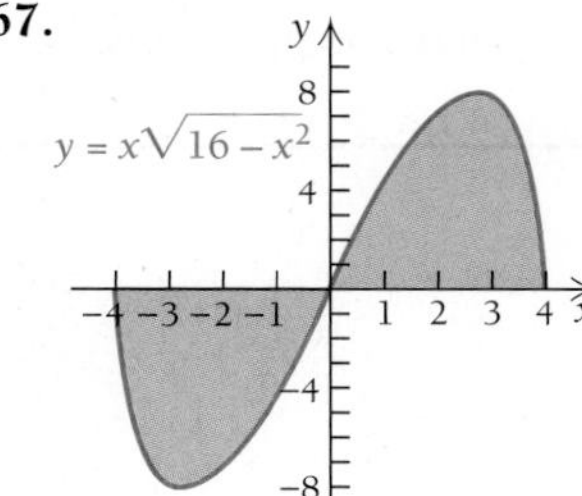

Evaluate. Assume $u > 0$ when $\ln u$ appears.

68. $\int \frac{dx}{ax + b}$

69. $\int 5x\sqrt{1 - 4x^2}\, dx$

70. $\int \frac{e^{\sqrt{t}}}{\sqrt{t}}\, dt$

71. $\int \frac{x^2}{e^{x^3}}\, dx$

72. $\int \frac{(\ln x)^{99}}{x}\, dx$

73. $\int \frac{e^{1/t}}{t^2}\, dt$

74. $\int (e^t + 2)e^t\, dt$

75. $\int \frac{dx}{x(\ln x)^4}$

76. $\int \frac{t^2}{\sqrt[4]{2 + t^3}}\, dt$

77. $\int x^2\sqrt{x^3 + 1}\, dx$

78. $\int \frac{[(\ln x)^2 + 3(\ln x) + 4]}{x}\, dx$

79. $\int \frac{x - 3}{(x^2 - 6x)^{1/3}}\, dx$

80. $\int \frac{t^3 \ln (t^4 + 8)}{t^4 + 8}\, dt$

81. $\int \frac{t^2 + 2t}{(t + 1)^2}\, dt$

$$\left(\textit{Hint: } \frac{t^2 + 2t}{(t + 1)^2} = \frac{t^2 + 2t + 1 - 1}{t^2 + 2t + 1} = 1 - \frac{1}{(t + 1)^2}.\right)$$

82. $\int \frac{x^2 + 6x}{(x + 3)^2}\, dx$ (*Hint*: See Exercise 81.)

83. $\int \frac{x + 3}{x + 1}\, dx$ $\left(\textit{Hint: } \frac{x + 3}{x + 1} = 1 + \frac{2}{x + 1}.\right)$

84. $\int \frac{t - 5}{t - 4}\, dt$

85. $\int \frac{dx}{x(\ln x)^n}$, $n \neq -1$

86. $\int \frac{dx}{e^x + 1}$ $\left(\textit{Hint: } \frac{1}{e^x + 1} = \frac{e^{-x}}{1 + e^{-x}}.\right)$

87. $\int \frac{e^x - e^{-x}}{e^x + e^{-x}}\, dx$

88. $\int \frac{(\ln x)^n}{x}\, dx$, $n \neq -1$

89. $\int \frac{dx}{x \ln x\, [\ln (\ln x)]}$

90. $\int \frac{e^{-mx}}{1 - ae^{-mx}}\, dx$

91. $\int 9x(7x^2 + 9)^n\, dx$, $n \neq -1$

92. $\int 5x^2(2x^3 - 7)^n\, dx$, $n \neq -1$

TW **93.** Determine whether the following is a theorem:

$$\int [2f(x)]\, dx = [f(x)]^2 + C.$$

4.6 Integration Techniques: Integration by Parts

OBJECTIVES

- Evaluate integrals using the formula for integration by parts.
- Solve applied problems involving integration by parts.

Recall the Product Rule for derivatives:

$$\frac{d}{dx}(uv) = u\frac{dv}{dx} + v\frac{du}{dx}.$$

Integrating both sides with respect to x, we get

$$uv = \int u\frac{dv}{dx}\, dx + \int v\frac{du}{dx}\, dx$$
$$= \int u\, dv + \int v\, du.$$

Solving for $\int u\, dv$, we get the following theorem.

THEOREM 7

The Integration-by-Parts Formula

$$\int u\,dv = uv - \int v\,du$$

This equation can be used as a formula for integrating in certain situations—that is, situations in which an integrand is a product of two functions, and one of the functions can be integrated using the techniques we have already developed. For example,

$$\int xe^x\,dx$$

can be considered as

$$\int x(e^x\,dx) = \int u\,dv,$$

where we let

$$u = x \quad \text{and} \quad dv = e^x\,dx.$$

In this case, differentiating u gives

$$du = dx,$$

and integrating dv gives

$$v = e^x. \qquad \text{We select } C = 0 \text{ to obtain the simplest antiderivative.}$$

Then the Integration-by-Parts Formula gives us

$$\begin{aligned}\int \overset{u}{(x)}\overset{dv}{(e^x\,dx)} &= \overset{u}{(x)}\overset{v}{(e^x)} - \int \overset{v}{(e^x)}\overset{du}{(dx)}\\ &= xe^x - e^x + C.\end{aligned}$$

This method of integrating is called **integration by parts.** As always, to check, we can simply differentiate. This check is left to the student.

Note that integration by parts, like substitution, is a trial-and-error process. In the preceding example, suppose that we had reversed the roles of x and e^x. We would have obtained

$$u = e^x, \qquad dv = x\,dx,$$

$$du = e^x\,dx, \qquad v = \frac{x^2}{2},$$

and

$$\int \overset{u}{(e^x)}\overset{dv}{(x\,dx)} = \overset{u}{(e^x)}\overset{v}{\left(\frac{x^2}{2}\right)} - \int \overset{v}{\left(\frac{x^2}{2}\right)}\overset{du}{(e^x\,dx)}.$$

Now the integrand on the right is more difficult to integrate than the one with which we began. When we can integrate *both* factors of an integrand, and thus have a choice as to how to apply the Integration-by-Parts Formula, it can happen that only one (or maybe none) of the possibilities will work.

Tips on Using Integration by Parts

1. If you have had no success using substitution, try integration by parts.
2. Use integration by parts when an integral is of the form

$$\int f(x)\, g(x)\, dx.$$

Match it with an integral of the form

$$\int u\, dv$$

by choosing a function to be $u = f(x)$, where $f(x)$ can be differentiated, and the remaining factor to be $dv = g(x)\, dx$, where $g(x)$ can be integrated.

3. Find du by differentiating and v by integrating.
4. If the resulting integral is more complicated than the original, make some other choice for $u = f(x)$ and $dv = g(x)\, dx$.
5. To check your solution, differentiate.

Let's consider some additional examples.

EXAMPLE 1 Evaluate: $\int \ln x\, dx.$

Solution Note that $\int (dx/x) = \ln x + C$, but we do not yet know how to find $\int \ln x\, dx$ since we have not yet found an antiderivative of $\ln x$. Since we can differentiate $\ln x$, we let

$$u = \ln x \quad \text{and} \quad dv = dx.$$

Then

$$du = \frac{1}{x}\, dx \quad \text{and} \quad v = x.$$

Using the Integration-by-Parts Formula gives

$$\begin{aligned}\int \overset{u}{(\ln x)}\overset{dv}{(dx)} &= \overset{u}{(\ln x)}\overset{v}{x} - \int \overset{v}{x}\overset{du}{\left(\frac{1}{x}\, dx\right)} \\ &= x \ln x - \int dx \\ &= x \ln x - x + C.\end{aligned}$$

✦

EXAMPLE 2 Evaluate: $\int x \ln x\, dx.$

Solution Let's examine several choices, as follows.

Attempt 1: We let

$$u = 1 \quad \text{and} \quad dv = x \ln x\, dx.$$

This will not work because we do not know how to integrate $dv = x \ln x\, dx$.

Attempt 2: We let

$$u = x \ln x \qquad \text{and} \qquad dv = dx.$$

Then $du = \left[x\left(\frac{1}{x}\right) + (\ln x)1 \right] dx$ and $v = x$

$= (1 + \ln x)\, dx.$

Using the Integration-by-Parts Formula, we have

$$\int \overset{u}{(x \ln x)}\, \overset{dv}{dx} = \overset{u}{(x \ln x)}\overset{v}{x} - \int \overset{v}{x}\overset{du}{((1 + \ln x)\, dx)}$$

$$= x^2 \ln x - \int (x + x \ln x)\, dx.$$

This integral seems more complicated than the original, but we will reconsider it in Example 6.

Attempt 3: We let

$$u = \ln x \qquad \text{and} \qquad dv = x\, dx.$$

Then

$$du = \frac{1}{x}\, dx \qquad \text{and} \qquad v = \frac{x^2}{2}.$$

Using the Integration-by-Parts Formula, we have

$$\int \overset{u}{x} \overset{dv}{\ln x\, dx} = \overset{u}{\ln x} \cdot \overset{v}{\frac{x^2}{2}} - \int \overset{v}{\frac{x^2}{2}} \overset{du}{\left(\frac{1}{x}\, dx\right)} = \frac{x^2}{2} \ln x - \frac{1}{2} \int x\, dx$$

$$= \frac{x^2}{2} \ln x - \frac{x^2}{4} + C.$$

This choice of u and dv allows us to evaluate the integral. ✦

EXAMPLE 3 Evaluate $\int x\sqrt{5x + 1}\, dx.$

Solution We let

$$u = x \qquad \text{and} \qquad dv = (5x + 1)^{1/2}\, dx.$$

Then

$$du = dx \qquad \text{and} \qquad v = \tfrac{2}{15}(5x + 1)^{3/2}.$$

Note that we had to use substitution in order to integrate dv. We see this as follows:

$$\int (5x + 1)^{1/2}\, dx = \frac{1}{5} \int (5x + 1)^{1/2}\, 5\, dx = \frac{1}{5} \int w^{1/2}\, dw \quad \underline{\text{Substitution}} \quad w = 5x + 1,\ dw = 5\, dx$$

$$v = \frac{1}{5} \cdot \frac{w^{1/2+1}}{\frac{1}{2} + 1} = \frac{2}{15} w^{3/2} = \frac{2}{15}(5x + 1)^{3/2}.$$

Using the Integration-by-Parts Formula gives us

$$\int \overset{u}{x}\left(\overset{dv}{\sqrt{5x+1}\,dx}\right) = \overset{u}{x} \cdot \overset{v}{\tfrac{2}{15}(5x+1)^{3/2}} - \int \overset{v}{\tfrac{2}{15}(5x+1)^{3/2}}\,\overset{du}{dx}$$

$$= \tfrac{2}{15}x(5x+1)^{3/2} - \tfrac{2}{15}\cdot\tfrac{2}{25}(5x+1)^{5/2} + C$$

$$= \tfrac{2}{15}x(5x+1)^{3/2} - \tfrac{4}{375}(5x+1)^{5/2} + C.$$ ✦

EXAMPLE 4 Evaluate: $\displaystyle\int_1^2 \ln x\,dx.$

Solution First, we find the indefinite integral (see Example 1). Next, we evaluate the definite integral:

$$\int_1^2 \ln x\,dx = [x\ln x - x]_1^2$$

$$= (2\ln 2 - 2) - (1\cdot\ln 1 - 1)$$

$$= 2\ln 2 - 2 + 1$$

$$= 2\ln 2 - 1.$$ ✦

Repeated Integration by Parts

In some cases, we may need to apply the Integration-by-Parts Formula more than once.

EXAMPLE 5 Evaluate $\displaystyle\int_0^7 x^2e^{-x}\,dx$ to find the area of the shaded region shown to the right.

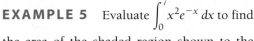

Solution We first let

$$u = x^2 \quad\text{and}\quad dv = e^{-x}\,dx.$$

Then

$$du = 2x\,dx \quad\text{and}\quad v = -e^{-x}.$$

Using the Integration-by-Parts Formula gives

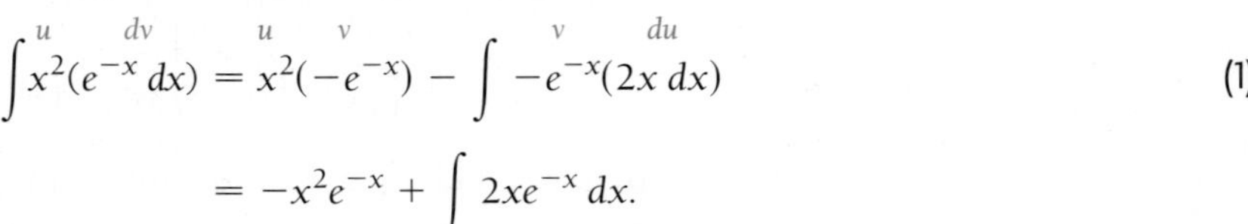

$$\int \overset{u}{x^2}(\overset{dv}{e^{-x}\,dx}) = \overset{u}{x^2}(\overset{v}{-e^{-x}}) - \int \overset{v}{-e^{-x}}(\overset{du}{2x\,dx}) \tag{1}$$

$$= -x^2e^{-x} + \int 2xe^{-x}\,dx.$$

EXERCISE

1. Use a calculator to evaluate
$$\int_0^7 x^2e^{-x}\,dx.$$

To evaluate the integral on the right, we can apply integration by parts again, as follows. We let

$$u = 2x \quad\text{and}\quad dv = e^{-x}\,dx.$$

Then

$$du = 2\,dx \quad\text{and}\quad v = -e^{-x}.$$

Using the Integration-by-Parts Formula once again, we get

$$\int \overset{u}{2x}(\overset{dv}{e^{-x}\,dx}) = \overset{u}{2x}(\overset{v}{-e^{-x}}) - \int \overset{v}{-e^{-x}}(\overset{du}{2\,dx})$$

$$= -2xe^{-x} - 2e^{-x} + C. \tag{2}$$

When we substitute equation (2) into (1), the original integral becomes

$$\int x^2e^{-x}\,dx = -x^2e^{-x} - 2xe^{-x} - 2e^{-x} + C$$

$$= -e^{-x}(x^2 + 2x + 2) + C. \quad \text{Factoring simplifies the next step.}$$

We now evaluate the definite integral:

$$\int_0^7 x^2e^{-x}\,dx = [-e^{-x}(x^2 + 2x + 2)]_0^7$$

$$= [-e^{-7}(7^2 + 2(7) + 2)] - [-e^{-0}(0^2 + 2(0) + 2)]$$

$$= -65e^{-7} + 2 \approx 1.94.$$ ✦

Recognizing a Reoccurrence

Occasionally integration by parts yields an integral of the form $\int v\,du$ that is identical to the original integral. If we are alert and notice this when it occurs, we can find a solution of the original integral algebraically. Let's use this approach and reconsider Example 2.

EXAMPLE 6 Evaluate $\int x \ln x\,dx$ using the result of the second attempt in Example 2.

Solution For the second attempt in Example 2, we let

$$u = x \ln x \quad \text{and} \quad dv = dx,$$

so that

$$du = (1 + \ln x)\,dx \quad \text{and} \quad v = x.$$

Let's now work further with the result we abandoned earlier:

$$\int (\overset{u}{x \ln x})\,\overset{dv}{dx} = (\overset{u}{x \ln x})\overset{v}{x} - \int \overset{v}{x}(\overset{du}{(1 + \ln x)\,dx})$$

$$= x^2 \ln x - \int (x + x \ln x)\,dx$$

$$= x^2 \ln x - \int x\,dx - \int x \ln x\,dx \quad \text{Substituting the integral of each term in the sum}$$

$$= x^2 \ln x - \tfrac{1}{2}x^2 - \int x \ln x\,dx \quad \text{The original integral is duplicated.}$$

$$2\int x \ln x\,dx = x^2 \ln x - \tfrac{1}{2}x^2 \quad \text{Adding } \int x \ln x\,dx \text{ to both sides}$$

$$\int x \ln x\,dx = \tfrac{1}{2}x^2 \ln x - \tfrac{1}{4}x^2 + C. \quad \text{Dividing both sides by 2}$$ ✦

Tabular Integration by Parts

In situations like that in Example 5, we have an integral,

$$\int f(x)\, g(x)\, dx,$$

for which $f(x)$ can be repeatedly differentiated easily to a derivative that is eventually 0. The function $g(x)$ can also be repeatedly integrated easily. In such cases, we can use integration by parts more than once to evaluate the integral.

EXAMPLE 7 Evaluate: $\int x^3 e^x\, dx$.

Solution We use integration by parts repeatedly, watching for patterns:

$$\int \overset{u}{x^3}\, \overset{dv}{e^x\, dx} = \overset{u}{x^3}\overset{v}{e^x} - \int \overset{v}{e^x}\overset{du}{3x^2\, dx}$$

$$= x^3 e^x - \int 3x^2 e^x\, dx. \quad \text{This integral is simpler than the original.} \tag{1}$$

To solve $\int 3x^2 e^x\, dx$, we select $u = 3x^2$ and $dv = e^x\, dx$, so

$$du = 6x\, dx \quad \text{and} \quad v = e^x,$$

and

$$\int 3x^2 e^x\, dx = \overset{u}{3x^2}\overset{v}{e^x} - \int \overset{v}{e^x}\overset{du}{6x\, dx}$$

$$= 3x^2 e^x - \int 6x e^x\, dx. \quad \text{This integral is the simplest so far.} \tag{2}$$

To solve $\int 6xe^x\, dx$, we select $u = 6x$ and $dv = e^x\, dx$, so

$$du = 6\, dx \quad \text{and} \quad v = e^x,$$

and

$$\int 6xe^x\, dx = \overset{u}{6x}\overset{v}{e^x} - \int \overset{v}{e^x}\overset{du}{6\, dx}$$

$$= 6xe^x - 6\int e^x\, dx$$

$$= 6xe^x - 6e^x + C. \tag{3}$$

Combining equations (1), (2), and (3), we have

$$\int x^3 e^x\, dx = x^3 e^x - \left(3x^2 e^x - \int 6xe^x\, dx\right) \quad \text{Substituting equation (2) into equation (1)}$$

$$= x^3 e^x - 3x^2 e^x + \int 6xe^x\, dx$$

$$= x^3 e^x - 3x^2 e^x + 6xe^x - 6e^x + C. \quad \text{Substituting equation (3)}$$

✦

As you can see, this approach can get complicated. Using **tabular integration**, as shown in the following table, can greatly simplify our work.

$f(x)$ and Repeated Derivatives	Sign of Product	$g(x)$ and Repeated Integrals
x^3	$(+)$	e^x
$3x^2$	$(-)$	e^x
$6x$	$(+)$	e^x
6	$(-)$	e^x
0		e^x

We then add products along the arrows, making the alternating sign changes to obtain

$$\int x^3 e^x\,dx = x^3e^x - 3x^2e^x + 6xe^x - 6e^x + C.$$

Exercise Set 4.6

Evaluate using integration by parts or substitution. Check by differentiating.

1. $\int 4xe^{4x}\,dx$

2. $\int 3xe^{3x}\,dx$

3. $\int x^3(3x^2)\,dx$

4. $\int x^2(2x)\,dx$

5. $\int xe^{5x}\,dx$

6. $\int 2xe^{4x}\,dx$

7. $\int xe^{-2x}\,dx$

8. $\int xe^{-x}\,dx$

9. $\int x^2 \ln x\,dx$

10. $\int x^3 \ln x\,dx$

11. $\int x \ln \sqrt{x}\,dx$

12. $\int x^2 \ln x^3\,dx$

13. $\int \ln (x + 5)\,dx$

14. $\int \ln (x + 4)\,dx$

15. $\int (x + 2) \ln x\,dx$

16. $\int (x + 1) \ln x\,dx$

17. $\int (x - 1) \ln x\,dx$

18. $\int (x - 2) \ln x\,dx$

19. $\int x\sqrt{x + 2}\,dx$

20. $\int x\sqrt{x + 5}\,dx$

21. $\int x^3 \ln (2x)\,dx$

22. $\int x^2 \ln (5x)\,dx$

23. $\int x^2e^x\,dx$

24. $\int (\ln x)^2\,dx$

25. $\int x^2e^{2x}\,dx$

26. $\int x^{-5} \ln x\,dx$

27. $\int x^3e^{-2x}\,dx$

28. $\int x^5e^{4x}\,dx$

29. $\int (x^4 + 4)e^{3x}\,dx$

30. $\int (x^3 - x + 1)e^{-x}\,dx$

Evaluate using integration by parts.

31. $\int_1^2 x^2 \ln x\,dx$

32. $\int_1^2 x^3 \ln x\,dx$

33. $\int_2^6 \ln (x + 8)\,dx$

34. $\int_0^5 \ln (x + 7)\,dx$

35. $\int_0^1 xe^x\,dx$

36. $\int_0^1 (x^3 + 2x^2 + 3)e^{-2x}\,dx$

APPLICATIONS

Business and Economics

37. Cost from marginal cost. A company determines that its marginal-cost function is given by

$$C'(x) = 4x\sqrt{x + 3}.$$

Find the total cost given that $C(13) = \$1126.40$.

38. **Profit from marginal profit.** A firm determines that its marginal-profit function is given by

$$P'(x) = 1000x^2e^{-0.2x}.$$

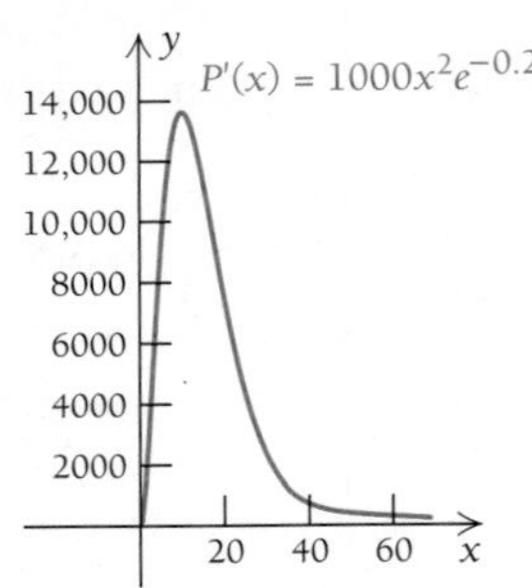

Find the total profit given that $P = -\$2000$ when $x = 0$.

Life and Physical Sciences

39. **Electrical energy use.** The rate of electrical energy used by the Ortiz family, in kilowatt-hours (kW-h) per day, is given by

$$K(t) = 10te^{-t},$$

where t is time, in hours. That is, t is in the interval $[0, 24]$.

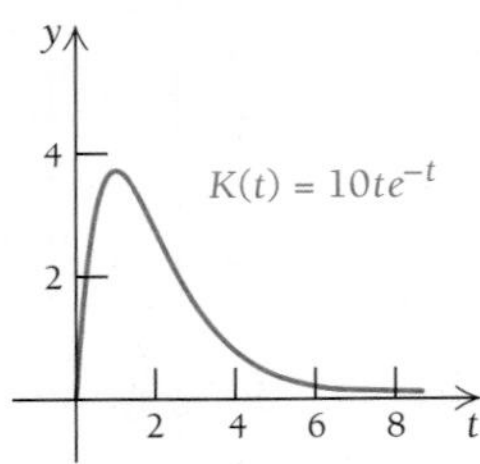

a) How many kilowatt-hours does the family use in the first T hours of a day ($t = 0$ to $t = T$)?

b) How many kilowatt-hours does the family use in the first 4 hours of the day?

40. **Drug dosage.** Suppose that an oral dose of a drug is taken. Over time, the drug is assimilated in the body and excreted through the urine. The total amount of the drug that has passed through the body in time T is given by

$$\int_0^T E(t)\,dt,$$

where E is the rate of excretion of the drug. A typical rate-of-excretion function is

$$E(t) = te^{-kt},$$

where $k > 0$ and t is the time, in hours.

a) Find a formula for

$$\int_0^T E(t)\,dt.$$

b) Find

$$\int_0^{10} E(t)\,dt, \quad \text{when } k = 0.2 \text{ mg/hr}.$$

SYNTHESIS

Evaluate using integration by parts.

41. $\int \sqrt{x} \ln x\,dx$

42. $\int \frac{te^t}{(t+1)^2}\,dt$

43. $\int \frac{\ln x}{\sqrt{x}}\,dx$

44. $\int \frac{13t^2 - 48}{\sqrt[5]{4t+7}}\,dt$

45. $\int (27x^3 + 83x - 2)\sqrt[6]{3x+8}\,dx$

46. $\int x^2 (\ln x)^2\,dx$

47. $\int x^n (\ln x)^2\,dx, \quad n \neq -1$

48. $\int x^n \ln x\,dx, \quad n \neq -1$

49. Verify that for any positive integer n,

$$\int x^n e^x\,dx = x^n e^x - n\int x^{n-1}e^x\,dx.$$

50. Verify that for any positive integer n,

$$\int (\ln x)^n\,dx = x(\ln x)^n - n\int (\ln x)^{n-1}\,dx.$$

TW 51. Determine whether the following is a theorem:

$$\int f(x)\,g(x)\,dx = \int f(x)\,dx \cdot \int g(x)\,dx.$$

Explain.

TW 52. Compare the procedures of differentiation and integration. Which seems to be the most complicated or difficult and why?

TECHNOLOGY CONNECTION

53. Use a graphing calculator to evaluate

$$\int_1^{10} x^5 \ln x\,dx.$$

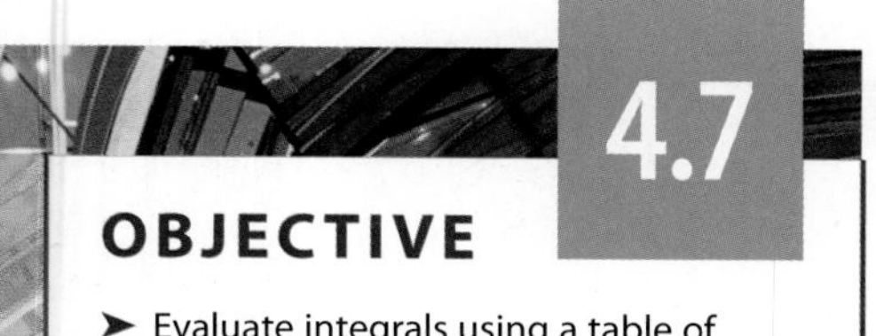

OBJECTIVE

- Evaluate integrals using a table of integration formulas.

Integration Techniques: Tables

Tables of Integration Formulas

You have probably noticed that, generally speaking, integration is more difficult and "tricky" than differentiation. Because of this, integral formulas that are reasonable and/or important have been gathered into tables. Table 1, shown below and at the back of the book on p. 613, is a brief example of such a table. Entire books of integration formulas are available in libraries, and lengthy tables are also available online. Such tables are usually classified by the form of the integrand. The idea is to properly match the integral in question with a formula in the table. Sometimes some algebra or a technique such as substitution or integration by parts may be needed as well as a table.

TABLE 1 Integration Formulas

1. $\int x^n\,dx = \frac{x^{n+1}}{n+1} + C, \quad n \neq -1$
2. $\int \frac{dx}{x} = \ln x + C, \quad x > 0$
3. $\int u\,dv = uv - \int v\,du$
4. $\int e^x\,dx = e^x + C$
5. $\int e^{ax}\,dx = \frac{1}{a} \cdot e^{ax} + C$
6. $\int xe^{ax}\,dx = \frac{1}{a^2} \cdot e^{ax}(ax - 1) + C$
7. $\int x^n e^{ax}\,dx = \frac{x^n e^{ax}}{a} - \frac{n}{a}\int x^{n-1}e^{ax}\,dx + C$
8. $\int \ln x\,dx = x\ln x - x + C$
9. $\int (\ln x)^n\,dx = x(\ln x)^n - n\int (\ln x)^{n-1}\,dx + C, \quad n \neq -1$
10. $\int x^n \ln x\,dx = x^{n+1}\left[\frac{\ln x}{n+1} - \frac{1}{(n+1)^2}\right] + C, \quad n \neq -1$
11. $\int a^x\,dx = \frac{a^x}{\ln a} + C, \quad a > 0, a \neq 1$
12. $\int \frac{1}{\sqrt{x^2 + a^2}}\,dx = \ln|x + \sqrt{x^2 + a^2}| + C$
13. $\int \frac{1}{\sqrt{x^2 - a^2}}\,dx = \ln|x\sqrt{x^2 - a^2}| + C$

(continued)

TABLE 1 *(continued)*

14. $\displaystyle\int \frac{1}{x^2 - a^2}\,dx = \frac{1}{2a}\ln\left|\frac{x-a}{x+a}\right| + C$

15. $\displaystyle\int \frac{1}{a^2 - x^2}\,dx = \frac{1}{2a}\ln\left|\frac{a+x}{a-x}\right| + C$

16. $\displaystyle\int \frac{1}{x\sqrt{a^2 + x^2}}\,dx = -\frac{1}{a}\ln\left|\frac{a+\sqrt{a^2+x^2}}{x}\right| + C$

17. $\displaystyle\int \frac{1}{x\sqrt{a^2 - x^2}}\,dx = -\frac{1}{a}\ln\left|\frac{a+\sqrt{a^2-x^2}}{x}\right| + C$

18. $\displaystyle\int \frac{x}{a + bx}\,dx = \frac{a}{b^2} + \frac{x}{b} - \frac{a}{b^2}\ln|a + bx| + C$

19. $\displaystyle\int \frac{x}{(a + bx)^2}\,dx = \frac{a}{b^2(a + bx)} + \frac{1}{b^2}\ln|a + bx| + C$

20. $\displaystyle\int \frac{1}{x(a + bx)}\,dx = \frac{1}{a}\ln\left|\frac{x}{a+bx}\right| + C$

21. $\displaystyle\int \frac{1}{x(a + bx)^2}\,dx = \frac{1}{a(a + bx)} + \frac{1}{a^2}\ln\left|\frac{x}{a+bx}\right| + C$

22. $\displaystyle\int \sqrt{x^2 \pm a^2}\,dx = \tfrac{1}{2}\left[x\sqrt{x^2 \pm a^2} \pm a^2\ln\left|x + \sqrt{x^2 \pm a^2}\right|\right] + C$

23. $\displaystyle\int x\sqrt{a + bx}\,dx = \frac{2}{15b^2}(3bx - 2a)(a + bx)^{3/2} + C$

24. $\displaystyle\int x^2\sqrt{a + bx}\,dx = \frac{2}{105b^3}(15b^2x^2 - 12abx + 8a^2)(a + bx)^{3/2} + C$

25. $\displaystyle\int \frac{x\,dx}{\sqrt{a + bx}} = \frac{2}{3b^2}(bx - 2a)\sqrt{a + bx} + C$

26. $\displaystyle\int \frac{x^2\,dx}{\sqrt{a + bx}} = \frac{2}{15b^3}(3b^2x^2 - 4abx + 8a^2)\sqrt{a + bx} + C$

EXAMPLE 1 Evaluate: $\displaystyle\int \frac{dx}{x(3 - x)}$.

Solution The integral $\displaystyle\int \frac{dx}{x(3 - x)}$ fits *formula 20* in Table 1:

$$\int \frac{1}{x(a + bx)}\,dx = \frac{1}{a}\ln\left|\frac{x}{a+bx}\right| + C.$$

In the given integral, $a = 3$ and $b = -1$, so we have, by the formula,

$$\begin{aligned}\int \frac{dx}{x(3 - x)} &= \frac{1}{3}\ln\left(\frac{x}{3 + (-1)x}\right) + C \\ &= \frac{1}{3}\ln\left(\frac{x}{3 - x}\right) + C.\end{aligned}$$

◆

EXAMPLE 2 Evaluate: $\int \frac{5x}{7x - 8}\,dx.$

Solution We first factor 5 out of the integral. The integral then fits *formula 18* in Table 1:

$$\int \frac{x}{a + bx}\,dx = \frac{a}{b^2} + \frac{x}{b} - \frac{a}{b^2}\ln|a + bx| + C.$$

In the given integral, $a = -8$ and $b = 7$, so we have, by the formula,

$$\begin{aligned}\int \frac{5x}{7x - 8}\,dx &= 5\int \frac{x}{-8 + 7x}\,dx \\ &= 5\left[\frac{-8}{7^2} + \frac{x}{7} - \frac{-8}{7^2}\ln|-8 + 7x|\right] + C \\ &= 5\left[\frac{-8}{49} + \frac{x}{7} + \frac{8}{49}\ln|7x - 8|\right] + C \\ &= -\frac{40}{49} + \frac{5x}{7} + \frac{40}{49}\ln|7x - 8| + C.\end{aligned}$$

◆

EXAMPLE 3 Evaluate: $\int \sqrt{16x^2 + 3}\,dx.$

Solution This integral *almost* fits *formula 22* in Table 1:

$$\int \sqrt{x^2 \pm a^2}\,dx = \tfrac{1}{2}\left[x\sqrt{x^2 \pm a^2} \pm a^2 \ln\left|x + \sqrt{x^2 \pm a^2}\right|\right] + C.$$

But the coefficient of x^2 needs to be 1. To achieve this, we first factor out 16. Then we apply *formula 22:*

$$\begin{aligned}\int \sqrt{16x^2 + 3}\,dx &= \int \sqrt{16\left(x^2 + \tfrac{3}{16}\right)}\,dx && \text{Factoring} \\ &= \int 4\sqrt{x^2 + \tfrac{3}{16}}\,dx && \text{Using the properties of radicals; } \sqrt{16} = 4 \\ &= 4\int \sqrt{x^2 + \tfrac{3}{16}}\,dx && \text{We have } a^2 = \tfrac{3}{16} \text{ in formula 22.} \\ &= 4 \cdot \tfrac{1}{2}\left[x\sqrt{x^2 + \tfrac{3}{16}} + \tfrac{3}{16}\ln\left|x + \sqrt{x^2 + \tfrac{3}{16}}\right|\right] + C \\ &= 2\left[x\sqrt{x^2 + \tfrac{3}{16}} + \tfrac{3}{16}\ln\left|x + \sqrt{x^2 + \tfrac{3}{16}}\right|\right] + C.\end{aligned}$$

In the given integral, $a^2 = 3/16$ and $a = \sqrt{3}/4$, though we did not need to use a in this form when applying the formula. ◆

EXAMPLE 4 Evaluate: $\int \frac{dx}{x^2 - 25}.$

Solution This integral fits *formula 14* in Table 1:

$$\int \frac{1}{x^2 - a^2}\,dx = \frac{1}{2a}\ln\left|\frac{x - a}{x + a}\right| + C.$$

In the given integral, $a^2 = 25$, so $a = 5$. We have, by the formula,

$$\int \frac{dx}{x^2 - 25} = \frac{1}{10} \ln \left(\frac{x-5}{x+5} \right) + C.$$ ✦

EXAMPLE 5 Evaluate: $\int (\ln x)^3 \, dx$.

Solution This integral fits *formula 9* in Table 1:

$$\int (\ln x)^n \, dx = x(\ln x)^n - n \int (\ln x)^{n-1} \, dx + C, \quad n \neq -1.$$

We must apply the formula three times:

$$\int (\ln x)^3 \, dx = x(\ln x)^3 - 3 \int (\ln x)^2 \, dx + C$$ Formula 9, with $n = 3$

$$= x(\ln x)^3 - 3\left[x(\ln x)^2 - 2 \int \ln x \, dx \right] + C$$ Applying formula 9 again, with $n = 2$

$$= x(\ln x)^3 - 3\left[x(\ln x)^2 - 2\left(x \ln x - \int dx \right) \right] + C$$ Applying formula 9 for the third time, with $n = 1$

$$= x(\ln x)^3 - 3x(\ln x)^2 + 6x \ln x - 6x + C.$$ ✦

The Web site www.integrals.com can be used to find integrals. If you have access to the Internet, use this Web site to check Examples 1–5 or to do the exercises in the following set.

Exercise Set 4.7

Evaluate using Table 1.

1. $\int xe^{-3x} \, dx$

2. $\int 2xe^{3x} \, dx$

3. $\int 6^x \, dx$

4. $\int \frac{1}{\sqrt{x^2 - 9}} \, dx$

5. $\int \frac{1}{25 - x^2} \, dx$

6. $\int \frac{1}{x\sqrt{4 + x^2}} \, dx$

7. $\int \frac{x}{3 - x} \, dx$

8. $\int \frac{x}{(1 - x)^2} \, dx$

9. $\int \frac{1}{x(8 - x)^2} \, dx$

10. $\int \sqrt{x^2 + 9} \, dx$

11. $\int \ln (3x) \, dx$

12. $\int \ln \left(\frac{4}{5} x \right) dx$

13. $\int x^4 \ln x \, dx$

14. $\int x^3 e^{-2x} \, dx$

15. $\int x^3 \ln x \, dx$

16. $\int 5x^4 \ln x \, dx$

17. $\int \frac{dx}{\sqrt{x^2 + 7}}$

18. $\int \frac{3 \, dx}{x\sqrt{1 - x^2}}$

19. $\int \frac{10 \, dx}{x(5 - 7x)^2}$

20. $\int \frac{2}{5x(7x + 2)} \, dx$

21. $\int \frac{-5}{4x^2 - 1} \, dx$

22. $\int \sqrt{9t^2 - 1} \, dt$

23. $\int \sqrt{4m^2 + 16} \, dm$

24. $\int \frac{3 \ln x}{x^2} \, dx$

25. $\int \frac{-5 \ln x}{x^3} \, dx$

26. $\int (\ln x)^4 \, dx$

27. $\int \frac{e^x}{x^{-3}}\,dx$

28. $\int \frac{3}{\sqrt{4x^2 + 100}}\,dx$

29. $\int x\sqrt{1 + 2x}\,dx$

30. $\int x\sqrt{2 + 3x}\,dx$

APPLICATIONS

Business and Economics

31. **Supply from marginal supply.** A lawn machinery company introduces a new kind of lawn seeder. It finds that its marginal supply for the seeder satisfies the function

$$S'(x) = \frac{100x}{(20 - x)^2}, \quad 0 \le x \le 19,$$

where S is the quantity purchased when the price is x thousand dollars per seeder. Find the supply function, $S(x)$, given that the company will sell 2000 seeders when the price is 19 thousand dollars.

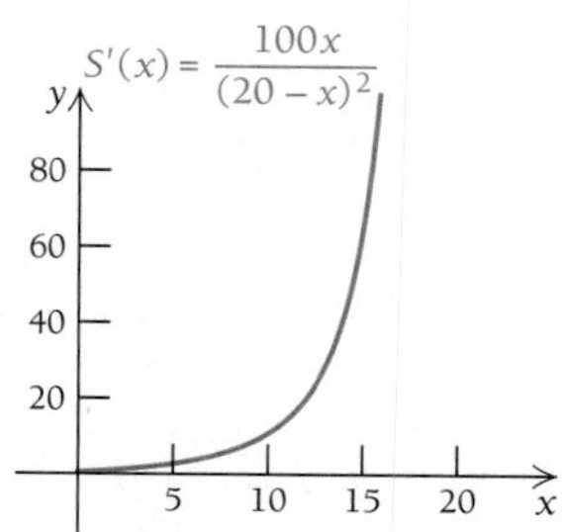

Social Sciences

32. **Learning rate.** The rate of change of the probability that an employee learns a task on a new assembly line is given by

$$p'(t) = \frac{1}{t(2 + t)^2},$$

where $p(t)$ is the probability of learning the task after t months. Find $p(t)$ given that $p = 0.8267$ when $t = 2$.

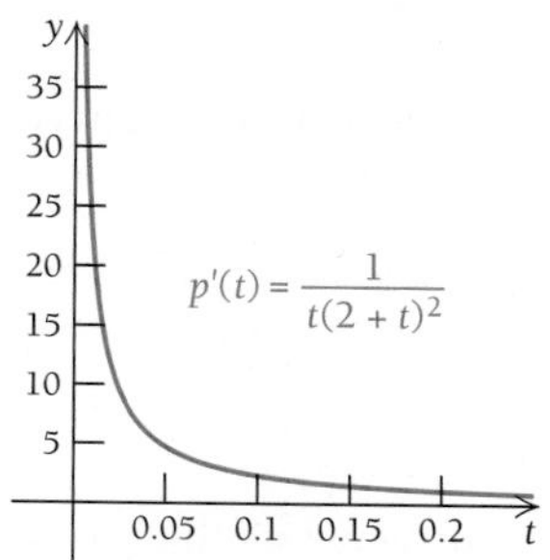

SYNTHESIS

Evaluate using Table 1 or the Web site www.integrals.com.

33. $\int \frac{8}{3x^2 - 2x}\,dx$

34. $\int \frac{x\,dx}{4x^2 - 12x + 9}$

35. $\int \frac{dx}{x^3 - 4x^2 + 4x}$

36. $\int e^x\sqrt{e^{2x} + 1}\,dx$

37. $\int \frac{-e^{-2x}\,dx}{9 - 6e^{-x} + e^{-2x}}$

38. $\int \frac{\sqrt{(\ln x)^2 + 49}}{2x}\,dx$

Chapter Summary

As an introduction to integral calculus, this chapter discussed how to approximate the area under a curve, how to calculate the area under a curve exactly, and how to find antiderivatives. Various applications of integrals were also considered.

The Area under a Curve

To approximate the area under a curve, over an interval $[a, b]$, we divide the interval into subintervals. In each subinterval, we select an x-value, x_i (in this text, we use left endpoints). We then calculate the area of the rectangle extending from the subinterval to the graph, $f(x_i)\,\Delta x$, where Δx is the width of the subinterval. We use **summation notation** to express this approximation of the area, which is called a **Riemann sum.** (pp. 394–395)

EXAMPLE Approximate the area under the graph of $f(x) = -\frac{1}{3}x^2 + 3x$ over the interval [1, 9] using 4 subintervals.

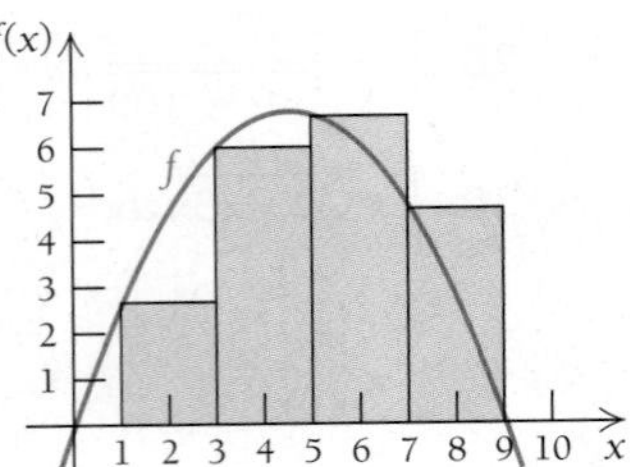

Solution We divide [1, 9] into 4 subintervals of size $\Delta x = (9 - 1)/4 = 2$, with x_i ranging from $x_1 = 1$ to $x_4 = 7$.

Thus,

$$\begin{aligned}\text{Area under curve over } [1, 9] &\approx \sum_{i=1}^{4} f(x_i)\,\Delta x \\ &= f(1)\cdot 2 + f(3)\cdot 2 + f(5)\cdot 2 + f(7)\cdot 2 \\ &= \frac{8}{3}\cdot 2 + 6\cdot 2 + \frac{20}{3}\cdot 2 + \frac{14}{3}\cdot 2 \\ &= 40.\end{aligned}$$

✦

In general, the more subintervals used, the more accurate the approximation of area is. The exact area is the limit of the Riemann sum:

$$A = \lim_{n\to\infty} \sum_{i=1}^{n} f(x_i)\,\Delta x. \qquad \text{(p. 401)}$$

A remarkable result is that if $A(x)$ is the area under the graph of f from 0 to x, then $A'(x) = f(x)$. (p. 404)

THEOREM 1

Let f be a nonnegative continuous function over an interval $[0, b]$, and let $A(x)$ be the area between the graph of f and the x-axis over the interval $[0, x]$, with $0 < x < b$. Then $A(x)$ is a differentiable function of x and $A'(x) = f(x)$.

In Theorem 1, $A(x)$ is an example of an **antiderivative** of $f(x)$, since the derivative of $A(x)$ is $f(x)$. Theorem 2 states that all of a function's antiderivatives differ by, at most, a constant. (p. 405)

THEOREM 2

If two functions F and G have the same derivative over an interval, then

$$F(x) = G(x) + C, \quad \text{where } C \text{ is a constant.}$$

Finding an antiderivative is often called **integration**. To indicate that the antiderivative of x^2 is $\frac{x^3}{3} + C$, we write

$$\int x^2\,dx = \frac{x^3}{3} + C.$$

The constant C is the *constant of integration* and can have any fixed value. Thus, the integral in the above equation is an example of an *indefinite integral*. The function being integrated—x^2 in the example—is called the *integrand*. (p. 405)

Some important formulas and rules for integration follow.

1. $\int k\,dx = kx + C$
2. $\int x^r\,dx = \frac{x^{r+1}}{r+1} + C$
3. $\int x^{-1}\,dx = \ln x + C, \quad x > 0$

   $\int x^{-1}\,dx = \ln|x| + C, \quad x < 0$
4. $\int be^{ax}\,dx = \frac{b}{a}e^{ax} + C$
5. $\int kf(x)\,dx = k\int f(x)\,dx$
6. $\int [f(x) \pm g(x)]\,dx = \int f(x)\,dx \pm \int g(x)\,dx$

(pp. 406–407)

EXAMPLE Evaluate: $\int \left(2 + \frac{5}{x} - \frac{1}{x^6}\right) dx$, for $x > 0$.

Solution

$$\int \left(2 + \frac{5}{x} - \frac{1}{x^6}\right) dx = \int 2\,dx + 5\int \frac{1}{x}\,dx - \int x^{-6}\,dx$$
$$= 2x + 5\ln x + \frac{x^{-5}}{5} + C$$

✦

Sometimes *initial conditions*, or *boundary conditions*, determine the value that C, the constant of integration, has. (p. 408)

EXAMPLE Find f such that $f'(x) = x^3$ and $f(2) = 7$.

Solution We have $f(x) = \int x^3\,dx = \frac{x^4}{4} + C$.

Since $\quad f(2) = \frac{2^4}{4} + C,$

we have $\quad 7 = 4 + C,$

and $\quad 3 = C.$

Thus, $f(x) = \frac{x^4}{4} + 3$.

✦

The definite integral represents the link between antiderivatives and area. (p. 413)

Let f be any continuous function over the interval $[a, b]$ and let F be any antiderivative of f. Then the **definite integral** of f from a to b is

$$\int_a^b f(x)\,dx = F(b) - F(a).$$ (p. 414)

EXAMPLE Evaluate: $\int_1^4 (x + 1)\,dx$.

Solution

$$\int_1^4 (x + 1)\,dx = \left[\frac{x^2}{2} + x\right]_1^4$$
$$= \left(\frac{4^2}{2} + 4\right) - \left(\frac{1^2}{2} + 1\right)$$
$$= 12 - \frac{3}{2} = 10\frac{1}{2}$$

✦

The Fundamental Theorem of Integral Calculus

If a continuous function f has an antiderivative F over $[a, b]$, then

$$\lim_{n\to\infty} \sum_{i=1}^{n} f(x_i)\,\Delta x = \int_a^b f(x)\,dx = F(b) - F(a). \qquad \text{(p. 415)}$$

EXAMPLE Find the area under $y = \sqrt[3]{x}$ over the interval $[1, 8]$.

Solution We have

$$\text{Area} = \int_1^8 \sqrt[3]{x}\,dx = \int_1^8 x^{1/3}\,dx = \left[\frac{3}{4}x^{4/3}\right]_1^8$$
$$= \frac{3}{4}\cdot 8^{4/3} - \frac{3}{4}\cdot 1^{4/3}$$
$$= 12 - \frac{3}{4} = 11\frac{1}{4}.$$

✦

One important application of the definite integral occurs when objects are in motion. (p. 419)

EXAMPLE Suppose that a vehicle's velocity, in miles per hour, is given by $v(t) = t^2 + t$, where t is the number of hours since travel began. How far does the vehicle travel from $t = 1$ to $t = 3$ (during hours 2 and 3)?

Solution Note that the distance, or position, function is an antiderivative of the velocity function. Thus,

$$\text{Distance traveled during hours 2 and 3} = \int_1^3 v(t)\,dt$$
$$= \int_1^3 (t^2 + t)\,dt$$
$$= \left[\frac{t^3}{3} + \frac{t^2}{2}\right]_1^3$$
$$= \left(\frac{3^3}{3} + \frac{3^2}{2}\right) - \left(\frac{1^3}{3} + \frac{1^2}{2}\right)$$
$$= 13\frac{1}{2} - \frac{5}{6} = 12\frac{2}{3}\text{ mi.}$$

✦

There are several ways in which definite integrals can be visualized:

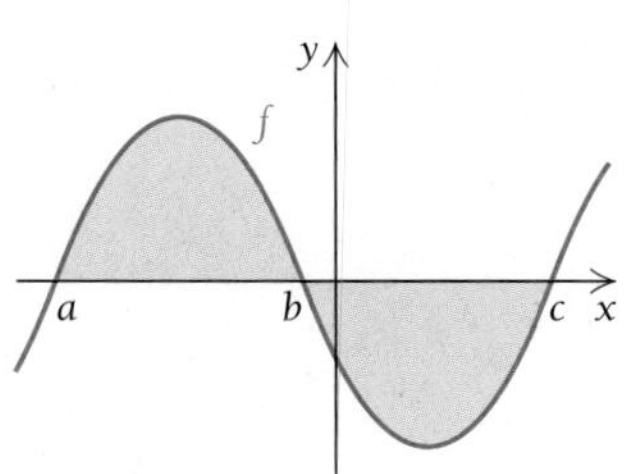

$$\int_a^b f(x)\,dx = -\int_b^c f(x)\,dx$$

(p. 417)

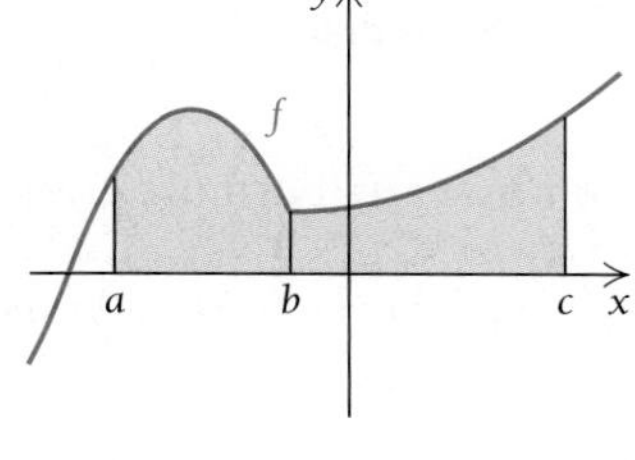

$$\int_a^c f(x)\,dx = \int_a^b f(x)\,dx + \int_b^c f(x)\,dx$$

(p. 426)

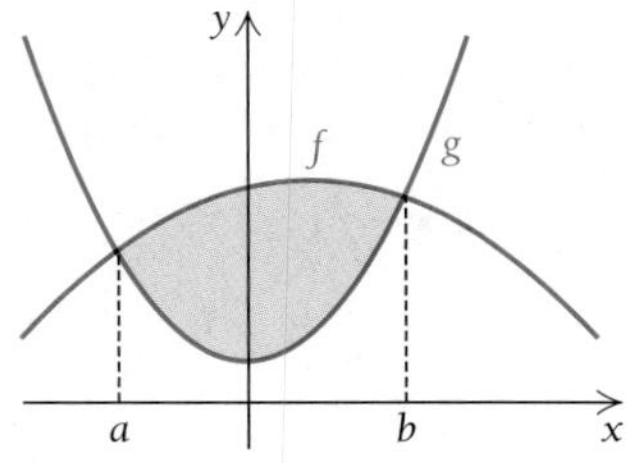

$$\text{Area} = \int_a^b [f(x) - g(x)]\,dx$$

(p. 427)

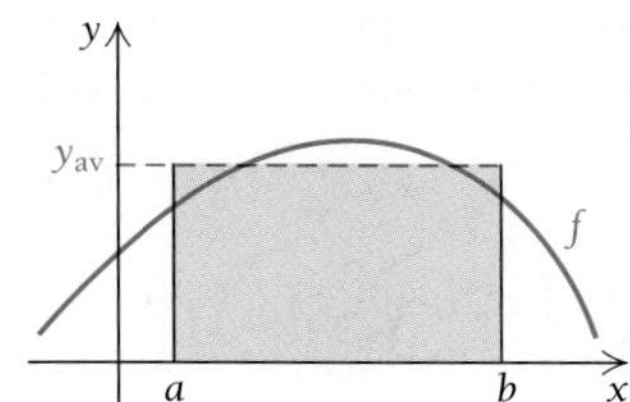

Average value, y_{av}, over $[a, b]$ $= \dfrac{1}{b-a}\displaystyle\int_a^b f(x)\,dx$

(p. 432)

Two important methods for finding certain antiderivatives are **substitution** (p. 437) and **integration by parts** (p. 445).

EXAMPLE Use substitution to evaluate $\displaystyle\int \frac{x^2\,dx}{(6+x^3)^5}$.

Solution

$$\int \frac{x^2\,dx}{(6+x^3)^5} = \int (6+x^3)^{-5}\,x^2\,dx$$

$$= \frac{1}{3}\int (6+x^3)^{-5}\,3x^2\,dx$$

$$= \frac{1}{3}\int u^{-5}\,du \qquad \text{Substitution} \quad u = 6 + x^3,\ du = 3x^2\,dx$$

$$= \frac{1}{3}\cdot\frac{u^{-4}}{-4} + C$$

$$= -\frac{1}{12}(6+x^3)^{-4} + C$$

✦

EXAMPLE Use integration by parts to evaluate $\int 4x\sqrt[3]{2x+1}\,dx$.

Solution We let $u = 4x$ and $dv = (2x+1)^{1/3}\,dx$.

Then $du = 4\,dx$ and $v = \frac{3}{4}\cdot\frac{1}{2}(2x+1)^{4/3} = \frac{3}{8}(2x+1)^{4/3}$.

The Integration-by-Parts Formula gives

$$\int u\,dv = uv - \int v\,du$$

$$\begin{aligned}\int 4x\sqrt[3]{2x+1}\,dx &= 4x\cdot\frac{3}{8}(2x+1)^{4/3} - \int\frac{3}{8}(2x+1)^{4/3}\,4\,dx\\ &= \frac{3}{2}x(2x+1)^{4/3} - \frac{12}{8}\cdot\frac{3}{7}\cdot\frac{1}{2}(2x+1)^{7/3} + C\\ &= \frac{3}{2}x(2x+1)^{4/3} - \frac{9}{28}(2x+1)^{7/3} + C.\end{aligned}$$ ✦

To evaluate some integrals, it may be necessary to use integration by parts more than once (p. 448). In some cases, a table of integration formulas can be helpful (p. 451).

Chapter Review Exercises

These review exercises are for test preparation. They can also be used as a practice test. Answers are at the back of the book. The bracketed section references tell you what part(s) of the chapter to restudy if your answer is incorrect.

CONCEPT REINFORCEMENT

Classify each statement as either true or false.

1. Riemann sums are a way of approximating the area under a curve by using rectangles. [4.1]
2. If a and b are both negative, then $\int_a^b f(x)\,dx$ is negative. [4.3]
3. For any continuous function f defined over $[-1, 7]$, it follows that
$$\int_{-1}^{2} f(x)\,dx + \int_{2}^{7} f(x)\,dx = \int_{-1}^{7} f(x)\,dx. \qquad [4.4]$$
4. Every integral can be solved using integration by parts. [4.6]

Match each integral in column A with the corresponding antiderivative in column B. [4.3, 4.5]

Column A	Column B
5. $\int \frac{1}{\sqrt{x}}\,dx$	**a)** $\ln x + C$
6. $\int (1+2x)^{-2}\,dx$	**b)** $-x^{-1} + C$
7. $\int \frac{1}{x}\,dx,\quad x > 0$	**c)** $-(1+x^2)^{-1} + C$
8. $\int \frac{2x}{1+x^2}\,dx$	**d)** $-\frac{1}{2}(1+2x)^{-1} + C$
9. $\int \frac{1}{x^2}\,dx$	**e)** $2x^{1/2} + C$
10. $\int \frac{2x}{(1+x^2)^2}\,dx$	**f)** $\ln(1+x^2) + C$

REVIEW EXERCISES

11. Business: total cost. The marginal cost, in dollars, of producing the xth car stereo is given by

$$C'(x) = 0.004x^2 - 2x + 500.$$

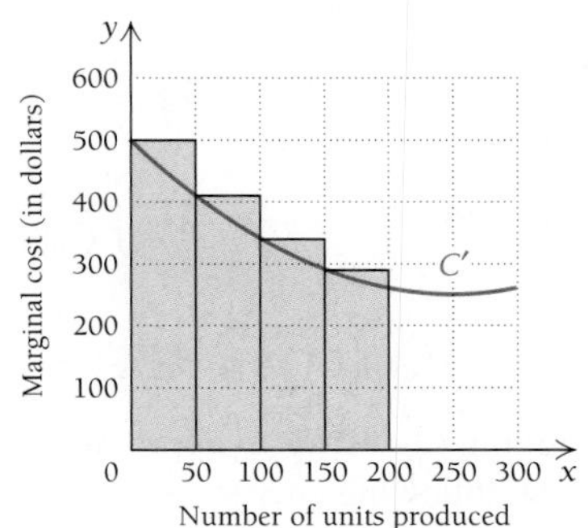

Approximate the total cost of producing 200 car stereos by computing the sum

$$\sum_{i=1}^{4} C'(x_i)\,\Delta x, \quad \text{with } \Delta x = 50. \qquad [4.1]$$

Evaluate. [4.2, 4.3]

12. $\int 20x^4\,dx$

13. $\int (3e^x + 2)\,dx$

14. $\int \left(3t^2 + 5t + \frac{1}{t}\right) dt$ (assume $t > 0$)

Find the area under the curve over the indicated interval. [4.3]

15. $y = 4 - x^2$; $[-2, 1]$

16. $y = x^2 + 2x + 1$; $[0, 3]$

In each case, give an interpretation of the shaded region. [4.3]

17.

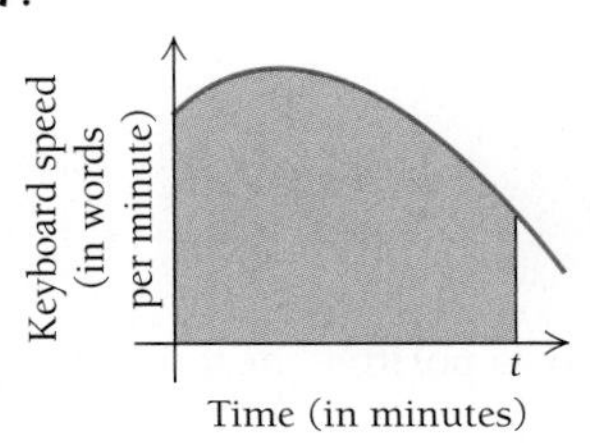

18.

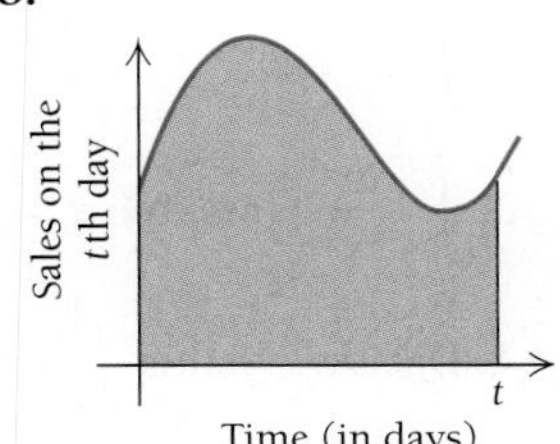

Evaluate. [4.3]

19. $\int_a^b x^5\,dx$

20. $\int_{-1}^{1} (x^3 - x^4)\,dx$

21. $\int_0^1 (e^x + x)\,dx$

22. $\int_1^4 \frac{2}{x}\,dx$

Decide whether $\int_a^b f(x)\,dx$ *is positive, negative, or zero.* [4.3]

23.

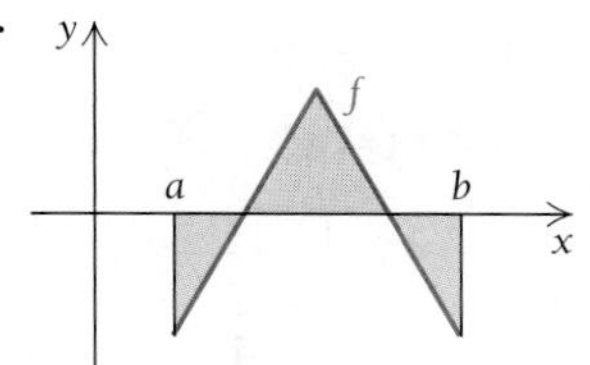

24.

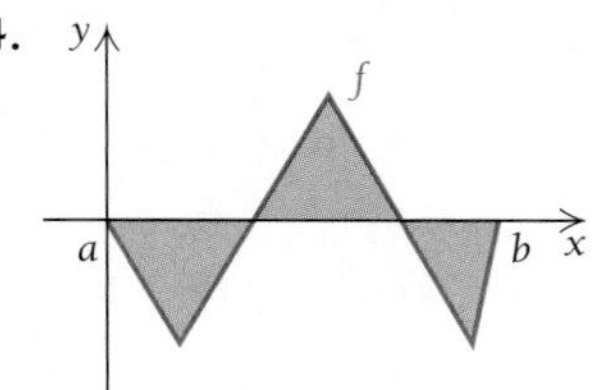

25.

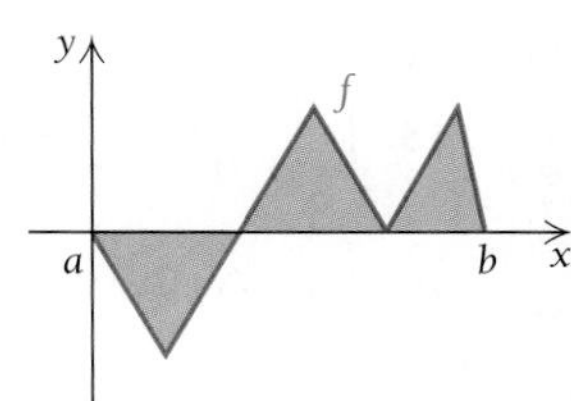

26. Find the area of the region bounded by $y = 3x^2$ and $y = 9x$. [4.4]

Evaluate using substitution. Do not use Table 1. [4.5]

27. $\int x^3 e^{x^4}\,dx$

28. $\int \frac{24t^5}{4t^6 + 3}\,dt$

29. $\int \frac{\ln(4x)}{2x}\,dx$

30. $\int 2e^{-3x}\,dx$

Evaluate using integration by parts. Do not use Table 1. [4.6]

31. $\int 3xe^{3x}\,dx$

32. $\int \ln \sqrt[3]{x^2}\,dx$

33. $\int 3x^2 \ln x\,dx$

Evaluate using Table 1. [4.7]

34. $\int \frac{1}{49 - x^2}\,dx$

35. $\int x^2 e^{5x}\,dx$

36. $\int \frac{x}{7x + 1}\,dx$

37. $\int \frac{dx}{\sqrt{x^2 - 36}}$

38. $\int x^6 \ln x\,dx$

39. $\int xe^{8x}\,dx$

40. Business: total cost. Refer to Exercise 11. Calculate the total cost of producing 200 car stereos. [4.4]

41. Find the average value of $y = xe^{-x}$ over $[0, 2]$. [4.4]

42. A particle starts out from the origin. Its velocity in mph after t hours is given by $v(t) = 3t^2 + 2t$. Find the distance that the particle travels during the first 4 hr (from $t = 0$ to $t = 4$). [4.3]

43. **Business: total revenue.** A company estimates that its revenue will grow continuously at a rate given by the function $S'(t) = 3e^{3t}$, where $S'(t)$ is the rate at which revenue is increasing on the tth day. Find the accumulated revenue for the first 4 days. [4.3]

Integrate using any method. [4.3–4.6]

44. $\int x^3 e^{0.1x}\,dx$

45. $\int \frac{12t^2}{4t^3 + 7}\,dt$

46. $\int \frac{x\,dx}{\sqrt{4 + 5x}}$

47. $\int 5x^4 e^{x^5}\,dx$

48. $\int \frac{dx}{x + 9}$ (assume $x > -9$)

49. $\int t^7 (t^8 + 3)^{11}\,dt$

50. $\int \ln (7x)\,dx$

51. $\int x \ln (8x)\,dx$

SYNTHESIS

Evaluate. [4.5–4.7]

52. $\int \frac{t^4 \ln (t^5 + 3)}{t^5 + 3}\,dt$

53. $\int \frac{dx}{e^x + 2}$

54. $\int \frac{\ln \sqrt{x}}{x}\,dx$

55. $\int x^{91} \ln x\,dx$

56. $\int \ln \left(\frac{x - 3}{x - 4}\right) dx$

57. $\int \frac{dx}{x (\ln x)^4}$

TECHNOLOGY CONNECTION

58. Use a graphing calculator to approximate the area between the following curves:

$y = 2x^2 - 2x, \quad y = 12x^2 - 12x^3.$ [4.4]

Chapter 4 Test

1. Approximate

$$\int_0^5 (25 - x^2)\,dx$$

by computing the area of each rectangle and adding.

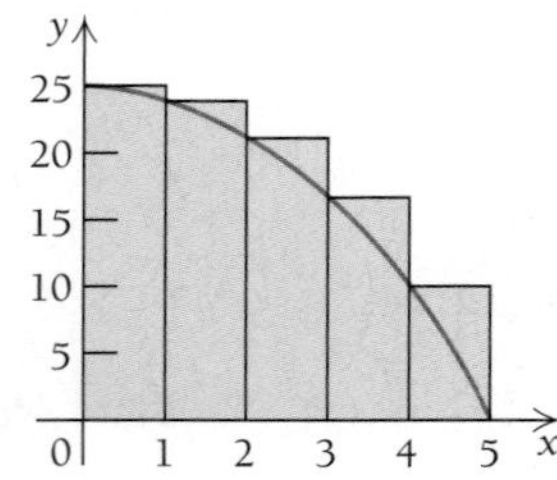

Evaluate.

2. $\int \sqrt{3x}\,dx$

3. $\int 1000x^5\,dx$

4. $\int \left(e^x + \frac{1}{x} + x^{3/8}\right) dx$ (assume $x > 0$)

Find the area under the curve over the indicated interval.

5. $y = x - x^2$; $[0, 1]$

6. $y = \frac{4}{x}$; $[1, 3]$

7. Give an interpretation of the shaded area.

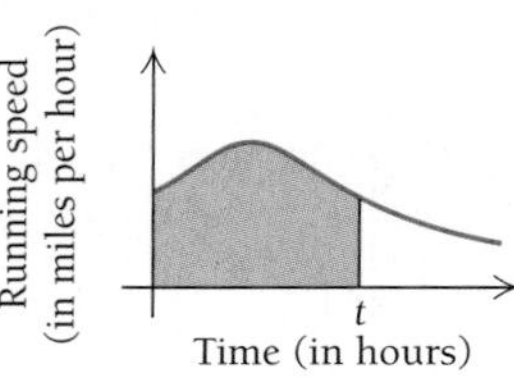

Evaluate.

8. $\int_{-1}^{2} (2x + 3x^2)\,dx$

9. $\int_0^1 e^{-2x}\,dx$

10. $\int_a^{a^2} \frac{dx}{x}$ (assume $a > 0$)

11. Decide whether $\int_a^b f(x)\,dx$ is positive, negative, or zero.

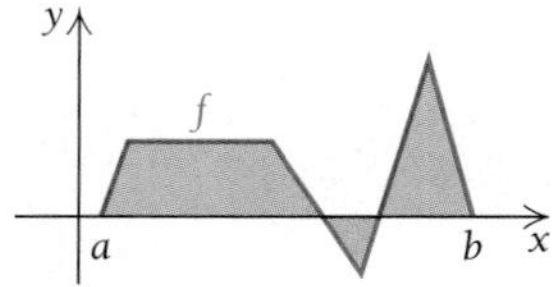

Evaluate using substitution. Assume $u > 0$ when $\ln u$ appears. Do not use Table 1.

12. $\int \frac{dx}{x + 12}$

13. $\int e^{-0.5x}\, dx$

14. $\int t^3(t^4 + 3)^9\, dt$

Evaluate using integration by parts. Do not use Table 1.

15. $\int xe^{5x}\, dx$

16. $\int x^3 \ln x^4\, dx$

Evaluate using Table 1.

17. $\int 2^x\, dx$

18. $\int \frac{dx}{x(7 - x)}$

19. Find the average value of $y = 4t^3 + 2t$ over $[-1, 2]$.

20. Find the area of the region in the first quadrant bounded by $y = x$ and $y = x^5$.

21. Business: cost from marginal cost. An air conditioning company determines that the marginal cost, in dollars, for the xth air conditioner is given by

$$C'(x) = -0.2x + 500, \quad C(0) = 0.$$

Find the total cost of producing 100 air conditioners.

22. Social science: learning curve. A translator's speed over a 4-min interval is given by

$$W(t) = -6t^2 + 12t + 90, \quad t \text{ in } [0, 4],$$

where $W(t)$ is the speed, in words per minute, at time t. How many words are translated during the second minute (from $t = 1$ to $t = 2$)?

23. A robot leaving a spacecraft has velocity given by $v(t) = -0.4t^2 + 2t$, where $v(t)$ is in kilometers per hour and t is the number of hours since the robot left the spacecraft. Find the total distance traveled during the first 3 hr.

Integrate using any method. Assume $u > 0$ when $\ln u$ appears.

24. $\int \frac{6}{5 + 7x}\, dx$

25. $\int x^5 e^x\, dx$

26. $\int x^5 e^{x^6}\, dx$

27. $\int \sqrt{x} \ln x\, dx$

28. $\int \frac{dx}{64 - x^2}$

29. $\int x^4 e^{-0.1x}\, dx$

30. $\int x \ln (13x)\, dx$

SYNTHESIS

Evaluate using any method.

31. $\int x^3\sqrt{x^2 + 4}\, dx$

32. $\int \frac{[(\ln x)^3 - 4(\ln x)^2 + 5]}{x}\, dx$

33. $\int \ln \left(\frac{x + 3}{x + 5}\right) dx$

34. $\int \frac{8x^3 + 10}{\sqrt[3]{5x - 4}}\, dx$

35. Evaluate $\int 5^x\, dx$ without using Table 1.

(*Hint:* $5 = e^{\ln 5}$.)

TECHNOLOGY CONNECTION

36. Use a calculator to approximate the area between the following curves:

$$y = 3x - x^2, \quad y = 2x^3 - x^2 - 5x.$$

EXTENDED TECHNOLOGY APPLICATION

Financial Predictions for Starbucks, Lowe's, The Gap, and Intel

One of the many tasks of corporate leaders is the analysis of all kinds of data regarding their businesses. The data might be total revenue, total costs, operating expenses, operating income, costs of sales and marketing, and so on.

One very important use of such data can be the prediction, or forecasting, of the future. In this application, you will analyze factual data for several actual companies. You will be asked to fit curves, to find models and make predictions.

EXERCISES

Starbucks is a national chain specializing in gourmet coffee drinks. The table at the right lists the total sales of Starbucks for several years.

YEAR	x	TOTAL SALES (IN MILLIONS OF DOLLARS)
1993	0	177
1994	1	285
1995	2	465
1996	3	696
1997	4	967
1998	5	1309
1999	6	1687
2000	7	2178
2001	8	2649
2002	9	3289
2003	10	4076
2004	11	5294
2005	12	6369

(*Source:* Starbucks annual financial reports.)

1. Make a scatterplot of the data. What kind of function—linear, exponential, or logarithmic—seems to best fit the data?
2. Use REGRESSION to fit the chosen function to the data.
3. Use the function to predict the total sales of Starbucks in 2008, 2012, and 2015.
4. Use integration to predict the total sales of the company from 1993 through 2020.

Lowe's is a retailer of home improvement products and services. The following table lists the total sales of Lowe's Companies, Inc., for several years.

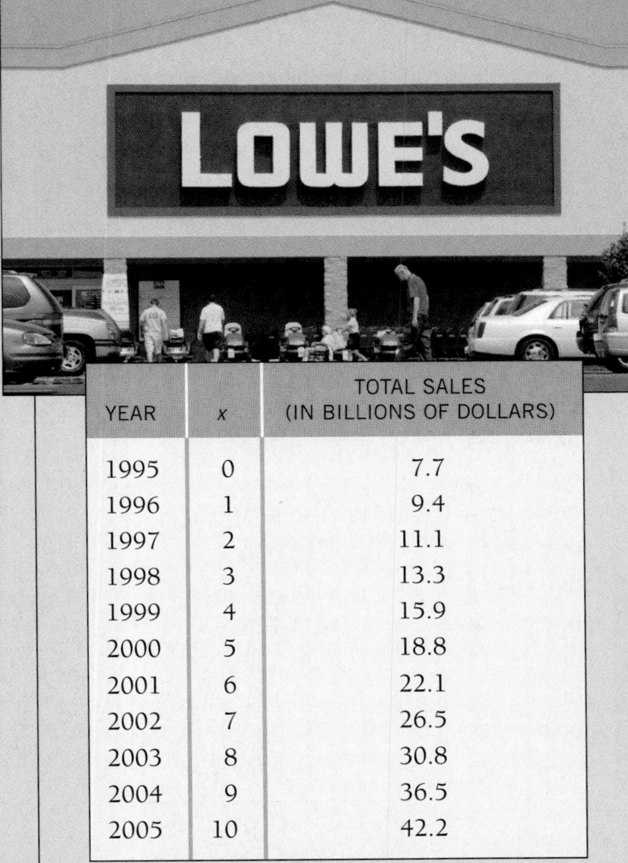

YEAR	x	TOTAL SALES (IN BILLIONS OF DOLLARS)
1995	0	7.7
1996	1	9.4
1997	2	11.1
1998	3	13.3
1999	4	15.9
2000	5	18.8
2001	6	22.1
2002	7	26.5
2003	8	30.8
2004	9	36.5
2005	10	42.2

(*Source:* Lowe's annual financial reports.)

5. Make a scatterplot of the data. What kind of function—linear, quadratic, cubic polynomial, quartic polynomial, exponential, or logarithmic—seems to best fit the data?

6. Use REGRESSION to fit the chosen function to the data.

7. Use the function to predict Lowe's total sales in 2009, 2012, and 2015.

8. Use integration to predict the total sales of the company from 1995 through 2015.

The Gap is a specialty retailer that sells casual apparel under private-label brand names. The following table lists the total sales of The Gap for several years.

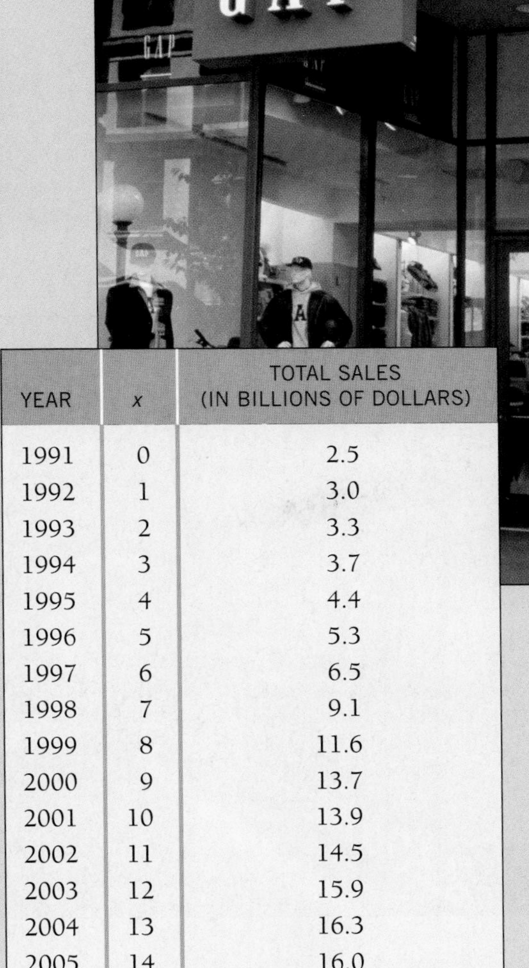

YEAR	x	TOTAL SALES (IN BILLIONS OF DOLLARS)
1991	0	2.5
1992	1	3.0
1993	2	3.3
1994	3	3.7
1995	4	4.4
1996	5	5.3
1997	6	6.5
1998	7	9.1
1999	8	11.6
2000	9	13.7
2001	10	13.9
2002	11	14.5
2003	12	15.9
2004	13	16.3
2005	14	16.0

(*Source:* The Gap annual financial reports.)

(*continued*)

9. Make a scatterplot of the data. What kind of function—linear, quadratic, cubic polynomial, quartic polynomial, exponential, logarithmic, or logistic—seems to best fit the data?
10. Use REGRESSION to fit the chosen function to the data.
11. Use the function to predict the total sales of The Gap in 2009, 2013, and 2017.
12. Use integration to predict the total sales of the company from 1991 through 2020.

13. Make a scatterplot of the data. What kind of function—linear, quadratic, cubic polynomial, quartic polynomial, exponential, logarithmic, or logistic—seems to best fit the data?
14. Use REGRESSION to fit the chosen function to the data.
15. Use the function to predict the total sales of Intel in 2009, 2013, and 2017.
16. Use integration to predict the total sales of the company from 1990 through 2020.

Intel is a company that specializes in microprocessors and other semiconductor products such as computer chips. The growth of personal computer sales is a top priority of the company. The table at the right lists the total sales of Intel for several years.

YEAR	x	TOTAL SALES (IN MILLIONS OF DOLLARS)
1986	0	1,265
1988	2	2,875
1990	4	3,921
1993	7	8,782
1995	9	16,202
1996	10	20,847
1997	11	25,070
1999	13	29,389
2000	14	33,726
2001	15	26,539
2003	17	30,141
2004	18	34,209
2005	19	38,826

(*Source:* Intel annual financial reports.)

Applications of Integration

5

APPLICATION Between 2006 and 2010, the annual world demand for oil is projected to increase from 30.8 billion barrels to 34.5 billion barrels. Assuming exponential growth in demand and given that world reserves will remain at the 2006 estimate of 1293 billion barrels, when will the reserves be depleted? (*Source:* Based on data from U.S. Department of Energy and *Oil and Gas Journal,* Jan. 1, 2006.)

This problem appears as Exercise 29 in Exercise Set 5.2.

INTRODUCTION

In this chapter, we study a wide variety of applications of integration. We first consider an economics application—finding consumer surplus and producer surplus. Then we study the integration of functions involved in exponential growth and decay. We will explore applications not only to business but also to such environmental concerns as the depletion of resources and the buildup of radioactivity in the atmosphere. Integration has applications in probability and statistics, in finding volume, and in many situations involving the solution of differential equations. We will examine these topics as well.

Topics in this chapter can be chosen to fit the needs of the student and the course.

An Economics Application: Consumer Surplus and Producer Surplus

OBJECTIVE

➤ Given demand and supply functions, find the consumer surplus and the producer surplus at the equilibrium point.

It has been convenient to think of demand and supply as quantities that are functions of price. For purposes of this section, we will find it convenient to think of them as prices that are functions of quantity: $p = D(x)$ and $p = S(x)$. Indeed, such an interpretation is common in economics.

The consumer's demand curve, $p = D(x)$, gives the price per unit that the consumer is willing to pay for x units. It is usually a decreasing function. The producer's supply curve, $p = S(x)$, gives the price per unit that the seller is willing to accept when supplying x units. It is usually an increasing function. The equilibrium point, (x_E, p_E), is the intersection of the two curves.

TECHNOLOGY CONNECTION

EXERCISE

1. Graph the demand and supply functions

$D(x) = (x - 5)^2$ and

$S(x) = x^2 + x + 3$

using the viewing window $[0, 5, 0, 30]$, with Yscl $= 5$. Find the equilibrium point using the INTERSECT feature.

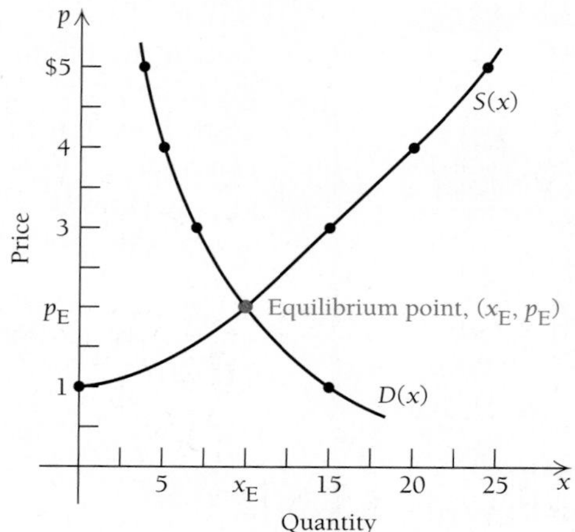

Utility is a function often considered in economics. When a consumer receives x units of a product, a certain amount of pleasure, or utility, U, is derived from them (see Exercise 27 in Exercise Set 1.3). For example, the number of movies that you see in a month gives you a certain utility. If you see four movies (unless they are not entertaining), you get more utility than if you see no movies. The same notion applies to having a meal in a restaurant or paying your heating bill to warm your home.

To help to explain the concepts of *consumer surplus* and *producer surplus*, we will consider the utility of seeing movies over a fixed amount of time, say, 1 month. We are also going to make the assumption that the movies seen are of about the same quality.

Suppose that the graphs in Figs. 1 and 2 show demand curves of a college student named Samantha for movies. We want to examine the utility she receives from going to the movies. Samantha goes to 0 movies per month if the price is $10. She will go to 1 movie per month if the price is $8.80. Suppose that she goes to 1 movie. Her total expenditure is then $8.80 · 1, or $8.80, as shown in Fig. 1.

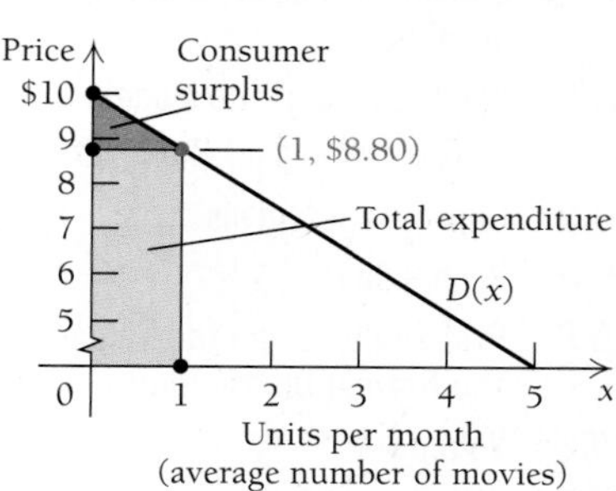

FIGURE 1

The area of the blue region in Fig. 1 represents her total expenditure, \$8.80. Look at the area of the orange region. It is $\frac{1}{2}(1)(\$1.20)$, or \$0.60. The total area under the curve, \$9.40, is what going to 1 movie a month is worth to Samantha. The area of the orange region, \$0.60, is a measure of the enjoyment Samantha gets, but for which she does not have to pay. Economists define this amount as *consumer surplus.* It is the extra utility that consumers enjoy when prices decrease as more units are purchased.

Suppose that Samantha goes to 2 movies per month if the price is \$7.50 (Fig. 2). Her total expenditure is then \$7.50(2), or \$15, which is the area of the blue region in Fig. 2. The total area under the curve, \$17.50, is what going to 2 movies a month is worth to Samantha. Look at the area of the orange region. It (the consumer surplus) is $\frac{1}{2}(2)(\$2.50)$, or \$2.50. It measures the utility Samantha received, but for which she did not have to pay.

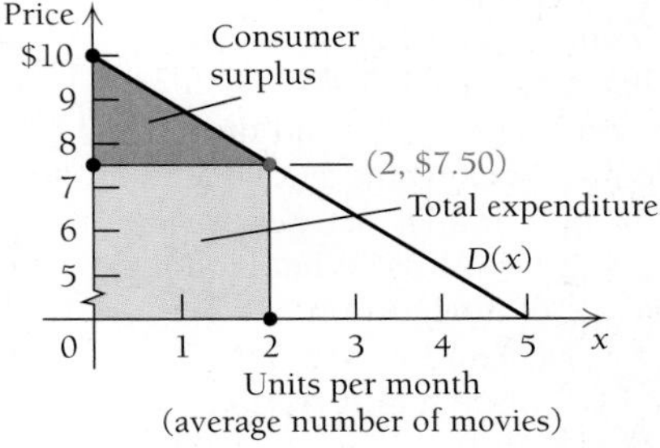

FIGURE 2

Suppose that the graph of the demand function is a curve, as shown below.

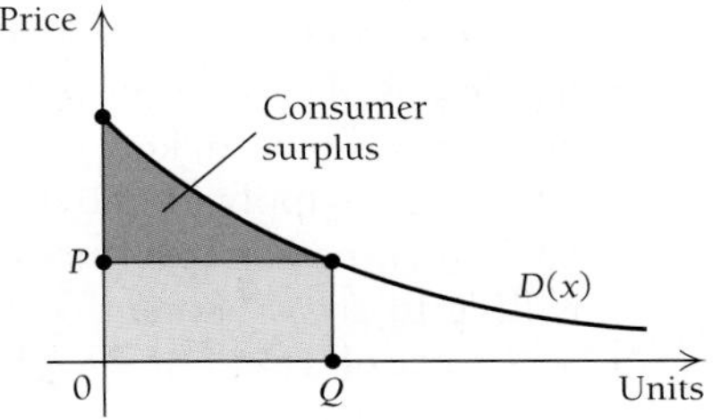

If Samantha goes to Q movies when the price is P, then her total expenditure is QP. The total area under the curve is the total utility, or the total enjoyment received, and is

$$\int_0^Q D(x)\,dx.$$

The *consumer surplus* is the total area under the curve minus the total expenditure. This surplus is the total utility minus the total cost and is given by

$$\int_0^Q D(x)\,dx - QP.$$

DEFINITION

Suppose that $p = D(x)$ describes the demand function for a commodity. Then the **consumer surplus** is defined for the point (Q, P) as

$$\int_0^Q D(x)\,dx - QP.$$

TECHNOLOGY CONNECTION

Exploratory

Graph $D(x) = (x - 5)^2$, the demand function in Example 1, using the viewing window [0, 5, 0, 30], with Yscl = 5. To find the consumer surplus at $x = 3$, we first find $D(3)$. Then we graph $y = D(3)$. What is the point of intersection of $y = D(x)$ and $y = D(3)$?

From the intersection, use DRAW to create a vertical line down to the x-axis. What does the area of the resulting rectangle represent? What does the area above the horizontal line and below the curve represent?

EXAMPLE 1 Find the consumer surplus for the demand function given by $D(x) = (x - 5)^2$ when $x = 3$.

Solution When $x = 3$, we have $D(3) = (3 - 5)^2 = (-2)^2 = 4$. Then

$$\begin{aligned}\text{Consumer surplus} &= \int_0^3 (x-5)^2\,dx - 3\cdot 4\\ &= \int_0^3 (x^2 - 10x + 25)\,dx - 12\\ &= \left[\frac{x^3}{3} - 5x^2 + 25x\right]_0^3 - 12\\ &= \left[\left(\frac{3^3}{3} - 5(3)^2 + 25(3)\right) - \left(\frac{0^3}{3} - 5(0)^2 + 25(0)\right)\right] - 12\\ &= (9 - 45 + 75) - 0 - 12\\ &= \$27.\end{aligned}$$

Let's now look at a *supply* curve for movies, as shown in Figs. 3 and 4. At a price of \$0 per movie ticket, the movie theater is willing to supply 0 movies. At a price of \$5.75 per person per movie, the theater is willing to show Samantha 1 movie and take in a total of 1(\$5.75), or \$5.75. The area of the beige triangle in Fig. 3 represents the total per-person cost to the theater of showing 1 movie, which is $\frac{1}{2}(1)(\$5.75)$, or \$2.875. The area of the green triangle is also \$2.875 and represents the surplus over cost. It is a contribution to profit. Economists call this number the *producer surplus*. It is the benefit a producer receives when supplying more units at a price that is higher than a price at which fewer units would be supplied. It is extra revenue that the producer receives as a result of not being forced to sell fewer units at a lower price.

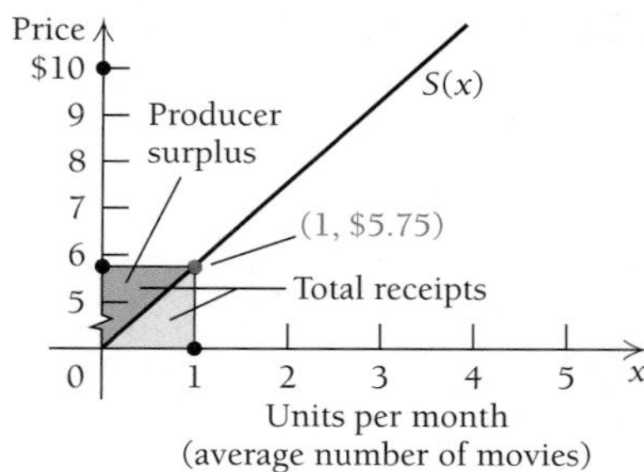

FIGURE 3

At a price of \$7.50, the theater will show Samantha 2 movies and collect total receipts of 2(\$7.50), or \$15. The area of the beige triangle in Fig. 4 represents the total cost to the theater of showing Samantha 2 movies, and is $\frac{1}{2}(2)(\$7.50)$, or \$7.50. The area of the green triangle is also \$7.50 and is the producer's surplus. It is a contribution to the theater's profit.

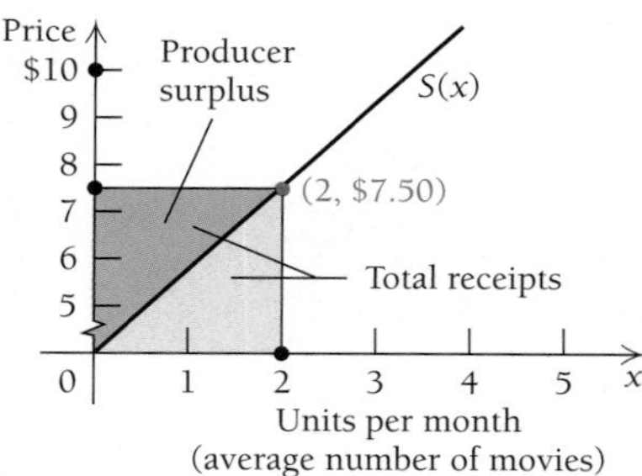

FIGURE 4

Suppose that the graph of the supply function is a curve, as shown at the right. If the theater shows Samantha Q movies when the price is P, the total receipts are QP. The *producer surplus* is the total receipts minus the area under the curve and is given by

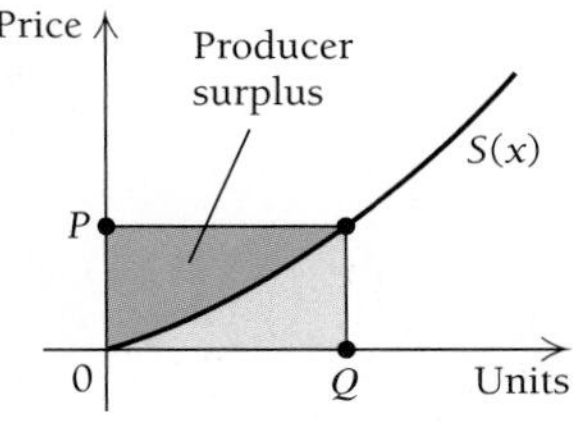

$$QP - \int_0^Q S(x)\,dx.$$

DEFINITION

Suppose that $p = S(x)$ is the supply function for a commodity. Then the **producer surplus** is defined for the point (Q, P) as

$$QP - \int_0^Q S(x)\,dx.$$

EXAMPLE 2 Find the producer surplus for $S(x) = x^2 + x + 3$ when $x = 3$.

Solution When $x = 3$, $S(3) = 3^2 + 3 + 3 = 15$. Then

$$\begin{aligned}\text{Producer surplus} &= 3 \cdot 15 - \int_0^3 (x^2 + x + 3)\,dx \\ &= 45 - \left[\frac{x^3}{3} + \frac{x^2}{2} + 3x\right]_0^3 \\ &= 45 - \left[\left(\frac{3^3}{3} + \frac{3^2}{2} + 3(3)\right) - \left(\frac{0^3}{3} + \frac{0^2}{2} + 3(0)\right)\right] \\ &= 45 - \left(9 + \frac{9}{2} + 9 - 0\right) \\ &= \$22.50.\end{aligned}$$

The **equilibrium point,** (x_E, p_E), in Fig. 5 is the point at which the supply and demand curves intersect. It is that point at which sellers and buyers come together and purchases and sales actually occur. In Fig. 6, we see the equilibrium point and the consumer and producer surpluses for the movie curves.

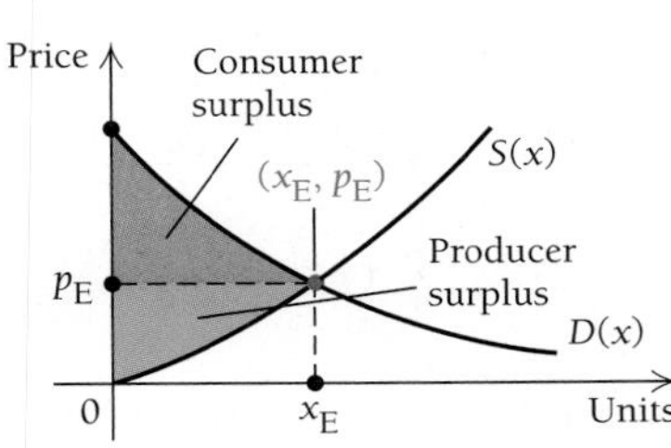

FIGURE 5

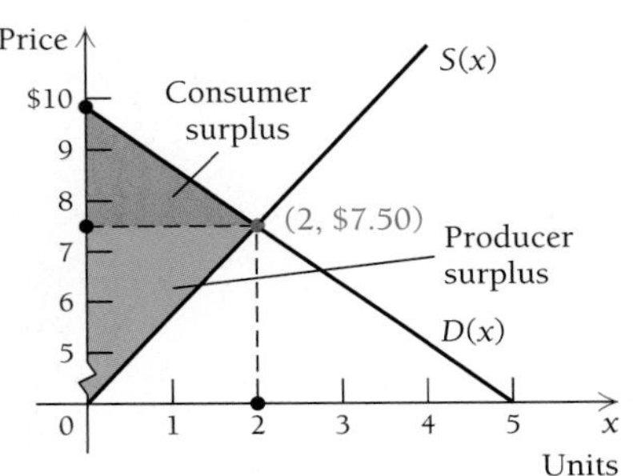

FIGURE 6

EXAMPLE 3 Given

$$D(x) = (x - 5)^2 \quad \text{and} \quad S(x) = x^2 + x + 3,$$

find each of the following.

a) The equilibrium point

b) The consumer surplus at the equilibrium point

c) The producer surplus at the equilibrium point

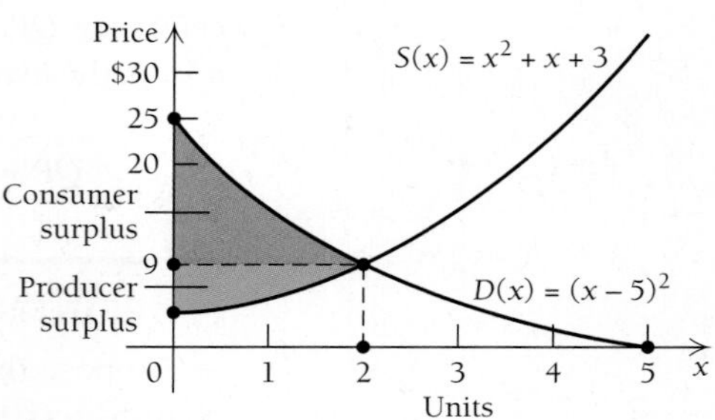

Solution

a) To find the equilibrium point, we set $D(x) = S(x)$ and solve:

$$\begin{aligned}(x - 5)^2 &= x^2 + x + 3\\ x^2 - 10x + 25 &= x^2 + x + 3\\ -10x + 25 &= x + 3\\ 22 &= 11x\\ 2 &= x.\end{aligned}$$

Thus, $x_E = 2$. To find p_E, we substitute x_E into either $D(x)$ or $S(x)$. If we choose $D(x)$, we have

$$\begin{aligned}p_E = D(x_E) &= D(2)\\ &= (2 - 5)^2\\ &= (-3)^2\\ &= \$9.\end{aligned}$$

Thus, the equilibrium point is (2, \$9).

b) The consumer surplus at the equilibrium point is

$$\int_0^{x_E} D(x)\,dx - x_E p_E,$$

or

$$\begin{aligned}\int_0^2 (x - 5)^2\,dx - 2 \cdot 9 &= \left[\frac{(x - 5)^3}{3}\right]_0^2 - 18\\ &= \left[\frac{(2 - 5)^3}{3} - \frac{(0 - 5)^3}{3}\right] - 18\\ &= \frac{(-3)^3}{3} - \frac{(-5)^3}{3} - 18 = -\frac{27}{3} + \frac{125}{3} - \frac{54}{3}\\ &= \frac{44}{3} \approx \$14.67.\end{aligned}$$

c) The producer surplus at the equilibrium point is

$$x_E p_E - \int_0^{x_E} S(x)\,dx,$$

or

$$2 \cdot 9 - \int_0^2 (x^2 + x + 3)\,dx = 2 \cdot 9 - \left[\frac{x^3}{3} + \frac{x^2}{2} + 3x\right]_0^2$$

$$= 18 - \left[\left(\frac{2^3}{3} + \frac{2^2}{2} + 3 \cdot 2\right) - \left(\frac{0^3}{3} + \frac{0^2}{2} + 3 \cdot 0\right)\right]$$

$$= 18 - \left(\frac{8}{3} + 2 + 6\right)$$

$$= \frac{22}{3} \approx \$7.33.$$

Exercise Set 5.1

In each of Exercises 1–14, D(x) is the price, in dollars per unit, that consumers are willing to pay for x units of an item, and S(x) is the price, in dollars per unit, that producers are willing to accept for x units. Find **(a)** *the equilibrium point,* **(b)** *the consumer surplus at the equilibrium point, and* **(c)** *the producer surplus at the equilibrium point.*

1. $D(x) = -\frac{5}{6}x + 9, \quad S(x) = \frac{1}{2}x + 1$

2. $D(x) = -3x + 7, \quad S(x) = 2x + 2$

3. $D(x) = (x - 4)^2, \quad S(x) = x^2 + 2x + 6$

4. $D(x) = (x - 3)^2, \quad S(x) = x^2 + 2x + 1$

5. $D(x) = (x - 6)^2, \quad S(x) = x^2$

6. $D(x) = (x - 8)^2, \quad S(x) = x^2$

7. $D(x) = 1000 - 10x, \quad S(x) = 250 + 5x$

8. $D(x) = 8800 - 30x, \quad S(x) = 7000 + 15x$

9. $D(x) = 5 - x$, for $0 \le x \le 5$; $\quad S(x) = \sqrt{x + 7}$

10. $D(x) = 7 - x$, for $0 \le x \le 7$; $\quad S(x) = 2\sqrt{x + 1}$

11. $D(x) = \frac{100}{\sqrt{x}}, \quad S(x) = \sqrt{x}$

12. $D(x) = \frac{1800}{\sqrt{x + 1}}, \quad S(x) = 2\sqrt{x + 1}$

13. $D(x) = (x - 4)^2, \quad S(x) = x^2 + 2x + 8$

14. $D(x) = 13 - x$, for $0 \le x \le 13$; $\quad S(x) = \sqrt{x + 17}$

SYNTHESIS

For Exercises 15 and 16, follow the directions given for Exercises 1–14.

15. $D(x) = e^{-x+4.5}, \quad S(x) = e^{x-5.5}$

16. $D(x) = \sqrt{56 - x}, \quad S(x) = x$

TW **17.** Explain why both consumers and producers feel good when consumer and producer surpluses exist.

TW **18.** Do some research on consumer and producer surpluses in an economics book. Write a brief description.

TECHNOLOGY CONNECTION

For Exercises 19 and 20, graph each pair of demand and supply functions. Then:

a) *Find the equilibrium point using the* INTERSECT *feature or another feature that will allow you to find this point of intersection.*

b) *Graph $y = D(x_E)$ and determine the regions of both consumer and producer surpluses.*

c) *Find the consumer surplus.*

d) *Find the producer surplus.*

19. $D(x) = \frac{x + 8}{x + 1}, \quad S(x) = \frac{x^2 + 4}{20}$

20. $D(x) = 15 - \frac{1}{3}x, \quad S(x) = 2\sqrt[3]{x}$

21. Bungee jumping. Reggie loves to go bungee jumping. The table shows the number of half-hours that Reggie is willing to go bungee jumping at various prices.

Time Spent (in half-hours per month)	Price (per half-hour)
8	\$ 2.50
7	5.00
6	7.50
5	10.00
4	12.50
3	15.00
2	17.50
1	20.00

a) Make a scatterplot of the data, and determine the type of function that you think fits best.

b) Fit that function to the data using REGRESSION.

c) If Reggie goes bungee jumping for 6 half-hours per month, what is his consumer surplus?

d) At a price of \$11.50 per half-hour, what is Reggie's consumer surplus?

5.2 Applications of the Models $\int_0^T P_0e^{kt}\,dt$ and $\int_0^T P_0e^{-kt}\,dt$

OBJECTIVES

- Perform computations involving interest compounded continuously and continuous money flow.
- Calculate the total consumption of a natural resource.
- Find the present value of an investment.
- Find the accumulated present value of an investment.

Recall the basic model of exponential growth (Section 3.3):

$$P'(t) = kP(t), \quad \text{or} \quad \frac{dP}{dt} = kP.$$

The function that satisfies the equation is

$$P(t) = P_0e^{kt}, \quad \text{where } P = P_0 \text{ when } t = 0. \tag{1}$$

One application of equation (1) is to compute the balance in a savings account t years after an initial amount, P_0, is invested at interest rate k, compounded continuously.

EXAMPLE 1 Business: Growth in an Investment. Find the balance in a savings account 3 yr after \$1000 is invested at 8%, compounded continuously.

TECHNOLOGY CONNECTION

EXERCISE

1. Graph the function $f(x) = 1000e^{0.08x}$ of Example 1 using the viewing window [0, 3, 0, 1500], with Xscl = 1 and Yscl = 100. Then find function values for $x = 0, 1, 2, 3$.

Solution Using equation (1) with $k = 0.08$, $t = 3$, and $P_0 = \$1000$, we get

$$\begin{aligned} P(3) &= 1000e^{0.08(3)} \\ &= 1000e^{0.24} \\ &= 1000(1.271249) \\ &\approx \$1271.25. \end{aligned}$$

The balance in the account after 3 yr will be approximately \$1271.25. ✦

In the remainder of this section, we examine applications that involve variations of the basic exponential growth model described by equation (1).

Continuous Money Flow

Suppose that a parking space in an airport lot generates a profit of \$3650 per year, or \$10 every day, and the money is invested in a savings account at 5%, compounded continuously, for 4 yr. We saw in Section 3.3 that the value of the first day's deposit will grow to

$$\frac{3650}{365}e^{0.05(4)}, \quad \text{or} \quad 10e^{0.05(4)},$$

at the end of the 4-yr period. The value of the money from the *second* day will grow in the 4-yr period to

$$\frac{3650}{365}e^{0.05(4-1/365)}, \quad \text{or} \quad 10e^{0.05(4-1/365)},$$

since it will be invested for 1 day, or 1/365 yr, less than the full 4 yr. The value of the money from the *third* day will grow to

$$10e^{0.05(4-2/365)},$$

and so on, for every day in the 4-yr period. Assuming that all deposits are made into the same account, the total value of that account after 4 yr is

$$10e^{0.05(4)} + 10e^{0.05(4-1/365)} + 10e^{0.05(4-2/365)} + \cdots + 10e^{0.05(2/365)} + 10e^{0.05(1/365)}.$$

Reversing the order of the terms in this sum, we have

$$10e^{0.05(1/365)} + 10e^{0.05(2/365)} + \cdots + 10e^{0.05(4-2/365)} + 10e^{0.05(4-1/365)} + 10e^{0.05(4)}.$$

If we express 10 as $3650 \cdot \Delta t$, with $\Delta t = \dfrac{1}{365}$, this sum can be written as

$$\sum_{i=1}^{4\cdot 365} 3650e^{0.05(i/365)}\,\Delta t.$$

This last expression is a Riemann sum. If the \$3650 flows into the account in a continuous manner, rather than through daily deposits, we have $\Delta t \to 0$, and the total value of the account becomes

$$\int_0^4 3650e^{0.05t}\,dt.$$

TECHNOLOGY CONNECTION

EXERCISE

1. Evaluate the integral

$$\int_0^3 \$1000e^{0.08x}\,dx.$$

Explain what the answer represents.

Before we generalize this result, let's complete the calculation:

$$\int_0^4 3650e^{0.05t}\,dt = \left[\frac{3650}{0.05}\cdot e^{0.05t}\right]_0^4$$
$$= \frac{3650}{0.05}(e^{0.05(4)} - 1)$$
$$\approx \$16{,}162.40.$$

Economists call a sum like \$16,162.40 the *future value of a continuous money flow*. In this case, the money is flowing according to a constant function $R(t) = \$3650$.

The situation we just considered can be generalized as follows.

FUTURE VALUE OF A CONTINUOUS MONEY FLOW

If the yearly flow of money into an investment is given by some constant function $R(t)$, then the **future value of the continuous money flow** at interest rate k, compounded continuously, over T years, is given by

$$\int_0^T R(t)e^{kt}\,dt.$$

TECHNOLOGY CONNECTION

Exploratory

Graph $f(x) = 1000e^{0.08x}$ using the viewing window [1, 15, 0, 5000], with Xscl = 2 and Yscl = 1000. Then graph the line $y = f(0)$. Consider the rectangle formed by the lines $y = f(0)$, $x = 15$, the y-axis, and the x-axis. Graph it if you can. What does the area of the rectangle represent? What does the area above the rectangle and below the graph of f represent? What does the entire area under the curve represent?

EXAMPLE 2 Business: Future Value of a Continuous Money Flow. Find the future value of the continuous money flow if \$1000 per year flows at a constant rate into an account paying 8%, compounded continuously, for 15 yr.

Solution

$$\int_0^{15} \$1000e^{0.08t}\,dt = \left[\frac{1000}{0.08}e^{0.08t}\right]_0^{15}$$
$$= \frac{1000}{0.08}(e^{1.2} - 1)$$
$$\approx \$29{,}001.46$$ ◆

Sometimes we might want to know how much money should be flowing into an investment so that we end up with a specified amount.

EXAMPLE 3 Business: Continuous Money Flow. Consider a continuous flow of money into an investment at the constant rate of P_0 dollars per year. What should P_0 be so that the amount of a continuous money flow over 20 yr, at an interest rate of 8%, compounded continuously, will be \$10,000?

Solution We find P_0 such that

$$10{,}000 = \int_0^{20} P_0e^{0.08t}\,dt.$$

We solve the following equation:

$$\begin{aligned}
10{,}000 &= P_0\left[\frac{1}{0.08}e^{0.08t}\right]_0^{20} \\
10{,}000 &= P_0[12.5e^{1.6} - 12.5] \\
10{,}000 &\approx 49.4129053P_0 \qquad \text{Using a calculator} \\
\$202.38 &\approx P_0.
\end{aligned}$$

A continuous money flow of \$202.38 per year, invested at 8%, compounded continuously for 20 years, will yield \$10,000. ✦

Life and Physical Sciences: Consumption of Natural Resources

Another application of the integral of exponential growth uses

$$P(t) = P_0e^{kt}$$

as a model of the demand for natural resources. Suppose that P_0 represents the annual amount of a natural resource (such as coal or oil) used at time $t = 0$ and that the growth rate for the use of this resource is k. Then, assuming exponential growth in demand (which is the case for the use of many resources), the amount used annually t years in the future is $P(t)$, given by

$$P(t) = P_0e^{kt}.$$

The total amount used during an interval $[0, T]$ is then given by

$$\int_0^T P(t)\,dt = \int_0^T P_0e^{kt}\,dt = \left[\frac{P_0}{k}e^{kt}\right]_0^T = \frac{P_0}{k}(e^{kT} - 1).$$

Consumption of a Natural Resource

Suppose that $P(t)$ is the annual consumption of a natural resource in year t. If consumption of the resource is growing exponentially at growth rate k, then the total consumption of the resource after T years is given by

$$\int_0^T P_0e^{kt}\,dt = \frac{P_0}{k}(e^{kT} - 1),$$

where P_0 is the annual consumption at time $t = 0$.

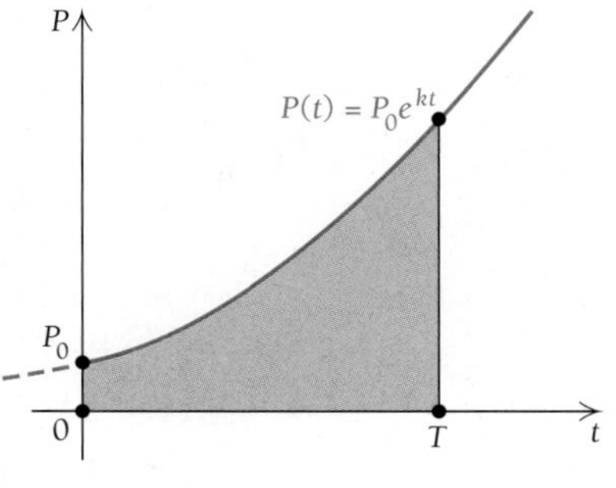

EXAMPLE 4 Physical Science: Gold Mining. In 2000 ($t = 0$), world gold production was 2547 metric tons, and it was growing exponentially at the rate of 0.6% per year. (*Source:* Based on data from the U.S. Geological Survey, U.S. Dept. of the Interior, 6/20/06.) If the growth continues at this rate, how many tons of gold will be produced from 2000 to 2013?

Solution Using the equation for consumption of a natural resource, we have

$$\begin{aligned}\int_0^{13} 2547e^{0.006t}\,dt &= \frac{2547}{0.006}(e^{0.006(13)} - 1)\\ &= 424{,}500(e^{0.078} - 1)\\ &\approx 424{,}500(1.081122659 - 1)\\ &\approx 34{,}437\end{aligned}$$

From 2000 to 2013, approximately 34,437 metric tons of gold will be produced. ✦

EXAMPLE 5 Physical Science: Depletion of Gold Reserves. The world reserves of gold in 2000 were estimated to be 77,000 metric tons. (*Source:* Based on data from the U.S. Geological Survey, U.S. Dept. of the Interior, 6/20/06. Data exclude China.) Assuming that the growth rate for production given in Example 4 continues and that no new reserves are discovered, when will the world reserves of gold be depleted?

Solution Using the equation for consumption of a natural resource, we want to find T such that

$$77{,}000 = \frac{2547}{0.006}(e^{0.006T} - 1).$$

We solve for T as follows:

$$\begin{aligned}77{,}000 &= 424{,}500(e^{0.006T} - 1)\\ 0.1814 &\approx e^{0.006T} - 1 && \text{Dividing both sides by 424,500}\\ 1.1814 &\approx e^{0.006T}\\ \ln 1.1814 &\approx \ln e^{0.006T} && \text{Taking the natural logarithm of each side}\\ \ln 1.1814 &\approx 0.006T && \text{Recall that } \ln e^k = k.\\ 28 &\approx T. && \text{Dividing both sides by 0.006 and rounding}\end{aligned}$$

Thus, assuming that world production of gold continues to increase at 0.6% per year and no new reserves are found (and disregarding Chinese reserves, for which data are unavailable), the world reserves of gold will be depleted 28 yr from 2000, in 2028. ✦

Applications of the Model $\int_0^T P_0e^{-kt}\,dt$

Recall the discussion of present value in Section 3.4. If you did not study that topic there, you may want to do so now. The *present value*, P_0, of an amount P, when P_0 is invested at interest rate k, compounded continuously, and due t years later, is found by solving the following equation for P_0:

$$\begin{aligned}P_0e^{kt} &= P\\ P_0 &= \frac{P}{e^{kt}} = Pe^{-kt}.\end{aligned}$$

DEFINITION

The **present value, P_0,** of an amount P, when P_0 is invested at interest rate k, compounded continuously, and due t years later, is given by

$$P_0 = Pe^{-kt}.$$

Note that this model can be interpreted as exponential decay from the future back to the present.

EXAMPLE 6 Business: Present Value. Find the present value of \$200,000 due 25 yr from now, at 8.7%, compounded continuously.

Solution We substitute 200,000 for P, 0.087 for k, and 25 for t in the equation for present value:

$$P_0 = 200{,}000e^{-0.087(25)} \approx 22{,}721.63.$$

Thus the present value is \$22,721.63. ✦

Suppose that a continuous flow of money will go into an account at the constant rate of P dollars per year, from now until some time T years in the future. We have seen that the future value of such an investment is given by

$$\int_0^T Pe^{kt}\,dt.$$

If we wished instead to deposit A dollars for T years at interest rate k, compounded continuously, so that the deposit had the same future value, that number of dollars must be such that

$$\begin{aligned} Ae^{kT} &= \int_0^T Pe^{kt}\,dt = \left[\frac{P}{k}e^{kt}\right]_0^T \\ &= \frac{P}{k}(e^{kT} - 1). \end{aligned}$$

Dividing both sides by e^{kT}, we have

$$\begin{aligned} A &= \frac{Pe^{kT}}{ke^{kT}} - \frac{P}{ke^{kT}} \\ &= \frac{P}{k}(1 - e^{-kT}) && \text{Simplifying} \\ &= \frac{P}{-k}(e^{-kT} - 1) && \text{Factoring out } -1 \\ &= \int_0^T Pe^{-kt}\,dt && \text{Recognizing the definite integral; check this by evaluating the integral.} \end{aligned}$$

Economists refer to A as the *accumulated present value* of a continuous money flow.

In the following definition, we drop the requirement that the investment be made at a constant rate.

DEFINITION

The **accumulated present value, A,** of a continuous money flow into an investment at a rate of $R(t)$ dollars per year from now until T years in the future is given by

$$A = \int_0^T R(t)e^{-kt}\,dt,$$

where k is the interest rate, and interest is compounded continuously.

EXAMPLE 7 Business: Accumulated Present Value. Find the accumulated present value of an investment over a 5-yr period if there is a continuous money flow of \$2400 per year and the interest rate is 14%, compounded continuously.

Solution The accumulated present value is

$$\begin{aligned}\int_0^5 \$2400e^{-0.14t}\,dt &= \left[\frac{2400}{-0.14}e^{-0.14t}\right]_0^5 \approx -17{,}142.86(e^{-0.14\cdot 5} - e^0)\\ &\approx -17{,}142.86(e^{-0.7} - 1)\\ &\approx \$8629.97.\end{aligned}$$

✦

Example 7 is an application of the model

$$\int_0^T Pe^{-kt}\,dt = \frac{P}{-k}(e^{-kT} - 1) = \frac{P}{k}(1 - e^{-kT}).$$

This model can be applied to a calculation of the buildup of a specific amount of radioactive material leaking into the atmosphere annually. Some of the material decays, but more continues to be released. The amount present at time T is given by the integral above.

Exercise Set 5.2

APPLICATIONS

Business and Economics

1. Find the amount in a savings account 3 yr after an initial investment of \$700 at interest rate 9%, compounded continuously.

2. Find the amount in a savings account 4 yr after an initial investment of \$700 at interest rate 10%, compounded continuously.

3. Find the future value of a continuous money flow when \$2000 per year is being invested at a constant rate, compounded continuously, at 6%, for 8 yr.

4. Find the future value of a continuous money flow when \$800 per year is being invested at a constant rate, compounded continuously, at 7%, for 10 yr.

5. Find the future value of a continuous money flow when \$1500 per year is being invested at a constant rate, compounded continuously, at 5%, for 20 yr.

6. Find the future value of a continuous money flow when \$1200 per year is being invested at a constant rate, compounded continuously, at 6%, for 15 yr.

7. What should P_0 be so that the amount of a continuous money flow, at a constant rate over 20 yr at 8.5% compounded continuously, will be \$50,000?

8. What should P_0 be so that the amount of a continuous money flow, at a constant rate over 20 yr at 7.5% compounded continuously, will be \$50,000?

9. What should P_0 be so that the amount of a continuous money flow, at a constant rate over 30 yr at 9% compounded continuously, will be \$40,000?

10. What should P_0 be so that the amount of a continuous money flow, at a constant rate over 30 yr at 10% compounded continuously, will be \$40,000?

11. Following the birth of their grandchild, Dori and Dwayne wish to make an initial investment, P_0, that will grow to \$50,000 by the child's 20th birthday. Interest is compounded continuously at 7%. What should the initial investment be?

12. Following the birth of a child, Addie and Wesley wish to make an initial investment, P_0, that will grow to \$60,000 by the child's 20th birthday. Interest is compounded continuously at 6%. What should the initial investment be?

13. Find the present value of \$100,000 due in 10 years, at 5%, compounded continuously.

14. Find the present value of \$50,000 due 16 yr from now, at 7.4%, compounded continuously.

15. Find the accumulated present value of an investment over a 10-yr period if there is a continuous money flow of \$2700 per year and the interest rate is 9%, compounded continuously.

16. Find the accumulated present value of an investment over a 10-yr period if there is a continuous money flow of \$2700 per year and the interest rate is 10%, compounded continuously.

17. At age 35, Rochelle accepts a position as vice president of a company. Assuming retirement at age 65 and an annual salary of \$95,000 that is paid as a continuous money flow, what is the accumulated present value of that flow? Assume an interest rate of 6%, compounded continuously.

18. Chris accepts a job as a truck driver at the age of 25. Assuming retirement at age 65 and an annual salary of \$41,000 that is paid as a continuous money flow, what is the accumulated present value of that flow? Assume an interest rate of 7%, compounded continuously.

Capitalized cost. *The capitalized cost, c, of an asset over its lifetime is the total of the initial cost and the present value of all maintenance that will occur in the future. It is computed by the formula*

$$c = c_0 + \int_0^L m(t)e^{-rt}\,dt,$$

where c_0 is the initial cost of the asset, L is the lifetime (in years), r is the interest rate (compounded continuously), and $m(t)$ is the annual cost of maintenance. Find the capitalized cost under each set of assumptions.

19. $c_0 = \$500{,}000$, $r = 5\%$, $m(t) = \$20{,}000$, $L = 20$

20. $c_0 = \$400{,}000$, $r = 5.5\%$, $m(t) = \$10{,}000$, $L = 25$

21. $c_0 = \$600{,}000$, $r = 4\%$, $m(t) = \$40{,}000 + \$1000e^{0.01t}$, $L = 40$

22. $c_0 = \$300{,}000$, $r = 5\%$, $m(t) = \$30{,}000 + \$500t$, $L = 20$

Life and Physical Sciences

23. The demand for natural gas. In 2006 ($t = 0$), the world consumption of natural gas was approximately 101.4 trillion cubic feet and was growing exponentially at about 2.6% per year. (***Source:*** *International Energy Outlook 2005,* U.S. Energy Information Administration, U.S. Department of Energy.) If the demand continues to grow at this rate, how many cubic feet of natural gas will the world use from 2006 to 2020?

24. The demand for aluminum ore (bauxite). In 2005 ($t = 0$), bauxite production was approximately 153 million metric tons, and the demand was growing exponentially at a rate of 2.5% per year. (***Source:*** Based on data from U.S. Energy Information Administration.) If the demand continues to grow at this rate, how many tons of bauxite will the world use from 2005 to 2030?

25. The depletion of natural gas. The world reserves of natural gas were approximately 6112 trillion cubic feet in 2006. (***Source:*** Estimate from *Oil and Gas Journal,* Jan. 1, 2006.) Assuming the growth described in Exercise 23 continues and that no new reserves are found, when will the world reserves of natural gas be depleted?

26. **The depletion of aluminum ore (bauxite).** In 2005, the world reserves of bauxite were about 23 billion metric tons. (*Source:* Based on data from U.S. Geological Survey summaries, Jan. 2005.) Assuming that the growth described in Exercise 24 continues and that no new reserves are discovered, when will the world reserves of bauxite be depleted?

27. **Radioactive buildup.** Plutonium-239 has a decay rate of approximately 0.003% per year. Suppose that plutonium-239 is released into the atmosphere each year for 20 yr at a constant rate of 1 lb per year. How much plutonium-239 will remain in the atmosphere after 20 yr?

28. **Radioactive buildup.** Cesium-137 has a decay rate of 2.3% per year. Suppose cesium-137 is released into the atmosphere each year for 20 yr at the rate of 1 lb per year. How much cesium-137 will remain in the atmosphere after 20 yr?

29. **Demand and depletion of oil.** Between 2006 and 2010, the annual world demand for oil is projected to increase from approximately 30.8 billion barrels to 34.5 billion barrels. (*Source:* Based on data from U.S. Department of Energy and *Oil and Gas Journal*, Jan. 1, 2006.)

a) Assuming an exponential growth model, compute the growth rate of demand.
b) Predict the demand in 2015.
c) The world reserves of crude oil in 2006 were estimated at 1293 billion barrels. Assuming that no new oil is found, when will the reserves be depleted?

SYNTHESIS

Let $R(t)$ be the rate of any continuous money flow, at interest rate k, compounded continuously, over T years. The future value of this money flow is then given by

$$\int_0^T R(t)e^{k(T-t)}\,dt.$$

Find the future value of a continuous money flow for each of the following.

30. $R(t) = t^2$, $k = 7\%$, and $T = 40$ yr

31. $R(t) = 2000t + 7$, $k = 8\%$, and $T = 30$ yr

*32. $R(t) = 100e^{0.2t}$, $k = 5\%$, and $T = 8$ yr

*33. $R(t) = 3000$, $k = 3\%$, and $T = 12$ yr

*34. $R(t) = 50t + 110$, $k = 4\%$, and $T = 10$ yr

35. **Charitable giving.** A foundation wants to fund a scholarship and is considering establishing either an account into which \$2500 per year flows at a constant rate for 15 yr or an account with a single initial investment of \$20,000. If both accounts earn 6% interest, compounded continuously, which account has the greatest future value?

TW 36. Describe the idea of present value to a friend who is not a business major. Then describe accumulated present value.

TW 37. Look up some data on rate of use and current world reserves of a natural resource not considered in this section. Predict when the world reserves for that resource will be depleted.

TECHNOLOGY CONNECTION

38. Graph the area under the curve represented by the demand for aluminum ore (bauxite) in Exercise 24. Then compute the area, and compare this result with the answer to the exercise.

39. Graph the area under the curve represented by the continuous money flow in Exercise 3. Then compute the area, and compare this result with the answer to the exercise.

* Exercise is new to this printing.

5.3 Improper Integrals

OBJECTIVES

- Determine whether an improper integral is convergent or divergent.
- Solve applied problems involving improper integrals.

Let's try to find the area of the region under the graph of $y = 1/x^2$ over the interval $[1, \infty)$.

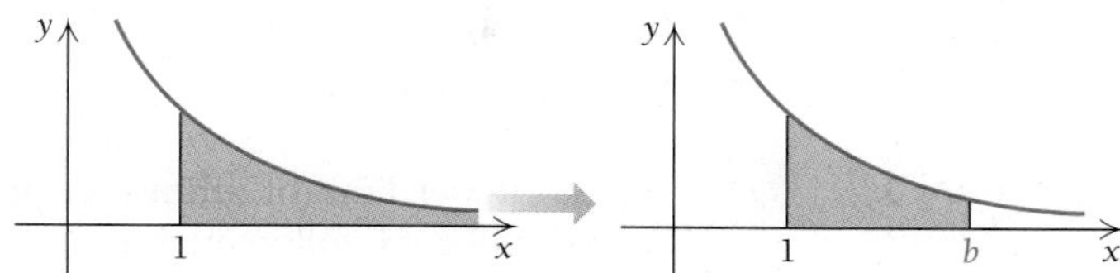

Note that this region is of infinite extent. We have not yet considered how to find the area of such a region. Let's find the area under the curve over the interval from 1 to b, and then see what happens as b gets very large. The area under the graph over $[1, b]$ is

$$\int_1^b \frac{dx}{x^2} = \left[-\frac{1}{x}\right]_1^b$$
$$= \left(-\frac{1}{b}\right) - \left(-\frac{1}{1}\right)$$
$$= -\frac{1}{b} + 1$$
$$= 1 - \frac{1}{b}.$$

Then

$$\lim_{b\to\infty} (\text{area from 1 to } b) = \lim_{b\to\infty}\left(1 - \frac{1}{b}\right) = 1.$$

We *define* the area from 1 to infinity to be this limit. Here we have an example of an infinitely long region with a finite area.

Such areas may not always be finite. Let's try to find the area of the region under the graph of $y = 1/x$ over the interval $[1, \infty)$.

TECHNOLOGY CONNECTION

EXPLORATORY EXERCISES

1. Using a calculator, find $\int_1^{10} \frac{dx}{x}$, $\int_1^{100} \frac{dx}{x}$, and $\int_1^{1000} \frac{dx}{x}$.
2. Predict the value of $\int_1^{\infty} \frac{dx}{x}$.

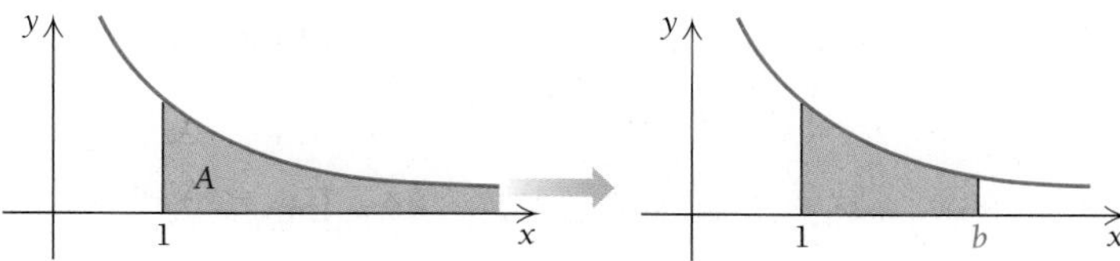

By definition, the area A from 1 to infinity is the limit as b approaches ∞ of the area from 1 to b, so

$$A = \lim_{b\to\infty} \int_1^b \frac{dx}{x} = \lim_{b\to\infty} [\ln x]_1^b$$
$$= \lim_{b\to\infty} (\ln b - \ln 1)$$
$$= \lim_{b\to\infty} \ln b.$$

In Section 3.2, we graphed $y = \ln x$ and saw that the function is always increasing. Therefore, the limit $\lim_{b\to\infty} \ln b$ does not exist and we have an infinitely long region with an infinite area.

Note that the graphs of $y = 1/x^2$ and $y = 1/x$ have similar shapes, but the region under one of them has a finite area and the other does not.

An integral such as

$$\int_a^\infty f(x)\,dx,$$

with an upper limit of infinity, is an example of an **improper integral.** Its value is defined to be the following limit.

DEFINITION

$$\int_a^\infty f(x)\,dx = \lim_{b\to\infty} \int_a^b f(x)\,dx$$

If the limit exists, then we say that the improper integral **converges,** or is **convergent.** If the limit does not exist, then we say that the improper integral **diverges,** or is **divergent.** Thus,

$$\int_1^\infty \frac{dx}{x^2} = 1 \ \textit{converges}, \quad \text{and} \quad \int_1^\infty \frac{dx}{x} \ \textit{diverges}.$$

EXAMPLE 1 Determine whether the following integral is convergent or divergent, and calculate its value if it is convergent:

$$\int_0^\infty 4e^{-2x}\,dx.$$

Solution We have

$$\begin{aligned}
\int_0^\infty 4e^{-2x}\,dx &= \lim_{b\to\infty} \int_0^b 4e^{-2x}\,dx \\
&= \lim_{b\to\infty} \left[\frac{4}{-2}e^{-2x}\right]_0^b \\
&= \lim_{b\to\infty} [-2e^{-2x}]_0^b \\
&= \lim_{b\to\infty} [-2e^{-2b} - (-2e^{-2\cdot 0})] \\
&= \lim_{b\to\infty} (-2e^{-2b} + 2) \\
&= \lim_{b\to\infty} \left(2 - \frac{2}{e^{2b}}\right).
\end{aligned}$$

As b approaches ∞, we know that e^{2b} approaches ∞ (see the graphs of $y = a^x$ in Chapter 3), so

$$\frac{2}{e^{2b}} \to 0 \quad \text{and} \quad \left(2 - \frac{2}{e^{2b}}\right) \to 2.$$

Thus,

$$\int_0^\infty 4e^{-2x}\,dx = \lim_{b\to\infty}\left(2 - \frac{2}{e^{2b}}\right) = 2.$$

The integral is convergent. ✦

Following are definitions of two other types of improper integrals.

DEFINITIONS

1. $\displaystyle\int_{-\infty}^{b} f(x)\,dx = \lim_{a\to-\infty}\int_a^b f(x)\,dx$

2. $\displaystyle\int_{-\infty}^{\infty} f(x)\,dx = \int_{-\infty}^{c} f(x)\,dx + \int_c^{\infty} f(x)\,dx,$
where c can be any real number.

In order for $\int_{-\infty}^{\infty} f(x)\,dx$ to converge, both integrals on the right in the second part of the definition must converge.

Applications of Improper Integrals

In Section 5.2, we learned that the accumulated present value of a continuous money flow of P dollars per year, at a constant rate, from now until T years in the future is given by

$$\int_0^T Pe^{-kt}\,dt = \frac{P}{k}(1 - e^{-kT}),$$

where k is the interest rate and interest is compounded continuously. Suppose that the money flow is to continue perpetually (forever). Under this assumption, the accumulated present value of the money flow is

$$\begin{aligned}\int_0^\infty Pe^{-kt}\,dt &= \lim_{T\to\infty}\int_0^T Pe^{-kt}\,dt\\ &= \lim_{T\to\infty}\frac{P}{k}(1 - e^{-kT})\\ &= \lim_{T\to\infty}\frac{P}{k}\left(1 - \frac{1}{e^{kT}}\right) = \frac{P}{k}.\end{aligned}$$

THEOREM 2

The **accumulated present value** of a continuous money flow into an investment at the rate of P dollars per year perpetually is given by

$$\int_0^\infty Pe^{-kt}\,dt = \frac{P}{k},$$

where k is the interest rate and interest is compounded continuously.

TECHNOLOGY CONNECTION

Exploratory

We can explore the situation of Example 2 with a calculator. To evaluate $\int_0^\infty 2000e^{-0.08x}\,dx$, we first consider

$$\int_0^t 2000e^{-0.08x}\,dx = \frac{2000}{0.08}(1 - e^{-0.08t}).$$

Then we examine what happens as t gets large. Graph

$$f(x) = \frac{2000}{0.08}(1 - e^{-0.08x})$$

using the window [0, 10, 0, 30000], with Xscl = 1 and Yscl = 5000. On the same set of axes, graph $y = 25{,}000$. Then change the viewing window to [0, 50, 0, 30000], with Xscl = 10 and Yscl = 5000, and finally to [0, 100, 0, 30000]. What happens as x gets larger? What is the significance of 25,000?

EXAMPLE 2 Business: Accumulated Present Value. Find the accumulated present value of an investment for which there is a perpetual continuous money flow of \$2000 per year. Assume that the interest rate is 8%, compounded continuously.

Solution The accumulated present value is 2000/0.08, or \$25,000. ◆

When an amount P of radioactive material is being released into the atmosphere annually, the total amount released at time T is given by

$$\int_0^T Pe^{-kt}\,dt = \frac{P}{k}(1 - e^{-kT}).$$

As T approaches ∞ (the radioactive material is to be released forever), the buildup of radioactive material approaches a limiting value P/k. It is no wonder that scientists and environmentalists are so concerned about radioactive waste. The radioactivity is "here to stay."

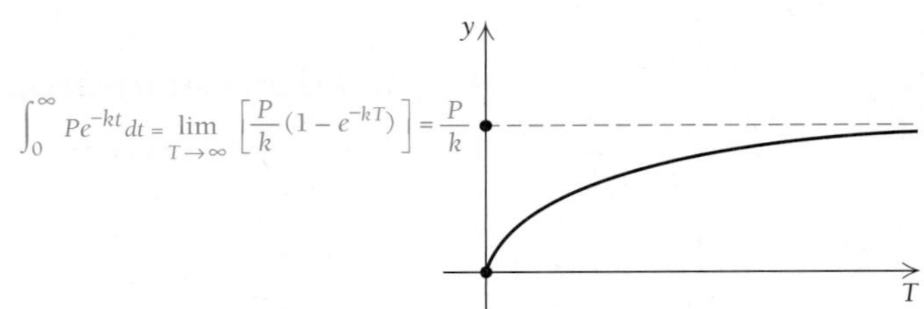

5.3 Exercise Set

Determine whether each improper integral is convergent or divergent, and calculate its value if it is convergent.

1. $\int_2^\infty \frac{dx}{x^2}$

2. $\int_4^\infty \frac{dx}{x^2}$

3. $\int_3^\infty \frac{dx}{x}$

4. $\int_4^\infty \frac{dx}{x}$

5. $\int_0^\infty 3e^{-3x}\,dx$

6. $\int_0^\infty 4e^{-4x}\,dx$

7. $\int_1^\infty \frac{dx}{x^3}$

8. $\int_1^\infty \frac{dx}{x^4}$

9. $\int_0^\infty \frac{dx}{2 + x}$

10. $\int_0^\infty \frac{4\,dx}{3 + x}$

11. $\int_2^\infty 4x^{-2}\,dx$

12. $\int_2^\infty 7x^{-2}\,dx$

13. $\int_0^\infty e^x\,dx$

14. $\int_0^\infty e^{2x}\,dx$

15. $\int_3^\infty x^2\,dx$

16. $\int_5^\infty x^4\,dx$

17. $\int_0^\infty xe^x\,dx$

18. $\int_1^\infty \ln x\,dx$

19. $\int_0^{\infty} me^{-mx}\,dx,\ m > 0$

20. $\int_0^{\infty} Qe^{-kt}\,dt,\ k > 0$

21. $\int_{\pi}^{\infty} \frac{dt}{t^{1.001}}$

22. $\int_1^{\infty} \frac{2t}{t^2 + 1}\,dt$

23. $\int_{-\infty}^{\infty} t\,dt$

24. $\int_1^{\infty} \frac{3x^2}{(x^3 + 1)^2}\,dx$

25. Find the area, if it is finite, of the region under the graph of $y = 1/x^2$ over the interval $[2, \infty)$.

26. Find the area, if it is finite, of the region under the graph of $y = 1/x$ over the interval $[2, \infty)$.

27. Find the area, if it is finite, of the region bounded by $y = 2xe^{-x^2}$, $x = 0$, and $[0, \infty)$.

28. Find the area, if it is finite, of the region bounded by $y = 1/\sqrt{(3x - 2)^3}$, $x = 6$, and $[6, \infty)$.

APPLICATIONS

Business and Economics

29. Accumulated present value. Find the accumulated present value of an investment for which there is a perpetual continuous money flow of $3600 per year at an interest rate of 7%, compounded continuously.

30. Accumulated present value. Find the accumulated present value of an investment for which there is a perpetual continuous money flow of $3500 per year at an interest rate of 6%, compounded continuously.

31. Total profit from marginal profit. A firm is able to determine that its marginal profit, in dollars, from producing x units of an item is given by

$$P'(x) = 200e^{-0.032x}.$$

Suppose that it were possible for the firm to make infinitely many units of this item. What would its total profit be?

32. Total profit from marginal profit. Find the total profit in Exercise 31 if

$$P'(x) = 200x^{-1.032}, \quad \text{where } x \geq 1.$$

33. Total cost from marginal cost. A company determines that its marginal cost, in dollars, for producing x units of a product is given by

$$C'(x) = 3600x^{-1.8}, \quad \text{where } x \geq 1.$$

Suppose that it were possible for the company to make infinitely many units of this product. What would the total cost be?

34. Total production. A firm determines that it can produce tires at a rate of

$$r(t) = 2000e^{-0.42t},$$

where t is the time, in years. Assuming that the firm endures forever (it never gets tired), how many tires can it make?

35. Accumulated present value. Find the accumulated present value of an investment for which there is a perpetual continuous money flow of $5000 per year, assuming continuously compounded interest at a rate of 8%.

36. Accumulated present value. Find the accumulated present value of an investment for which there is a perpetual continuous money flow of $\$2000e^{-0.01t}$ per year, assuming continuously compounded interest at a rate of 7%.

Capitalized cost. *The capitalized cost, c, of an asset for an unlimited lifetime is the total of the initial cost and the present value of all maintenance that will occur in the future. It is computed by the formula*

$$c = c_0 + \int_0^{\infty} m(t)e^{-rt}\,dt,$$

where c_0 is the initial cost of the asset, r is the interest rate (compounded continuously), and $m(t)$ is the annual cost of maintenance. Find the capitalized cost under each set of assumptions.

37. $c_0 = \$500{,}000, \quad r = 5\%, \quad m(t) = \$20{,}000$

38. $c_0 = \$700{,}000, \quad r = 5\%, \quad m(t) = \$30{,}000$

Life and Physical Sciences

39. Radioactive buildup. Plutonium has a decay rate of 0.003% per year. Suppose that a nuclear accident causes plutonium to be released into the atmosphere perpetually at the rate of 1 lb each year. What is the limiting value of the radioactive buildup?

40. Radioactive buildup. Cesium-137 has a decay rate of 2.3% per year. Suppose that a nuclear accident causes cesium-137 to be released into the atmosphere perpetually at the rate of 1 lb each year. What is the limiting value of the radioactive buildup?

Radioactive implant treatments. *In the treatment of prostate cancer, radioactive implants are often used. The implants are left in the patient and never removed. The amount of energy that is transmitted to the body from the implant is measured in rem units and is given by*

$$E = \int_0^a P_0 e^{-kt}\, dt,$$

where k is the decay constant for the radioactive material, a is the number of years since the implant, and P_0 is the initial rate at which energy is transmitted. Use this information for Exercises 41 and 42.

41. Suppose that the treatment uses iodine-125, which has a half-life of 60.1 days.

a) Find the decay rate, k, of iodine-125.

b) How much energy (measured in rems) is transmitted in the first month if the initial rate of transmission is 10 rem per year?

c) What is the total amount of energy that the implant will transmit to the body?

42. Suppose that the treatment uses palladium-103, which has a half-life of 16.99 days.

a) Find the decay rate, k, of palladium-103.

b) How much energy (measured in rems) is transmitted in the first month if the initial rate of transmission is 10 rem per year?

c) What is the total amount of energy that the implant will transmit to the body?

SYNTHESIS

Determine whether each improper integral is convergent or divergent, and calculate its value if it is convergent.

43. $\int_0^\infty \frac{dx}{x^{2/3}}$

44. $\int_1^\infty \frac{dx}{\sqrt{x}}$

45. $\int_0^\infty \frac{dx}{(x+1)^{3/2}}$

46. $\int_{-\infty}^0 e^{2x}\, dx$

47. $\int_0^\infty xe^{-x^2}\, dx$

48. $\int_{-\infty}^\infty xe^{-x^2}\, dx$

Life science: drug dosage. *Suppose that an oral dose of a drug is taken. Over time, the drug is assimilated in the body and excreted through the urine. The total amount of the drug that has passed through the body in time T is given by*

$$\int_0^T E(t)\, dt,$$

where $E(t)$ is the rate of excretion of the drug. A typical rate-of-excretion function is $E(t) = te^{-kt}$, where $k > 0$ and t is the time, in hours. Use this information for Exercises 49 and 50.

49. Find $\int_0^\infty E(t)\, dt$, and interpret the answer. That is, what does the integral represent?

50. A physician prescribes a dosage of 100 mg. Find k.

TW **51.** Consider the functions

$$y = \frac{1}{x^2} \quad \text{and} \quad y = \frac{1}{x}.$$

Suppose that you go to a paint store to buy paint to cover the region under each graph over the interval $[1, \infty)$. Discuss whether you could be successful and why or why not.

TW **52.** Suppose that you are the owner of a building that yields a continuous series of rental payments and you decide to sell the building. Explain how you would use the concept of the accumulated present value of a perpetual continuous money flow to determine a fair selling price.

TECHNOLOGY CONNECTION

53. Graph the function E and shade the area under the curve for each situation in Exercises 49 and 50.

Approximate each integral.

54. $\int_1^\infty \frac{4}{1+x^2}\, dx$

55. $\int_1^\infty \frac{6}{5+e^x}\, dx$

OBJECTIVES

- Verify certain properties of probability density functions.
- Solve applied problems involving probability density functions.

Probability

A number between 0 and 1 that represents the likelihood that an event will occur is referred to as the event's **probability.** There are two types of probability, *experimental* and *theoretical.*

Experimental and Theoretical Probability

If we toss a coin a great number of times—say, 1000—and count the number of times we get heads, we can determine the probability of the coin landing heads up. If it lands heads up 503 times, we calculate the probability of it landing heads up to be

$$\frac{503}{1000}, \quad \text{or} \quad 0.503.$$

This is an **experimental** determination of probability. Such a determination of probability is discovered by the observation and study of data and is quite common and very useful. Here, for example, are some probabilities that have been determined *experimentally:*

1. If a person has a heart attack, the probability he or she will die is $\frac{1}{3}$.
2. If you kiss someone who has a cold, the probability of catching the cold is 0.07.
3. A person who has just been released from prison has an 80% probability of returning to prison.

If we consider tossing a coin and reason that we are just as likely to get heads as tails, we would calculate the probability of it landing heads up to be $\frac{1}{2}$, or 0.5. This is a **theoretical** determination of probability. Here, for example, are some probabilities that have been determined *theoretically*, using mathematics:

1. If there are 30 people in a room, the probability that two of them have the same birthday (excluding year) is 0.706.
2. While on a trip, you meet someone and, after a period of conversation, discover that you have a common acquaintance. "It's a small world!" is actually not a very appropriate reaction in this situation, because the probability of such an occurrence is surprisingly high—just over 22%.

In summary, experimental probabilities are determined by making observations and gathering data. Theoretical probabilities are determined by reasoning mathematically. Examples of experimental and theoretical probabilities like those above, especially those we do not expect, lead us to see the value of a study of probability. You might ask, "What is *true* probability?" In fact, there is none. Experimentally, we can determine probabilities within certain limits. These may or may not agree with the probabilities that we obtain theoretically. There are situations in which it is much easier to determine one type of probability than the other. For example, it would be quite difficult to determine the theoretical probability of catching a cold.

In the discussion that follows, we will consider primarily theoretical probability. Eventually, calculus will come to bear on our considerations.

EXAMPLE 1 What is the probability of drawing an ace from a well-shuffled deck of cards?

A desire to calculate odds in games of chance gave rise to the theory of probability.

Solution Since there are 52 possible outcomes, and each card has the same chance of being drawn, and since there are 4 aces, the probability of drawing an ace is $\frac{4}{52}$ or $\frac{1}{13}$, or about 7.7%.

In practice, we may not draw an ace 7.7% of the time, but in a large number of trials, after shuffling the cards and drawing one, replacing the card, and shuffling the cards and drawing one, we would expect to draw an ace about 7.7% of the time. That is, the more draws we make, the closer we expect to get to 7.7%. ✦

EXAMPLE 2 A jar contains 7 black balls, 6 yellow balls, 4 green balls, and 3 red balls, all the same size and weight. The jar is shaken well, and you remove 1 ball without looking. What is the probability that the ball is red? That it is white?

Solution There are 20 balls altogether and of these 3 are red, so the probability of drawing a red ball is $\frac{3}{20}$. There are no white balls, so the probability of drawing a white one is $\frac{0}{20}$, or 0. ✦

Below is a table of probabilities for the situation in Example 2. Note that the sum of these probabilities is 1. We are certain that we will draw either a black, yellow, green, or red ball. The probability of that event is 1. Let's arrange these data from the table into what is called a **relative frequency graph,** or **histogram,** which shows the proportion of times that each event occurs (the probability of each event). If we assign a width of 1 to each rectangle in this graph, then the sum of the areas of

Color	Probability
Black (B)	$\frac{7}{20}$
Yellow (Y)	$\frac{6}{20}$
Green (G)	$\frac{4}{20}$
Red (R)	$\frac{3}{20}$

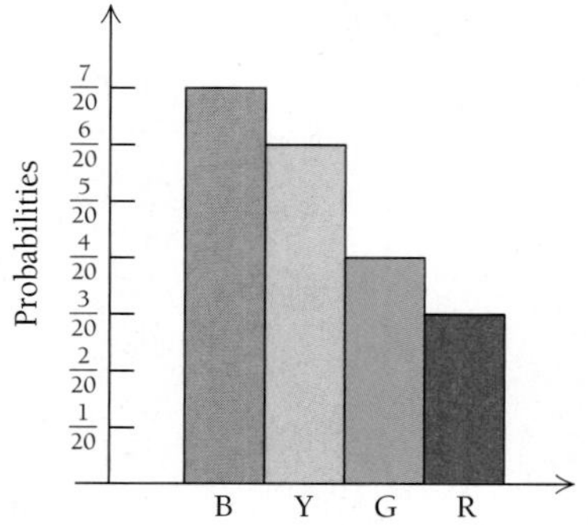

the rectangles is 1. That is, it is certain that you will draw a ball of one of these colors.

Continuous Random Variables

Suppose that we throw a dart at a number line in such a way that it always lands in the interval $[1, 3]$. Let x be the number that the dart hits. There is an infinite number of possibilities for x. Note that x is a quantity that can be observed (or measured) repeatedly and whose possible values comprise an interval of real numbers. Such a variable is called a **continuous random variable.**

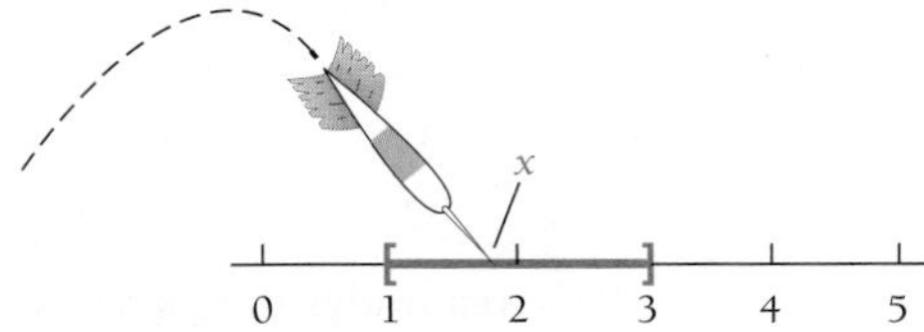

Suppose that we throw the dart a large number of times and it lands 43% of the time in the subinterval $[1.6, 2.8]$ of the main interval $[1, 3]$. The probability, then, that the dart lands in the interval $[1.6, 2.8]$ is 0.43.

Let's consider some other examples of continuous random variables.

EXAMPLE 3 Suppose that buses traveling from Philadelphia to New York City require at least 2 hr and at most 5 hr for the trip. If x is the number of hours a bus takes to make the trip, then x is a continuous random variable distributed over the interval $[2, 5]$.

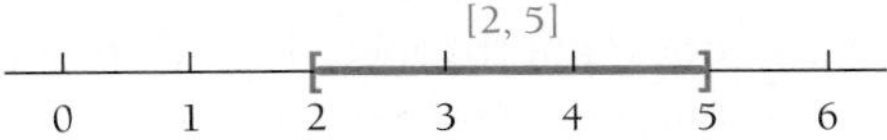

✦

EXAMPLE 4 Suppose that x is the corn acreage of any farm in the United States or Canada. The interval is $[0, a]$, where a is the highest acreage. Not knowing what the highest acreage is, we could use $[0, \infty)$ to allow for all possibilities. *Note*: It might be argued that there is a value in $[0, a]$ or $[0, \infty)$ for which no farm has that acreage, but for practical convenience, all values are included in our consideration.

Then x is a continuous random variable distributed over the interval $[0, a]$ or $[0, \infty)$.

✦

Considering Example 3 on the travel times of buses, suppose that we want to know the probability that a bus will take between 4 hr and 5 hr, as represented by

$$P([4, 5]), \quad \text{or} \quad P(4 \leq x \leq 5).$$

There may be a function $y = f(x)$ such that the area under the graph over a subinterval of $[2, 5]$ gives the probability that a particular trip time appears in the subinterval. For example, suppose that we have a constant function $f(x) = \frac{1}{3}$ that gives us these probabilities. Look at its graph.

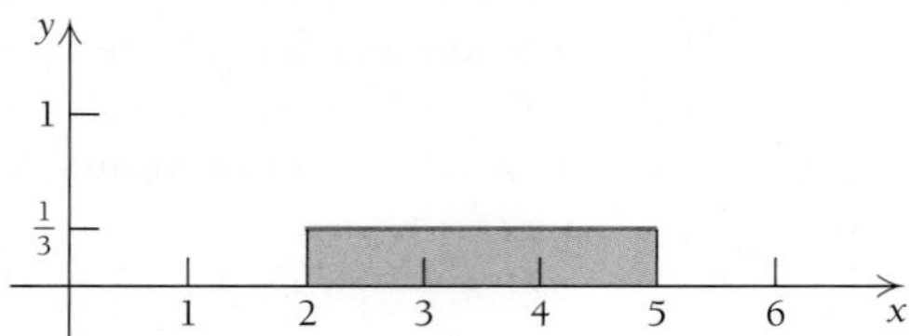

The area under the curve is $3 \cdot \frac{1}{3}$, or 1. The probability that a trip takes between 4 hr and 5 hr is the area that lies over the interval $[4, 5]$. That is,

$$P([4, 5]) = \tfrac{1}{3} = 33\tfrac{1}{3}\%.$$

The probability that a trip takes between 2 hr and 4.5 hr is $\frac{5}{6}$, or $83\frac{1}{3}\%$. This is the area of the rectangle over $[2, 4.5]$.

Note that when $f(x) = \frac{1}{3}$, any interval between the numbers 2 and 5 of width 1 has probability $\frac{1}{3}$. This does not happen for all functions. Suppose instead that

$$f(x) = \tfrac{3}{117}x^2.$$

As shown in the graph, the area under the graph of f from 4 to 5 is given by the definite integral over the interval $[4, 5]$ and yields the probability that a trip takes between 4 hr and 5 hr.

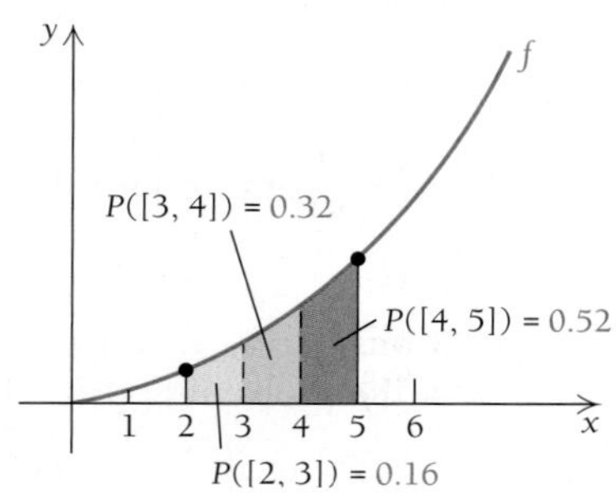

We have

$$P([4, 5]) = \int_4^5 f(x)\,dx$$
$$= \int_4^5 \frac{3}{117}x^2\,dx$$
$$= \frac{3}{117}\left[\frac{x^3}{3}\right]_4^5$$
$$= \frac{1}{117}\left[x^3\right]_4^5$$
$$= \frac{1}{117}(5^3 - 4^3) = \frac{61}{117} \approx 0.52.$$

Thus, according to this model, there is a probability of 0.52 that a bus trip takes betweeen 4 hr and 5 hr. The function f is called a *probability density function.* Its integral over *any* subinterval gives the probability that x "lands" in that subinterval.

Similar calculations for the bus trip example are shown in the following table.

Trip Time	Probability That a Trip Time Occurs during the Interval
2 hr to 3 hr	$\int_2^3 \frac{3}{117}x^2\,dx = 0.16 = P([2, 3])$
3 hr to 4 hr	$\int_3^4 \frac{3}{117}x^2\,dx = 0.32 = P([3, 4])$
4 hr to 5 hr	$\int_4^5 \frac{3}{117}x^2\,dx = 0.52 = P([4, 5])$
2 hr to 5 hr	$\int_2^5 \frac{3}{117}x^2\,dx = 1.00 = P([2, 5])$

TECHNOLOGY CONNECTION

Exploratory

Graph the function

$$f(x) = \tfrac{3}{117}x^2$$

using a viewing window of [0, 5, 0, 1]. Then successively evaluate each of the following integrals, shading the appropriate area if possible:

$$\int_2^3 \tfrac{3}{117}x^2\,dx,$$

$$\int_3^4 \tfrac{3}{117}x^2\,dx,$$

and

$$\int_4^5 \tfrac{3}{117}x^2\,dx.$$

Add your results, and explain the meaning of the total.

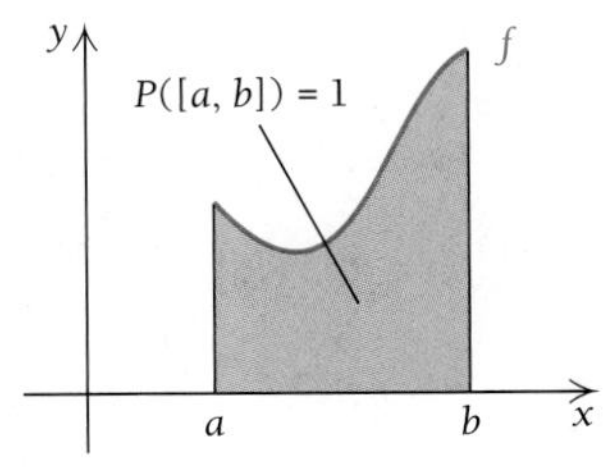

FIGURE 1

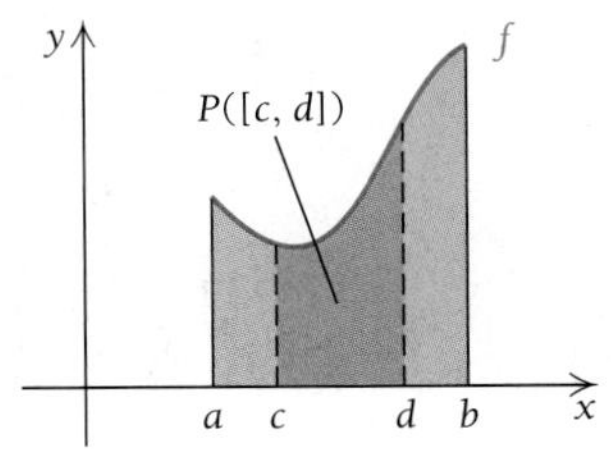

FIGURE 2

The results in the table lead us to the following definition of a probability density function.

DEFINITION

Let x be a continuous random variable. A function f is said to be a **probability density function** for x if:

1. For all x in the domain of f, we have $0 \leq f(x)$.
2. The area under the graph of f is 1 (see Fig. 1).
3. For any subinterval $[c, d]$ in the domain of f (see Fig. 2), the probability that x will be in that subinterval is given by

$$P([c, d]) = \int_c^d f(x)\,dx.$$

EXAMPLE 5 Verify Property 2 of the above definition for

$$f(x) = \frac{3}{117}x^2, \quad \text{for } 2 \leq x \leq 5.$$

Solution

$$\int_2^5 \frac{3}{117}x^2\,dx = \frac{3}{117}\left[\frac{1}{3}x^3\right]_2^5$$
$$= \frac{1}{117}\left[x^3\right]_2^5$$
$$= \frac{1}{117}(5^3 - 2^3)$$
$$= \frac{117}{117} = 1$$

✦

EXAMPLE 6 Business: Life of a Product. A company that produces fluorescent bulbs determines that the life t of a bulb is from 3 to 6 yr and that the probability density function for t is given by

$$f(t) = \frac{24}{t^3}, \quad \text{for } 3 \leq t \leq 6.$$

a) Verify Property 2 of the definition of a probability density function.

b) Find the probability that a bulb will last no more than 4 yr.

c) Find the probability that a bulb will last at least 4 yr and at most 5 yr.

Solution

a) We want to show that $\int_3^6 f(t)\,dt = 1$. We have

$$\int_3^6 \frac{24}{t^3}\,dt = 24\int_3^6 t^{-3}\,dt = 24\left[\frac{t^{-2}}{-2}\right]_3^6$$
$$= -12\left[\frac{1}{t^2}\right]_3^6 = -12\left(\frac{1}{6^2} - \frac{1}{3^2}\right)$$
$$= -12\left(\frac{1}{36} - \frac{1}{9}\right) = -12\left(-\frac{3}{36}\right) = 1.$$

b) The probability that a bulb will last no more than 4 yr is

$$P(3 \leq t \leq 4) = \int_3^4 \frac{24}{t^3}\,dt = 24\int_3^4 t^{-3}\,dt$$
$$= 24\left[\frac{t^{-2}}{-2}\right]_3^4 = -12\left[\frac{1}{t^2}\right]_3^4$$
$$= -12\left(\frac{1}{4^2} - \frac{1}{3^2}\right) = -12\left(\frac{1}{16} - \frac{1}{9}\right)$$
$$= -12\left(-\frac{7}{144}\right) = \frac{7}{12} \approx 0.58.$$

c) The probability that a bulb will last at least 4 yr and at most 5 yr is

$$\begin{aligned} P(4 \le t \le 5) &= \int_4^5 \frac{24}{t^3}\,dt = 24\int_4^5 t^{-3}\,dt \\ &= 24\left[\frac{t^{-2}}{-2}\right]_4^5 = -12\left[\frac{1}{t^2}\right]_4^5 \\ &= -12\left(\frac{1}{5^2} - \frac{1}{4^2}\right) = -12\left(\frac{1}{25} - \frac{1}{16}\right) \\ &= -12\left(-\frac{9}{400}\right) = \frac{27}{100} = 0.27. \end{aligned}$$

✦

Constructing Probability Density Functions

Suppose that you have an arbitrary nonnegative function $f(x)$ whose definite integral over some interval $[a, b]$ is K. Then

$$\int_a^b f(x)\,dx = K.$$

Multiplying on both sides by $1/K$ gives us

$$\frac{1}{K}\int_a^b f(x)\,dx = \frac{1}{K}\cdot K = 1, \quad \text{or} \quad \int_a^b \frac{1}{K}\cdot f(x)\,dx = 1.$$

Thus, when we multiply the function $f(x)$ by $1/K$, we have a function whose area over the given interval is 1.

EXAMPLE 7 Find k such that

$$f(x) = kx^2$$

is a probability density function over the interval $[1, 4]$. Then write the probability density function.

Solution We have

$$\begin{aligned} \int_1^4 x^2\,dx &= \left[\frac{x^3}{3}\right]_1^4 \\ &= \frac{4^3}{3} - \frac{1^3}{3} = \frac{64}{3} - \frac{1}{3} = \frac{63}{3} = 21. \end{aligned}$$

Thus, for f to be a probability density function, we must have $\int_1^4 kx^2\,dx = 1$, or $k\int_1^4 x^2\,dx = 1$, and

$$k = \tfrac{1}{21}.$$

The probability density function is

$$f(x) = \tfrac{1}{21}x^2.$$

✦

Uniform Distributions

Suppose that the probability density function of a continuous random variable is constant. How is it described? Consider the graph shown below.

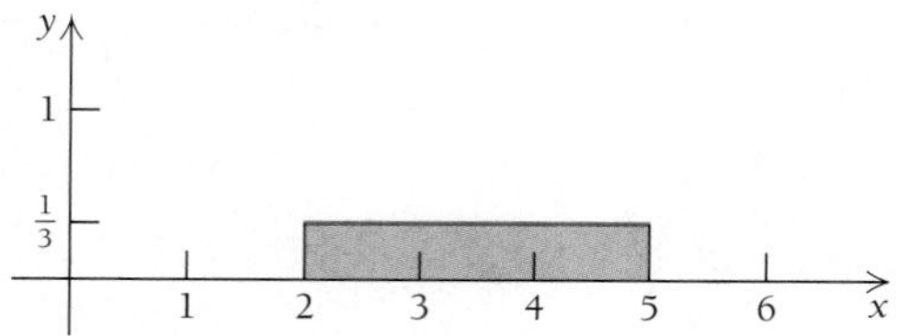

The length of the shaded rectangle is the length of the interval $[2, 5]$, which is 3. In order for the shaded area to be 1, the height of the rectangle must be $\frac{1}{3}$. Thus, $f(x) = \frac{1}{3}$.

The length of the shaded rectangle shown below is the length of the interval $[a, b]$, which is $b - a$. In order for the shaded area to be 1, the height of the rectangle must be $1/(b - a)$. Thus, $f(x) = 1/(b - a)$.

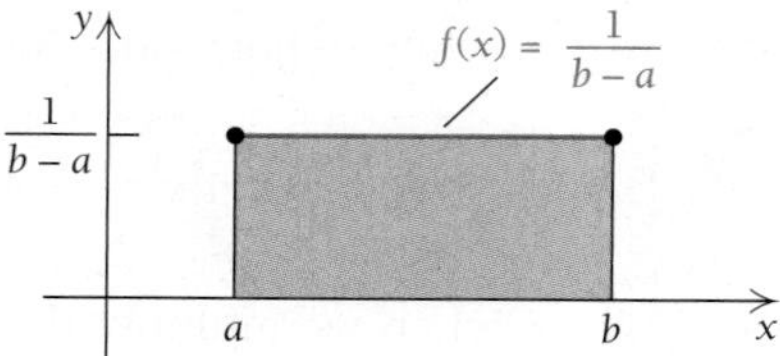

DEFINITION

A continuous random variable x is said to be **uniformly distributed** over an interval $[a, b]$ if it has a probability density function f given by

$$f(x) = \frac{1}{b - a}, \quad \text{for } a \leq x \leq b.$$

EXAMPLE 8 A number x is selected at random from the interval $[40, 50]$. The probability density function for x is given by

$$f(x) = \tfrac{1}{10}, \quad \text{for } 40 \leq x \leq 50.$$

Find the probability that a number selected is in the subinterval $[42, 48]$.

Solution The probability is

$$P(42 \leq x \leq 48) = \int_{42}^{48} \tfrac{1}{10}\, dx = \tfrac{1}{10}[x]_{42}^{48}$$

$$= \tfrac{1}{10}(48 - 42) = \tfrac{6}{10} = 0.6.$$

✦

EXAMPLE 9 Business: Quality Control. A company produces sirens used for tornado warnings. The maximum loudness, L, of the sirens ranges from 70 to 100 decibels. The probability density function for L is

$$f(L) = \tfrac{1}{30}, \quad \text{for } 70 \leq L \leq 100.$$

A siren is selected at random off the assembly line. Find the probability that its maximum loudness is from 70 to 92 decibels.

Solution The probability is

$$P(70 \leq L \leq 92) = \int_{70}^{92} \tfrac{1}{30}\, dL = \tfrac{1}{30}[L]_{70}^{92}$$

$$= \tfrac{1}{30}(92 - 70) = \tfrac{22}{30} = \tfrac{11}{15} \approx 0.73.$$

Exponential Distributions

The duration of a phone call, the distance between successive cars on a highway, and the amount of time required to learn a task are all examples of *exponentially distributed* random variables. That is, their probability density functions are exponential.

DEFINITION

A continuous random variable is **exponentially distributed** if it has a probability density function of the form

$$f(x) = ke^{-kx}, \quad \text{over the interval } [0, \infty).$$

To see that $f(x) = 2e^{-2x}$ is such a probability density function, note that

$$\int_0^{\infty} 2e^{-2x}\, dx = \lim_{b \to \infty} \int_0^b 2e^{-2x}\, dx = \lim_{b \to \infty} [-e^{-2x}]_0^b = \lim_{b \to \infty} \left(\frac{-1}{e^{2b}} - (-1) \right) = 1.$$

The general case,

$$\int_0^{\infty} ke^{-kx}\, dx = 1,$$

can be verified in a similar way.

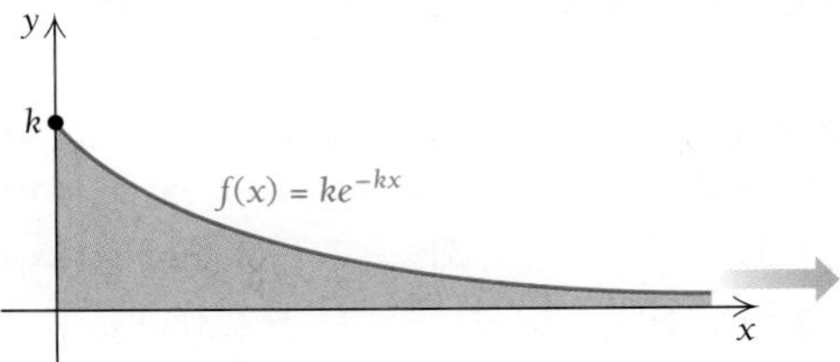

Why is it reasonable to assume that the distance between cars is exponentially distributed? Part of the reason is that there are many more cases in which distances are small, though we can find other distributions that are "skewed" in this manner. The same argument holds for the duration of a phone call. That is, there are more short calls than long ones. The rest of the reason might lie in an analysis of the data involving such distances or phone calls.

A transportation planner can determine the probabilities that cars are certain distances apart.

EXAMPLE 10 Business: Transportation Planning. The distance x, in feet, between successive cars on a certain stretch of highway has a probability density function

$$f(x) = ke^{-kx}, \quad \text{for } 0 \le x < \infty,$$

where $k = 1/a$ and a is the average distance between successive cars over some period of time.

A transportation planner determines that the average distance between cars on a certain stretch of highway is 166 ft. What is the probability that the distance between two successive cars, chosen at random, is 50 ft or less?

Solution We first determine k:

$$k = \frac{1}{166} \approx 0.006024.$$

The probability density function for x is

$$f(x) = 0.006024e^{-0.006024x}, \quad \text{for } 0 \le x < \infty.$$

The probability that the distance between the cars is 50 ft or less is

$$\begin{aligned} P(0 \le x \le 50) &= \int_0^{50} 0.006024e^{-0.006024x}\,dx \\ &= \left[\frac{0.006024}{-0.006024}e^{-0.006024x}\right]_0^{50} \\ &= \left[-e^{-0.006024x}\right]_0^{50} \\ &= (-e^{-0.006024\cdot 50}) - (-e^{-0.006024\cdot 0}) \\ &= -e^{-0.301200} + 1 \\ &= 1 - e^{-0.301200} \\ &= 1 - 0.739930 \approx 0.260. \end{aligned}$$

Exercise Set 5.4

In Exercises 1–12, verify Property 2 of the definition of a probability density function over the given interval.

1. $f(x) = \frac{1}{4}x, \quad [1, 3]$

2. $f(x) = 2x, \quad [0, 1]$

3. $f(x) = 3, \quad \left[0, \frac{1}{3}\right]$

4. $f(x) = \frac{1}{5}, \quad [3, 8]$

5. $f(x) = \frac{3}{64}x^2, \quad [0, 4]$

6. $f(x) = \frac{3}{26}x^2, \quad [1, 3]$

7. $f(x) = \dfrac{1}{x}, \quad [1, e]$

8. $f(x) = \dfrac{1}{e-1}e^x, \quad [0, 1]$

9. $f(x) = \frac{3}{2}x^2, \quad [-1, 1]$

10. $f(x) = \frac{1}{3}x^2, \quad [-2, 1]$

11. $f(x) = 3e^{-3x}, \quad [0, \infty)$

12. $f(x) = 4e^{-4x}, \quad [0, \infty)$

Find k such that each function is a probability density function over the given interval. Then write the probability density function.

13. $f(x) = kx, \quad [2, 5]$

14. $f(x) = kx, \quad [1, 4]$

15. $f(x) = kx^2, \quad [-1, 1]$

16. $f(x) = kx^2, \quad [-2, 2]$

17. $f(x) = k, \quad [1, 7]$

18. $f(x) = k, \quad [3, 9]$

19. $f(x) = k(2 - x), \quad [0, 2]$

20. $f(x) = k(4 - x), \quad [0, 4]$

21. $f(x) = \dfrac{k}{x}, \quad [1, 3]$

22. $f(x) = \dfrac{k}{x}, \quad [1, 2]$

23. $f(x) = ke^x, \quad [0, 3]$

24. $f(x) = ke^x, \quad [0, 2]$

25. A dart is thrown at a number line in such a way that it always lands in the interval $[0, 10]$. Let x represent the number that the dart hits. Suppose that the probability density function for x is given by

$$f(x) = \tfrac{1}{50}x, \quad \text{for } 0 \le x \le 10.$$

a) Find $P(2 \le x \le 6)$, the probability that the dart lands in $[2, 6]$.

TW **b)** Interpret your answer to part (a).

26. In Exercise 25, suppose that the dart always lands in the interval $[0, 5]$, and that the probability density function for x is given by

$$f(x) = \tfrac{3}{125}x^2, \quad \text{for } 0 \le x \le 5.$$

a) Find $P(1 \le x \le 4)$, the probability that the dart lands in $[1, 4]$.

TW **b)** Interpret your answer to part (a).

27. A number x is selected at random from the interval $[4, 20]$. The probability density function for x is given by

$$f(x) = \tfrac{1}{16}, \quad \text{for } 4 \le x \le 20.$$

Find the probability that a number selected is in the subinterval $[9, 20]$.

28. A number x is selected at random from the interval $[5, 29]$. The probability density function for x is given by

$$f(x) = \tfrac{1}{24}, \quad \text{for } 5 \le x \le 29.$$

Find the probability that a number selected is in the subinterval $[14, 29]$.

APPLICATIONS

Business and Economics

29. Transportation planning. Refer to Example 10. A transportation planner determines that the average distance between cars on a certain highway is 100 ft. What is the probability that the distance between two successive cars, chosen at random, is 40 ft or less?

30. Transportation planning. Refer to Example 10. A transportation planner determines that the average distance between cars on a certain highway is 200 ft. What is the probability that the distance between two successive cars, chosen at random, is 10 ft or less?

31. Duration of a phone call. A telephone company determines that the duration t, in minutes, of a phone call is an exponentially distributed random variable with a probability density function

$$f(t) = 2e^{-2t}, \quad 0 \le t < \infty.$$

Find the probability that a phone call will last no more than 5 min.

32. Duration of a phone call. Referring to Exercise 31, find the probability that a phone call will last no more than 2 min.

33. Time to failure. The *time to failure, t*, in hours, of a machine is often exponentially distributed with a probability density function

$$f(t) = ke^{-kt}, \quad 0 \le t < \infty,$$

where $k = 1/a$ and a is the average amount of time that will pass before a failure occurs. Suppose that the average amount of time that will pass before a failure occurs is 100 hr. What is the probability that a failure will occur in 50 hr or less?

34. Reliability of a machine. The *reliability* of the machine (the probability that it will work) in Exercise 33 is defined as

$$R(T) = 1 - \int_0^T 0.01e^{-0.01t}\,dt,$$

where $R(T)$ is the reliability at time T. Write $R(T)$ without using an integral.

Life and Physical Sciences

35. Mortality rate. For every 100,000 females in the United States of any age x between 1 and 85, the number that died in 2003 can be approximated by the function

$$f(x) = 8.1305e^{0.074x}.$$

(*Source:* Based on data from the Centers for Disease Control.)

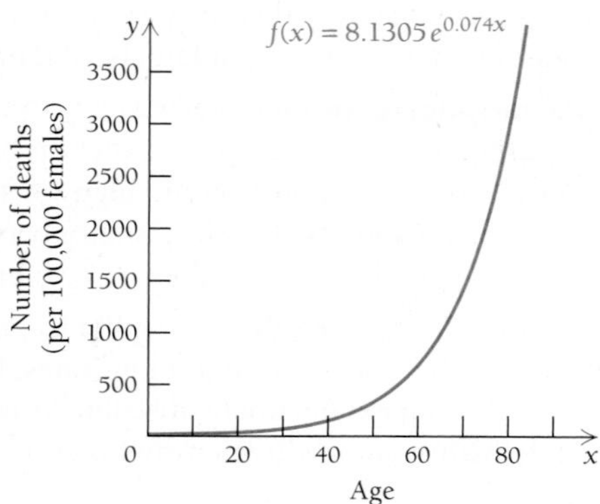

a) Find a value k to make the function $kf(x)$ a probability density function.

b) Use your answer from part (a) to approximate the probability that a female aged 1 through 85 who died in 2003 was between 25 and 40 years old.

36. Mortality rate. For every 100,000 males in the United States of any age x between 1 and 85, the number that died in 2003 can be approximated by the function

$$f(x) = 17.359e^{0.067x}.$$

(*Source:* Based on data from the Centers for Disease Control.)

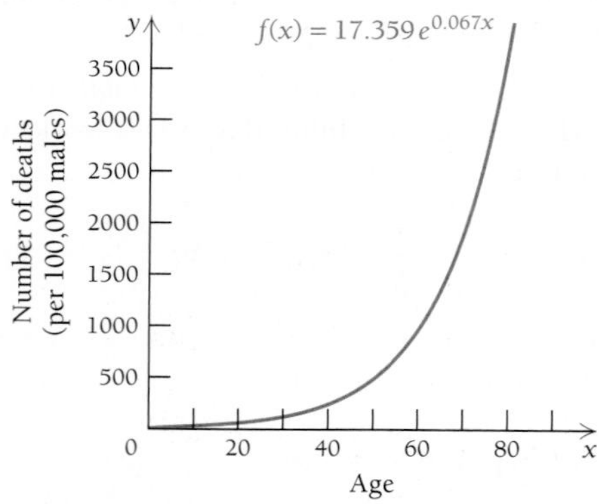

a) Find a value k to make the function $kf(x)$ a probability density function.

b) Use your answer from part (a) to approximate the probability that a male aged 1 through 85 who died in 2003 was between 20 and 30 years old.

Social Sciences

37. Time in a maze. In a psychology experiment, the time t, in seconds, that it takes a rat to learn its way through a maze is an exponentially distributed random variable with the probability density function

$$f(t) = 0.02e^{-0.02t}, \quad 0 \le t < \infty.$$

Find the probability that a rat will learn its way through a maze in 150 sec or less.

The time that it takes a rat to learn its way through a maze is an exponentially distributed random variable.

38. Time in a maze. Using the situation and the equation in Exercise 37, find the probability that a rat will learn its way through the maze in 50 sec or less.

SYNTHESIS

39. The function $f(x) = x^3$ is a probability density function over $[0, b]$. What is b?

40. The function $f(x) = 12x^2$ is a probability density function over $[-a, a]$. What is a?

41. Use your answer to Exercise 37 to find the probability that a rat requires more than 150 sec to learn its way through the maze.

TW **42.** List five examples of the use of probability in daily life.

TECHNOLOGY CONNECTION

43–54. Verify Property 2 of the definition of a probability density function for each of the functions in Exercises 1–12.

5.5 Probability: Expected Value; The Normal Distribution

OBJECTIVES

- Find $E(x)$, $E(x^2)$, the mean, the variance, and the standard deviation.
- Evaluate normal distribution probabilities using a table.

Expected Value

Let's again consider throwing a dart at a number line in such a way that it always lands in the interval $[1, 3]$. This time we assume a uniform distribution so that it is equally likely that the dart will land anywhere in the interval.

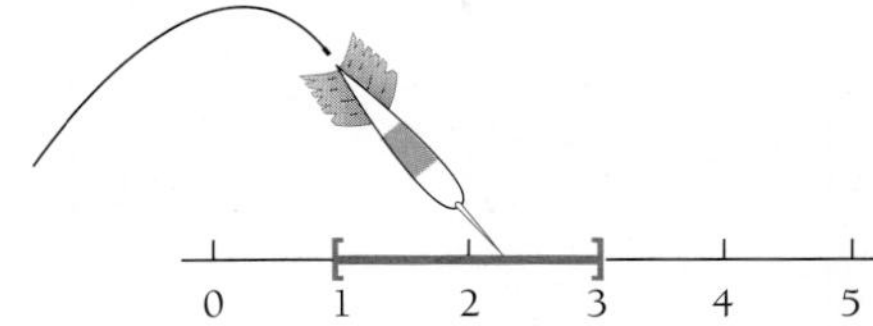

Suppose that we throw the dart at the line 100 times and keep track of the numbers it hits. Then we calculate the arithmetic mean (or average) $\overline{x}$ of all these numbers:

$$\overline{x} = \frac{x_1 + x_2 + x_3 + \cdots + x_{100}}{100} = \frac{\sum_{i=1}^{100} x_i}{100} = \sum_{i=1}^{100} x_i \cdot \frac{1}{100}.$$

$f(x) = \frac{1}{2}$, $1 \le x \le 3$

Assuming that the x_i's are uniformly distributed over $[1, 3]$, as shown to the left,

$$\sum_{i=1}^{n} x_i \cdot \frac{1}{n}, \quad \text{or} \quad \sum_{i=1}^{n} \left(x_i \cdot \frac{1}{2}\right)\frac{2}{n},$$

Note that the interval width is 2.

is analogous to

$$\int_1^3 x \cdot f(x)\, dx,$$

where $f(x) = 1/2$ is a probability density function for x. Because the width of $[1, 3]$ is 2, we can regard $2/n$ as Δx. The probability density function gives a "weight" to x. We add all the values of

$$\left(x_i \cdot \frac{1}{2}\right)\left(\frac{2}{n}\right)$$

when we find $\sum_{i=1}^{n} \left(x_i \cdot \frac{1}{2}\right)\left(\frac{2}{n}\right)$. Similarly, we add all the values of

$$(x \cdot f(x))(\Delta x)$$

when we find $\int_1^3 x \cdot f(x)\, dx$:

$$\int_1^3 x \cdot \frac{1}{2}\, dx = \frac{1}{2}\left[\frac{x^2}{2}\right]_1^3 = \frac{1}{4}[3^2 - 1^2] = 2.$$

This result, representing the average value of the distribution, is not surprising, since the distribution is uniform and centered around 2. Although the average of 100 dart throws may not be 2, as $n \to \infty$, we will have $\bar{x} \to 2$.

Suppose that we use the probability density function $f(x) = \frac{1}{4}x$ over the interval $[1, 3]$. As we can see from the graph below, this function gives more "weight" to the right side of the interval than to the left. Perhaps more points are awarded if a dart lands on the right. Then

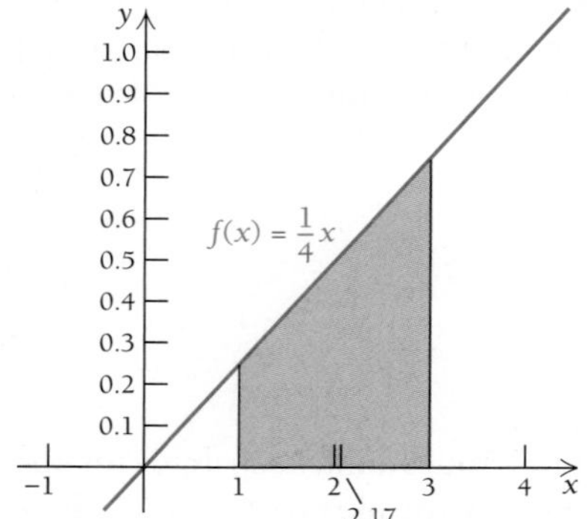

$$\begin{aligned}\int_1^3 x \cdot f(x)\,dx &= \int_1^3 x \cdot \frac{1}{4}x\,dx \\ &= \frac{1}{4}\int_1^3 x^2\,dx \\ &= \frac{1}{4}\left[\frac{x^3}{3}\right]_1^3 \\ &= \frac{1}{12}(3^3 - 1^3) = \frac{26}{12} \approx 2.17.\end{aligned}$$

Suppose that we continue to throw the dart and compute averages. The more times we throw the dart, the closer we expect the averages to come to 2.17.

DEFINITION

Let x be a continuous random variable over the interval $[a, b]$ with probability density function f. The **expected value** of x is defined by

$$E(x) = \int_a^b x \cdot f(x)\,dx.$$

The concept of expected value of a random variable can be generalized to functions of the random variable. Suppose that $y = g(x)$ is a function of the random variable x. Then we have the following.

DEFINITION

The **expected value** of $g(x)$ is defined by

$$E(g(x)) = \int_a^b g(x) \cdot f(x)\,dx,$$

where f is a probability density function for x.

For example,

$$E(x) = \int_a^b xf(x)\,dx,$$

$$E(x^2) = \int_a^b x^2f(x)\,dx,$$

$$E(e^x) = \int_a^b e^xf(x)\,dx,$$

and

$$E(2x + 3) = \int_a^b (2x + 3)f(x)\,dx.$$

EXAMPLE 1 Given the probability density function

$$f(x) = \tfrac{1}{2}x, \quad \text{over } [0, 2],$$

find $E(x)$ and $E(x^2)$.

Solution

$$\begin{aligned} E(x) &= \int_0^2 x \cdot \frac{1}{2}x\,dx = \int_0^2 \frac{1}{2}x^2\,dx \\ &= \frac{1}{2}\left[\frac{x^3}{3}\right]_0^2 = \frac{1}{6}\left[x^3\right]_0^2 \\ &= \frac{1}{6}(2^3 - 0^3) \\ &= \frac{1}{6} \cdot 8 = \frac{4}{3}; \\ E(x^2) &= \int_0^2 x^2 \cdot \frac{1}{2}x\,dx = \int_0^2 \frac{1}{2}x^3\,dx = \frac{1}{2}\left[\frac{x^4}{4}\right]_0^2 \\ &= \frac{1}{8}\left[x^4\right]_0^2 = \frac{1}{8}(2^4 - 0^4) \\ &= \frac{1}{8} \cdot 16 = 2 \end{aligned}$$

✦

DEFINITION

The **mean, μ,** of a continuous random variable x is defined to be $E(x)$. That is,

$$\mu = E(x) = \int_a^b xf(x)\,dx,$$

where f is a probability density function for x defined over $[a, b]$. (The symbol μ is the lowercase Greek letter *mu*.)

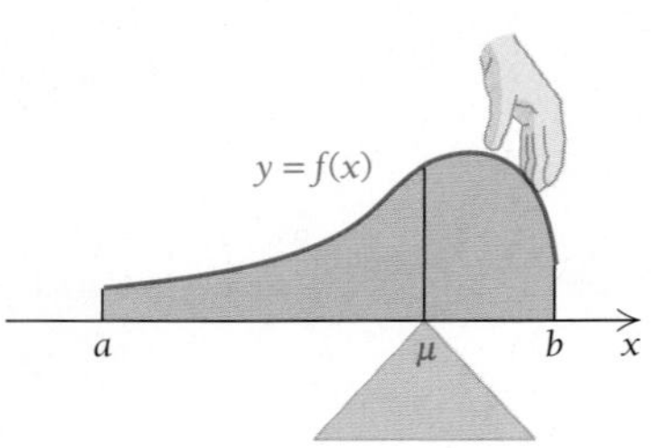

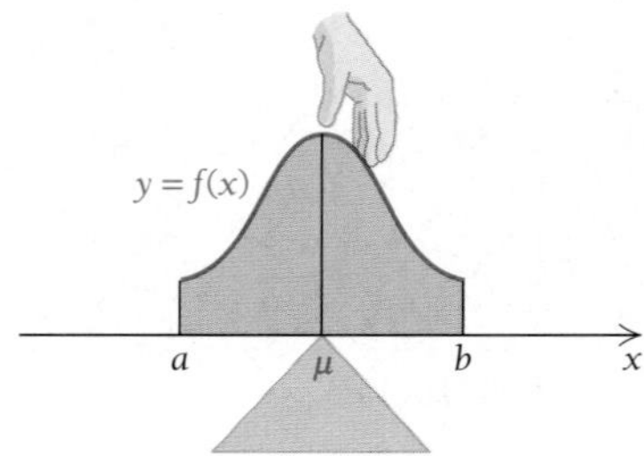

We can get a physical idea of the mean of a random variable by pasting the graph of the probability density function on cardboard and cutting out the area under the curve over the interval $[a, b]$. Then we try to find a balance point on the x-axis. That balance point is the mean, μ.

Variance and Standard Deviation

Because two very different distributions can have the same mean, it is useful to have a second statistic that serves as a measure of how the data in a distribution are spread out. Statistics that provide such a measure are the *variance* and (especially) the *standard deviation* of a distribution. A full derivation of how these statistics were developed is beyond the scope of this book.

DEFINITION

The **variance, σ^2,** of a continuous random variable x, defined on $[a, b]$, with probability density function f, is

$$\begin{aligned}\sigma^2 &= E(x^2) - \mu^2 \\ &= E(x^2) - [E(x)]^2 \\ &= \int_a^b x^2 f(x)\,dx - \left[\int_a^b xf(x)\,dx\right]^2.\end{aligned}$$

The **standard deviation, σ,** of a continuous random variable is defined as

$$\sigma = \sqrt{\text{variance}}.$$

(The symbol σ is the lowercase Greek letter *sigma*.)

EXAMPLE 2 Given the probability density function

$$f(x) = \tfrac{1}{2}x, \quad \text{over } [0, 2],$$

find the mean, the variance, and the standard deviation.

Solution From Example 1, we have $E(x) = \frac{4}{3}$ and $E(x^2) = 2$. Thus,

$$\begin{aligned}\textit{Mean} &= \mu = E(x) = \tfrac{4}{3}; \\ \textit{Variance} &= \sigma^2 = E(x^2) - [E(x)]^2 \\ &= 2 - \left(\tfrac{4}{3}\right)^2 = 2 - \tfrac{16}{9} \\ &= \tfrac{18}{9} - \tfrac{16}{9} = \tfrac{2}{9}; \\ \textit{Standard deviation} &= \sigma = \sqrt{\tfrac{2}{9}} \\ &= \tfrac{1}{3}\sqrt{2} \approx 0.47.\end{aligned}$$

✦

Loosely speaking, the standard deviation is a measure of how closely bunched the graph of f is, that is, how far the points on f are, on average, from the line $x = \mu$, as indicated below.

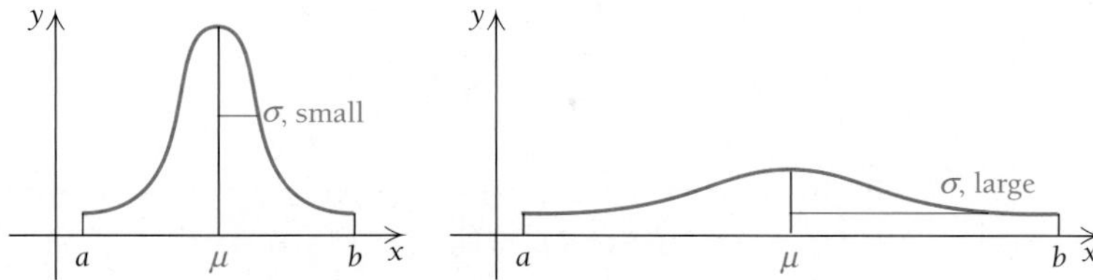

The Normal Distribution

Suppose that the average score on a test is 70. Usually there are about as many scores above the average as there are below the average; and the farther away from the average a particular score is, the fewer people there are who get that score. On this test, it is probable that more people scored in the 80s than in the 90s, and more people scored in the 60s than in the 50s. Test scores, heights of human beings, and weights of human beings are all examples of random variables that are often *normally* distributed.

Consider the function

$$g(x) = e^{-x^2/2}, \quad \text{over the interval } (-\infty, \infty).$$

This function has the entire set of real numbers as its domain. Its graph is the bell-shaped curve shown below. We can find function values by using a calculator:

TECHNOLOGY CONNECTION

Exploratory

To confirm that

$$\int_{-\infty}^{\infty} e^{-x^2/2}\,dx = \sqrt{2\pi},$$

let y1 = e^(−x^2/2).

1. Using the window [−5, 5, −0.5, 1.5] and the integral option of the CALC menu, set the limits of integration at −2 and 2.
2. Repeat, setting the limits at −3 and 3.
3. Compare your results with $\sqrt{2\pi} \approx 2.507$.

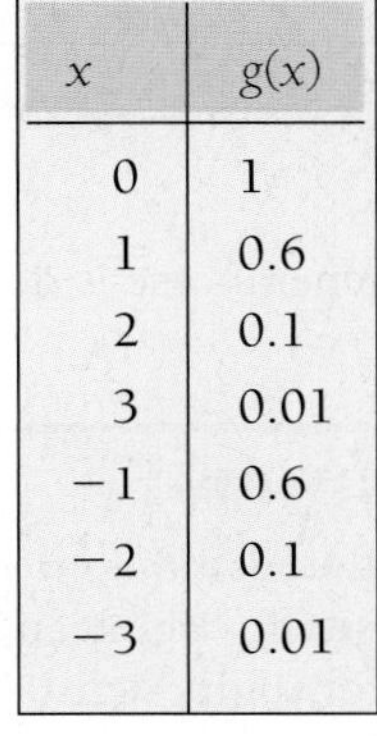

x	$g(x)$
0	1
1	0.6
2	0.1
3	0.01
−1	0.6
−2	0.1
−3	0.01

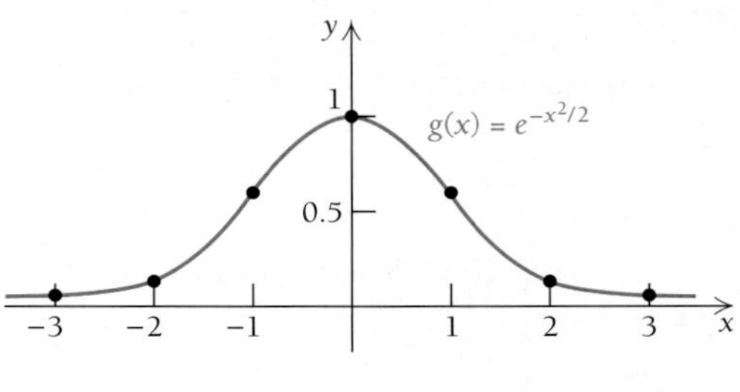

This function has an antiderivative, but that antiderivative has no basic integration formula. Nevertheless, it can be shown that the improper integral converges over the interval $(-\infty, \infty)$ to a number given by

$$\int_{-\infty}^{\infty} e^{-x^2/2}\,dx = \sqrt{2\pi}.$$

That is, although an elementary expression for the antiderivative cannot be found, there is a numerical value for the improper integral evaluated over the set of real numbers. Note that since the area is not 1, the function g is not a probability density function, but the following function is:

$$f(x) = \frac{1}{\sqrt{2\pi}} e^{-x^2/2}.$$

DEFINITION

A continuous random variable x has a **standard normal distribution** if its probability density function is

$$f(x) = \frac{1}{\sqrt{2\pi}} e^{-x^2/2}, \quad \text{over } (-\infty, \infty).$$

This standard normal distribution has a mean of 0 and a standard deviation of 1. Its graph follows.

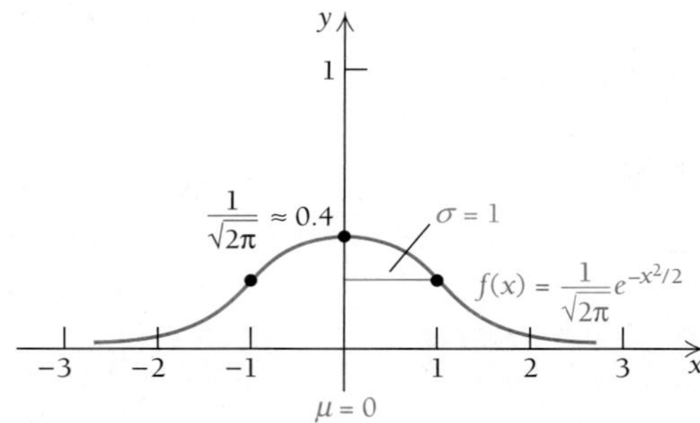

TECHNOLOGY CONNECTION

Exploratory

Use a graphing calculator to approximate

$$\int_{-b}^{b} \frac{1}{\sqrt{2\pi}} e^{-x^2/2}\, dx$$

for $b = 10$, 100, and 1000. What does this suggest about

$$\int_{-\infty}^{\infty} \frac{1}{\sqrt{2\pi}} e^{-x^2/2}\, dx?$$

This is a way to verify part of the assertion that

$$f(x) = \frac{1}{\sqrt{2\pi}} e^{-x^2/2}$$

is a probability density function. Use a similar approximation procedure to show that the mean is 0 and the standard deviation is 1.

The general case is defined as follows.

DEFINITION

A continuous random variable x is **normally distributed** with mean μ and standard deviation σ if its probability density function is given by

$$f(x) = \frac{1}{\sigma\sqrt{2\pi}} e^{-(1/2)[(x-\mu)/\sigma]^2}, \quad \text{over } (-\infty, \infty).$$

The graph of any normal distribution is a transformation of the graph of the standard normal distribution. This can be shown by translating the graph of any normal distribution along the x-axis and adjusting how tightly clustered the graph is about the mean. Some examples follow.

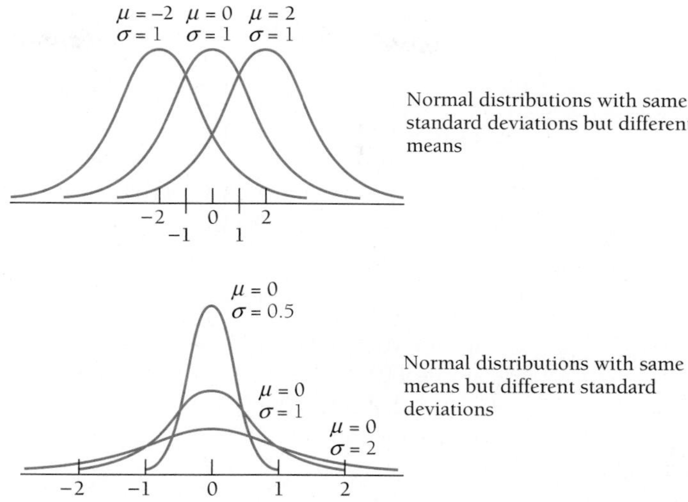

The normal distribution is extremely important in statistics; it underlies much of the research in the behavioral and social sciences. Because of this, tables of approximate values of the definite integral of the standard normal distribution have been prepared using numerical approximation methods like the Trapezoidal Rule given in Exercise Set 4.1. Table 2 at the back of the book is such a table. It contains values of

$$P(0 \leq x \leq z) = \int_0^z \frac{1}{\sqrt{2\pi}} e^{-x^2/2}\, dx.$$

The symmetry of the graph of this function about the mean allows many types of probabilities to be computed from the table. Some involve addition or subtraction of areas.

EXAMPLE 3 Let x be a continuous random variable with a standard normal distribution. Using Table 2 at the back of the book, find each of the following.

a) $P(0 \leq x \leq 1.68)$ **b)** $P(-0.97 \leq x \leq 0)$

c) $P(-2.43 \leq x \leq 1.01)$ **d)** $P(1.90 \leq x \leq 2.74)$

e) $P(-2.98 \leq x \leq -0.42)$ **f)** $P(x \geq 0.61)$

Solution

a) $P(0 \leq x \leq 1.68)$ is the area bounded by the standard normal curve and the lines $x = 0$ and $x = 1.68$. We look this up in Table 2 by going down the left column to 1.6, then moving to the right to the column headed 0.08. There we read 0.4535. Thus,

$$P(0 \leq x \leq 1.68) = 0.4535.$$

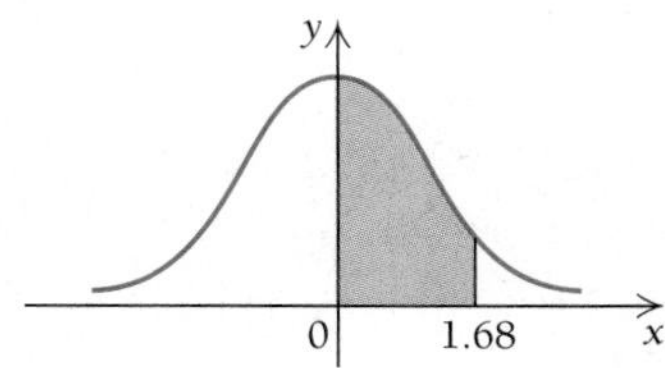

b) Because of the symmetry of the graph,

$$P(-0.97 \le x \le 0)$$
$$= P(0 \le x \le 0.97)$$
$$= 0.3340.$$

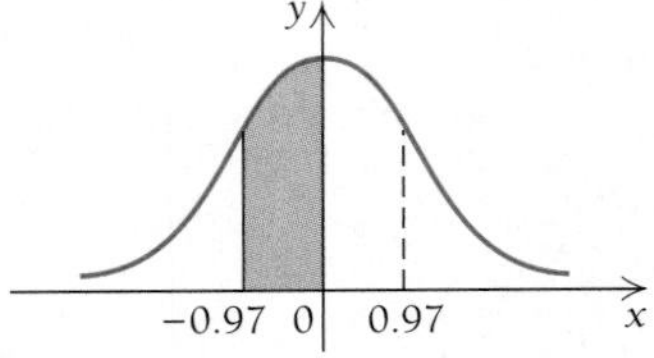

c) $P(-2.43 \le x \le 1.01)$
$= P(-2.43 \le x \le 0) + P(0 \le x \le 1.01)$
$= P(0 \le x \le 2.43) + P(0 \le x \le 1.01)$
$= 0.4925 + 0.3438$
$= 0.8363$

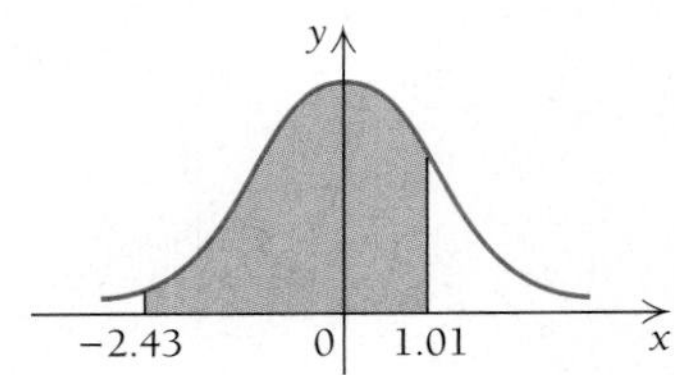

d) $P(1.90 \le x \le 2.74)$
$= P(0 \le x \le 2.74) - P(0 \le x \le 1.90)$
$= 0.4969 - 0.4713$
$= 0.0256$

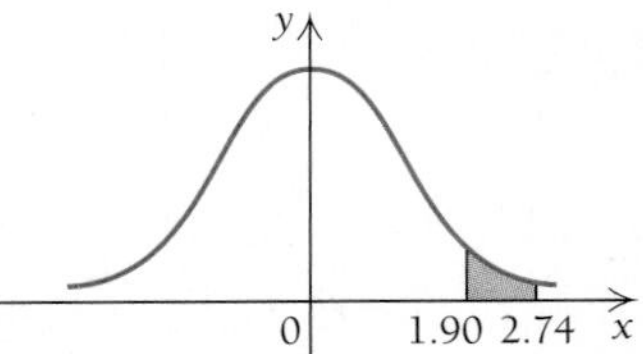

e) $P(-2.98 \le x \le -0.42)$
$= P(0.42 \le x \le 2.98)$
$= P(0 \le x \le 2.98) - P(0 \le x \le 0.42)$
$= 0.4986 - 0.1628$
$= 0.3358$

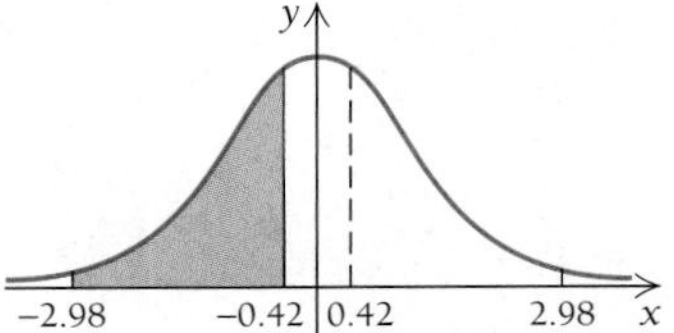

f) $P(x \ge 0.61)$
$= P(x \ge 0) - P(0 \le x \le 0.61)$
$= 0.5000 - 0.2291$ Because of the symmetry about the line $x = 0$, half the area is on each side of the line, and since the entire area is 1, we have $P(x \ge 0) = 0.5000$.
$= 0.2709$ ✦

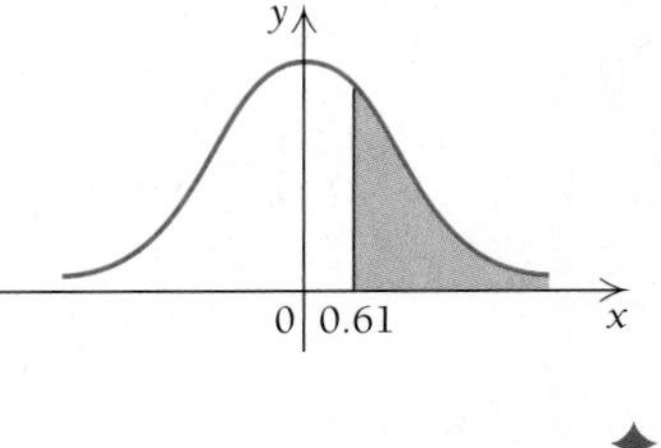

For most normal distributions, $\mu \ne 0$ and $\sigma \ne 1$. It would be a hopeless task to make tables for all values of the mean μ and the standard deviation σ. For any normal distribution, the transformation

$$z = \frac{x - \mu}{\sigma}$$

standardizes the distribution, since $(x - \mu)/\sigma$ is a measure of how many standard deviations x is from μ. Subtracting μ from all x-values and then dividing by σ preserves the order of the x-values while permitting the use of Table 2 at the back of this book. Such converted values are called *z-scores* or *z-values*.

$$P(a \le x \le b) = P\left(\frac{a - \mu}{\sigma} \le z \le \frac{b - \mu}{\sigma}\right),$$

and this last probability can be found using Table 2.

EXAMPLE 4 The weights, w, of the students in a calculus class are normally distributed with a mean, μ, of 150 lb and a standard deviation, σ, of 25 lb. Find the probability that the weight of a student in the class is from 160 lb to 180 lb.

Solution We first standardize the weights:

$$180 \text{ is standardized to } \frac{b-\mu}{\sigma} = \frac{180-150}{25} = 1.2;$$

$$160 \text{ is standardized to } \frac{a-\mu}{\sigma} = \frac{160-150}{25} = 0.4.$$

These z-values measure the distance of w from μ in terms of σ.

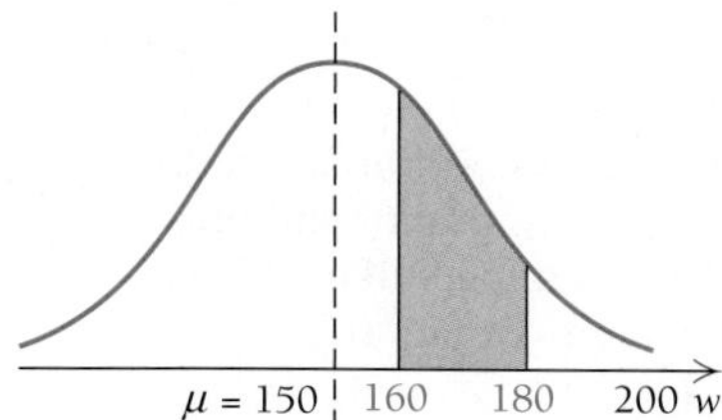

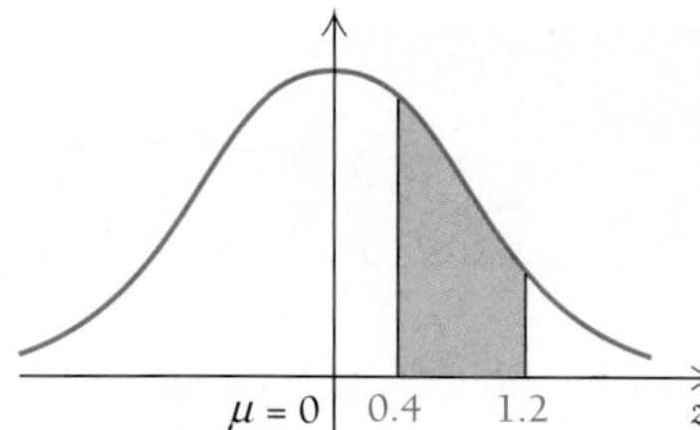

Then we have

$$\begin{aligned} P(160 \le w \le 180) &= P(0.4 \le z \le 1.2) \\ &= P(0 \le z \le 1.2) - P(0 \le z \le 0.4) \\ &= 0.3849 - 0.1554 \\ &= 0.2295. \end{aligned}$$

Now we can use Table 2.

Thus, the probability that the weight of a student in that class is 160 lb to 180 lb is 0.2295. That is, about 23% of the students weigh from 160 lb to 180 lb. ✦

TECHNOLOGY CONNECTION

Statistics on a Calculator

It is possible to use a TI-84 or TI-83 Plus to make an approximation of the probability in Example 4 without performing a standard conversion, using Table 2, or entering the normal probability density function. We first select an appropriate window, [0, 300, −0.002, 0.02], with Xscl = 50 and Yscl = 0.01. Next, we use the ShadeNorm command, which we find by pressing 2ND DISTR ▷ ENTER. We enter the values as shown

(continued)

Statistics on a Calculator (continued)

and press ENTER. (If necessary, the ClearDraw option on the DRAW menu can be used to clear the graph.)

Left endpoint of interval Right endpoint of interval

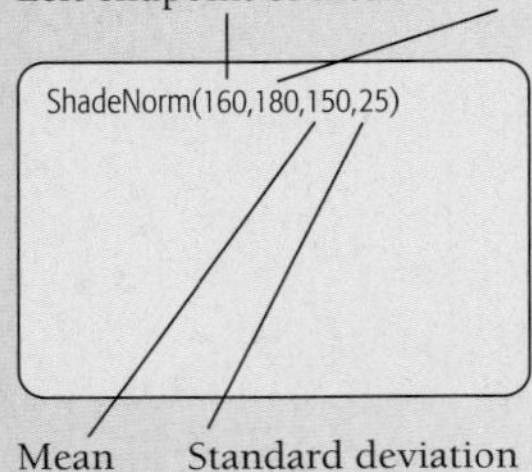

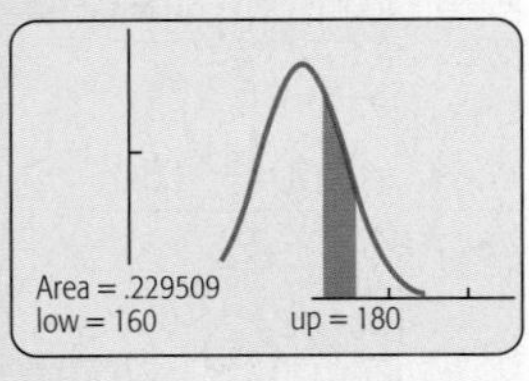

Mean Standard deviation

The area is shaded and given as 0.229509, or about 23%. If we are entering intervals involving infinity without endpoints, using extremely large numbers like 1 E 99 or −1 E 99 will allow adequate approximations.

EXERCISES

1. The weights of the students in a calculus class are normally distributed with mean $\mu = 150$ lb and standard deviation $\sigma = 25$ lb.
   a) What is the probability that a student's weight is from 125 lb to 170 lb?
   b) What is the probability that a student's weight is greater than 200 lb?
2. *SAT Scores.* Combined SAT reading and math scores are normally distributed with mean $\mu = 1020$ and standard deviation $\sigma = 140$.
   a) What percentage of the scores are between 400 and 900? (Note that 400 is the minimum score possible.)
   b) What percentage of the scores are above 840? (Note that 1600 is the maximum score.)

5.5 Exercise Set

For each probability density function, over the given interval, find $E(x)$, $E(x^2)$, the mean, the variance, and the standard deviation.

1. $f(x) = \frac{1}{4}$, $[3, 7]$

2. $f(x) = \frac{1}{5}$, $[3, 8]$

3. $f(x) = \frac{1}{8}x$, $[0, 4]$

4. $f(x) = \frac{2}{9}x$, $[0, 3]$

5. $f(x) = \frac{1}{4}x$, $[1, 3]$

6. $f(x) = \frac{2}{3}x$, $[1, 2]$

7. $f(x) = \frac{3}{2}x^2$, $[-1, 1]$

8. $f(x) = \frac{1}{3}x^2$, $[-2, 1]$

9. $f(x) = \dfrac{1}{\ln 5} \cdot \dfrac{1}{x}$, $[1.5, 7.5]$

10. $f(x) = \dfrac{1}{\ln 4} \cdot \dfrac{1}{x}$, $[0.8, 3.2]$

Let x be a continuous random variable with a standard normal distribution. Using Table 2, find each of the following.

11. $P(0 \le x \le 2.13)$

12. $P(0 \le x \le 0.36)$

13. $P(-1.37 \le x \le 0)$

14. $P(-2.01 \le x \le 0)$

15. $P(-1.89 \le x \le 0.45)$

16. $P(-2.94 \le x \le 2.00)$

17. $P(1.35 \le x \le 1.45)$

18. $P(0.76 \le x \le 1.45)$

19. $P(-1.27 \le x \le -0.58)$

20. $P(-2.45 \le x \le -1.24)$

21. $P(x \ge 3.01)$

22. $P(x \ge 1.01)$

23. a) $P(-1 \le x \le 1)$
b) What percentage of the area is from -1 to 1?

24. a) $P(-2 \le x \le 2)$
b) What percentage of the area is from -2 to 2?

Let x be a continuous random variable that is normally distributed with mean $\mu = 22$ and standard deviation $\sigma = 5$. Using Table 2, find each of the following.

25. $P(24 \le x \le 30)$

26. $P(22 \le x \le 27)$

27. $P(19 \le x \le 25)$

28. $P(18 \le x \le 26)$

29–46. *Use a graphing calculator to do Exercises 11–28.*

APPLICATIONS

Business and Economics

47. Mail orders. The number of orders, N, received daily by an online vendor of used CDs is normally distributed with mean 250 and standard deviation 20. The company has to hire extra help or pay overtime on those days when the number of orders received is 300 or higher. What percentage of days will the company have to hire extra help or pay overtime?

48. Bread baking. The number of loaves of bread, N, baked each day by Fireside Bakers is normally distributed with mean 1000 and standard deviation 50. The bakery pays bonuses to its employees on those days when at least 1100 loaves are baked. What percentage of days will the bakery have to pay a bonus?

Manufacturing. *In an automotive body-welding line, delays encountered during the manufacturing process can be modeled by various probability distributions. (**Source:** R. R. Inman, "Empirical Evaluation of Exponential and Independence Assumptions in Queueing Models of Manufacturing Systems," Production and Operations Management, Vol. 8, 409–432 (1999).)*

49. The processing time for the robogate has a normal distribution with mean 38.6 sec and standard deviation 1.729 sec. Find the probability that the next operation of the robogate will take 40 sec or less.

50. The processing time for the automatic piercing station has a normal distribution with mean 36.2 sec and standard deviation 2.108 sec. Find the probability that the next operation of the piercing station will take between 35 and 40 sec.

Social Sciences

51. Test score distribution. The scores on a biology test are normally distributed with mean 65 and standard deviation 20. A score from 80 to 89 is a B. What is the probability of getting a B?

Percentiles. *Let x be a continuous random variable. The pth percentile of its distribution is defined to be the number z for which the probability that x is less than or equal to z is $p\%$:*

$$P(x \leq z) = \frac{p}{100}.$$

For example, if x has a standard normal distribution, Table 2 tells us that

$$P(x \leq 1.28) \approx 90\%.$$

Therefore, we say that 1.28 is the 90th percentile for a standard normal distribution. Use the preceding information for Exercises 52 and 53.

52. Find the following percentiles for a standard normal distribution.

a) 30th percentile
b) 50th percentile
c) 95th percentile

53. Combined SAT reading and math scores are normally distributed with mean 1020 and standard deviation 140. Find the SAT scores that correspond to these percentiles.

a) 35th percentile
b) 60th percentile
c) 92nd percentile

General Interest

54. Bowling scores. At the time this book was written, the bowling scores, S, of author Marv Bittinger (shown below) were normally distributed with mean 201 and standard deviation 23.

a) Find the probability that a score is from 185 to 215, and interpret your results.
b) Find the probability that a score is from 160 to 175, and interpret your results.
c) Find the probability that a score is greater than 200, and interpret your results.

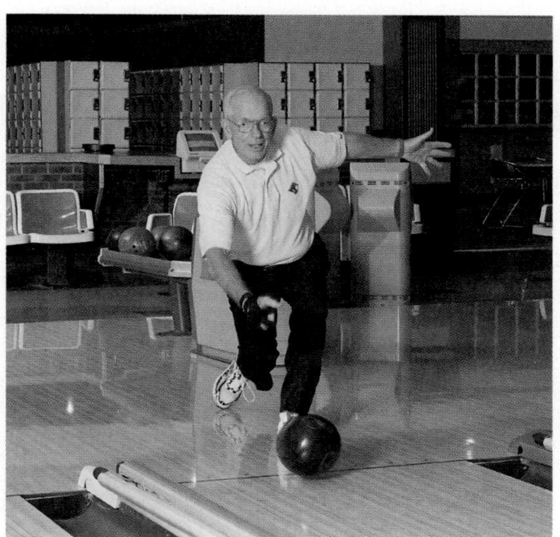

SYNTHESIS

For each probability density function, over the given interval, find $E(x)$, $E(x^2)$, the mean, the variance, and the standard deviation.

55. $f(x) = \dfrac{1}{b - a}$, over $[a, b]$

56. $f(x) = \dfrac{3a^3}{x^4}$, over $[a, \infty)$

Median. *Let x be a continuous random variable over $[a, b]$ with probability density function f. Then the median of the x-values is that number m for which*

$$\int_a^m f(x)\,dx = \tfrac{1}{2}.$$

Find the median.

57. $f(x) = \frac{1}{2}x$, $[0, 2]$

58. $f(x) = \frac{3}{2}x^2$, $[-1, 1]$

59. $f(x) = ke^{-kx}$, $[0, \infty)$

60. Business: coffee production. Suppose that the amount of coffee beans loaded into a vacuum-packed bag has a mean weight of μ ounces, which can be adjusted on the filling machine. Suppose that the amount dispensed is normally distributed with $\sigma = 0.2$ oz. What should μ be set at to ensure that only 1 bag in 50 will have less than 16 oz?

61. Business: Does thy cup overflow? Suppose that the mean amount of cappuccino, μ, dispensed by a vending machine can be set. If a cup holds 8.5 oz and the amount dispensed is normally distributed with $\sigma = 0.3$ oz, what should μ be set at to ensure that only 1 cup in 100 will overflow?

TW **62.** Explain the uses of integration in the study of probability.

TW **63.** You are told that "The antiderivative of the function $f(x) = e^{-x^2/2}$ has no basic integration formula." Make some guesses of functions that might seem reasonable to you as antiderivatives and show why they are not.

TECHNOLOGY CONNECTION

64. Approximate the integral

$$\int_{-\infty}^{\infty} e^{-x^2}\,dx.$$

5.6 Volume

OBJECTIVE

- Find the volume of a solid of revolution.

Consider the graph of $y = f(x)$ in Fig. 1. If the upper half-plane is rotated about the x-axis, then each point on the graph has a circular path, and the whole graph sweeps out a certain surface, called a *surface of revolution*.

The plane region bounded by the graph, the x-axis, $x = a$, and $x = b$ sweeps out a *solid of revolution*. To calculate the volume of this solid, we first approximate it as a finite sum of thin right circular cylinders, or disks (Fig. 2). We divide the interval $[a, b]$ into equal subintervals, each of length Δx. Thus, the height h of each disk is Δx (Fig. 3). The radius of each disk is $f(x_i)$, where x_i is the right-hand endpoint of the subinterval that determines that disk. If $f(x_i)$ is negative, we can use $|f(x_i)|$.

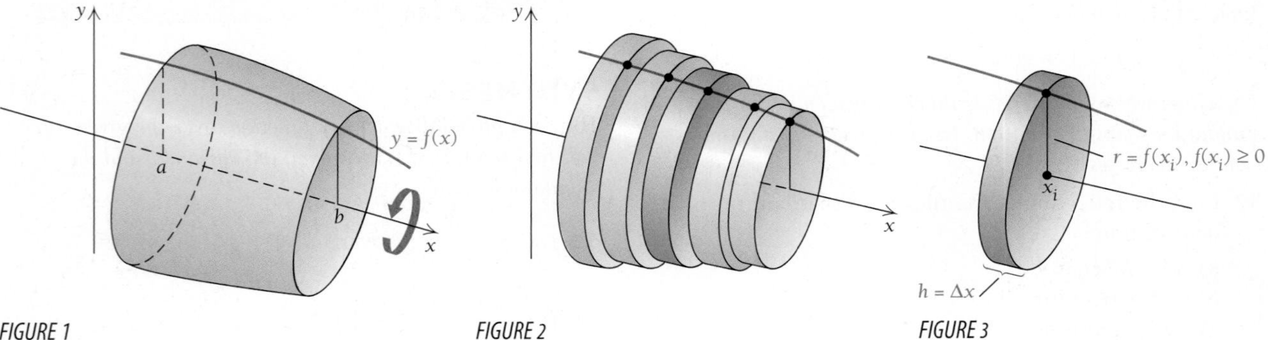

FIGURE 1 FIGURE 2 FIGURE 3

Since the volume of a right circular cylinder is given by

$$V = \pi r^2 h, \quad \text{or} \quad \text{Volume} = \text{area of the base} \cdot \text{height},$$

each of the approximating disks has volume

$$\pi|f(x_i)|^2\, \Delta x = \pi[f(x_i)]^2\, \Delta x. \qquad \text{Squaring makes use of the absolute value unnecessary.}$$

The volume of the solid of revolution is approximated by the sum of the volumes of all the cylinders:

$$V \approx \sum_{i=1}^{n} \pi[f(x_i)]^2\, \Delta x.$$

The actual volume is the limit as the thickness of the disks approaches zero, or the number of disks approaches infinity:

$$V = \lim_{n\to\infty} \sum_{i=1}^{n} \pi[f(x_i)]^2\, \Delta x = \int_a^b \pi[f(x)]^2\, dx.$$

(See Section 4.3.) That is, the volume is the value of the definite integral of the function $y = \pi[f(x)]^2$ from a to b.

THEOREM 3

For a continuous function f defined on $[a, b]$, the **volume, V, of the solid of revolution** obtained by rotating about the x-axis the area between the graph of f and $[a, b]$ is given by

$$V = \int_a^b \pi[f(x)]^2\, dx.$$

EXAMPLE 1 Find the volume of the solid of revolution generated by rotating about the x-axis the region under the graph of $y = \sqrt{x}$ from $x = 0$ to $x = 1$.

Solution

$$\begin{aligned} V &= \int_0^1 \pi[f(x)]^2\, dx \\ &= \int_0^1 \pi[\sqrt{x}]^2\, dx \\ &= \int_0^1 \pi x\, dx \\ &= \pi\left[\frac{x^2}{2}\right]_0^1 \\ &= \frac{\pi}{2}\left[x^2\right]_0^1 \\ &= \frac{\pi}{2}(1^2 - 0^2) = \frac{\pi}{2} \end{aligned}$$

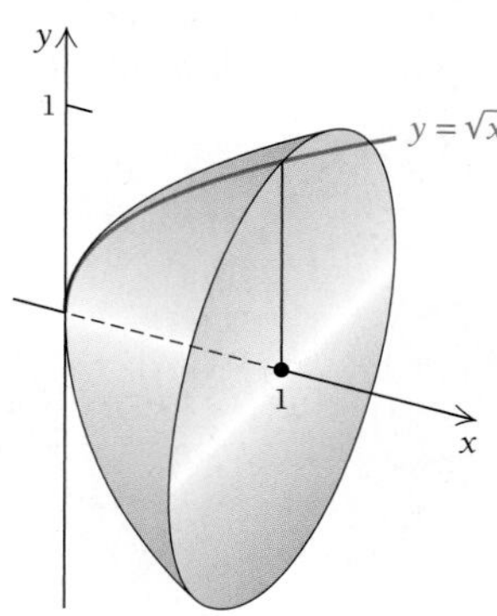

✦

Explain how this could be interpreted as a solid of revolution.

EXAMPLE 2 Find the volume of the solid of revolution generated by rotating about the x-axis the region under the graph of

$$y = e^x$$

from $x = -1$ to $x = 2$.

Solution

$$V = \int_{-1}^{2} \pi[f(x)]^2\,dx$$

$$= \int_{-1}^{2} \pi[e^x]^2\,dx$$

$$= \int_{-1}^{2} \pi e^{2x}\,dx$$

$$= \left[\frac{\pi}{2}e^{2x}\right]_{-1}^{2}$$

$$= \frac{\pi}{2}\left[e^{2x}\right]_{-1}^{2}$$

$$= \frac{\pi}{2}(e^{2\cdot 2} - e^{2(-1)})$$

$$= \frac{\pi}{2}(e^4 - e^{-2}) \approx 85.55$$

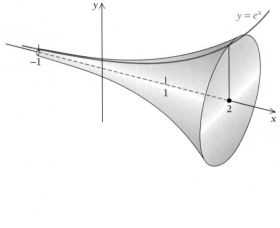

Exercise Set 5.6

Find the volume generated by rotating about the x-axis the regions bounded by the graphs of each set of equations.

1. $y = x, x = 0, x = 1$

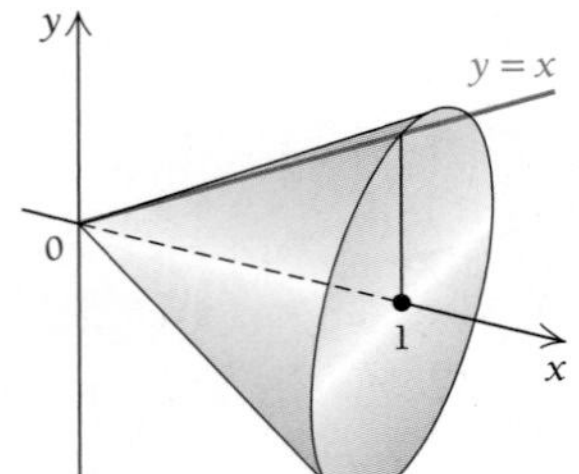

2. $y = x, x = 0, x = 2$

3. $y = \sqrt{x}, x = 1, x = 4$

4. $y = 2x, x = 1, x = 3$

5. $y = e^x, x = -2, x = 5$

6. $y = e^x, x = -3, x = 2$

7. $y = \dfrac{1}{x}, x = 1, x = 3$

8. $y = \dfrac{1}{x}, x = 1, x = 4$

9. $y = \dfrac{2}{\sqrt{x}}, x = 4, x = 9$

10. $y = \dfrac{1}{\sqrt{x}}, x = 1, x = 4$

11. $y = 4, x = 1, x = 3$

12. $y = 5, x = 1, x = 3$

13. $y = x^2, x = 0, x = 2$

14. $y = x + 1, x = -1, x = 2$

15. $y = \sqrt{1 + x}, x = 2, x = 10$

16. $y = 2\sqrt{x}, x = 1, x = 2$

17. $y = \sqrt{4 - x^2}, x = -2, x = 2$

18. $y = \sqrt{r^2 - x^2}, x = -r, x = r$ (assume $r > 0$)

SYNTHESIS

19. Graph $y = \sqrt{4 - x^2}$ and $y = \sqrt{r^2 - x^2}$, with $r > 0$, and explain how the results can be used to calculate the volume of a common shape. (See Exercises 17 and 18.)

Find the volume generated by rotating about the x-axis the regions bounded by the graphs of each set of equations.

20. $y = \sqrt{\ln x}, x = e, x = e^3$

21. $y = \sqrt{xe^{-x}}, x = 1, x = 2$

22. Consider the function $y = 1/x$ over the interval $[1, \infty)$. We showed in Section 5.3 that the area under the curve does not exist; that is,

$$\int_1^\infty \frac{1}{x}\,dx$$

diverges. Find the volume of the solid of revolution formed by rotating about the x-axis the region under the graph of $y = 1/x$ over the interval $[1, \infty)$. That is, find

$$\int_1^\infty \pi\left[\frac{1}{x}\right]^2 dx.$$

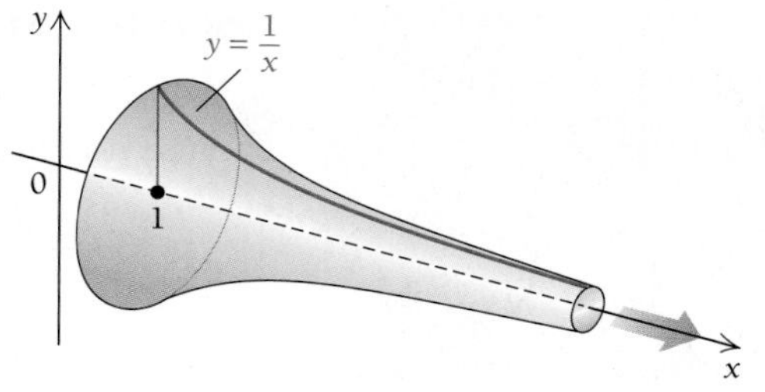

This solid is sometimes referred to as *Gabriel's horn.*

TECHNOLOGY CONNECTION

23. Paradox of Gabriel's horn or the infinite paint can. Though we cannot prove it here, the surface area of Gabriel's horn (see Exercise 22) is given by

$$S = \int_1^\infty \frac{2\pi}{x}\sqrt{1 + \frac{1}{x^4}}\,dx.$$

Show that the surface area of Gabriel's horn does not exist. The paradox is that the volume of the horn exists, but the surface area does not. This is like a can of paint that has a finite volume but, when full, does not hold enough paint to paint the outside of the can.

5.7 Differential Equations

OBJECTIVES

- Solve differential equations.
- Verify that a given function is a solution of a differential equation.
- Solve differential equations using separation of variables.
- Solve applications involving differential equations.

A **differential equation** is an equation that involves derivatives, or differentials. In Chapter 3, we studied one very important differential equation,

$$\frac{dP}{dt} = kP, \quad \text{or} \quad P'(t) = k \cdot P(t),$$

where P, or $P(t)$, is the population at time t. This equation is a model of uninhibited population growth. Its solution is the function

$$P(t) = P_0e^{kt},$$

where the constant P_0 is the size of the population at $t = 0$. As this example illustrates, differential equations are rich in applications and have solutions that are functions.

Waves can be represented by differential equations.

Solving Certain Differential Equations

In this section, we will frequently use the notation y' for a derivative—mainly because it is simple. Thus, if $y = f(x)$, then

$$y' = \frac{dy}{dx} = f'(x).$$

We actually find solutions of certain differential equations when we find their antiderivatives or indefinite integrals. The differential equation

$$\frac{dy}{dx} = g(x), \quad \text{or} \quad y' = g(x),$$

has the solution

$$y = \int g(x)\, dx.$$

EXAMPLE 1 Solve: $y' = 2x$.

Solution

$$y = \int 2x\, dx$$
$$= x^2 + C$$

TECHNOLOGY CONNECTION

Exploratory

For $y' = 3x^2$, write the general solution. Then graph the particular solutions for $C = -2$, $C = 0$, and $C = 1$.

Look again at the solution of Example 1. Note the constant of integration, C. This solution is called a *general solution* because taking all values of C gives *all* the solutions. Taking specific values of C gives *particular solutions*. For example, the following are particular solutions of $y' = 2x$:

$$y = x^2 + 3,$$
$$y = x^2,$$
$$y = x^2 - 3.$$

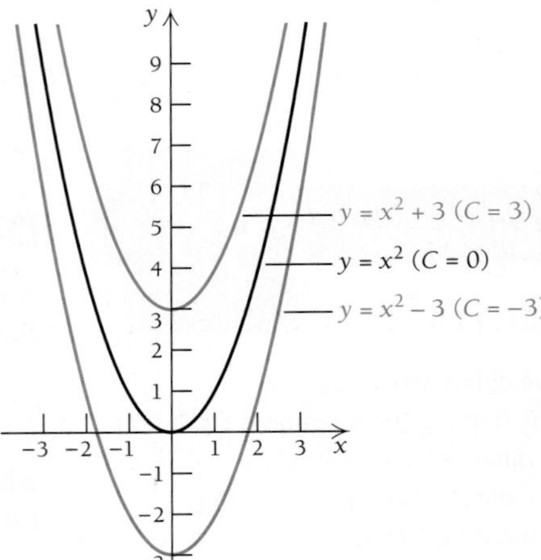

The graph shows the curves of these few particular solutions. The general solution can be regarded as the set of all particular solutions, a *family* of curves.

Knowing the value of a function at a particular point may allow us to select a particular solution from the general solution.

EXAMPLE 2 Solve

$$f'(x) = e^x + 5x - x^{1/2},$$

given that $f(0) = 8$.

Solution

We first find the general solution:

$$f(x) = \int f'(x)\,dx = \int (e^x + 5x - x^{1/2})\,dx$$
$$= e^x + \tfrac{5}{2}x^2 - \tfrac{2}{3}x^{3/2} + C.$$

Next, since $f(0) = 8$, we substitute to find C:

$$8 = e^0 + \tfrac{5}{2}\cdot 0^2 - \tfrac{2}{3}\cdot 0^{3/2} + C$$
$$8 = 1 + C$$
$$7 = C.$$

Thus the particular solution is

$$f(x) = e^x + \tfrac{5}{2}x^2 - \tfrac{2}{3}x^{3/2} + 7.$$ ✦

Verifying Solutions

To verify that a function is a solution to a differential equation, we find the necessary derivatives and substitute.

EXAMPLE 3 Show that $y = 4e^x + 5e^{3x}$ is a solution of $y'' - 4y' + 3y = 0$.

Solution

We first find y' and y'':

$$y' = \frac{d}{dx}(4e^x + 5e^{3x}) = 4e^x + 15e^{3x};$$
$$y'' = \frac{d}{dx}y' = 4e^x + 45e^{3x}.$$

Then we substitute in the differential equation, as follows:

$y'' - 4y' + 3y$	$= 0$
$(4e^x + 45e^{3x}) - 4(4e^x + 15e^{3x}) + 3(4e^x + 5e^{3x})$	0
$4e^x + 45e^{3x} - 16e^x - 60e^{3x} + 12e^x + 15e^{3x}$	
0	0

Since a true equation, $0 = 0$, results, we know that $y = 4e^x + 5e^{3x}$ is a solution of the differential equation. ✦

Separation of Variables

Consider the differential equation

$$\frac{dy}{dx} = 2xy. \quad (1)$$

We treat dy/dx as a quotient, as we did in Sections 2.6 and 4.5. Multiplying equation (1) by dx and then by $1/y$, we get

$$\frac{dy}{y} = 2x\,dx, \quad y \neq 0. \quad (2)$$

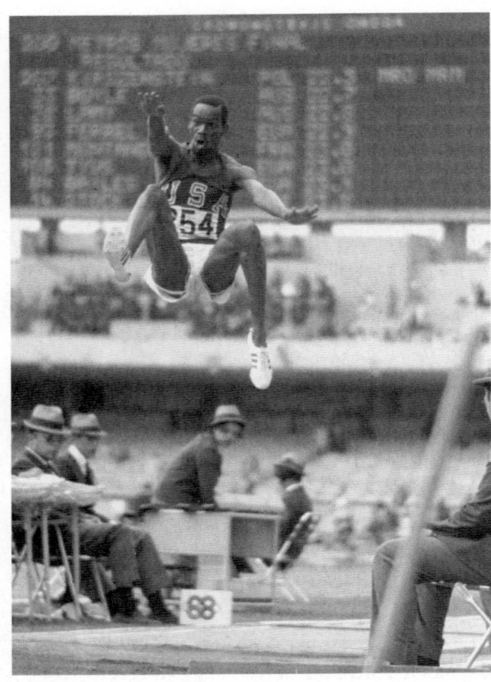

At the 1968 Olympic Games in Mexico City, Bob Beamon made a miraculous long jump of 29 ft, $2\frac{1}{2}$ in. Many believed that the jump's record length was due to the altitude, which was 7400 ft. Using differential equations for analysis, M. N. Bearley refuted the altitude theory in "The Long Jump Miracle of Mexico City" (Mathematics Magazine, *Vol. 45, 241–246 (November 1972)). Bearley argues that the world-record jump was a result of Beamon's exceptional speed (9.5 sec in the 100-yd dash) and the fact that he hit the take-off board in perfect position.*

We have now **separated the variables,** meaning that all the expressions involving y are on one side and all those involving x are on the other. We then integrate both sides of equation (2):

$$\int \frac{dy}{y} = \int 2x\,dx$$
$$\ln|y| = x^2 + C.$$

We use only one constant because the two antiderivatives differ by, at most, a constant. Recall that the definition of logarithms says that if $\log_a b = t$, then $b = a^t$. We have $\ln|y| = \log_e|y| = x^2 + C$, so

$$|y| = e^{x^2+C}, \quad \text{or} \quad y = \pm e^{x^2} \cdot e^C.$$

Thus, the solution to differential equation (1) is

$$y = C_1 e^{x^2}, \quad \text{where } C_1 = \pm e^C.$$

In fact, C_1 is still an arbitrary constant.

EXAMPLE 4 Solve:

$$3y^2 \frac{dy}{dx} + x = 0, \quad \text{where } y = 5 \text{ when } x = 0.$$

Solution

We first separate the variables as follows:

$$3y^2 \frac{dy}{dx} = -x \qquad \text{Adding } -x \text{ to both sides}$$
$$3y^2\,dy = -x\,dx. \qquad \text{Multiplying both sides by } dx$$

We then integrate both sides:

$$\int 3y^2\,dy = \int -x\,dx$$
$$y^3 = -\frac{x^2}{2} + C$$
$$y^3 = C - \frac{x^2}{2}$$
$$y = \sqrt[3]{C - \frac{x^2}{2}}. \qquad \text{Taking the cube root of both sides}$$

Since $y = 5$ when $x = 0$, we substitute to find C:

$$5 = \sqrt[3]{C - \frac{0^2}{2}} \qquad \text{Substituting 5 for } y \text{ and 0 for } x$$
$$5 = \sqrt[3]{C}$$
$$125 = C. \qquad \text{Cubing both sides}$$

The particular solution is

$$y = \sqrt[3]{125 - \frac{x^2}{2}}.$$

✦

EXAMPLE 5 Solve:

$$\frac{dy}{dx} = \frac{x}{y}.$$

Solution We first separate the variables:

$$y\frac{dy}{dx} = x$$
$$y\,dy = x\,dx.$$

We then integrate both sides:

$$\int y\,dy = \int x\,dx$$
$$\frac{y^2}{2} = \frac{x^2}{2} + C$$
$$y^2 = x^2 + 2C$$
$$y^2 = x^2 + C_1,$$

where $C_1 = 2C$. We make this substitution in order to simplify the equation. We then obtain the solutions

$$y = \sqrt{x^2 + C_1} \quad \text{and} \quad y = -\sqrt{x^2 + C_1}.$$ ✦

EXAMPLE 6 Solve: $y' = x - xy$.

Solution Before we separate the variables, we replace y' with dy/dx:

$$\frac{dy}{dx} = x - xy.$$

Then we separate the variables:

$$dy = (x - xy)\,dx$$
$$dy = x(1 - y)\,dx$$
$$\frac{dy}{1 - y} = x\,dx.$$

Next, we integrate both sides:

$$\int \frac{dy}{1 - y} = \int x\,dx$$
$$-\ln|1 - y| = \frac{x^2}{2} + C$$
$$\ln|1 - y| = -\frac{x^2}{2} - C$$
$$|1 - y| = e^{-x^2/2 - C}$$
$$1 - y = \pm e^{-x^2/2 - C}$$
$$1 - y = \pm e^{-x^2/2}e^{-C}$$

TECHNOLOGY CONNECTION

Exploratory

Solve $y' = 2x + xy$. Graph the particular solutions for $C_1 = -2$, $C_1 = 0$, and $C_1 = 1$.

Since C is an arbitrary constant, $\pm e^{-C}$ is an arbitrary constant. Thus, we can replace $\pm e^{-C}$ with C_1:

$$1 - y = C_1 e^{-x^2/2}$$

$$-y = C_1 e^{-x^2/2} - 1$$

$$y = 1 - C_1 e^{-x^2/2}. \quad \text{Multiplying both sides by } -1$$

Or, if we choose to replace $-C_1$ with C_2, we have

$$y = 1 + C_2 e^{-x^2/2}. \quad \text{Addition is generally regarded as "simpler" than subtraction.}$$

✦

An Application to Economics: Elasticity

EXAMPLE 7 Suppose that for a certain product, the elasticity of demand is 1 for all prices $x > 0$. That is, $E(x) = 1$ for $x > 0$. Find the demand function $q = D(x)$. (See Section 3.6. Note that this use of the symbol E is unrelated to our earlier work on expected value.)

Solution Since $E(x) = 1$ for all $x > 0$,

$$1 = E(x) = -\frac{x D'(x)}{D(x)} \quad \text{Substituting}$$

$$= -\frac{x}{q} \cdot \frac{dq}{dx}.$$

Then

$$-\frac{q}{x} = \frac{dq}{dx}.$$

Separating the variables, we get

$$\frac{dx}{x} = -\frac{dq}{q}.$$

Now we integrate both sides:

$$\int \frac{dx}{x} = -\int \frac{dq}{q}$$

$$\ln x = -\ln q + C. \quad \text{Note that both the price, } x\text{, and the quantity, } q\text{, can be assumed to be positive.}$$

Then

$$\ln x + \ln q = C$$

$$\ln (xq) = C \quad \text{Using a property of logarithms}$$

$$xq = e^C. \quad \text{Writing an equivalent exponential equation}$$

We let $C_1 = e^C = xq$. Then

$$q = \frac{C_1}{x}, \quad \text{or} \quad x = \frac{C_1}{q}.$$

This result characterizes those demand functions for which the elasticity is always 1.

✦

An Application to Psychology: Reaction to a Stimulus

The Weber–Fechner Law

In psychology, one model of stimulus–response asserts that the rate of change dR/dS of the reaction R with respect to a stimulus S is inversely proportional to the intensity of the stimulus. That is,

$$\frac{dR}{dS} = \frac{k}{S},$$

where k is some positive constant.

To solve this equation, we first separate the variables:

$$dR = k \cdot \frac{dS}{S}.$$

We then integrate both sides:

$$\int dR = \int k \cdot \frac{dS}{S}$$

$$R = k \ln S + C. \quad \text{We assume } S > 0. \tag{1}$$

Now suppose that we let S_0 be the lowest level of the stimulus that can be detected. This is the *threshold value*, or the *detection threshold*. For example, the lowest level of sound that can be consistently detected is the tick of a watch from 20 ft away, under very quiet conditions. If S_0 is the lowest level of stimulus that can be detected, it seems reasonable that $R(S_0) = 0$. Substituting this condition into equation (1), we get

$$0 = k \ln S_0 + C,$$

or

$$-k \ln S_0 = C.$$

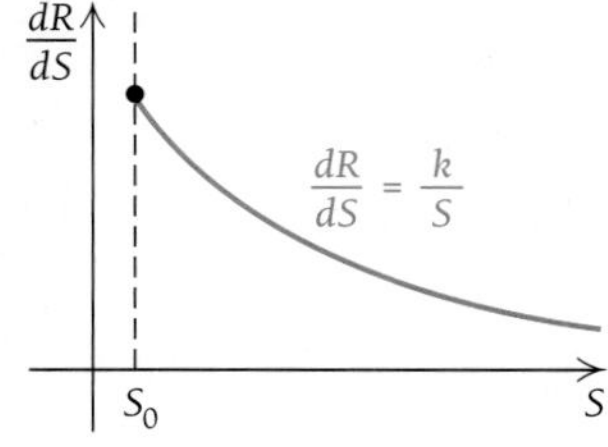

Replacing C in equation (1) with $-k \ln S_0$ gives us

$$R = k \ln S - k \ln S_0 \quad \text{As a check, note that } \frac{dR}{dS} = \frac{k}{S}.$$

$$= k(\ln S - \ln S_0).$$

Using a property of logarithms, we have

$$R = k \cdot \ln \frac{S}{S_0}.$$

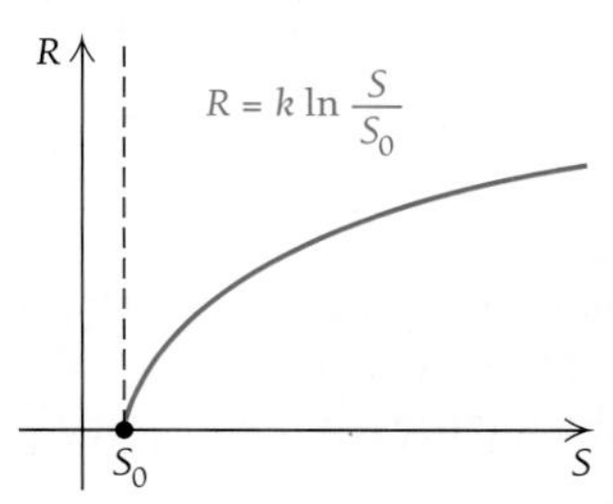

Look at the graphs of dR/dS and R on the left. Note that as the stimulus gets larger, the rate of change decreases; that is, the change in reaction becomes smaller as the stimulation received becomes stronger. For example, suppose that a lamp has a 50-watt bulb in it. If the bulb were suddenly changed to 100 watts, you would probably be very aware of the difference. That is, your reaction would be strong. If the bulb were then changed to 150 watts, your reaction would not be as great as it was to the change from 50 to 100 watts. A change from a 150- to a 200-watt bulb would cause even less reaction, and so on.

For your interest, here are some other detection thresholds.

Stimulus	Detection Threshold
Light	The flame of a candle 30 miles away on a dark night
Taste	Water diluted with sugar in the ratio of 1 teaspoon to 2 gallons
Smell	One drop of perfume diffused into the volume of three average-size rooms
Touch	The wing of a bee dropped on your cheek at a distance of 1 centimeter (about $\frac{3}{8}$ of an inch)

Exercise Set 5.7

Find the general solution and three particular solutions.

1. $y' = 5x^4$

2. $y' = 6x^5$

3. $y' = e^{2x} + x$

4. $y' = e^{4x} - x + 2$

5. $y' = \frac{8}{x} - x^2 + x^5$

6. $y' = \frac{3}{x} + x^2 - x^4$

Find the particular solution determined by the given condition.

7. $y' = x^2 + 2x - 3;\quad y = 4$ when $x = 0$

8. $y' = 3x^2 - x + 5;\quad y = 6$ when $x = 0$

9. $f'(x) = x^{2/3} - x;\quad f(1) = -6$

10. $f'(x) = x^{2/5} + x;\quad f(1) = -7$

11. Show that $y = x \ln x + 3x - 2$ is a solution of

$$y'' - \frac{1}{x} = 0.$$

12. Show that $y = x \ln x - 5x + 7$ is a solution of

$$y'' - \frac{1}{x} = 0.$$

13. Show that $y = e^x + 3xe^x$ is a solution of

$$y'' - 2y' + y = 0.$$

14. Show that $y = -2e^x + xe^x$ is a solution of

$$y'' - 2y' + y = 0.$$

Solve.

15. $\frac{dy}{dx} = 4x^3y$

16. $\frac{dy}{dx} = 5x^4y$

17. $3y^2\frac{dy}{dx} = 8x$

18. $3y^2\frac{dy}{dx} = 5x$

19. $\frac{dy}{dx} = \frac{2x}{y}$

20. $\frac{dy}{dx} = \frac{x}{2y}$

21. $\frac{dy}{dx} = \frac{6}{y}$

22. $\frac{dy}{dx} = \frac{7}{y^2}$

23. $y' = 3x + xy;\quad y = 5$ when $x = 0$

24. $y' = 2x - xy;\quad y = 9$ when $x = 0$

25. $y' = 5y^{-2};\quad y = 3$ when $x = 2$

26. $y' = 7y^{-2};\quad y = 3$ when $x = 1$

27. $\frac{dy}{dx} = 3y$

28. $\frac{dy}{dx} = 4y$

29. $\frac{dP}{dt} = 2P$

30. $\frac{dP}{dt} = 4P$

31. Solve

$$f'(x) = \frac{1}{x} - 4x + \sqrt{x},$$

given that $f(1) = \frac{23}{3}$.

APPLICATIONS

Business and Economics

32. Total revenue from marginal revenue. The marginal revenue for a certain product is given by $R'(x) = 300 - 2x$. Find the total-revenue function, $R(x)$, assuming that $R(0) = 0$.

33. **Total cost from marginal cost.** The marginal cost for a certain product is given by $C'(x) = 2.6 - 0.02x$. Find the total-cost function, $C(x)$, and the average cost, $A(x)$, assuming that fixed costs are \$120; that is, $C(0) = \$120$.

34. **Capital expansion.** *Domar's capital expansion model* is

$$\frac{dI}{dt} = hkI,$$

where I is the investment, h is the investment productivity (constant), k is the marginal productivity to the consumer (constant), and t is the time.

a) Use separation of variables to solve the differential equation.

b) Rewrite the solution in terms of the condition $I_0 = I(0)$.

35. **Total profit from marginal profit.** A firm's marginal profit, P, as a function of its total cost, C, is given by

$$\frac{dP}{dC} = \frac{-200}{(C+3)^{3/2}}.$$

a) Find the profit function, $P(C)$, if $P = \$10$ when $C = \$61$.

b) At what cost will the firm break even ($P = 0$)?

36. **Stock growth.** The growth rate of a certain stock, in dollars, can be modeled by

$$\frac{dV}{dt} = k(L - V),$$

where V is the value of the stock, per share, after t months; k is a constant; $L = \$24.81$, the *limiting value* of the stock; and $V(0) = 20$. Find the solution of the differential equation in terms of t and k.

37. **Utility.** The reaction R in pleasure units by a consumer receiving S units of a product can be modeled by the differential equation

$$\frac{dR}{dS} = \frac{k}{S+1},$$

where k is a positive constant.

a) Use separation of variables to solve the differential equation.

b) Rewrite the solution in terms of the initial condition $R(0) = 0$.

TW c) Explain why the condition $R(0) = 0$ is reasonable.

Elasticity. *Find the demand function $q = D(x)$, given each set of elasticity conditions.*

38. $E(x) = \dfrac{4}{x}$; $q = e$ when $x = 4$

39. $E(x) = \dfrac{x}{200 - x}$; $q = 190$ when $x = 10$

40. $E(x) = 2$, for all $x > 0$

41. $E(x) = n$, for some constant n and all $x > 0$

Life and Physical Sciences

42. **Exponential growth.**

a) Use separation of variables to solve the differential-equation model of uninhibited growth,

$$\frac{dP}{dt} = kP.$$

b) Rewrite the solution of part (a) in terms of the condition $P_0 = P(0)$.

Social Sciences

43. **The Brentano–Stevens Law.** The validity of the Weber–Fechner Law has been the subject of great debate among psychologists. An alternative model,

$$\frac{dR}{dS} = k \cdot \frac{R}{S},$$

where k is a positive constant, has been proposed. Find the general solution of this equation. (This model has also been referred to as the *Power Law of Stimulus–Response.*)

SYNTHESIS

Solve.

44. $\dfrac{dy}{dx} = 5x^4y^2 + x^3y^2$

45. $e^{-1/x} \cdot \dfrac{dy}{dx} = x^{-2} \cdot y^2$

TW 46. Discuss as many applications as you can of the use of integration in this chapter

TW 47. In Example 6 in this section, it is stated that "Since C is an arbitrary constant, $\pm e^{-C}$ is an arbitrary constant." Explain why the $\pm$ is necessary.

TECHNOLOGY CONNECTION

48. Solve $dy/dx = 5/y$. Graph the particular solutions for $C_1 = 5$, $C_1 = -200$, and $C_1 = 100$.

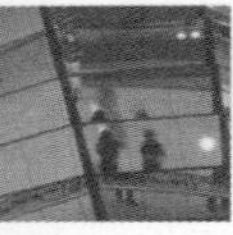

Chapter Summary

This chapter covered a variety of topics and applications involving integrals.

When market forces allow a producer and consumers to settle on an equilibrium point, there is agreement on both price and the quantity of the item bought at that price. Usually some consumers would have been willing to pay more per unit for the item. To the extent that they benefit from the equilibrium price being lower than what they were willing to pay, these consumers enjoy a *consumer surplus*.

Suppose that $p = D(x)$ describes the demand function for a commodity. Then the **consumer surplus** is defined for the point (Q, P) as

$$\int_0^Q D(x)\,dx - QP.$$ (p. 471)

The equilibrium point usually benefits the producer as well. To the extent that a producer benefits from selling more units at the equilibrium price, which is usually greater than the price the producer would charge for fewer units, a *producer surplus* exists.

Suppose that $p = S(x)$ is the supply function for a commodity. Then the **producer surplus** is defined for the point (Q, P) as

$$QP - \int_0^Q S(x)\,dx.$$ (p. 473)

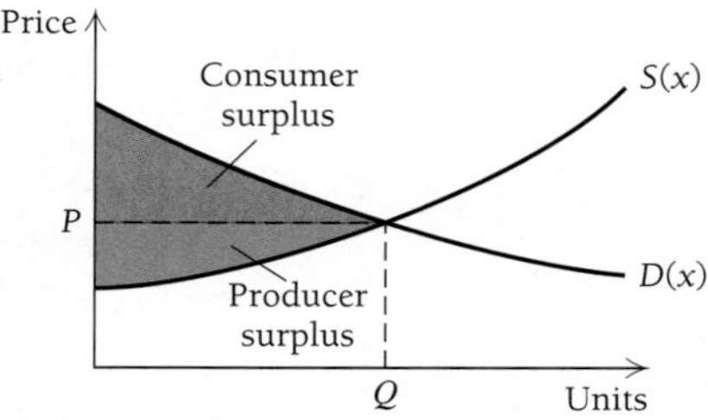

If money is continuously being invested in an account that is earning interest compounded continuously, a *continuous money flow* exists.

FUTURE VALUE OF A CONTINUOUS MONEY FLOW

If the yearly flow of money into an investment is given by some constant function $R(t)$, then the **future value of the continuous money flow** at interest rate k, compounded continuously, over T years, is given by

$$\int_0^T R(t)e^{kt}\,dt. \qquad \text{(p. 478)}$$

EXAMPLE If \$600 per year is flowing at a constant rate for 10 years into an account paying 5%, compounded continuously, the future value of the money flow is

$$\int_0^{10} 600e^{0.05t}\,dt = \left[\frac{600}{0.05}e^{0.05t}\right]_0^{10}$$
$$= \frac{600}{0.05}(e^{0.5} - 1) \approx \$7784.66.$$ ✦

Integrals of the form $\int_0^T P_0e^{kt}\,dt$ also can be used to determine the consumption of a natural resource. (p. 479)

Just as the model $P = P_0e^{kt}$ can be used to find the **present value, P_0,** for exponential growth or continuous compounding (p. 481), the model for a continuous money flow can be used to determine the single amount that, invested with continuous compounding, will yield the same future value. This amount is the *accumulated present value*.

The **accumulated present value** of a continuous money flow into an investment at a rate of $R(t)$ dollars per year from now until T years in the future is given by

$$\int_0^T R(t)e^{-kt}\,dt,$$

where k is the interest rate, and interest is compounded continuously. (p. 482)

EXAMPLE The accumulated present value of a continuous money flow of \$700 per year, over 8 yr, with 7% interest compounded continuously, is

$$\int_0^8 700e^{-0.07t}\,dt = \left[\frac{700}{-0.07}e^{-0.07t}\right]_0^8$$
$$= -10{,}000(e^{-0.56} - 1) \approx \$4287.91.$$ ✦

When a continuous money flow continues perpetually, we have an integral of the form

$$\int_0^\infty f(t)\,dt.$$

Such an expression, in which at least one limit of integration is either $-\infty$ or ∞, is called an **improper integral**. Limits are used to evaluate such integrals:

$$\int_a^{\infty} f(x)\,dx = \lim_{b\to\infty}\int_a^b f(x)\,dx;$$

$$\int_{-\infty}^{b} f(x)\,dx = \lim_{a\to-\infty}\int_a^b f(x)\,dx;$$

and $\displaystyle\int_{-\infty}^{\infty} f(x)\,dx = \int_{-\infty}^{c} f(x)\,dx + \int_c^{\infty} f(x)\,dx,$ where c can be any real number.

If the limit of an improper integral exists, we say that the integral is **convergent**. If a limit does not exist, the integral is **divergent**. (pp. 486–487)

EXAMPLE $\displaystyle\int_1^{\infty} e^{-x}\,dx$ is convergent, since the following limit exists:

$$\lim_{b\to\infty}\int_1^b e^{-x}\,dx = \lim_{b\to\infty}\left[-e^{-x}\right]_1^b$$

$$= \lim_{b\to\infty}\left(-\frac{1}{e^b} - \left(-\frac{1}{e}\right)\right) = \frac{1}{e}.$$

✦

The **accumulated present value** of a continuous money flow into an investment at a rate of P dollars per year perpetually is given by

$$\int_0^{\infty} Pe^{-kt}\,dt = \frac{P}{k},$$

where k is the interest rate and interest is compounded continuously. (p. 487)

Another use of integrals is to calculate the probabilities of certain events. To use an integral for this purpose, the variable of integration must represent a **continuous random variable**—a quantity that can be observed or measured repeatedly and whose possible values consist of an entire interval of real numbers. (p. 493)

A *probability density function* is a function whose integral over any subinterval for which a random variable exists gives the probability that the variable lands in that subinterval. Let x be a continuous random variable. A function f is said to be a **probability density function** for x if:

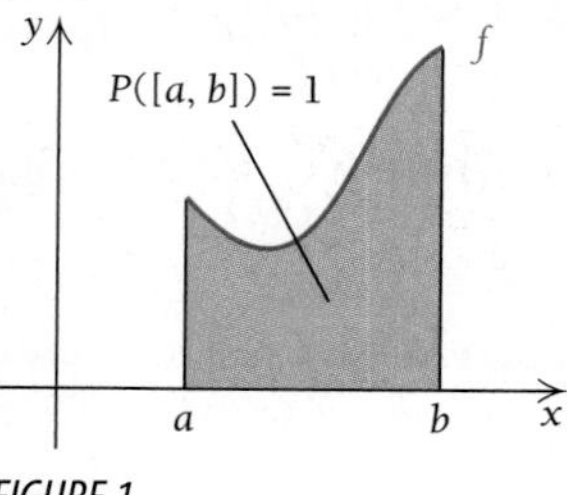

FIGURE 1

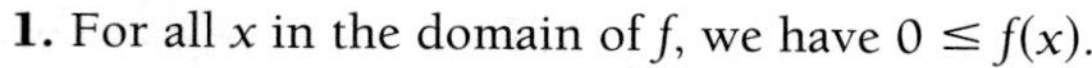

1. For all x in the domain of f, we have $0 \le f(x)$.
2. The area under the graph of f is 1 (see Fig. 1).
3. For any subinterval $[c, d]$ in the domain of f (see Fig. 2), the probability that x will be in that subinterval is given by

$$P([c, d]) = \int_c^d f(x)\,dx.$$ (p. 495)

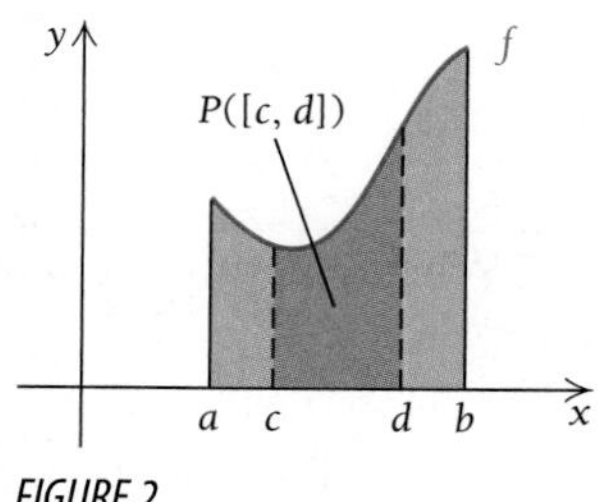

FIGURE 2

EXAMPLE To verify Property 2 for the function given by

$$f(x) = \frac{2}{35}x, \quad \text{for } 1 \leq x \leq 6,$$

we note that

$$\int_1^6 \frac{2}{35}x\,dx = \left[\frac{1}{35}x^2\right]_1^6 = \frac{36}{35} - \frac{1}{35} = 1.$$

✦

Three important types of probability distributions occur in a wide variety of applications.

A continuous random variable x is said to be **uniformly distributed** over an interval $[a, b]$ if it has a probability density function f given by

$$f(x) = \frac{1}{b-a}, \quad \text{for } a \leq x \leq b. \qquad \text{(p. 498)}$$

A continuous random variable is **exponentially distributed** if it has a probability density function given by

$$f(x) = ke^{-kx}, \quad \text{over the interval } [0, \infty). \qquad \text{(p. 499)}$$

A continuous random variable x has a **standard normal distribution** if its probability density function is

$$f(x) = \frac{1}{\sqrt{2\pi}}e^{-x^2/2}. \qquad \text{(p. 508)}$$

The data in any distribution can be characterized using numbers that represent a measure of central tendency (*mean*) and the spread (*variance* and, especially, *standard deviation*).

The **mean, μ,** of a continuous random variable x is defined to be $E(x)$, the expected value of x. That is,

$$\mu = E(x) = \int_a^b xf(x)\,dx,$$

where f is a probability density function for x. (p. 505)

The **variance, σ^2,** of a continuous random variable x is defined as

$$\begin{aligned}\sigma^2 &= E(x^2) - \mu^2 \\ &= E(x^2) - [E(x)]^2 \\ &= \int_a^b x^2f(x)\,dx - \left[\int_a^b xf(x)\,dx\right]^2. \qquad \text{(p. 506)}\end{aligned}$$

The **standard deviation, σ,** of a continuous random variable is defined as

$$\sigma = \sqrt{\text{variance}}. \qquad \text{(p. 506)}$$

The standard normal distribution has a mean of 0 and a standard deviation of 1. Its graph follows.

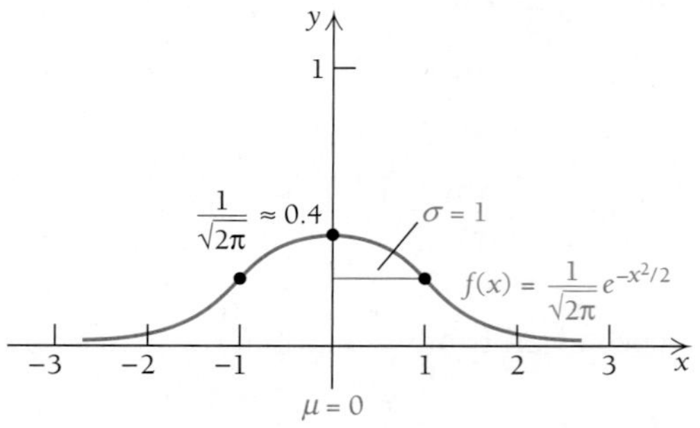

The general case is defined as follows: A continuous random variable x is **normally distributed** with mean μ and standard deviation σ if its probability density function is given by

$$f(x) = \frac{1}{\sigma\sqrt{2\pi}} e^{-(1/2)[(x-\mu)/\sigma]^2}, \quad \text{over } (-\infty, \infty). \quad \text{(p. 508)}$$

To calculate areas for the standard normal distribution, tables of values have been developed using numerical methods. Such a table allows for calculating probabilities for any normal distribution in which the mean μ and standard deviation σ are both known. (p. 509)

EXAMPLE For $\mu = 45$ and $\sigma = 5$, we find $P(43 \le x \le 49)$ by standardizing the scores and viewing them in terms of intervals, as required for Table 2:

$$\begin{aligned} P(43 \le x \le 49) &= P\left(\frac{43 - 45}{5} \le z \le \frac{49 - 45}{5}\right) \\ &= P(-0.4 \le z \le 0.8) \\ &= P(0 \le z \le 0.4) + P(0 \le z \le 0.8) \\ &= 0.1554 + 0.2881 = 0.4435. \end{aligned}$$

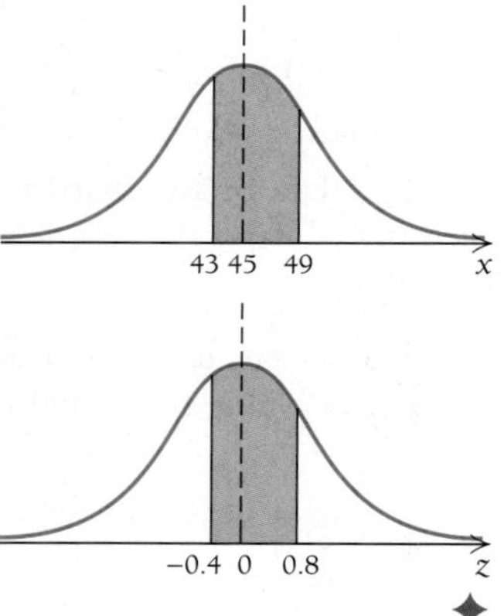

Integrals can also be used to calculate the volume when an area below a curve is rotated about the x-axis.

For a continuous function f defined on $[a, b]$, the **volume, V, of the solid of revolution** obtained by rotating about the x-axis the area between the graph of f and $[a, b]$ is given by

$$V = \int_a^b \pi[f(x)]^2\, dx. \quad \text{(p. 515)}$$

EXAMPLE The volume of the solid of revolution generated by rotating about the x-axis the region under the graph of $y = \frac{1}{4}x^2$ from $x = -1$ to $x = 3$ is

$$\begin{aligned} V &= \int_{-1}^{3} \pi[f(x)]^2\,dx = \int_{-1}^{3} \pi\left(\frac{1}{4}x^2\right)^2 dx \\ &= \int_{-1}^{3} \frac{\pi}{16}x^4\,dx = \left[\frac{\pi}{16}\cdot\frac{x^5}{5}\right]_{-1}^{3} = \frac{\pi}{80}(3^5 - (-1)^5) \\ &= \frac{\pi}{80}(243 - (-1)) = \frac{\pi}{80}(244) = \frac{61\pi}{20}. \end{aligned}$$

✦

The final topic of this chapter, **differential equations,** was first studied in Chapter 3. Solutions of differential equations are functions. When a constant of integration is included, the solution is a **general solution;** when information that leads to a specific choice of that constant is given, we have a **particular solution**. (p. 518)

EXAMPLE Given that $f'(x) = 2e^x + 4x^2$ and $f(0) = 7$, the general solution is

$$f(x) = 2e^x + \frac{4}{3}x^3 + C.$$

Since $f(0) = 2e^0 + \frac{4}{3}\cdot 0^3 + C = 7$, it follows that

$$2\cdot 1 + C = 7, \text{ and } C = 5.$$

Thus, the particular solution is $f(x) = 2e^x + \frac{4}{3}x^3 + 5$. ✦

Some differential equations are most easily solved by **separating the variables.** (p. 520)

EXAMPLE To solve $5y^4\frac{dy}{dx} - x = 0$, we proceed as follows:

$$\begin{aligned} 5y^4\frac{dy}{dx} &= x \\ 5y^4\,dy &= x\,dx \\ \int 5y^4\,dy &= \int x\,dx \\ y^5 &= \frac{x^2}{2} + C \\ y &= \sqrt[5]{\frac{x^2}{2} + C}. \end{aligned}$$

✦

Chapter Review Exercises

These review exercises are for test preparation. They can also be used as a practice test. Answers are at the back of the book. The bracketed section references tell you what part(s) of the chapter to restudy if your answer is incorrect.

CONCEPT REINFORCEMENT

Match each term in column A with the most appropriate graph in column B.

Column A — **Column B**

1. Consumer surplus [5.1]

a)

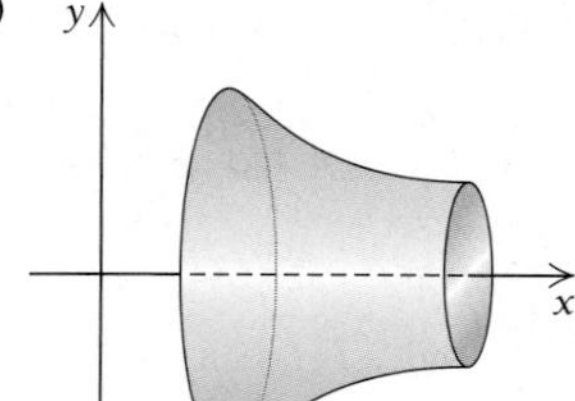

2. Producer surplus [5.1]

b)

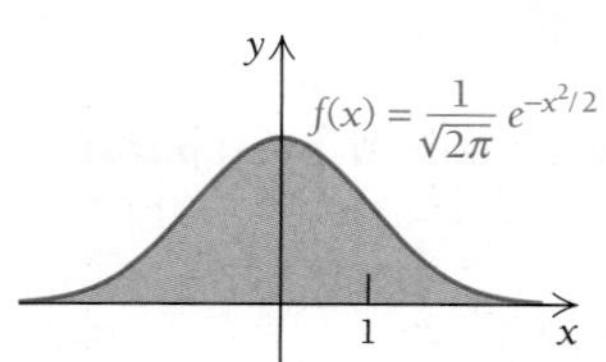

3. Exponential distribution [5.4]

c)

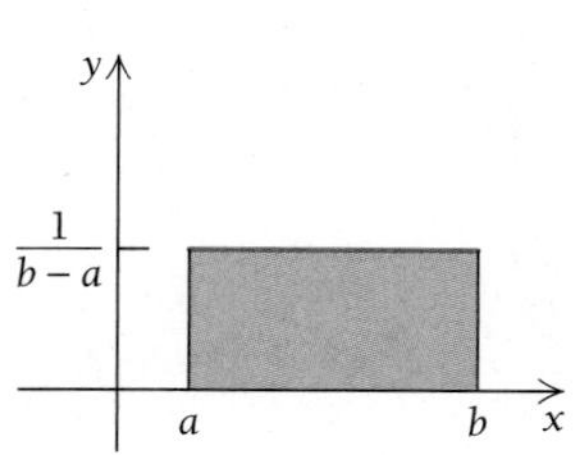

4. Standard normal distribution [5.5]

d)

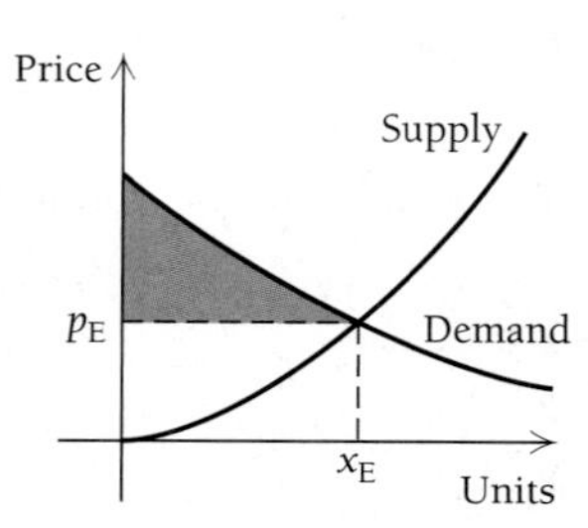

5. Uniform distribution [5.4]

e)

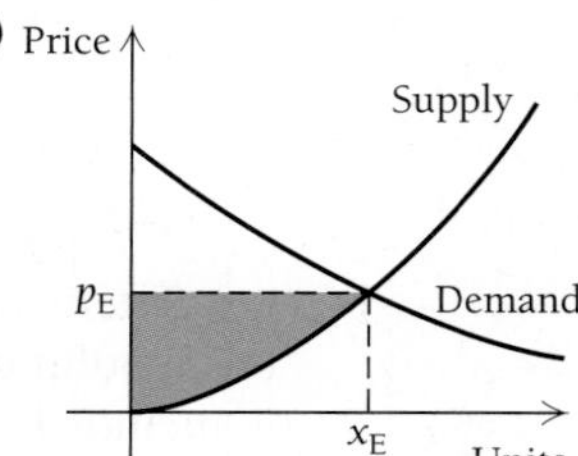

6. Solid of revolution [5.6]

f)

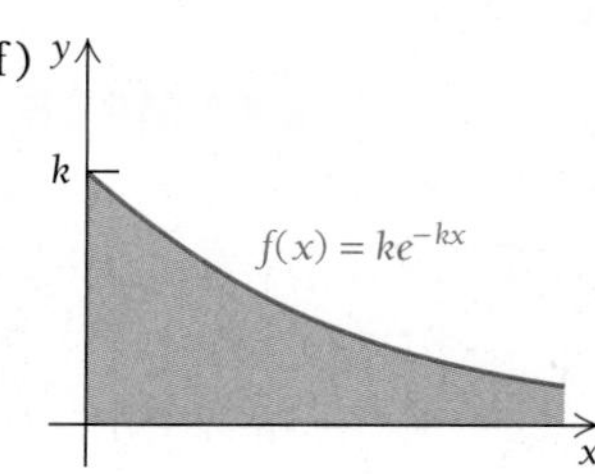

Classify each statement as either true or false.

7. The accumulated present value of an investment is the value of the investment as a tax-deductible present to a nonprofit charity. [5.2]

8. If an integral has $-\infty$ or ∞ as one of the limits of integration, it is an improper integral. [5.3]

9. If f is a probability density function over $[a, b]$, then $f(x) \geq 0$ for all x in $[a, b]$. [5.4]

10. If f is a probability density function over $[a, b]$ and x is a continuous random variable over $[a, b]$, then the mean value of $f(x)$ is $(b - a)/2$. [5.5]

11. To find the volume of the solid of revolution obtained by rotating the graph of $y = f(x)$ about the x-axis, we must have $f(x) \geq 0$. [5.6]

12. If $y = f(x)$ is a solution of $y'' + y' = 5$, then $y = f(x) + C$ is also a solution. [5.7]

REVIEW EXERCISES

Let $D(x) = (x - 6)^2$ be the price, in dollars per unit, that consumers are willing to pay for x units of an item, and $S(x) = x^2 + 12$ be the price, in dollars per unit, that producers are willing to accept for x units.

13. Find the equilibrium point. [5.1]

14. Find the consumer surplus at the equilibrium point. [5.1]

15. Find the producer surplus at the equilibrium point. [5.1]

16. Business: future value of a continuous money flow. Find the future value of a continuous money flow when \$2400 per year is being invested at a constant rate, compounded continuously at 6%, for 15 yr. [5.2]

17. Business: continuous money flow. What should P_0 be so that the amount of a continuous money flow, at P_0 per year over 25 yr, at 7% compounded continuously, will be \$30,000? [5.2]

18. Business: accumulated present value. A grandparent's will stipulates that for a span of 15 yr, each grandchild is to receive \$1200 per yr, paid as a continuous money flow at a constant rate into an account earning 5% compounded continuously. What is the accumulated present value of such an account at the start of the 15-yr period? [5.2]

19. Business: accumulated present value. Find the accumulated present value of one of the accounts described in Exercise 18 if the continuous money flow does not end after 15 years, but instead continues perpetually. [5.3]

20. Business: present value. Find the present value of \$100,000 due 50 yr later, at 12%, compounded continuously. [5.2]

21. Physical science: iron ore consumption. In 2005 ($t = 0$), the world production of iron ore was estimated at 1.23 billion metric tons, and production was growing exponentially at the rate of 3% per year. (*Source*: Based on data from the U.S. Energy Information Administration.) If the production continues to grow at this rate, how much iron ore will be produced from 2005 to 2016? [5.2]

22. Physical science: depletion of iron ore. The world reserves of iron ore in 2005 were estimated to be 160 billion metric tons. (*Source*: Based on data from the U.S. Geological Survey.) Assuming that the growth rate in Exercise 21 continues and no new reserves are discovered, when will the world reserves of iron ore be depleted? [5.2]

Determine whether each improper integral is convergent or divergent, and calculate its value if it is convergent. [5.3]

23. $\int_1^\infty \frac{1}{x^2}\,dx$

24. $\int_1^\infty e^{4x}\,dx$

25. $\int_0^\infty e^{-2x}\,dx$

26. Find k such that $f(x) = k/x^3$ is a probability density function over the interval $[1, 2]$. Then write the probability density function. [5.4]

27. Business: waiting time. A person arrives at a random time at a bus stop where the waiting time t for a bus is no more than 25 min. The probability density function for t is $f(t) = \frac{1}{25}$, for $0 \le t \le 25$. Find the probability that a person will have to wait no more than 15 min for a bus. [5.4]

Given the probability density function

$$f(x) = 6x(1 - x), \quad \text{over } [0, 1]$$

find each of the following. [5.5]

28. $E(x^2)$

29. $E(x)$

30. The mean

31. The variance

32. The standard deviation

Let x be a continuous random variable with a standard normal distribution. Using Table 2, find each of the following. [5.5]

33. $P(0 \le x \le 1.85)$

34. $P(-1.74 \le x \le 1.43)$

35. $P(-2.08 \le x \le -1.18)$

36. $P(x \ge 0)$

37. Business: pizza sales. The number of pizzas sold daily at Benito's Pizzeria is normally distributed with mean $\mu = 90$ and standard deviation $\sigma = 20$. What is the probability that at least 100 pizzas are sold during a day? [5.5]

Find the volume generated by rotating about the x-axis the region bounded by the graphs of the given equations. [5.6]

38. $y = x^3, x = 1, x = 2$

39. $y = \frac{1}{x + 2}, x = 0, x = 1$

Solve each differential equation. [5.7]

40. $\frac{dy}{dx} = 11x^{10}y$

41. $\frac{dy}{dx} = \frac{2}{y}$

42. $\frac{dy}{dx} = 4y; \quad y = 5$ when $x = 0$

43. $\frac{dv}{dt} = 5v^{-2}; \quad v = 4$ when $t = 3$

44. $y' = \frac{3x}{y}$

45. $y' = 8x - xy$

46. **Economics: elasticity.** Find the demand function $q = D(x)$, given the elasticity condition

$$E(x) = \frac{x}{100 - x}; \quad q = 70 \text{ when } x = 30. \quad [5.7]$$

47. **Business: stock growth.** The growth rate of a stock, in dollars per month, can be modeled by

$$\frac{dV}{dt} = k(L - V),$$

where V is the value of a share, in dollars, after t months; k is a constant; $L = \$36.37$, the limiting value of the stock; and $V(0) = 30$. Find the solution of the differential equation in terms of t and k. [5.7]

SYNTHESIS

48. The function $f(x) = x^8$ is a probability density function over the interval $[-c, c]$ Find c. [5.4]

Determine whether each improper integral is convergent or divergent, and calculate its value if it is convergent. [5.3]

49. $\displaystyle\int_{-\infty}^{0} x^4 e^{-x^5}\, dx$

50. $\displaystyle\int_{0}^{\infty} \frac{dx}{(x + 1)^{4/3}}$

TECHNOLOGY CONNECTION

51. Approximate the integral

$$\int_1^{\infty} \frac{\ln x}{x^2}\, dx. \quad [5.3]$$

Chapter 5 Test

Let $D(x) = (x - 7)^2$ be the price, in dollars per unit, that consumers are willing to pay for x units of an item, and let $S(x) = x^2 + x + 4$ be the price, in dollars per unit, that producers are willing to accept for x units. Find:

1. The equilibrium point
2. The consumer surplus at the equilibrium point
3. The producer surplus at the equilibrium point
4. **Business: amount of a continuous money flow.** Find the future value of a continuous money flow when \$1400 per year is being invested at a constant rate, compounded continuously at 5% for 10 yr.
5. **Business: continuous money flow.** Consider a continuous money flow into an investment at the rate of P_0 dollars per year. What should P_0 be so that the amount of a continuous money flow at a constant rate over 12 yr, at 6% compounded continuously, will be \$20,000?
6. **Physical science: demand for potash.** In 2004 ($t = 0$), the world production of potash was approximately 49.9 million metric tons, and demand was increasing at the rate of 9.9% a year. (*Source*: Based on data from the U.S. Energy Information Administration.) If the demand continues to grow at this rate, how much potash will be produced from 2004 to 2016?
7. **Physical science: depletion of potash.** See Exercise 6. The world reserves of potash in 2004 were approximately 8300 million metric tons. (*Source*: U.S. Geological Survey.) Assuming the demand for potash continues to grow at the rate of 9.9% per year and no new reserves are discovered, when will the world reserves be depleted?
8. **Business: present value.** Following the birth of a child, a parent wants to make an initial investment P_0 that will grow to \$10,000 by the child's 20th birthday. Interest is compounded continuously at 7%. What should the initial investment be?
9. **Business: accumulated present value.** Find the accumulated present value of an investment over a 20-yr period if there is a continuous money flow of \$3800 per year and the interest rate is 11%.
10. **Business: accumulated present value.** Find the accumulated present value of an investment for which the continuous money flow in Exercise 9 is perpetual.

Determine whether each improper integral is convergent or divergent, and calculate its value if it is convergent.

11. $\displaystyle\int_1^{\infty} \frac{dx}{x^5}$

12. $\displaystyle\int_0^{\infty} \frac{4}{1 + 3x}\, dx$

13. Find k such that $f(x) = kx^3$ is a probability density function over the interval $[0, 2]$. Then find the probability density function.

14. Business: times of telephone calls. A telephone company determines that the length of a phone call, t, in minutes, is an exponentially distributed random variable with probability density function

$$f(t) = 2e^{-2t}, \quad 0 \leq t < \infty.$$

Find the probability that a phone call will last no more than 3 min.

Given the probability density function $f(x) = \frac{1}{4}x$ *over* $[1, 3]$, *find each of the following.*

15. $E(x)$

16. $E(x^2)$

17. The mean

18. The variance

19. The standard deviation

Let x be a continuous random variable with a standard normal distribution. Using Table 2, find each of the following.

20. $P(0 \leq x \leq 1.3)$

21. $P(-2.31 \leq x \leq -1.05)$

22. $P(-1.61 \leq x \leq 1.76)$

23. The price per pound p of wild salmon at various stores in a certain city is normally distributed with mean $\mu = \$12$ and standard deviation $\sigma = \$2.50$. What is the probability that the price at a randomly selected store is at least \$13.25 per pound?

Find the volume generated by rotating about the x-axis the regions bounded by the following.

24. $y = \dfrac{1}{\sqrt{x}}, \quad x = 1, x = 5$

25. $y = \sqrt{2 + x}, \quad x = 0, x = 1$

Solve each differential equation.

26. $\dfrac{dy}{dx} = 8x^7y$

27. $\dfrac{dy}{dx} = \dfrac{9}{y}$

28. $\dfrac{dy}{dt} = 6y; \quad y = 11$ when $t = 0$

29. $y' = 5x^2 - x^2y$

30. $\dfrac{dv}{dt} = 2v^{-3}$

31. $y' = 4y + xy$

32. Economics: elasticity. Find the demand function $q = D(x)$, given the elasticity condition

$$E(x) = 4 \quad \text{for all } x > 0.$$

33. Business: stock growth. The growth rate of Fabric Industries stock, in dollars per month, can be modeled by

$$\frac{dV}{dt} = k(L - V),$$

where V is the value of a share, in dollars, after t months; $L = \$36$, the limiting value of the stock; k is a constant; and $V(0) = 0$.

a) Write the solution $V(t)$ in terms of L and k.

b) If $V(6) = 18$, determine k to the nearest hundredth.

c) Rewrite $V(t)$ in terms of t and k using the value of k found in part (b).

d) Use the equation in part (c) to find $V(12)$, the value of the stock after 12 months.

e) In how many months will the value be \$30?

SYNTHESIS

34. The function $f(x) = x^3$ is a probability density function over the interval $[0, b]$. What is b?

35. Determine whether the following improper integral is convergent or divergent, and calculate its value if it is convergent:

$$\int_{-\infty}^{0} x^3 e^{-x^4}\, dx.$$

TECHNOLOGY CONNECTION

36. Approximate the integral

$$\int_{-\infty}^{\infty} \frac{1}{1 + x^2}\, dx.$$

EXTENDED TECHNOLOGY APPLICATION

Curve Fitting and the Volume of a Bottle of Soda

Consider the urn or vase below. How could we estimate the volume? One way would be to simply fill the container with a liquid and then pour the liquid into a measuring device.

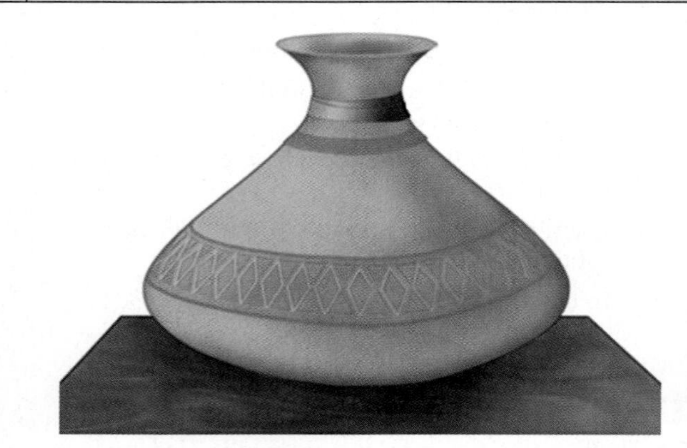

Another way, using calculus and the curve-fitting or REGRESSION feature of a graphing calculator, would be to turn the urn on its side, as shown below, take a series of vertical measurements from the center to the top, use REGRESSION, and then integrate (either by hand or with the aid of the calculator).

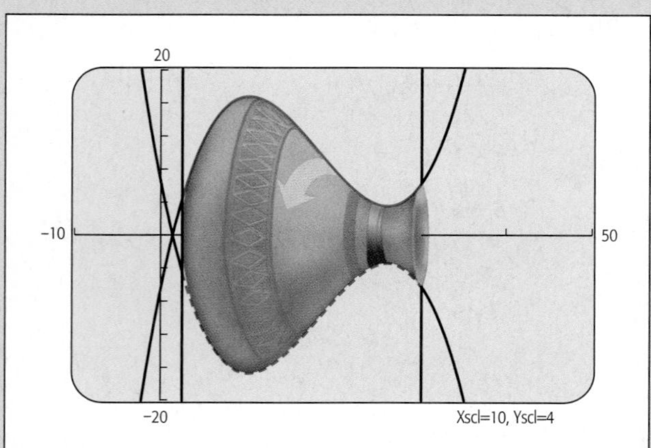

The following table is a table of values for the red curve.

x (in centimeters)	y (in centimeters)
3	4
5	10
10	17
15	16
20	10
25	5
28	3.5
34	7

EXERCISES

1. Using REGRESSION, fit a cubic polynomial function to the data.
2. Using the function found in Exercise 1, integrate over the interval [3, 34] to find the volume of the urn. *Hint*: If the function in Exercise 1 is Y1, find the volume by using the VARS key to enter πY1^2 as Y2. Then use the CALC option to integrate.

Now consider the soda bottle pictured to the right. To find its volume in a similar manner, we can turn the bottle on its side, use a measuring device to take vertical measurements, and proceed as we did with the urn.

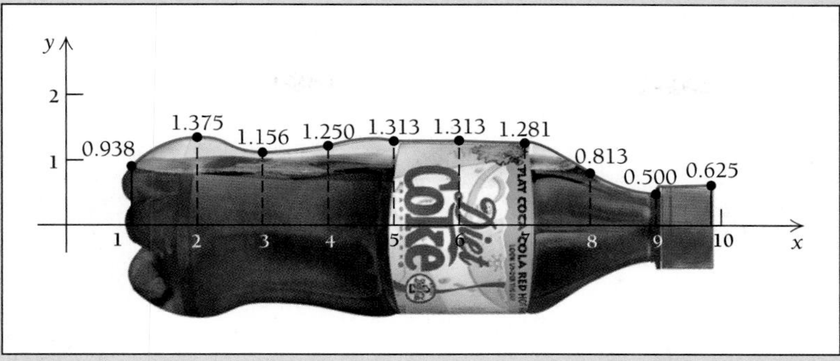

The table of measurements is as follows.

x (in inches)	y (in inches)
1	0.938
2	1.375
3	1.156
4	1.250
5	1.313
6	1.313
7	1.281
8	0.813
9	0.500
9.875	0.625

EXERCISES

3. Using REGRESSION, fit a quartic polynomial function to the data.

4. Using the function found in Exercise 3, integrate to find the volume of the bottle. Your answer will be in cubic inches. Convert it to fluid ounces using the fact that 1 $\text{in}^3 =$ 0.55424 fluid ounce.

5. The bottle in question holds 20 oz. How good was our curve-fitting procedure for making the volume estimate?

6. Find a curve that gives a better estimate of the volume. What is the curve and what is the estimated volume?

Functions of Several Variables

APPLICATION The minimum hourly wage in the United States has grown over the years, as shown in the table below.

Number of Years, x, since 1980	Minimum Hourly Wage
0	$3.10
1	3.35
10	3.80
11	4.25
16	4.75
17	5.15

(*Source:* www.workworld.org.) Find the regression line, and use it to predict the minimum hourly wage in 2010 and 2015.

This problem appears as Exercise 5 in Section 6.4.

INTRODUCTION

Functions that have more than one input are called *functions of several variables.* We introduce these functions in this chapter and learn to differentiate them to find *partial derivatives.* Then we use these functions and their partial derivatives to find regression lines and solve maximum–minimum problems. Finally, we consider the integration of such functions.

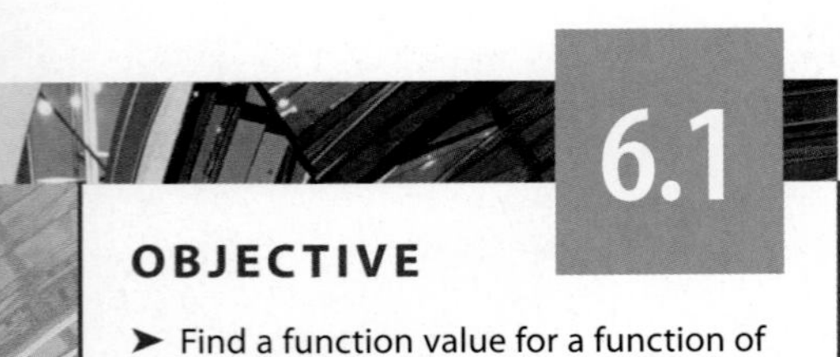

OBJECTIVE

- Find a function value for a function of several variables.

6.1 Functions of Several Variables

Suppose that a one-product firm produces x units of its product at a profit of \$4 per unit. Then its total profit $P(x)$ is given by

$$P(x) = 4x.$$

This is a function of one variable.

Suppose that a two-product firm produces x units of one product at a profit of \$4 per unit and y units of a second product at a profit of \$6 per unit. Then its total profit P is a function of the *two* variables x and y, and is given by

$$P(x, y) = 4x + 6y.$$

This function assigns to the input pair (x, y) a unique output number, $4x + 6y$.

DEFINITION

A **function of two variables** assigns to each input pair, (x, y), exactly one output number, $f(x, y)$.

We can regard a function of two variables as a machine that has two inputs. Thus, the domain is a set of pairs (x, y) in the plane. When such a function is given by a formula, the domain normally consists of all ordered pairs (x, y) that are meaningful replacements in the formula.

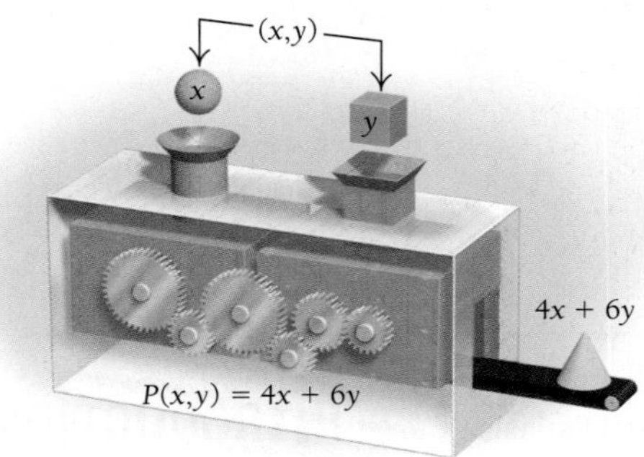

EXAMPLE 1 For the above profit function, $P(x, y) = 4x + 6y$, find $P(25, 10)$.

Solution $P(25, 10)$ is defined to be the value of the function found by substituting 25 for x and 10 for y:

$$\begin{aligned} P(25, 10) &= 4 \cdot 25 + 6 \cdot 10 \\ &= 100 + 60 \\ &= \$160. \end{aligned}$$

This result means that by selling 25 units of the first product and 10 of the second, the two-product firm will make a profit of \$160. ✦

The following are examples of **functions of several variables,** that is, functions of two or more variables. If there are n variables, then there are n inputs for such a function.

EXAMPLE 2 Business: Total Cost. The total cost to a company, in thousands of dollars, is given by

$$C(x, y, z, w) = 4x^2 + 5y + z - \ln(w + 1),$$

where x dollars is spent for labor, y dollars for raw materials, z dollars for advertising, and w dollars for machinery. This is a function of four variables (all in thousands of dollars). Find $C(3, 2, 0, 10)$.

Solution We substitute 3 for x, 2 for y, 0 for z, and 10 for w:

$$\begin{aligned} C(3, 2, 0, 10) &= 4 \cdot 3^2 + 5 \cdot 2 + 0 - \ln(10 + 1) \\ &= 4 \cdot 9 + 10 + 0 - 2.397895 \\ &\approx \$43.6 \text{ thousand, or } \$43{,}600. \end{aligned}$$ ✦

EXAMPLE 3 Business: Cost of Storage Equipment. A business purchases a piece of storage equipment that costs C_1 dollars and has capacity V_1. Later it wishes to replace the original with a new piece of equipment that costs C_2 dollars and has capacity V_2. Industrial economists have found that in such cases, the cost of the new piece of equipment can be estimated by the function of three variables

$$C_2 = \left(\frac{V_2}{V_1}\right)^{0.6} C_1.$$

For \$45,000, a beverage company buys a manufacturing tank that has a capacity of 10,000 gallons. Later it decides to buy a tank with double the capacity of the original. Estimate the cost of the new tank.

Solution We substitute 20,000 for V_2, 10,000 for V_1, and 45,000 for C_1:

$$\begin{aligned} C_2 &= \left(\frac{20{,}000}{10{,}000}\right)^{0.6}(45{,}000) \\ &= 2^{0.6}(45{,}000) \\ &\approx \$68{,}207.25. \end{aligned}$$

Note that a 100% increase in capacity was achieved by about a 52% increase in cost. This is independent of any increase in the costs of labor, management, or other equipment resulting from the purchase of the tank. ✦

EXAMPLE 4 Social Science: The Gravity Model. As the populations of two cities grow, the number of telephone calls between the cities increases, much like the gravitational pull will increase between two growing objects in space. The average number of telephone calls per day between two cities is given by

$$N(d, P_1, P_2) = \frac{2.8P_1P_2}{d^{2.4}},$$

where d is the distance, in miles, between the cities and P_1 and P_2 are their populations. The cities of Dallas and Fort Worth are 30 mi apart and have populations of 1,213,825 and 624,067, respectively. (*Sources:* Population Division, U.S. Census Bureau, June 2006, and Rand McNally.) Find the average number of calls per day between the two cities.

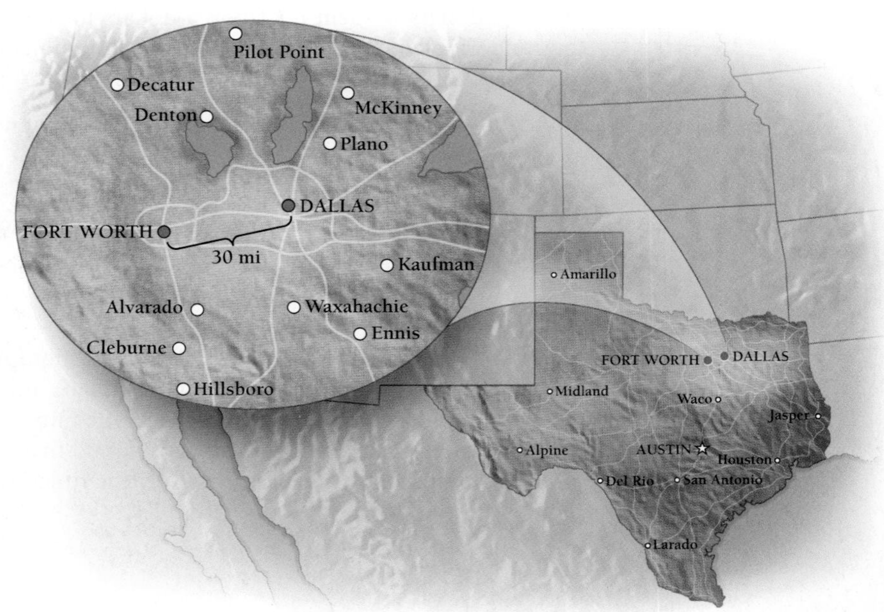

Solution We evaluate the function with the aid of a calculator:

$$N(30, 1{,}213{,}825, 624{,}067) = \frac{2.8(1{,}213{,}825)(624{,}067)}{30^{2.4}}$$
$$\approx 604{,}580{,}752.$$

✦

It is also possible for a function of two or more variables to be a constant function.

EXAMPLE 5 For $f(x, y) = -3$, find $f(5, 7)$ and $f(-2, 0)$.

Solution Since this is a constant function, it has the value -3 for any values of x and y. Thus,

$$f(5, 7) = -3 \quad \text{and} \quad f(-2, 0) = -3.$$

✦

Geometric Interpretations

Visually, a function of two variables,

$$z = f(x, y),$$

can be thought of as matching a point (x_1, y_1) in the xy-plane with the number z_1 on a number line.

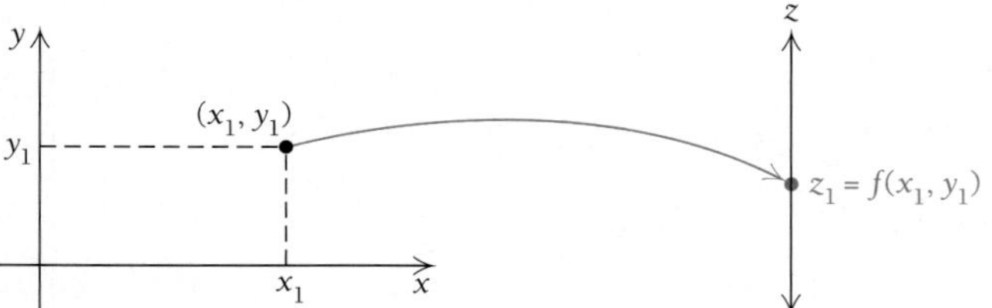

Thus, to graph a function of two variables, we need a three-dimensional coordinate system. The axes are generally placed as shown to the right. The line z, called the z-axis, is placed perpendicular to the xy-plane at the origin.

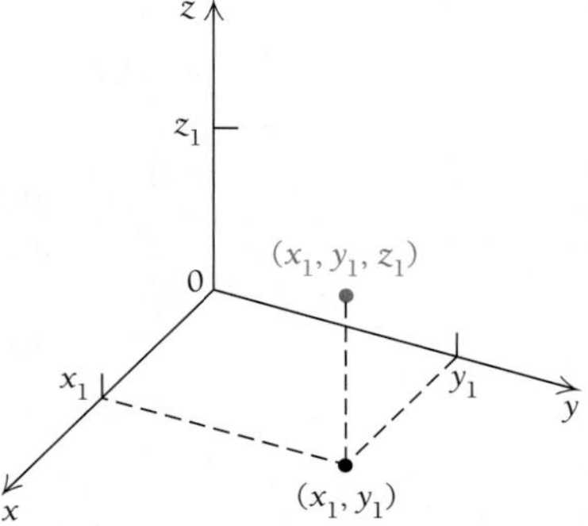

To help visualize this, think of looking into the corner of a room, where the floor is the xy-plane and the z-axis is the intersection of the two walls. To plot a point (x_1, y_1, z_1), we locate the point (x_1, y_1) in the xy-plane and move up or down in space according to the value of z_1.

EXAMPLE 6 Plot these points:

$P_1(2, 3, 5)$,
$P_2(2, -2, -4)$,
$P_3(0, 5, 2)$, and
$P_4(2, 3, 0)$.

Solution The solution is shown at the right.

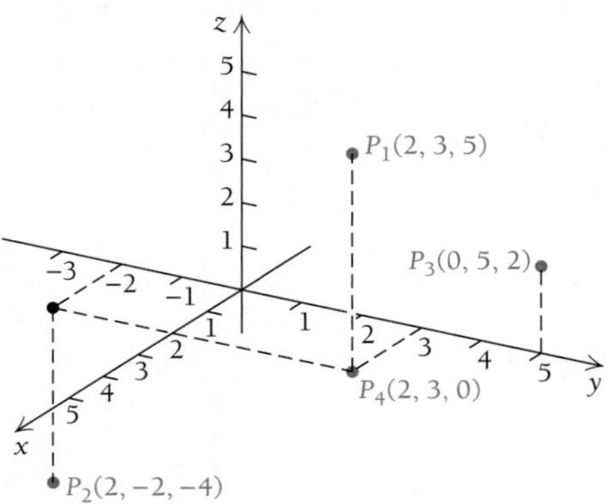

◆

The *graph* of a function of two variables,

$$z = f(x, y),$$

consists of ordered triples (x_1, y_1, z_1), where $z_1 = f(x_1, y_1)$. The domain of f is a region D in the xy-plane, and the graph of f is a surface S.

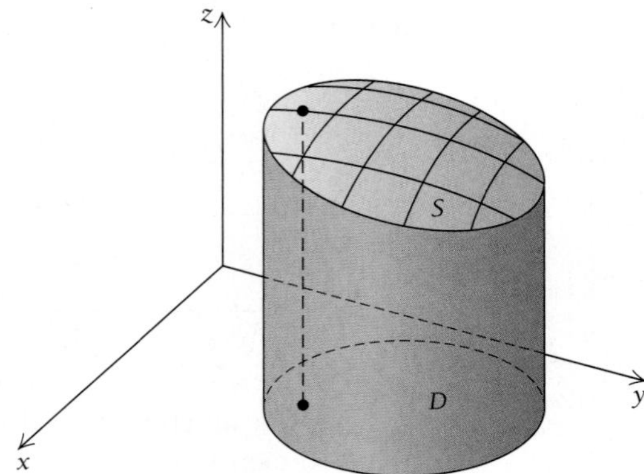

The following graphs have been generated from their equations by the computer graphics program called *Mathematica*. There are many excellent graphics programs for generating graphs of functions of two variables. The graphs can be generated with very few keystrokes. Seeing one of these graphs often provides more insight about the behavior of the function than does its formula.

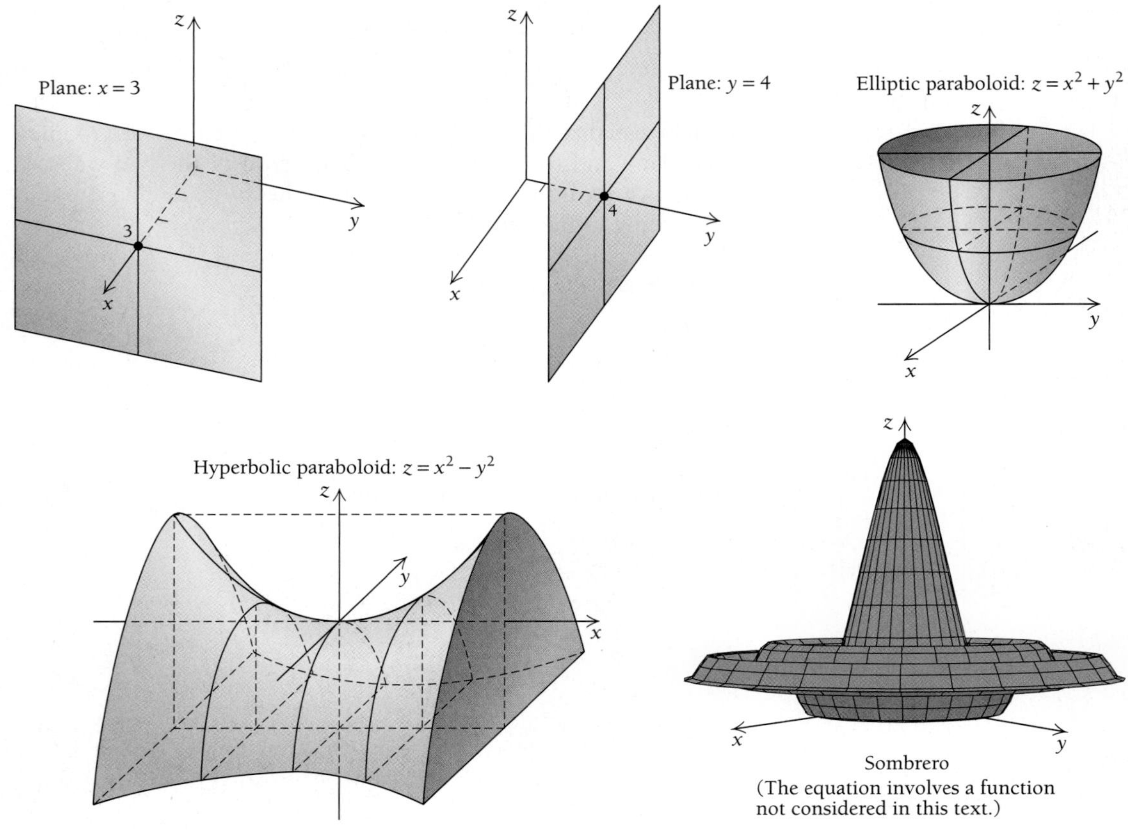

Sombrero
(The equation involves a function not considered in this text.)

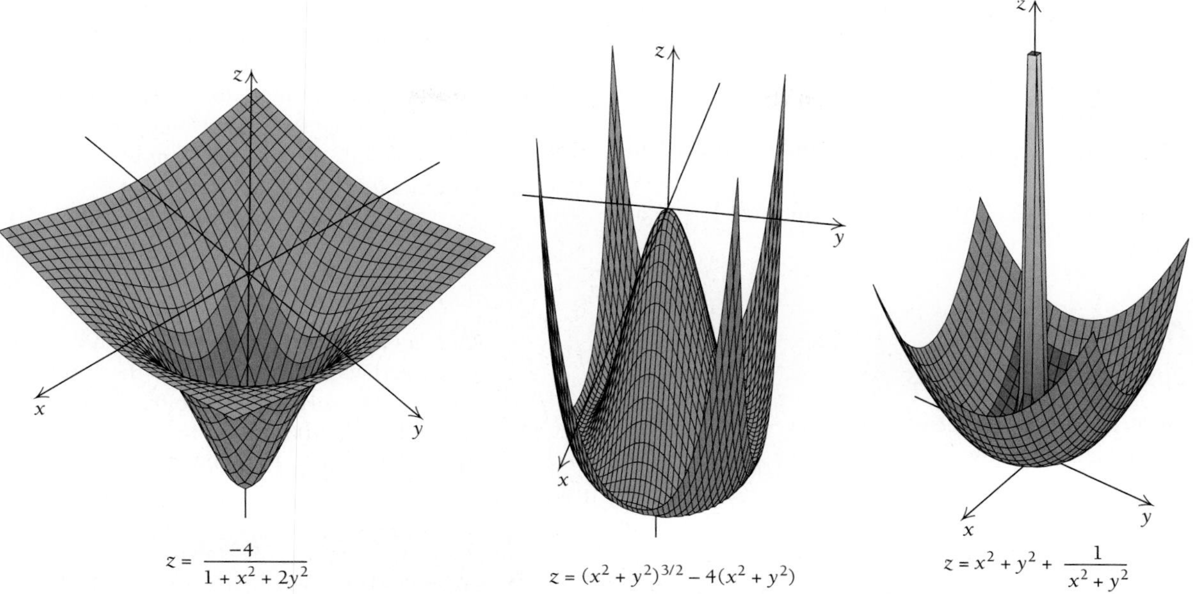

TECHNOLOGY CONNECTION

Exploratory

Many computer programs and some calculators are capable of generating graphs of functions of two variables. Some software, such as Graph3D, is free and can be downloaded into a graphing calculator. Use such software to verify the graphs shown here. What problems do you have with viewing windows?

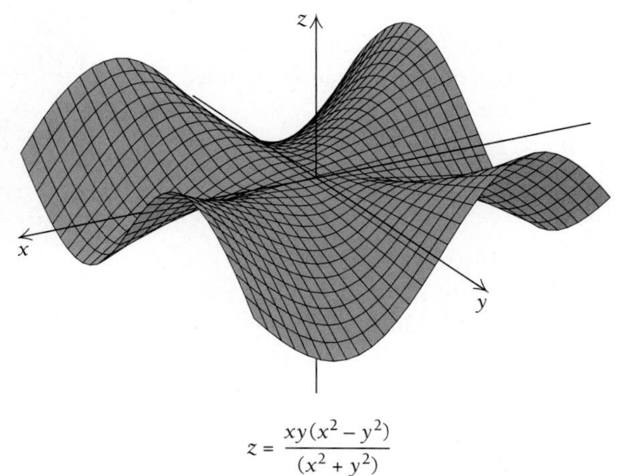

Exercise Set 6.1

1. For $f(x, y) = x^2 - 3xy$, find $f(0, -2)$, $f(2, 3)$, and $f(10, -5)$.
2. For $f(x, y) = (y^2 + 2xy)^3$, find $f(-2, 0)$, $f(3, 2)$, and $f(-5, 10)$.
3. For $f(x, y) = 3^x + 7xy$, find $f(0, -2)$, $f(-2, 1)$, and $f(2, 1)$.
4. For $f(x, y) = \log_{10}(x + y) + 3x^2$, find $f(3, 7)$, $f(1, 99)$, and $f(2, -1)$.

5. For $f(x, y) = \ln x + y^3$, find $f(e, 2)$, $f(e^2, 4)$, and $f(e^3, 5)$.
6. For $f(x, y) = 2^x - 3^y$, find $f(0, 2)$, $f(3, 1)$, and $f(2, 3)$.
7. For $f(x, y, z) = x^2 - y^2 + z^2$, find $f(-1, 2, 3)$ and $f(2, -1, 3)$.
8. For $f(x, y, z) = 2^x + 5zy - x$, find $f(0, 1, -3)$ and $f(1, 0, -3)$.

APPLICATIONS

Business and Economics

9. **Price–earnings ratio.** The *price–earnings ratio* of a stock is given by

$$R(P, E) = \frac{P}{E},$$

where P is the price of the stock and E is the earnings per share. Recently, the price per share of Hewlett-Packard stock was \$32.03 and the earnings per share were \$1.25. (*Source:* yahoo.finance.com.) Find the price–earnings ratio. Use decimal notation rounded to the nearest hundredth.

10. **Yield.** The *yield* of a stock is given by

$$Y(D, P) = \frac{D}{P},$$

where D is the dividends per share of stock and P is the price per share. Recently, the price per share of Texas Instruments stock was \$30, and the dividends per share were \$0.12. (*Source:* yahoo.finance.com.) Find the yield. Use percent notation rounded to the nearest hundredth of a percent.

11. **Cost of storage equipment.** Consider the cost model in Example 3. For \$100,000, a company buys a storage tank that has a capacity of 80,000 gal. Later it replaces the tank with a new tank with double the capacity of the original. Estimate the cost of the new tank.

Life and Physical Sciences

12. **Poiseuille's Law.** The speed of blood in a vessel is given by

$$V(L, p, R, r, v) = \frac{p}{4Lv}(R^2 - r^2),$$

where R is the radius of the vessel, r is the distance of the blood from the center of the vessel, L is the length of the blood vessel, p is the pressure, and v is the viscosity. Find $V(1, 100, 0.0075, 0.0025, 0.05)$.

13. **Wind speed of a tornado.** Under certain conditions, the *wind speed* S, in miles per hour, of a tornado at a distance d feet from its center can be approximated by the function

$$S(a, d, V) = \frac{aV}{0.51d^2},$$

where a is a constant that depends on certain atmospheric conditions and V is the approximate volume of the tornado, in cubic feet. Approximate the wind speed 100 ft from the center of a tornado when its volume is 1,600,000 ft^3 and $a = 0.78$.

14. **Body surface area.** The Mosteller formula for approximating the surface area S, in square meters (m^2), of a human is given by

$$S(h, w) = \frac{\sqrt{hw}}{60},$$

where h is the person's height in centimeters and w is the person's weight in kilograms. (*Source:* www.halls.md.) Use the Mosteller approximation to estimate the surface area of a person whose height is 165 cm and whose weight is 80 kg.

15. **Body surface area.** The Haycock formula for approximating the surface area S, in square meters (m^2), of a human is given by

$$S(h, w) = 0.024265h^{0.3964}w^{0.5378},$$

where h is the person's height in centimeters and w is the person's weight in kilograms. (*Source:* www.halls.md.) Use the Haycock approximation to estimate the surface area of a person whose height is 165 cm and whose weight is 80 kg.

Social Sciences

16. **Intelligence quotient.** In psychology, the *intelligence quotient* is given by

$$Q(m, c) = 100 \cdot \frac{m}{c},$$

where m is a person's mental age and c is his or her chronological, or actual, age. Find $Q(21, 20)$ and $Q(19, 20)$.

SYNTHESIS

17. For the tornado described in Exercise 13, if the wind speed measures 200 mph, how far from the center was the measurement taken?

18. According to the Mosteller formula in Exercise 14, if a person's weight drops 19%, by what percentage does his or her surface area change?

TW 19. Explain the difference between a function of two variables and a function of one variable.

TW 20. Find some examples of functions of several variables not considered in the text, even some that may not have formulas.

TECHNOLOGY CONNECTION

General Interest

Wind chill temperature. *Because wind speed enhances the loss of heat from the skin, we feel colder when there is wind than when there is not. The* wind chill temperature *is what the temperature would have to be with no wind in order to give the same chilling effect. The wind chill temperature, W, is given by*

$$W(v, T) = 91.4 - \frac{(10.45 + 6.68\sqrt{v} - 0.447v)(457 - 5T)}{110},$$

where T is the actual temperature measured by a thermometer, in degrees Fahrenheit, and v is the speed of the wind, in miles per hour. Find the wind chill temperature in each case. Round to the nearest degree.

21. $T = 30°\text{F}, v = 25$ mph

22. $T = 20°\text{F}, v = 20$ mph

23. $T = 20°\text{F}, v = 40$ mph

24. $T = -10°\text{F}, v = 30$ mph

Use a 3D graphics program to generate the graph of each function.

25. $f(x, y) = y^2$

26. $f(x, y) = x^2 + y^2$

27. $f(x, y) = (x^4 - 16x^2)e^{-y^2}$

28. $f(x, y) = 4(x^2 + y^2) - (x^2 + y^2)^2$

29. $f(x, y) = x^3 - 3xy^2$

30. $f(x, y) = \dfrac{1}{x^2 + 4y^2}$

6.2 Partial Derivatives

OBJECTIVES

- Find the partial derivatives of a given function.
- Evaluate partial derivatives.
- Find the four second-order partial derivatives of a function in two variables.

Finding Partial Derivatives

Consider the function f given by

$$z = f(x, y) = x^2y^3 + xy + 4y^2.$$

Suppose for the moment that we fix y at 3. Then

$$f(x, 3) = x^2(3^3) + x(3) + 4(3^2) = 27x^2 + 3x + 36.$$

Note that we now have a function of only one variable. Taking the first derivative with respect to x, we have

$$54x + 3.$$

In general, without replacing y with a specific number, we can consider y fixed. Then f becomes a function of x alone, and we can calculate its derivative with respect to x. This derivative is called the *partial derivative of f with respect to x*. Notation for this partial derivative is

$$\frac{\partial f}{\partial x} \quad \text{or} \quad \frac{\partial z}{\partial x}.$$

Now, let's again consider the function

$$z = f(x, y) = x^2y^3 + xy + 4y^2.$$

The color blue indicates the variable x when we fix y and treat it as a constant. The expressions y^3, y, and y^2 are then also treated as constants. We have

$$\frac{\partial f}{\partial x} = \frac{\partial z}{\partial x} = 2xy^3 + y.$$

Similarly, we find $\partial f/\partial y$ or $\partial z/\partial y$ by fixing x (treating it as a constant) and calculating the derivative with respect to y. From

$$z = f(x, y) = x^2y^3 + xy + 4y^2, \qquad \text{The color blue indicates the variable.}$$

we get

$$\frac{\partial f}{\partial y} = \frac{\partial z}{\partial y} = 3x^2y^2 + x + 8y.$$

A definition of partial derivatives is as follows.

DEFINITION

For $z = f(x, y)$, the **partial derivatives with respect to x and y** are

$$\frac{\partial z}{\partial x} = \lim_{h \to 0} \frac{f(x + h, y) - f(x, y)}{h},$$

and $$\frac{\partial z}{\partial y} = \lim_{k \to 0} \frac{f(x, y + k) - f(x, y)}{k}.$$

We can find partial derivatives of functions of any number of variables. Because of our earlier theorems for finding derivatives, we will rarely need to use the definition to find a partial derivative.

EXAMPLE 1 For $w = x^2 - xy + y^2 + 2yz + 2z^2 + z$, find

$$\frac{\partial w}{\partial x}, \quad \frac{\partial w}{\partial y}, \quad \text{and} \quad \frac{\partial w}{\partial z}.$$

Solution In order to find $\partial w/\partial x$, we regard x as the variable and y and z as constants. From

$$w = x^2 - xy + y^2 + 2yz + 2z^2 + z,$$

we get

$$\frac{\partial w}{\partial x} = 2x - y.$$

From

$$w = x^2 - xy + y^2 + 2yz + 2z^2 + z,$$

we get

$$\frac{\partial w}{\partial y} = -x + 2y + 2z;$$

and from

$$w = x^2 - xy + y^2 + 2yz + 2z^2 + z,$$

we get

$$\frac{\partial w}{\partial z} = 2y + 4z + 1.$$

✦

We will often make use of a simpler notation f_x for the partial derivative of f with respect to x and f_y for the partial derivative of f with respect to y. Similarly, if $z = f(x, y)$, then z_x represents the partial derivative of z with respect to x, and z_y represents the partial derivative of z with respect to y.

EXAMPLE 2 For $f(x, y) = 3x^2y + xy$, find f_x and f_y.

Solution We have

$f_x = 6xy + y,$ Treating y as a constant

$f_y = 3x^2 + x.$ Treating x^2 and x as constants

✦

For the function in Example 2, let's evaluate f_x at $(2, -3)$:

$$f_x(2, -3) = 6 \cdot 2 \cdot (-3) + (-3)$$
$$= -39.$$

If we use the notation $\partial f/\partial x = 6xy + y$, where $f = 3x^2y + xy$, the value of the partial derivative at $(2, -3)$ is given by

$$\left.\frac{\partial f}{\partial x}\right|_{(2,\,-3)} = 6 \cdot 2 \cdot (-3) + (-3)$$
$$= -39.$$

However, this notation is not quite as convenient as $f_x(2, -3)$.

TECHNOLOGY CONNECTION

Exploratory

Consider finding values of a partial derivative of $f(x, y) = 3x^3y + 2xy$ using a calculator that finds derivatives of functions of one variable. How can you find $f_x(-4, 1)$? Then how can you find $f_y(2, 6)$?

EXAMPLE 3 For $f(x, y) = e^{xy} + y \ln x$, find f_x and f_y.

Solution

$$f_x = y \cdot e^{xy} + y \cdot \frac{1}{x}$$
$$= ye^{xy} + \frac{y}{x},$$
$$f_y = x \cdot e^{xy} + 1 \cdot \ln x$$
$$= xe^{xy} + \ln x$$

◆

The Geometric Interpretation of Partial Derivatives

The graph of a function of two variables $z = f(x, y)$ is a surface S, which might have a graph similar to the one shown to the right, where each input pair (x, y) in the domain D has only one output, $z = f(x, y)$.

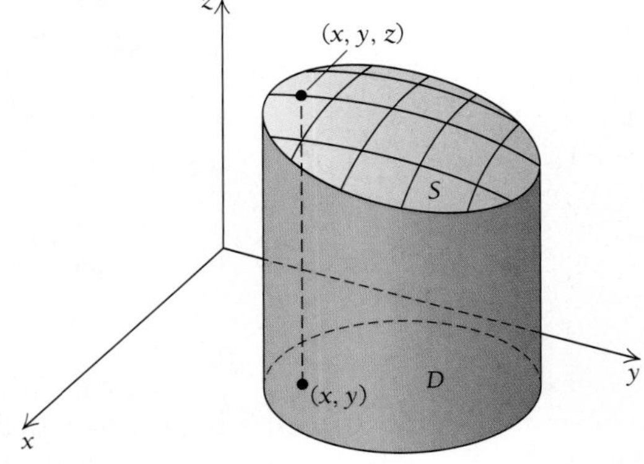

Now suppose that we hold x fixed, say, at the value a. The set of all points for which $x = a$ is a plane parallel to the yz-plane; thus, when x is fixed at a, y and z vary along that plane, as shown to the right. The plane in the figure cuts the surface along the curve C_1. The partial derivative f_y gives the slope of tangent lines to this curve.

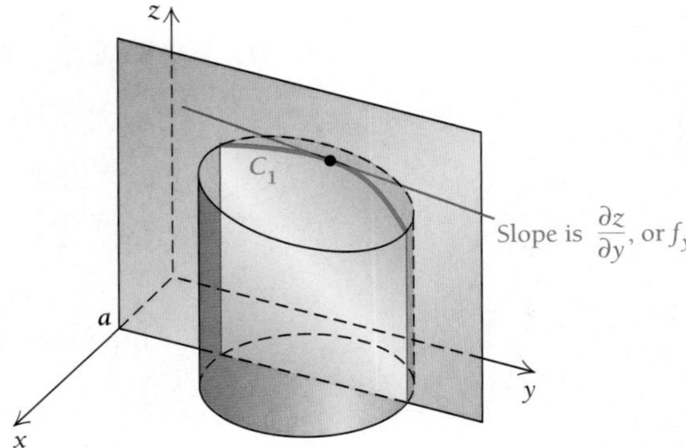

Similarly, if we hold y fixed, say, at the value b, we obtain a curve C_2, as shown. The partial derivative f_x gives the slope of tangent lines to this curve.

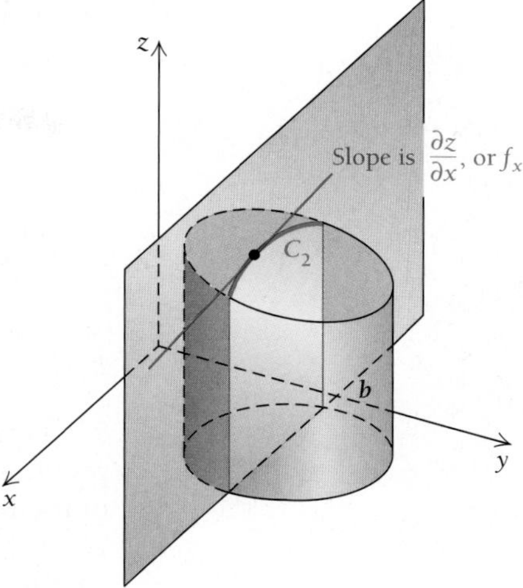

An Economics Application: The Cobb–Douglas Production Function

One model of production that is frequently considered in business and economics is the *Cobb–Douglas production function:*

$$p(x, y) = Ax^a y^{1-a}, \quad \text{for} \quad A > 0 \quad \text{and} \quad 0 < a < 1,$$

where p is the number of units produced with x units of labor and y units of capital. (Capital is the cost of machinery, buildings, tools, and other supplies.) The partial derivatives

$$\frac{\partial p}{\partial x} \quad \text{and} \quad \frac{\partial p}{\partial y}$$

are called, respectively, the *marginal productivity of labor* and the *marginal productivity of capital.*

EXAMPLE 4 A cellular phone company has the following production function for a certain product:

$$p(x, y) = 50x^{2/3}y^{1/3},$$

where p is the number of units produced with x units of labor and y units of capital.

a) Find the number of units produced with 125 units of labor and 64 units of capital.

b) Find the marginal productivities.

c) Evaluate the marginal productivities at $x = 125$ and $y = 64$.

Solution

a) $p(125, 64) = 50(125)^{2/3}(64)^{1/3} = 50(25)(4) = 5000$ units

b) Marginal productivity of labor $= \dfrac{\partial p}{\partial x} = p_x = 50\left(\dfrac{2}{3}\right)x^{-1/3}y^{1/3} = \dfrac{100y^{1/3}}{3x^{1/3}}$

Marginal productivity of capital $= \dfrac{\partial p}{\partial y} = p_y = 50\left(\dfrac{1}{3}\right)x^{2/3}y^{-2/3} = \dfrac{50x^{2/3}}{3y^{2/3}}$

c) For 125 units of labor and 64 units of capital, we have

$$\text{Marginal productivity of labor} = p_x(125, 64) = \frac{100(64)^{1/3}}{3(125)^{1/3}} = \frac{100(4)}{3(5)} = 26\tfrac{2}{3},$$

and

$$\text{Marginal productivity of capital} = p_y(125, 64) = \frac{50(125)^{2/3}}{3(64)^{2/3}} = \frac{50(25)}{3(16)} = 26\tfrac{1}{24}.$$

✦

Let's interpret the marginal productivities of Example 4. To visualize marginal productivity of labor, suppose that capital is fixed at 64 units. Then a one-unit change in labor, from 125 to 126, will cause production to increase by about $26\frac{2}{3}$ units. To visualize the marginal productivity of capital, suppose that the amount of labor is fixed at 125 units. Then a one-unit change in capital from 64 to 65 will cause production to increase by about $26\frac{1}{24}$ units.

A Cobb–Douglas production function is consistent with the law of diminishing returns. That is, if one input (either labor or capital) is held fixed while the other increases infinitely, then production will eventually increase at a decreasing rate. With such functions, it also turns out that if a certain maximum production is possible, then the expense of more labor, for example, may be required for that maximum output to be attainable.

Higher-Order Partial Derivatives

Consider

$$z = f(x, y) = 3xy^2 + 2xy + x^2.$$

Then

$$\frac{\partial z}{\partial x} = \frac{\partial f}{\partial x} = 3y^2 + 2y + 2x.$$

Suppose that we continue and find the first partial derivative of $\partial z/\partial x$ with respect to y. This will be a **second-order partial derivative** of the original function z. Its notation is as follows:

$$\frac{\partial}{\partial y}\left(\frac{\partial z}{\partial x}\right) = \frac{\partial}{\partial y}\left(\frac{\partial f}{\partial x}\right) = \frac{\partial}{\partial y}(3y^2 + 2y + 2x) = 6y + 2.$$

The notation $\frac{\partial}{\partial y}\left(\frac{\partial z}{\partial x}\right)$ is often expressed as

$$\frac{\partial^2 z}{\partial y\, \partial x} \quad \text{or} \quad \frac{\partial^2 f}{\partial y\, \partial x}.$$

We could also denote the preceding partial derivative using the notation f_{xy}:

$$f_{xy} = 6y + 2.$$

Note that in the notation f_{xy}, x and y are in the order (left to right) in which the differentiation is done, but in

$$\frac{\partial^2 f}{\partial y\,\partial x},$$

the order of x and y is reversed. In each case, the differentiation with respect to x is done first, followed by differentiation with respect to y.

Notation for the four second-order partial derivatives is as follows.

DEFINITION

Second-Order Partial Derivatives

1. $\dfrac{\partial^2 z}{\partial x\,\partial x} = \dfrac{\partial^2 f}{\partial x\,\partial x} = \dfrac{\partial^2 z}{\partial x^2} = \dfrac{\partial^2 f}{\partial x^2} = f_{xx}$ Take the partial with respect to x, and then with respect to x again.

2. $\dfrac{\partial^2 z}{\partial y\,\partial x} = \dfrac{\partial^2 f}{\partial y\,\partial x} = f_{xy}$ Take the partial with respect to x, and then with respect to y.

3. $\dfrac{\partial^2 z}{\partial x\,\partial y} = \dfrac{\partial^2 f}{\partial x\,\partial y} = f_{yx}$ Take the partial with respect to y, and then with respect to x.

4. $\dfrac{\partial^2 z}{\partial y\,\partial y} = \dfrac{\partial^2 f}{\partial y\,\partial y} = \dfrac{\partial^2 z}{\partial y^2} = \dfrac{\partial^2 f}{\partial y^2} = f_{yy}$ Take the partial with respect to y, and then with respect to y again.

EXAMPLE 5 For

$$z = f(x, y) = x^2y^3 + x^4y + xe^y,$$

find the four second-order partial derivatives.

Solution

a) $$\frac{\partial^2 f}{\partial x^2} = f_{xx} = \frac{\partial}{\partial x}(2xy^3 + 4x^3y + e^y) = 2y^3 + 12x^2y$$

Differentiate twice with respect to x.

b) $$\frac{\partial^2 f}{\partial y\,\partial x} = f_{xy} = \frac{\partial}{\partial y}(2xy^3 + 4x^3y + e^y) = 6xy^2 + 4x^3 + e^y$$

Differentiate with respect to x and then with respect to y.

c) $$\frac{\partial^2 f}{\partial x\,\partial y} = f_{yx} = \frac{\partial}{\partial x}(3x^2y^2 + x^4 + xe^y) = 6xy^2 + 4x^3 + e^y$$

Differentiate with respect to y and then with respect to x.

d) $$\frac{\partial^2 f}{\partial y^2} = f_{yy} = \frac{\partial}{\partial y}(3x^2y^2 + x^4 + xe^y) = 6x^2y + xe^y$$

Differentiate twice with respect to y. ✦

We see by comparing parts (b) and (c) of Example 5 that

$$\frac{\partial^2 f}{\partial y\,\partial x} = \frac{\partial^2 f}{\partial x\,\partial y} \quad \text{and} \quad f_{xy} = f_{yx}.$$

Although this will be true for virtually all functions that we consider in this text, it is *not* true for all functions. One function for which it is not true is given in Exercise 69.

In Section 6.3, we will see how higher-order partial derivatives are used in applications to find extrema for functions of two variables.

Exercise Set 6.2

Find $\frac{\partial z}{\partial x}, \frac{\partial z}{\partial y}, \left.\frac{\partial z}{\partial x}\right|_{(-2,-3)}$, *and* $\left.\frac{\partial z}{\partial y}\right|_{(0,-5)}$.

1. $z = 2x - 3y$

2. $z = 7x - 5y$

3. $z = 3x^2 - 2xy + y$

4. $z = 2x^3 + 3xy - x$

Find $f_x, f_y, f_x(-2, 4)$, *and* $f_y(4, -3)$.

5. $f(x, y) = 2x - 5xy$

6. $f(x, y) = 5x + 7y$

Find $f_x, f_y, f_x(-2, 1)$, *and* $f_y(-3, -2)$.

7. $f(x, y) = \sqrt{x^2 + y^2}$

8. $f(x, y) = \sqrt{x^2 - y^2}$

Find f_x *and* f_y.

9. $f(x, y) = e^{2x-y}$

10. $f(x, y) = e^{3x-2y}$

11. $f(x, y) = e^{xy}$

12. $f(x, y) = e^{2xy}$

13. $f(x, y) = y \ln (x + 2y)$

14. $f(x, y) = x \ln (x - y)$

15. $f(x, y) = x \ln (xy)$

16. $f(x, y) = y \ln (xy)$

17. $f(x, y) = \frac{x}{y} - \frac{y}{3x}$

18. $f(x, y) = \frac{x}{y} + \frac{y}{5x}$

19. $f(x, y) = 3(2x + y - 5)^2$

20. $f(x, y) = 4(3x + y - 8)^2$

Find $\frac{\partial f}{\partial b}$ *and* $\frac{\partial f}{\partial m}$.

21. $f(b, m) = m^3 + 4m^2b - b^2 + (2m + b - 5)^2 + (3m + b - 6)^2$

22. $f(b, m) = 5m^2 - mb^2 - 3b + (2m + b - 8)^2 + (3m + b - 9)^2$

Find f_x, f_y, *and* f_λ. *(The symbol* λ *is the Greek letter* lambda.)

23. $f(x, y, \lambda) = 5xy - \lambda(2x + y - 8)$

24. $f(x, y, \lambda) = 9xy - \lambda(3x - y + 7)$

25. $f(x, y, \lambda) = x^2 + y^2 - \lambda(10x + 2y - 4)$

26. $f(x, y, \lambda) = x^2 - y^2 - \lambda(4x - 7y - 10)$

Find the four second-order partial derivatives.

27. $f(x, y) = 5xy$

28. $f(x, y) = 2xy$

29. $f(x, y) = 7xy^2 + 5xy - 2y$

30. $f(x, y) = 3x^2y - 2xy + 4y$

31. $f(x, y) = x^5y^4 + x^3y^2$

32. $f(x, y) = x^4y^3 - x^2y^3$

Find f_{xx}, f_{xy}, f_{yx}, *and* f_{yy}. *(Remember,* f_{yx} *means to differentiate with respect to y and then with respect to x.)*

33. $f(x, y) = 2x - 3y$

34. $f(x, y) = 3x + 5y$

35. $f(x, y) = e^{2xy}$

36. $f(x, y) = e^{xy}$

37. $f(x, y) = x + e^y$

38. $f(x, y) = y - e^x$

39. $f(x, y) = y \ln x$

40. $f(x, y) = x \ln y$

APPLICATIONS

Business and Economics

41. The Cobb–Douglas model. Lincolnville Sporting Goods has the following production function for a certain product:

$$p(x, y) = 2400x^{2/5}y^{3/5},$$

where p is the number of units produced with x units of labor and y units of capital.

a) Find the production from 32 units of labor and 1024 units of capital.
b) Find the marginal productivities.
c) Evaluate the marginal productivities at $x = 32$ and $y = 1024$.
TW **d)** Interpret the meanings of the marginal productivities found in part (c).

42. The Cobb–Douglas model. Riverside Appliances has the following production function for a certain product:

$$p(x, y) = 1800x^{0.621}y^{0.379},$$

where p is the number of units produced with x units of labor and y units of capital.

a) Find the production from 2500 units of labor and 1700 units of capital.
b) Find the marginal productivities.
c) Evaluate the marginal productivities at $x = 2500$ and $y = 1700$.
TW **d)** Interpret the meanings of the marginal productivities found in part (c).

Nursing facilities. *A study of Texas nursing homes found that the annual profit P (in dollars) of profit-seeking, independent nursing homes in urban locations follows the function*

$$P(w, r, s, t) = 0.007955w^{-0.638}r^{1.038}s^{0.873}t^{2.468}.$$

In this function, w is the average hourly wage of nurses and aides (in dollars), r is the occupancy rate (as a percentage), s is the total square footage of the facility, and t is the Texas Index of Level of Effort (TILE), a number between 1 and 11 that measures state Medicaid reimbursement. (***Source:*** *K. J. Knox, E. C. Blankmeyer, and J. R. Stutzman, "Relative Economic Efficiency in Texas Nursing Facilities,"* Journal of Economics and Finance, *Vol. 23, 199–213 (1999).) Use the preceding information for Exercises 43 and 44.*

43. A profit-seeking, independent Texas nursing home in an urban setting has nurses and aides with an average hourly wage of \$20 an hour, a TILE of 8, an occupancy rate of 70%, and 400,000 ft^2 of space.

a) Estimate the nursing home's annual profit.
b) Find the four partial derivatives of P.
TW **c)** Interpret the meaning of the partial derivatives found in part (b).

44. The change in P due to a change in w when the other variables are held constant is approximately

$$\Delta P \approx \frac{\partial P}{\partial w}\Delta w.$$

Use the values of w, r, s, and t in Exercise 43 and assume that the nursing home gives its nurses and aides a small raise so that the average hourly wage is now \$20.25 an hour. By approximately how much does the profit change?

Life and Physical Sciences

Temperature–humidity heat index. *In the summer, humidity interacts with the outdoor temperature, making a person feel hotter due to a reduced heat loss from the skin caused by higher humidity. The* temperature–humidity index, T_h, *is what the temperature would have to be with no humidity in order to give the same heat effect. One index often used is given by*

$$T_h = 1.98T - 1.09(1 - H)(T - 58) - 56.9,$$

where T is the air temperature, in degrees Fahrenheit, and H is the relative humidity, expressed as a decimal. Find the temperature–humidity index in each case. Round to the nearest tenth of a degree.

45. $T = 85°$ and $H = 60\%$

46. $T = 90°$ and $H = 90\%$

47. $T = 90°$ and $H = 100\%$

48. $T = 78°$ and $H = 100\%$

TW **49.** Find $\dfrac{\partial T_h}{\partial H}$, and interpret its meaning.

TW **50.** Find $\dfrac{\partial T_h}{\partial T}$, and interpret its meaning.

51. Body surface area. The Mosteller formula for approximating the surface area, S, in m^2, of a human is given by

$$S = \frac{\sqrt{hw}}{60},$$

where h is the person's height in centimeters and w is the person's weight in kilograms. (***Source:*** www.halls.md.)

a) Compute $\frac{\partial S}{\partial h}$.

b) Compute $\frac{\partial S}{\partial w}$.

c) The change in S due to a change in w when h is constant is approximately

$$\Delta S \approx \frac{\partial S}{\partial w}\Delta w.$$

Use this formula to approximate the change in someone's surface area given that the person is 170 cm tall, weighs 80 kg, and loses 2 kg.

52. Body surface area. The Haycock formula for approximating the surface area, S, in m^2, of a human is given by

$$S = 0.024265h^{0.3964}w^{0.5378},$$

where h is the person's height in centimeters and w is the person's weight in kilograms. (*Source:* www.halls.md.)

a) Compute $\frac{\partial S}{\partial h}$.

b) Compute $\frac{\partial S}{\partial w}$.

c) The change in S due to a change in w when h is constant is approximately

$$\Delta S \approx \frac{\partial S}{\partial w}\Delta w.$$

Use this formula to approximate the change in someone's surface area given that the person is 170 cm tall, weighs 80 kg, and loses 2 kg.

Social Sciences

Reading ease. *The following formula is used by psychologists and educators to predict the reading ease, E, of a passage of words:*

$$E = 206.835 - 0.846w - 1.015s,$$

where w is the number of syllables in a 100-word section and s is the average number of words per sentence. Find the reading ease in each case.

53. $w = 146$ and $s = 5$

54. $w = 180$ and $s = 6$

55. Find $\frac{\partial E}{\partial w}$.

56. Find $\frac{\partial E}{\partial s}$.

SYNTHESIS

Find f_x and f_t.

57. $f(x, t) = \frac{x^2 + t^2}{x^2 - t^2}$

58. $f(x, t) = \frac{x^2 - t}{x^3 + t}$

59. $f(x, t) = \frac{2\sqrt{x} - 2\sqrt{t}}{1 + 2\sqrt{t}}$

60. $f(x, t) = \sqrt[4]{x^3t^5}$

61. $f(x, t) = 6x^{2/3} - 8x^{1/4}t^{1/2} - 12x^{-1/2}t^{3/2}$

62. $f(x, t) = \left(\frac{x^2 + t^2}{x^2 - t^2}\right)^5$

Find f_{xx}, f_{xy}, f_{yx}, and f_{yy}.

63. $f(x, y) = \frac{x}{y^2} - \frac{y}{x^2}$

64. $f(x, y) = \frac{xy}{x - y}$

TW **65.** Do some research on the Cobb–Douglas production function, and explain how it was developed.

TW **66.** Explain the meaning of the first partial derivatives of a function of two variables in terms of slopes of tangent lines.

67. Consider $f(x, y) = \ln(x^2 + y^2)$. Show that f is a solution to the partial differential equation

$$\frac{\partial^2 f}{\partial x^2} + \frac{\partial^2 f}{\partial y^2} = 0.$$

68. Consider $f(x, y) = x^3 - 5xy^2$. Show that f is a solution to the partial differential equation

$$xf_{xy} - f_y = 0.$$

69. Consider the function f defined as follows:

$$f(x, y) = \begin{cases} \dfrac{xy(x^2 - y^2)}{x^2 + y^2}, & \text{for } (x, y) \neq (0, 0), \\ 0, & \text{for } (x, y) = (0, 0). \end{cases}$$

a) Find $f_x(0, y)$ by evaluating the limit

$$\lim_{h \to 0} \frac{f(h, y) - f(0, y)}{h}.$$

b) Find $f_y(x, 0)$ by evaluating the limit

$$\lim_{h \to 0} \frac{f(x, h) - f(x, 0)}{h}.$$

c) Now find and compare $f_{yx}(0, 0)$ and $f_{xy}(0, 0)$.

6.3 Maximum – Minimum Problems

OBJECTIVE

- Find relative extrema of a function of two variables.

We will now find maximum and minimum values of functions of two variables.

DEFINITION

A function f of two variables:

1. has a **relative maximum** at (a, b) if

   $$f(x, y) \leq f(a, b)$$

   for all points in a rectangular region containing (a, b);

2. has a **relative minimum** at (a, b) if

   $$f(x, y) \geq f(a, b)$$

   for all points in a rectangular region containing (a, b).

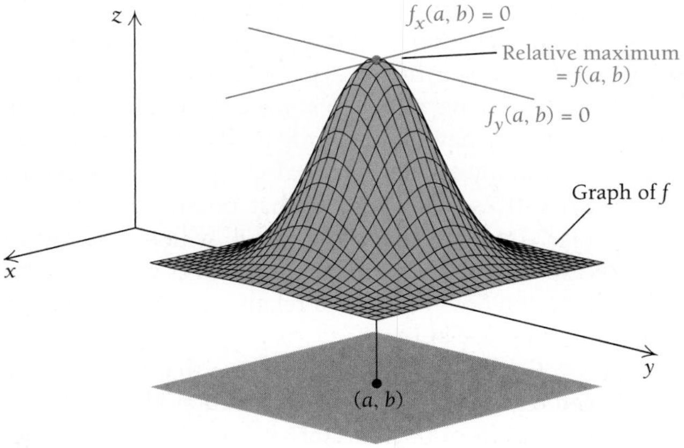

FIGURE 1

This definition is illustrated in Figs. 1 and 2. A relative maximum (or minimum) may not be an "absolute" maximum (or minimum), as illustrated in Fig. 3.

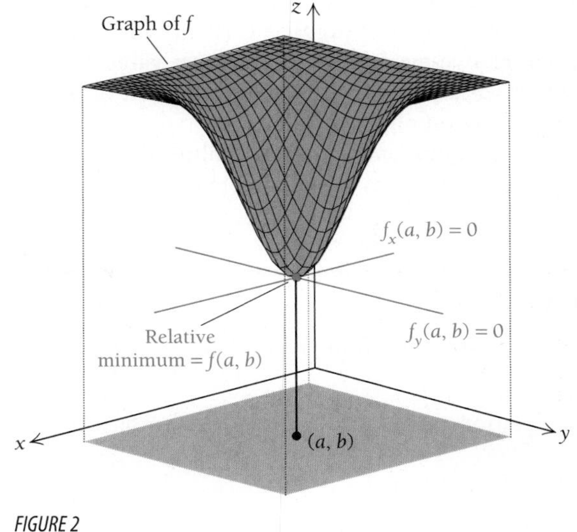

FIGURE 2

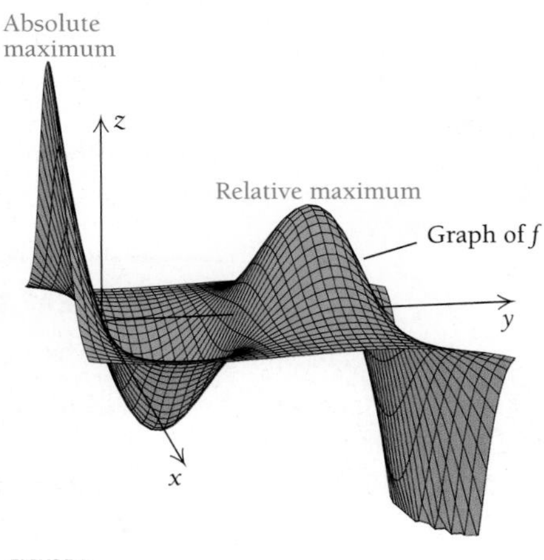

FIGURE 3

Determining Maximum and Minimum Values

Suppose that a function f has a relative maximum or minimum value at some point (a, b) inside its domain. (We assume that f and its partial derivatives exist and are "continuous" inside its domain, though we will not formally define continuity.) If we fix y at the value b, then $f(x, b)$ can be regarded as a function of x. Because a relative maximum or minimum occurs at (a, b), we know that $f(x, b)$ achieves a maximum or minimum at (a, b) and $f_x = 0$. Similarly, if we fix x at a, then $f(a, y)$ can be regarded as a function of y that achieves a relative extremum at (a, b), and thus $f_y = 0$. In short, since an extremum exists at (a, b), we must have

$$f_x = 0 \quad \text{and} \quad f_y = 0. \tag{1}$$

We call a point (a, b) at which both partial derivatives are 0 a *critical point*. This concept of a critical value is comparable to that for functions of one variable. Thus, one strategy for finding relative maximum or minimum values is to solve a system of equations like (1) to find critical points. Just as for functions of one variable, this strategy does *not* guarantee that we will have a relative maximum or minimum value. We have argued only that *if* f has a maximum or minimum value at (a, b), *then* both its partial derivatives must be 0 at that point. Look back at Figs. 1 and 2. Then note Fig. 4, which illustrates a case in which the partial derivatives are 0 but the function does not have a relative maximum or minimum value at (a, b).

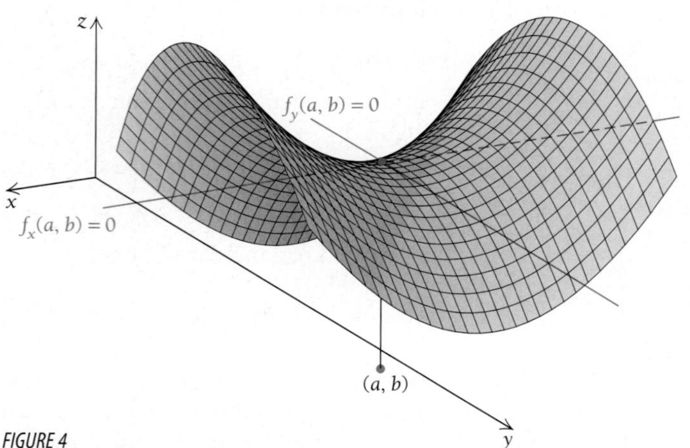

FIGURE 4

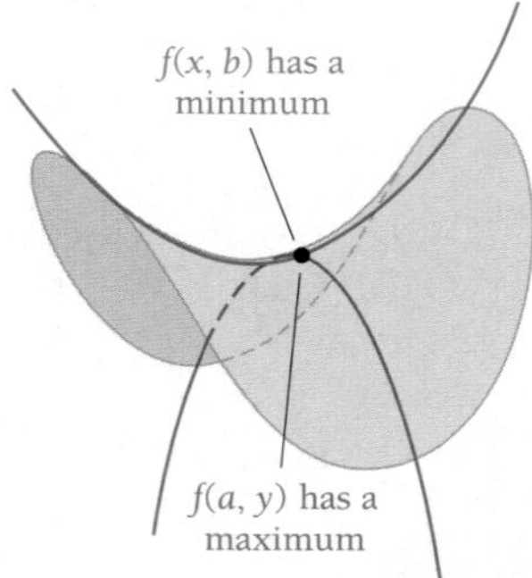

Considering Fig. 4, suppose that we fix y at a value b. Then $f(x, b)$, considered as the output of a function of one variable x, has a minimum at a, but f does not. Similarly, if we fix x at a, then $f(a, y)$, considered as the output of a function of one variable y, has a maximum at b, but f does not. The point $f(a, b)$ is called a **saddle point.** In other words, $f_x(a, b) = 0$ and $f_y(a, b) = 0$ [the point (a, b) is a critical point], but f does not attain a relative maximum or minimum value at (a, b). A saddle point for a function of two variables is comparable to a point of inflection that occurs at a critical value for a function of one variable.

A test for finding relative maximum and minimum values that involves the use of first- and second-order partial derivatives is stated below. We will not prove this theorem.

THEOREM 1 The *D*-Test

To find the relative maximum and minimum values of f:

1. Find f_x, f_y, f_{xx}, f_{yy}, and f_{xy}.
2. Solve the system of equations $f_x = 0, f_y = 0$. Let (a, b) represent a solution.
3. Evaluate D, where $D = f_{xx}(a, b) \cdot f_{yy}(a, b) - [f_{xy}(a, b)]^2$.
4. Then:

   a) f has a maximum at (a, b) if $D > 0$ and $f_{xx}(a, b) < 0$.
   b) f has a minimum at (a, b) if $D > 0$ and $f_{xx}(a, b) > 0$.
   c) f has neither a maximum nor a minimum at (a, b) if $D < 0$. The function has a *saddle point* at (a, b). See Fig. 4.
   d) This test is not applicable if $D = 0$.

The D-test is somewhat analogous to the Second Derivative Test (Section 2.2) for functions of one variable. Saddle points are analogous to critical values at which concavity changes and there are no relative maximum or minimum values.

A relative maximum or minimum *may or may not be an absolute maximum or minimum value.* Tests for absolute maximum or minimum values are rather complicated. We will restrict our attention to finding *relative* maximum or minimum values. Fortunately, in most of our applications, relative maximum or minimum values turn out to be absolute as well.

The shape of a perfect tent. *To give a tent roof the maximum strength possible, designers draw the fabric into a series of three-dimensional shapes that, viewed in profile, resemble a horse's saddle and that mathematicians call an* anticlastic curve. *Two people with a stretchy piece of fabric such as Lycra Spandex can duplicate the shape, as shown above. One person pulls up and out on two diagonal corners; the other person pulls down and out on the other two corners. The opposing tensions draw each point of the fabric's surface into rigid equilibrium. The more pronounced the curve, the stiffer the surface.*

EXAMPLE 1 Find the relative maximum or minimum values of

$$f(x, y) = x^2 + xy + y^2 - 3x.$$

Solution

1. Find f_x, f_y, f_{xx}, f_{yy}, and f_{xy}:

$$f_x = 2x + y - 3, \qquad f_y = x + 2y,$$
$$f_{xx} = 2; \qquad f_{yy} = 2;$$
$$f_{xy} = 1.$$

2. Solve the system of equations $f_x = 0, f_y = 0$:

$$2x + y - 3 = 0, \tag{1}$$
$$x + 2y = 0. \tag{2}$$

Solving equation (2) for x, we get $x = -2y$. Substituting $-2y$ for x in equation (1) and solving, we get

$$\begin{aligned} 2(-2y) + y - 3 &= 0 \\ -4y + y - 3 &= 0 \\ -3y &= 3 \\ y &= -1. \end{aligned}$$

To find x when $y = -1$, we substitute -1 for y in equation (1) or equation (2). We choose equation (2):

$$\begin{aligned} x + 2(-1) &= 0 \\ x &= 2. \end{aligned}$$

Thus, $(2, -1)$ is the only critical point, and $f(2, -1)$ is our candidate for a maximum or minimum value.

3. We must check to see whether $f(2, -1)$ is a maximum or minimum value:

$$\begin{aligned} D &= f_{xx}(2, -1) \cdot f_{yy}(2, -1) - [f_{xy}(2, -1)]^2 \\ &= 2 \cdot 2 - [1]^2 \quad \text{Using Step 1 above} \\ &= 3. \end{aligned}$$

4. Thus, $D = 3$ and $f_{xx}(2, -1) = 2$. Since $D > 0$ and $f_{xx}(2, -1) > 0$, it follows from the D-test that f has a relative minimum at $(2, -1)$. That minimum value is found as follows:

$$\begin{aligned} f(2, -1) &= 2^2 + 2(-1) + (-1)^2 - 3 \cdot 2 \\ &= 4 - 2 + 1 - 6 \\ &= -3. \quad \text{This is the relative minimum.} \end{aligned}$$

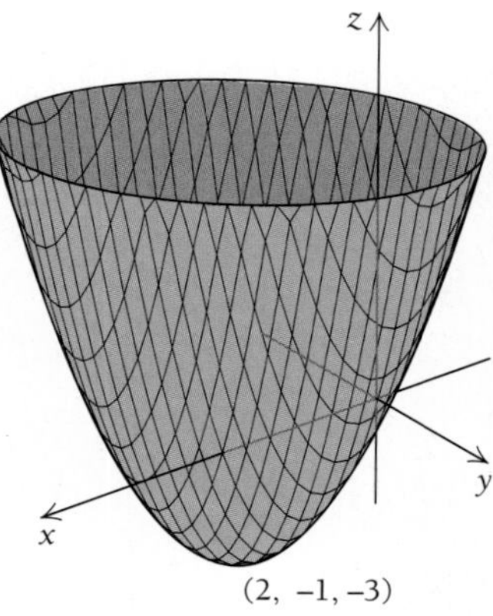

$z = f(x, y) = x^2 + xy + y^2 - 3x$

✦

EXAMPLE 2 Find the relative maximum and minimum values of

$$f(x, y) = xy - x^3 - y^2.$$

Solution

1. Find f_x, f_y, f_{xx}, f_{yy}, and f_{xy}:

$$f_x = y - 3x^2, \qquad f_y = x - 2y,$$
$$f_{xx} = -6x; \qquad f_{yy} = -2;$$
$$f_{xy} = 1.$$

2. Solve the system of equations $f_x = 0$, $f_y = 0$:

$$y - 3x^2 = 0, \quad (1)$$
$$x - 2y = 0. \quad (2)$$

Solving equation (1) for y, we get $y = 3x^2$. Substituting $3x^2$ for y in equation (2) and solving, we get

$$\begin{aligned} x - 2(3x^2) &= 0 \\ x - 6x^2 &= 0 \\ x(1 - 6x) &= 0. \quad \text{Factoring} \end{aligned}$$

Setting each factor equal to 0 and solving, we have

$$\begin{aligned} x = 0 \quad &or \quad 1 - 6x = 0 \\ x = 0 \quad &or \quad x = \tfrac{1}{6}. \end{aligned}$$

To find y when $x = 0$, we substitute 0 for x in equation (1) or equation (2). We choose equation (2):

$$\begin{aligned} 0 - 2y &= 0 \\ -2y &= 0 \\ y &= 0. \end{aligned}$$

Thus, $(0, 0)$ is a critical point, and $f(0, 0)$ is one candidate for a maximum or minimum value. To find the other, we substitute $\frac{1}{6}$ for x in either equation (1) or equation (2). We choose equation (2):

$$\begin{aligned} \tfrac{1}{6} - 2y &= 0 \\ -2y &= -\tfrac{1}{6} \\ y &= \tfrac{1}{12}. \end{aligned}$$

Thus, $\left(\frac{1}{6}, \frac{1}{12}\right)$ is another critical point, and $f\left(\frac{1}{6}, \frac{1}{12}\right)$ is another candidate for a maximum or minimum value.

3–4. We must check both $(0, 0)$ and $\left(\frac{1}{6}, \frac{1}{12}\right)$ to see whether they yield maximum or minimum values.

$$\text{For } (0, 0): \quad \begin{aligned} D &= f_{xx}(0, 0) \cdot f_{yy}(0, 0) - [f_{xy}(0, 0)]^2 \\ &= (-6 \cdot 0) \cdot (-2) - [1]^2 \quad \text{Using Step 1 above} \\ &= -1. \end{aligned}$$

Since $D < 0$, it follows that $f(0, 0)$ is neither a maximum nor a minimum value, but a saddle point.

$$\text{For } \left(\tfrac{1}{6}, \tfrac{1}{12}\right): \quad \begin{aligned} D &= f_{xx}\left(\tfrac{1}{6}, \tfrac{1}{12}\right) \cdot f_{yy}\left(\tfrac{1}{6}, \tfrac{1}{12}\right) - \left[f_{xy}\left(\tfrac{1}{6}, \tfrac{1}{12}\right)\right]^2 \\ &= \left(-6 \cdot \tfrac{1}{6}\right) \cdot (-2) - [1]^2 \\ &= -1(-2) - 1 \quad \text{Using Step 1 above} \\ &= 1. \end{aligned}$$

Thus, $D = 1$ and $f_{xx}\left(\frac{1}{6}, \frac{1}{12}\right) = -1$. Since $D > 0$ and $f_{xx}\left(\frac{1}{6}, \frac{1}{12}\right) < 0$, it follows that f has a relative maximum at $\left(\frac{1}{6}, \frac{1}{12}\right)$; that maximum value is

$$\begin{aligned} f\left(\tfrac{1}{6}, \tfrac{1}{12}\right) &= \tfrac{1}{6} \cdot \tfrac{1}{12} - \left(\tfrac{1}{6}\right)^3 - \left(\tfrac{1}{12}\right)^2 \\ &= \tfrac{1}{72} - \tfrac{1}{216} - \tfrac{1}{144} = \tfrac{1}{432}. \end{aligned}$$

This is the relative maximum. ✦

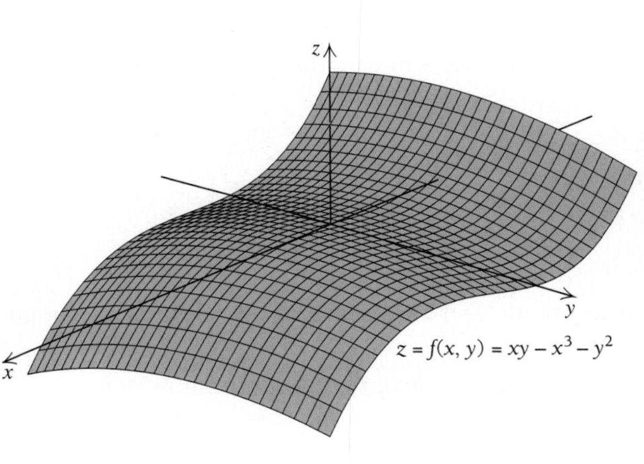

EXAMPLE 3 Business: Maximizing Profit. A firm produces two kinds of golf ball, one that sells for \$3 and one priced at \$2. The total revenue, in thousands of dollars, from the sale of x thousand balls at \$3 each and y thousand at \$2 each is given by

$$R(x, y) = 3x + 2y.$$

The company determines that the total cost, in thousands of dollars, of producing x thousand of the \$3 ball and y thousand of the \$2 ball is given by

$$C(x, y) = 2x^2 - 2xy + y^2 - 9x + 6y + 7.$$

How many balls of each type must be produced and sold in order to maximize profit?

Solution The total profit $P(x, y)$ is given by

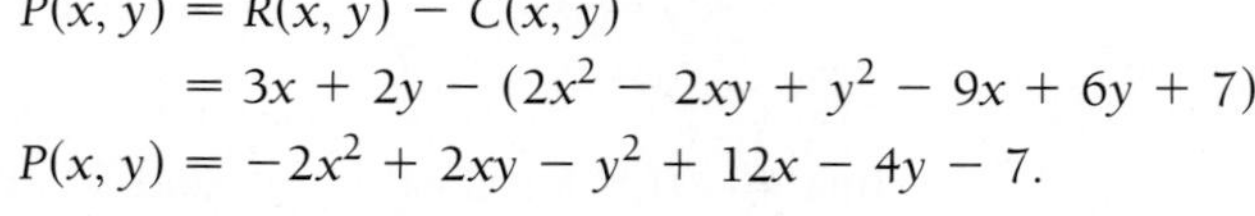

$$\begin{aligned} P(x, y) &= R(x, y) - C(x, y) \\ &= 3x + 2y - (2x^2 - 2xy + y^2 - 9x + 6y + 7) \\ P(x, y) &= -2x^2 + 2xy - y^2 + 12x - 4y - 7. \end{aligned}$$

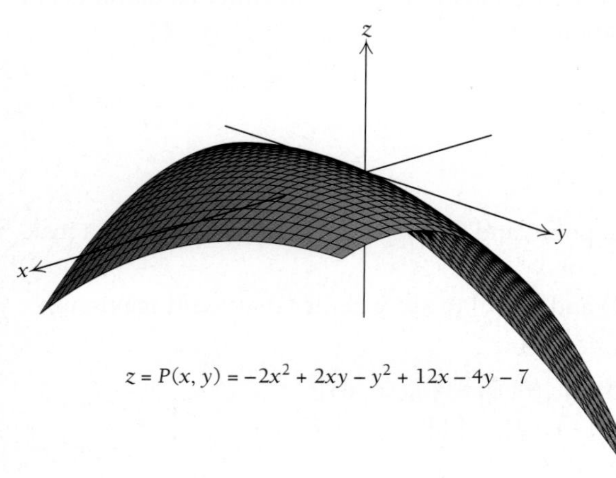

1. Find P_x, P_y, P_{xx}, P_{yy}, and P_{xy}:

$$P_x = -4x + 2y + 12, \qquad P_y = 2x - 2y - 4,$$
$$P_{xx} = -4; \qquad P_{yy} = -2;$$
$$P_{xy} = 2.$$

2. Solve the system of equations $P_x = 0$, $P_y = 0$:

$$-4x + 2y + 12 = 0, \qquad (1)$$
$$2x - 2y - 4 = 0. \qquad (2)$$

Adding these equations, we get

$$-2x + 8 = 0.$$

Then

$$\begin{aligned} -2x &= -8 \\ x &= 4. \end{aligned}$$

To find y when $x = 4$, we substitute 4 for x in equation (1) or equation (2). We choose equation (2):

$$\begin{aligned} 2 \cdot 4 - 2y - 4 &= 0 \\ -2y + 4 &= 0 \\ -2y &= -4 \\ y &= 2. \end{aligned}$$

Thus, $(4, 2)$ is the only critical point, and $P(4, 2)$ is a candidate for a maximum or minimum value.

3. We must check to see whether $P(4, 2)$ is a maximum or minimum value:

$$\begin{aligned} D &= P_{xx}(4, 2) \cdot P_{yy}(4, 2) - [P_{xy}(4, 2)]^2 \\ &= (-4)(-2) - 2^2 \qquad \text{Using Step 1 above} \\ &= 4. \end{aligned}$$

4. Thus, $D = 4$ and $P_{xx}(4, 2) = -4$. Since $D > 0$ and $P_{xx}(4, 2) < 0$, it follows that P has a relative maximum at (4, 2). So, in order to maximize profit, the company must produce and sell 4 thousand of the \$3 golf balls and 2 thousand of the \$2 golf balls. The maximum profit will be

$$P(4, 2) = -2 \cdot 4^2 + 2 \cdot 4 \cdot 2 - 2^2 + 12 \cdot 4 - 4 \cdot 2 - 7 = 13,$$

or \$13 thousand. ✦

Exercise Set 6.3

Find the relative maximum and minimum values.

1. $f(x, y) = x^2 + xy + y^2 - y$

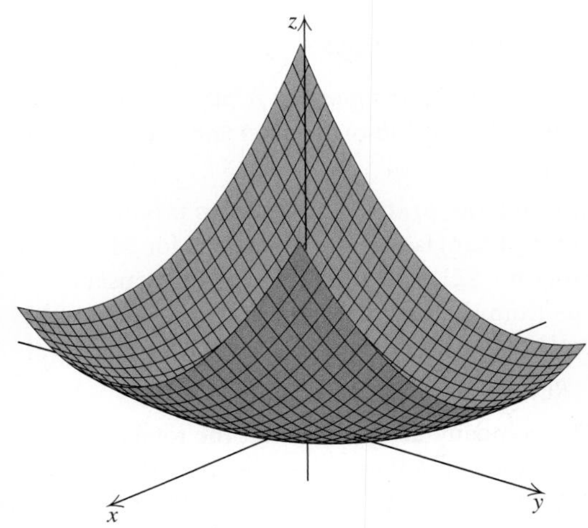

2. $f(x, y) = x^2 + xy + y^2 - 5y$

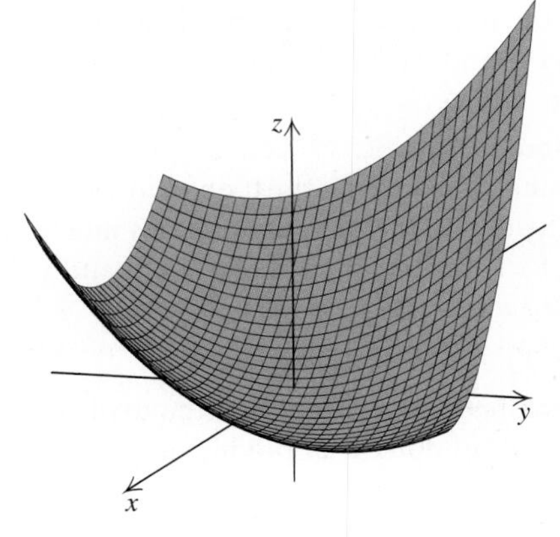

3. $f(x, y) = 2xy - x^3 - y^2$

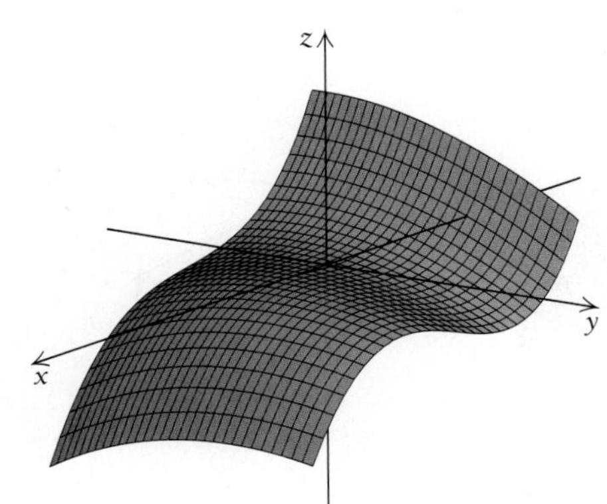

4. $f(x, y) = 4xy - x^3 - y^2$

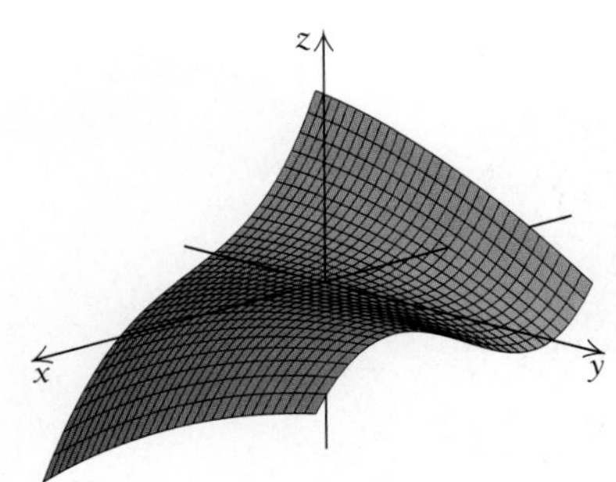

5. $f(x, y) = x^3 + y^3 - 3xy$
6. $f(x, y) = x^3 + y^3 - 6xy$
7. $f(x, y) = x^2 + y^2 - 2x + 4y - 2$
8. $f(x, y) = x^2 + 2xy + 2y^2 - 6y + 2$

9. $f(x, y) = x^2 + y^2 + 2x - 4y$

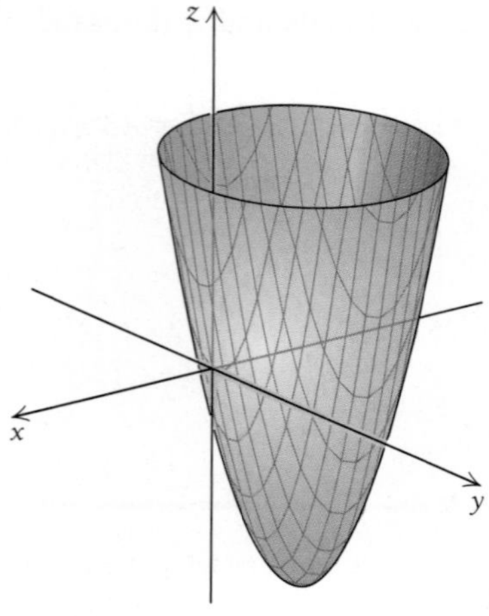

10. $f(x, y) = 4y + 6x - x^2 - y^2$

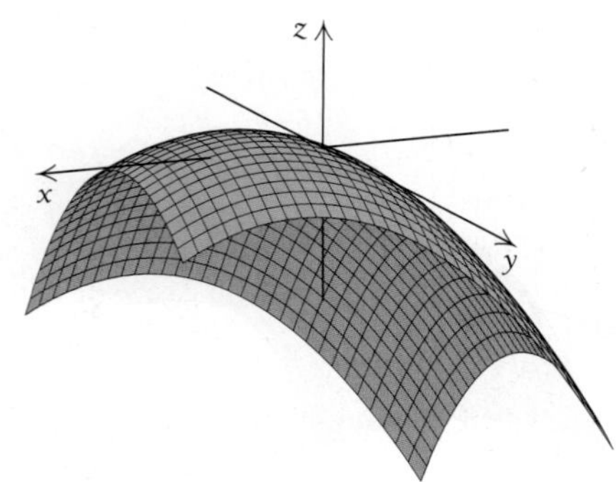

11. $f(x, y) = 4x^2 - y^2$

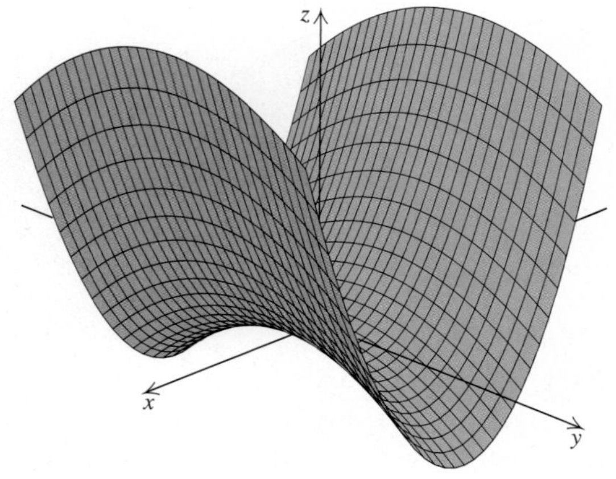

12. $f(x, y) = x^2 - y^2$

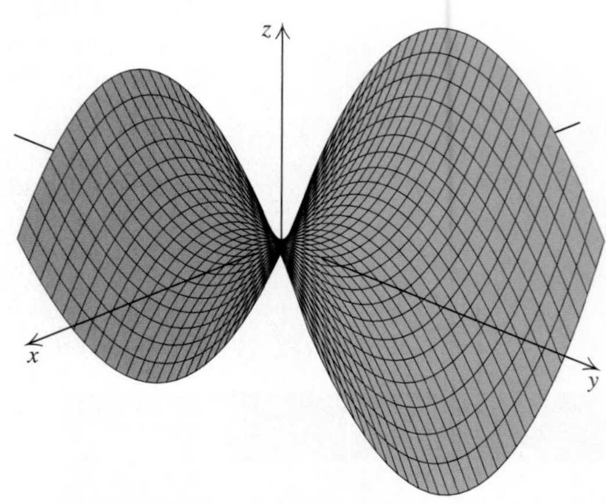

13. $f(x, y) = e^{x^2+y^2+1}$

14. $f(x, y) = e^{x^2-2x+y^2-4y+2}$

APPLICATIONS

Business and Economics

In Exercises 15–22, assume that relative maximum and minimum values are absolute maximum and minimum values.

15. **Maximizing profit.** Safe Shades produces two kinds of sunglasses; one kind sells for \$17, and the other for \$21. The total revenue in thousands of dollars from the sale of x thousand sunglasses at \$17 each and y thousand at \$21 each is given by

$$R(x, y) = 17x + 21y.$$

The company determines that the total cost, in thousands of dollars, of producing x thousand of the \$17 sunglasses and y thousand of the \$21 sunglasses is given by

$$C(x, y) = 4x^2 - 4xy + 2y^2 - 11x + 25y - 3.$$

Find the number of each type of sunglasses that must be produced and sold in order to maximize profit.

16. **Maximizing profit.** A concert promoter produces two kinds of souvenir shirt; one kind sells for \$18, and the other for \$25. The total revenue from the sale of x thousand shirts at \$18 each and y thousand at \$25 each is given by

$$R(x, y) = 18x + 25y.$$

The company determines that the total cost, in thousands of dollars, of producing x thousand of the \$18 shirt and y thousand of the \$25 shirt is given by

$$C(x, y) = 4x^2 - 6xy + 3y^2 + 20x + 19y - 12.$$

How many of each type of shirt must be produced and sold in order to maximize profit?

17. Maximizing profit. A one-product company finds that its profit, P, in millions of dollars, is given by

$$P(a, p) = 2ap + 80p - 15p^2 - \tfrac{1}{10}a^2p - 80,$$

where a is the amount spent on advertising, in millions of dollars, and p is the price charged per item of the product, in dollars. Find the maximum value of P and the values of a and p at which it is attained.

18. Maximizing profit. A one-product company finds that its profit, P, in millions of dollars, is given by

$$P(a, n) = -5a^2 - 3n^2 + 48a - 4n + 2an + 290,$$

where a is the amount spent on advertising, in millions of dollars, and n is the number of items sold, in thousands. Find the maximum value of P and the values of a and n at which it is attained.

19. Minimizing the cost of a container. A trash company is designing an open-top, rectangular container that will have a volume of 320 ft^3. The cost of making the bottom of the container is \$5 per square foot, and the cost of the sides is \$4 per square foot. Find the dimensions of the container that will minimize total cost. (*Hint:* Make a substitution using the formula for volume.)

20. Two-variable revenue maximization. Boxowitz, Inc., a computer firm, markets two kinds of calculator that compete with one another. Their demand functions are expressed by the following relationships:

$$q_1 = 78 - 6p_1 - 3p_2, \quad (1)$$
$$q_2 = 66 - 3p_1 - 6p_2, \quad (2)$$

where p_1 and p_2 are the prices of the calculators, in multiples of \$10, and q_1 and q_2 are the quantities of the calculators demanded, in hundreds of units.

a) Find a formula for the total-revenue function, R, in terms of the variables p_1 and p_2. [*Hint:* $R = p_1q_1 + p_2q_2$; then substitute expressions from equations (1) and (2) to find $R(p_1, p_2)$.]

b) What prices p_1 and p_2 should be charged for each product in order to maximize total revenue?

c) How many units will be demanded?

d) What is the maximum total revenue?

21. Two-variable revenue maximization. Repeat Exercise 20, using

$$q_1 = 64 - 4p_1 - 2p_2$$

and

$$q_2 = 56 - 2p_1 - 4p_2.$$

Life and Physical Sciences

22. Temperature. A flat metal plate is located on a coordinate plane. The temperature of the plate, in degrees Fahrenheit, at point (x, y) is given by

$$T(x, y) = x^2 + 2y^2 - 8x + 4y.$$

Find the minimum temperature and where it occurs. Is there a maximum temperature?

SYNTHESIS

Find the relative maximum and minimum values and the saddle points.

23. $f(x, y) = e^x + e^y - e^{x+y}$

24. $f(x, y) = xy + \dfrac{2}{x} + \dfrac{4}{y}$

25. $f(x, y) = 2y^2 + x^2 - x^2y$

26. $S(b, m) = (m + b - 72)^2 + (2m + b - 73)^2 + (3m + b - 75)^2$

TW **27.** Is a cross-section of an anticlastic curve always a parabola? Why or why not?

TW **28.** Explain the difference between a relative minimum and an absolute minimum of a function of two variables.

TECHNOLOGY CONNECTION

Use a 3D graphics program to graph each of the following functions. Then estimate any relative extrema.

29. $f(x, y) = \dfrac{-5}{x^2 + 2y^2 + 1}$

30. $f(x, y) = x^3 + y^3 + 3xy$

31. $f(x, y) = \dfrac{3xy(x^2 - y^2)}{(x^2 + y^2)}$

32. $f(x, y) = \dfrac{y + x^2y^2 - 8x}{xy}$

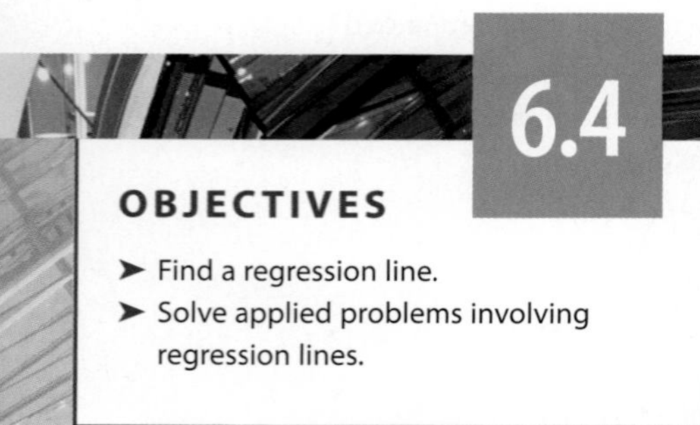

OBJECTIVES

- Find a regression line.
- Solve applied problems involving regression lines.

An Application: The Least-Squares Technique

The problem of fitting an equation to a set of data occurs frequently. We first considered this in Section R.6. Such an equation provides a model of the phenomenon that the data measure, from which predictions can be made. For example, in business, one might want to predict future sales on the basis of past data. In ecology, one might want to predict future demand for natural gas on the basis of past usage. Suppose that we wish to find a linear equation,

$$y = mx + b,$$

to fit some data. To determine this equation is to determine the values of m and b. But how? Let's consider some factual data.

Suppose that a car rental company that offers hybrid (gas–electric) vehicles charts its revenue as shown in Fig. 1 and the table that follows. How best could we predict the company's revenue for the year 2011?

YEARLY REVENUE OF SKY BLUE CAR RENTALS

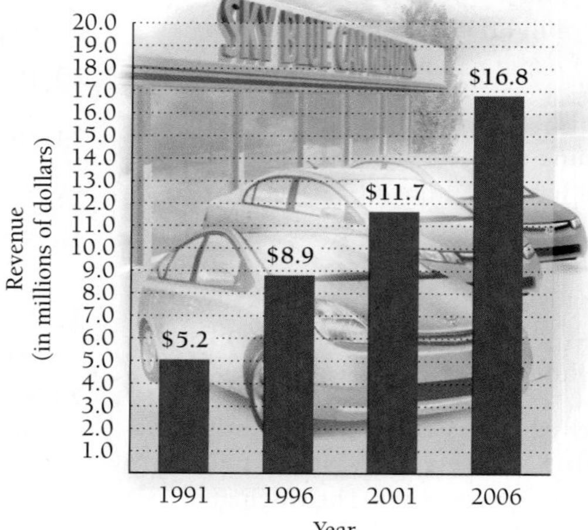

FIGURE 1

Year, x	1991	1996	2001	2006	2011
Yearly Revenue, y (in millions of dollars)	5.2	8.9	11.7	16.8	?

Suppose that we plot these points and try to draw a line through them that fits. Note that there are several ways in which this might be done (see Figs. 2

and 3). Each would give a different estimate of the company's total revenue for 2011.

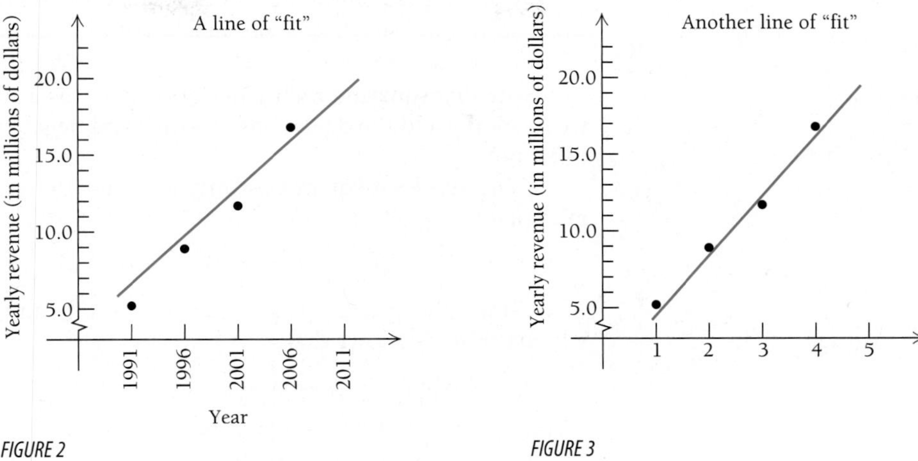

FIGURE 2

FIGURE 3

Note that the years given follow 5-yr increments. Thus, computations can be simplified if we use the data points (1, 5.2), (2, 8.9), (3, 11.7), and (4, 16.8), as plotted in Fig. 3, where each horizontal unit represents 5 years.

To determine the equation of the line that "best" fits the data, we note that for each data point there will be a deviation, or error, between the y-value at that point and the y-value of the point on the line that is directly above or below the point. Those deviations, in this case, $y_1 - 5.2$, $y_2 - 8.9$, $y_3 - 11.7$, and $y_4 - 16.8$, will be positive or negative, depending on the location of the line.

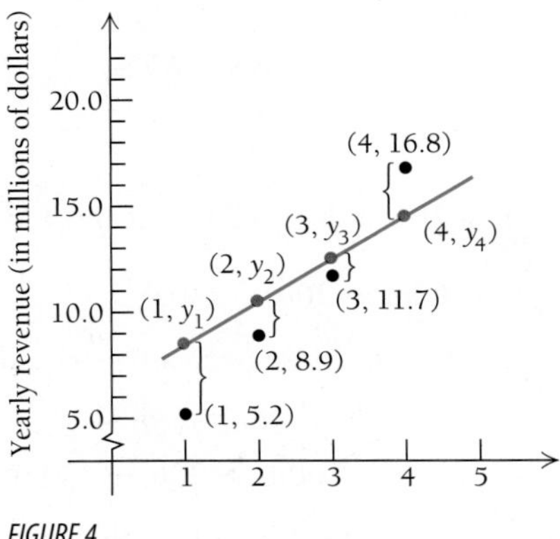

FIGURE 4

We wish to fit these data points with a line,

$$y = mx + b,$$

that uses values of m and b that, somehow, minimize the y-deviations in order to have a good fit. One way of minimizing the deviations is based on the *least-squares assumption*.

The Least-Squares Assumption

The line of best fit is the line for which the sum of the squares of the y-deviations is a minimum. This is called the **regression line.**

Note that squaring each y-deviation gives us a sum of nonnegative terms. Were we to simply add the deviations, positive and negative deviations would cancel each other out.

Using the least-squares assumption with the yearly revenue data, we want to minimize

$$(y_1 - 5.2)^2 + (y_2 - 8.9)^2 + (y_3 - 11.7)^2 + (y_4 - 16.8)^2. \qquad (1)$$

Also, since the points $(1, y_1)$, $(2, y_2)$, $(3, y_3)$, and $(4, y_4)$ must be solutions of $y = mx + b$, it follows that

$$\begin{aligned} y_1 &= m(1) + b = m + b, \\ y_2 &= m(2) + b = 2m + b, \\ y_3 &= m(3) + b = 3m + b, \\ y_4 &= m(4) + b = 4m + b. \end{aligned}$$

Substituting $m + b$ for y_1, $2m + b$ for y_2, $3m + b$ for y_3, and $4m + b$ for y_4 in equation (1), we now have a function of two variables:

$$S(m, b) = (m + b - 5.2)^2 + (2m + b - 8.9)^2 + (3m + b - 11.7)^2 + (4m + b - 16.8)^2.$$

Thus, to find the regression line for the given set of data, we must find the values of m and b that minimize the function S given by the sum in this last equation.

To apply the D-test, we first find the partial derivatives $\partial S/\partial b$ and $\partial S/\partial m$:

$$\begin{aligned} \frac{\partial S}{\partial b} &= 2(m + b - 5.2) + 2(2m + b - 8.9) + 2(3m + b - 11.7) + 2(4m + b - 16.8) \\ &= 20m + 8b - 85.2, \end{aligned}$$

and

$$\begin{aligned} \frac{\partial S}{\partial m} &= 2(m + b - 5.2) + 2(2m + b - 8.9)2 + 2(3m + b - 11.7)3 + 2(4m + b - 16.8)4 \\ &= 60m + 20b - 250.6. \end{aligned}$$

We set these derivatives equal to 0 and solve the resulting system:

$$\begin{aligned} 20m + 8b - 85.2 &= 0, \\ 60m + 20b - 250.6 &= 0; \end{aligned} \qquad \text{or} \qquad \begin{aligned} 5m + 2b &= 21.3, \\ 15m + 5b &= 62.65. \end{aligned}$$

It can be shown that the solution of this system is

$$b = 1.25, \qquad m = 3.76.$$

We leave it to the student to complete the D-test to verify that $(1.25, 3.76)$ does, in fact, yield the minimum of S. There is no need to compute $S(1.25, 3.76)$.

The values of m and b are all we need to determine $y = mx + b$. The regression line is

$$y = 3.76x + 1.25. \qquad \text{Substituting for } m \text{ and } b$$

The graph of this "best-fit" regression line together with the data points is shown in Fig. 5. Compare it to Figs. 1, 2, and 3.

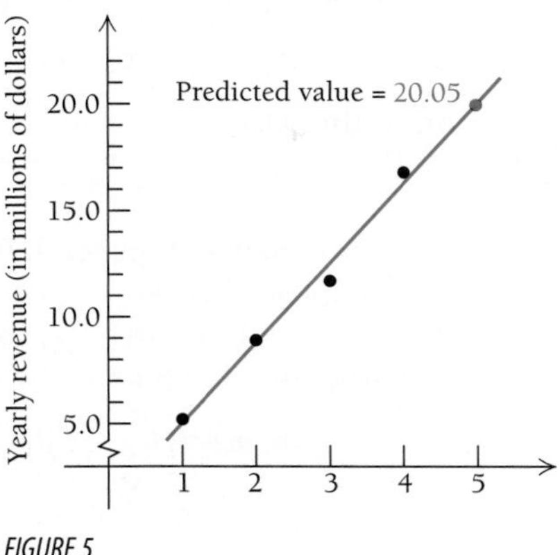

FIGURE 5

We can now extrapolate from the data to predict the car rental company's yearly revenue in 2011:

$$y = 3.76(5) + 1.25 = 20.05.$$

The yearly revenue in 2011 is predicted to be about \$20.05 million. How might you check this prediction?

The method of least squares is a statistical process illustrated here with only four data points in order to simplify the explanation. Most statistical researchers would warn that many more than four data points should be used to get a "good" regression line. Furthermore, making predictions too far in the future from any mathematical model may not be valid. The further into the future a prediction is made, the more dubious one should be about the prediction.

*The Regression Line for an Arbitrary Collection of Data Points $(c_1, d_1), (c_2, d_2), \ldots, (c_n, d_n)$

Look again at the regression line

$$y = 3.76x + 1.25$$

for the data points (1, 5.2), (2, 8.9), (3, 11.7), and (4, 16.8). Let's consider the arithmetic averages, or means, of the x-coordinates, denoted $\overline{x}$, and of the y-coordinates, denoted $\overline{y}$:

$$\overline{x} = \frac{1 + 2 + 3 + 4}{4} = 2.5,$$

$$\overline{y} = \frac{5.2 + 8.9 + 11.7 + 16.8}{4} = 10.65.$$

---

*This subsection is considered optional and can be omitted without loss of continuity.

It turns out that the point $(\bar{x}, \bar{y})$, or (2.5, 10.65), is on the regression line since

$$10.65 = 3.76(2.5) + 1.25.$$

Thus, the equation for the regression line can be written

$$y - \bar{y} = m(x - \bar{x}),$$

or, in this case,

$$y - 10.65 = m(x - 2.5).$$

All that remains, in general, is to determine m.

Suppose that we want to find the regression line for an arbitrary number of points $(c_1, d_1), (c_2, d_2), \dots, (c_n, d_n)$. To do so, we find the values m and b that minimize the function S given by

$$S(b, m) = (y_1 - d_1)^2 + (y_2 - d_2)^2 + \cdots + (y_n - d_n)^2 = \sum_{i=1}^{n} (y_i - d_i)^2,$$

where $y_i = mc_i + b$.

Using a procedure like the one we used earlier to minimize S, we can show that $y = mx + b$ takes the form

$$y - \bar{y} = m(x - \bar{x}),$$

where $\bar{x} = \dfrac{\sum_{i=1}^{n} c_i}{n}$, $\bar{y} = \dfrac{\sum_{i=1}^{n} d_i}{n}$, and $m = \dfrac{\sum_{i=1}^{n}(c_i - \bar{x})(d_i - \bar{y})}{\sum_{i=1}^{n}(c_i - \bar{x})^2}$.

Let's see how this works out for the yearly revenue data from our earlier example.

c_i	d_i	$c_i - \bar{x}$	$(c_i - \bar{x})^2$	$(d_i - \bar{y})$	$(c_i - \bar{x})(d_i - \bar{y})$
1	5.2	−1.5	2.25	−5.45	8.175
2	8.9	−0.5	0.25	−1.75	0.875
3	11.7	0.5	0.25	1.05	0.525
4	16.8	1.5	2.25	6.15	9.225
$\sum_{i=1}^{4} c_i = 10$	$\sum_{i=1}^{4} d_i = 42.6$		$\sum_{i=1}^{4} (c_i - \bar{x})^2 = 5$		$\sum_{i=1}^{4} (c_i - \bar{x})(d_i - \bar{y}) = 18.8$
$\bar{x} = 2.5$	$\bar{y} = 10.65$			$m = \dfrac{18.8}{5} = 3.76$	

Thus, the regression line is

$$y - 10.65 = 3.76(x - 2.5),$$

which simplifies to

$$y = 3.76x + 1.25.$$

TECHNOLOGY CONNECTION

As we have seen in Section R.6 and in other parts of the book, graphing calculators can perform linear regression, as well as quadratic, exponential, and logarithmic regression. Use such a calculator now to fit a linear equation to the yearly revenue data.

With some calculators, you will also obtain a number r, called the **coefficient of correlation.** Although we cannot develop that concept in detail in this text, keep in mind that r is used to describe the strength of the linear relationship between x and y. The closer $|r|$ is to 1, the better the correlation.

For the yearly revenue data, $r \approx 0.993$, which indicates a fairly good linear relationship. Keep in mind that a high linear correlation does not necessarily indicate a "cause-and-effect" connection between the variables.

Exercise Set 6.4

For each data set, find the regression line without using a calculator.

1.

x	1	2	4	5
y	1	3	3	4

2.

x	1	3	5
y	2	4	7

3.

x	1	2	3	5
y	0	1	3	4

4.

x	1	2	4
y	3	5	8

All of the following exercises can be done with a graphing calculator if your instructor so directs. The calculator can also be used to check your work.

APPLICATIONS

Business and Economics

5. **Labor force.** The minimum hourly wage in the United States has grown over the years, as shown in the table below.

Number of Years, x, since 1980	Minimum Hourly Wage
0	\$3.10
1	3.35
10	3.80
11	4.25
16	4.75
17	5.15

(*Source*: www.workworld.org.)

a) For the data in the table on p. 571, find the regression line, $y = mx + b$.
b) Use the regression line to predict the minimum hourly wage in 2010 and 2015.

6. **Football ticket prices.** Ticket prices for NFL football games have experienced steady growth, as shown in the following table.

Number of Years, x, since 1999 Season	Average Ticket Price (dollars)
0	$45.03
1	49.35
2	47.49
3	50.02
4	52.95
5	54.75
6	58.95

(*Source*: Team Marketing Report.)

a) Find the regression line, $y = mx + b$.
b) Use the regression line to predict the average ticket price for an NFL game in 2009 and in 2015.

Life and Physical Sciences

7. **Life expectancy of women.** Consider the data in the following table showing the average life expectancy of women in various years. Note that x represents the actual year.

Year, x	Life Expectancy of Women, y (years)
1950	71.1
1960	73.1
1970	74.7
1980	77.4
1990	78.8
2000	79.5
2003	80.1

(*Source*: Centers for Disease Control, June, 2006.)

a) Find the regression line, $y = mx + b$.
b) Use the regression line to predict the life expectancy of women in 2010 and 2015.

8. **Life expectancy of men.** Consider the following data showing the average life expectancy of men in various years. Note that x represents the actual year.

Year, x	Life Expectancy of Men, y (years)
1950	65.6
1960	66.6
1970	67.1
1980	70.0
1990	71.8
2000	74.1
2003	74.8

(*Source*: Centers for Disease Control, June, 2006.)

a) Find the regression line, $y = mx + b$.
b) Use the regression line to predict the life expectancy of men in 2010 and 2015.

General Interest

9. **Grade predictions.** A professor wants to predict students' final examination scores on the basis of their midterm test scores. An equation was determined on the basis of data on the scores of three students who took the same course with the same

instructor the previous semester (see the following table).

Midterm Score, x	Final Exam Score, y
70%	75%
60	62
85	89

a) Find the regression line, $y = mx + b$. (*Hint:* The y-deviations are $70m + b - 75$, $60m + b - 62$, and so on.)

b) The midterm score of a student was 81%. Use the regression line to predict the student's final exam score.

10. Predicting the world record in the high jump. It has been established that most world records in track and field can be modeled by a linear function. The table below shows world high-jump records for various years. Note that x represents the actual year.

Year, x	World Record in High Jump, y (in inches)
1912 (George Horme)	78.0
1956 (Charles Dumas)	84.5
1973 (Dwight Stones)	90.5
1989 (Javier Sotomayer)	96.0
1993 (Javier Sotomayer)	96.5

(*Source*: wikipedia.org.)

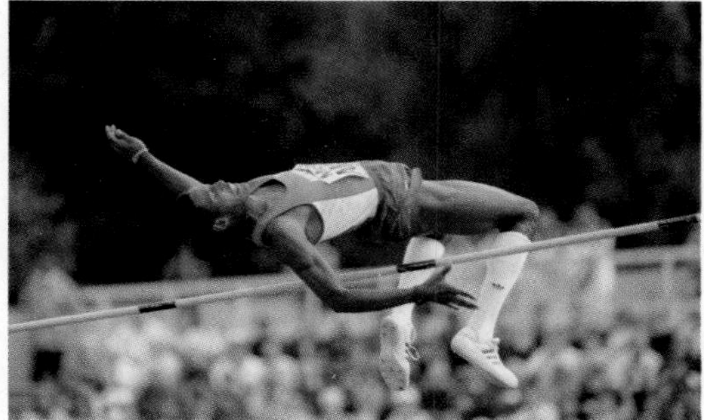

a) Find the regression line, $y = mx + b$.

b) Use the regression line to predict the world record in the high jump in 2010 and in 2050.

TW **c)** Does your answer in part (b) for 2050 seem realistic? Explain why extrapolating so far into the future could be a problem.

TW **11.** How would you explain the concept of linear regression to a friend?

TW **12.** Discuss the idea of linear regression with a professor from another discipline in which regression is used. Explain how it is used in that field.

TECHNOLOGY CONNECTION

13. General interest: predicting the world record for running the mile. Note that x represents the actual year in following table.

Year, x	World Record, y (in minutes:seconds)
1875 (Walter Slade)	4:24.5
1894 (Fred Bacon)	4:18.2
1923 (Paavo Nurmi)	4:10.4
1937 (Sidney Wooderson)	4:06.4
1942 (Gunder Hägg)	4:06.2
1945 (Gunder Hägg)	4:01.4
1954 (Roger Bannister)	3:59.6
1964 (Peter Snell)	3:54.1
1967 (Jim Ryun)	3:51.1
1975 (John Walker)	3:49.4
1979 (Sebastian Coe)	3:49.0
1980 (Steve Ovett)	3:48.40
1985 (Steve Cram)	3:46.31
1993 (Noureddine Morceli)	3:44.39

(*Source*: USA Track & Field and infoplease.com.)

a) Find the regression line, $y = mx + b$, that fits the data in the table. $\left(\textit{Hint}\text{: Convert each time to decimal notation; for instance, } 4{:}24.5 = 4\frac{24.5}{60} = 4.4083.\right)$

b) Use the regression line to predict the world record in the mile in 2010 and in 2015.

c) In July 1999, Hicham El Guerrouj set the current (as of December 2006) world record of 3:43.13 for the mile. (***Source:*** USA Track & Field and infoplease.com.) How does this compare with what is predicted by the regression line?

Constrained Maximum and Minimum Values: Lagrange Multipliers

OBJECTIVES

- Find maximum and minimum values using Lagrange multipliers.
- Solve applied problems involving Lagrange multipliers.

Before we proceed in detail, let's return to a problem we considered in Chapter 2: A hobby store has 20 ft of fencing to fence off a rectangular electric-train area in one corner of its display room. The two sides up against the wall require no fence. What dimensions of the rectangle will maximize the area?

We maximize the function

$$A = xy$$

subject to the condition, or *constraint*, $x + y = 20$. Note that A is a function of two variables.

When we solved this earlier, we first solved the constraint for y:

$$y = 20 - x.$$

We then substituted $20 - x$ for y to obtain

$$\begin{aligned} A(x, y) &= x(20 - x) \\ &= 20x - x^2, \end{aligned}$$

which is a function of one variable. Next, we found a maximum value using Maximum–Minimum Principle 1 (see Section 2.4). By itself, the function of two variables

$$A(x, y) = xy$$

has no maximum value. This can be checked using the D-test. With the constraint $x + y = 20$, however, the function does have a maximum. We see this in the following graph.

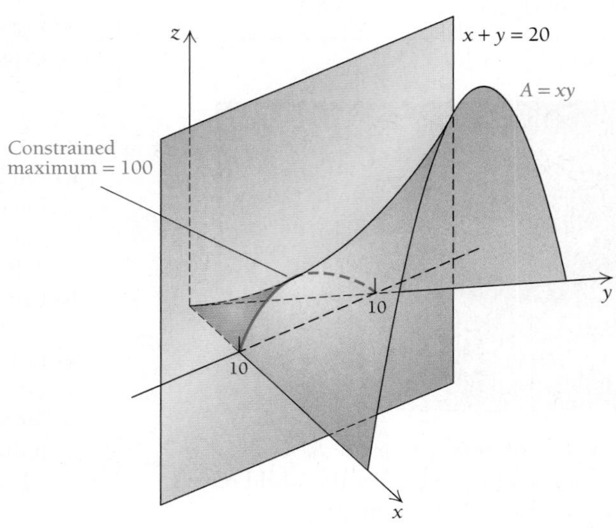

It may be quite difficult to solve a constraint for one variable. The procedure outlined below allows us to proceed without doing so.

The Method of Lagrange Multipliers

To find a maximum or minimum value of a function $f(x, y)$ subject to the constraint $g(x, y) = 0$:

1. Form a new function:

$$F(x, y, \lambda) = f(x, y) - \lambda g(x, y).$$

The variable λ (lambda) is called a **Lagrange multiplier.**

2. Find the first partial derivatives F_x, F_y, and F_λ.

3. Solve the system

$$F_x = 0, \qquad F_y = 0, \quad \text{and} \quad F_\lambda = 0.$$

Let (a, b, λ) represent a solution of this system. We normally must determine whether (a, b, λ) yields a maximum or minimum of the function f. For the problems in this text, we will specify that a maximum or minimum exists.

The method of Lagrange multipliers can be extended to functions of three (or more) variables.

We can illustrate the method of Lagrange multipliers by resolving the electric-train area problem.

EXAMPLE 1 Find the maximum value of

$$A(x, y) = xy$$

subject to the constraint $x + y = 20$.

Solution Note first that $x + y = 20$ is equivalent to $x + y - 20 = 0$.

1. We form the new function F, given by

$$F(x, y, \lambda) = xy - \lambda \cdot (x + y - 20).$$

2. We find the first partial derivatives:

$$F_x = y - \lambda,$$
$$F_y = x - \lambda,$$
$$F_\lambda = -(x + y - 20).$$

3. We set each derivative equal to 0 and solve the resulting system:

$$y - \lambda = 0, \tag{1}$$
$$x - \lambda = 0, \tag{2}$$
$$-(x + y - 20) = 0, \quad \text{or} \quad x + y - 20 = 0. \tag{3}$$

From equations (1) and (2), it follows that

$$x = y = \lambda.$$

Substituting x for y in equation (3), we get

$$\begin{aligned} x + x - 20 &= 0 \\ 2x &= 20 \\ x &= 10. \end{aligned}$$

Thus, $y = x = 10$. The maximum value of A subject to the constraint occurs at $(10, 10)$ and is

$$\begin{aligned} A(10, 10) &= 10 \cdot 10 \\ &= 100. \end{aligned}$$

✦

EXAMPLE 2 Find the maximum value of

$$f(x, y) = 3xy$$

subject to the constraint

$$2x + y = 8.$$

Note: f might be interpreted, for example, as a production function with budget constraint $2x + y = 8$.

Solution Note that first we express $2x + y = 8$ as $2x + y - 8 = 0$.

1. We form the new function F, given by

$$F(x, y, \lambda) = 3xy - \lambda(2x + y - 8).$$

2. We find the first partial derivatives:

$$\begin{aligned} F_x &= 3y - 2\lambda, \\ F_y &= 3x - \lambda, \\ F_\lambda &= -(2x + y - 8). \end{aligned}$$

3. We set each derivative equal to 0 and solve the resulting system:

$$3y - 2\lambda = 0, \quad (1)$$

$$3x - \lambda = 0, \quad (2)$$

$$-(2x + y - 8) = 0, \quad \text{or} \quad 2x + y - 8 = 0. \quad (3)$$

Solving equation (2) for λ, we get

$$\lambda = 3x.$$

Substituting in equation (1) for λ, we get

$$3y - 2 \cdot 3x = 0, \quad \text{or} \quad 3y = 6x, \quad \text{or} \quad y = 2x. \quad (4)$$

Substituting $2x$ for y in equation (3), we get

$$\begin{aligned} 2x + 2x - 8 &= 0 \\ 4x &= 8 \\ x &= 2. \end{aligned}$$

> **TECHNOLOGY CONNECTION**
>
> Use a 3D graphics program to graph both $f(x, y) = 3xy$ and $2x + y = 8$. Then check the results of Example 2 visually.

Then, using equation (4), we have

$$y = 2 \cdot 2 = 4.$$

The maximum value of f subject to the constraint occurs at $(2, 4)$ and is

$$f(2, 4) = 3 \cdot 2 \cdot 4 = 24. \quad ◆$$

EXAMPLE 3 Business: The Beverage-Can Problem. The standard beverage can has a volume of 12 fl. oz, or 21.66 in^3. What dimensions yield the minimum surface area? Find the minimum surface area. (Assume that the shape of the can is a right circular cylinder.)

Solution We want to minimize the function s, given by

$$s(h, r) = 2\pi rh + 2\pi r^2,$$

subject to the volume constraint

$$\pi r^2 h = 21.66,$$

or $$\pi r^2 h - 21.66 = 0.$$

Note that s does not have a minimum without the constraint.

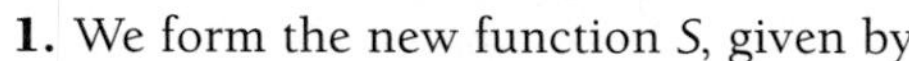

1. We form the new function S, given by

$$S(h, r, \lambda) = 2\pi rh + 2\pi r^2 - \lambda(\pi r^2 h - 21.66).$$

2. We find the first partial derivatives:

$$\frac{\partial S}{\partial h} = 2\pi r - \lambda\pi r^2,$$

$$\frac{\partial S}{\partial r} = 2\pi h + 4\pi r - 2\lambda\pi rh,$$

$$\frac{\partial S}{\partial \lambda} = -(\pi r^2 h - 21.66).$$

3. We set each derivative equal to 0 and solve the resulting system:

$$2\pi r - \lambda\pi r^2 = 0, \quad (1)$$

$$2\pi h + 4\pi r - 2\lambda\pi rh = 0, \quad (2)$$

$$-(\pi r^2 h - 21.66) = 0, \quad \text{or} \quad \pi r^2 h - 21.66 = 0. \quad (3)$$

Note that, since π is a constant, we can solve equation (1) for r:

$$\pi r(2 - \lambda r) = 0$$

$$\pi r = 0 \quad \text{or} \quad 2 - \lambda r = 0$$

$$r = 0 \quad \text{or} \quad r = \frac{2}{\lambda}. \qquad \text{We assume } \lambda \neq 0.$$

Since $r = 0$ cannot be a solution to the original problem, we continue by substituting $2/\lambda$ for r in equation (2):

$$2\pi h + 4\pi \cdot \frac{2}{\lambda} - 2\lambda\pi \cdot \frac{2}{\lambda} \cdot h = 0$$

$$2\pi h + \frac{8\pi}{\lambda} - 4\pi h = 0$$

$$\frac{8\pi}{\lambda} - 2\pi h = 0$$

$$-2\pi h = -\frac{8\pi}{\lambda},$$

so

$$h = \frac{4}{\lambda}.$$

Since $h = 4/\lambda$ and $r = 2/\lambda$, it follows that $h = 2r$. Substituting $2r$ for h in equation (3) yields

$$\pi r^2(2r) - 21.66 = 0$$

$$2\pi r^3 - 21.66 = 0$$

$$2\pi r^3 = 21.66$$

$$\pi r^3 = 10.83$$

$$r^3 = \frac{10.83}{\pi}$$

$$r = \sqrt[3]{\frac{10.83}{\pi}} \approx 1.51 \text{ in.}$$

Thus, when $r = 1.51$ in., we have $h = 3.02$ in. The surface area is then a minimum and is approximately

$$2\pi(1.51)(3.02) + 2\pi(1.51)^2, \quad \text{or about } 42.98 \text{ in}^2.$$ ✦

The actual dimensions of a standard-sized 12-oz beverage can are $r = 1.25$ in. and $h = 4.875$ in. A natural question arising from the solution of Example 3 is, "Why don't beverage companies make cans using the dimensions found in that example?" To do this would mean an enormous cost for retooling. New can-making machines and new beverage-filling machines would have to be designed and purchased. Vending machines would no longer be the correct size. A partial response to the desire to save aluminum has been found in recycling and in manufacturing cans with bevelled edges. These cans require less aluminum. As a result of many engineering advances, the amount of aluminum required to make 1000 cans has been reduced over the years from 36.5 lb to 28.1 lb. Consumer preference is another very important factor affecting the shape of the can. Market research has shown that a can with the dimensions found in Example 3 is not as comfortable to hold and might not be accepted by consumers.

Exercise Set 6.5

Find the maximum value of f subject to the given constraint.

1. $f(x, y) = xy;\quad 3x + y = 10$

2. $f(x, y) = 2xy;\quad 4x + y = 16$

3. $f(x, y) = 4 - x^2 - y^2;\quad x + 2y = 10$

4. $f(x, y) = 3 - x^2 - y^2;\quad x + 6y = 37$

Find the minimum value of f subject to the given constraint.

5. $f(x, y) = x^2 + y^2;\quad 2x + y = 10$

6. $f(x, y) = x^2 + y^2;\quad x + 4y = 17$

7. $f(x, y) = 2y^2 - 6x^2;\quad 2x + y = 4$

8. $f(x, y) = 2x^2 + y^2 - xy;\quad x + y = 8$

9. $f(x, y, z) = x^2 + y^2 + z^2;\quad y + 2x - z = 3$

10. $f(x, y, z) = x^2 + y^2 + z^2;\quad x + y + z = 2$

Use the method of Lagrange multipliers to solve each of the following.

11. Of all numbers whose sum is 50, find the two that have the maximum product.

12. Of all numbers whose sum is 70, find the two that have the maximum product.

13. Of all numbers whose difference is 6, find the two that have the minimum product.

14. Of all numbers whose difference is 4, find the two that have the minimum product.

15. Of all points (x, y, z) that satisfy $x + 2y + 3z = 13$, find the one that minimizes $(x - 1)^2 + (y - 1)^2 + (z - 1)^2$.

16. Of all points (x, y, z) that satisfy $3x + 4y + 2z = 52$, find the one that minimizes $(x - 1)^2 + (y - 4)^2 + (z - 2)^2$.

APPLICATIONS

Business and Economics

17. Maximizing typing area. A standard piece of printer paper has a perimeter of 39 in. Find the dimensions of the paper that will give the most area. What is that area? Does standard $8\frac{1}{2} \times 11$ in. paper have maximum area?

18. Maximizing room area. A carpenter is building a rectangular room with a fixed perimeter of 80 ft. What are the dimensions of the largest room that can be built? What is its area?

19. Minimizing surface area. An oil drum of standard size has a volume of 200 gal, or 27 ft^3. What dimensions yield the minimum surface area? Find the minimum surface area.

Do these drums appear to be made in such a way as to minimize surface area?

20. Juice-can problem. A standard-sized juice can has a volume of 99 in^3. What dimensions yield the minimum surface area? Find the minimum surface area.

21. Maximizing total sales. The total sales, S, of a one-product firm are given by

$$S(L, M) = ML - L^2,$$

where M is the cost of materials and L is the cost of labor. Find the maximum value of this function subject to the budget constraint

$$M + L = 90.$$

22. Maximizing total sales. The total sales, S, of a one-product firm are given by

$$S(L, M) = ML - L^2,$$

where M is the cost of materials and L is the cost of labor. Find the maximum value of this function subject to the budget constraint

$$M + L = 70.$$

23. Minimizing construction costs. A company is planning to construct a warehouse whose interior volume is to be 252,000 ft^3. Construction costs per square foot are estimated to be as follows:

Walls: $3.00

Floor: $4.00

Ceiling: $3.00

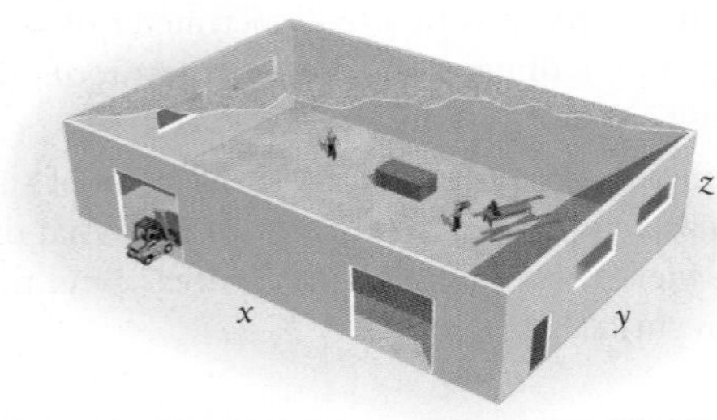

a) The total cost of the building is a function $C(x, y, z)$, where x is the length, y is the width, and z is the height. Find a formula for $C(x, y, z)$.

b) What dimensions of the building will minimize the total cost? What is the minimum cost?

24. Minimizing the costs of container construction. A container company is going to construct a shipping crate of volume 12 ft^3 with a square bottom and top. The cost of the top and the sides is $2 per square foot, and the cost for the bottom is $3 per square foot. What dimensions will minimize the cost of the crate?

25. Minimizing total cost. Each unit of a product can be made on either machine A or machine B. The nature of the machines makes their cost functions differ:

Machine A: $C(x) = 10 + \frac{x^2}{6}$,

Machine B: $C(y) = 200 + \frac{y^3}{9}$.

Total cost is given by $C(x, y) = C(x) + C(y)$. How many units should be made on each machine in order to minimize total costs if $x + y = 10{,}100$ units are required?

SYNTHESIS

Find the indicated maximum or minimum values of f subject to the given constraint.

26. Minimum: $f(x, y) = xy$; $x^2 + y^2 = 9$

27. Minimum: $f(x, y) = 2x^2 + y^2 + 2xy + 3x + 2y$; $y^2 = x + 1$

28. Maximum: $f(x, y, z) = x + y + z$; $x^2 + y^2 + z^2 = 1$

29. Maximum: $f(x, y, z) = x^2y^2z^2$; $x^2 + y^2 + z^2 = 2$

30. Maximum: $f(x, y, z) = x + 2y - 2z$; $x^2 + y^2 + z^2 = 4$

31. Maximum: $f(x, y, z, t) = x + y + z + t$; $x^2 + y^2 + z^2 + t^2 = 1$

32. Minimum: $f(x, y, z) = x^2 + y^2 + z^2$; $x - 2y + 5z = 1$

33. Economics: the Law of Equimarginal Productivity. Suppose that $p(x, y)$ represents the production of a two-product firm. The company produces x units of the first product at a cost of c_1 each and y units of the second product at a cost of c_2 each. The budget constraint, B, is a constant given by

$$B = c_1x + c_2y.$$

Use the method of Lagrange multipliers to find the value of λ in terms of p_x, p_y, c_1, and c_2. The resulting equation holds for any production function p and is called the *Law of Equimarginal Productivity.*

34. Business: maximizing production. A computer company has the following Cobb–Douglas production function for a certain product:

$$p(x, y) = 800x^{3/4}y^{1/4},$$

where x is the labor, measured in dollars, and y is the capital, measured in dollars. Suppose that the company can make a total investment in labor and capital of $1,000,000. How should it allocate the investment between labor and capital in order to maximize production?

TW **35.** Discuss the difference between solving a maximum–minimum problem using the method of Lagrange multipliers and the method of Section 6.3. Describe the difference graphically.

TW **36.** Write a brief report on the life and work of the mathematician Joseph Louis Lagrange (1736–1813).

TECHNOLOGY CONNECTION

37–44. Use a 3D graphics program to graph both equations in each of Exercises 1–8. Then visually check the results that you found analytically.

6.6

OBJECTIVE

➤ Evaluate a double integral.

Double Integrals

The following is an example of a *double integral:*

$$\int_3^6 \int_{-1}^2 10xy^2\, dx\, dy, \quad \text{or} \quad \int_3^6 \left(\int_{-1}^2 10xy^2\, dx \right) dy.$$

Evaluating a double integral is somewhat similar to "undoing" a second partial derivative. We first evaluate the inside integral, indicated by the innermost differential (here dx), and treat the other variable(s) (here y) as constant(s):

$$\int_{-1}^2 10xy^2\, dx = 10y^2 \left[\frac{x^2}{2} \right]_{-1}^2 = 5y^2[x^2]_{-1}^2 = 5y^2[2^2 - (-1)^2] = 15y^2.$$

Color indicates the variable. All else is constant.

Then we evaluate the outside integral, associated with the differential dy:

$$\begin{aligned}\int_3^6 15y^2\, dy &= 15\left[\frac{y^3}{3}\right]_3^6 \\ &= 5[y^3]_3^6 \\ &= 5(6^3 - 3^3) \\ &= 945.\end{aligned}$$

More precisely, the given double integral is called a **double iterated integral.** The word "iterate" means "to do again."

If dx and dy, as well as the limits of integration, are interchanged, we have

$$\int_{-1}^2 \int_3^6 10xy^2\, dy\, dx.$$

We first evaluate the inside, y-integral, treating x as a constant:

$$\begin{aligned}\int_3^6 10xy^2\, dy &= 10x\left[\frac{y^3}{3}\right]_3^6 \\ &= \frac{10x}{3}\left[y^3\right]_3^6 \\ &= \frac{10}{3}x(6^3 - 3^3) = 630x.\end{aligned}$$

Then we evaluate the outside, x-integral:

$$\begin{aligned}\int_{-1}^2 630x\, dx &= 630\left[\frac{x^2}{2}\right]_{-1}^2 \\ &= 315[x^2]_{-1}^2 \\ &= 315[2^2 - (-1)^2] = 945.\end{aligned}$$

Note that we get the same result.

DEFINITION

If $f(x, y)$ is defined over the rectangular region R bounded by $x = a$, $x = b$, $y = c$, and $y = d$, then the *double integral* of $f(x, y)$ over R is given by

$$\int_c^d \int_a^b f(x, y)\, dx\, dy \quad \text{or} \quad \int_a^b \int_c^d f(x, y)\, dy\, dx.$$

In a more technical definition of double integral, Riemann sums are used. However, for the functions in this text, the above definition is sufficient.

Sometimes variables occur as limits of integration.

EXAMPLE 1 Evaluate

$$\int_0^1 \int_{x^2}^x xy^2\, dy\, dx.$$

Solution We first evaluate the y-integral, treating x as a constant:

$$\begin{aligned}\int_{x^2}^x xy^2\, dy &= x\left[\frac{y^3}{3}\right]_{x^2}^x \\ &= \frac{1}{3}x[x^3 - (x^2)^3] \\ &= \frac{1}{3}(x^4 - x^7).\end{aligned}$$

Then we evaluate the outside integral:

$$\begin{aligned}\frac{1}{3}\int_0^1 (x^4 - x^7)\, dx &= \frac{1}{3}\left[\frac{x^5}{5} - \frac{x^8}{8}\right]_0^1 \\ &= \frac{1}{3}\left[\left(\frac{1^5}{5} - \frac{1^8}{8}\right) - \left(\frac{0^5}{5} - \frac{0^8}{8}\right)\right] = \frac{1}{40}.\end{aligned}$$

Thus, $\displaystyle\int_0^1 \int_{x^2}^x xy^2\, dy\, dx = \frac{1}{40}.$ ✦

The Geometric Interpretation of Multiple Integrals

Suppose that the region D in the xy-plane is bounded by the functions $y_1 = g(x)$ and $y_2 = h(x)$ and the lines $x_1 = a$ and $x_2 = b$. We want the volume, V, of the solid above D and under the surface $z = f(x, y)$. We can think of the solid as composed of many vertical columns, one of which is shown in Fig. 1 in red. The volume of this column can be thought of as $l \cdot w \cdot h$, or $z \cdot \Delta y \cdot \Delta x$. Integrating such columns in the y-direction, we obtain

$$\int_{y_1}^{y_2} z\, dy, \quad \text{so that} \quad \left[\int_{y_1}^{y_2} z\, dy\right] \Delta x$$

can be pictured as a "slab," or slice. Then integrating such slices in the x-direction, we obtain the entire volume:

$$V = \int_a^b \left[\int_{y_1}^{y_2} z \, dy \right] dx,$$

This can be thought of as a collection of slices that fills the volume.

or

$$V = \int_a^b \int_{g(x)}^{h(x)} z \, dy \, dx,$$

where $z = f(x, y)$.

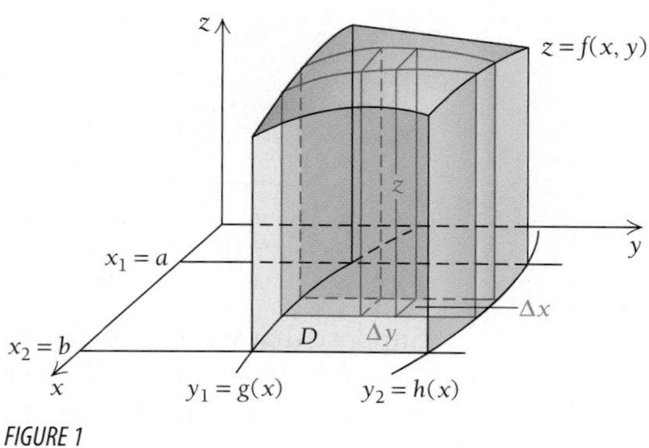

FIGURE 1

In Example 1, the region of integration D is the plane region between the graphs of $y = x^2$ and $y = x$, as shown in Figs. 2 and 3.

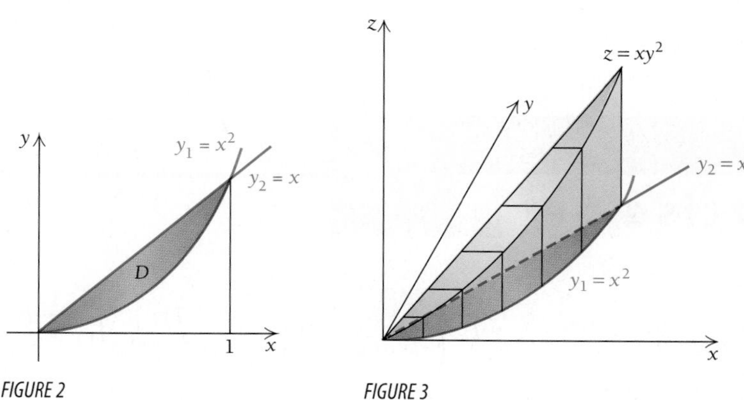

FIGURE 2

FIGURE 3

When we evaluated the double integral in Example 1, we found the volume of the solid based on D and capped by the surface $z = xy^2$, as shown in Fig. 3.

An Application to Probability

Suppose that we throw a dart at a region R in a plane. We assume that the dart lands on a point (x, y) in R (see Fig. 4 on the next page). We can think of (x, y) as a continuous random variable whose coordinates can be those of any ordered pair in region R. A function f is said to be a **joint probability density function** if

$$f(x, y) \geq 0, \quad \text{for all } (x, y) \text{ in } R,$$

and

$$\iint_R f(x, y)\, dx\, dy = 1,$$

where $\iint_R$ refers to the double integral evaluated over the region R.

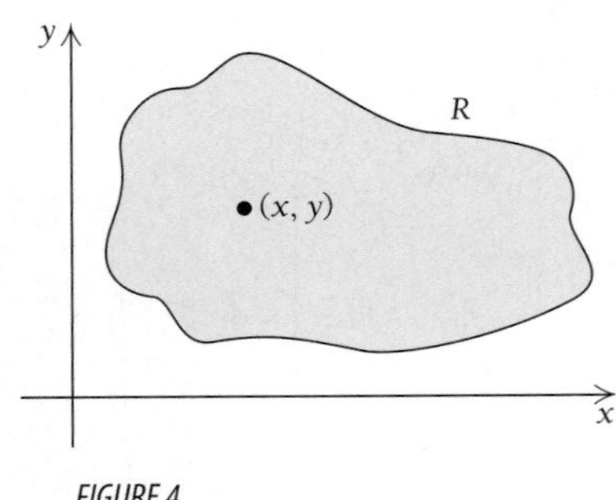

FIGURE 4

FIGURE 5

Suppose that we want to know the probability that the dart hits a point (x, y) in a rectangular subregion G of R, where G is the set of points for which $a \leq x \leq b$ and $c \leq y \leq d$ (Fig. 5). This probability is given by

$$\iint_G f(x, y)\, dx\, dy = \int_c^d \int_a^b f(x, y)\, dx\, dy.$$

Exercise Set 6.6

Evaluate

1. $\displaystyle\int_0^3 \int_0^1 2y\, dx\, dy$

2. $\displaystyle\int_0^1 \int_0^4 3x\, dx\, dy$

3. $\displaystyle\int_{-1}^3 \int_1^2 x^2 y\, dy\, dx$

4. $\displaystyle\int_1^4 \int_{-2}^1 x^3 y\, dy\, dx$

5. $\displaystyle\int_0^5 \int_{-2}^{-1} (3x + y)\, dx\, dy$

6. $\displaystyle\int_{-4}^{-1} \int_1^3 (x + 5y)\, dx\, dy$

7. $\displaystyle\int_{-1}^1 \int_x^1 xy\, dy\, dx$

8. $\displaystyle\int_{-1}^1 \int_x^2 (x + y)\, dy\, dx$

9. $\displaystyle\int_0^1 \int_{x^2}^x (x + y)\, dy\, dx$

10. $\displaystyle\int_0^2 \int_0^x e^{x+y}\, dy\, dx$

11. $\displaystyle\int_0^1 \int_1^{e^x} \frac{1}{y}\, dy\, dx$

12. $\displaystyle\int_0^1 \int_{-1}^x (x^2 + y^2)\, dy\, dx$

13. $\displaystyle\int_0^2 \int_0^x (x + y^2)\, dy\, dx$

14. $\displaystyle\int_1^3 \int_0^x 2e^{x^2}\, dy\, dx$

15. Find the volume of the solid capped by the surface $z = 1 - y - x^2$ over the region bounded on the xy-plane by $y = 1 - x^2$, $y = 0$, $x = 0$, and $x = 1$, by evaluating the integral

$$\int_0^1 \int_0^{1-x^2} (1 - y - x^2)\, dy\, dx.$$

16. Find the volume of the solid capped by the surface $z = x + y$ over the region bounded on the xy-plane by $y = 1 - x$, $y = 0$, $x = 0$, and $x = 1$, by evaluating the integral

$$\int_0^1 \int_0^{1-x} (x + y)\, dy\, dx.$$

For Exercises 17 and 18, suppose that a continuous random variable has a joint probability density function given by

$$f(x, y) = x^2 + \tfrac{1}{3}xy,$$
$$0 \le x \le 1, \quad 0 \le y \le 2.$$

17. Find

$$\int_0^2 \int_0^1 f(x, y)\, dx\, dy.$$

18. Find the probability that a point (x, y) is in the region bounded by $0 \le x \le \frac{1}{2}$, $1 \le y \le 2$, by evaluating the integral

$$\int_1^2 \int_0^{1/2} f(x, y)\, dx\, dy.$$

For Exercises 19 and 20, suppose that a continuous random variable has a joint probability density function given by

$$f(x, y) = x^2 - 3x + \tfrac{1}{3}xy - \tfrac{1}{3}y + 2,$$
$$1 \le x \le 2, \quad 3 \le y \le 5$$

19. Find the probability that a point is in the region bounded by $1 \le x \le 2$ and $3 \le y \le 4$.

20. Find the probability that a point is in the region bounded by $1 \le x \le 2$ and $4 \le y \le 5$.

SYNTHESIS

A triple iterated integral such as

$$\int_r^s \int_c^d \int_a^b f(x, y, z)\, dx\, dy\, dz$$

is evaluated in much the same way as a double iterated integral. We first evaluate the inside x-integral, treating y and z as constants. Then we evaluate the middle y-integral, treating z as a constant. Finally, we evaluate the outside z-integral. Evaluate these triple integrals.

21. $\int_0^1 \int_1^3 \int_{-1}^2 (2x + 3y - z)\, dx\, dy\, dz$

22. $\int_0^2 \int_1^4 \int_{-1}^2 (8x - 2y + z)\, dx\, dy\, dz$

23. $\int_0^1 \int_0^{1-x} \int_0^{2-x} xyz\, dz\, dy\, dx$

24. $\int_0^2 \int_{2-y}^{6-2y} \int_0^{\sqrt{4-y^2}} z\, dz\, dx\, dy$

TW 25. Describe the geometric meaning of the double integral of a function of two variables.

TW 26. Explain how Exercise 17 can be answered without performing any calculations or finding any antiderivatives.

TECHNOLOGY CONNECTION

27. Use a calculator that does multiple integration to evaluate some double integrals found in this exercise set.

Chapter Summary

This chapter covered a variety of topics that all make use of functions of two or more variables. Most of the functions are functions of two variables:

A **function of two variables** assigns to each input pair, (x, y), exactly one output number, $f(x, y)$. (p. 540)

Thus, if $f(x, y) = 2x + y^2$, then $f(1, 3) = 2 \cdot 1 + 3^2 = 11$.

In much the same way that a function given by $y = f(x)$ has a rate of change with respect to x, so does a function given by $z = f(x, y)$. When this rate of change is instantaneous and all variables except x are treated as constants, it is called the **partial derivative of f with respect to x**, denoted by

$$\frac{\partial f}{\partial x} \quad \text{or} \quad \frac{\partial z}{\partial x} \quad \text{or} \quad f_x. \qquad \text{(p. 548)}$$

Similarly, the **partial derivative of f with respect to y** is denoted by

$$\frac{\partial f}{\partial y} \quad \text{or} \quad \frac{\partial z}{\partial y} \quad \text{or} \quad f_y. \qquad \text{(p. 548)}$$

Thus, if $f(x, y) = 5x^2y^3$, then $f_x = 10xy^3$ and $f_y = 15x^2y^2$.

Partial derivatives have many real-world applications. Among those are the Cobb–Douglas production function (p. 551), maximum–minimum problems (pp. 557–563), and applications in probability and statistics (pp. 566–571).

Similar to the second- and higher-order derivatives studied in Chapter 2, a **second-order partial derivative** is a partial derivative of a partial derivative (p. 552). For example, if $f(x, y) = 5x^2y^3$, then

$$\frac{\partial f}{\partial x} = 10xy^3$$

and

$$\frac{\partial}{\partial y}\left(\frac{\partial f}{\partial x}\right) = \frac{\partial^2 f}{\partial y\,\partial x} = f_{xy} = 30\,xy^2.$$

It is important to notice that each notation indicates which partial derivative is found first.

Second-Order Partial Derivatives

1. $\dfrac{\partial^2 z}{\partial x\,\partial x} = \dfrac{\partial^2 f}{\partial x\,\partial x} = \dfrac{\partial^2 z}{\partial x^2} = \dfrac{\partial^2 f}{\partial x^2} = f_{xx}$ Take the partial derivative with respect to x, and then with respect to x again.

2. $\dfrac{\partial^2 z}{\partial y\,\partial x} = \dfrac{\partial^2 f}{\partial y\,\partial x} = f_{xy}$ Take the partial derivative with respect to x, and then with respect to y.

3. $\dfrac{\partial^2 z}{\partial x\,\partial y} = \dfrac{\partial^2 f}{\partial x\,\partial y} = f_{yx}$ Take the partial derivative with respect to y, and then with respect to x.

4. $\dfrac{\partial^2 z}{\partial y\,\partial y} = \dfrac{\partial^2 f}{\partial y\,\partial y} = \dfrac{\partial^2 z}{\partial y^2} = \dfrac{\partial^2 f}{\partial y^2} = f_{yy}$ Take the partial derivative with respect to y, and then with respect to y again. (p. 553)

One reason for learning about higher-order partial derivatives is that they are useful for finding extrema for functions of two variables. A function f of two variables:

1. has a **relative maximum** at (a, b) if

   $f(x, y) \le f(a, b)$

   for all points in a rectangular region containing (a, b);
2. has a **relative minimum** at (a, b) if

   $f(x, y) \ge f(a, b)$

   for all points in a rectangular region containing (a, b). (p. 557)

To identify points that may be extrema, we use the D-test.

The *D*-Test

To find the relative maximum and minimum values of f:

1. Find f_x, f_y, f_{xx}, f_{yy}, and f_{xy}.
2. Solve the system of equations $f_x = 0, f_y = 0$. Let (a, b) represent a solution.
3. Evaluate D, where $D = f_{xx}(a, b) \cdot f_{yy}(a, b) - [f_{xy}(a, b)]^2$.
4. Then:
   a) f has a maximum at (a, b) if $D > 0$ and $f_{xx}(a, b) < 0$.
   b) f has a minimum at (a, b) if $D > 0$ and $f_{xx}(a, b) > 0$.
   c) f has neither a maximum nor a minimum at (a, b) if $D < 0$. The function has a **saddle point** at (a, b).
   d) This test is not applicable if $D = 0$. (p. 558)

EXAMPLE For $f(x, y) = x^2 + xy + y^2 - x$, find the relative maximum or minimum values.

Solution

1. $f_x = 2x + y - 1, \qquad f_y = x + 2y,$

   $f_{xx} = 2; \qquad f_{yy} = 2;$

   $f_{xy} = 1.$
2. We solve the system

$$2x + y - 1 = 0,$$
$$x + 2y = 0,$$

by solving the second equation for x and substituting in the other equation:

$$x = -2y,$$
$$2(-2y) + y - 1 = 0.$$

Solving for y, we find that $y = -\frac{1}{3}$. Substituting $-\frac{1}{3}$ into $x = -2y$, we find that $x = \frac{2}{3}$.

Thus, $\left(\frac{2}{3}, -\frac{1}{3}\right)$ is a critical point and $f\left(\frac{2}{3}, -\frac{1}{3}\right)$ is a possible maximum or minimum value.

3. The D-test yields

$$\begin{aligned} D &= f_{xx}\left(\tfrac{2}{3}, -\tfrac{1}{3}\right) \cdot f_{yy}\left(\tfrac{2}{3}, -\tfrac{1}{3}\right) - \left[f_{xy}\left(\tfrac{2}{3}, -\tfrac{1}{3}\right)\right]^2 \\ &= 2 \cdot 2 - 1^2 = 3. \end{aligned}$$

4. Since $D > 0$ and $f_{xx}\left(\tfrac{2}{3}, -\tfrac{1}{3}\right) > 1$, it follows that $f\left(\tfrac{2}{3}, -\tfrac{1}{3}\right)$ is a minimum. Thus, the relative minimum value of f is $f\left(\tfrac{2}{3}, -\tfrac{1}{3}\right) = -\tfrac{1}{3}$. ✦

Partial derivatives are also used when fitting a regression line to a set of data points. This is accomplished using the **least-squares assumption**:

The line of best fit is the line for which the sum of the squares of the y-deviations is a minimum. This is called the **regression line**. (p. 568)

EXAMPLE A researcher's data consists of the points (1, 3), (4, 5), and (6, 7). Find the regression line.

Solution Using the least-squares assumption, we must minimize

$$(y_1 - 3)^2 + (y_2 - 5)^2 + (y_3 - 7)^2, \tag{1}$$

where
$$\begin{aligned} y_1 &= m(1) + b = m + b, \\ y_2 &= m(4) + b = 4m + b, \\ y_3 &= m(6) + b = 6m + b. \end{aligned}$$

After substituting into equation (1), we can find m and b for the regression line by minimizing S, where

$$S(m, b) = (m + b - 3)^2 + (4m + b - 5)^2 + (6m + b - 7)^2.$$

To do so, we find $\dfrac{\partial S}{\partial m}$ and $\dfrac{\partial S}{\partial b}$ and apply the D-test:

$$\frac{\partial S}{\partial m} = 2(m + b - 3) + 2(4m + b - 5)4 + 2(6m + b - 7)6 = 106m + 22b - 130,$$

$$\frac{\partial S}{\partial b} = 2(m + b - 3) + 2(4m + b - 5) + 2(6m + b - 7) = 22m + 6b - 30.$$

Setting the partial derivatives equal to 0 and solving, we find that the solution of the system is

$$m = \frac{15}{19}, \quad b = \frac{40}{19}.$$ You can confirm this.

The D-test, which the student can perform, confirms that $\left(\tfrac{15}{19}, \tfrac{40}{19}\right)$ does, in fact, minimize S. Thus, the equation of the regression line is $y = \tfrac{15}{19}x + \tfrac{40}{19}$. ✦

Partial derivatives also play an important role in determining a maximum or minimum value of a function of two variables that is subject to a *constraint*. (p. 574)

The Method of Lagrange Multipliers

To find a maximum or minimum value of a function $f(x, y)$ subject to the constraint $g(x, y) = 0$:

1. Form a new function:

   $$F(x, y, \lambda) = f(x, y) - \lambda g(x, y).$$

   The variable λ (lambda) is called a **Lagrange multiplier.**
2. Find the first partial derivatives F_x, F_y, and F_λ.
3. Solve the system

   $$F_x = 0, \quad F_y = 0, \quad \text{and} \quad F_\lambda = 0.$$

   Let (a, b, λ) represent a solution of this system. We normally must determine whether (a, b, λ) yields a maximum or minimum of the function f. For the problems in this text, we will specify that a maximum or minimum exists. (p. 575)

Reversing the process of finding partial derivatives gives an antiderivative in which one variable is used and any others are treated as constants. In a **double iterated integral**, this process occurs more than once. To evaluate such an integral, we first evaluate the innermost integral, indicated by the innermost differential. (p. 581)

If $f(x, y)$ is defined over the rectangular region R bounded by $x = a$, $x = b$, $y = c$, and $y = d$, then the *double integral* of $f(x, y)$ over R is given by

$$\int_c^d \int_a^b f(x, y)\,dx\,dy \quad \text{or} \quad \int_a^b \int_c^d f(x, y)\,dy\,dx.$$ (p. 582)

Chapter Review Exercises

These review exercises are for test preparation. They can also be used as a practice test. Answers are at the back of the book. The bracketed section references tell you what part(s) of the chapter to restudy if your answer is incorrect.

CONCEPT REINFORCEMENT

Match each expression in column A with an equivalent expression in column B. Assume that $z = f(x, y)$. [6.2, 6.6]

Column A	Column B
1. $\frac{\partial z}{\partial x}$	**a)** $\int_2^3 \frac{1}{2}x\,dx$
2. $\frac{\partial z}{\partial y}$	**b)** f_{yx}
3. $\frac{\partial}{\partial x}(5x^3y^7)$	**c)** f_{xy}
4. $\frac{\partial}{\partial y}(5x^3y^7)$	**d)** $\int_2^3 y^3\,dy$
5. $\frac{\partial^2 z}{\partial x\,\partial y}$	**e)** f_x
6. $\frac{\partial^2 z}{\partial y\,\partial x}$	**f)** $15x^2y^7$
7. $\int_2^3\int_0^1 2xy^3\,dx\,dy$	**g)** $35x^3y^6$
8. $\int_2^3\int_0^1 2xy^3\,dy\,dx$	**h)** f_y

REVIEW EXERCISES

Given $f(x, y) = e^y + 3xy^3 + 2y$, find each of the following. [6.1, 6.2]

9. $f(2, 0)$ **10.** f_x **11.** f_y
12. f_{xy} **13.** f_{yx} **14.** f_{xx}
15. f_{yy}

Given $z = 2x^3 \ln y + xy^2$, find each of the following. [6.2]

16. $\frac{\partial z}{\partial x}$ **17.** $\frac{\partial z}{\partial y}$ **18.** $\frac{\partial^2 z}{\partial x\,\partial y}$
19. $\frac{\partial^2 z}{\partial y\,\partial x}$ **20.** $\frac{\partial^2 z}{\partial x^2}$ **21.** $\frac{\partial^2 z}{\partial y^2}$

Find the relative maximum and minimum values. [6.3]

22. $f(x, y) = x^3 - 6xy + y^2 + 6x + 3y - \frac{1}{5}$
23. $f(x, y) = x^2 - xy + y^2 - 2x + 4y$
24. $f(x, y) = 3x - 6y - x^2 - y^2$
25. $f(x, y) = x^4 + y^4 + 4x - 32y + 80$

26. Consider the data in the following table regarding enrollment in colleges and universities during a recent three-year period. [6.4]

Year, x	Enrollment, y (in millions)
1	7.2
2	8.0
3	8.4

a) Find the regression line, $y = mx + b$.
b) Use the regression line to predict enrollment in the fourth year.

27. Consider the data in the table below regarding workers' average monthly out-of-pocket premium for health insurance for a family. [6.4]

a) Find the regression line, $y = mx + b$.
b) Use the regression line to predict workers' average monthly out-of-pocket premium for health insurance for a family in 2012.

Year, x	Workers' average monthly out-of-pocket premium for health insurance for a family
0 (1999)	\$129
2 (2001)	149
4 (2003)	201
5 (2004)	222
6 (2005)	226

(*Source*: Kaiser Family Foundation and *The New York Times*, 10/23/05.)

28. Find the minimum value of

$$f(x, y) = x^2 - 2xy + 2y^2 + 20$$

subject to the constraint $2x - 6y = 15$. [6.5]

29. Find the maximum value of

$$f(x, y) = 6xy$$

subject to the constraint $2x + y = 20$. [6.5]

Evaluate. [6.6]

30. $\int_0^1 \int_1^2 x^2 y^3 \, dy \, dx$

31. $\int_0^1 \int_{x^2}^x (x - y) \, dy \, dx$

SYNTHESIS

32. Evaluate

$$\int_0^2 \int_{1-2x}^{1-x} \int_0^{\sqrt{2-x^2}} z \, dz \, dy \, dx. \quad [6.6]$$

33. **Business: minimizing surface area.** Suppose that beverages could be packaged in either a cylindrical container or a rectangular container with a square top and bottom. Each container is designed to be of minimum surface area for its shape. If we assume a volume of 26 in^3, which container would have the smaller surface area? [6.3, 6.5]

TECHNOLOGY CONNECTION

34. Use a 3D graphics program to graph

$$f(x, y) = x^2 + 4y^2. \quad [6.1]$$

Chapter 6 Test

Given $f(x, y) = e^x + 2x^3y + y$, find each of the following.

1. $f(-1, 2)$
2. $\dfrac{\partial f}{\partial x}$
3. $\dfrac{\partial f}{\partial y}$
4. $\dfrac{\partial^2 f}{\partial x^2}$
5. $\dfrac{\partial^2 f}{\partial x \, \partial y}$
6. $\dfrac{\partial^2 f}{\partial y \, \partial x}$
7. $\dfrac{\partial^2 f}{\partial y^2}$

Find the relative maximum and minimum values.

8. $f(x, y) = x^2 - xy + y^3 - x$
9. $f(x, y) = 4y^2 - x^2$
10. **Business: predicting total sales.** Consider the data in the following table regarding the total sales of a company during the first three years of operation.

Year, x	Sales, y (in millions)
1	\$10
2	15
3	19

a) Find the regression line, $y = mx + b$.
b) Use the regression line to predict sales in the fourth year.

11. Find the maximum value of

$$f(x, y) = 6xy - 4x^2 - 3y^2$$

subject to the constraint $x + 3y = 19$.

12. Evaluate

$$\int_0^3 \int_1^3 4x^3y^2 \, dx \, dy.$$

SYNTHESIS

13. **Business: maximizing production.** Southwest Appliances has the following Cobb–Douglas production function for a certain product:

$$p(x, y) = 50x^{2/3}y^{1/3},$$

where x is labor, measured in dollars, and y is capital, measured in dollars. Suppose that Southwest can make a total investment in labor and capital of \$600,000. How should it allocate the investment between labor and capital in order to maximize production?

14. Find f_x and f_t:

$$f(x, t) = \frac{x^2 - 2t}{x^3 + 2t}.$$

TECHNOLOGY CONNECTION

15. Use a 3D graphics program to graph

$$f(x, y) = x - \tfrac{1}{2}y^2 - \tfrac{1}{3}x^3.$$

EXTENDED TECHNOLOGY APPLICATION

Minimizing Employees' Travel Time in a Building

If employees spend considerable time moving between offices, designing a building to minimize travel time can reap enormous savings for a company.

For a multilevel building with a square base, one design concern is minimizing travel time between the most remote points. We will make use of Lagrange multipliers to help design such a building.

Let's assume that each floor has a square grid of hallways, as shown in the figure at the upper right. Suppose that you are standing at point P in the top northeast corner of the twelfth floor of this building. How long will it take to reach the most remote point at the southwest corner on the first floor—that is, point Q?

Let's call the time t. We find a formula for t in two steps:

1. You are to go from the twelfth floor to the first floor. This is a move in a vertical direction.
2. You need to cross horizontally from one corner of the building to the other.

The vertical time is h, the height of point P from the ground, divided by a, the speed at which you can travel in a vertical direction (elevator speed). Thus, vertical time is given by h/a.

The horizontal time is the time it takes to go across one level, by way of the square grid of hallways (from R to Q in the figure). If each floor is a square with side of length k, then the distance from R to Q is $2k$. If the walking speed is b, then the horizontal time is given by $2k/b$.

Thus, the time it will take to go from P to Q is a function of two variables, h and k, given by

$$t(h, k) = \text{vertical time} + \text{horizontal time}$$
$$= \frac{h}{a} + \frac{2k}{b},$$

where a and b, the elevator speed and walking speed, are constants.

What happens if we must choose between two (or more) building plans with the same floor area, but with different dimensions? Will the travel time be the same? Or will it be different for the two buildings? First, what is the total floor area of a given building? Suppose that the building has n floors, each a square of side k. Then the total floor area is given by

$$A = nk^2.$$

Note that the area of the roof is not included.

If h is the height of point P and c is the height of each floor—that is, the distance from the carpeting on one floor to the carpeting on the floor above—then $n = 1 + h/c$, with

$$A = (1 + h/c)\,k^2.$$

Let's return to the problem of two buildings with the same total floor area, but with different dimensions, and see what happens to $t(h, k)$.

EXERCISES

1. Use the TABLE feature on your calculator or spreadsheet software to complete the table below. For each case in the table, let the elevator speed $a = 10$ ft/sec, the walking speed $b = 4$ ft/sec, and the height of each floor $c = 15$ ft. Each case in the table covers two situations, though the floor area stays essentially the same for a particular case.

CASE	BUILDING	n	k	A	h	$t(h, k)$
1	B1	2	40	3200	15	21.5
	B2	3	32.66	3200	30	19.4
2	B1	2	60	7200		
	B2	3	48.99			
3	B1	4	40			
	B2	5	35.777			
4	B1	5	60			
	B2	10	42.426			
5	B1	5	150			
	B2	10	106.066			
6	B1	10	40			
	B2	17	30.679			
7	B1	10	80			
	B2	17	61.357			
8	B1	17	40			
	B2	26	32.344			
9	B1	17	50			
	B2	26	40.43			
10	B1	26	77			
	B2	50	55.525			

2. Do different dimensions, with a fixed floor area, yield different travel times?

In Exercises 3–5, assume that you are finding the dimensions of a multilevel building with a square base that will minimize travel time t between the most remote points in the building. Each floor has a square grid of hallways. The height of point P is h, and the length of a side of each floor is k. The elevator speed is 10 ft/sec and the average speed of a person walking is 4 ft/sec. The total floor area of the building is 40,000 ft^2. The height of each floor is 12 ft.

3. Use the information given to find a formula for the function $t(h, k)$.

4. Find a formula for the constraint.

5. Use the method of Lagrange multipliers to find the dimensions of the building that will minimize travel time t between the most remote points in the building.

6. Use a 3D graphics program to graph both equations in Exercises 3 and 4. Then visually check the results you found analytically.

Cumulative Review

1. Write an equation of the line with slope -4 and containing the point $(-7, 1)$.
2. For $f(x) = x^2 - 5$, find $f(x + h)$.
3. a) Graph:

$$f(x) = \begin{cases} 5 - x, & \text{for } x \neq 2, \\ -3, & \text{for } x = 2. \end{cases}$$

   b) Find $\lim_{x \to 2} f(x)$.

   c) Find $f(2)$.

   d) Is f continuous at 2?

Find each limit, if it exists. If a limit does not exist, state that fact.

4. $\lim_{x \to -4} \dfrac{x^2 - 16}{x + 4}$
5. $\lim_{x \to 1} \sqrt{x^3 + 8}$
6. $\lim_{x \to 3} \dfrac{4}{x - 3}$
7. $\lim_{x \to \infty} \dfrac{12x - 7}{3x + 2}$
8. $\lim_{x \to \infty} \dfrac{2x^3 - x}{8x^5 - x^2 + 1}$
9. If $f(x) = x^2 + 3$, find $f'(x)$ by determining $\lim_{h \to 0} \dfrac{f(x + h) - f(x)}{h}$.

Differentiate.

10. $y = -9x + 3$
11. $y = x^2 - 7x + 3$
12. $y = x^{1/4}$
13. $f(x) = x^{-6}$
14. $f(x) = \sqrt[3]{2x^5 - 8}$
15. $f(x) = \dfrac{5x^3 + 4}{2x - 1}$
16. $y = \ln (x^2 + 5)$
17. $y = e^{\ln x}$
18. $y = e^{3x} + x^2$
19. $y = e^{\sqrt{x - 3}}$
20. $f(x) = \ln (e^x - 4)$
21. For $y = x^2 - \dfrac{2}{x}$, find d^2y/dx^2.
22. **Business: average cost.** Doubletake Clothing finds that the cost, in dollars, of producing x pairs of jeans is given by $C(x) = 320 + 9\sqrt{x}$. Find the rate at which the average cost is changing when 100 pairs of jeans have been produced.
23. Differentiate implicitly to find dy/dx if $x^3 + x/y = 7$.
24. Find an equation of the tangent line to the graph of $y = e^x - x^2 - 3$ at the point $(0, -2)$.

Sketch the graph of each function. List and label the coordinates of any extrema and points of inflection. State where the function is increasing or decreasing, where it is concave up or concave down, and where any asymptotes occur.

25. $f(x) = x^3 - 3x + 1$
26. $f(x) = 2x^2 - x^4 - 3$
27. $f(x) = \dfrac{8x}{x^2 + 1}$
28. $f(x) = \dfrac{8}{x^2 - 4}$

Find the absolute maximum and minimum values, if they exist, over the indicated interval. If no interval is indicated, consider the entire real number line.

29. $f(x) = 3x^2 - 6x - 4$
30. $f(x) = -5x + 1$
31. $f(x) = \frac{1}{3}x^3 - x^2 - 3x + 5$; $[-2, 0]$

32. Business: maximizing profit. For custom sweatshirts, Detailed Clothing's total revenue and total cost, in dollars, are given by

$$R(x) = 4x^2 + 11x + 110,$$

$$C(x) = 4.2x^2 + 5x + 10.$$

Find the number of sweatshirts, x, that must be produced and sold in order to maximize profit.

33. Business: minimizing inventory costs. An appliance store sells 450 MP3 players each year. It costs \$4 to store a player for a year. When placing an order, there is a fixed cost of \$1 plus \$0.75 for each player. How many times per year should the store reorder MP3 players, and in what lot size, in order to minimize inventory costs?

34. For $f(x) = 3x^2 - 7$, $x = 5$, and $\Delta x = 0.1$, find Δy and $f'(x)\,\Delta x$.

35. Business: exponential growth. A national frozen yogurt firm is experiencing growth of 10% per year in the number, N, of franchises that it owns; that is,

$$\frac{dN}{dt} = 0.1N,$$

where N is the number of franchises and t is the time, in years, from 2001.

a) Given that there were 8000 franchises in 2001, find the solution of the equation, assuming that $N_0 = 8000$ and $k = 0.1$.

b) How many franchises will there be in 2009?

c) What is the doubling time of the number of franchises?

36. Economics: elasticity of demand. Consider the demand function

$$q = D(x) = 240 - 20x,$$

where q is the quantity of coffee mugs demanded at a price of x dollars.

a) Find the elasticity.

b) Find the elasticity at $x = \$2$, and state whether the demand is elastic or inelastic.

c) Find the elasticity at $x = \$9$, and state whether the demand is elastic or inelastic.

d) At a price of \$2, will a small increase in price cause total revenue to increase or decrease?

e) Find the value of x for which the total revenue is a maximum.

Evaluate.

37. $\int 3x^5\,dx$

38. $\int_{-1}^{0} (2e^x + 1)\,dx$

39. $\int \frac{x}{(7-3x)^2}\,dx$ (Use Table 1.)

40. $\int x^3 e^{x^4}\,dx$ (Do not use Table 1.)

41. $\int (x+3)\ln x\,dx$

42. $\int \frac{75}{x}\,dx$

43. $\int_0^1 3\sqrt{x}\,dx$

44. Find the area under the graph of $y = x^2 + 3x$ over the interval $[1, 5]$.

45. Business: present value. Find the present value of \$250,000 due in 30 yr at 6%, compounded continuously.

46. Business: accumulated present value. Find the accumulated present value of an investment made over a 50-yr period during which there is a continuous money flow of \$4500 and the current interest rate is 7.2%, compounded continuously.

47. Determine whether the following improper integral is convergent or divergent, and calculate its value if it is convergent:

$$\int_3^{\infty} \frac{1}{x^7}\,dx.$$

48. Given the probability density function

$$f(x) = \frac{3}{2x^2} \quad \text{over } [1, 3],$$

find $E(x)$.

49. Let x be a continuous random variable that is normally distributed with mean $\mu = 3$ and standard deviation $\sigma = 5$. Using Table 2, find $P(-2 \le x \le 8)$.

50. Economics: supply and demand. Demand and supply functions are given by

$$p = D(x) = (x - 20)^2$$

and

$$p = S(x) = x^2 + 10x + 50,$$

where p is the price per unit, in dollars, when x units are sold. Find the equilibrium point and the consumer's surplus.

51. Find the volume of the solid of revolution generated by rotating the region under the graph of

$$y = e^{-x}, \quad \text{from } x = 0 \text{ to } x = 5,$$

about the x-axis.

52. Solve the differential equation $dy/dx = xy$.

53. Consider the data in the following table.

Age of Business (in years)	1	3	5
Profit (in tens of thousands of dollars)	4	7	9

a) Find the regression line, $y = mx + b$.

b) Use the regression line to predict the profit when the business is 10 years old.

Given $f(x, y) = e^y + 4x^2y^3 + 3x$, *find each of the following.*

54. f_x

55. f_{yy}

56. Find the relative maximum and minimum values of $f(x, y) = 8x^2 - y^2$.

57. Maximize $f(x, y) = 4x + 2y - x^2 - y^2 + 4$, subject to the constraint $x + 2y = 9$.

58. Evaluate

$$\int_0^3 \int_{-1}^2 e^x \, dy \, dx.$$

Appendix

OBJECTIVES

- Manipulate exponential expressions.
- Multiply and factor algebraic expressions.
- Solve equations, inequalities, and applied problems.

Review of Basic Algebra

This appendix covers most of the algebraic topics essential to a study of calculus. It might be used in conjunction with Chapter R or as the need for certain skills arises throughout the book.

Exponential Notation

Let's review the meaning of an expression

$$a^n,$$

where a is any real number and n is an integer; that is, n is a number in the set $\{\ldots, -3, -2, -1, 0, 1, 2, 3, \ldots\}$. The number a is called the **base** and n is called the **exponent.** If n is greater than 1, then

$$a^n = \underbrace{a \cdot a \cdot a \cdots a.}_{n \text{ factors}}$$

In other words, a^n is the product of n **factors,** each of which is a.

EXAMPLE 1 Express each of the following without exponents:

a) 4^3 **b)** $(-2)^5$ **c)** $(-2)^4$ **d)** -2^4 **e)** $(1.08)^2$ **f)** $\left(\frac{1}{2}\right)^3$

Solution

a) $4^3 = 4 \cdot 4 \cdot 4 = 64$

b) $(-2)^5 = (-2)(-2)(-2)(-2)(-2) = -32$

c) $(-2)^4 = (-2)(-2)(-2)(-2) = 16$

d) $-2^4 = -(2^4) = -(2)(2)(2)(2) = -16$ The base is 2, not -2.

e) $(1.08)^2 = 1.08 \times 1.08 = 1.1664$

f) $\left(\frac{1}{2}\right)^3 = \frac{1}{2} \cdot \frac{1}{2} \cdot \frac{1}{2} = \frac{1}{8}$ ✦

We define an exponent of 1 as follows:

> $a^1 = a$, for any real number a.

In other words, any real number to the first power is that number itself.

We define an exponent of 0 as follows:

> $a^0 = 1$, for any nonzero real number a.

That is, any nonzero real number a to the zero power is 1.

EXAMPLE 2 Express without exponents:

a) $(-2x)^0$ **b)** $(-2x)^1$ **c)** $\left(\frac{1}{2}\right)^0$ **d)** e^0 **e)** e^1 **f)** $\left(\frac{1}{2}\right)^1$

Solution

a) $(-2x)^0 = 1$

b) $(-2x)^1 = -2x$

c) $\left(\frac{1}{2}\right)^0 = 1$

d) $e^0 = 1$

e) $e^1 = e$

f) $\left(\frac{1}{2}\right)^1 = \frac{1}{2}$ ✦

The meaning of a negative integer as an exponent is as follows:

> $a^{-n} = \frac{1}{a^n} = \left(\frac{1}{a}\right)^n$, for any nonzero real number a.

That is, any nonzero real number a to the $-n$ power is the reciprocal of a^n, or equivalently, (the reciprocal of $a)^n$.

EXAMPLE 3 Express without negative exponents:

a) 2^{-5} **b)** 10^{-3} **c)** $\left(\frac{1}{4}\right)^{-2}$ **d)** x^{-5} **e)** e^{-k} **f)** t^{-1}

Solution

a) $2^{-5} = \dfrac{1}{2^5} = \dfrac{1}{2 \cdot 2 \cdot 2 \cdot 2 \cdot 2} = \dfrac{1}{32}$

b) $10^{-3} = \dfrac{1}{10^3} = \dfrac{1}{10 \cdot 10 \cdot 10} = \dfrac{1}{1000}$, or 0.001

c) $\left(\dfrac{1}{4}\right)^{-2} = \left(\dfrac{4}{1}\right)^2 = 4^2 = 16$

d) $x^{-5} = \dfrac{1}{x^5}$

e) $e^{-k} = \dfrac{1}{e^k}$

f) $t^{-1} = \dfrac{1}{t^1} = \dfrac{1}{t}$ ✦

Properties of Exponents

Note the following:

$$\begin{aligned} b^5 \cdot b^{-3} &= (b \cdot b \cdot b \cdot b \cdot b) \cdot \frac{1}{b \cdot b \cdot b} \\ &= \frac{b \cdot b \cdot b}{b \cdot b \cdot b} \cdot b \cdot b \\ &= 1 \cdot b \cdot b = b^2. \end{aligned}$$

We can obtain the same result by adding the exponents. This is true in general.

THEOREM 1

For any nonzero real number a and any integers n and m,

$$a^n \cdot a^m = a^{n+m}.$$

(To multiply when the bases are the same, add the exponents.)

EXAMPLE 4 Multiply:

a) $x^5 \cdot x^6$ **b)** $x^{-5} \cdot x^6$ **c)** $2x^{-3} \cdot 5x^{-4}$ **d)** $r^2 \cdot r$

Solution

a) $x^5 \cdot x^6 = x^{5+6} = x^{11}$

b) $x^{-5} \cdot x^6 = x^{-5+6} = x$

c) $2x^{-3} \cdot 5x^{-4} = 10x^{-3+(-4)} = 10x^{-7}$, or $\dfrac{10}{x^7}$

d) $r^2 \cdot r = r^{2+1} = r^3$ ✦

Note the following:

$$\begin{aligned} b^5 \div b^2 &= \frac{b^5}{b^2} = \frac{b \cdot b \cdot b \cdot b \cdot b}{b \cdot b} \\ &= \frac{b \cdot b}{b \cdot b} \cdot b \cdot b \cdot b \\ &= 1 \cdot b \cdot b \cdot b = b^3. \end{aligned}$$

We can obtain the same result by subtracting the exponents. This is true in general.

THEOREM 2

For any nonzero real number a and any integers n and m,

$$\frac{a^n}{a^m} = a^{n-m}.$$

(To divide when the bases are the same, subtract the exponent in the denominator from the exponent in the numerator.)

EXAMPLE 5 Divide:

a) $\dfrac{a^3}{a^2}$ **b)** $\dfrac{x^7}{x^7}$ **c)** $\dfrac{e^3}{e^{-4}}$ **d)** $\dfrac{e^{-4}}{e^{-1}}$

Solution

a) $\dfrac{a^3}{a^2} = a^{3-2} = a^1 = a$ **b)** $\dfrac{x^7}{x^7} = x^{7-7} = x^0 = 1$

c) $\dfrac{e^3}{e^{-4}} = e^{3-(-4)} = e^{3+4} = e^7$ **d)** $\dfrac{e^{-4}}{e^{-1}} = e^{-4-(-1)} = e^{-4+1} = e^{-3}$, or $\dfrac{1}{e^3}$

◆

Note the following:

$$(b^2)^3 = b^2 \cdot b^2 \cdot b^2 = b^{2+2+2} = b^6.$$

We can obtain the same result by multiplying the exponents. The other results in Theorem 3 can be similarly motivated.

THEOREM 3

For any nonzero real numbers a and b, and any integers n and m,

$$(a^n)^m = a^{nm}, \qquad (ab)^n = a^n b^n, \quad \text{and} \quad \left(\frac{a}{b}\right)^n = \frac{a^n}{b^n}.$$

EXAMPLE 6 Simplify:

a) $(x^{-2})^3$ **b)** $(e^x)^2$ **c)** $(2x^4y^{-5}z^3)^{-3}$ **d)** $\left(\dfrac{x^2}{p^4q^5}\right)^3$

Solution

a) $(x^{-2})^3 = x^{-2 \cdot 3} = x^{-6}$, or $\frac{1}{x^6}$

b) $(e^x)^2 = e^{2x}$

c) $(2x^4y^{-5}z^3)^{-3} = 2^{-3}(x^4)^{-3}(y^{-5})^{-3}(z^3)^{-3}$

$$= \frac{1}{2^3}x^{-12}y^{15}z^{-9}, \quad \text{or} \quad \frac{y^{15}}{8x^{12}z^9}$$

d) $\left(\frac{x^2}{p^4q^5}\right)^3 = \frac{(x^2)^3}{(p^4q^5)^3} = \frac{x^6}{(p^4)^3(q^5)^3} = \frac{x^6}{p^{12}q^{15}}$ ✦

Multiplication

The distributive law is important when multiplying. This law is as follows.

The Distributive Law

For any numbers A, B, and C,

$$A(B + C) = AB + AC.$$

Because subtraction can be regarded as addition of an additive inverse, it follows that

$$A(B - C) = AB - AC.$$

EXAMPLE 7 Multiply:

a) $3(x - 5)$ **b)** $P(1 + i)$ **c)** $(x - 5)(x + 3)$ **d)** $(a + b)(a + b)$

Solution

a) $3(x - 5) = 3 \cdot x - 3 \cdot 5 = 3x - 15$

b) $P(1 + i) = P \cdot 1 + P \cdot i = P + Pi$

c) $(x - 5)(x + 3) = (x - 5)x + (x - 5)3$

$$= x \cdot x - 5x + 3x - 5 \cdot 3$$
$$= x^2 - 2x - 15$$

d) $(a + b)(a + b) = (a + b)a + (a + b)b$

$$= a \cdot a + ba + ab + b \cdot b$$
$$= a^2 + 2ab + b^2$$ ✦

The following formulas, which are obtained using the distributive law, are also useful when multiplying. All three are used in Example 8, which follows.

$$(A + B)^2 = A^2 + 2AB + B^2$$
$$(A - B)^2 = A^2 - 2AB + B^2$$
$$(A - B)(A + B) = A^2 - B^2$$

EXAMPLE 8 Multiply:

a) $(x + h)^2$ **b)** $(2x - t)^2$ **c)** $(3c + d)(3c - d)$

Solution

a) $(x + h)^2 = x^2 + 2xh + h^2$

b) $(2x - t)^2 = (2x)^2 - 2(2x)t + t^2 = 4x^2 - 4xt + t^2$

c) $(3c + d)(3c - d) = (3c)^2 - d^2 = 9c^2 - d^2$ ✦

Factoring

Factoring is the reverse of multiplication. That is, to factor an expression, we find an equivalent expression that is a product. Always remember to look first for a common factor.

EXAMPLE 9 Factor:

a) $P + Pi$ **b)** $2xh + h^2$ **c)** $x^2 - 6xy + 9y^2$

d) $x^2 - 5x - 14$ **e)** $6x^2 + 7x - 5$ **f)** $x^2 - 9t^2$

Solution

a) $P + Pi = P \cdot 1 + P \cdot i = P(1 + i)$ We used the distributive law.

b) $2xh + h^2 = h(2x + h)$

c) $x^2 - 6xy + 9y^2 = (x - 3y)^2$

d) $x^2 - 5x - 14 = (x - 7)(x + 2)$ We looked for factors of -14 whose sum is -5.

e) $6x^2 + 7x - 5 = (2x - 1)(3x + 5)$ We first considered ways of factoring the first coefficient—for example, $(2x \quad)(3x \quad)$. Then we looked for factors of -5 such that when we multiply, we obtain the given expression.

f) $x^2 - 9t^2 = (x - 3t)(x + 3t)$ We used the formula $(A - B)(A + B) = A^2 - B^2$. ✦

Some expressions with four terms can be factored by first looking for a common binomial factor. This is called **factoring by grouping.**

EXAMPLE 10 Factor:

a) $t^3 + 6t^2 - 2t - 12$ **b)** $x^3 - 7x^2 - 4x + 28$

Solution

a) $t^3 + 6t^2 - 2t - 12 = t^2(t + 6) - 2(t + 6)$ Factoring the first two terms and then the second two terms

$= (t^2 - 2)(t + 6)$ Factoring out the common binomial factor, $t + 6$

b) $x^3 - 7x^2 - 4x + 28 = x^2(x - 7) - 4(x - 7)$ Factoring the first two terms and then the second two terms

$= (x - 7)(x^2 - 4)$ Factoring out the common binomial factor, $x - 7$

$= (x - 7)(x - 2)(x + 2)$ Using $(A - B)(A + B) = A^2 - B^2$

✦

Solving Equations

Basic to the solution of many equations are the *Addition Principle* and the *Multiplication Principle*. We can add (or subtract) the same number on both sides of an equation and obtain an equivalent equation, that is, a new equation that has the same solutions as the original equation. We can also multiply (or divide) by a nonzero number on both sides of an equation and obtain an equivalent equation.

The Addition Principle

For any real numbers a, b, and c,

$$a = b \quad \text{is equivalent to} \quad a + c = b + c.$$

The Multiplication Principle

For any real numbers a, b, and c, with $c \neq 0$,

$$a = b \quad \text{is equivalent to} \quad a \cdot c = b \cdot c.$$

When solving a linear equation, we use these equation-solving principles and other properties of real numbers to get the variable alone on one side. Then it is easy to determine the solution.

EXAMPLE 11 Solve: $-\frac{5}{6}x + 10 = \frac{1}{2}x + 2$.

Solution We first multiply by 6 on both sides to clear the fractions:

$$6\left(-\tfrac{5}{6}x + 10\right) = 6\left(\tfrac{1}{2}x + 2\right)$$ Using the Multiplication Principle

$$6\left(-\tfrac{5}{6}x\right) + 6 \cdot 10 = 6\left(\tfrac{1}{2}x\right) + 6 \cdot 2$$ Using the distributive law

$$-5x + 60 = 3x + 12$$ Simplifying

$$60 = 8x + 12$$ Using the Addition Principle: We add $5x$ on both sides.

$$48 = 8x$$ Adding -12 on both sides

$$\tfrac{1}{8} \cdot 48 = \tfrac{1}{8} \cdot 8x$$ Multiplying by $\frac{1}{8}$ on both sides

$$6 = x.$$

The variable is now alone on one side, and we see that 6 is the solution. We can check by substituting 6 into the original equation. ✦

The third principle for solving equations is the *Principle of Zero Products*.

The Principle of Zero Products

For any numbers a and b, if $ab = 0$, then $a = 0$ or $b = 0$; and if $a = 0$ or $b = 0$, then $ab = 0$.

To solve an equation using this principle, we must have a 0 on one side and a product on the other. The solutions are then obtained by setting each factor equal to 0 and solving the resulting equations.

EXAMPLE 12 Solve: $3x(x - 2)(5x + 4) = 0$.

Solution We have

$$3x(x - 2)(5x + 4) = 0$$

$$3x = 0 \quad or \quad x - 2 = 0 \quad or \quad 5x + 4 = 0 \qquad \text{Using the Principle of Zero Products}$$

$$\tfrac{1}{3} \cdot 3x = \tfrac{1}{3} \cdot 0 \quad or \quad x = 2 \quad or \quad 5x = -4 \qquad \text{Solving each separately}$$

$$x = 0 \quad or \quad x = 2 \quad or \quad x = -\tfrac{4}{5}.$$

The solutions are 0, 2, and $-\frac{4}{5}$. ✦

Note that the Principle of Zero Products applies *only* when a product is 0. For example, although we may know that $ab = 8$, *we do not know* that $a = 8$ or $b = 8$.

EXAMPLE 13 Solve: $4x^3 = x$.

Solution We have

$$4x^3 = x$$

$$4x^3 - x = 0 \qquad \text{Adding } -x \text{ to both sides}$$

$$x(4x^2 - 1) = 0$$

$$x(2x - 1)(2x + 1) = 0 \qquad \text{Factoring}$$

$$x = 0 \quad or \quad 2x - 1 = 0 \quad or \quad 2x + 1 = 0 \qquad \text{Using the Principle of Zero Products}$$

$$x = 0 \quad or \quad 2x = 1 \quad or \quad 2x = -1$$

$$x = 0 \quad or \quad x = \tfrac{1}{2} \quad or \quad x = -\tfrac{1}{2}.$$

The solutions are 0, $\frac{1}{2}$, and $-\frac{1}{2}$. ✦

Rational Equations

Expressions like the following are polynomials in one variable:

$$x^2 - 4, \qquad x^3 + 7x^2 - 8x + 9, \qquad t - 19.$$

The **least common multiple, LCM,** of two polynomials is found by factoring and using each factor the greatest number of times that it occurs in any one factorization.

EXAMPLE 14 Find the LCM: $x^2 + 2x + 1$, $5x^2 - 5x$, and $x^2 - 1$.

Solution

$$\left.\begin{aligned} x^2 + 2x + 1 &= (x + 1)(x + 1); \\ 5x^2 - 5x &= 5x(x - 1); \\ x^2 - 1 &= (x + 1)(x - 1) \end{aligned}\right\} \quad \text{Factoring}$$

$$\text{LCM} = 5x(x + 1)(x + 1)(x - 1)$$

✦

A **rational expression** is a ratio of polynomials. Each of the following is a rational expression:

$$\frac{x^2 - 6x + 9}{x^2 - 4}, \quad \frac{x - 2}{x - 3}, \quad \frac{a + 7}{a^2 - 16}, \quad \frac{5}{5t - 15}.$$

A **rational equation** is an equation containing one or more rational expressions. Here are some examples:

$$\frac{2}{3} - \frac{5}{6} = \frac{1}{x}, \qquad x + \frac{6}{x} = 5, \qquad \frac{2x}{x - 3} - \frac{6}{x} = \frac{18}{x^2 - 3x}.$$

To solve a rational equation, we first clear the equation of fractions by multiplying on both sides by the LCM of all the denominators. The resulting equation might have solutions that are *not* solutions of the original equation. Thus, we must check all possible solutions in the original equation.

EXAMPLE 15 Solve: $\dfrac{2x}{x - 3} - \dfrac{6}{x} = \dfrac{18}{x^2 - 3x}$.

Solution Note that $x^2 - 3x = x(x - 3)$. The LCM of the denominators is $x(x - 3)$. We multiply by $x(x - 3)$.

$$x(x - 3)\left(\frac{2x}{x - 3} - \frac{6}{x}\right) = x(x - 3)\left(\frac{18}{x^2 - 3x}\right) \qquad \text{Multiplying by the LCM on both sides}$$

$$x(x - 3) \cdot \frac{2x}{x - 3} - x(x - 3) \cdot \frac{6}{x} = x(x - 3)\left(\frac{18}{x^2 - 3x}\right) \qquad \text{Using the distributive law}$$

$$2x^2 - 6(x - 3) = 18 \qquad \text{Simplifying}$$

$$2x^2 - 6x + 18 = 18$$

$$2x^2 - 6x = 0$$

$$2x(x - 3) = 0$$

$$2x = 0 \quad or \quad x - 3 = 0$$

$$x = 0 \quad or \quad x = 3$$

The numbers 0 and 3 are possible solutions. We look at the original equation and see that each makes a denominator 0. We can also carry out a check, as follows.

Check

For 0:

$\dfrac{2x}{x - 3} - \dfrac{6}{x}$	$= \dfrac{18}{x^2 - 3x}$	
$\dfrac{2(0)}{0 - 3} - \dfrac{6}{0}$	$\overset{?}{}\ \dfrac{18}{0^2 - 3(0)}$	
$0 - \dfrac{6}{0}$	$\dfrac{18}{0}$	UNDEFINED; FALSE

For 3:

$\dfrac{2x}{x - 3} - \dfrac{6}{x}$	$= \dfrac{18}{x^2 - 3x}$	
$\dfrac{2(3)}{3 - 3} - \dfrac{6}{3}$	$\overset{?}{}\ \dfrac{18}{3^2 - 3(3)}$	
$\dfrac{6}{0} - 2$	$\dfrac{18}{0}$	UNDEFINED; FALSE

The equation has *no solution*. ✦

EXAMPLE 16 Solve: $\dfrac{x^2}{x-2} = \dfrac{4}{x-2}$.

Solution The LCM of the denominators is $x - 2$. We multiply by $x - 2$.

$$(x-2) \cdot \frac{x^2}{x-2} = (x-2) \cdot \frac{4}{x-2}$$

$$x^2 = 4 \qquad \text{Simplifying}$$

$$x^2 - 4 = 0$$

$$(x+2)(x-2) = 0$$

$$x = -2 \quad or \quad x = 2 \qquad \text{Using the Principle of Zero Products}$$

Check

For 2:

$$\frac{x^2}{x-2} = \frac{4}{x-2}$$

$\frac{2^2}{2-2}$?	$\frac{4}{2-2}$	
$\frac{4}{0}$	$\frac{4}{0}$	UNDEFINED; FALSE

For −2:

$$\frac{x^2}{x-2} = \frac{4}{x-2}$$

$\frac{(-2)^2}{-2-2}$?	$\frac{4}{-2-2}$	
$\frac{4}{-4}$	$\frac{4}{-4}$	
-1	-1	TRUE

The number -2 is a solution, but 2 is not (it results in division by 0). ✦

Solving Inequalities

Two inequalities are **equivalent** if they have the same solutions. For example, the inequalities $x > 4$ and $4 < x$ are equivalent. Principles for solving inequalities are similar to those for solving equations. We can add the same number to both sides of an inequality. We can also multiply on both sides by the same nonzero number, but if that number is negative, we must reverse the inequality sign. The following are the inequality-solving principles.

The Inequality-Solving Principles

For any real numbers a, b, and c,

$$a < b \quad \text{is equivalent to} \quad a + c < b + c.$$

For any real numbers a, b, and any *positive* number c,

$$a < b \quad \text{is equivalent to} \quad ac < bc.$$

For any real numbers a, b, and any *negative* number c,

$$a < b \quad \text{is equivalent to} \quad ac > bc.$$

Similar statements hold for $\leq$ and $\geq$.

EXAMPLE 17 Solve: $17 - 8x \geq 5x - 4$.

Solution We have

$$\begin{aligned} 17 - 8x &\geq 5x - 4 \\ -8x &\geq 5x - 21 && \text{Adding } -17 \text{ to both sides} \\ -13x &\geq -21 && \text{Adding } -5x \text{ to both sides} \\ -\tfrac{1}{13}(-13x) &\leq -\tfrac{1}{13}(-21) && \text{Multiplying both sides by } -\tfrac{1}{13} \text{ and } \textit{reversing} \text{ the inequality sign} \\ x &\leq \tfrac{21}{13}. \end{aligned}$$

Any number less than or equal to $\frac{21}{13}$ is a solution. ✦

Applications

To solve applied problems, we first translate to mathematical language, usually an equation. Then we solve the equation and check to see whether the solution to the equation is a solution to the problem.

EXAMPLE 18 *Life Science: Weight Gain.* After a 5% gain in weight, a grizzly bear weighs 693 lb. What was its original weight?

Solution We first translate to an equation:

$$\underbrace{(\textit{Original weight})}_{w} + 5\%\underbrace{(\textit{Original weight})}_{w} = 693$$

$$w + 5\% \quad w = 693.$$

Now we solve the equation:

$$\begin{aligned} w + 5\%w &= 693 \\ 1 \cdot w + 0.05w &= 693 \\ (1 + 0.05)w &= 693 \\ 1.05w &= 693 \\ w &= \frac{693}{1.05} = 660. \end{aligned}$$

Check: $600 + 5\% \cdot 660 = 660 + 0.05 \cdot 660 = 660 + 33 = 693.$

The original weight of the bear was 660 lb. ✦

EXAMPLE 19 *Business: Total Sales.* Raggs, Ltd., a clothing firm, determines that its total revenue, in dollars, from the sale of x suits is given by

$$200x + 50.$$

Determine the number of suits that the firm must sell to ensure that its total revenue will be more than \$70,050.

Solution We translate to an inequality and solve:

$$\begin{aligned} 200x + 50 &> 70{,}050 \\ 200x &> 70{,}000 && \text{Adding } -50 \text{ to both sides} \\ x &> 350. && \text{Multiplying both sides by } \tfrac{1}{200} \end{aligned}$$

Thus the company's total revenue will exceed \$70,050 when it sells more than 350 suits. ✦

Exercise Set A

Express as an equivalent expression without exponents.

1. 5^3
2. 7^2
3. $(-7)^2$
4. $(-5)^3$
5. $(1.01)^2$
6. $(1.01)^3$
7. $\left(\frac{1}{2}\right)^4$
8. $\left(\frac{1}{4}\right)^3$
9. $(6x)^0$
10. $(6x)^1$
11. t^1
12. t^0
13. $\left(\frac{1}{3}\right)^0$
14. $\left(\frac{1}{3}\right)^1$

Express as an equivalent expression without negative exponents.

15. 3^{-2}
16. 4^{-2}
17. $\left(\frac{1}{2}\right)^{-3}$
18. $\left(\frac{1}{2}\right)^{-2}$
19. 10^{-1}
20. 10^{-4}
21. e^{-b}
22. t^{-k}
23. b^{-1}
24. h^{-1}

Multiply.

25. $x^2 \cdot x^3$
26. $t^3 \cdot t^4$
27. $x^{-7} \cdot x$
28. $x^5 \cdot x$
29. $5x^2 \cdot 7x^3$
30. $4t^3 \cdot 2t^4$
31. $x^{-4} \cdot x^7 \cdot x$
32. $x^{-3} \cdot x \cdot x^3$
33. $e^{-t} \cdot e^t$
34. $e^k \cdot e^{-k}$

Divide.

35. $\frac{x^5}{x^2}$
36. $\frac{x^7}{x^3}$
37. $\frac{x^2}{x^5}$
38. $\frac{x^3}{x^7}$
39. $\frac{e^k}{e^k}$
40. $\frac{t^k}{t^k}$
41. $\frac{e^t}{e^4}$
42. $\frac{e^k}{e^3}$
43. $\frac{t^6}{t^{-8}}$
44. $\frac{t^5}{t^{-7}}$
45. $\frac{t^{-9}}{t^{-11}}$
46. $\frac{t^{-11}}{t^{-7}}$
47. $\frac{ab(a^2b)^3}{ab^{-1}}$
48. $\frac{x^2y^3(xy^3)^2}{x^{-3}y^2}$

Simplify.

49. $(t^{-2})^3$
50. $(t^{-3})^4$
51. $(e^x)^4$
52. $(e^x)^5$
53. $(2x^2y^4)^3$
54. $(2x^2y^4)^5$
55. $(3x^{-2}y^{-5}z^4)^{-4}$
56. $(5x^3y^{-7}z^{-5})^{-3}$
57. $(-3x^{-8}y^7z^2)^2$
58. $(-5x^4y^{-5}z^{-3})^4$
59. $\left(\frac{cd^3}{2q^2}\right)^4$
60. $\left(\frac{4x^2y}{a^3b^3}\right)^3$

Multiply.

61. $5(x-7)$
62. $x(1+t)$
63. $(x-5)(x-2)$
64. $(x-4)(x-3)$
65. $(a-b)(a^2+ab+b^2)$
66. $(x^2-xy+y^2)(x+y)$
67. $(2x+5)(x-1)$
68. $(3x+4)(x-1)$
69. $(a-2)(a+2)$
70. $(3x-1)(3x+1)$
71. $(5x+2)(5x-2)$
72. $(t-1)(t+1)$
73. $(a-h)^2$
74. $(a+h)^2$
75. $(5x+t)^2$
76. $(7a-c)^2$
77. $5x(x^2+3)^2$
78. $-3x^2(x^2-4)(x^2+4)$

Use the following equation for Exercises 79–82.

$$\begin{aligned}(x+h)^3 &= (x+h)(x+h)^2\\ &= (x+h)(x^2+2xh+h^2)\\ &= (x+h)x^2+(x+h)2xh+(x+h)h^2\\ &= x^3+x^2h+2x^2h+2xh^2+xh^2+h^3\\ &= x^3+3x^2h+3xh^2+h^3\end{aligned}$$

79. $(a+b)^3$
80. $(a-b)^3$
81. $(x-5)^3$
82. $(2x+3)^3$

Factor.

83. $x-xt$
84. $x+xh$
85. $x^2+6xy+9y^2$
86. $x^2-10xy+25y^2$
87. $x^2-2x-15$
88. $x^2+8x+15$
89. x^2-x-20
90. $x^2-9x-10$
91. $49x^2-t^2$
92. $9x^2-b^2$
93. $36t^2-16m^2$
94. $25y^2-9z^2$
95. $a^3b-16ab^3$
96. $2x^4-32$
97. a^8-b^8
98. $36y^2+12y-35$
99. $10a^2x-40b^2x$
100. $x^3y-25xy^3$
101. $2-32x^4$
102. $2xy^2-50x$
103. $9x^2+17x-2$
104. $6x^2-23x+20$
105. x^3+8 (*Hint:* See Exercise 66.)
106. a^3-27 (*Hint:* See Exercise 65.)
107. y^3-64t^3
108. $m^3+1000p^3$
109. $3x^3-6x^2-x+2$
110. $5y^3+2y^2-10y-4$

111. $x^3 - 5x^2 - 9x + 45$

112. $t^3 + 3t^2 - 25t - 75$

Solve.

113. $-7x + 10 = 5x - 11$

114. $-8x + 9 = 4x - 70$

115. $5x - 17 - 2x = 6x - 1 - x$

116. $5x - 2 + 3x = 2x + 6 - 4x$

117. $x + 0.8x = 216$

118. $x + 0.5x = 210$

119. $x + 0.08x = 216$

120. $x + 0.05x = 210$

121. $2x(x + 3)(5x - 4) = 0$

122. $7x(x - 2)(2x + 3) = 0$

123. $x^2 + 1 = 2x + 1$

124. $2t^2 = 9 + t^2$

125. $t^2 - 2t = t$

126. $6x - x^2 = x$

127. $6x - x^2 = -x$

128. $2x - x^2 = -x$

129. $9x^3 = x$

130. $16x^3 = x$

131. $(x - 3)^2 = x^2 + 2x + 1$

132. $(x - 5)^2 = x^2 + x + 3$

133. $\frac{4x}{x + 5} + \frac{20}{x} = \frac{100}{x^2 + 5x}$

134. $\frac{x}{x + 1} + \frac{3x + 5}{x^2 + 4x + 3} = \frac{2}{x + 3}$

135. $\frac{50}{x} - \frac{50}{x - 2} = \frac{4}{x}$

136. $\frac{60}{x} = \frac{60}{x - 5} + \frac{2}{x}$

137. $0 = 2x - \frac{250}{x^2}$

138. $5 - \frac{35}{x^2} = 0$

139. $3 - x \le 4x + 7$

140. $x + 6 \le 5x - 6$

141. $5x - 5 + x > 2 - 6x - 8$

142. $3x - 3 + 3x > 1 - 7x - 9$

143. $-7x < 4$

144. $-5x \ge 6$

145. $5x + 2x \le -21$

146. $9x + 3x \ge -24$

147. $2x - 7 < 5x - 9$

148. $10x - 3 \ge 13x - 8$

149. $8x - 9 < 3x - 11$

150. $11x - 2 \ge 15x - 7$

151. $8 < 3x + 2 < 14$

152. $2 < 5x - 8 \le 12$

153. $3 \le 4x - 3 \le 19$

154. $9 \le 5x + 3 < 19$

155. $-7 \le 5x - 2 \le 12$

156. $-11 \le 2x - 1 < -5$

APPLICATIONS

Business and Economics

157. Investment increase. An investment is made at $8\frac{1}{2}\%$, compounded annually. It grows to \$705.25 at the end of 1 yr. How much was invested originally?

158. Investment increase. An investment is made at 7%, compounded annually. It grows to \$856 at the end of 1 yr. How much was invested originally?

159. Total revenue. Sunshine Products determines that the total revenue, in dollars, from the sale of x flowerpots is

$3x + 1000.$

Determine the number of flowerpots that must be sold so that the total revenue will be more than \$22,000.

160. Total revenue. Beeswax Inc. determines that the total revenue, in dollars, from the sale of x candles is

$5x + 1000.$

Determine the number of candles that must be sold so that the total revenue will be more than \$22,000.

Life and Physical Sciences

161. Weight gain. After a 6% gain in weight, an elk weighs 508.8 lb. What was its original weight?

162. Weight gain. After a 7% gain in weight, a deer weighs 363.8 lb. What was its original weight?

Social Sciences

163. Population increase. After a 2% increase, the population of a city is 826,200. What was the former population?

164. Population increase. After a 3% increase, the population of a city is 741,600. What was the former population?

General Interest

165. Grade average. To get a B in a course, a student's average must be greater than or equal to 80% (at least 80%) and less than 90%. On the first three tests, Claudia scores 78%, 90%, and 92%. Determine the scores on the fourth test that will guarantee her a B.

166. Grade average. To get a C in a course, a student's average must be greater than or equal to 70% and less than 80%. On the first three tests, Horace scores 65%, 83%, and 82%. Determine the scores on the fourth test that will guarantee him a C.

Tables

TABLE 1 Integration Formulas (see also: www.integrals.com)

1. $\int x^n \, dx = \frac{x^{n+1}}{n+1} + C, \quad n \neq -1$
2. $\int \frac{dx}{x} = \ln x + C, \quad x > 0$
3. $\int u \, dv = uv - \int v \, du$
4. $\int e^x \, dx = e^x + C$
5. $\int e^{ax} \, dx = \frac{1}{a} \cdot e^{ax} + C$
6. $\int xe^{ax} \, dx = \frac{1}{a^2} \cdot e^{ax}(ax - 1) + C$
7. $\int x^n e^{ax} \, dx = \frac{x^n e^{ax}}{a} - \frac{n}{a} \int x^{n-1} e^{ax} \, dx + C$
8. $\int \ln x \, dx = x \ln x - x + C$
9. $\int (\ln x)^n \, dx = x(\ln x)^n - n \int (\ln x)^{n-1} \, dx + C, \quad n \neq -1$
10. $\int x^n \ln x \, dx = x^{n+1} \left[\frac{\ln x}{n+1} - \frac{1}{(n+1)^2} \right] + C, \quad n \neq -1$
11. $\int a^x \, dx = \frac{a^x}{\ln a} + C, \quad a > 0, a \neq 1$
12. $\int \frac{1}{\sqrt{x^2 + a^2}} \, dx = \ln |x + \sqrt{x^2 + a^2}| + C$
13. $\int \frac{1}{\sqrt{x^2 - a^2}} \, dx = \ln |x + \sqrt{x^2 - a^2}| + C$

(continued)

TABLE 1 **Integration Formulas** *(continued)*

14. $\displaystyle\int \frac{1}{x^2 - a^2}\,dx = \frac{1}{2a}\ln\left|\frac{x-a}{x+a}\right| + C$

15. $\displaystyle\int \frac{1}{a^2 - x^2}\,dx = \frac{1}{2a}\ln\left|\frac{a+x}{a-x}\right| + C$

16. $\displaystyle\int \frac{1}{x\sqrt{a^2 + x^2}}\,dx = -\frac{1}{a}\ln\left|\frac{a+\sqrt{a^2+x^2}}{x}\right| + C$

17. $\displaystyle\int \frac{1}{x\sqrt{a^2 - x^2}}\,dx = -\frac{1}{a}\ln\left|\frac{a+\sqrt{a^2-x^2}}{x}\right| + C$

18. $\displaystyle\int \frac{x}{a + bx}\,dx = \frac{a}{b^2} + \frac{x}{b} - \frac{a}{b^2}\ln|a+bx| + C$

19. $\displaystyle\int \frac{x}{(a + bx)^2}\,dx = \frac{a}{b^2(a+bx)} + \frac{1}{b^2}\ln|a+bx| + C$

20. $\displaystyle\int \frac{1}{x(a + bx)}\,dx = \frac{1}{a}\ln\left|\frac{x}{a+bx}\right| + C$

21. $\displaystyle\int \frac{1}{x(a + bx)^2}\,dx = \frac{1}{a(a+bx)} + \frac{1}{a^2}\ln\left|\frac{x}{a+bx}\right| + C$

22. $\displaystyle\int \sqrt{x^2 \pm a^2}\,dx = \tfrac{1}{2}\left[x\sqrt{x^2 \pm a^2} \pm a^2\ln\left|x + \sqrt{x^2 \pm a^2}\right|\right] + C$

23. $\displaystyle\int x\sqrt{a + bx}\,dx = \frac{2}{15b^2}(3bx - 2a)(a+bx)^{3/2} + C$

24. $\displaystyle\int x^2\sqrt{a + bx}\,dx = \frac{2}{105b^3}(15b^2x^2 - 12abx + 8a^2)(a+bx)^{3/2} + C$

25. $\displaystyle\int \frac{x\,dx}{\sqrt{a + bx}} = \frac{2}{3b^2}(bx - 2a)\sqrt{a+bx} + C$

26. $\displaystyle\int \frac{x^2\,dx}{\sqrt{a + bx}} = \frac{2}{15b^3}(3b^2x^2 - 4abx + 8a^2)\sqrt{a+bx} + C$

TABLE 2 Areas for a Standard Normal Distribution

Entries in the table represent area under the curve between $z = 0$ and a positive value of z. Because of the symmetry of the curve, area under the curve between $z = 0$ and a negative value of z would be found in a similar manner.

$$\text{Area} = \text{Probability} = P(0 \le x \le z) = \int_0^z \frac{1}{\sqrt{2\pi}} e^{-x^2/2}\, dx$$

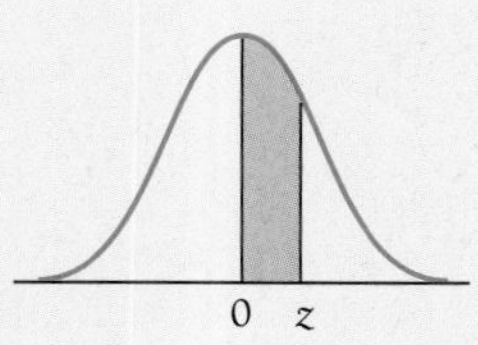

z	0.00	0.01	0.02	0.03	0.04	0.05	0.06	0.07	0.08	0.09
0.0	.0000	.0040	.0080	.0120	.0160	.0199	.0239	.0279	.0319	.0359
0.1	.0398	.0438	.0478	.0517	.0557	.0596	.0636	.0675	.0714	.0753
0.2	.0793	.0832	.0871	.0910	.0948	.0987	.1026	.1064	.1103	.1141
0.3	.1179	.1217	.1255	.1293	.1331	.1368	.1406	.1443	.1480	.1517
0.4	.1554	.1591	.1628	.1664	.1700	.1736	.1772	.1808	.1844	.1879
0.5	.1915	.1950	.1985	.2019	.2054	.2088	.2123	.2157	.2190	.2224
0.6	.2257	.2291	.2324	.2357	.2389	.2422	.2454	.2486	.2517	.2549
0.7	.2580	.2611	.2642	.2673	.2704	.2734	.2764	.2794	.2823	.2852
0.8	.2881	.2910	.2939	.2967	.2995	.3023	.3051	.3078	.3106	.3133
0.9	.3159	.3186	.3212	.3238	.3264	.3289	.3315	.3340	.3365	.3389
1.0	.3413	.3438	.3461	.3485	.3508	.3531	.3554	.3577	.3599	.3621
1.1	.3643	.3665	.3686	.3708	.3729	.3749	.3770	.3790	.3810	.3830
1.2	.3849	.3869	.3888	.3907	.3925	.3944	.3962	.3980	.3997	.4015
1.3	.4032	.4049	.4066	.4082	.4099	.4115	.4131	.4147	.4162	.4177
1.4	.4192	.4207	.4222	.4236	.4251	.4265	.4279	.4292	.4306	.4319
1.5	.4332	.4345	.4357	.4370	.4382	.4394	.4406	.4418	.4429	.4441
1.6	.4452	.4463	.4474	.4484	.4495	.4505	.4515	.4525	.4535	.4545
1.7	.4554	.4564	.4573	.4582	.4591	.4599	.4608	.4616	.4625	.4633
1.8	.4641	.4649	.4656	.4664	.4671	.4678	.4686	.4693	.4699	.4706
1.9	.4713	.4719	.4726	.4732	.4738	.4744	.4750	.4756	.4761	.4767
2.0	.4772	.4778	.4783	.4788	.4793	.4798	.4803	.4808	.4812	.4817
2.1	.4821	.4826	.4830	.4834	.4838	.4842	.4846	.4850	.4854	.4857
2.2	.4861	.4864	.4868	.4871	.4875	.4878	.4881	.4884	.4887	.4890
2.3	.4893	.4896	.4898	.4901	.4904	.4906	.4909	.4911	.4913	.4916
2.4	.4918	.4920	.4922	.4925	.4927	.4929	.4931	.4932	.4934	.4936
2.5	.4938	.4940	.4941	.4943	.4945	.4946	.4948	.4949	.4951	.4952
2.6	.4953	.4955	.4956	.4957	.4959	.4960	.4961	.4962	.4963	.4964
2.7	.4965	.4966	.4967	.4968	.4969	.4970	.4971	.4972	.4973	.4974
2.8	.4974	.4975	.4976	.4977	.4977	.4978	.4979	.4979	.4980	.4981
2.9	.4981	.4982	.4982	.4983	.4984	.4984	.4985	.4985	.4986	.4986
3.0	.4987	.4987	.4987	.4988	.4988	.4989	.4989	.4989	.4990	.4990

Photo Credits

1, © Royalty-Free/Corbis **12,** © Matt Inden, Corbis **42,** © Michael T. Sedam/ CORBIS **46,** © Skyscan/Corbis **53,** © Randy Faris/CORBIS **67,** digital vision, Getty Images **80,** PureStock, Getty Images **97,** © LUCY NICHOLSON/Reuters/ Corbis **99,** CMSP/Getty Images **118,** © William Manning/CORBIS **124,** Corbis/ Digital Stock **136,** © Phil Schermeister/CORBIS **170,** Brian Spurlock **194,** © Chuck Savage/CORBIS **196,** Bob Schutz, AP/Feature Photo **197,** © NASA/Corbis **212 left,** Chad Ehlers/Photographer's Choice/Getty Images **212 right,** © Paul A. Souders/CORBIS **229,** © Royalty-Free/Corbis4 **245,** © DAVE KAUP/Reuters/Corbis **257,** © Axel Koester/Corbis **266,** © Mark Peterson/CORBIS **273,** © Dana Hoff/Beateworks/Corbis **287,** David Ellenbogen **289,** © NASA/Corbis **290,** Custom Medical Stock Photo **301,** © Royalty-Free/ Corbis **302 top,** © Tom Brakefield/CORBIS **302 bottom,** © DLILLC/Corbis **303,** © Erik Freeland/Corbis **316,** © Erik Freeland/Corbis **338,** © Kapoor Baldev/Sygma/CORBIS **339,** © Najlah Feanny/Corbis **340,** © Royalty-Free/ Corbis **342,** © Bob Daemmrich, Photo Edit **344,** Photonica/Getty Images **347,** © Royalty-Free/Corbis **348,** AFP/Getty Images **350 left,** © Andy Rouse/ Corbis **350 right,** © Royalty-Free/Corbis **355,** © Les Stone/ Sygma/Corbis **360,** © Royalty-Free/Corbis **364,** © Royalty-Free/Corbis **369,** © Roger Ressmeyer/CORBIS **372,** Associated Press, AP **387,** © Mambo Inc./ZUMA/Corbis **388 left,** © Nancy Kaszerman/ZUMA/Corbis **388 right,** © Lucas Jackson/ Reuters/Corbis **389,** © Reuters/CORBIS **391,** © Royalty-Free/Corbis **392,** © Chuck Savage/CORBIS **394,** © Tim Street-Porter/Beateworks/Corbis **410,** © Royalty-Free/Corbis **411,** PhotoDisc **425,** © Royalty-Free/Corbis **465,** © CORBIS **466,** Associated Press, AP **467 left,** Associated Press, AP **467 right,** © Gail Mooney/CORBIS **468,** © Christopher J. Morris/CORBIS **469,** © Ashley Cooper/Corbis **480,** Digital Vision **484,** © Ashley Cooper/ Corbis **500,** Stockbyte/Getty Images **513,** Michael St-Andre **516,** © CORBIS **518,** © Sean Daveys/Australian Picture Library/Corbis **520,** Tony Duffy/Getty Images **539,** © Ariel Skelley/CORBIS **541,** © Macduff Everton/CORBIS **546,** PhotoDisc **547,** © David Trood/Stone/Getty Images **555,** © Reuters/ CORBIS **559,** Rick Hanston, Latent Images **572,** © Ariel Skelley/CORBIS **573,** Gray Mortimore/Allsport/Getty Images **578,** Copyright © Tony Freeman/ Photo Edit **579,** PhotoDisc **592,** © C. Devan/zefa/Corbis

Answers

CHAPTER R

Technology Connection, p. 8

1–20. Left to the student.

Exercise Set R.1, p. 11

1.

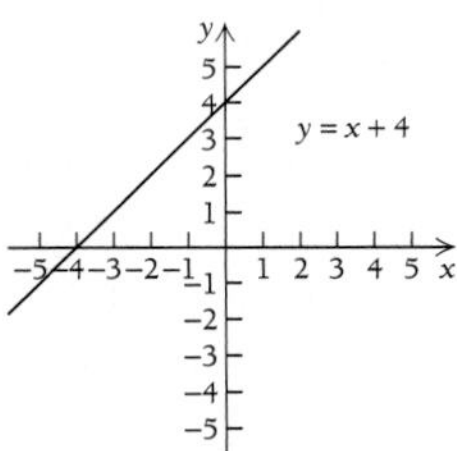

3.

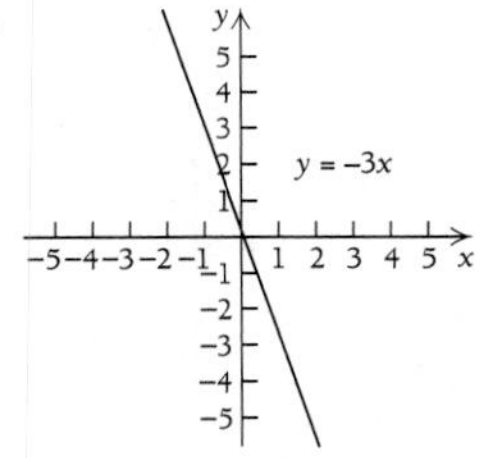

5.

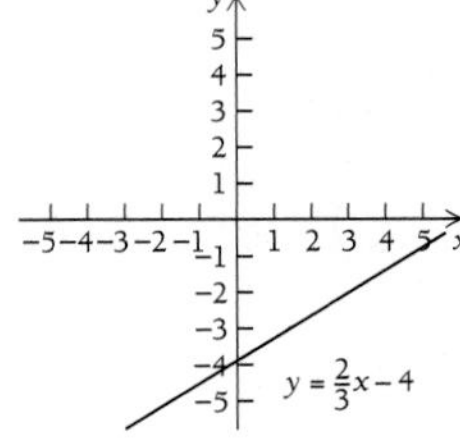

7.

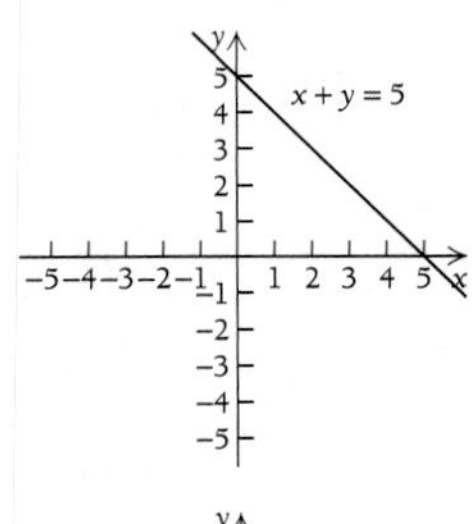

9.

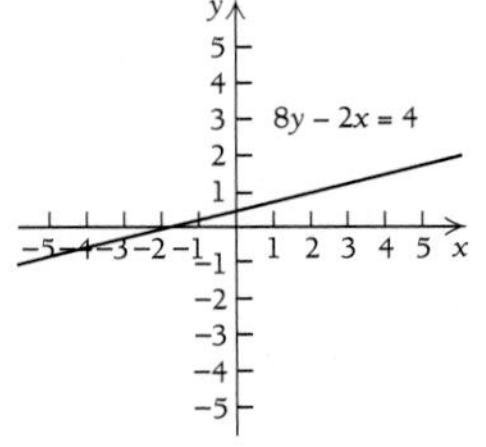

11.

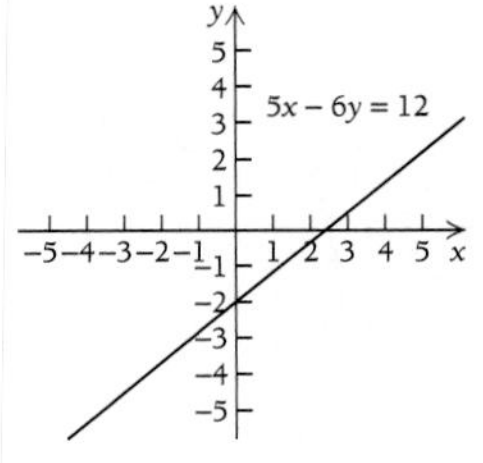

13.

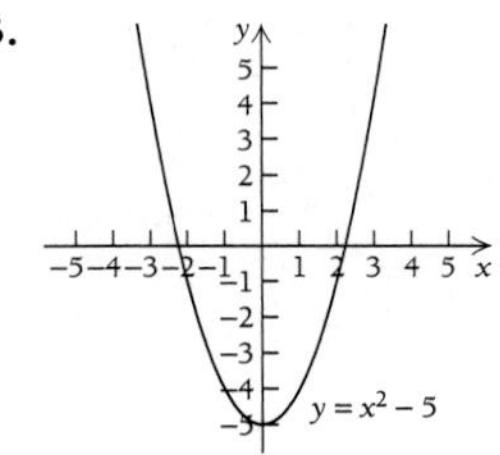

15.

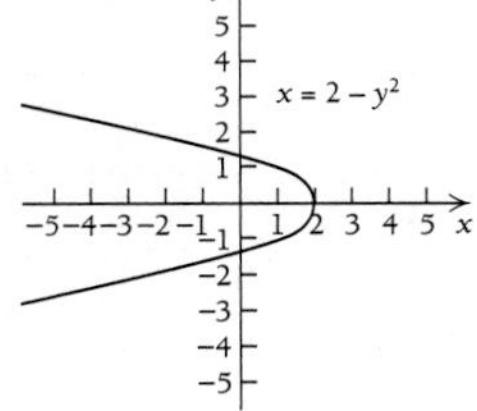

17.

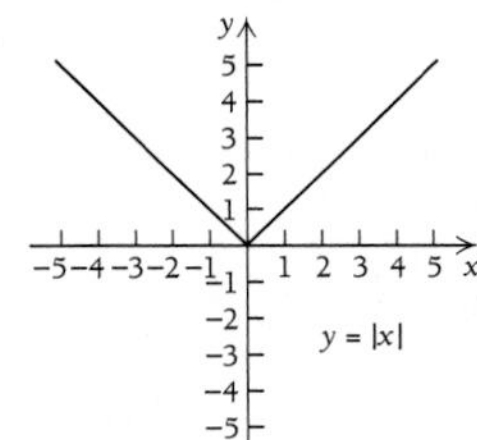

19.

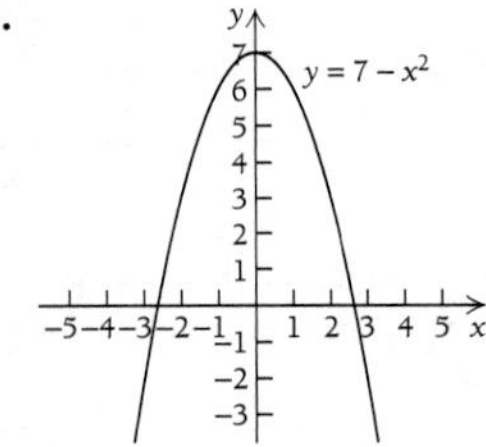

21.

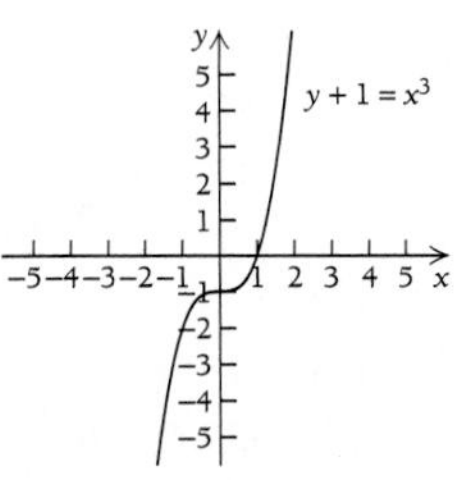

23. 3.98 min (1954), 3.66 min (2008) **25.** About 25.5 mi/hr, or mph **27. (a)** 1.8 million, 3.7 million, 4.4 million, 4.5 million; **(b)** 44 and 67; **(c)** about 58; **(d)** TW **29. (a)** \$1060; **(b)** \$1060.90; **(c)** \$1061.36; **(d)** \$1061.83, assuming 365 days in a year; **(e)** \$1061.84 **31. (a)** \$2249.73; **(b)** \$2252.32; **(c)** \$2253.65;

(d) \$2254.98; **(e)** \$2254.99 **33.** \$578.70
35. \$112,351 **37.** **(a)** 1800–1858, 1862–1888, 1900–1937, 1975–1984, 1987–2000; **(b)** 1937–1975, 1984–1987; **(c)** 1858–1862, 1888–1900

39.

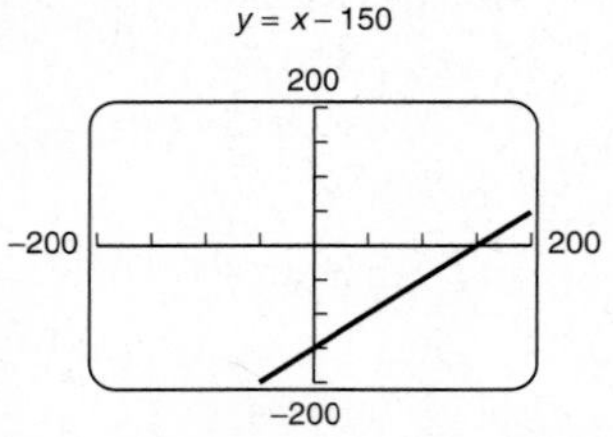
y = x − 150
200
−200
200
−200

41.

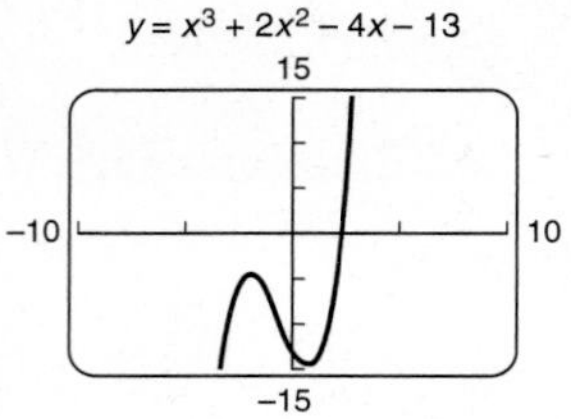
y = x³ + 2x² − 4x − 13
15
−10
10
−15

43.

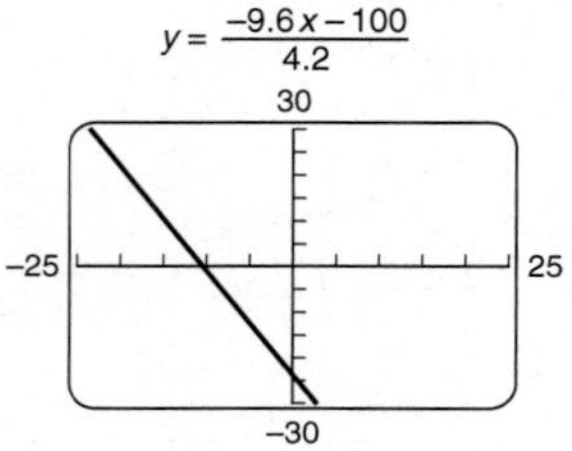
y = (−9.6x − 100)/4.2
30
−25
25
−30

45.

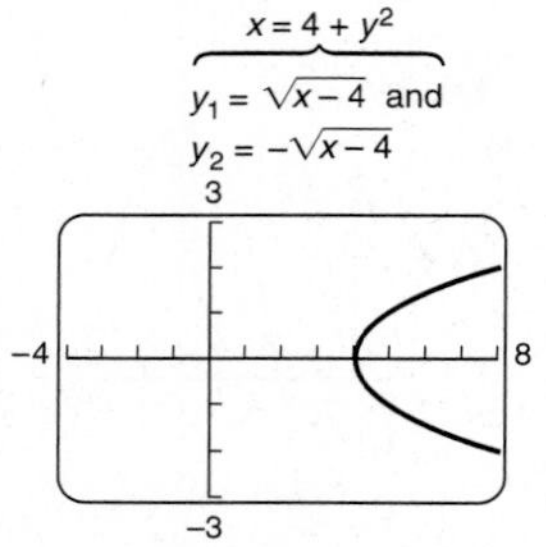
x = 4 + y²
y₁ = √(x − 4) and
y₂ = −√(x − 4)
3
−4
8
−3

Technology Connection, p. 18

1. 951; 42,701 **2.** 21,813

Technology Connection, p. 19

1. 6; 3.99; 150; $-1.\overline{5}$, or $-\frac{14}{9}$ **2.** -21.3; -18.39; -117.3; $3.2\overline{5}$, or $\frac{293}{90}$ **3.** -75; -65.466; -420.6; $1.6\overline{8}$, or $\frac{76}{45}$

Technology Connection, p. 23

1–3. Left to the student

Exercise Set R.2, p. 24

1. Yes **3.** No **5.** Yes **7.** Yes **9.** Yes
11. Yes **13.** Yes **15.** No **17.** Yes

19. **(a)**

x	4.1	4.01	4.001	4
$f(x)$	14.3	14.03	14.003	14

(b) $f(5) = 17, f(-1) = -1, f(k) = 3k + 2,$
$f(1 + t) = 3t + 5, f(x + h) = 3x + 3h + 2$
21. $g(-1) = -2, g(0) = -3, g(1) = -2, g(5) = 22,$
$g(u) = u^2 - 3, g(a + h) = a^2 + 2ah + h^2 - 3,$ and
$$\frac{g(a + h) - g(a)}{h} = 2a + h, h \neq 0$$

23. **(a)** $f(4) = \frac{1}{49}$, $f(-3)$ is undefined, $f(0) = \frac{1}{9}$,
$$f(a) = \frac{1}{(a + 3)^2},\ f(t + 4) = \frac{1}{(t + 7)^2},$$
$$f(x + h) = \frac{1}{(x + h + 3)^2},\text{ and}$$
$$\frac{f(x + h) - f(x)}{h} = \frac{-2x - h - 6}{(x + h + 3)^2(x + 3)^2},\ h \neq 0$$
(b) Take an input, square it, add six times the input, add 9, and then take the reciprocal of the result.

25.

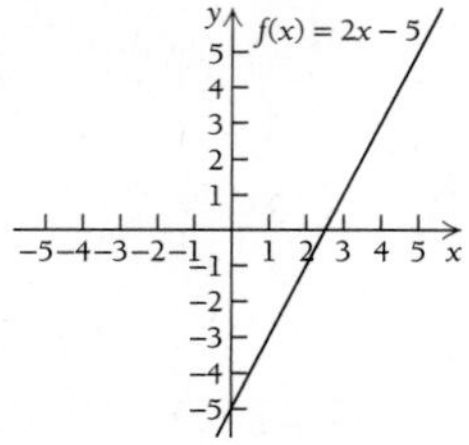
f(x) = 2x − 5

27.

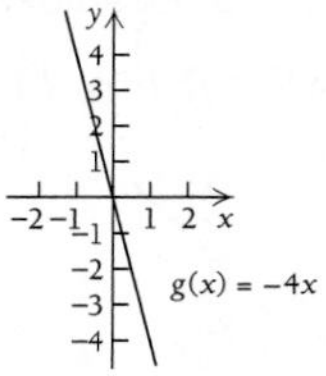
g(x) = −4x

29.

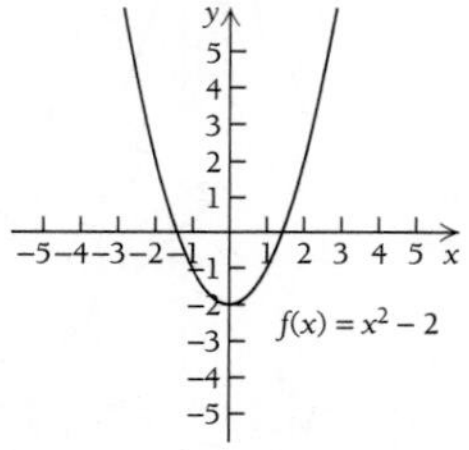
f(x) = x² − 2

31.

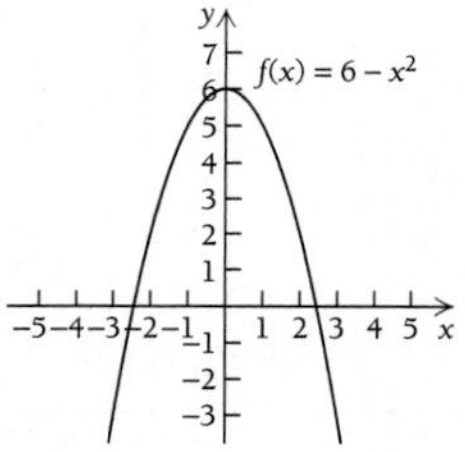
f(x) = 6 − x²

33.

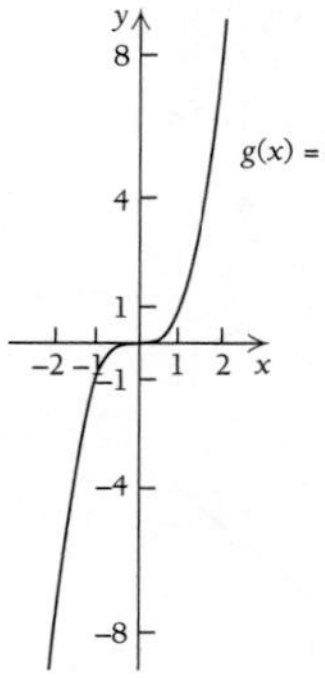
g(x) = x³

35. Yes **37.** Yes **39.** No **41.** No **43.** Yes

45. Yes **47. (a)**

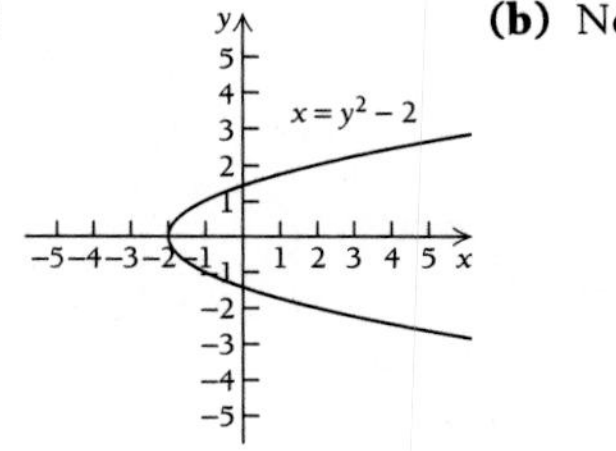

(b) No

49. $\dfrac{f(x+h) - f(x)}{h} = 2x + h - 3, h \neq 0$

51. $f(-1) = 3, f(1) = -2$

53. $f(0) = 17, f(10) = 6$

55.

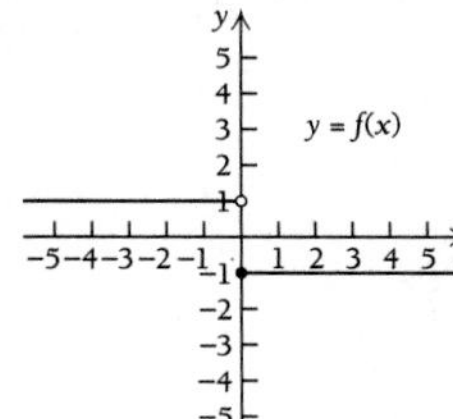

57.

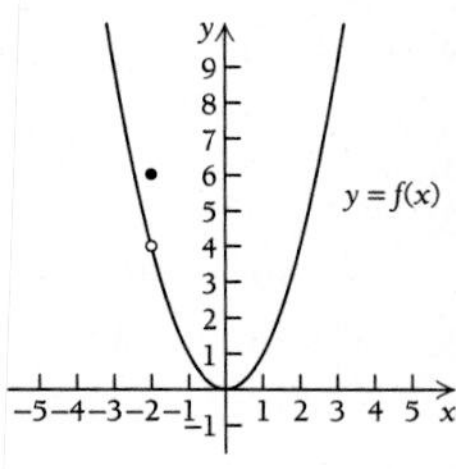

59.

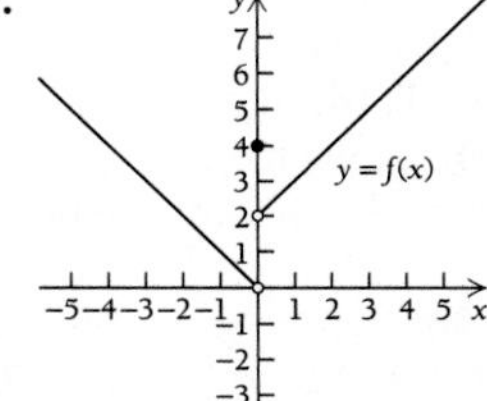

61.

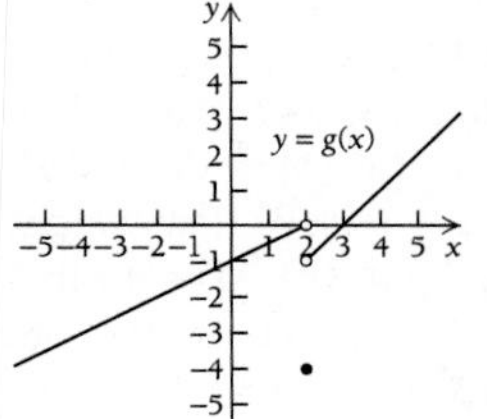

63.

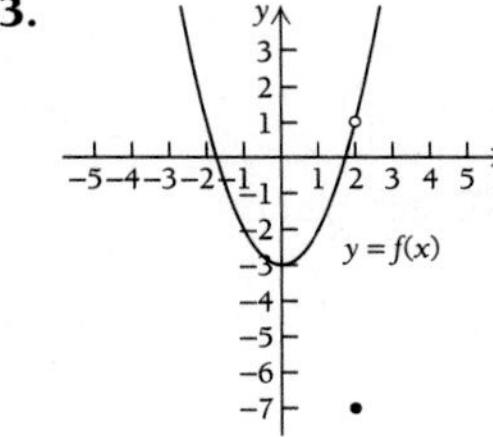

65. (a) About 600 million; **(b)** 2001 **67. (a)** Yes; **(b)** about $1050 billion; **(c)** 1982 or 1983 **69.** $563.25 **71. (a)** 1.818 m^2; **(b)** 2.173 m^2; **(c)** 1.537 m^2 **73. (a)** Yes; a unique "scale of impact" number is assigned to each event. **(b)** The inputs are the events; the outputs are the scale of impact numbers.

75. $y = \pm\sqrt{\dfrac{x+5}{2}}$; this is *not* a function.

77. $y = 2 \cdot \sqrt[3]{x}$; this is a function.

79. TW **81.**

X	Y1
-3	.6
-2	ERR:
-1	-1
0	-.75
1	-1
2	ERR:
3	.6
X = -3	

83. Left to the student

Technology Connection, p. 33

1.

X	Y1
-3	.2
-2.5	.4444
-2	ERR:
-1.5	-.5714
-1	-.3333
-.5	-.2667
0	-.25
X = -3	

2. Answers may vary.

Technology Connection, p. 34

1. Domain = ℝ; range = $[-4, \infty)$
2. Domain = ℝ; range = ℝ
3. Domain = $\{x \mid x$ is a real number and $x \neq 0\}$; range = $\{x \mid x$ is a real number and $x \neq 0\}$
4. Domain = ℝ; range = $[-8, \infty)$
5. Domain = $[-4, \infty)$; range = $[0, \infty)$
6. Domain = $[-3, 3]$; range = $[0, 3]$
7. Domain = $[-3, 3]$; range = $[-3, 0]$
8. Domain = ℝ; range = ℝ

Exercise Set R.3, p. 36

1. $[-2, 4]$ **3.** $(0, 5)$ **5.** $[-9, -4)$ **7.** $[x, x+h]$
9. (p, ∞) **11.** $[-2, 2]$
13. $[-4, -1)$
15. $(-\infty, -2]$
17. $(-2, 3]$
19. $(-\infty, 12.5)$
21. (a) 3; **(b)** $\{-3, -1, 1, 3, 5\}$; **(c)** 3; **(d)** $\{-2, 0, 2, 3, 4\}$ **23. (a)** 4; **(b)** $\{-5, -3, 1, 2, 3, 4, 5\}$; **(c)** $\{-5, -3, 4\}$; **(d)** $\{-3, 2, 4, 5\}$ **25. (a)** -1; **(b)** $[-2, 4]$; **(c)** 3; **(d)** $[-3, 3]$ **27. (a)** -2; **(b)** $[-4, 2]$; **(c)** -2; **(d)** $[-3, 3]$ **29. (a)** 3; **(b)** $[-3, 3]$; **(c)** about -1.4 and 1.4; **(d)** $[-5, 4]$ **31. (a)** 1; **(b)** $[-5, 5)$; **(c)** $[3, 5)$; **(d)** $\{-2, -1, 0, 1, 2\}$ **33.** $\{x \mid x$ is a real number and $x \neq 2\}$ **35.** $\{x \mid x \geq 0\}$ **37.** ℝ **39.** $\{x \mid x$ is a real number and $x \neq 2\}$ **41.** ℝ **43.** $\{x \mid x$ is a real number and $x \neq 3.5\}$ **45.** $\{x \mid x \geq -\frac{4}{5}\}$ **47.** ℝ **49.** $\{x \mid x$ is a real number and $x \neq 5, x \neq -5\}$ **51.** ℝ **53.** $\{x \mid x$ is a real number and $x \neq 5, x \neq 1\}$ **55.** $[-1, 2]$

57. (a) $A(t) = 5000\left(1 + \dfrac{0.08}{2}\right)^{2t}$; **(b)** $\{t \mid t \geq 0\}$

59. (a) $[0, 84.7]$; **(b)** approximately $[0, 4{,}600{,}000]$; **(c)** TW **61. (a)** $[0, 65]$; **(b)** $[8, 75]$ **63.** TW **65.** TW **67.** $(-\infty, 0) \cup (0, \infty)$; $[0, \infty)$; ℝ; $[1, \infty)$; ℝ

Technology Connection, p. 41

1. The line will slant up from left to right, will intersect the y-axis at $(0, 1)$, and will be steeper than $y = 10x + 1$.
2. The line will slant up from left to right, will pass through the origin, and will be less steep than $y = \frac{2}{31}x$.

3. The line will slant down from left to right, will pass through the origin, and will be steeper than $y = -10x$. **4.** The line will slant down from left to right, will intersect the y-axis at $(0, -1)$, and will be less steep than $y = -\frac{5}{32}x - 1$.

Technology Connection, p. 43

1. The graph of y_2 is a shift 3 units up of the graph of y_1, and y_2 has y-intercept $(0, 3)$. The graph of y_3 is a shift 4 units down of the graph of y_1, and y_3 has y-intercept $(0, -4)$. The graph of $y = x - 5$ is a shift 5 units down of the graph of $y = x$, and $y = x - 5$ has y-intercept $(0, -5)$. All lines are parallel. **2.** For any x-value, the y_2-value is 3 more than the y_1-value and the y_3-value is 4 less than the y_1-value.

Exercise Set R.4, p. 51

1. $x = 3$

3. $y = -2$

5. $x = -4.5$

7.

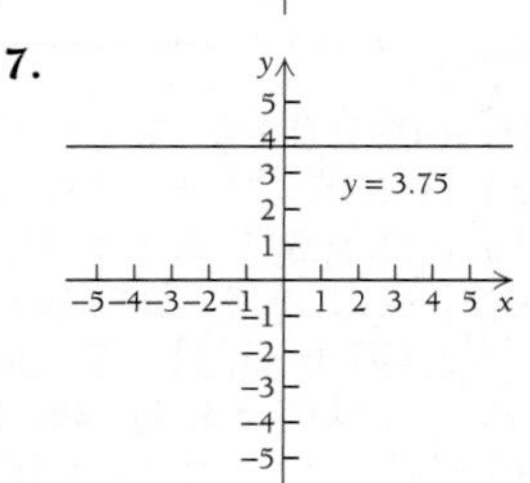

9.

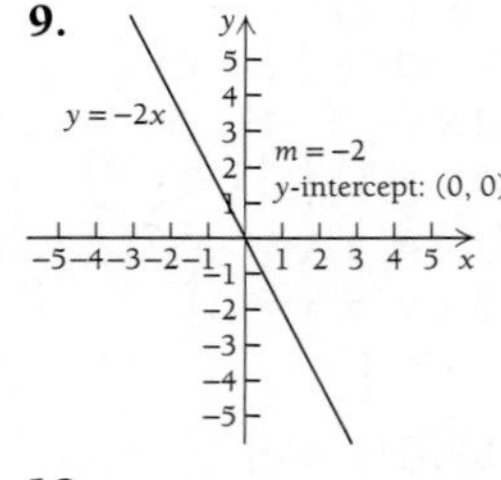

11.

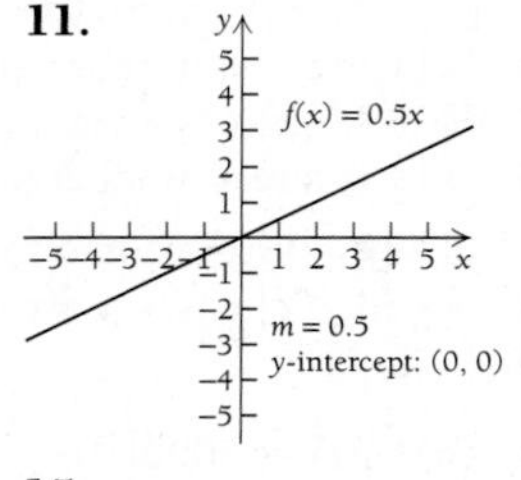

13.

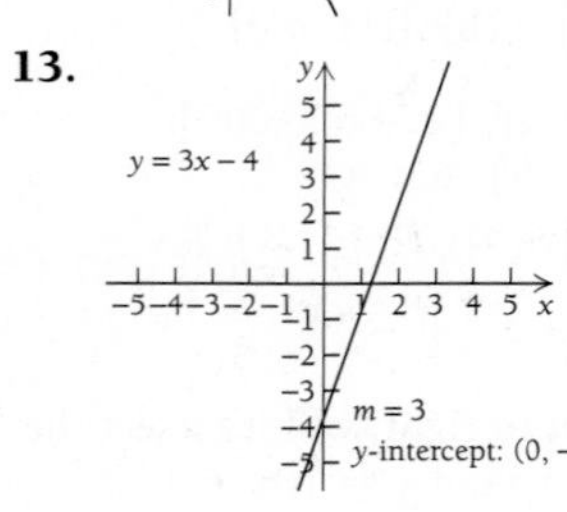

15.

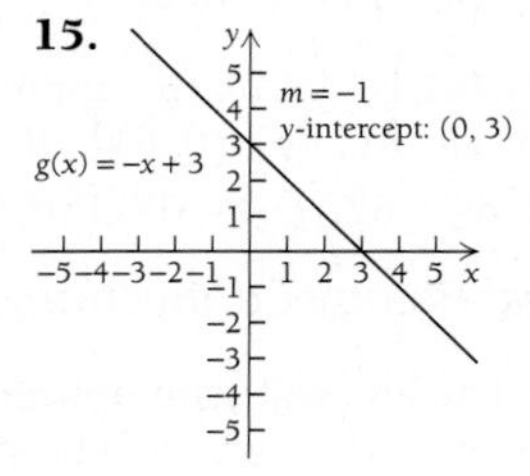

17. $y = 7$; $m = 0$; y-intercept: $(0, 7)$

19. $m = 3$, y-intercept: $(0, 6)$
21. $m = -2$, y-intercept: $(0, 3)$
23. $m = -1$, y-intercept: $(0, -4)$
25. $m = \frac{1}{3}$, y-intercept: $\left(0, -\frac{7}{3}\right)$
27. $y + 3 = -5(x + 2)$, or $y = -5x - 13$
29. $y - 3 = -2(x - 2)$, or $y = -2x + 7$
31. $y - 0 = 2(x - 3)$, or $y = 2x - 6$
33. $y + 6 = \frac{1}{2}(x - 0)$, or $y = \frac{1}{2}x - 6$
35. $y - 3 = 0 \cdot (x - 2)$, or $y = 3$
37. $-\frac{4}{7}$ **39.** $\frac{1}{3}$ **41.** Undefined slope
43. $-\frac{34}{3}$ **45.** 0 **47.** 3 **49.** 2
51. $y = -\frac{4}{7}x - \frac{1}{7}$ **53.** $y = \frac{1}{3}x - \frac{11}{3}$
55. $x = 3$ **57.** $y = -\frac{34}{3}x + \frac{91}{15}$ **59.** $y = 3$
61. $\frac{2}{5}$ **63.** 3.5% **65.** About \$888 per year
67. \$53 per year **69. (a)** $R = 4.17T$; **(b)** $R \approx 25$
71. (a) $M = 0.4W$; **(b)** $0.4 = 40\%$, so we have $M = 40\%W$. The weight of the muscles is 40% of the body weight. **(c)** $M = 48$ lb
73. (a) $T(w) = 0.0008w$; **(b)** \$2.96
75. **(a)** $C(x) = 20x + 100{,}000$; **(b)** $R(x) = 45x$; **(c)** $P(x) = R(x) - C(x) = 25x - 100{,}000$; **(d)** a profit of \$3,650,000; **(e)** 4000 calculators

77. (a) $V(t) = 5200 - 512.5t$;
(b) $V(0) = \$5200$, $V(1) = \$4687.50$, $V(2) = \$4175$, $V(3) = \$3662.50$, $V(4) = \$3150$, $V(7) = \$1612.50$, $V(8) = \$1100$
79. About \$743,590 **81. (a)** The number -700 indicates that the value of the photocopier decreases by \$700 per year, and the number 3500 indicates that the original value of the copier was \$3500. **(b)** 5 years. **(c)** TW
83. (a) $D(5) = 6$ ft, $D(10) = 11.5$ ft, $D(20) = 22.5$ ft, $D(50) = 55.5$ ft, $D(65) = 72$ ft
(b)

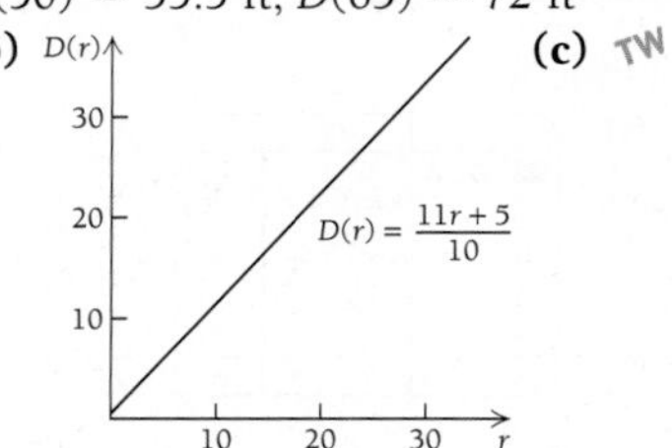

(c) TW

85. **(a)** Approximately $y = 178x - 353{,}952$; **(b)** Approximately 2048 manatees; **(c)** TW **87.** **(a)** $N = 1.02P$; **(b)** $N = 204{,}000$ people; **(c)** $P = 360{,}000$ people **89.** TW **91.** **(a)** Graph III; **(b)** graph IV; **(c)** graph I; **(d)** graph II

Technology Connection, p. 57

1–2. Left to the student

Technology Connection, p. 59

1. **(a)** 2 and 4; **(b)** 2 and -5; **(c)** and **(d)** left to the student

Technology Connection, p. 61

1. -5 and 2 **2.** -4 and 6 **3.** -2 and 1 **4.** 0, -1.414, and 1.414 (approx.) **5.** 0 and 700 **6.** -2.079, 0.463, and 3.116 (approx.) **7.** -3.096, -0.646, 0.646, and 3.096 (approx.) **8.** -1 and 1 **9.** -0.387 and 1.721 **10.** 6.133 **11.** -2, -1.414, 1, and 1.414 **12.** -3, -1, 2, and 3

Technology Connection, p. 63

1. Left to the student **2.** For $x = 3$, y_1 is ERR and y_2 is 6.

Technology Connection, p. 64

1–8. Left to the student

Technology Connection, p. 65

1–2. Left to the student

Technology Connection, p. 70

1. Approximately (14, 266)

Exercise Set R.5, p. 70

1.

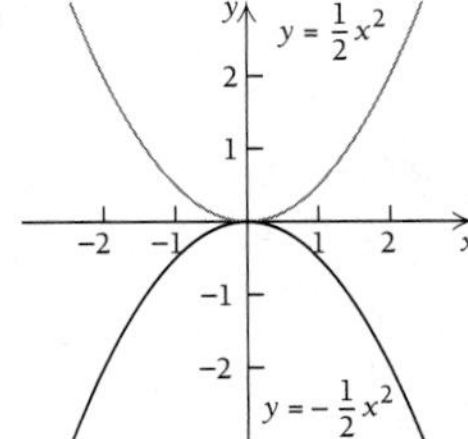

3.

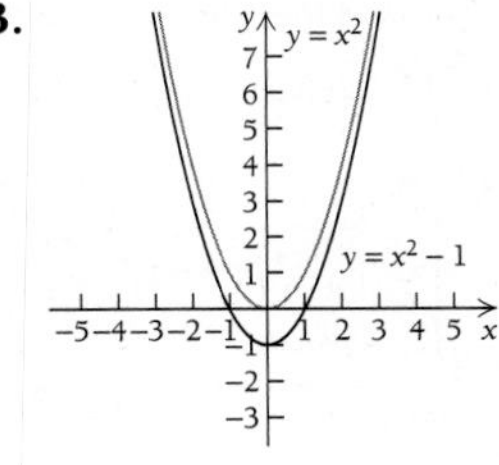

5.

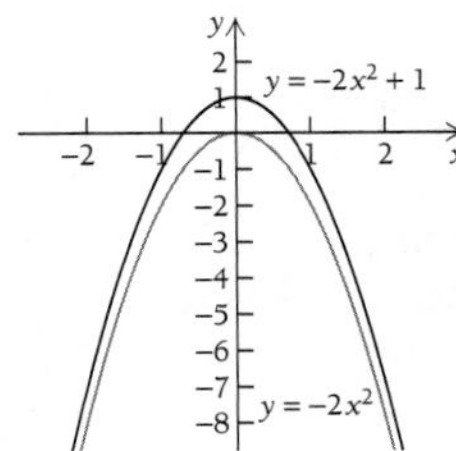

7.

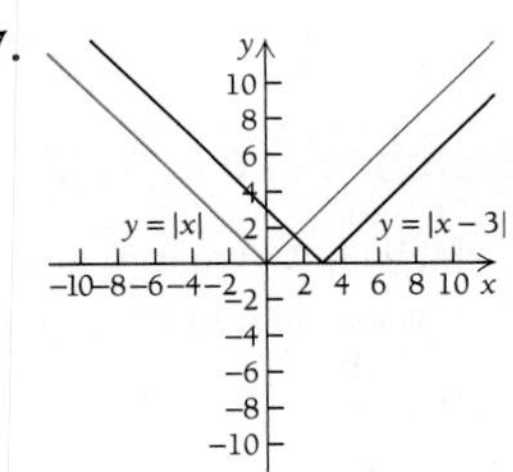

9.

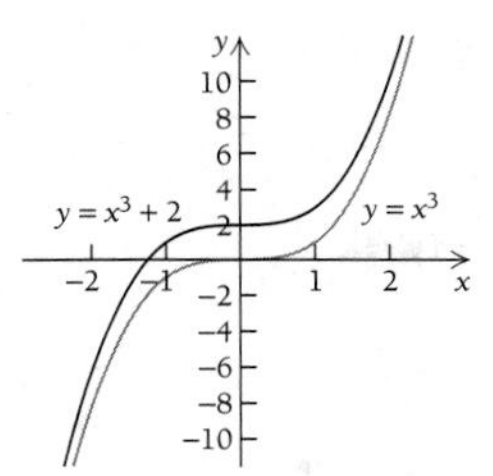

11.

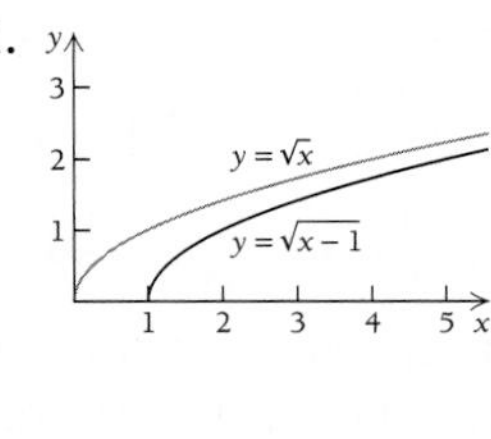

13. Parabola with vertex at $(-2, -11)$
15. Not a parabola

17.

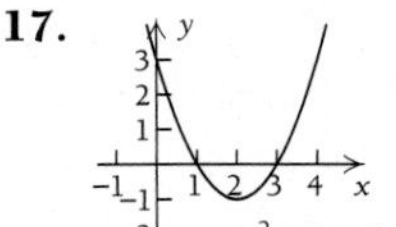

19.

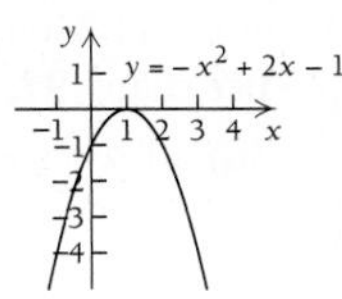

21.

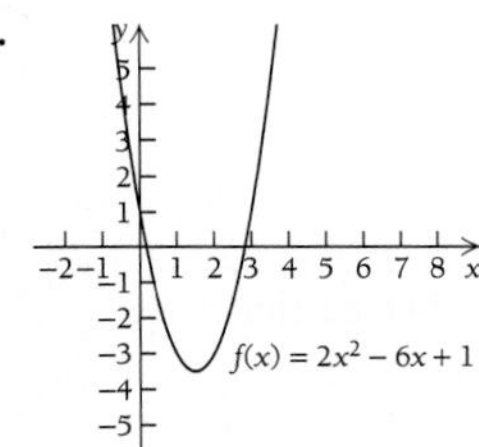

23.

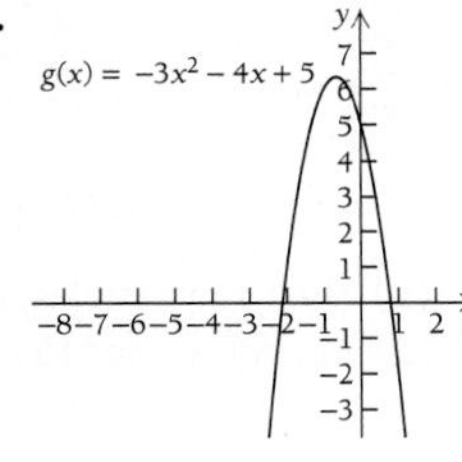

25.

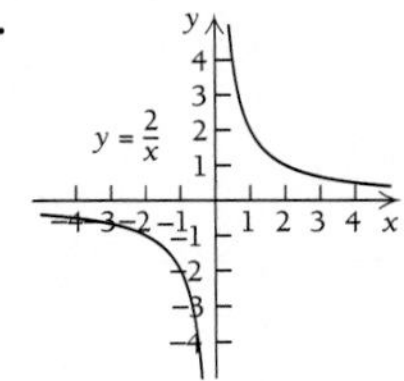

27.

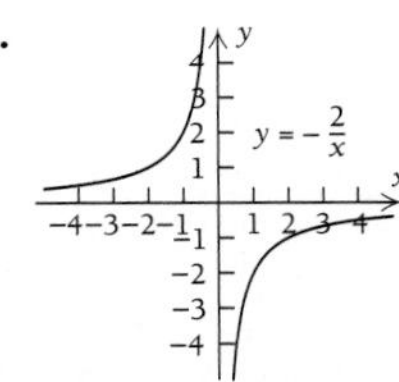

29.

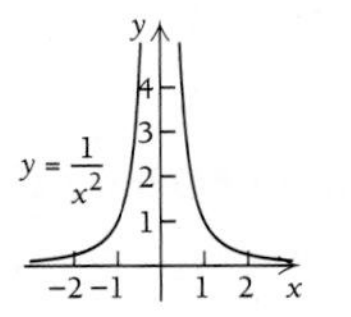

31.

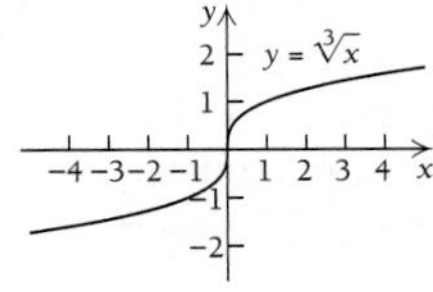

33.

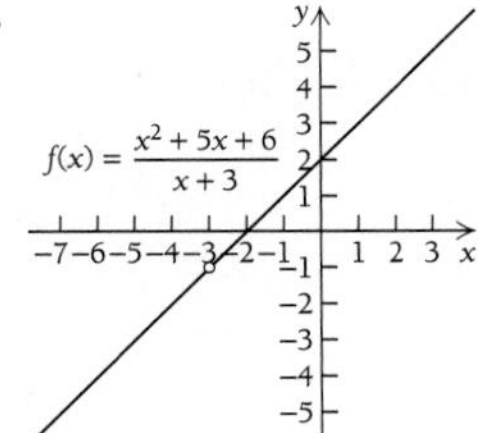

35.

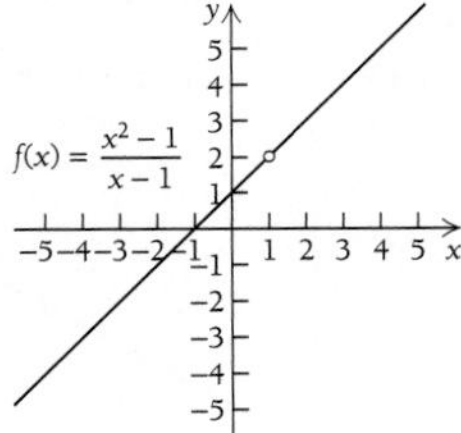

37. $1 \pm \sqrt{3}$, or -0.732, 2.732
39. $-3 \pm \sqrt{10}$, or -6.162, 0.162
41. $\dfrac{1 \pm \sqrt{2}}{2}$, or -0.207, 1.207
43. $\dfrac{-4 \pm \sqrt{10}}{3}$, or -2.387, -0.279
45. $\dfrac{-7 \pm \sqrt{13}}{2}$, or -5.303, -1.697

47. $x^{3/2}$ **49.** $a^{3/5}$ **51.** $t^{1/7}$ **53.** x^3 **55.** $t^{-5/2}$

57. $(x^2 + 7)^{-1/2}$ **59.** $\sqrt[5]{x}$ **61.** $\sqrt[3]{y^2}$ **63.** $\frac{1}{\sqrt[5]{t^2}}$

65. $\frac{1}{\sqrt[3]{b}}$ **67.** $\frac{1}{\sqrt[6]{e^{17}}}$ **69.** $\frac{1}{\sqrt{x^2 - 3}}$ **71.** $\frac{1}{\sqrt[3]{t^2}}$

73. 27 **75.** 16 **77.** 8 **79.** $\{x \mid x \neq 5\}$

81. $\{x \mid x \neq 2, x \neq 3\}$ **83.** $\{x \mid x \geq -\frac{4}{5}\}$

85. $\{x \mid x \leq 7\}$ **87.** (50, 500); $x = \$50$, $q = 500$ items

89. (5, 1); price is \$500, and quantity is 1000.

91. (1, 4); price is \$1, and quantity is 400.

93. (2, 3); price is \$2000, and quantity is 3000.

95. \$78.20 per share **97.** **(a)** 166 mi, 176 mi, 184 mi;
(b)

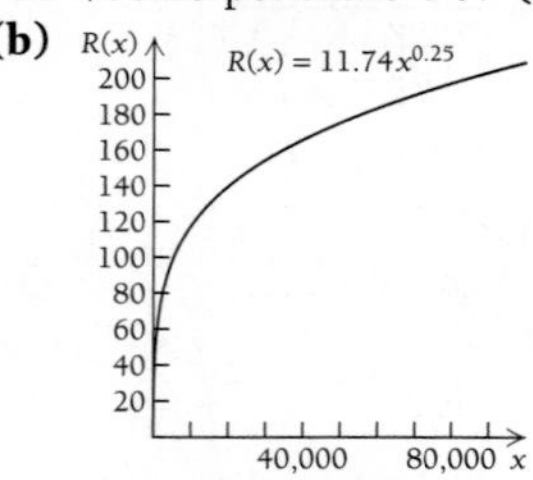

99. **(a)** 99,130 particles/cm³, 108,347 particles/cm³, 127,322 particles/cm³ **(b)**

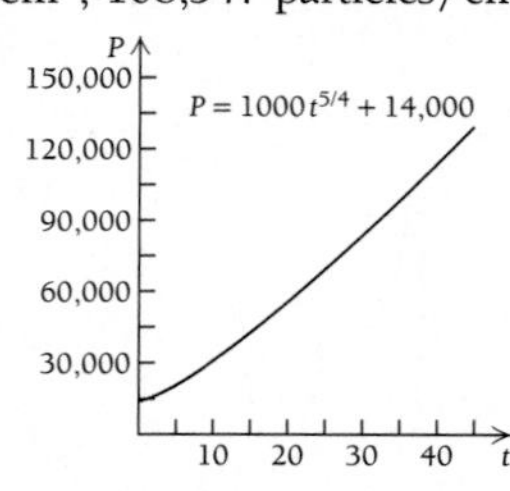

101. 15 cities; 30 cities **103.** TW

105. −1.831, −0.856, 3.188

107. 1.489, 5.673

109. −2, 3 **111.** [−1, 2]

113. Approximately (75.11, 7893); produce 7893 units at a price of \$75.11 each.

Technology Connection, p. 79

1. **(a)** 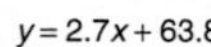

(b) 93.5; **(c)** TW

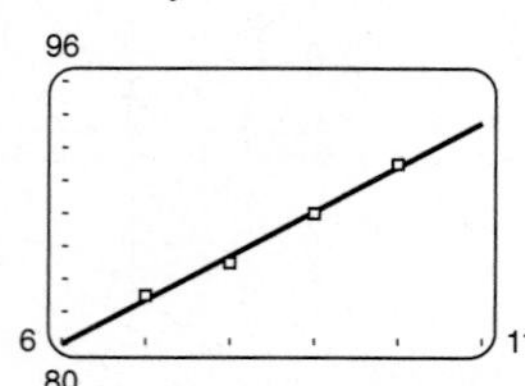

Technology Connection, p. 82

1. **(a)** $y = -0.00005368295x^4 + 0.037566680x^3 - 3.4791715x^2 + 105.81080x - 916.68952$

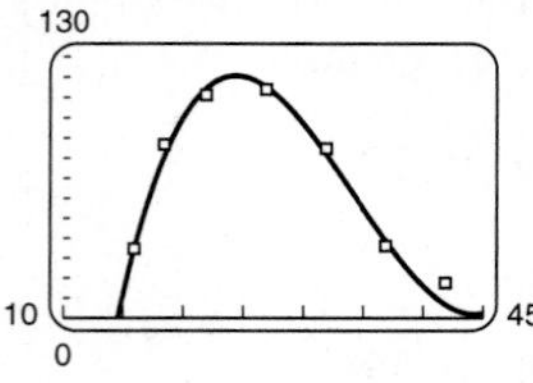

(b) TW

2. **(a)** $y = -62.8327x^2 + 5417.8404x - 57264.7856$

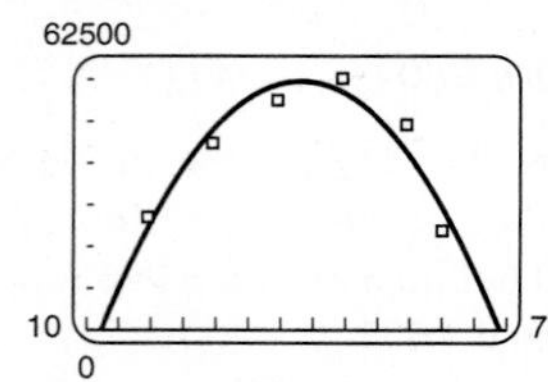

(b) $y = -1.6519x^3 + 145.6606x^2 - 2658.3088x + 36491.7730$

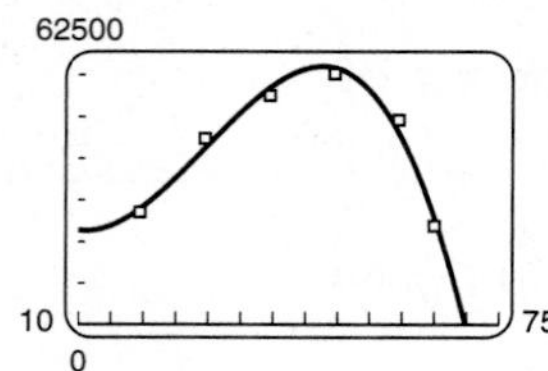

(c) $y = -0.0771x^4 + 11.3952x^3 - 639.2276x^2 - 17037.1915x - 135483.9938$

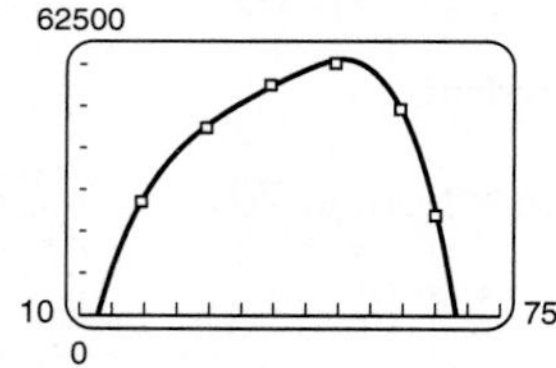

(d) quartic function; **(e)** \$38,853, \$58,887

3. $y = 93.2857x^2 - 1336x + 5460.8286$

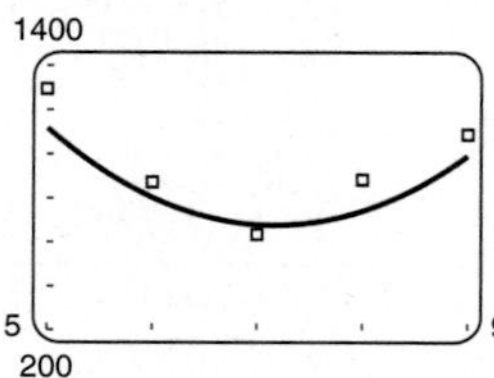

Exercise Set R.6, p. 83

1. Linear **3.** Quadratic, $a < 0$ **5.** Linear

7. Polynomial, neither linear nor quadratic **9.** Linear

11. **(a)** $y = 2x + 13$; **(b)**

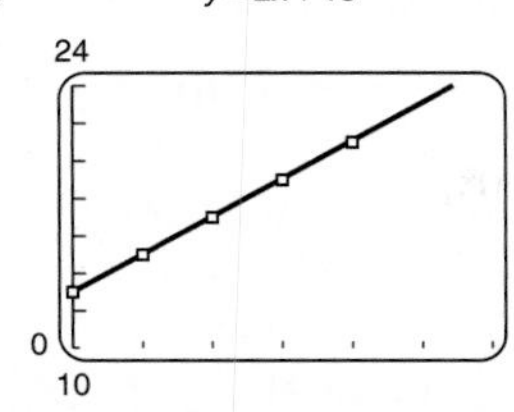

(c) \$33

13. **(a)** $y = 0.144x^2 - 4.63x + 60$; **(b)** 188.5 ft; **(c)** TW **15.** Answers will vary. **17.** TW **19.** TW **21.** **(a)** $y = 0.172x + 4.98$; **(b)** 6.012%. **(c)** Both are about the same. **(d)** $y = -0.000486x^3 + 0.011x^2 + 0.097x + 5.106$; 5.98%. **(e)** TW **23.** **(a)** $y = 0.00027x^3 - 0.1382x^2 + 4.056x + 42.178$; **(b)** \$72.8 billion in 2006, \$70.2 billion in 2010; **(c)** TW

Chapter Review Exercises, p. 91

1. (d) **2.** (b) **3.** (f) **4.** (a) **5.** (e) **6.** (c) **7.** True **8.** False **9.** True **10.** True **11.** False **12.** False **13.** True **14.** True **15.** **(a)** About 56 per 1000 women; **(b)** 18, 30; **(c)** [15, 45]; this interval covers typical human child-bearing ages. **16.** \$1340.24 **17.** \$5017.60 **18.** Not a function. One input, Richard, has three outputs. **19.** **(a)** $f(3) = -6$; **(b)** $f(-5) = -30$; **(c)** $f(a) = -a^2 + a$; **(d)** $f(x + h) = -x^2 - 2xh - h^2 + x + h$

20.

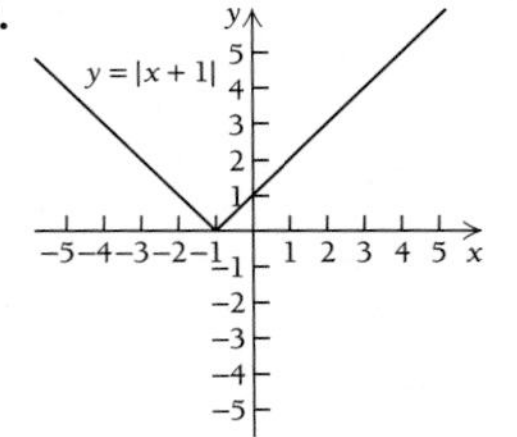

21.

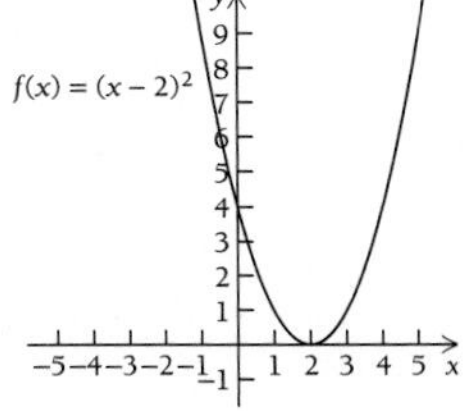

22.

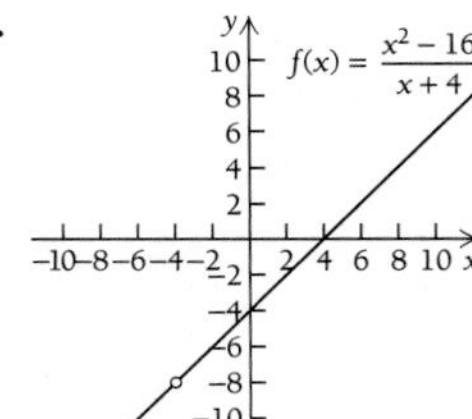

23.

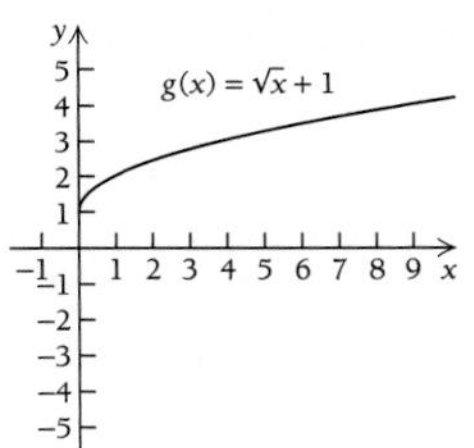

24. Not a function **25.** Function **26.** Function **27.** Not a function **28.** **(a)** $f(2) = 1$; **(b)** $[-4, 4]$; **(c)** $x = -3$; **(d)** $[-1, 3]$ **29.** **(a)** $f(-1) = 1$, $f(1.5) = 4$, $f(6) = 3$ **(b)**

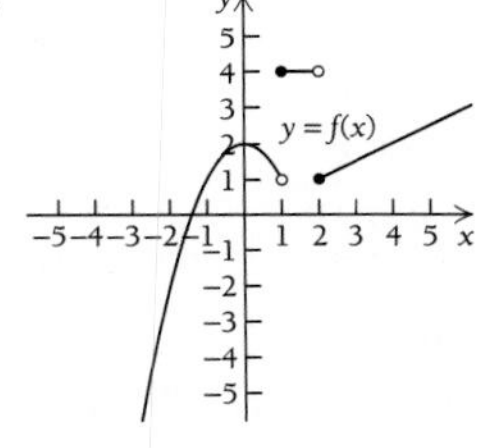

30. **(a)** $[-2, 5]$; **(b)** $(-1, 3]$; **(c)** $(-\infty, a)$ **31.** **(a)** $[-4, 5)$; **(b)** $(2, \infty)$; **32.** **(a)** $f(-3) = -2$; **(b)** $\{-3, -2, -1, 0, 1, 2, 3\}$; **(c)** $-1, 3$; **(d)** $\{-2, 1, 2, 3, 4\}$ **33.** **(a)** $(-\infty, 5) \cup (5, \infty)$; **(b)** $[-6, \infty)$ **34.** Slope, -3; y-intercept, 2 **35.** $y + 5 = \frac{1}{4}(x - 8)$, or $y = \frac{1}{4}x - 7$ **36.** -3 **37.** $-\$350$ per year **38.** 75 pages per day **39.** $A = \dfrac{7}{200}V$ **40.** **(a)** $C(x) = 0.50x + 4000$; **(b)** $R(x) = 10x$; **(c)** $P(x) = 9.5x - 4000$

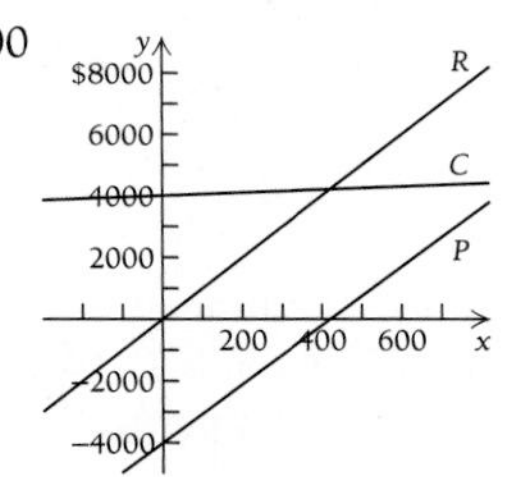

(d) 422 CDs

41. **(a)** **(b)**

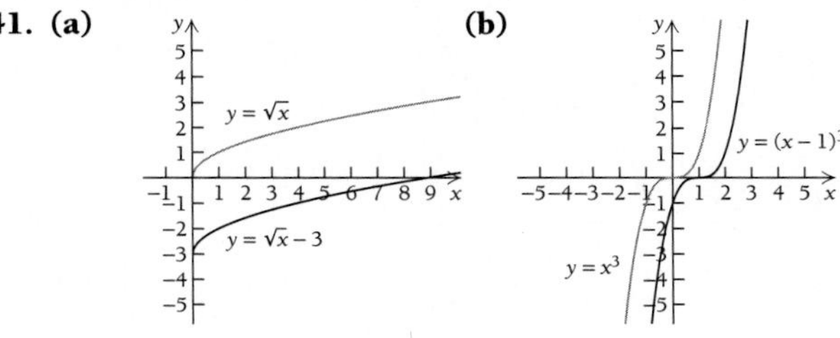

42. **(a)**

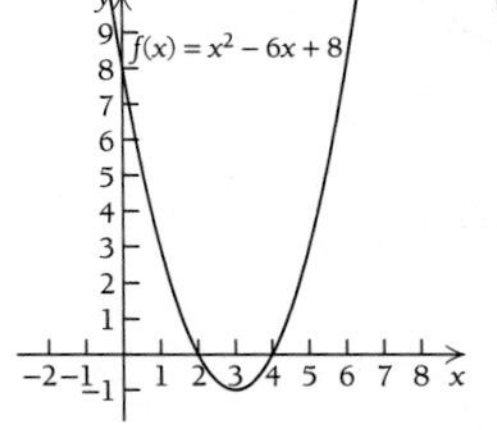

Vertex at $(3, -1)$

(b)

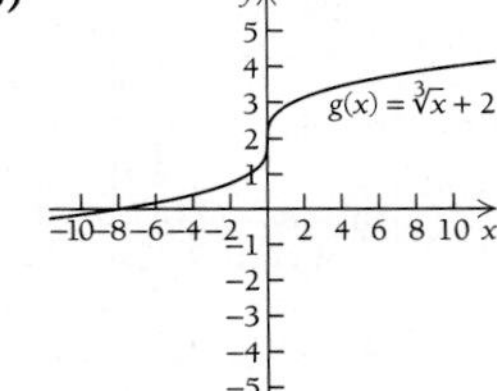

(c)

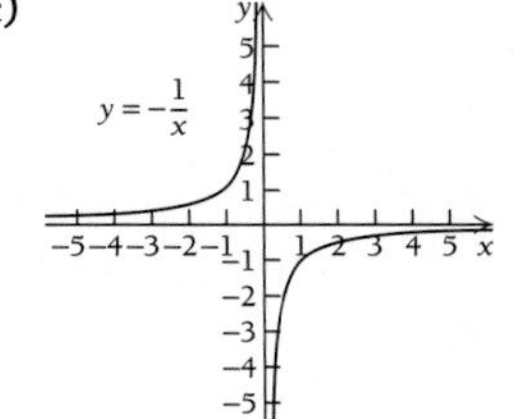

(d)

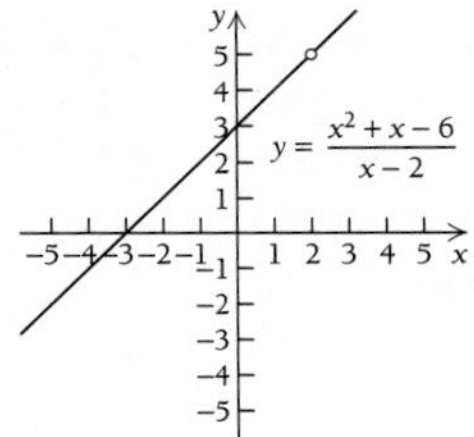

43. **(a)** $x = 1, x = 3$; **(b)** $x = \dfrac{2 \pm \sqrt{10}}{2}$ **44.** **(a)** $x^{4/5}$; **(b)** t^4; **(c)** $m^{-2/3}$; **(d)** $(x^2 - 9)^{-1/2}$ **45.** **(a)** $\sqrt[5]{x^2}$ **(b)** $\dfrac{1}{\sqrt[5]{m^3}}$; **(c)** $\sqrt{x^2 - 5}$; **(d)** $\sqrt[3]{t}$ **46.** $\left[\frac{9}{2}, \infty\right)$ **47.** (3, 16); price = \$3, quantity =1600 units

48. About 3.3 hr
49. **(a)** $y = 0.2x + 160$ **(b)**

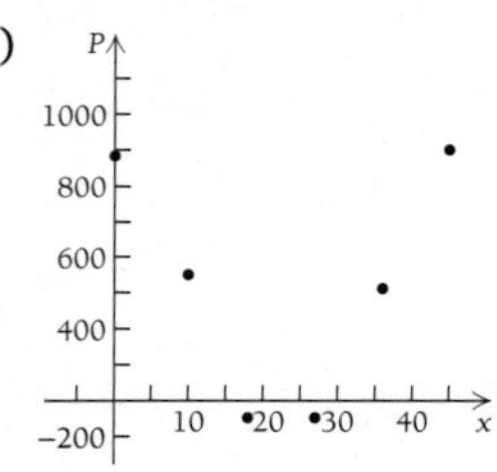

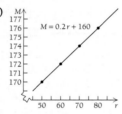

(c) 173.4 beats/min **50.** **(a)**

(b) Data fit a quadratic function.
(c) $y = 2.0x^2 - 89.8\bar{4}x + 870$; **(d)** About $-\$25.33$;
(e) TW **51.** **(a)** 525,375 lb; **(b)** \$4.31/lb
52.

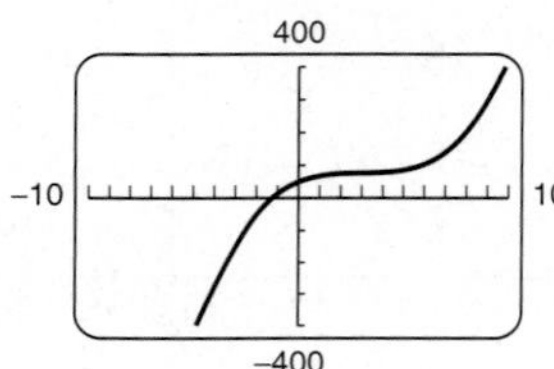

Zero: $x = -1.25$; domain: $\mathbb{R}$; range: $\mathbb{R}$

53.

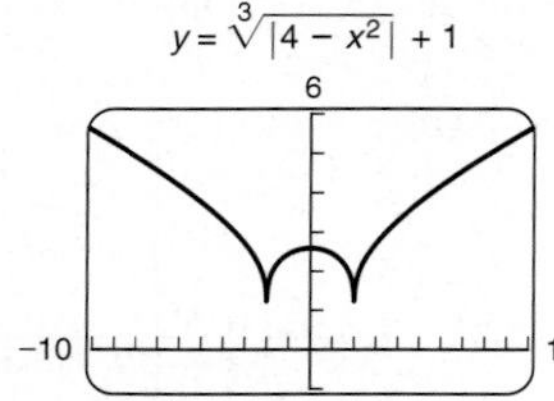

Zero: none; domain: $\mathbb{R}$; range: $[1, \infty)$

54. $(-1.21, 2.36)$ **55.** **(a)** $y = 0.2x + 160$;
(b) 173.4 beats/min; **(c)** TW
56. **(a)** $y = 1.86x^2 - 84.18x + 943.86$; **(b)** \$95.46;
(c) TW **57.** **(a)** $y = 37.58x + 294.48$;
$y = -0.59x^2 + 74.61x - 117.72$;
$y = 0.02x^3 - 2.60x^2 + 125.71x - 439.65$;
$y = 0.003x^4 - 0.324x^3 + 11.46x^2 - 88.51x + 507.84$
(b)

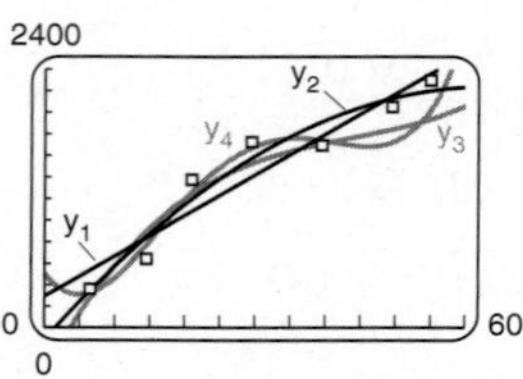

(c) TW

Chapter R Test, p. 95

1. **(a)** [R.1] \$29,500; **(b)** [R.1] 1995;
(c) [R.3] $[25{,}000, 42{,}167]$
2. [R.1] \$750 **3.** [R.2] **(a)** $f(-3) = -4$;
(b) $f(a + h) = -a^2 - 2ah - h^2 + 5$ **4.** [R.4] Slope, $\frac{4}{5}$; y-intercept, $-\frac{2}{3}$ **5.** [R.4] $y - 7 = \frac{1}{4}(x + 3)$, or $y = \frac{1}{4}x + \frac{31}{4}$ **6.** [R.4] $-\frac{1}{2}$ **7.** [R.4] $-\$700$/yr
8. [R.4] $\frac{1}{2}$ lb/bag **9.** [R.4] $F = \frac{2}{3}W$
10. [R.4] **(a)** $C(x) = 0.08x + 8000$; **(b)** $R(x) = 0.50x$;
(c) $P(x) = 0.42x - 8000$; **(d)** 19,048 cards
11. [R.5] (3, 16); $x = \$3$, $q = 16$ thousand units
12. [R.2] Yes **13.** [R.2] No
14. [R.3] **(a)** $f(1) = -4$; **(b)** $\mathbb{R}$ **(c)** $x = \pm 3$;
(d) $[-5, \infty)$ **15.** [R.5]

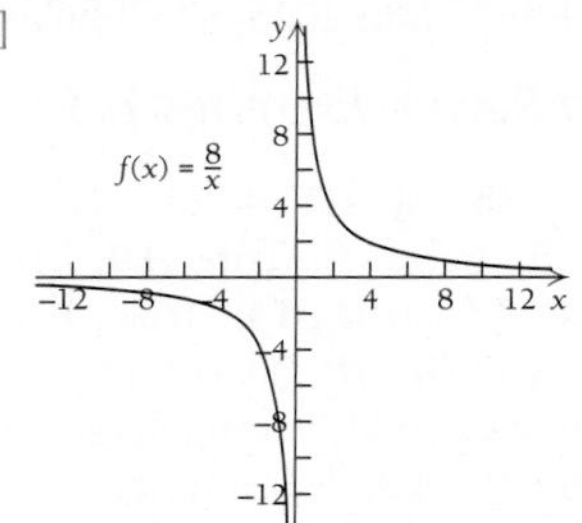

16. [R.5] $t^{-1/2}$ **17.** [R.5] $\dfrac{1}{\sqrt[5]{t^3}}$
18. [R.5]

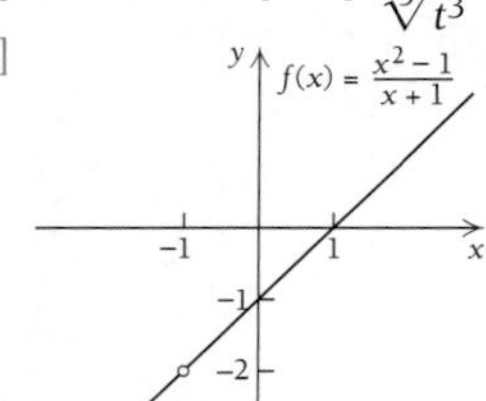

19. [R.5] $(-\infty, -7) \cup (-7, 2) \cup (2, \infty)$
20. [R.5] $(-2, \infty)$ **21.** [R.3] $[c, d)$
22. [R.2]

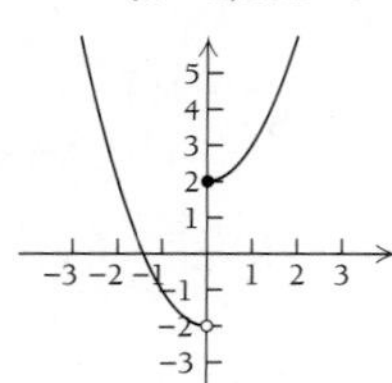

f(x) = { x² + 2, for x ≥ 0; x² − 2, for x < 0 }

23. [R.6] **(a)** **(b)** yes;
(c) $y = -1.94x^2 + 102.74x + 1253.49$;
(d) 2589.9 calories; **(e)** TW **24.** [R.5] $\frac{1}{16}$

25. [R.5] Domain: $\left(-\infty, \frac{5}{3}\right]$; zero: $x = -798\frac{2}{3}$
26. [R.5] Answers will vary. One possibility is $(x + 3)(x - 1)(x - 4) = 0$. **27.** [R.4] $\frac{51}{7}$
28. [R.5] Zeros: $\pm\sqrt{8} \approx \pm 2.828$, $\pm\sqrt{10} \approx \pm 3.162$; domain: $\mathbb{R}$; range: $[-1, \infty)$

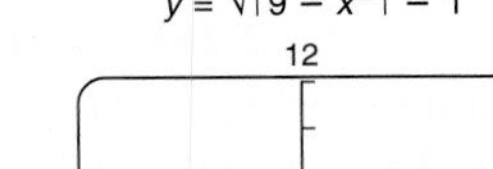

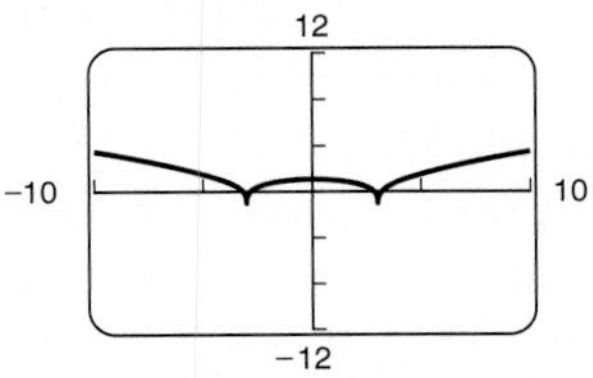

29. [R.6] **(a)** $y = -1.51x^2 + 79.98x + 1436.93$; **(b)** 2480.4 calories; **(c)** TW

Extended Technology Application, p. 97

1. **(a)** $y = 3.221899135x - 3.207929395$;
(b)

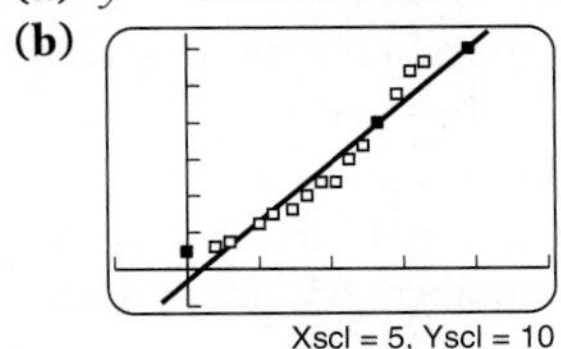

(c) \$139,697, \$54.786 million; the first estimate seems too low, but the second estimate is reasonable.
(d) \$74.118 million, \$202.994 million;
(e) 2017–2018 season
2. **(a)** $y = 0.0937284172x^2 + 1.484547034x + 2.300978582$; **(b)**

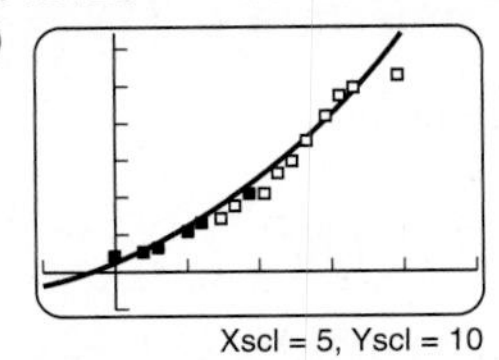

(c) \$3.879 million, \$59.391 million; the first estimate seems a bit small, and the second a bit large, but both estimates are reasonable; **(d)** \$91.918 million, \$481.224 million; **(e)** 2010–2011 season
3. **(a)** $y = -0.0130089164x^3 + 0.4643709669x^2 - 1.282274685x + 6.178257497$;
(b)

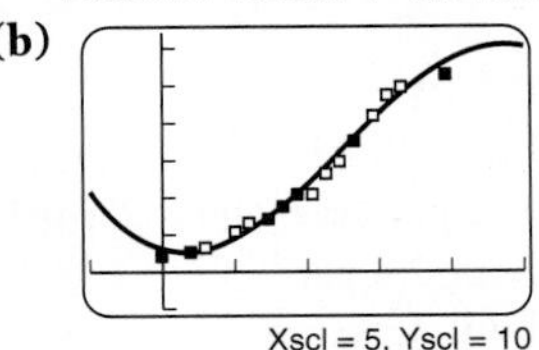

(c) \$5.347 million, \$57.686 million; both estimates are reasonable; **(d)** \$63.046 million, −\$1584 million (clearly not possible); **(e)** 1973–1974 season (clearly unrealistic)
4. **(a)** $y = -0.0018786318x^4 + 0.0590047461x^3 - 0.4069318808x^2 + 2.189955552x + 3.902016562$;
(b)

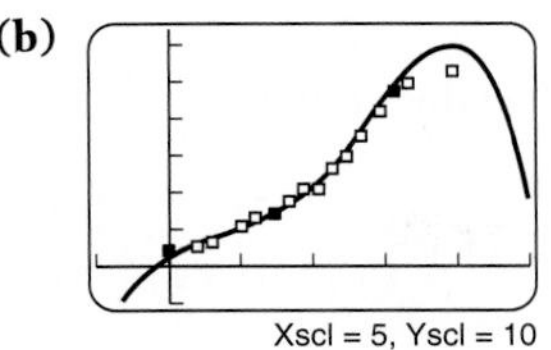

(c) \$5.742 million, \$58.380 million; both estimates are reasonable; **(d)** \$14.465 million, −\$17.573 billion (clearly not possible); **(e)** this function's value never exceeds \$59 million; so \$100 million is not in its range.
5. Left to the student

CHAPTER 1

Technology Connection, p. 102

1. 5 **2.** −4 **3.** $g(x) = 327, 456.95, 475.24, 492.1, 493.81, 494.19, 495.9, 513.24, 573.9, 685.17$ **4.** 494 **5.** −1

Exercise Set 1.1, p. 110

1. 11 **3.** −2 **5.** The limit, as x approaches 4, of $f(x)$ **7.** The limit, as x approaches 5 from the left, of $F(x)$ **9.** $\lim_{x \to 2^+}$ **11.** 2 **13.** −3 **15.** Does not exist **17.** 3 **19.** 4 **21.** −1 **23.** −1 **25.** 0 **27.** 5 **29.** Does not exist **31.** 2 **33.** 2 **35.** 1 **37.** Does not exist **39.** 0 **41.** 3 **43.** 1

45.

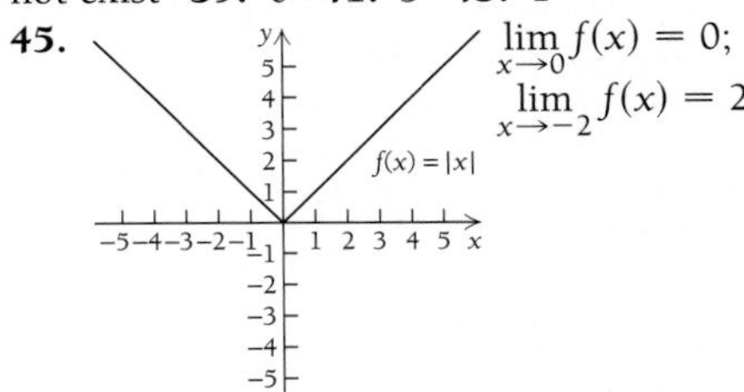

$\lim_{x \to 0} f(x) = 0$;
$\lim_{x \to -2} f(x) = 2$

47.

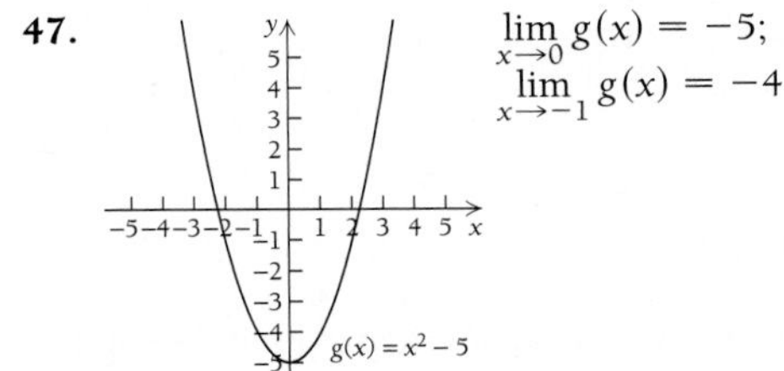

$\lim_{x \to 0} g(x) = -5$;
$\lim_{x \to -1} g(x) = -4$

49.

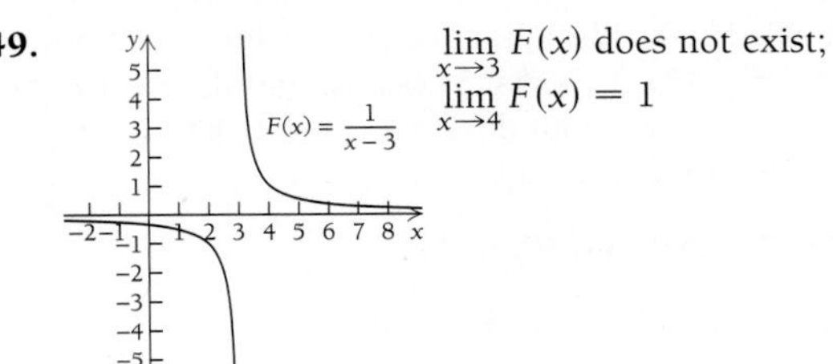

$\lim_{x \to 3} F(x)$ does not exist;
$\lim_{x \to 4} F(x) = 1$

51.

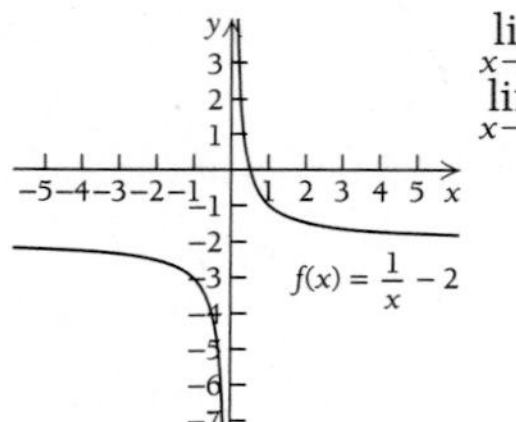

$\lim_{x\to\infty} f(x) = -2$; $\lim_{x\to 0} f(x)$ does not exist

53.

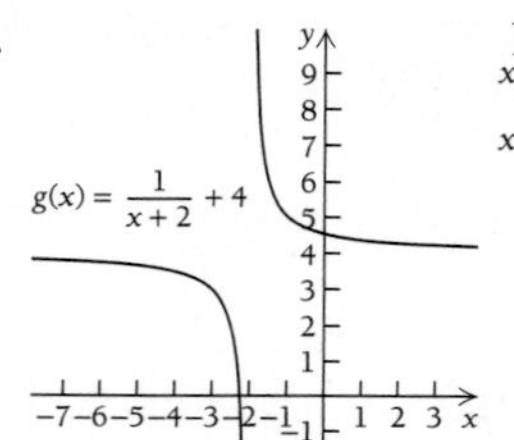

$\lim_{x\to\infty} g(x) = 4$; $\lim_{x\to -2} g(x)$ does not exist

55.

$\lim_{x\to 1^-} F(x) = 3$; $\lim_{x\to 1^+} F(x) = 1$; $\lim_{x\to 1} F(x)$ does not exist

57.

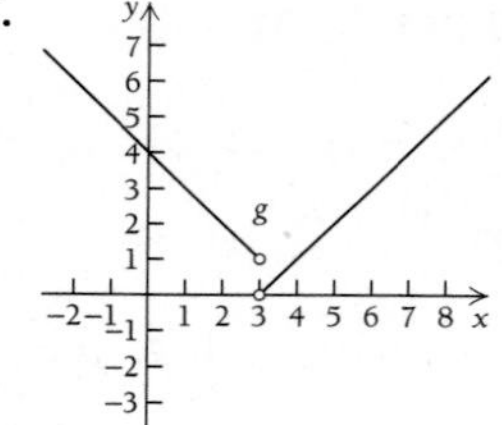

$\lim_{x\to 3^-} g(x) = 1$; $\lim_{x\to 3^+} g(x) = 0$; $\lim_{x\to 3} g(x)$ does not exist

59. $\lim_{x\to -1} G(x) = 1$

61. \$3.30; \$3.30; \$3.30 **63.** \$3.70; \$4.10; limit does not exist **65.** \$0.63; \$0.87; limit does not exist **67.** Limit does not exist. **69.** 1100 deer; 1200 deer; limit does not exist **71.** TW **73.** 35 bears; 34 bears; limit does not exist **75.** 3 **77.** -1 **79.** Limit does not exist; 2 **81.** Limit does not exist; limit does not exist.

Technology Connection, p. 117

1. 53 **2.** 2.8284 **3.** 0.25 **4.** 0.16667

Exercise Set 1.2, p. 121

1. True **3.** False **5.** True **7.** False **9.** 5 **11.** -3 **13.** 4 **15.** 15 **17.** 1 **19.** 6 **21.** $\sqrt{7}$ **23.** Limit does not exist. **25.** 0 **27.** Not continuous **29.** Not continuous **31.** Not continuous **33. (a)** $-2, -2, -2$; **(b)** -2; **(c)** yes, $\lim_{x\to 1} g(x) = g(1)$; **(d)** limit does not exist; **(e)** -3; **(f)** no, the limit does not exist **35. (a)** 2; **(b)** 2; **(c)** yes, $\lim_{x\to 1} h(x) = h(1)$; **(d)** 0; **(e)** 0; **(f)** yes, $\lim_{x\to -2} h(x) = h(-2)$ **37. (a)** 3; **(b)** 1; **(c)** limit does not exist; **(d)** 1; **(e)** no, $\lim_{x\to 3} G(x)$ does not exist; **(f)** yes, $\lim_{x\to 0} G(x) = G(0)$; **(g)** yes, $\lim_{x\to 2.9} G(x) = G(2.9)$ **39.** Yes; $\lim_{x\to 5} f(x) = f(5)$ **41.** No; $\lim_{x\to 0} G(x)$ does not exist and $G(0)$ does not exist. **43.** Yes; $\lim_{x\to 3} g(x) = g(3)$ **45.** No; $\lim_{x\to 3} F(x)$ does not exist. **47.** Yes; $\lim_{x\to 3} f(x) = f(3)$ **49.** No; $\lim_{x\to 2} G(x)$ does not equal $G(2)$. **51.** Yes; $\lim_{x\to 5} f(x) = f(5)$ **53.** No; $g(5)$ does not exist and $\lim_{x\to 5} g(x)$ does not exist. **55.** Yes; $\lim_{x\to 4} F(x) = F(4)$ **57.** Yes; $g(x)$ is continuous at each point on $(-4, 4)$. **59.** No; $f(x)$ is not continuous at $x = 0$. **61.** Yes, since $g(x)$ is continuous at each real number **63.** Limit does not exist. **65.** 6 **67.** -0.2887, or $-\dfrac{1}{2\sqrt{3}}$

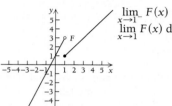

69. 0.75 **71.** 0.25

Technology Connection, p. 128

1–2. Left to the student

Exercise Set 1.3, p. 130

1. (a) $8x + 4h$; **(b)** 48, 44, 40.4, 40.04
3. (a) $-8x - 4h$; **(b)** $-48, -44, -40.4, -40.04$
5. (a) $2x + h + 1$; **(b)** 13, 12, 11.1, 11.01
7. (a) $\dfrac{-2}{x \cdot (x + h)}$; **(b)** $-\frac{2}{35}, -\frac{1}{15}, -\frac{4}{51}, -\frac{40}{501}$
9. (a) -2; **(b)** $-2, -2, -2, -2$
11. (a) $-3x^2 - 3xh - h^2$; **(b)** $-109, -91, -76.51, 75.1501$
13. (a) $2x + h - 3$; **(b)** 9, 8, 7.1, 7.01
15. (a) $2x + h + 4$; **(b)** 16, 15, 14.1, 14.01
17. About $3\frac{1}{3}$% per year; about 0.4% per year; about 2% per year **19.** About 6.5% per year; about 2% per year; about 4.5% per year **21.** About $2\frac{5}{6}$% per year; about 2.2% per year; about 2.5% per year **23.** About $-1\frac{2}{3}$% per year; about 1.2% per year; about -0.4% per year **25.** 2.3 quadrillion BTUs/yr; 0.6 quadrillion BTUs/yr; 1.1 quadrillion BTUs/yr; **27. (a)** 70 pleasure units/unit of product, 39 pleasure units/unit of product, 29 pleasure units/unit of product, 23

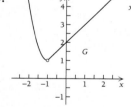

pleasure units/unit of product; **(b)** TW **29. (a)** \$11.35; **(b)** \$22.35; **(c)** \$11.00; **(d)** \$1.10 is the average annual increase in ticket price from the 4th to the 14th year. **31.** \$909.72 is the annual increase in the debt from the 2nd to the 3rd year. **33.** \$19.95 is the cost to produce the 301st unit. **35. (a)** 1.0 lb/month; **(b)** 0.54 lb/month; **(c)** 0.77 lb/month; **(d)** 0.67 lb/month; **(e)** growth rate is greatest in the first 3 months. **37. (a)** Approximately 1.49 hectares/g; **(b)** 1.09 represents the average growth rate, in hectares/g, of home range with respect to body weight when the mammal grows from 200 to 300 g **39. (a)** 1.25 words/min, 1.25 words/min, 0.625 words/min, 0 words/min, 0 words/min; **(b)** TW **41. (a)** 256 ft; **(b)** 128 ft/sec **43. (a)** 125 million people/yr for both countries; **(b)** TW **(c)** *A*: 290 million people/year, −40 million people/year, −50 million people/year, 300 million people/year, *B*: 125 million people/yr in all intervals; **(d)** TW **45. (a)** 1985–86; **(b)** 1975–76, 2003–04, and 2004–05; **(c)** about \$2472 for public and about \$5356 for private **47.** $2ax + ah + b$

49. $\dfrac{1}{\sqrt{x+h}+\sqrt{x}}$

51. $\dfrac{-2x-h}{x^2(x+h)^2}$

53. $\dfrac{1}{(x+1)(x+1+h)}$

55. $\dfrac{-1}{\sqrt{x}\sqrt{x+h}\left(\sqrt{x}+\sqrt{x+h}\right)}$

Technology Connection, p. 141

1. $f'(x) = -\dfrac{3}{x^2}$; $f'(-2) = -\frac{3}{4}$; $f'\left(-\frac{1}{2}\right) = -12$

2. $y = -\frac{3}{4}x - 3$; $y = -12x - 12$

3. Left to the student

Technology Connection, p. 144

Graphs are left to the student.

1. 60, 26, 0, −80 **2.** −96, −11, 24, −11, −56 **3.** −36, 0, 12, 0, −43.47 **4.** Derivative does not exist, 0.41, 1.81, 2.00, 1.15, derivative does not exist

5. TW

Exercise Set 1.4, p. 145

1. (a) and **(b)**

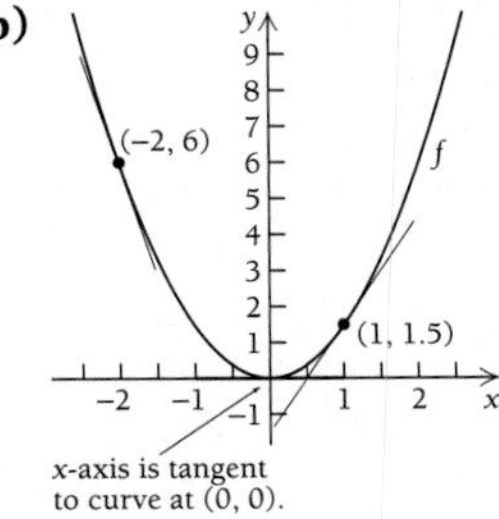

(c) $f'(x) = 3x$; **(d)** −6, 0, 3

3. (a) and **(b)**

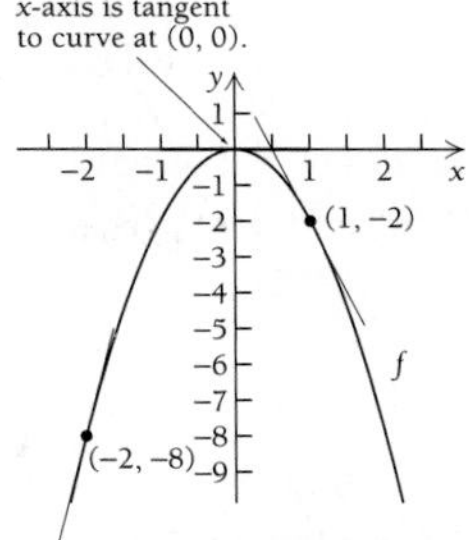

(c) $f'(x) = -4x$; **(d)** 8, 0, −4

5. (a) and **(b)**

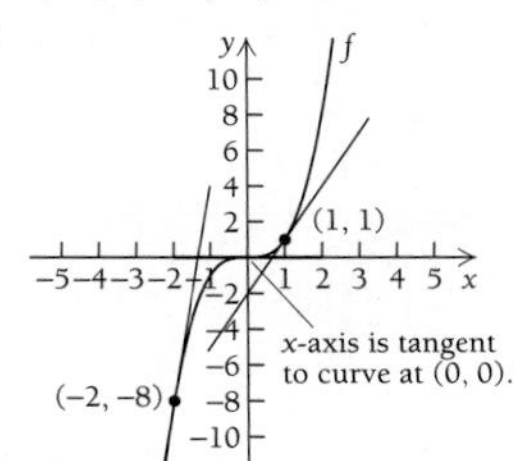

(c) $f'(x) = 3x^2$; **(d)** 12, 0, 3

7. (a) and **(b)** All tangent lines are identical to the graph of the original function.

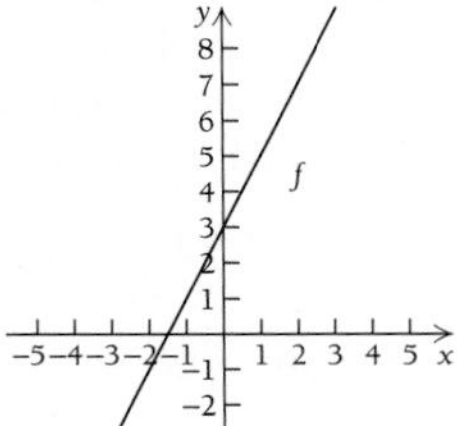

(c) $f'(x) = 2$; **(d)** 2, 2, 2

9. (a) and **(b)** All tangent lines are identical to the graph of the original function.

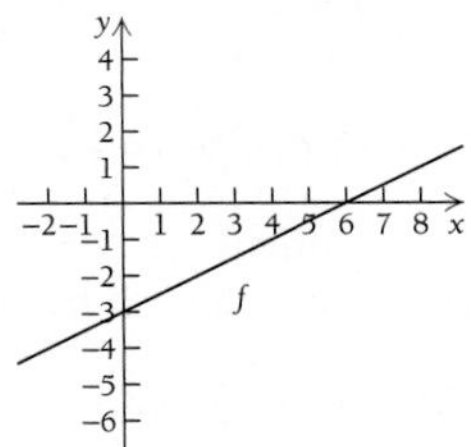

(c) $f'(x) = \dfrac{1}{2}$; **(d)** $\dfrac{1}{2}, \dfrac{1}{2}, \dfrac{1}{2}$

11. (a) and **(b)**

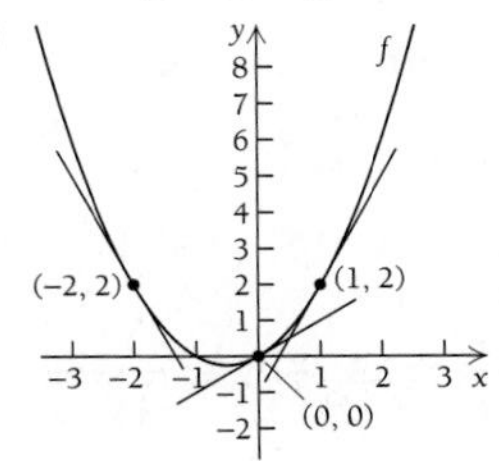

(c) $f'(x) = 2x + 1$; **(d)** −3, 1, 3

13. **(a)** and **(b)**

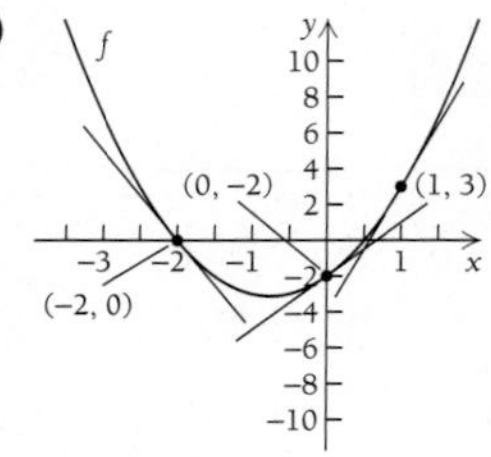

(c) $f'(x) = 4x + 3$; **(d)** $-5, 3, 7$
15. **(a)** and **(b)** There is no tangent line for $x = 0$.

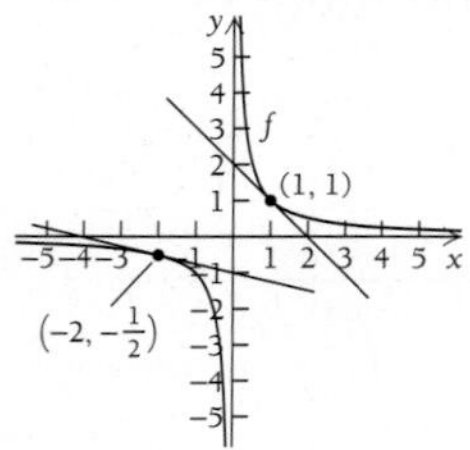

(c) $f'(x) = -\frac{1}{x^2}$; **(d)** $-\frac{1}{4}$, does not exist, -1
17. **(a)** $y = 6x - 9$; **(b)** $y = -2x - 1$; **(c)** $y = 20x - 100$ **19.** **(a)** $y = -2x + 4$; **(b)** $y = -2x - 4$; **(c)** $y = -0.0002x + 0.04$
21. **(a)** $y = 2x + 5$; **(b)** $y = 4$; **(c)** $y = -10x + 29$
23. $f'(x) = m$ **25.** $x_0, x_3, x_4, x_6, x_{12}$ **27.** Answers will vary. **29.** 1, 2, 3, 4, . . . **31.** The function is differentiable at all values. **33.** TW **35.** $f'(x) = -\frac{2}{x^3}$
37. $f'(x) = \frac{1}{(1 + x)^2}$ **39.** $f'(x) = \frac{1}{2\sqrt{x}}$
41. **(a)** -3; **(b)** TW
43–47. Left to the student
49. $f'(x)$ does not exist for $x = 5$.

Technology Connection, p. 149

Graphs are left to the student. **1.** 152, −76, −100, −180 **2.** 36, 0, 12, 0, 43.47 **3.** −2.31, 3.69, 0.81

Technology Connection, p. 155

1. Graph is left to the student. Tangent line is horizontal at $\left(2, \frac{8}{3}\right)$.

Exercise Set 1.5, p. 156

1. $7x^6$ **3.** -3 **5.** 0 **7.** $30x^{14}$ **9.** $-6x^{-7}$
11. $-8x^{-3}$ **13.** $3x^2 + 6x$ **15.** $\frac{4}{\sqrt{x}}$ **17.** $0.9x^{-0.1}$
19. $\frac{2}{5}x^{-1/5}$ **21.** $-\frac{21}{x^4}$ **23.** $\frac{4}{5}$ **25.** $\frac{1}{4\sqrt[4]{x^3}} + \frac{3}{x^2}$
27. $\frac{1}{2\sqrt{x}} + \frac{1}{x\sqrt{x}}$ **29.** $-\frac{10}{3}\sqrt[3]{x^2}$ **31.** $10x - 7$
33. $0.9x^{0.5}$ **35.** $\frac{2}{3}$ **37.** $-\frac{12}{7x^4}$ **39.** $-\frac{5}{x^2} - \frac{2}{3}x^{-1/3}$
41. 4 **43.** $\frac{1}{3}x^{1/3}$ **45.** $-0.02x - 0.5$
47. $-2x^{-5/3} + \frac{3}{4}x^{-1/4} + \frac{6}{5}x^{1/5} - \frac{24}{x^4}$
49. $-\frac{2}{x^2} - \frac{1}{2}$ **51.** 24 **53.** 1
55. **(a)** $y = 10x - 15$; **(b)** $y = x + 3$; **(c)** $y = -2x + 1$ **57.** $(0, -3)$ **59.** $(0, 1)$
61. $\left(\frac{5}{6}, \frac{23}{12}\right)$ **63.** $(-25, 76.25)$ **65.** None **67.** The tangent line is horizontal at *all* points of the graph.
69. $(-1, -4), \left(\frac{5}{3}, 5\frac{13}{27}\right)$ **71.** $(\sqrt{3}, 2 - 2\sqrt{3})$, or approximately $(1.73, -1.46)$; $(-\sqrt{3}, 2 + 2\sqrt{3})$, or approximately $(-1.73, 5.46)$ **73.** $(0, -2), \left(-1, -\frac{11}{6}\right)$
75. $(9.5, 99.75)$ **77.** $(60, 150)$
79. $\left(-2 + \sqrt{3}, \frac{4}{3} - \sqrt{3}\right)$, or approximately $(-0.27, -0.40)$; $\left(-2 - \sqrt{3}, \frac{4}{3} + \sqrt{3}\right)$, or approximately $(-3.73, 3.07)$ **81.** **(a)** $A'(r) = 6.28r$; **(b)** TW
83. **(a)** $w'(t) = 1.82 - 0.1192t + 0.002274t^2$; **(b)** about 21 lb; **(c)** about 0.86 lb/month
85. **(a)** $R'(v) = -\frac{6000}{v^2}$; **(b)** 75 beats/min; **(c)** −0.94 beat/min per mL **87.** **(a)** $\frac{dP}{dt} = 4000t$; **(b)** 300,000 people; **(c)** 40,000 people/yr; **(d)** TW
89. **(a)** $V' = \frac{0.61}{\sqrt{h}}$; **(b)** 244 mi; **(c)** 0.0031 mi/ft; **(d)** TW **91.** $(2, \infty)$ **93.** $(-\infty, -1)$ and $(3, \infty)$
95. $(0, -2), \left(\sqrt{\frac{1}{3}}, -\frac{55}{27}\right), \left(-\sqrt{\frac{1}{3}}, -\frac{55}{27}\right)$ **97.** $2x$
99. $\frac{5}{4}x - 1$ **101.** $3x^2 - 6x - \frac{2}{x^2} - \frac{8}{x^3}$
103. $\frac{\sqrt{7}}{2\sqrt{x}}$ **105.** $2x - 6$ **107.** $1 + \frac{5}{3}x^{-1/6} + \frac{2}{3}x^{-1/3}$
109. $\frac{d}{dx}1 = \frac{d}{dx}x^0 = 0x^{-1} = 0$ **111.** TW
113.

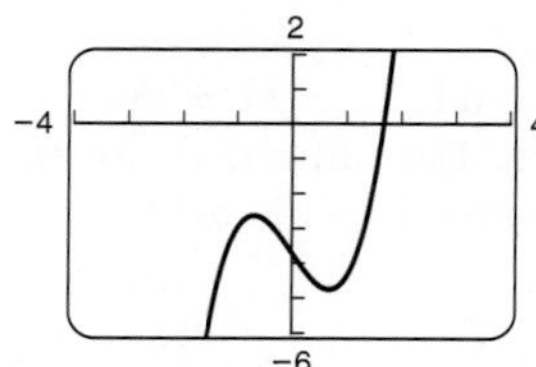

$(-0.692, -2.639), (0.692, -4.761)$
115.

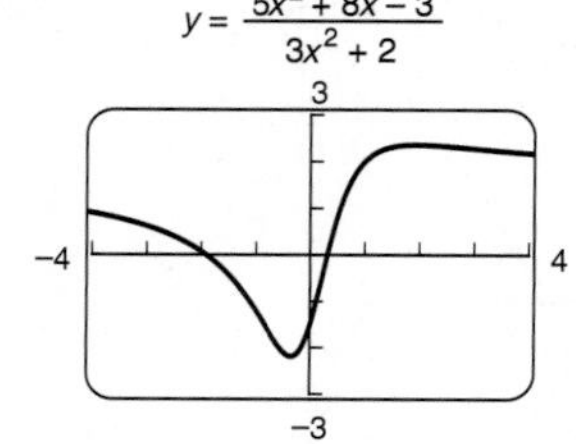

$(-0.346, -2.191), (1.929, 2.358)$

117.

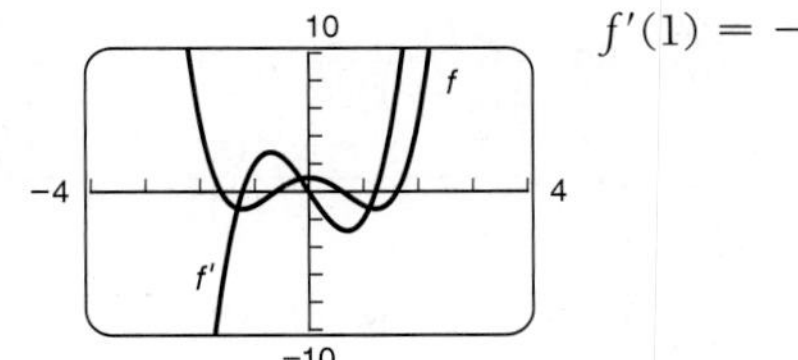

$f'(1) = -2$

119.

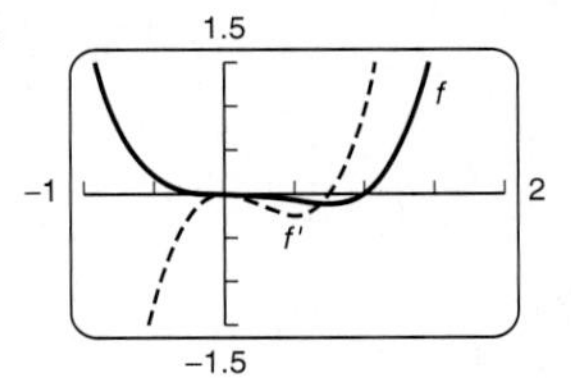

$f'(1) = 1$

121.

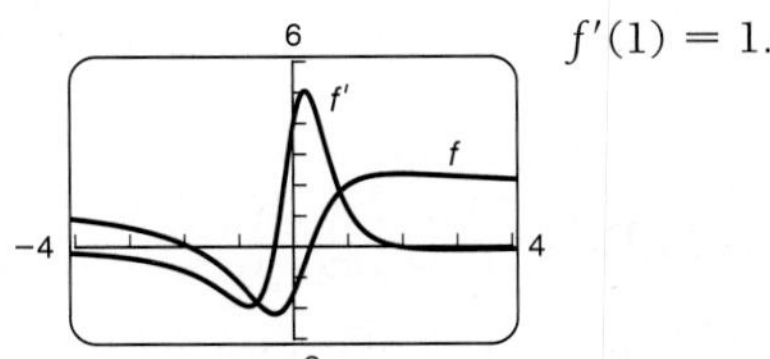

$f'(1) = 1.2$

Technology Connection, p. 163

1. (c) **2–5.** Left to the student

Technology Connection, p. 164

1–2. Left to the student

Exercise Set 1.6, p. 165

1. $11x^{10}$ **3.** $12x + 7$
5. $20x^4 + 60x^2$ **7.** $\frac{15}{2}x^{3/2} + 4x$ **9.** $24x^2 + 12x + 11$
11. $\frac{9\sqrt{t}}{2} - \frac{1}{2\sqrt{t}} + 2$ **13.** $4x^3$, for $x \neq 0$
15. $8x^3 + 1$, for $x \neq 0$ **17.** $8x + 2$, for $x \neq \frac{1}{2}$
19. 1, for $t \neq -4$ **21.** $48x^3 + 3x^2 + 22x + 17$
23. $\frac{-2x(5x^3 - 3x - 15)}{(2x^3 + 3)^2}$
25. $120x^2 + \frac{25}{2}x^{3/2} + \frac{3}{2}x^{-1/2} + 24$
27. $\frac{3}{(3 - t)^2} + 15t^2$
29. $2x + 6$ **31.** $2x(x^2 - 4)(3x^2 - 4)$
33. $5 - 100x^{-3} + 30x^{-4}$ **35.** $3t^2 - 1 + \frac{6}{t^2}$
37. $\frac{-x^4 - 3x^2 - 2x}{(x^3 - 1)^2} - 10x$
39. $\frac{(x^{1/2} + 3)\left(\frac{1}{3}x^{-2/3}\right) - (x^{1/3} - 7)\left(\frac{1}{2}x^{-1/2}\right)}{(\sqrt{x} + 3)^2}$, or $\frac{6 - \sqrt{x} + 21x^{1/6}}{6x^{2/3}(\sqrt{x} + 3)^2}$
41. $\frac{2x^{-1} + 1}{(x^{-1} + 1)^2}$, or $\frac{x(x + 2)}{(x + 1)^2}$, for $x \neq 0$
43. $\frac{-1}{(t - 4)^2}$ **45.** $\frac{-2(x^2 - 3x - 1)}{(x^2 + 1)^2}$
47. $\frac{(t^2 - 2t + 4)(-2t + 3) - (-t^2 + 3t + 5)(2t - 2)}{(t^2 - 2t + 4)^2}$, or $\frac{-t^2 - 18t + 22}{(t^2 - 2t + 4)^2}$ **49–95.** Left to the student
97. **(a)** $y = 2$; **(b)** $y = \frac{1}{2}x + 2$ **99.** **(a)** $y = x + 5$; **(b)** $y = \frac{21}{4}x - \frac{21}{4}$ **101.** $-\$0.006875$/jacket
103. $-\$0.0053$/jacket **105.** About \$0.0016/jacket
107. \$1.64/vase **109.** **(a)** $P'(t) = 57.6t^{0.6} - 104$; **(b)** \$461.4 billion/yr; **(c)** TW
111. **(a)** $T'(t) = \frac{-4(t^2 - 1)}{(t^2 + 1)^2}$; **(b)** 100.2 °F; **(c)** −0.48 °F/hr **113.** $30t^2 + 10t - 15$
115. $3x^2\left(\frac{x^2 + 1}{x^2 - 1}\right) + \frac{-4x}{(x^2 - 1)^2}(x^3 - 8)$, or $\frac{3x^6 - 4x^4 - 3x^2 + 32x}{(x^2 - 1)^2}$ **117.** $\frac{-x^6 + 4x^3 - 24x^2}{(x^4 - 3x^3 - 5)^2}$
119. **(a)** $f'(x) = \frac{-2x}{(x^2 - 1)^2}$; **(b)** $g'(x) = \frac{-2x}{(x^2 - 1)^2}$; **(c)** TW **121.** TW **123.** **(a)** Definition of derivative; **(b)** adding and subtracting the same quantity is the same as adding 0; **(c)** the limit of a sum is the sum of the limits; **(d)** factoring; **(e)** the limit of a product is the product of the limits and $\lim_{h \to 0} f(x + h) = f(x)$; **(f)** definition of derivative; **(g)** using Leibniz notation
125. \$2.58/jacket; \$0.014/jacket at $x = 184$
127.

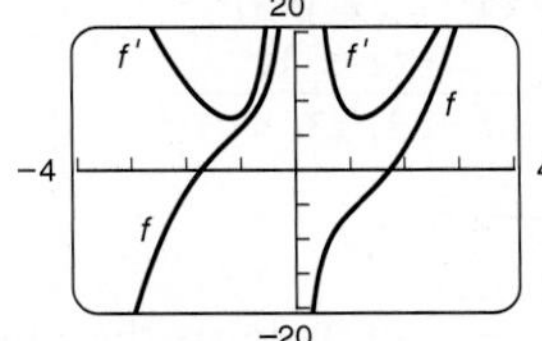

There are no points at which the tangent line is horizontal.
129.

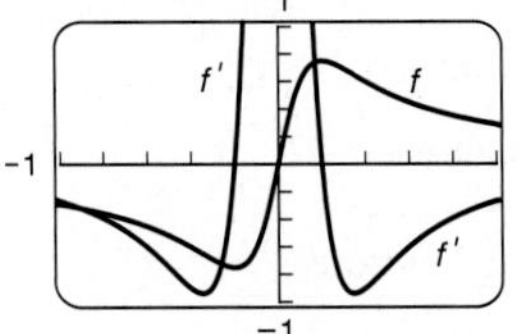

$(-0.2, -0.75)$, $(0.2, 0.75)$
131.

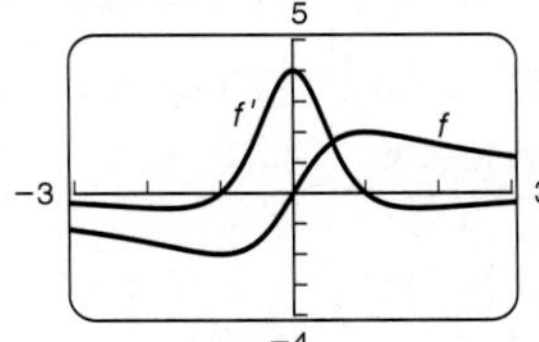

$(-1, -2)$, $(1, 2)$

Exercise Set 1.7, p. 174

1. $8x + 4$ **3.** $-55(7 - x)^{54}$ **5.** $\frac{4}{\sqrt{1 + 8x}}$

7. $\frac{3x}{\sqrt{3x^2 - 4}}$ **9.** $-640x(8x^2 - 6)^{-41}$

11. $4(x - 4)^7(2x + 3)^5(7x - 6)$

13. $\frac{-6}{(3x + 8)^3}$ **15.** $\frac{4x(5x + 14)}{(7 - 5x)^4}$

17. $9x^2(1 + x^3)^2 - 32x^7(2 + x^8)^3$

19. $4x - 400$ **21.** $\frac{1}{2\sqrt{x}} + 3(x - 3)^2$

23. $-5(2x - 3)^3(10x - 3)$

25. $(3x - 1)^6(2x + 1)^4(72x + 11)$

27. $\frac{2x(5x - 1)}{\sqrt{4x - 1}}$ **29.** $\frac{5x^4 + 6}{3\sqrt[3]{(x^5 + 6x)^2}}$, or $\frac{5x^4 + 6}{3x^{2/3}(x^4 + 6)^{2/3}}$

31. $\frac{44(3x - 1)^3}{(5x + 2)^5}$ **33.** $\frac{-7}{2(x + 3)^{3/2}(4 - x)^{1/2}}$

35. $200(2x^3 - 3x^2 + 4x + 1)^{99}(3x^2 - 3x + 2)$

37. $\frac{68(5x - 1)^3}{(2x + 3)^5}$ **39.** $\frac{-1}{(x - 1)^{3/2}(x + 1)^{1/2}}$

41. $\frac{-(2x + 3)^3(6x + 61)}{(3x - 2)^6}$ **43.** $\frac{(3x - 4)^{1/4}(138x - 19)}{(2x + 1)^{1/3}}$

45. $\frac{1}{2\sqrt{u}}, 2x, \frac{x}{\sqrt{x^2 - 1}}$

47. $50u^{49}, 12x^2 - 4x, 50(4x^3 - 2x^2)^{49}(12x^2 - 4x)$

49. $2u + 1, 3x^2 - 2, (2x^3 - 4x + 1)(3x^2 - 2)$

51. $3x^2(10x^3 + 13)$ **53.** $\frac{2(2x - 1)}{3(2x^2 - 2x + 5)^{2/3}}$

55. $\frac{3(-6t - 11)}{(5 + 3t)^2(6 + 3t)^2}$, or $\frac{-6t - 11}{3(t + 2)^2(3t + 5)^2}$

57. $y = \frac{5}{4}x + \frac{3}{4}$ **59.** $y = 4x - 3$

61. (a) $\frac{2x - 3x^2}{(1 + x)^6}$; **(b)** $\frac{2x - 3x^2}{(1 + x)^6}$; **(c)** They are the same.

63. $f(x) = x^5, g(x) = 3x^2 - 7$

65. $f(x) = \frac{x + 1}{x - 1}, g(x) = x^3$

67. -216 **69.** $4(13)^{-2/3}$, or about 0.72

71. \$1,000,000/item

73. $P'(x) = \frac{500(2x - 0.1)}{\sqrt{x^2 - 0.1x}} - \frac{4000x}{3(x^2 + 2)^{2/3}}$

75. (a) $0.84x^3 - 17.76x^2 + 101.06x - 18.92$; **(b)** TW; **(c)** \$336 billion/yr **77. (a)** $dA/di = 3000(1 + i)^2$; **(b)** TW **79. (a)** $D(t) = \frac{80{,}000}{1.6t + 9}$; **(b)** -4.482 units/day

81. (a) $D(c) = 4.25c + 106.25$, $c(w) = \frac{95w}{43.2} \approx 2.199w$ **(b)** 4.25 mg/unit of creatine clearance; **(c)** 2.199 units of creatine clearance/kg; **(d)** 9.35 mg/kg; **(e)** TW

83. $\frac{6x^7 + 32x^5 + 5x^4}{(x^3 + 6x + 1)^{2/3}}$ **85.** $\frac{3x^2(x - 2)}{2(x - 1)^{5/2}}$

87. $\frac{1}{(1 - x)\sqrt{1 - x^2}}$ **89.** $\frac{3(x^2 - x - 1)^2(x^2 + 4x - 1)}{(x^2 + 1)^4}$

91. $\frac{6\sqrt{t} + 1}{4\sqrt{t}\sqrt{3t + \sqrt{t}}}$ **93.** TW

95.

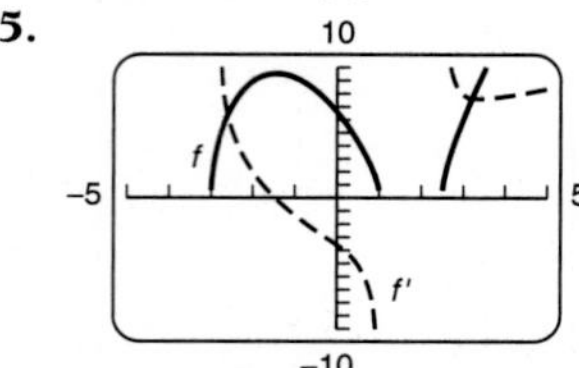

$(-1.47481, 9.4878)$

97. $\frac{2x - 40}{(x - 10)^{3/2}}$

Exercise Set 1.8, p. 181

1. $20x^3$ **3.** $24x^2$ **5.** 8 **7.** 0 **9.** $\frac{6}{x^4}$ **11.** $\frac{-1}{4x^{3/2}}$

13. $12x^2 + \frac{6}{x^3}$ **15.** $\frac{-4}{25x^{9/5}}$ **17.** $\frac{48}{x^5}$

19. $14(x^2 + 3x)^5(13x^2 + 39x + 27)$

21. $10(2x^2 - 3x + 1)^8(152x^2 - 228x + 85)$

23. $\frac{3(x^2 + 2)}{4(x^2 + 1)^{5/4}}$ **25.** $\frac{-2}{9x^{4/3}}$ **27.** $\frac{45x^4 - 54x^2 - 3}{16(x^3 - x)^{5/4}}$

29. $\frac{5}{8}x^{-3/4} - \frac{1}{4}x^{-3/2}$

31. $24x^{-5} + 6x^{-4}$ **33.** $24x - 2$

35. $\frac{44}{(2x - 3)^3}$ **37.** 24 **39.** $720x$

41. $120x^{-6} + \frac{15}{16}x^{-7/2}$

43. 0 **45. (a)** $v(t) = 3t^2 + 1$; **(b)** $a(t) = 6t$; **(c)** $v(4) = 49$ ft/sec, $a(4) = 24$ ft/sec^2

47. (a) $v(t) = 3$; **(b)** $a(t) = 0$; **(c)** $v(2) = 3$ mi/hr, $a(2) = 0$ mi/hr^2 **(d)** TW

49. (a) 144 ft; **(b)** 96 ft/sec; **(c)** 32 ft/sec^2

51. $v(2) = 19.62$ m/sec, $a(2) = 9.81$ m/sec^2

53. (a) The velocity at $t = 20$ sec is greater, since the slope of a tangent line is greater there. **(b)** The acceleration is positive, since the velocity (slope of a tangent line) is increasing over time. **55. (a)** \$146,000/month, \$84,000/month, $-$\$4000/month; **(b)** $-$\$68,000/month^2, $-$\$56,000/month^2, $-$\$32,000/month^2; **(c)** TW

57. $\frac{6}{(1 - x)^4}$ **59.** $\frac{-15}{(2x + 1)^{7/2}}$ **61.** $\frac{3x^{1/2} - 1}{2x^{3/2}(x^{1/2} - 1)^3}$

63. $k(k - 1)(k - 2)(k - 3)(k - 4)x^{k-5}$

65. $f'(x) = \frac{3}{(x + 2)^2}, f''(x) = \frac{-6}{(x + 2)^3}, f'''(x) = \frac{18}{(x + 2)^4},$ $f^4(x) = \frac{-72}{(x + 2)^5}$ **67.** 2.29 sec

69.

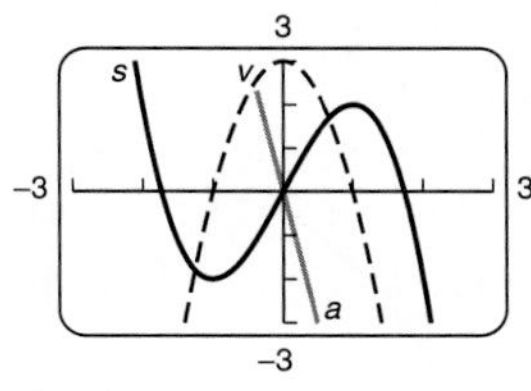

$v(t)$ switches at $t = 0$.

71.

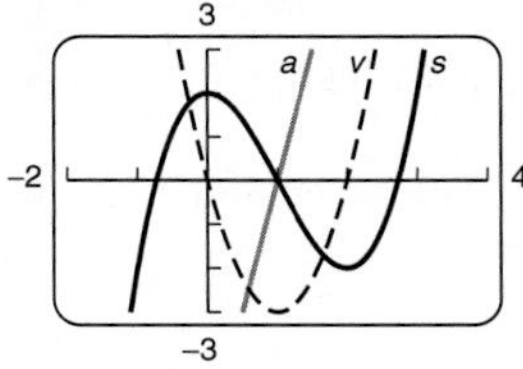

$v(t)$ switches at $t = 1$.

Chapter Review Exercises, p. 190

1. False **2.** False **3.** True **4.** False **5.** True **6.** True **7.** False **8.** True **9.** (e) **10.** (c) **11.** (a) **12.** (f) **13.** (b) **14.** (d)

15. (a)

$x \to -7^-$	$f(x)$
−8	−11
−7.5	−10.5
−7.1	−10.1
−7.01	−10.01
−7.001	−10.001
−7.0001	−10.0001

$x \to -7^+$	$f(x)$
−6	−9
−6.5	−9.5
−6.9	−9.9
−6.99	−9.99
−6.999	−9.999
−6.9999	−9.9999

(b) $\lim_{x\to -7^-} f(x) = -10$
$\lim_{x\to -7^+} f(x) = -10$
$\lim_{x\to -7} f(x) = -10$

16.

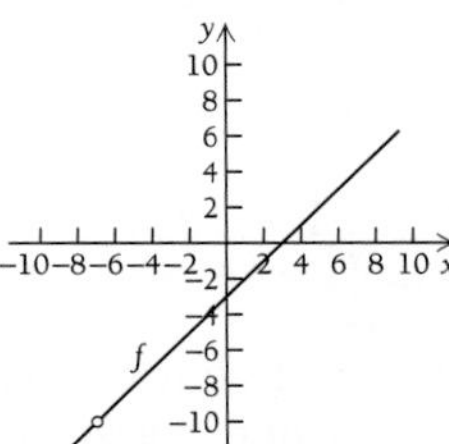

17. $\lim_{x\to -7} \dfrac{x^2 + 4x - 21}{x + 7} = \lim_{x\to -7} \dfrac{(x + 7)(x - 3)}{x + 7} = \lim_{x\to -7} (x - 3) = -10$ **18.** −4 **19.** 10 **20.** −12 **21.** 3

22. Not continuous, since $\lim_{x\to -2} g(x)$ does not exist **23.** Continuous **24.** −4 **25.** −4 **26.** Continuous, since $\lim_{x\to 1} g(x) = g(1)$ **27.** Does not exist **28.** −2 **29.** Not continuous, since $\lim_{x\to -2} g(x)$ does not exist **30.** 2 **31.** −3 **32.** $4x + 2h$ **33.** $y = x - 1$ **34.** (4, 5) **35.** (5, −108) **36.** $45x^4$ **37.** $\frac{8}{3}x^{-2/3}$ **38.** $\frac{24}{x^9}$ **39.** $6x^{-3/5}$ **40.** $0.7x^6 - 12x^3 - 3x^2$ **41.** $\frac{5}{2}x^5 + 32x^3 - 2$ **42.** $2x, x \neq 0$ **43.** $\dfrac{-x^2 + 16x + 8}{(8 - x)^2}$ **44.** $2(5 - x)(2x - 1)^4(-7x + 26)$ **45.** $35x^4(x^5 - 3)^6$ **46.** $\dfrac{x(11x + 4)}{(4x + 2)^{1/4}}$ **47.** $-48x^{-5}$ **48.** $3x^5 - 60x + 26$

49. (a) $v(t) = 1 + 4t^3$; **(b)** $a(t) = 12t^2$; **(c)** $v(2) = 33$ ft/sec, $a(2) = 48$ ft/sec^2

50. (a) $A_C(x) = 5x^{-1/2} + 100x^{-1}$, $A_R(x) = 40$, $A_P(x) = 40 - 5x^{-1/2} - 100x^{-1}$; **(b)** average cost is dropping at approximately \$1.33 per item.

51. (a) $P'(t) = 100t$; **(b)** 30,000; **(c)** 2000/yr

52. $(f \circ g)(x) = 4x^2 - 4x + 6$; $(g \circ f)(x) = -2x^2 - 9$

53. $\dfrac{-9x^4 - 4x^3 + 9x + 2}{2\sqrt{1 + 3x}(1 + x^3)^2}$ **54.** −0.25 **55.** $\frac{1}{6}$

56.

(−1.7137, 37.445), (0, 0), (1.7137, −37.445)

Chapter 1 Test, p. 192

1. [1.1] **(a)**

$x \to 6^-$	$f(x)$
5	11
5.7	11.7
5.9	11.9
5.99	11.99
5.999	11.999
5.9999	11.9999

$x \to 6^+$	$f(x)$
7	13
6.5	12.5
6.1	12.1
6.01	12.01
6.001	12.001
6.0001	12.0001

(b) $\lim_{x \to 6^-} f(x) = 12$
$\lim_{x \to 6^+} f(x) = 12$
$\lim_{x \to 6} f(x) = 12$

2. [1.1]

3. [1.2] $\lim_{x \to 6} \frac{x^2 - 36}{x - 6} = \lim_{x \to 6} \frac{(x + 6)(x - 6)}{x - 6} = \lim_{x \to 6} (x + 6) = 12$

4. [1.1] Does not exist **5.** [1.1] 0 **6.** [1.1] Does not exist **7.** [1.1] 2 **8.** [1.1] 4 **9.** [1.1] 1 **10.** [1.1] 1 **11.** [1.1] 1 **12.** [1.2] Continuous **13.** [1.2] Not continuous, since $\lim_{x \to 3} f(x)$ does not exist **14.** [1.1, 1.2] Does not exist **15.** [1.1, 1.2] 1 **16.** [1.1, 1.2] No **17.** [1.1, 1.2] 3 **18.** [1.1, 1.2] 3 **19.** [1.1, 1.2] Yes **20.** [1.1, 1.2] 6 **21.** [1.1, 1.2] $\frac{1}{8}$ **22.** [1.1, 1.2] Does not exist, since $\lim_{x \to 0^-} \frac{1}{x} \neq \lim_{x \to 0^+} \frac{1}{x}$

23. [1.3] $4x + 3 + 2h$ **24.** [1.4] $y = \frac{3}{4}x + 2$ **25.** [1.5] $(0, 0)$, $(2, -4)$ **26.** [1.5] $23x^{22}$ **27.** [1.5] $\frac{4}{3}x^{-2/3} + \frac{5}{2}x^{-1/2}$ **28.** [1.5] $\frac{10}{x^2}$ **29.** [1.5] $\frac{5}{4}x^{1/4}$ **30.** [1.5] $-1.0x + 0.61$ **31.** [1.5] $x^2 - 2x + 2$ **32.** [1.6] $\frac{-6(x - 2)}{x^4}$

33. [1.6] $\frac{5}{(5 - x)^2}$ **34.** [1.7] $(x + 3)^3(7 - x)^4(-9x + 13)$ **35.** [1.7] $-5(x^5 - 4x^3 + x)^{-6}(5x^4 - 12x^2 + 1)$ **36.** [1.6, 1.7] $\frac{2x^2 + 5}{\sqrt{x^2 + 5}}$ **37.** [1.8] $24x$

38. [1.6] **(a)** $A_R = 50$, $A_C = x^{-1/3} + 750x^{-1}$, $A_P = 50 - x^{-1/3} - 750x^{-1}$; **(b)** average cost is dropping at approximately $11.74 per item.
39. [1.5] **(a)** $M'(t) = -0.003t^2 + 0.2t$; **(b)** 9; **(c)** 1.7 words/min
40. [1.7] $(f \circ g)(x) = 4x^6 - 2x^3$; $(g \circ f)(x) = 2(x^2 - x)^3$
41. [1.6, 1.7] $\frac{-1 - 9x}{2(1 - 3x)^{2/3}(1 + 3x)^{5/6}}$
42. [1.2] 27 **43.** [1.5]

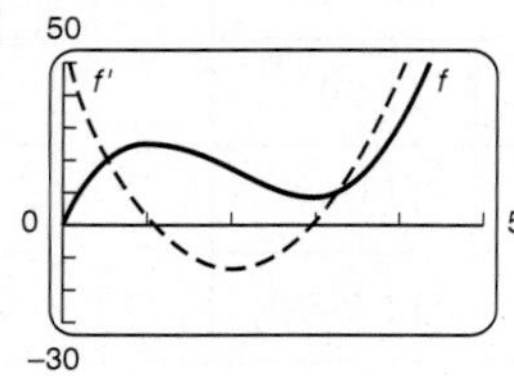

(1.0836, 25.1029) and (2.9503, 8.6247) **44.** [1.5] 0.5

Extended Technology Application, p. 195

1.

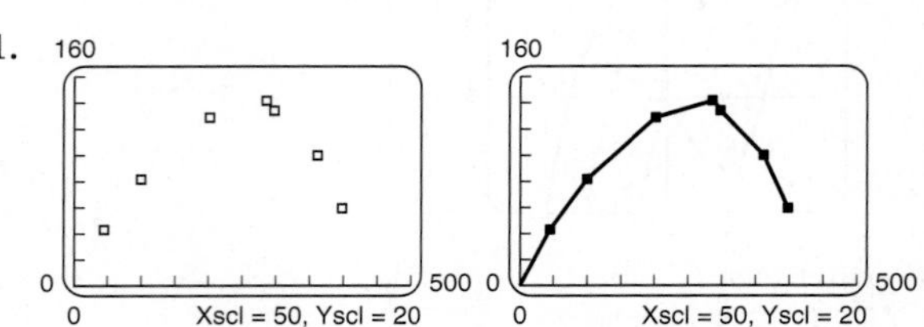

2. (a) $y = -0.0000045x^3 + 0.000204x^2 + 0.7806x + 4.6048$;
(b)
(c) acceptable fit;
(d) about 441 ft;
(e) $dy/dx = -0.0000135x^2 + 0.000408x + 0.7806$;
(f) approximately (256, 142); at about 256 ft from home plate, the ball reached its maximum height of approximately 142 ft.
3. (a) $y = -0.0000000024x^4 - 0.0000026x^3 - 0.00026x^2 + 0.8150x + 4.3026$;
(b)
(c) acceptable fit;
(d) about 440 ft;
(e) $dy/dx = -0.0000000096x^3 - 0.0000078x^2 - 0.00053x + 0.815$; **(f)** approximately (257, 142); at about 257 ft from home plate, the ball reached its maximum height of approximately 142 ft.
4. (a)
(b) 450 ft; **(c)** $\frac{dy}{dx} = \frac{303.75 - 0.003x^2}{\sqrt{202{,}500 - x^2}}$;
(d) approximately (318, 152); at about 318 ft from home plate, the ball reached its maximum height of approximately 152 ft. **5.** The two models are very similar. The main difference seems to be that the maximum height is reached further from home plate with the model in Exercise 4.
6. 466 ft, 442 ft, 430 ft **7.** The estimate of the reporters is way off. Even with a low trajectory, the ball would at best have traveled a horizontal distance of about 526 ft.

CHAPTER 2

Technology Connection, p. 208

1.

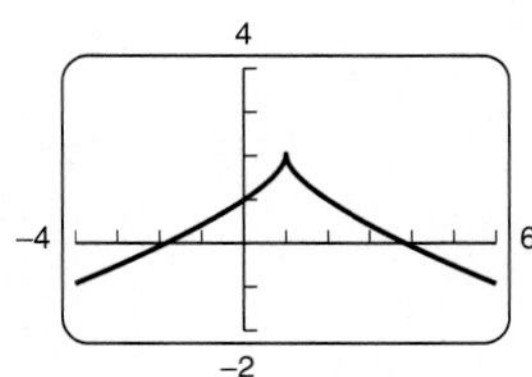

2.

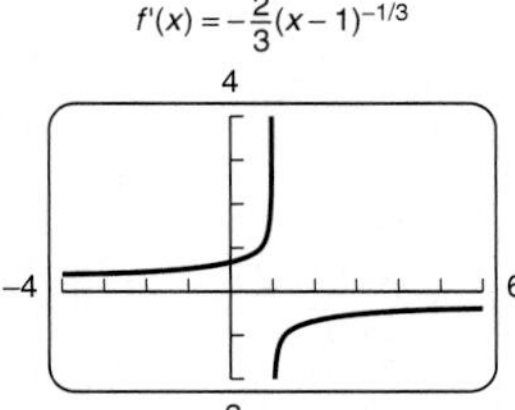

The derivative is not defined at $(1, 2)$.

Technology Connection, p. 211

1. Relative maximum at $(-1, 19)$; relative minimum at $(2, -8)$

Exercise Set 2.1, p. 211

1. Relative minimum at $(-2, 1)$

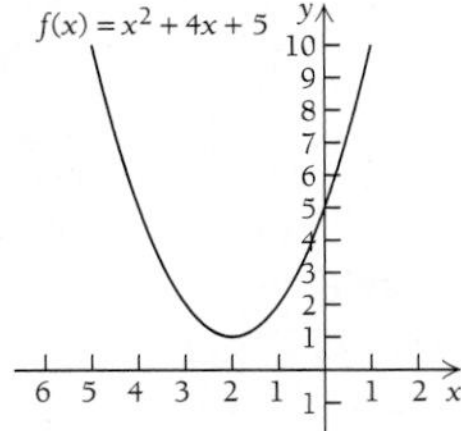

3. Relative maximum at $\left(-\frac{1}{2}, \frac{21}{4}\right)$

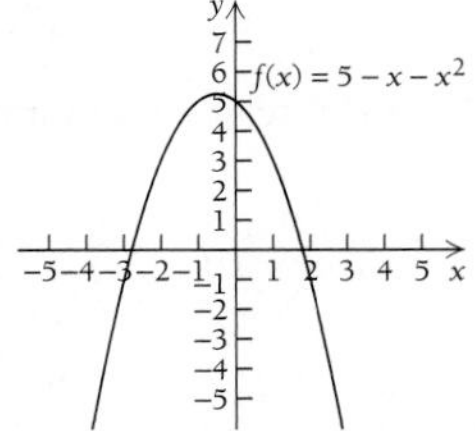

5. Relative minimum at $(-1, -2)$

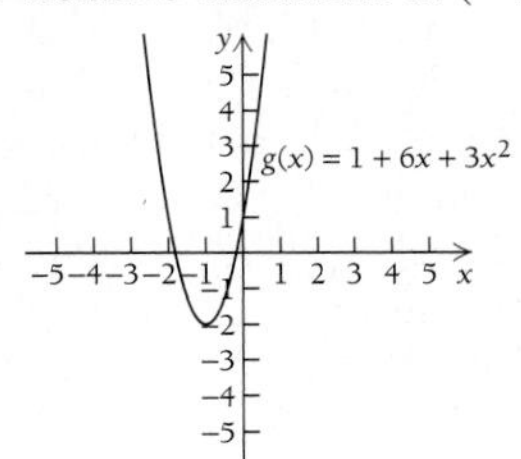

7. Relative minimum at $(1, 1)$; relative maximum at $\left(-\frac{1}{3}, \frac{59}{27}\right)$

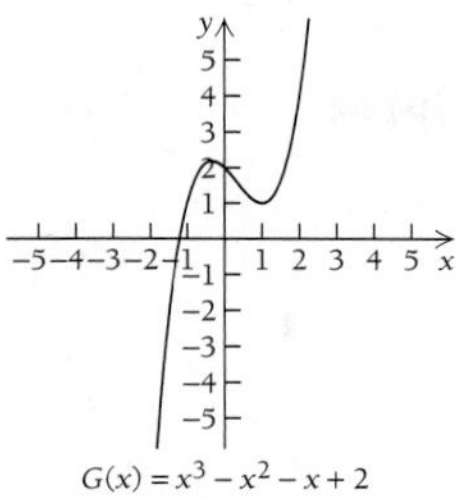

9. Relative minimum at $(1, 4)$; relative maximum at $(-1, 8)$

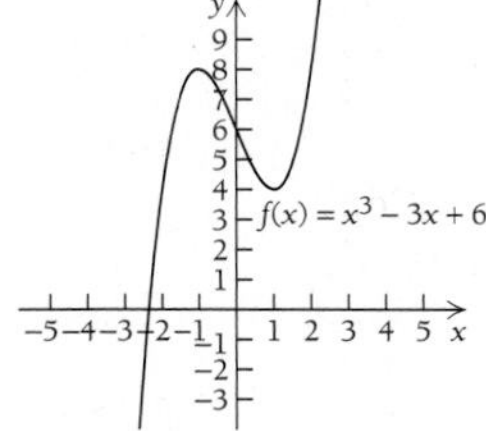

11. Relative minimum at $(0, 0)$; relative maximum at $(-1, 1)$

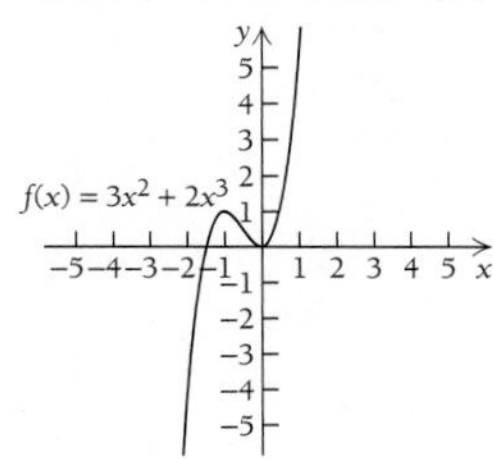

13. No relative extrema exist.

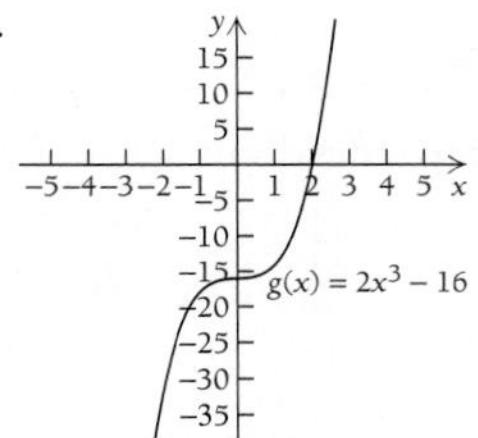

15. Relative minimum at $(4, -22)$; relative maximum at $(0, 10)$

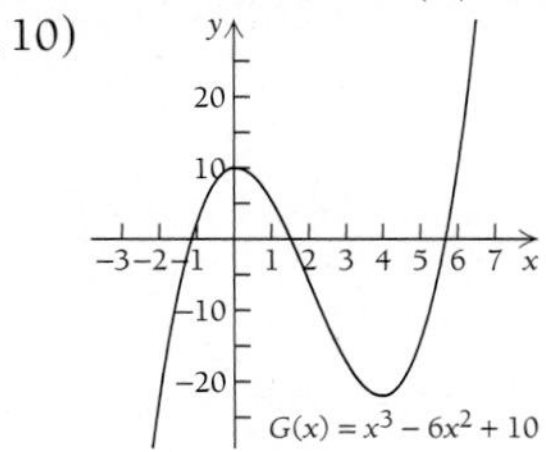

17. Relative maximum at $\left(\frac{3}{4}, \frac{27}{256}\right)$

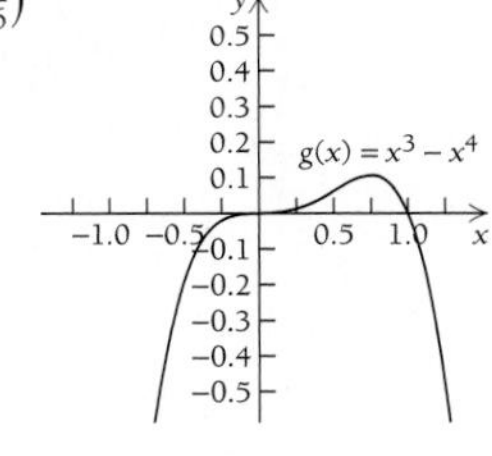

19. No relative extrema exist.

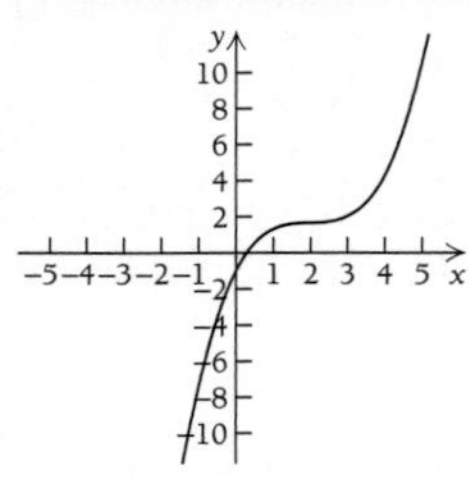

$f(x) = \frac{1}{3}x^3 - 2x^2 + 4x - 1$

21. Relative minima at $(-\sqrt{5}, -32)$ and $(\sqrt{5}, -32)$; relative maximum at $(0, 18)$

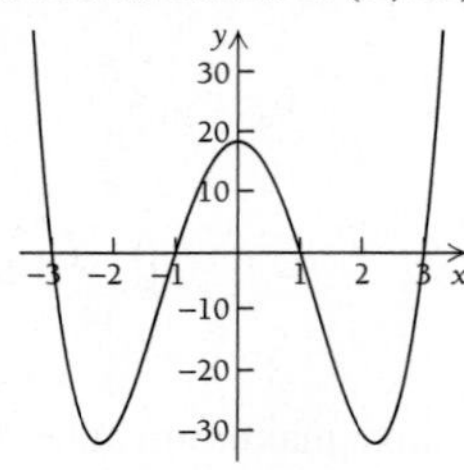

$g(x) = 2x^4 - 20x^2 + 18$

23. No relative extrema exist.

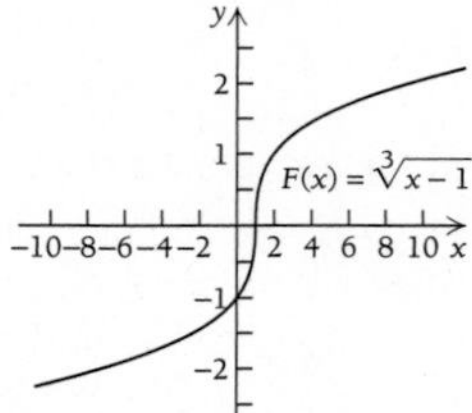

25. Relative maximum at $(0, 1)$

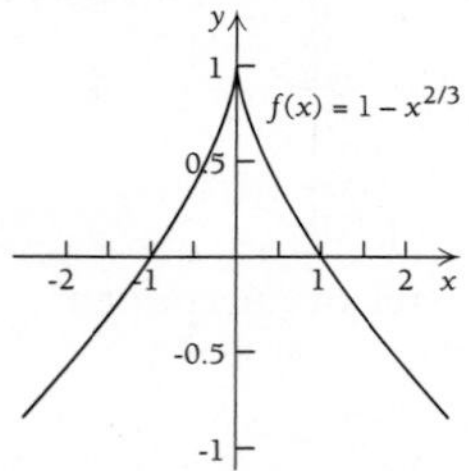

27. Relative minimum at $(0, -8)$

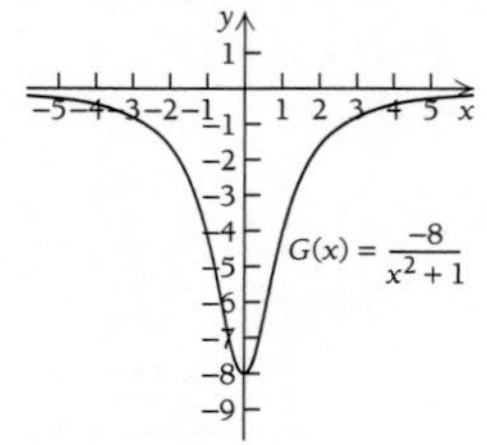

29. Relative minimum at $(-1, -2)$; relative maximum at $(1, 2)$

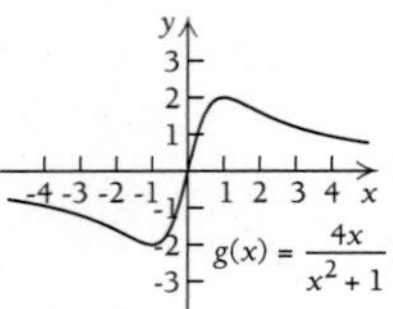

31. No relative extrema exist.

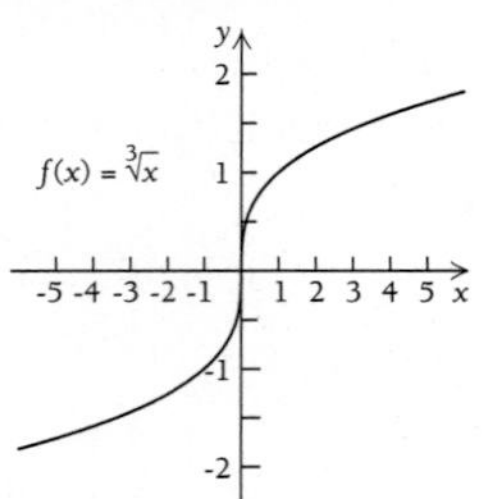

33. Relative minimum at $(-1, 2)$

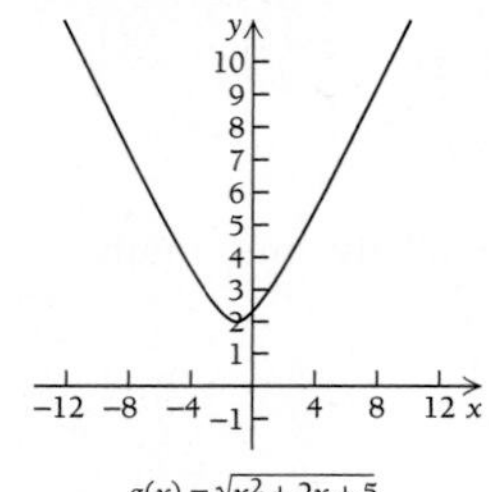

35–77. Left to the student.

79. Relative maximum at $(6, 102.2)$

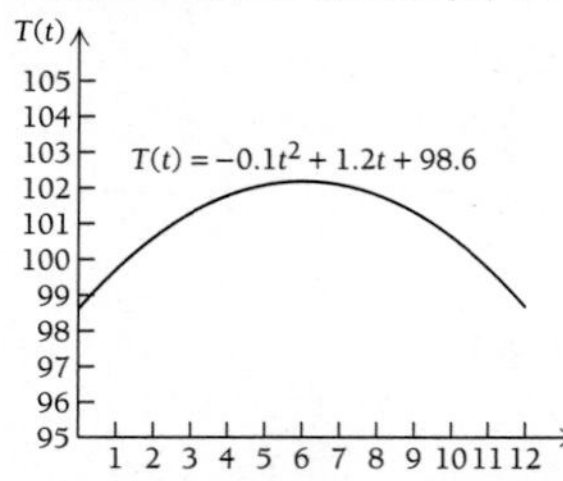

81. Relative maximum at $(2.5553, 9.5436)$; relative minimum at $(8.8409, 6.2531)$

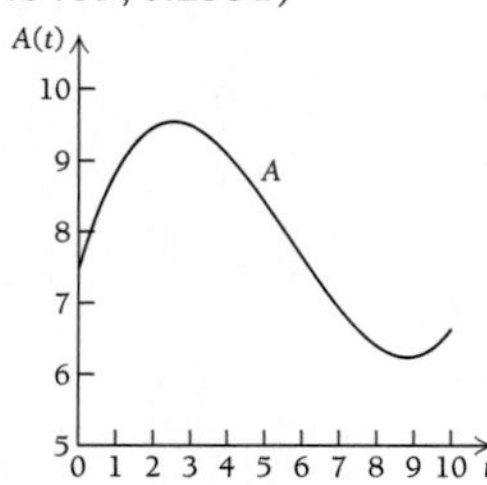

83. TW **85.** Increasing on $[-1, \infty)$, decreasing on $(-\infty, -1]$; relative minimum at $x = -1$. Graph is left to the student. **87.** Increasing on $(-\infty, 1]$, decreasing on $[1, \infty)$; relative maximum at $x = 1$. Graph is left to the student. **89.** Increasing on $[-4, 2]$, decreasing on $(-\infty, -4]$ and $[2, \infty)$; relative minimum at $x = -4$, relative maximum at $x = 2$. Graph is left to the student.

91. 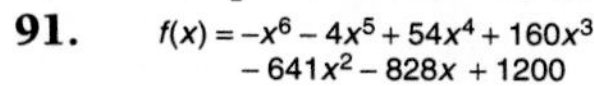

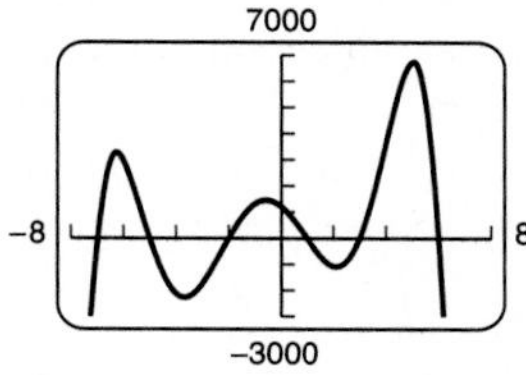

Relative minima at $(-3.683, -2288.03)$ and $(2.116, -1083.08)$; relative maxima at $(-6.262, 3213.8)$ and $(-0.559, 1440.06)$ and $(5.054, 6674.12)$

93. $f(x) = \sqrt[3]{|4 - x^2|} + 1$

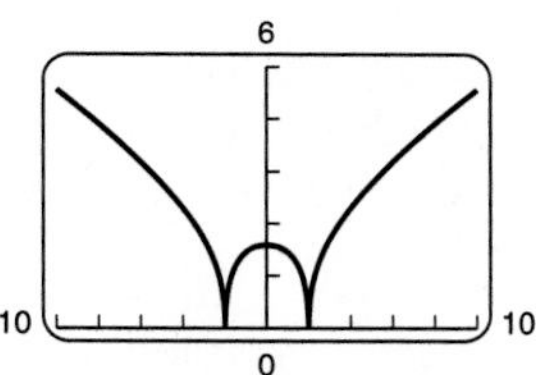

Relative minima at $(-2, 1)$ and $(2, 1)$; relative maximum at $(0, 2.587)$ **95.** TW **97.** TW

Technology Connection, p. 224

1. Relative minimum at $(1, -1)$; inflection points at $(0, 0)$, $(0.553, -0.512)$, $(1.447, -0.512)$, and $(2, 0)$

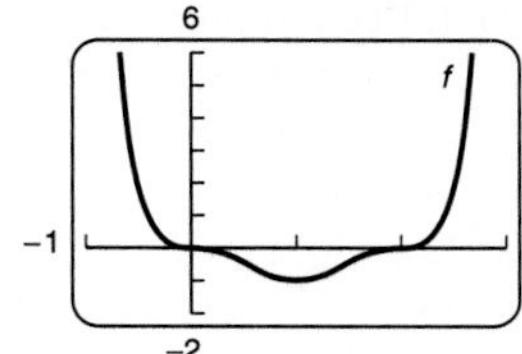

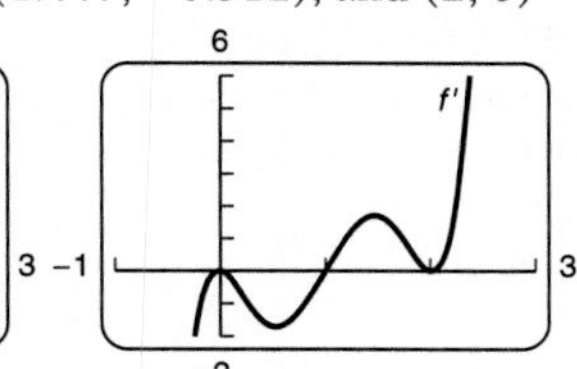

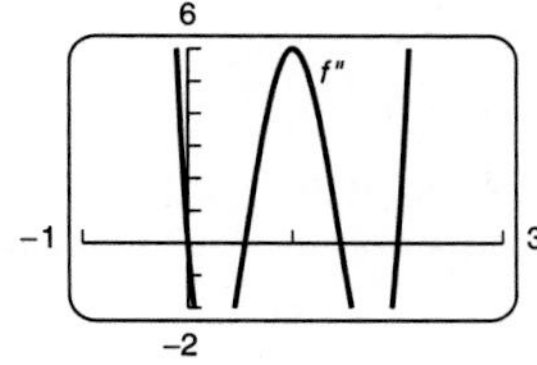

Technology Connection, p. 228

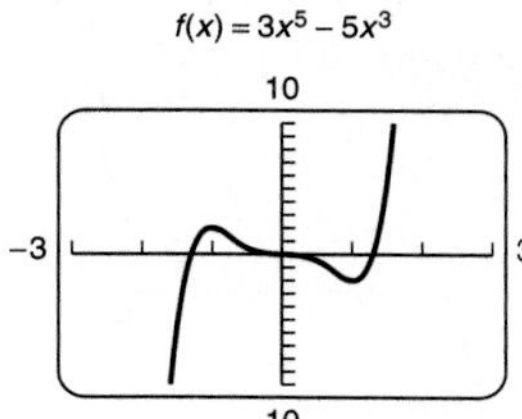

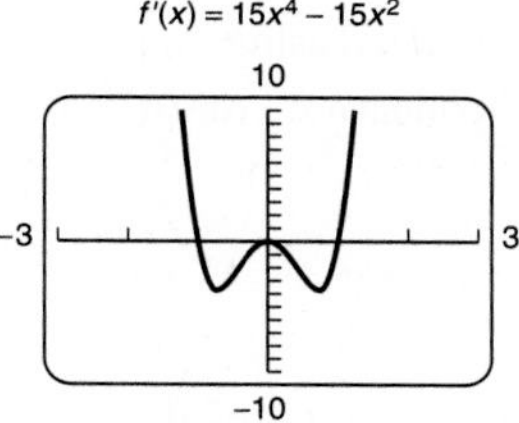

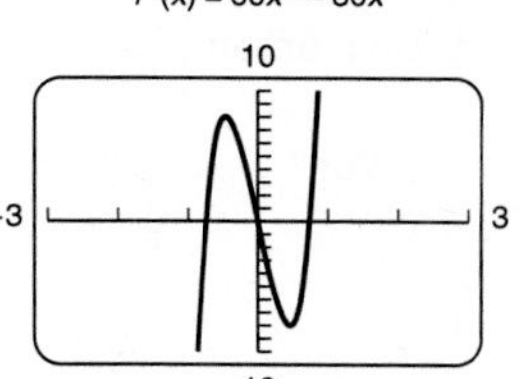

1. Critical values: -1, 0, and 1
2. Inflection points at -0.707, 0, and 0.707

Exercise Set 2.2, p. 228

1. Relative maximum is $f(0) = 5$.
3. Relative minimum is $f\left(\frac{1}{2}\right) = -\frac{1}{4}$.
5. Relative maximum is $f\left(\frac{4}{5}\right) = -\frac{19}{5}$.
7. Relative minimum is $f\left(\frac{1}{2}\right) = -1$; relative maximum is $f\left(-\frac{1}{2}\right) = 3$. **9.** Relative minimum at $(2, -16)$, relative maximum at $(-2, 16)$; inflection point at $(0, 0)$; increasing on $(-\infty, -2]$ and $[2, \infty)$, decreasing on $[-2, 2]$; concave down on $(-\infty, 0)$, concave up on $(0, \infty)$

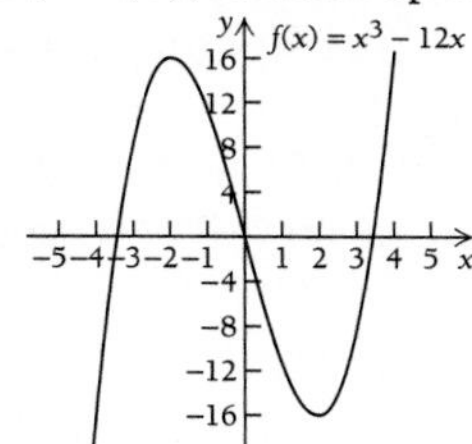

11. Relative minimum at $(2, -51)$, relative maximum at $(-2, 45)$; inflection point at $(0, -3)$; increasing on $(-\infty, -2]$ and $[2, \infty)$, decreasing on $[-2, 2]$; concave down on $(-\infty, 0)$, concave up on $(0, \infty)$

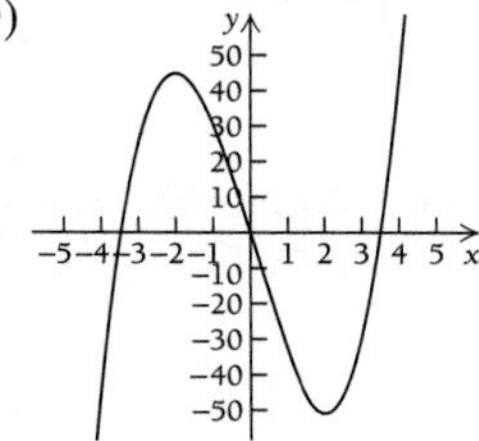

13. Relative minimum at $\left(\frac{1}{2}, -\frac{1}{3}\right)$, relative maximum at $\left(-\frac{1}{2}, 1\right)$; inflection point at $\left(0, \frac{1}{3}\right)$; increasing on $\left(-\infty, -\frac{1}{2}\right]$ and $\left[\frac{1}{2}, \infty\right)$, decreasing on $\left[-\frac{1}{2}, \frac{1}{2}\right]$; concave down on $(-\infty, 0)$, concave up on $(0, \infty)$

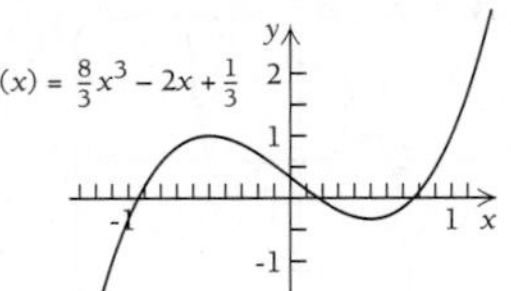

15. Relative minimum at $(0, -4)$, relative maximum at $(2, 0)$; inflection point at $(1, -2)$; increasing on $[0, 2]$, decreasing on $(-\infty, 0]$ and $[2, \infty)$; concave up on $(-\infty, 1)$, concave down on $(1, \infty)$

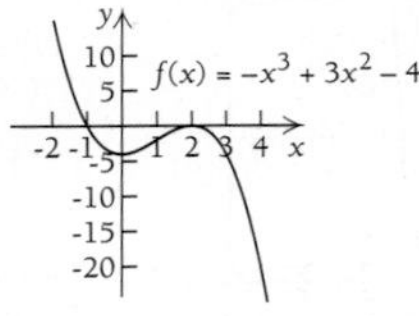

17. Relative minima at $(0, 0)$, and $(3, -27)$, relative maximum at $(1, 5)$; inflection points at $(0.451, 2.321)$ and $(2.215, -13.358)$; increasing on $[0, 1]$, and $[3, \infty)$, decreasing on $(-\infty, 0]$ and $[1, 3]$; concave up on $(-\infty, 0.451)$ and $(2.215, \infty)$, concave down on $(0.451, 2.215)$

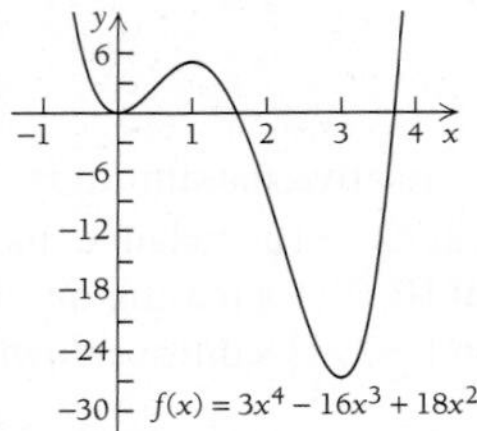

19. Relative minima at $\left(-\sqrt{3}, -9\right)$ and $\left(\sqrt{3}, -9\right)$, relative maximum at $(0, 0)$; inflection points at $(-1, -5)$ and $(1, -5)$; increasing on $\left[-\sqrt{3}, 0\right]$ and $\left[\sqrt{3}, \infty\right)$, decreasing on $\left(-\infty, -\sqrt{3}\right]$ and $\left[0, \sqrt{3}\right]$; concave up on $(-\infty, -1)$ and $(1, \infty)$, concave down on $(-1, 1)$

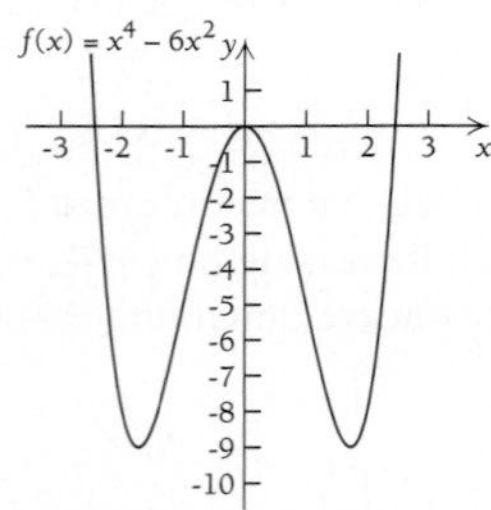

21. Relative minimum at $(2, -5)$, relative maximum at $\left(-\frac{2}{3}, \frac{121}{27}\right)$; inflection point at $\left(\frac{2}{3}, -\frac{7}{27}\right)$; increasing on $\left(-\infty, -\frac{2}{3}\right]$ and $[2, \infty)$, decreasing on $\left[-\frac{2}{3}, 2\right]$; concave down on $\left(-\infty, \frac{2}{3}\right)$, concave up on $\left(\frac{2}{3}, \infty\right)$

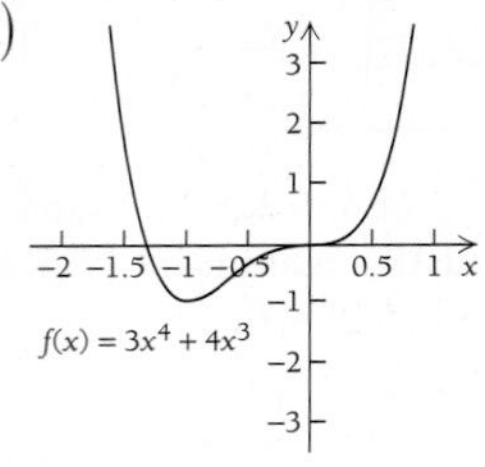

23. Relative minimum at $(-1, -1)$; inflection points at $\left(-\frac{2}{3}, -\frac{16}{27}\right)$ and $(0, 0)$; increasing on $[-1, \infty)$, decreasing on $(-\infty, -1]$; concave up on $\left(-\infty, -\frac{2}{3}\right)$ and $(0, \infty)$, concave down on $\left(-\frac{2}{3}, 0\right)$

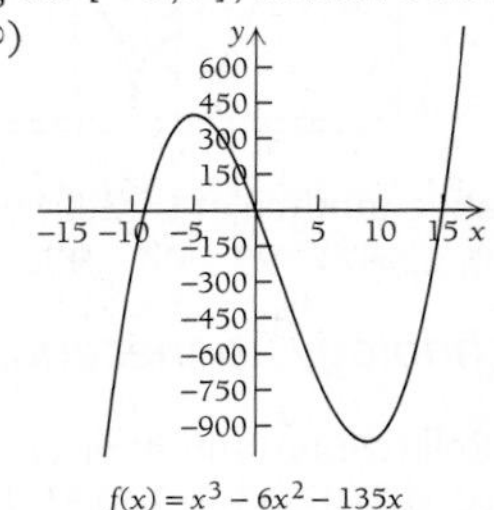

25. Relative minimum at $(9, -972)$, relative maximum at $(-5, 400)$; inflection point at $(2, -286)$; increasing on $(-\infty, -5]$ and $[9, \infty)$, decreasing on $[-5, 9]$; concave down on $(-\infty, 2)$, concave up on $(2, \infty)$

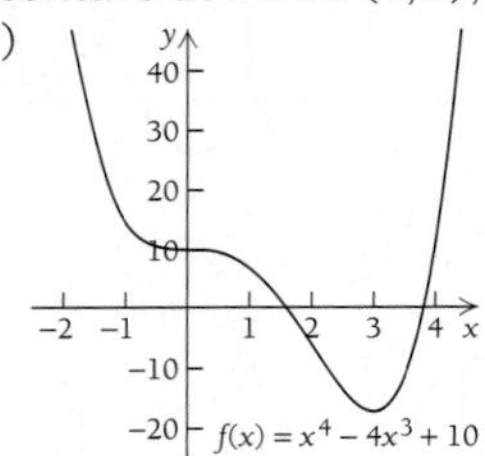

27. Relative minimum at $(3, -17)$; inflection points at $(0, 10)$ and $(2, -6)$; increasing on $[3, \infty)$, decreasing on $(-\infty, 3]$; concave down on $(0, 2)$, concave up on $(-\infty, 0)$ and $(2, \infty)$

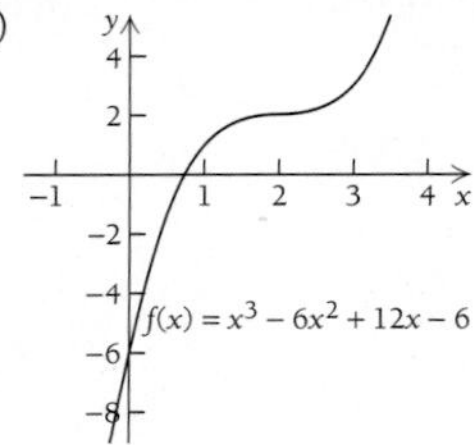

29. No relative extrema; inflection point at $(2, 2)$; increasing on $(-\infty, \infty)$; concave down on $(-\infty, 2)$, concave up on $(2, \infty)$

31. Relative minimum at $(-1,-2)$, relative maximum at $(1,2)$; inflection points at $(-0.707,-1.237)$, $(0,0)$, and $(0.707,1.237)$; increasing on $[-1,1]$, decreasing on $(-\infty,-1]$ and $[1,\infty)$; concave down on $(-0.707,0)$ and $(0.707,\infty)$, concave up on $(-\infty,-0.707)$ and $(0,0.707)$

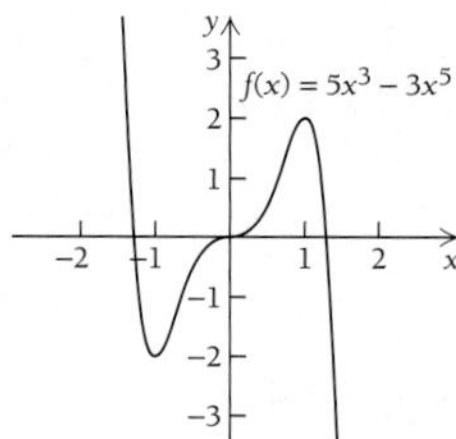

33. Relative minima at $(0,0)$ and $(3,0)$, relative maximum at $\left(\frac{3}{2},\frac{81}{16}\right)$; inflection points at $(0.634, 2.25)$ and $(2.366, 2.25)$; increasing on $\left[0,\frac{3}{2}\right]$ and $[3,\infty)$, decreasing on $(-\infty,0]$ and $\left[\frac{3}{2},3\right]$; concave down on $(0.634, 2.366)$, concave up on $(-\infty, 0.634)$ and $(2.366,\infty)$

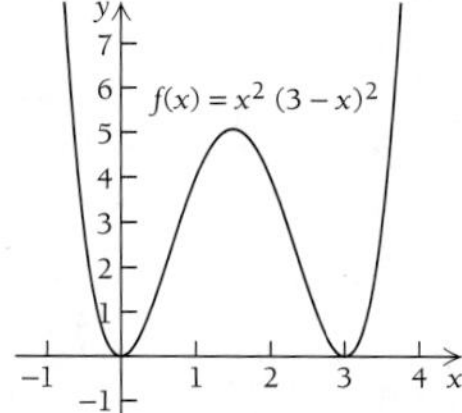

35. Relative minimum at $(-1,0)$; no inflection points; increasing on $[-1,\infty)$, decreasing on $(-\infty,-1]$; concave down on $(-\infty,-1)$ and $(-1,\infty)$

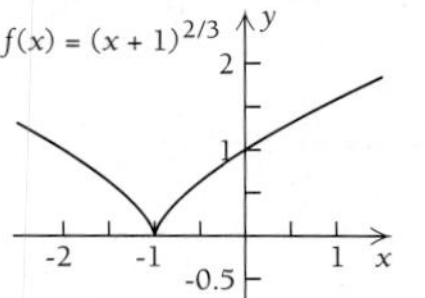

37. No relative extrema; inflection point at $(3,-1)$; increasing on $(-\infty,\infty)$; concave up on $(-\infty,3)$, concave down on $(3,\infty)$

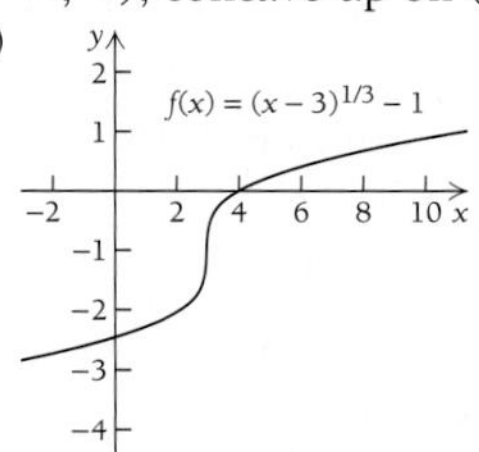

39. Relative maximum at $(4,5)$; no inflection points; increasing on $(-\infty,4]$, decreasing on $[4,\infty)$; concave up on $(-\infty, 4)$ and $(4,\infty)$

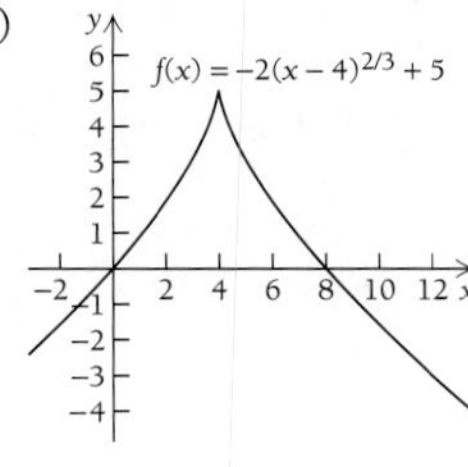

41. Relative minimum at $\left(-\sqrt{2},-2\right)$, relative maximum at $\left(\sqrt{2},2\right)$; inflection point at $(0,0)$; increasing on $\left[-\sqrt{2},\sqrt{2}\right]$, decreasing on $\left[-2,-\sqrt{2}\right]$ and $\left[\sqrt{2},2\right]$; concave up on $(-2,0)$, concave down on $(0,2)$

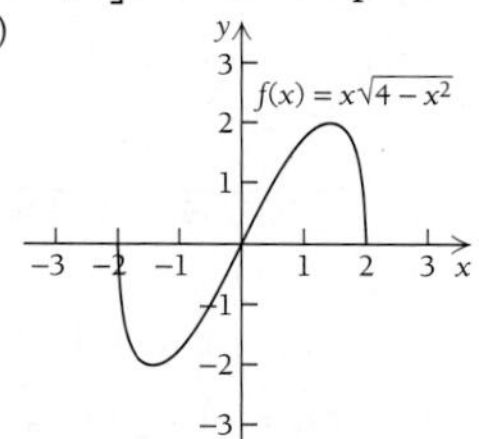

43. Relative minimum at $\left(-1,-\dfrac{1}{2}\right)$, relative maximum at $\left(1,\dfrac{1}{2}\right)$; inflection points at $\left(-\sqrt{3},-\dfrac{\sqrt{3}}{4}\right)$ and $(0,0)$ and $\left(\sqrt{3},\dfrac{\sqrt{3}}{4}\right)$; increasing on $[-1,1]$, decreasing on $(-\infty,-1]$ and $[1,\infty)$; concave up on $\left(-\sqrt{3},0\right)$ and $\left(\sqrt{3},\infty\right)$, concave down on $\left(-\infty,-\sqrt{3}\right)$ and $\left(0,\sqrt{3}\right)$

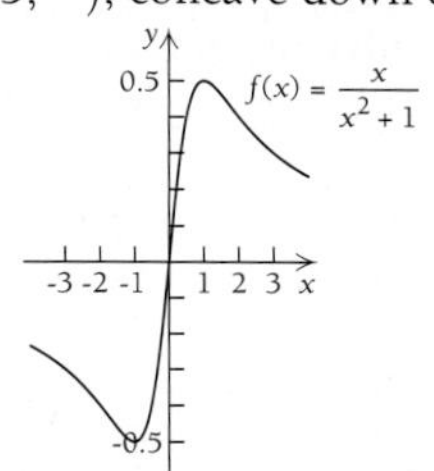

45. Relative maximum at $(0,3)$; inflection points at $\left(-\sqrt{\dfrac{1}{3}},\dfrac{9}{4}\right)$ and $\left(\sqrt{\dfrac{1}{3}},\dfrac{9}{4}\right)$; increasing on $(-\infty,0]$, decreasing on $[0,\infty)$; concave up on $\left(-\infty,-\sqrt{\dfrac{1}{3}}\right)$ and $\left(\sqrt{\dfrac{1}{3}},\infty\right)$, concave down on $\left(-\sqrt{\dfrac{1}{3}},\sqrt{\dfrac{1}{3}}\right)$

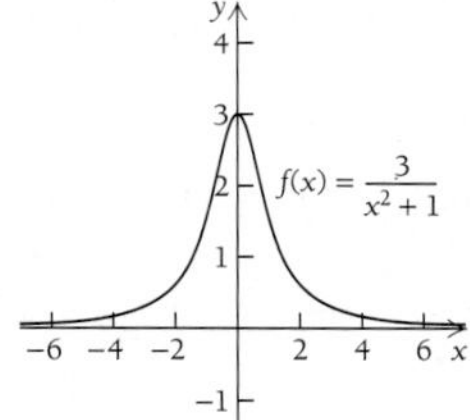

47–99. Left to the student

101.

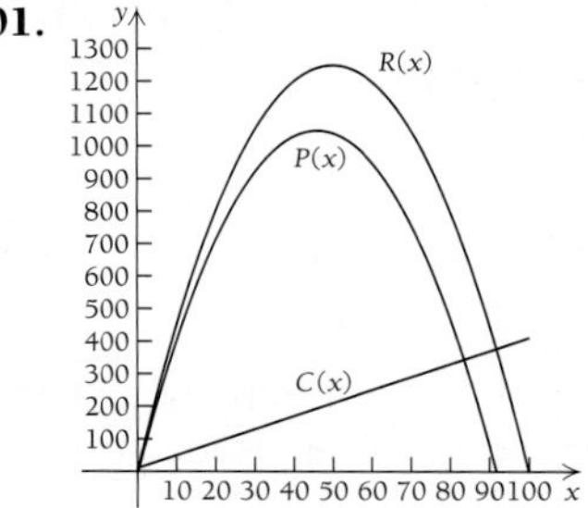

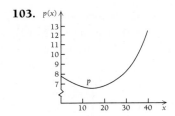

103.

105. Radius $= \frac{40}{3}$, or $13\frac{1}{3}$ mm

107. TW **109.** TW **111.** Left to the student

113. Relative maximum at $(0, 0)$; relative minimum at $(1, -2)$

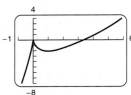

$f(x) = 4x - 6x^{2/3}$

115. Relative minimum at $(0, 0)$; relative maximum at $(0.4, 0.035)$

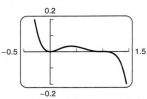

$f(x) = x^2(1 - x)^3$

117. Relative maximum at $(-1, 1.587)$; relative minimum at $(1, -1.587)$

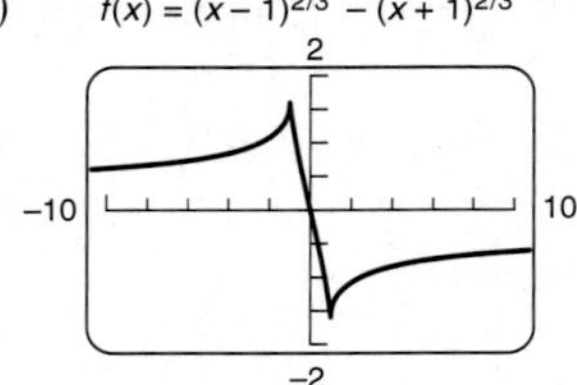

Technology Connection, p. 233

1. Vertical asymptotes: $x = -7$ and $x = 4$ **2.** Vertical asymptotes: $x = 0$, $x = 3$, and $x = -2$

Technology Connection, p. 234

1. and **2.** Left to the student **3.** 2 **4.** Left to the student

Technology Connection, p. 235

1. and **2.** Left to the student

Technology Connection, p. 236

Graphs are left to the student.
1. Horizontal asymptote: $y = 0$ **2.** Horizontal asymptote: $y = 3$ **3.** Horizontal asymptote: $y = 0$ **4.** Horizontal asymptote: $y = \frac{1}{2}$

Technology Connection, p. 237

Graphs are left to the student.
1. $y = 3x - 1$ **2.** $y = 5x$

Technology Connection, p. 237

1. x-intercepts: $(0, 0)$, $(3, 0)$, and $(-5, 0)$; y-intercept: $(0, 0)$
2. x-intercepts: $(0, 0)$, $(1, 0)$, and $(-3, 0)$; y-intercept: $(0, 0)$

Exercise Set 2.3, p. 243

1. $x = 5$ **3.** $x = -3$ and $x = 3$ **5.** $x = 0$, $x = 2$, and $x = 4$ **7.** $x = -1$ **9.** No vertical asymptotes **11.** $y = \frac{3}{4}$ **13.** $y = 0$ **15.** $y = 5$ **17.** No horizontal asymptotes **19.** $y = 0$ **21.** $y = \frac{1}{2}$ **23.**

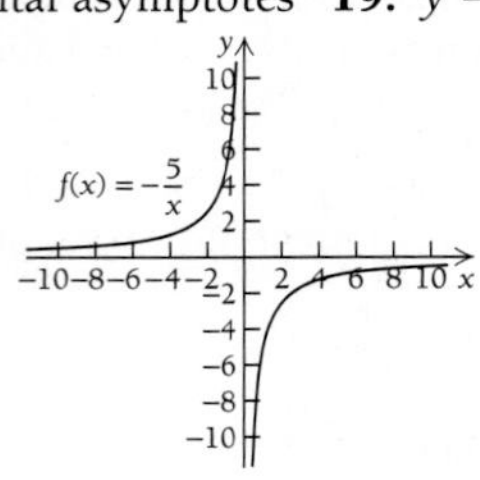

Increasing on $(-\infty, 0)$ and $(0, \infty)$
No relative extrema
Asymptotes: $x = 0$ and $y = 0$
Concave up on $(-\infty, 0)$; concave down on $(0, \infty)$
No intercepts

25.

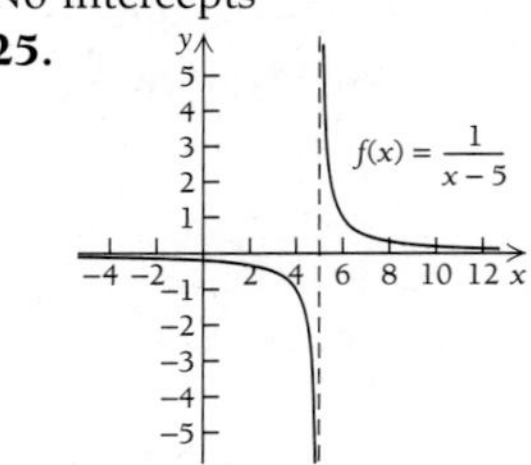

Decreasing on $(-\infty, 5)$ and $(5, \infty)$
No relative extrema
Asymptotes: $x = 5$ and $y = 0$
Concave down on $(-\infty, 5)$; concave up on $(5, \infty)$
y-intercept: $\left(0, -\frac{1}{5}\right)$

27.

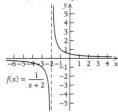

$f(x) = \frac{1}{x+2}$

Decreasing on $(-\infty, -2)$ and $(-2, \infty)$
No relative extrema
Asymptotes: $x = -2$ and $y = 0$
Concave down on $(-\infty, -2)$; concave up on $(-2, \infty)$
y-intercept: $\left(0, \frac{1}{2}\right)$

29.

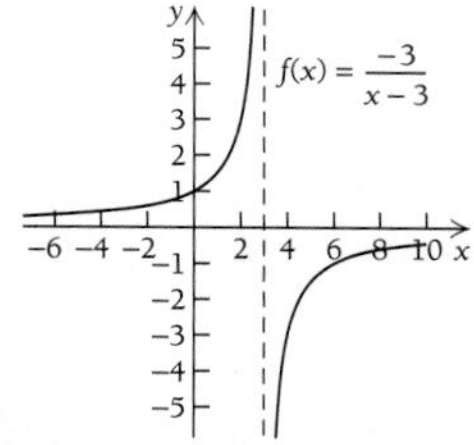

Increasing on $(-\infty, 3)$ and $(3, \infty)$
No relative extrema
Asymptotes: $x = 3$ and $y = 0$
Concave up on $(-\infty, 3)$; concave down on $(3, \infty)$
y-intercept: $(0, 1)$

31.

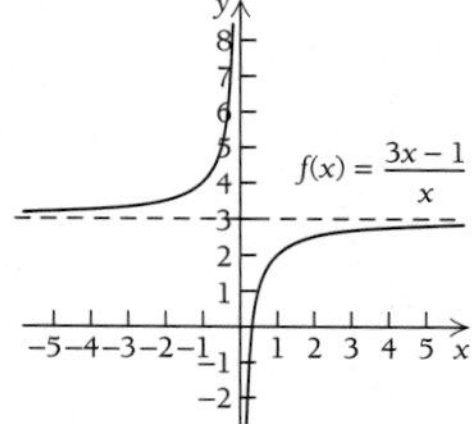

Increasing on $(-\infty, 0)$ and $(0, \infty)$
No relative extrema
Asymptotes: $x = 0$ and $y = 3$
Concave up on $(-\infty, 0)$; concave down on $(0, \infty)$
x-intercept: $\left(\frac{1}{3}, 0\right)$

33.

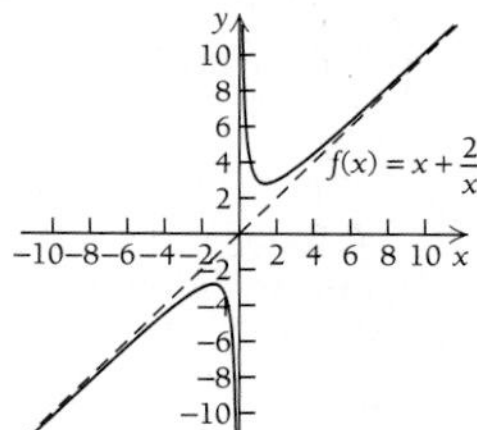

Increasing on $\left(-\infty, -\sqrt{2}\right]$ and $\left[\sqrt{2}, \infty\right)$; decreasing on $\left[-\sqrt{2}, 0\right)$ and $\left(0, \sqrt{2}\right]$
Relative minimum at $\left(\sqrt{2}, 2\sqrt{2}\right)$; relative maximum at $\left(-\sqrt{2}, -2\sqrt{2}\right)$
Asymptotes: $x = 0$ and $y = x$
Concave down on $(-\infty, 0)$; concave up on $(0, \infty)$
No intercepts

35.

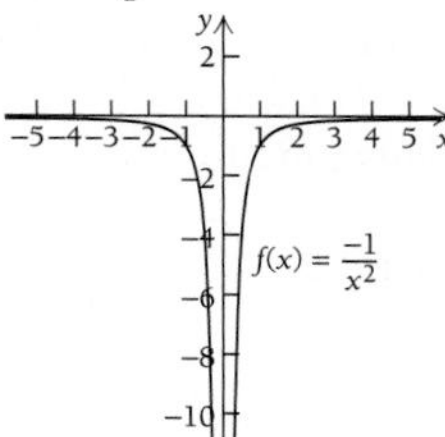

Decreasing on $(-\infty, 0)$; increasing on $(0, \infty)$
No relative extrema
Asymptotes: $x = 0$ and $y = 0$
Concave down on $(-\infty, 0)$ and $(0, \infty)$
No intercepts

37.

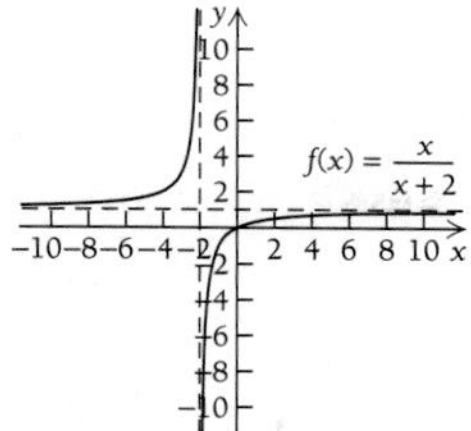

Increasing on $(-\infty, -2)$ and $(-2, \infty)$
No relative extrema
Asymptotes: $x = -2$ and $y = 1$
Concave up on $(-\infty, -2)$; concave down on $(-2, \infty)$
x- and y-intercept: $(0, 0)$

39.

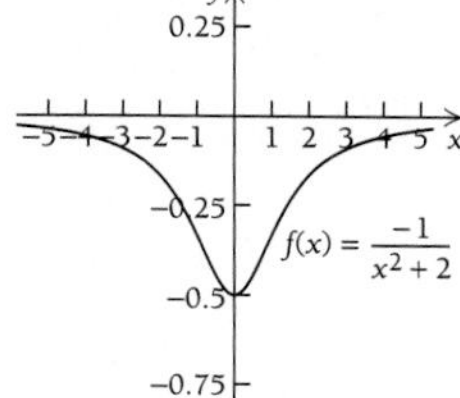

Decreasing on $(-\infty, 0]$; increasing on $[0, \infty)$
Relative minimum at $\left(0, -\frac{1}{2}\right)$
Asymptote: $y = 0$
Concave up on $\left(-\sqrt{\frac{2}{3}}, \sqrt{\frac{2}{3}}\right)$; concave down on $\left(-\infty, -\sqrt{\frac{2}{3}}\right)$ and $\left(\sqrt{\frac{2}{3}}, \infty\right)$
Inflection points: $\left(-\sqrt{\frac{2}{3}}, -\frac{3}{8}\right)$ and $\left(\sqrt{\frac{2}{3}}, -\frac{3}{8}\right)$
y-intercept: $\left(0, -\frac{1}{2}\right)$

41.

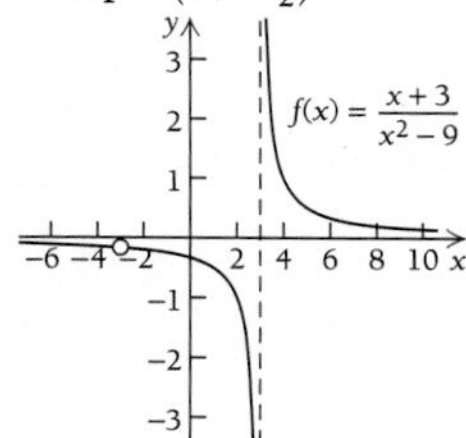

Decreasing on $(-\infty, -3)$, $(-3, 3)$, and $(3, \infty)$
No relative extrema
Asymptotes: $x = 3$ and $y = 0$
Concave down on $(-\infty, -3)$ and $(-3, 3)$; concave up on $(3, \infty)$
y-intercept: $\left(0, -\frac{1}{3}\right)$

43.

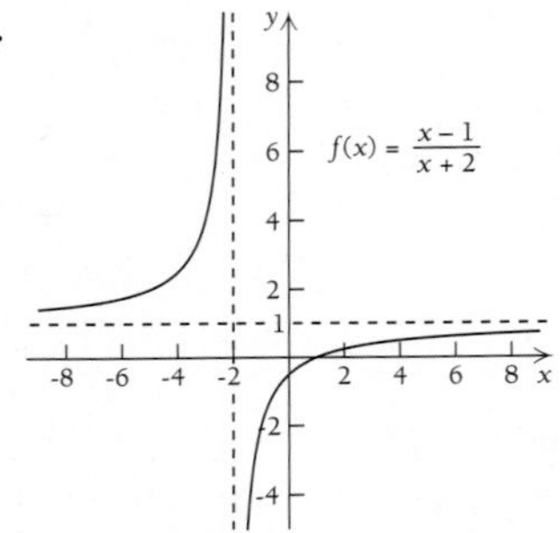

Increasing on $(-\infty, -2)$ and $(-2, \infty)$
No relative extrema
Asymptotes: $x = -2$ and $y = 1$
Concave up on $(-\infty, -2)$; concave down on $(-2, \infty)$
x-intercept: $(1, 0)$; y-intercept: $\left(0, -\frac{1}{2}\right)$

45.

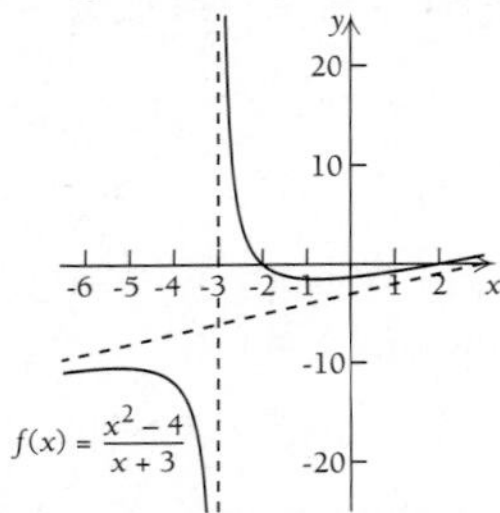

Increasing on $(-\infty, -3 - \sqrt{5})$ and $[-3 + \sqrt{5}, \infty)$, or approximately $(-\infty, -5.236]$ and $[-0.764, \infty)$; decreasing on $[-3 - \sqrt{5}, 3)$ and $(-3, -3 + \sqrt{5}]$, or approximately $[-5.236, -3)$ and $(-3, -0.764]$
Relative maximum at $(-3 - \sqrt{5}, -6 - 2\sqrt{5})$ or approximately $(-5.236, -10.472)$; relative minimum at $(-3 + \sqrt{5}, -6 + 2\sqrt{5})$, or approximately $(-0.764, -1.528)$
Asymptotes: $x = -3$ and $y = x - 3$
Concave down on $(-\infty, -3)$; concave up on $(-3, \infty)$
x-intercepts: $(-2, 0)$, $(2, 0)$; y-intercept: $\left(0, -\frac{4}{3}\right)$

47.

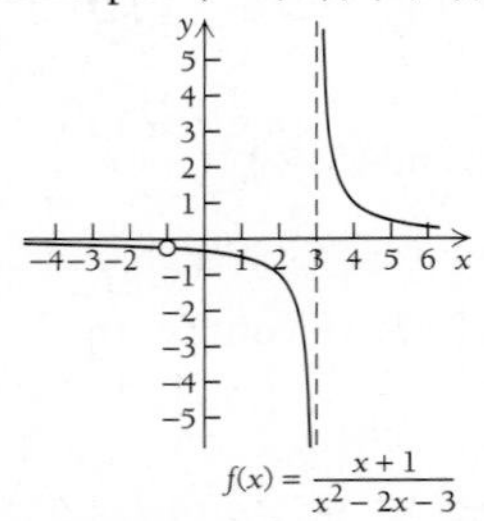

Decreasing on $(-\infty, -1)$, $(-1, 3)$, and $(3, \infty)$
No relative extrema
Asymptotes: $x = 3$ and $y = 0$
Concave down on $(-\infty, -1)$ and $(-1, 3)$; concave up on $(3, \infty)$
y-intercept: $\left(0, -\frac{1}{3}\right)$

49.

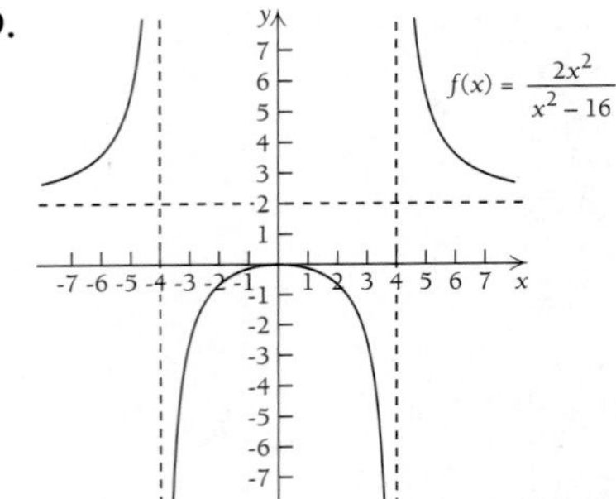

Increasing on $(-\infty, -4)$ and $(-4, 0]$; decreasing on $[0, 4)$ and $(4, \infty)$
Relative maximum at $(0, 0)$
Asymptotes: $x = -4$, $x = 4$, and $y = 2$
Concave up on $(-\infty, -4)$ and $(4, \infty)$; concave down on $(-4, 4)$
x- and y-intercept: $(0, 0)$

51.

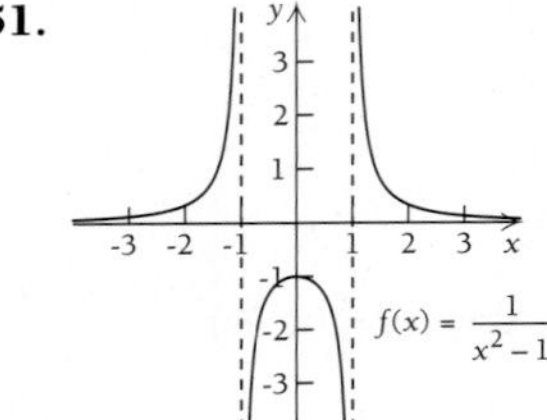

Increasing on $(-\infty, -1)$ and $(-1, 0]$; decreasing on $[0, 1)$ and $(1, \infty)$
Relative maximum at $(0, -1)$
Asymptotes: $x = -1$, $x = 1$, and $y = 0$
Concave up on $(-\infty, -1)$ and $(1, \infty)$; concave down on $(-1, 1)$
y-intercept: $(0, -1)$

53.

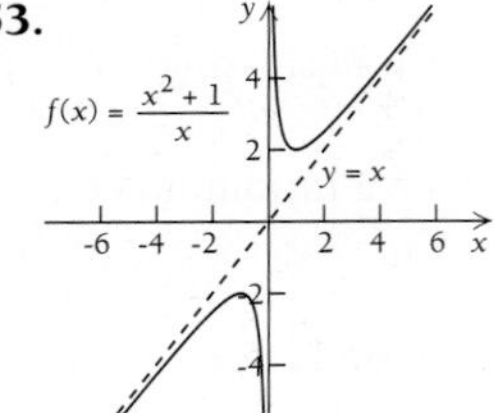

Increasing on $(-\infty, -1]$ and $[1, \infty)$; decreasing on $[-1, 0)$ and $(0,1]$
Relative maximum at $(-1, -2)$; relative minimum at $(1, 2)$
Asymptotes: $x = 0$ and $y = x$
Concave down on $(-\infty, 0)$; concave up on $(0, \infty)$
No intercepts

55.

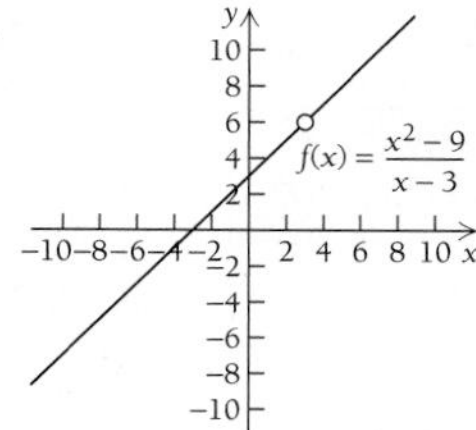

Increasing on $(-\infty, 3)$ and $(3, \infty)$
No relative extrema
No asymptotes
No concavity
x-intercept: $(-3, 0)$; y-intercept: $(0, 3)$
57. (a) \$50, \$37.24, \$32.64, \$26.37; **(b)** maximum $= 50$ at $t = 0$; **(c)**

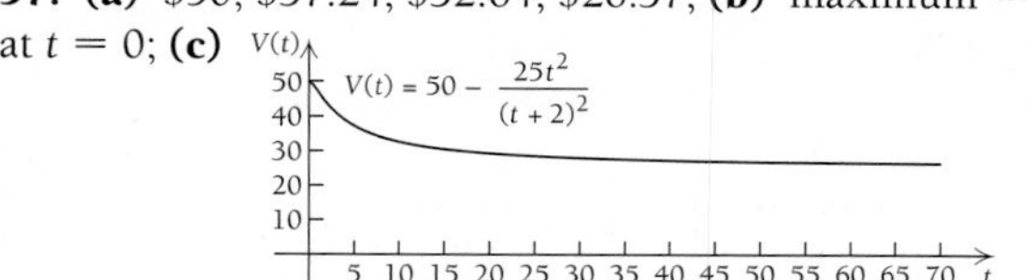

(d) TW
59. (a) \$480, \$600, \$2400, \$4800; **(b)** $[0, 100)$
(c)

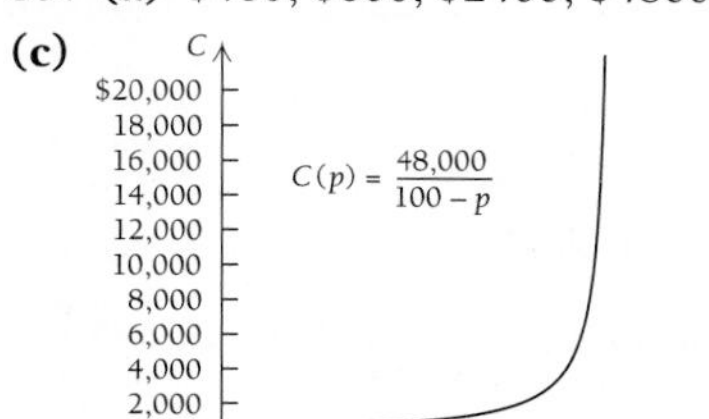

(d) TW

61. (a) \$1.22, \$0.79, \$0.47; **(b)** 36.8 yr after 1970; **(c)** 0 **63. (a)**

n	9	8	7	6	5	4	3	2	1	$\frac{2}{3}$	$\frac{1}{3}$
E	4.00	4.50	5.14	6.00	7.20	9.00	12.00	18.00	36.00	54.00	108.00

(b) $\lim_{n \to 0} E(n) = \infty$ **65.** TW **67.** Does not exist -3
69. $-\infty$ **71.** $\frac{3}{2}$ **73.** $-\infty$
75.

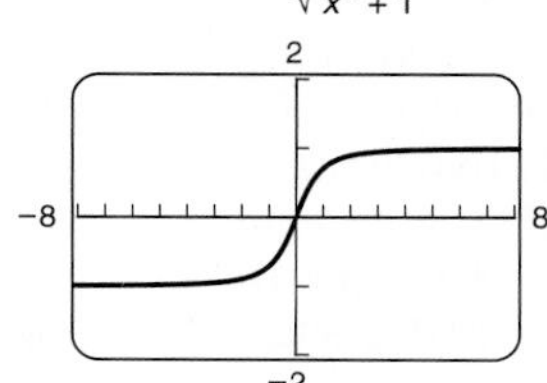

77.

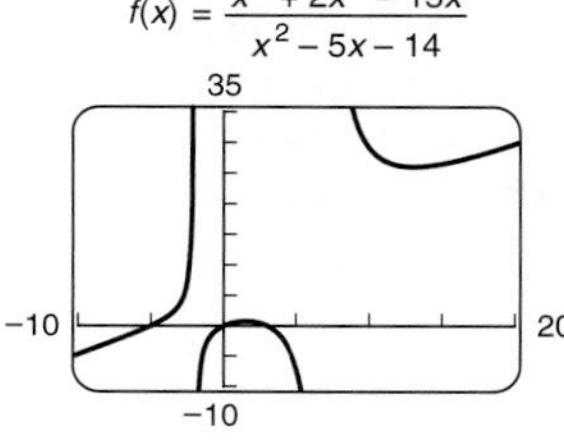

79.

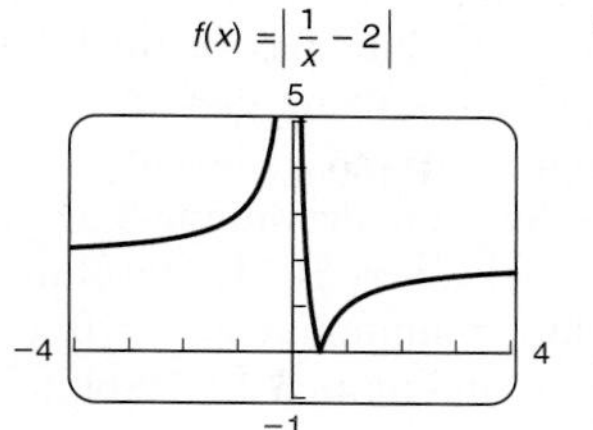

81. TW

Technology Connection, p. 249

1. On $[-2, 1]$, absolute minimum is -8 at $x = -2$, and absolute maximum is 2.185 at $x = -0.333$; on $[-1, 2]$, absolute minimum is 1 at $x = -1$ and $x = 1$, and absolute maximum is 4 at $x = 2$

Technology Connection, p. 250

1. Absolute minimum: -4 at $x = 2$; no absolute maximum

Technology Connection, p. 254

1. No absolute maximum; absolute minimum: 6.325 at $x = 0.316$

Exercise Set 2.4, p. 254

1. (a) 55 mph **(b)** 5 mph **(c)** 25 mpg **3.** Absolute maximum: $5\frac{1}{4}$ at $x = \frac{1}{2}$; absolute minimum: 3 at $x = 2$
5. Absolute maximum: 4 at $x = 2$; absolute minimum: 1 at $x = -1$ and $x = 1$ **7.** Absolute maximum: $\frac{86}{27}$ at $x = -\frac{1}{3}$; absolute minimum: 2 at $x = -1$ **9.** Absolute maximum: 8 at $x = 3$; absolute minimum: -17 at $x = -2$
11. Absolute maximum: 15 at $x = -2$; absolute minimum: -13 at $x = 5$ **13.** Absolute maximum: -5 for $-1 \leq x \leq 1$; absolute minimum: -5 for $-1 \leq x \leq 1$ **15.** Absolute maximum: 4 at $x = -1$; absolute minimum: -12 at $x = 3$
17. Absolute maximum: $\frac{16}{5}$ at $x = -\frac{1}{5}$; absolute minimum: -48 at $x = 3$ **19.** Absolute maximum: 50 at $x = 5$; absolute minimum: -4 at $x = 2$ **21.** Absolute maximum: 2 at $x = -1$; absolute minimum: -110 at $x = -5$
23. Absolute maximum: 513 at $x = -8$; absolute minimum: -511 at $x = 8$ **25.** Absolute maximum: 17 at $x = 1$; absolute minimum: -15 at $x = -3$ **27.** Absolute maximum: 32 at $x = -2$; absolute minimum: $-\frac{27}{16}$ at $x = \frac{3}{2}$
29. Absolute maximum: 13 at $x = -2$ and $x = 2$; absolute minimum: 4 at $x = -1$ and $x = 1$ **31.** Absolute maximum: -1 at $x = 5$; absolute minimum: -5 at $x = -3$
33. Absolute maximum: $20\frac{1}{20}$ at $x = 20$; absolute minimum: 2 at $x = 1$ **35.** Absolute maximum: $\frac{4}{5}$ at $x = -2$ and $x = 2$; absolute minimum: 0 at $x = 0$ **37.** Absolute maximum: 3 at $x = 26$; absolute minimum: -1 at $x = -2$ **39–47.** Left to the student **49.** Absolute maximum: 36 at $x = 6$
51. Absolute minimum: 70 at $x = 10$ **53.** Absolute maximum: $\frac{1}{3}$ at $x = \frac{1}{2}$ **55.** Absolute maximum: 900 at $x = 30$ **57.** Absolute maximum: $2\sqrt{3}$ at $x = -\sqrt{3}$; absolute minimum: $-2\sqrt{3}$ at $x = \sqrt{3}$

59. Absolute maximum: 5700 at $x = 2400$ **61.** Absolute minimum: $-55\frac{1}{3}$ at $x = 1$ **63.** Absolute maximum: 2000 at $x = 20$; absolute minimum: 0 at $x = 0$ and $x = 30$ **65.** Absolute minimum: 24 at $x = 6$ **67.** Absolute minimum: 108 at $x = 6$ **69.** Absolute maximum: 3 at $x = -1$; absolute minimum: $-\frac{3}{8}$ at $x = \frac{1}{2}$ **71.** Absolute maximum: 2 at $x = 8$; absolute minimum: 0 at $x = 0$ **73.** No absolute maximum or minimum **75.** Absolute maximum: -1 at $x = 1$; absolute minimum: -5 at $x = -1$ **77.** No absolute maximum; absolute minimum: -5 at $x = -1$ **79.** Absolute maximum: 1 at $x = -1$ and $x = 1$; absolute minimum: 0 at $x = 0$ **81.** No absolute maximum or minimum **83.** Absolute maximum: $-\frac{10}{3} + 2\sqrt{3}$ at $x = 2-\sqrt{3}$; absolute minimum: $-\frac{10}{3} - 2\sqrt{3}$ at $x = 2 + \sqrt{3}$ **85.** No absolute maximum; absolute minimum: -1 at $x = -1$ and $x = 1$ **87–95.** Left to the student **97.** 1430 units; 25 yr of service **99.** 1986 **101.** 1999; 37.4 billion barrels **103.** **(a)** $P(x) = -\frac{1}{2}x^2 + 400x - 5000$; **(b)** 400 items **105.** About 1.26 at $x = \frac{1}{9}$ cc, or about 0.11 cc **107.** Absolute maximum: $3\sqrt{6}$ at $x = 3$; absolute minimum: -2 at $x = -2$ **109.** Minimum: \$20,000 at $x = 7$ "quality units" **111.** TW **113.** **(a)** 2.755 billion barrels in 1981; **(b)** 0.0765 billion barrels / yr **115.** No absolute maximum; absolute minimum: 0 at $x = 1$ **117.** **(a)** $P(t) = t + 8.857$; $P(7) = 15.857$ mm Hg; **(b)** $P(t) = 0.117t^4 - 1.520t^3 + 6.193t^2 - 7.018t + 10.009$; $P(7) = 24.86$ mm Hg; $P(0.765) = 7.62$ mm Hg is the smallest contraction.

Technology Connection, p. 260

1.

x	$y = 20 - x$	$A = x(20 - x)$
0	20	0
4	16	64
6.5	13.5	87.75
8	12	96
10	10	100
12	8	96
13.2	6.8	89.76
20	0	0

2. Left to the student **3.** Maximum: 100 at $x = 10$

Technology Connection, p. 261

1.

x	$8 - 2x$	$4x^3 - 32x^2 + 64x$
0	8	0
0.5	7	24.5
1.0	6	36
1.5	5	37.5
2.0	4	32
2.5	3	22.5
3.0	2	12
3.5	1	3.5
4.0	0	0

2. Left to the student **3.** Maximum: about 37.9 at $x \approx 1.33$

Technology Connection, p. 268

1. Left to the student **2.** Minimum: \$23,500 at $x = 100$; yes

Exercise Set 2.5, p. 270

1. Maximum $Q = 625$; $x = 25$, $y = 25$ **3.** TW **5.** Minimum product $= -4$; $x = 2$, $y = -2$ **7.** Maximum $Q = \frac{1}{4}$; $x = \frac{1}{2}$, $y = \sqrt{\frac{1}{2}}$ **9.** Minimum $Q = 30$; $x = 3$, $y = 2$ **11.** Maximum $Q = 21\frac{1}{3}$; $x = 2$, $y = 10\frac{2}{3}$ **13.** Maximum area $= 4050$ yd^2; width is 45 yd, and length (parallel to beach) is 90 yd **15.** $x = 13.5$ ft, $y = 13.5$ ft; maximum area $= 182.25$ ft^2 **17.** Dimensions: $33\frac{1}{3}$ cm by $33\frac{1}{3}$ cm by $8\frac{1}{3}$ cm; maximum volume $= 9259\frac{7}{27}$ cm^3 **19.** Dimensions: 5 in. by 5 in. by 2.5 in.; minimum surface area $= 75$ in^2 **21.** Dimensions: 2.08 yd by 4.16 yd by 1.387 yd **23.** \$1048; 46 units **25.** \$19; 70 units **27.** \$5481; 1667 units **29.** **(a)** $R(x) = x(150 - 0.5x)$; **(b)** $P(x) = -0.75x^2 + 150x - 4000$; **(c)** 100 suits; **(d)** \$3500; **(e)** \$100/suit **31.** \$12.75/ticket; 57,500 people **33.** 25 trees/acre **35.** **(a)** $q(x) = 3.13 - 0.04x$; **(b)** \$39.13 **37.** 4 ft by 4 ft by 20 ft **39.** Order 5 times/yr; lot size is 20. **41.** Order 12 times/yr; lot size is 60. **43.** Order 8 times/yr; lot size is 32. **45.** 14 in. by 14 in. by 28 in. **47.** $x \approx 3.36$ ft, $y \approx 3.36$ ft **49.** $\sqrt[3]{0.1}$, or approximately 0.4642 **51.** 9% **53.** S is 3.25 mi downshore from A.

55. $x = \frac{bp}{a+b}$ **57. (a)** $A'(x) = \frac{x \cdot C'(x) - C(x)}{x^2}$; **(b)** $A'(x_0) = 0 = \frac{x_0 \cdot C'(x_0) - C(x_0)}{x_0^2}$; solving for $C'(x_0)$, we get $C'(x_0) = \frac{C(x_0)}{x_0} = A(x_0)$. **59.** $x = -\sqrt{2}, y = 0$, $Q = -3\sqrt{2} \approx -4.24$ **61.** Order 25 times; lot size: 100 units

Technology Connection, p. 276

1. $P(x) = -3 + 40x - 0.5x^2$; $R(40) = 1200$, $C(40) = 403$, $P(40) = 797$, $R'(40) = 10$, $C'(40) = 10$, $P'(40) = 0$; marginal cost is constant. Graphs are left to the student.

Technology Connection, p. 277

1. $P'(50) = \$140/\text{unit}$; $P(51) - P(50) = \$217/\text{unit}$

Exercise Set 2.6, p. 280

1. (a) $P(x) = -0.001x^2 + 3.8x - 60$; **(b)** $R(100) = \$500$, $C(100) = \$190$, $P(100) = \$310$; **(c)** $R'(x) = 5$; $C'(x) = 0.002x + 1.2$; $P'(x) = -0.002x + 3.8$; **(d)** $R'(100) = \$5$, $C'(100) = \$1.40$, $P'(100) = \$3.60$; **(e)** TW **3. (a)** \$1234.38; **(b)** \$24.52; **(c)** \$24.38; **(d)** \$48.75; **(e)** \$1283.13 **5. (a)** \$1799; **(b)** \$235.88; **(c)** \$75.40; **(d)** $R(71) = \$1874.40$, $R(72) = \$1949.80$, $R(73) = \$2025.20$ **7. (a)** \$4572.78; **(b)** \$594.03; **(c)** \$593.63; **(d)** \$5166.41 **9.** If the price increases from \$1000 to \$1001, sales will decrease by 100 units. **11.** \$2.01; \$2.00 **13.** \$2; \$2 **15. (a)** $P(x) = -0.01x^2 + 1.4x - 30$; **(b)** −\$0.01; \$0 **17. (a)** $dD/dp = 0.021p^2 - p + 150$; **(b)** 3547 units; **(c)** TW; **(d)** TW **19.** −\$0.01 **21.** \$491.03 billion **23.** TW **25.** About \$0.21 paid in taxes per dollar earned **27.** 0.0401; 0.04 **29.** 0.2816; 0.28 **31.** −0.556; −1 **33.** 6; 6 **35.** 5.1 **37.** 10.1 **39.** 10.017 **41.** $\frac{1}{2\sqrt{x+1}}\,dx$ **43.** $9x^2\sqrt{2x^3+1}\,dx$ **45.** $\frac{1}{5(x+27)^{4/5}}\,dx$ **47.** $(4x^3 - 6x^2 + 10x + 3)\,dx$ **49.** 3.1 **51.** 7.2 **53.** 657.00 **55.** $-0.01345\ \text{m}^2$ **57.** The concentration changes more from 1 hr to 1.1 hr. **59.** $\frac{10}{2\pi} = 1.59$ ft **61.** $R'(x) = 100 - \frac{3\sqrt{x}}{2}$ **63.** $R'(x) = 500 - 2x$ **65.** $R'(x) = 5$ **67.** TW

Exercise Set 2.7, p. 288

1. $\frac{-x^2}{2y^2}; -2$ **3.** $\frac{4x}{9y^2}; -\frac{8}{9}$ **5.** $\frac{x}{y}; \sqrt{\frac{3}{2}}$ **7.** $\frac{-y}{2x}; \frac{1}{4}$ **9.** $\frac{3x - 2y^2}{2xy}; -\frac{1}{12}$ **11.** $\frac{1-y}{x+2}; -\frac{1}{9}$ **13.** $\frac{6x^2 - 2xy}{x^2 - 3y^2}; -\frac{36}{23}$ **15.** $\frac{-y}{x}$ **17.** $\frac{x}{y}$ **19.** $\frac{3x^2}{5y^4}$ **21.** $\frac{-3xy^2 - 2y}{4x^2y + 3x}$ **23.** $\frac{3}{3p^2 + 1}$ **25.** $\frac{-p}{3x}$ **27.** $\frac{2-p}{x-2}$ **29.** $\frac{-p-4}{x+3}$ **31.** $-\frac{3}{4}$

33. \$400/day, \$80/day, \$320/day

35. \$16/day, \$8/day, \$8/day

37. −1.18 sales/day

39. $-21{,}830\ \text{mi}^2/\text{yr}$

41. Decreasing by $0.0256\ \text{m}^2/\text{month}$

43. (a) $\frac{dV}{dt} = 952.38R\,\frac{dR}{dt}$; **(b)** $0.0143\ \text{mm/sec}^2$

45. $-2\frac{1}{12}$ ft/sec **47.** $494.8\ \text{cm}^3/\text{week}$ **49.** $\frac{-y^3}{x^3}$

51. $\frac{2x}{y(x^2+1)^2}$, or $\frac{x(1-y^2)}{y(1+x^2)}$

53. $\frac{5x^4 - 3(x-y)^2 - 3(x+y)^2}{3(x+y)^2 - 3(x-y)^2 - 5y^4}$, or $\frac{6y^2 - 5x^4 + 6x^2}{y(5y^3 - 12x)}$

55. $\frac{-6(y^2 - xy + x^2)}{(2y-x)^3}$ **57.** $\frac{2x(y^3 - x^3)}{y^5}$ **59.** TW

61.

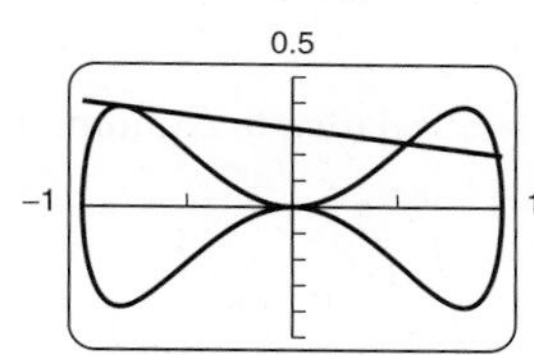

63.

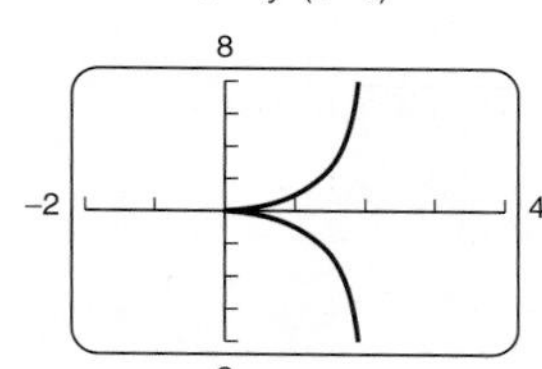

Chapter Review Exercises, p. 297

1. (g) **2.** (e) **3.** (f) **4.** (a) **5.** (b) **6.** (d) **7.** (c) **8.** False **9.** False **10.** True **11.** False **12.** False **13.** True **14.** Relative maximum: $\frac{25}{4}$ at $x = -\frac{3}{2}$

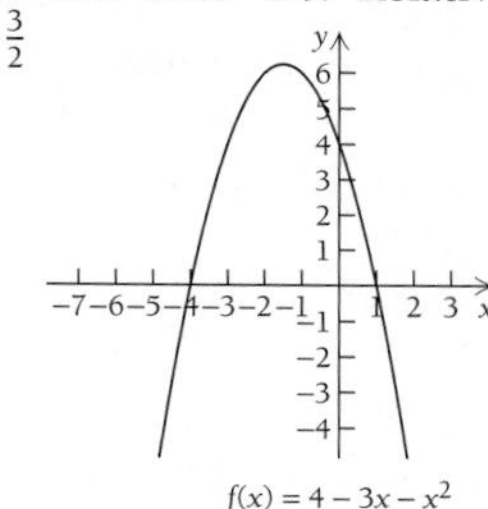

15. Relative minima: 2 at $x = -1$ and 2 at $x = 1$; relative maximum: 3 at $x = 0$

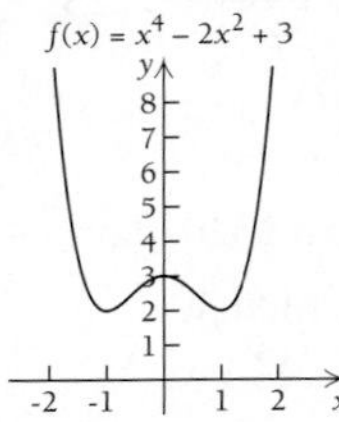

16. Relative minimum: -4 at $x = 1$; relative maximum: 4 at $x = -1$

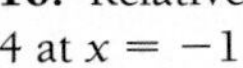

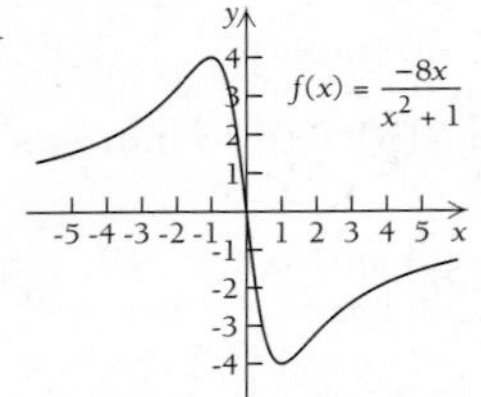

17. No relative extrema

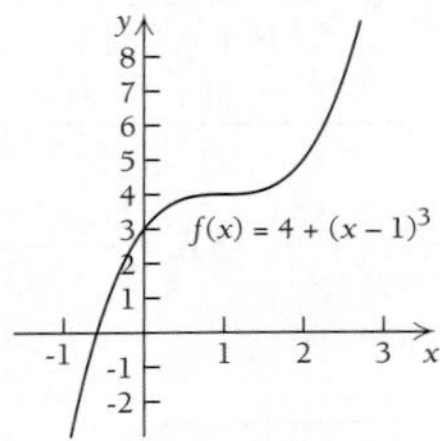

18. Relative minimum: $\frac{76}{27}$ at $x = \frac{1}{3}$; relative maximum: 4 at $x = -1$

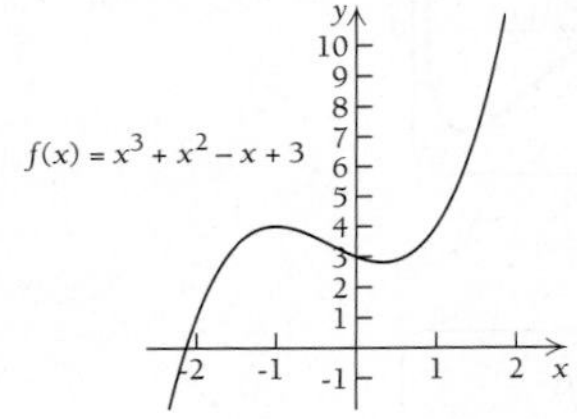

19. Relative minimum: 0 at $x = 0$

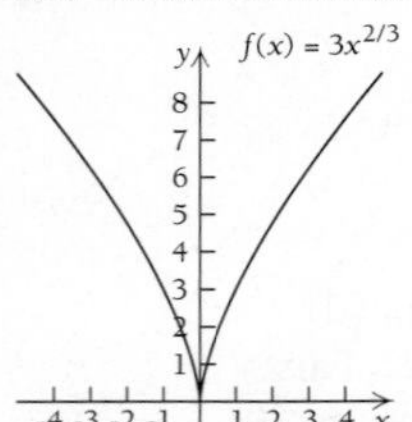

20. Relative maximum: 17 at $x = -1$; relative minimum: -10 at $x = 2$

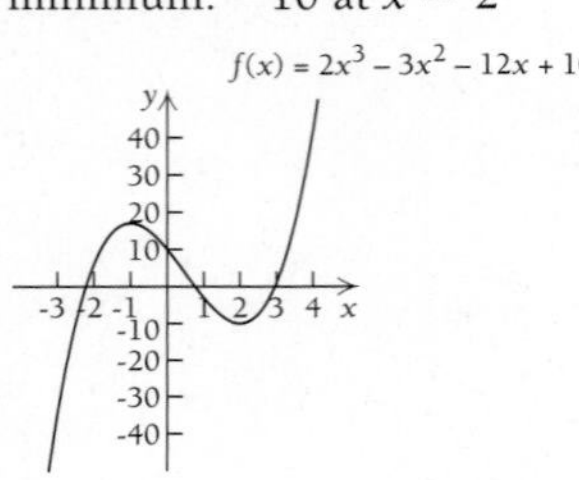

21. Relative maximum: 4 at $x = -1$; relative minimum: 0 at $x = 1$

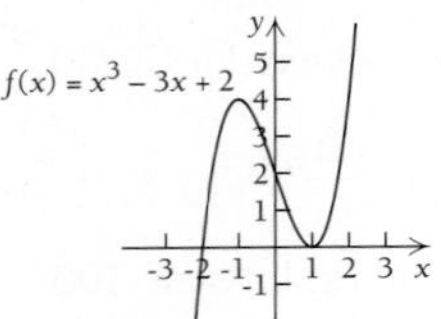

22.

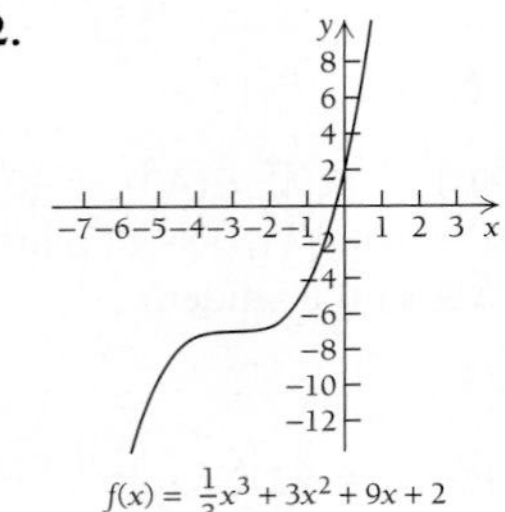

No relative extrema
Inflection point at $(-3, -7)$
Increasing on $(-\infty, \infty)$
Concave down on $(-\infty, -3)$; concave up on $(-3, \infty)$

23.

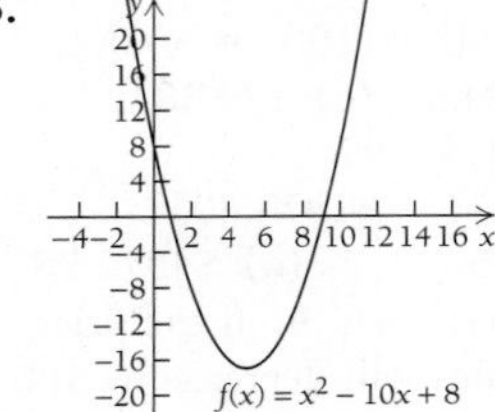

Relative minimum: -17 at $x = 5$
Decreasing on $(-\infty, 5]$; increasing on $[5, \infty)$
Concave up on $(-\infty, \infty)$

24.

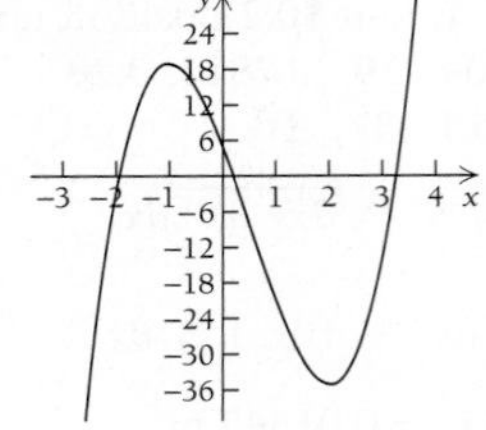

Relative minimum: -35 at $x = 2$; relative maximum: 19 at $x = -1$.
Inflection point at $\left(\frac{1}{2}, -8\right)$
Increasing on $(-\infty, -1]$ and $[2, \infty)$; decreasing on $[-1, 2]$
Concave down on $\left(-\infty, \frac{1}{2}\right)$; concave up on $\left(\frac{1}{2}, \infty\right)$

25.

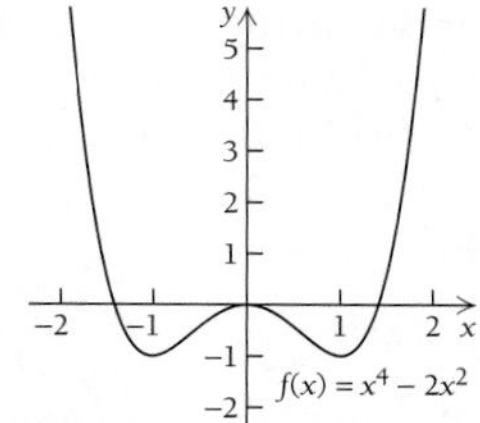

Relative minima: -1 at $x = -1$ and -1 at $x = 1$; relative maximum: 0 at $x = 0$

Inflection points at $\left(-\sqrt{\frac{1}{3}}, -\frac{5}{9}\right)$ and $\left(\sqrt{\frac{1}{3}}, -\frac{5}{9}\right)$

Increasing on $[-1, 0]$ and $[1, \infty)$; decreasing on $(-\infty, -1]$ and $[0, 1]$

Concave up on $\left(-\infty, -\sqrt{\frac{1}{3}}\right)$ and $\left(\sqrt{\frac{1}{3}}, \infty\right)$; concave down on $\left(-\sqrt{\frac{1}{3}}, \sqrt{\frac{1}{3}}\right)$

26.

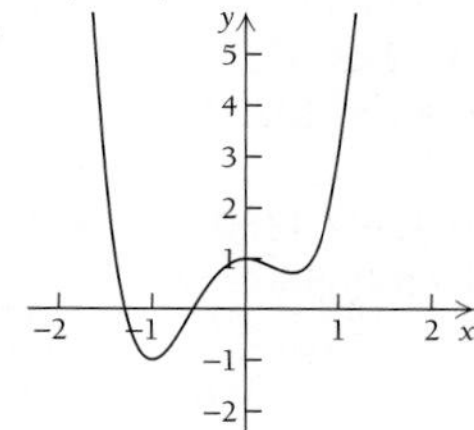

$f(x) = 3x^4 + 2x^3 - 3x^2 + 1$

Relative minima: -1 at $x = -1$ and $\frac{11}{16}$ at $x = \frac{1}{2}$; relative maximum: 1 at $x = 0$

Inflection points at $(-0.608, -0.147)$ and $(0.274, 0.833)$

Increasing on $[-1, 0]$ and $\left[\frac{1}{2}, \infty\right)$; decreasing on $(-\infty, -1]$ and $\left[0, \frac{1}{2}\right]$

Concave down on $(-0.608, 0.274)$; concave up on $(-\infty, -0.608)$ and $(0.274, \infty)$

27.

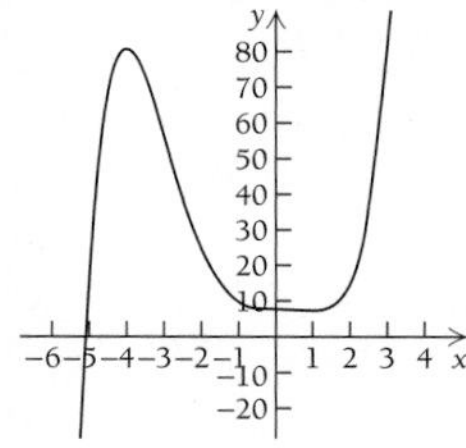

$f(x) = \frac{1}{5}x^5 + \frac{3}{4}x^4 - \frac{4}{3}x^3 + 8$

Relative minimum: $\frac{457}{60}$ at $x = 1$; relative maximum: $\frac{1208}{15}$ at $x = -4$

Inflection points at $(-2.932, 53.701)$, $(0, 8)$, and $(0.682, 7.769)$

Increasing on $(-\infty, -4]$ and $[1, \infty)$; decreasing on $[-4, 1]$

Concave down on $(-\infty, -2.932)$ and $(0, 0.682)$; concave up on $(-2.932, 0)$ and $(0.682, \infty)$

28.

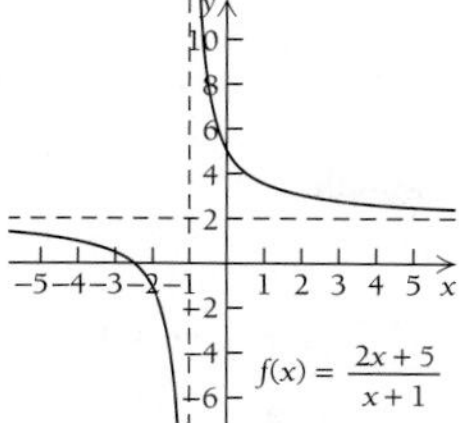

No relative extrema

Decreasing on $(-\infty, -1)$ and $(-1, \infty)$

Concave down on $(-\infty, -1)$; concave up on $(-1, \infty)$

Asymptotes: $x = -1$ and $y = 2$

x-intercept: $\left(-\frac{5}{2}, 0\right)$; y-intercept: $(0, 5)$

29.

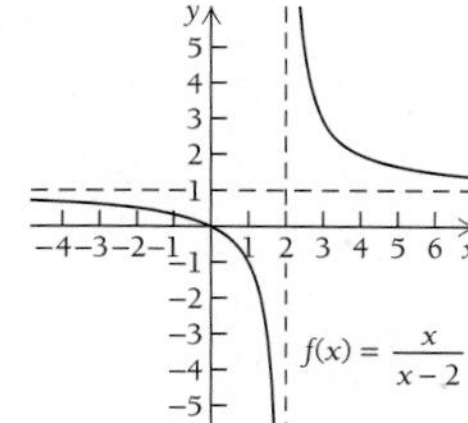

No relative extrema

Decreasing on $(-\infty, 2)$ and $(2, \infty)$

Concave down on $(-\infty, 2)$; concave up on $(2, \infty)$

Asymptotes: $x = 2$ and $y = 1$

x-intercept: $(0, 0)$; y-intercept: $(0, 0)$

30.

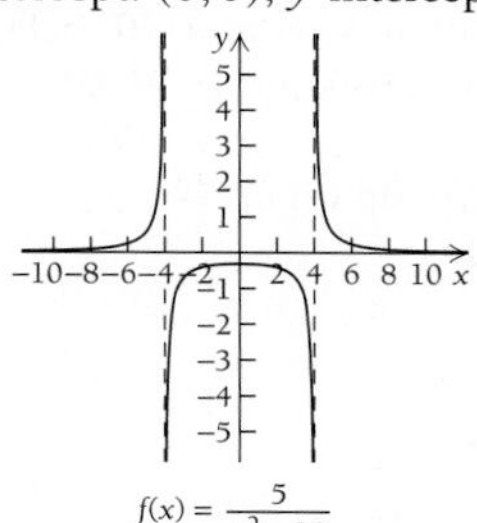

$f(x) = \frac{5}{x^2 - 16}$

Relative maximum at $\left(0, -\frac{5}{16}\right)$

Decreasing on $(0, 4)$ and $(4, \infty)$; increasing on $(-\infty, -4)$ and $(-4, 0)$

Concave down on $(-4, 4)$; concave up on $(-\infty, -4)$ and $(4, \infty)$

Asymptotes: $x = -4$, $x = 4$, and $y = 0$

y-intercept: $\left(0, -\frac{5}{16}\right)$

31.

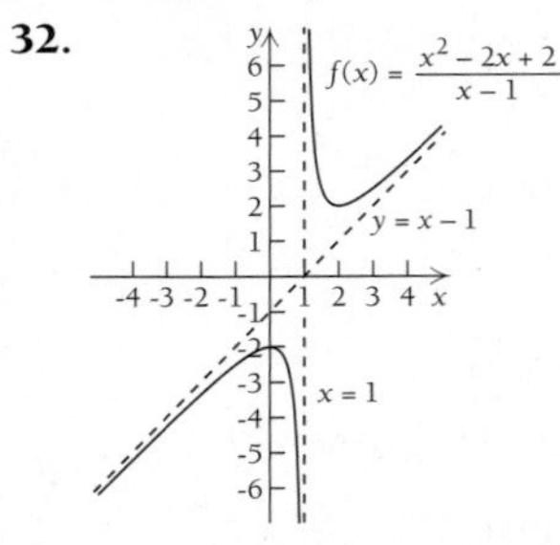

No relative extrema
Increasing on $(-\infty, -1)$, $(-1, 2)$, and $(2, \infty)$
Concave up on $(-\infty, -1)$ and $(-1, 2)$; concave down on $(2, \infty)$
Asymptotes: $x = 2$ and $y = 0$
y-intercept: $(0, \frac{1}{2})$

32.

Relative minimum at (2, 2); relative maximum at $(0, -2)$
Decreasing on $[0, 1)$ and $(1, 2]$; increasing on $(-\infty, 0]$ and $[2, \infty)$
Concave down on $(-\infty, 1)$; concave up on $(1, \infty)$
Asymptotes: $x = 1$ and $y = x - 1$
y-intercept: $(0, -2)$

33.

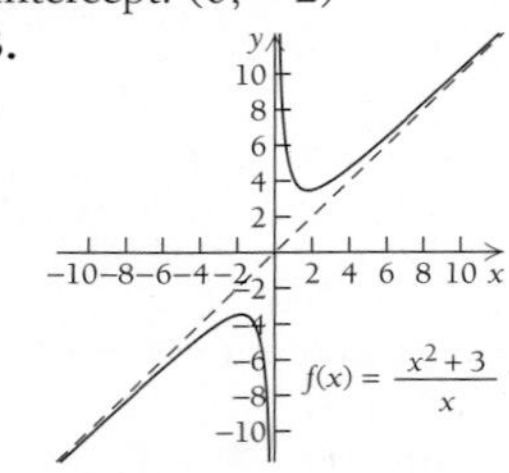

Relative minimum at $(\sqrt{3}, 2\sqrt{3})$; relative maximum at $(-\sqrt{3}, -2\sqrt{3})$
Decreasing on $[-\sqrt{3}, 0)$ and $(0, \sqrt{3}]$; increasing on $(-\infty, -\sqrt{3}]$ and $[\sqrt{3}, \infty)$
Concave down on $(-\infty, 0)$; concave up on $(0, \infty)$
Asymptotes: $x = 0$ and $y = x$
No intercepts
34. Absolute maximum: 66 at $x = 3$; absolute minimum: 2 at $x = 1$ **35.** Absolute maximum: $75\frac{23}{27}$ at $x = \frac{16}{3}$; absolute minima: 0 at $x = 0$ and $x = 8$ **36.** No absolute maxima; absolute minimum: $10\sqrt{2}$ at $x = 5\sqrt{2}$ **37.** No absolute maxima; absolute minima: 0 at $x = -1$ and $x = 1$ **38.** 30 and 30 **39.** $Q = -1$ when $x = -1$ and $y = -1$ **40.** Maximum profit is \$451 when 30 units are produced and sold. **41.** 10 ft by 10 ft by 25 ft **42.** Order 12 times per year with a lot size of 30 **43.** **(a)** \$108; **(b)** \$1/dinner; **(c)** \$109 **44.** $\Delta y = -0.335$, $dy = -0.35$ **45.** **(a)** $(6x^2 + 1)\,dx$; **(b)** 0.25 **46.** 9.111 **47.** $\frac{-3y - 2x^2}{2y^2 + 3x}$; $\frac{4}{5}$ **48.** -1.75 ft/sec **49.** \$600/day, \$450/day, \$150/day **50.** No maximum; absolute minimum: 0 at $x = 3$ **51.** $\frac{3x^5 - 2(x-y)^3 - 2(x+y)^3}{2(x+y)^3 - 2(x-y)^3 - 3y^5}$ **52.** Relative maximum at (0, 0); relative minima at $(-9, -9477)$ and $(15, -37{,}125)$ **53.** Relative maxima at $(-1.714, 37.445)$; relative minimum at $(1.714, -37.445)$ **54.** Relative maximum at $(0, 1.08)$; relative minima at $(-3, -1)$ and $(3, -1)$
55. **(a)** Linear: $y = 6.998187602x - 124.6183581$
Quadratic: $y = 0.0439274846x^2 + 2.881202838x - 53.51475166$
Cubic: $y = -0.0033441547x^3 + 0.4795643605x^2 - 11.35931622x + 5.276985809$
Quartic: $y = -0.00005539834x^4 + 0.0067192294x^3 - 0.0996735857x^2 - 0.8409991942x - 0.246072967$
(b) The quartic function best fits the data.
(c) The domain is $[26, 102]$. Very few women outside of the age range from 26 to 102 years old develop breast cancer.
(d) Maximum: 466 per 100,000 women at $x = 79.0$ years old

Chapter 2 Test, p. 299

1. [2.1, 2.2] Relative minimum: -9 at $x = 2$
Decreasing on $(-\infty, 2]$; increasing on $[2, \infty)$

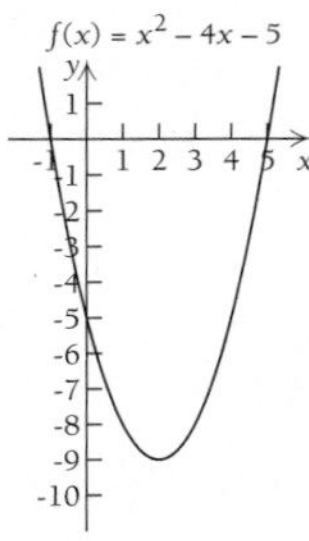

2. [2.1, 2.2] Relative minimum: 2 at $x = -1$; relative maximum: 6 at $x = 1$
Decreasing on $(-\infty, -1]$ and $[1, \infty)$; increasing on $[-1, 1]$

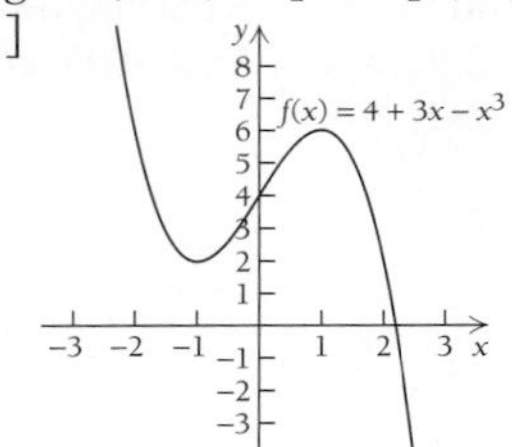

3. [2.1, 2.2] Relative minimum: -4 at $x = 2$
Decreasing on $(-\infty, 2]$; increasing on $[2, \infty)$

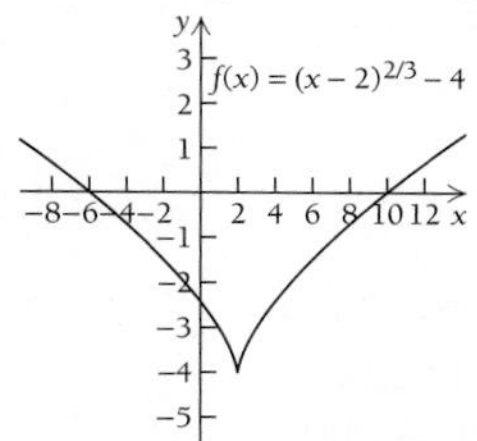

4. [2.1, 2.2] Relative maximum: 4 at $x = 0$
Increasing on $(-\infty, 0]$; decreasing on $[0, \infty)$

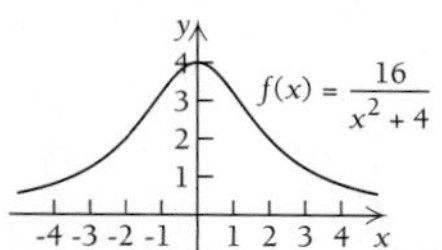

5. [2.3]

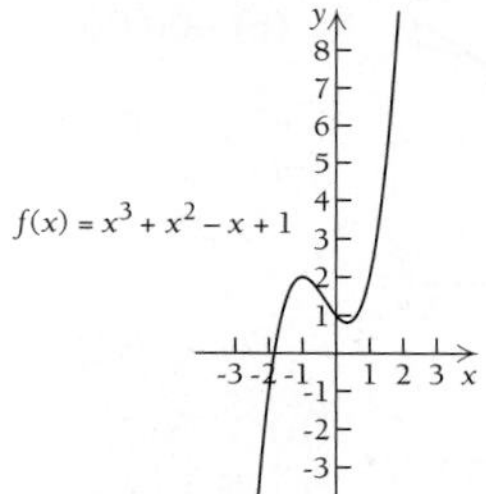

Relative maximum: 2 at $x = -1$; relative minimum: $\frac{22}{27}$ at $x = \frac{1}{3}$
Inflection point: $\left(-\frac{1}{3}, \frac{38}{27}\right)$

6. [2.3]

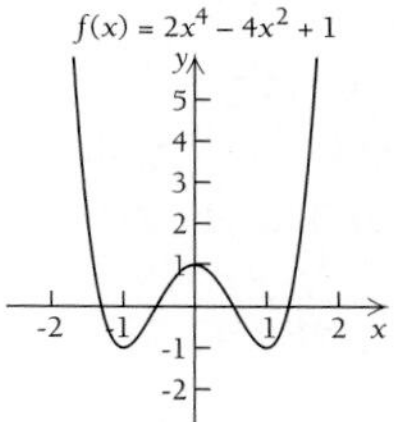

Relative maximum: 1 at $x = 0$; relative minima: -1 at $x = -1$ and $x = 1$
Inflection points: $\left(-\sqrt{\frac{1}{3}}, -\frac{1}{9}\right)$ and $\left(\sqrt{\frac{1}{3}}, -\frac{1}{9}\right)$

7. [2.3]

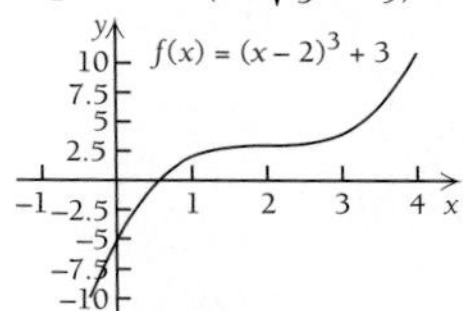

No relative extrema
Inflection point: $(2, 3)$

8. [2.3]

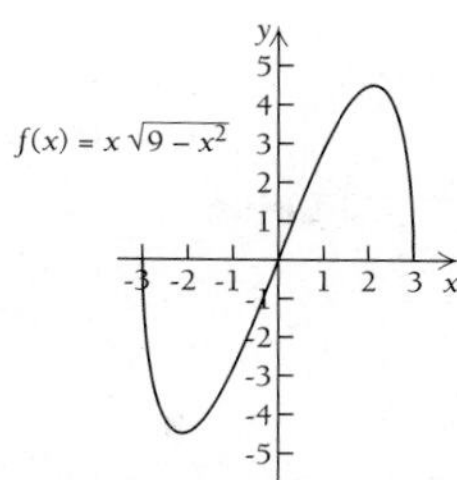

Relative maximum: $\frac{9}{2}$ at $x = \sqrt{\frac{9}{2}}$; relative minimum: $-\frac{9}{2}$ at $x = -\sqrt{\frac{9}{2}}$
Inflection point: $(0, 0)$

9. [2.3]

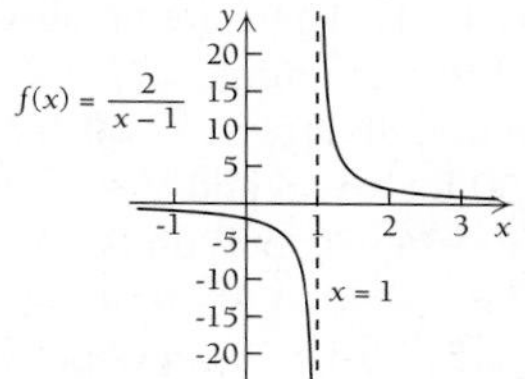

No relative extrema
Asymptotes: $x = 1$ and $y = 0$

10. [2.3]

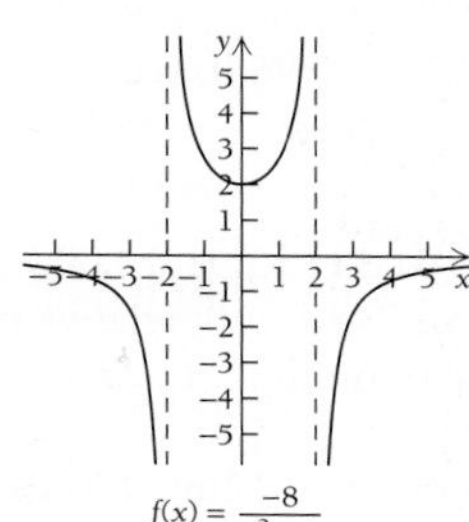

Relative minimum: 2 at $x = 0$
Asymptotes: $x = -2$, $x = 2$, and $y = 0$

11. [2.3]

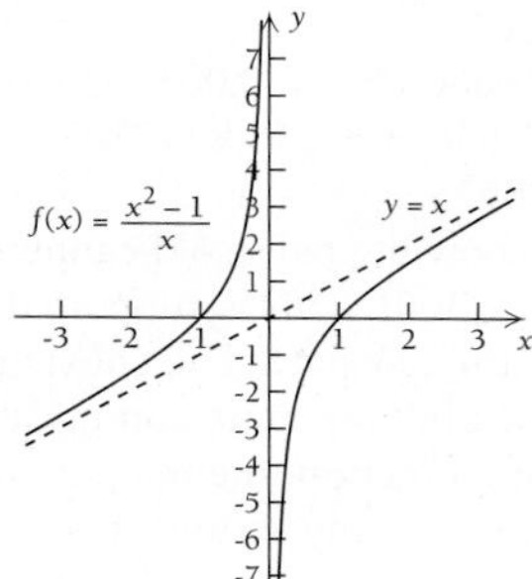

No relative extrema
Asymptotes: $x = 0$ and $y = x$

12. [2.3]

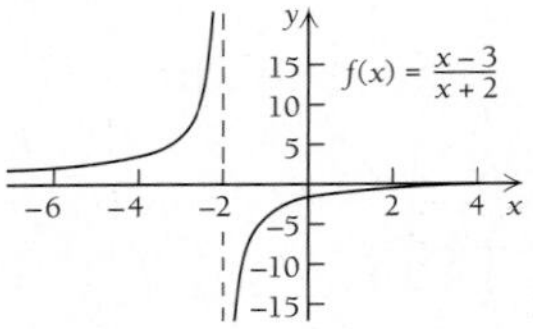

No relative extrema
Asymptotes: $x = -2$ and $y = 1$
13. [2.4] Absolute maximum: 9 at $x = 3$; no absolute minimum **14.** [2.4] Absolute maximum: 2 at $x = -1$; absolute minimum: -1 at $x = -2$ **15.** [2.4] Absolute maximum: 28.49 at $x = 4.3$; no absolute minimum
16. [2.4] Absolute maximum: 7 at $x = -1$; absolute minimum: 3 at $x = 1$ **17.** [2.4] There are no absolute extrema. **18.** [2.4] Absolute minimum: $-\frac{13}{12}$ at $x = \frac{1}{6}$
19. [2.4] Absolute minimum: 48 at $x = 4$ **20.** [2.5] 4 and -4 **21.** [2.5] $Q = 50$ for $x = 5$ and $y = -5$
22. [2.5] Maximum profit: \$24,980; 500 units.
23. [2.5] Dimensions: 40 in. by 40 in. by 10 in; maximum volume: 16,000 in^3 **24.** [2.5] Order 35 times per year; lot size, 35 **25.** [2.6] $\Delta y = 1.01$; $f'(x)\,\Delta x = 1$ **26.** [2.6] 7.0714 **27.** [2.6] **(a)** $\dfrac{x}{\sqrt{x^2+3}}\,dx$; **(b)** 0.00756
28. [2.7] $\dfrac{-x^2}{y^2}$; $-\dfrac{1}{4}$ **29.** [2.7] -0.96 ft/sec
30. [2.4] Absolute maximum: $\dfrac{2^{2/3}}{3} \approx 0.529$ at $x = \sqrt[3]{2}$; absolute minimum: 0 at $x = 0$ **31.** [2.5] 10,000 units
32. [2.4] Absolute minimum: 0 at $x = 0$; relative maximum: 25.103 at $x = 1.084$; relative minimum: 8.625 at $x = 2.95$
33. [2.1, 2.2]
(a) Linear: $y = -0.7707142857x + 12691.60714$
Quadratic: $y = -0.9998904762x^2 + 299.1964286x + 192.9761905$
Cubic: $y = 0.000084x^3 - 1.037690476x^2 + 303.3964286x + 129.9761905$
Quartic: $y = -0.000001966061x^4 + 0.0012636364x^3 - 1.256063636x^2 + 315.8247403x + 66.78138528$
(b) Since the number of bowling balls sold cannot be negative, the domain is [0, 300]. This is supported by both the quadratic model and the raw data. The cubic and quartic models can also be used but are more complicated.
(c) Based on the quadratic function, the maximum value is 22,575 bowling balls. The company should spend \$150,000 on advertising.

Extended Technology Application, p. 302

1. (a)

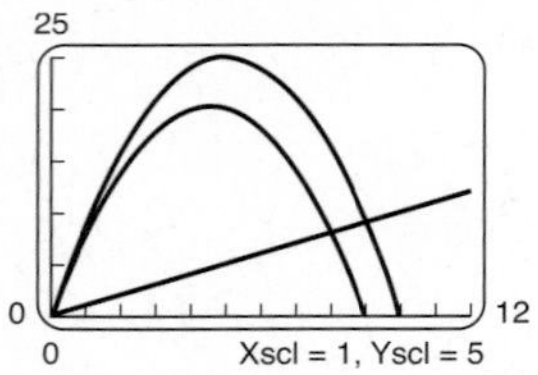

(b) 4500; **(c)** 20,250

2. (a)

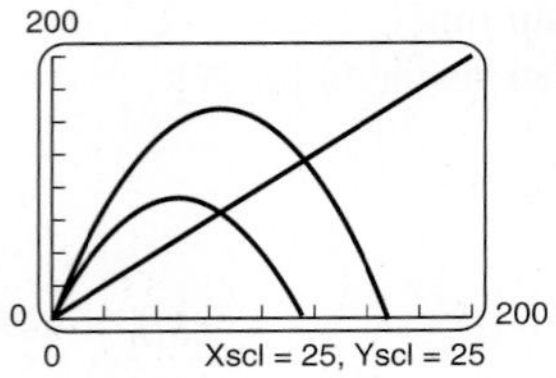

(b) 60,000; **(c)** 90,000

3. (a)

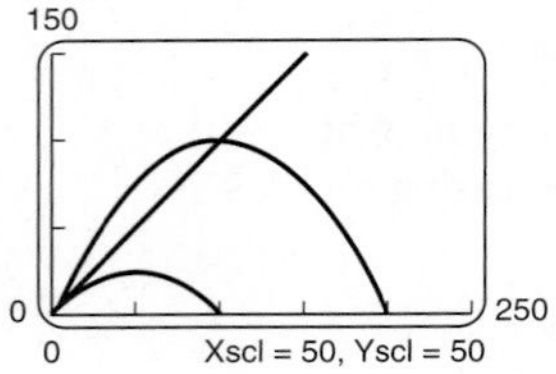

(b) 50,000; **(c)** 25,000

4. (a)

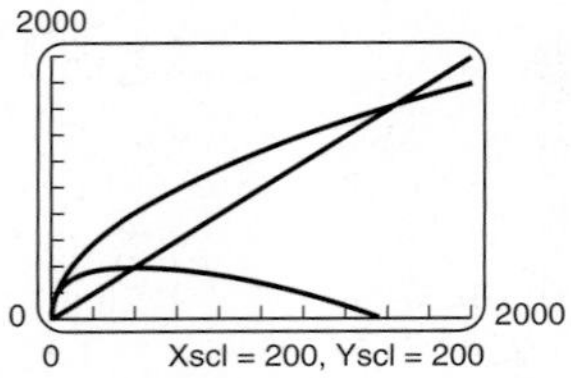

(b) 400,000; **(c)** 400,000

5. (a)

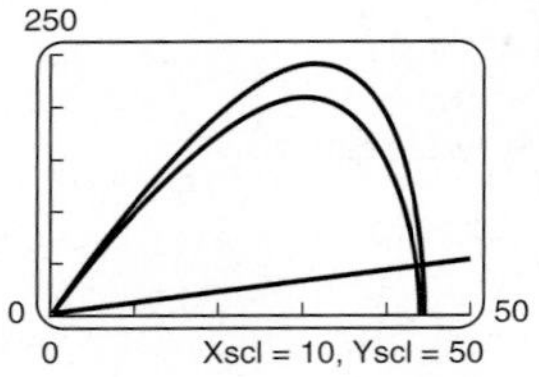

(b) 30,513; **(c)** 205,923

6. (a) $y = -0.0011P^3 + 0.0715P^2 - 0.0338P + 4$
(b)

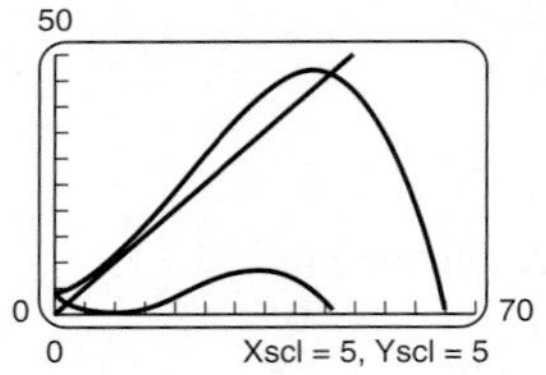

(c) 33,841

CHAPTER 3

Technology Connection, p. 305

1. 156.993 **2.** 16.242 **3.** 0.064 **4.** 0.000114

Technology Connection, p. 309

Left to the student

Exercise Set 3.1, p. 315

1.
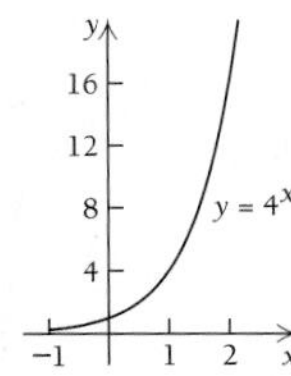

3.
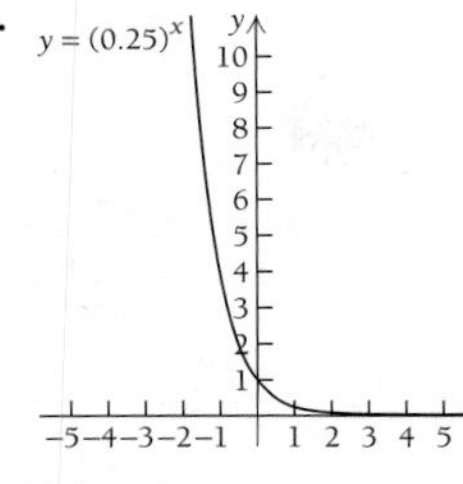

5.
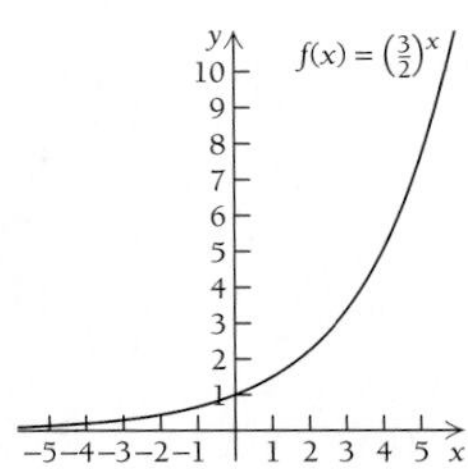

7.
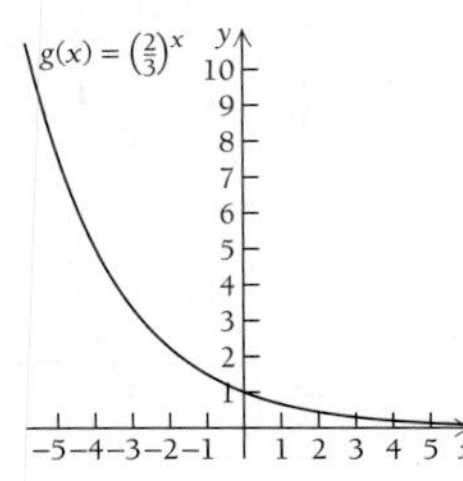

9.
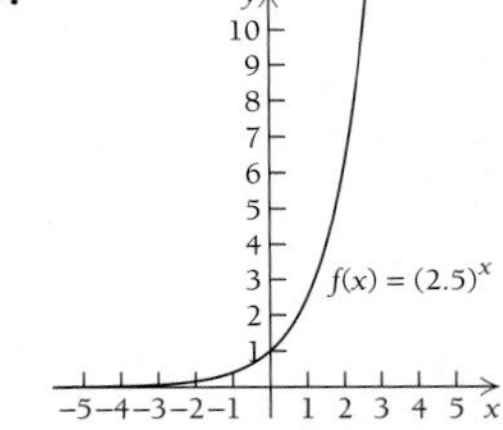

11. e^x **13.** $2e^{2x}$ **15.** $6e^x$ **17.** $-7e^{-7x}$ **19.** $8e^{4x}$

21. $3e^{-x}$ **23.** $-\frac{5}{2}e^{-5x}$ **25.** $-\dfrac{4x}{3} \cdot e^{x^2}$ **27.** $15e^{5x}$

29. $5x^4 - 12e^{6x}$ **31.** $5x^4 \cdot e^{2x} + 2x^5 \cdot e^{2x}$

33. $\dfrac{2e^{2x}(x-2)}{x^5}$ **35.** $e^x(x^2 + 5x - 6)$

37. $\dfrac{e^x(x-4)}{x^5}$ **39.** $(-2x + 7)e^{-x^2+7x}$ **41.** $-xe^{-x^2/2}$

43. $\dfrac{e^{\sqrt{x-7}}}{2\sqrt{x-7}}$ **45.** $\dfrac{e^x}{2\sqrt{e^x - 1}}$

47. $-2xe^{-2x} + e^{-2x} - e^{-x} + 3x^2$ **49.** e^{-x} **51.** ke^{-kx}

53. $(4x^2 + 3x)e^{x^2-7x}(2x - 7) + (8x + 3)e^{x^2-7x}$, or $(8x^3 - 22x^2 - 13x + 3)e^{x^2-7x}$

55.
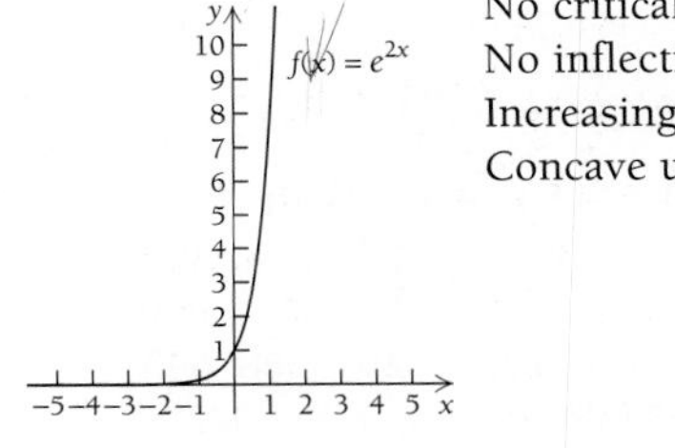

No critical values
No inflection points
Increasing on $(-\infty, \infty)$
Concave up on $(-\infty, \infty)$

57.
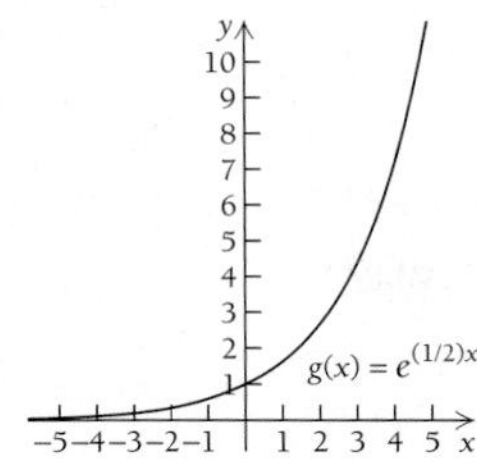

No critical values
No inflection points
Increasing on $(-\infty, \infty)$
Concave up on $(-\infty, \infty)$

59.
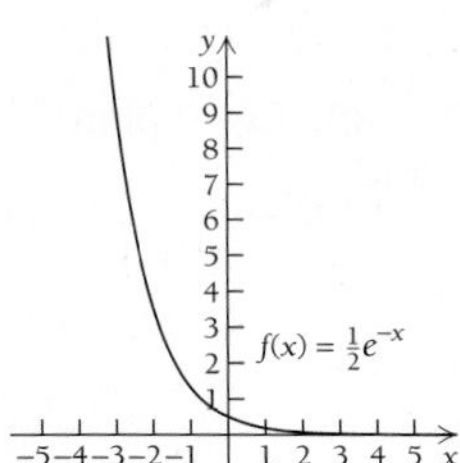

No critical values
No inflection points
Decreasing on $(-\infty, \infty)$
Concave up on $(-\infty, \infty)$

61.
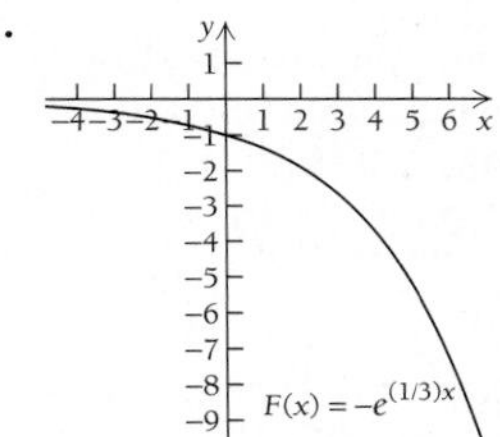

No critical values
No inflection points
Decreasing on $(-\infty, \infty)$
Concave down on $(-\infty, \infty)$

63.
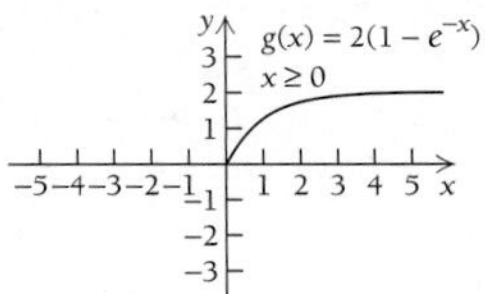

No critical values on $(0, \infty)$
No inflection points
Increasing on $[0, \infty)$
Concave down on $(0, \infty)$

65–73. Left to the student **75.** 1

77. $y = -x + 1$ **79.** Left to the student

81. (a) \$30.20 billion; **(b)** \$3.17 billion/yr

83. (a) 5.162 million, 5.199 million; **(b)** 5200; in the long run, eventually nearly 5.2 million cell phones will be in use in Israel.

85. (a) $C'(t) = 50e^{-t}$; **(b)** \$50 million/yr;

(c) \$916,000/yr; **(d)** TW

87. **(a)** 113,000; **(b)**

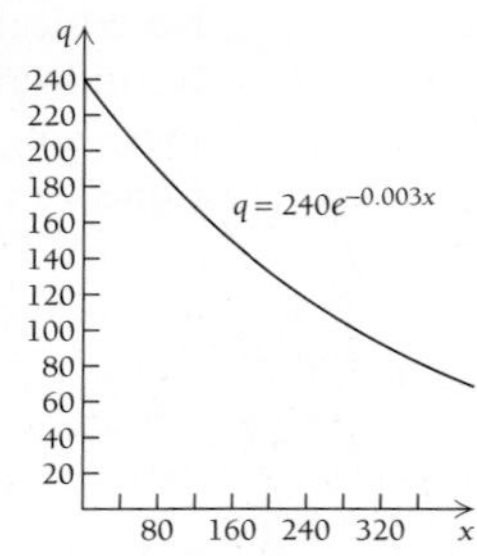

(c) $q'(x) = -0.72e^{-0.003x}$; **(d)** TW **89.** **(a)** 0 ppm, 3.7 ppm, 5.4 ppm, 4.5 ppm, 0.05 ppm;
(b)

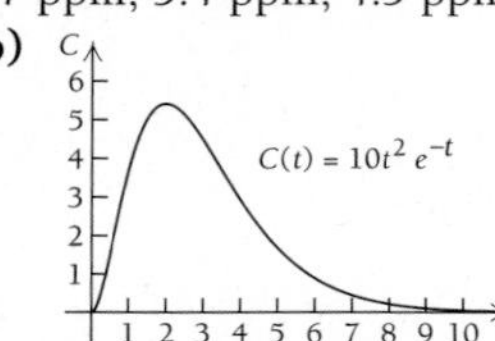

(c) $C'(t) = 10te^{-t}(2 - t)$; **(d)** 5.4 ppm at $t = 2$ hr; **(e)** TW

91. **(a)** $D'(t) = \dfrac{-64.6e^{-0.072t}}{(1 + 29.44e^{-0.072t})^2}$ represents the change in the death rate, in deaths/thousand people per yr, at t years since 1900; **(b)** declining at a rate of 0.025 deaths/thousand people per yr **93.** $15e^{3x}(e^{3x} + 1)^4$

95. $-e^{-t} - 3e^{3t}$ **97.** $\dfrac{(x^2 - 2x + 1)e^x}{(x^2 + 1)^2}$

99. $\dfrac{e^{\sqrt{x}}}{2\sqrt{x}} + \dfrac{1}{2}e^{x/2}$ **101.** $e^{x/2}\left(\dfrac{x}{2\sqrt{x - 1}}\right)$

103. $\dfrac{4}{(e^x + e^{-x})^2}$ **105.** 2; 2.25; 2.48832; 2.59374; 2.71692

107. $4e^{-2} \approx 0.5413$, for $x = 2$ **109.** TW

111.

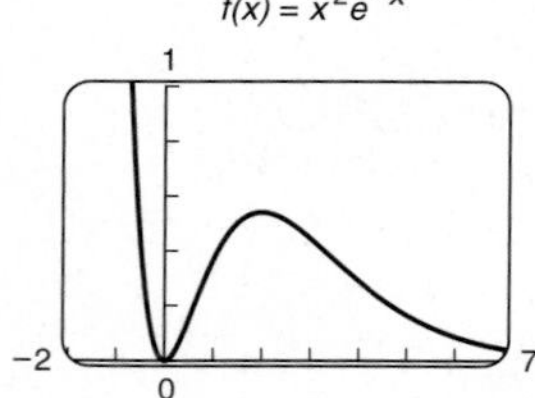

Relative minimum at $(0, 0)$; relative maximum at $(2, 0.5413)$

113.

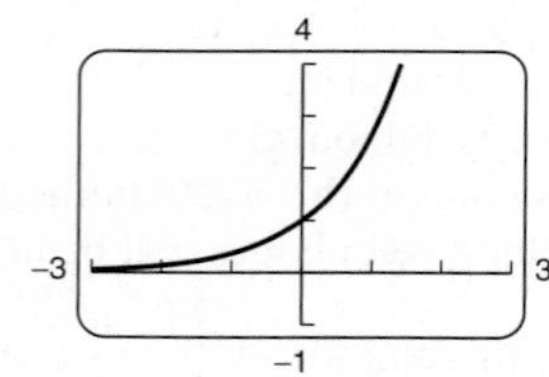

115.

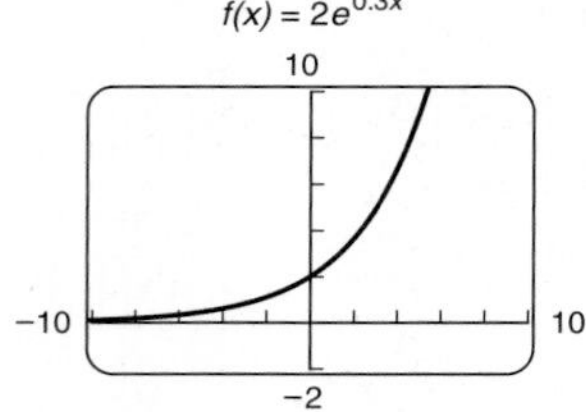

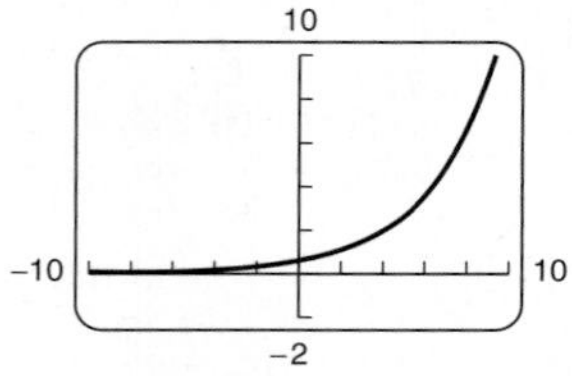

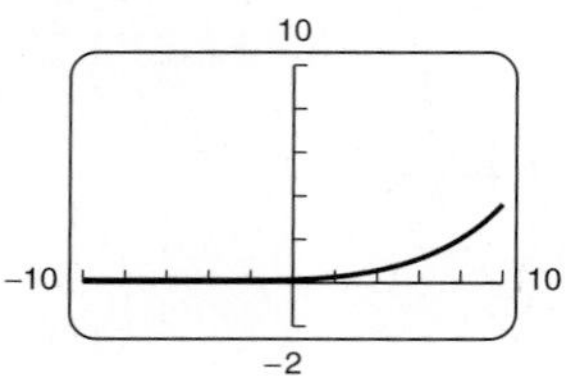

117.

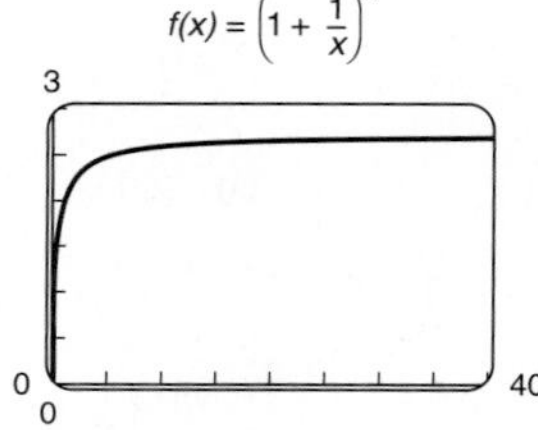

Technology Connection, p. 320

1–4. Left to the student

Technology Connection, p. 323

1. Graph is left to the student; function values are 1000, 5, 0.699, 3. **2.** Left to the student ($y = \log x/\log 2$)

Technology Connection, p. 325

1. $t = 6.9$ **2.** $x = -4.1$ **3.** $t = 74.9$ **4.** $x = 46.2$ **5.** $x = 38.7$

Exercise Set 3.2, p. 332

1. $2^3 = 8$ **3.** $8^{1/3} = 2$ **5.** $a^J = K$ **7.** $10^{-p} = h$ **9.** $\ln b = M$ **11.** $\log_{10} 100 = 2$ **13.** $\log_{10} 0.1 = -1$ **15.** $\log_M V = p$ **17.** 0.51 **19.** 2.708 **21.** 2.609 **23.** 2.9957 **25.** -1.6094 **27.** 4 **29.** 8.681690 **31.** -4.006334 **33.** 8.999619 **35.** $t \approx 4.382$ **37.** $t \approx 3.454$ **39.** $t \approx 2.303$ **41.** $t \approx 140.671$

43. $-\frac{8}{x}$ **45.** $x^3 + 4(\ln x)x^3 - x$ **47.** $\frac{1}{x}$
49. $x + 2x\ln(7x)$ **51.** $\frac{1 - 4\ln x}{x^5}$ **53.** $\frac{2}{x}$
55. $\frac{2(3x+1)}{3x^2 + 2x - 1}$ **57.** $\frac{x^2+7}{x(x^2-7)}$ **59.** $\frac{2e^x}{x} + 2e^x \ln x$
61. $\frac{e^x}{e^x + 1}$ **63.** $\frac{4(\ln x)^3}{x}$ **65.** $\frac{1}{x\ln(8x)}$
67. $\frac{\ln(5x) + \ln(3x)}{x}$ **69.** $y = 8.455x - 11.94$
71. $y = 0.732x - 0.990$ **73. (a)** 2000 units;
(b) $N'(a) = \frac{500}{a}$, $N'(10) = 50$ units per \$1000 spent on advertising; **(c)** minimum is 2000 units; **(d)** TW
75. 58 days **77. (a)** \$58.69, \$78.00;
(b) $V'(t) = 63.8e^{-1.1t}$; **(c)** 2.7 months; **(d)** TW
79. (a) $P'(x) = 1.7 - 0.3\ln x$; **(b)** TW; **(c)** 289.069
81. $A\ln\left(\frac{A}{B}\right) - A$ **83. (a)** 78%; **(b)** 53.9%;
(c) 29.7%; **(d)** 37.5%; **(e)** $S'(t) = \frac{-15}{t+1}$;
(f) maximum = 78; **(g)** TW **85. (a)** 26 words/min, 91 words/min; **(b)** $W'(t) = 30e^{-0.3t}$; **(c)** 10 weeks;
(d) TW **87.** $t = \frac{\ln(P/P_0)}{k}$ **89.** $\frac{7(2t-1)}{t(t-1)}$
91. $\frac{1}{x\ln(3x)\cdot\ln(\ln(3x))}$ **93.** $\frac{-1}{1-t} - \frac{1}{1+t}$, or $\frac{-2}{(1-t)(1+t)}$ **95.** $\frac{1}{x\ln 5}$ **97.** $\frac{x}{x^2+5}$ **99.** $x^4 \ln x$
101. $\frac{1}{\sqrt{x}(1-\sqrt{x})(1+\sqrt{x})}$, or $\frac{1}{\sqrt{x}(1-x)}$
103. Definition of logarithm; Product Rule for exponents; definition of logarithm; substitution **105.** Definition of logarithm; if $a = b$, then $a^c = b^c$; Power Rule for exponents; definition of logarithm; substitution and the commutative law for multiplication **107.** 1 **109.** e^{π} **111.** 0
113–115. Left to the student **117.** Minimum: $-e^{-1} \approx -0.368$

Technology Connection, p. 341

1. 6.7401 billion **2.** 7.2893 billion **3.** 8.2626 billion
4. 13.22 billion **5.** $y = 4262.5311 \cdot 1.057932922^x$; $y = 4262.5311e^{0.0563169307x}$; exponential growth rate = 5.632% **6.** \$13,147; \$24,427; \$84,323

Exercise Set 3.3, p. 347

1. $f(x) = ce^{4x}$ **3.** $A(t) = ce^{-9t}$ **5.** $Q(t) = ce^{kt}$
7. (a) $N(t) = 112{,}000e^{0.046t}$; **(b)** $N(30) = 445{,}189$;
(c) 15.1 yr **9. (a)** $P(t) = P_0e^{0.065t}$; **(b)** \$1067.16, \$1138.83; **(c)** 10.7 yr **11. (a)** $G(t) = 4.7e^{0.093t}$;
(b) 11.91 billion gallons; **(c)** 7.5 yr **13.** 4.62%
15. 6.9 yr after 2006 **17.** 11.2 yr; \$102,256.88
19. \$7,500; 8.3 yr **21. (a)** $k = 0.151$, or 15.1%, $V(t) = 30{,}000e^{0.151t}$; **(b)** \$258,124,520; **(c)** 4.6 yr;
(d) 69 yr **23. (a)** $I(t) = 2277e^{0.0607t}$;
(b) $I(60) = \$86{,}908$; **(c)** about 62.3 yr after 1960;
(d) 11.4 yr after 1960
25. (a) $y = 136.3939183 \cdot 1.071842825^x$, $y = 136.3939183e^{0.0693794334x}$, and exponential growth rate = 0.069, or 6.9%; **(b)** 444 million, 628 million;
(c) 18.7 yr; **(d)** 10 yr **27.** Approximately \$5.2 billion
29. 12.85%; \$5,002,484; \$18,082,319 **31. (a)** 2%;
(b) 3.8%, 7%, 21.6%, 50.2%, 93.1%, 98%;
(c) $P'(x) = \frac{637e^{-0.13x}}{(1 + 49e^{-0.13x})^2}$;
(d)

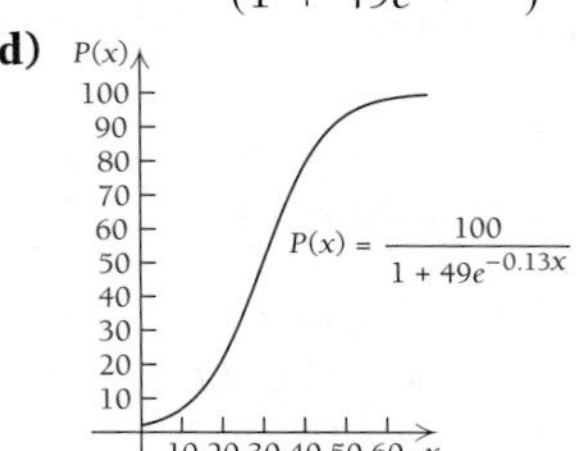

33. 19.8 yr **35.** 10%/yr **37.** 24.8 yr
39. $B(t) = 190e^{0.035t}$; 770 bears **41.** 0.205%
43. (a) 1000, 1375, 1836, 3510, 5315, 5771;
(b) $P'(t) = \frac{11{,}051.36e^{-0.4t}}{(1 + 4.78e^{-0.4t})^2}$;
(c)

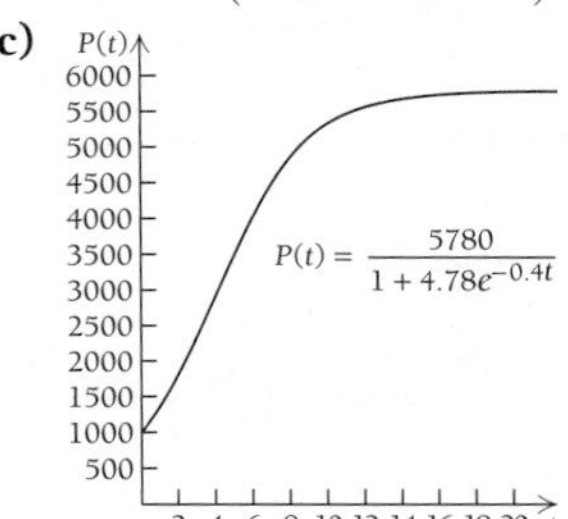

45. $N(t) = 48{,}869e^{0.0378t}$, where $t_0 = 1930$; exponential growth rate = 3.78% **47. (a)** 0%, 33%, 55%, 70%, 86%, 99.2%, 99.8%; **(b)** at 7 months, the percentage of doctors who are prescribing the medication is growing by 2.4% per month;
(c)

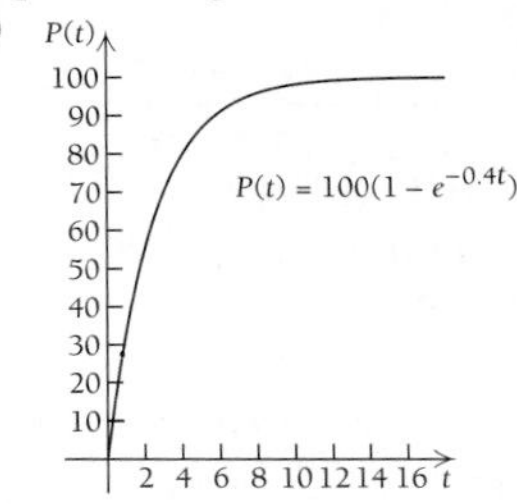

49–55. TW **57.** $\ln 4 = kT_4$ **59.** 2 yr **61.** 7.57%
63. 9% **65.** $k = \frac{\ln(y_2/y_1)}{t_2 - t_1}$ **66–67.** TW

Exercise Set 3.4, p. 360

1. **(a)** $N(t) = N_0 e^{-0.096t}$; **(b)** 341 g; **(c)** 7.2 days **3.** **(a)** $A(t) = A_0 e^{-kt}$; **(b)** 11 hr **5.** 23.1%/min **7.** 22 yr **9.** 42.9 g **11.** 4239 yr **13.** 25 days **15.** 3980 yr **17.** \$13,858.23 **19.** \$6,393,134 **21.** \$27,195 **23.** **(a)** \$40,000; **(b)** \$5413.41 **(c)** TW **25.** **(a)** $V(t) = 34{,}001.78697 \cdot 0.6702977719^x$; $V(t) = 34{,}001.78697e^{-0.4000332297t}$; **(b)** \$2067.20, \$622.56; **(c)** 8.8 yr; **(d)** 1.7 yr; **(e)** TW **27.** **(a)** $k = 0.016$, $B(t) = 80e^{-0.016t}$; **(b)** 62.9 lb per person; **(c)** 86.6 yr after 1985 **29.** **(a)** $k = 0.0096$, $P(t) = 51.9e^{-0.0096t}$; **(b)** 42.8 million people; **(c)** 1851 yr after 1995 **31.** **(a)** 27; **(b)** 0.05878; **(c)** 83°; **(d)** 28.7 min; **(e)** TW **33.** The murder was committed at 7 P.M. **35.** **(a)** 145 lb; **(b)** -1.2 lb/day **37.** **(a)** 11.2 W; **(b)** 173 days; **(c)** 402 days; **(d)** 50 W; **(e)** TW **39.** (a) **41.** (c) **43.** (d) **45.** (f) **47.** (b) **49.** TW **51.** Left to the student **53.** 2084

Exercise Set 3.5, p. 368

1. $(\ln 7)7^x$ **3.** $(\ln 8)8^x$ **5.** $x^3 \cdot \ln 5.4 \cdot (5.4)^x + 3x^2(5.4)^x$ **7.** $(\ln 7) \cdot 7^{x^4+2} \cdot 4x^3$ **9.** $8e^{8x}$ **11.** $(\ln 3) \cdot 3^{x^4+1} \cdot (4x^3)$ **13.** $\dfrac{1}{x \cdot \ln 4}$ **15.** $\dfrac{1}{x \cdot \ln 17}$ **17.** $\dfrac{5}{(5x+1)\ln 6}$ **19.** $\dfrac{6}{(6x-7)\ln 10}$ **21.** $\dfrac{3x^2+1}{(x^3+x)\ln 8}$ **23.** $\dfrac{2}{(x - 2\sqrt{x})\ln 7}$ **25.** $\dfrac{6^x}{x \cdot \ln 7} + 6^x \cdot \ln 6 \cdot \log_7 x$ **27.** $5(\log_{12} x)^4\left(\dfrac{1}{x \ln 12}\right)$ **29.** $\dfrac{(4x+1)7^x \cdot \ln 7 - 4 \cdot 7^x}{(4x+1)^2}$ **31.** $\dfrac{6 \cdot 5^{2x^3-1}}{(6x+5)(\ln 10)} + (\ln 5)5^{2x^3-1} \cdot 6x^2 \cdot \log(6x+5)$ **33.** $(\ln 7)7^x \cdot (\log_4 x)^9 + \dfrac{7^x \cdot 9 \cdot (\log_4 x)^8}{x \cdot \ln 4}$ **35.** $5(3x^5+x)^4(15x^4+1) \cdot (\log_3 x) + \dfrac{(3x^5+x)^5}{\ln 3 \cdot x}$ **37.** **(a)** $V'(t) = 5200(\ln 0.80)(0.80)^t$; **(b)** TW **39.** **(a)** \$19.84 trillion; **(b)** \$905 billion/yr; **(c)** TW **41.** **(a)** 0.82; **(b)** -0.015/yr; **(c)** TW **43.** 6.9 **45.** **(a)** $I = I_0 10^{10}$; **(b)** $I = I_0 10$; **(c)** the power mower is 10^9 times louder than a just audible sound; **(d)** $dI/dL = I_0 10^{0.1L}(\ln 10)(0.1)$; **(e)** TW **47.** **(a)** $\dfrac{dL}{dI} = \dfrac{10}{(\ln 10)I}$; **(b)** TW **49.** $\ln 3 \approx 1.0986$ **51.** $(\ln 2)2^{x^4} \cdot 4x^3$ **53.** $\dfrac{1}{\ln 3 \cdot \log x \cdot \ln 10 \cdot x}$ **55.** $\ln a \cdot a^{f(x)} \cdot f'(x)$ **57.** $\left(\dfrac{g(x) \cdot f'(x)}{f(x)} + g'(x) \cdot \ln(f(x))\right) \cdot [f(x)]^{g(x)}$ **59.** TW

Technology Connection, p. 373

1. $E(x) = \dfrac{x}{300 - x}$; $R(x) = 300x - x^2$ **2.** Left to the student **3.** \$150

Exercise Set 3.6, p. 375

1. **(a)** $E(x) = \dfrac{x}{400 - x}$; **(b)** $\dfrac{5}{11}$, inelastic; **(c)** \$200 **3.** **(a)** $E(x) = \dfrac{x}{50 - x}$; **(b)** 11.5, elastic; **(c)** \$25 **5.** **(a)** $E(x) = 1$; **(b)** 1, unit elasticity; **(c)** total revenue is independent of x. **7.** **(a)** $E(x) = \dfrac{x}{2(600 - x)}$; **(b)** 0.10, inelastic; **(c)** \$400 **9.** **(a)** $E(x) = 0.25x$; **(b)** 2.5, elastic; **(c)** \$4 **11.** **(a)** $E(x) = \dfrac{2x}{x + 3}$; **(b)** 0.5, inelastic; **(c)** \$3 **13.** **(a)** $E(x) = \dfrac{25x}{967 - 25x}$; **(b)** approximately 19¢; **(c)** prices greater than 19¢; **(d)** prices less than 19¢; **(e)** approximately 19¢; **(f)** decrease **15.** **(a)** $E(x) = \dfrac{3x^3}{2(200 - x^3)}$; **(b)** $\dfrac{81}{346}$; **(c)** increase **17.** **(a)** $E(x) = n$; **(b)** no; **(c)** yes, at $n = 1$ **19.** $E(x) = (-x)\,L'(x)$ **21.** TW

Chapter Review Exercises, p. 382

1. (b) **2.** (e) **3.** (f) **4.** (c) **5.** (a) **6.** (d) **7.** False **8.** True **9.** True **10.** False **11.** True **12.** False **13.** True **14.** False **15.** True **16.** $\dfrac{1}{x}$ **17.** e^x **18.** $\dfrac{4x^3}{x^4 + 5}$ **19.** $\dfrac{e^{2\sqrt{x}}}{\sqrt{x}}$ **20.** $\dfrac{1}{2x}$ **21.** $3x^4e^{3x} + 4x^3e^{3x}$ **22.** $\dfrac{1 - 3\ln x}{x^4}$ **23.** $2xe^{x^2}(\ln 4x) + \dfrac{e^{x^2}}{x}$ **24.** $4e^{4x} - \dfrac{1}{x}$ **25.** $8x^7 - \dfrac{8}{x}$ **26.** $\dfrac{1 - x}{e^x}$ **27.** $(\ln 9)9^x$ **28.** $\dfrac{1}{(\ln 2)\,x}$ **29.** $3^x(\ln 3)(\log_4(2x + 1)) + \dfrac{(3^x)2}{(2x + 1)(\ln 4)}$

30.

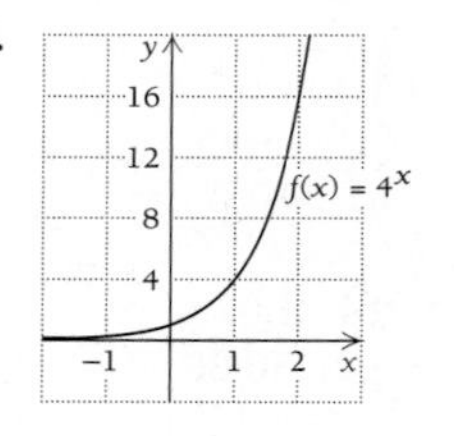

31.

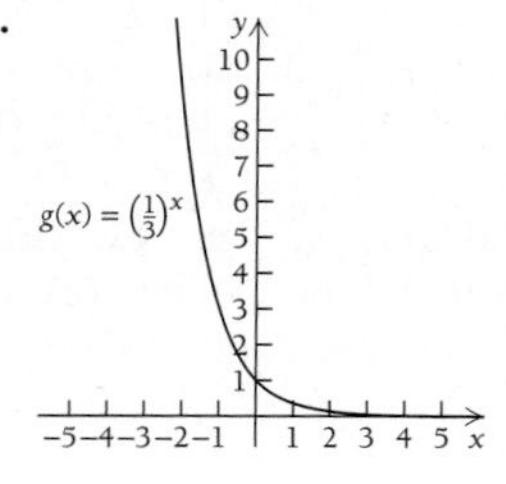

32. 6.93 **33.** −3.2698 **34.** 8.7601
35. 3.2698 **36.** 2.54995 **37.** −3.6602
38. $Q(t) = 25e^{7t}$ **39.** 4.3% **40.** 10.2 yr
41. **(a)** 0.051, $C(t) = 4.65e^{0.051t}$; **(b)** \$46.15, \$59.55
42. **(a)** $N(t) = 60e^{0.12t}$; **(b)** 123 franchises;
(c) 5.8 yr after 2007 **43.** 5.3 yr **44.** 18.2%
45. **(a)** $A(t) = 800e^{-0.07t}$; **(b)** 197 g; **(c)** 9.9 days
46. **(a)** 0.50, 0.75, 0.97, 0.999, 0.9999;
(b) $p'(t) = 0.7e^{-0.7t}$; **(c)** TW
(d)

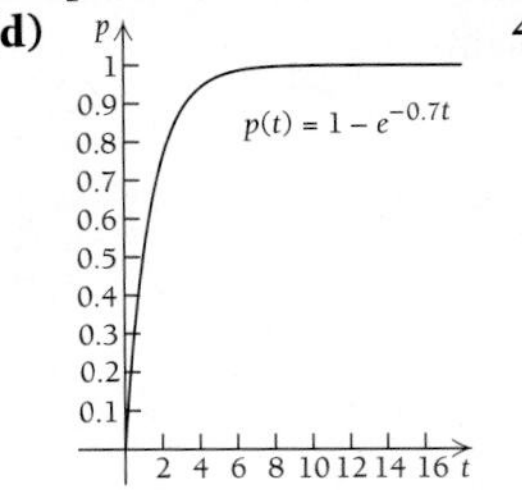

47. \$34,735.26

48. **(a)** $E(x) = \frac{2x}{x+4}$; **(b)** 0.4, inelastic;
(c) 1.5, elastic; **(d)** decrease; **(e)** \$4

49. $\frac{-8}{(e^{2x} - e^{-2x})^2}$ **50.** $-\frac{1}{1024e} \approx 0$

51. $f(x) = \frac{e^{1/x}}{(1+e^{1/x})^2}$

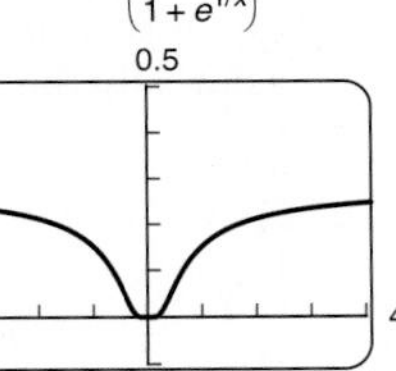

52. 0

53. **(a)** $y = 9.033880177 \cdot 1.431864118^x$,
$y = 9.033880177e^{0.3589771744x}$, 0.3589771744;
(b) \$671.0 billion, \$24.3 trillion; **(c)** 10.56 yr;
(d) 1.93 yr

Chapter 3 Test, p. 385

1. [3.1] $6e^{3x}$ **2.** [3.2] $\frac{4(\ln x)^3}{x}$ **3.** [3.1] $-2xe^{-x^2}$

4. [3.2] $\frac{1}{x}$ **5.** [3.1] $e^x - 15x^2$

6. [3.1, 3.2] $\frac{3e^x}{x} + 3e^x \cdot \ln x$

7. [3.5] $(\ln 7)7^x + (\ln 3)3^x$ **8.** [3.5] $\frac{1}{(\ln 14)x}$

9. [3.2] 1.0674 **10.** [3.2] 0.5554 **11.** [3.2] 0.4057
12. [3.3] $M(t) = 2e^{6t}$ **13.** [3.3] 23.1%
14. [3.3] 10.0 yr **15.** [3.3] **(a)** 2.7%, $C(t) = 0.54e^{0.027t}$;
(b) \$3.30, \$3.98 **16.** [3.4] **(a)** $A(t) = 3e^{-0.1t}$;
(b) 1.1 cc; **(c)** 6.9 hr **17.** [3.4] About 16.47 centuries, or
1647 yr **18.** [3.4] 4.0773%/sec **19.** [3.3] **(a)** 4%;
(b) 5.2%, 14.5%, 40.7%, 73.5%, 91.8%, 99.5%, 99.9%;
(c) $P'(t) = \frac{672e^{-0.28t}}{(1 + 24e^{-0.28t})^2}$; **(d)** TW;
(e)

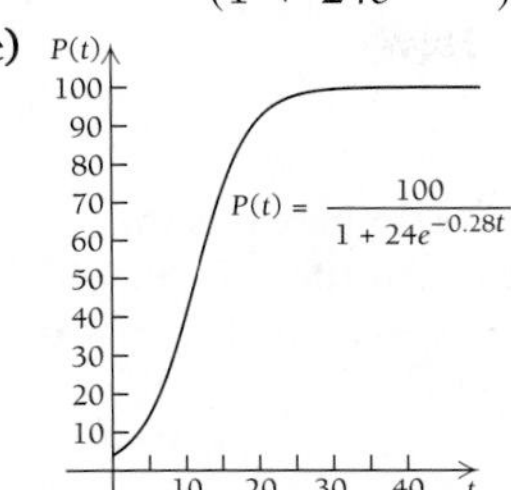

20. [3.4] \$10,945,729 **21.** [3.6] **(a)** $E(x) = 0.2x$;
(b) 0.6, inelastic; **(c)** 3.6, elastic; **(d)** increase;
(e) \$5 **22.** [3.2] $(\ln x)^2$ **23.** [3.1] Maximum is
$\frac{256}{e^4} \approx 4.689$; minimum is 0

24. [3.1]

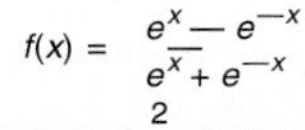

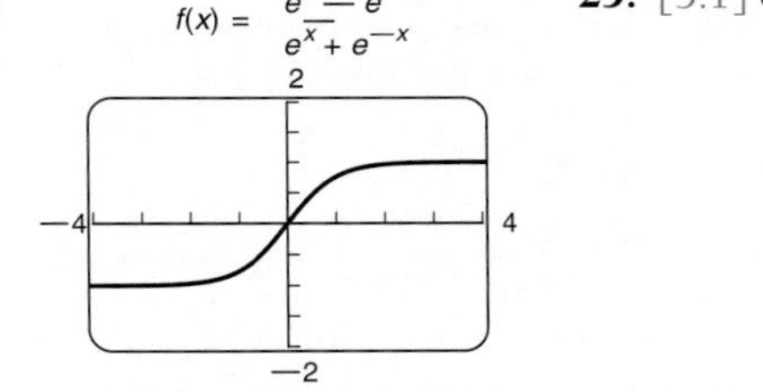

25. [3.1] 0

26. [3.3] **(a)** $y = 686996.582 \cdot 1.086747476^x$,
$y = 686996.582 \cdot e^{0.0831892684x}$; **(b)** \$3,630,000,
\$8,330,000; **(c)** 87.5 yr; **(d)** 8.3 yr

Extended Technology Application, p. 387

1.

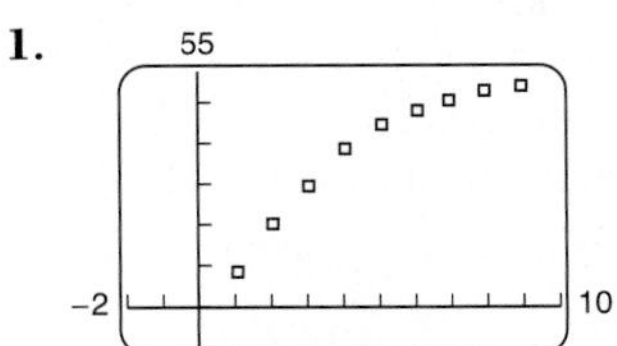

A logistic equation seems to fit the data points, because there appears to be a limiting value for R.

2. $R(t) = \frac{48.8515}{1 + 7.512e^{-0.7543t}}$

3. $R'(t) = \frac{276.81e^{-0.7543t}}{(1 + 7.512e^{-0.7543t})^2}$, which represents the rate of change of the total revenue; $\lim_{t \to \infty} R'(t) = 0$, which means that eventually the total revenue does not change.

4.

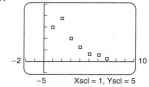

Answers will vary. Domain is $[0, a]$, where a is the longest time that a movie is shown at theaters. Rarely will a exceed 12 weeks, so $[0, 12]$ can usually serve as the domain.

5. Answers will vary.

6.

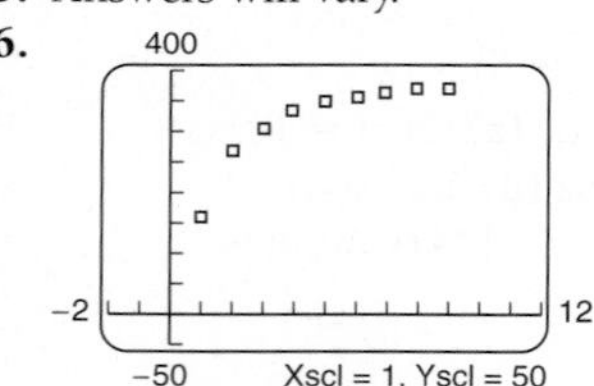

A logistic equation seems to fit the data points, because there appears to be a limiting value for R.

7. $R(t) = \dfrac{365.1}{1 + 3.094e^{-0.9704t}}$

8. $R'(t) = \dfrac{1096.2e^{-0.9704t}}{(1 + 3.094e^{-0.9704t})^2}$, which represents the rate of change of the total revenue; $\lim_{t\to\infty} R'(t) = 0$, which means that eventually the total revenue does not change.

9.

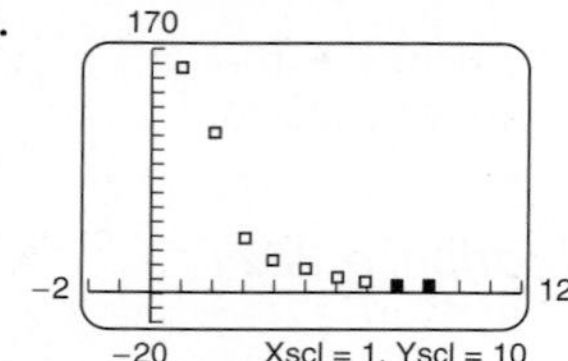

Answers will vary. Domain is $[0, a]$, where a is the number of weeks the film was showing in theaters. Thus, $[0, 12]$ probably suffices here.

10. Answers will vary.

11.

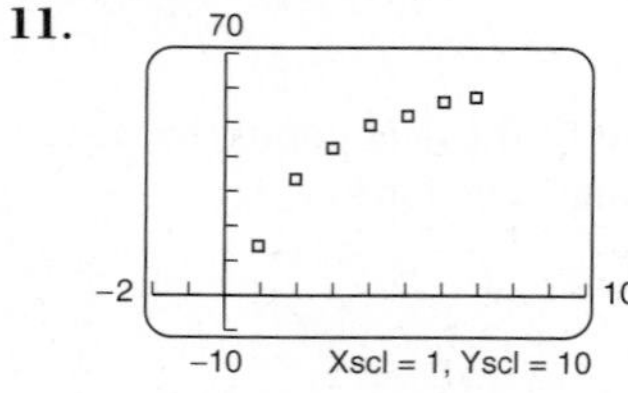

A logistic equation seems to fit the data points, because there appears to be a limiting value for R.

12. $R(t) = \dfrac{55.36}{1 + 7.057e^{-1.099t}}$

13. $R'(t) = \dfrac{429.4e^{-1.099t}}{(1 + 7.057e^{-1.099t})^2}$, which represents the rate of change of the total revenue; $\lim_{t\to\infty} R'(t) = 0$, which means that eventually the total revenue does not change.

14.

Answers will vary. Domain is $[0, a]$, where a is the number of weeks the film is shown in theaters. Thus, $[0, 12]$ may be a suitable domain.

15. Answers will vary.

16.

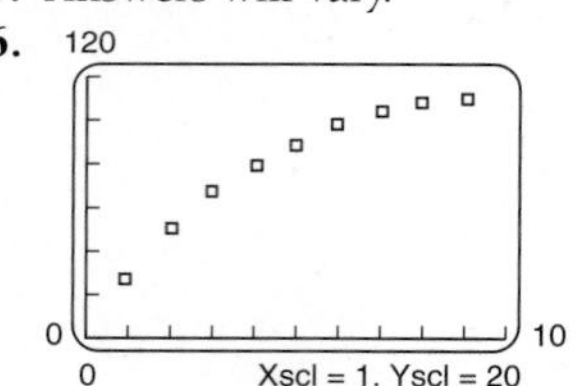

A logistic equation seems to fit the data points, because there appears to be a limiting value for R.

17. $R(t) = \dfrac{110.6770}{1 + 4.763e^{-0.635t}}$

18. $R'(t) = \dfrac{334.743e^{-0.635t}}{(1 + 4.763e^{-0.635t})^2}$, which represents the rate of change of the total revenue; $\lim_{t\to\infty} R'(t) = 0$, which means that eventually the total revenue does not change.

19.

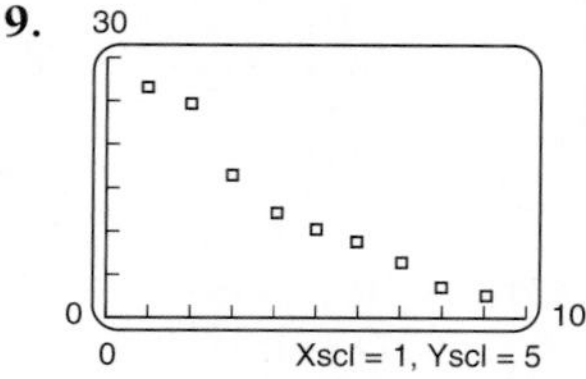

Answers will vary. Domain is $[0, a]$, where a is the number of weeks the film appears in theaters. Thus, $[0, 12]$ is probably a suitable domain.

20. Answers will vary. **21.** Yes; eventually a movie will generate little additional income, and its total revenue will approach a limiting value. **22.** Answers will vary.

CHAPTER 4

Exercise Set 4.1, p. 398

1. \$1060 **3.** 46,800¢, or \$468 **5.** −\$255,000 **7.** \$8400 **9.** 23302.4¢, or \$233.02 **11.** \$471.96 **13.** $\sum_{i=1}^{6} 3i$ **15.** $\sum_{i=1}^{4} f(x_i)$ **17.** $\sum_{i=1}^{15} G(x_i)$ **19.** $2^1 + 2^2 + 2^3 + 2^4$, or 30 **21.** $f(x_1) + f(x_2) + f(x_3) + f(x_4) + f(x_5)$ **23. (a)** 1.4914; **(b)** 1.1418 **25.** 3,166,250¢, or \$31,662.50 **27.** 247.68 **29.** 124

31. $\sum_{x=1}^{4} kf(x_i) = kf(x_1) + kf(x_2) + kf(x_3) + kf(x_4)$

$= k[f(x_1) + f(x_2) + f(x_3) + f(x_4)]$

$= k \sum_{x=1}^{4} f(x_i);$

$\sum_{i=1}^{n} kf(x_i) = kf(x_1) + kf(x_2) + kf(x_3) + \cdots + kf(x_n)$

$= k[f(x_1) + f(x_2) + f(x_3) + \cdots + f(x_n)]$

$= k \sum_{i=1}^{n} f(x_i)$

33. 47.5 **35.** 75.42 (Exact area is $\frac{1}{2} \cdot 49\pi$.)

Exploratory Exercises, p. 402

1. (a) $A(x) = 5x$; **(b)** 5, 10, 35; **(c)** $A(x) = f(x) \cdot x$

2. (a) $A(x) = \frac{3}{2}x^2$; **(b)** $\frac{3}{2}$, 6, 18.375;

(c) $A(x) = \frac{x}{2} \cdot f(x)$

Exercise Set 4.2, p. 409

1. $\frac{x^7}{7} + C$ **3.** $2x + C$ **5.** $\frac{4}{5}x^{5/4} + C$

7. $\frac{1}{3}x^3 + \frac{1}{2}x^2 - x + C$ **9.** $\frac{2}{3}t^3 + \frac{5}{2}t^2 - 3t + C$

11. $-\frac{x^{-2}}{2} + C$ **13.** $\frac{3}{4}x^{4/3} + C$ **15.** $\frac{2}{7}x^{7/2} + C$

17. $-\frac{x^{-3}}{3} + C$ **19.** $\ln x + C$ **21.** $3 \ln x - 5x^{-1} + C$

23. $-21x^{1/3} + C$ **25.** $e^{2x} + C$ **27.** $\frac{1}{3}e^{3x} + C$

29. $\frac{1}{7}e^{7x} + C$ **31.** $\frac{5}{3}e^{3x} + C$ **33.** $\frac{3}{4}e^{8x} + C$

35. $-\frac{2}{27}e^{-9x} + C$ **37.** $\frac{5}{3}x^3 - \frac{2}{7}e^{7x} + C$

39. $\frac{x^3}{3} - x^{3/2} - 3x^{-1/3} + C$ **41.** $3x^3 + 6x^2 + 4x + C$

43. $3 \ln x - \frac{5}{2}e^{2x} + \frac{2}{9}x^{9/2} + C$

45. $14x^{1/2} - \frac{2}{15}e^{5x} - 8 \ln x + C$

47. $f(x) = \frac{1}{2}x^2 - 3x + 13$ **49.** $f(x) = \frac{1}{3}x^3 - 4x + 7$

51. $f(x) = \frac{5}{3}x^3 + \frac{3}{2}x^2 - 7x + 9$

53. $f(x) = x^3 - \frac{5}{2}x^2 + x + 4$

55. $f(x) = \frac{5}{2}e^{2x} - 2$ **57.** $f(x) = 8x^{1/2} - 13$

59. $D(t) = 17{,}198 + 1975t - 595t^2 + 199t^3 - 17.825t^4$

61. $C(x) = \frac{x^4}{4} - x^2 + 7000$

63. (a) $R(x) = \frac{x^3}{3} - 3x$; **(b)** TW

65. $D(x) = \frac{4000}{x} + 3$ **67. (a)** $E(t) = 32 + 30t - 5t^2$;
(b) $E(3) = 77\%$, $E(5) = 57\%$

69. (a) $I(t) = 32.31e^{0.1049t} - 32.31$;
(b) about 516 people; **(c)** about 1111 people;
(d) about 595 people **71.** $f(t) = \frac{2}{3}t^{3/2} + 2t^{1/2} - \frac{28}{3}$

73. $\frac{25}{7}t^7 + \frac{20}{3}t^6 + \frac{16}{5}t^5 + C$ **75.** $\frac{2}{3}t^{3/2} - \frac{2}{5}t^{5/2} + C$

77. $\frac{x^2}{2} - 6 \ln x + \frac{7}{2}x^{-2} + C$

79. $\frac{1}{\ln 10} \cdot \ln x + C$, or $\log x + C$

81. $3x^4 - \frac{8}{3}x^3 - \frac{17}{2}x^2 - 5x + C$ **83.** $\frac{x^2}{2} - x + C$

85. TW

Technology Connection, p. 418

1. 0 **2.** 13.75 **3.** 0.535 **4.** 27.972 **5.** −260

Exercise Set 4.3, p. 421

1. 8 **3.** 8 **5.** $41\frac{2}{3}$ **7.** $\frac{1}{4}$ **9.** $10\frac{2}{3}$

11. $e^3 - 1 \approx 19.086$ **13.** $3 \ln 6 \approx 5.375$
15. Total cost, in dollars, for t days **17.** Total number of kilowatts used in t hours **19.** Total revenue, in dollars, for x units produced **21.** Total amount of the drug, in milligrams, in v cubic centimeters of blood **23.** Total number of words memorized in t minutes **25.** 4

27. $9\frac{5}{6}$ **29.** 12 **31.** $e^5 - e^{-1}$, or approximately 148.045 **33.** TW **35.** 0; the area above the x-axis is the same as the area below the x-axis. **37.** $-\frac{4}{15}$; all the area is below the x-axis. **39–41.** Left to the student

43. 40 **45.** $\frac{5}{3}$ **47.** $\frac{637}{6}$ **49.** $e^2 - e^{-5}$, or approximately 7.382 **51.** $\frac{b^3 - a^3}{6}$ **53.** $\frac{e^{2b} - e^{2a}}{2}$

55. $\frac{e^2 + 1}{2}$, or approximately 4.195 **57.** $\frac{8}{3}$

59. \$628.56 **61.** \$29.13 **63. (a)** \$2948.26;
(b) \$2913.90 **65.** \$7627.28 billion **67.** 18.69 hr; 20.12 hr **69.** 7 words **71.** about 5 words
73. $s(t) = t^3 + 4$ **75.** $v(t) = 2t^2 + 20$

77. $s(t) = -\frac{t^3}{3} + 3t^2 + 6t + 10$ **79. (a)** 104.17 m;
(b) 229.17 m **81. (a)** 60 mph; **(b)** $\frac{1}{8}$ mi

83. **(a)** 13.33 km/hr; **(b)** 0.1875 km
85. $s(t) = -16t^2 + v_0t + s_0$ **87.** $\frac{1}{4}$ mi **89.** 148 mi
91. On the 10th day **93.** 3.5 **95.** $359\frac{7}{15}$ **97.** 6.75
99. 30 **101.** $5\frac{1}{3}$ **103.** $14\frac{2}{3}$ **105.** TW
107. 4068.789 **109.** 7.571 **111.** 9.524
113. 10.987

Technology Connection, p. 429

1. $\frac{4}{3}$

Technology Connection, p. 432

1.

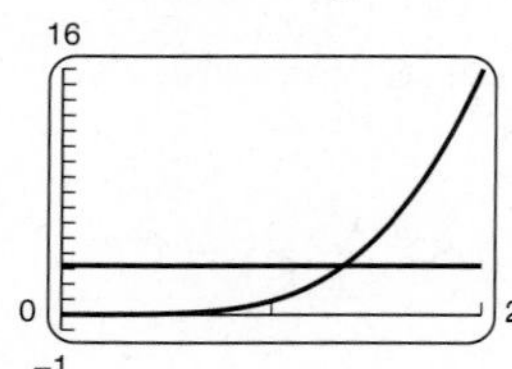

Over the interval [0, 2], the areas under $f(x) = x^4$ and $y_{av} = 3.2$ are equal.

Exercise Set 4.4, p. 433

1. 22 **3.** $18\frac{1}{6}$ **5.** $89\frac{11}{12}$ **7.** $x = -3$ and $x = 3$
9. $x = \dfrac{3 \pm \sqrt{29}}{2}$, or approximately $x = -1.193$ and $x = 4.193$ **11.** $x = -3$ and $x = 5$ **13.** $\frac{3}{2}$ **15.** 62.5
17. $\frac{1}{4}$ **19.** $4\frac{1}{2}$ **21.** $20\frac{5}{6}$ **23.** $4\frac{1}{2}$ **25.** $\frac{3}{10}$ **27.** $41\frac{2}{3}$
29. $10\frac{2}{3}$ **31.** 3 **33.** $85\frac{1}{3}$ **35.** $\frac{8}{3}$ **37.** $-e^{-1} + 1$, or approximately 0.632 **39.** $\frac{16}{3}$ **41.** $2a + 5$
43. $\dfrac{2^{n+1} - 1}{n + 1}$ **45.** **(a)** \$2,201,556.58; **(b)** \$220,155.66
47. \$26,534.37 **49.** \$32,781.35 **51.** **(a)** Ben; **(b)** 2 more words; **(c)** 0.7 word per minute; **(d)** 0.9 word per minute **53.** **(a)** 90 words per minute; **(b)** 96 words per minute, at $t = 1$ min; **(c)** 70 words per minute **55.** **(a)** 42.03 μg/mL; **(b)** 22.44 μg/mL
57. **(a)** 31.7°; **(b)** −10°; **(c)** 46.25° **59.** $40\frac{8}{15}$
61. 16 **63.** 6 **65.** 4 **67.** 4 **69.** 5.886
71. 0.237

Technology Connection, p. 441

1. 4.673

Exercise Set 4.5, p. 442

1. $\frac{1}{6}(8 + x^3)^6 + C$ **3.** $\frac{1}{16}(x^2 - 6)^8 + C$
5. $\frac{1}{24}(3t^4 + 2)^2 + C$ **7.** $\ln(2x + 1) + C$
9. $\frac{1}{4}(\ln x)^4 + C$ **11.** $\frac{1}{3}e^{3x} + C$ **13.** $3e^{x/3} + C$
15. $\frac{1}{5}e^{x^5} + C$ **17.** $-\frac{1}{2}e^{-t^2} + C$
19. $\frac{1}{2}\ln(5 + 2x) + C$ **21.** $\frac{1}{3}\ln(12 + 3x) + C$
23. $-\ln(1 - x) + C$ **25.** $\frac{1}{12}(t^2 - 1)^6 + C$
27. $\frac{1}{8}(x^4 + x^3 + x^2)^8 + C$ **29.** $\ln(4 + e^x) + C$
31. $(\ln x)^2 + C$ **33.** $\ln(\ln x) + C$
35. $\dfrac{1}{3a}(ax^2 + b)^{3/2} + C$ **37.** $\dfrac{P_0}{k}e^{kt} + C$
39. $\dfrac{1}{24(2 - x^4)^6} + C$ **41.** $\frac{5}{6}(1 + 6x^2)^{6/5} + C$
43. $e - 1$ **45.** $\frac{21}{4}$ **47.** ln 5 **49.** ln 19
51. $1 - e^{-b}$ **53.** $1 - e^{-mb}$ **55.** $\frac{208}{3}$ **57.** $\frac{1640}{6561}$
59. $\frac{315}{8}$ **61.** Left to the student
63. **(a)** $P(t) = 2000t^2 - 50{,}000 - 25{,}000e^{-0.1t}$; **(b)** \$82,042.95 **65.** **(a)** About 51,218,297; **(b)** about 27,059,930 **67.** $42\frac{2}{3}$ **69.** $-\frac{5}{12}(1 - 4x^2)^{3/2} + C$
71. $-\frac{1}{3}e^{-x^3} + C$ **73.** $-e^{1/t} + C$ **75.** $-\frac{1}{3}(\ln x)^{-3} + C$
77. $\frac{2}{9}(x^3 + 1)^{3/2} + C$ **79.** $\frac{3}{4}(x^2 - 6x)^{2/3} + C$
81. $t + \dfrac{1}{t + 1} + C$ **83.** $x + 1 + 2\ln(x + 1) + C$, or $x + 2\ln(x + 1) + K$, where $K = 1 + C$
85. $\dfrac{(\ln x)^{-n+1}}{-n + 1} + C$ **87.** $\ln(e^x + e^{-x}) + C$
89. $\ln[\ln(\ln x)] + C$
91. $\dfrac{9}{14(n + 1)}(7x^2 + 9)^{n+1} + C$
93. TW

Technology Connection, p. 448

1. 1.941

Exercise Set 4.6, p. 451

1. $xe^{4x} - \frac{1}{4}e^{4x} + C$ **3.** $\dfrac{x^6}{2} + C$ **5.** $\frac{1}{5}xe^{5x} - \frac{1}{25}e^{5x} + C$
7. $-\frac{1}{2}xe^{-2x} - \frac{1}{4}e^{-2x} + C$ **9.** $\dfrac{x^3 \ln x}{3} - \dfrac{x^3}{9} + C$
11. $\frac{1}{4}x^2 \ln x - \frac{1}{8}x^2 + C$
13. $(x + 5)\ln(x + 5) - x + C$

15. $\left(\frac{x^2}{2} + 2x\right) \ln x - \frac{x^2}{4} - 2x + C$

17. $\left(\frac{x^2}{2} - x\right) \ln x - \frac{x^2}{4} + x + C$

19. $\frac{2}{3}x(x + 2)^{3/2} - \frac{4}{15}(x + 2)^{5/2} + C$

21. $\frac{x^4}{4} \ln (2x) - \frac{x^4}{16} + C$, or $\frac{x^4 \ln 2}{4} - \frac{x^4 \ln x}{4} + \frac{x^4}{16} + C$

23. $x^2e^x - 2xe^x + 2e^x + C$

25. $\frac{1}{2}x^2e^{2x} - \frac{1}{2}xe^{2x} + \frac{1}{4}e^{2x} + C$

27. $-\frac{1}{2}x^3e^{-2x} - \frac{3}{4}x^2e^{-2x} - \frac{3}{4}xe^{-2x} - \frac{3}{8}e^{-2x} + C$

29. $\frac{1}{3}(x^4 + 4)e^{3x} - \frac{4}{9}x^3e^{3x} + \frac{4}{9}x^2e^{3x} - \frac{8}{27}xe^{3x} + \frac{8}{81}e^{3x} + C$ **31.** $\frac{8}{3}\ln 2 - \frac{7}{9}$

33. $14 \ln 14 - 10 \ln 10 - 4$ **35.** 1

37. $C(x) = \frac{8}{3}x(x + 3)^{3/2} - \frac{16}{15}(x + 3)^{5/2}$

39. **(a)** $-10Te^{-T} - 10e^{-T} + 10$; **(b)** about 9.084 kW-h

41. $\frac{2}{3}x^{3/2} \ln x - \frac{4}{9}x^{3/2} + C$ **43.** $2\sqrt{x}(\ln x) - 4\sqrt{x} + C$

45. $\frac{2}{7}(27x^3 + 83x - 2)(3x + 8)^{7/6} - \frac{4}{91}(81x^2 + 83)(3x + 8)^{13/6} + \frac{1296}{1729}x(3x + 8)^{19/6} - \frac{2592}{43,225}(3x + 8)^{25/6} + C$

47. $\frac{x^{n+1}}{n + 1}(\ln x)^2 - \frac{2x^{n+1}}{(n + 1)^2}\ln x + \frac{2x^{n+1}}{(n + 1)^3} + C$

49. Let $u = x^n$ and $dv = e^x\,dx$. Then $du = nx^{n-1}\,dx$ and $v = e^x$. Next, use integration by parts. **51.** TW **53.** About 355,986

Exercise Set 4.7, p. 456

1. $-\frac{1}{9}e^{-3x}(3x + 1) + C$ **3.** $\frac{6^x}{\ln 6} + C$

5. $\frac{1}{10}\ln\left|\frac{5 + x}{5 - x}\right| + C$ **7.** $3 - x - 3 \ln |3 - x| + C$

9. $\frac{1}{8(8 - x)} + \frac{1}{64}\ln\left|\frac{x}{8 - x}\right| + C$

11. $(\ln 3)x + x \ln x - x + C$ **13.** $\frac{x^5}{5}(\ln x) - \frac{x^5}{25} + C$

15. $\frac{x^4}{4}(\ln x) - \frac{x^4}{16} + C$ **17.** $\ln |x + \sqrt{x^2 + 7}| + C$

19. $\frac{2}{5 - 7x} + \frac{2}{5}\ln\left|\frac{x}{5 - 7x}\right| + C$

21. $-\frac{5}{4}\ln\left|\frac{x - 1/2}{x + 1/2}\right| + C$

23. $m\sqrt{m^2 + 4} + 4 \ln |m + \sqrt{m^2 + 4}| + C$

25. $\frac{5}{2x^2}(\ln x) + \frac{5}{4x^2} + C$

27. $x^3e^x - 3x^2e^x + 6xe^x - 6e^x + C$

29. $\frac{1}{15}(3x - 1)(1 + 2x)^{3/2} + C$

31. $S(x) = 100\left[\frac{20}{20 - x} + \ln(20 - x)\right]$

33. $-4 \ln\left|\frac{x}{3x - 2}\right| + C$

35. $\frac{-1}{2(x - 2)} + \frac{1}{4}\ln\left|\frac{x}{x - 2}\right| + C$

37. $\frac{-3}{e^{-x} - 3} + \ln |e^{-x} - 3| + C$

Chapter Review Exercises, p. 462

1. True **2.** False **3.** True **4.** False **5.** (e) **6.** (d) **7.** (a) **8.** (f) **9.** (b) **10.** (c) **11.** \$77,000 **12.** $4x^5 + C$ **13.** $3e^x + 2x + C$ **14.** $t^3 + \frac{5}{2}t^2 + \ln t + C$ **15.** 9 **16.** 21 **17.** Total number of words keyboarded in t minutes **18.** Total sales in t days **19.** $\frac{b^6 - a^6}{6}$ **20.** $-\frac{2}{5}$ **21.** $e - \frac{1}{2}$

22. $2 \ln 4$, or $4 \ln 2$ **23.** Zero **24.** Negative **25.** Positive **26.** $13\frac{1}{2}$ **27.** $\frac{1}{4}e^{x^4} + C$

28. $\ln(4t^6 + 3) + C$ **29.** $\frac{1}{4}(\ln 4x)^2 + C$

30. $-\frac{2}{3}e^{-3x} + C$ **31.** $xe^{3x} - \frac{1}{3}e^{3x} + C$

32. $-\frac{2x}{3} + x \ln x^{2/3} + C$ **33.** $x^3 \ln x - \frac{x^3}{3} + C$

34. $\frac{1}{14}\ln\left|\frac{7 + x}{7 - x}\right| + C$

35. $\frac{1}{5}x^2e^{5x} - \frac{2}{25}xe^{5x} + \frac{2}{125}e^{5x} + C$

36. $\frac{1}{49} + \frac{x}{7} - \frac{1}{49}\ln |7x + 1| + C$

37. $\ln |x + \sqrt{x^2 - 36}| + C$ **38.** $x^7\left(\frac{\ln x}{7} - \frac{1}{49}\right) + C$

39. $\frac{1}{64}e^{8x}(8x - 1) + C$ **40.** About \$70,666.67

41. $\frac{1}{2}(1 - 3e^{-2})$, or approximately 0.297 **42.** 80 mi

43. About \$162,753.79 **44.** $10x^3e^{0.1x} - 300x^2e^{0.1x} + 6000xe^{0.1x} - 60{,}000e^{0.1x} + C$

45. $\ln|4t^3 + 7| + C$
46. $\frac{2}{75}(5x - 8)\sqrt{4 + 5x} + C$
47. $e^{x^5} + C$ **48.** $\ln(x + 9) + C$
49. $\frac{1}{96}(t^8 + 3)^{12} + C$ **50.** $x\ln(7x) - x + C$
51. $\frac{x^2}{2}\ln(8x) - \frac{x^2}{4} + C$ **52.** $\frac{1}{10}\left[\ln|t^5 + 3|\right]^2 + C$
53. $-\frac{1}{2}\ln(1 + 2e^{-x}) + C$
54. $(\ln\sqrt{x})^2 + C$, or $\frac{1}{4}(\ln x)^2 + C$
55. $x^{92}\left(\frac{\ln x}{92} - \frac{1}{8464}\right) + C$
56. $(x - 3)\ln(x - 3) - (x - 4)\ln(x - 4) + C$
57. $-\frac{1}{3(\ln x)^3} + C$ **58.** 1.343

Chapter 4 Test, p. 464

1. [4.1] 95 **2.** [4.2] $\frac{2\sqrt{3}}{3}x^{3/2} + C$
3. [4.2] $\frac{500}{3}x^6 + C$ **4.** [4.2] $e^x + \ln x + \frac{8}{11}x^{11/8} + C$
5. [4.3] $\frac{1}{6}$ **6.** [4.3] $4\ln 3$ **7.** [4.3] Total miles run in t hours **8.** [4.3] 12 **9.** [4.3] $\frac{1 - e^{-2}}{2}$
10. [4.3] $\ln a$ **11.** [4.3] Positive
12. [4.5] $\ln(x + 12) + C$ **13.** [4.5] $-2e^{-0.5x} + C$
14. [4.5] $\frac{(t^4 + 3)^{10}}{40} + C$ **15.** [4.6] $\frac{1}{5}xe^{5x} - \frac{1}{25}e^{5x} + C$
16. [4.6] $\frac{x^4}{4}\ln x^4 - \frac{x^4}{4} + C$, or $x^4\ln|x| - \frac{x^4}{4} + C$
17. [4.7] $\frac{2^x}{\ln 2} + C$ **18.** [4.7] $\frac{1}{7}\ln\left|\frac{x}{7 - x}\right| + C$
19. [4.4] 6 **20.** [4.4] $\frac{1}{3}$ **21.** [4.4] \$49,000
22. [4.3] 94 words **23.** [4.3] 5.4 km
24. [4.5] $\frac{6}{7}\ln(5 + 7x) + C$
25. [4.6] $x^5e^x - 5x^4e^x + 20x^3e^x - 60x^2e^x + 120xe^x - 120e^x + C$
26. [4.5] $\frac{1}{6}e^{x^6} + C$
27. [4.6, 4.7] $\frac{2}{3}x^{3/2}(\ln x) - \frac{4}{9}x^{3/2} + C$
28. [4.7] $\frac{1}{16}\ln\left(\frac{8 + x}{8 - x}\right) + C$
29. [4.6, 4.7] $-10x^4e^{-0.1x} - 400x^3e^{-0.1x} - 12{,}000x^2e^{-0.1x} - 240{,}000xe^{-0.1x} - 2{,}400{,}000e^{-0.1x} + C$
30. [4.6] $\frac{x^2}{2}\ln(13x) - \frac{x^2}{4} + C$
31. [4.6] $\frac{1}{15}(3x^2 - 8)(x^2 + 4)^{3/2} + C$
32. [4.5] $\frac{(\ln x)^4}{4} - \frac{4}{3}(\ln x)^3 + 5\ln x + C$
33. [4.6] $(x + 3)\ln(x + 3) - (x + 5)\ln(x + 5) + C$
34. [4.6, 4.7] $\frac{3}{10}(8x^3 + 10)(5x - 4)^{2/3} - \frac{108}{125}x^2(5x - 4)^{5/3} + \frac{81}{625}x(5x - 4)^{8/3} - \frac{243}{34{,}375}(5x - 4)^{11/3} + C$
35. [4.5] $\frac{1}{\ln 5}e^{(\ln 5)x} + C$, or $\frac{5^x}{\ln 5} + C$ **36.** [4.4] 16

Extended Technology Application, p. 466

1. Exponential

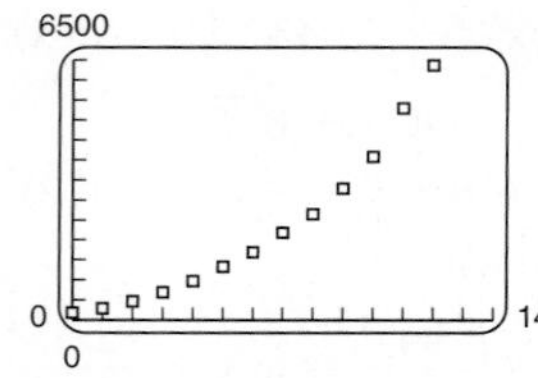

2. $S(x) = 254.9923987(1.330525137)^x$
3. $S(15) \approx \$18{,}487$ million; $S(19) \approx \$57{,}938$ million; $S(22) \approx \$136{,}469$ million
4. About \$1,991,763 million
5. Exponential

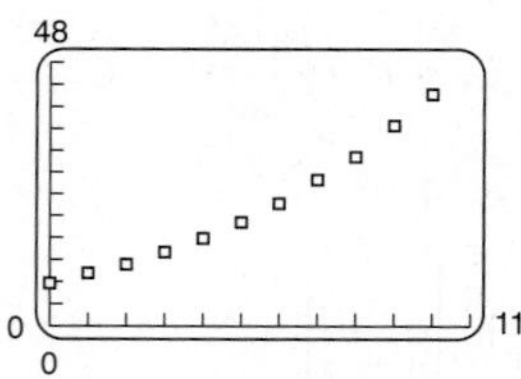

6. $S(x) = 7.921736972(1.185327894)^x$
7. $S(14) \approx \$85.6$ billion; $S(17) \approx \$142.6$ billion; $S(20) \approx \$237.5$ billion
8. About \$1350.1 billion **9.** Logistic

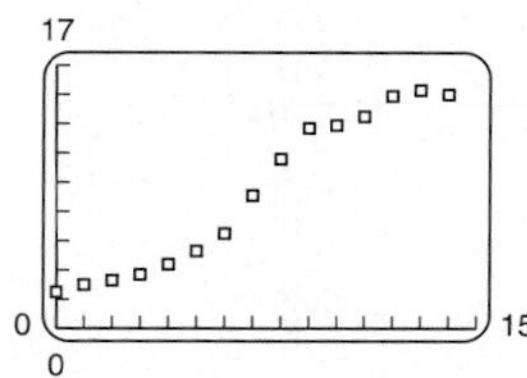

10. $S(x) =$
$17.980785389535/(1 + 11.426350559698e^{-0.35984217275628x})$
11. $S(18) \approx \$17.7$ billion; $S(22) \approx \$17.9$ billion; $S(26) \approx \$18.0$ billion
12. Approximately \$395.5 billion
13. Quartic polynomial

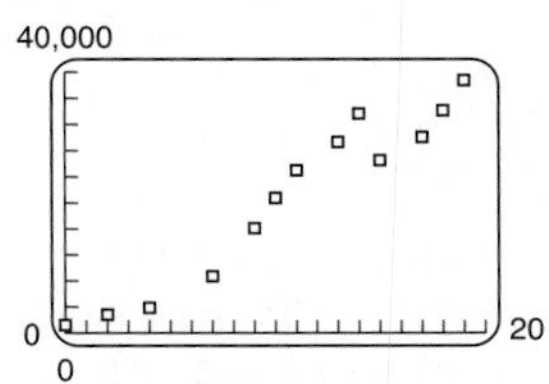

14. $S(x) = 1.519037653x^4 - 68.85343134x^3 + 1005.625846x^2 - 2866.336375x + 2605.964972$
15. $S(23) \approx \$56{,}006$ million; $S(27) \approx \$110{,}353$ million; $S(31) \approx \$231{,}807$ million **16.** About \$2,402,843 million

CHAPTER 5

Technology Connection, p. 470

1.

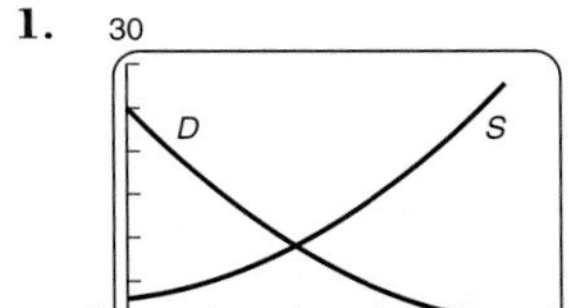

The point of intersection is (2, 9); this is the equilibrium point.

Exercise Set 5.1, p. 475

1. (a) (6, \$4); **(b)** \$15; **(c)** \$9 **3. (a)** (1, \$9); **(b)** \$3.33; **(c)** \$1.67 **5. (a)** (3, \$9); **(b)** \$36; **(c)** \$18 **7. (a)** (50, \$500); **(b)** \$12,500; **(c)** \$6250 **9. (a)** (2, \$3); **(b)** \$2; **(c)** \$0.35 **11. (a)** (100, \$10); **(b)** \$1000; **(c)** \$333.33 **13. (a)** (0.8, \$10.24); **(b)** \$2.22; **(c)** \$0.98 **15. (a)** (5, \$0.61); **(b)** \$86.36; **(c)** \$2.45 **17.** TW
19. (a) (6, \$2); **(b)**

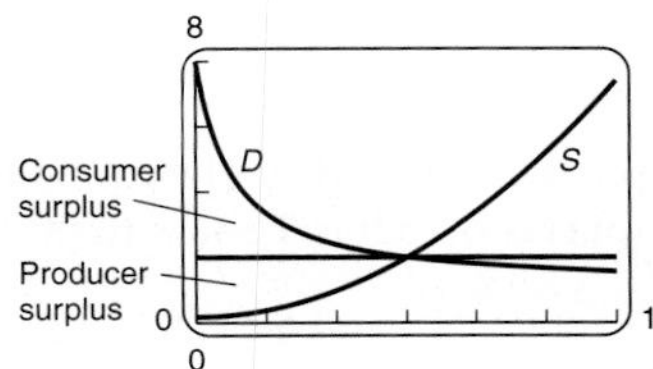

(c) \$7.62; **(d)** \$7.20
21. (a) Linear

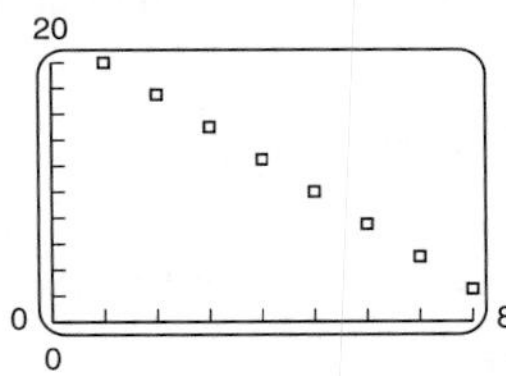

(b) $y = -2.5x + 22.5$; **(c)** \$45; **(d)** \$24.20

Technology Connection, p. 477

1.

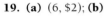

$f(0) = 1000, f(1) = 1083.287, f(2) = 1173.511,$
$f(3) = 1271.249$

Technology Connection, p. 478

1. \$3390.61; the future value of the continuous money flow if \$1000 per year flows at a constant rate into an account paying 8%, compounded continuously, for 3 yr

Exercise Set 5.2, p. 482

1. \$916.98 **3.** \$20,535.81 **5.** \$51,548.45 **7.** \$949.94 **9.** \$259.37 **11.** \$12,329.85 **13.** \$60,653.07 **15.** \$17,802.91 **17.** \$1,321,610.09 **19.** \$752,848 **21.** \$1,421,397 **23.** 1712.4 trillion cubic feet **25.** 2042 **27.** 19.994 lb **29. (a)** 0.0284, or 2.84%; **(b)** 39.76 billion barrels **(c)** 27.7 yr after 2006 **31.** \$2,383,119.65 **33.** \$43,332.94 **35.** The account with the continuous money flow has a future value of \$60,816.80. The single initial investment has a future value of \$49,192.06. **37.** TW
39. $y = 2000\,e^{0.06x}$

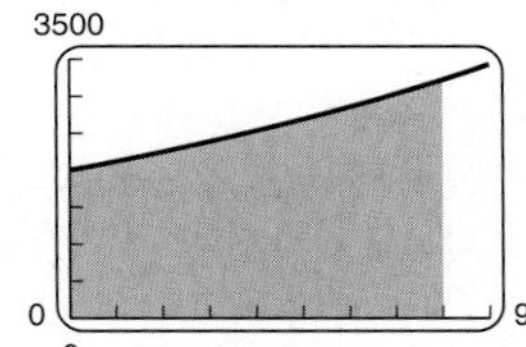

The area and the future value of the continuous money flow are both \$20,535.81.

Exercise Set 5.3, p. 488

1. Convergent; $\frac{1}{2}$ **3.** Divergent **5.** Convergent; 1 **7.** Convergent; $\frac{1}{2}$ **9.** Divergent **11.** Convergent; 2 **13.** Divergent **15.** Divergent **17.** Divergent **19.** Convergent; 1 **21.** Convergent; $\frac{1000}{\pi^{0.001}}$ **23.** Divergent **25.** $\frac{1}{2}$ **27.** 1 **29.** \$51,428.57 **31.** \$6250 **33.** \$4500 **35.** \$62,500 **37.** \$900,000 **39.** $33{,}333\frac{1}{3}$ lb **41. (a)** 4.20963; **(b)** 0.702858 rem; **(c)** 2.37551 rems **43.** Divergent **45.** Convergent; 2 **47.** Convergent; $\frac{1}{2}$ **49.** $\frac{1}{k^2}$ is the total dose of the drug.
51. TW

53.

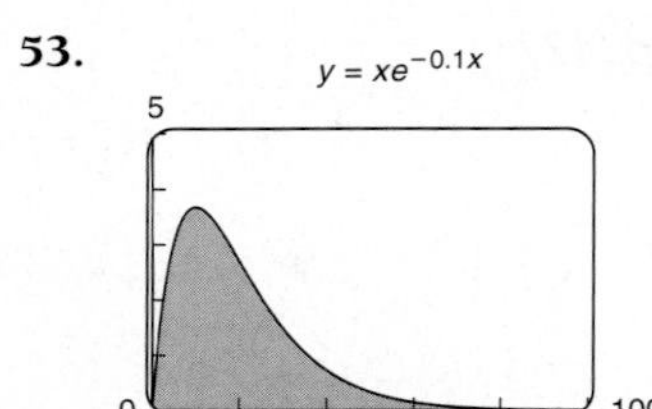

55. About 1.252

Exercise Set 5.4, p. 500

1. $\int_1^3 \frac{1}{4}x\,dx = \left[\frac{x^2}{8}\right]_1^3 = \frac{9}{8} - \frac{1}{8} = 1$

3. $\int_0^{1/3} 3\,dx = [3x]_0^{1/3} = 3\left(\frac{1}{3} - 0\right) = 1$

5. $\int_0^4 \frac{3}{64}x^2\,dx = \left[\frac{x^3}{64}\right]_0^4 = \frac{1}{64}(4^3 - 0^3) = 1$

7. $\int_1^e \frac{1}{x}\,dx = [\ln x]_1^e = \ln e - \ln 1 = 1 - 0 = 1$

9. $\int_{-1}^1 \frac{3}{2}x^2\,dx = \left[\frac{x^3}{2}\right]_{-1}^1 = \frac{1^3}{2} - \frac{(-1)^3}{2} = \frac{1}{2} + \frac{1}{2} = 1$

11. $\int_0^\infty 3e^{-3x}\,dx = \lim_{b\to\infty}\int_0^b 3e^{-3x}\,dx = \lim_{b\to\infty}[-e^{-3x}]_0^b$

$= \lim_{b\to\infty}[-e^{-3b} - (-e^0)] = \lim_{b\to\infty}\left(-\frac{1}{e^{3b}} + 1\right) = 1$

13. $\frac{2}{21}$; $f(x) = \frac{2}{21}x$ **15.** $\frac{3}{2}$; $f(x) = \frac{3}{2}x^2$ **17.** $\frac{1}{6}$; $f(x) = \frac{1}{6}$

19. $\frac{1}{2}$; $f(x) = \frac{2 - x}{2}$ **21.** $\frac{1}{\ln 3}$; $f(x) = \frac{1}{x \ln 3}$

23. $\frac{1}{e^3 - 1}$; $f(x) = \frac{e^x}{e^3 - 1}$ **25.** **(a)** $\frac{8}{25}$, or 0.32; **(b)** TW

27. $\frac{11}{16}$, or 0.6875 **29.** 0.3297 **31.** 0.999955

33. 0.3935 **35.** **(a)** $k \approx \frac{1}{59{,}119.3} \approx 0.000017$; **(b)** 0.02404 **37.** 0.950213 **39.** $\sqrt{2}$ **41.** 0.049787 **43–53.** Check using the answers to Exercises 1–11.

Technology Connection, p. 512

1. **(a)** 0.6295, or 62.95%; **(b)** 0.0228, or 2.28%
2. **(a)** About 19.6%; **(b)** about 90.1%

Exercise Set 5.5, p. 512

1. $\mu = E(x) = 5$, $E(x^2) = \frac{79}{3}$, $\sigma^2 = \frac{4}{3}$, $\sigma = \frac{2}{\sqrt{3}}$

3. $\mu = E(x) = \frac{8}{3}$, $E(x^2) = 8$, $\sigma^2 = \frac{8}{9}$, $\sigma = \frac{2\sqrt{2}}{3}$

5. $\mu = E(x) = \frac{13}{6}$, $E(x^2) = 5$, $\sigma^2 = \frac{11}{36}$, $\sigma = \frac{\sqrt{11}}{6}$

7. $\mu = E(x) = 0$, $E(x^2) = \frac{3}{5}$, $\sigma^2 = \frac{3}{5}$, $\sigma = \sqrt{\frac{3}{5}}$

9. $\mu = E(x) = \frac{6}{\ln 5}$, $E(x^2) = \frac{27}{\ln 5}$, $\sigma^2 = \frac{27 \ln 5 - 36}{(\ln 5)^2}$,

$\sigma = \frac{\sqrt{27 \ln 5 - 36}}{\ln 5}$

11. 0.4834 **13.** 0.4147 **15.** 0.6442 **17.** 0.0150 **19.** 0.1790 **21.** 0.0013 **23.** **(a)** 0.6826; **(b)** 68.26% **25.** 0.2898 **27.** 0.4514 **29–45.** Check using the answers to Exercises 11–27. **47.** 0.62% **49.** 0.7910 **51.** 0.1115 **53.** **(a)** 966; **(b)** 1055; **(c)** 1217

55. $\mu = E(x) = \frac{b + a}{2}$, $E(x^2) = \frac{b^2 + ab + a^2}{3}$,

$\sigma^2 = \frac{(b - a)^2}{12}$, $\sigma = \frac{b - a}{2\sqrt{3}}$

57. $\sqrt{2}$ **59.** $\frac{\ln 2}{k}$ **61.** 7.801 oz **63.** TW

Exercise Set 5.6, p. 516

1. $\frac{\pi}{3}$, or about 1.05 **3.** $\frac{15\pi}{2}$, or about 23.56

5. $\frac{\pi}{2}(e^{10} - e^{-4})$, or about 34,599.06

7. $\frac{2\pi}{3}$, or about 2.09 **9.** $4\pi \ln \frac{9}{4}$, or about 10.19

11. 32π, or about 100.53 **13.** $\frac{32\pi}{5}$, or about 20.11

15. 56π, or about 175.93 **17.** $\frac{32\pi}{3}$, or about 33.51

19.

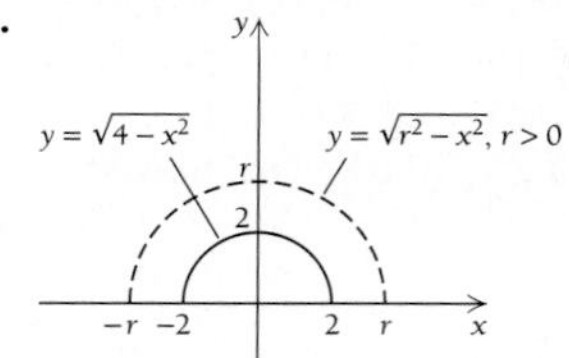

The graphs are semicircles. Their rotation about the x-axis creates spheres of radius 2 and r, respectively.

21. $\pi\left(\frac{-3 + 2e}{e^2}\right)$, or about 1.04 **23.** Using the fnInt feature on a graphing calculator, with successively larger values for the upper limit, you will find that the integral diverges.

Exercise Set 5.7, p. 524

1. $y = x^5 + C$; $y = x^5$, $y = x^5 - 1$, $y = x^5 + \pi$ (answers may vary)

3. $y = \frac{1}{2}e^{2x} + \frac{1}{2}x^2 + C$; $y = \frac{1}{2}e^{2x} + \frac{1}{2}x^2$, $y = \frac{1}{2}e^{2x} + \frac{1}{2}x^2 - 3$, $y = \frac{1}{2}e^{2x} + \frac{1}{2}x^2 + 3$ (answers may vary)
5. $y = 8 \ln x - \frac{1}{3}x^3 + \frac{1}{6}x^6 + C$; $y = 8 \ln x - \frac{1}{3}x^3 + \frac{1}{6}x^6$, $y = 8 \ln x - \frac{1}{3}x^3 + \frac{1}{6}x^6 + 5$, $y = 8 \ln x - \frac{1}{3}x^3 + \frac{1}{6}x^6 - 17$ (answers may vary)
7. $y = \frac{1}{3}x^3 + x^2 - 3x + 4$ **9.** $f(x) = \frac{3}{5}x^{5/3} - \frac{1}{2}x^2 - \frac{61}{10}$
11. $y' = \ln x + 3$; $y'' = 1/x$

$$\begin{array}{c|c} y'' - \dfrac{1}{x} & 0 \\ \hline \dfrac{1}{x} - \dfrac{1}{x} & 0 \\ 0 & 0 \end{array}$$

13. $y' = 4e^x + 3xe^x$; $y'' = 7e^x + 3xe^x$

$$\begin{array}{c|c} y'' \quad - \quad 2y' \quad + \quad y & 0 \\ \hline (7e^x + 3xe^x) - 2(4e^x + 3xe^x) + (e^x + 3xe^x) & 0 \\ (7e^x - 8e^x + e^x) + (3xe^x - 6xe^x + 3xe^x) & \\ 0 & 0 \end{array}$$

15. $y = C_1 e^{x^4}$, where $C_1 = \pm e^C$
17. $y = \sqrt[3]{4x^2 + C}$
19. $y = \sqrt{2x^2 + C_1}$, $y = -\sqrt{2x^2 + C_1}$, where $C_1 = 2C$
21. $y = \sqrt{12x + C_1}$ and $y = -\sqrt{12x + C_1}$, where $C_1 = 2C$
23. $y = 8e^{x^2/2} - 3$ **25.** $y = \sqrt[3]{15x - 3}$
27. $y = C_1 e^{3x}$, where $C_1 = \pm e^C$
29. $P = C_1 e^{2t}$, where $C_1 = \pm e^C$
31. $f(x) = \ln x - 2x^2 + \frac{2}{3}x^{3/2} + 9$
33. $C(x) = 2.6x - 0.01x^2 + 120$, $A(x) = 2.6 - 0.01x + \dfrac{120}{x}$
35. **(a)** $P(C) = \dfrac{400}{\sqrt{C + 3}} - 40$; **(b)** \$97
37. **(a)** $R = k \ln(S + 1) + C$; **(b)** $R = k \ln(S + 1)$; **(c)** TW **39.** $q = 200 - x$
41. $q = \dfrac{C_1}{x^n}$, where $C_1 = e^C$
43. $R = C_1 S^k$, where $C_1 = e^C$
45. $y = \dfrac{1}{e^{1/x} - C}$ **47.** TW

Chapter Review Exercises, p. 532

1. (d) **2.** (e) **3.** (f) **4.** (b) **5.** (c) **6.** (a) **7.** False **8.** True **9.** True **10.** False **11.** False **12.** True **13.** (2, \$16) **14.** \$18.67 **15.** \$5.33 **16.** \$58,384.12 **17.** \$441.68 **18.** \$12,663.20 **19.** \$24,000 **20.** \$247.88 **21.** 16.03 billion metric tons **22.** 53.0 yr from 2005 **23.** Convergent; 1 **24.** Divergent **25.** Convergent; $\frac{1}{2}$ **26.** $k = \frac{8}{3}$; $f(x) = \dfrac{8}{3x^3}$ **27.** 0.6 **28.** $\frac{3}{10}$ **29.** $\frac{1}{2}$ **30.** $\frac{1}{2}$ **31.** $\frac{1}{20}$ **32.** $\dfrac{1}{2\sqrt{5}}$ **33.** 0.4678 **34.** 0.8827 **35.** 0.1002 **36.** 0.5000 **37.** 0.3085 **38.** $\dfrac{127\pi}{7}$ **39.** $\dfrac{\pi}{6}$
40. $y = C_1 e^{x^{11}}$, where $C_1 = \pm e^C$ **41.** $y = \sqrt{4x + C_1}$ and $y = -\sqrt{4x + C_1}$, where $C_1 = 2C$ **42.** $y = 5e^{4x}$
43. $v = \sqrt[3]{15t + 19}$ **44.** $y = \sqrt{3x^2 + C_1}$ and $y = -\sqrt{3x^2 + C_1}$, where $C_1 = 2C$
45. $y = C_1 e^{-x^2/2} + 8$, where $C_1 = \pm e^{-C}$
46. $q = 100 - x$ **47.** $V = -6.37e^{-kt} + 36.37$ **48.** $\sqrt[9]{4.5}$
49. Divergent **50.** Convergent; 3 **51.** 1

Chapter 5 Test, p. 534

1. [5.1] (3, \$16) **2.** [5.1] \$45 **3.** [5.1] \$22.50 **4.** [5.2] \$18,164.20 **5.** [5.2] \$1138.05 **6.** [5.2] 1149.5 million metric tons **7.** [5.2] 28.89 yr after 2004 **8.** [5.2] \$2465.97 **9.** [5.2] \$30,717.71 **10.** [5.3] \$34,545.45 **11.** [5.3] Convergent; $\frac{1}{4}$ **12.** [5.3] Divergent **13.** [5.4] $\frac{1}{4}$, $f(x) = \frac{1}{4}x^3$ **14.** [5.4] 0.9975 **15.** [5.5] $E(x) = \frac{13}{6}$ **16.** [5.5] $E(x^2) = 5$ **17.** [5.5] $\mu = \frac{13}{6}$ **18.** [5.5] $\sigma^2 = \frac{11}{36}$ **19.** [5.5] $\sigma = \dfrac{\sqrt{11}}{6}$ **20.** [5.5] 0.4032 **21.** [5.5] 0.1365 **22.** [5.5] 0.9071 **23.** [5.5] 0.3085 **24.** [5.6] $\pi \ln 5$ **25.** [5.6] $\dfrac{5\pi}{2}$
26. [5.7] $y = C_1 e^{x^8}$, where $C_1 = e^C$
27. [5.7] $y = \sqrt{18x + C_1}$ and $y = -\sqrt{18x + C_1}$, where $C_1 = 2C$ **28.** [5.7] $y = 11e^{6t}$
29. [5.7] $y = \pm C_1 e^{-x^3/3} + 5$, where $C_1 = \pm e^{-C}$
30. [5.7] $v = \pm\sqrt[4]{8t + C_1}$, where $C_1 = 4C$
31. [5.7] $y = C_1 e^{4x + x^2/2}$, where $C_1 = \pm e^C$
32. [5.7] $q = \dfrac{C_1}{x^4}$, where $C_1 = e^{-C}$
33. [5.7] **(a)** $V(t) = 36(1 - e^{-kt})$; **(b)** $k = 0.12$; **(c)** $V(t) = 36(1 - e^{-0.12t})$; **(d)** $V(12) = \$27.47$; **(e)** $t \approx 14.9$ months
34. [5.4] $\sqrt[4]{4}$, or $\sqrt{2}$ **35.** [5.3] Convergent; $-\frac{1}{4}$
36. [5.3] π

Extended Technology Application, p. 536

1. $y = 0.00525488582427x^3 - 0.31949926791313x^2 + 5.2617546608767x - 8.994864719578$
2. 12,348.28 cm^3

3. $y = 0.00109000713314x^4 - 0.02219272885861x^3 + 0.11944088382992x^2 - 0.11585042438606x + 1.0466143607372$ **4.** 35.170535 in^3, 19.493 fl. oz **5.** It seems good since the volume estimate was 19.493 fl. oz. **6.** The cubic function $y = 0.00149729174531x^3 - 0.05193696325118x^2 + 0.34763221627001x + 0.68805240705899$ yields an estimated volume of 35.4635 in^3, or 19.655 fl. oz, which is a better estimate.

CHAPTER 6

Exercise Set 6.1, p. 545

1. 0; −14; 250 **3.** 1; $-\frac{125}{9}$; 23 **5.** 9; 66; 128 **7.** 6; 12 **9.** 25.62 **11.** \$151,571.66 **13.** 244.7 mph **15.** 1.939 m^2 **17.** 110.6 ft **19.** TW **21.** 0°F **23.** −22°F

25.

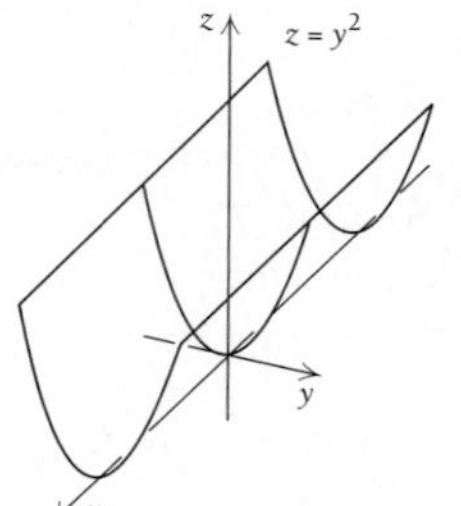

27.

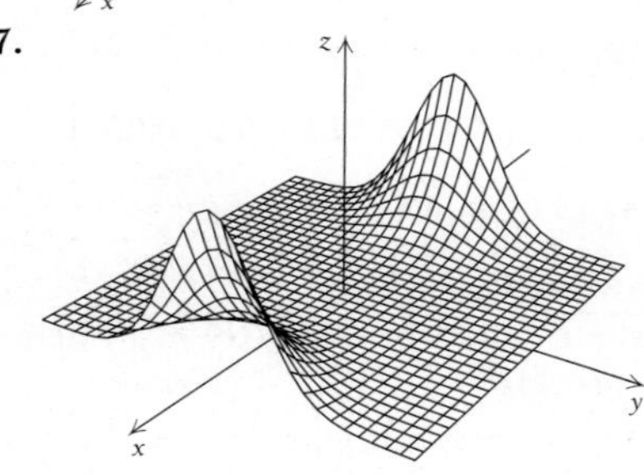

$z = (x^4 - 16x^2)e^{-y^2}$

29.

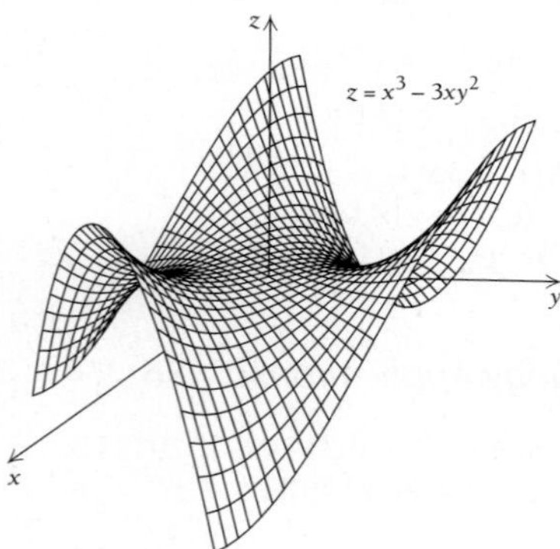

Exercise Set 6.2, p. 554

1. 2; −3; 2; −3 **3.** $6x - 2y$; $-2x + 1$; −6; 1 **5.** $2 - 5y$; $-5x$; −18; −20 **7.** $\frac{x}{\sqrt{x^2+y^2}}$; $\frac{y}{\sqrt{x^2+y^2}}$; $\frac{-2}{\sqrt{5}}$; $\frac{-2}{\sqrt{13}}$ **9.** $2e^{2x-y}$; $-e^{2x-y}$ **11.** ye^{xy}; xe^{xy} **13.** $\frac{y}{x+2y}$; $\frac{2y}{x+2y} + \ln(x+2y)$ **15.** $1 + \ln(xy)$; $\frac{x}{y}$ **17.** $\frac{1}{y} + \frac{y}{3x^2}$; $-\frac{x}{y^2} - \frac{1}{3x}$ **19.** $12(2x + y - 5)$; $6(2x + y - 5)$ **21.** $4m^2 + 10m + 2b - 22$; $3m^2 + 8mb + 10b + 26m - 56$ **23.** $5y - 2\lambda$; $5x - \lambda$; $-(2x + y - 8)$ **25.** $2x - 10\lambda$; $2y - 2\lambda$; $-(10x + 2y - 4)$ **27.** $f_{xx} = 0$; $f_{yy} = 0$; $f_{xy} = 5$; $f_{yx} = 5$ **29.** $f_{xx} = 0$; $f_{yy} = 14x$; $f_{xy} = 14y + 5$; $f_{yx} = 14y + 5$ **31.** $f_{xx} = 20x^3y^4 + 6xy^2$; $f_{yy} = 12x^5y^2 + 2x^3$; $f_{xy} = 20x^4y^3 + 6x^2y$; $f_{yx} = 20x^4y^3 + 6x^2y$ **33.** 0; 0; 0; 0 **35.** $4y^2e^{2xy}$; $4xye^{2xy} + 2e^{2xy}$; $4xye^{2xy} + 2e^{2xy}$; $4x^2e^{2xy}$ **37.** 0; 0; 0; e^y **39.** $\frac{-y}{x^2}$; $\frac{1}{x}$; $\frac{1}{x}$; 0 **41. (a)** 614,400 units **(b)** $\frac{\partial p}{\partial x} = 960\left(\frac{y}{x}\right)^{3/5}$, $\frac{\partial p}{\partial y} = 1440\left(\frac{x}{y}\right)^{2/5}$; **(c)** $\left.\frac{\partial p}{\partial x}\right|_{(32,\,1024)} = 7680$, $\left.\frac{\partial p}{\partial y}\right|_{(32,\,1024)} = 360$; **(d)** TW **43. (a)** \$1.274 million; **(b)** $\frac{\partial P}{\partial w} = -0.005075w^{-1.638}r^{1.038}s^{0.873}t^{2.468}$, $\frac{\partial P}{\partial r} = 0.008257w^{-0.638}r^{0.038}s^{0.873}t^{2.468}$, $\frac{\partial P}{\partial s} = 0.006945w^{-0.638}r^{1.038}s^{-0.127}t^{2.468}$, $\frac{\partial P}{\partial t} = 0.019633w^{-0.638}r^{1.038}s^{0.873}t^{1.468}$; **(c)** TW **45.** 99.6° **47.** 121.3° **49.** TW **51. (a)** $\frac{\sqrt{w}}{120\sqrt{h}}$; **(b)** $\frac{\sqrt{h}}{120\sqrt{w}}$; **(c)** −0.0243 m^2 **53.** 78.244 **55.** −0.846 **57.** $f_x = \frac{-4xt^2}{(x^2 - t^2)^2}$; $f_t = \frac{4x^2t}{(x^2 - t^2)^2}$ **59.** $f_x = \frac{1}{\sqrt{x}(1 + 2\sqrt{t})}$; $f_t = \frac{-1 - 2\sqrt{x}}{\sqrt{t}(1 + 2\sqrt{t})^2}$ **61.** $f_x = 4x^{-1/3} - 2x^{-3/4}t^{1/2} + 6x^{-3/2}t^{3/2}$; $f_t = -4x^{1/4}t^{-1/2} - 18x^{-1/2}t^{1/2}$ **63.** $f_{xx} = \frac{-6y}{x^4}$; $f_{xy} = \frac{-2}{y^3} + \frac{2}{x^3}$; $f_{yx} = \frac{-2}{y^3} + \frac{2}{x^3}$; $f_{yy} = \frac{6x}{y^4}$

65. TW

67. $f_{xx} = \dfrac{-2x^2 + 2y^2}{(x^2 + y^2)^2}$ and $f_{yy} = \dfrac{2x^2 - 2y^2}{(x^2 + y^2)^2}$, so $f_{xx} + f_{yy} = 0$

69. **(a)** $\lim_{h \to 0} \dfrac{y(h^2 - y^2)}{h^2 + y^2} = -y$;
(b) $\lim_{h \to 0} \dfrac{x(x^2 - h^2)}{x^2 + h^2} = x$; **(c)** $f_{yx}(0, 0) = 1$ and $f_{xy}(0, 0) = -1$; at $(0, 0)$, the mixed partials are *not* equal.

Exercise Set 6.3, p. 563

1. Relative minimum $= -\frac{1}{3}$ at $\left(-\frac{1}{3}, \frac{2}{3}\right)$
3. Relative maximum $= \frac{4}{27}$ at $\left(\frac{2}{3}, \frac{2}{3}\right)$
5. Relative minimum $= -1$ at $(1, 1)$
7. Relative minimum $= -7$ at $(1, -2)$
9. Relative minimum $= -5$ at $(-1, 2)$
11. No relative extrema
13. Relative minimum $= e$ at $(0, 0)$
15. 6 thousand of the \$17 sunglasses and 5 thousand of the \$21 sunglasses **17.** Maximum value of $P =$ \$55 million when $a =$ \$10 million and $p =$ \$3 **19.** The bottom measures 8 ft by 8 ft, and the height is 5 ft.
21. **(a)** $R = 64p_1 - 4p_1^2 - 4p_1p_2 + 56p_2 - 4p_2^2$;
(b) $p_1 = 6$, or \$60, $p_2 = 4$, or \$40;
(c) $q_1 = 32$, or 3200 units, $q_2 = 28$, or 2800 units;
(d) \$304,000 **23.** No relative extrema; saddle point at $(0, 0)$ **25.** Relative minimum $= 0$ at $(0, 0)$; saddle points at $(2, 1)$ and $(-2, 1)$ **27.** TW **29.** Relative minimum $= -5$ at $(0, 0)$ **31.** No relative extrema

Exercise Set 6.4, p. 571

1. $y = 0.6x + 0.95$
3. $y = \frac{36}{35}x - \frac{29}{35}$
5. **(a)** $y = 0.1064679772x + 3.090710209$;
(b) \$6.28, \$6.82
7. **(a)** $y = 0.1705202312x - 261.0738233$;
(b) 81.7 yr, 82.5 yr
9. **(a)** $y = 1.068421053x - 1.236842105$;
(b) 85% **11.** TW
13. **(a)** $y = -0.0059379586x + 15.57191398$;
(b) 3:38.2, 3:36.4; **(c)** the value predicted is 3:42.1, about a second less than the actual record.

Exercise Set 6.5, p. 579

1. Maximum $= \frac{25}{3}$ at $\left(\frac{5}{3}, 5\right)$
3. Maximum $= -16$ at $(2, 4)$
5. Minimum $= 20$ at $(4, 2)$
7. Minimum $= -96$ at $(8, -12)$
9. Minimum $= \frac{3}{2}$ at $\left(1, \frac{1}{2}, -\frac{1}{2}\right)$
11. 25 and 25 **13.** 3 and -3 **15.** $\left(\frac{3}{2}, 2, \frac{5}{2}\right)$
17. $9\frac{3}{4}$ in. by $9\frac{3}{4}$ in.; $95\frac{1}{16}$ in^2; no

19. $r = \sqrt[3]{\dfrac{27}{2\pi}} \approx 1.6$ ft, $h = 2r \approx 3.2$ ft; about 48.3 ft^2
21. Maximum value of S is 1012.5 at $L = 22.5$, $M = 67.5$
23. **(a)** $C(x, y, z) = 7xy + 6yz + 6xz$;
(b) $x = 60$ ft, $y = 60$ ft, $z = 70$ ft; minimum cost is \$75,600
25. 10,000 units on A, 100 units on B
27. Minimum $= -\frac{155}{128}$ at $\left(-\frac{7}{16}, -\frac{3}{4}\right)$
29. Maximum $= \frac{8}{27}$ at $\left(\pm\sqrt{\frac{2}{3}}, \pm\sqrt{\frac{2}{3}}, \pm\sqrt{\frac{2}{3}}\right)$
31. Maximum $= 2$ at $\left(\frac{1}{2}, \frac{1}{2}, \frac{1}{2}, \frac{1}{2}\right)$
33. $\lambda = \dfrac{p_x}{c_1} = \dfrac{p_y}{c_2}$ **35.** TW **37–43.** Left to the student

Exercise Set 6.6, p. 584

1. 9 **3.** 14 **5.** -10 **7.** 0 **9.** $\frac{3}{20}$ **11.** $\frac{1}{2}$ **13.** 4
15. $\frac{4}{15}$ **17.** 1 **19.** $\frac{5}{12}$ **21.** 39 **23.** $\frac{13}{240}$ **25.** TW **27.** Left to the student

Chapter Review Exercises, p. 590

1. (e) **2.** (h) **3.** (f) **4.** (g) **5.** (b) **6.** (c)
7. (d) **8.** (a) **9.** 1 **10.** $3y^3$ **11.** $e^y + 9xy^2 + 2$
12. $9y^2$ **13.** $9y^2$ **14.** 0 **15.** $e^y + 18xy$
16. $6x^2 \ln y + y^2$ **17.** $\dfrac{2x^3}{y} + 2xy$ **18.** $\dfrac{6x^2}{y} + 2y$
19. $\dfrac{6x^2}{y} + 2y$ **20.** $12x \ln y$ **21.** $\dfrac{-2x^3}{y^2} + 2x$
22. Relative minimum $= -\frac{549}{20}$ at $\left(5, \frac{27}{2}\right)$; relative minimum $= \frac{121}{20}$ at $\left(1, \frac{3}{2}\right)$
23. Relative minimum $= -4$ at $(0, -2)$
24. Relative maximum $= \frac{45}{4}$ at $\left(\frac{3}{2}, -3\right)$
25. Relative minimum $= 29$ at $(-1, 2)$
26. **(a)** $y = \dfrac{3}{5}x + \dfrac{20}{3}$; **(b)** 9.1 million
27. **(a)** $y = 17.93965517x + 124.4051724$; **(b)** \$358
28. Minimum $= \frac{125}{4}$ at $\left(-\frac{3}{2}, -3\right)$
29. Maximum $= 300$ at $(5, 10)$ **30.** $\frac{5}{4}$ **31.** $\frac{1}{60}$
32. 0 **33.** The cylindrical container

34.

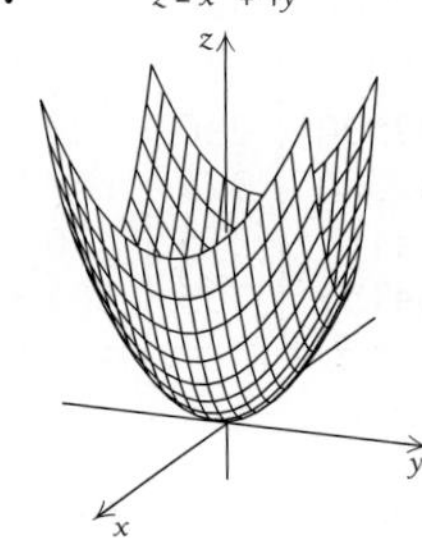

Chapter 6 Test, p. 591

1. [6.1] $e^{-1} - 2$ **2.** [6.2] $e^x + 6x^2y$ **3.** [6.2] $2x^3 + 1$
4. [6.2] $e^x + 12xy$ **5.** [6.2] $6x^2$ **6.** [6.2] $6x^2$
7. [6.2] 0 **8.** [6.3] Minimum $= -\frac{7}{16}$ at $\left(\frac{3}{4}, \frac{1}{2}\right)$
9. [6.3] None **10.** [6.4] **(a)** $y = \frac{9}{2}x + \frac{17}{3}$;
(b) \$23.67 million
11. [6.5] Maximum $= -19$ at $(4, 5)$ **12.** [6.6] 720
13. [6.5] \$400,000 for labor, \$200,000 for capital
14. [6.2] $f_x = \dfrac{-x^4 + 4xt + 6x^2t}{(x^3 + 2t)^2}$, $f_t = \dfrac{-2x^3 - 2x^2}{(x^3 + 2t)^2}$
15. [6.1]

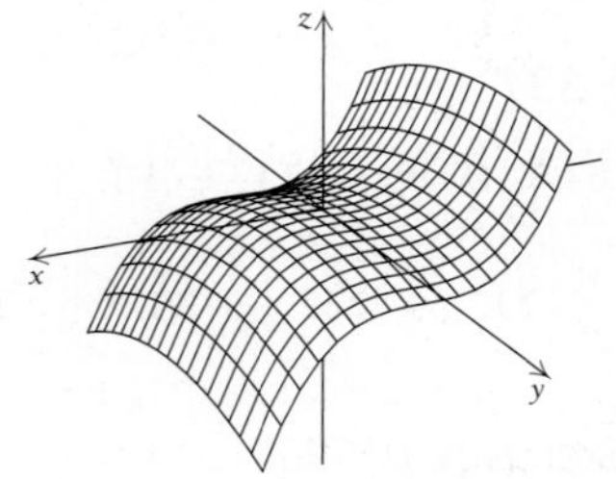

Extended Technology Application, p. 592

1.

Case	Bldg.	n	k	A	h	t(h, k)
1	B1	2	40	3200	15	21.5
	B2	3	32.66	3200	30	19.3
2	B1	2	60	7200	15	31.5
	B2	3	48.99	7200	30	27.5
3	B1	4	40	6400	45	24.5
	B2	5	35.777	6400	60	23.9
4	B1	5	60	18000	60	36
	B2	10	42.426	18000	135	34.7
5	B1	5	150	112500	60	81
	B2	10	106.066	112500	135	66.5
6	B1	10	40	16000	135	33.5
	B2	17	30.679	16000	240	39.3
7	B1	10	80	64000	135	53.5
	B2	17	61.357	64000	240	54.7
8	B1	17	40	27200	240	44
	B2	26	32.344	27200	375	53.7
9	B1	17	50	42500	240	49
	B2	26	40.43	42500	375	57.7
10	B1	26	77	154154	375	76
	B2	50	55.525	154152	735	101.3

2. Yes **3.** $t(h, k) = \dfrac{h}{10} + \dfrac{k}{2}$
4. $A = \left(1 + \dfrac{h}{12}\right)k^2 = 40{,}000$
5. About 57.7 ft by 57.7 ft by 132.2 ft (11 floors)
6. Left to the student

CUMULATIVE REVIEW, P. 595

1. [R.4] $y = -4x - 27$ **2.** [R.2] $x^2 + 2xh + h^2 - 5$
3. [1.2] **(a)**

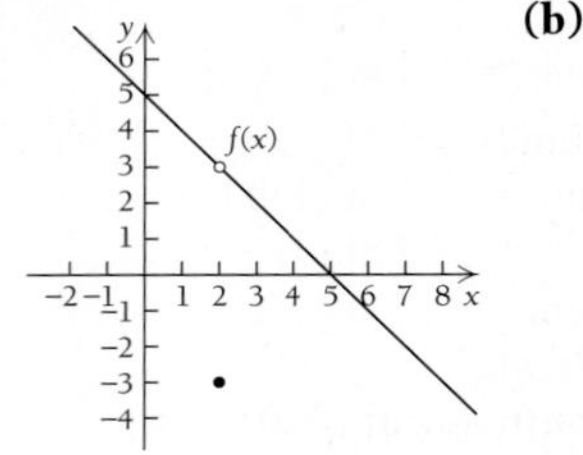

(b) 3; **(c)** -3; **(d)** no **4.** [1.2] -8 **5.** [1.2] 3 **6.** [1.2] Does not exist **7.** [2.3] 4 **8.** [2.3] 0 **9.** [1.4] $f'(x) = 2x$
10. [1.5] -9 **11.** [1.5] $2x - 7$ **12.** [1.5] $\frac{1}{4}x^{-3/4}$
13. [1.5] $-6x^{-7}$ **14.** [1.7] $\frac{10}{3}x^4(2x^5 - 8)^{-2/3}$
15. [1.6] $\dfrac{20x^3 - 15x^2 - 8}{(2x - 1)^2}$ **16.** [3.2] $\dfrac{2x}{x^2 + 5}$
17. [3.2] 1 **18.** [3.1] $3e^{3x} + 2x$
19. [3.1] $\dfrac{e^{\sqrt{x-3}}}{2\sqrt{x - 3}}$ **20.** [3.2] $\dfrac{e^x}{e^x - 4}$
21. [1.8] $2 - 4x^{-3}$ **22.** [2.5] $-\$0.04$/pair
23. [2.7] $3xy^2 + \dfrac{y}{x}$ **24.** [3.1] $y = x - 2$
25. [2.2] Relative maximum at $(-1, 3)$, relative minimum at $(1, -1)$; point of inflection at $(0, 1)$; increasing on $(-\infty, -1]$ and $[1, \infty)$, decreasing on $[-1, 1]$; concave down on $(-\infty, 0)$, concave up on $(0, \infty)$

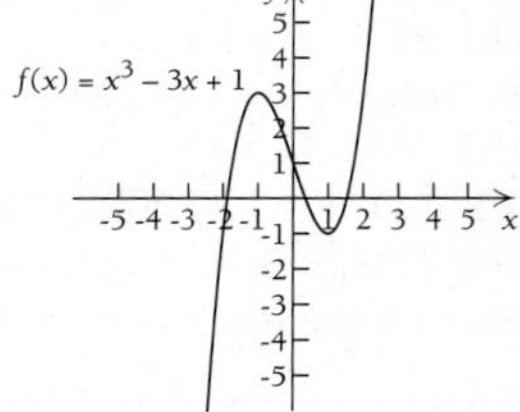

26. [2.2] Relative maxima at $(-1, -2)$ and $(1, -2)$, relative minimum at $(0, -3)$; points of inflection at $\left(-\frac{1}{\sqrt{3}}, -\frac{22}{9}\right)$ and $\left(\frac{1}{\sqrt{3}}, -\frac{22}{9}\right)$; increasing on $(-\infty, -1]$ and $[0, 1]$, decreasing on $[-1, 0]$ and $[1, \infty)$; concave down on $\left(-\infty, -\frac{1}{\sqrt{3}}\right)$ and $\left(\frac{1}{\sqrt{3}}, \infty\right)$, concave up on $\left(-\frac{1}{\sqrt{3}}, \frac{1}{\sqrt{3}}\right)$

$f(x) = 2x^2 - x^4 - 3$

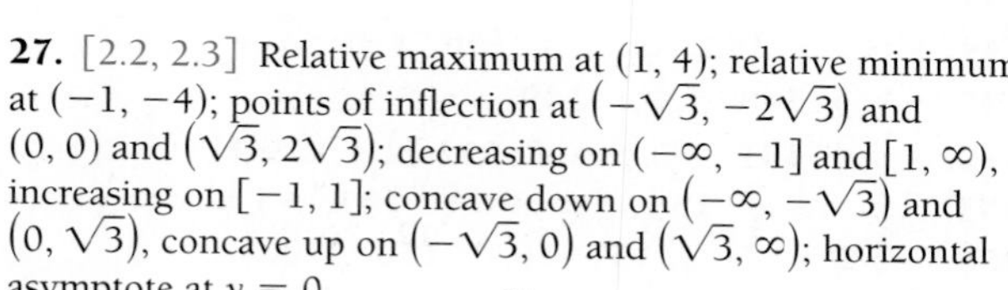

27. [2.2, 2.3] Relative maximum at $(1, 4)$; relative minimum at $(-1, -4)$; points of inflection at $(-\sqrt{3}, -2\sqrt{3})$ and $(0, 0)$ and $(\sqrt{3}, 2\sqrt{3})$; decreasing on $(-\infty, -1]$ and $[1, \infty)$, increasing on $[-1, 1]$; concave down on $(-\infty, -\sqrt{3})$ and $(0, \sqrt{3})$, concave up on $(-\sqrt{3}, 0)$ and $(\sqrt{3}, \infty)$; horizontal asymptote at $y = 0$

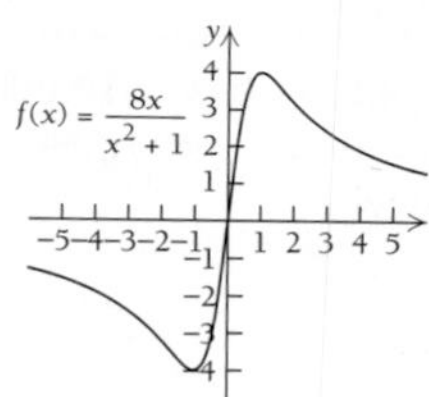

28. [2.3] Relative maximum at $(0, -2)$; no points of inflection; vertical asymptotes at $x = -2$ and $x = 2$; horizontal asymptote at $y = 0$; increasing on $(-\infty, -2)$ and $(-2, 0]$, decreasing on $[0, 2)$ and $(2, \infty)$; concave up on $(-\infty, -2)$ and $(2, \infty)$, concave down on $(-2, 2)$

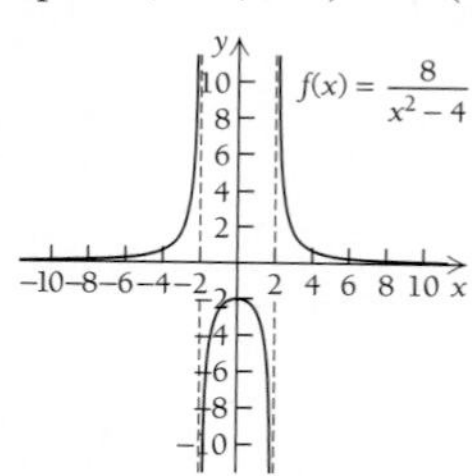

29. [2.4] Minimum $= -7$ at $x = 1$ **30.** [2.4] No absolute extrema **31.** [2.4] Maximum $= 6\frac{2}{3}$ at $x = -1$; minimum $= 4\frac{1}{3}$ at $x = -2$ **32.** [2.5] 15 sweatshirts **33.** [2.5] 30 times; lot size of 15 **34.** [2.6] $\Delta y = 3.03$; $f'(x)\Delta x = 3$ **35.** [3.3] **(a)** $N(t) = 8000e^{0.1t}$; **(b)** 17,804; **(c)** 6.9 yr

36. [3.6] **(a)** $E(x) = \frac{x}{12 - x}$; **(b)** $E(2) = \frac{1}{5}$, inelastic; **(c)** $E(9) = 3$, elastic; **(d)** increase; **(e)** $x = \$6$

37. [4.2] $\frac{1}{2}x^6 + C$ **38.** [4.3] $3 - \frac{2}{e}$

39. [4.7] $\frac{7}{9(7 - 3x)} + \frac{1}{9}\ln|7 - 3x| + C$

40. [4.5] $\frac{1}{4}e^{x^4} + C$

41. [4.6] $\left(\frac{x^2}{2} + 3x\right)\ln x - \frac{x^2}{4} - 3x + C$

42. [4.2] $75 \ln x + C$ **43.** [4.3] 2

44. [4.3] $\frac{232}{3}$, or $77\frac{1}{3}$ **45.** [5.2] \$41,324.72

46. [5.2] \$60,792.27 **47.** [5.3] Convergent, $\frac{1}{4374}$

48. [5.5] $\frac{3}{2}\ln 3$ **49.** [5.5] 0.6826

50. [5.1] (7, \$169); \$751.33

51. [5.6] $-\frac{\pi}{2}\left(\frac{1}{e^{10}} - 1\right)$, or approximately 1.571

52. [5.7] $y = C_1 e^{x^2/2}$, where $C_1 = e^C$

53. [6.4] **(a)** $y = \frac{5}{4}x + \frac{35}{12}$; **(b)** \$154,167

54. [6.2] $8xy^3 + 3$ **55.** [6.2] $e^y + 24x^2y$

56. [6.3] No relative extrema **57.** [6.5] Maximum = 4 at (3, 3) **58.** [6.6] $3(e^3 - 1)$

APPENDIX

Exercise Set A, p. 610

1. $5 \cdot 5 \cdot 5$, or 125 **2.** $7 \cdot 7$, or 49 **3.** $(-7)(-7)$, or 49 **4.** $(-5)(-5)(-5)$, or -125 **5.** 1.0201 **6.** 1.030301 **7.** $\frac{1}{16}$ **8.** $\frac{1}{64}$ **9.** 1 **10.** $6x$ **11.** t

12. 1 **13.** 1 **14.** $\frac{1}{3}$ **15.** $\frac{1}{3^2}$, or $\frac{1}{9}$ **16.** $\frac{1}{4^2}$, or $\frac{1}{16}$

17. 8 **18.** 4 **19.** 0.1 **20.** 0.0001 **21.** $\frac{1}{e^b}$

22. $\frac{1}{t^k}$ **23.** $\frac{1}{b}$ **24.** $\frac{1}{h}$ **25.** x^5 **26.** t^7

27. x^{-6}, or $\frac{1}{x^6}$ **28.** x^6 **29.** $35x^5$ **30.** $8t^7$ **31.** x^4

32. x **33.** 1 **34.** 1 **35.** x^3 **36.** x^4

37. x^{-3}, or $\frac{1}{x^3}$ **38.** x^{-4}, or $\frac{1}{x^4}$ **39.** 1 **40.** 1

41. e^{t-4} **42.** e^{k-3} **43.** t^{14} **44.** t^{12} **45.** t^2

46. t^{-4}, or $\frac{1}{t^4}$ **47.** a^6b^5 **48.** x^7y^7 **49.** t^{-6}, or $\frac{1}{t^6}$

50. t^{-12}, or $\frac{1}{t^{12}}$ **51.** e^{4x} **52.** e^{5x} **53.** $8x^6y^{12}$

54. $32x^{10}y^{20}$ **55.** $\frac{1}{81}x^8y^{20}z^{-16}$, or $\frac{x^8y^{20}}{81z^{16}}$
56. $\frac{1}{125}x^{-9}y^{21}z^{15}$, or $\frac{y^{21}z^{15}}{125x^9}$ **57.** $9x^{-16}y^{14}z^4$, or $\frac{9y^{14}z^4}{x^{16}}$
58. $625x^{16}y^{-20}z^{-12}$, or $\frac{625x^{16}}{y^{20}z^{12}}$ **59.** $\frac{c^4d^{12}}{16q^8}$
60. $\frac{64x^6y^3}{a^9b^9}$ **61.** $5x - 35$ **62.** $x + xt$
63. $x^2 - 7x + 10$ **64.** $x^2 - 7x + 12$ **65.** $a^3 - b^3$
66. $x^3 + y^3$ **67.** $2x^2 + 3x - 5$ **68.** $3x^2 + x - 4$
69. $a^2 - 4$ **70.** $9x^2 - 1$ **71.** $25x^2 - 4$
72. $t^2 - 1$ **73.** $a^2 - 2ah + h^2$ **74.** $a^2 + 2ah + h^2$
75. $25x^2 + 10xt + t^2$ **76.** $49a^2 - 14ac + c^2$
77. $5x^5 + 30x^3 + 45x$ **78.** $-3x^6 + 48x^2$
79. $a^3 + 3a^2b + 3ab^2 + b^3$ **80.** $a^3 - 3a^2b + 3ab^2 - b^3$
81. $x^3 - 15x^2 + 75x - 125$
82. $8x^3 + 36x^2 + 54x + 27$ **83.** $x(1 - t)$
84. $x(1 + h)$ **85.** $(x + 3y)^2$ **86.** $(x - 5y)^2$
87. $(x - 5)(x + 3)$ **88.** $(x + 5)(x + 3)$
89. $(x - 5)(x + 4)$ **90.** $(x - 10)(x + 1)$
91. $(7x - t)(7x + t)$ **92.** $(3x - b)(3x + b)$
93. $4(3t - 2m)(3t + 2m)$ **94.** $(5y - 3z)(5y + 3z)$
95. $ab(a + 4b)(a - 4b)$ **96.** $2(x^2 + 4)(x + 2)(x - 2)$
97. $(a^4 + b^4)(a^2 + b^2)(a + b)(a - b)$
98. $(6y - 5)(6y + 7)$ **99.** $10x(a + 2b)(a - 2b)$
100. $xy(x + 5y)(x - 5y)$
101. $2(1 + 4x^2)(1 + 2x)(1 - 2x)$
102. $2x(y + 5)(y - 5)$ **103.** $(9x - 1)(x + 2)$
104. $(3x - 4)(2x - 5)$ **105.** $(x + 2)(x^2 - 2x + 4)$
106. $(a - 3)(a^2 + 3a + 9)$
107. $(y - 4t)(y^2 + 4yt + 16t^2)$
108. $(m + 10p)(m^2 - 10mp + 100p^2)$
109. $(3x^2 - 1)(x - 2)$ **110.** $(y^2 - 2)(5y + 2)$
111. $(x - 3)(x + 3)(x - 5)$
112. $(t - 5)(t + 5)(t + 3)$ **113.** $\frac{7}{4}$ **114.** $\frac{79}{12}$
115. -8 **116.** $\frac{4}{5}$ **117.** 120 **118.** 140 **119.** 200
120. 200 **121.** $0, -3, \frac{4}{5}$ **122.** $0, 2 -\frac{3}{2}$ **123.** 0, 2
124. $3, -3,$ **125.** 0, 3 **126.** 0, 5 **127.** 0, 7
128. 0, 3 **129.** $0, \frac{1}{3}, -\frac{1}{3}$ **130.** $0, \frac{1}{4}, -\frac{1}{4}$ **131.** 1
132. 2 **133.** No solution **134.** No solution
135. -23 **136.** -145 **137.** 5
138. $-\sqrt{7}, \sqrt{7}$ **139.** $x \geq -\frac{4}{5}$ **140.** $x \geq 3$
141. $x > -\frac{1}{12}$ **142.** $x > -\frac{5}{13}$ **143.** $x > -\frac{4}{7}$
144. $x \leq -\frac{6}{5}$ **145.** $x \leq -3$ **146.** $x \geq -2$
147. $x > \frac{2}{3}$ **148.** $x \leq \frac{5}{3}$ **149.** $x < -\frac{2}{5}$
150. $x \leq \frac{5}{4}$ **151.** $2 < x < 4$ **152.** $2 < x \leq 4$
153. $\frac{3}{2} \leq x \leq \frac{11}{2}$ **154.** $\frac{6}{5} \leq x < \frac{16}{5}$ **155.** $-1 \leq x \leq \frac{14}{5}$
156. $-5 \leq x < -2$ **157.** \$650 **158.** \$800
159. More than 7000 units **160.** More than 4200 units
161. 480 lb **162.** 340 lb **163.** 810,000
164. 720,000 **165.** $60\% \leq x < 100\%$
166. $50\% \leq x < 90\%$

Index

See also Index of Applications and Index of Technology Connections at the beginning of the book.

N

O

P

Q

R

S

SUMMARY OF IMPORTANT FORMULAS FOR DIFFERENTIATION

1. **Power Rule.** For any real number k, $\frac{d}{dx} x^k = kx^{k-1}$.
2. **Derivative of a Constant Function.** If $F(x) = c$, then $F'(x) = 0$.
3. **Derivative of a Constant Times a Function.** If $F(x) = cf(x)$, then $F'(x) = cf'(x)$.
4. **Derivative of a Sum.** If $F(x) = f(x) + g(x)$, then $F'(x) = f'(x) + g'(x)$.
5. **Derivative of a Difference.** If $F(x) = f(x) - g(x)$, then $F'(x) = f'(x) - g'(x)$.
6. **Derivative of a Product.** If $F(x) = f(x)g(x)$, then $F'(x) = f(x)g'(x) + g(x)f'(x)$.
7. **Derivative of a Quotient.** If $F(x) = \frac{f(x)}{g(x)}$, then
$$F'(x) = \frac{g(x)f'(x) - f(x)g'(x)}{[g(x)]^2}.$$
8. **Extended Power Rule.** If $F(x) = [g(x)]^k$, then $F'(x) = k[g(x)]^{k-1}g'(x)$.
9. **Chain Rule.** If $F(x) = f[g(x)]$, then $F'(x) = f'[g(x)]g'(x)$.
Or, if $y = f(u)$ and $u = g(x)$, then
$$\frac{dy}{dx} = \frac{dy}{du} \cdot \frac{du}{dx}.$$

SUMMARY OF IMPORTANT FORMULAS FOR DIFFERENTIATION

(continued)

10. $\dfrac{d}{dx}e^x = e^x$

11. $\dfrac{d}{dx}e^{f(x)} = e^{f(x)} \cdot f'(x)$

12. $\dfrac{d}{dx}\ln x = \dfrac{1}{x}, \quad x > 0$

13. $\dfrac{d}{dx}\ln f(x) = \dfrac{f'(x)}{f(x)}, \quad f(x) > 0$

14. $\dfrac{d}{dx}\ln |x| = \dfrac{1}{x}, \quad x \neq 0$

15. $\dfrac{d}{dx}\ln |f(x)| = \dfrac{f'(x)}{f(x)}, \quad f(x) \neq 0$

16. $\dfrac{d}{dx}a^x = (\ln a)a^x$

17. $\dfrac{d}{dx}\log_a x = \dfrac{1}{\ln a} \cdot \dfrac{1}{x}, \quad x > 0$

18. $\dfrac{d}{dx}\log_a |x| = \dfrac{1}{\ln a} \cdot \dfrac{1}{x}, \quad x \neq 0$